Cell Biology

Cell Biology

Channarayappa

Professor and Head, Department of Biotechnology
MS Ramaiah Institute of Technology (MSRIT)
Bengaluru, Karnataka

Universities Press

CELL BIOLOGY

UNIVERSITIES PRESS (INDIA) PRIVATE LIMITED

Registered Office
3-6-747/1/A & 3-6-754/1, Himayatnagar, Hyderabad 500 029, Telangana, India
info@universitiespress.com; www.universitiespress.com

Distributed by
Orient Blackswan Private Limited

Registered Office
3-6-752 Himayatnagar, Hyderabad 500 029, Telangana, India

Other Offices
Bengaluru / Chennai / Guwahati / Hyderabad / Kolkata
Mumbai / New Delhi / Noida / Patna

Reprinted 2014, 2015, 2024

ISBN: 978 81 7371 716 1

501995

Typeset in Times New Roman 11/13 *by*
MacroTex Solutions,
Chennai 600 088

Printed at
The Print Park, Chennai 600 117

Published by
Universities Press (India) Private Limited
3-6-747/1/A & 3-6-754/1, Himayatnagar, Hyderabad 500 029, Telangana, India

Preface

I am happy to present this book to my readers, particularly teachers and students. This book covers one of the most fundamental and elaborately studied areas of biology, and provides an overview of the structural and functional aspects of the cell. The cell is undoubtedly the basic unit of life and has all the structural and functional properties required for life. Further, it has also been experimentally demonstrated that any genetic modification desired in an organism has to be made at the cellular level only. Therefore, understanding the cell in the context of its structure and function is essential and inevitable.

The book has been structured such that undergraduates, postgraduates and others can use it, by taking the information suitable for their level. This book can serve as a basic textbook for students of molecular biology, genetics, biochemistry, agriculture and biotechnology, or as a reference book for those interested in learning the fundamentals of cell biology, in particular, the origin, organization and functions of subcellular components and cell types.

The architecture of the cell is unique and there is no equivalent to it. The cell is the only system that has the ability to harvest the physical and chemical properties of atoms and create a unique biological phenomenon—life. In the endeavour to understand this miracle, a plethora of information has been accumulated. Condensing this information into a concise book, but without compromising on the essential details, has been a hard but deeply satisfying task. Wherever possible, I have also made an attempt to blend theology and sociology with biology, to convey that all fields are interconnected and cannot be studied in isolation.

The book has been divided into 20 chapters—beginning with the origin of biological systems and ending with tools for the study of cells. At the end of each chapter there is a summary to aid quick revision, model questions to test understanding, and references for further reading. Every effort has been made to include the most recent information. Each chapter is provided with an adequate number of illustrations generated by myself. However, the concept and information needed for drawings and text were extracted from various sources based on pedagogy. The information obtained from other sources is acknowledged and sufficiently updated to suit the text and recent developments.

Most cell biology books are concise and lack sufficient detail. It is essential that students of life sciences, particularly those majoring in cell biology and wanting to pursue a career in the field of cell and molecular biology, learn the subject in a comprehensive manner. A clear understanding of the subject helps one to appreciate nature's ingenuity in the creation of life and apply the knowledge in the laboratory and field. Textbooks should not merely be the means to prepare for examinations or interviews. A textbook must expose students to the essence and essentials of the cell and its interactive environment, beginning from the perplexing origin of life. Understanding the cell is like understanding ourselves; I have therefore made a sincere attempt to provide relevant information. I welcome suggestions from readers to further improve the book.

I would like to thank a number of people for the encouragement and inspiration they have given me while writing this book. I want to thank Dr. Tong-man Ong, Chief, Microbiology Section, National Institute for Occupational Safety and Health (NIOSH), Morgantown, West Virginia, USA, and Dr J Nath, Professor, West Virginia University (WVU), Morgantown, West Virginia, USA, for their guidance, and for providing research facilities during my PhD and postdoctoral program in USA.

I acknowledge with deep gratitude the suggestions and encouragement given by Dr MR Jayaram, Chairman, Gokula Education Foundation (GEF), MR Seetharam, Vice-Chairman, GEF and Director, MSRIT, MR Kodandaram, Director, GEF, Dr K Rajanikanth, Principal, MS Ramaiah Institute of Technology (MSRIT), Bengaluru, Dr KV Devaraj, Former Vice-Chancellor, University of Agricultural Sciences (UAS), Dr M Udaya Kumar, Professor, Department of Crop Physiology, UAS, Dr MV Nagaraj Setty, UAS, and Mr Veerabadraiah, UAS, Bengaluru.

I am grateful to the staff and management of Universities Press and Orient Blackswan for their editorial assistance and the fine production of this book. I especially thank Ms Jayanthi Singaram ad Ms Ramaa Kishore for their careful editing, and MacroTex Solutions, Chennai, for the typesetting.

Channarayappa
crayappa@rediffmail.com

To

my parents

Shri Venkataswamy and Smt Nanjamma

CONTENTS

List of Colour Plates

1 ORIGIN OF BIOLOGICAL SYSTEMS

Learning objectives

After reading this chapter, you will be able to:

- Appreciate the various theories on the origin of life on earth
- List the systematic methods of identification of the biological code
- Identify fundamental features of living systems

1.1 INTRODUCTION

What is life? Where does it come from? Where is it heading? What is its purpose? Early efforts to answer these questions on the origin of life were largely based on religious beliefs. While the biologists attempted an objective explanation, they came up with a description of life as a collection of functions. So far, they have succeeded in identifying the molecular basis of life forms.

With rapid progress in the 19th century, scientists have developed sophisticated techniques to probe the rich diversity of living organisms. The main features of biological systems are:

- they are cellular in nature
- they are hierarchically organized and self-sustaining
- they adapt and evolve frequently but very slowly
- they are complex and dynamic
- they are fundamentally information-based.

1.2 THEORIES OF THE ORIGIN AND HISTORY OF LIFE

Origin of life studies is a limited field of research, despite its profound impact on biology and human understanding of the natural world. Progress in this area is generally slow, though it still draws a lot of attention. Many attempts have been made on a scientific basis to understand how life originated. This scientific inquisition progressed in many and different directions and consequently, many new theories were proposed. However, these theories are diverse and uncertain. The major theories accounting for the origin of life on earth are:

Special creation: Life caused by a supernatural being at a particular time

Spontaneous generation: Life arose from non-living matter on numerous occasions by a process of spontaneous generation.

Steady-state: Life has no origin; it has remained so forever.

Cosmozoan: Life arrived on this planet from elsewhere.

Biochemical evolution: Life arose according to chemical and physical laws.

All the above theories were critically debated and no single theory can categorically answer all the questions related to the origin of life. A discussion of these theories is beyond the scope of this chapter. A brief note on the Hindu perspective on the origin of life is given in the following box.

Hindu perspective of the origin of life

Different schools of Indian philosophy and Hindu religion have elaborated on the origin of life. Ancient Hindu scripts conceptualize the world as having been created by the three cosmic forces: Parameshwara, Narayana and Brahma. It could be said that *Parmeshwara* (or Shiva, the destroyer) represents the proton, which is positively charged; *Narayana* (or Vishnu, the conservator), represents the neutron, which is neutral in charge, since it contains both the positively charged proton and the negatively charged electron; and *Brahma* (evolutionary or creator) represents the electron which is negatively charged. The electrons help in the formation of new molecules and matter by breaking and making bonds, like Brahma the creator is believed to do so. It has been suggested that the three subatomic particles are made up of even more elementary particles called *quarks* (most basic forms of matter). It is also possible to conceptualize that the quarks are the equivalent of *Audibrahma*, the root creator of the three divine forces. As explained in nuclear physics, when a neutron is bombarded by a positron it generates an electron, since a neutron contains both negetron and positron. Similarly, the ancient scripts say that Brahma (equivalent of electron) is born out of Vishnu (equivalent of neutron). As shown in **Fig. 1.1**, these three divine powers can be equated with the three subatomic particles (*proton, neutron* and *electron*) present within the atom.

To explain the evolution of higher forms of life, according to the scripts, Vishnu *reincarnates* (*avathara*) into different life forms on earth to harmonize life (restoration of order) by saving the good and punishing the evil. In close parallel the three subatomic particles are the basis for the formation of diverse and dynamic elements with a myriad of physical and chemical properties

Parameshwara the proton (p), Narayana the neutron (n), and the evolutionary Bramha the electron (e) Atma within the Atom

Fig. 1.1 Three divine forces of universe manifest within the atom:.

essential for life; only the neutron can give rise to all three forms of subatomic particles. When neutron is bombarded with a positron, it replaces the electron (negatron), and a proton (with two positrons) and an electron are generated. By analogy, this could be the reason why only Vishnu has reincarnations. More surprisingly, at the organism level, the evolution of different life forms, as hypothesized by biologists on scientific facts and interpreted in the form of cladograms, also correlate with the series of reincarnations of Vishnu. The different *avathars,* listed chronologically are: Matsya (fish), Kurma (tortoise), Narsimha (lion/cat-headed), Varha (boar headed), Rama (perfect human forms) (**Fig. 1.2**).

1.2.1 Biochemical evolution

The theory of molecular evolution, is increasingly gaining credibility, because it is based on scientific hypotheses, which are capable of being tested. Moreover, the information and the results obtained from molecular experiments are consistent and reliable. All living forms are made up of molecular structures controlled by physical and chemical laws.

Most biologists, astronomers and geologists accept that the earth is around 4.5–5 billion years old. Accordingly, the present form of the earth bears little semblance to the original state. The primordial earth (**Fig. 1.3**) was hot (about 4000–8000°C) and as it cooled, carbon and the less volatile metals condensed and formed the earth's core; the surface was probably barren and rugged as volcanic activity, continuous earth movements and contraction on cooling, folded

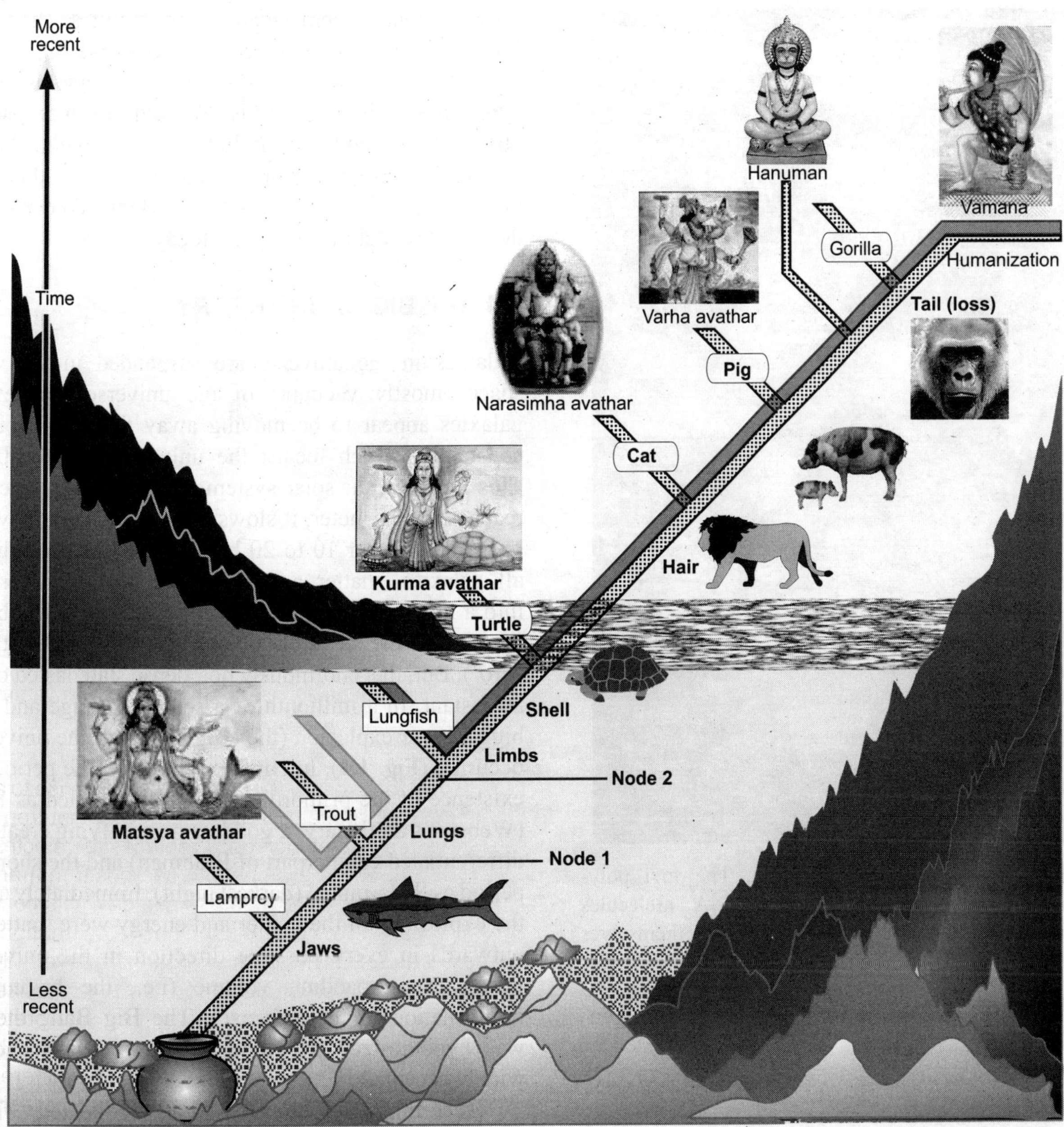

Fig. 1.2 Constructing a cladogram: To see how phylogenetic reconstruction works in practice, a group of vertebrates were compared. For each of these animals, six different traits, shown in the figure have been compared. It allows us to make inferences about the common ancestor of the organism being considered. Interestingly, this also correlates with the order and forms of Vishnu's incarnations.

and fractured the surface. The atmosphere is believed to have been totally different in those days. The lighter gases such as hydrogen, helium, nitrogen, oxygen and argon would have escaped because the gravitational field of the partially condensed planet could not contain them. However, simple compounds containing these elements (amongst others) such as water, ammonia, carbon dioxide and methane would have been retained, and until the earth cooled below 100°C all water would have existed as vapour. The atmosphere was 'reduced atmosphere', as indicated by the presence of metals in their reduced form (such as, iron II) in the oldest rocks of the earth. But more recent rocks contain metals in their oxidized form (for example, iron III). The lack of oxygen in the atmosphere was probably necessary, since it would have been easier to generate organic

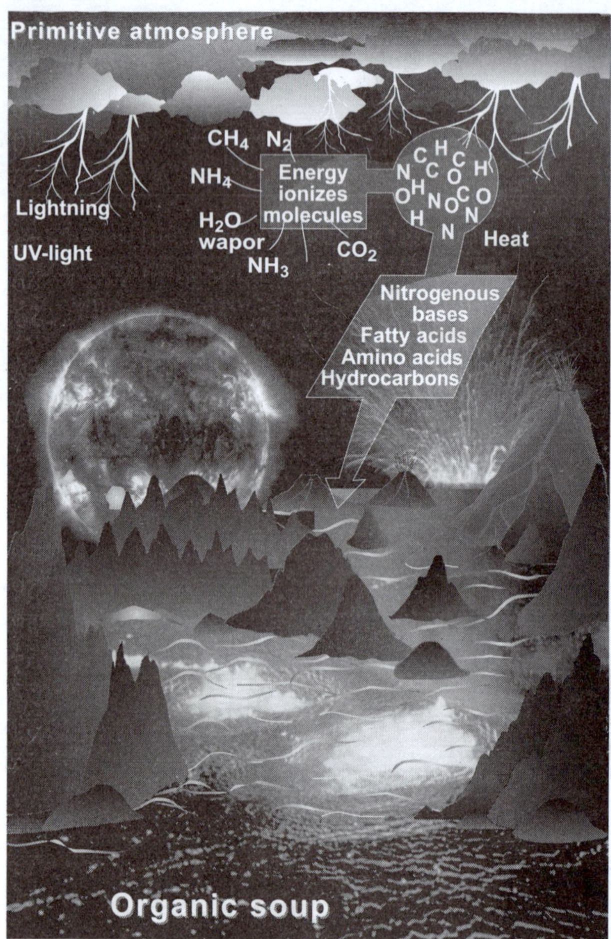

Fig. 1.3 Prebiotic replication of RNA. The first polynucleotide sequence may have been RNA molecules that polymerized abiotically and replicated themselves spontaneously.

molecules in a reducing atmosphere than in an oxygen-rich atmosphere.

1.3 THE UNIVERSE

The rise of geology in the 18th and 19th centuries facilitated ascertaining the age of the earth in scientific terms. Accumulation of data and development of new techniques in other sciences, in particular, astronomy and biochemistry, helped to develop scientific interpretations that explain the geological timescale of the origin of the earth and even the universe. The prerequisite for understanding the process of formation of life is to know the origin of the raw materials (matter, elements, stars, solar systems and planets). The field, which deals with the study of the creation of the **cosmos**, is called **cosmology**. The Universe or cosmos is an aggregation of diverse celestial bodies, which are organized into systems such as Milky Ways, nebulas, galaxies, solar systems, planets and satellites. The term 'universe' refers here to the visible universe.

According to cosmology, the Universe is ever expanding. This result is derived from redshift observations and quantified by Hubble's Law. The most probable theory based on the fact that the universe is expanding, is the 'Big Bang' theory—the dominant model in cosmology today.

1.4 THE BIG BANG THEORY

Galaxies in the universe are suspended in the vast space (mostly vacuum) of the universe. However, galaxies appear to be moving away from each other constantly, which means the universe is expanding. This includes our solar system also. Probably, once it reaches its perimeter, it slows down and starts moving backwards. After 10 to 20 billion years of rewinding, all galaxies, matter and space, become compressed into a hot, dense, volume (a primordial fireball) about the size of the sun (this point considered to be 'time zero'). But, this enormously hot, dense state lasted only an instant. In a millionth of a second, a huge and the biggest ever explosion (the 'Big bang') in the universe occurred (**Fig. 1.4**). In Hindu mythology, the period of existence of the primordial fireball was called as Kali (Webster's dictionary: a goddess personifying creative, differentiated counterpart of Brahman) and the shortest period as 'kala-rathri' (darkest night). Immediately after the explosion, all the matter and energy were scattered outwards in every possible direction in the universe in an ever-expanding volume (i.e., the beginning of formation of the universe). The Big Bang theory was hypothesized by Abbe Lemaitre (1950) and is widely accepted.

According to the evidence collected from astrophysics and cosmology, the 'Big Bang' might have occurred about 13.7 billion years ago, with a margin of error of about 1%.

1.5 THE SOLAR SYSTEM

Our solar system was once a part of a massive cloud of interstellar gases and cosmic dust (**Fig. 1.5**) that extended trillions of kilometres in space. As the particles of matter were pulled together by gravitational attraction, the vast cloud condensed into a gigantic spinning flat disk about 4.6 billion years ago. Nearly 90% of the matter gravitated to the centre, causing a dense, hot centre—the solar system. In addition, the cloud of solar dust gave birth to one or more planets

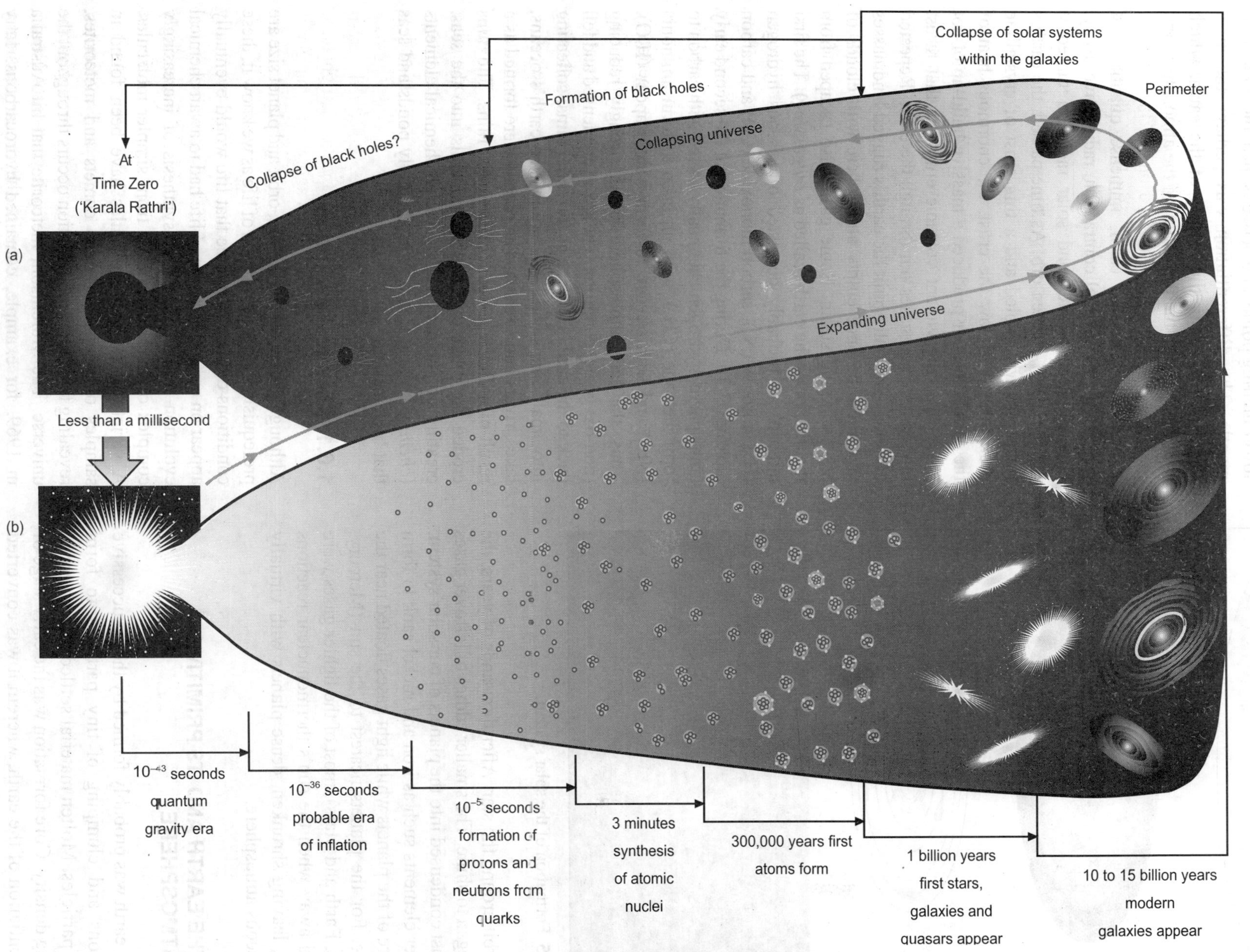

Fig. 1.4 Artistic view of the Big Bang: (a) A dark universal matter at just a fraction of a second before the explosion; (b) Cosmic timeline showing the evolution of our universe from the Big Bang to the present day.

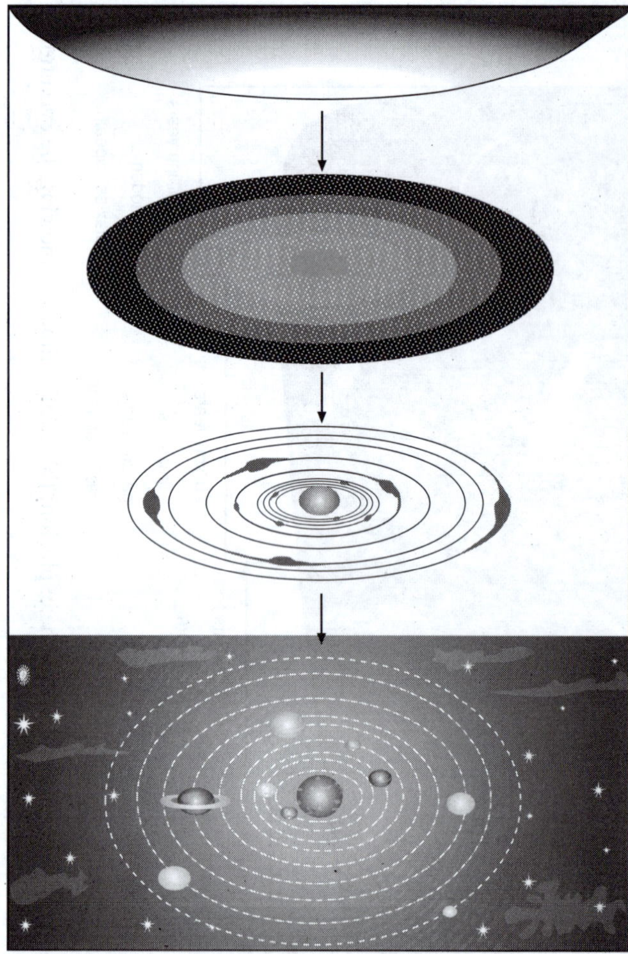

Fig. 1.5 Formation of the solar system.

that orbit around the star. After thousands of years, the sun began to shine. The smaller eddies of leftover gases and dust condensed into the planets of our solar system. Heavier elements such as iron and nickel sank to form the core of the planets while light gases floated near the surface. For the planets nearest to the sun (Mercury, Venus, Earth and Mars), most of the lighter gases were blasted away when the sun's thermonuclear reactions began, leaving shrunken, dense planets with virtually no gaseous atmosphere.

1.6 THE EARTH AND ITS PRIMITIVE ATMOSPHERE

Planet earth was probably formed by these successive collisions and clumping of tiny particles to form larger particles. Molten material settled into layers of varying density. Core formation was the early stage of differentiation of the earth, wherein it was converted from a homogeneous body to a zoned, layered or shelled body with a dense nickel and iron core. The surface or crust consisted of lighter material with lower melting points and between them the remaining mantle. The differentiation was perhaps the most significant event in the history of the earth, which eventually led to the formation of the atmosphere and the oceans.

On the primitive earth, immense quantities of heat from gravitational contraction, radioactive decay, impacts of meteorites and solar radiation turned it into a red hot, molten orb. As time passed, the earth's surface slowly cooled and a thin crust of crystalline rock formed. Below the crust, the enormous heat of the earth's interior produced a massive build-up of hot gases, sparking violent volcanic eruptions that thrust molten rock and gases out through the crust. Repeated eruptions gradually built the earth's rugged landmasses and filled the once empty atmosphere with clouds of hot gases and steam. Some gases also escaped from the interior of the earth (called as *outgassing*). The first atmosphere probably consisted of gaseous hydrogen (H_2), nitrogen (N_2), carbon monoxide (CO) and carbon dioxide (CO_2). During the formation of the second early atmosphere, hydrogen chemically united with carbon to form methane (CH_4), with nitrogen to form ammonia (NH_3) and with oxygen to form water vapour (H_2O). Thus, the early atmosphere had a strongly reducing (non-oxygenic) condition. The steam condensed and fell back to the earth's surface as torrential rains, collecting into small ponds and later forming the earth's oceans, lakes, rivers and streams. The oceans were formed as a result of millions of years of torrential rains. The rains eroded the rocks and washed minerals into the seas, creating a combination of the varied chemical elements (*chemical soup*). It was from the early ponds and seas that life emerged.

1.6.1 Chemical evolution

Although distance from the sun, orbit and planet size are prerequisites for the evolution of life as we know it, these conditions did not guarantee that life would eventually appear on earth. The dawn of life had to await chemical evolution—the spontaneous synthesis of increasingly complex organic compounds from simpler molecules. A variety of organic molecules have been found in samples of interstellar dust, comets and meteorites, revealing that chemical evolution occurs throughout the universe. The Murchisom meteorite that hit Australia in 1969, for example, contained hydrocarbons, fatty acids and over 100 amino acids. Organic compounds are constructed primarily of four, highly reactive atoms (carbon, hydrogen, oxygen and nitrogen); a source of energy causes these reactive gases, if present together, to spontaneously assemble into complex organic compounds. These observations indicate that organic molecules must have been synthesized abiotically. Therefore, abiotic synthesis of organic molecules could have occurred on the primeval earth. The most significant outcomes of chemical evolution are the origin of biologically important macromolecules such as proteins and nucleic acids (RNA and DNA) and formation of an environment to sustain life.

1.6.2 Spontaneous formation of organic molecules

Scientific inquiries into the mechanisms of abiogenesis began with Charles Darwin, who stated that life might have arisen in a 'warm little pond,' which he speculated contained ammonia, phosphate and other molecules, so that a protein was chemically formed ready to undergo more complex changes. In 1924, two biochemists, JBS Haldane in Scotland and AP Oparin in Russia, demonstrated virtually identical hypotheses (i.e., **abiotic synthesis**) for how organic compounds could have formed on the primitive earth and simulated the conditions most likely to have existed in the earth's early atmosphere. Instead of the nitrogen and oxygen that comprises most of today's atmosphere, Haldane and Oparin speculated that the earth's early atmosphere was a reducing atmosphere (non-oxidizing) containing a mixture of carbon monoxide (CO), carbon dioxide (CO_2), water vapour (H_2O), nitrogen (N_2) and large amounts of hydrogen (H_2). If oxygen had been present, as it does today, atoms and compounds would have oxidized before they could assemble into organic molecules. Life could therefore not emerge in an atmosphere that contained free oxygen. The early atmosphere would have also contained ammonia (NH_3), methane (CH_4) and hydrogen sulphide (H_2S), all of which are complexes of hydrogen and other gases. Energy from various sources could have supported the required endergonic reactions. Such sources could have been lightning, solar radiation (e.g., ultraviolet rays), heat from lava flows, radioactive decay of various elements in the rocky crust of the earth and collisions with meteorites, which might have ionized these gases briefly, freeing carbon, hydrogen, nitrogen and oxygen. These ions immediately reacted with each other to form amino acids, nucleotides, hydrocarbons, lipids and fatty acids—the organic building blocks of life. Oparin further argued that if one considered the multitude of simple molecules present in the oceans, the surface area of the earth, the energy available and the time scale, it was conceivable that oceans would gradually accumulate organic molecules to produce a '**primeval soup/organic soup**', in which life could have arisen.

Experimental evidence

Experiments conducted by Stanley Miller and Harold Urey in 1953 reinforced Haldane and Oparin's hypothesis of abiotic synthesis. In a closed reaction vessel, Miller and Urey repeatedly jolted a mixture of hydrogen, ammonia, methane and water vapour with electricity to simulate lightning (**Fig. 1.6**). The results of the week-long experiment proved significant in several respects. They found a number of organic compounds in the reaction vessel, including eight amino acids of which four were commonly found in organisms—glycine, alanine, aspartic acid and

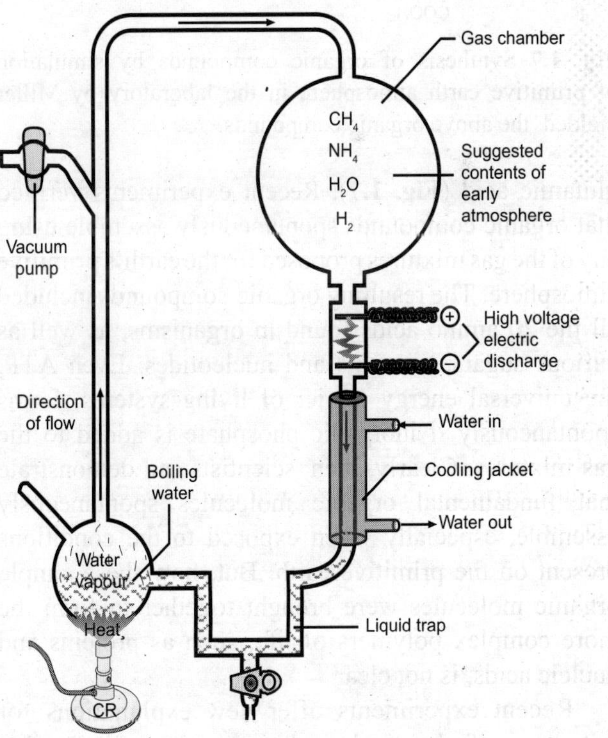

Fig 1.6 Stanley Miller's apparatus simulating conditions on primitive earth. He synthesized organic compounds from gases under conditions thought to have been present in the primeval atmosphere. The gases and water vapour were circulated under pressure and exposed to high voltage for one week. Finally, the liquid products in the trap were analyzed.

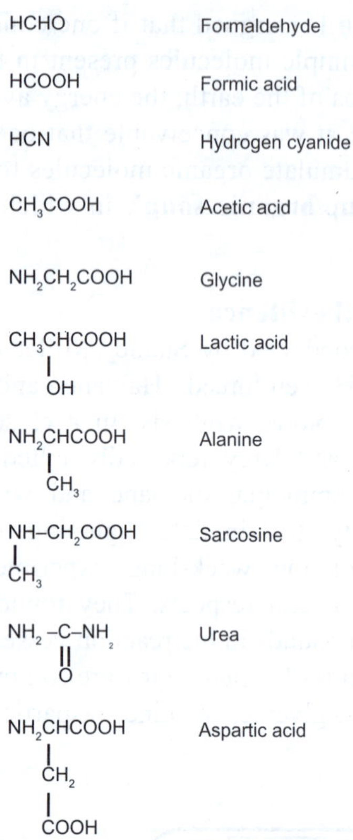

Fig. 1.7 Synthesis of organic compounds by simulation of primitive earth atmosphere in the laboratory by Miller yielded the above organic compounds.

glutamic acid (**Fig. 1.7**). Recent experiments verified that organic compounds spontaneously assemble using any of the gas mixtures proposed for the earth's primitive atmosphere. The resulting organic compounds included all the 20 amino acids found in organisms, as well as various sugars, vitamins and nucleotides. Even ATP, the universal energy carrier of living systems, forms spontaneously if inorganic phosphate is added to the gas mixtures. Clearly, then, scientists can demonstrate that fundamental organic molecules spontaneously assemble, especially when exposed to the conditions present on the primitive earth. But, how these simple organic molecules were brought together to form the more complex polymers of life, such as proteins and nucleic acids, is not clear.

Recent experiments offer new explanations for this stage of chemical evolution as scientist's edge closer to restructuring the chemical events that led to the origin of life. Three major models of self-contained systems enclosed within a selectively permeable boundary have been proposed:

- the ***coacervate model*** of Alexander Oparin
- the ***proteinoid microsphere model*** of Sidney Fox
- the ***lipid bilayer model*** proposed by Richard Goldacre

In experiments performed in the 1960s, Sydney Fox demonstrated that if amino acids were dry and hot, they joined together to form **proteinoids**—complex polymers made up of hundreds of amino acids. On the primitive earth, dry, hot amino acids may have formed along the edges of ponds near hot volcanoes or in hot evaporated pools. The dry, hot amino acids then combined into proteinoids. Scientists also speculate that it was probably also along the edges of ponds or in evaporated pools that spontaneously assembled nucleotides became concentrated (**see box**)), allowing for the formation of nucleic acids (DNA and RNA).

Eigen's hypothesis

In the early 1970s, a team of scientists at the Max Planck Institute tried to examine the transient stages between the molecular chaos in prebiotic soup and the transient stages of a self-replicating hypercycle, between the molecular chaos in prebiotic soup and simple macromolecular self-reproducing systems. In a hypercycle, the information storing system (RNA) produces an enzyme, which catalyzes the formation of the first information system. Hypercycles could create quasispecies, which through natural selection entered into a form of Darwinian evolution. The significance of hypercycle theory was the discovery that RNA, in certain circumstances transforms itself into ribozymes, a form of RNA enzyme.

1.7 ORIGIN OF BIOLOGICAL SYSTEMS (LIFE FORMS)

The earth is a small planet, circulating a very ordinary star (sun). This seemingly unremarkable planet is home to millions of different life forms. What conditions on a planet enable it to support life? The following conditions must be available:

- Water sustains life. A planet must orbit the star at the centre of the planetary system it belongs to at the right distance, so that temperatures will not continually exceed 100°C or remain below 0°C. These are the temperature extremes at which water exists as a liquid, unless under very high pressure.
- The planet's orbit must be nearly circular, maintaining a constant distance from the star and therefore a somewhat stable temperature range within the limits that permit life to occur.

- The planet must be large enough for its gravity to hold a gaseous atmosphere, yet not so large that its gravity creates too dense an atmosphere for sunlight to penetrate.

Within our solar system, only the earth appears to meet all these criteria.

About 4.1 billion years ago, when the Earth's crust began to solidify and 3.5 billion years ago, when the planet inhabited by bacteria advanced to build **stromatolites** (ancient prokaryotes found in rocks), the organisms came into being. But the question is what was their origin? The information obtained from the analysis of geological, biological and chemical remnants have allowed scientists to develop an overview of the history of life. However, the precise mechanisms by which life originated is still the subject of debate. Two approaches (or basic strategies) in origin of life studies have been used: the top-down and the bottom-up approaches. In the first approach, the evolutionary history of modern organisms has been traced back through time with phylogenetic analysis, i.e., by investigating the similarities and differences among organisms that are clues to their evolutionary past. Alternatively, in the bottom-up approach, researchers have focussed on reconstructing the processes by which inanimate matter was transformed on primitive earth into the first living organisms, a process called **abiogenesis** (see box). This will help in ascertaining the physical and chemical conditions that facilitated the origin of life. They also analyzed biomolecules in modern species that are considered to be a vestige from the prebiotic world. As of 2007, no one has yet synthesized a **protocell** using basic components which would have the necessary properties of life (the 'bottom-up-approach').

Abiogenesis

It is interesting to see the unravelling of the exact mechanism by which biological life emerged from the physical and chemical conditions that prevailed in the primordial era. It is concerned with how organic molecules like sugars and proteins were formed and how they linked up to form large chains containing information. There is still no conclusive theory, although a lot of evidence is being collected.

1.7.1 The first life forms

The fact that life evolved relatively "soon" after the earth formed has led many biologists to conclude that life may not be, after all, such an unusual phenomenon in the universe. Living things are curious, intricately organized chemical factories that take in matter from their surroundings and use these raw materials to generate copies of themselves. Nevertheless, the leap from organic soup to the simplest living thing is a big one, for even the simplest cell is highly organized and carries out complicated metabolic reactions. Before the first cell could have evolved, one or more types of simpler **precells** would have formed. It is estimated that there are more than a million living species on earth today. Each species is different and reproduces itself faithfully, yielding progeny that belong to the same species with very little or no difference. Cellular life as we know it has three basic components:

- A boundary that separates the living system from its environment
- Metabolism or coordinated sets of chemical reactions catalyzed by enzymes;
- A set of genes that stores information that can be replicated and transmitted to descendents like itself.

With the incorporation of a genetic system, protobionts would have evolved into **eubionts**, or true life forms.

Precells would be able to concentrate organic molecules, allowing them to react more frequently. Within the precell, catalysts would direct the synthesis of specific organic polymers; in other words, the precell would be conducting metabolism. As metabolism produced new organic molecules, the precell would grow. The eventual evolution of a genetic code would enable the precell to pass on naturally selected codes for metabolism, allowing the precell to reproduce itself. At this point, the precell would possess three of the basic characteristics of life—metabolism, growth and reproduction—crossing the line between precell and living cell.

1.7.2 Geological record and timescale of life on earth

The first living cell could have appeared on the earth probably between 3.5 and 4.6 billion years ago, approximately 750 million years after the earth was formed (**Fig. 1.8**). Life remained unicellular for 3 billion years. For the first 2 billion years, only simple

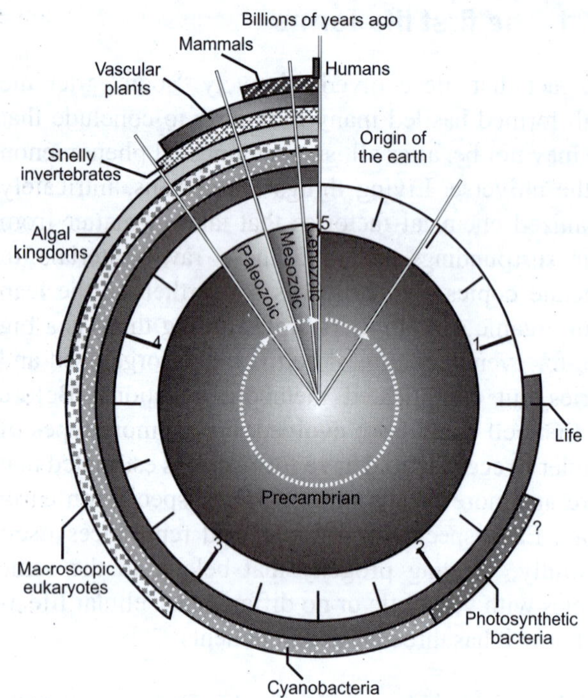

Fig. 1.8 Earth's biological clock: A diagram of the past five billion years of the earth's history, illustrating the period of evolution of the major groups of organisms.

prokaryotes existed. The evolution of single-celled eukaryotes paved the way for the evolution of the first multicellular organisms one billion years later. The multicellular plan was a tremendous success, leading to a proliferation of complex life forms, some of which moved onto land and eventually took to the air. Today, about 2–30 million species of organisms populate the earth.

Geologists divide the earth's 4.6-billion-year lifespan into four great eras: the **Proterozoic** (a period that produced few fossils), **Paleozoic**, **Mesozoic** and **Cenozoic** eras. With the exception of the Proterozoic era, the others are divided into periods and epochs on the basis of geological formations and the fossil record. Chemical evolution and transformation of the precell into a living cell might have occurred during the Proterozoic era between 0.5 and 4.6 billion years ago.

1.8 SIMPLE INORGANIC MOLECULES TO COMPLEX FREE LIVING CELLS

What is life? ***It may be defined as the coordinated behaviour of molecules within cells or between cells in the case of higher organisms.*** This calls for the right substance in the right proportions and in the right arrangement. The substances are water, certain salts and carbon compounds. The latter are called organic compounds and consist of four types of atoms: carbon, oxygen, nitrogen and hydrogen. These four atoms together constitute about 99% in life forms. Organic molecules, therefore, form a large and formidable array, endless in variety and of the most bewildering complexity. Besides being available in adequate amounts and proper proportions, they also need to be placed in the right arrangement.

What is a cell? ***It is a self-contained, self-reproducing functional and structural unit of life.*** Cells may exist independently as free living organisms like prokaryotic bacteria or as constituent functional units in multicellular eukaryotes such as fungi, animals and plants (the human body, for example, consists of about 100 trillion cells). In the construction of cells, a molecular hierarchy exists in which a few types of small molecules are utilized as building blocks for a large variety of macromolecules of several major classes. These macromolecules in turn are assembled into complexes and organelles, which are further organized into the units that we call living cells. The molecules inherit the physical and chemical properties of atoms from which they are constituted and the same physical and chemical laws apply throughout the universe. They attract, repel and form aggregates with each other. They combine to form new stable compounds with novel size, structure and chemical properties. Life's mystery lies in the programmed coordination of the myriad of chemical reactions necessary for the metabolic activities of cells. To maintain the cellular integrity and growth and to carry out its unique mission, each cell regulates its own internal environment and draws selectively from the external one the raw materials it needs. A cell comes into being and formulates its own identity as it develops from some pre-existing living entity, carries out unique functions, procreates and eventually disappears forever. Although we now know many of the molecular interactions and essential chemical processes required for the creation of a living cell, we simply cannot yet create "life in the test tube".

1.8.1 Development of complex chemical systems

Simple organic molecules such as amino acids and nucleotides can associate to form **polymers**. The linear polymers known as polypeptides and polynucleotides are formed by the process of polymerization. One amino acid can join with another by forming a peptide bond and two nucleotides can join together by a phosphodiester bond. The repetition of these reactions leads to linear polymers known as polypeptides (proteins) and polynucleotides (nucleic acids, DNA and RNA), respectively. In living cells, proteins and nucleic acids are commonly viewed as the most important constituents. A restricted set of 20 amino acids constitutes the universal building blocks of the proteins, while RNA and DNA molecules are constructed from just four types of nucleotides each. The earliest polymers may have formed in any one of the several ways (e.g., heating of dry organic compounds or by the catalytic activity of high concentrations of inorganic compounds or other crude mineral catalysts). Once a polymer has formed (**Fig. 1.9**), however, it can itself influence subsequent chemical reactions by acting as a catalyst.

The origin of life requires an assortment of molecules, which must have possessed, if only to a small extent, this crucial property: the ability to catalyze reactions that lead, directly or indirectly, to production of more molecules of the catalyst itself. Thus, one can envisage the gradual development of an increasingly complex chemical system of organic monomers and polymers that function together to generate more molecules of the same type, fuelled by a supply of simple raw materials in the environment. Such an **autocatalytic system** would have some of the properties we think of as characteristics of living matter (reproduction, competition and maintenance of chemical equilibrium).

1.9 ORIGIN OF GENETIC INFORMATION

A cell stores its genetic information as DNA, transcribes the information into RNA and then translates the messages into specific enzymes and other proteins. Instructions are transmitted by the replication of DNA when a cell divides. However, even before DNA, some primitive mechanism may have existed for aligning amino acids along strands of RNA that could have replicated within them. Short polymers of ribonucleotides have been produced abiotically in test tube experiments. If RNA is added to a test tube solution containing monomers for making more RNA, sequences about five to ten nucleotides long are copied from the template according to the base-pairing rules. And if zinc is added as a catalyst, sequences up to 40 nucleotides long are synthesized with less than 1% error. These results support the idea that RNA strands might have self-replicated without the help of elaborate enzymes.

1.9.1 Polynucleotides are capable of reproduction

Once the necessary chemical building blocks were available, living systems began to evolve. Before the appearance of life, simple molecular systems could

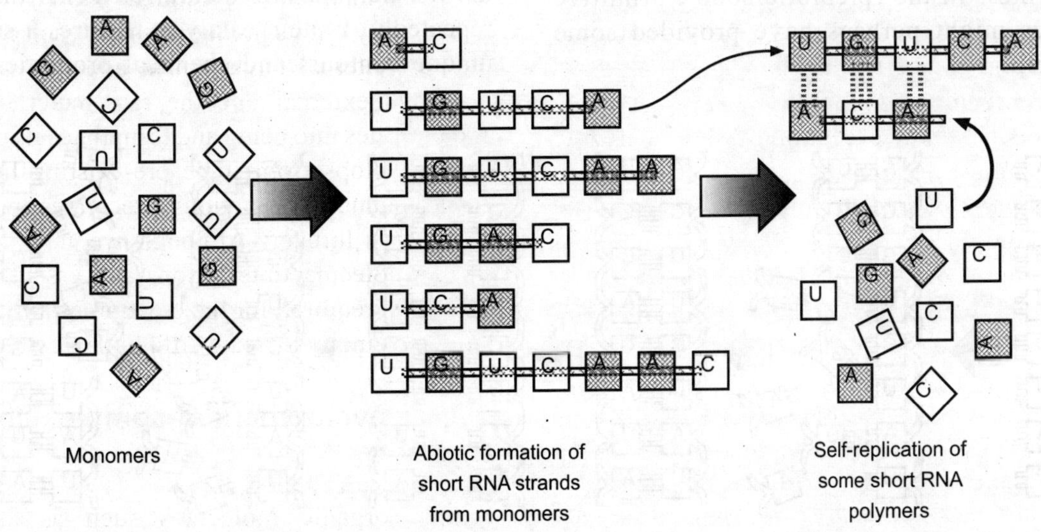

Fig. 1.9 Prebiotic replication of RNA. The first polynuceotide sequence may have been RNA molecules that polymerized abiotically and replicated themselves spontaneously while bound to clay surfaces.

have existed, which subsequently evolved into the complex chemical systems that are characteristic of organisms. To address how this evolution occurred, we need to consider first, the most fundamental property of evolving systems—their ability to replicate or reproduce. Without this ability, each species of molecules might have become extinct as soon as all its individual molecules degraded. However, molecules that can replicate continue to be represented in the population, even if the lifetime of each individual molecule remains short.

To understand the mechanisms that make life possible, one must understand the structure of the molecules that store the genetic information. Another important process that needs to be elucidated is the expression of hereditary information, which today requires extraordinarily complex machinery and proceeds from DNA to protein through an RNA intermediate. How does this intricate but highly efficient machinery arise? Most organisms use DNA as the genetic material and are made up of polynucleotides. They have limited capabilities as catalysts, but can directly guide the formation of exact copies of their own sequence. This property depends on **complementary pairing** of nucleotide subunits, which enables one polynucleotide to act as a **template** for the formation of another (**Fig. 1.10**). Templating mechanisms, however, require additional catalysts to promote polymerization; otherwise, the process is slow. Today, the catalytic functions that polymerize nucleotides are provided by highly specialized catalytic proteins called enzymes. In the 'prebiotic soup', primitive polypeptides might perhaps have provided some catalytic help.

1.9.2 RNA molecules act as catalysts

The development of systems beyond simple replication required the generation of specific catalysts. Prior to the 1980s, all biological catalysts, termed enzymes, were considered to be proteins. Tom Cech and Sidney Altman independently discovered that certain RNA molecules could be effective catalysts. The discovery of ribozymes suggested the possibility that catalytic RNA molecules could have played fundamental roles early in the evolution of life. One hypothesis is that an RNA world existed on earth before modern cells arose. According to this concept, relatively short RNA molecules, that were capable of catalyzing their own continuing replication, could have spontaneously formed. RNA molecules could have stored both genetic information and catalyzed the chemical reactions in primitive cells. Only later in evolution did DNA take over as the genetic material and proteins become the major catalyst and structural component of cells. RNA molecules with an appropriate nucleotide sequence can act as a catalyst for the accurate replication of another RNA molecule—the template—whose sequence can be arbitrary. This versatility of the RNA molecules is thought to have played a central role in the origin of life.

As previously mentioned, the complementary base pairing and other types of hydrogen bonds can occur between nucleotides in the same chain, causing an RNA molecule to fold up in a unique way determined by its nucleotide sequence. Comparison of many RNA structures has revealed conserved motifs, i.e., short structural elements that are used repeatedly. Protein catalysis requires a surface with unique contours and chemical properties on which

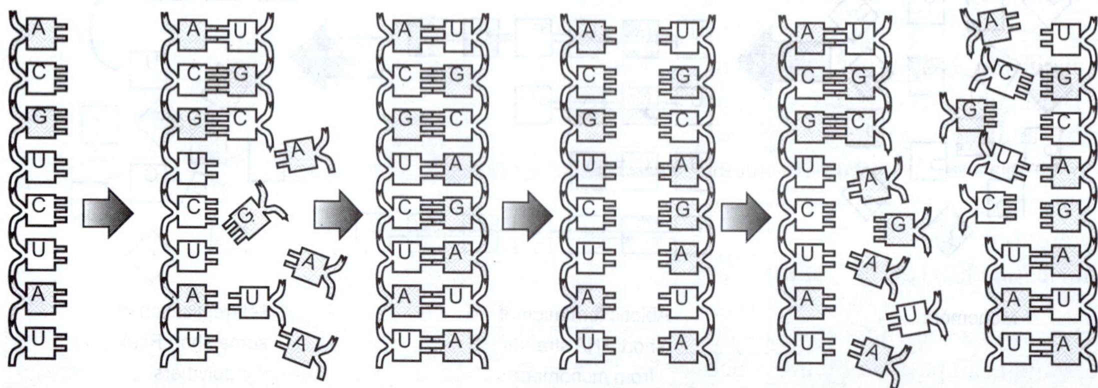

Fig. 1.10 Self-replication of RNA. Complementary base pairing between nucleotides (adenine [A] with uracil [U] and guanine [G] with cytosine [C]) allows one strand of RNA to serve as a template for the synthesis of a new strand with the complementary sequence.

a given set of substrates can react. Similarly, only an RNA molecule with an appropriately folded shape can serve as an enzyme. In vitro creations of various catalytic RNAs have shown to catalyze a wide variety of biochemical reactions. Like proteins, RNAs can also undergo allosteric conformational changes, either in response to small molecules or to other RNAs. One artificially synthesized ribozyme can exist in two entirely different conformations, each with a different catalytic activity.

Although RNA has enzymatic and self-replicative capabilities, some biochemical constraints indicate that it may not be the first molecule that might have existed. For example, the length and order of nucleotides in RNA molecules may be difficult to happen purely by non-enzymatic means. Presumably, the first molecules that possessed both catalytic activity and information storage capabilities might have been polymers that resemble RNA but are much simpler in form.

1.9.3 Self-replicating molecules undergo variation and natural selection

Variation and competition are important fundamental factors in evolution. Natural selection on the molecular level works on diverse populations so that replicating molecules must carry changes. Evolution cannot occur on the same type of molecules. Polynucleotides not only carry information but also have chemical personalities that affect their behaviour (definite 3D folding of RNA in solution by hydrogen bonding). The 3D-folded structure of a polynucleotide affects its stability, its actions on other molecules and its ability to replicate. Therefore, certain polynucleotides will be especially successful in any self-replicating mixture. Self-replicating molecules compete with one another for available resources such as chemical precursors and the competition allows the process of **evolution by natural selection** to occur (**Fig. 1.11**). Those molecules that are stable and are best able to survive and replicate themselves will increase in relative concentration. Thus, new molecules that are better able to replicate under the conditions of their environment are formed.

A polynucleotide has two unique characteristics:

- It can carry information encoded in its nucleotide sequence, which can pass on by the process of replication.
- It has a specific folded structure that enables it to interact selectively with other molecules and also determines how it will respond to ambient conditions.

The molecule thus has a **genotype** (its nucleotide sequence) and a **phenotype** (its conformation, which interacts with surrounding molecules in specific ways). These special features, one informational and the other functional, are the two properties essential for evolution. The nucleotide sequence of an RNA molecule is analogous to the genotype—the hereditary information—of an organism. The folded 3D structure is analogous to the phenotype—the expression of the genetic information on which natural selection operates.

1.9.4 Flow of information from polynucleotides to polypeptides

Probably between 3.5 to 4 billion years ago, self-replicating systems of RNA molecules, mixed with other organic molecules, including simple polypeptides, began the process of molecular evolution. Molecules containing groups of polymers competed for the available precursor materials to construct copies of themselves. Success depended on the accuracy, speed and stability of the copies made. The structure of polynucleotides was well suited for information storage and replication, but lacked catalytic ability. Thus, at the origin of life any polypeptide that helped guide the synthesis of a useful polypeptide in its environment would have had a great advantage in the evolutionary struggle for survival. These polynucleotides must also have the ability of catalysis to join selected amino acids together. Thus, the first steps toward the replication and translation of genetic information may have been taken by molecular evolution even before RNA and polypeptides became packaged within membranes.

One important question to be answered, however, is how did protein synthesis evolve? It is conceptually difficult to imagine how protein synthesis evolved, because it is now performed by a complex interlocking system of protein and RNA molecules. The proteins could not have existed until an early version of the translation apparatus was already in place, but now we can only speculate on the origins of protein synthesis and the genetic code by several experimental approaches. In comparison to the present protein synthesis events, it seems likely, then, that RNA guided the primordial synthesis of proteins, perhaps in a clumsy and primitive fashion.

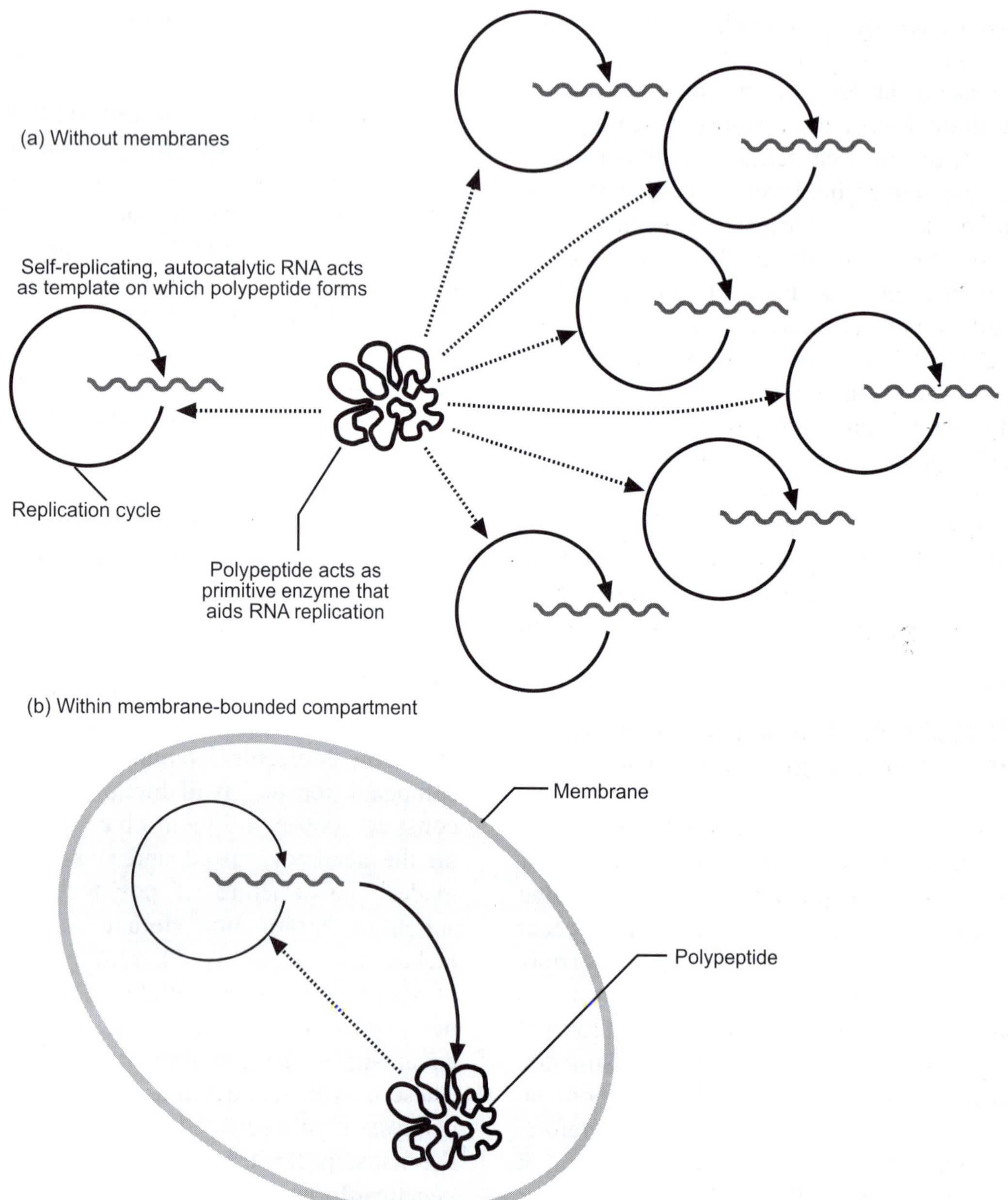

Fig. 1.11 The beginning of molecular cooperation: (a) An RNA strand enhances its own replication if it orders amino acids into a polypeptide that in turn functions as an enzyme helping the RNA strand replicate. Such cooperation may have been the rudiments of the transition of genetic information into protein structure. The benefits of this protein, however, whould have been shared by competing RNA molecules not yet segregated within protobiots. (b) Once genes were in membrane-enclosed compartments, they would have benefits exclusively from their protein products.

In this way, RNA was able to create tools—in the form of proteins—for more efficient biosynthesis and some of these could have been put to use in the replication of RNA and in the process of tool production itself. The synthesis of specific proteins under the guidance of RNA required the evolution of a code by which the polynucleotide sequence specifies the amino acid sequence that makes up the protein. This code—the genetic code—is spelt out in a 'dictionary' of three-letter words: different triplets of nucleotides encode specific amino acids. The code seems to have been selected arbitrarily, yet it was virtually the same in all living organisms. This strongly suggests that all present-day cells have descended from a single line of primitive cells that have evolved the mechanism of protein synthesis.

1.9.5 DNA as genetic material

The cell forms of the RNA world must presumably have been simple and less efficient in reproducing themselves, since catalytic activity of RNA is less efficient than that of proteins. Primitive cells might have existed as a set of self-replicating molecules enclosed in a membrane and with a few other simple components. However, all the present-day cells use double-stranded DNA (dsDNA) as genetic material to store hereditary information. Both the types of polynucleotides (DNA and RNA) are found in present-day cells; they function differently but in a collaborative manner. Each has evolved to perform specialized tasks. Minor chemical differences in these two molecules provide distinct functions.

- The 2-hydroxyl group in the ribose unit of the RNA backbone is replaced in DNA by a hydrogen atom. This group makes the RNA susceptible to base-catalyzed hydrolysis (about 100-fold).
- The covalent structures of RNA and DNA differ in another way: RNA contains uracil whereas DNA contains a methylated uracil derivative termed thymine. This modification also serves to protect the integrity and stability of the genetic material. The methyl group present in thymine facilitates the repair of damaged DNA (conversion of cytosine to uracil by abnormal deamination), providing an additional selective advantage.
- DNA, unlike RNA, is found in cells principally in a double-stranded form, composed of a pair of complementary but antiparallel polynucleotide molecules.
- The double-stranded structure makes DNA in cells more robust and stable than RNA. It also makes DNA relatively easy to replicate and permits a repair mechanism to use the intact strand as a template for the correction or repair of the associated damaged strand.

Although DNA replaced RNA in the role of storing genetic information, the latter maintained many of its other functions. DNA guides the synthesis of specific RNA molecules on the basis of complementary base pairing. This pairing now takes place between slightly different types of nucleotides. The resulting ssRNA molecules then perform two crucial functions: they direct protein synthesis both as coding RNA molecules (mRNA) and as RNA catalysts (ribosomal and other non-messenger RNAs). This implies that RNA preceded DNA in evolution, having both genetic and catalytic properties. Eventually, DNA took over the primary genetic function and proteins became the major catalysts, while RNA remained primarily as the intermediary connecting the two.

1.9.6 Origin of progenote

An undefined and transient stage of organization has been termed **progenote**. The nature of progenote is uncertain and it might have been just a functional, slow replicating, independently evolving organism, or it might have been an amorphous group of primitive transcription units plus a primitive transcription and translation apparatus from which many different cell lineages evolved. The progenote might have contained DNA, genes with introns, slow and heterotrophic nutrition with a few genes, a few proteins and a limiting plasma membrane. These multiplied rapidly in an environment with a copious supply of dissolved organic substances and continued until the store of organic molecules lasted. Eventually, living systems were forced to evolve with the ability to synthesize their own organic requirements from simple inorganic substances. This led to the evolution of autotrophs, which were able to manufacture organic nutrients from simpler ones.

1.10 THE FIRST CELL

The crucial event that led to the formation of the first cell could have been the development of an outer **cell membrane** (plasma membrane). This container acts as a selective barrier that enables the cell to concentrate molecules synthesized inside and nutrients gathered from its environment for its own use, while excreting waste products. For example, the proteins synthesized under the direction of certain RNA molecules would not perpetuate unless they remained in the neighbourhood of the RNA. Moreover, as long as these proteins were free to diffuse among the population of replicating RNA molecules, this RNA could not have a selective advantage over others. Selection of molecules according to the quality of the proteins they generated could not occur efficiently until some form of compartment evolved to contain the proteins made by an RNA molecule and thereby make them available only to the RNA that had generated them (**Fig. 1.12**).

The need for containment is easily fulfilled by another class of molecules that has the simple physicochemical property of being **amphipathic**,

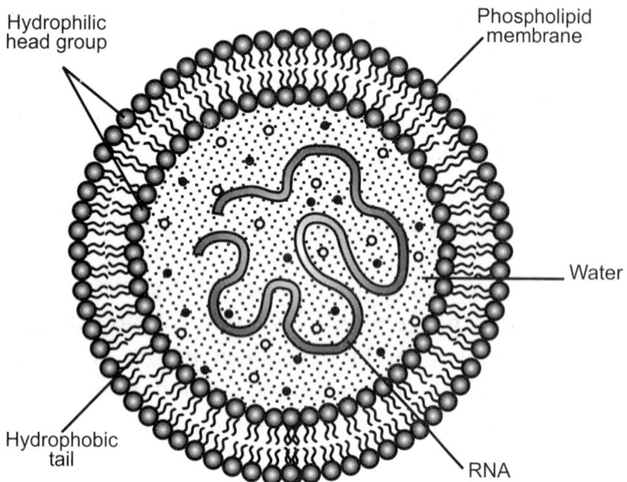

Fig. 1.12 Enclosure of self-replicating RNA in phospholipid membrane.

i.e., consisting of both hydrophobic and hydrophilic parts. In aqueous conditions, amphipathic molecules of appropriate shape spontaneously aggregate to form **bilayers**, creating small closed vesicles whose aqueous contents are isolated from the external medium. Although the chemical details vary, the hydrophobic tails of the predominant membrane molecules in all cells are hydrocarbon polymers. At some stage in evolution, sufficient quantities of appropriate amphipathic molecules must have accumulated from biosynthetic or other processes to allow some nucleic acids to become entrapped and cell-like organisms to form. Presumably, the first membrane-bound cells were formed by spontaneous assembly of phospholipid molecules from the prebiotic soup, enclosing a self-replicating mixture of RNA and other molecules. According to the **bubble theory**, waves breaking on the shore create foam composed of bubbles made up of phospholipids. It is possible that organic molecules were concentrated on the shorelines in much the same way. Oily bubbles happen to be much more stable, lending more time to the particular bubble to perform these crucial experiments. Once RNA molecules were sealed within a closed membrane, they began to evolve as carriers of genetic instructions: they could be selected not merely on the basis of their own structure, but also according to their effect on other molecules in the same compartment. The nucleotide sequences of the RNA molecules could now be expressed in the characteristic of a unitary living cell. The main advantage of compartmentalization is that the products of enzymatic reactions do not diffuse away into the environment.

Despite its many advantages, the membrane compartments can be affected by osmosis. When macromolecules are concentrated inside a compartment surrounded by such a semipermeable membrane (permeable to water and small non-polar molecules, but impermeable to macromolecules), the cells have evolved two distinct mechanisms for overcoming these osmotic forces: introduction of an additional structure such as a **cell wall** to toughen the cell membrane and the use of **energy-dependent ion pumps**. These pumps can lower the concentration of ions inside a cell relative to the outside, facilitating the flow of water from inside to outside. The resulting unequal distribution of ions across an inherently impermeable membrane is called an **ion gradient**. It can prevent osmotic crises, but they require energy. An ATP-driven proton pump was the first existing component of the machinery for generating an ion gradient.

1.11 ORIGIN OF INFORMATION SYSTEMS

Cellular life stores its genetic information in the form of a triplet code in linear or circular DNA molecules. The code itself is essentially universal, although some exceptions are observed in mitochondrial genes. These observations indicate the common ancestry of all life forms. It has also been suggested that an RNA information storage system may have existed earlier and have been replaced later by genetic DNA. Probably the codons that specify the amino acids share an initial doublet and differ only in the third base. In half of the 16 codon groupings, all four members of the codon family specify the same amino acids (Leu, Val, Ser, Pro, Thr, Ala, Arg and Gly). By virtue of the anticodon wobble base in the 5 position, a single tRNA may recognize and interact with two related codons. Once a doublet code was established, it expanded to a triplet code; however, changing of reading frame from doublet to triplet was unlikely. The existence of common initial doublets in codons and of wobble in the tRNA anticodons suggests that the original triplet code may have utilized only the first two bases to specify an amino acid and that the third base may have been irrelevant in coding but important for other reasons.

Why are only 20 amino acids included in the genetic code and many others are not? According to the 'frozen accident' hypothesis, the original allocation of codons was a matter of chance. As life forms became more complex, natural selection would have favoured maintenance of the present code. Too many disturbances in the code might cause changes in structure and metabolism that depend on specific proteins; minor modifications were tolerated, but large-scale changes were not adaptive.

1.12 CELLULAR ENERGY SOURCES

Living organisms obtain their free energy in different ways. Some organisms, for example, lower animals, fungi and the bacteria that live in the human gut, get it by feeding on other living things or the organic chemicals they produce, which are called **organotrophic** (Greek *trophe* for 'food'). Others get their energy directly from the non-living world. These fall into·two classes: those that harvest the energy of sunlight (**phototrophic**, feeding on sunlight) and those that capture their energy from energy-rich systems of inorganic chemicals in the environment (**lithotrophic**, feeding on rock). Organotrophic organisms could not exist without these primary energy converts, which constitute the largest mass of living matter on earth.

1.13 EVOLUTION OF METABOLIC DIVERSITY

The earliest living organisms were simple cells that contained nucleic acids (RNA), enzymes and ribosomes. Because their genetic information was not complex enough to enable them to manufacture all the organic molecules required for metabolism and reproduction, these early **chemical heterotrophs** or **chemotrophs** were totally dependent on abiotic nutrients: the spontaneously synthesized amino acids, sugars, nucleotides and ATP that already existed in the surrounding organic soup. With an abundant supply of these nutrients at first, these early heterotrophs proliferated at a rapid rate. Eventually they consumed the organic molecules faster than abiotic synthesis could replace them, causing competition for limited nutrients. This self-limiting situation did not last long, so cells needed to evolve their own mechanisms of generating energy and synthesizing the molecules necessary for their replication. Some cells evolved the ability to manufacture their own ATP through fermentation, giving them a competitive edge over other heterotrophs that relied solely on spontaneously formed ATP in the soup. But even these fermentative heterotrophs depended on spontaneously formed organic compounds. The only organisms that could be freed from the limitations of abiotically synthesized organic molecules were those that could manufacture their own organic nutrients and energy source and those were the earth's first autotrophs.

Evolution of autotrophs

If autotrophs had not emerged, evolution of life might have ended about 3.5 billion years ago. The first autotrophs not only have supplied themselves with nutrients, but also became a food source for the evolution of the heterotrophs. Without autotrophs, life would have remained simple and sparse, limited by the meagre supply of abiotically synthesized organic food. The earth's first autotrophs were probably very much like present-day anerobic photosynthetic bacteria. They used the energy of sunlight to remove hydrogen atoms from hydrogen gas, hydrogen sulphide, ethanol, or lactic acid. They combined the liberated hydrogen with carbon dioxide to produce organic compounds. These compounds could then be used in glycolysis and fermentation to produce ATP.

But the metabolic success of anerobic photosynthesizers soon gave way to another group of organisms that evolved the ability to extract hydrogen by splitting water, releasing oxygen as a waste product. These new autotrophs were the first **cyanobacteria**, a 3 billion-year-old group of photosynthetic bacteria that continue to flourish even today. Because water was plentiful, these ancient cyanobacteria flourished, tinting the ponds and seas green with their light-capturing photosynthetic pigments (chlorophyll) and releasing vast amounts of oxygen into the water and air. Most other organisms were ill equipped to deal with oxygen's tendency to break down organic compounds and were either driven to extinction or forced into very limited, oxygen-free habitats. Cyanobacteria, thus, have a selective advantage over other organisms by having the ability to utilize carbon and nitrogen atoms directly from the atmosphere.

1.14 EVOLUTION OF METABOLIC REACTIONS

As mentioned earlier, cells are presumed to have originated in a sea of organic molecules (**organic soup**) and were able to obtain food and energy directly from their environment. But such a situation was self-limiting and lasted only for a short period. Therefore, primitive cells were under enormous selection pressure to evolve their own mechanisms for generating energy and synthesizing the molecules necessary for their replication. The generation and efficient harnessing of metabolic energy is central to all cell activities, which explains why the principal pathways of energy metabolism are highly conserved in present-day cells. All cells, for example, use ATP as their source of metabolic energy to drive the synthesis of cell constituents and for other cellular activities. The mechanisms used by cells for the generation of ATP are believed to have evolved in three stages, corresponding to the evolution of

- glycolysis
  $C_6H_{12}O_6 \rightarrow 2C_3H_6O_3 + 2ATP$
- photosynthesis
  $6CO_2 + 6H_2O \rightarrow C_6H_{12}O_6 + 6O_2$
- oxidative metabolism
  $C_6H_{12}O_6 + 6O_2 \rightarrow 6CO_2 + 6H_2O + 36–38\ ATP$.

The development of these metabolic pathways changed the earth's atmosphere significantly, thereby altering the course of further evolution.

A bacterium growing in a simple salt solution containing a single type of carbon source (e.g., glucose) must carry out a large number of chemical reactions to convert glucose into many other organic molecules that the cell requires. These reactions are catalyzed by hundreds of enzymes working in reaction chains (metabolic pathways) so that the product of one reaction is the substrate for the next. However, primitive autotrophs would have required a whole array of enzymes to direct a multistep chain of metabolic reactions as in the synthesis of a protein. Indeed, it would be too much to expect that all the necessary enzymes evolved at the same time. Norman H Horowitz (1959) proposed that the chain of steps in a complicated chemical pathway evolved backward (**retrograde evolution**). According to this concept, an organism might acquire the enzymes required in a biosynthetic pathway in the reverse order from that in which they normally appear in a sequence, by successive mutations. In other words, evolution began with the end product of the pathway and worked backwards, one step at a time, towards the beginning of the reaction chain. The complete pathway became constructed when the organism had tapped from its surroundings all of the intermediate precursors of the reaction.

The first simple autotrophy arose in an anerobic world—one in which little free oxygen was available. Since the early atmosphere on earth was anerobic, therefore, the first energy-generating reactions presumably involved the breakdown of organic molecules (**fermentation**). This anerobic reaction, in addition to providing available energy in the form of ATP, also provided reducing power in the form of NADH that could be used to derive a biosynthetic reaction. This is likely to have been a form of present-day glycolysis (the anerobic breakdown of glucose to lactic acid plus the net energy gain of two ATP). Glycolysis provided a mechanism by which the energy in preformed organic molecules (e.g., glucose) could be converted to ATP, which could then be used as a source of energy to drive other metabolic reactions. The metabolic processes of the anerobic autotrophs resulted in the liberation of a large amount of carbon dioxide (CO_2) into the atmosphere. Once this occurred, the way was paved for the evolution of organisms that could use CO_2 as the sole source of carbon in synthesizing organic compounds and could use sunlight as the only source of energy.

1.15 EVOLUTION OF PHOTOSYNTHESIS AND AEROBIC RESPIRATION

One of the first sunlight-driven reactions was probably the generation of 'reducing power'. The carbon and nitrogen atoms in the atmosphere—CO_2 and N_2—are in an oxidized and inert state. One way to make them more reactive, so that they can participate in biosynthetic reactions, is to reduce them, i.e., to give them a larger number of electrons. This is achieved in several steps. In the first step, electrons are removed from a poor electron donor and transferred to a strong electron donor by chlorophyll in a reaction that is driven by sunlight. The strong electron donor is then used to reduce CO_2 or N_2. Comparison of the mechanisms of photosynthesis in various present-day bacteria suggests that one of the first sources of electrons was H_2S, from which the primary waste product would have been elemental sulphur. Later, the more difficult but ultimately more rewarding process of obtaining electrons from water was accomplished and oxygen was released in large amounts as a waste product.

During its evolution from anerobic heterotrophy, primitive life forms developed modified electron transport pathways that were of utmost significance for the subsequent appearance of eukaryotic organisms. Initially, electron transport systems were modified, particularly, in three important aspects:

- The incorporation of a second photosystem by which water served as an electron donor for photosynthesis
- The release of large amounts of molecular oxygen into the air and water when water molecules were

processed to release their electrons in expanded photosynthetic activity
- The development of a respiratory electron transport pathway in which molecular oxygen behaved as the final electron acceptor.

Modern photosynthesis provides inexhaustible food supply for all of life from green cells and it continuously replenishes atmospheric oxygen for aerobic life, which now predominates on the earth. By aerobic respiration, large amounts of chemical energy become available from a smaller food requirement in aerobes than in anerobes and sufficient energy is conserved in mitochondrial reactions and stored in ATP to subsidize the large organisms and vigorous activity that characterize eukaryotic life.

Respiration evolved independently in other bacterial lineages as well. The appearance of aerobic respiration involving O_2 as the final electron acceptor could have occurred only if there were adequate supplies of dissolved oxygen in water and of gaseous oxygen in the atmosphere. The transition from an anerobic to an aerobic environment must have been dependent largely on the photosynthetic activities of prokaryotic blue green and green algae, which derive electrons from H_2O and release O_2 as a by product. From fossil records, it seems that prokaryotic blue green algal forms existed almost 2.5 billion years ago, which might have been preceded by anerobic bacterial photosynthesizers. Photosynthesis thus evolved early in the history of life on earth. With the change to an oxygenic atmosphere, an ozone layer developed in the upper atmosphere. Ozone (O_3) is produced from oxygen in the presence of the longer wavelengths of ultraviolet (UV) radiation from the sun and from space. The blanket of ozone absorbs much of the germicidal and mutagenic short-wavelength UV radiation, which is harmful to living systems. It is interesting to note in this regard that virtually all organisms have more than one system by which UV-damaged DNA can be repaired, implying these repair systems were adaptive inherited features of ancient life forms that received large amounts of germicidal UV radiation.

1.16 CELLS FIXING NITROGEN, CARBON DIOXIDE AND OTHERS

Living cells require matter in useable form and free energy. Most molecules in living cells contain just six elements: hydrogen, carbon, nitrogen, oxygen, sulphur and phosphorus. These are abundant in the earth's crest, water and atmosphere, but not in chemically available forms that allow easy incorporation into biological molecules. Atmospheric N_2 and CO_2, in particular, are extremely unreactive and a large amount of free energy is required to drive the reactions that use these inorganic molecules to make the organic compounds needed for further biosynthesis. Many types of living cells lack the biochemical machinery to achieve this fixation and rely on other classes of cells for this purpose. Plants can fix CO_2 from the atmosphere, but lack the ability to fix atmospheric nitrogen and they depend in part on nitrogen-fixing bacteria to supply their need for nitrogen compounds. However, plants can harbour symbiotic nitrogen-fixing bacteria in nodules in their roots, which depend on plants mainly for their carbon source. Living things therefore differed widely on the most basic aspects of their biochemistry and developed close associations with others on complementary needs and capabilities. These associations have evolved into diverse forms to the point where the partners have lost their separate identities altogether. These complementary and capable forces have joined to form a single composite cell.

1.17 FROM PROKARYOTIC TO EUKARYOTIC CELLS

The period between 2 and 4 billion years ago is referred to as the age of prokaryotes and three evolutionary lines emerged during this time. One line, the eukaryotes, gave rise to a new type of cell—the **eukaryotic cell**—which set the stage for the evolution of multicellular life (see Chapter 5). **Prokaryotic cells** (see Chapter 4) do not have a true nucleus in which the DNA is covered with a membrane. In contrast, eukaryotic cells contain a **true nucleus** which is enclosed in a membranous envelope. These membrane enclosures might have protected genes and protein products from competitors. With specialized organelles and a nucleus that sequestered genetic material, eukaryotic cells were more organized than their prokaryotic ancestors and were capable of conducting more complex functions. Meiosis is one such complex function performed by eukaryotic cells. It leads to the formation of gametes for sexual reproduction, a process that accelerates the rate of evolution by producing new gene combinations and spreading them quickly through a population. The ability of eukaryotes to reproduce sexually facilitated quick proliferation.

Early eukaryotes left no fossilized remains; therefore, speculations on how eukaryotic cells evolved from simpler prokaryotic ancestors are based on the characteristics and lifestyles of modern unicellular eukaryotes. There are a few modern theories that could explain how this happened.

The general agreement scientists have had for decades is that prokaryotic cells arose before eukaryotic cells and they gave rise to eukaryotic cells. Until the last quarter of the 19th century, it was believed that eukaryotic cells evolved from prokaryotic ones by a process of gradual evolution. However, acceptance of this concept dramatically changed when Lynn Margulis proposed that the mitochondria and chloroplasts had evolved from smaller prokaryotic cells that had taken up residence in the cytoplasm of a larger host cell. This hypothesis is referred to as the **endosymbiont theory**. An alternative mechanism proposed was the **membrane invagination** theory.

1.17.1 Endosymbiosis theory

The endosymbiosis theory (Margulis's theory) proposes that some organelles were once free-living prokaryotic cells that took up residence inside a larger prokaryotic cell, forming a union that was beneficial to both (**Fig. 1.13**). There are plenty of examples in nature: eukaryotic cells carrying mitochondria, resemble aerobic bacteria (i.e. *Paracoccus*) in size and structure. In such mitochondrial structures, the inner membrane is like a bacterial plasma membrane and replicates independently. It has its own DNA and contains genes for coding a few proteins required for specialized tasks. Another example of the endosymbiotic theory are the chloroplasts. Chloroplasts may have descended from aerobic eubacteria that engage in oxygen-producing photosynthesis (e.g., blue-green algae). Such cells may have been engulfed and may have resisted digestion. By providing their respiring host cell with oxygen, their endosymbiotic existence would have been favoured. And they could have evolved into chloroplast eventually. In metabolism and overall nucleic acid sequence, chloroplasts resemble some eubacteria. They are self-dividing and have their own DNA. The endosymbiosis theory is supported by strong evidence:

- Some bacteria form symbiotic partnerships with eukaryotic cells, which can be seen even today. These bacteria resemble mitochondria in structure and carry out the same steps of aerobic respiration as mitochondrial.
- Mitochondria and plastids contain their own nucleic acids, including a circular strand of DNA that closely resembles the chromosomes of prokaryotic cells.
- Mitochondria and plastids can 'divide' independently in the host cell. Because these organelles contain their own genetic information, replication is not solely under the control of the nucleus as it is for other organelles.
- Like an independent cell, a mitochondria and chloroplast contain their own ribosomes. These ribosomes resemble those of prokaryotic bacteria more than those found in eukaryotic cells.
- Chloroplasts have the identical arrangement of pigments as found in some photosynthetic prokaryotes. Electron transport chains in chloroplasts are also identical to those of photosynthetic bacteria.

Along with host cell reproduction, the endosymbiont also reproduces and forms a colony of these composite cells. Over many generations, endosymbionts have lost many of the traits that were no longer required for survival and the once-independent oxygen-respiring microbes evolved into precursors of the modern-day mitochondria. Similarly, the free-living photosynthetic bacteria, cyanobacterium, could have converted an early heterotrophic eukaryote into an ancestor of photosynthetic eukaryotes: the green algae and plants which contain chloroplast in the cells. Therefore, biologists accept the endosymbiosis theory as the explanation for the origin of possibly all the organelles in eukaryotic cells. Even eukaryotic flagella (also cilia) show evidence that they were once independent prokaryotic organisms. The ancestors of present-day cells, through the sequence of symbiotic events have given rise to a line of cells that evolved other basic characteristics of the eukaryotic cells. It includes a system of membranes (a nuclear envelope, endoplasmic reticulum, Golgi complex, lysosomes), a complex cytoskeleton and a mitotic type of cell division. All these subcellular structures are proposed to have arisen by gradual evolution.

1.17.2 Membrane invagination theory

The membrane invagination theory proposes that the plasma membrane of the ancestral prokaryotic cells gradually folded in on itself, forming pockets that isolated specific metabolic reactions (**Fig. 1.14**). Some of these pockets pinched off, partitioning function into membrane-enclosed organelles. Over long periods,

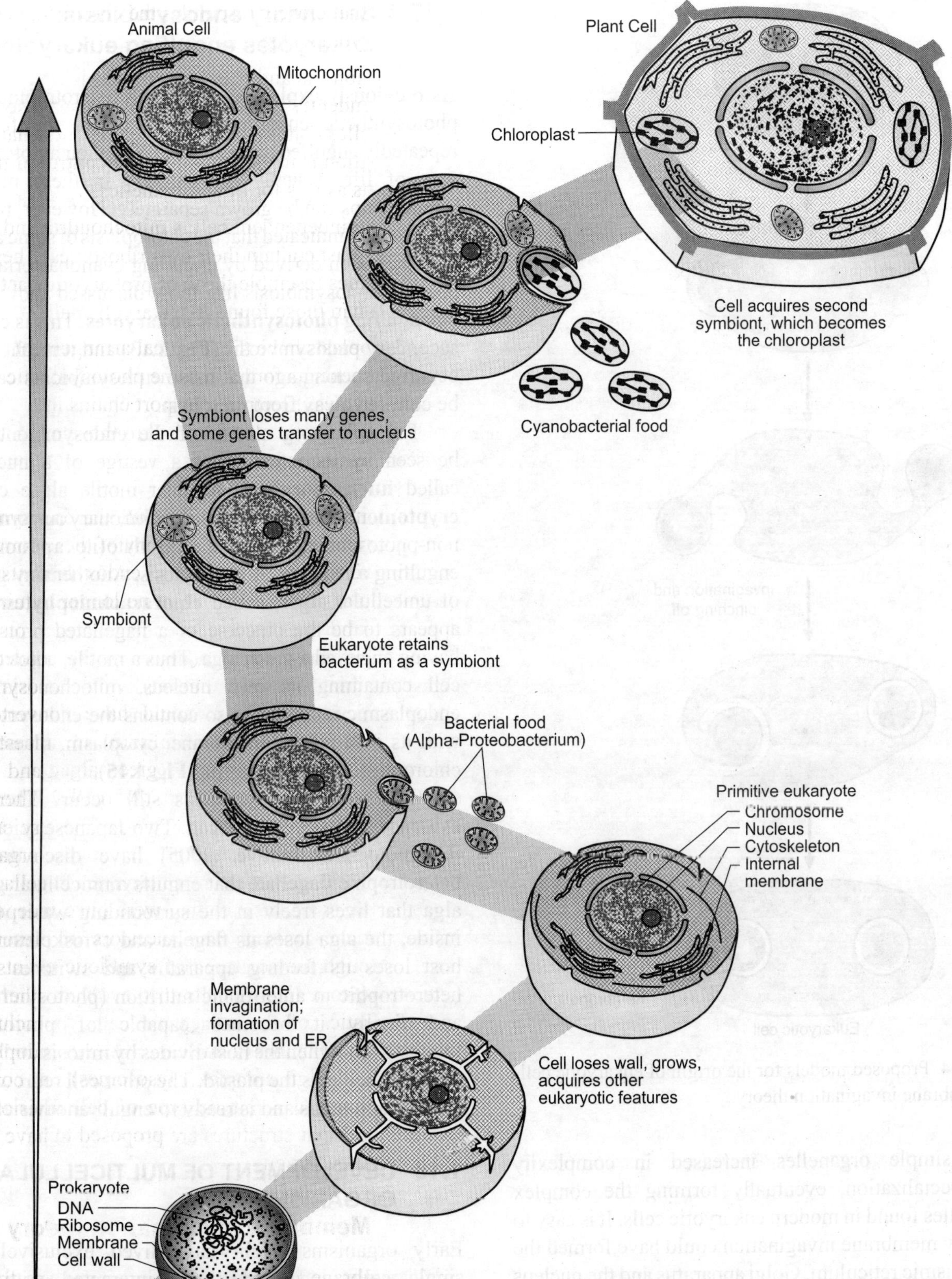

Fig. 1.13 Proposed models for the origin of eukaryotic cells by endosymbiosis theory.

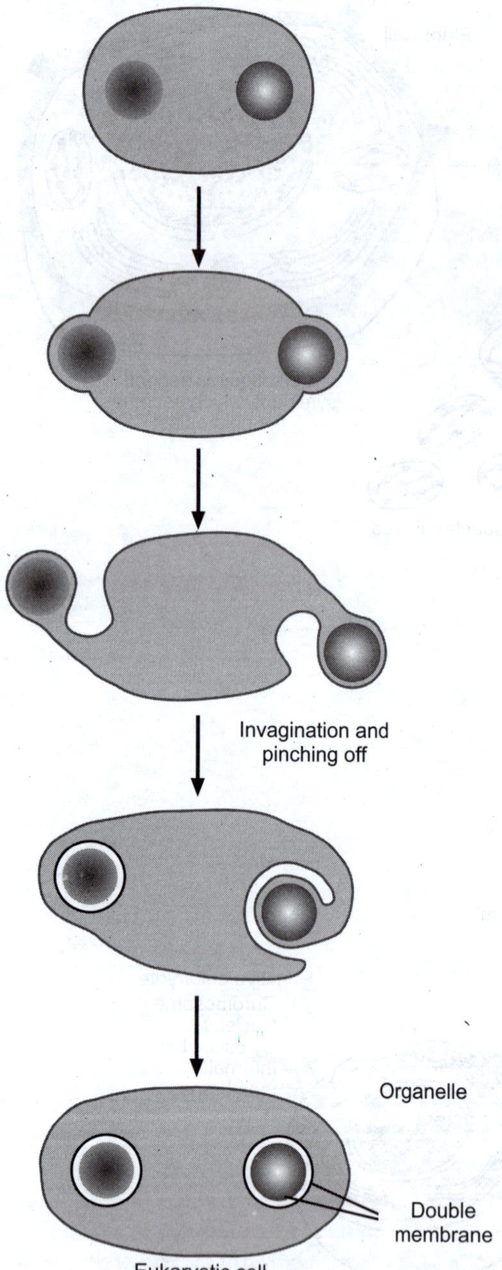

Fig. 1.14 Proposed models for the origin of eukaryotic cells by membrane invagination theory.

these simple organelles increased in complexity and specialization, eventually forming the complex organelles found in modern eukaryotic cells. It is easy to see how membrane invagination could have formed the endoplasmic reticulum, Golgi apparatus and the nucleus by the same infolding process. The nuclear membrane of the eukaryotic cell is often connected to the cell membrane. The evolution of complex organelles such as chloroplasts and mitochondria, however, is usually explained by another theory, the endosymbiosis theory.

1.17.3 Secondary endosymbiosis: Eukaryotes engulfing eukaryotes

As previously explained, once both heterotrophic and photosynthetic eukaryotes had evolved, the former repeatedly engulfed the latter to exploit their autotrophic way of life. Usually the partners in these mutual relationships can be grown separately. However, recent studies have indicated that the chloroplasts of some algae have not been derived by engulfing cyanobacteria in a primary endosymbiosis like those discussed above, but by engulfing **photosynthetic eukaryotes**. This is called secondary endosymbiosis (**Fig. 1.13**) and it might have occurred so long ago that these endosymbionts cannot be cultured away from their host.

The eukaryotic nature of the endosymbiont can be seen by its retention of a vestige of a nucleus, called **nucleomorph**. Unicellular motile algae called **cryptomonads** appear to be the evolutionary outcome of a non-photosynthetic eukaryotic flagellate (i.e., a protozoan) engulfing a red alga by endocytosis. Another tiny group of unicellular algae, called **chlorarachniophytes**, also appears to be the outcome of a flagellated protozoan having engulfed a green alga. Thus a motile, autotrophic cell containing its own nucleus, mitochondria and endoplasmic reticulum, also contains the endosymbiont with its own plasma membrane, cytoplasm, ribosomes, chloroplast and nucleomorph (**Fig. 1.15**).

Can secondary symbiosis still occur? There is evidence to believe that it can. Two Japanese scientists (Okamoto and Inouye, 2005) have discovered a heterotrophic flagellate that engulfs a unicellular green alga that lives freely in the surrounding water. Once inside, the alga loses its flagella and cytoskeleton, the host loses its feeding apparatus and switches from heterotrophic to autotrophic nutrition (photosynthesis) and finally it becomes capable of phototaxis. Interestingly, when the host divides by mitosis, only one daughter cell gets the plastid. The other cell regrows the feeding apparatus and is ready to engulf another alga.

1.18 DEVELOPMENT OF MULTICELLULAR ORGANISMS

Early organisms might have lived exclusively as single cells and could have interacted with one another indirectly by competing for resources in their environments. Some of these organisms, however, developed the ability to form colonies comprising many interacting cells. In such groups, the environment of a cell is dominated by the presence of surrounding cells,

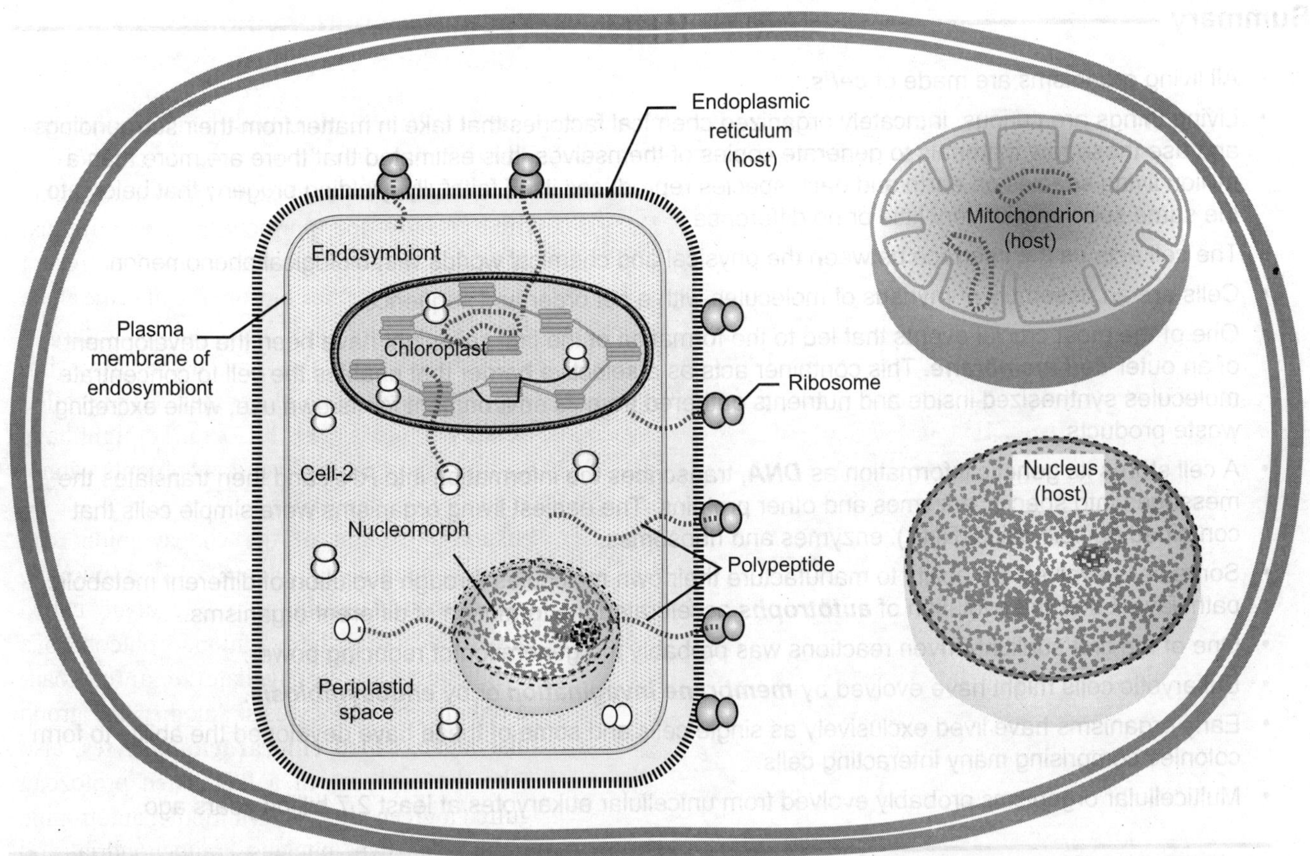

Fig. 1.15 Secondary endosymbiosis: Eukaryotes engulfing eukaryotes. The cell contains four genomes: its own nuclear genome, own mitochondria, chloroplast and the nucleomorph.

which may be in direct contact with one another. These cells communicate with one another by a variety of signalling mechanisms and may respond to signals by altering enzyme activity of levels of gene expression. One event may be a process of **cell differentiation**. Many organisms are able to switch back and forth from existence as independent single cells to existence as multicellular colonies of differentiated cells.

Multicellular organisms probably evolved from unicellular eukaryotes at least 2.7 billion years ago. Some unicellular eukaryotes form multicellular aggregates that appear to represent an evolutionary transition from single cells to multicellular organisms. For instance, the cells of many algae (e.g., *Volvox*, green algae) associate with each other to form multicellular colonies, which are believed to have been the evolutionary precursors of present-day plants. Increasing cell specialization then led to the transition from colonial aggregates to truly multicellular organisms. Continuing cell specialization and division of labour among the cells of an organism have led to the complexity and diversity.

Another example is the slime mold, *Dictyostelium*. In favourable environments, it lives as individual cells; under conditions of starvation, however, the cells come together to form a cell aggregate. This aggregate, sometimes called a **slug**, can move as a unit to a more favourable environment where it then forms a multicellular structure, termed a **fruiting body**, that rises substantially above the surface on which the cells are growing. Wind may carry cells released from the top of the fruiting body to sites where the food supply is plentiful. The transition from unicellular to multicellular growth is triggered by cell–cell communication and reveals much about signalling processes between and within cells. Under starvation conditions, *Dictyostelium* cells release the signal molecule, cyclic AMP or cAMP. This molecule signals by binding to a membrane-bound protein receptor on the cell surface. The binding of cAMP to these receptors triggers several responses, including movement in the direction of higher cAMP and aggregation of cells. The differentiation and diversity of plant and animal cells have been discussed in Chapters 12–17.

Summary

- All living organisms are made of ***cells***.
- Living things are curious, intricately organized chemical factories that take in matter from their surroundings and use these raw materials to generate copies of themselves. It is estimated that there are more than a million living species on earth and each species reproduces itself faithfully, yielding progeny that belong to the same species with very little or no difference.
- The cell acts as the interface between the physical and chemical worlds with biological phenomenon.
- Cells are an assembly of myriads of molecules with a set order and sequence.
- One of the most crucial events that led to the formation of the first cell might have been the development of an outer ***cell membrane***. This container acts as a selective barrier that enables the cell to concentrate molecules synthesized inside and nutrients gathered from its environment for its own use, while excreting waste products.
- A cell stores its genetic information as ***DNA***, transcribes the information into RNA and then translates the messages into specific enzymes and other proteins. The earliest living organisms were simple cells that contained nucleic acids (RNA), enzymes and ribosomes.
- Some cells evolved the ability to manufacture their own molecules through evolution of different metabolic pathways. Later the evolution of ***autotrophs*** accelerated the formation of different organisms.
- One of the first sunlight-driven reactions was probably the generation of reducing power.
- Eukaryotic cells might have evolved by ***membrane invagination*** or by ***endosymbiosis***.
- Early organisms have lived exclusively as single cells and some of these have developed the ability to form colonies comprising many interacting cells.
- Multicellular organisms probably evolved from unicellular eukaryotes at least 2.7 billion years ago.

Exercises

1. What are the different theories that have been developed to explain the origin of life?
2. Explain the 'Big Bang' theory.
3. How do scientists visualize primitive earth's atmosphere?
4. Explain Miller's experimental evidence of chemical evolution with a diagram.
5. Give a brief outline of the origin of genetic information.
6. The membrane defined the first cell. Explain with diagrams.
7. Explain the mechanism of cell organelle formation by endosymbiosis.

Further Reading

Books

Anonymous (1985), *Life–How did it get here? By evolution or by creation?* Watch Tower Bible and Tract Society of New York, Inc., International Bible Students Association, Brooklyn, New York, USA.

Campbell NA (1990), *Biology*, 2nd edition, The Benjamin/Cummings Publishing Company, Inc.

Cooper GM and Hausman RE (2007), *The Cell: A Molecular Approach*, 4th edition, ASM Press and Sinauer Associates, Inc., USA.

Davis P (1999), *The Fifth Miracle: The Search for the Origin and Meaning of Life*, Simon and Schuster, New York.

Hartwell L, Hood L, Goldberg ML Reynolds A, Silver LM and Veres RC, (2000), *Genetics: from Genes to Genomes*, McGraw-Hill, Boston.

Margulis L (2000), *What is Life?*, University of California Press, Berkeley.

Margulis L and Sagan D (1997), *Microcosmos: Four Billion Years of Evolution from Our Microbial Ancestors*, University of California Press, Berkeley.

McKee T and McKee JR (2003), *Biochemistry: The Molecular Basis of Life*, 3rd edition, McGraw Hill Companies, Inc.

Wilson EO (1992), *The Diversity of Life*, The Belknap Press and Harvard University Press, Cambridge, Massachusetts.

References

Avery DT, MacLeod CM and McCarty M (1944), Studies on the chemical nature of the substance inducing transformation of pneumococcal types. Induction of transformation by a deoxyribonucleic acid fraction isolated from pneumococcus type III, *Jour. Exp. Med.*, 79:137–258.

Baldauf SL (2003), The deep roots of eukaryotes, *Science*, 300:1703–6.

Bartel DP and Unrau PJ (1999), Constructing an RNA world. *Trends Cell Biol.*, 9:M9–M13.

Benton MJ and Ayala FJ (2003), Dating the tree of life, *Science*, 300:1698–1700.

Cech T (1986), RNA as an enzyme, *Sci. Am.*, 255:64–75.

Crick FHC (1970), Central dogma of molecular biology, *Nature*, 227:561–63.

Des Marais DJ and Walter MR (1999), Astrobiology exploring the origins, evolution and distribution of life in the universe, *Annu. Rev. Ecology and Systematics*, 30:397–420.

Doolittle WF (2000), Uprooting the tree of life, *Sci. Am.*, 282:72–77 (Feb.).

Fraenkel–Conrat H and Singer B (1957), Virus reconstitution: Combination of protein and nucleic acid from different stains, *Biochem. Et Biophys. Acta*, 24:540–48.

Gadgil M (1996), Life: Complexity and diversity, *Resonance*, 1:14–19.

Griffith F (1928), The significance of pneumococcal types, *Jour. Hygiene*, 127:113–59.

Hershey AD and Chase M (1952), Independent functions of viral protein and nucleic acid in growth of bacteriophage, *Jour. Gen. Physiol.*, 36:39–56.

Kasting JF and JL Siefert (2002), Life and the evolution of earth's atmosphere, *Science*, 296: 1066–68.

Miller SL (1953), A production of amino acids under possible primitive earth conditions, *Science*, 117:528–29.

Orgel LE (1994), The origin of life on the earth, *Sci. Am.*, 271:77–83.

Orgel LE (1998), The origin of life: A review of facts and speculations, *Trends Biochem. Sci.*, 23:491–95.

Orgel LE (2000), Origin of life. A simpler nucleic acid,. *Science*, 290:1306–7.

Rees M (1999), Exploring our universe and others, *Sci. Am.*, 44–49 (Dec.).

Szathmary E (1999), The origin of the genetic code: Amino acids as cofactors in an RNA world, *Trends Genet.*, 15:223–29.

Wilson AC, Carlson SS and White TJ (1977), Biochemical evolution, *Annu. Rev. Biochem.*, 46:573–639.

2

ATOMIC BASIS OF LIFE

Learning objectives

After reading this chapter, you will be able to:

- Understand how the atomic structure and subatomic components influence biological processes
- List the various atomic reactions that underly biological reactions

2.1 INTRODUCTION

The basic chemical and physical properties and the drive to formulate biological information exist within the atom. The biological phenomenon is purely a hierarchical manifestation of inherent properties of different atoms. The properties of cells and their organelles derive directly from the activities of the molecules of which they are composed and their respective atoms. Thus, probing the molecular basis of life is the first step in understanding biology. It reveals how different atoms can be arranged and rearranged to form an unlimited variety of organisms and provide the driving force required for all living organisms on Earth.

2.2 FINE STRUCTURE OF ATOMS

The units of matter are called **atoms**, a term derived from Greek, which means 'indivisible'. Each element consists of a unique atom, which is different from the atoms of any other element. The symbols for the atoms are mostly derived from the name of the element. Thus, C stands for both the element carbon and a single atom of it. An atom is the simplest possible representation of an element. Beyond doubt atoms are the basis for all the complex forms of matter.

All life forms are composed of various kinds of atoms. Remarkably, every atom in the universe consists of three types of subatomic particles:

- ***Protons***, which are positively charged
- ***Neutrons***, which carry neutral or no charge
- ***Electrons***, which are negatively charged.

These three kinds of minute particles combine to form atoms, the fundamental units of matter that can enter into chemical reactions (**Fig. 2.1**). Every atom has a central nucleus that contains one or more protons. However, the atom will be unstable if the nucleus has only protons in its nucleus, because the protons repel each other. To make the atom stable, neutrons are present in the nucleus between the protons. The neutrons are electrically neutral (uncharged) and keep protons apart from each other, thus preventing proton repulsion. Since neutrons are uncharged, the atomic nucleus, therefore always has a net positive charge and carries one unit of positive charge for each proton present. Smaller but, negatively charged electrons, orbit around the nucleus. The protons and neutrons are also called **baryons** and they are collectively known as **nucleons**. Recently, it has been suggested that baryons are made up of even more elementary particles called **quarks** (**Table 2.1**). Quarks are of six types—Up, Down, Strangeness, Charm, Bottom or Beauty and Top or Truth—and are also called **flavours**. But only three pairs of quarks can exist in each baryon. Interestingly, all six quarks have their anti-quarks with charges opposite in value to their quark counterparts. In an atom, particles with opposite charges of equal magnitude neutralize each other (one positive charge

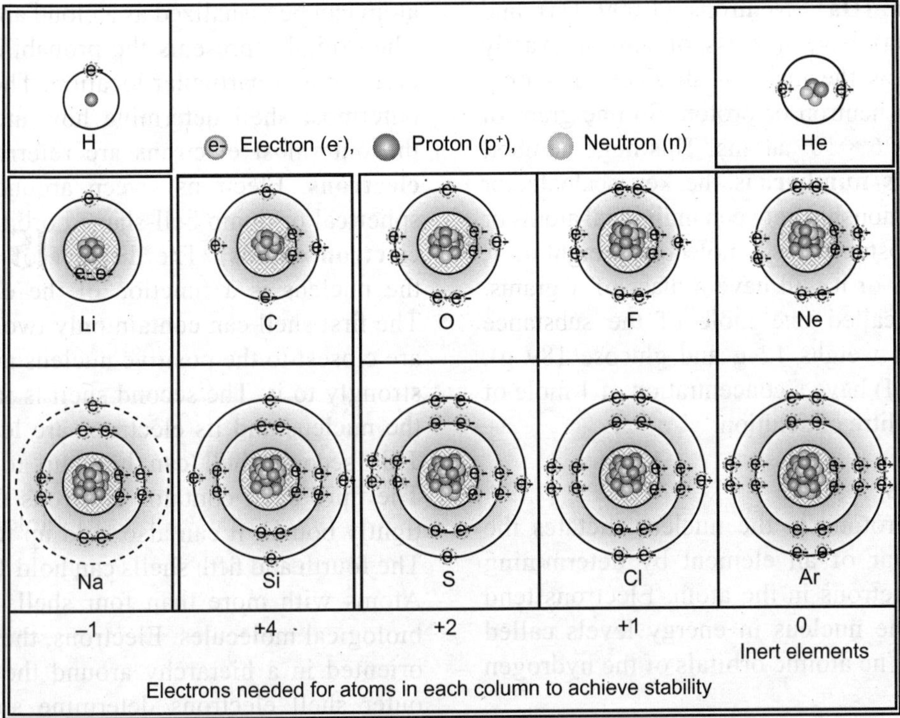

Fig.2.1 Representation of the arrangement of electrons in a number of common atoms.

Table 2.1 Elementary particles or subatomic particles

	Baryons /Nucleons		
Property	**Proton**	**Neutron**	**Electron**
Symbol	p^+	n	e^-
Mass (kg)	1.673×10^{-27}	1.675×10^{-27}	9.109×10^{-31}
Mass (MeV)	938.2	939.6	0.51
Electric charge	+1	0	–1
Quark triplets	2 Up (u) + Down (d)	2 Down (d) + Up (u)	
Quark charges	+(2/3) + (2/3) –(1/3) = +1	–(1/3) –(1/3) +(2/3) = 0	

Note: 1 MeV = 1 million electron volts = 1.6×10^{-13} J = 1.8×10^{-30} kg

of proton cancels one negative charge of the electron). Atoms have equal numbers of protons and electrons and are electrically neutral (inert) and stable.

The number of protons in the atom is called atomic number, which determines its identity and its chemical properties. Atomic numbers of naturally occurring atoms range from one (for hydrogen, the smallest) to 92 (for uranium, the largest). There are also 14 more elements (93 through 106) that have been synthesized. Each element also has a characteristic atomic weight that is roughly equivalent to the total number of protons and neutrons in the nucleus. Electrons have little effect on the total weight of an atom (since 1840 electrons are needed to add one unit of mass). Two or more elements can combine in a fixed ratio to produce a compound. Table salt, for example, is actually sodium chloride (NaCl), a compound composed of the elements sodium (Na) and chlorine (Cl), but O_2, H_2 and N_2 are not compounds, since they contain only a single element.

Atoms are so small that it is not possible even to imagine their size. One proton or a neutron weighs approximately $1/(6 \times 10^{23})$ g. Grams and other conventional units are not very useful for describing the mass of miniscule objects. Thus, for atoms and subatomic particles, scientists use a unit of measurement called the **dalton (Da)**. Neutrons (1.009 Da) and protons (1.007 Da) have a mass of almost exactly 1 Da each, whereas the mass of an electron is only about 1/2000 of a neutron or proton. So one gram of hydrogen contains 6×10^{23} atoms. This huge number, called **Avogadro's number,** is the key scale factor describing the relationship between individual atoms or molecules. If a substance has a molecular weight of *X*, 6×10^{23} molecules of it will have a mass of *X* grams. This quantity is called one mole of the substance (1 mole of carbon weighs 12 g and glucose 189 g). Molar solutions (M) have a concentration of 1 mole of the substance in 1 litre of solution.

Electron orbitals

The number of protons in the nucleus dictates the chemical behaviour of an element by determining the number of electrons in the atom. Electrons tend to orbit around the nucleus in energy levels called **shells (Fig. 2.2)**. The atomic orbitals of the hydrogen

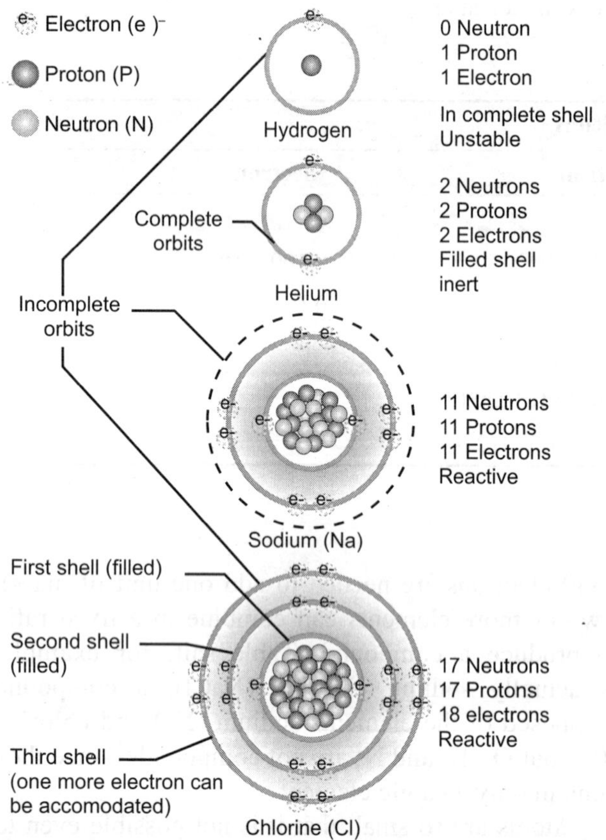

Fig.2.2 Filled and unfilled electron shells in some common elements. Elements with unfilled outermost shells are only found in living organisms and can participate in chemical reactions.

atom can be visualized as a cloud around the nucleus. The orbital represents the probability of finding the electron at a particular location. The electrons in the outermost shell determine how atoms interact and the outermost electrons are referred to as **valence electrons**. Electrons sweep around the nucleus in spherical or dumb bell-shaped orbits (often called as **electron clouds**). The distance of an electron from the nucleus is a function of the electron's energy. The first shell can contain only two electrons; which are closest to the positive nucleus and attracted most strongly to it. The second shell is farther away from the nucleus and its electrons are less tightly bound. This second shell can hold up to eight electrons. The third shell contains electrons that are even less tightly bound; it can also hold up to eight electrons. The fourth and fifth shells can hold 18 electrons each. Atoms with more than four shells are very rare in biological molecules. Electrons, therefore, tend to be oriented in a hierarchy around the nucleus and the outer shell electrons determine some fundamental chemical properties of the atom. The shells of atoms of certain elements are filled completely and these elements do not enter into any chemical reactions; they are called **inert elements**. They have completely filled up outer shells. Hence, they are stable and do not enter into reactions easily. For example, neon is an inert element because its outer shell contains the maximum of eight electrons. By contrast, a carbon atom, with six electrons has two electrons in its first shell and only four in the second and outer shell (**Fig. 2.3**). For this reason carbon is extremely active chemically and forms innumerable combinations with other atoms.

2.3 ORGANIZATION OF MATTER

The three forms of matter—solids, liquids and gases—are due to the physical and chemical properties of atoms and the amount of energy absorbed from the surroundings. The electronic arrangement of atoms determines how they combine and the chemical properties of the resultant compound. In fact, a compound has characteristics quite different from those of the combining elements.

Solids have a definite volume and rigid properties, mainly because the atoms are held together by strong chemical bonds arranged in a three-dimensional network. These forces are weaker in fluids and hence liquids and gases have no fixed shape but only definite volume as they fill up the available space.

Atomic number	Element	Electron distribution			
		Shell I	Shell II	Shell III	Shell IV
1	Hydrogen	●			
2	Helium	○○			
6	Carbon	○○	●●●●		
7	Nitrogen	○○	●●●●●		
8	Oxygen	○○	●●●●●●		
10	Neon	○○	○○○○○○○○		
11	Sodium	○○	○○○○○○○○	●	
12	Magnesium	○○	○○○○○○○○	●●	
15	Phosphorous	○○	○○○○○○○○	●●●●●	
16	Sulphur	○○	○○○○○○○○	●●●●●●	
17	Chlorine	○○	○○○○○○○○	●●●●●●●	
19	Potassium	○○	○○○○○○○○	○○○○○○○○	●
20	Calcium	○○	○○○○○○○○	○○○○○○○○	●●

Fig. 2.3 Filled and unfilled electron shells in some common elements. Elements with unfilled outermost shells (showing filled circles), only found in living organisms, can participate in chemical reactions. The number of circles (both filled and unfilled) in each row represents the total number of electrons of an element. Helium and neon are inert elements (shaded rows), since their shells are completely filled with electrons.

2.4 ATOMIC INTERACTIONS

There are two types of atomic interactions—between different atoms (inter-atomic) and within the same atom (intra-atomic). Neutrons with neutral charge provide half the atomic mass and the positively charged protons make up the other half. Protons also regulate the number and position of the negatively charged electrons. It is the electrons that influence bond formation. The overall charge of an atom depends on the number of protons and number of electrons and the ratio between the protons and electrons determines whether an atom is negatively charged or positively charged or neutral. Only those atoms that have imbalance in charge due to the differences in the number of protons and electrons and also the variation in position or distribution of electrons, participate in evolution of higher order molecules. This also reveals that the vital force required for the seed of life to be generated is primarily from intra-atomic interactions.

As mentioned, **inter-atomic interactions** are the basis for the formation of molecules and their interactive properties (inter-atomic interactions)with other molecules. These interactions determine the structural and functional ability of a molecule. The structural organization and functions of all the biomolecules including water, amino acids, proteins, nucleic acids, sugars and lipids are determined by intermolecular interactions. Therefore to understand, for example, how the protein structure might catalyze a reaction, we need to understand enough about chemical reaction mechanisms to develop a hypothesis about how the reaction might work, given the shape of the protein and the location of various amino acids with specific intermolecular interactions. Protein association is often mediated by the electrostatic (**intra-atomic interactions**) properties of the protein structure; interactive molecules can be drawn together over considerable distances by strong electrostatic potentials. Within protein structures, hydrogen bonds and other inter-atomic interactions confer structural stability (**Fig. 2.4**). A study of inter-atomic forces such as covalent bonds, hydrogen bonds, hydrophobic and hydrophilic interactions, van der Waals forces, charge–charge, charge–dipole and dipole–dipole interactions, repulsive forces, relative strength of inter-atomic forces, helps in understanding the complexity and functions of proteins.

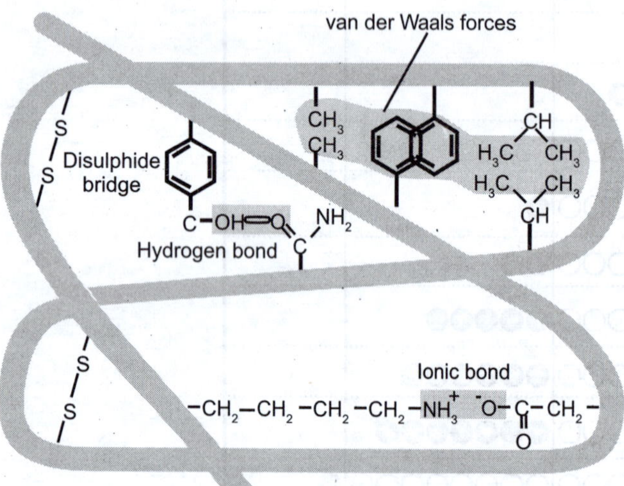

Fig. 2.4 Types of non-covalent bonds maintaining the conformation of proteins.

2.5 ELEMENTS THAT CONSTITUTE LIFE

The earth's crust contains approximately 100 elements and yet only 16 of these are essential for life (**Table 2.2**) and about 30 of them are present in living cells. Most of the elements in living mater have relatively low atomic numbers. ***The four common elements in living organisms account for more than 96% of the mass (dry weight)***. The biological importance of H, O, N and C is largely due to their having valencies of 1, 2, 3 and 4 respectively and their ability to form more stable covalent bonds than any other elements with these valencies. These limited numbers of atoms combine to form biomolecules, the building blocks of life. These molecules vary enormously in size from simple molecules such as carbon dioxide and water to macromolecules such as proteins. The larger molecules tend to be used for storage, structural and information use. The smaller molecules are soluble, easily transported and frequently enter into the general chemical activity of cells known as metabolism. Of the smaller molecules, water is the most abundant, typically making up between 60%–95% of the fresh mass of an organism.

2.6 CHEMICAL REACTIONS AND MOLECULAR FORCES

Life is a chemical process involving hundreds and thousands of reactions occurring in an organized manner, which are called **metabolic reactions** (collectively called as **metabolism**). Most atoms tend to establish stable partnerships with other atoms, forming larger, multiatomic complexes called **molecules**. Two or more atoms can team up to form a molecule that is more stable (less reactive) than the individual atoms. The formation of a molecule is governed by a fundamental principle—an atom is most stable when its outermost electron shell is filled. Atoms join together by sharing electrons or by a complete transfer of electrons. This sharing or transfer creates the chemical bond that keeps the atoms together.

Chemical bonds are characterized in several ways. An obvious characteristic of a bond is its strength (**Table 2.3**). There are two kinds of association between atoms and molecules: strong covalent bonds and weaker non-covalent bonds. Strong bonds

Table 2.2 Elements found in living organisms

Major elements	%*	Ions of elements	%*	Trace elements	
Hydrogen (H)	8	Sodium (Na^+)	0.03	Iodine (I)	Boron (B)
Carbon (C)	50	Magnesium (Mg^{2+})	0.01	Iron (Fe)	Aluminum (Al)
Nitrogen (N)	14	Chloride (Cl^-)	0.08	Cobalt (Co)	Silicon (Si)
Oxygen (O)	20	Potassium (K^+)	0.06	Copper (Cu)	Vanadium (V)
Phosphorus (P)	3	Calcium (Ca^{2+})	0.31	Zinc (Zn)	Molybdenum (Mo)
Sulphur (S)	1			Manganese (Mn)	Selenium (Se)
				Fluorine (F)	

***Dry weight**

Table 2.3 Bonds and forces

Type of bond	Example	Comments
Covalent bond	–C–C–, –C=C–, N=N bonds in many carbohydrates, proteins and in fatty acids	Very strong; bond energy of carbon–carbon bond is around 350 kJ/mol
Ionic bond	$–CO_2^-$ on glutamyl residue and $–NH_3–$ on lysyl residue	Potentially very strong, depending upon the dielectric constant
Hydrogen bond	C=O and NH in adjacent peptide bonds	Bond energy is about 20 kJ/mol; forms because oxygen is electronegative creating dipoles
Hydrophobic effects	Tendency of hydrophobic molecules to keep away from aqueous phase	Placing hydrocarbons in an aqueous medium requires energy to change water structure
Van der Waal's forces	All atoms have a weak tendency to stick together over short ranges	Results from electrostatic attraction of positively charged nucleus of one atom and negatively charged electrons of the other; weak and bond energy is about 4 kJ/mol

almost never fall apart at physiological temperatures. Covalent bonds hold the atoms within an individual molecule together. Weak bonds are easily broken. ***Non-covalent bonds determine the three-dimensional architecture of large biological molecules and molecular complexes through cooperation: although a single non-covalent bond is not strong, the effect of many weak bonds functioning together can be very powerful***. Non-covalent bonds are more easily broken, making them the basis of many dynamic biological processes. The bond energy—the energy required to break the bond—is comparative. The strength of a bond is correlated with its length. All chemical bonds, weak as well as strong, are based on electrostatic forces. The number of covalent bonds that an atom can form is called its **valency**. For example, hydrogen has a valency of one and oxygen has a valency of two. An angle between two bonds originating from a single atom is called the **bond angle**. Bonds also differ in the **freedom of rotation** they allow. Single covalent bonds permit free rotation of bound atoms, whereas double and triple bonds are quite rigid (**Fig. 2.5**).

The spontaneous formation of a bond between two atoms always involves the release of some of the internal energy of the unbounded atoms and its conversion to another energy form. It is called the **bond energy**. The stronger the bond, the greater the amount of energy released upon its formation. The amount of energy that must be added to break a bond is exactly equal to the amount that was released during the formation of the bond. This equivalence follows from the first law of thermodynamics, which states that energy can be neither made nor destroyed. Every bond is thus a result of the combined actions of bond-making and bond-breaking forces.

$$K_{eq} = \frac{[AB]}{[A] \times [B]}$$

Where K_{eq} is the *equilibrium constant* and [*A*], [*B*] and [*AB*] are molar concentrations of the substrate and the product.

2.6.1 Chemical bonds

In most cases, atoms can achieve stability by sharing pairs of their valence electrons through a process called covalent bonding (forming a linkage). In a covalent bond, two atoms are held close together, because, in their outermost orbital electrons are shared by both the atoms. For example, the hydrogen molecule consists of two hydrogen atoms held together by a covalent chemical bond (H–H). Atoms can also be joined by double covalent bonds (–C=C–) on even triple covalent bonds (N≡N). The majority of bonds in organic compounds are covalent bonds. Some atoms can form more than four covalent bonds, but with different atoms nitrogen and phosphorous each have five electrons in their outer shells, which can hold up to eight electrons. In ammonia (NH_3), the nitrogen atom forms three covalent bonds. In the ammonium ion (NH_4^+), the nitrogen atom forms four covalent bonds. Phosphorous can form up to five covalent bonds, as in phosphoric acid (H_3PO_4).

Esters of phosphoric acid form the backbone of nucleic acids. Phosphates play a key role in cellular energetics and in the regulation of cell function. The difference between the bonding patterns of nitrogen and phosphorus is primarily due to the

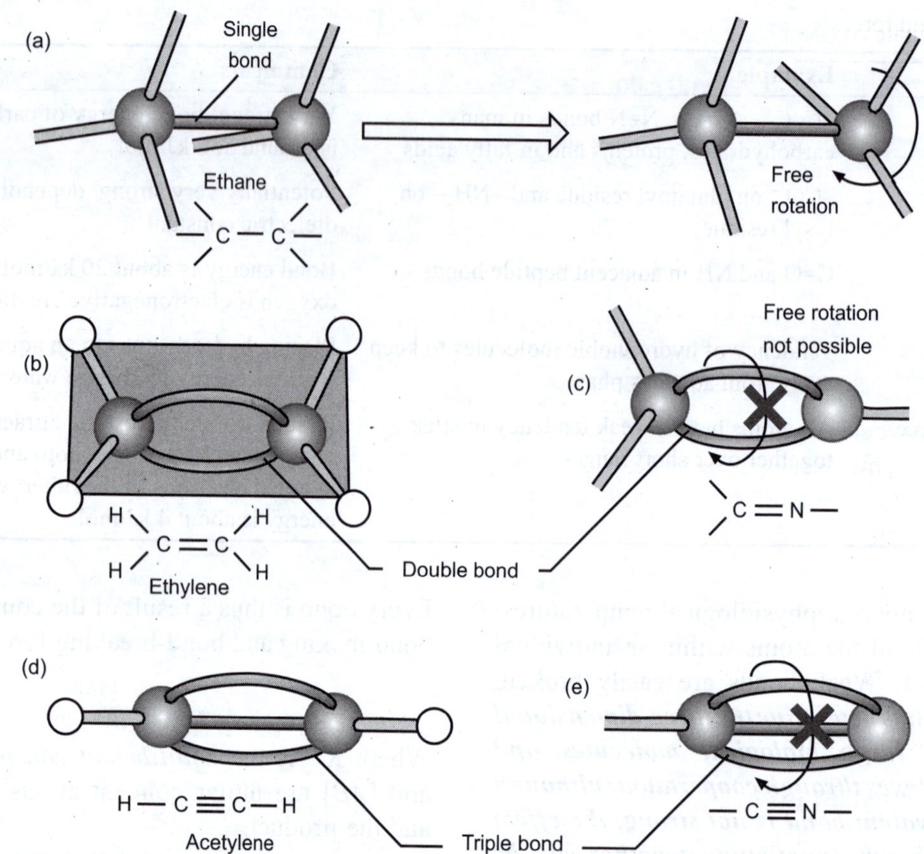

Fig. 2.5 Covalent bonds: (a) Single covalent bonds rotate freely, (b) and (c) Double covalent bonds and (d) and (e) Triple covalent bonds are rigid and the bonded atoms cannot rotate freely.

relative sizes of the two atoms. The smaller nitrogen atom has only enough space for four bonding pairs of electrons to be arranged without creating destructive repulsions between them, while the larger sphere of the phosphorus atom allows more electron pairs to be arranged around it without the pairs being too close together. Both oxygen and sulphur contain six electrons in their outermost orbitals. However, an atom of oxygen usually forms only two covalent bonds, as in molecular oxygen (O_2). Sulphur can form as few as two covalent bonds, as in hydrogen sulphide (H_2S), or as many as six, as in sulphur trioxide (SO_3) or sulphuric acid (H_2SO_4). The esters of sulphuric acid are important constituents of the proteoglycans that form part of the matrix surrounding most animal cells.

Energy changes in covalent bonds

Covalent bonds establish very stable links between atoms. The energy required to break a covalent bond is much greater than the thermal energy available at room or body temperatures (25ºC or 37ºC). Thermal energy at 25ºC is less than 1 kcal/mol while, the energy required for breaking a C–C bond in ethane is about 83 kcal/mol. The bond strength is measured by the amount of energy that must be supplied to break that bond. This is often expressed in units of kilocalories per mole (kcal/mole)

$$H_3C\text{—}CH_3 \rightarrow H_3\dot{C} + \dot{C}H_3 \qquad \Delta E = +\,83\ \text{kcal/mol}$$

where, ΔE represents the difference between the total energy of all of the bonds in the reactants and in the products. The positive value indicates that an input of energy is needed to cause the reaction and that the products contain more energy than the reactants. The high energy needed for breaking of the ethane bond means that, at room temperature (25°C), well under 1 in 10^{12} ethane molecules exists as a pair of $\bullet CH_3$ radicals. The covalent bonds in the biological molecules (**Table 2.4**) have ΔE values similar to that of the C–C bond in ethane.

In the course of a reaction, molecules usually adopt temporary states of much higher energy, called **transition states** or **activated complexes**. The high energy requirement for achieving these states, often in the range of 100 kcal/mol, means that many chemical reactions, even though favourable in terms of the

Table 2.4 Approximate covalent bond energies

Bond	Energy (kcal/mol)
C–C	83
C–H	99
C–O	84
C=O	174
O–H	111
O=O	118

energy differential between reactants and products, will not occur over a measurable time scale. That is the reason why covalent structures of molecules remain intact over very long periods. This higher energy input needed before the bonds will break is called the **activation energy** and is discussed in detail later. Specific biological catalysts (**Fig. 2.6**) and the enzymes (**Fig. 2.7**) act by reducing the activation energy requirements of particular reactions and thus channel chemical reactions in specific directions.

Bond angles in covalent molecules

When two or more atoms form covalent bonds with another central atom, these bonds are oriented at precise angles to one another. The angles are determined by the mutual repulsion of the outer electron orbitals of the central atom. In methane, for example, the central carbon atom is bonded to four hydrogen atoms; these angle between neighbouring C–H bonds is 109.5° (**Fig. 2.8**). In a double bond (C_2H_4), the carbon atom and all three atoms linked to it lie in the same plane (**Fig. 2.9**). Atoms connected by a double bond cannot rotate freely about the bond axis. The rigid planarity imposed by double bonds has enormous significance for the shape of large biological molecules such as proteins and nucleic acids. Triple bonds, where two atoms share six electrons are rare in biological molecules.

The outer electron orbitals not involved in covalent bond formation also contribute to molecular configurations. For example, the outer shell of the oxygen atom has two such pairs of electrons. In a water molecule, the orbitals of these non-bonding electrons have a specific spatial orientation and a high electron density that compresses the angle between the covalent H–O–H bonds to 104.5° (**Fig. 2.10**).

Unequal sharing of electrons

In certain cases, the bonded atoms exert different attractions for the electrons of the bond, resulting

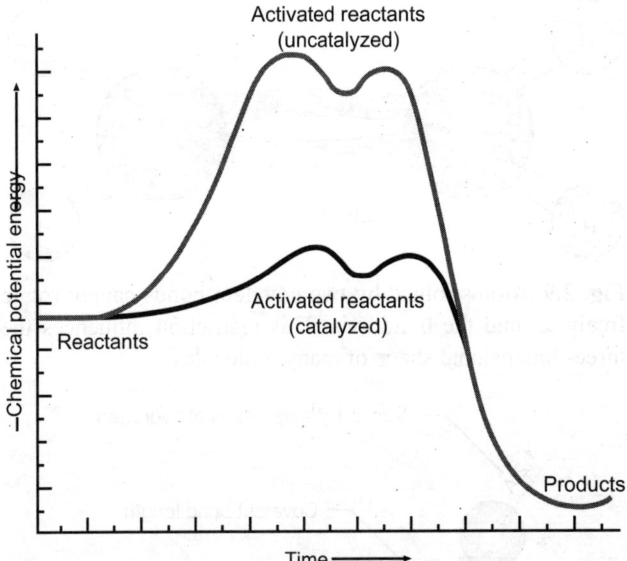

Fig. 2.6 A catalyst accelerates the rate of a reaction by reducing the activation energy required.

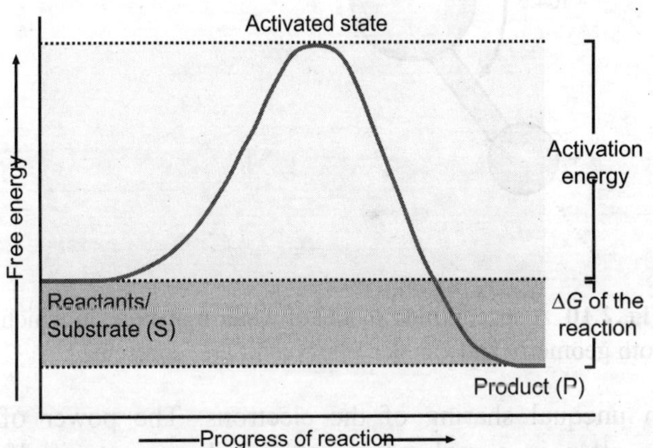

Fig. 2.7 Energy diagrams for activation of a chemical reaction.

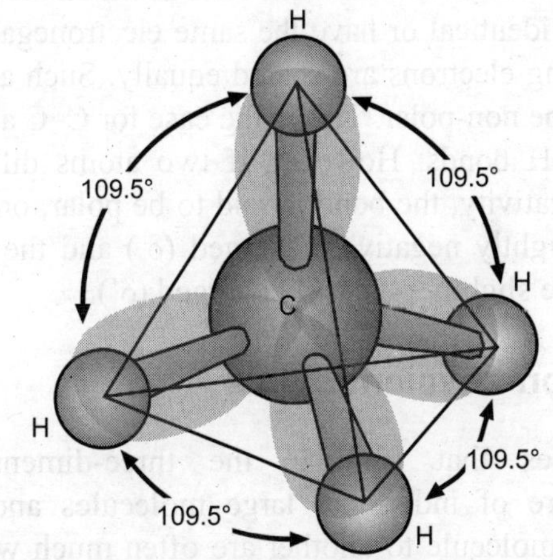

Fig. 2.8 The shape of methane: The four hydrogens are at the corners of an imaginary tetrahedron.

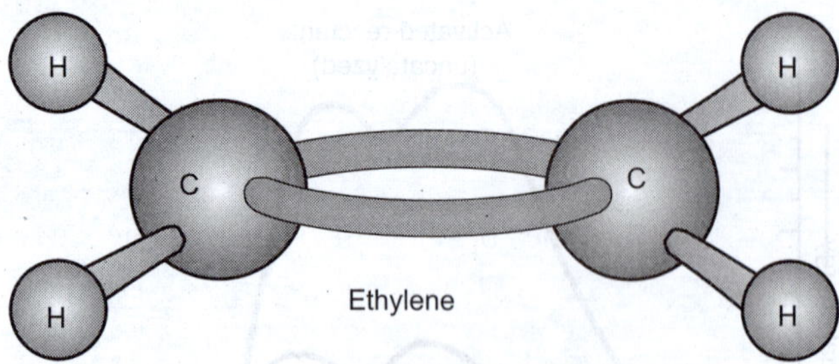

Fig. 2.9 Atoms joined by two covalent bonds cannot rotate freely around the bond axis. This restriction influences the three-dimensional shape of many molecules.

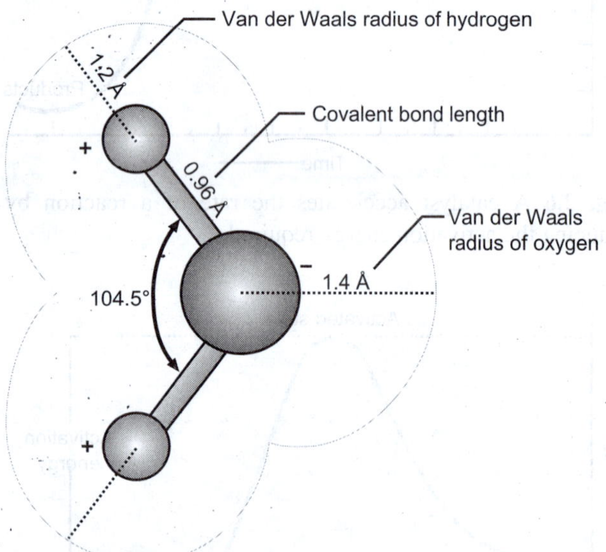

Fig. 2.10 A space filling model of water molecule, in which both geometry and van der Waals radii are represented.

in unequal sharing of the electrons. The power of an atom in a molecule to attract electrons to itself, called **electronegativity**, is measured on a scale from 4.0 (for fluorine, the most electronegative atom) to a hypothetical zero. In a covalent bond in which the atoms either are identical or have the same electronegativity, the bonding electrons are shared equally. Such a bond is said to be non-polar. This is the case for C–C as well as for C–H bonds. However, if two atoms differ in electronegativity, the bond is said to be polar, one end will be slightly negatively charged (δ^-) and the other end will be slightly positively charged (δ^+).

2.6.2 Non-covalent bonds

The forces that stabilize the three-dimensional architecture of individual large molecules and that bind one molecule to another are often much weaker and thus they are sometimes called **interactions** rather than bonds. The energy released in the formation of non-covalent bonds is only 1–5 kcal/mol. Because the average kinetic energy of molecules at room temperature (25°C) is about 0.6 kcal/mol, many molecules have enough energy to break these weak bonds. At physiological temperatures (25°C–37°C), weak bonds have a transient existence, but many can act together to produce highly stable structures. The four main types of non-covalent bonds are the hydrogen bond, the ionic interaction, van der Waals interactions and the hydrophobic bond.

Hydrogen bonds

It is formed when two molecules share an atom of hydrogen. Positively charged hydrogen of one polar molecule aligns next to the negative portions of another molecule (electronegative atom). In biological systems, both donors and acceptors are usually nitrogen or oxygen atoms, especially those atoms in amino ($-NH_2$) and hydroxyl (-OH) groups. Because all covalent N–H and O–H bonds are polar, their H atoms can, participate in hydrogen bond formations. By contrast, C–H bonds are non-polar and so these H atoms are never involved in a hydrogen bond.

An important feature of all hydrogen bonds is directionality. In the strongest hydrogen bonds, the donor, the hydrogen and the acceptor atoms all lie in a straight line. Non-linear hydrogen bonds are weaker than linear ones. ***Hydrogen bonds are important stabilizing forces in nucleic acids as well as in proteins. They stabilize giant protein molecules and nucleic acids.*** However, the strengths of these hydrogen bonds are only 1 or 2 kcal/mol, much weaker than the 5 kcal/mol characteristic of the hydrogen bonds between water molecules. The strength of a hydrogen bond in water, in turn, is much weaker than a covalent H–O bond. The strengths of other hydrogen bonds are similar to that of the hydrogen bond in water. Hydrogen bonding is not limited to water alone but is common among polar molecules in general.

Ionic bonds

Ionic bonds form as a result of the attraction between ions of opposite charge. Some atoms are so electron-attractive that they can capture electrons from other atoms—those that have a weaker hold on their one or two outer-shell electrons. The bonding electrons are never shared; these electrons are always found around the more electronegative atom. The atom that captures the extra electron acquires a negative charge (Cl^-), whereas the electron donor becomes positively charged (Na^+). Such electrically charged atoms are called **ions**. Oppositely charged ions attract one another in much the same way as do protons and electrons, forming a chemical linkage called an ionic bond (as in NaCl).

Because the electrons are not shared, the bonds in such compounds cannot be considered covalent. They are ionic interactions that result from the attraction of a positively charged ion—a **cation**—for a negatively charged ion—**anion**. Unlike covalent or hydrogen bonds, ionic bonds do not have fixed or specific geometric orientations, because the electrostatic field around an ion—its attraction for an opposite charge—is uniform in all directions. However, they do have very regular structures because that is the energetically most favourable way of packing together positive and negative ions. In aqueous solutions, simple ions of biological significance, such as Na^+, K^+, Ca^{2+}, Mg^{2+} and Cl^-, do not exist as free, isolated entities. Instead, each of these ions is surrounded by a stable and tightly held shell of water molecules. An ionic interaction occurs between the ion and the oppositely charged end of the water dipole. Most ionic compounds are soluble in water because a large amount of energy is released when ions bind water molecules. This is known as the **energy of hydration**.

Similar bonds may also form between charged organic molecules. For example, an ionic bond will form between a negatively charged carboxyl ion ($R–CO_2^-$) and a positively charged amino group ($^+H_3N–R'$). ***Organisms capitalize on the mutual attraction of oppositely charged ions to accomplish a number of biological tasks. For example, ionic movements through membranes are essential for the conduction of nerve impulses and for the stimulation of muscle contraction.*** One way channels accommodate some ions and exclude others is by size and an estimation of the size of an ion should include its shell of water molecules.

Van der Waals interactions

When one atom is more electron-attractive than its partner, the shift of electron distribution concentrates negative and positive charges resulting in a polar molecule, with distinct positive and negative regions—a **dipole**. This polarity induces an opposite polarity in neighbouring molecules such that a temporary negative charge in one atom may be attracted to a temporary positive charge in ano ther. Such forces, called van der Waals' forces, are weaker than a covalent bond and are non-specific. When two molecules have complementary shapes, hundreds or thousands of van der Waals' forces may form between them. Many biological interactions are governed by complementary shapes and van der Waals' force.

The van der Waals attraction decreases rapidly with increasing distance and is effective only when atoms are quite close to one another. However, if atoms get too close together, they become repelled by the negative charges in their outer electron shells. ***Van der Waals forces are among the interactions that take place between an antibody molecule and its specific antigen and also between many enzymes and their specifically bound substrates.***

Hydrophobic interactions

Polar molecules are attracted to water molecules, that is, they are hydrophilic (*hydro* = water, *philia* = loving). Hydrophilic substances readily dissolve in water. Non-polar substances, on the other hand, are insoluble in water or hydrophobic (*phobia* = fearing) **(Fig. 2.11)**. This explains why fats and oils, both of which are hydrophobic substances, form globules in water. ***Hydrophobic forces hold together cell membranes and many other biologically important structures.***

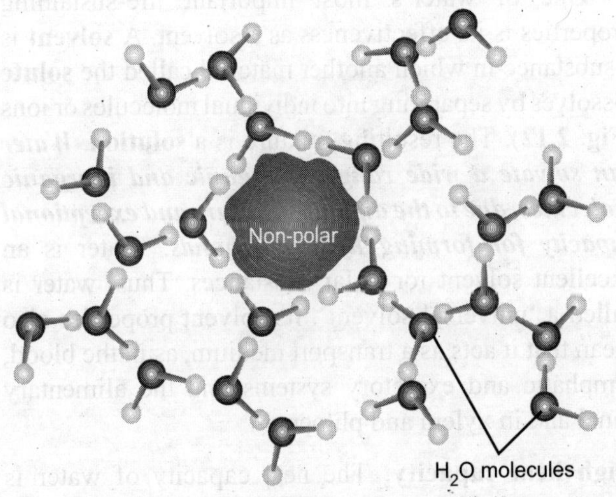

Fig. 2.11 Hydrophobic interactions between water and a non-polar substance.

2.6.3 Molecular collisions

Chemical reactions occur only when the reactants physically come into contact with one another. Atoms and molecules are in constant random motion and inevitably collide with each other. While some collisions simply change the direction and movement of the particles, others have strong forces to trigger a chemical reaction. Heat, for example, can increase the rate of chemical reactions by increasing molecular velocity. The minimum collision energy needed to produce a chemical reaction is the activation energy for that reaction.

2.7 WATER

In addition to energy, life on earth depends on water. Water is the predominant chemical component of living organisms and it makes up about 70% of most organisms. Water is important for two reasons:

- It is a vital chemical constituent of living cells.
- It provides an environment for those organisms that live in water.

These properties are rather unusual and due mostly to its small size, its polarity and the hydrogen bonding between its molecules. Water is an excellent **nucleophile** (a nucleophile is an electron-rich molecule, whereas an electron-poor atoms is called an **electrophile**) and is a reactant or product in many metabolic reactions. It has a slight propensity to dissociate into hydroxide ions and protons.

One of water's most important life-sustaining properties is its effectiveness as a solvent. A **solvent** is a substance in which another material called the **solute** dissolves by separating into individual molecules or ions (**Fig. 2.12**). The resulting mixture is a **solution**. ***Water can solvate a wide range of organic and inorganic molecules, due to the dipolar structure and exceptional capacity for forming hydrogen bonds***. Water is an excellent solvent for polar substances. Thus, water is called a 'universal solvent'. Its solvent properties also mean that it acts as a transport medium, as in the blood, lymphatic and excretory systems and the alimentary canal and in xylem and phloem.

High heat capacity: The heat capacity of water is the amount of heat required to rise the temperature of 1 kg of water by 1°C. Water has a high heat capacity.

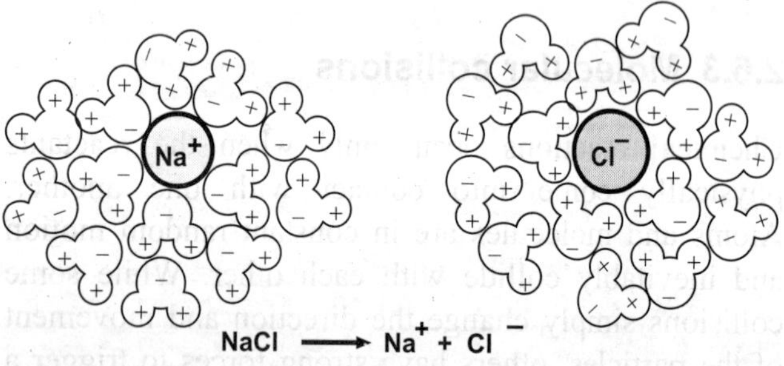

Fig. 2.12 Water molecules separate ions in solution. Each ion has shells of water molecules.

This means that a large increase in heat energy results in a relatively small rise in temperature. This is because much of the energy is used to break the hydrogen bonds, which restrict the movement of the molecules.

High heat of vaporization: Latent heat of vaporization is a measure of the heat energy required to vaporize a liquid, i.e., to overcome the attractive forces between its molecules so that they can escape as a gas.

High heat of fusion: Latent heat of fusion is a measure of the heat energy required to melt a solid, in this case ice. With its high heat capacity, water requires relatively large amounts of heat energy to thaw it. Conversely, liquid water must lose a relatively large amount of heat energy to freeze.

Density and freezing properties: The density of water decreases below 4°C and ice therefore tends to float. It is the only substance whose solid form is less dense than its liquid form.

High surface tension and cohesion: The attraction between water molecules creates high **surface tension**, that is, the liquid's surface resists being broken. Water has a higher surface tension than any other liquid. The high cohesion of water molecules is important in cells and in translocation of water through xylem.

Water as a reagent: Water is biologically significant as an essential metabolite, that is, it participates in the chemical reactions of metabolism.

2.7.1 Weak interactions in aqueous systems

Hydrogen bonds between water molecules provide the cohesive forces that make water a liquid at room temperature and favour the extreme ordering of molecules that is typical of crystalline water (ice). Hydrogen bonding profoundly influences the physical properties of water and accounts for its exceptionally high viscosity, surface tension and boiling point. Polar biomolecules dissolve readily in water because they can replace water–water interactions. Hydrogen bonds, ionic, hydrophobic and van der Waals interactions are individually weak, but collectively they have a significant influence on the three-dimensional structures of proteins, nucleic acids, polysaccharides and membrane lipids (**Fig. 2.13**).

Water is a polar solvent. It readily dissolves most biomolecules, which are generally charged or polar

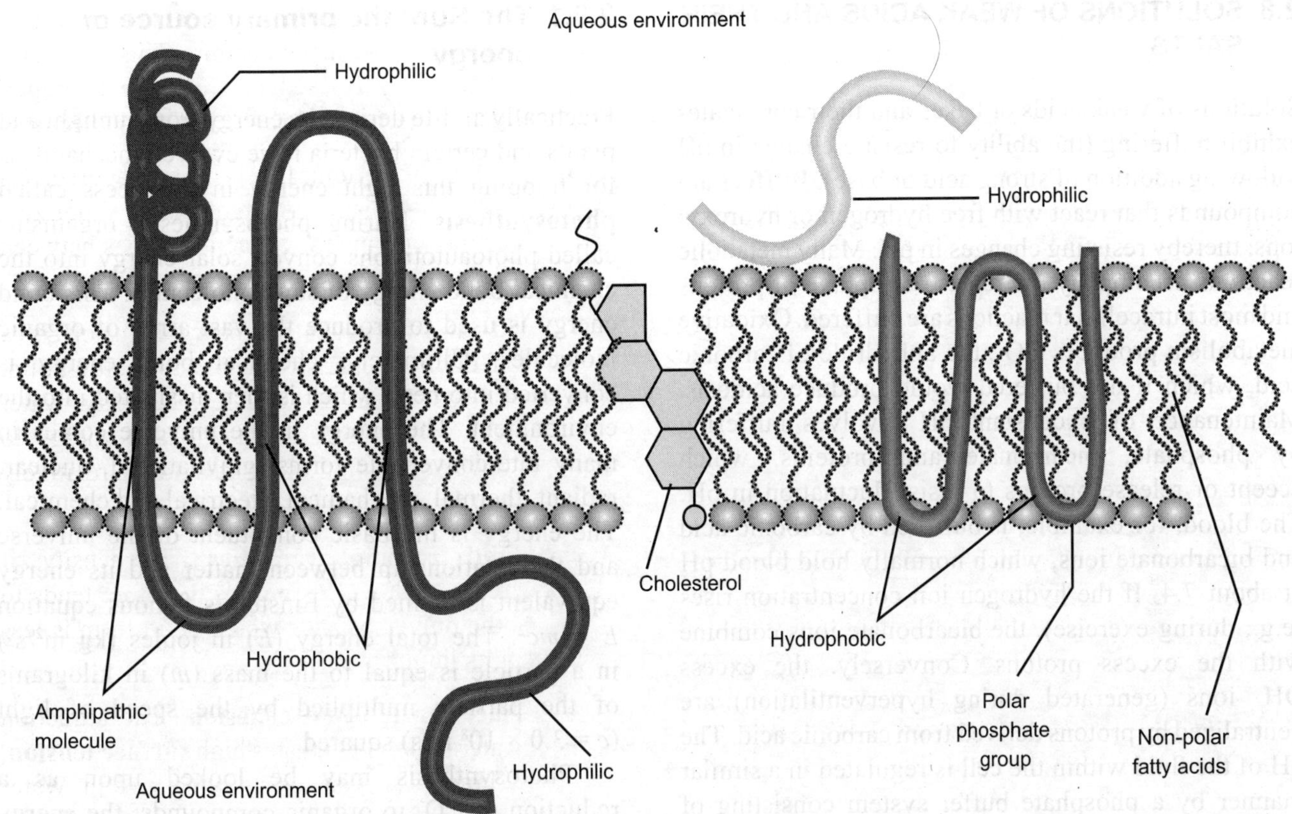

Fig. 2.13 Water stabilizes different molecules in a region based on the type of charges they have. Hydrophobic molecules group together in a membrane and hydrophilic molecules expose out into aqueous phase.

compounds. Sugars dissolve readily in water because of the stabilizing effect of hydrogen bonds between the hydroxyl groups or carbonyl oxygen of the sugar and the polar water molecules.

2.7.2 Interaction with water and structure of biomolecules

Weak interactions are crucial to macromolecular structure, stability and function. Most biomolecules are **amphipathic**, that is, they possess regions rich in charged or polar functional groups as well as regions with hydrophobic character. Macromolecules such as proteins, DNA and RNA contain so many sites of potential hydrogen bonding or ionic, van der Waals, or hydrophobic interactions, that the cumulative effect of the many small binding forces is enormous. This pattern maximizes the opportunities for the formation of energetically favourable charge–dipole, dipole–dipole and hydrogen bonding interactions between polar groups on the biomolecule and water. It also minimizes energetically unfavourable contact between water and hydrophobic groups. For macromolecules, the most stable (**native**) structure is usually that in which weak-bonding possibilities are maximized. The binding of an antigen to a specific antibody and the binding of a hormone or a neurotransmitter to its cellular receptor protein is the result of weak interactions.

2.7.3 Dissociative tendency of water molecules

The ability of water to ionize is of central importance for life. Since water can act both as an acid and as a base, its ionization may be represented as an intermolecular proton transfer that forms a hydronium ion (H_3O^+) and a hydroxide ion (OH^-). Since hydronium and hydroxide ions continuously recombine to form water molecules, hydrogen or oxygen cannot be stated to be present as an ion or as part of a water molecule. Since 1 g of water contains 3.46 x 10^{-22} molecules, the ionization of water can be described statistically. The actual probability of a hydrogen atom in pure water existing as a hydrogen ion is approximately 1.8 x 10^{-9}. The hydrogen and hydroxyl ions contribute significantly to the properties of water.

$$2H_2O \rightleftharpoons H_3O^+ + OH^-$$

2.8 SOLUTIONS OF WEAK ACIDS AND THEIR SALTS

Solutions of weak acids or bases and their conjugates exhibit buffering (the ability to resist a change in pH following addition of strong acid or base). **Buffers** are compounds that react with free hydrogen or hydroxyl ions, thereby resisting changes in pH. Many metabolic reactions accompany the release or uptake of protons and most intracellular reactions are buffered. Oxidative metabolism produces CO_2, the anhydride of carbonic acid, which if not buffered would produce acidosis. Maintenance of a constant pH involves buffering by phosphate, bicarbonate and proteins, which accept or release protons to resist fluctuation in pH. The blood, for example, is buffered by carbonic acid and bicarbonate ions, which normally hold blood pH at about 7.4. If the hydrogen ion concentration rises (e.g., during exercise), the bicarbonate ions combine with the excess protons. Conversely, the excess OH^- ions (generated during hyperventilation) are neutralized by protons derived from carbonic acid. The pH of the fluid within the cell is regulated in a similar manner by a phosphate buffer system consisting of $H_2PO_4^-$ and HPO_4^{2-}.

2.9 BIOENERGETICS

The production of energy, its storage and its utilisation are fundamental to cellular activities. A living cell displays millions of biological activities such as assembling of raw materials (e.g., synthesis of glucose) production and excretion of waste products, flow of genetic instructions from the nucleus to the cytoplasm, vesicular movement: ions are pumped across cell membranes. To maintain such a high level of activity, a cell must acquire and expend energy. The study that deals with the various types of energy transformations that occur in living organisms is referred to as bioenergetics.

Energy is defined as the capacity to do work, that is, to move matter and to drive the reaction. Life activities such as muscular contraction, orientating of leaves towards light, beating of flagella, conduction of nerve impulses and building new living matter, are all energy-consuming processes. Organisms not only need to obtain energy to drive these processes, but also have the necessary mechanism to transform and to store it. ***A large portion of the cell's activities is, therefore, devoted to obtaining energy from the environment and using that energy to drive energy-requiring reactions.***

2.9.1 The Sun: the primary source of energy

Practically all life derives its energy from sunlight and plants and certain bacteria have evolved mechanisms for trapping this light energy in a process called **photosynthesis**. During photosynthesis, organisms called photoautotrophs convert solar energy into the chemical bond energy of sugar molecules. This bond energy is used to produce the vast array of organic molecules. Ultimately, chemical bond energy is converted into heat, which is then dissipated into the environment. The energy in the universe comes in many interconvertible forms: gravitational, nuclear, radiant, thermal, mechanical, electrical and chemical. The energy is the basic constituent of the universe and the relationship between matter and its energy equivalent is defined by Einstein's famous equation $E = mc^2$. The total energy (E) in joules (kg m^2/s^2) in a particle is equal to the mass (m) in kilograms of the particle multiplied by the speed of light ($c = 3.0 \times 10^8$ m/s) squared.

Photosynthesis may be looked upon as a reduction of CO_2 to organic compounds; the energy required to drive this reduction being supplied by sunlight. All other organisms (animals and non-photosynthetic microorganisms and fungi) survive by taking in molecules synthesized by plants and breaking them down to release the energy contained in them. It matters little whether these organisms feed on plant matter directly or obtain pre-formed organic molecules by consuming animals that have eaten plants.

The process of breaking down organic molecules is called **catabolism**. Catabolic processes are mostly oxidations and release energy. They may reach completion and oxidize the organic material to CO_2 and H_2O, or occur incompletely, the products being used as building blocks for biosynthetic activities. Organisms also store energy. Therefore, it is appropriate to consider a few basic principles here.

Forms of energy

The principles of thermodynamics are used to evaluate the flow and interchange of matter and energy. There are two principal forms of energy: **kinetic** and **potential**. Kinetic energy is derived from particles in motion and includes thermal, acoustical, electrical, radiant and mechanical. Various forms of potential energy are biologically significant. But the main source to biological activities is the **chemical**

potential energy, the energy stored in the bonds connecting atoms in molecules. Other important sources of potential energy are **concentration gradient** and **electrical potential**.

Cells transform energy from one form to another

The laws of thermodynamics govern chemical equilibrium and determine the energetically favourable direction of all chemical reactions. Many of the reactions that take place within cells are in energetically unfavourable conditions; therefore reactions are able to proceed only at the cost of additional energy input. Consequently, cells must constantly expend energy derived from the environment. The generation and utilization of metabolic energy is thus fundamental to cell biology.

2.9.2 Fundamentals of biochemical energetics

Biological systems are generally held at constant temperature and pressure. Therefore, it is possible to predict the direction of a chemical reaction by using a measure of potential energy called **free energy**, or *G* (after the American chemist Josiah Willard Gibbs, 1939–1903). Free energy is that portion of a system's energy that can be used to do work when temperature is uniform throughout the system, as it is in the living cell. Gibbs showed that under conditions of constant pressure and temperature, as generally found in biological systems, 'all systems change in such a way that free energy is minimized'.

Free energy is the energy available to do work under conditions of constant temperature and pressure. It is the free energy released in the breakdown of organic molecules that is of importance to biochemical systems. Many of the mechanical devices we use (steam and petrol engines) are heat engines which work by making heat move from a region of high temperature to one of low temperature. However, organisms are uniform in temperature and cannot use temperature gradients to drive their reactions.

Knowing the state of free energy when molecules or molecular configurations change is one of the important aspects of biochemical energetics. In particular, with the difference between the values before and after the change, written as ΔG, where Δ stands for difference. Therefore:

$$\Delta G = G_{prod} - G_{reac}$$

Exergonic and endergonic reactions

Based on their free energy changes, reactions can be classified as either exergonic (meaning 'energy outward') or endergonic (meaning 'energy inward'). An exergonic reaction proceeds with a net release of free energy. Since the chemical mixture loses free energy, ΔG is negative for an exergonic reaction. In other words, exergonic reactions are those that occur spontaneously. The magnitude of ΔG for an exergonic reaction is the maximum amount of work the reaction can perform. For example, cellular respiration is an exergonic reaction.

$$C_6H_{12}O_6 + 6O_2 \rightarrow 6CO_2 + 6H_2O \quad \Delta G = -686 \text{ kcal/mol}$$

For each mole (180 g) of glucose broken down by respiration, 686 kcal of energy are made available for work.

An endergonic reaction is one that absorbs free energy from its surroundings. Because this kind of reaction stores more free energy in the molecules, ΔG is positive. Such reactions are non-spontaneous and the magnitude of ΔG is the minimum quantity of work required to drive the reaction.

If a chemical process is exergonic in one direction, then the reverse process must be endergonic. If $\Delta G = -686$ kcal/mol for respiration, then for photosynthesis, $\Delta G = +686$ kcal/mol.

Generation of ATP from glucose

The breakdown of carbohydrates, particularly glucose, is a major source of cellular energy. The complete oxidative breakdown of glucose to CO_2 and H_2O, plus energy can be written as follows:

$$C_6H_{12}O_6 + 6O_2 \rightarrow 6CO_2^- + 6H_2O + 38ATP$$

The reaction yields a large amount of free energy: $\Delta G = -686$ kcal/mol. To harness this free energy in usable form, glucose is oxidized within cells in a series of steps coupled to the synthesis of ATP.

Glycolysis, the breakdown of glucose in the cytoplasm (the initial stage of glucose metabolism), is common to almost all cells. It is an anerobic process and occurs in the absence of oxygen and can provide all the metabolic energy of anerobic organisms. In aerobic cells, however, glycolysis is only the first stage in glucose degradation. During this process glucose breaks down into pyruvate with the net gain of two molecules of ATP. In addition to producing ATP, glycolysis converts two molecules of the coenzyme NAD^+ to NADH. In this reaction, NAD^+ acts as an oxidizing agent that accepts electrons from glyceraldehyde-3-phosphate. The NADH will recycle by serving as a donor of electrons for other oxidation–reduction reactions within the cell. In anerobic conditions, the NADH reoxidizes to NAD^+ by the conversion of pyruvate to lactate or ethanol. In aerobic organisms, however, the NADH serves as an additional source of energy by donating its electrons to the electron transport chain, where they are ultimately used to reduce O_2 to H_2O, coupled to the generation of additional ATPs. Glycolysis in eukaryotic cells takes place in the cytosol and generates ATP, NADH and pyruvate similar to an anerobic glucose breakdown.

In the presence of oxygen, pyruvate formed in glycolysis is transported into the mitochondria, where it is oxidized by O_2 to CO_2 in a series of oxidation reactions collectively termed as **cellular respiration**. These reactions generate an estimated 28 additional ATPs per glucose molecule. Pyruvate is first converted to acetyl CoA with the formation of NADH (**Fig. 2.14**). Fatty acids are also converted to acetyl CoA with the formation of NADH and FADH. In the citric acid cycle of reaction, the acetyl group in acetyl CoA is oxidized to CO_2 and H_2O and produces most of the ATPs, that is, in a series of reactions containing various intermediate molecules (**Fig. 2.15**). In eukaryotes all these reactions take place in the mitochondria, the organelle to which pyruvate and fatty acids are directed for producing acetyl CoA.

(a)

Glucose — ATP ATP — NADH ATP ATP — NADH ATP ATP — Two molecules of pyruvate

(b)

Pyruvate + CoA–SH + NAD^+ → CO_2 + CoA–S–C(=O)–CH_3 + NADH

CoA–SH

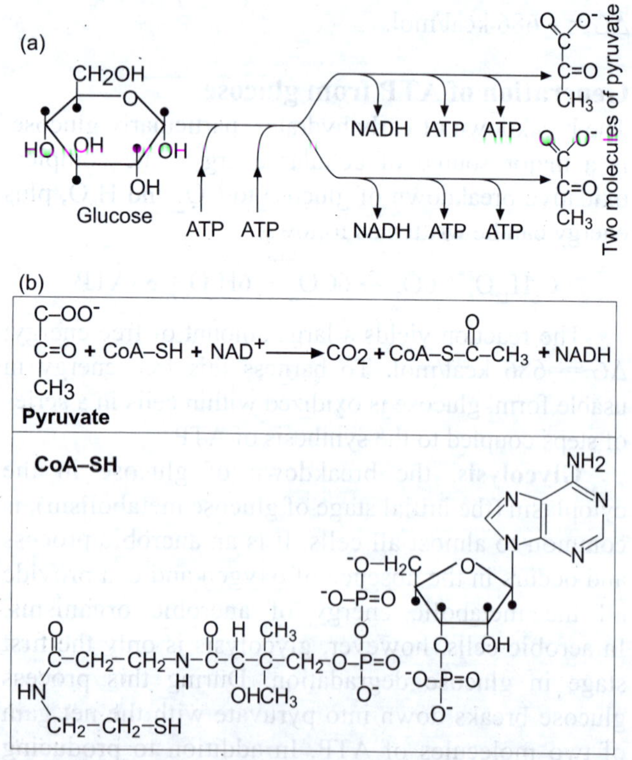

Fig. 2.14 Oxidative decarboxylation of pyruvate: Pyruvate is converted to CO_2 and acetyl CoA and one molecule of NADH is produced in the process.

Photosynthesis

The energy required for the synthesis of organic molecules is ultimately derived from sunlight, which is harvested from the plants and photosynthetic bacteria. In the first stage of photosynthesis, energy from sunlight is used to drive the synthesis of ATP and NADPH, coupled to the oxidation of H_2O to O_2. The ATP and NADPH produced by these reactions are then used to synthesize glucose from CO_2 and H_2O.

2.10 METABOLIC REACTIONS

Metabolism is the collection of biochemical reactions that occur within the cell, which include a variety of molecular conversions. The compounds formed in each step along the pathway are **metabolic intermediates** (or **metabolites**) that lead ultimately to the formation of an **end product**. The overall process is made up of reactions that produce energy (**catabolism**) and those that require energy (**anabolism**). Catabolic pathways lead to the disassembly of complex molecules to form simpler products. They serve two functions: make available the raw material for synthesis of various other molecules and provide chemical energy required for the many activities of a cell. Anabolic pathways lead to the synthesis of more complex compounds from simpler starting materials. They require energy and utilize chemical energy released by the exergonic catabolic reactions.

Metabolism builds all the molecules that collectively comprise the organism. Many of these molecules are macromolecules, polymers of high molecular weight assembled from relatively simple precursors. Macromolecules themselves may be further assembled into supramolecular complexes. Most of these molecular giants share one common characteristic with even the smallest compounds of life—they all have chemical 'skeletons' composed of carbon. ***Although the total number of reactions that occur in a given cell of a given organism can run into thousands, all these reaction types forms the basis of our understanding of biochemical process.***

2.10.1 Types of metabolic reactions

The metabolic reactions in our body fall into five broad types and studying of these reactions will make it easier for understanding the important metabolic reactions that happen in organisms.

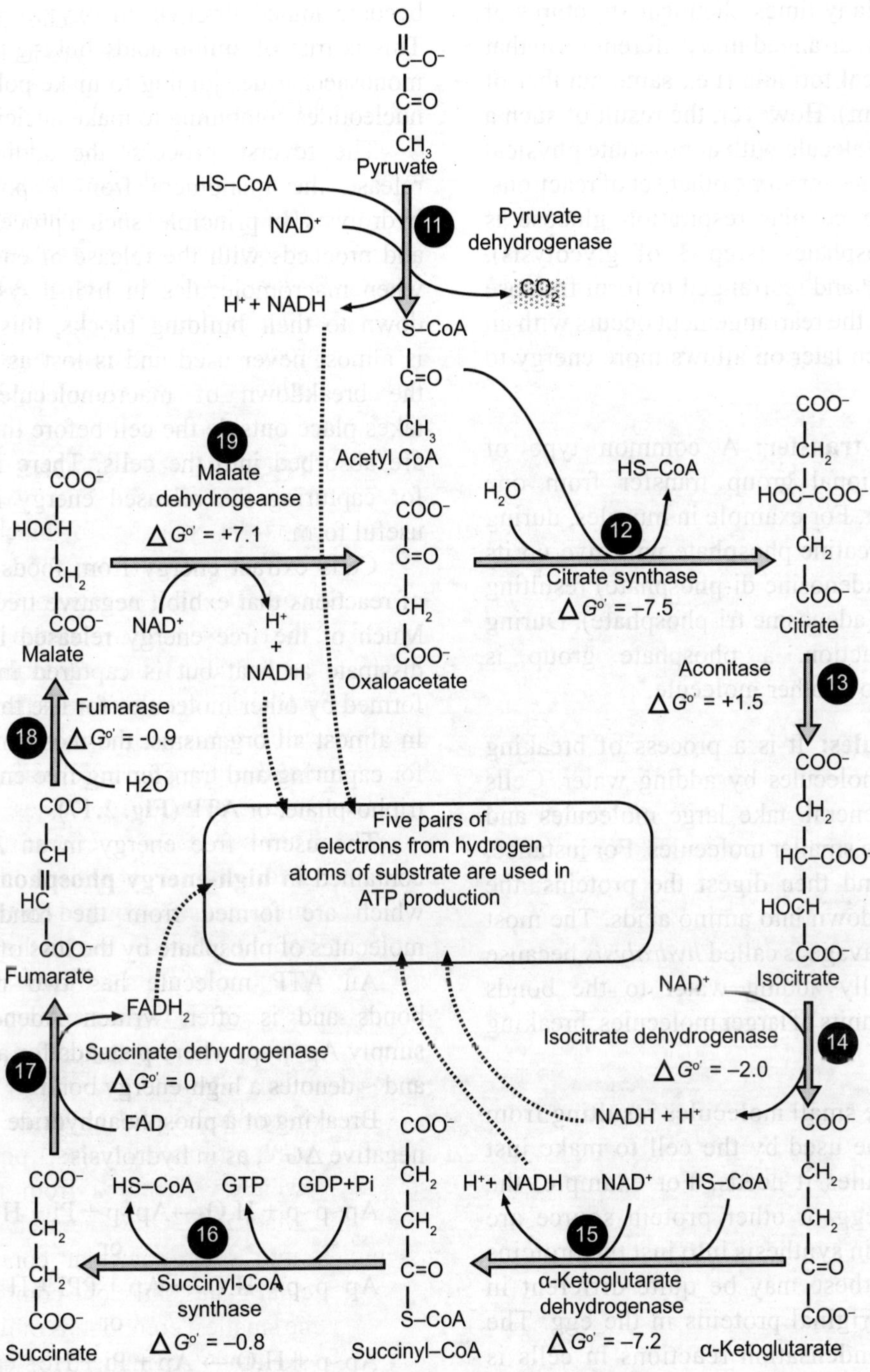

Fig. 2.15 The tricarboxylic acid (TCA) cycle or Krebs cycle.

Electron transfer: Electrons carry energy and this energy is released by transferring electrons from one substance to another by oxidation-reduction reactions. For example, from an electron carrier to some type of protein system that can transport electrons and use the electron's energy to do work. For instance, the NADH (reduced) an electron carrier being broken down into NAD^+ (oxidized) and H^+ (oxidized) along with two electrons. These electrons are picked up by a system of proteins in the plasma membrane of certain organelles. These proteins then pass the electrons along from one to the other resulting in energy being released to do work. Often this work involves making ATP or to move (work) the organism around its environment (in case of bacteria) (refer Chapters 7 and 8).

Rearrangement: Many times chemical structures of a molecule's will be rearranged into different form that has the same empirical formula (i.e., same number of atoms of each element). However, the result of such a rearrangement is a molecule with appropriate physical and chemical properties for some other set of reactions. For instance, during cellular respiration glucose is combined with phosphates (step 3 of glycolysis), transferred from ATP and rearranged to form fructose 1-6 biphosphate. But the rearrangement occurs with an input of energy which later on allows more energy to be harvested.

Functional group transfer: A common type of reaction is a functional group transfer from one compound to another. For example in muscles, during muscular activity, creatine phosphate will give up its phosphate to ADP (adenosine di-phosphate) resulting in creatine and ATP (adenosine tri-phosphate). During phosphorylation reaction, a phosphate group is covalently coupled to another molecule.

Cleavage of molecules: It is a process of breaking covalent bonds of molecules by adding water. Cells and organisms in general, take large molecules and break them down into smaller molecules. For instance, when we eat egg and then digest the proteins; the proteins are broken down into amino acids. The most common form of cleavage is called *hydrolysis* because it involves essentially adding water to the bonds joining the small subunits of larger molecules, breaking those bonds.

Condensation: The small molecules resulting from cleavage can then be used by the cell to make just the complex molecules it needs. For example, the amino acids from egg or other protein source are recombined in protein synthesis into just the proteins the cell needs and these may be quite different in function from the original proteins in the egg. The common type of condensation reactions in cells is called *dehydration synthesis* because it involved removing a hydroxyl group from one molecule and a hydrogen from the other molecule when the two molecules are joined resulting in water as a by-product.

Condensation and hydrolysis

The joining together of monomers to producer macromolecules almost always takes place by a process called condensation. The two molecules that condense lose the elements of a molecule of water and become joined through an oxygen atom (**Fig. 2.16**). This is true of amino acids linking to make proteins, monosaccharides joining to make polysaccharides and nucleotides combining to make nucleic acids.

The reverse process, the addition of water to release the monomers from a polymer, is called hydrolysis. In principle, such a process is spontaneous and proceeds with the release of energy. In practice, when macromolecules in living systems are broken down to their building blocks, this released energy is almost never used and is lost as heat. Most often the breakdown of macromolecules to monomers takes place outside the cell before the building blocks are absorbed into the cells. There is no mechanism for capturing the released energy in a biologically useful form.

Cells extract energy from foods through a series of reactions that exhibit negative free energy changes. Much of the free energy released is not allowed to dissipate as heat but is captured in chemical bonds formed by other molecules for use throughout the cell. In almost all organisms, the most important molecule for capturing and transferring free energy is adenosine triphosphate, or ATP (**Fig. 2.17**).

The useful free energy in an ATP molecule is contained in **high-energy phosphoanhydride bonds**, which are formed from the condensation of two molecules of phosphate by the loss of water.

An ATP molecule has two phosphoanhydride bonds and is often written Adenosine—p~p~p or simply Ap~p~p, where p stands for a phosphate group and ~ denotes a high-energy bond.

Breaking of a phosphoanhydride bond has a highly negative $\Delta G^{\circ\prime}$, as in hydrolysis:

$$\text{Ap}{\sim}\text{p}{\sim}\text{p} + H_2O \rightarrow \text{Ap}{\sim}\text{p} + \text{Pi} + H^+$$

or

$$\text{Ap}{\sim}\text{p}{\sim}\text{p} + H_2O \rightarrow \text{Ap} + \text{PPi} + H^+$$

or

$$\text{Ap}{\sim}\text{p} + H_2O \rightarrow \text{Ap} + \text{Pi} + H^+$$

where, Pi stands for inorganic phosphate and PPi for inorganic pyrophosphate, two phosphate groups linked by a phosphoanhydride bond. As the above two reactions show, the removal of a phosphate or a pyrophosphate group from ATP leaves adenosine diphosphate (ADP) or adenosine monophosphate (AMP), respectively.

Phosphorylation

There are many other reactions that go on in living systems and they have two major objectives. One is

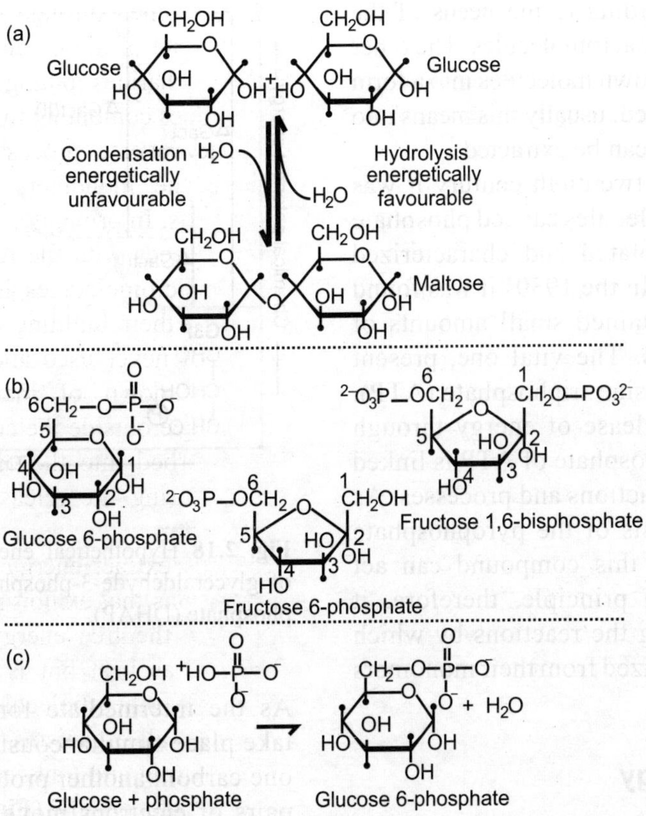

Fig. 2.16 (a) Condensation and hydrolysis are opposite reactions: Condensation is removal of water and the formation of a link; hydrolysis is addition of water and breakage of linkage. The formation of the disaccharide, maltose, from two glucose molecules is an example. (b) Structures of some sugar phosphate esters. (c) Formation of a sugar-phosphate ester between glucose and phosphate.

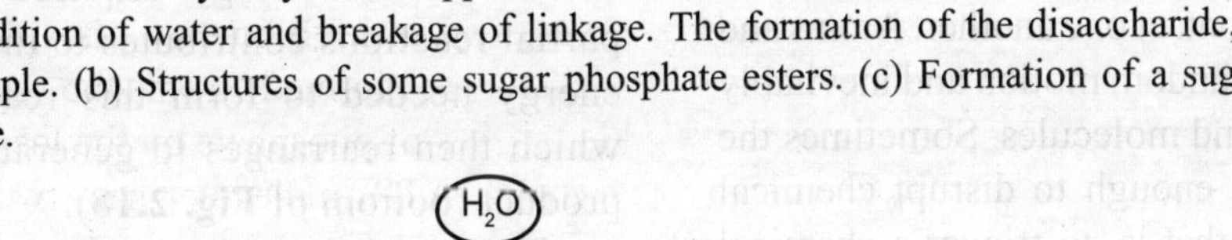

Fig. 2.17 A hydrolysis reaction. The hydrolysis of ATP drives many energy-requiring biochemical reactions.

to distribute molecules according to the needs of the organisms for building their macromolecules. The other is to get the partially broken down molecules into a form such that energy can be released: usually this means into a form from which electrons can be extracted.

In the early part of the twentieth century it was discovered that many biomolecules carried phosphate groups. The first to be isolated and characterized were the sugar-phosphates. In the 1930s it was found that all living systems contained small amounts of **pyrophosphate** compounds. The vital one, present in all organisms, was adenosine triphosphate (ATP). In a variety of ways the release of energy through the hydrolysis of the pyrophosphate of ATP is linked to other energy-requiring reactions and processes. As a simple example, hydrolysis of the pyrophosphate bonds of ATP means that this compound can act as a dehydrating agent. In principle, therefore, it has the potential for driving the reactions by which macromolecules are synthesized from their monomers through condensation.

2.10.2 Activation energy

Chemical reactions occur only when the reactants (atoms or molecules) physically contact one another. Atoms and molecules are in constant random motion and inevitably collide with other atoms and molecules. Sometimes the force of collision is great enough to disrupt chemical bonds or form new ones, that is, to trigger a chemical reaction. Heat increases the rate of chemical reactions by increasing molecular velocity. The minimum collision energy needed to produce a chemical reaction is the activation energy for that reaction.

All biochemical reactions with a negative ΔG (such as the formation of H_2O from H_2 and O_2) have activation energy; they proceed through one or more intermediate transition states whose content of free energy is greater than that of either the reactants or products. In some reactions, certain covalent bonds are moved to a strained position in the transition state and an input of energy is essential for this to happen. In other reactions, to form the transition state electrons must be excited, requiring an input of energy; only then can they pair up in a covalent bond in the product. In still others, molecules need only enough energy to overcome the mutual repulsion of their electron clouds to get close enough to react. For example, the conversion of glyceraldehyde-3-phosphate (G3P) to dihydroxyacetone phosphate (DHAP) involves at least one intermediate (**Fig. 2.18**).

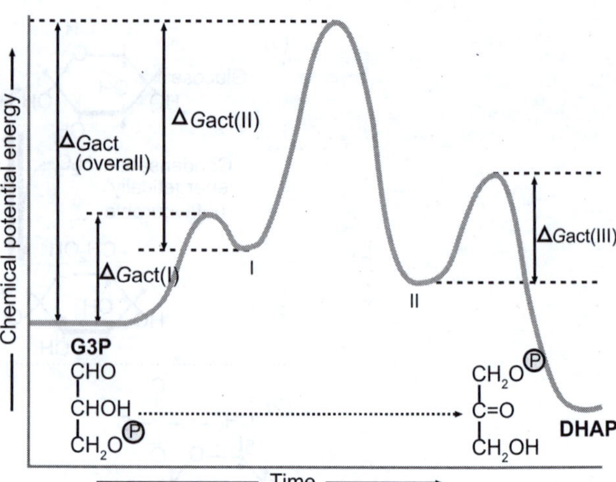

Fig. 2.18 Hypothetical energy changes in the conversion of glyceraldehyde-3-phosphate (G3P) to dihydroxyacetone phosphate (DHAP).

As the intermediate forms, the following changes take place simultaneously: a proton is removed from one carbon, another proton is donated to oxygen and pairs of electrons move from one bond to another. The activation energy required by each of these partial reactions contributes to the overall activation energy needed to form this reaction intermediate, which then rearranges to generate the final reaction product (bottom of **Fig. 2.18**).

Thus each stage in a multistep reaction has its own activation energy. For the overall reaction to proceed, the highest activation energy must be achieved. Because biochemical reactions occur at moderate temperatures, the kinetic energy of colliding molecules is generally insufficient to provide the necessary activation energy; in most cases, enzymes reduce the activation energy requirement, allowing specific reactions to proceed (**Fig. 2.19**).

Synthesis of biopolymers

The cellular dry mass constitutes many biomolecules, which are made from subunits (or monomers) that are linked together in condensation reaction. The nucleic acids (DNA and RNA), proteins and polysaccharides are all polymers that are produced by the repeated addition of subunits onto one end of a growing chain. The synthesis reactions for these macromolecules are outlined in the later chapters. However, it is interesting to note that the polymerization reactions that produce macromolecules can be oriented in one of two ways (**Fig. 2.20**), giving rise to either the head

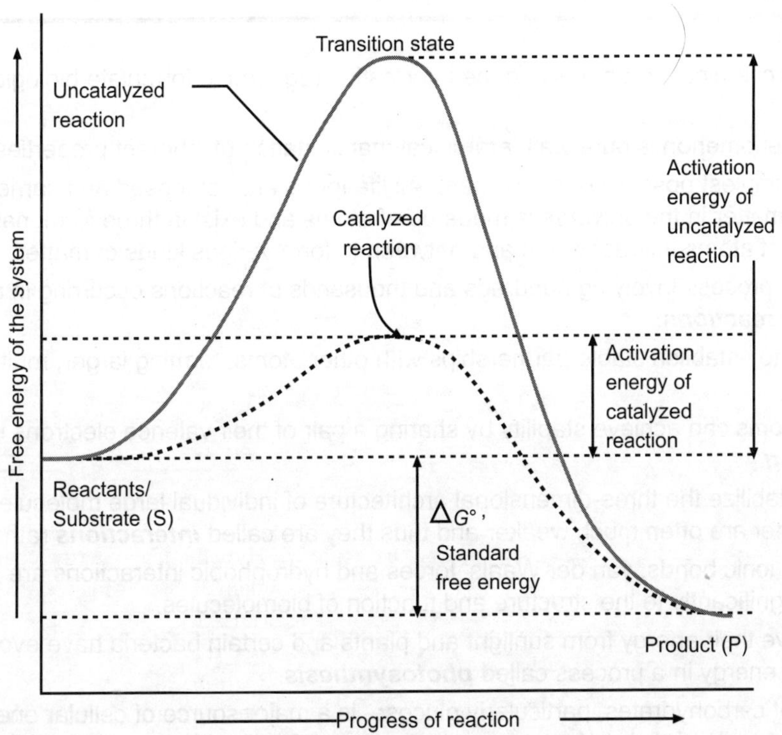

Fig. 2.19 Enzymes lower activation energ and thus increase the rate of the reaction.

polymerization or the tail polymerization of monomers. In head polymerization, the reactive bond required for the condensation reaction is carried on the end of the growing polymer and it must therefore be regenerated each time that a monomer is added. In this case, each monomer brings with it the reactive bond that will be used in adding the next monomer in the series. In tail polymerization, the reactive bond carried by each monomer is instead used immediately for its own addition.

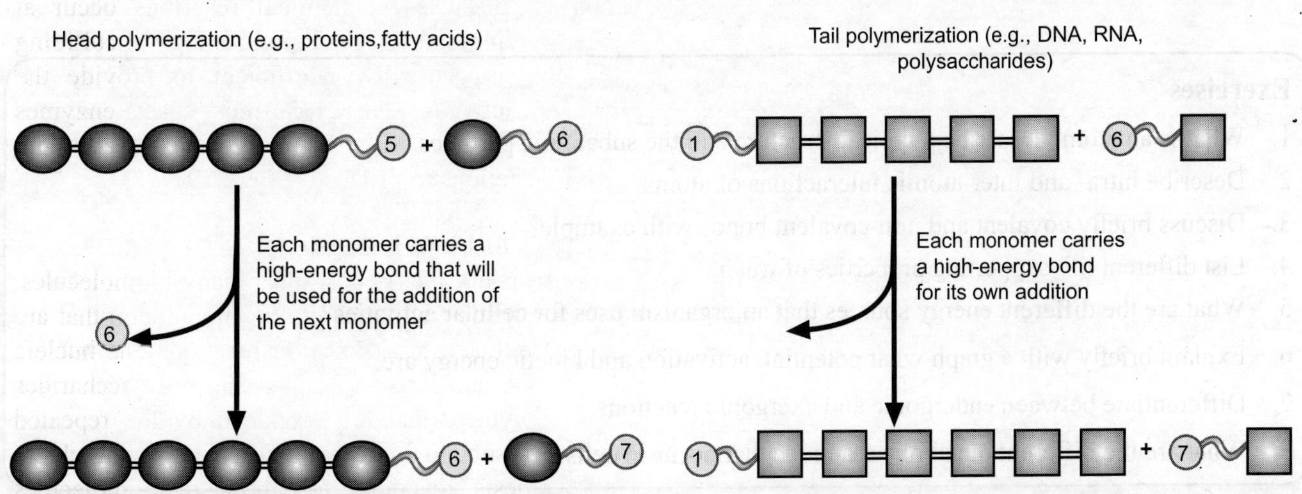

Fig. 2.20 The orientation of the active intermediates in biological polymerization reactions, which generates various biological macromolecules (Source: Alberts et al, 2002).

Summary

- The physical and chemical properties and the information required to formulate biological molecules exist within the atom.
- Any biological phenomenon is purely a hierarchical manifestation of inherent properties of different atoms.
- An ***entity*** is the simplest possible of an element. All life forms are composed and carried out by various kinds of atoms. All the matter in the universe is made up of atoms and exist in three form: namely, solids, liquids and gase. Different atoms interact within and between to form various kinds of matter.
- Life is a chemical process involving hundreds and thousands of reactions occurring in an organized manner, called ***metabolic reactions***.
- Most atoms tend to establish stable partnerships with other atoms, forming larger, multiatomic complexes called *molecules*.
- In most cases, atoms can achieve stability by sharing a pair of their valence electrons by a process called ***covalent bonding***.
- The forces that stabilize the three-dimensional architecture of individual large molecules and that bind one molecule to another are often much weaker and thus they are called ***interactions*** rather than bonds.
- Hydrogen bonds, ionic bonds, van der Waals' forces and hydrophobic interactions are non-covalent bonds; they contribute significantly to the structure and function of biomolecules.
- All life forms derive their energy from sunlight and plants and certain bacteria have evolved mechanisms for trapping this light energy in a process called ***photosynthesis***.
- The breakdown of carbohydrates, particularly glucose, is a major source of cellular energy. The breakdown of glucose in the cytoplasm (***glycolysis***) is common to almost all cells.
- ***Glycolysis*** is only the first stage in glucose degradation in aerobic organisms. In the presence of oxygen, pyruvate formed in glycolysis is transported into mitochondria and oxidized to O_2 to CO_2 in a series of oxidation reactions which generate an estimated 28 additional ATPs per glucose molecule.
- During cellular metabolism, cells use **ATP** as an energy source and build various biomolecules.
- Chemical reactions occur only when reactants come in contact with one another.
- The minimum collision energy needed to produce a chemical reaction is the ***activation energy*** and in a multistep reaction each step has its own activation energy. For the overall reaction to proceed, the highest activation energy must be achieved.
- In most cases, ***enzymes*** reduce the activation energy requirement, allowing specific reactions to proceed.

Exercises

1. What is an atom? Describe a typical structure and the subatomic particles.
2. Describe intra- and inter-atomic interactions of atoms.
3. Discuss briefly covalent and non-covalent bonds with examples.
4. List different life supporting properties of water.
5. What are the different energy sources that an organism uses for cellular activities?
6. Explain briefly with a graph what potential, activation and kinetic energy are.
7. Differentiate between endergonic and exergonic reactions.
8. What are the different metabolic reactions that occur in cells.

Further Reading

Books

Albert B, Bray D, Johnson A, Lewis J, Raff M, Roberts K and Walter P (1998), *Essential Cell Biology: An Introduction to the Molecular Biology of the Cell*, Garland, New York

Alberty RA and Silbey RJ (2000), *Physical Chemistry*, 3rd edition, Wiley.

Atkins PW (2000), *The Elements of Physical Chemistry*, 3rd edition, WH Freeman and Company.

Berg JM, Tymoczko and Stryer L (2002), *Biochemistry*, 5th edition, WH Freeman and Company.

Bergethon PR (1998), *The Physical Basis of Biochemistry: The Foundations of Molecular Biophysics*, Springer, New York.

Gennis RB (1989), *Biomembranes: Molecular Structure and Function*, Springer–Verlag, New York, Inc.

Hill TJ (1977), *Free Energy Transduction in Biology*, Academic Press.

Mathews CK, van Holde KE and Ahern KG (2000), *Biochemistry*, 3rd edition, Benjamin Cummings, California.

Murray RK, Granner DK, Mayes PA and Rodwell VW (2003), *Harper's Illustrated Biochemistry*, 26th edition, McGraw Hill.

Nelson DL and Cox MM (2000), *Lehninger: Principles of Biochemistry*, 3rd edition, Worth Publishing, New York.

Nicholls DG and SJ Fergusion (2002), *Bioenergetics*, 3rd edition, Academic Press, London.

Stryer L (1995), *Biochemistry*, 4th edition, Freeman, New York.

Tinoco I, Sauer K and Wang J (2001), *Physical Chemistry—Principles and Applications in Biological Sciences*, 4th edition, Prentice Hall.

Voet D, Voet JG and Pratt C (2002), *Fundamentals of Biochemistry*, John Wiley and Sons, New York.

References

Beinert H, Holm RH and Munck E (1997), Iron–sulphur clusters: Nature's modular, multipurpose structures, *Science*, 277:653–59.

Deisenhofer J and Michel H (1991) Structures of bacterial photosynthetic reaction centres, *Annu. Rev. Cell Biol.*, 7:1–23.

Krebs HA and Kornberg HL (1957), A survey of the energy transformation in living material, *Ergeb. Physiol. Biol. Chem. Exp. Pharmakol.*, 49:213.

Narlikar GJ and Herschlag (1997), Mechanistic aspects of enzymatic catalysis: Lesions from comparison of RNA and protein enzymes, *Annu. Rev. Biochem.*, 66:19–59.

Saraste M (1999), Oxidative phosphorylation at the fin de siècle, *Science*, 283:1488–93.

Wiggins PM (1990), Role of water in some biological processes, *Microbial Rev.*, 54:433.

3 BIOMOLECULES

Learning objectives

After reading this chapter, you will be able to:

- Identify the basic components of biological molecules
- Understand the various molecules and processes by which they combine to give different molecules
- List the features of specific biomolecules such as carbohydrates, lipids and proteins

3.1 INTRODUCTION

In the previous chapter we discussed the structure, physical and chemical properties of atoms and their role in creation of matter. In this chapter we will discuss how these atoms are linked in highly structured groups called **molecules**. Molecules range in size from two to many millions of atoms. They are, therefore, extraordinarily diverse and generate molecular information required for creation of various macromolecules with diverse biological roles (**biomolecules**). A typical cell contains 10^4–10^5 different kinds of molecules, in which roughly 50% are small inorganic and organic compounds and the remaining are polymers called **macromolecules**. Fundamentally, the biological behaviour of living forms—form, growth, movement, the metabolic functions, response to stimuli from the environment and reproduction—are due to the physical and chemical behaviour of molecules.

3.2 ELEMENTS ESSENTIAL FOR LIFE

Living organisms are composed of thousands of different kinds of inorganic and organic molecules. However, chemical analysis of organisms (microorganisms, animals, or plants) on earth shows that only a few elements (**Table 3.1**) from the periodic table are present in all living things. About 25 of the 92 natural elements

Table 3.1 Elemental composition of animals, plants and microorganisms

Element	Humans	Green plants	Bacterium
	% wet weight		
Oxygen (O)	62.81	77.80	74.68
Carbon (C)	19.37	11.34	12.14
Hydrogen (H)	9.31	8.72	9.94
Nitrogen (N)	5.14	0.83	4.04
Phosphorus (P)	0.63	0.71	0.60
Sulphur (S)	0.64	0.10	0.32
Ions			
Sodium (Na)	0.2		
Potassium (K)	0.4		
Magnesium (Mg)	0.1		
Calcium (Ca)	1.5		
Trace elements (less than 0.01%)			
Aluminium (Al)	Copper (Cu)	Manganese (Mn)	Silicon (Si)
Boron (B)	Fluorine (F)	Molybdenum (Mo)	Tin (Sn)
Chromium (Cr)	Iodine (I)	Nickel (Ni)	Vanadium (Va)
Cobalt (Co)	Iron (Fe)	Selenium (Se)	Zinc (Zn)

are known to be essential to life. What is more, these are always the same few elements and they are almost always found in similar proportions (more than 95% of living matter constitutes C, H, O and N). Although these six elements are combined in many different ways, chemical analysis of any organism again shows that, at the molecular level, it consists of relatively few types of compounds. Water, an inorganic molecule, makes up about 70% and of the remainder, the major compounds are carbohydrates, fats (or lipids), amino acids, mostly combined to form proteins and nucleotides, many of them in combination as nucleic acids.

3.3 CARBON SKELETON

The most important molecules found in cells are those based on a skeleton of covalently bonded carbon atoms. The simplest organic molecules are **hydrocarbons,** made up of just carbon and hydrogen. Different functional groups attached to the carbon give a molecule its unique chemical properties. They also affect the shape (3D), structure and charge distribution.

Nearly every molecule that contains carbon atoms belongs to the special category of organic molecules. These are molecules based on a skeleton of carbon atoms linked by covalent bonds and have immense biological importance. The chemistry of compounds containing carbon is called **organic chemistry**. Originally, the term 'organic' referred to molecules that could be manufactured only by living organisms. Chemists eventually referred to all compounds containing carbon as 'organic compounds'.

Why is **carbon** the central atom in biological systems? Carbon's chemical properties make it ideal to be the central element on which life is based. Perhaps the most important of these properties is its ability to form long chains with itself. Two properties of carbon merit particular attention:

- the ability to form covalent bonds with itself
- the tetrahedral nature of the four covalent bonds about a carbon atom.

Together these properties hold the potential for an incredible variety of linear, branched and cyclic compounds of carbon whose diversity is further multiplied by the possibilities for including N, O and H atoms in these compounds. Without these chains, the complex molecules of life could not exist. Chemical chains are the backbone of biochemicals. They can be linear, branched, or cyclic, or with more complex formations (**Fig. 3.1**).

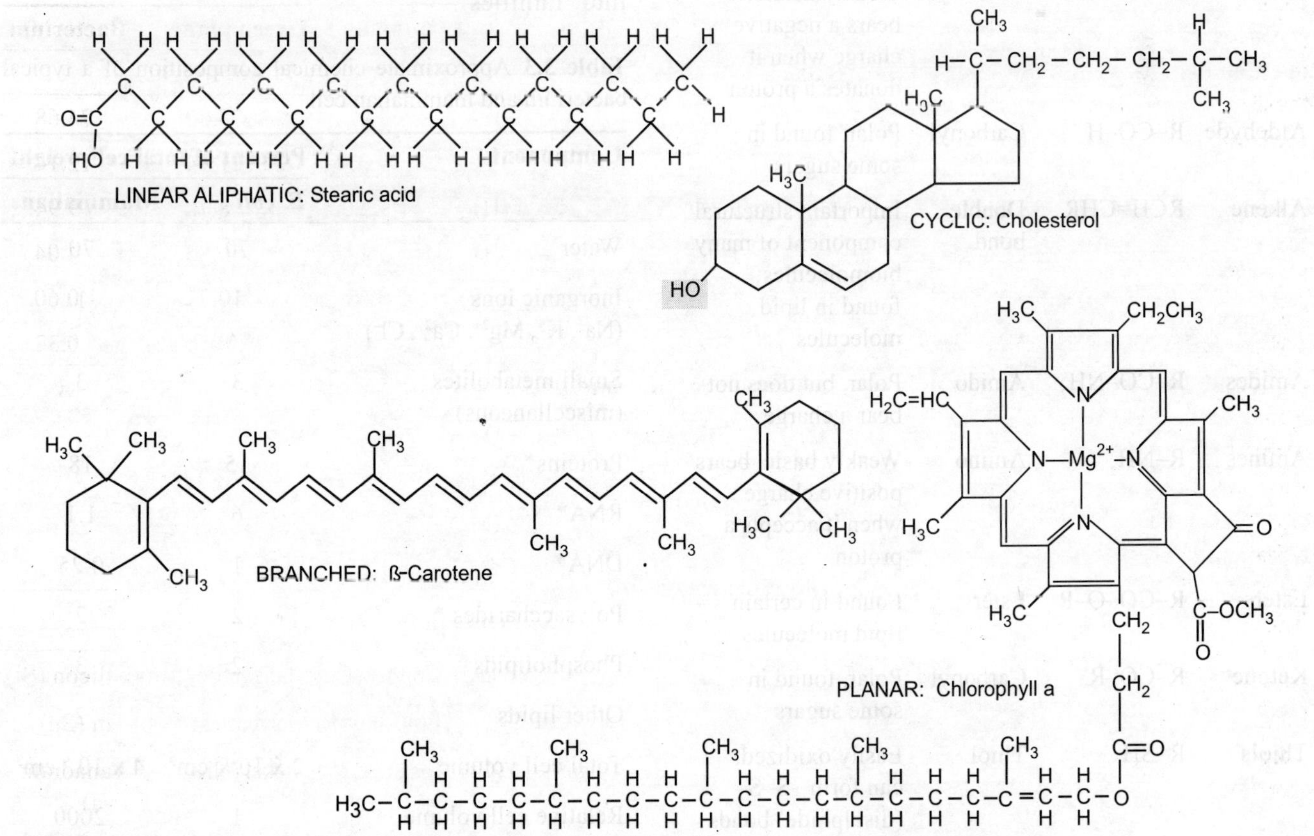

Fig. 3.1 Examples of the versatilty of C–C bonds in complex structures: Linear, cyclic, branched and planar.

Even after forming one or two bonds with other carbons in a chain, each carbon still has at least two free bonding sites to which additional chemical groups and elements (e.g., H, O, N, P or more carbons) can attach. A carbon atom may also be bonded to a chemical group composed of several atoms rather than to a single element. These accessory chemical entities, called **functional groups (Table 3.2)**, help determine the identity and chemical properties of various functional groups.

3.4 FUNCTIONAL GROUPS OF BIOMOLECULES

Most biomolecules contain more than one functional group. For example, many simple sugar molecules have several hydroxyl groups and an aldehyde group. Nucleotides, the building block molecules of nucleic acids, have amino and carboxyl groups. Similarly, amino acids, the building block molecules of proteins, have both an amino group and a carboxyl group. The distinct chemical properties of each functional group contribute to the behaviour of the molecule containing (**Table 3.2**).

Table 3.2 Important functional groups of biomolecules

Family name	Group structure	Group name	Significance
Alcohol	R–OH	Hydroxyl	Polar, forms hydrogen bonds
Acids	R–COOH	Carboxyl	Weakly acidic, bears a negative charge when it donates a proton
Aldehyde	R–CO–H	Carbonyl	Polar, found in some sugars
Alkene	RCH=CHR	Double bond	Important structural component of many biomolecules; found in lipid molecules
Amides	R–CO–NH_2	Amido	Polar, but does not bear a charge
Amines	R–NH_2	Amino	Weakly basic, bears positive charge when it accepts a proton
Esters	R–CO–O–R	Ester	Found in certain lipid molecules
Ketone	R–CO–R'	Carbonyl	Polar, found in some sugars
Thiols	R–SH	Thiol	Easily oxidized; can form –S–S– (disulphide) bonds readily

3.5 CLASSES OF BIOMOLECULES

Organisms must synthesize all the chemicals needed to operate, maintain and reproduce cells. The various biological polymers are either repetitive or non-repetitive in structure. Repetitive biological polymers contain one kind of monomeric subunit: distinctions between different types of the same polymers are due to differences in molecular weight and the degree of branching of the polymer chains. The major function of repetitive polymers in the cell is to provide structures, chemical inertness, permeability and nutrient storage. Non-repetitive polymers may contain up to 20 different monomer species. Further, each of these biological polymers has a fixed molecular weight and monomer composition (**Table 3.3**). There are two different classes of molecules. micromolecules and macromolecules. Many macromolecules are further grouped into 'families'.

Table 3.3 Approximate chemical composition of a typical bacterium and mammalian cell

Component	Percent of total cell weight	
	E. coli	Mammalian
Water	70	70
Inorganic ions (Na^+, K^+, Mg^{2+}, Ca^{2+}, Cl^-)	1	1
Small metabolites (miscellaneous)	3	3
Proteins*	15	18
RNA*	6	1.1
DNA*	1	0.25
Polysaccharides *	2	2
Phospholipids	2	3
Other lipids	–	2
Total cell volume	2 x 10^{-12} cm^3	4 x 10^{-9} cm^3
Relative cell volume	1	2000

*macromolecules

Micromolecules

Astonishingly, all living cells are composed of the same chemical elements as the physical world, but their proportions and organization in living tissues differ. Elements exist in cells as part of chemical compounds or as ions. Micromolecules are the basic ingredients of the cell. Some important micromolecules are water (70%), simple sugars, amino acids, fatty acids, nucleotides, vitamins, steroids, ATP and cAMP, urea and so on. At least 750 different types of small molecules are found in virtually every cell. This is obviously a broad category of molecules but not as large as macromolecules.

Macromolecules

Macromolecules are polymers of chains constructed by assembling together small molecular subunits and each subunit is called a monomer. The chains may be branched or unbranched. If the units are identical, the result is a **homopolymer**. If the units differ, the product is a **heteropolymer**. The majority of the biomolecules that enable organisms to function are polymers. There are about 40 different types of monomers from which all polymers, and therefore all organisms on earth, are constructed.

The major precursors for the formation of biomolecules are water, carbon dioxide and three inorganic nitrogen entities—ammonium (NH_4^+), nitrate (NO_3^-) and dinitrogen (N_2). Metabolic processes assimilate and transform these inorganic precursors into complex biomolecular products. In the first step, precursors are converted to **metabolites**, simple organic compounds that are intermediates in cellular energy transformation and in the biosynthesis of various sets of building blocks: amino acids, sugars, nucleotides, fatty acids and glycerol. These building blocks construct proteins, polysaccharides, polynucleotides (DNA and RNA) and lipids (**Table 3.4**). Interactions among macromolecules lead to the next level of structural organization, **supramolecular complexes** (e.g., enzyme complexes, ribosomes, chromosomes and cytoskeletal elements). Every macromolecule falls into one of the four fundamental families of organic compounds: carbohydrates, lipids, proteins and nucleic acids.

Due to the three-dimensional orientation of molecules two compounds can be mirror images of one another. Although, two such compounds have similar chemical and physical properties, their behaviour in biological systems is usually quite distinct. This is because biological systems work by recognizing shapes, or molecular configurations—the precise positions of atoms and groups of atoms relative to each other.

3.6 CARBOHYDRATES

Carbohydrates are carbon compounds that contain large quantities of hydroxyl groups and are the most abundant biomolecules on earth. These biomolecules are the source of solar energy stored in the chemical bond energy of living organisms. More than half of all 'organic' carbon on the earth is found in carbohydrates. Each year, photosynthesis converts more than 100 billion metric tons of CO_2 and H_2O into cellulose and other plant products. Carbohydrates are substances which contain the element carbon, hydrogen and oxygen and have the

Table 3.4 Major macromolecules found in living systems

Macromolecule	Class of monomers	Some major functions and examples
Carbohydrates	Sugars	Energy storage, structural elements, e.g., starch, glycogen, cellulose
Lipids		
Triglycerides	Fatty acids and glycerol, phosphate and an R group	Energy storage; thermal insulation; shock absorption, e.g., fat, oil
Phospholipids	Fatty acids and glycerol, phosphate and an R group	Foundation for membranes, e.g., plasma membrane
Waxes	Fatty acids and long-chain alcohols	Water proofing; protection against infection, e.g., cutin, suberin, beeswax
Steroids	four-ringed structure	Membrane stability hormones, e.g., cholesterol, testosterone, estrogen
Nucleic acids	Ribonucleotides; deoxyribonucleotides	Inheritance of genetic information; ultimate director of metabolism, e.g., DNA, RNA
Protein	Amino acids	Catalysis for metabolic reactions; hormones; oxygen transport; physical structure, e.g., hormones, haemoglobin, enzymes.

basic formula $(CH_2O)_n$ or $C_x(H_2O)_y$, where x, y and n are variable numbers. The formula $(CH_2O)_n$ is sufficiently accurate to be useful in calculating overall elemental balances and energy release in cellular reactions. In addition they also have properties such as all are aldehydes (termed **polyhydroxyaldehydes**) or ketones (**polyhydroxyketones**), all contain several hydroxyl groups and these groups determine the chemistry of carbohydrates. The name 'carbohydrate' reflects this ratio of carbon and water (carbo = carbon, C and hydrate = addition of water, H_2O). Carbohydrate moieties also occur as components of other biomolecules, such as proteins, nucleic acids and lipids. Recent investigations have revealed that carbohydrates, in addition to the above roles, are also involved in logical processes such as signal transduction, cell–cell interactions and endocytosis.

3.6.1 Biosynthesis of carbohydrates

In most organisms, in addition to being obtained directly from food or generated by photosynthesis, glucose can be synthesized from other organic molecules. In animal cells, glucose synthesis (**gluconeogenesis**) usually starts with lactate (produced by anerobic glycolysis), amino acids (derived from the breakdown of proteins) or glycerol (produced by the breakdown of lipids). Plants (but not animals) are also able to synthesize glucose from fatty acids—a process that is particularly important during the germination of seeds, when energy stored as fats must be converted to carbohydrates to support growth of the plant. In both animal and plant cells, simple sugars are polymerized and stored as polysaccharides.

3.6.2 Nomenclature and classification of carbohydrates

The predominant carbohydrates encountered in the body are structurally related to the aldotriose glyceraldehyde and to the ketotriose dihydroxyacetone (**Table 3.5**). All carbohydrates contain at least one asymmetrical (chiral) carbon and are, therefore, optically active. In addition, carbohydrates can exist in either of two conformations, as determined by the orientation of the hydroxyl group about the asymmetric carbon farthest from the carbonyl. With a few exceptions, those carbohydrates that are of physiological significance exist in the D-conformation. The mirror image conformations, called **enantiomers**, are in the L conformation. Carbohydrates are divided into three main classes, monosaccharides, oligosaccharides and polysaccharides (Latin, *saccharum* = sugar, Sanskrit, *sarkara* = sugar) and also a number of carbohydrate derivative molecules (heterosaccharides).

The **monosaccharides** are single sugar units, which cannot be hydrolyzed further into simpler carbohydrates. The families of monosaccharides are **aldoses** (sugars with an aldehyde functional group) and **ketoses** (sugars with a ketone functional group). They can be classified based on the number of carbon atoms they contain: trioses (3C), tetroses (4C), pentoses (5C), hexoses (6C) and heptoses (7C) and as aldoses or ketoses depending upon whether they have an aldehyde or ketone group. In aqueous solution, all monosaccharides with four or more carbon atoms in the backbone occur predominantly as cyclic or **ring** structures.

Disaccharides are the most abundant carbohydrates; they contain two monosaccharide units. When they are linked together by glycosidic bonds, monosaccharides form a variety of molecules that perform diverse biological functions. Disaccharides are formed when two monosaccharides, usually hexoses, combine by means of a chemical reaction known as **condensation,** or removal of water. The most common disaccharides are maltose (glucose + glucose), lactose (glucose + galactose), sucrose (glucose + fructose) and cellobiose a degradation product of cellulose (glucose + glucose).

Oligosaccharides (*oligo* = few) consist of short chains of monosaccharide units or residues (two to ten monosaccharides) joined by characteristic linkages called **glycosidic bonds**. Most often such chains are found covalently attached to lipids and proteins, converting them into glycolipids and glycoproteins, respectively. These molecules may be composed of many different combinations of sugar units; they play an informational role.

Most carbohydrates found in nature occur as **polysaccharides (Fig. 3.2)**, as polymers of medium to high molecular weight. Polysaccharides function mainly as nutritional sugar stores and as structural materials. Polysaccharides can be broadly divided into two classes: **homopolysaccharides**, which are composed of one type of monosaccharide and **heteropolysaccharides**, which contain two or more types of monosaccharides. The important homopolysaccharides are starch, glycogen, cellulose and chitin. The most important polysaccharides are starch in plant cells and glycogen in animal cells. Both occur intracellularly as large clusters or granules. Cellulose and starch are glucose polymers, but differ in the way the glucose monomers are linked.

Heteropolysaccharides are high-molecular weight carbohydrate polymers that contain more than one kind

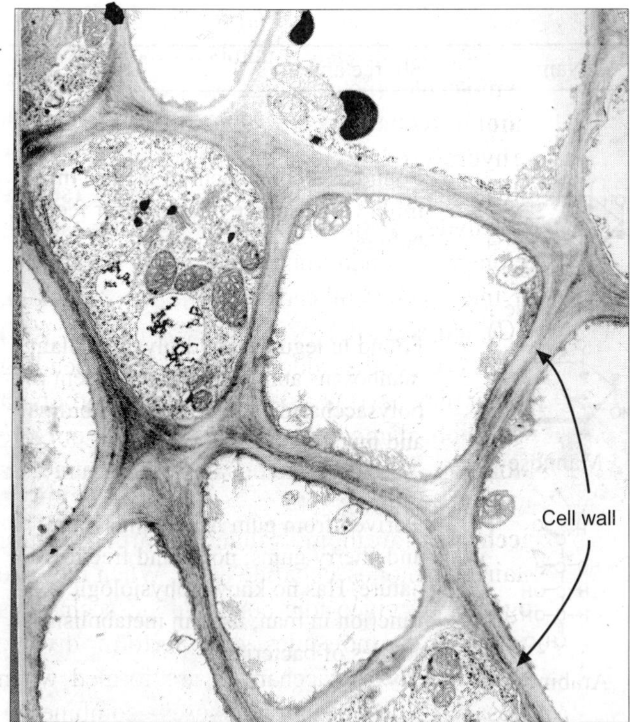

Fig. 3.2 TEM image showing the cell wall of tomato leaf cell. Polysaccharides are the principal components of the cell wall and form its main structural framework (Channarayappa, 1992).

of monosaccharide. Important examples include the glycosaminoglycans (GAGs), the principal components of proteoglycans and murein (peptidoglycan), a major component of bacterial cell walls.

Cellulose is a linear polymer with β (1→4) linkages and is a major structural component of the plant cell wall. Cellulose is synthesized by an enzyme complex associated with the plasma membrane. The substrate for cellulose is UDP-glucose, which can be formed from hexose phosphates by UDP-glucose phosphorylase and from sucrose by sucrose synthase. A plasma membrane-bound isoenzyme of sucrose synthase forms a complex with cellulose synthase, channelling UDP-glucose from sucrose catabolism directly into cell wall synthesis (**Fig. 3.3**).

Other forms of polysaccharides are starch, glycogen and chitin. **Starch** is a sugar store and is found in large intracellular granules, which can be hydrolyzed quickly to release glucose for metabolism. It contains two components: α-amylose, a linear polymer with α(1→4) linkages and amylopectin, branched polymer with additional α(1→6) linkages. **Glycogen** is the main carbohydrate storage molecule in vertebrates. It is a polymer of α(1→4) linked subunits of glucose, with α(1→6) linked branches. **Chitin** is found in fungal cell walls and in the exoskeleton of insects and crustacean. The monomeric units (*N*-acetylglucosamine) are linked in unbranched chains by β(1,4) glycosidic bonds.

3.7 LIPIDS

Lipids are a heterogeneous group of compounds that are preferentially soluble in a non-aqueous solvent

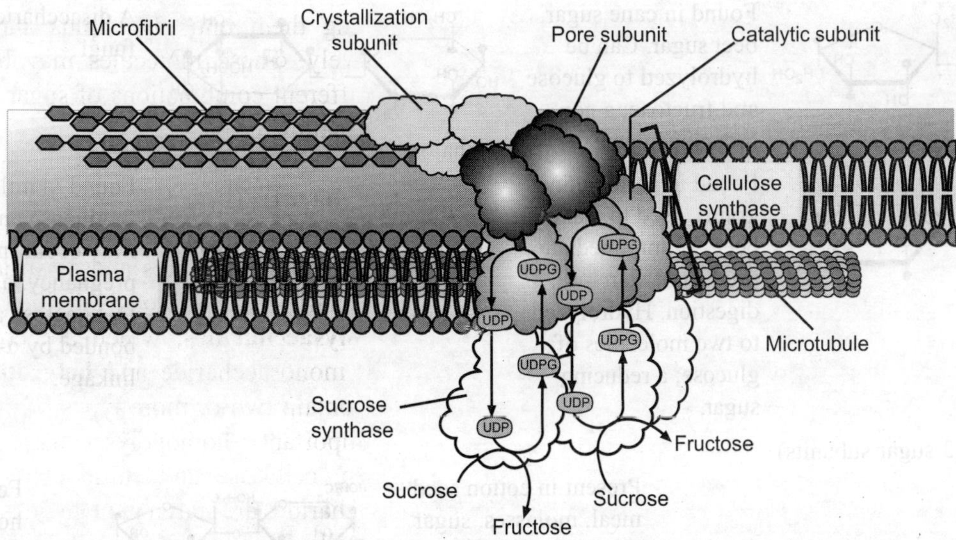

Fig. 3.3 Model of cellulose synthase complex. Cellulose is formed outside the cell by the addition of glucose units to the growing cellulose molecule. The glucose units come from UDP-glucose (UDPG), which might be supplied by sucrose synthase associated with the complex that synthesizes cellulose. This would solve the problem of transferring UDP-glucose across the plasma membrane of the cell.

Table 3.5 Common types of carbohydrates: Classification and sources

Name	Source and uses	Name	Source and uses
A. Monosaccharides: simple sugars			
Hexoses			
Glucose	Found in sweet wine; grape sugar, blood sugar, dextrose. Glucose is the major substrate for glycolysis and energy source.	Galactose	Found as a component of lactose in milk. Good source of sugars for feeding babies.
Fructose	It is also known as levulose, found in fruits and honey; sweetest sugar.	Mannose	Found in legumes; hydrolysis of plant mannosans and gums. A constituent of polysaccharide of albumins, globulins and mucoids.
Pentoses			
Ribose	Derived from nucleic acid of meats and tissues. Structural element of nucleic acids, ATP and coenzymes (NAD and FAD). Ribose and Deoxyribose are found in the backbone structure of RNA and DNA, respectively.	Arabinose	Derived from gum Arabic plus plum and cherry gums; not found free in nature. Has no known physiologic function in man; used in metabolism studies of bacteria.
Deoxyribose			
Ribulose	Formed in metabolic processes. Intermediate in direct oxidative pathway of glucose breakdown.	D-xylose and L-xylose	Found in wood gums, corn cobs and peanut shells; not found free in nature. Poorly digested and has no known physiologic function; used medicinally as a diabetic food.
B. Oligosaccharides (2–10 sugar units)			
Disaccharides **contain two monosaccharides**			
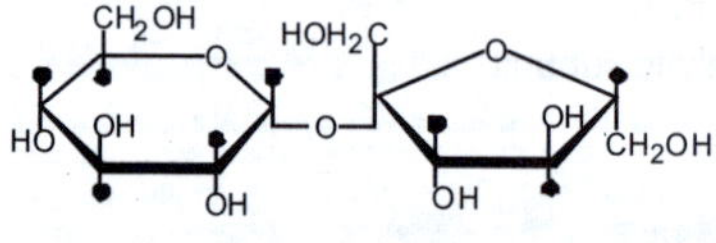 Sucrose (table sugar)	Found in cane sugar, beet sugar. Can be hydrolyzed to glucose and fructose; a non-reducing sugar.	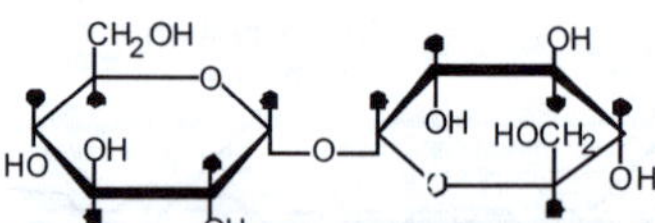 Trehalose	A disaccharide produced in fungi
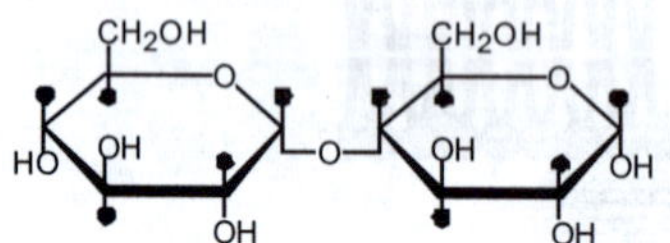 Maltose	Found in germinating grains, used to make beer. An intermediate product of starch digestion. Hydrolyzed to two molecules of glucose; a reducing sugar.	Lactose	Found in milk. Hydrolyzed to glucose and galactose. May occur in urine during pregnancy; a reducing sugar. D-galactose and D-glucose bonded by α-1,4 glycosidic linkage.
Trisaccharides (3 sugar subunits)			
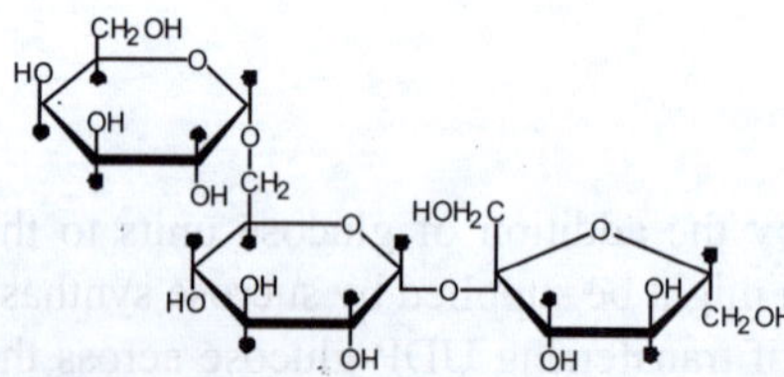 Raffinose	Present in cotton seed meal, molasses, sugar beets and stems. Can be hydrolyzed by enzymes of intestinal bacteria to glucose, fructose and galactose.	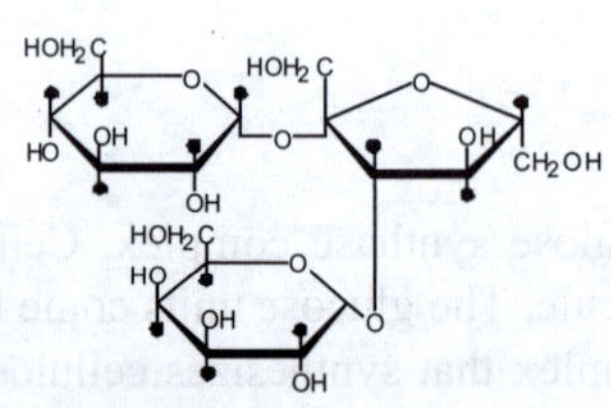 Melizitose	Found in honey, poplars and conifers. Composed of one fructose unit and two glucose units.

Table 3.5 *Continued*

Name	Source and uses
C. Polysaccharides (more than 10 sugar units)	
Glycogen	Rich in meat products and sea foods; polysaccharides of the animal body, often called animal starch; storage form of carbohydrates in body, mainly in liver and muscle.
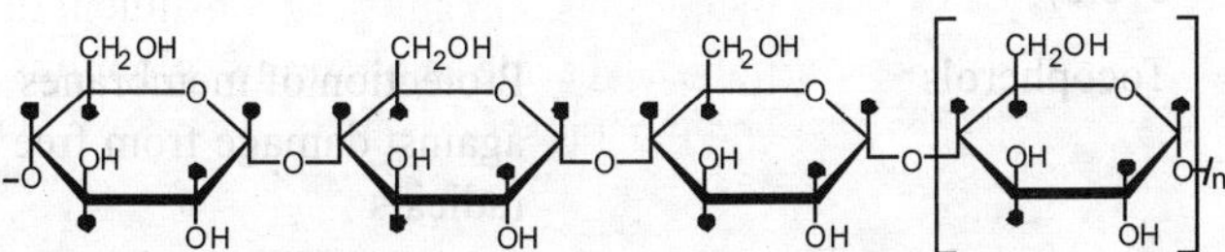 Starch (amylase)	Plants store glucose as a polysaccharide starch. The cereal grains (wheat, rice, corn, oats and barley) as well as tubers such as potatoes are rich in starch. Starch is an important food source of carbohydrates; composed chiefly of amylose and amylopectin; hydrolyzed to glucose.
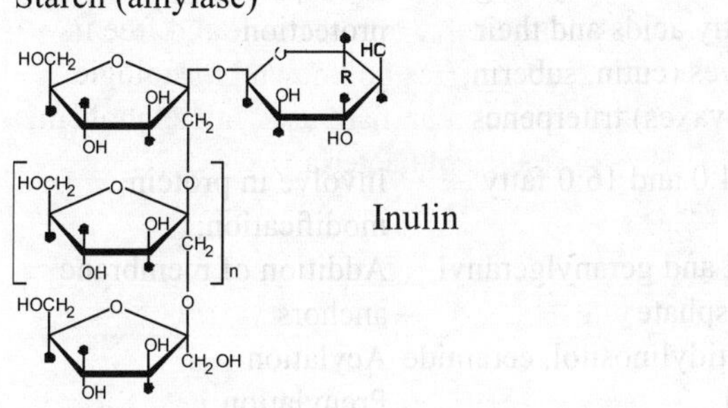 Inulin	Found in tubers and roots of dahlias, artichokes, dandelions, onions and garlic. Hydrolyzable to fructose; used in physiologic investigation for determination of glomerular filtration rate.
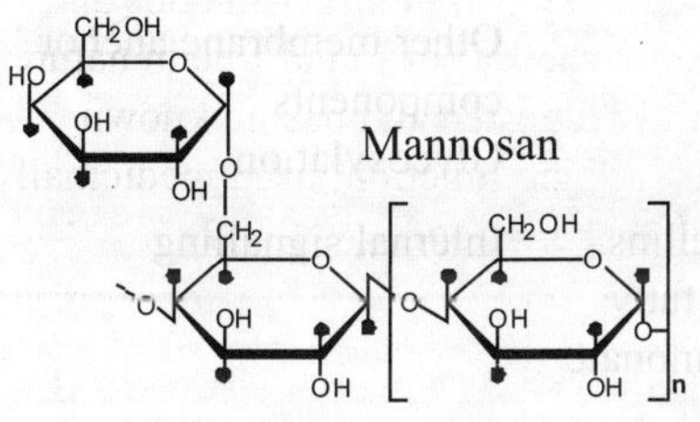 Mannosan	Found in legumes and plant gums. Hydrolyzable to mannose but digestion incomplete; further splitting by bacteria may occur in large bowel.*
Cellulose	The major component in the rigid cell walls in plants and is a linear polysaccharide polymer with many glucose monosaccharide units.
Hemicelluloses	Woody fibres and leaves. Less polymerized than cellulose; may be digested to some extent by microbial enzymes, yielding xylose.
Pectin (Polygacaturonic acid)	Pectin forms a gel with sugar. For this reason, pectin is used in combination with sugar as a thickening agent in the food industry. Commercially pectin is made from apple- and orange-peel.
Chitin	Chitin is an unbranched polymer of *N*-acetyl-D-glucosamine. It is found in fungi and is the principal component of arthropod and lower animal exoskeletons, e.g., insect, crab, and shrimp shells. It is a derivative of cellulose in which the hydroxyl groups of the second carbon of each glucose unit have been replaced with acetamido [-NH(C=O)CH_3] groups.

such as chloroform. Lipids are dominated by carbon and hydrogen atoms. They include fats, oils, steroids, phospholipids, waxes, carotenoids and related compounds; They resemble each other more by their physical than by their chemical properties.

Lipids are vital for biological systems being important energy storage molecules and the major constituent of cell membranes. As the major components of biological membranes, they form a hydrophobic barrier that is critical to life. Lipids are important dietary constituents because of their high energy value and contain fat-soluble vitamins and the essential fatty acids (**Table 3.6**). Lipids have three major roles in cells:

- They provide an important form of energy storage.
- They are the major components of cell membranes.
- They play an important role in cell signalling (e.g., steroids and messenger molecules).

Lipids are synthesized from acetyl CoA, which is formed from the breakdown of carbohydrates, in a series of reactions that resemble the reverse of fatty acid oxidation. Fatty acid biosynthesis in plants takes place within plastids (**Fig. 3.4**).

Lipids are a diverse group of organic compounds and may be classified broadly into three groups:

- simple lipids (e.g., fats and waxes)
- complex lipids (e.g., phospholipids, glycolipids and other complex lipids)
- precursor and derived lipids (e.g., fatty acids, glycerol, steroids, other alcohols, fatty aldehydes and ketone bodies, hydrocarbons, lipid-soluble vitamins and hormones).

The best known lipids are the **fats**, found in small quantities in free state, but in large quantities involved in ester linkages. Fats and oils are the principal long-term energy storage forms of many organisms. The biological functions of the lipids are as diverse as their chemistry. Lipids can be best explained based on structural characteristics.

Biological membranes

Lipids are important components of biological membranes. Membranes define the external boundaries of cells and regulate the molecular traffic across the membrane boundary. In eukaryotic cells, membranes divide the internal space into discrete compartments to segregate processes and components. They organize complex reaction sequences and are central to both biological energy conservation and cell-to-cell communication and structure. When placed in an aqueous environment, phospholipids and sphingolipids naturally form a lipid bilayer with the polar groups on the outside and the non-polar hydrocarbon chains on the inside. However, the precise lipid composition in cell membranes varies from cell to cell and from organelle to organelle.

Table 3.6 Functions of lipid molecules in higher eukaryotes

Lipid types	Major functions
Glycerolipids, sphingolipids, sterols	Membrane structural components
Triacylglycerols, waxes	Storage compounds
Chlorophyll and other pigments, ubiquinone, plastoquinone	Compounds active in electron transfer reactions
Carotenoids (xanthophyll cycle)	Photoprotection
Tocopherols	Protection of membranes against damage from free radicals
Long-chain and very long-chain fatty acids and their derivatives (cutin, suberin, surface waxes) triterpenes	Waterproofing and surface protection
Many 14:0 and 16:0 fatty acids Farnesyl and geranylgeranyl pyrophosphate Phosphatidylinositol, ceramide Dolichol	Involve in protein modification: Addition of membrane anchors Acylation Prenylation Other membrane anchor components Glycosylation
Abscisic acid, gibberellins, brassinosteroids 18:3 fatty acid precursors of jasmonate Inositol phosphates Diacylglycerols	Internal signalling
Jasmonate Volatile insect attractants	External signalling
Essential oils Latex components (rubber) Resin components (terpenes)	Defence and anti-feeding compounds

3.8 PROTEINS

The term 'protein' is derived from a Greek word meaning 'primary' or 'holding first place'. Proteins are the

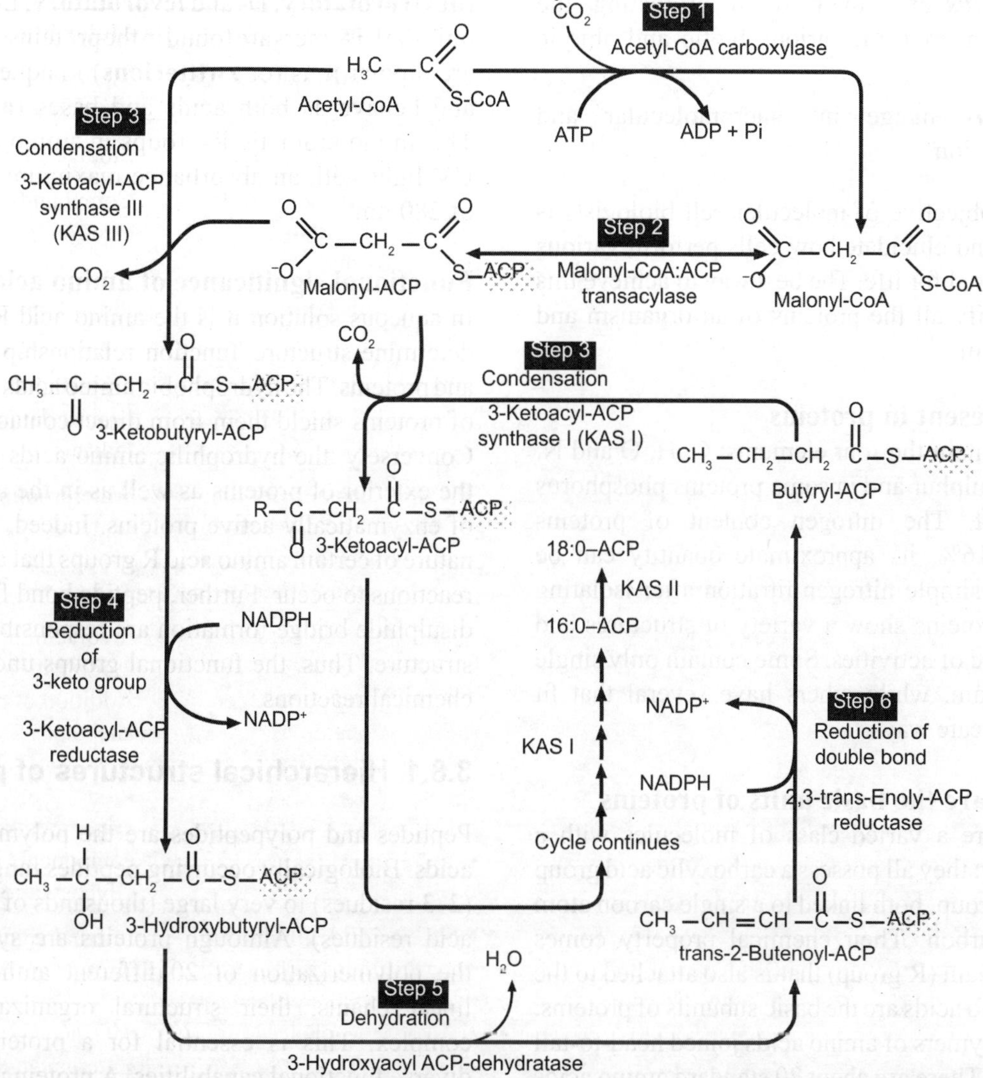

Fig. 3.4 Overview of fatty acid synthesis. Fatty acids grow by addition of two-carbon units. ACP: acetyl-carrier protein; KSA: ketoacyl-ACP synthase.

most abundant biological macromolecules, occurring in all cells and all parts of cells. More than 50% of the dried remains of an organism consist of proteins. Proteins occur in great variety and exhibit enormous diversity of biological function and are the most important final products of the information pathways. Proteins are the molecular instruments through which genetic information is expressed and executed. All proteins are polymers of α-amino acids linked together, in linear and unbranched chain, by peptide bonds. Proteins, whether from the most ancient lines of bacteria or from the most complex forms of life, are constructed from the same ubiquitous set of 20 amino acids. Several other amino acids are found in the body in free or in combined states. Most protein molecules exist in one or other of two general conformations: they can be extended, relatively rigid rods (**fibrous**) or compact and **globular** in shape.

Proteins exit in various forms with diverse functions in nature. Therefore, they have been classified variously. However, they can be grouped into different classes based on their major functions as below:

Structural proteins provide shape and structural rigidity to the cell.

Transporter proteins control the flow of materials across cellular membranes.

Regulatory proteins act as sensors and switches to control protein activity and gene function.

Signalling proteins include cell surface receptors and other proteins that transmit external signals to the cell interior.

Defence proteins are involved in protecting the cell or an organism from various biotic and abiotic invasions.

Motor proteins engage in macromolecular and cellular locomotion.

The main objective of molecular cell biologists is to understand and elucidate how cells perform various processes essential for life. The best way to achieve this goal is to identify all the proteins of an organism and characterize them.

Chemicals present in proteins

All proteins contain the four elements: C, H, O and N. Many contain sulphur and in some proteins phosphorus is also present. The nitrogen content of proteins being around 16%, its approximate quantity can be estimated by a simple nitrogen titration after isolating the proteins. Proteins show a variety of structures and carry out a range of activities. Some contain only single polypeptide chain, while others have several that fit together in intricate ways.

Amino acids are the basic units of proteins

Amino acids are a varied class of molecules with a unique property: they all possess a carboxylic acid group and an amino group, both linked to a single carbon atom called the α-carbon. Their chemical property comes from the side chain (R group) that is also attached to the α-carbon. Amino acids are the basic subunits of proteins. Proteins are polymers of amino acids joined head-to-tail in a long chain. There are about 20 standard amino acids and many numbers of nonstandard amino acids.

Proteins are dehydration polymers of L-amino acids. Apart from proline all other 20 amino acids found in proteins have a common structure (**Fig. 3.5**). In an amino acid, a carbon atom (α-carbon) linked to a carboxyl group, a primary amino group and a side chain (R), which is different in each amino acid. As indicated by their name, amino acids contain one acidic group and one primary amino group in the α position with respect to the carboxyl group. Except in glycine, the α-carbon atom is **chiral** (asymmetric)—it has four chemically active stereoisomers

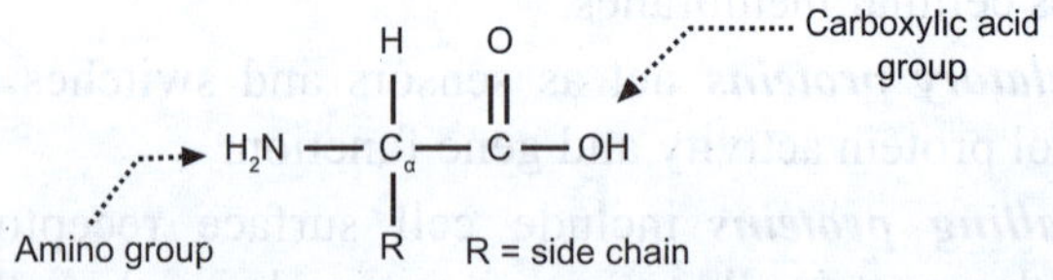

Fig. 3.5 General structure of amino acids.

(**dextrorotatory**, D- and **levorotatory**, L-). However, only the L isomers are found in the proteins. Amino acids are dipolar ions (or **zwitterions**) in aqueous solution and behave as both acids and bases (**amphoteric**). The amino aromatic R-groups in amino acids absorb UV light with an absorbance maximum in the range of 280 nm.

Functional significance of amino acid R-groups

In aqueous solution it is the amino acid R-groups that determine structure–function relationships of peptides and proteins. The hydrophobic amino acids in the interior of proteins shield them from direct contact with water. Conversely, the hydrophilic amino acids are found on the exterior of proteins as well as in the active centres of enzymatically active proteins. Indeed, it is the very nature of certain amino acid R groups that allow enzyme reactions to occur. Further, peptide bond formation and disulphide bridge formation are responsible for protein structure. Thus, the functional groups undergo various chemical reactions.

3.8.1 Hierarchical structures of proteins

Peptides and polypeptides are the polymers of amino acids. Biologically occurring peptides range from small (2–3 residues) to very large (thousands of linked amino acid residues). Although proteins are synthesized by the polymerization of 20 different amino acids into linear chains, their structural organization is very complex. This is essential for a protein to achieve diverse functional capabilities. A protein is one or more polypeptides with a specific 3D shape and can exist in linear form or it can fold into a unique shape that is stabilized by non-covalent interactions between regions in the linear sequence of amino acids. The special organization of a protein into its 3D shape is the key to its unique function.

Two broad classes of large proteins may be distinguished. **Globular proteins** are folded compactly and behave in solution more or less as spherical particles; most enzymes, are globular in nature. **Fibrous proteins** have very high axial ratios (length/width) and are often important structural proteins (e.g., silk fibroin and keratin in hair and wool). Molecular masses can range from a few thousand Da to at least 5 million Da. Some proteins contain bound non-protein material, either in the form of small prosthetic groups, which may act as co-factors in enzyme reactions, or as large associations (e.g., lipoproteins, glycoproteins).

Traditionally, four levels of protein structures are commonly defined (**Fig. 3.6**): primary, secondary, tertiary and quaternary structures (**Table 3.6**). But all levels of proteins have a primary structure, which is simply the sequence of amino acids in the polypeptide, plus any disulphide bridges.

Primary structure

A polypeptide chain consists of a large number of amino acids linked together (20 different amino acids). ***The primary structure of a protein is the linear arrangement (the sequence) of the amino acids in a polypeptide chain***. Polypeptides that have similar amino acid sequences and functions are said to be **homologous**. Sequence comparisons among homologous polypeptides have been used to trace the genetic relationships of different species. The amino acid residues that are identical in all homologues of a protein, referred to as **invariant**, are presumed to be essential for the protein's function.

Table 3.6 Hierarchical structures of proteins

I	Primary	Linear sequence	1D
II	Secondary	Local, repetitive spatial arrangements	2D
III	Tertiary	Three dimensional structure of native fold	3D
IV	Quaternary	Non-covalent oligomerization of subunits (single polypeptides) into protein complexes	4D

The crucial feature of amino acids that allows them to polymerize to form peptides and proteins is the existence of their two identifying chemical groups: the amino ($-NH_3^+$) and carboxyl ($-COO^-$) groups. The α-carboxyl group of one amino acid covalently linked to the α-amino group of the next amino acid by condensation and the bond formed is an amide bond, commonly known as a **peptide bond** when in proteins. When two amino acid residues are linked in head-to-tail fashion the product is a **dipeptide**. When a few amino acids are joined in this fashion, the structure called as an **oligopeptide**. If many amino acids are linked by peptide bonds they form a **polypeptide** (generally have molecular weights below 10,000 Da).

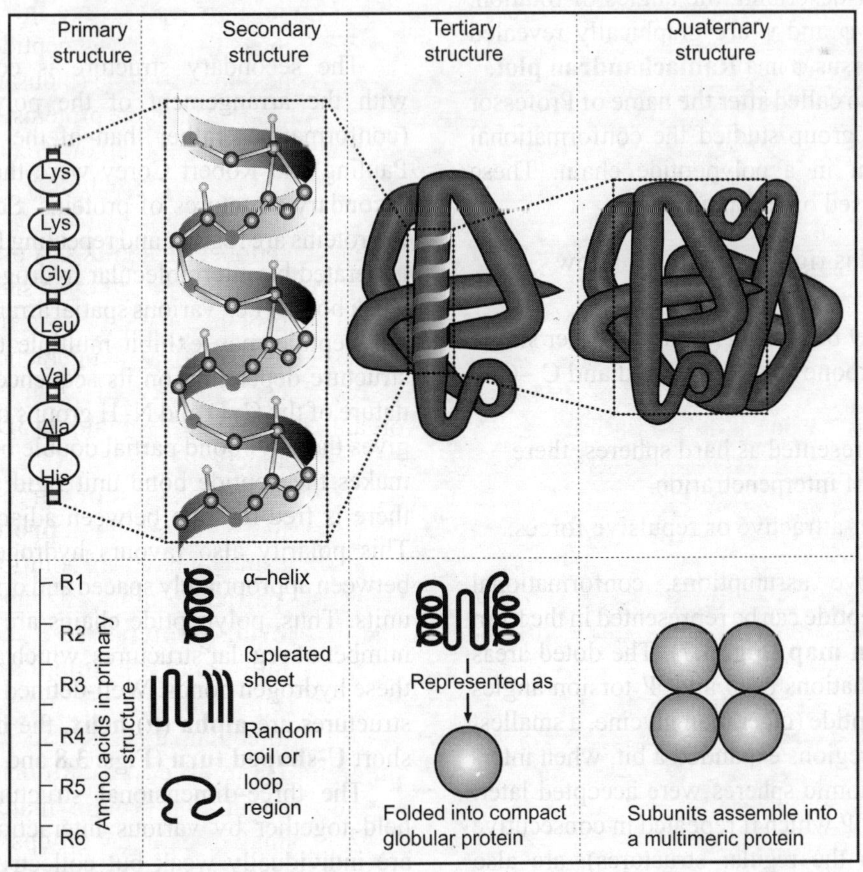

Fig. 3.6 Diagrammatic illustration of levels of structures in proteins.

Secondary structure

When proteins are immersed in water a polypeptide chain folds up according to the polarity of the side chains it contains and the rotation of the peptide backbone; the bond angles are largely determined by van der Waals radii of side chains. To understand protein structures we can measure two torsion angles (also called **dihedral angles**) in the backbone, which define the tilt between two neighbouring amide planes (the plane of peptide bond) with the C_α at the centre of rotation. Thus, we can calculate the allowed backbone conformations of a peptide through interplay of rotation around the bonds defined by the torsion angles and the steric hindrance of side groups determined by their van der Waals radii.

The bond angles resulting from rotations at C_α are labelled φ for the N–C_α bond and ψ for the C_α–C bond. Both φ and ψ are defined as 180° when the polypeptide is in its fully extended conformation and all peptide groups are in the same plane. The φ and ψ angles can have any value between –180° and +180°, but many values are prohibited by steric interference between atoms in the polypeptide backbone and amino acid side chains. The conformation in which both φ and ψ are 0° is prohibited for this reason; this is used merely as a reference point for describing the angles of rotation. Allowed values for φ and ψ are graphically revealed when ψ is plotted versus φ in a **Ramachandran plot**.

These plots are so called after the name of Professor Ramachandran. His group studied the conformational possibilities inherent in a polypeptide chain. These possibilities were based on the assumption:

- The peptide bond is rigid and will not allow rotation.
- The molecule may be rotated about the other single bonds in the backbone (the N–C_α bond and C_α–C' bonds).
- All atoms are represented as hard spheres, there being no chance of interpenetration.
- There are no other attractive or repulsive forces.

Based on the above assumptions, conformational possibilities of a dipeptide can be represented in the form of a **Ramachandran map (Fig. 3.7)**. The doted areas represent the combinations of ϕ and Ψ torsion angles allowed for any dipeptide (other than glycine, a smallest amino acid). These regions expanded a bit, when inter-penetration of the atomic spheres were accepted later. The values of ϕ and Ψ which if repeated in consecutive residues (leading to the regular structures), are also marked in the map (helix α- and β- sheets).

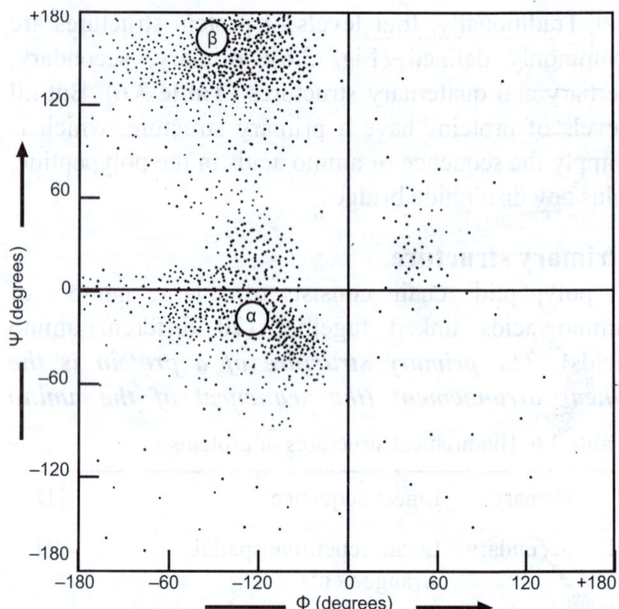

Fig. 3.7 Diagramatic illustration of Ramachandran plot. The conformation of the main-chain atoms in a protein is determined by one pair of φ and ψ angles for each amino acid; because of steric collisions between atoms within each amino acid, most pairs of φ and ψ angles do not occur. In this so-called Ramachandran plot, each dot represents an observed pair of angles in a protein.

The secondary structure is concerned primarily with the arrangement of the polypeptide backbone (conformation) rather than of the side chains. Linus Pauling and Robert Corey were the first to study the secondary structures of proteins. Secondary structures in proteins are regular and repeating local configurations generated by intermolecular hydrogen bonds and other weak bonds (i.e., various spatial arrangements). A single polypeptide may exhibit multiple types of secondary structure depending on its sequence. The highly polar nature of the C=O and N–H groups of the peptide bonds gives the C–N bond partial double bond character. This makes the peptide bond unit rigid and planar, though there is free rotation between adjacent peptide bonds. This polarity also favours hydrogen bond formation between appropriately spaced and oriented peptide bond units. Thus, polypeptide chains are able to fold into a number of regular structures, which are held together by these hydrogen bonds. Well-defined periodic secondary structures are **alpha** (α) **helix**, the **beta** (ß) **sheet**, or a short **U-shaped turn (Figs. 3.8** and **3.9**).

The three-dimensional structure of a protein is held together by various interactions most of which are individually weak but collectively produce stable molecules. These interactions involve the amino acid

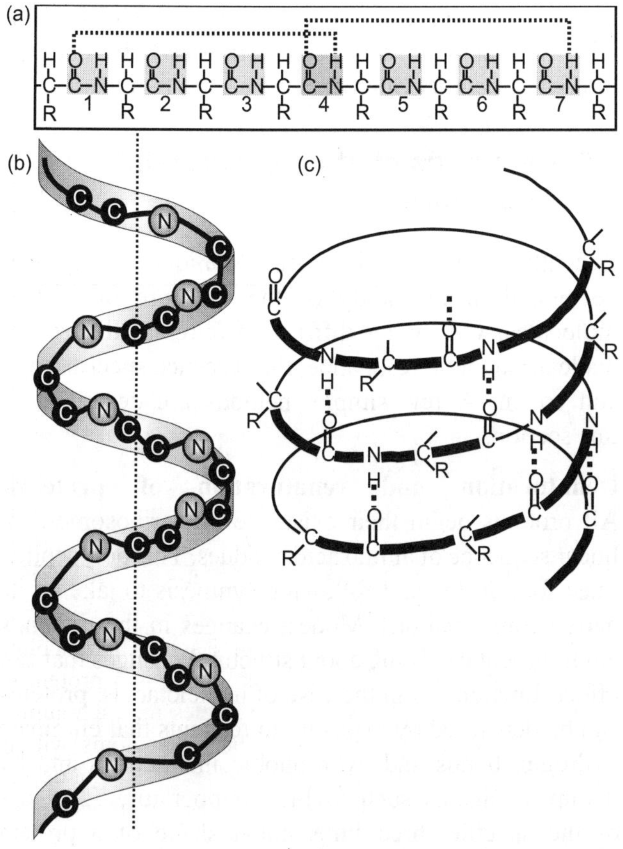

Fig. 3.8 Properties of an α-helix: (a) Two hydrogen bonds in which peptide group is engaged. (b) Formation of a right-handed α-helix. The planes of the rigid peptide bonds are parallel to the long axis of the helix. (c) An α-helix drawn in three dimensions, showing how the hydrogen bonds stabilize the structure.

Parallel   Antiparallel

Fig. 3.9 Two forms of β-pleated sheet: parallel and antiparallel. Dotted lines represents hydrogen bonds.

side chains that interact with one another, with the backbone, with the water molecules and with a prosthetic group or cofactor. These interactions may be due to one or more of hydrogen bonds, electrostatic interactions, van der Waals forces, hydrophobic interactions and disulphide bonds.

Tertiary structure

It refers to the overall conformation of a single polypeptide chain—that is, the 3D arrangement of all its amino acid residues (a native structure). It is stabilized by an array of non-covalent bonds between the diverse side chains of the polypeptide. The native fold is defined as the **active conformation**. This structure of a protein often has protrusions, clefts, or grooves on the surface where particular amino acids are positioned to form sites that bind ligands and in the case of enzymes, catalyze reactions within or between ligands. Most proteins can be categorized on the basis of their overall conformation as being either **fibrous proteins**, which have a highly elongated shape, or **globular proteins**, which have a compact shape. The tertiary structure of a protein also reflects the packaging together (biologically active) of secondary and super-secondary structures to form compact globular domains.

Quaternary structure

This structure means the arrangement of multiple folded protein molecules in a multi-subunit complex. Many polypeptides function in isolation, others assemble into **oligomeric** or multimeric complexes containing either copies of the same polypeptide (**homomultimers**, e.g., ß-galactosidase) or different polypeptides (**heteromultimers**, e.g., hemoglobin) or composed of more than two polypeptide chains (trimeric, tetrameric and so on).

A further level of structural organization is seen when proteins interact with non-protein molecules: cofactors, ligands and substrates. Some cofactors however, are covalently joined to their cognate proteins, in which case they are termed **prosthetic groups**. Many of the cofactors are involved in enzyme-catalyzed reactions (e.g., nicotinamide adenine dinucleotide, NAD^+, in many dehydrogenases.).

Amino acids can be roughly grouped into those with hydrophobic (non-polar) side chains and those with hydrophilic (polar or charged) side chains. In a protein, it is thermodynamically unacceptable to expose predominantly hydrophobic residues to water, therefore soluble proteins posses a **hydrophobic core** where the non-polar residues (Ala, Val, Leu, Ile, Phe) are sequestered and a polar or charged surface exposed to the solvent. Very few charged residues are buried in the interior of a protein, but polar residues are not excluded and the backbone, which is also polar must run through the core. Polar atoms generally make hydrogen bonds with water, but where this is not possible, as in the hydrophobic core of a protein, the hydrogen bonding potential must be taken up by the secondary structure. Otherwise, the protein can be denatured. The formation of secondary structures, predominantly α-helices and ß-sheets, neutralizes the polar atoms of the backbone and polar side chains can also be neutralized by hydrogen bonding with the backbone and with each other. Van der Waals interaction also stabilizes the hydrophobic core.

3.8.2 Binding sites of proteins

Every protein in a cell carries out a particular function, which may be structural, enzymatic or regulatory. The structural proteins are generally fibrous, whereas the other functions generally performed by globular proteins are utilizing highly specific interactions between the protein molecule and one or more intracellular molecules. These interactions occur on the surface of the protein, primarily with the amino acid side chains. Because of the folding of the polypeptide chain, the side chains produce a unique surface conformation of charged and hydrophobic groups. This interacting surface is called a **binding site**. In an enzyme, the binding site is the catalytically active region and is called the **active site**. Binding can occur in any of the following ways:

- Two charged groups in the protein can be arranged to match with the two other charged groups on a small molecule.
- A hydrophobic cluster of side chains of protein can provide a non-polar surface onto which a molecule having a large non-polar group can adsorb.
- Hydrogen bonding groups in the protein can be brought closer to facilitate hydrogen bonds to complementary groups on a small molecule.
- An electronic conformation is formed on the protein surface in an arrangement that allows attractive van der Waals interactions between the protein and many portions of the molecule being bound (***steric fit***).

Binding a small molecule or a part of a bigger molecule to the binding site of a protein may change the shape of a protein. This alteration in shape can improve contact between the active site and the substrate molecule on which an enzyme acts (called **induced fit**) and the biological activity of an enzyme can be regulated.

3.8.3 Methods of studying protein structures

Recently, many sophisticated techniques have been developed in the study of the structure of protein molecules, viz., x-ray diffraction, circular dichroism, nuclear magnetic resonance, fluorescence spectroscopy and so on. A few simple methods are described in this section.

Denaturation and renaturation of proteins: All proteins begin their existence on a ribosome as a linear sequence of amino acid residues. This polypeptide must fold during and following synthesis to take up its native conformations. Modest changes in the protein's environment can bring about structural changes that can affect function. As in the case of nucleic acids, proteins can be denatured by exposure to reagents that eliminate hydrogen bonds and hydrophobic interactions and by disruptive agents such as high temperature. The loss of the specific three-dimensional shape of a protein molecule is called *denaturation*. The change may be temporary or permanent, but the amino acid sequence of the protein remains unaffected. Some of the agents that may cause denaturation are: heat or radiation, strong acids and alkalis, heavy metals, organic solvents and detergents, reducing agents, salt concentration and mechanical stress.

Sometimes, proteins spontaneously refold into the original structure after denaturation, provided suitable conditions,exist. This is called renaturation. This process explains how a polypeptide chain folds or refolds. For example, heat-denatured proteins can be brought back to native form by decreasing temperature to the original level. Removal of chemical denaturing agents can also renature the proteins. A polypeptide chain guided by side chain interactions spontaneously folds and refolds until it achieves the conformation having minimum free energy.

Hydrolysis of proteins: Proteins can be hydrolyzed to free amino acids by addition of 1 M HCl at high temperature. Numerous hydrolytic enzymes have been identified whose role is in digestion or in destruction of bacteria and viruses. These enzymes are called *proteolytic enzymes* or *proteases*. These enzymes are very useful not only in biological systems but also in the laboratory.

3.8.4 Functions of proteins

Of all the molecules encountered in living organisms, proteins are undoubtedly the most functionally diverse biomolecules. Primarily, the function of a protein depends on its amino acid sequence, whereas the secondary, tertiary, quaternary structures and protein modifications provide additional functions to the proteins. The bacterium (e.g., *E. coli*) produces more than 3000 different proteins; a human being produces 50,000 to 100,000. In both cases, each type of protein has a unique 3D structure and this structure confes a unique function. Proteins with different amino acid sequences have different functions and alterations in a single amino acid and can alter the function of a protein. Many proteins have sequence flexibility for similar function. An estimated 20%–30% of the proteins in humans are **polymorphic**, having amino acid sequence variants in the human populations. Proteins often contain crucial regions within their amino acid sequence that are essential to their biological functions. The fraction of the sequence that is critical varies from protein to protein, complicating the task of relating sequence to 3D structure and structure to function. Some major functions of proteins in cells are discussed below.

Catalysis

Enzymes catalyze almost all of the biochemical reactions; their presence enables cellular growth through the life-supporting chemical reactions. They make reactions proceed faster and generate specific products from specific substrates.

Enzymes are biological polymers that direct and accelerate the biochemical reactions in such processes as digestion, energy capture and biosynthesis. Most enzymes work in some compartment of the cell. Enzymes are the most important determinants of biochemical reaction rates (catalysts). The increase in reaction rates can be 10^6- to 10^{12}- fold and like all catalysts enzymes cause reactions to reach equilibrium faster. A catalyst accelerates the rates of forward and reverse reactions by the same factor. It does not alter the change in free energy or equilibrium constant. Enzymes are also highly specific: each catalyzes only a single reaction or a set of closely related reactions.

The enzyme–substrate complex*:* Most enzymes are proteins, with the exception of a small group of catalytically active RNA molecules. To achieve this, an enzyme binds either a single substrate (the molecules on which the enzyme acts) or a set of similar substrates. The molecules acted upon by enzymes are generally called reaction *substrates* rather than ligand and the ligand binding site is called the catalytic site or active site.

Enzyme binding the substrate*:* Two models have been proposed to explain the enzyme binding the substrate (**Fig. 3.10**) In the '*lock and key*' model the substrate fits neatly into the rigid shape of the active site, helped by an appropriate distribution of charges. Some enzymes work this way and a good example is lysozyme, found in tears, saliva and egg white. It digests the cell walls of bacteria, allowing them to burst. The '*induced fit*' model is a little more complex. As the substrate binds, the enzyme changes conformation (or shape). For example, carboxypeptidase is a digestive enzyme. It nibbles away at the carboxyl ends of polypeptides, removing amino acids one at a time. Induced fit is an important mechanism of some enzymes, which bind two substrates. The enzyme has to have two coupled binding sites, one for each substrate. When one substrate molecule binds, the enzyme changes shape, making it easier for the second to bind. Hexokinase is a well known example. This enzyme, catalyses the transfer of a phosphoryl group from ATP to glucose. Both ATP and glucose must bind before the reaction can take place.

Structure of enzymes: Enzymes are proteins and can have molecular weights ranging from 12,000 Da to over 1 million Da. An enzyme may consist of a

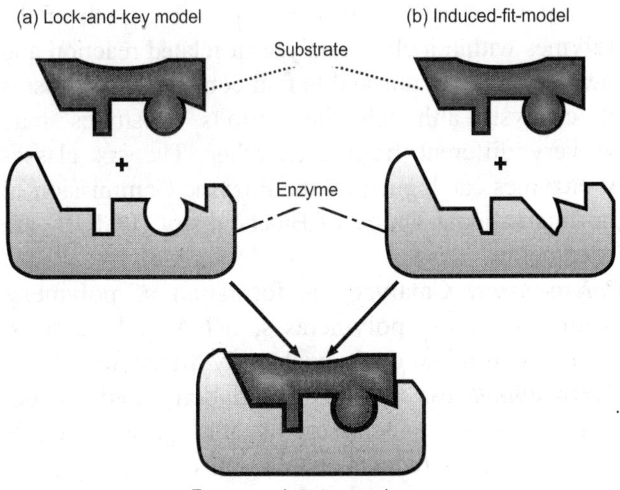

Fig. 3.10 Two models for enzyme substrate binding: (a) Lock and key model: The active site of the enzyme by itself is complementary in shape to that of the substrate. (b) The induced fit model: The enzyme changes shape upon binding a substrate. The active site has a shape complementary to that of the substrate only after the substrate is bound.

single polypeptide chain or multiple subunits, or may form part of a multi-enzyme complex. Some enzymes require no chemical groups for activity; others require an additional chemical component called a *cofactor* (inorganic ions or a complex organic or metalloorganic molecules, called a *coenzyme*). A coenzyme or metal ion that is very tightly bound to the enzyme protein is called a prosthetic group. A complete, catalytically active enzyme, together with its bound coenzyme is called as *holoenzyme*. The protein part of such an enzyme is called as the *apoenzyme* or *apoprotein*. Many enzyme proteins are modified covalently by phosphorylation, glycosylation and other process.

Factors affecting the rate of reaction: Several factors affect the rate of enzyme-mediated reactions. Temperature, pH, the concentration of molecules involved in the reaction (substrate), concentration of the enzyme and the presence of other molecules, which may activate or inhibit the enzyme, all have an effect on the reaction rate.

Metabolic pathways: A number of important biological molecules are rarely made by a single enzymatic reaction. A series of reactions in sequence, with an enzyme for each step, convert a substrate to the final product. These production lines are called metabolic pathways. The pathway as a whole needs to be regulated, to make sure that there is always enough product (or optimal quantities), but never too much.

Classification of enzymes: It is useful in identifying the types of reactions that are catalyzed. Each of the enzymes within a class catalyzes a related reaction and therefore it is not unusual to find common mechanisms of catalysis although the various substrates may be very different from each other. The six classes of enzymes catalogued by the Enzyme Commission of the International Union of Biochemistry (ECIUB) are listed below.

Polymerases: Catalyze the formation of polymers; examples: DNA polymerases, RNA polymerases, primases, terminal deoxynucleotidyl transferase.

Oxidoreductases: Catalyze oxidations and reduction reactions; subclasses of this group include dehydrogenases, reductases, oxygenases, oxidases, peroxidases; and hydroxylases. Examples are: NAD-oxidoreductase, ferricytochrome-c-oxidoreductase, glucose oxidase.

Transferases: Catalyze transfer of groups from one molecule to another, such as methyl, phosphoryl, amino, carboxyl, carbonyl, acyl and glycosyl groups from a donor to an acceptor molecule; examples: Methyl transferases, hydroxy methyl transferases, carboxyl transferases and carbamoyl transferases, acyl transferases, glycosyl transferases, alkyl transferase, amino transferases and phosphoryl transferases.

Hydrolases: Catalyze the hydrolytic cleavage (cleavage of bonds is done by adding water) of C–C, C–O, C–N, P–O and certain other bonds, including acid anhydride bonds. This class includes the esterases, phosphatases and peptidases. Examples are: Lipase or glycerol-ester hydrolase, alkaline phosphatase, ribonulcease and deoxyribonucleases, glucosidases, aminopeptidase, carboxypeptidase A and trypsin.

Lyases: Catalyze cleavage of C–C, C–O, C–N and other bonds by elimination, leaving double bonds and also add groups to double bonds. Examples are: Decarboxylases, hydratases, dehydratases, deaminases and synthases.

Isomerase: A heterogeneous group of enzymes that catalyzes several types of intramolecular rearrangements, geometric or structural changes within a single molecule. The epimerases catalyze the inversion of asymmetric carbon atoms. Mutases catalyze the intramolecular transfer of functional groups. Examples are: Alanine racemase, *cis–trans* isomerases, triosephosphate isomerase, L-methyl malonyl-coA-carbonyl mutase.

Ligases or synthetases: Catalyze bond formation (joining together) between two molecules, coupled in ATP or a similar nucleoside triphosphate as an energy source. Examples are: Synthetase and several other carboxylases.

Signalling

Receptor proteins in cell membranes can bind ligands from the extracellular medium and, by virtue of the resulting conformational change, initiate reactions within the cell in response to that ligand. Ligand binding is similar to substrate binding but the ligand usually remains unchanged. Some hormones are small proteins, such as insulin and growth hormone.

Transport

Proteins function to transport specific substances from one place to another. Transport proteins include haemoglobin, which transports oxygen in the red blood cells and the lipoproteins (LDL and HDL), which transport lipids from the liver and intestines to other organs. Transferrin and ceruloplasmin are serum proteins, which transport iron and copper, respectively, to the liver. Dietary fats are carried in the blood by lipoproteins. The transport of fatty acids from adipose tissue to various organs is by the blood protein **serum albumin**. A very different type of transport of metabolites across permeability barriers such as cell membranes occurs, as mediated by specific membrane proteins. These membrane transport proteins take up metabolite molecules on one side of a membrane and release them on the other side. Examples of membrane proteins include the Na^+–K^+ ATPase and the glucose transporter.

Storage

Proteins that serve as a reservoir of an essential nutrient are called **storage proteins**. For example, **ovalbumin**, the protein of egg white, provides the developing bird embryo with a source of nitrogen during its isolation within the egg. Casein is the most abundant protein in milk and thus the major nitrogen source for mammalian infants. The seeds of the higher plants often contain as much as 60% storage protein to make the germinating seed nitrogen-sufficient during this crucial period of plant development (e.g., zeins in corn, phaseolin in peas, gliadins and glutenin in wheat). Many other molecules and ions are transported and stored in a protein-bound form. This can enhance solubility and reduce reactivity until they are required. Thus, they serve as a source of food.

3.8.5 Structural proteins

Structural proteins provide strength and protection to cells and tissues. Monomeric units of structural proteins typically polymerize to generate long fibres or protective sheets of fibrous arrays. For example, collagen (the major components of connective tissues) is the major protein in skin, bone and connective tissue, fibroin (silk protein) and keratins (a major component of hair) have significant mechanical strength. Elastin, the rubber-like protein found in elastic fibres is found in several tissues in the body (e.g., blood vessels and skin) that must be elastic to function properly. There are also many other structural proteins within the cell, for example, cytoskeleton.

3.8.6 Contractile and motile proteins

Proteins are involved in all cell movements. Certain proteins, for example, actin, tubulin and other proteins that comprise the cytoskeleton, endow cells with unique capabilities for movement. Cell division, muscle contraction and cell motility represent some of the ways in which cells execute motion. The major muscle proteins actin and myosin form sliding filaments which are the basis of muscle contraction. Another class of proteins involved in movement includes dynein and kinesin, the so called **motor proteins** that drive the movement of vesicles (endocytosis or exocytosis), granules and organelles along microtubules serving as established cytoskeletal 'tracks'. The cytoskeletal proteins are also active in ameboid movement of white blood cells.

3.8.7 Defence

A wide variety of proteins are protective (e.g., keratin, fibrinogen and thrombin, immunoglobulin). Prominent among the protective proteins are immunoglobulins (or antibodies) which recognize and bind to bacteria, viruses and other foreign material (the antigen). Another important group of protective proteins is the blood clotting proteins, thrombin and fibrinogen, which prevent the loss of blood when the circulatory system is damaged. Some fishes have antifreeze proteins to protect their blood against freezing in the subzero temperatures. In addition, various proteins serve defensive roles for organisms, including the lytic and neurotoxic proteins of snake and bee venoms and toxic plant proteins, such as ricin.

3.8.8 Exotic proteins

Some proteins for example, **monellin**, a protein found in an African plant, has a very sweet taste and is being considered as an artificial sweetener for human consumption. **Resilin**, a protein having exceptional elastic properties, is found in the hinges of insect wings. Certain marine organisms such as mussels secrete **glue proteins**, allowing them to attach firmly to hard surfaces.

3.8.9 Regulation

Many proteins modify the functions of other molecules by binding to them. Binding of hormone molecule or growth factor to cognate receptors on the target cell changes cellular function. Examples of peptide hormones include insulin and glucagon, both of which regulate blood glucose levels. Growth hormones (are also proteins) stimulate cell growth and division; for example, pituitary **somatotropin** and **thyrotropin** stimulate the thyroid gland. Growth factors (e.g., PDGF and EGF) are polypeptides that control

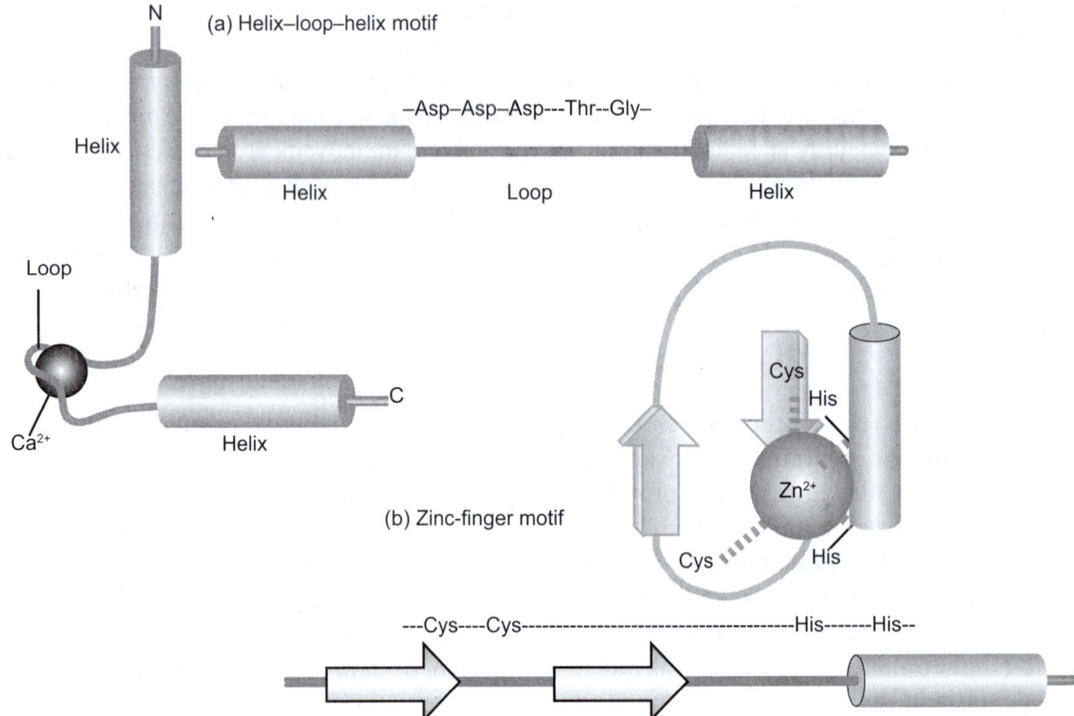

Fig. 3.11 Secondary structure motifs: (a) Helix–loop–helix motif is a characteristic feature of many calcium-binding proteins; (b) Zinc-finger motif is present in many proteins that bind nucleic acids.

animal cell division and differentiation. Another group of regulatory proteins is involved in regulation of gene expression. Examples include **repressors** (e.g., *lac* repressor), which block transcription and in contrast, the mammalian AP1 activates gene expression of the ß-globin gene. Transcription factors bind to and modulate the function of DNA. These various DNA-binding regulatory proteins often posses characteristic structural features, such as helix-turn-helix, leucine zipper and zinc finger motifs (**Fig. 3.11**).

3.8.10 Nucleic acid–binding proteins

Proteins interacting with nucleic acids are very important because they control some of the most fundamental biological processes, including replication, recombination, DNA repair, transcription, RNA processing and protein synthesis. Protein–nucleic acid interactions can be divided into four major categories.

Structural and packaging roles: Includes histones in chromatin, HU in the bacterial nucleoid, viral capsid proteins.

Transport and localization roles: Includes DNA segregation and localization in the nucleus, RNA export and localization and plasmid transfer.

Metabolism and rearrangement roles: Includes DNA and RNA polymerases, nucleases, helicases, DNA repair enzymes, recombinases, topoisomerases.

Gene expression: Includes RNA polymerases, transcription factors, ribosomes and initiation factors, the RNA splicing apparatus, amino acyl tRNA synthetases.

3.8.11 Stress response

Certain proteins mediate the ability of an organism to survive a variety of abiotic and biotic stresses. For example, cytochrome P450, a diverse group of enzymes found in animals and plants, converts a variety of toxic organic contaminants to less toxic; metallothionein, a cysteine-rich intracellular protein found in virtually all mammalian cell binds to and sequesters toxic metals such as cadmium, mercury and silver. High temperatures and other stresses result in the synthesis of a class of proteins called the **heat shock proteins** (hsps) that promote the correct refolding of damaged proteins. If protein damage is severe, hsps proteins promote their degradation. During radiation stress, cells are protected from radiation by DNA repair enzymes.

3.8.12 Complex macromolecules with diverse functions

Many macromolecules contain covalent or non-covalent associations of more than one of the major classes of large biomolecules. This can greatly increase the functionality or structural capabilities of the resulting complex. For example, nearly all enzymes are proteins, but some have a non-covalently attached RNA component, which is essential for catalytic activity. Associations of nucleic acid and protein are known as nucleoproteins. (e.g., ribonuclease, telomerase, ribozyme).

Glycoproteins

Glycoproteins (**Fig. 3.12**) contain both protein and carbohydrate components (one to several oligosaccharides joined covalently to protein).

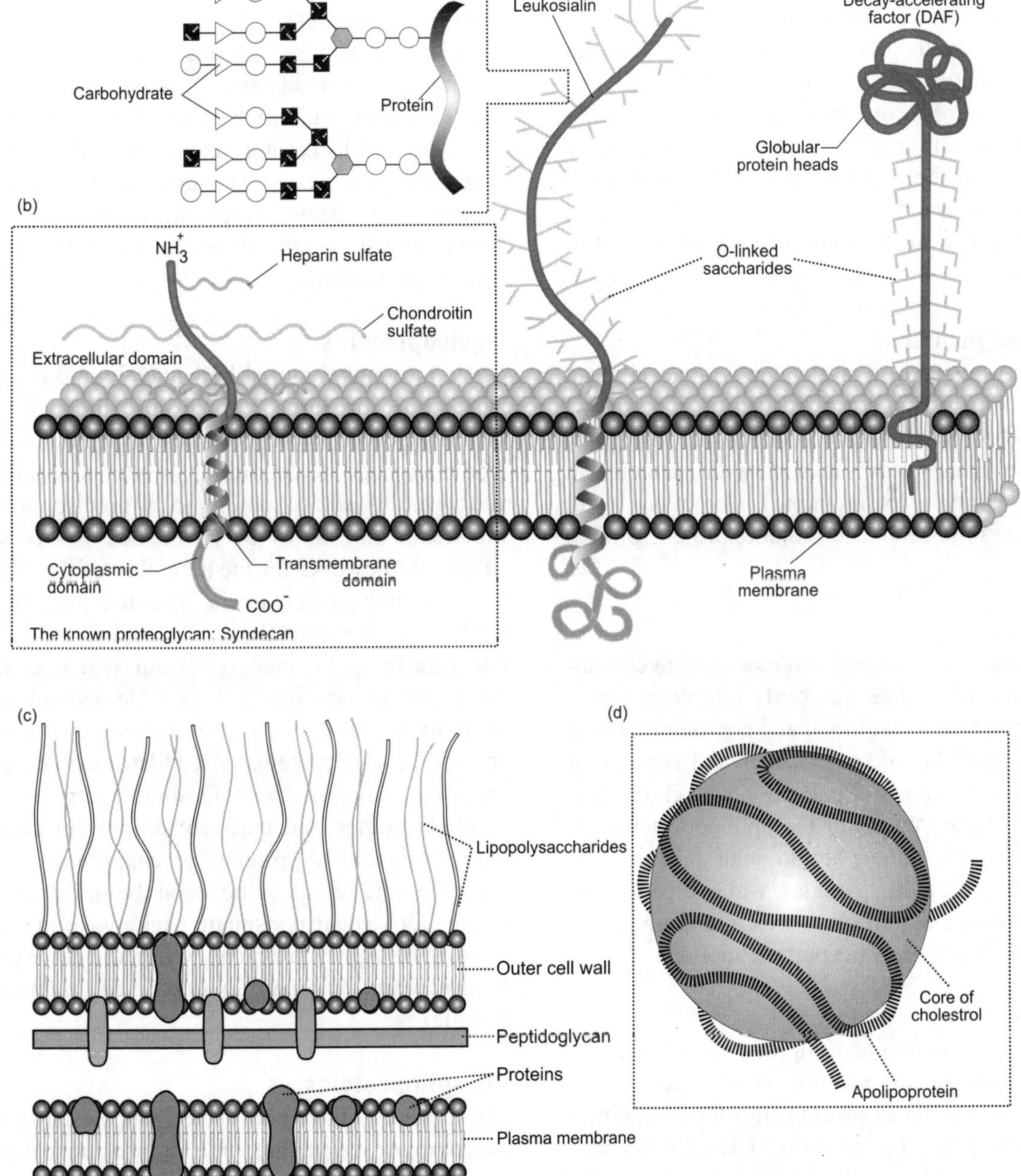

Figure 3.12 (a) Glycoprotein structures (leukosialin and decay-accelerating factor (DAF), different monosaccharide units (different symbols) linked to give oligosaccharides attached to protein; (b) Proteoglycan syndecan; (c) Lipopolysaccharide (LPS) coats the outer membrane of Gram-negative bacteria; (d) A typical lipoprotein (e.g., apolipoprotein).

The carbohydrate is always covalently attached to the surface of the protein, never the interior and is often variable in composition, causing microheterogeneity. Glycoproteins have functions that span the entire range of protein activities and are usually found extracellularly (extracellular matrix). They are important components of cell membranes and mediate cell–cell recognition.

Proteoglycans

Proteoglycans (mucoproteins) are large complexes (>10^7 Da) of protein and mucopolysaccharide found in bacterial cell walls and in the extracellular matrix space in connective tissue. The basic proteoglycan unit consists of a 'core protein' with covalently attached glycosaminoglycan(s). Many proteoglycans are secreted into the extracellular matrix, but some are integral membrane proteins. Proteoglycans act as lubricants and shock absorbers in extracellular spaces.

Lipid-linked proteins

Lipid-linked proteins have a covalently attached lipid component. This is usually a fatty acyl (e.g., myristoyl or palmitoyl) or isoprenoid (e.g., farensyl or geranyl) group. These groups serve to anchor the proteins in membranes through hydrophobic interactions with the membrane lipids and also promote protein–protein associations.

Lipoproteins

In lipoproteins, the lipids and proteins are linked non-covalently. Because lipids are poorly soluble in water, they are transported in the blood as lipoproteins. These are basically particles of triglycerides and cholesterol esters coated with a layer of phospholipids, cholesterol and protein. The structures of the apolipoproteins are such that their hydrophobic amino acids face towards the lipid interior of the particles while the charged and polar amino acids face outwards into the aqueous environment. This renders the particles soluble.

Glycolipids

Glycolipids are membrane lipids in which the hydrophilic head groups are oligosaccharides, which as in glycoproteins, act as specific sites for recognition by carbohydrate-binding proteins. Glycolipids such as cerebrosides and gangliosides, have covalently linked lipid and carbohydrate components and are especially abundant in the membranes of nerve cells.

Lipopolysaccharides

Lipopolysaccharides are the dominant surface feature of the outer membrane of Gram-negative bacteria such as *E. coli* and *Salmonella typhimurium*. They are prime targets of the antibodies produced by the vertebrate immune system in response to bacterial infection and are therefore important determinants of the serotype of bacterial strains. The lipopolysaccharides of some bacteria are toxic to humans and other animals (it lowers blood pressure).

Protein complexes

Cellular architecture contains many large complexes of the different classes of macromolecules with themselves or with each other. Many of the major structural and locomotory elements of the cell consist of protein complexes. The **cytoskeleton** is an array of protein filaments, which organizes the shape and motion of cells and the intracellular distribution of subcellular organelles.

Nucleoproteins

Nucleoproteins comprise both nucleic acid and protein. Ribosomes are large cytoplasmic ribonucleoprotein complexes, which are the sites of protein synthesis. Under appropriate conditions, mixtures of the rRNAs and proteins self-assemble in a precise order into functional ribosomes *in vitro*. The RNAs are not simply frameworks for the assembly of the ribosomal proteins, but participate in both the binding of the mRNA and in the catalysis of peptide bond synthesis. **Chromatin** is the material from which eukaryotic chromosomes are made. It is a deoxynucleoprotein complex made up of roughly equal amounts of DNA and small, basic proteins called **histones**. These form a repeating unit called a nucleosome. Correct assembly of nucleosomes and many other protein complexes requires assembly proteins, or chaperones. Histones neutralize the repulsion between the negative charges of the DNA sugar phosphate backbone and allow the DNA to be tightly packaged within the chromosomes. Viruses are another example of nucleoprotein complexes.

Phosphoproteins

These proteins have phosphate groups esterified to the hydroxyls of serine, threonine, or tyrosine residues. For example, casein, the major protein in milk, contains many phosphates and serves to bring essential phosphorus to the growing infants. Glycogen phosphorylase is a well-studied example.

Metalloproteins

Metalloproteins are either metal storage forms, as in the case of ferritin, or enzymes in which the metal atom participates in a catalytically important manner. Hemoproteins are a subclass of metalloproteins because their prosthetic group is **heme.** Because heme-containing proteins have many prominent biological functions, they are often considered as a special class of metalloproteins.

Flavoproteins

Flavin is an essential substance for the activity of a number of important oxidoreductases. For example, FMN and FAD are involved in electron transport and oxidative phosphorylation.

Chromoproteins

Chromoproteins are coloured proteins. Their prosthetic group often contains a metal element (Fe, Cu, Mg). These compounds are generally classified on the basis of their prosthetic group having, or not having, the tetrapyrrol structure called porphyrin and according to their respiratory or non-respiratory physiological role.

Porphyrin chromoproteins include:

- Respiratory proteins: hemoglobins, cytochromes, cruorins and heme enzymes (e.g., catalase and peroxidases)
- Non-respiratory pigments: chloroplastins, phycobilins

Non-porphyrin chromoproteins include:

- Respiratory pigments: hemerythrins and hemocyanins
- Non-respiratory pigments: carotenoproteins and flavoproteins

Metalloproteins include:

- Proteins containing iron: ferritin and siderophilin
- Proteins containing copper: hemocuprein and ceruleoplasmin

Extracellular matrix proteins

Other than the globular proteins discussed above, there are many elongated fibrous proteins: proteins of extracellular matrix. The dense connective tissue (bone and tendons) for example, predominantly contains **collagen**, the most plentiful protein in the mammalian body. A layer of protein fibres known as the basal lamina supports the loose connective tissue found under all epithelial layers.

3.9 NUCLEIC ACIDS

The nucleic acids are among the largest molecules found in organisms. Nucleotides have a variety of roles in cellular metabolism. They are the energy currency in metabolic transactions; the essential chemical links in the response of cells to hormones and other extracellular stimuli; and the structural components of an array of enzyme cofactors and metabolic intermediates. Mainly, nucleic acids are the molecular repositories of genetic information. The structure of every protein and thus of every biomolecule and cellular component, is a product of information programmed into the nucleotide sequence of a cell's nucleic acids. The ability to store and transmit genetic information from one generation to the next is a fundamental condition for life. Two types of nucleic acids function in all living things:

- deoxyribonucleic acid (DNA)
- ribonucleic acid (RNA).

DNA and RNA are polymers of nucleotides, which are consist of a nitrogenous base, a pentose sugar and phosphoric acid.

3.9.1 Clinical significance

Clinical problems associated with nucleotide metabolism in humans are predominantly the result of abnormal catabolism of the purines. Abnormal purine metabolism arises from the insolubility of the degradation byproduct, uric acid. Excess accumulation of uric acid leads to hyperuricemia, more commonly known as gout. This condition results from the precipitation of sodium urate crystals in the synovial fluid of the joints, leading to severe inflammation and arthritis. Most forms of gout are the result of excess purine or of a partial deficiency in the salvage enzyme HGPRT. Two severe disorders are associated with defects in purine metabolism: Lesch–Nyhan syndrome and severe combined immunodeficiency disease (SCID). This deficiency selectively leads to a destruction of B and T lymphocytes. Another glycogen storage disease von Gierke's disease also leads to excessive uric acid production. This disorder results from a deficiency in glucose 6-phosphatase activity.

Summary

- Molecules are extraordinarily diverse and generate molecular information required for creation of various macromolecules with diverse biological roles (***biomolecules***).
- Cells contain thousands of different molecules made up of only a few elements. However, the most important molecules found in cells are those based on a skeleton of covalently bonded carbon atoms, called ***hydrocarbons***, which have immense biological importance.
- Most biomolecules contain more than one ***functional group***. The distinct chemical properties of each functional group contribute to the behaviour of any molecule that contains it.
- Elements exist in cells as small chemical compounds or ions and macromolecules.
- ***Macromolecules*** are polymers of chains constructed by assembling together small molecular subunits, which include: Carbohydrates, Lipids, Proteins and Nucleic acids.
- ***Carbohydrates*** contain carbon, hydrogen and oxygen and have the basic formula $(CH_2O)_n$.
- ***Lipids*** are a heterogeneous group of compounds that are insoluble in a aqueous solvent. They include fats, oils, steroids, phospholipids, waxes, carotenoids and related compounds; they resemble each other more by their physical than by their chemical properties. Lipids are vital molecules in biological systems and are important energy storage molecules and the major constituent of cell membranes.
- ***Membranes*** define the external boundaries of cells and regulate the molecular traffic across the membrane boundary. Membranes also divide the internal space into discrete compartments to segregate processes and components.
- ***Proteins*** are polymers of α-amino acids linked together, in linear and unbranched chains, by peptide bonds. They exist in various forms with diverse functions. Traditionally, four levels of protein structures are commonly defined: primary, secondary, tertiary and quaternary structures.
- The three-dimensional structure of a protein is held together by various interactions, most of which are individually weak but collectively produce stable molecules. These interactions may be due to one or more hydrogen bonds, electrostatic interactions, van der Waals forces, hydrophobic interactions and disulphide bonds.
- The functions of proteins depend on its amino acid sequence. Secondary, tertiary, quaternary structures and protein modifications provide additional functions to the proteins.
- ***Enzymes*** catalyze reactions or make them faster and generate specific products from specific substrates.
- ***Receptor proteins*** can bind ligands and initiate reactions. Their function is to transport specific substances from one place to another.
- Proteins act as a reservoir of essential nutrients called ***storage proteins***. The seeds of the higher plants often contain as much as 60% storage proteins.
- ***Structural proteins*** provide strength and protection to cells and tissues and are involved in cellular movement. A wide variety of proteins are involved in protection.
- Proteins that interact with nucleic acids control some of the most fundamental biological processes, including replication, recombination, DNA repair, transcription, RNA processing and protein synthesis.
- ***Nucleotides*** have a variety of roles in cellular metabolism. They are the energy currency in metabolic transactions; the essential chemical links in the response of cells to hormones and other extra–cellular stimuli; and the structural components of an array of enzyme cofactors and metabolic intermediates.
- ***Nucleic acids*** are the molecular repositories of genetic information. The structure of every protein and ultimately of every biomolecule and cellular component, is a product of information programmed into the nucleotide sequence of a cell's nucleic acids. The ability to store and transmit genetic information from one generation to the next is a fundamental condition for life.
- Two types of nucleic acids function in all living things: deoxyribonucleic acid (DNA) and ribonucleic acid (RNA). ***DNA*** and ***RNA*** are polymers of nucleotides, which consist of a nitrogenous base, a pentose sugar and phosphoric acid.

Exercises

1. How many types of elements do living organisms generally contain?
2. Explain the role of carbon in organic matter.
3. What are the different types of functional groups that are involved in molecular interactions?
4. Briefly explain the different classes of biomolecules found in cells.
5. Describe the biosynthesis of carbohydrates and their classification.
6. List the biological significance of carbohydrates
7. What are lipids? What is the basis of classification of lipids?
8. Describe the main functions of lipids in biological systems.
9. Describe the structure of amino acids and their classification.
10. Describe different levels of protein organization.
11. Briefly discuss the role of proteins in cellular functions.
12. List the different classes of enzymes with an example.
13. Describe the major steps involved in the biosynthesis of ribonucleotides..
14. Explain the mechanism of formation of deoxynucleotides.
15. How is nucleotide biosynthesis regulated?

Further Reading

Books

Bailey JE and Ollis DF (1986), *Biochemical Engineering Fundamentals,* McGraw-Hill Book Company, New York, New Delhi.

Branden C, Tooze J (1991), *Introduction to Protein Structure,* Garland.

Channarayappa (1993), Resistance mechanism of *Lycopersicon* species to the vector *Bemisia tabaci* and ultrastructural changes in leaf curl virus infected tomato leaves, PhD thesis submitted to the University of Agricultural Sciences, GKVK, Bangalore.

Cullis PR, Fenske DB, Hope MJ (1996), Physical properties and functional roles of lipids in membranes, In: Vance DE, Vance JE (editors), *Biochemistry of Lipids, Lipoproteins and Membranes*, Elsevier.

Garrett RH and Grisham CM (1999), *Biochemistry*, 2nd edition, Saunders College Publishing, San Diego.

Gurr MI, Hardwood JL (1991), *Lipid Biochemistry: An Introduction*, 4th edition, Chapman and Hall.

Murray RK, Granner DK, Mayes PA and Rodwell VW (2003), *Harper's Illustrated Biochemistry*, 26th edition, McGraw-Hill.

Smith CA and Wood EJ (1991), *Biological Molecules*, Chapman and Hall, London.

Starr C and Taggart R (1995), *Biology, The Unity and Diversity of Life*, Wadsworth Publishing Company, A Division of International Thomson Publishing Inc.

Weil JH (1997), *General Biochemistry*, New Age International Ltd.

References

Chothia C (1984), Principles that determine the structure of proteins, *Annu. Rev. Biochem*, 53:537–72.

Goodsell DS (1991), Inside a living cell, *Trends Biochem. Sci.*, 16:203–6.

Kendrew JC (1961), The three-dimensional structure of a protein molecule, *Sci. Am.*, 205(6):96–111.

Lindahl U, Hook M (1978), Glycosaminoglycans and their binding to biological macromolecules, *Annu. Rev. Biochem.*, 47:385.

Narlikar GJ and Herschlag D (1997), Mechanistic aspects of enzymatic catalysis: Lesions from comparisǒn of RNA and protein enzymes, *Annu. Rev. Biochem.*, 66:19–59.

Schramm VL (1998), Enzymatic transition states and transition state analog design, *Annu. Rev. Biochem.*, 67:693–720.

4 Prokaryotic Cells

Learning objectives

After reading this chapter, you will be able to:

- List the different types of prokaryotes
- Describe the composition and structure of a bacterial cell–cell membrane and cell wall
- Understand the structure, composition and applications of nanobacteria

4.1 INTRODUCTION

The building blocks of all life forms are cells. They occur singly as a great variety of unicellular organisms and as the subunits of organs and tissues in plants, animals and other multicellular organisms. Thus, the cell is life's basic unit of structure and function. The term, 'cellula' was first used by Robert Hooke in 1665, which means the small cavities surrounded by walls that he saw in cork. Anton van Leeuwenhoek (1632–1723) was the first to observe freeliving cells in soil and water. He described bacteria (bacilli, cocci, spirilla and Monera), protozoa, rotiferas and hydra. He was also the first to describe the movement of blood cells and sperm cells of humans and many other animals and insects. Later, the protoplasm—the substance within the cell––was discovered (Hanstein 1880). In 1831, Robert Brown discovered the **nucleus** in an epidermal cell of an orchid plant. In 1862, Kolliker introduced the term **cytoplasm**. Cells are tiny expandable factories that simultaneously synthesize several thousand different molecules. ***The essence of life is ability to grow and divide to produce progeny cells, which are likewise capable of generating new cellular molecules and replicating themselves.***

Schleiden and Schwann were the first to postulate that the cell is the basic unit of structure and function. This simple but fundamental biological generalization is known as the **cell theory**. It states that:

- All living systems are composed of cells and their products.
- All cells arise from other cells or pre-existing cells.
- All the metabolic reactions of a living organism take place within cells.

The above postulates allow us to first study the component parts of the cells and expand the conclusions to the complete organism, because cells from a wide variety of organisms share many common features in structure and function. By concentrating on the apparently universal features of cellular function, a basic framework for understanding all living systems can be established. Cell and molecular biology is a rapidly growing field of research and is fundamental not only for advancement in basic sciences, but also finds application in medicine, agriculture and biotechnology.

Evolutionary tree of life

The classification of living things has traditionally depended on comparisons of their outward appearances (morphological characteristics). The relative closeness of traits among the organisms can be interpreted in terms of evolution from common ancestors.

When the differences between them became wider, this method of classification became inadequate. In prokaryotes which are organisms without a membrane-bound nucleus, the task becomes more complex due to close similarities between the species. Therefore, in prokaryotes, classification was traditionally based on the biochemical and nutritional requirements. However, recent methods of genome analysis have simplified the process and are more accinate determining evolutionary relationships.

The first to appear in the evolutionary stage are **protistans**. With their efficient metabolic strategies, the early protistans underwent rapid divergences and adaptive radiations. Somewhere in this evolutionary tree, some of their descendents gave rise to the kingdoms of animals, fungi and plants (**Fig. 4.1**).

4.2 FIVE KINGDOM APPROACH OF CLASSIFICATION

Grouping organisms on the basis of common features is called **classification**. The science of classification is called **taxonomy**. Modern biologists feel that a five kingdom approach to a classification of living organisms is superior to the other classifications. The biological nomenclature is based on the binomial system (Carl Linnaeus, 1707–78). In this system each organism has two Latin names: a generic name beginning with a capital letter and a specific name beginning with a lower case letter (e.g., humans are named *Homo sapiens*).

The extended binomial system is arranged in a hierarchy with the largest group, the kingdom, at the

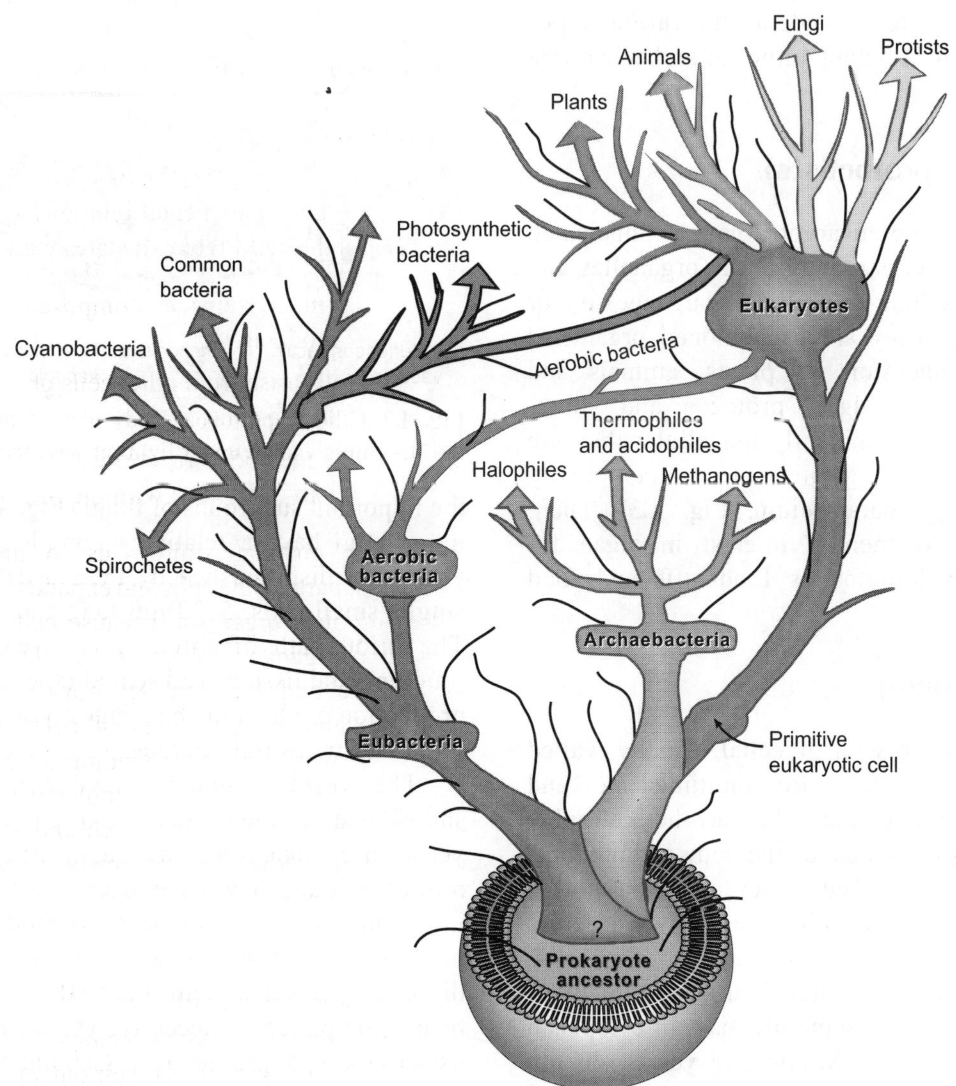

Fig. 4.1 The evolutionary tree for all modern groups of organisms. Molecular analysis now indicates that all life forms investigated so far can be classified into three domains.

top of the hierarchy: Kingdom, phylum, class, order, family, genus and species.

4.2.1 Monera (prokaryotes)

Prokaryotes are microscopically small, single-celled organisms that live nearly everywhere (no organized nucleus or cellular organelles). They are divided into two major taxonomic groups: bacteria (or *eubacteria*) and archaea (*archaebacteria*). Prokaryotes usually exist alone, without associating with other cells. According to current opinion, members of archaea are more closely related to eukaryotes than they are to other prokaryotes. Prokaryotes, such as mycoplasmas, bacteria and cyanobacteria (blue-green algae) are autotrophic or heterotrophic, motile or non-motile and predominantly unicellular. Present-day prokaryotes, which include all the various types of bacteria (**Fig. 4.2**) belong either to archaebacteria or eubacteria.

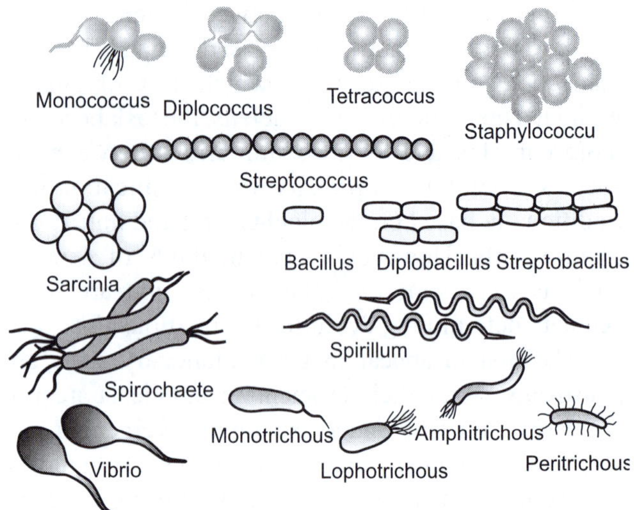

Fig. 4.2 Different forms of bacteria.

4.2.2 Protista (protoctists)

These relatively large eukaryotes have sophisticated and highly organized structures, true organelles and nucleus, are unicellular or multicellular, ingestive or photoautotrophic. They are trichomonad organisms resembling the ancestors of plants, animals and fungi. They include algae, protozoa and slime-molds. Although predominantly unicellular they are much larger and have greater internal complexity than bacteria (e.g., paramecium, **Fig. 4.3**). There is considerable commercial interest in algae for their possible exploitation as foodstuffs and food supplements.

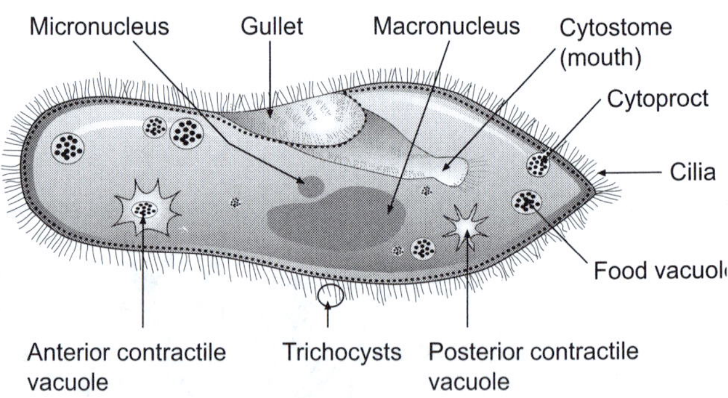

Fig. 4.3 Ciliated protozoa. Body plan of paramecium and the structures visible under light microscope.

4.2.3 Fungi (fungi)

Fungi are eukaryotic with mycelial, usually walled and septate cells; they are multinucleate and predominantly multicellular. They are heterotrophic so that food is digested outside the body and products of digestion are absorbed. They are non-motile. Molds are higher fungi with a vegetative structure called **mycelium**. As illustrated in **Fig. 4.4**, the mycelium is a highly branched system of tubes and within these tubes is a mobile mass of cytoplasm containing many nuclei. Molds like yeasts, do not contain chlorophyll. Reproduction may be sexual or asexual. Spore properties form an important component in fungi classification. Yeasts form one of the important subgroups of fungi (**Fig. 4.5**). Although most fungi have a relatively complex morphology, yeasts are distinguished by their usual existence as single, small cells, 5–30 µm long and 1–5 µm wide. The various paths of reproduction of yeasts are asexual (budding and fission) and sexual (sporulation). In the production of alcoholic beverages, yeasts are the only important industrial microbes.

The yeast is another favourable model that shares many of the advantages of bacteria. However, yeasts are eukaryotes, having a true nucleus and mitochondria. They have been used for centuries in producing wine and beer. In modern molecular biology, mutant strains of yeast are often used to discover genes that control growth, division and cell behaviour patterns. Recently, yeast cells have been used as tools for producing large numbers of copies of human chromosome fragments (called 'yeast artificial chromosomes'). This feature has immense potential in genetic engineering.

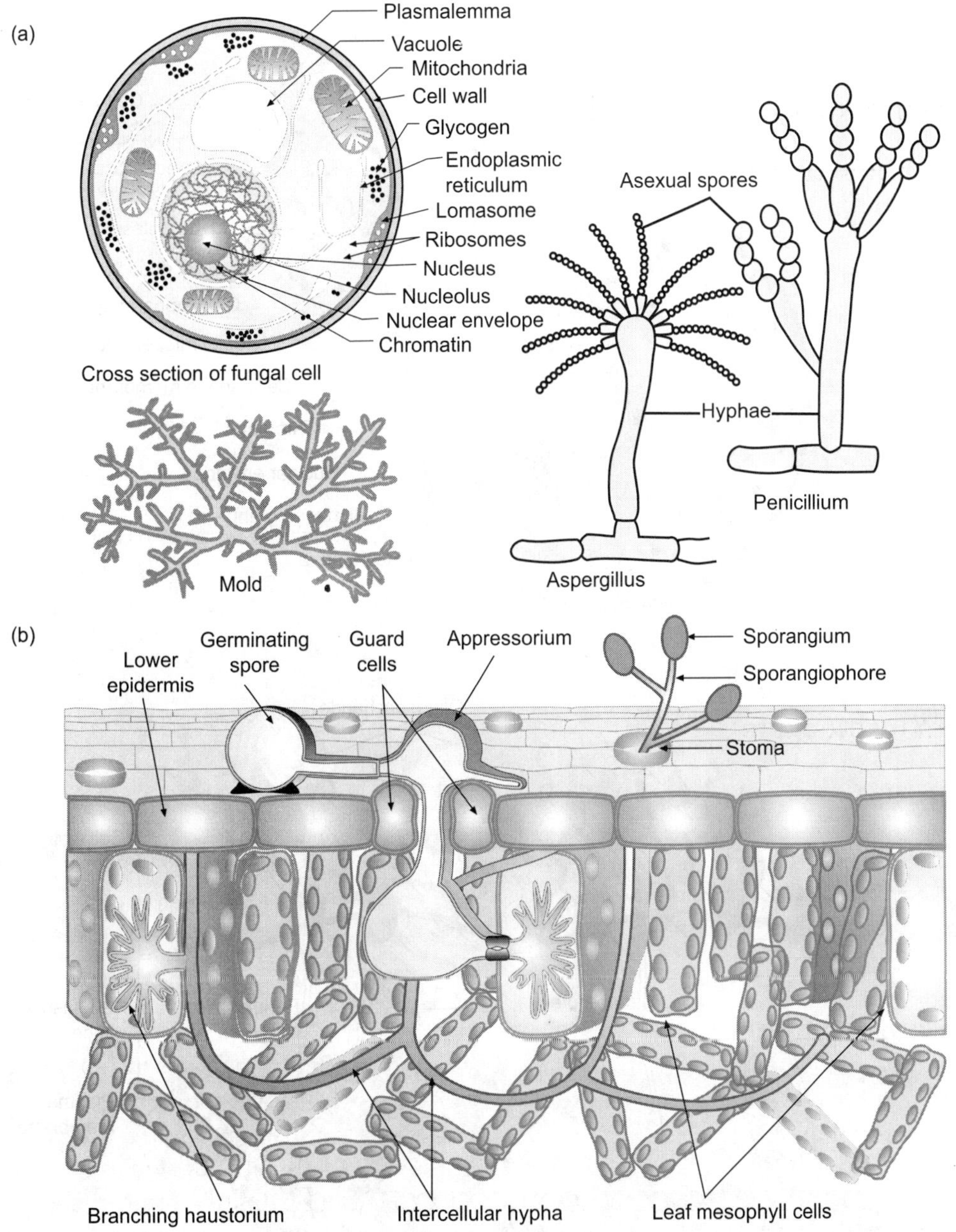

Fig. 4.4 Fungi: (a) The mycelial structures of molds and cross section of a typical fungal cell; (b) *Phytophthora infestans* growing in a diseased potato leaf, with sporangiophores emerging from the underside of the leaf.

4.2.4 Animalia (animals)

They have eukaryotic cells and are predominantly multicellular but wall-less. They are aerobic and capable of complex differentiation. They are heterotrophic (they cannot produce their own food; they depend on other organisms for it), food is ingested before digestion. They generally move about far more than other kinds of organisms. Humans are one of the important groups of the animal kingdom.

In vitro animal cell culture has been used for the study of structure, function and development. Many types of cells, including several kinds of human cells, can be cultured using simple in vitro techniques. Primary cell cultures represent normal animal tissue that is usually derived from either skin cells or embryos. Due to their short life in cultures, many immortalized cell lines have been developed and used in several molecular biology experiments. **Embryonic stem cells** or cells from early-stage mammalian embryos

(a)

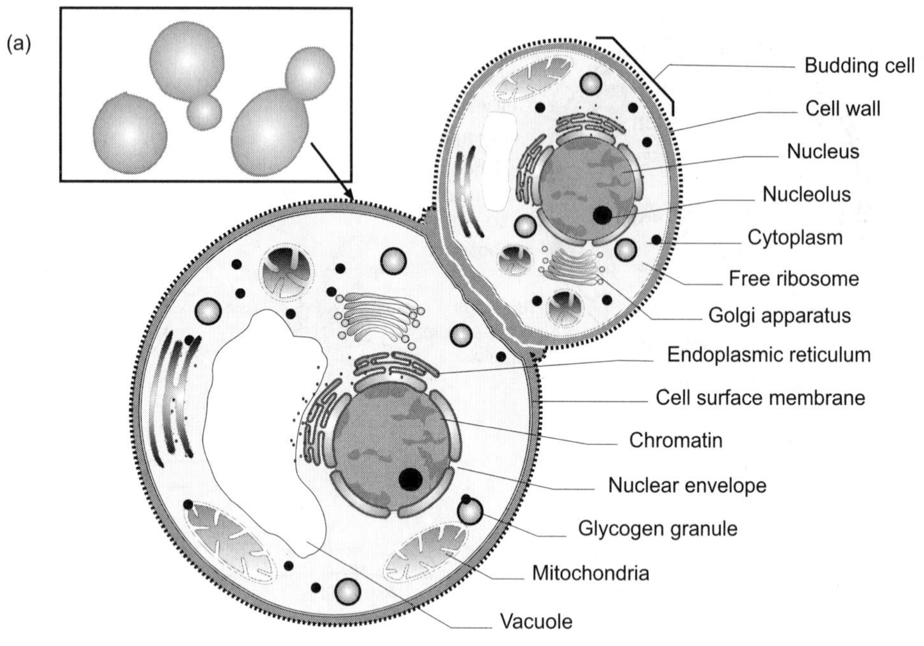

(b)

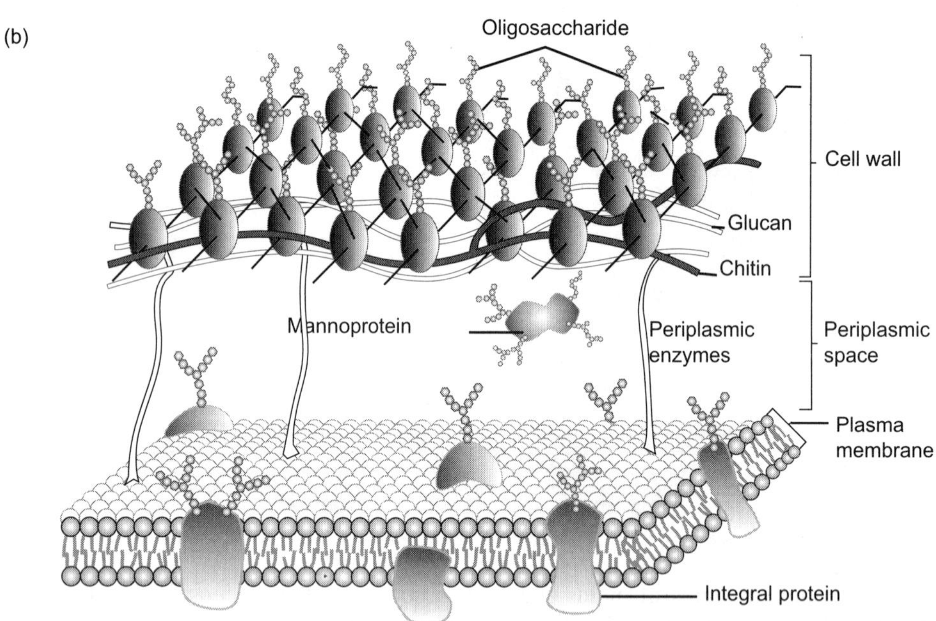

Fig. 4.5 Yeast: (a) Fine structure of yeast (saccharomyces) visible under an electron microscope and (b) the yeast cell surface.

provide an especially versatile model. Under specific conditions, they can differentiate into mature cells, which can be conveniently studied by molecular biologists. This area of research has great promise for the production of tissues and organs for replacement therapy. Recently, molecular biologists were successful in transforming many cells by introducing foreign genes and expressing the inserted recombinant gene(s). From these transgenic embryonic cells, **transgenic animals** have been generated.

4.2.5 Plantae (Plants)

Plants are eukaryotes predominantly multicellular cell walled primarily photoautotrophic nutrition (produce their own food through photosynthesis) aerobic usually differentiated and non-motile. The discovery that many plant cells have totipotency (a single cell can give rise to complete plant in cell culture environment) stimulated molecular biologists to standardize tissue culture protocols for many plant species. Using these cell culture techniques, mechanism of action of many plant hormones (e.g., auxins, gibberelins, cytokines.) have been well studied. Recently, a small plant, the common wall cress (*Arabidopsis*), has been used as a tool for molecular genetics.

4.3 CELLULAR DIVERSITY WITHIN AN ORGANISM

The cell is the fundamental unit of life; all living organisms consist of cells, which contain either membrane-bound nuclei or comparable structures without membranes. Elucidation of the similarities and differences between cells is particularly important to understanding cell biology. This is because, on the one hand, all cells share common fundamental properties that have been conserved throughout evolution (e.g., all cells use DNA as their genetic material, are surrounded by plasma membranes and have similar mechanisms for energy metabolism). On the other hand, present-day cells have evolved a variety of different lifestyles. The universe of the cell is complex and diverse. However, it is well observed that where there is diversity, there is also similarity. A closer examination of the complexity and organization of living systems reveals that there are basically three levels of complexity: **viral**, **prokaryotic** and **eukaryotic**. Organisms such as bacteria, protozoa and yeasts consist of single cells that are capable of independent self-replication. However, the more complex organisms are composed of collections of cells that function in a coordinated manner, with different cells specialized to perform particular tasks. The fundamental similarities between different types of cells provide a unifying theme to cell biology, allowing the basic principles learned from experiments with one kind of cell to be extrapolated and generalized to other cell types.

Cells differ greatly in size (e.g., 0.2–5 μm in bacteria to 6 inches in an ostrich egg). Cell number enormously varies from a single cell in bacteria and other unicellular organisms to trillions of cells in higher organisms, e.g., 6×10^{12} cells in humans. The shape of the cell is also highly variable all forms of cell shapes exist from fixed: to irregular structures. Cellular activities (particularly in eukaryotes) are highly diverse and specialized.

4.4 ARCHAEBACTERIA

This prokaryotic group is probably a modern descendent of a very ancient type of prokaryotes. The molecular properties of archaeans, when compared to those of bacteria and eukaryotes, revealed that archaeans are in many ways closer to the eukaryotes than to the outwardly similar bacteria. For example, the archaean system for synthesizing proteins is more like that of eukaryotes. Archaea, although clearly a prokaryotic (no nuclear membrane), differ from bacteria by lack of peptidoglycan in their cell wall and the presence of ether-linked lipids bonded to glycerol in their membranes. These prokaryotes have adaptability to extreme conditions and are often referred to as **extremophiles**. Extremophiles can be classified according to the type of exotic conditions in which they live: very high (thermophile) or low temperatures (psychrophile), low (acidophile) or high salt concentrations (alkalophile), or high pressure (barophile).

The genomes of many archaeal species have now been sequenced and have been found to have circular, compact chromosomes of about 1.5 to 3 Mbp. Chromosomes are organized by nucleosome-like entities. DNA replication appears to be mediated by polymerases that are similar to those of eukaryotes. Transcription machinery of archaea is also similar to those of eukaryotes. No sigma factors are found in archaea, which are the key players in bacterial transcription. Researchers are very interested in the **extremozymes**, enzymes (protein catalysts) that work under noxious conditions.

4.5 BACTERIA

Bacteria are diverse in their habitats and nutritional capacities. Various species possess traits that allow them to exploit virtually every conceivable energy source, nutrient and habitat. Many bacterial species are essential to human life and life on earth; they are of great interest to humans. Although they are notorious for their role in causing human diseases, a number of species are essential to good health and foods such as cheese, sourdough bread and sauerkraut, are manufactured with the aid of certain bacteria. Other types of bacteria such as actinomycetes produce antibiotics.

4.5.1 Composition of the bacterial cell

There is no 'typical cell', which represents basic characteristics of all the cells. However, when a bacterium is compared with an eukaryotic cell, five universal structural features can be observed in living cells: (i) chromosome (DNA), (ii) ribosomes, (iii) cell membrane, (iv) cell wall and (v) a surface layer which may be an inherent part of the cell wall.

The chemical composition of these structures is made up of macromolecules such as nucleic acids (DNA and RNA), proteins, polysaccharides, phospholipids, or some combination thereof. The macromolecules are made up of primary subunits such as nucleotides, amino acids or sugars (**Table 4.1**). The arrangement or sequence, in which the subunits are put together, called the primary structure of the molecule, determines the exact properties that the macromolecule will have. Thus, at a molecular level, the primary structure of a macromolecule determines its function or role in the cell and the functional aspects of bacteria are related directly to the structure and organization of the macromolecules in a cell. The variations present within the primary structure of these molecules is responsible for the diversity that exists among prokaryotes.

The overall chemical composition of a bacterial cell may be inferred by chemical analysis of a bacterium such as *E. coli* (**Table 4.2**). The macromolecules comprise about 96% of the dry weight of the cell (dry weight is the remaining material after all water is removed). Small molecules and inorganic ions, which are constituents of the cytoplasm, comprise the remaining 4%.

4.5.2 General structure of prokaryotic cells

The prokaryotes are an immense and heterogenous group that includes bacteria and archaea. The *prokaryotic* cells (*pro* = before, *karyon* = nucleus) do not have a true nucleus. The development of the electron microscope in the 1950s revealed the anatomical structure of bacteria and further confirmed the lack of a nuclear membrane in the cell. Most prokaryotes are similar in their external appearance. Prokaryotic cells are comparatively small, seldom more than a few micrometres.

Morphologically, just as in higher animals, there are different shapes among bacterial cells. Cell shape is generally characteristic of a given bacterial species, but can vary depending on growth conditions. The most common shapes are (**Fig. 4.6**) spherical (coccioid), rods (bacilli) and spirals (spirilli). Bacterial cells can also form chains (streptococcus) or clusters (staphylococcus) of cells. However, some bacteria have complex lifecycles involving the production of stalks and appendages and some produce elaborate structures bearing reproductive spores.

The most obvious structural characteristic of bacteria is their small size. For example, *E. coli*, an average sized bacterium with average cell length of 1 μm has a cell volume of approximately 1–2 μm^3. This corresponds to a wet mass of ca. 1 pg, assuming that the cell contains sufficient amount of water. The dry mass of a single cell can be estimated as 20% of the wet mass, amounting to 0.2 pg. About half of the dry mass of a bacterial cell consists of carbon and also about half of it can be attributed to proteins. Therefore, a fully-grown 1-litre culture of *E. coli* (at an optical density of 1.0, corresponding to ca. 10^9 cells/ml) yields ca. 1 g wet cell mass. The small size allows for a large surface area-to-volume ratio, which facilitates rapid uptake and intracellular distribution of nutrients and excretion of wastes.

Table 4.1 Macromolecules of cell material

Macromolecules	Primary subunits	Location in the cell
Proteins	Amino acids	Flagella, pili, cell walls, cytoplasmic membranes, ribosomes, cytoplasm
Polysaccharides	Sugars (carbohydrates)	Capsules, inclusions (storage), cell wall
Phospholipids	Fatty acids	Membranes
Nucleic acids (DNA/RNA)	Nucleotides	DNA: nucleoid (chromosome), plasmids; rRNA, ribosomes; mRNA, tRNA, cytoplasm

Table 4.2 Chemical composition of *E. coli* under balanced growth conditions

Molecules	Dry weight (%)
Protein	55
Total RNA	20.5
DNA	3.1
Phospholipids	9.1
Lipopolysaccharides	3.4
Peptidoglycan	2.5
Glycogen	2.5
Small molecules: precursors, metabolites, vitamins	2.9
Inorganic ions	1.0
Total dry weight	100.0

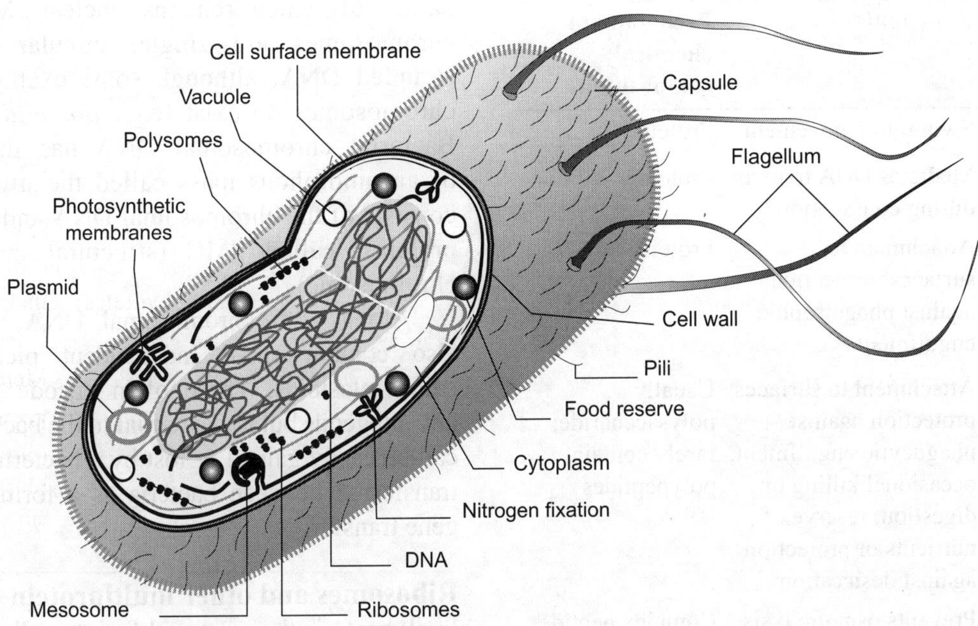

Fig. 4.6 Structure of a generalised rod-shaped bacterium (a typical prokaryotic cell).

Structurally, a bacterial cell has three distinct architectural regions:

- **Appendages** in the form of flagella and pili (or fimbriae);
- A **cell envelope** consisting of a capsule, cell wall and plasma membrane;
- A **cytoplasmic region** that contains the cell chromosome (DNA), ribosomes and various inclusions.

Different structures have different functions and specific chemical compositions (**Table 4.3**). Many bacteria are capable of relatively rapid movement generated by the action of thread-like structures—the **cilia** and **flagella** that protrude from the cell surface.

Prokaryotes have thousands of species. Most are identified on the basis of more subtle characteristics, such as nutritional requirements, energy sources, chemical compositions, biochemical capacities and staining with a specific dye. Mycoplasms are the primitive form of prokaryotes, which do not have a cell wall (**Fig. 4.7**).

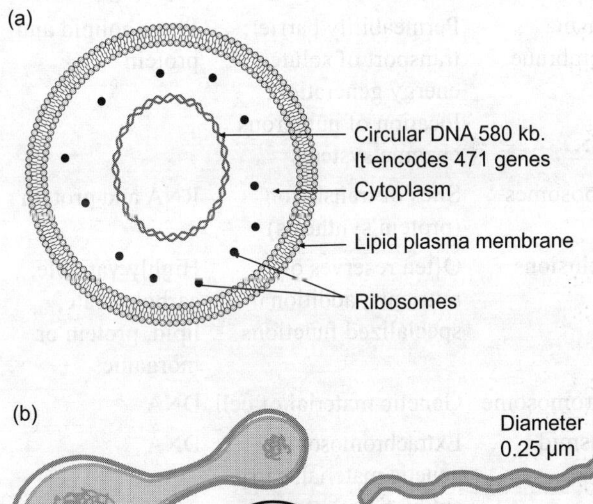

Fig. 4.7 Primitive bacteria: (a) Mycoplasma (*Mycoplasma genitalium*) and (b) MLO, *Sirochaeta citri*.

4.6 INTRACELLULAR STRUCTURES OF BACTERIAL CELL

In comparison to an eukaryotic cell, the intracellular features of the bacterial cell are extremely simple. Bacteria do not contain cell organelles, instead, the chromosome and perhaps ribosomes are the only easily observable intracellular structures found in all bacteria. However, specialized groups of bacteria that contain more complex intracellular structures do exist.

Table 4.3 Characteristics of typical bacterial cell structures

Structure	Function(s)	Predominant chemical composition
Flagella	Swimming movement	Protein
Pili sex pilus	Mediates DNA transfer during conjugation	Protein
Common pili or fimbriae	Attachment to surfaces; protection against phagotrophic engulfment	Protein
Capsules (includes slime layers and glycocalyx)	Attachment to surfaces; protection against phagocytic engulfment, occasional killing or digestion; reserve of nutrients or protection against desiccation	Usually polysaccharide; rarely contain polypeptides
Cell wall	Prevents osmotic lysis of cell protoplast and confers rigidity and shape on cells	Complex peptide and carbohydrate polymers (called peptidoglycan)
Plasma membrane	Permeability barrier; transport of solutes; energy generation; location of numerous enzyme systems	Phospholipid and protein
Ribosomes	Sites of translation (protein synthesis)	RNA and protein
Inclusions	Often reserves of nutrients; additional specialized functions	Highly variable; carbohydrate, lipid, protein or inorganics
Chromosome	Genetic material of cell	DNA
Plasmid	Extrachromosomal genetic material, carry antibiotic-resistant genes	DNA

Bacterial chromosome and plasmids

The bacterial chromosome is not enclosed inside a membrane-bound nucleus, but resides freely inside the cytoplasm. This means that the transfer of cellular information through the processes of transcription, translation and DNA replication all occur within the same compartment and can interact with other cytoplasmic structures readily. The bacterial chromosome is not packaged using histones to form chromatin as in eukaryotes but exists as a highly compact supercoiled structure, the precise nature of which remains unclear. Most bacterial chromosomes are single, circular and double-stranded DNA, although some examples of linear chromosomes do exist (e.g., *Borrelia burgdorferi*). Bacterial chromosomal DNA has the appearance of an amorphous mass called the **nucleoid**, which consists of the chromosomal DNA and its associated proteins, called SMC (structural maintenance of chromosome).

Along with chromosomal DNA, most bacteria also contain small independent pieces of DNA called plasmids. These often encode traits that are advantageous but not essential to the bacteria. Plasmids can be easily gained or lost by a bacterium and can be transferred between bacteria as a form of horizontal gene transfer.

Ribosomes and other multiprotein complexes

In all bacteria the most prolific intracellular structure is the ribosome, the site of protein synthesis. Ribosomes are present in all living organisms. All prokaryotes have 70 S (where S = Svedberg units) ribosomes while eukaryotes contain 80 S ribosomes in their cytoplasm. The 70S ribosome is made up of 50 S and 30 S subunits. The 50S subunit contains the 23 S and 5 S rRNA, while the 30S subunit contains the 16 S rRNA. These rRNA molecules differ in size in eukaryotes and are complexed with a large number of ribosomal proteins, the number and type of which can vary slightly between organisms. The ribosome is the most commonly observed intracellular multiprotein-complex in bacteria.

Intracellular membranes

Although it is not a common feature of all bacteria, many microbes contain intracellular membranes in addition to (or as extensions of) their cytoplasmic membranes. These membranes normally function to maximize the amount of membrane available for use by membrane-bound energy-yielding processes, much the same as they do in mitochondria and chloroplasts. Examples of bacteria containing intracellular membranes are phototrophs, nitrifying bacteria and methane-oxidizing bacteria. Intracellular membranes are also found in bacteria belonging to the poorly studied Planctomycetes group.

Cytoskeleton

The prokaryotic cytoskeleton is not well developed as much as seen in the eukaryotic cells. It is the collective name for all the structural filaments in prokaryotes. Recent advances in staining and visualization technology and structure determination have confirmed that filaments indeed exist in these cells. Homologues for all major cytoskeletal proteins in eukaryotes have been found in prokaryotes. The cytoskeletal elements play an essential role in cell division, protection, locomotion and determination of shape and polarity in various prokaryotes.

Nutrient storage structures

Most bacterial habitats do not contain large amounts of essential nutrients at all times and the food supply is not secured. To accommodate these varying levels of nutrients, bacteria have evolved several ways of storing nutrients in times of plenty for use in times of want. For example, many bacteria store excess carbon in the form of polyhydroxyalkanoates or glycogen. Some microbes store soluble nutrients such as nitrate in vacuoles. Sulphur is most often stored as elemental granules which can be deposited either intra or extracellularly. Sulphur granules are especially common in bacteria that use hydrogen sulphide as an electron source. Most of these structures are visible under a microscope and appear surrounded by a thin non-unit membrane to separate them from the cytoplasm.

Gas vesicles

Gas vesicles are spindle-shaped structures found in some planktonic bacteria and provide buoyancy to these cells by decreasing their overall cell density. The gas vesicles are made up of a protein coat that is impermeable to solvents such as water but permeable to most gases. By adjusting the amount of gas present in their gas vesicles, bacteria can increase or decrease their overall cell density and thereby move up or down within the water column to maintain their position in an environment optimal for growth.

Carboxysomes

Carboxysomes are intracellular structures found in many autotrophic bacteria such as cyanobacteria, knallgas bacteria, nitrosobacteria and nitrobacteria. They are proteinaceous structures resembling phage heads in their morphology and contain the enzymes for carbon dioxide fixation, especially ribulose bisphosphate carboxylase/oxygenase, RuBisCO and carbonic anhydrase. It is thought that the high local concentration of the enzymes along with the fast conversion of bicarbonate to CO_2 by carbonic anhydrase allows faster and more efficient CO_2 fixation than possible inside the cytoplasm.

Similar structures are known to harbour the coenzyme B_{12}-containing glycerol dehydratase, the key enzyme of glycerol fermentation to 1,3-propanediol, in some enterobacteriaceae (e.g., *Salmonella*).

Magnetosomes

Magnetosomes are intracellular structures found in magnetotactic bacteria that allow them to sense and align themselves along a magnetic field (magnetotaxis). The ecological role of magnetotaxis is unknown but it is hypothesized to be involved in the determination of optimal oxygen concentrations. Magnetosomes are composed of the mineral magnetite and are surrounded by a non-unit membrane. The morphology of magnetosomes is species-specific.

4.7 BACTERIAL CELL COMPONENTS

4.7.1 Cytoplasm

Prokaryotic cells, even though lacking in internal membranes, appear to have functional compartments. The most obvious of these is a large, ill-defined region called the **nuclear zone** (or **nucleoid**), which is the control centre for cell operation. The chromosome of bacteria is typically one large circular molecule of DNA, more or less free in the cytoplasm. The grainy dark spots apparent in the cell interior are the ribosomes. These (small spherical bodies about 15 nm in diameter), crowd the cytoplasm and are the sites of protein synthesis. Prokaryotic ribosomes are 70S in size, being composed of 30S and 50S subunits. Few prokaryotes contain vacuoles, vesicles and reserve deposits of complex sugars or inorganic materials. Many bacteria also contain additional small circular DNA molecules called plasmids that exist separately from the cell's chromosome.

The cytoplasm of bacterial cells consists of an aqueous solution of three groups of molecules:

- macromolecules such as proteins (enzymes), DNA, mRNA and tRNA
- small molecules that are energy sources, precursors of macromolecules, metabolites or vitamins
- various inorganic ions and cofactors.

The cytoplasm of prokaryotes is more gel-like than that of eukaryotes and the processes of cytoplasmic streaming, which are evident in eukaryotes, do not occur.

4.7.2 Inclusions

The cytoplasm of prokaryotic cells contains different types of inclusion granules. They are usually reserve materials which may occupy a substantial part of the cytoplasm (**Table 4.4**). For example, carbon and energy reserves may be stored in the form of glycogen (a polymer of glucose) or as polybetahydroxybutyric acid (a type of fat) granules. Polyphosphate inclusions are reserves of PO_4 and possibly energy; elemental sulphur (sulphur globules) is stored by some phototrophic and some lithotrophic prokaryotes as reserves of energy or electrons. Some inclusion bodies are actually membranous vesicles or intrusions into the cytoplasm, which contain photosynthetic pigments or enzymes (**Fig. 4.8**).

Prokaryotes grow rapidly when plenty of nutrients and optimal conditions available and punctuated by

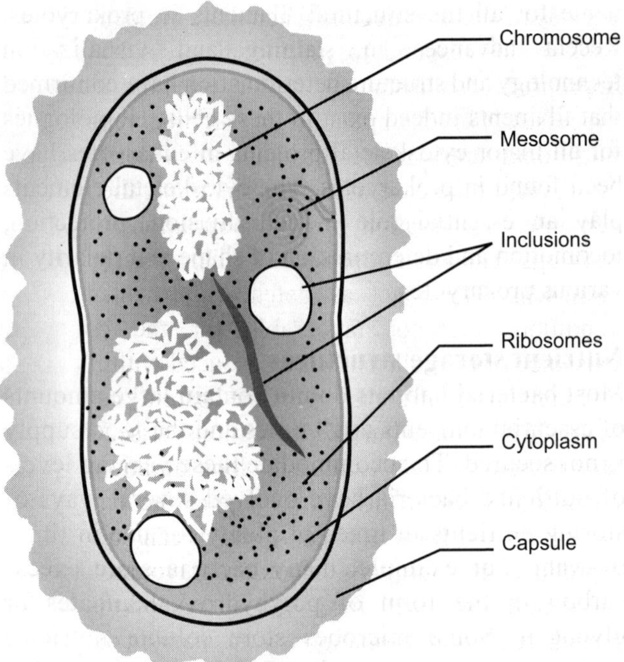

Fig. 4.8 Bacterial cell containing inclusions, capsule and mesosome.

Table 4.4 Commonly found inclusions in bacterial cells

Cytoplasmic inclusions	Found in	Composition	Functions
Glycogen	Many bacteria, e.g., *E. coli*	Polyglucose	Reserve carbon and energy source
Polybetahydroxybutyric acid (PHB)	Many bacteria, e.g., Pseudomonas	Polymerized hydroxy butyrate	Reserve carbon and energy source
Polyphosphate (volutin granules)	Many bacteria, e.g., Corynebacterium	Linear or cyclical polymers of PO_4	Reserve phosphate
Sulphur globules	Phototrophic purple and green sulphur bacteria and lithotrophic colourless sulphur bacteria	Elemental sulphur	Reserve of electrons (reducing source) in phototrophs and reserve energy source in lithotrophs
Gas vesicles	Especially in aquatic bacteria	Protein hulls or shells inflated with gases	Buoyancy in the vertical water column
Parasporal crystals	Endospore-forming bacilli	Protein	Not known but toxic to certain insects
Magnetosomes	Certain aquatic bacteria	Magnetite (iron oxide) Fe_3O_4	Orienting and migrating along geo-magnetic field lines
Carboxysomes	Many autotrophic bacteria	Enzymes for autotrophic CO_2 fixation	Site of CO_2 fixation
Phycobilisomes	Cyanobacteria	Phycobiliproteins	Light-harvesting pigments
Chlorosomes	Green bacteria	Lipid, protein and bacteriochlorophyll	Light-harvesting pigments and antennae

periods of stasis or decline when there are deficiencies. Prokaryotes respond to nutritional stress by making appropriate readjustments to the relevant metabolic processes. For example, in *E. coli*, the response to nutritional stress is the formation of a dormant cell. However, in the case of *Bacillus subtilus*, the ultimate response to starvation lies in sporulation, the spore develops as an **endospore** within the bacterium (**Fig. 4.9**). These act as defence mechanism when conditions are not favourable for the continued existence of the cell. They are produced within the cell and contain all of the DNA of the 'parent' cell within them. When unfavourable conditions arise, they are able to hold the DNA intact for long durations. When conditions become favourable again, the endospores germinate and form new cells, which can survive and reproduce. If conditions become unfavourable again, these new cells will form endospores of their own. Hence, endospore formation is a mechanism of survival rather than a method of reproduction.

Endospores exhibit no signs of life (**cryptobiotic**); they are highly resistant to environmental stresses such as high temperature (even on boiling for hours they still retain their viability), irradiation, strong acids or disinfectants (**Table 4.5**). Endospores are probably the most durable cell types, because when appropriate environmental conditions resume, they germinate back into vegetative cells. The trigger for endospore formation from vegetative cells comes from environmental signals that indicate a limiting factor for vegetative growth, such as exhaustion of an essential nutrient.

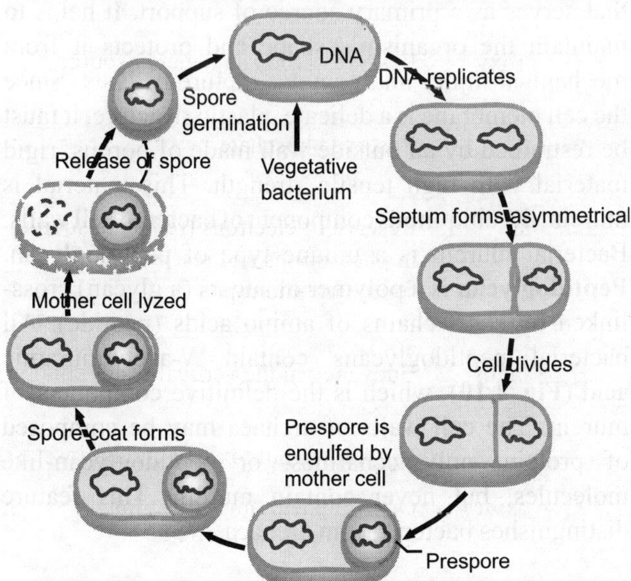

Fig. 4.9 *Bacillus subtilis* endospore formation. The entire process involves formation of a prespore that is engulfed by the mother cell. After formation of the spore coat, the mother cell will be lyzed and the mature spore is released into the environment.

Table 4.5 Differences between endospores and vegetative cells

Property	Vegetative cells	Endospores
Surface coats	Typical Gram-positive murein cell wall polymer	Thick spore coat, cortex and peptidoglycan core wall
Microscopic appearance	Non-refractile	Refractile
Calcium dipicolinic acid	Absent	Present in core
Cytoplasmic water activity	High	Very low
Enzymatic activity	Present	Absent
Macromolecular synthesis	Present	Absent
Heat resistance	Low	Very high
Resistance to chemicals and acids	Low	Very high
Radiation resistance	Low	Very high
Sensitivity to lysozyme	Sensitive	Resistant
Sensitivity to dyes and staining	Sensitive	Resistant

4.7.3 Cytoplasmic cell membrane

Inside the wall and tightly pressed against it, is the outer membrane of the prokaryotic cell, the **plasma membrane** or **plasmalemma**. Prokaryotic cell membranes are the extremely thin layers of material, consisting mostly of lipids and protein, that separate cell and most cell parts, from their surroundings. The bacterial membrane is composed of 40% phospholipid and 60% protein. Most prokaryotic cells have only the surface membrane that surrounds each cell. Cell membranes may be smooth or have infoldings that extend into the cell, forming structures called **mesosomes**.

The plasma membrane is the most dynamic structure in the cell and performs various functions (**Table 4.6**). In addition to acting as a selective barrier, the bacterial membrane possesses receptor proteins that detect nutrients and toxins in their environment. Bacteria have a variety of transport systems, which can be used alternatively in different environmental situations. Besides controlling molecular trafficking, membranes also have enzymatic reactions including photosynthesis and respiration and the role in cell replication. Unlike eukaryotes, bacteria do not have intracellular organelles for processes producing energy such as respiration or photosynthesis. Instead, the cytoplasmic membrane carries out these functions. The membrane is the location of electron transport systems (ETS) used to produce energy during photosynthesis and respiration and it is the location of an enzyme called ATP synthetase (ATPase) which is used to synthesize ATP. Unlike eukaryotes, bacteria can have a wide variety of fatty acids within their membranes. Along with typical saturated and unsaturated fatty acids, bacteria contain fatty acids with additional methyl, hydroxy or even cyclic groups. The relative proportions of these fatty acids can be modulated by the bacterium to maintain optimum fluidity of the membrane in response to a change in environmental factors (e.g., following temperature change).

Since plasma membrane is a phospholipid bilayer, the lipid portion of the outer membrane is impermeable to charged molecules. However, channels called **porins** are present in the outer membrane that allows for passive transport of many ions, sugars and amino acids across the outer membrane. These molecules are therefore present in the periplasm, the region between the cytoplasmic and outer membranes. The periplasm contains the peptidoglycan layer and many proteins responsible for substrate binding or hydrolysis and reception of extracellular signals. The periplasm, because of its location between the cytoplasmic and outer membranes, signals received from outside and substrates bound on membrane are available to be transported across the cytoplasmic membrane using transport and signalling proteins imbedded there.

The composition of membranes in the archaea is remarkably different from both bacteria and the eukaryotes. Instead of the straight-chain fatty acids linked to glycerol through ester bonds, the hydrocarbon chains in the membranes of the archaea have ether linkages. In addition, archaean membrane lipids also contain some branched chain hydrocarbons.

Table 4.6 Functions of the prokaryotic plasma membrane

- Regulates osmotic pressure or functions as permeability barrier
- Location of transport systems for trafficking specific solutes (nutrients and ions)
- A site for energy-generating functions, involving respiratory and photosynthetic electron transport systems, establishment of proton motive force and ATP-synthesizing ATPase
- Biosynthesis of membrane lipids (including lipopolysaccharide in Gram-negative cells)
- Synthesis of murein (cell wall peptidoglycan)
- Facilitates DNA replication and segregation with septum formation and cell division
- A specialized structure, where enzyme systems are located, such as for photosynthesis, CO_2 fixation and nitrogen fixation

4.7.4 Cell wall

Most prokaryotes have a rigid cell wall (10–20 nm). They usually lack cellulose and are hence chemically different from the cell walls of higher plants. The prokaryotic cell wall is a semi-rigid structure that serves as a primary source of support. It helps to maintain the organism's shape and protects it from mechanical injury and osmotic rupture or lysis. Since the cell membrane is a delicate, plastic structure, it must be restrained by an outside wall made of porous, rigid material with high tensile strength. This material is murein, the ubiquitous component of bacterial cell walls. Bacterial murein is a unique type of peptidoglycan. **Peptidoglycan** is a polymer of sugars (a glycan) cross-linked by short chains of amino acids (peptide). All bacterial peptidoglycans contain *N*-acetylmuramic acid (**Fig. 4.10**), which is the definitive component of murein. The cell walls of archaea may be composed of protein, polysaccharides, or peptidoglycan-like molecules, but never contain murein. This feature distinguishes bacteria from archaea.

Gram-positive cell wall

In Gram-positive bacteria there are many different peptide arrangements among peptidoglycans, which are responsible for the retention of the crystal violet dye during staining. The best studied is the murein of *Staphylococcus aureus*. The most significant

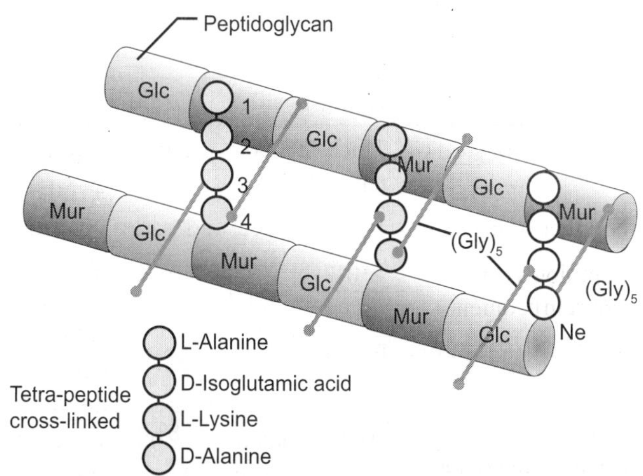

Fig. 4.10 Bacterial cell walls. The peptidoglycan layer of Gram-positive bacteria *Staphylococcus aureus*. (Glc: glucosamine; Mur: muramic acid).

difference in the Gram-positive wall is the occurrence of an interpeptide bridge of amino acids that connects nearby side chains to one another. Embedded in the Gram-positive cell wall are polyalcohols called teichoic acids, some of which are lipid-linked to form lipoteichoic acid (**Fig. 4.11**). Since the lipoteichoic acids are covalently linked to lipids within the cytoplasmic membrane, they are responsible for linking the peptidoglycan to the cytoplasmic membrane. Teichoic acids give the Gram-positive cell wall an overall negative charge due to the presence of phosphodiester bonds between teichoic acid monomers. Assembly of the interpeptide bridge in Gram-positive murein is inhibited by the beta lactam antibiotics in the same manner as the interpeptide bond in Gram-negative murein. Gram-positive bacteria are more sensitive to penicillin than Gram-negative bacteria because the peptidoglycan is not protected by an outer membrane and it is a more abundant molecule.

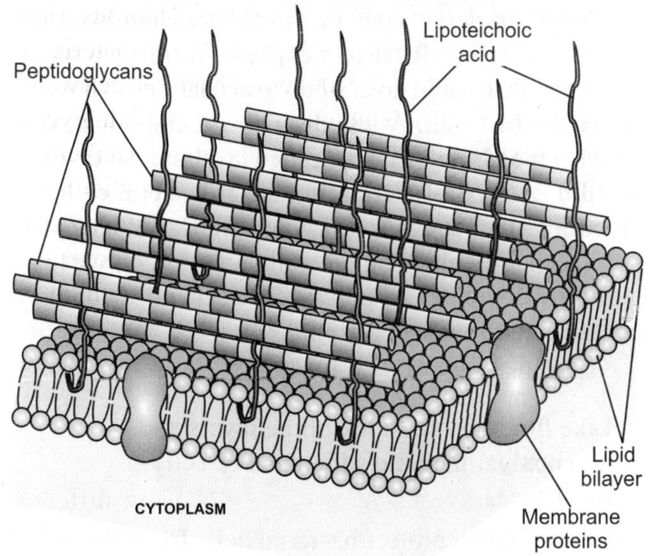

Fig. 4.11 Schematic drawing of the bacterial cell wall structure of Gram-positive *Staphylococcus aureus*.

Gram-negative cell wall

In the Gram-negative bacteria (which do not retain the crystal violet in the Gram-stain procedure), the cell wall is relatively thin (10 nm) and is composed of a single layer of peptidoglycan surrounded by a membranous structure called the outer membrane. The outer membrane of Gram-negative bacteria contains an additional outer membrane composed by phospholipids and lipopolysaccharides (LPS), which face into the external environment (**Fig. 4.12**). LPS or endotoxin, is toxic to animals and is usually considered a part of the cell wall. In *E. coli* and other Gram-negative bacteria, the glycan backbone of peptodoglycan is made up of alternating molecules of *N*-acetylglucosamine (G) and *N*-acetylmuramic acid (M). The *N*-acetylmuramic acid (M) attaches a peptide side chain that contains the amino acids L-alanine, (L-ala), D-glutamate (D-glu), diaminopimelic acid (DAP) and D-alanine (D-ala). The glycan backbone of the peptidoglycan molecule can be cleaved by the enzyme lysozyme that is present in animal serum, tissues and secretions and in phagocyte granules.

The toxic component of endotoxin (LPS) is Lipid A. The O-specific polysaccharide may provide for

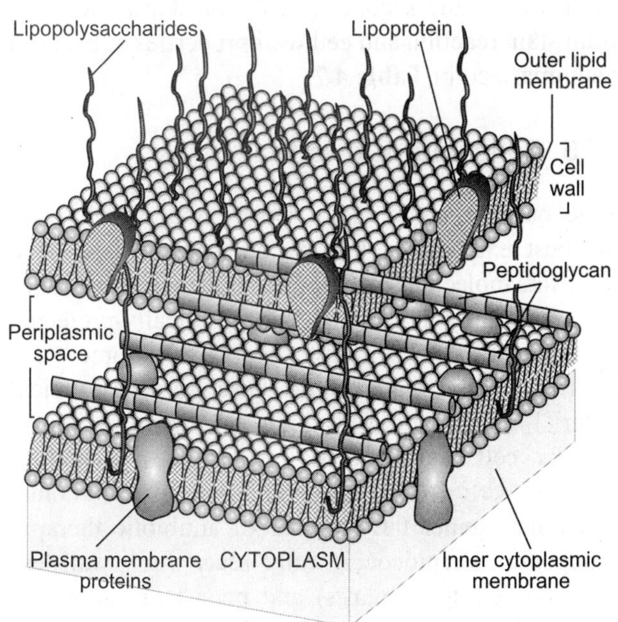

Fig. 4.12 Schematic drawing of the cell wall structure of Gram-negative bacteria *Escherichia coli*.

adherence or resistance to phagocytosis, in the same manner as fimbriae and capsules. The O-polysaccharide (also referred to as the O-antigen) also accounts for multiple antigenic types (serotypes) among Gram-negative bacterial pathogens.

4.7.5 Gram staining of bacteria

Since there are many different types of bacteria there are differences in the composition of these cells. One of the primary differences is the amount of certain substances found in the cell wall. The difference in the amount of murein found in these cell walls gives us a differential staining technique known as the Gram stain. In the Gram stain, cells are subjected to two different colours of dyes and a fixing agent. During the de-staining step cells are washed with an acetone and alcohol mixture. The lipid content of the Gram-negative wall probably affects the outcome of this step so that Gram-positive cells retain a primary stain while Gram-negative cells are de-stained. Those cells that contain a large quantity of murein retain the colour of the first dye, which is purple, because of the action of the fixing agent. The cells that do not contain a high quantity of murein do not retain the colour of the first dye, but are counterstained with the second dye, which is pink in colour. Thus, when a Gram stain is performed, we can conclude that the microbe is Gram-positive because of the purple colour that it retains, versus Gram-negative organisms, which are pink in colour due to the second dye. A correlation between Gram stain reaction and cell wall properties of bacteria is summarized in **Table 4.7**.

4.7.6 Cells without walls

Some bacteria are able to live without a cell wall. The best example is the mycoplasmas. These have sterol-like molecules incorporated into their membranes and are usually inhabitants of osmotically-protected environments (live in host cytoplasm). Some such as *Mycoplasma pneumoniae* cause primary atypical bacterial pneumonia ("walking pneumonia"). Since it lacks cell wall-synthesizing enzymes, penicillin is ineffective in treatment of this type of pneumonia. Sometimes, under the pressure of antibiotic therapy, pathogenic streptococci can revert to wall-less forms (called spheroplasts) and persist or survive in osmotically-protected tissues. When the antibiotic is withdrawn from therapy, the organisms may regrow their cell walls and reinfect unprotected tissues.

Table 4.7 Correlation of the Gram stain with properties of bacterial cell wall

Property	Gram-positive	Gram-negative
Thickness of wall	Thick (20–80 nm)	Thin (10 nm)
Number of layers	1	2
Peptidoglycan (murein) content	>50%	10%–20%
Teichoic acids in wall	Present	Absent
Lipid and lipoprotein content	zero to 3%	58%
Protein content	0	9%
Lipopolysaccharide content	0	13%
Sensitivity to penicillin G	Yes	No (1)
Sensitivity to lysozyme	Yes	No (2)

4.7.7 The cell envelope

The bacterial cell envelope consists of several layers of material that enclose the protoplasm of the cell. The cell protoplasm (cytoplasm) is surrounded by a plasma membrane, cell wall and capsule. Outside the cell wall, foremost as a surface structure, may be a polysaccharide capsule or glycocalyx.

Most bacteria contain a polysaccharide layer outside the cell wall called a **capsule**. A true capsule is a discrete detectable layer of polysaccharides deposited above the cell wall. A less discrete structure or matrix which embeds the cell is called a slime layer or a **biofilm**. A type of capsule found in bacteria called a glycocalyx or microcapsule is a very thin layer of tangled polysaccharide fibres on the cell surface. Capsules are generally composed of polysaccharides; rarely do they contain amino sugars or peptides.

Capsules have several functions:

- Like fimbriae, capsules, slime layers and glycocalyx, mediate adherence of cells to surfaces.
- Capsules also protect bacterial cells from engulfment by protozoa or white blood cells (phagocytes), or from attack by antimicrobial agents of plant or animal origin.

- Capsules in certain soil bacteria protect cells from perennial effects of drying or desiccation.
- Capsular materials (e.g., dextrans) may be overproduced when bacteria are fed sugars to become reserves of carbohydrates for subsequent metabolism.

Some bacteria produce slime materials to adhere and float as colonial masses in their environments. Other bacteria do so to attach themselves to a surface or substrate. Bacteria may attach themselves to the surface, produce slime, divide and produce microcolonies within the slime layer and construct a **biofilm**, which becomes an enriched and protected environment. Structurally biofilms consist of microcolonies with a complex web of water channels that extend throughout the structure and provide its residents with constant nutrient supply. The individual organisms within a biofilm are embedded in a protective polysaccharide matrix. The sticky polymeric matrix surrounding the biofilm helps to concentrate carbon, nitrogen, phosphate and other nutrients within the biofilm. It also protects microbes from antibiotics, sheer stress and host defences. An example of biofilm construction is the formation of dental plaque mediated by the oral bacterium, *Streptococcus mutans*. The bacteria adhere to the surface of the tooth and then grow and synthesize a polysaccharide capsule, which binds them to the enamel and forms a biofilm which is 300–500 microbial cells in thickness. Bacteria convert sugars in the diet into dextran that forms plaque and cements the bacteria to tooth enamel. Within the plaque, the bacteria convert sugars to lactic acid which then lead to dental caries or bacterial infection of the tooth.

Another important characteristic of capsules is their ability to block some steps in the phagocytic process and thereby prevent the bacterial cells from being engulfed or destroyed by phagocytes (white blood cells). For example, the human pathogens *Streptococcus pneumoniae* (lobar pneumonia), *Bacillus anthracis* (anthrax) and *Neisseria meningitidis* (meningitis) are able to resist attack by phagocytes because they cover their surfaces with a type of capsular material.

4.7.8 Pili

Many bacterial cells have external appendages, mainly used for locomotion, attachment and transfer of genetic material (e.g., pili and flagella). Fimbriae and pili (singular, pilus) are interchangeable terms used to designate short, hair-like structures on the surfaces of prokaryotic cells. Fimbriae are shorter and stiffer than flagella and slightly smaller in diameter (**Table 4.8**). Generally, fimbriae have no role in bacterial movement. But they are common in Gram-negative bacteria and also occur in some archaea and Gram-positive bacteria. It is mainly involved in adherence of bacteria to surfaces, substrates and other cells or tissues in nature. In *E. coli*, a specialized type of pilus, the F or sex pilus, mediates the transfer of DNA between mating bacteria during conjugation (**Fig. 4.13**), but the function of the smaller, more numerous common pili is quite different.

Table 4.8 Properties of pili and fimbriae

Bacterial species	Per cell	Distribution	Functions
E. coli (F or sex pilus)	1–4	Uniform	Mediates DNA transfer during conjugation
E. coli (common pili or Type 1 fimbriae)	100–200	Uniform	Surface adherence to epithelial cells of the GI tract
Neisseria gonorrhoeae	100–200	Uniform	Surface adherence to epithelial cells of the urogenital tract
Streptococcus pyogenes (fimbriae plus the M protein)	NA	Uniform	Adherence, resistance to phagocytosis; antigenic variability
Pseudomonas aeruginosa	10–20	Polar	Surface adherence

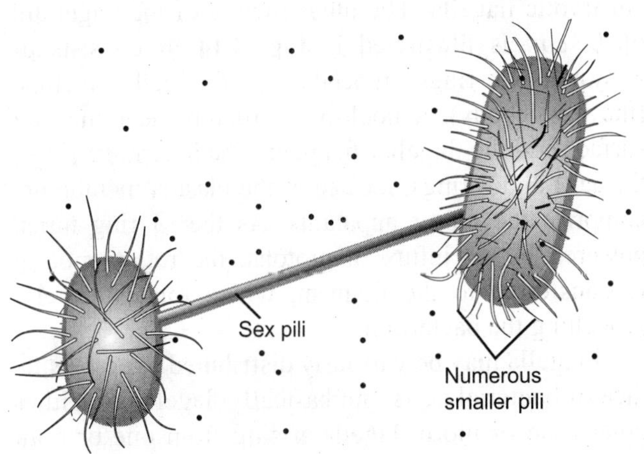

Fig. 4.13 Bacterial pili. A sex pilus that connects two conjugating *E. coli*.

Common pili are usually involved in specific adherence (attachment) of prokaryotes to surfaces. In medical conditions, they are major determinants of bacterial virulence because they allow pathogens to colonize tissues and resist attacks by phagocytic blood cells. For example, pathogenic *Neisseria gonorrhoeae* adheres specifically to the human cervical or urethral epithelium by means of its fimbriae. Enterotoxigenic strains of *E. coli* adhere to the mucosal epithelium of the intestines by means of specific fimbriae and the M protein and associated fimbriae of *Streptococcus pyogenes* are involved in both adherence and resistance to engulfment by phagocytes.

4.7.9 S layers

The S layer is a cell surface protein layer found in many bacteria and in some archaea, where it serves as the cell wall. It is made up of a two-dimensional array of proteins and has crystalline appearance, the symmetry of which differs between species. The exact function of the S layers is unknown, but it has been suggested that it may act as a partial permeability barrier for large substrates. For example, an S layer could conceivably keep extracellular proteins near the cell membrane by preventing their diffusion away from the cell. In some pathogenic species, an S layer may help to facilitate survival within the host by conferring protection against host defence mechanisms.

4.7.10 Flagella

Flagellum is a flexible corkscrew-shaped protein filament that is used for locomotion. The prokaryotic flagellum (about 20 nm) is much thinner than the eukaryotic flagella. The ultrastructure of the flagellum of *E. coli* is illustrated in **Fig. 4.14**. It consists of a system of rings imbedded in the cell envelope (the basal body), a hook-like structure near the cell surface and the flagellar filament. The innermost rings, the aaM and S rings, located in the plasma membrane, comprise the motor apparatus. As the M ring turns, powered by an influx of protons, the rotary motion is transferred to the filament, which rotates thereby propelling the bacterium.

Flagella may be variously distributed over the surface of bacterial cells, but basically flagella are either polar (one or more flagella arising from one or both poles of the cell) or peritrichous (lateral flagella distributed over the entire cell surface). Flagellar distribution is a genetically distinct trait that is occasionally used to characterize or distinguish bacteria. For example, among Gram-negative rods, pseudomonads have polar flagella to distinguish them from enteric bacteria, which have peritrichous flagella.

Prokaryotes are known to exhibit various types of tactic behaviour, i.e., the ability to move (swim) in response to environmental stimuli. The flagellar filament is rotated by a motor apparatus in the plasma membrane allowing the cell to swim in fluid environments. The forward motion occurs when flagella rotate in a counter-clockwise direction. In contrast, clockwise rotation results in a stop-and-tumble motion, allowing the cell to reorient for a subsequent forward run. The filament of the flagellum is anchored into the cell by a protein complex. The important one is the motor protein that converts chemical energy into rotational motion.

4.7.11 Quorum sensing

The recent finding that bacteria can communicate with each other has intensified the investigation on many single cell organisms inhabiting our world. The term 'Quorum Sensing' (QS) has been used to describe the phenomenon of accumulation of signalling molecules that enable a single cell to sense the number of bacteria

Bacterial growth curve

The reproduction of microbes occurs in a fashion that can be plotted graphically. Changes in parameters can affect this growth curve, which is generally the same for all microbes. The curve consists of four stages or phases.

Lag phase: In the lag phase, growth and reproduction are just beginning to occur. This is ideally the phase in which to destroy the organisms, since there are relatively few of them.

Log or exponential phase: During the log phase, reproduction is occurring at an exponential rate. This exponential growth occurs until the environmental surroundings and food supply cannot support any more exponential growth.

Stationary phase: In this phase, the number of microbes has reached its limit. While reproduction still goes on, the number of microbes does not increase with the speed of the log phase.

Death phase: Finally, when all of the nutrients have been exhausted, the microbes enter the death phase. In this phase, the population begins to die off, thus decreasing the numbers rapidly.

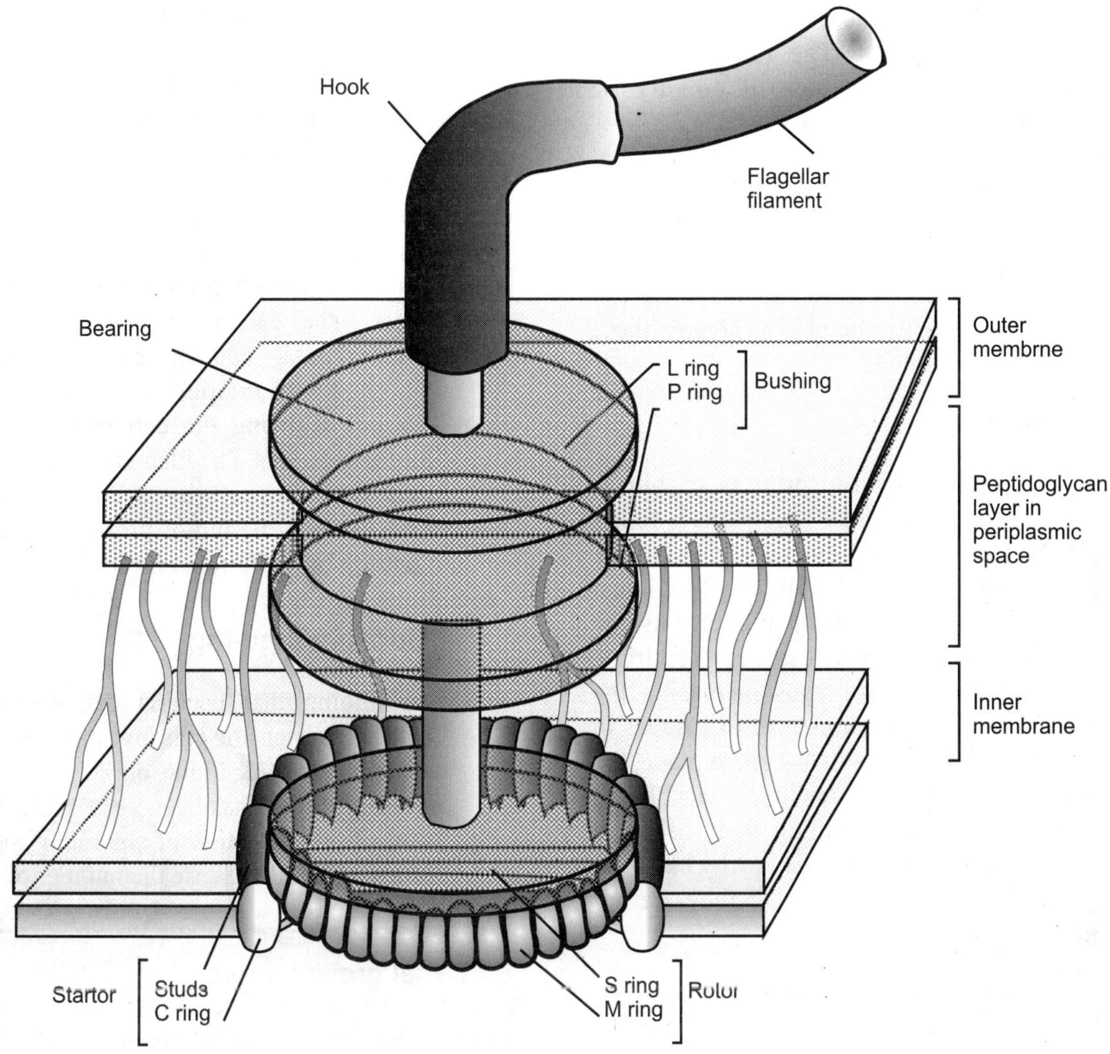

Fig. 4.14 Schematic drawing of the flagellar rotatory 'motor' of *E. coli.*

(cell density). Prokaryotes communicate by secretion of diffusible signalling molecules that are sensed by nearby microbes. Signal molecules called **auto-inducers** allow for coordinated gene expression. However, the autoinducers are sensed only when they are present in large quantities. Therefore, it requires a large number of microbes to secrete enough autoinducers to cause gene expression changes.

Recently, several quorum sensing systems were intensively studied in various organisms such as marine bacteria and several pathogenic bacteria. QS enables bacteria to coordinate their behaviour. As environmental conditions often change rapidly, bacteria need to respond quickly in order to survive. These responses include adaptation to availability of nutrients, defence against other microorganisms which may compete for the same nutrients and the avoidance of toxic compounds potentially dangerous for the bacteria. It is very important for pathogenic bacteria during infection of a host to coordinate their virulence in order to escape the immune response.

Different bacterial species use different molecules to communicate. There are several different classes of signalling molecules. In some cases a single bacterial species can have more than one QS system and therefore use more than one signal molecule. The bacterium may respond to each molecule in a different way. In this sense, the signal molecules can be thought of as words of a language. There is evidence that interspecies communication via QS can occur, which is referred to as crosstalk. This has implications in many areas of microbiology as in nature bacteria almost always exist in mixed species populations such as biofilms. The formation of a biofilm represents the advancement of prokaryotic community behaviour.

Quorum sensing research has many potential applications: most of these involve controlling bacteria by interfering with their signalling systems. For example, many bacteria rely on QS to control the expression of the genes which cause disease. If we can block the QS systems we may be able to prevent these bacteria from being dangerous.

4.7.12 Reproduction

Most bacteria reproduce by the process of binary fission. The genetic code for these organisms is found in a single molecule of DNA, which contains all the information that the cell needs for the biological processes that take place in the cell. In binary fission (**Fig. 4.15**), the DNA strand is duplicated in its entirety. The cell also increases in size, thus increasing the amount of cytoplasm and the number of organelles in the original cell. Once this growth has taken place, the cell divides into two separate 'daughter' cells. These cells are then capable of continuing their lifecycle and are also capable of dividing. This is why bacterial infections can easily get out of hand if they are not treated promptly.

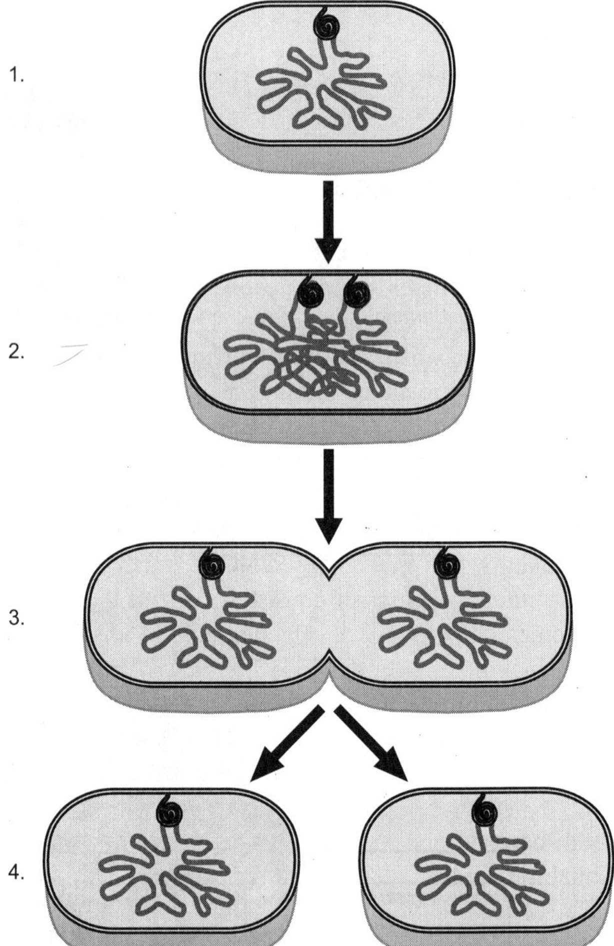

Fig. 4.15 Cell division in prokaryotes. Bacteria use a simple form of cell division called binary fission.
Step 1: Shows a bacterial cell containing cell wall, cell membrane, single loop of chromosome and the cytoplasm. The spiral structure is a membrane invagination (called mesosome) where chromosome attach to the cell membrane.
Step 2: Duplication of DNA: During this step the chromosome replicates leading to two identical chromosomes attached to separate points of attachment.
Step 3: Cell begins to divide: Each daughter chromosome moves apart with the help of a sliding mesosome.
Step 4: Cell divides into two cells, each having one chromosome.

4.7.13 Prokaryotic cell division

Bacteria divide precisely at mid-cell to produce two identical daughter cells. After chromosome segregation, the cell undergoes cytokinesis. It involves the annular (ring-like) ingrowth of all of the cell envelope layers. During DNA replication, each strand of the replicating bacterial DNA attaches to proteins at what will become the cell division plane. As the bacterium grows to full size, the newly replicated chromosomes become separated. In the centre of the bacterium, a group of proteins called **Fts proteins** interact to form a ring at the cell division plane. These proteins form the cell division apparatus known as the divisome. Proteins at the divisome synthesize the peptidoglycan and new membrane material that both splits the bacterium into two daughter cells and subsequently enables each to grow to full size.

A number of genes required for division was isolated and characterization of Fts (filamentous temperature-sensitive) mutations. About 8 Fts genes are present in most bacteria. The FtsZ protein forms a ring-like structure just under the cell membrane at the site of impending division. The other division proteins are then recruited to this 'Z ring' in the sequential order. The functions of several Fts proteins have been identified:

- **FtsZ**, similar to tubulin in eukaryotic cells, forms a constricting ring at the division site.
- **FtsA**, similar to actin—along with FtsZ—is the driving force behind membrane invagination at the division site.

- **FtsK** helps in separating the replicated bacterial chromosome.
- **FtsI** plays a role in the later stages of peptidoglycan synthesis.

4.8 NANOBACTERIA

Researchers have found a strange and tiny new group of microbes called nanobacteria or calcifying nanoparticles (CNPs) or **nanobes**. They are the smallest cellular organisms recently discovered. The organisms are less than 400 nm across and seem unable to survive alone. The environment causes drastic changes in their unit size: under unfavourable conditions they form very large multicellular units. The relationship between these tiny microbes and their hosts is unclear. It is unknown if they do any damage and their hosts do fine without them. The microbe has one of the smallest genomes known with only half a million DNA letters. Sequencing its genome may point to the minimum number of genes needed for an organism to sustain itself. The new microbes are called **Nanoarchaeota**, nanobacteria being an ancestor for a group of bacteria, the α_2 subgroup of *Proteobacteria*, which include both bacteria present in the environment and bacteria inhabiting mammalian blood and tissues. The nearest relatives are phyllobacteria found in soils causing tropical plant diseases. However, these bacteria do not produce apatite and differ much from nanobacteria.

4.8.1 Structure

Nanobacteria are cell-walled microorganisms with a diameter well below the generally accepted lower limit (about 200 nm) for bacteria. They pass through sterile filters and endure γ irradiation (1 megarad not effective), like a virus. Their size is between that of a virus and cell-walled bacteria. The average diameter of nanobacteria has been measured with electron microscopic techniques. The smallest units of nanobacteria capable of starting replication in culture, possibly as aggregates of several nanobacteria (sizes of about 0.05 μm), was detected based upon filtration and electron microscopic results (**Fig. 4.16**). The theoretical minimum diameter of a cell, based on the size of those macromolecules now considered necessary for a living cell, has been calculated to be about 0.14 μm.

Nanobacteria produce a slimy biomatrix that forms carbonate apatite mineral around them in culture. This bizarre new form of life seems to have adapted to living inside the host—an ecologically free but hostile niche.

4.8.2 Discovery of nanobacteria

Torella and Morita (1981) described very small cells called ultra-microbacteria. By 1982 MacDonell and Hood found that some particles could pass through a 200 nm membrane. Early in 1989, geologist Robert L Folk found what he later identified as nanobacteria from the hot springs of Viterbo, Italy. Initially, searching for a bacterial cause for travertine deposition, scanning electron microscope examination of the mineral, where no bacteria were detectable, revealed extremely small objects, which appeared to be biological. He proposed that nanobacteria are the principal agents of precipitation of all minerals and crystals formed in water. They also cause oxidation of metals and are abundant in many biological specimens. In 2004, a team from Mayo Clinic reported having isolated nanobacteria in diseased human arteries and kidney stones. Working with particles less than 0.2 um in size, they found indirect evidence that the particles had self-replicated and had a cell-like appearance under an electron microscope. They also believed that the particles could produce RNA, since they absorbed one of its building blocks, uridine, in greater quantities than would be expected in the case of pure absorption (by crystals such as apatite). Using an antibody produced by the Finnish researchers, the particles were found to bind to diseased arterial tissue and to the same sites to which a DNA stain bound. The researchers now plan to isolate RNA and DNA from the nanobacteria.

4.8.3 Growth and composition of nanobacteria

Nanobacteria can be cultured and have doubling time of around 3 days. They can be passaged apparently forever. Their growth can be prevented with small concentrations of tetracycline antibiotics, or with high concentrations of aminoglycoside antibiotics. Both stop bacteria protein synthesis at the ribosomal level. Small concentrations of cytosine arabinoside or fluorouracil (anti-metabolites) also inhibit growth of nanobacteria by preventing nucleic acid synthesis. Nanobacteria have unique strategies for social behaviour and for multiplication, including communities, budding and fragmentation.

Nanobacteria have a very simplistic lifestyle. When nutrients are available they replicate, produce protective mineral layer to 'hibernate' as nutrient pool diminishes

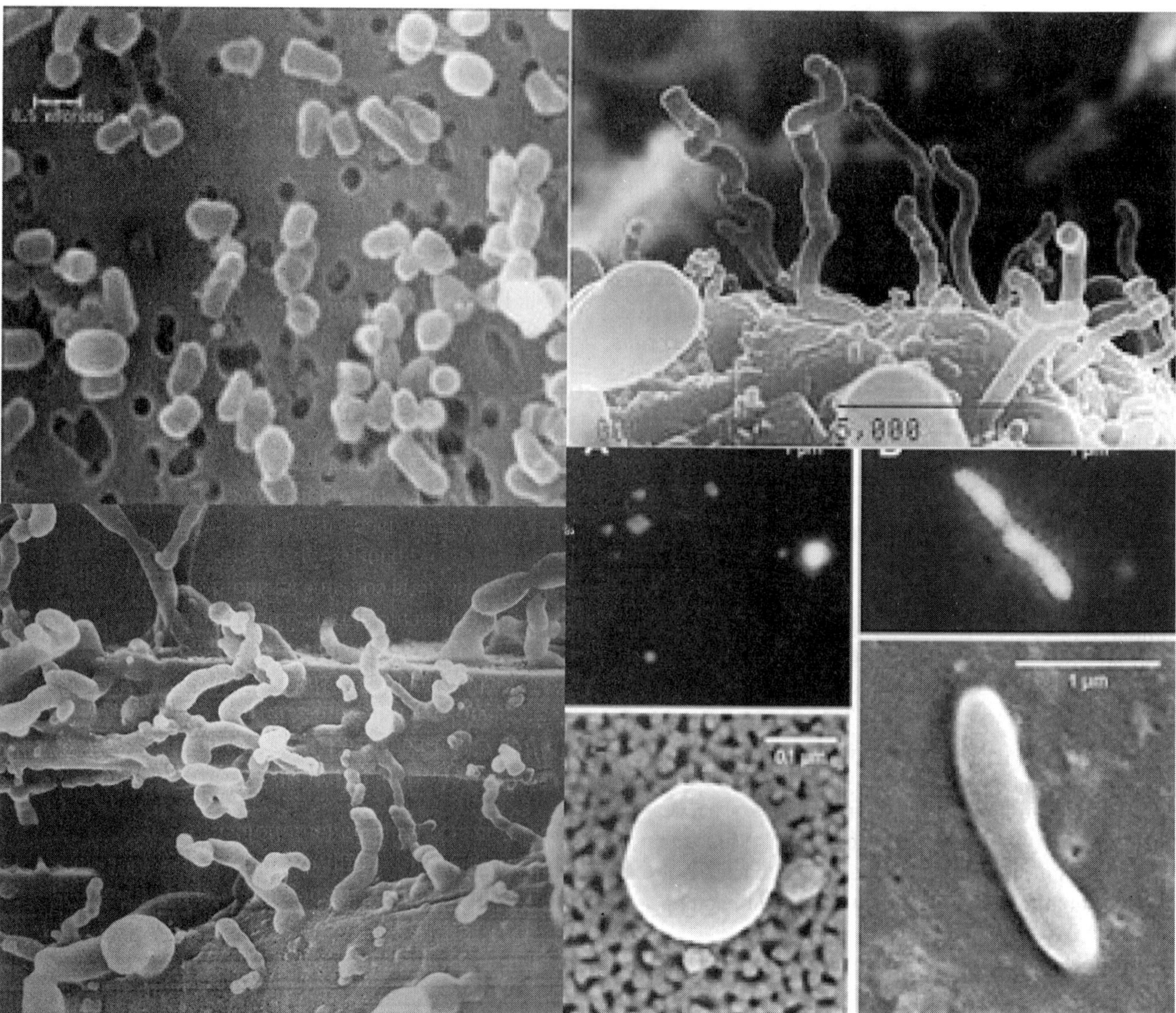

Fig. 4.16 Nanobacteria (collected from various sources)

and wait for a new cycle of nutrient availability. Nanobacteria absorb nutrients by diffusion and Brownian movements. They take only readily available amino acids from the surroundings, preferably, large amounts of Gln, Asn and Arg from the medium for synthesis of structural components, energy production and mineralization process. They can use ready fatty acids from the medium and when fatty acids are scarce, nanobacteria protected by replacing membrane lipids partly with apatite. They react to stress by becoming social and form communities. They can hibernate for extensive periods waiting for optimal conditions. Nanobacteria have very low internal pressure, when compared to normal bacteria (3–5 bars). Because of this nanobacteria may show high resistance to near-boiling temperatures. Nanobacteria may form and shed units resembling viruses that could spread even via tiny pores or cracks in rocks. The survival strategy of nanobacteria appears to be 'being small is efficient'.

Nanobacterial mineral is biogenic

All carbonate apatite produced in the human body is biogenic. Nanobacterial mineral formation is also due to a specific biogenic process, since, mineral grows directly on the nanobacteria, forming parts of the cell envelope. Without nanobacterial presence there will not be any mineralization in the culture medium. Mineral growth is dependent on a biomatrix made by the nanobacteria. The mineral layer is under active remodelling of its size and shape. No significant mineralization takes place if nanobacteria are killed with γ irradiation. Mineralization is an active process that can bring phosphate levels to zero. Mineralization occurs as in a biomatrix comparable to that of pearls and mineral crystallization is under biocontrol with serum factors, similar to bone development.

Composition

Nanobacteria produce biomass at a rate of about 0.0001 times that of *E. coli*. Their biomass contains novel proteins (about 30 proteins have been detected by SDS-PAGE) and 'tough' polysaccharides. Amino terminal sequences are available for six different proteins and one of them is a functional porin protein. Porins are a hallmark of Gram-negative bacteria. Porins are normally located in their outer membrane and are involved in the trafficking of small molecules. In nanobacteria, porins seem to be located in mineral layers. Polyamines are now considered essential for cell proliferation. Normal bacteria may contain puterscine, spermidine and many other 30 di- and polyamines. However, nanobacteria do not have these polyamines, but contain a compound having similar mobility with cadaverine, a special polyamine used by several eubacteria, a peptidoglycan component. The parasitic bacteria acquire their polyamines from their hosts. The environment provides compensation for the loss of many genes. Muramic acid, a major component of peptidoglycan, has also been detected.

Genetic material

Nanobacteria have apparently small genomes. Hoescht 33258 staining indicated that nanobacteria should have DNA amounts between that of mycoplasmas and mitochondria. They can be stained with DNA fluorochromes similar to mitochondrial staining. Nanobacteria contain modified nucleic acids detectable specifically with fluorescent staining and spectroscopy techniques. What is the smallest genetic size for life? The suggested minimum number of genes, 256 genes, may be still too high a number for the simplest genome. The primordial soup might have provided nutrients for supporting organisms with fewer genes than are necessary to survive in present-day environments.

4.8.4 Nanobacteria in disease causation

Nanobacterium sanguineum was proposed in 1998 as an explanation of certain kinds of pathologic calcification (apatite in kidney stones) by Finnish researcher Olavi Kajander and Turkish researcher Neva Ciftcioglu, University of Kuopio, Finland. According to the researchers, the particles self-replicated in microbiological culture. Researchers further reported having identified DNA in these structures by staining. They have developed medical diagnostic kits for identifying nanobacteria and are working towards developing prescription medical treatment for calcification-associated diseases. A strong correlation has been found between antibodies to nanobacteria and coronary artery calcification (associated with increased risk of coronary artery disease).

Kajander and his colleagues were trying to find out why certain cultures of mammalian cells in the lab would die no matter how carefully they prepared them. When they took a closer look at their old cultures under an electron microscope (EM), they saw the particles like bacteria, but an astonishingly 100 times smaller and seemed to be thriving inside the dying cells. Believing them to be a possible new form of life, Kajander named the particles 'nanobacteria'. A steadily increasing number of studies are linking nanobacteria to serious health problems, including kidney stones, aneurysms and ovarian cancer. The studies show that nanobacteria can infect humans. Kajander and Neva Çiftçioglu (1998) had observed, through an EM, nanobacteria particles building shells of calcium phosphate around themselves. They began to investigate whether such particles played a role in causing kidney stones, which are also made of calcium compounds. A breakthrough came in 2003 when researchers discovered nanobacteria in the calcified debris found in tissue samples from ovarian cancer patients and in the samples of calcified arteries. Researchers also found that nanobacteria do indeed self-replicate and endorsed the idea that the particles are life forms. Patients with chronic pelvic pain—thought to be linked to urinary stones and prostate calcification—reported 'significant improvement' after using an experimental treatment for nanobacteria.

Now it appears that nanobacteria may build essential parts of larger organisms. Biomineralization could result in the formation of bones, shells, teeth and kidney stones and arterial plates.

4.8.5 Nanobacteria in space research

Nanobacteria research team at NASA announced the results of an experiment in which a bioreactor chamber that simulates conditions of space travel was used to culture nanobacteria suspected of rapidly forming kidney stones in astronauts. In this microgravity environment, they were found to multiply five times faster than in normal earth gravity. NASA also determined that nanobacteria were shown to be a possible infection risk for crew-members living in close quarters.

Summary

- The building blocks of all life forms are ***cells***, which occur singly and as the subunits of multicellular organisms. Schleiden and Schwann were the first to postulate that the cell is the basic unit of structure and function.
- Broadly all the organisms on earth have been classified into ***five kingdoms***: prokaryotes, protista, fungi, animal and plant.
- The ***prokaryotes*** are an immense and heterogeneous group that includes bacteria and archaea.
- ***Arcahea***, probably a modern descendent of ancient prokaryotes, appear in many ways closer to the eukaryotes than to the bacteria and have adaptability to highly extreme conditions and often referred to as extremophiles.
- ***Bacteria*** are diverse in their habitats and nutritional capacities.
- Structurally, a bacterial cell has three distinct architectural regions: (i) appendages in the form of ***flagella*** and ***pili*** (or fimbriae), (ii) a cell envelope consisting of a ***capsule***, ***cell wall*** and ***plasma membrane*** and (iii) a ***cytoplasmic region*** that contains the cell chromosome (DNA), ribosomes and various sorts of inclusions.
- Most prokaryotes have a rigid cell wall that serves as a primary source of support and helps to maintain the organism's shape and protects it from mechanical injury and osmotic rupture or lysis. Gram-positive bacteria are more sensitive to penicillin than gram-negative bacteria, because the peptidoglycan is not protected by an outer membrane.
- Bacteria also have many other protective layers such as ***fimbriae***, ***capsules***, ***slime layers*** and ***glycocalyx***, which mediate adherence of cells to surfaces.
- Most bacteria reproduce by the process of ***binary fission***: the DNA strand is duplicated in its entirety and divide precisely at mid-cell to produce two identical daughter cells.
- Bacteria have evolved a sophisticated mechanism to overcome nutritional stress by converting vegetative phase to ***spores*** (dormant cells). They also form ***biofilms*** that provide them protection against toxins and stress and facilitate absorption of nutrients.
- Researchers have found a strange and tiny new group of microbes called ***nanobacteria*** or calcifying nanoparticles (CNPs). They are the smallest cellular organisms recently discovered.
- Nanobacteria are cell-walled microorganisms with a diameter well below 200 nm.
- Nanobacteria have a very simplistic lifestyle. When nutrients are available they replicate, produce protective mineral layer to 'hibernate' as nutrient pool seizes and wait for a new cycle of nutrients availability.

Exercises

1. Discuss briefly, with examples, the five kingdom-approach system of classification.
2. Explain diagrammatically the structure of a typical prokaryotic cell.
3. Discuss briefly the different intracellular components of bacteria.
4. Explain the complex nature of the bacterial cell wall.
5. How does the Gram stain help to differentiate two groups of bacteria?
6. Explain bacterial cell division.
7. What is biofilm and why do bacteria produce it?
8. Mention the important contrasting characteristics between prokaryotic and eukaryotic cells.
9. What are nanobacteria? How can they harm humans?

Further Reading

Books

Adams RLP, Knowler JT and Leader DP (1992), *The Biochemistry of the Nucleic Acids,* Chapman and Hall, London.

Becker WM, Kleinsmith LJ and Hardin J (2000), *The World of the Cell*, 4th edition, Benjamin Cummings, New York.

Cappuccino JG and Sherman N (1999), *Microbiology: A Laboratory Manual,* Benjamin Cummings.

De Duve C (1984), *A Guided Tour of the Living Cell,* Scientific American Library, New York.

Madigan MT, Martinko JM (2005), *Brock Biology of Microorganisms*, 11th edition, Pearson Prentice Hall, New Jersey.

Starr C and Taggart R (1995), *Biology, the Unity and Diversity of Life*, Wadsworth Publishing Company, A division of International Thomson Publishing Inc.

References

Brimacombe R, Stoffler G and Willman HG (1978), Ribosome structure, *Annu. Rev. Biochem.*, 47:217.

De Duve C (1996), The birth of complex cells, *Sci. Am.*, 274(4):50–7.

Gitai Z (2005), The new bacterial cell biology: moving parts and subcellular architecture, *Cell*,120(5):577–86. PMID 15766522.

Harowtz HJ (1987), *Mayonnaise and the Origin of Life*, Berkeley Publications, New York.

Martinko JM and Parker J (1997), *Brock Biology of Microorganisms*, Prentice Hall International, Inc.

Liljas A (1991), Comparative biochemistry and biophysics of ribosomal proteins, *Int. Rev. Cyto.*, 124:103.

Nomura M (1986), The control of ribosome synthesis, *Sci. Am.*, 250:102.

Shih YL, Rothfield L (2006), The bacterial cytoskeleton, *Microbiol. Mol. Biol. Rev.*, 70(3):729–54. PMID 16959967.

Von Heijne G (1994), Membrane proteins: from sequence to structure, *Annu. Rev. Biophys. Biomolec. Struc.*, 23:167–92.

5 EUKARYOTIC CELLS

Learning objectives

After reading this chapter, you will be able to:

- Identify the important components and characteristics of a eukaryotic cell
- Understand functional specialization and organization of plants and animals

5.1 INTRODUCTION

Based on the organization of their cellular structures, all living cells can be divided into two groups: prokaryotic and eukaryotic. Eukaryotic cells are generally larger and more complex than prokaryotic ones and contain a variety of cellular bodies called **organelles** (**Table 5.1**). These organelles function in the activities of the cell and are compartments for metabolic functions. Protozoa, unicellular algae and fungi are microscopic eukaryotes, whereas, plants, animals and humans are multicellular (macroscopic eukaryotic cells) organisms.

As discussed in the previous chapter, eukaryotic cells have a distinctive nucleus, composed primarily of protein and nucleic acids (DNA and RNA). The DNA is organized into linear units called **chromosomes** and the functional segments of the chromosomes are referred to as **genes**. The nuclear proteins belong to a class of proteins called histones, which provides a supportive framework for the DNA in the chromosomes. DNA replicates in eukaryotic cells by mitosis. Some examples of organelles are the endoplasmic reticulum or ER, (a series of membranes that extend throughout the cytoplasm), Golgi body, lysosomes, mitochondria, chloroplast, vacuoles, microbodies and cytoskeleton. Many specialized eukaryotic cells contain flagella and cilia to facilitate cellular locomotion. The cell wall is very prominent in many species of eukaryotes, such as fungi and plants. The eukaryotic cell membrane conforms to the **fluid-mosaic model** and is a dynamic structure governing passage of dissolved molecules and particles into and out of the cytoplasm, communication with other cells and responds to the environmental changes.

5.2 STRUCTURE OF EUKARYOTIC CELL

Eukaryotic (*eu* = true, *karyon* = nucleus) make up the other major class of cell types. In order to meet the many specialized needs, eukaryotic cells exist in many different forms. By coexisting and interacting in a cooperative manner in a higher organism, these cells do not need the biochemical flexibility and adaptability, which is essential to prokaryotes. The internal structure of eukaryotes is considerably more complex than that of prokaryotic cells, as can be seen in **Figs. 5.1–3**. Earlier, the eukaryotic cell was divided into two important compartments: (i) the nucleus and (ii) the cytoplasm. With the development of high resolution microscopes including the electron microscope, it became possible to study the cells in greater detail (**Fig. 5.3**). Since there is a great diversity in cell size (cell diameter varies 10–30 μm), shape and subcellular structures, there is no 'typical cell.' The features of generalized plant and animal cells are described here. A plant cell is divided into the cell wall and protoplast. The cell wall was, for long regarded as a non-living excretion of living cell matter, but recently, more and more evidence has shown that the protoplast and the wall have organic unity. The protoplast contains two sets of subcomponents.

Table 5.1 Main differences between prokaryotic and eukaryotic cells

Features	Prokaryotic cell	Eukaryotic cell
Evolution	Most primitive organisms	Evolved from prokaryotic cells and highly advanced
Typical organisms	Bacteria, cyanobacteria and archaea	Protists, fungi, animals and plants
Cellular organization	Usually single cells	Single cells, colonies, higher multicellular organisms with specialized cells
Typical size	~1–10 μm	~10–100 μm (apart from sperm and neurons)
Cell wall	Yes	Absent in animals and present in plants
Peptidoglycan	Present	Absent
Cytoplasmatic structure	Very few structures	Highly structured and compartmentalized
Endomenbrane system	Absent	Present extensively in endoplasmic membrane and in other cytoplasmic vesicles
Cell organelles	None	Mitochondria (present in all eukaryotes) and chloroplasts present in photosynthetic organisms
Centrioles	No	Yes, all animals and some lower plant forms contains
Golgi complex	No	Yes
Lysosomes	No	Common in animals, not present in plants
Peroxisomes	No	Yes
Ribosomes	70 S (50 S + 30 S)	80 S (60 S + 40 S)
Plasma membrane	Yes	Yes
Cytoskeletal filaments	Absent or very few	Present, microtubules, actin/microfilaments and intermediate filaments
Cilia and flagella	Yes, complex	Yes, simple
Nucleus	Nucleoid region, no real nucleus	True nucleus with nuclear envelope
Chromosomes	DNA strand (usually circular) and no nucleosomes	Several chromosomes (linear molecules associated with histone proteins)
DNA amount	Low (up to 1.4×10^{-2} pg/ cell)	High (up to 1.6×10^{-2} – 96 pg/ cell)
Genes	Polycistronic (operons), contain no introns	Monocistronic (contain introns in genes, split genes)
RNA/protein synthesis	Coupled in cytoplasm	RNA synthesis occurs in nucleus and protein synthesis in cytoplasm
Plasmids	Present in cytoplasm	absent
Recombination	Conjugation	Meiosis and syngami
Cell division	Binary fission, take place in less than an hour	Mitosis and meiosis, it need many hours to complete one division
Death	Cells die only by injury	Cells die by many ways: injury, ageing and senescence, cancer, suicide, etc.

Note: This list does not contain every conceivable difference, but focusses on the main features that distinguish prokaryotic from eukaryotic cells.

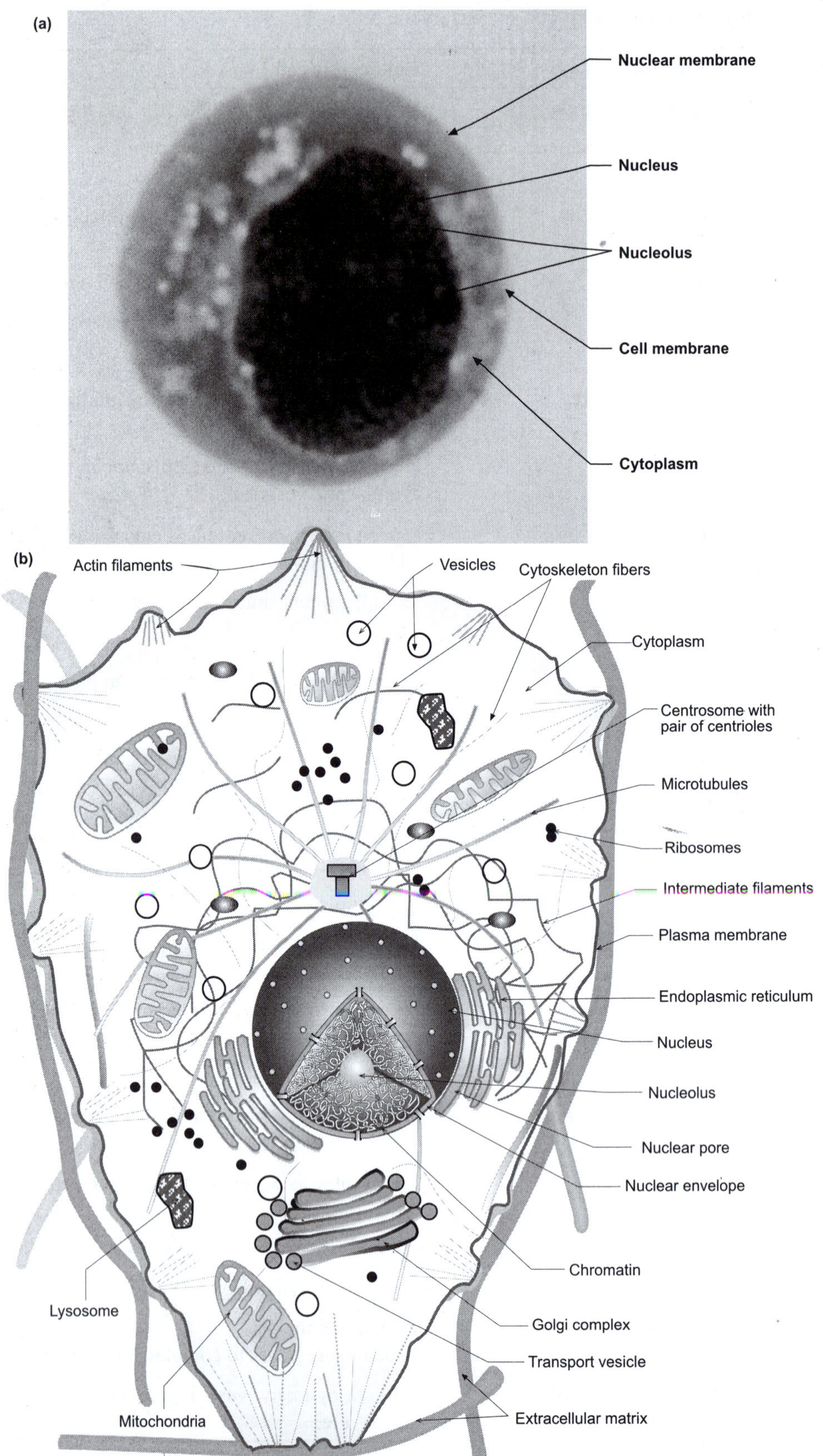

Fig. 5.1 Animal cell: (a) Human blood lymphocyte (interphase) stained with Giemsa and observed under light microscope at oil immersion (Channarayappa, 1990); (b) The diagrammatic representation of the major features of a generalized animal cell.

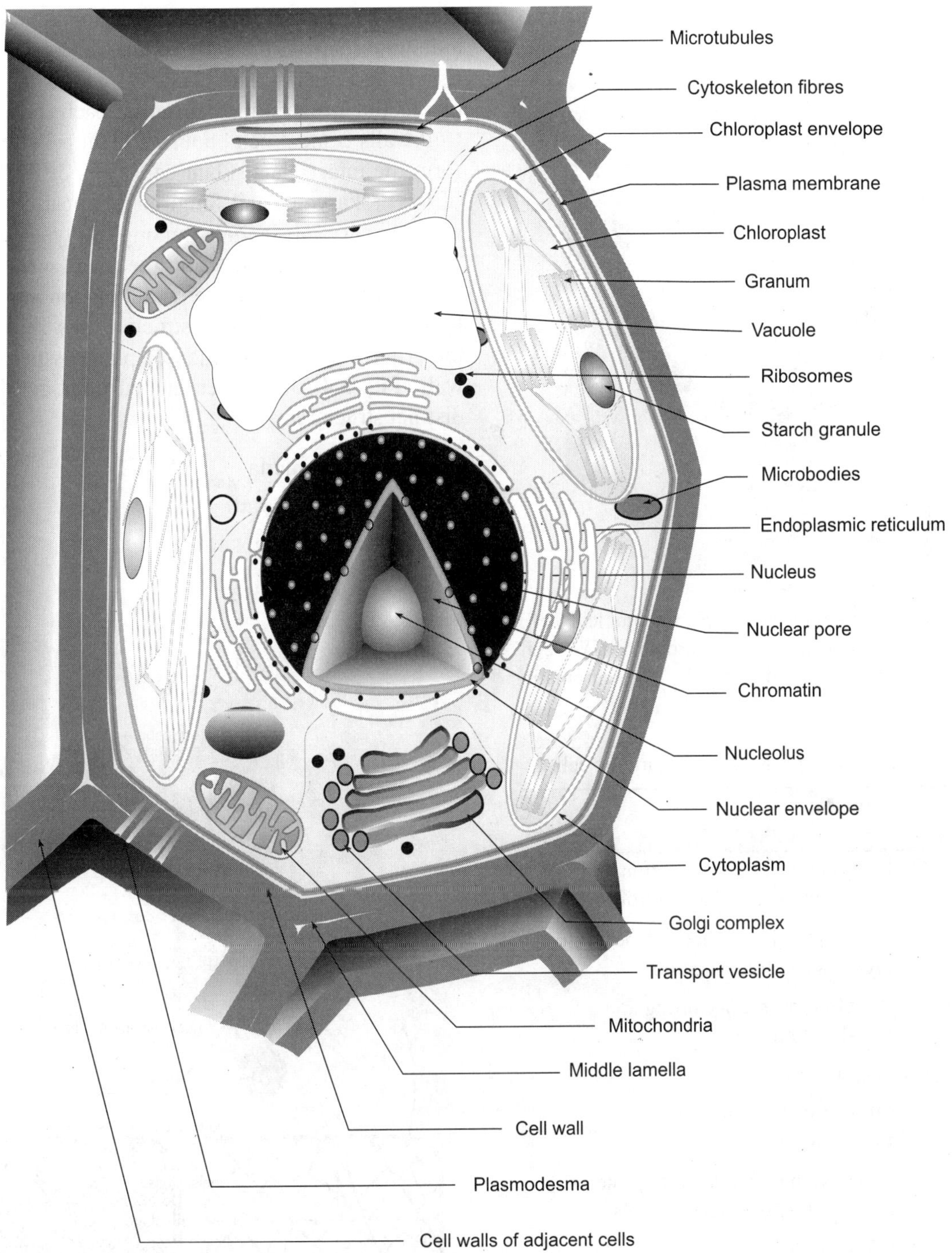

Fig. 5.2 Ultrastructure of a generalized plant cell as seen with an electron microscope.

Protoplasmic components include the cytoplasm and organelles (**Table 5.2**), such as nucleus, plastids and mitochondria. The second group consists of vacuoles, which are inclusions surrounded by the cytoplasm and ergastic substances, viz., reserve materials such as starch grains, oil droplets and aleurone grains and other products of metabolism such as various crystals.

5.3 SUBCELLULAR STRUCTURES OF EUKARYOTES

5.3.1 Cell wall

Plant cells, prokaryotes and fungi are surrounded by a relatively rigid wall, which is secreted by the living cell

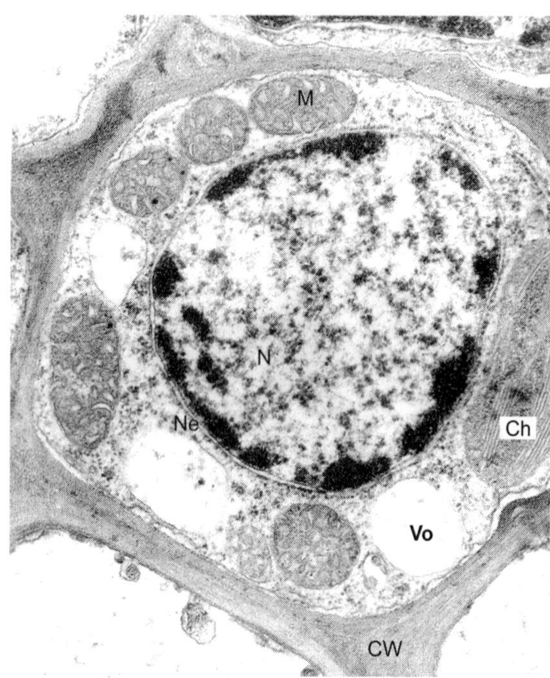

Fig. 5.3 TEM image showing a typical tomato leaf cell. N: nucleus; M: mitochondrion; Ch: chloroplast; CW: cell wall; Ne: nuclear envelope; Vo: vacuole; Ch: chromatin (Channarayappa, 1992).

Table 5.2 Subcellular structures found in eukaryotic cells

Subcellular structures	Functions
Nucleus	Carrying genetic information and regulation of cellular activities
Endoplasmic reticulum	Site of protein synthesis, processing proteins and translocation
Golgi apparatus	Protein processing, sorting and translocation
Secretory vesicles	Secretion
Lysosomes	Intracellular breakdown
Mitochondria	Energy production
Plastids	Photosynthesis; storage of proteins, oil, starch; pigments to tissues
Peroxisomes	Generate H_2O_2
Vacuoles	Storage, regulation of cell volume
Endosomes	Intracellular traffic; sorting of ligands and receptors
Ribosomes	Protein synthesis
Centrosomes	Cell division
Hydrogenosome	Energy generation
Cell wall	Provides definite shape and mechanical support and strength to tissues and organs

within. The main function of the cell wall is to provide a definite shape to the cells and mechanical strength to tissues and organs. The plant cell wall is an elaborate extracellular matrix that surrounds each cell in a plant. Cell walls cement neighbouring plant cells together to form the intact plant. Plant cell walls differ from that of the prokaryotes and fungi in chemical composition and also vary substantially among cell types of a plant. The pattern of deposition on the inner surface of cell walls shows considerable variation among different tissues. In fact, cell wall structure has been used as the criteria for identification of plant tissues. The cell wall (**Fig. 5.4**) of a differentiated plant cell has three distinct structures:

- middle lamella
- primary cell wall
- secondary cell wall

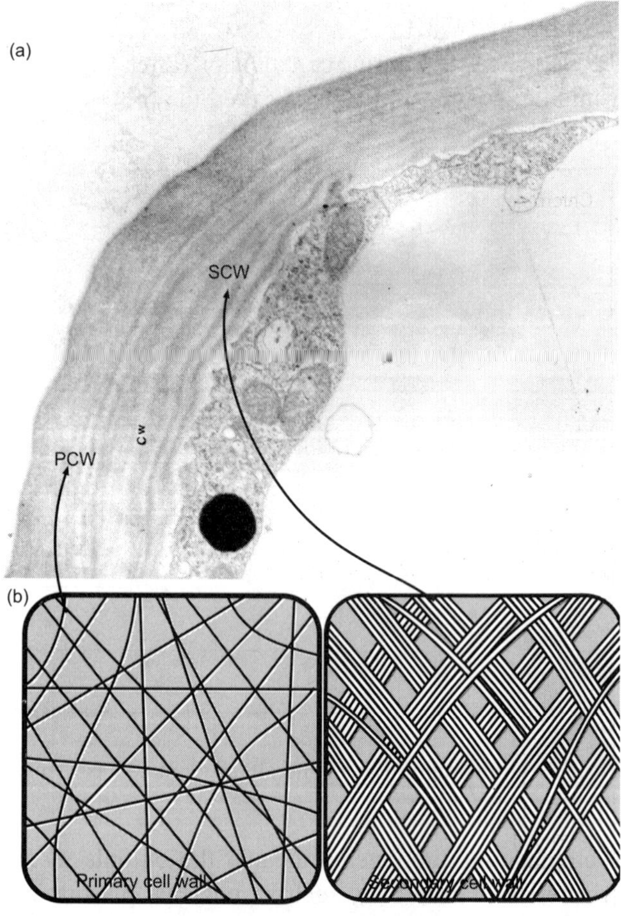

Fig. 5.4 The plant cell wall: (a) Electron micrograph of tomato cell wall and (b) arrangement of cellulose microfibrils in plant cell wall. In primary cell walls (PCW) microfibrils are dispersed, but in secondary cell walls (SCW) cellulose microfibrils are arranged in parallel (Channarayappa, 1992).

Middle lamella

In plants, the walls of adjacent cells are joined by middle lamella, which is a thin layer composed mainly of pectic substances (calcium and magnesium pectates) that hold the neighbouring cell walls together (**Fig. 5.5**). The middle lamella is the first to originate following cell division. During the cytokinesis of mitotic cell division (at telophase), three types of membranous structures—small cisterns of ER origin, spherical vesicles of 20 μm diameter and phragmosomes of about 250 μm) assemble at the equatorial plate of a cell. The cisterns and vesicles fuse with each other to form the cell plate. The 20 μm vesicles (produced by Golgi) are filled with pectin, which is deposited between the two newly formed membranes; this pectin gives rise to the middle lamella. Phragmosomes contain a dense substance which is mainly a polysaccharide used for the formation of the plasma-lemma. Phragmosomes disappear after the cell plate has formed.

Primary cell wall

The wall laid down during cell division (after formation of middle lamella) in plants is called the primary wall and lies between the middle lamella and plasmalemma. This rigid but extensible primary cell wall contains cellulose microfibrils running through a matrix of other complex polysaccharides. Its main constituents are hemicellulose (53%), cellulose (30%), pectin (5%), protein (5%) and lipid (7%). Cellulose molecules are like fibres; each molecule is made up of up to 3,000 glucose residues. Several cellulose molecules associate to form a cellulose microfibril (a single microfibril contains up to 2000 cellulose molecules arranged parallel to each other). Cellulose microfibrils are the units of the structural organization of the cell wall. They are present in a dispersed and loose arrangement (**Fig. 5.6**). Primary walls are pliable, so the surface area of cells can enlarge under incoming water pressure.

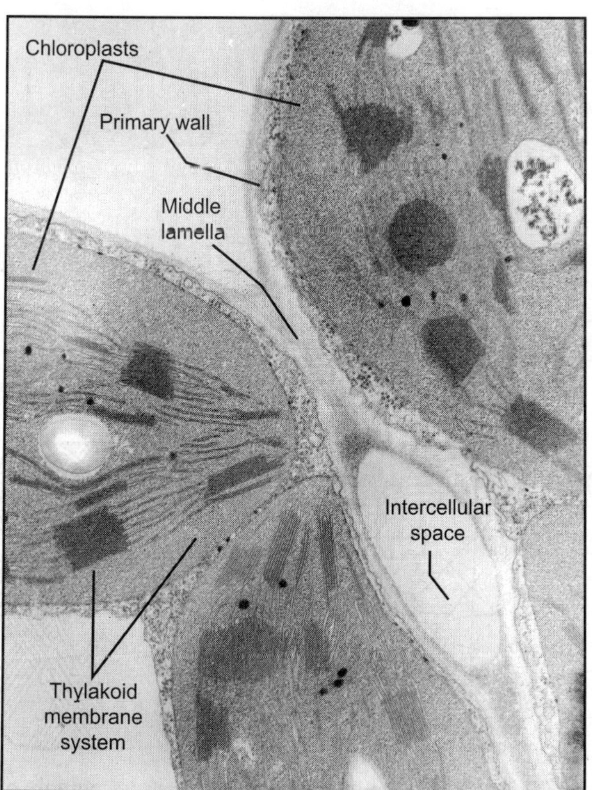

Fig. 5.5 Transmission electron micrograph showing the middle lamella primary cell wall, chloroplasts, intercellular space and thylakoid membrane system (Channarayappa, 1992).

Secondary cell wall

Later, in many cell types, more layers were deposited inside the first wall. They form a rigid, secondary wall that helps to maintain the cell's shape. Secondary walls are often impregnated with a hardening substance called **lignin**. It is a complex network of phenolic compounds found in the walls of the xylem vessels and fibre cells of woody tissues. Lignified secondary wall layers are responsible for the strength of wood and natural fibres. Some secondary walls also contain **suberin** or **cutin**, which are waxes that restrict water movement. Deposits of **pectin** cement the adjacent walls together. In secondary cell walls, cellulose microfibrils are relatively more closely packed and they are arranged more or less parallel to each other (**Fig. 5.6**). Water can move freely through free space in the cell wall. Trees invest substantial amount of their energy in the cellulose and lignin that comprise the bulk of their biomass.

Functions of plant cell wall

The extracellular matrix in animals can fulfil several functions such as cell–cell adhesion, networking for cell growth, strength, rigidity and compressibility of tissues. Plant cell walls serve many of the same functions, even though they are composed of entirely different macromolecules. Some of the important functions of the cell wall are listed on the next page.

1. Provides mechanical strength and skeletal support to individual cells (protoplast) and to the whole plant; it also provides chemically buffered environment, which allows the cell to perform homeostatis upon itself and live in a natural and contained environment. The cell wall allows for

Structure of the cell wall

Middle lamella

- Joins adjacent cells
- Composed of pectates of calcium and magnesium
- First to arise following cell division

Primary cell wall

- Arises after middle lamella
- Lies between middle lamella and plasmalemma
- Contains hemicellulose, cellulose, pectin, protein and lipid

Secondary cell wall

- Contains lignin, the hardening substance
- Provides the characteristic strength to wood and natural fibres
- More closely packed cells
- Last to arise

the circulation and distribution of water, minerals and other small nutrient molecules into and out of the cell.

2. Provides turgidity to resist expansion;
3. Provides rigid building blocks from which structures such as leaves and stems can be produced; this gives the cell a stable area and self-contained environment.
4. Limits the orientation of cellulose microfibrils and helps to control cell growth and shape.
5. Develops a coating of waxy cutin, the cuticle, on exposed epidermal surfaces reducing water loss and risk of infection.
6. Forms a barrier to water movement due to the suberin impregnated in the wall of root epidermal cells.
7. Some cell walls, modified as food reserves (hemicellulose in seeds), provide a storage site of regulatory molecules that sense the presence of pathogenic microbes and control the development

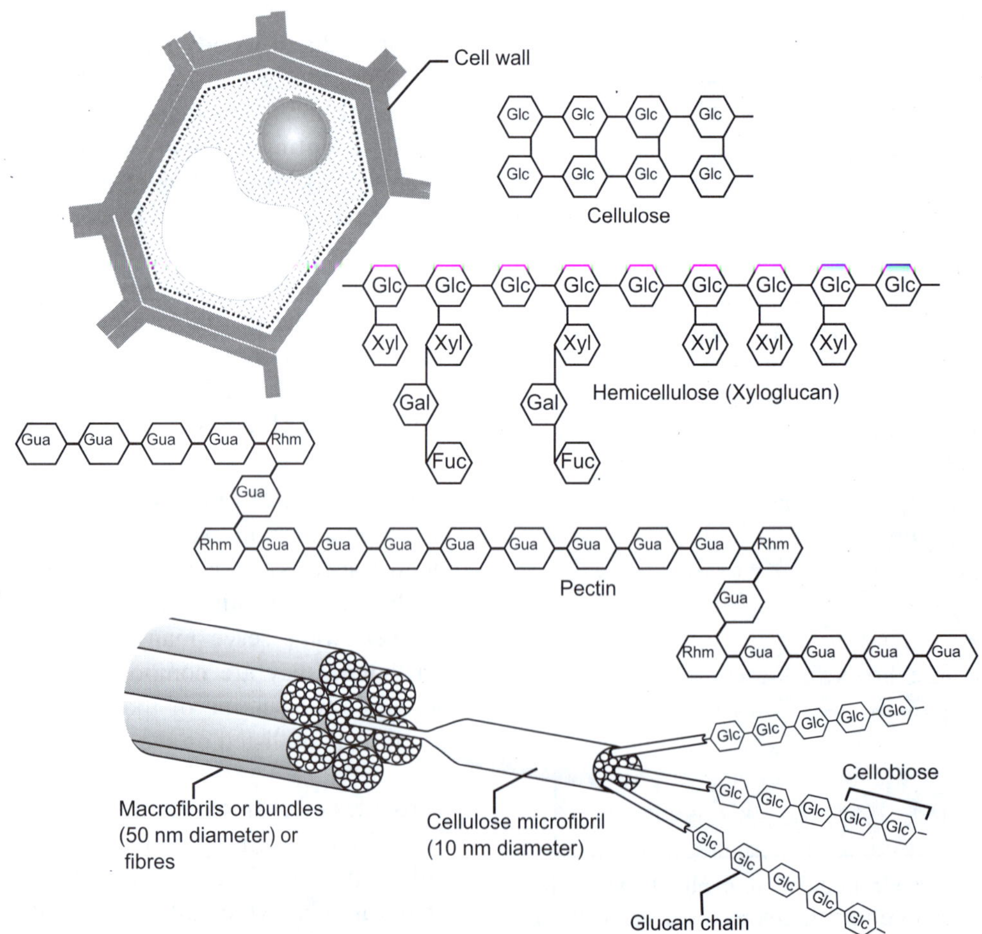

Fig. 5.6 Structure of the cell wall and the polysaccharide constituents.

of tissues within the cell. The secondary walls of the cotyledon and the endosperm of developing seeds contain single non-cellulosic polysaccharide. They provide a strong wall to protect the embryo or impose mechanical dormancy; they contain specialized storage carbohydrates that are digested during germination; converted to sucrose and transported to the growing seedling.

8. Develops increased surface area in transfer cells.
9. Helps to distinguish between plant cell from the animal cell.

Composition

All plant walls contain fibres of **cellulose** (in the form of organized microfibrils), a polysaccharide made up of thousands of glucose units. Like collagen in tendons, cellulose provides the tensile strength of the wall. Cellulose fibres are embedded in and linked to a matrix containing two other types of polysaccharides: **hemicellulose** and **pectin** (crosslinking glycans) and also a group of proteins, the most prominent of which is a family of hydroxyproline-rich fibrous glycoproteins called **extensins**. The composition and structure of the cell wall varies in different parts of a plant and in different types of plants. Other polymers that contain several different saccharides are arabinose, xylose, fucose, rhamnose, galactose and galacturonic acid. The best-studied constituents are those of the primary cell wall produced by cultured sycamore cells (**Fig. 5.7**).

Hemicellulose is a highly branched polysaccharide with a backbone of about 50 ß(1→4)-linked sugars of a single type. It is linked by hydrogen bonds to the surface of cellulose microfibrils. The hemicellulose branches help to bind the microfibrils to each other and to other matrix components, particularly the pectins. Pectins, like hyaluronan, contain multiple negatively charged saccharides rich in galacturonic acid. They bind cations such as Ca^{2+} and like hyaluronan, are highly hydrated. When purified, they bind water and form a gel and are hence used in many processed foods. Pectins are often crosslinked to hemicellulose, thus participating in forming a complex network of all the principal wall components. This interlinked network helps bind adjacent cells together. Pectins are abundant in the middle lamella, the layer between the cell walls of adjacent cells. Treatment of tissues with pectinase or other enzymes that degrade pectin frequently causes cells with their walls to separate from one another.

The primary cell wall can be thought of as a gel-like matrix in which cellulose microfibrils are embedded. Such walls are more impermeable; only water and ions diffuse freely. This is one of the reasons why plant hormones are small, water-soluble molecules. The secondary wall of the mature cell may have several layers (**Fig. 5.8**); within each layer the cellulose fibrils

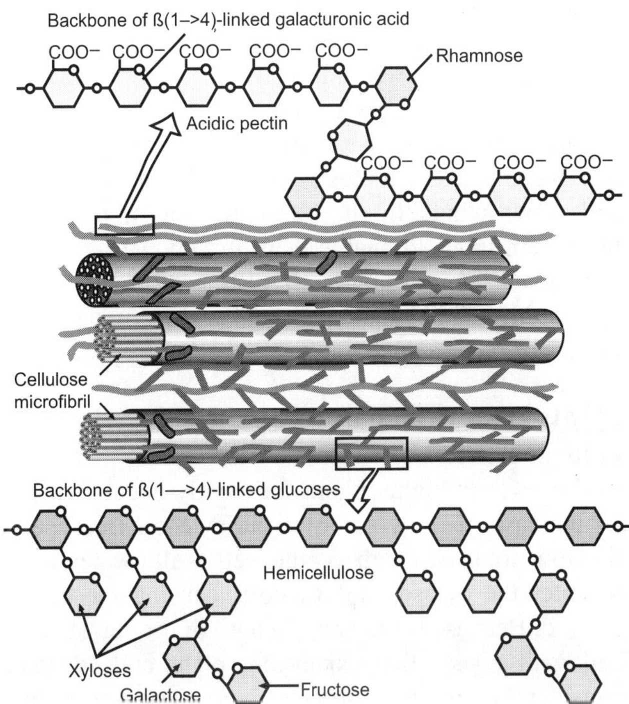

Fig. 5.7 The structure and interconnections of major polymers in the primary cell wall. Hemicellulose molecules are hydrogen bonded to the surface of cellulose microfibrils.

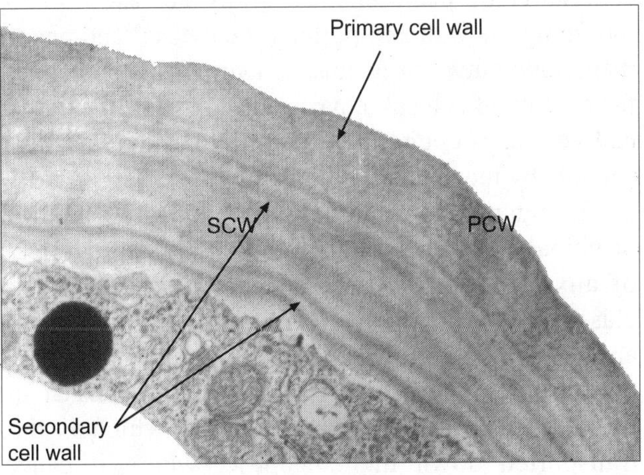

Fig. 5.8 TEM micrograph shows deposition of many layers of secondary cell wall (SCW) and a single layer of primary cell wall (PCW) (Channarayappa, 1992).

are parallel to each other, but the orientation differs in adjacent layers.

More than 15% of the primary cell wall is composed of extensin, a glycoprotein made up of roughly 300 amino acids. Extensins are incorporated into the insoluble polysaccharide network and are believed to have a structural role, forming the scaffolding upon which the cell wall architecture is formed.

Lignin, a complex insoluble polymer of phenolic residues, strengthens cell walls. It is particularly abundant in wood, where it accumulates in primary cell walls and in the secondary walls of the xylem. Lignin associates with cellulose, like cartilage proteoglycans. It resists compression forces on the matrix. Particularly for soil-grown plants, lignin is essential for strengthening the xylem tubes to enable them to conduct water and salts over long distances. It also protects the plant against invasion by pathogens and against predation by insects or other animals.

Plants grow primarily by auxin-induced cell enlargement

Cell growth in higher plants frequently occurs without an increase in the volume of the cytosol. Because of the low ionic strength of the cell wall, water tends to enter the cytosol and vacuole, causing the cell to expand. Because the tensile strength of the surrounding cell wall resists this expansion, plant cells develop considerable internal pressure, or **turgor pressure**. A localized loosening of the primary cell wall, induced by **auxin**, allows the cell to expand in a particular direction (**Fig. 5.9**). The amount and direction of enlargement determine the size and shape of a plant cell. Individual plant cells can rapidly increase in size by loosening the wall and pushing the cytosol and plasma membrane outwards against it. During this elongation, the amount of cytosol remains constant. The increase in cell volume is due to the expansion of the intracellular vacuole by the uptake of water.

Elongation of plant cells is induced by auxin, a principal plant growth hormone. The rapid effects of auxin on cell elongation were first observed in classical experiments on coleoptiles from grasses and oats. Coleoptiles like many plant parts are phototropic. The growing meristematic cells at the tip of a coleoptile synthesize auxin, which is then transported down the coleoptile, where it causes many effects. Light inhibits auxin production, so that more auxin and preferential cell elongation occur on the side away from the light, causing the coleoptile to bend towards light. Auxin is thought to induce cell elongation by causing the cell wall to soften at the 'growing' end of the cell. In some plants, this is believed to be due to the induction of localized proton secretion by auxin at the ends of the cell. Auxin apparently activates a proton pump bound to the plasma membrane, which lowers the pH of the cell wall region near the plasma membrane (pH 4.5). The low pH activates a class of wall proteins termed **expansins,** which disrupt the hydrogen bonding between cellulose microfibrils. As a consequence, the rigidity of the wall is reduced and the cell can elongate. Expansins are not enzymes that degrade cellulose and the expansin-triggered loosening of the wall is reversed when the pH is brought back to 7.0. Like auxin, fusicoccin induces rapid cell elongation and it triggers pumping out of protons by sensitive cells and localized loosening of the cell wall.

Plasmodesmata

Cells in higher plants are separated by walls containing cellulose. However, the cytoplasms of adjacent cells are interconnected by tube-like plasmodesmata that penetrate the wall. It is a fine cytoplasmic thread, linking the cytoplasm of two neighbouring cells through a fine pore in the cell walls, which acts as an intercellular cytoplasmic bridge to facilitate communication and transport of materials between

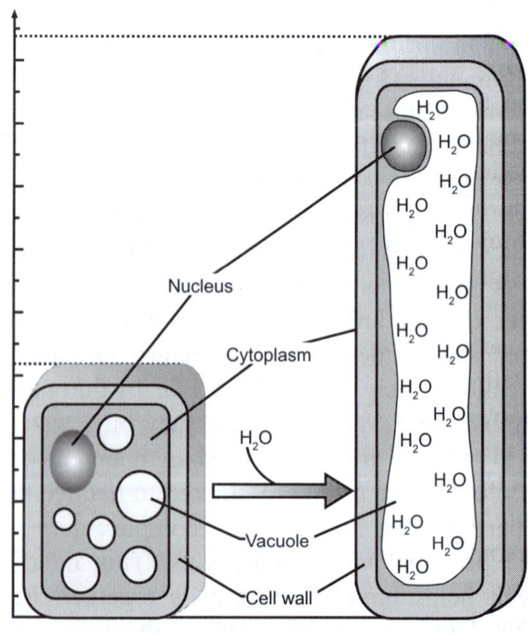

Fig. 5.9 Expansion of plant cells. Turgor pressure provides the force required to drive the expansion of plant cells by uptake of water, which is accumulated in a large central vacuole.

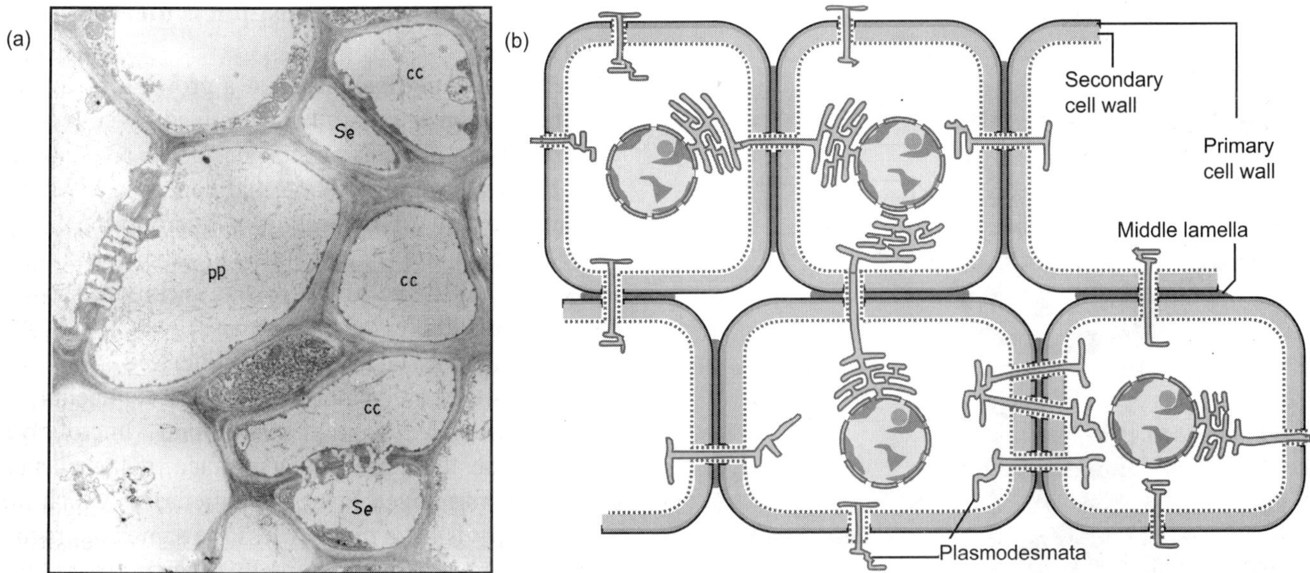

Fig. 5.10 The structure of the cell wall of a mature plant cell and intercellular macromolecule trafficking through plasmodesmata. (a) TEM photomicrograph of tomato leaf vein cross section (Channarayappa, 1992) and (b) Illustration of intercellular plasmodesmatal connections.

the cells (**Fig. 5.10**). Plasmodesmata are located in narrow areas of cell walls called primary pit fields and they are so dense in these areas (up to one million per square millimetre) that they make up 1% of the entire area of the cell wall.

Structure and composition: The pore is lined with the cell surface membrane and has a central tubular core, often associated at each end with ER (**Fig. 5.11**). This enables a continuous system of cytoplasm, the *symplast*, to be formed between neighbouring cells for the transport of substances between cells. Sieve plate pores of phloem sieve tubes are formed from plasmodesmata. Like gap junctions, plasmodesmata provide intercellular channels for molecules of up to 1000 molecular weight, including a variety of metabolic and signalling compounds. Electron micrographs reveal that in plasmodesmata the plasma membranes of the adjacent cells extend continuously through the cell wall (**Fig. 5.12**). The diameter of the channel is about 60 nm and plasmodesmata can traverse cell walls up to 90 nm thick. Within the plasmodesmal pore, a tightly wound cylinder of membrane termed the *desmotubule* runs the length of the plasmodesma. Thus, the desmotubule is essentially a tube within a larger tube bordered by the plasma membrane. The space between the plasmalemma and the desmotubule is the *cytoplasmic sleeve* or *cytoplasmic annulus* and transport through plasmodesmata has been proposed to occur either through the lipid portions of the desmotubule or this cytoplasmic sleeve (or both). Recent studies have suggested that there is an actin–myosin network associated with plasmodesmata, since these proteins were found to be associated with plasmodesmata using immunogold cytochemistry. The actin and myosin wrap around the central desmotubule is known to stabilize the desmotubule and consequently the entire plasmodesma.

Signalling and transport mechanisms: Plasmodesmata are involved in cell–cell communication. Many normal metabolic products, such as sucrose, spread across cells. Transport of such substances is proportional to the number of plasmodesmata and does not occur between cells not connected by such junctions. A small fluorescently-labelled probe was used to determine the passive transport mechanisms of plasmodesmata, that is, the *siz exclusion limits*, or the maximum size of a molecule that can pass through plasmodematal annulus passively. It was determined that the average width of channels allowing molecules to passively travel through a plasmodesmal annulus is 3 nm and thus the average size exclusion limit is molecules weighing approximately 800–1000 Da. Despite considerable structural differences between gap junctions and plasmodesmata, many of their functional aspects are remarkably similar. However, recent studies have shown that size exclusion limits vary between species and even cell types during passive transport. For example, in tobacco mesophyll cells, the size exclusion limit is only 1 kDa, whereas in trichome cells, the size exclusion limit is as large as 7 kDa.

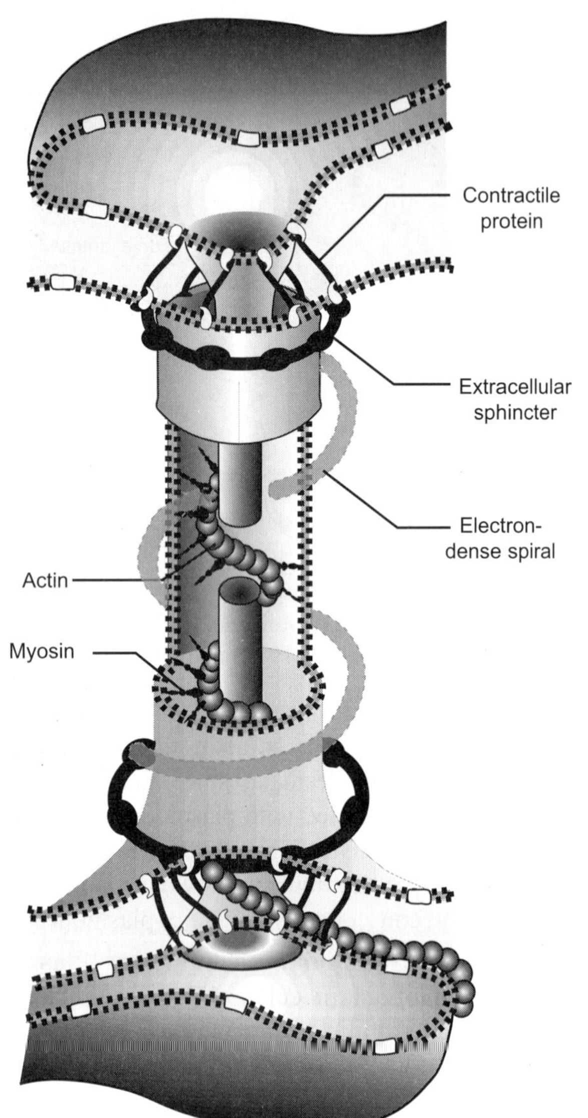

Fig. 5.11 Model of plasmodesmal structure, proposed by Overall and Blackman (1996).

Many other molecules destined for intercellular transport are shuttled through plasmodesmata via **active transport mechanisms**. Here, plasmodesmata can alter their dimensions such that they expand/dilate outward into an electron-lucent sleeve surrounding the normal-sized plasmodesmata. This expansion would allow for the transport of larger molecules through the cytoplasmic sleeve. The mechanism of this plasmodesmatal dilation is not clearly understood yet. Myosin has been found in several species of plants and is an ATP-dependent protein that generates directional movement. High ATPase activity has been demonstrated in plasmodesmata. Hence, myosin is a prime candidate for aiding active transport.

Direction of cell wall deposition is directed by microtubules

The characteristic final shape of a growing plant cell and the form of the plant is determined by controlled cell expansion. Cell expansion occurs in response to turgor pressure in a direction that depends in part on the arrangement of the cellulose microfibrils in the wall. The cellulose (like hyaluronan) is spun out from the surface of the cell by a plasma membrane-bound enzyme complex (cellulose synthase), which uses the sugar nucleotide UDP—glucose supplied from the cytosol. As they are synthesized, the nascent cellulose chains assemble spontaneously into microfibrils that form on the extracellular surface of the plasma membrane forming a layer, or lamella, in which all the microfibrils have more or less the same alignment. Each new lamella forms internally to the previous one. The cellulose microfibrils that are newly deposited in the cell wall will be in the same orientation of cortical microtubules that are closest to the cytosolic face of the plasma membrane. The congruent orientation of the cortical array of microtubules and cellulose microfibrils are observed in many types and shapes of plant cells.

Plant cells change their direction of expansion by a sudden change in the orientation of their cortical array of microtubules. Since plant cells are immobile, determination of the final morphology of a multicellular plants depends on the coordinated, highly patterned control of cortical microtubule orientations during plant development.

5.4 PLASMA MEMBRANE

Cells are separated and protected by a thin, fragile structure called the plasma membrane that is only 5–10 nm wide (**Fig. 5.13**). Cell membranes are important for a number of reasons. The primary function of membranes is to hold together the contents of the cell. They separate the contents of cells from the external environment, controlling exchange of materials (take in nutrients and exclude noxious materials and secrete finished products). All cells require a selectively permeable membrane. Even more fundamentally, cells can operate efficiently only by concentrating needed materials within their boundaries and keeping them there. Membranes also enable separate compartments to be formed inside cells in which specialized metabolic processes such as photosynthesis and aerobic respiration take place. They also act as receptor sites for recognizing hormones, neurotransmitters and other chemicals,

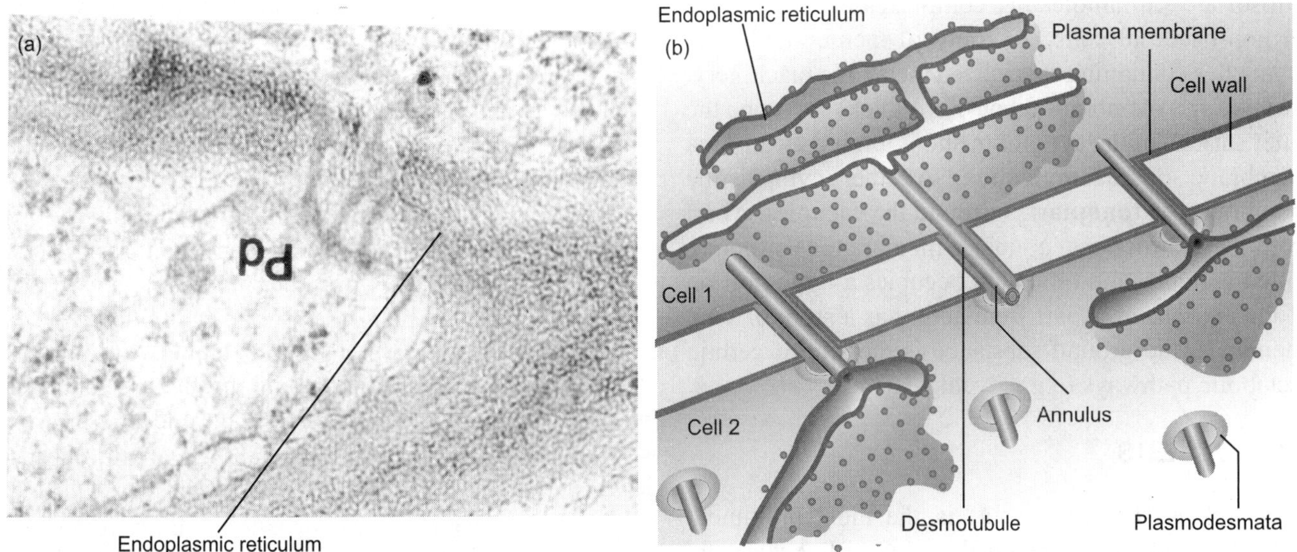

Fig. 5.12 (a) TEM photograph showing plasmodesmata (Pd) and (b) Drawing showing detailed structure and components of plasmodesmata (side view) (Channarayappa, 1992).

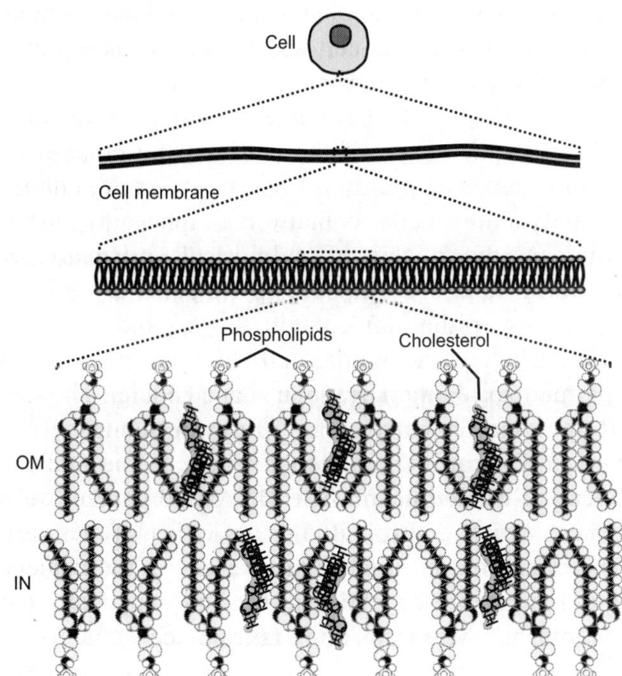

Fig. 5.13 Plasma membrane. Lipid bilayer structure at different resolutions (OM: outer membrane and IN: inner membrane).

either from the external environment or from other parts of the organism (see Chapter 6).

5.5 CYTOPLASM

The total cellular substance, except the nucleus, surrounded by the plasmalemma is known as cytoplasm (**cytoplasmic matrix**) and is a part of the protoplast (nucleus and cytoplasm together called **protoplasm**). The terms, cytoplasm and nucleoplasm, were introduced by Strasburger (1982). The internal volume bounded by the plasma membrane, the cytoplasm, is composed of an aqueous solution, the cytosol, containing a variety of insoluble suspended particles and a number of cell organelles. The soluble part of the cytoplasm forms the 'background material' or '**ground substance**' (cytoplasmic matrix) between the cell organelles. The portion of cytoplasm other than these structures is known as **hyaloplasm**. The cytosol is a highly concentrated aqueous solution with a complex composition and gel-like consistency (90% water). One of the important properties of the matrix is its ability to undergo reversible sol–gel transformation, which is attributed to the colloidal properties of the cell. This property also helps in viscosity changes, intracellular motion (**cyclosis**), spindle formation and cell cleavage and it is the site for many fibrillar differentiations such as microtubules and filaments. Earlier, it was believed that the matrix consists of homogeneous peripheral part known as the **ectoplasm** and the inner granular plasma known as the **endoplasm**.

The cytosol consists of many enzymes and RNA molecules that encode them; the monomeric subunits (amino acids and nucleotides) from which these macromolecules are assembled, large amounts of organic molecules called metabolites, intermediates in biosynthetic and degradative pathways, coenzymes and inorganic ions. Among the particles suspended in the cytosol are supramolecular complexes and, in almost all non-bacterial cells, a variety of membrane-bound organelles containing specialized metabolic machinery. A single unit membrane (about 80 Å thick) occurs on the outer surface of the cytoplasm and is termed the plasma membrane or **plasmalemma**. Another cytoplasmic membrane the **tonoplast**, surround the vacuoles. Apart from these structures, a tubular membranous structure called endoplasmic reticulum occupies a large portion of the cytoplasm. Apart from acting as a store of vital chemicals, the ground substance is the site of certain metabolic pathways (e.g., glycolysis).

5.6 NUCLEUS

The nucleus (L., *nux* = nut) is the most prominent structure and the functional centre of eukaryotic cells. Robert Brown (1833) discovered a prominent body within the cell and termed it nucleus. The role of the membrane in delimiting the nucleus was first demonstrated by O Hertwig in 1893. However, the role of the nucleus in heredity was firmly established by the grafting experiments of Hammerling (1953) with *Acetabularia*. The nuclear equivalent of a prokaryotic organism is known as **prokaryon** or more commonly as **nucleoid**. It does not have a nuclear envelope and does not divide by regular mitosis.

The nucleus is vital, houses the hereditary information of an organism and the prime control centre of a cell. This information includes all the genetic instructions for:

- governing the synthesis of chemicals for normal cell functions
- controlling the production of a new cell or an entire multicellular organism
- directing the development of genetically acquired characteristics.

DNA replication, transcription and RNA processing take place within the nucleus. The genetic information in the nucleus governs the synthesis of specific enzymes that in turn catalyze particular metabolic processes or contribute to the formation of a certain trait.

5.6.1 Nuclear morphology

In the late 19th century, Verworm, Balbiani and others showed that following microsurgery, nucleated halves of various protozoans survived and grew, whereas enucleated halves degenerated. Hammerling demonstrated by means of interspecific nuclear transplants, that morphological features, notably the shape of a cap, were determined by the nucleus.

The nucleus is the heart of all eukaryotic cells of plants and animals (**Fig. 5.14**). Generally the cells contain a single nucleus but the number of the nucleus can vary from cell to cell.

1. ***Mononucleate cells***: Most plant and animal cells contain a single nucleus, with somatic chromosome numbers.
2. ***Binucleate cells***: These contain two nuclei. Such cells occur in certain protozoans such as paramecium and cells of cartilage and liver.
3. ***Polynucleate cells***: Cells contain many (from 3 to 100) nuclei.

The polynucleate cells of the animals are termed as **syncytial cells**, while the polynucleate cells of the plants are known as **coenocytes**. The most common example of the syncytial cells is the osteoblast (polykaryocytes of the bone marrow), which contains about 100 nuclei per cell and striated muscle fibres each of which contain many hundred nuclei.

The shape of the nucleus normally remains related to the shape of the cell and the number of chromosomes or ploidy level. The spheroid, cuboid or polyhedral cells contain the spheroid nuclei, whereas, the nuclei of cylindrical, prismatic or fusiform cells are ellipsoid in shape. The cells of squamous epithelium contain the discoidal nuclei. The leucocytes, glandular cells of some insects and spermatozoa contain irregular shaped nuclei. The size of nuclei is typically about 10 μm in diameter (can vary 3–25 μm). The aqueous phase of the nucleus is called **nucleoplasm**. The DNA is organized into genes, which control all the activities of the cell. The transcription of RNA takes place in the nucleus and the processed RNA molecules pass into the cytoplasm where translation takes place.

5.6.2 Ultrastructure

The nucleus is composed of many prominent structures (**Fig. 5.15**) such as chromatin fibres, nucleolus, nucleoplasm, nuclear membrane or karyotheca or nuclear envelope and nuclear pores.

Nucleoplasm

The nucleoplasm (or **nuclear sap** or **karyolymph**) is the semifluid substance in the nucleus containing proteins, granules and fibrils, which is situated

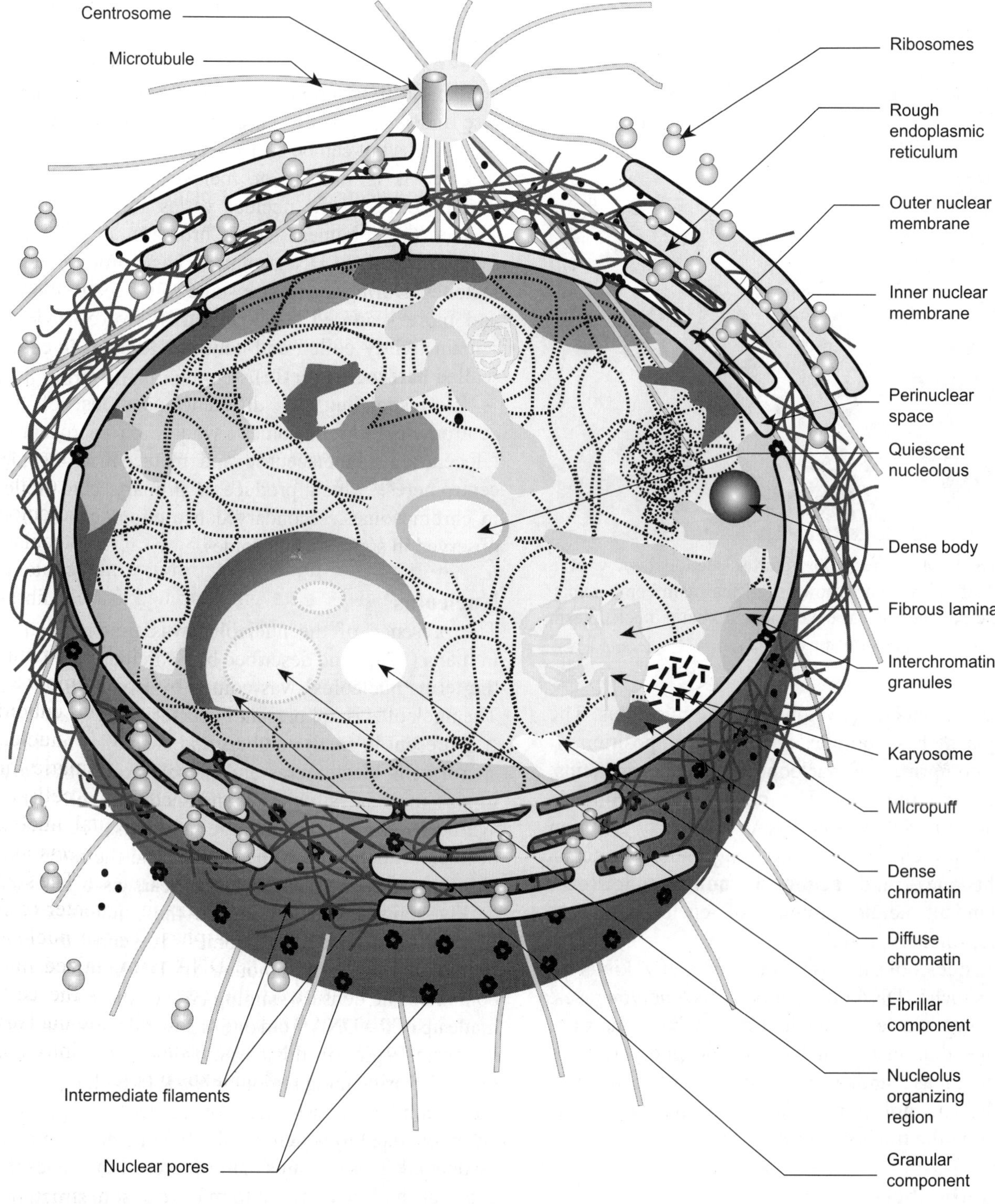

Fig. 5.14 A cross-sectional view of a typical nucleus showing all the main components **(See Plate 1)**.

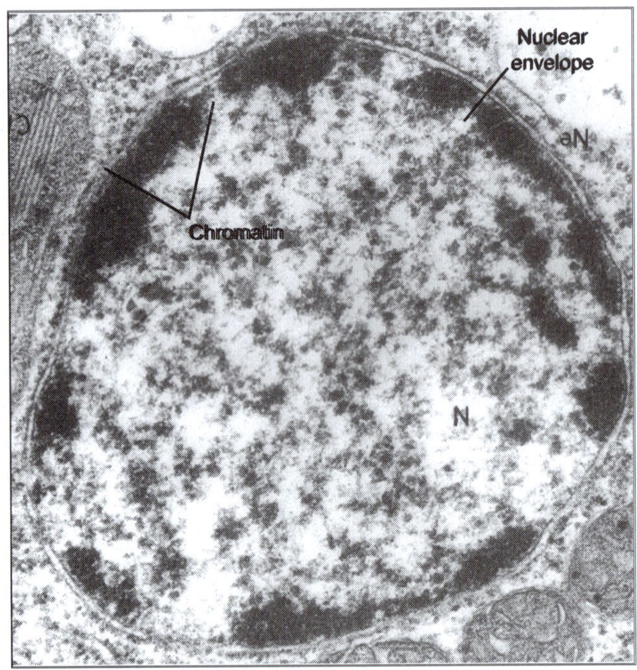

Fig. 5.15 TEM micrograph of cross-sectional view of a tomato leaf cell nucleus showing all the main components, which include the nuclear envelope, chromatin, nucleoplasm and nucleolus (Channarayappa, 1992).

between the nuclear envelope and the nucleolus. The nucleoplasm has complex chemical composition. It mainly comprises of nucleoproteins, nucleic acids, enzymes and minerals. The most common nucleic acids are the DNA and RNA. The two common types of proteins are: **basic** (e.g., nucleoproteins, nucleohistones) and **acidic** or **non-histone** (e.g., the abundant acidic proteins of euchromatin are phosphoproteins) proteins.

The nucleoplasm also contains many kinds of enzymes such as DNA polymerases, RNA polymerases, NAD synthetases, nucleotide triphosphatases, adenosine diaminase, nucleoside phosphorylase and pyruvate kinase and a small amount of lipids and minerals. The molecular 'skeleton' of the nucleus has been called the **nuclear matrix**.

Chromatin fibres

The nucleoplasm contains a number of thread-like coiled and elongated structures, called **chromatin fibres** and these can be observed only in interphase nucleus. Strasburger (1875) discovered thread-like structures, which appeared during cell division and he called them **chromosomes** (*chroma* = colour) due to their affinity for basic dyes. Chemically, chromatin consists of DNA and proteins and small quantities of RNA. The fibres are twisted, finely anastomosed and uniformly distributed in the nucleoplasm. In all types of higher organisms, the well-organized nucleus contains a fixed number of chromosomes of definite size and shape. These chromosomes are invisible in the nucleus but can be easily seen during cell division, whether it is mitosis or meiosis. The number of chromosomes in the nucleus varies according to the species (e.g., humans 46). Chromatin is composed mainly of coils of DNA bound to basic proteins called histones. During interphase some of it becomes looser and more scattered (**euchromatin**). However, some remain tightly coiled and continue to stain intensely (called **heterochromatin**).

Corresponding to different positions of the centromere, chromosomes would be acrocentric, telocentric, submetacentric and metacentric. Besides centromeres, which produce a primary constriction in chromosomes, secondary constrictions can also be observed in some chromosomes.

Nucleolus

The presence of the nucleolus was first noticed by Fontana (1781) and described by Schleiden (1938), but the term 'nucleolus' was coined by Bowman (1840). The nucleolus is not present in bacteria and yeast cells. It is present in small numbers and is smaller in size in sperm cells blastomeres and muscle cells. It is large in size in oocytes, neurons and secretory cells which synthesize the proteins in larger amounts. The number of nucleoli depends on the species and the number of chromosomes. The nucleolus appears as a rounded, darkly stained structure inside the nucleus and its function is to make ribosomes. It stains intensely because of the large amounts of DNA and RNA it contains. The densely staining core of the nucleolus is made up of the DNA from one or several chromosomes. A special region in this chromosome is known as the nucleolar organizing region or **NOR** (which represents the secondary constriction of nucleolar organizing chromosomes) to which usually the nucleolus remains associated. This contains many copies of the genes that code for the RNA needed to make ribosomes (rRNA). Nucleolus can be seen as a conspicuous structure in the interphase nucleus. It disappears during mitosis and reappears at the next interphase, but this is because the DNA disperses.

The nucleolus is not bounded by any limiting membrane (**Fig. 5.16**). It mainly contains protein (70%), RNA (30%) and a small amount of DNA. Calcium ions regulate maintenance of its integrity.

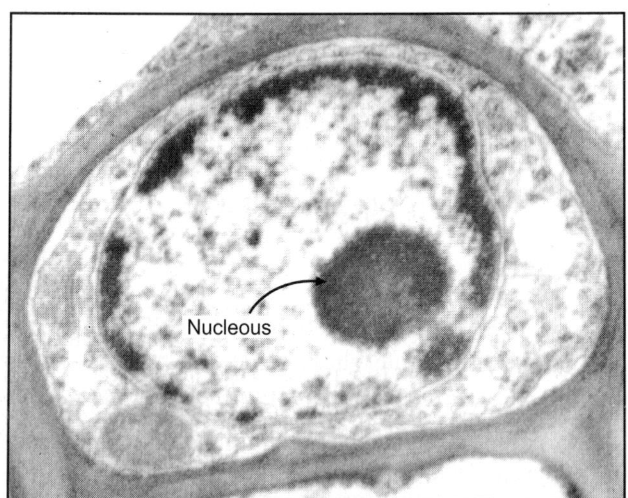

Fig. 5.16 Nucleolus. TEM micrograph of tomato leaf cell showing the very conspicuous nucleolus (Channarayappa, 1992).

It contains four types of rRNAs, more than 70 types of ribosomal proteins, RNA binding proteins (e.g., nucleolin) and RNA splicing nucleoproteins (U_1, U_2, ...U_{12}). It also contains phospholipids, orthophosphates, Ca^{2+} ions, some enzymes, which include acid phosphatase, nucleoside phosphorylase and NAD^+ synthesizing enzymes, nucleotides and ribosomal rRNA.

The ultrastructure of nucleoli reveals four regions:

- amorphous region
- granular region
- fibrillar region
- chromatin fibres.

The central region of a nucleolus is amorphous and is composed mainly of protein. The central amorphous region is surrounded by nucleolemma, which consists of a granular portion mostly containing ribosomal subunits of 150–200 Å diameter and a fibrillar region primarily containing RNA fibres of 50–150 Å diameter. In addition, chromatin fibres from the nucleolar organizer region also pass through nucleoli.

Functions: The main function of the nucleolus is the production and organization of ribosomes. It is the site where biogenesis of ribosomal subunits (i.e., 40 S and 60 S) takes place. Within a nucleolar region, three types of rRNAs, namely, 18 S, 58 S and 28 S are transcribed as part of a much longer precursor molecule (45 S transcript), which undergoes rRNA processing by the help of two types of proteins (such as nucleolin and U3 snRNP). However, the 5 S rRNA is transcribed on the chromosome existing outside the nucleolus and the 70 types of ribosomal proteins are synthesized in the cytoplasm. All these cytoplasmic components of the ribosomes migrate to the nucleolus, where they are assembled into two types of ribosomal subunits, which are eventually transported back to the cytoplasm.

Nuclear envelope

Nuclear envelope is not a mere membranous nuclear barrier; it is a complex and highly organized structure composed of various molecular components.

Structure

The nuclear boundary of the interphase and prophase nuclei is called the **nuclear membrane** and it separates the contents of the nucleus from the cytoplasm by forming a distinct biochemical compartment. The ultrastructure of the nuclear envelope pore complexes and nuclear lamina was studied by Kirschner and others. (1977) and Schatten and Thoman (1978). It consists of a lipid bilayer (double membrane) having two unit membranes (each 5–10 nm thick), called inner and outer nuclear membranes. The space between two unit membranes varies in width (10–50 nm) and is known as **perinuclear space**. It is a fluid-filled compartment, which is continuous with the ER lumen and may contain fibres, crystalline deposits and lipid droplets. Both the inner and outer membrane layers are similar in structure to the plasma membrane. The **outer nuclear membrane** is continuous with the endoplasmic reticulum (ER) and like the ER it may be covered with ribosomes carrying protein synthesis. Each unit membrane is 7.5 nm in diameter and perinuclear space may vary from regular 15 nm wide space to irregular cavities of several hundred times wide. The outer membrane of the nuclear envelope is smooth and interrupted by pores, which appear circular in surface view. The diameter of these pores is about 90 nm and allow passage of moderately large molecules (RNA and proteins). At the boundary of these pores, the outer membrane and inner unit membranes are joined. Nuclear pores allow exchange of substances between the nucleus and the cytoplasm. The outer membrane is surrounded by a network of intermediate filaments. The spherical **inner nuclear membrane** contains specific proteins that act as binding sites for the supporting fibrous sheath of intermediate filaments (IF), called **nuclear lamina**, which is composed of A, B1, B2 and C type of **lamins** (a specialized type of intermediate filament protein of 60–80 kDa). The nuclear lamins form dimers that have a rod-like domain and two globular heads at one end. The association of lamins with the inner nuclear membrane is facilitated by the post-translational addition of lipid (prenylation of *C*-terminal cysteine). The nuclear lamina is a dynamic structure. The transient phosphorylation of several serine residues on the lamins causes the lamina to reversibly disassemble into tetramers of hypophosphorylated lamin A and lamin C and membrane-associated lamin B. The lamins play an important role in the assembly of interphase nuclei.

Nuclear envelope assembly

Once a cell begins the mitotic division, the nuclear envelope disintegrates at the end of the prophase and is re-formed at the end of the nuclear division, i.e., it disassembles at the onset of mitosis and is re-assembled at the end of it. The mechanism of assembly involves the attachment of vesicles to the chromatin, followed by the fusion of vesicles to form a double membrane system (**Fig. 5.17**). The vesicles used for assembly of the nuclear envelope form a subset of ER-derived vesicles and are distinct from the majority of other ER-derived vesicles (e.g., COPII). Lamin deploymerization and polymerization during disassembly and re-assembly of the nuclear envelope also involves reversible phosphorylation.

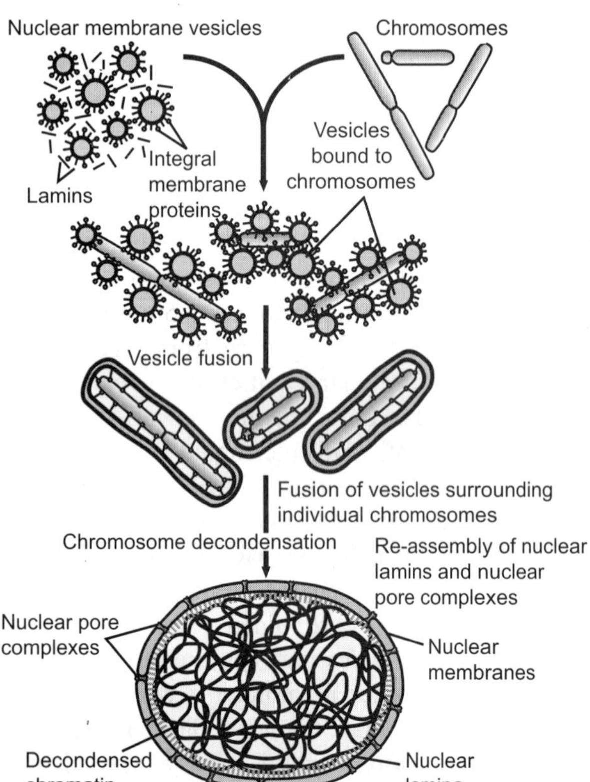

Fig. 5.17 Re-formation of the nuclear envelope and assembly of the nucleus.

5.6.3 Nuclear pore complex

Nuclear pores

A distinctive feature of the nuclear envelope is the presence of microscopically visible pores. Electron microscopy shows the pores to be membrane-lined channels in which the outer and inner membranes of the envelope are fused. The number of pores varies in cell types and species from 1 to 100 per square micrometre. Nuclear pores are created by a number of proteins, which are involved in the formation of a highly organized nuclear pore complex (**NPC**). The NPCs are the only channels through which small polar molecules, ions and macromolecules can travel between the nucleus and the cytoplasm. The nuclear pore occupies a central position among the major cellular structures, but still remains one of the least understood structures. In 1950, Callan and Tomlin first observed the nucleopores in the nuclei of amphibian oocytes. Recently, a number of genes and gene products involved in pore complex formation have been isolated. In vertebrates, the NPC is composed of 50 to 100 different pore proteins. Further, the pore has been reconstituted in vitro, a number of '**signal sequences**' and one or more '**signal sequence receptors**' have been identified and a new '**basket-like structure**' has been found attached to the inner side of the nuclear pore(**Fig. 5.18**).

Structure

Nuclear pores appear circular in surface view and have a diameter of 10–100 nm, i.e., about 30 times the size of the ribosome. The pore has a diameter of 120 nm and thickness of 50 nm and an eight-fold or octagonal symmetry. It consists of four separate elements:

- the scaffold
- the central hub or transporter
- short thick filaments attached to the cytoplasmic side of the pore
- a newly discovered basket attached to the nucleoplasmic side of the pore.

The pore complex consists of two '**rings**' (R or annuli) at its periphery with an inside diameter of

80 nm, a large particle that forms a **central plug** (**C**) and radial '**spokes**' (S) that extend from the plug to the rings (**Fig. 5.18**). **Particles** (**P**) are anchored to the cytoplasmic ring and are thought to be inactive ribosomes. The '**hole**' in the centre of the pore complex is an aqueous channel through which water-soluble molecules shuttle between the nucleus and the cytoplasm. This hole often appears to be plugged by a large central granule (**central plug**), which may consist of newly made ribosomes or other particles caught in transit. The number of pores can vary depending on the transcriptional activity of the cell (about 3–50 pores/ per micrometre.).

Functions of nuclear pore complex

The nuclear pore complexes are the sites for exchange of macromolecules (selective traffic of proteins and RNA) between the cytoplasm and the nucleus (i.e., **nucleo-cytoplasmic traffic**). NPCs have a mass of 125 megadaltons in higher eukaryotes and contain about 100 different polypeptides called **Nup** (**nuclear pore proteins**). The nuclear pore complex has a passive diffusion channel, 9 nm in diameter, permitting diffusion of many but not all the smaller proteins including cytochrome. Several small proteins like histones (<9 nm) and other larger proteins (>9 nm and up to 25 nm) pass through the NPC by an active transport process. The active process is facilitated by energy, signal sequence and saturability so that carrier molecules mediate the process.

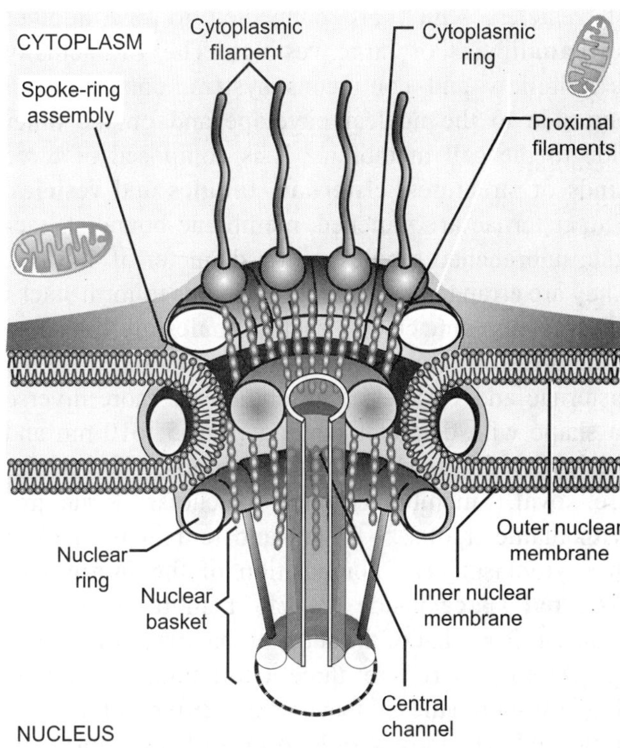

Fig. 5.18 Model of the nuclear pore complex **(See Plate 2).**

5.7 ENDOPLASMIC RETICULUM

The basic components of the **endomembrane system** (or cytoplasmic vacuolar systems) are the **nuclear envelope**, the endoplasmic reticulum (ER) and the **Golgi complex**. The ER (first identified by Porter in 1948) is an extensive membrane system within the cytoplasm and is continuous with the nuclear envelope (**Fig. 5.19**). Its membrane may account for about one-half of all cell membranes and the space enclosed by the ER (the lumen or cisternal space) may represent about 10% of the total cell volume. The spacc of thc ER opens into perinuclear space between the two nuclear membranes. This membranous organization travelling the matrix is often called **cytoplasmic vesicular system**. The other parts of this vesicular system are nuclear envelope, endosome, lysosome and Golgi complex. ER is present in both plants and animals and only in differentiated cells; it is not found in RBCs, eggs and in prokaryotic cells.

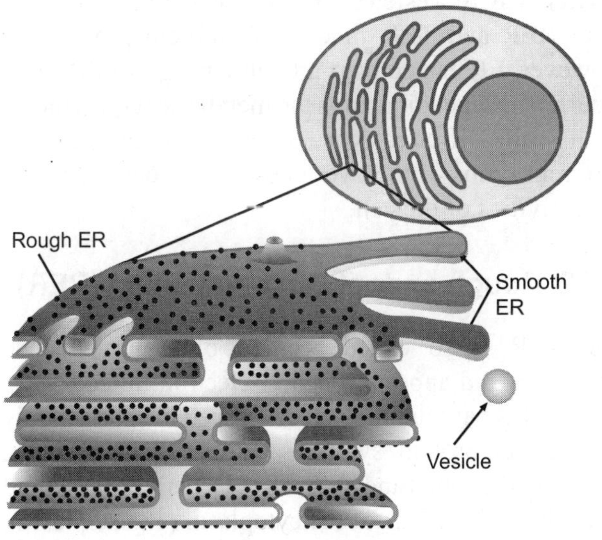

Fig. 5.19 Sketch of endomembranes. Endoplasmic reticulum.

5.7.1 Ultrastructure of ER

The membrane of ER is the same as that of a unit membrane. The body of ER appears as a labyrinth of branching tubules and flattened sacs with variable spaces. These strands form tubes (50–100 mμ diameter) and vesicles (25–500 mμ diameter) in the cytoplasm. Sometimes they may be found in chains of vesicles, which are connected to one another by **canalicules** or large vesicles. The ER actually is a hollow and continuous system, connected on one side to the nuclear envelope and on the other side to the cell membrane. It is composed of three kinds of structures: **cisternae**, **tubules** and **vesicles**. The cisternae are flattened, membrane-bounded, sac-like, unbranched tubules with a diameter of 4–5 nm. They are arranged in parallel in bundles to form stacks of lamellae connected to each. Tubules are branched structures forming the reticular system along with the cisternae and vesicles. The tubules are more diverse in shape with diameter ranging from 5 to10 nm and are involved in the synthesis of steroids. The vesicles are oval, membrane-bound vacuolar structures with diameter of 2.5–50 nm and remain isolated in the cytoplasm. The composition of the **luminal** or **cisternal space** inside the ER membranes is quite different from that of the surrounding **cytosolic space**. In general, there are three contiguous membrane domains within the ER that perform different functions in the cell. The most visible forms of ER in most cells are: **rough endoplasmic reticulum** (**RER**), **smooth endoplasmic reticulum** (**SER**) and **transitional ER** (**TER**), where vesicles exit to the Golgi apparatus. Both RER and TER function in protein processing. However, ER can convert into either RER or SER and vice versa, depending upon the metabolic requirements of the cell. Different types of cells contain markedly different ratios of the two types of ER, depending on the activities of the cell.

5.7.2 Rough endoplasmic reticulum (RER)

The RER is so called because of many attached ribosomes and appears rough in electron micrographs. The RER is particularly well developed in cells actively engaged in protein synthesis (e.g., pancreatic cells, plasma cells, goblet and liver cells). Generally, it is found in the basophilic part of the cytoplasm (ergastoplasm), a property that is attributed to attached ribosomes, which contain RNA.

The ribosomes are attached on the outer surface of RER membrane, as polysomes held together by mRNA. The ribosomes are always attached to the membrane by the large subunit. The ribosomes are bound by two transmembrane proteins: ribophorin I (65 kDa) and ribophorin II (64 kDa). These proteins are present in REM only. The ribophorins are involved in mediating the attachment of ribosomes to the ER membrane. These interact with each other and may form an intramembranous network, which could control the distribution of ribosome binding sites in the plane of the ER membrane.

The ribosomes on RER specially synthesize proteins intended for secretion (transport of proteins) by the cell, such as plasma or milk proteins, or those destined for the plasma membrane or certain organelles. The lipids and proteins (integral membrane proteins) synthesized on the RER are transported in specialized transport vesicles to the Golgi complex—a stock of flattened membrane vesicles which further modifies, sorts and directs them to their final destination.

5.7.3 Smooth endoplasmic reticulum (SER)

ER without ribosomes is called SER. It forms a continuous system with RER, but has a different morphology. The SER is more common in adipose cells, interstitial cells, liver cells, spermatocytes and leukocytes. The muscle cells are also rich in SER, known as sarcoplasmic reticulum. In the pigmented retinal cells it exists as tightly packed tubules known as **myeloid bodies**.

Functions of SER: The main function is the formation of *transport vesicles*, which carry membrane-bound enzymes, including those involved in the biosynthesis of certain lipids and oxidation and detoxification of foreign compounds such as drugs.

1. One of the chief functions of SER is *lipid synthesis*.
2. SER also makes *steroid* hormones (which are a type of lipid) in the endocrine cells of the gonad and adrenal cortex.
3. Detoxification in the liver of a wide variety of organic compounds, including barbiturates and ethanol, can lead to proliferation of SER in liver cells. Detoxification is carried out by a system of oxygen-transferring enzymes (oxygenases), including the cytochrome P450 family.
4. Release of glucose from glucose 6-phosphate in SER of liver cells by the enzyme glucose 6-phosphatase a step in glueoneogenesis. Large reserves of glycogen are stored in the liver as granules on the cytosolic side of SER membranes.
5. SER is highly developed in seeds. Sarcoplasmic reticulum, a type of SER in skeletal muscle cell, stores and releases calcium ions for muscle cells. The regulated release of Ca^{2+} from SER (or sarcoplasmic reticulum) triggers contraction, in the case of muscle cells.

5.8 GOLGI COMPLEX

Golgi apparatus was first described by Camillo Golgi (1822–1885) in nerve cells. Since originally these were known to be networks, they were also called **dictyosomes** (**Fig. 5.20a**), but now it is clear that there are no networks. The Golgi complex along with other organelles (ER and lysosomes) makes an extensive membrane system. It is a differentiated portion of the endomembrane system, which is morphologically similar in animal and plant cells. It is absent in the prokaryotic cells and eukaryotic cells of certain fungi, sperm cells of bryophytes and pteridophytes, cells of mature sieve tubes of plants, mature sperm and red blood cells of animals. It does not attach to ribosomes. This membrane system has a continuous lumen, which forms a distinct channel separated from the cytoplasm and the nucleoplasm of the cell. Within this channel the material can be synthesized, modified and transported (see chapter on Protein synthesis and targeting). Since it provides a connection with the nucleus, it plays an important role in the transport of materials from the central control site to the cytoplasm.

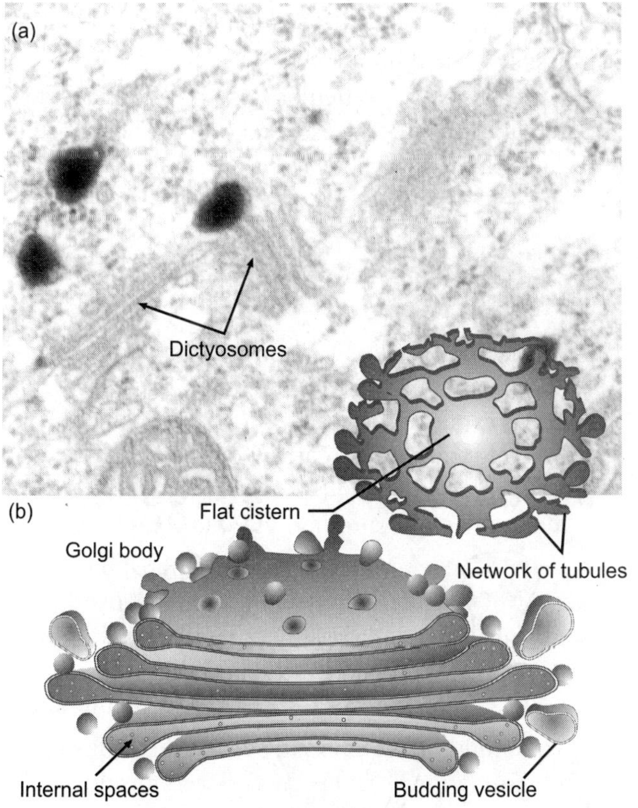

Fig. 5.20 (a) TEM showing the dictyosomes in a plant cell (Channarayappa, 1992). (b) Sketch of endomembranes. Complete cistern of a Golgi apparatus (top view) and cross section of Golgi body along with the vesicles.

5.8.1 Ultrastructure of Golgi complex

Nearly all eukaryotes have Golgi complexes (absent in prokaryotes), systems of membranous sacs, or **cisternae**, arranged as flattened stacks (**Fig. 5.20b**). It is generally, a single large structure, but many Golgi bodies can be seen in nerve cells, liver cells (about 50) and variable numbers in most plant cells (range from hundreds to several thousands). The Golgi complex varies in form from compact discrete granules or mass to a dispersed filamentous reticulum. The shape and function of this organelle varies with the metabolic state of the cell. In other words, the Golgi complex is **pleomorphic**. It consists of a series of double membranes, concentrically bent. These double membranes enclose submicroscopic spaces and are really flattened membrane-bound sacs, which are also referred to as cisternae, saccules or lamellar units. These individual vesicles (60 nm in diameter) are sometimes also referred to as **dictyosomes** or **golgiosomes**. The size and number of these cisternae vary greatly and usually range from 3 to 12. The cisternae of lamellae are interconnected, appearing as a system of branched tubules (300–500 Å). The number of Golgi bodies varies according to the type of cell, its function and stage of growth. Their relative lengths may differ from species to species and different stages of activity and ontogeny. Golgi bodies originate from ER elements, particularly from RER. The Golgi saccules proximal to ER are formed by the fusion of ER-derived vesicles, while distal saccules (mature Golgi) "give their all" to vesicle formation.

The cisternae of Golgi complex are stacked together and are found as **Golgi stacks**, each stack consisting of 4–6 cisternae. Each Golgi stack has two faces, a ***cis*-face** (entry face) and a ***trans*-face** (exit face). The *cis*-side faces the rough endoplasmic reticulum (which receives vesicles from ER) and the *trans*-side faces the plasma membrane (from where Golgi vesicles leave the Golgi stack); the **medial elements** lie between these. At one end of the stack new cisternae are constantly being formed by vesicles from the SER, while at the other end on the 'maturing' face, the cisternae break up into vesicles again. These vesicles are 40–80 nm in diameter and are of two types.

- ***Smooth vesicles or secretory vesicles*:** These are 20–80 μm in diameter, contain secretory material and are budded off from the ends of the cisternal tubules within the net.
- ***Clathrin-coated vesicles*:** These vesicles are spherical protuberances about 50 μm in diameter,

with a rough surface. They are found at the periphery of organelles and are morphologically quite distinct from the secretory vesicles.

5.8.2 Functions

The main functions of the Golgi apparatus are storage, packaging, transport and chemically modifying the lipids and proteins. Specific modifications allow them to be sorted out and packaged in vesicles for transport to specific locations. Thus, it is often referred to as the '**traffic police**' of the cell. They are involved in formation of primary lysosomes and other enzyme-containing cellular inclusions (e.g., melanin granules) and in formation of secretory granules (e.g., pancreas, pituitary and mammary glands and mucus secreting glands of the intestine). The Golgi apparatus is also a region of sorting of cellular secretory proteins and is sometimes involved in the secretion of carbohydrates.

In general the Golgi complex performs functions, such as reception, processing, sorting, packaging and dispatching of a variety of materials. To carry out these functions Golgi vesicles contain different sets of enzymes in different subcompartments. Recently, the division of labour in the subcompartments (*cis, medial* and *trans*) has been reported.

More specifically, in plants, the Golgi apparatus is mainly involved in the secretion of materials of primary and secondary cell walls, which include glycoproteins, lipids, pectins and monomers for cellulose, hemicellulose and lignin. During cytokinesis, the vesicles originating from the periphery of the Golgi apparatus coalesce in the phragmoplast area to form a semisolid layer, called **cell plate**. In later stages, the unit membrane of Golgi vesicles fuses during cell plate formation and becomes part of the plasma membrane of daughter cells.

In animals, depending on the cell types, the Golgi complex is involved in processing, packaging and exocytosis of various materials. Some important compounds produced by Golgi apparatus are:

- zymogen of exocrine pancreatic cells
- mucus secretion by goblet cells of the intestine
- lactoprotein (casein) by mammary gland cells (called **merocrine secretion**)
- thyroglobulins (thyroxine hormone) by thyroid cells
- tropocollagen and collagen
- melanin granules and other pigments
- yolk and vitelline membrane of growing primary oocytes.

It is also involved in the formation of certain cellular organelles such as plasma membrane, lysosomes, acrosome of spermatozoa and cortical granules of a variety of oocytes.

5.9 RIBOSOMES

Ribosomes were first observed by GE Palade (1952) under the EM. They appeared as dense particles or granules. They are tiny cellular particles (organelles), about 20 nm in diameter and found mainly on membranes of the ER and nucleus as also in the cytoplasm ('free floating'). In prokaryotes, they are present freely floating in the cytoplasm (**Fig. 5.21**). The ribosomes are oblate spheroid structures, 150–250 Å in diameter. In addition, ribosomes are also found freely in mitochondrial matrix and chloroplasts of eukaryotic cells. Ultracentrifugal analysis of ribosomes revealed that ribosomes sediment at discrete peaks in the 40–77 S range. When purified and resolved by gel electrophoresis, they are found to contain equal measures RNA and proteins.

In yeast cells, reticulocytes or lymphocytes, meristematic plant tissues, embryonic nerve cells and cancerous cells, ribosomes are present in large

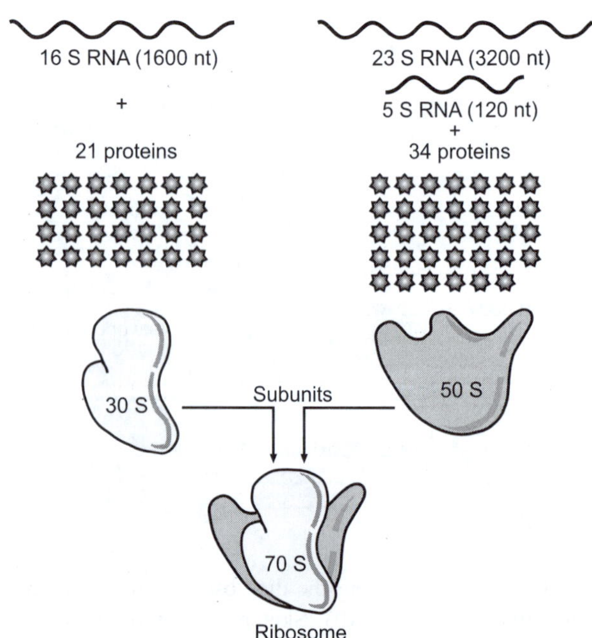

Fig. 5.21 Structure of a typical prokaryotic ribosome.

numbers. A typical bacterial cell contains about 10,000 ribosomes (25% of total cell mass), free in the cytoplasm, while eukaryotic cells possess many more (about ten millions per cell). They are made of roughly equal parts of protein (several kinds) and RNA (40%–60% of rRNA). Since these particles are rich in RNA, they are called ribosomes. Slightly smaller ribosomes are found in mitochondria (and in chloroplasts). Although there are several types of ribosomes, all have the same function—protein synthesis. Ribosomes act as chemical 'matchmakers', matching amino acids with the molecular instructions from the nucleus to construct a specific protein. Eukaryotic ribosomes are either free in the cytoplasmic matrix or attached to the membranes of the ER. Ribosomes may form **polysomes** (polyribosomes) that are collections of ribosomes strung along messenger RNA. The polysomes consist of those ribosomes which are actively engaged in protein synthesis and are held together by the RNA strand.

Prokaryotic and eukaryotic ribosomes can be distinguished on the basis of size and sedimentation coefficient (*S*), . The 70 S ribosomes of prokaryotes are comparatively small and have low sedimentation value. Similar types of ribosomes are also present in mitochondria and chloroplasts. The 80 S ribosomes occur in eukaryotic cells.

5.9.1 Ultrastructure of ribosomes

Each ribosome is porous and hydrated and consists of two subunits (**Fig. 5.22**). The association and dissociation of these subunits depend on the concentration of Mg^{2+} ion. If the ion concentration is lowered, the subunits fall apart, while the addition of Mg^{2+} ions will promote the association of subunits into complete ribosomes. In bacterial cells as well as the chloroplasts of higher organisms, the sedimentation coefficient of the complete ribosomes is 70 S (64 S–72 S). However, in mitochondria, the size of ribosomes varies from 60 S to 78 S. The 70 S subunit contains two subunits: the smaller subunit is 30 S and the bigger subunit is 50 S. The70 S ribosome contains 36%–37% proteins and about 63% rRNA. 30 S subunit contains 21 different proteins designated as S1–S21 and 16 S rRNA. The 50 S subunit contains 34 proteins designated as L1–L34 and 23 S and 5 S rRNA.

In higher organisms, the sedimentation coefficient of the ribosomes is 80 S and those of the subunits are 40 S and 60 S. The 80 S ribosomes contain less rRNA (40%–44%) than the proteins (60%–56%). The 40 S subunit contains one molecule of 18 S RNA and 33 proteins. The 60 S subunit, contains one molecule each of 5 S, 5.8 S, 28 S rRNAs and 49 proteins.

The fine structure of the ribosome is very complex. The RNA and proteins are intertwined and arranged in a complex manner in the two subunits. Negative staining of isolated ribosomes and their subunits has led to a better understanding of the fine structure of these organelles. It was shown that the smaller subunit consists of a **head**, a **base** and **platform**. The flat form separates head from the base with the help of a cleft. Similarly, the larger subunit consists of a **ridge**, a **central protuberance** and a **stalk**. The ridge and the central protuberance are separated with the help of a valley.

5.9.2 Chemical composition

The major components of the ribosomes are RNA and proteins. Ribosomes are strongly negative by binding cations and basic dyes. The 30 S and 50 S subunits may dissociate to give rise to RNA (16 S and 23 S respectively) and some core proteins. These RNAs in eukaryotes form 28 S and 18 S particles. Besides these core proteins, there are other proteins called **split proteins (SP)** in both subunits. These split proteins in 30 S and 50 S subunits are called SP30 and SP50 respectively and can be further fractionated into acidic

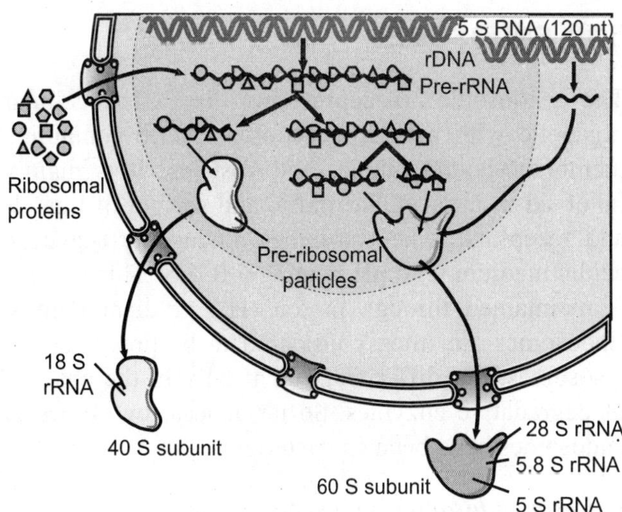

Fig. 5.22 Ribosome assembly in eukaryotes. Ribosomal proteins are imported in to the nucleolus from the cytoplasm and begin to assemble on the pre-rRNA prior to its cleavage. As the pre-rRNA is processed, additional ribosomal proteins and the 5 S rRNA assemble to form pre ribosomal particles. Finally the pre ribosomal particles export to the cytoplasm, yielding the 40 S and 60 S ribosomal subunits.

and basic proteins. All the 21 proteins have been isolated in smaller subunits (30 S) of *E. coli* ribosome and have been designated as S1 to S21. Similarly in larger subunits (50 S) 31 different proteins have been isolated and designated as L1 to L34. Similar organization of ribosomal RNAs and proteins are found in eukaryotic 80 S ribosomes.

5.9.3 Biogenesis

Ribosomes do not self-replicate. The complex organization of ribosomes involves different rRNA species and diverse proteins, suggesting that at least 50 genes may be involved in their formation. In bacteria, the genes for 5 S, 23 S and 16 S rRNAs are clustered in one region to form a single operon working as a functional unit. In eukaryotes, however, the biogenesis of ribosomes involves several regions of the cell. The 18 S, 5.8 S and 28 S rRNA molecules are synthesized as bigger molecules in the nucleolar organizer, which has multiple copies of DNA (ribosomal DNA repeat units) responsible for the synthesis of rRNA. The DNA responsible for 5 S rRNA also contains many copies but these are synthesized outside the nucleolus. The formation of rRNA is coupled with the formation of different ribosomal proteins that are synthesized in the cytoplasm. Most of the subunits of ribosomes are processed in the nucleolus and delivered to the cytoplasm, suggesting a key role for the nucleolus in the biogenesis of ribosomes.

5.10 ENDOSOMES

The endosome (**receptosome**) is a subcellular organelle with a heterogeneous structure containing membrane-bound tubules and vesicles. It is mainly involved in intracellular traffic and sorting of ligands and receptors. Like lysosomes, endosomes have an acidic medium with pH ranging 5.0–5.5. This low pH is maintained through proton (H^+) gradient pumps. Endosomes are often considered to be precursors of lysosomes, but differ from them only in the absence of degradative enzymes. So far at least two types of endosomes have been characterized:

- ***Early endosomes*** lie just beneath the plasma membrane and receive endocytosed proteins within a minute.
- ***Late endosomes*** are closer to the nucleus and receive endocytosed proteins within 5–10 min.

The early endosomes are actually involved in sorting proteins during the endocytic pathway. The relationship between the two types of endosomes and the lysosomes are not well understood.

5.11 VACUOLES

The vacuole is a fluid-filled sac bound by a single membrane. Animal cells contain relatively small vacuoles, such as phagocytic, food, autophagic and contractile vacuoles. However, plant cells notably mature parenchyma cells, have a large central vacuole with variable size and are surrounded by a unit membrane called **tonoplast** (**Fig. 5.23**). The fluid they contain is called **cell sap**. Plant cells do not have lysosomes, but their vacuoles carry out similar degradative reactions as well as other functions. Cell sap is a concentrated solution of various substances, such as mineral salts, sugars, phenols, pigments (e.g., anthocyanin), organic acids and hydrolytic enzymes. Apical meristem cells typically contain numerous small vacuoles, which coalesce into one or a few larger vacuoles as the cell matures and expands (up to 90% of the total cell volume). It also acts as a store house of various substances, including inorganic ions, organic acids, sugars, enzymes, storage proteins and many types of secondary metabolites

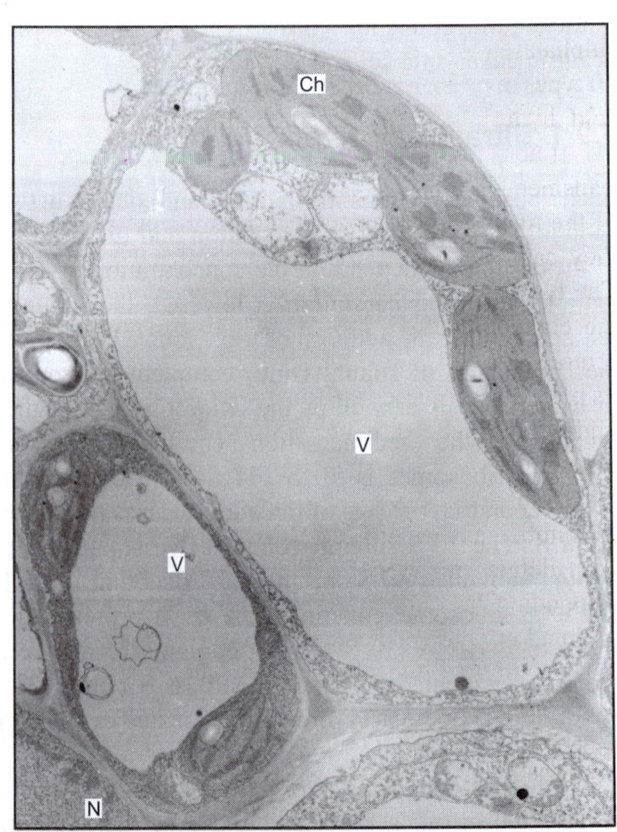

Fig. 5.23 TEM micrograph of plant cell vacuole:V (Channarayappa, 1992).

and also waste products. The resulting accumulation of solutes in the vacuole drives the osmotic uptake of water, producing the turgor pressure. It also functions in increase of cell size by using turgor pressure. The vacuole is also an important homeostatic device, so that it can maintain pH fluctuations.

Many plant cells contain at least two different vacuole systems, during some stage of development: neutral protein storage vacuoles and acidic lytic vacuoles. During development, storage products may be mobilized either by fusing the two vacuole types or by delivering specific lytic enzymes to the storage vacuoles.

5.12 LYSOSOMES

Lysosomes are found in most eukaryotic cells. These emerge from the Golgi membranes. In animal cells, lysosomes are usually spherical and 0.2–0.5 μm in diameter. In plant cells the large central vacuoles may act as lysosomes, although bodies similar to the lysosomes of animal cells are sometimes seen in the cytoplasm. Lysosomes are small single membrane-bound organelles and contain a variety of digestive (hydrolytic) enzymes capable of degrading proteins (proteases), nucleic acids (nucleases), lipids (lipases, phospholiphases) and carbohydrates (glycosidases), phosphatases and sulphatases (**Fig. 5.24**). Each lysosome contains about 40 types of hydrolytic enzymes. For the activity of these acid hydrolases, the lysosomes maintain an internal pH of about 5. The lysosome membrane also contains transmembrane proteins, which help in the transport of the final products of digestion to the cytosol, where they can be either excreted or re-utilized in the cell. The lysosomal membrane has H^+ pumps, which utilize the energy released due to ATP hydrolysis to pump protons into the lysosome, so that the lumen can lower its pH. Most of these lysosomal membrane proteins are highly glycosylated, which perhaps protects them from lysosomal proteases.

Lysosomes play diverse roles in cellular functions. A number of human genetic diseases are associated with defects in one or more of lysosomal enzymes. Depending upon the functions, lysosomes have been grouped into three categories.

1. ***Primary lysosome***, also known as starch granule, is a small body containing enzymes synthesized by ribosomes attached to ER.
2. ***Secondary lysosome***, also known as heterophagosome or digestive vacuole, results from phagocytosis or pinocytosis of foreign material by the cell. Hydrolytic enzymes digest the engulfed material and the products of digestion pass through the membrane of the lysosome and are incorporated into the cell.
3. ***Incomplete bodies*** may be due to the absence of some enzymes and may lead to accumulation of phospholipids or sphingolipids or in some cases, glycogen, which leads to severe pathologic disturbances.
4. ***Autophagic vacuole***, also known as *cytolysosome* or *autophagosome*, is a lysosome containing a part of the cell itself for digestion. These intracellular parts involved in digestion may include mitochondrion or ER and are formed during physiologic and pathologic processes. In general it acts as recycling centres for macromolecules brought in from outside the cell or from damaged organelles.

5.13 MICROBODIES

Peroxisomes and glyoxysomes are collectively known as microbodies (**Fig. 5.25**). They are spherical, ellipsoidal, or irregular organelles with

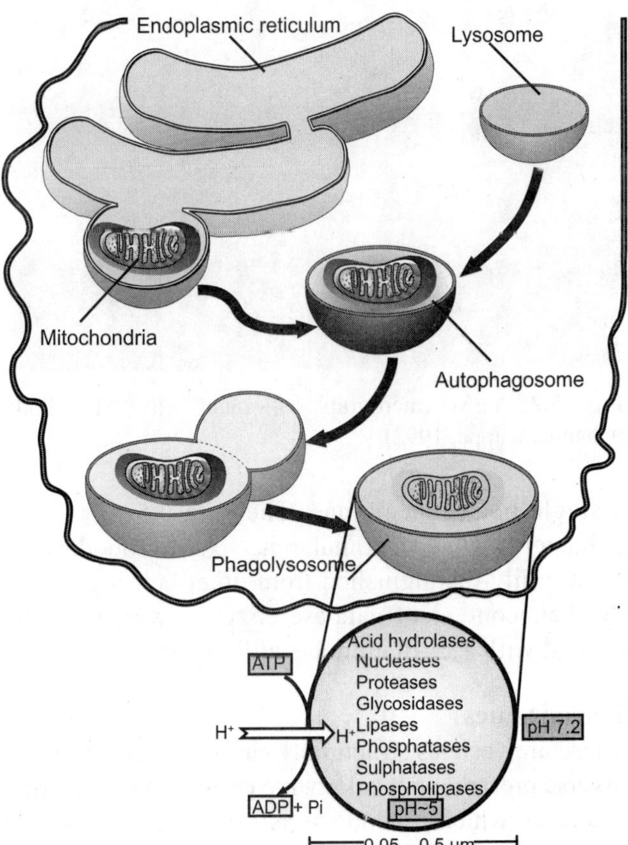

Fig. 5.24 The lysosome system. Lysosome contains various acid hydrolases.

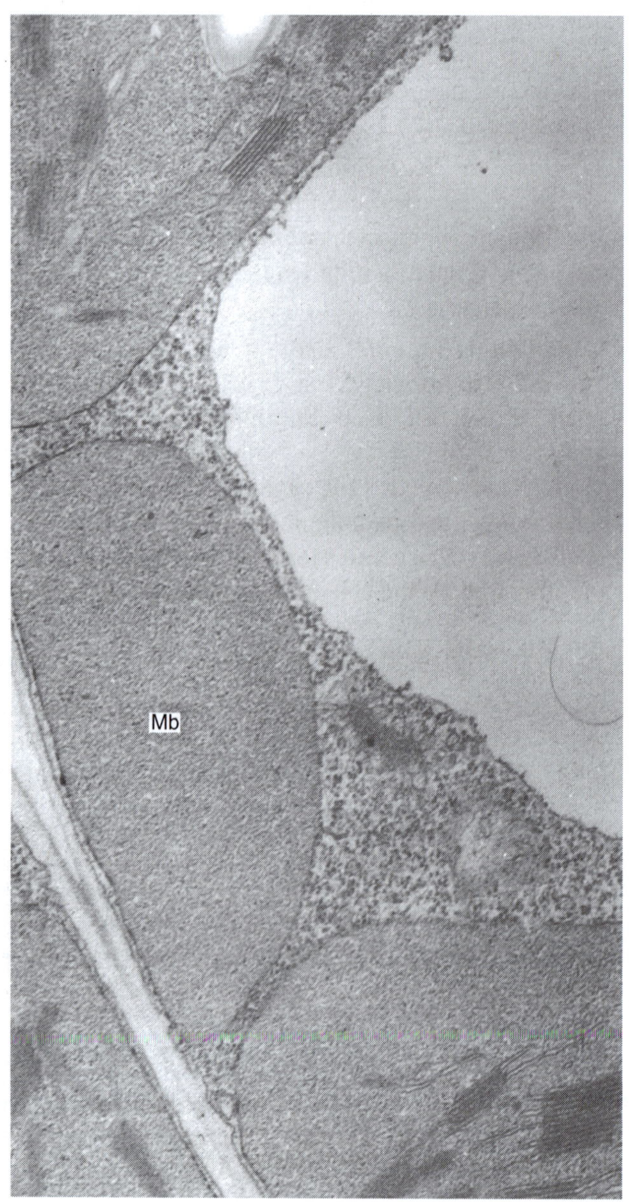

Fig. 5.25 TEM micrograph of plant Mb: Microbody (Channarayappa, 1992).

a single bound membrane. They are 0.5–1.5 μm in diameter and have a granular matrix. Microbodies are most easily distinguished from other cell organelles by their content of catalase enzyme, which can be stained with 3,3'-diaminobenzidine (DAB).

Peroxisomes

These are vesicles structurally simple but functionally diverse present in all eukaryotic cells. Some metabolic reactions, which generate highly reactive free radicals and hydrogen peroxide are confined within peroxisomes to prevent damage to the cellular components. Peroxisomes contain many oxidative enzymes, such as peroxidase, catalase, D-amino oxidase and urea oxidase. The name peroxidase was applied because this organelle is specifically involved in the formation and decomposition of hydrogen peroxide ($2H_2O_2 \rightarrow 2H_2O + O_2$). A second function of peroxisomes is the ß-oxidation of fatty acids.

Glyoxisomes

These are ovoid granules surrounded by a single membrane. They contain a fine granular substance condensing in the centre to form an opaque homogeneous core. In addition to glycolate oxidase and catalase, glyoxisomes also contain a number of enzymes not found in animal cells. These are prominent in plant seedlings, which rely upon stored fatty acid for energy and material to begin the formation of a new plant. Glyoxisomes are specialized structures in plants, which carry out reactions of the glyoxylate cycle. Their enzymes help convert stored fats and oils to the sugars necessary for early rapid growth. They are abundant in lipid-rich seeds, such as peanuts. Since the glyoxisomes resemble the mitochondrion in utilizing oxygen, it has been considered to be a vestige of an ancient organelle, which carried out oxygen metabolism in primitive cells without mitochondria.

Breakdown of fatty acid molecules by a process called **ß-oxidation** (alkyl chains of fatty acids converted into acetyl CoA) is another function of glyoxisomes. In plants, leaf-type peroxisomes help in photorespiration in which O_2 is used and CO_2 is liberated. But in germinating seeds, they convert fatty acids into sugars by a series of reactions know as glyoxylate cycle, so that these peroxisomes are called as **glyoxysomes**.

5.14 HYDROGENOSOME

The hydrogenosome has been identified only recently (1997). It has an endosymbiotic origin, yet lacks a genome and is entirely dependent on the nucleus for its genetic information. It occurs widely, in **trichomonads** (parasitic flagellated protists). These anaerobes lack mitochondria and peroxisomes, but possess hydrogenosomes, the unusual energy-generating organelles. Like mitochondria, hydrogenosomes have a double membrane envelope, divide autonomously by fission, import protein post-translationally and produce ATP by substrate-level phosphorylation. They use enzymes typical of anaerobes and produce large quantities of hydrogen, giving them the name hydrogenosome.

5.15 SPHEROSOME

These are vesicles of diameter 0.5–1 μm and mostly contain up to 98% lipid and some acid phosphatase. Spherosomes are found only in plants and absent in animal and prokaryotic cells. Spherosomes are surrounded by a single unit membrane. They are believed to aid lipid storage.

5.16 CENTROSOME

The term, 'centrosome' was coined by T Boveri in 1888 to represent a circumscribed portion of the clear cytoplasmic region close to the interphase nucleus. From this organelle, the spindle fibres radiate as an aster during mitosis in animal cells. However, in plant cells, it is generally invisible or absent. Each centrosome contains two centrioles (**diplosomes**) located at right angles to each other and looks like a cylinder, which is about 300–500 nm in length and 150–180 nm in diameter. The wall of each centriole contains nine triplet fibres arranged around a central axis, which may not have any fibre. Each triplet fibre consists of three secondary fibres or tubules and each tubule is about 20 nm in diameter. The three tubules are arranged in a line tilted at about 40° to the tangent of the centriole. In addition to the fibres, appendages sometimes extend radially from a group of centriole fibres, which end in a head (**Fig. 5.26**) known as satellite. All these fibres are interconnected and enclosed in a thick membrane. Another term 'centrosphere' is also used and refers to the cytoplasm at the poles of the spindle. The centrioles and the centrosphere together form the centrosome. There is evidence for the presence of DNA and RNA in centrioles. The information for the synthesis of centrioles is probably derived from a tiny DNA-containing unit with about 2×10^{-16} g double-stranded DNA.

The centrosome found in cells and certain lower plants has been implicated in the formation of basal bodies, spindle fibres, cilia and so on. Centrioles constitute a part of the centrosome and are involved in the formation of mitotic apparatus. They direct the migration of the separating chromosomes during nuclear division. At the start of cell division, the centrioles separate and move to opposite poles of the nucleus. The spindle is formed between them and the chromosomes get arranged at the equator. With the help of the spindle fibres attached to the kinetochore plate of the chromosomes, the latter are dragged towards the poles by contraction of the spindle fibres.

Centrioles multiply mainly through replication, which is semi-autonomous. The daughter centrioles arise as 'pro-centrioles' from the side of a pre-existing centriole close to one end and grow to mature size in late prophase, maintaining their close proximity to and orientation at right angles to the mother centriole. When, the interphase nuclei re-form at the end of a nuclear division, a centrosome containing two centrioles again forms beside each nucleus.

5.17 MITOCHONDRIA AND PLASTIDS

Cell organelles, such as mitochondria and plastids, are unique structures in the eukaryotic cells. These two organelles have their own genomes and protein synthesizing apparatus. Because of their genetic importance and evolutionary significance, they have been individually discussed in detail in Chapters 7 and 8. Only a brief explanation on the organelles is given here.

Mitochondria are present in all eukaryotic cells and are the major sites of aerobic respiration within cells. Cellular respiration (aerobic respiration), that is, the oxidation of nutrients to generate energy in the form of adenosine 5'-triphosphate (ATP), takes place in the mitochondria. The mitochondrion

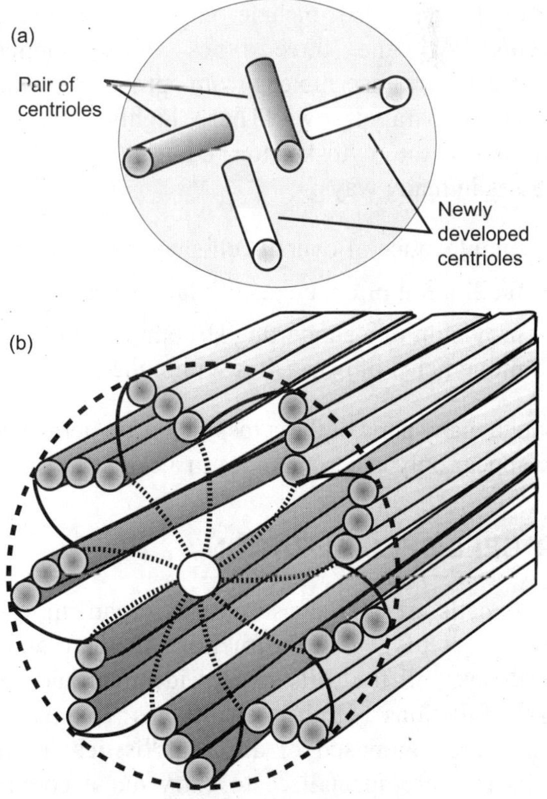

Fig. 5.26 Structure of (a) centrosome and (b) centriole.

is bound by an envelope of two membranes. They have a smooth outer membrane and a convoluted inner membrane that forms protrusions called cristae (**Fig. 5.27**). The genome of plant mitochondria is very large and it seems likely that most of the mitochondrial genome is non-coding DNA. Both animal and plant mitochondria encode their own ribosomal and transfer RNAs and a few ribosomes on which some mitochondrial proteins are synthesized. However, the majority of mitochondrial proteins are encoded by nuclear DNA and synthesized in the cytoplasm.

Plastids are found only in plant cells. Various types of plastids can be distinguished. These differ in structure and function, but develop from similar primordial organelles. But the most important plastids are **chloroplasts** (containing green pigments) that carry out photosynthesis by converting light energy into chemical energy (**Figs. 5.28** and **5.29**). The chloroplasts of plants are the site of photosynthesis.

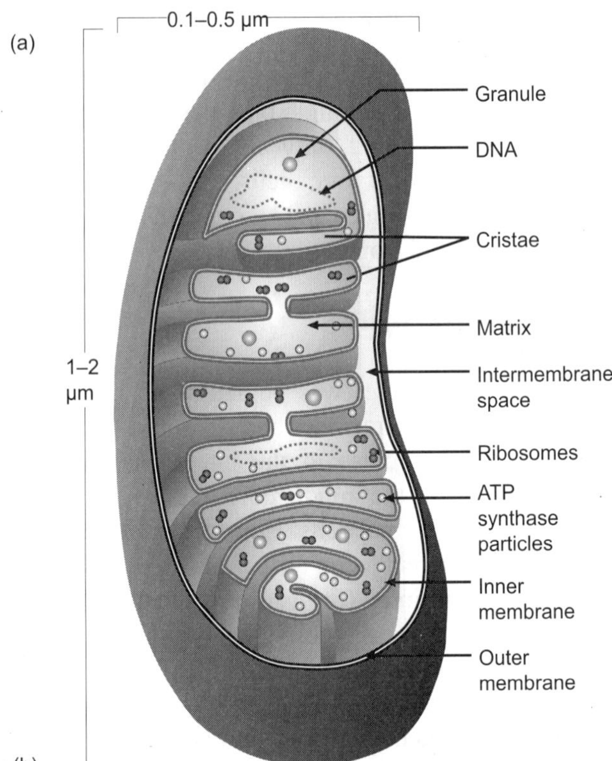

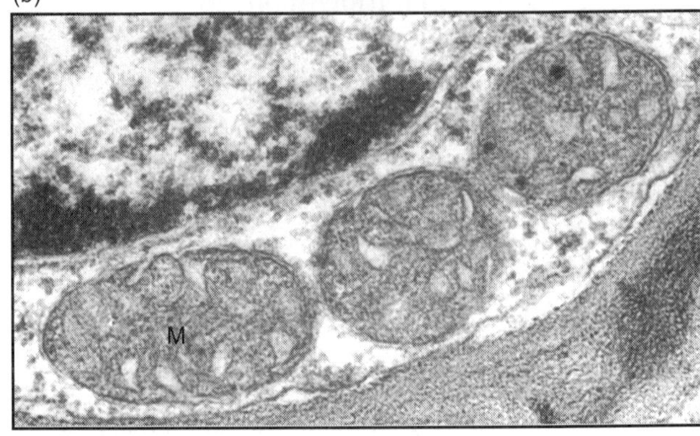

Fig. 5.27 Ultrastructure of mitochondria: (a) Longitudinal view and (b) TEM showing cross section of mitochodria (M). Inside of mitochondrion shows extensive folding of the inner membrane called cristae (Channaryappa, 1992).

5.18 THE CYTOSKELETON

It consists mainly of **microtubules**, **microfilaments** and **intermediate filaments**. The cytoskeletons of animal cells also include various intermediate filaments. All the three types of components (**Fig. 5.30**) are assembled from protein subunits arranged in parallel rows. These highly developed networks of the cytoskeleton contribute to living processes in many ways:

- maintenance of overall cell shape
- facilitation of coherent cellular movement
- provision of a supporting structure to guide the movements of organelles within the cell.

Although many parts of the cytoskeleton are permanent, some appear only at certain times in a cell's life.

5.19 CELL SPECIALIZATION

All the cells of that constitute an organism carry the same set of genetic instructions in their nuclei. Nevertheless, all the cells are not identical and have diverse functions. Higher eukaryotes (plants and animals) are composed of different **tissues** (groups of cells that are specialized to carry out a common function). This specialization occurs because different cell types read out different parts of the genomic DNA

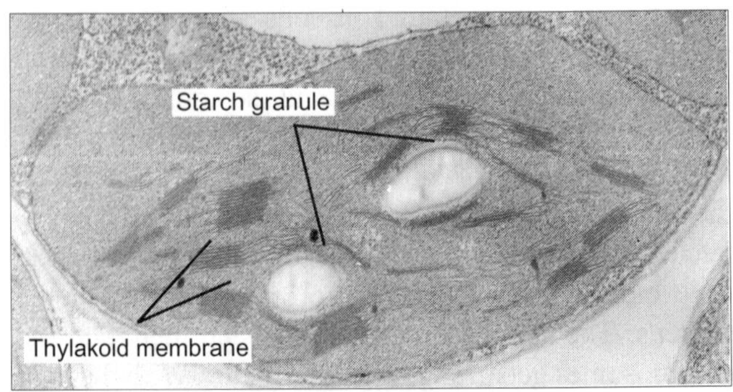

Fig. 5.28 TEM photomicrograph of tomato leaf spongy parenchyma cell filled with chloroplast containing starch granules (Channarayappa, 1992).

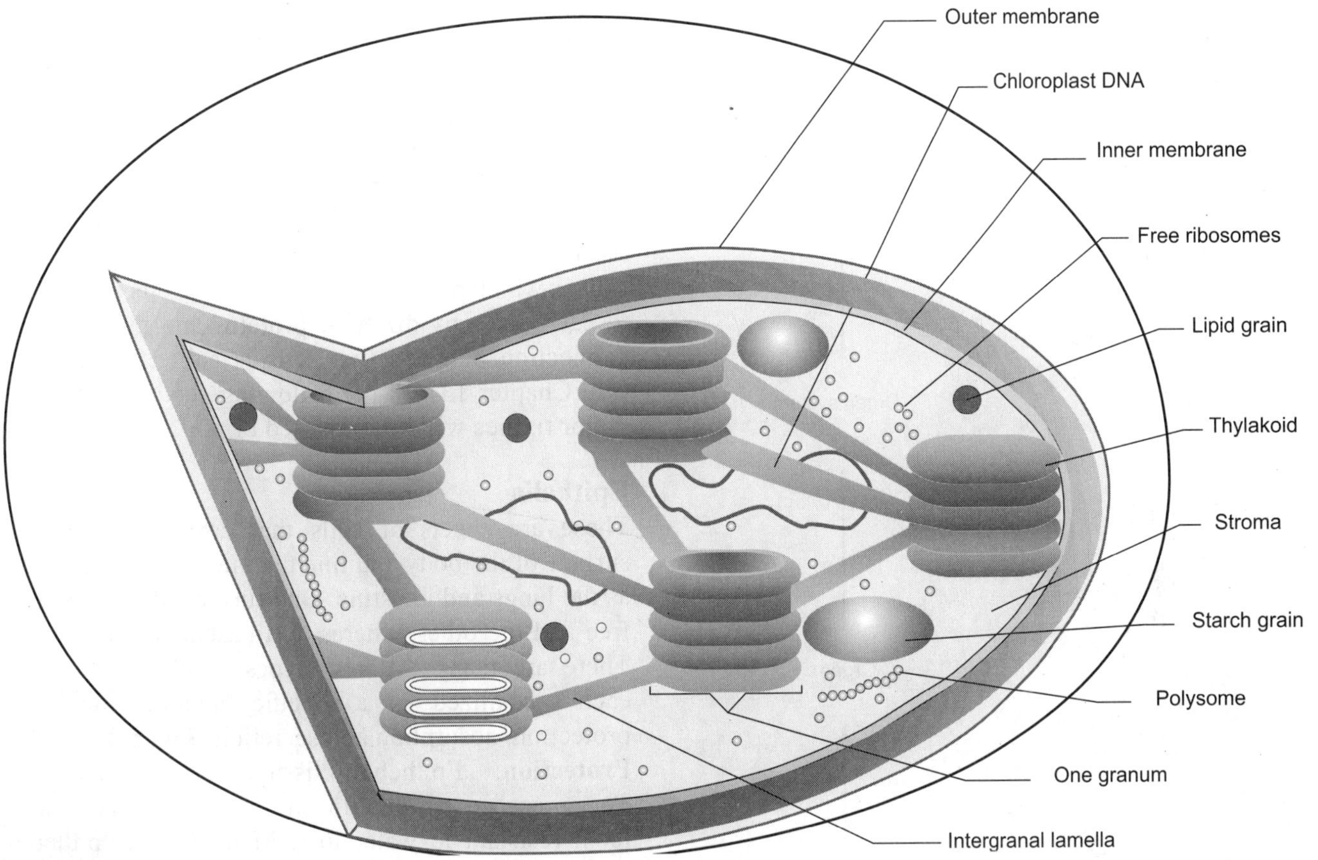

Fig. 5.29 Plant chloroplast structure (**See Plate 2**).

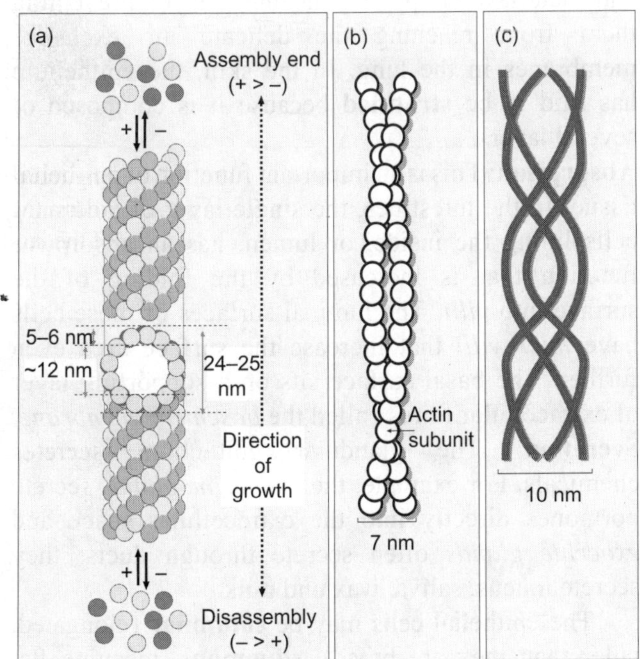

Fig. 5.30 Examples of cytoskeletons: (a) Microtubules, (b) Microfilaments and (c) Intermediate filaments.

and therefore make different proteins. The majority of these cells are diploid somatic cells and constitute most of the cells and tissues of the human body. In animals, there are four major tissue types: epithelium, connective, nerve and muscle. Examination of plant tissues reveals further distinct types. Despite wide differences in size and shape all cells share many common features.

5.19.1 Animal cell types

The cells found in animals are considerably more diverse than of plants. The human body, for example, is composed of more than 200 different kinds of cells. These cells are organized at different levels: tissue (embryonic and somatic tissues), organs, organ systems and cell junctions (**Fig. 5.31**). A tissue is composed of cells that function together in a specialized activity. They give rise to all the tissues, organs and organ systems in the body An **embryonic tissue** consists of the ectoderm (outer layer of skin and nervous system), mesoderm (muscles,

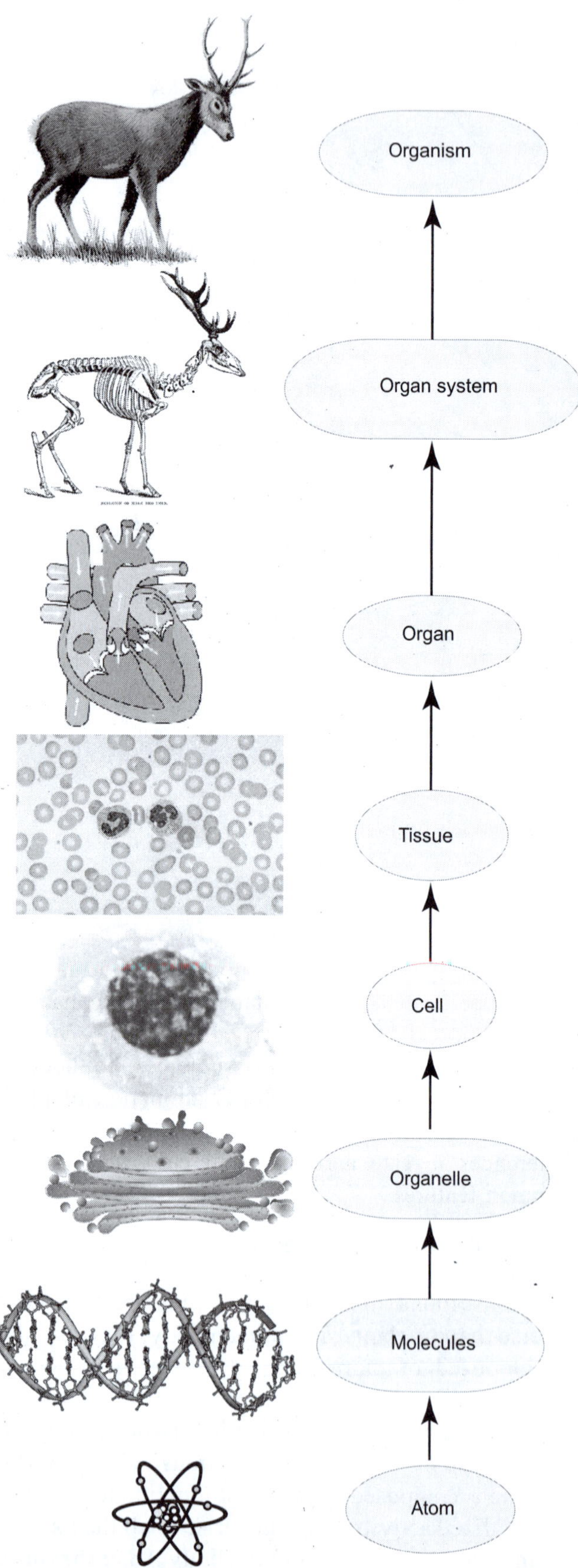

Fig. 5.31 Flowchart showing hierarchy in functional organization of living organisms.

connective tissue, skeleton, kidneys and circulatory and reproductive organs) and endoderm (lining of the gut, respiratory tract and urinary bladder and also the glands associated, with the gut and respiratory tract). **Organs** are composed of two or more tissues which function together to perform a common task. For example, the heart contains all the four types of tissues. An **organ system**, consists of two or more organs which perform a specific task. Cells are joined to each other by different proteins and the point of connection between two cells is called a **junction** (see Chapter 13). In the following sections only the major tissues will be discussed briefly.

Epithelia

These are sheets of cells that cover the external surface of the body and line the internal cavities such as the lungs and intestine. One surface of the tissue is free and the other adheres to a basement membrane. There are many different types of epithelial cells; each specialized for a specific function, including protection, absorption and secretion (**Fig. 5.32**).

Protection: Epithelial tissue forms the skin. The presence of keratin in the skin cells makes them resistant to water loss. Many of the epithelial cells of the airways—those lining the trachea and bronchioles—have **cilia** on their surfaces. These are hair-like appendages that actively beat back and forth moving a layer of mucus away from the lungs. Particles and bacteria are trapped in the mucus layer, preventing them from reaching the delicate air exchange membranes in the lung. In the skin, the epithelium has said to be **stratified** because it is composed of several layers.

Absorption: This is an important function of epithelial tissue. In the intestines, the single layer of columnar cells lining the inside, or lumen, has an absorptive function that is increased by the folding of the surface into *villi*. The luminal surfaces of these cells have *microvilli* that increase the surface area even further. The basal surface sits on a supporting layer of extracellular fibres called the *basement membrane*.

Secretion: The glandular epithelium secretes chemicals. For example, the *endocrine glands* secrete hormones directly into the extracellular space and *exocrine glands* often secrete through ducts; they secrete mucus, saliva, wax and milk.

The epithelial cells may be **columnar** (elongated, taller than they are broad), **squamous**, meaning flat and cube-shaped (cuboidal). These cells organize in different layers.

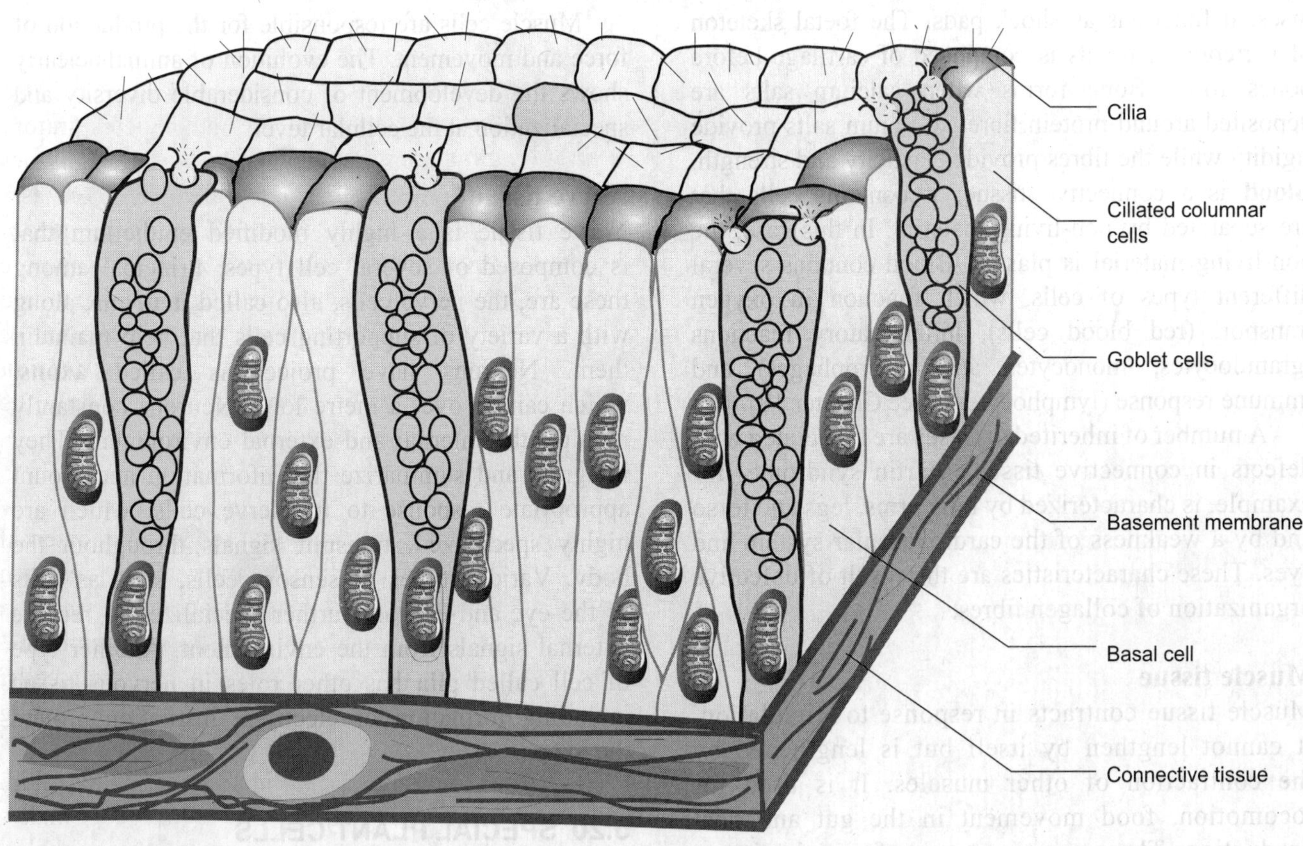

Fig. 5.32 The intestinal epithelial tissue containing various types of specialized functional cells.

- ***Simple epithelium:*** These cells are only one cell thick so that gasses can pass through quickly. They are present in respiratory surfaces such as the lining of the lungs or the skin of a frog.
- ***Striated epithelium:*** This has more than one layer, for example human skin contains many layers of cells.

Connective tissue

Connective tissues include bone, cartilage, adipose tissue and blood, each of which is formed by different types of cells (osteoblasts, chondriocytes and adipocytes). The cells of connective tissue are separated by non-living material. Connective tissue binds and supports body parts, protects, fills spaces, stores fat (for energy) and transports materials.

Loose connective tissue and **dense connective tissue** contain three kinds of fibres. Unlike other tissues, loose connective tissue contains relatively few cells within a large volume of **extracellular matrix** that consists of different types of fibre embedded in amorphous ground substances. The most abundant of the fibres is **collagen**, a protein with the tensile properties of steel that accounts for about one third of the protein of the human body. **Reticular fibres** are small and branched; they provide a support framework for organs such as the liver and lymph nodes. Of the many cell types found in connective tissue, two of the most important are **fibroblasts**—the loose connective tissue that underlies epithelial layers and fills the spaces between organs and tissues in the body—which secrete the ground substance and fibres and **macrophages**. **Tendons** connect muscle to bone and **ligaments** connect bone to bone.

Adipose tissue is a type of loose connective tissue. It has reduced matrix material and contains enlarged fibroblasts that store fat. Adipose tissue functions to store energy, insulate and provide padding, especially in the skin and around the kidneys and heart. The cells of **cartilage** are embedded in a protein-containing matrix that is strong but flexible. Cartilage contains collagen and elastic fibres and is resilient, resists compression and though flexible maintains its shape. It is found in the ends of bones where it prevents friction within the joints. In the nose, external ear and the walls of the trachea it supports the softer tissues. In intervertebral discs, it functions as shock pads. The foetal skeleton of vertebrate animals is composed of cartilage before bones form. Bone forms when calcium salts are deposited around protein fibres. Calcium salts provide rigidity while the fibres provide elasticity and strength. **Blood** is a connective tissue; it contains cells that are separated by non-living material. In this case, the non-living material is plasma. Blood contains several different types of cells, which function in oxygen transport (red blood cells), inflammatory reactions (granulocytes, monocytes and macrophages) and immune response (lymphocytes) (See Chapter 18).

A number of inherited diseases are associated with defects in connective tissue. Martin syndrome, for example, is characterized by long arms, legs and torso and by a weakness of the cardiovascular system and eyes. These characteristics are the result of defective organization of collagen fibres.

Muscle tissue

Muscle tissue contracts in response to stimulation. It cannot lengthen by itself but is lengthened by the contraction of other muscles. It is used for locomotion, food movement in the gut and heat production. There are two types of muscle tissue: smooth and striated. **Smooth muscle** cells are long and slender and are usually found in the walls of tubular organs such as the intestine and many blood vessels. In general, smooth muscle cells contract slowly and can maintain the contracted state for a long time. They are involuntary. They surround the blood vessels and control the distribution of blood. There is not enough blood in the body to fill all the blood vessels so some must contract while others get filled. For example, during eating, the blood vessels of the gut are opened while many of those in the skeletal muscles contract. There are two classes of **striated muscle**: cardiac and skeletal. **Cardiac muscle** cells make up the walls of the heart chambers. They are branched cells that are connected electrically by gap junctions and their automatic rhythmic contraction powers the beating of the heart. Each **skeletal muscle** is a bundle of hundreds or thousands of fibres, each fibre being a giant single cell with many nuclei. The cells are long, extending along the length of the muscle. This rather unusual situation is the result of an event that occurs in the embryo when the cells that give rise to the fibres fuse together, pooling their nuclei in a common cytoplasm. The mechanism of contraction of skeletal muscle will be described in Chapter 14.

Muscle cells are responsible for the production of force and movement. The evolution of animals clearly shows the development of considerable diversity and specialization at the cellular level.

Nerve tissue

Nerve tissue is a highly modified epithelium that is composed of several cell types. Principal among these are, the nerve cells, also called **neurons**, along with a variety of supporting cells that help maintain them. Neurons have projections called **axons**, which can be over a metre long. Neurons constantly monitor the internal and external environment. They integrate and summarize the information and mount appropriate response to it. Nerve cells, which are highly specialized, transmit signals throughout the body. Various types of sensory cells, such as cells of the eye and ear, are further specialized to receive external signals from the environment. Another type of cell called **glia** has other roles in nervous tissue including formation of electrical insulation around the axons.

5.20 SPECIAL PLANT CELLS

There are fewer cell types in plants than in animals, but each type of cell is specialized to perform specific tasks required by the organism as a whole (**Fig. 5.33**).

5.20.1 Parenchyma cells

These are the least specialized plant cells. They are large and thin-walled and have a large central vacuole. The cell walls are somewhat flexible. They are often partially separated from each other and usually stuffed with plastids. These are living at maturity and carry on most of the plant's metabolic functions. Generally, the size of the vacuole indicates the age of the cell and older cells contain bigger vacuoles. Most parenchyma cells have the ability to differentiate into other cell types under special conditions, particularly, during repair and replacement of organs after injury.

5.20.2 Collenchyma cells

Collenchyma cells have thicker primary cell walls (usually with uneven thickness) that are especially thicker at their corners and also living at maturity. They are most often found in areas of rapid growth and need to be strengthened. The **petiole** of leaves

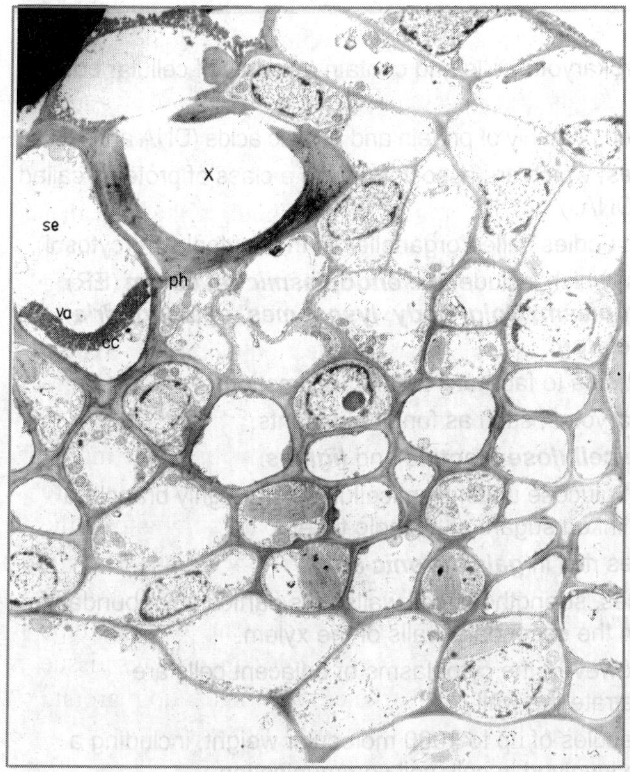

Fig. 5.33 TEM photomicrograph showing different plant cell types in a cross-section of tomato leaf vein (se: seive element; x: xylem vessel; ph: phloem cells; cc: companion cells) (Channarayappa, 1992).

is usually reinforced with collenchyma. They support herbaceous plants by developing into 'strings'.

5.20.3 Sclerenchyma cells

Sclerenchyma cells have thick secondary cell walls built up in a uniform layer around the entire margin of the cell. They are dead at maturity and therefore, cannot increase in length. They are present in two types of plant structures:

- ***Fibres:*** These are long, slender cells with a more or less regular secondary cell wall (e.g., hemp).
- ***Sclereids:*** These are shorter cells with an irregular shape (e.g., stone cells in mangoes and hard nut and seed shells).

5.20.4 Xylem

Xylem cells have thick secondary cell walls, often deposited unevenly in a coil-like pattern so that they may stretch. Cells are dead at maturity and are involved in conducting water and ions in the plant. Two major types of xylem cells are **tracheids** and **vessels**. Tracheids are long, slender cells connected to each other by pits and are found in all vascular plants. In woody plants, the older xylem ceases to participate in water transport and simply serves to give strength to the trunk. Wood is xylem. When counting the annual rings of a tree, it is the rings of xylem that are counted.

5.20.5 Phloem

Phloem cells are mainly involved in the transport of sucrose, other organic compounds and some ions from the place of synthesis to all other parts. These cells are living at functional maturity, but the protoplast may lack organelles and nucleus. The end walls connect to each other via a sieve plate. There are two types of cells in phloem cells: **sieve tube members** and **companion cells**. Sieve elements are so named because their end walls are perforated. This allows cytoplasmic connections between vertically stacked cells. Sieve tube members actually conduit sucrose transport. Sieve elements have no nucleus and only a spare collection of other organelles. They depend on adjacent companion cells for their functions.

5.20.6 Tissue organization in angiosperms

Plant cells are organized into three tissue systems: ground tissue, **dermal tissue and vascular tissue**.

Ground tissue makes up the bulk of the plant and contains parenchyma cells, which perform diverse functions and most of the metabolic reactions of the plant, including photosynthesis, storage and support. It also contains two specialized cell types (collenchyma and sclerenchyma cells) that are characterized by thick cell walls. They provide structural support to the plant.

Dermal tissue covers the surface of the plant and is generally composed of a single layer of epidermal cells (the skin of the plant). It consists primarily of parenchyma cells, which form a protective coat and allow the absorption of nutrients.

Vascular tissue constitutes several types of elongated cells (composed of xylem, phloem, parenchyma and schlerenchyma), which is responsible for the transport of water, ions, minerals and nutrients throughout the plant (**Fig. 5.33**). This also has a secondary role in support.

Summary

- ***Eukaryotic cells*** are larger and more complex than prokaryotic cells and contain a variety of cellular bodies called organelles.
- Eukaryotic cells have a distinct nucleus and are composed primarily of protein and nucleic acids (DNA and RNA).
- DNA is organized into linear units called ***chromosomes***, which are associated with a class of proteins called ***histones*** that provide a supportive framework for the DNA.
- Eukaryotic cells contain a large number of microscopic bodies called organelles in their cytoplasm (cytosol).
- Vital functions of the cell go on within these organelles, which include, the ***endoplasmic reticulum*** (ER), a series of membranes that extend throughout the ***cytoplasm***, ***Golgi body***, ***lysosomes***, ***mitochondria***, ***chloroplast***, ***vacuoles***, ***microbodies*** and the ***cytoskeleton***.
- Many specialized eukaryotic cells contain ***flagella*** and ***cilia*** to facilitate cellular locomotion.
- The cell wall is very prominent in many species of eukaryotes, such as fungi and plants.
- The walls of all plants contain fibres of ***cellulose***, ***hemicellulose***, ***pectins*** and ***lignins***.
- Cellulose is a polysaccharide made up of thousands of glucose units. Hemicellulose are highly branched polysaccharides with a backbone of about 50 $\beta(1 \rightarrow 4)$-linked sugars of a single type.
- Pectins contain multiple negatively charged saccharides rich in ***galacturonic acid***.
- Lignin, a complex insoluble polymer of phenolic residues, strengthens cell walls. It is particularly abundant in wood, where it accumulates in primary cell walls and in the secondary walls of the xylem.
- All cells in higher plants are separated by cell walls. However, the cytoplasms of adjacent cells are interconnected by tube-like ***plasmodesmata*** that penetrate the wall.
- Plasmodesmata provide intercellular channels for molecules of up to 1000 molecular weight, including a variety of metabolic and signalling compounds that are involved in cell–cell communication.
- Cells are separated and protected by a thin, fragile structure called the ***plasma membrane*** that is only 5–10 nm wide.
- The primary function of the plasma membrane is to hold together the contents of the cell. It separates the contents of the cells from the external environment and controlling exchange of materials.
- Cells can operate efficiently only by concentrating the required materials within their boundaries and keeping them there.
- Membranes also act as receptor sites for recognizing hormones, neurotransmitters and other chemicals, either from the external environment or from other parts of the organism. The internal volume bounded by the plasma membrane or the cytoplasm, is composed of an aqueous solution, the ***cytosol*** and a variety of insoluble suspended particles and contain a variety of cell organelles.
- The human body is composed of more than 200 different kinds of cells, which are generally considered to be components of five main types of tissues: epithelial, connective, blood, nervous and muscle. These cells are organized at different levels: tissue (embryonic and somatic tissues), organs, organ systems and cell junctions.

Further Reading

Books

Alberts B, Bray D, Johnson A, Lewis J, Raff M, Roberts K and Walter P (1998), *Essential Cell Biology: An Introduction to the Molecular Biology of the Cell*, Garland Press, New York.

Becker WM, Kleinsmith LJ and Hardin J (2000), *The World of the Cell*, 4th edition, Benjamin Cummings, New York.

de Duve C (1984), *A Guided Tour of the Living Cell*, Scientific American Books, New York, USA.

Fahn A (1974), *Plant Anatomy*, Pergamon Press, Oxford.

Freifelder D (1987), *Molecular Biology*, 2nd edition. Narosa Publishing House, New Delhi.

Gennis RB (1989), *Molecular Structure and Function*, Springer–Verlag, New York.

Lewin B, Cassimeris L, Lingappa VR and Plopper G (2007), *Cells*, Jones and Bartlett Publishers.

Lodish H, Baltimore D, Berk A, Zipursky SL, Matsudaira P and Darnell J (1995), *Molecular Cell Biology*, Scientific American Books, USA.

Miesfeld RL (1999), *Applied Molecular Genetics*, Wiley–Liss, John Wiley and Sons, Inc, Publication, New York.

Salisbury FB and Ross CW (1995), *Plant Physiology*, 3rd edition. CBS Publishers and Distributors, New Delhi.

Twyman RM (1998), *Advanced Molecular Biology*, Bios Scientific Publishers, VIVA Books Pvt. Ltd, New Delhi.

Exercises

1. Draw and describe the various parts of a typical eukaryotic cell.
2. List the different organelles in the eukaryotic cell and mention their functions.
3. Describe in detail the structure and organization of the cell wall.
4. Explain the composition of the cytoplasm.
5. Explain with a diagram the biological significance of the plasma membrane.
6. Give an account of the structure and composition of the nucleus.
7. Describe the structure and functions of the ER.
8. What are the main functions of the Golgi complex in a cell?
9. Explain the structure and organization of the ribosome.
10. How is the ribosome involved in protein synthesis?
11. What is a microbody? Explain its biological role.
12. How does the vacuole in the plant cell regulate various functions?
13. What are the major differences between animal and plant cells?
14. Explain the different functions of lysosomes.
15. List the different tissue types of tissue in animals and plants.

References

Brinkley BR (1985), Microtubule organizing centres, *Annu. Rev. Cell Biol.*, 1:145–72.

Barbara Mcclintock (1934), The relation of a particular chromosomol element to the development of nucleoli is zea mays with 21 fingers in the text and plates viii–xiv, *zeitschritt zellforschung and microskopische Anatomic* 21:294–328.

Carpita NC, Campbell M and Tierney M, (eds), (2001), Plant cell walls, *Plant Mol. Biol.*, 47:1–34. (special issue).

Channarayappa (1990), Cytogenetic analysis of clastogens and aneuploidogens by immunofluorescent and micro-nucleus methods, PhD thesis submitted to the West Virginia University, Morgantown, West Virginia, USA

Channarayappa (1992), Resistance mechanism of *Lycopersicon* species to the vector *Bemisia tabaci* and ultrastructural changes in leaf curl virus infected tomato plants, PhD Thesis submitted to University of Agricultural Sciences (UAS), Bangalore. India

Channarayappa, Muniyappa V, Schwegler–Berry D and Shivashankar G (1992), Ultrastructural changes in tomato infected with tomato leaf curl virus, a whitefly–transmitted geminivirus, *Can. J. of Botany*, 70:1747–53.

Channarayappa, Nath J and Ong T (1992), Immunofluorescent staining of kinetochores in micronucleus for detection of aneuploidy inducing agents, *J. of Tissue Cul. Meth.*, 14:125–32.

Channarayappa, Ong T and Nath J (1992), Cytogenetic effects of vincristine sulphate and ethylene dibromide in human peripheral lymphocytes: Micronucleus analysis, *Environ. and Mol. Mutag.*, 20:117–26.

Citovsky V and Zambryski P (1993), Transport of nucleic acids through membrane channels: snaking through small holes, *Annu. Rev. Microb.*, 47:167–97.

Collmer C and Howell S (1992), Role of satellite RNA in the expression of symptoms caused by plant viruses, *Annu. Rev. Phytopathol.,* 30:419–42.

Kirschner, Ruslim and TE Martin (1977), Characterization of the nuclear envelope, pore complexes and dense lamina & mouse liver nuclei by high resolution scanning electron microscopy, *J. Cell Biology*, 72:118–32.

Lake JA (1981), The ribosomes, *Sci. Am.* 245:84.

Liljas A (1991), Comparative biochemistry and biophysics of ribosomal proteins, *Int. Rev. Cyto.,* 124:103.

McIntosh JR (983), The centrosome as an organizer of the cytoskeleton, *Mod. Cell Biol.,* 2:115–42.

Moffat AS (1994), Mapping the sequence of disease resistance, *Science* , 265:1804–5.

Nomura M (1986), The control of ribosome synthesis, *Sci. Am.*, 250:102.

Porter KR and Thompson HP (1948), A particulate body associated with epithelial cells cultured from mammary carcinomas of mice of a milkfactor strain, *J. Exp. Med.*, 88:15.

Palade GE (1952), A study of fixation for electron microscopy, *J. Exp.Med.*, 95:285.

PopotJ-L and Engelman DM (2000), Helical membrane protein folding, stability and evolution. *Annu. Rev. Biochem.*, 69:881–922.

Schatten H and M Thoman (1978), Nuclear surface complex as observed with High-resolution scanning electron-microscope-visualization of membrane surface of nuclear-envelope and nuclear cortex from Xenopus lavis oocytes, *J.Cell Biology*, 77:517–35.

Von Heijne G (1994), Membrane proteins: from sequence to structure, *Annu. Rev. Biophys. Biomolec. Structure*, 23:167–92.

6 Biological Membranes

Learning objectives

After reading this chapter, you will be able to:

- Identify the different constituents of biological membranes
- Establish the functional details of the constituents
- List the various processes of interaction between the membranes

6.1 INTRODUCTION

Every living cell is surrounded by a complex and biologically active membrane that protects its delicate internal contents from a highly fluctuating and often harsh and inhospitable environment. Most living cells are separated from the external world by a thin, fragile protective layer (structure) called the **plasma membrane** that defines its boundaries. Eukaryotic cells have an elaborate internal membrane system: the membranes of endoplasmic reticulum, Golgi apparatus, mitochondria and other organelles maintain the characteristic differences between the contents of each organelle and the cytosol. In all cells, the plasma membrane also contains proteins that act as sensors of external signals and these sensors or receptors transfer information. Membranes also regulate influx and efflux of various solutes and synthesize ATP by establishing proton gradients.

Membranes are vital as they separate the cell from the outside world. They also serve in separating compartments within the cell to protect important processes and events. The term 'cell membrane' was coined by C Nageli and C Cramer (1855) and the term **plasmalemma** was given by JQ Plowe in 1931. Because of its thin nature (only 5–10 nm wide), its structural details were not known until the late 1950s. Early electron micrographs (JD Robertson of Duke University) well portrayed the plasma membrane as a three-layered structure, consisting of two darkly staining outer layers and a lightly staining middle layer. Later it was confirmed that all membranes (**Table 6.1**) whether plasma, nuclear, cytoplasmic, or organellar, showed the same ultrastructure.

Differences among various cytoplasmic membranes can be clearly seen with the electron microscope. There are marked differences in thickness and electron density (e.g., mitochondrial membrane is 50 Å, while plasma membrane is about 90–100 Å). All basic types of cellular membranes are inherited. Daughter cells inherit a complete set of membranes from their mother; each potential mother cell maintains a complete set of membranes. New membranes can be produced only by growth and fission of existing membranes; they cannot form de novo. In this chapter, the structure, composition and major biological functions of the cell membranes are described.

6.2 PLASMA MEMBRANE

The membrane enclosing the cytoplasm of a cell (protoplast) is known as plasmalemma or plasma membrane and constitutes about 5% of the total membranes present in a cell. It is the outermost boundary of the living cell and functions as an active interface between the cell and its environment.

Table 6.1 Basic types of membranes found in living cells

Plasma membrane
Nuclear membrane
Endoplasmic reticulum
Golgi cisternae (cis, medial, trans types)
Trans Golgi network/partially coated reticulum
Clathrin- and COP (coat protein)-coated transport vesicles
Endocytic vesicle membrane
Endosomal membrane
Multivesicular body/autophagic vacuole membrane
Tonoplast (vacuole membrane)
Lysosomal membrane
Peroxisomal membrane
Glyoxysomal membrane
Chloroplast envelope membranes (inner and outer)
Thylakoid membranes
Mitochondrial membranes (inner and outer)
Other membrane types

Cell membranes are not mechanically strong, in contrast to the rigid and tough cellulose walls of plants and bacteria. Their role is to provide a selective barrier, not a tough wall. Their fragile nature makes it difficult to investigate their structure and chemical nature.

The first proposal that cellular membranes might contain a lipid bilayer was made in 1925 by two Dutch scientists (E Gorter and F Grendel). These scientists extracted the lipid from human red blood cells and measured the amount of surface area the lipid would cover when spread over the surface of water. Since the red blood cells lack any internal lipid-containing structure, all the lipids extracted from the cell could be assumed to have resided in the cell plasma membranes. They calculated the ratio between the surface area of water covered by the extracted lipid to the surface area calculated for the red blood cells from which the lipid was extracted. This varied between 1.8 to 1 and 2.2 to 1, which suggests an average actual ratio of 2:1 and thus concluded that the plasma membrane contained a bimolecular layer of lipids, i.e., **lipid bilayer**. They also suggested that the polar groups of each molecular layer (or **leaflet**) were directed towards the aqueous environment, since this would be the thermodynamically favoured arrangement.

Fluid-mosaic model

In the first quarter of the 20th century, cell physiologists obtained evidence that the plasma membrane contains more than simply a lipid bilayer. In 1935, Hugh Davidson and James Danielli proposed that the plasma membrane is composed of a lipid bilayer that was lined on both its inner and outer surface by a layer of globular proteins. In 1972, Jonathan Singer and Garth Nicolson of the University of California proposed the new concept of membrane structure based on the fluid-mosaic model. It describes the molecular organization of lipids and proteins in cellular membranes and illustrates how a membrane's mechanical and physiological traits are defined by the physico-chemical characteristics of its various molecular components. It states that proteins occur as a 'mosaic' of discontinuous particles that penetrate the lipid sheet. Further, Daniel Branton used the freeze etching technique to examine membrane structure in a number of different cell types. He published electron micrographs that show the two outer surfaces of membranes, as well as their internal surfaces. The model proposed by Singer and Nicolson (1972) is the widely accepted one. According to this model, there is a continuous lipid bilayer in which the integral proteins are embedded to varying degrees (**Fig. 6.1**). Both these components are capable of translational diffusion within the bilayer.

6.2.1 Chemical composition of membranes

The early electron micrographs of tissues fixed with osmium tetraoxide, taken by Robertson showed the plasma membrane as three distinct layers. The two darkly-staining outer layers are the inner and outer polar surfaces of the bilayers, whereas the central lightly staining region is a hydrophobic fatty acid trail that resembles a railway track. The two outer layers (each 20 Å thick) are relatively dense and **osmiophilic** in nature and separated by a relatively light **osmiophobic layer** of about 35 Å thickness (**Fig. 6.2**). These layers together are known as **unit membrane** (coined by Robertson). It is about 75 Å thick, but thickness can vary depending on the tissue type.

The plasma membrane is a fluid bilayer of phospholipids in which proteins and a few other types of macromolecules are suspended and held together by non-covalent bonds. The lipid bilayer serves primarily as a structural backbone of the membrane and the proteins of the membrane carry

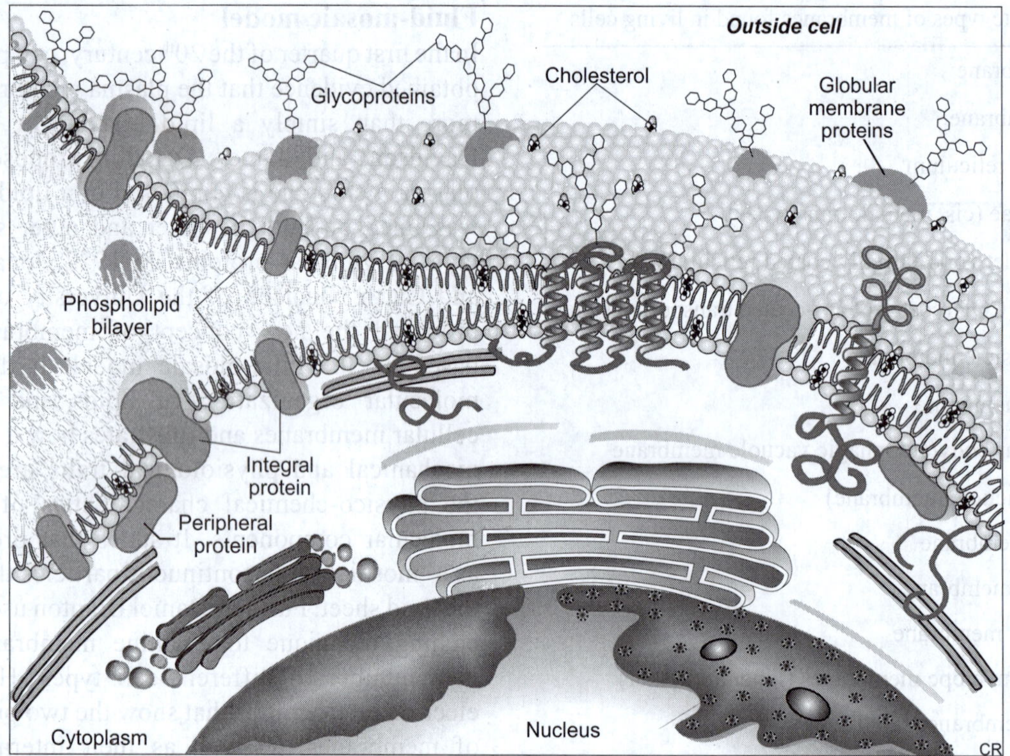

Fig. 6.1 Fluid-mosaic model for membrane structure. Simplified diagram of plasma membrane with different membrane proteins.

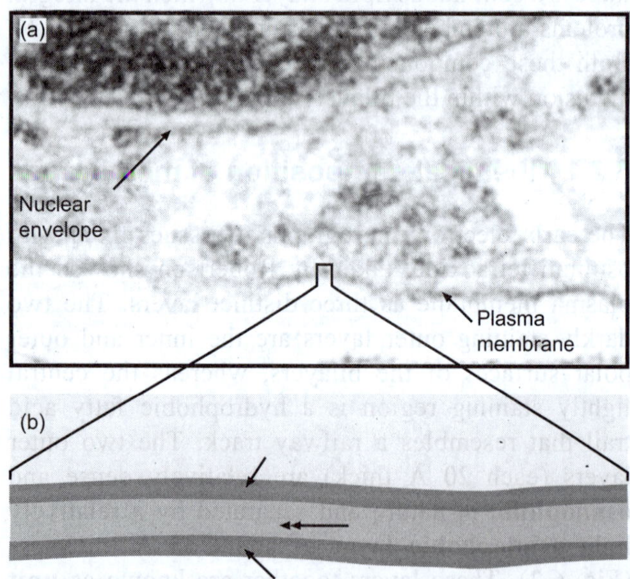

Fig. 6.2 Plasma membrane structure: (a) Electron micrograph showing the structure of the plasma membrane of tomato plant cell (Channarayappa, 1992). (b) Diagram showing a trilaminar appearance of plasma membrane. The single-headed arrows denote the dense inner and outer edges of the membrane, whereas the double-headed arrow denotes the lighter region of the membrane.

out most of the specific functions. Each type of differentiated cell contains a unique complement of membrane proteins, which contributes to the specialized activities of that cell type. The ratio of lipid to protein in a membrane varies depending on the type of cellular membrane (plasma membrane, ER, Golgi), the type of organism (bacterial, plant, animal) and the type of cell (liver cell, muscle cell). Membranes also contain carbohydrates, which are attached to the lipids and proteins.

Membrane lipids

Membranes contain a wide diversity of lipids, all of which are amphipathic, that is, they contain both hydrophobic (hydrocarbon non-polar tail) and hydrophilic regions (polar head group) (**Fig. 6.3**). This dimorphic behaviour of lipids have a number of molecular consequences, including the spontaneous formation (self-assembling) of surface monolayers, bilayers, micelles and vesicles by lipids in contact with water. However, exactly what kind of structural behaviour is exhibited when a lipid is in contact with water depends on the specific molecular structure of the hydrophillic and hydrophobic parts of that lipid.

All biological membranes contain lipids as major constituents. The three major classes of membrane-forming lipids are glycerophospholipids,

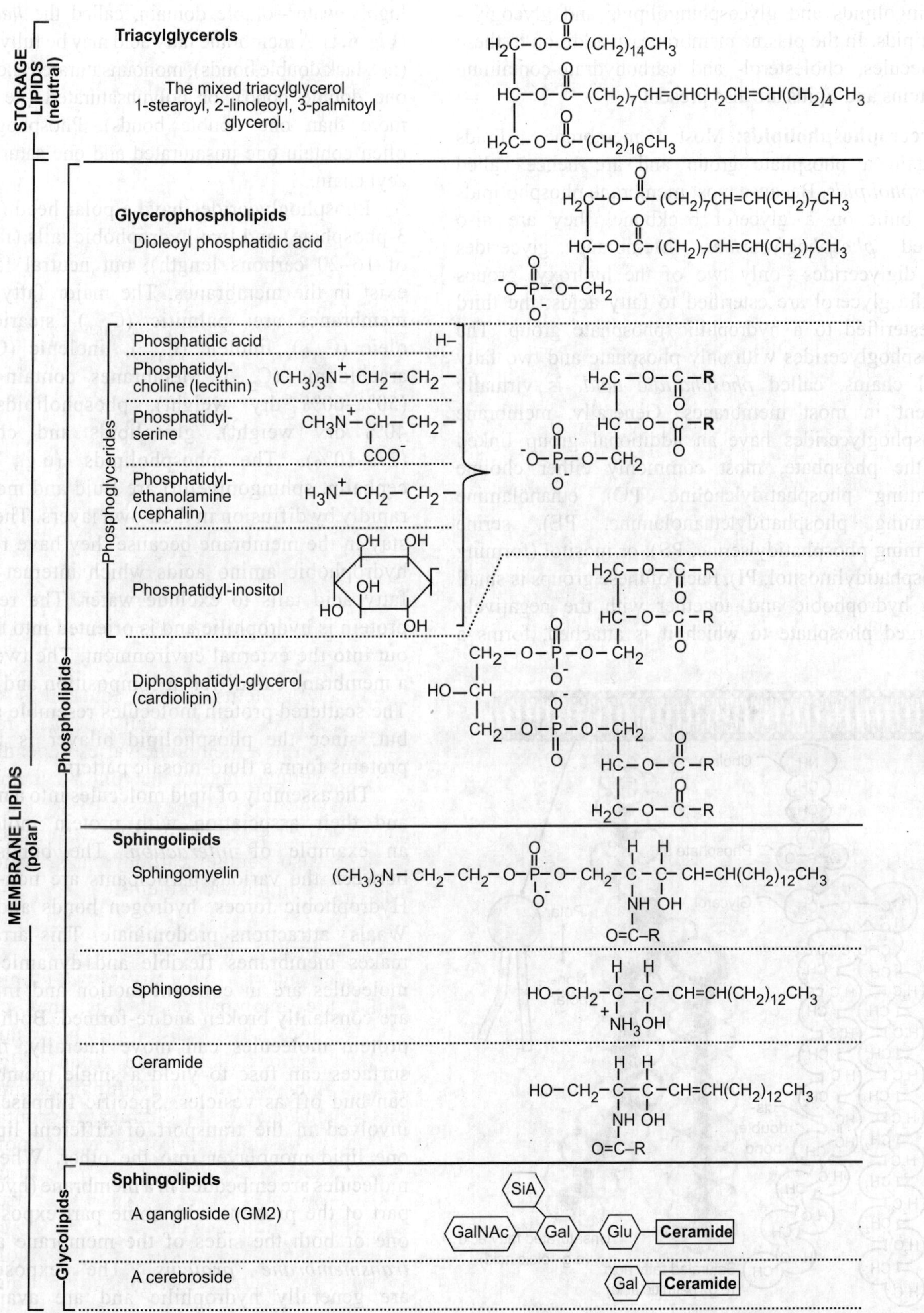

Fig. 6.3 The principal classes of storage and membrane lipids. The lipids shown here have either glycerol or sphingosine as the backbone. Sterol lipids are not included here.

sphingolipids and glycosphingolipids and glycoglycerolipids. In the plasma membrane, in addition to these molecules, cholesterol- and carbohydrate-containing proteins and lipids are also present.

Glycerophospholipids: Most membrane lipids contain a phosphate group and are hence called *phospholipids*. Because most membrane phospholipids are built on a glycerol backbone, they are also called *phosphoglycerides*. Membrane glycerides are diglycerides—only two of the hydroxyl groups of the glycerol are esterified to fatty acids; the third is esterified to a hydrophilic phosphate group. The phosphoglycerides with only phosphate and two fatty acyl chains, called *phosphatidic acid*, is virtually absent in most membranes. Generally, membrane phosphoglycerides have an additional group linked to the phosphate, most commonly either choline (forming phosphatidylcholine, PC), ethanolamine (forming phosphatidylethanolamine, PE), serine (forming phosphatidylserine, PS), or inositol (forming phosphatidylinositol, PI). Each of these groups is small and hydrophobic and, together with the negatively charged phosphate to which it is attached, forms a

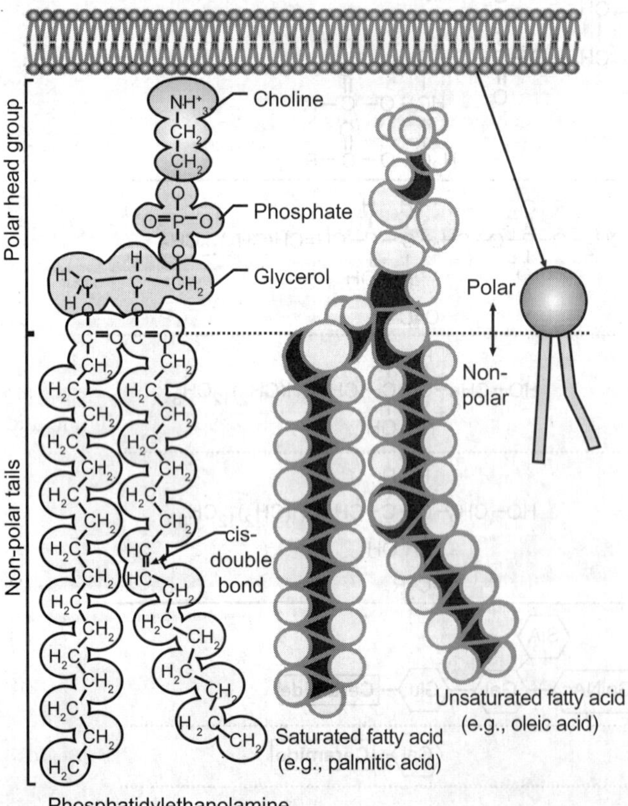

Fig. 6.4 Glycerophospholipids: Different chemical structures of phosphatidylethanolamine.

highly water-soluble domain, called the *head group* (**Fig. 6.4**). A membrane fatty acid may be fully saturated (i.e., lack double bonds), monounsaturated (i.e., possess one double bond), or polyunsaturated (i.e., possess more than one double bonds). Phosphoglycerides often contain one unsaturated and one saturated fatty acyl chain.

Phosphoglycerides have a polar head (glycerol-3-phosphate) and two hydrophobic tails (fatty acids of 16–20 carbons length), but neutral fats never exist in the membranes. The major fatty acids in membranes are: palmitic ($C_{16:0}$), stearic ($C_{18:0}$), oleic ($C_{18:1}$), linoleic ($C_{18:2}$), linolenic ($C_{18:3}$) and arachidonic ($C_{20:4}$). Membranes contain proteins (50%–60% dry weight), phospholipids (30%–40% dry weight), glycolipids and cholesterol (1%–10%). The phospholipids (e.g., lecithin, cephalin, sphingomyelin) are fluid and move about rapidly by diffusion in their own layers. The proteins stay in the membrane because they have regions of hydrophobic amino acids which interact with the fatty acid tails to exclude water. The rest of the protein is hydrophilic and is oriented into the cell or out into the external environment. The two sides of a membrane can differ in composition and function. The scattered protein molecules resemble a mosaic; but, since the phospholipid bilayer is fluid, the proteins form a fluid-mosaic pattern.

The assembly of lipid molecules into a membrane and their association with protein molecules is an example of *interaction*. The bonds formed between the various participants are non-covalent. Hydrophobic forces, hydrogen bonds and van der Waals' attractions predominate. This arrangement makes membranes flexible and dynamic. All the molecules are in constant motion and interactions are constantly broken and re-formed. Both lipid and protein molecules can move laterally, membrane surfaces can fuse to yield a single membrane and can bud off as vesicles. Specific flippases may be involved in the transport of different lipids from one lipid monolayer into the other. When protein molecules are embedded in a membrane (hydrophobic part of the protein) with some part exposed, either one or both the sides of the membrane are called *transmembrane proteins*. The exposed parts are generally hydrophilic and are available for various reactions.

Sphingolipids and glycosphingolipids: These are a less abundant class of membrane liquids.

Sphingolipids are the derivative of sphingosine, an amino alcohol that contains a long hydrocarbon chain. Shingolipids consist of sphingosine linked to a fatty acid by its amino group (**Fig. 6.5**). This molecule is a *ceramide*. Many shingosine-based lipids have additional groups esterified to the terminal alcohol of shingosine moiety; for example, *sphingomyelin* (substitution is phosphorylcholine), *glycolipid* (substitution is carbohydrate), *cerebroside* (substitution is simple sugar) and *ganglioside* (substitution is oligosaccharide). Glycolipids are important membrane components, though little is known. The nervous system is particularly rich in glycolipids; the myelin sheath is rich in a particular glycolipid, called galactocerebroside. The sites of synthesis of glycolipids are the target for a group of fungal toxins called *fumonisins*.

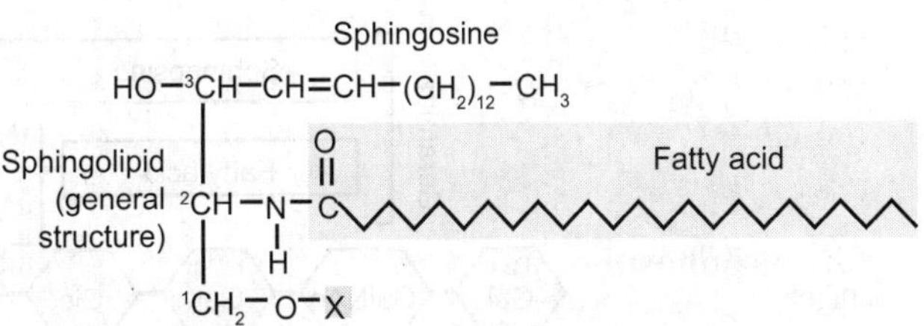

Fig. 6.5 Sphingolipids: Different types of sphingolipids differ in the polar head group (X) attached at C1.

Cholesterol: Another lipid component of certain membranes is *cholesterol*. It is a member of a large group of substances called steroids (**Fig. 6.6**). In fact, cholesterol is the precursor for the synthesis of many hormones. Other steroids are also found in membranes; for example, *lanosterol* is prominent in plant cell membranes. Plant plasma membranes also contain other sterols such as camposterol, sitosterol and stigmasterol. The amount of cholesterol may vary with the type of membrane—in certain cells it may constitute up to 50% of the lipid molecules in the plasma membrane, that is, nearly one cholesterol per phospholipid molecule. Cholesterol is absent from the plasma membrane of most plant and all bacterial cells. The orientation of cholesterol molecules is, similar to phospholipid molecules, with their small hydrophilic hydroxyl group oriented toward the membrane surface and the remainder of the molecule embedded in the lipid bilayer (a weakly amphipathic substance). The hydrophobic rings of a cholesterol molecule are flat and rigid and interfere with the movement of the fatty acid tails of the phospholipids.

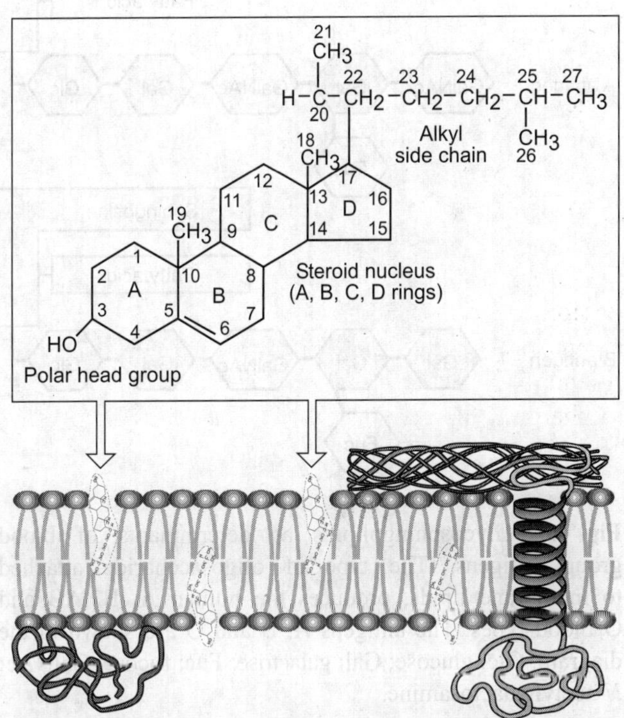

Fig. 6.6 Cholesterol, an amphipathic sterol present in the animal cell membrane. The steroid nuclear rings are labelled A through D and the carbons numbered from 1 to 27. The hydroxyl group on C3 represents the polar head group. Cholesterol head groups face outside the membrane and the hydrophobic (the steroid nucleus) part is inserted within the lipid bilayer.

Membrane carbohydrates

The plasma membranes of eukaryotic cells also contain carbohydrates of various types, which are covalently linked in both lipid (glycolipids) and protein components (glycoproteins). Depending on the cell type and species, the total carbohydrate content of the plasma membrane varies (between 2% and 10% by weight). All the carbohydrate molecules of the plasma membrane face outward into the extracellular space. The carbohydrates of internal cellular membranes also face away from the cytosol. The oligosaccharides attached to membrane proteins and lipids can display considerable variability in composition and structure. These carbohydrate projections play a vital role in mediating the interactions of a cell with its environment and in sorting membrane proteins to different cellular compartments. For example, carbohydrates of the glycolipids of the red blood cell plasma membrane determine A, B, AB, O blood group types (**Fig. 6.7**).

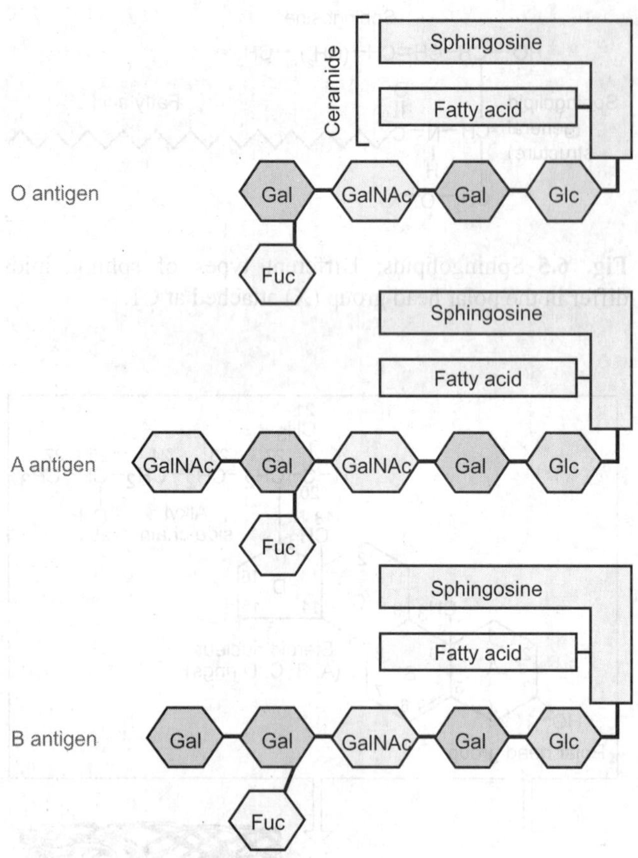

Fig. 6.7 Glycosphingolipids as determinants of blood group antigens. The type of oligosaccharide attached to membrane lipids, produces the human A, B, AB and O blood types. The antigens A, B and O are shown in the diagram. Glc: glucose; Gal: galactose; Fuc: fucose; GalNAc: *N*-acetylgalactosamine.

6.2.2 Structure of membrane proteins

The functions associated with biological membranes require the presence of different protein molecules (**Fig. 6.8**). Membrane proteins therefore give each type of membrane in the cell characteristic functional properties. Membranes contain hundreds of different proteins, which may vary depending on the cell type and organelle. For example, myelin contains only 20% protein, plasma membrane contains about 50% and the inner mitochondrial and chloroplast membranes contain 75%. Each membrane protein has a defined orientation relative to the cytoplasm, so that the properties of one surface of a membrane are very different from those of the others. This asymmetry is referred to as 'membrane sidedness'. Membrane proteins that interact with other cells or with extracellular ligands, such as hormones or growth factors, project outward into the extracellular space, whereas those parts of membrane proteins that interact with cytoplasmic molecules, such as G proteins or protein kinases, project into the cytosol. Membrane proteins are often classified by the function they perform: structural components,

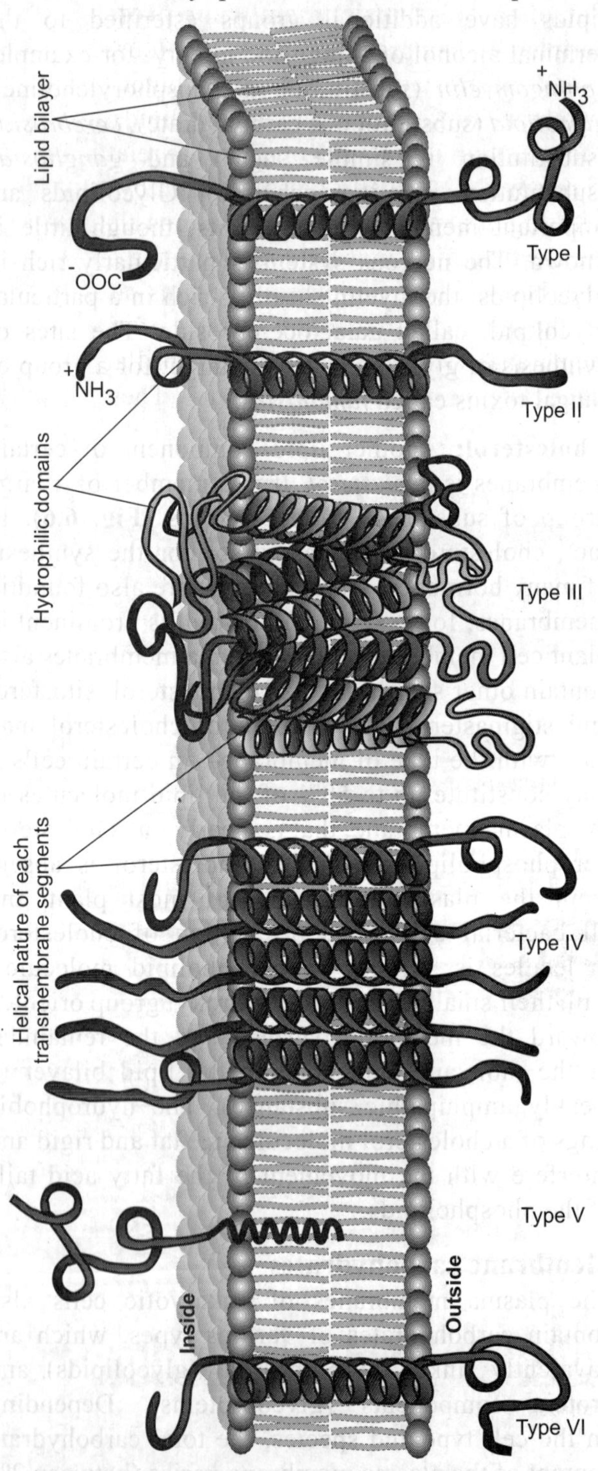

Fig. 6.8 Integral membrane proteins. For known integral membrane protein of the plasma membrane, the spatial relationships of protein domains to the lipid bilayer fall into six categories.

enzymes, hormone receptors or transport proteins. They are also classified according to the structural relationship to membrane: integral proteins, peripheral proteins and lipid-anchored proteins (**Fig. 6.9**).

Peripheral (extrinsic) membrane proteins

Peripheral membrane proteins are loosely bound (weak electrostatic bonds) to the outer and inner membrane surfaces or anchored via a lipid or glycosyl phosphatidylinositol anchor. They are relatively easy to remove, can usually be solubilized by extraction with high concentration salt solutions that weaken the electrostatic bonds holding them to a membrane. The best-studied peripheral proteins are located on the internal (cytosolic) surface of the plasma membrane, where they form a fibrillar network that acts as a membrane 'skeleton'. These proteins

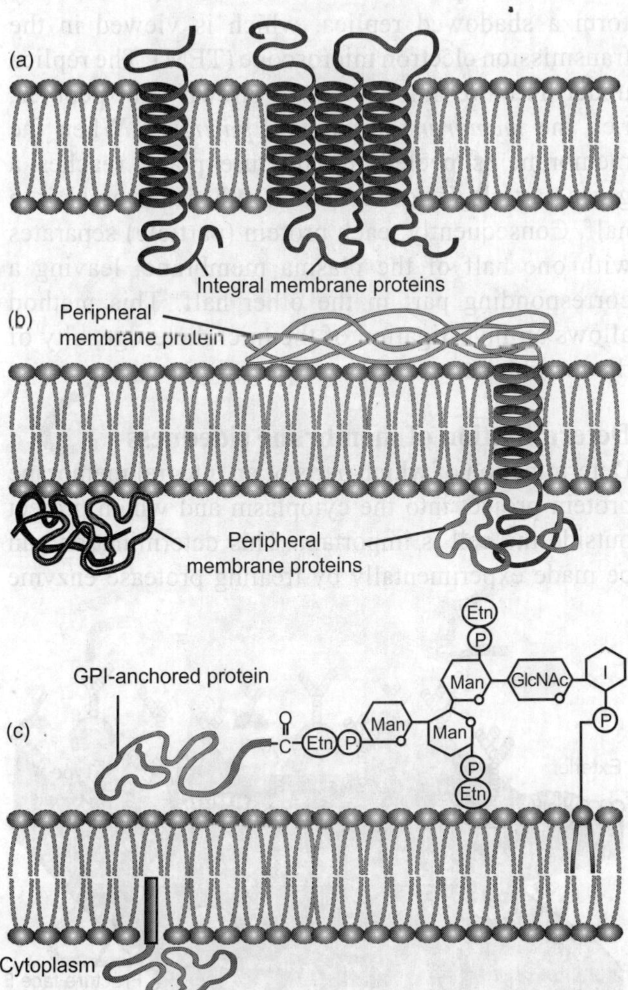

Fig. 6.9 Three classes of membrane proteins: (a) Integral proteins, (b) Peripheral proteins and (c) Lipid-anchored proteins. I: inositol; GlcNAc: *N*-acetylglucosamine; Man: mannose; Etn: ethanolamine.

provide mechanical support for the membrane and function as an anchor for integral membrane proteins. Many other peripheral proteins on the internal surface of the plasma membrane function as enzymes, specialized coats. Peripheral proteins associated with the external surface of the plasma membrane are typically part of the extracellular matrix (See Chapter 10).

Among the major types of peripheral proteins, spectrin, ankryn, actin and G3PD are the major ones (**Fig. 6.10**). The polypeptide chains of the spectrin polymers are located along the inner surface of the membrane. They are present as long, flexible rods of approximately 200 nm in length, composed of two pairs of subunits. The attachment of spectrin to the membrane requires the presence of other proteins such as ankyrin. Spectrin polymers play an important role in maintaining the shape and restraining the movement of the membrane's integral proteins. Another membrane proteins actin, exists usually in a filamentous form, but it can also be present as oligomeric clusters. The complex of spectrin, actin, ankyrin and other proteins form an association on the inner surface of the membrane. This structure maintains the shape of the membrance, provides strength, rigidity and pliability to the cell and is also responsible for endocytosis.

Integral (intrinsic) membrane proteins

Integral membrane proteins are embedded in the membrane and are difficult to remove without destroying the membrane. Like the phospholipids of the bilayer, integral membrane proteins are also amphipathic, having both hydrophilic and hydrophobic portions.

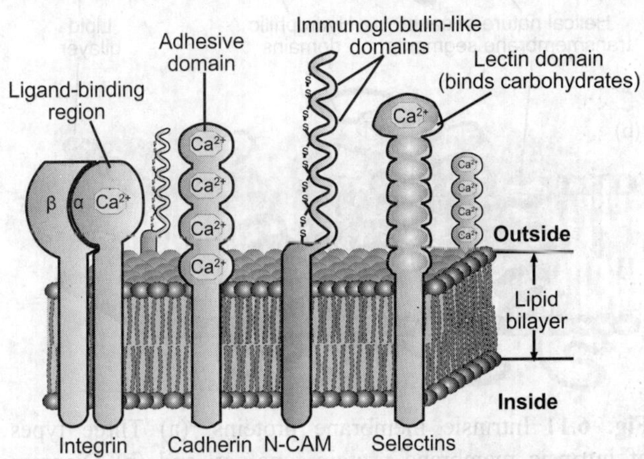

Fig. 6.10 Four examples of integral protein types that function in cell–cell interactions.

Some membrane proteins protrude from the outer or inner surface of the membrane while transmembrane proteins span the bilayer completely and have both extracellular domains. The three types (**Fig. 6.11a**) of intrinsic membrane protein topology are:

- ***Monotopic***—embedded in the lipid bilayer
- ***Bitopic***—traverses the bilayer once
- ***Polytopic***—traverse the bilayer more than once.

Several large families of membrane proteins contain an interior channel that provides an aqueous passageway through the lipid bilayer (**Fig. 6.11b**). For example, the two major integral proteins in the red blood cell membrane are glycophorin and the anion channel protein. Certain oligosaccharide groups on glycophorin constitute the ABO and MN blood group antigens. *Anion channel protein* is composed of two identical subunits, each consisting of 929 amino acids, and plays an important role in CO_2 transport in blood. The HCO_3^- ion formed from CO_2 with the aid of carbonic anhydrase diffuses into and out of the red blood cell through the anion channel exchanging with the chloride ion (Cl^-), called *chloride shift.*

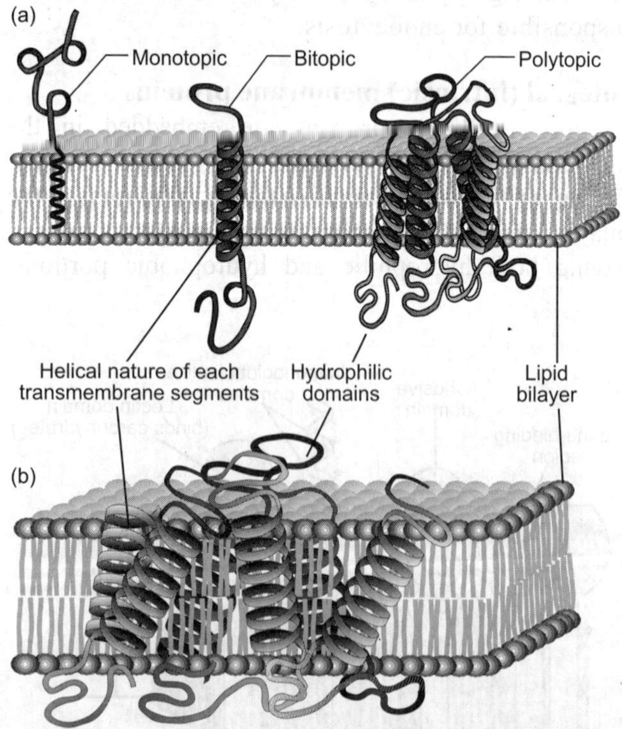

Fig. 6.11 Intrinsic membrane proteins: (a) Three types of intrinsic membrane protein topology and (b) Diagram of a membrane-spanning protein as it resides within the plasma membrane.

Structure and properties: Cytochemists study membranes either by first splitting them along the plane of the two lipid leaflets (freeze fracturing) between the lipid bilayer or by labelling transmembrane proteins and observe the dynamics of binding, membrane fluidity, or membrane uptake. Another approach is to study the specialized junctions (see Chapter 13) formed at key membrane sites.

Freeze fracture analysis: The property of proteins to integrate into the membrane bilayer was derived primarily from the results of a technique called freeze fracture replication. In this procedure, the tissue is frozen solid and then struck with a knife blade. The fracture plane often takes a path between the two leaflets of the lipid bilayer (**Fig. 6.12**). Once the membranes are split in this manner, metals are deposited on their exposed surfaces to form a shadowed replica, which is viewed in the transmission electron microscope (TEM). The replica under TEM resembles a road strewn with pebbles, i.e., the *membrane-associated particles.* When the membrane is peeled the fracture plane reaches a given particle, it goes around it rather than crack it in half. Consequently, each protein (particle) separates with one half of the plasma membrane, leaving a corresponding part in the other half. This method allows an investigation of the microheterogeneity of the membrane,

Determination of membrane sidedness

Determination of which parts of an integral membrane protein project into the cytoplasm and which project outside the cell is important. This determination can be made experimentally by treating protease enzyme

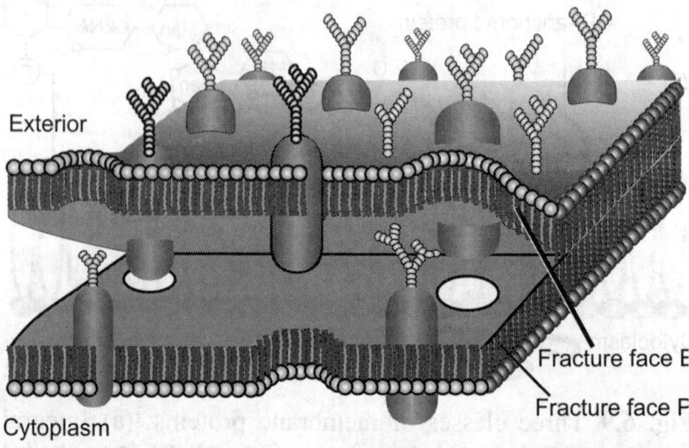

Fig. 6.12 Freeze fracture: Technique for investigating cell membrane structure.

sequentially. A preparation of intact cells is treated with a proteolytic enzyme, such as trypsin, which is too large to penetrate the plasma membrane; those parts of the membrane proteins that are situated on the external side would be subjected to digestion, while those parts within the bilayer or on the cytoplasmic face of the membrane would not be affected. To determine whether a portion of a protein projects from the cytoplasmic face of the membrane, the cells can be made permeable by treatment with non-ionic detergents or by osmotic shock. Under these conditions, the cytoplasmic portion of the protein are also subjected to digestion. The effect of protease treatment on membrane proteins can be monitored by extracting the proteins after each treatment and subjecting them to SDS–PAGE (**Fig. 6.13**). From the gel it is very clear that the pattern of bands obtained from each treatment is very different.

Identification of transmembrane domains: Identification of the segments of the polypeptide chain embedded in the lipid bilayer (transmembrane domain) is generally by prediction from the amino acid sequence of the protein, which is deduced from the nucleotide sequence of an isolated gene. Transmembrane domain generally contain a string of 20–30 predominantly non-polar amino acids that adopt an α-helical secondary structure. Of the 20 amino acids that make up the lone α-helix of glycophorin, all but three have hydrophobic side chains.

Lipid-anchored proteins

These are located outside the lipid layer, on either side of the extracellular or cytoplasmic surface, but are covalently linked to a lipid molecule that is situated within the bilayer. A number of lipid-anchored membrane proteins can be distinguished. The proteins present on the external face of the plasma membrane are bound to the membrane by a small, complex oligosaccharide linked to a molecule of phosphatidylinositol that is embedded in the outer leaflet of the lipid bilayer (**GPI-anchored proteins**).

6.2.3 Biological features of membranes

Lipids are largely responsible for several important features of biological membranes.

Membrane fluidity

Functioning biological membranes are not rigid. They are fluid and very flexible. The physical state of the lipids of a membrane is described by its fluidity (or viscosity). Many of the lipids and protein components are in constant motion. When human and mouse cells, each carrying a distinctive fluorescent marker in its plasma membrane, are fused together, the two kinds of markers gradually become completely intermixed, demonstrating the occurrence of lateral diffusion in the membrane

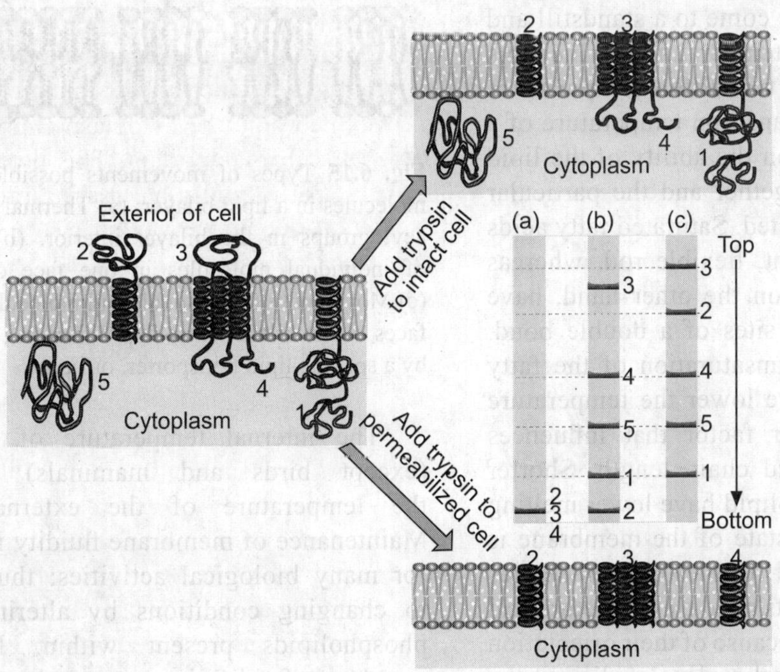

Fig. 6.13 Experimental procedure for determining the orientation of proteins within a plasma membrane. A hypothetical membrane with five distinct proteins being studied.

(**Fig. 6.14**). Lipid molecules in membranes rarely migrate from one lipid monolayer to monolayers of other lipid bilayers. This type of movement of lipids between monolayers is called **flip–flop** or **transbilayer movement** (**Fig. 6.15**). The rate of flip–flop movements of lipids vary depending on membrane type and conditions and facilitated by membrane-bound enzymes called phospholipid translocators (e.g., flipase). Rapid lateral movement of lipid molecules is apparently responsible for the proper functioning of many membrane proteins. Lipid molecules also readily exchange places with their neighbours within a monolayer (~10^7 times per second), resulting in rapid lateral diffusion. Individual lipid molecules rotate rapidly about their long axes and their hydrocarbon chains are flexible.

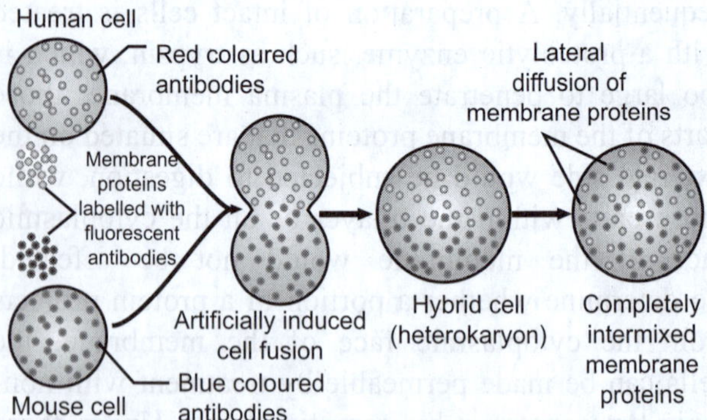

Fig. 6.14 An experimental demonstration of membrane fluidity: After fusion of mouse cells with human cells, the location of each type of membrane protein is determined by staining cells with species-specific antibodies. The anti-mouse and anti-human antibodies are specifically tagged with the molecules that fluoresce with different colours. The coloured antibodies are seen to mix on the surface of the hybrid cell within minutes after the two cell membranes fuse, indicating lateral diffusion of the membrane proteins throughout the lipid bilayer.

Role of membrane composition and temperature

The percentage of unsaturated fatty acids in the phospholipid molecules largely determines membrane fluidity. Similar to all fatty substances, membrane lipids exist in two different physical states: semicrystalline gel and fluid. Lipid or a mixture of lipids can be converted from gel to fluid—that is, melted—by a temperature increase. If the temperature is slowly lowered, a point is reached where the bilayer distinctly changes to frozen crystalline gel, which restricts the movement of the phospholipid fatty acid chains. Membrane activities come to a standstill and permeability increases. The temperature at which this change occurs is called the **transition temperature** or **phase transition**. The transition temperature of a particular bilayer depends on the ability of the lipid molecules to be packed together and the particular lipids of which it is constructed. Saturated fatty acids have the shape of a straight, flexible rod, whereas *cis*-unsaturated fatty acids on the other hand, have crooks in the chain at the sites of a double bond. The greater the degree of unsaturation of the fatty acids, lower the temperature lower the temperature for gel formation. Another factor that influences bilayer fluidity is fatty acid chain length. Shorter fatty acyl chains of phospholipid have lower melting temperature. The physical state of the membrane is also affected by cholesterol. Cholesterol moderates membrane stability, due to its rigid ring structure and flexible hydrocarbon tail. Because of their orientation within the bilayer, cholesterol molecules disrupt the close packaging of fatty acyl chains and interfere with their mobility.

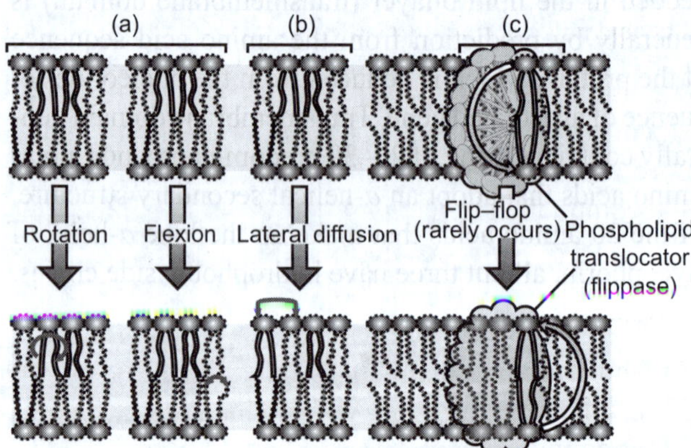

Fig. 6.15 Types of movements possible for phospholipid molecules in a lipid bilayer: (a) Thermal motion of the fatty acyl groups in the bilayer interior, (b) Lateral diffusion of individual molecules in one face of the bilayer and (c) Movement of individual lipid molecules between the two faces of the bilayer (flip–flop diffusion), which is catalyzed by a specific lipid transporter, or lipase.

The internal temperature of most organisms (except birds and mammals) fluctuates with the temperature of the external environment. Maintenance of membrane fluidity is very important for many biological activities; thus, cells respond to changing conditions by altering the types of phospholipids present within. Maintenance of membrane fluidity is an example of homeostasis at the cellular level. For example, at lower temperature, cells respond by altering their metabolic activity.

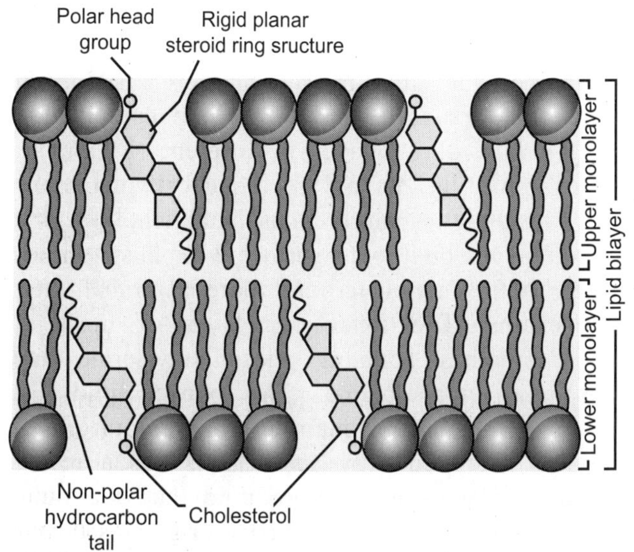

Fig. 6.16 The cholesterol molecules are inserted within the monolayer between the two phospholipid molecules. Cholesterol molecules of a membrane are oriented with their small hydrophilic end facing the external surface of the bilayer and the bulk of their structure packed in among the fatty acid tails of the phosphate.

The initial 'emergency' response is mediated by enzymes that remodel membranes, making the cell more cold-resistant. This membrane remodelling is accomplished by:

- desaturating single bonds in fatty acyl chains to form double bonds,
- reshuffling the chains between different phospholipid molecules to produce molecules that contain two unsaturated fatty acids, which greatly lower the melting temperature of the bilayer.

In addition, the cell also changes the types of phospholipids being synthesized in favour of those containing more unsaturated fatty acids. Maintenance of membrane fluidity by adjustments in fatty acyl composition has been found in a variety of organisms, including hibernating mammals, freshwater fish and bacteria living in hot water springs.

Eukaryotic plasma membranes are found to contain a large amount of cholesterol—up to one molecule for every phospholipid molecule. Cholesterol molecules orient themselves in the lipid bilayer in such a way that the hydroxyl groups remain close to the polar head and the rigid plate-like steroid rings interact with and partly immobilize regions of hydrocarbon chains that are closest to the polar head groups (**Fig. 6.16**). Cholesterol inhibits phase transition by preventing hydrocarbon chains from aggregating together and crystallizing. It also decreases membrane permeability to small water-soluble molecules and is known to enhance both the flexibility and the mechanical stability of the membrane.

Cold acclimation

Low temperature is most influential in limiting the productivity and distribution of plants. Plants that are able to withstand freezing temperatures have the ability to freeze-proof their cells by a process known as cold acclimation. This physiologic process involves altering the composition and physical properties of membranes, the cytoplasm and cell walls, to withstand not only subzero temperatures but also freeze-induced dehydration. One of the most cold-hardy woody species is the mulberry. The most significant and critical alterations that occur during cold acclimation are changes in lipid composition of plasma membranes. Cold acclimation in plants causes an increase in the proportion of phospholipids and a decrease in the proportion of glucocerebrosides.

Self-sealing character

The lipid bilayer is effectively a two-dimensional fluid. This gives the cells flexibility and a self-sealing potential that is essential for cell division. When lipid bilayers are disrupted, they immediately and spontaneously re-seal because a break in a lipid bilayer exposes the hydrophobic hydrocarbon chains to water. With their ability to break and reseal, two membranes can fuse, as in exocytosis and endocytosis (**Fig. 6.17**).

When secreting a substance such as a digestive enzyme, the vesicle containing the enzyme (synthesized inside the cell and enclosed within a membrane bounded vesicle) migrates to the plasma membrane, fuses with it and releases its contents (exocytosis). The ability of a substance to diffuse through a lipid bilayer is largely related to fat solubility. Small fat-soluble molecules penetrate easily, whereas strongly polar molecules traverse extremely slowly if at all. The majority of the molecules of biochemical interest are polar compounds and cannot pass through the lipid bilayer at rates commensurate with cellular needs. Surprisingly, water molecules, despite being polar, pass through the lipid bilayer with sufficient ease, presumably due to the small size of the molecule and the lack of charge. The lipid bilayer also readily allows gases in solution such as oxygen to diffuse through it.

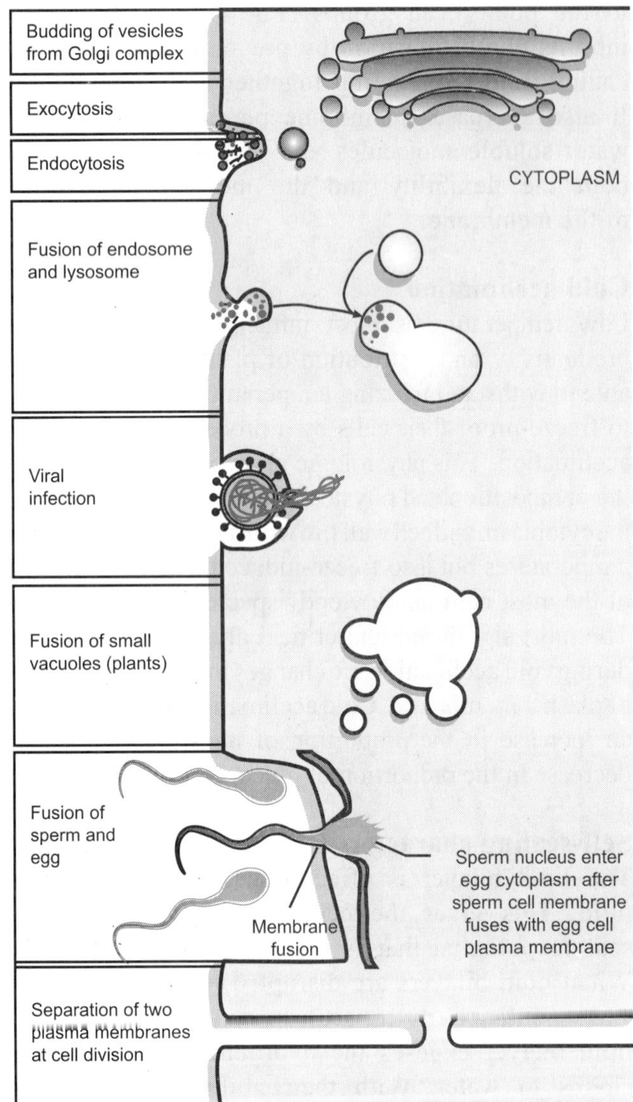

Fig. 6.17 Membrane fusion. The fusion of two membranes is central to a variety of cellular processes involving both organelles and plasma membrane.

Selective permeability

Because of their hydrophobic nature, the hydrocarbon chains in lipid bilayers provide a virtually impermeable barrier to the transport of ionic and polar substances (**Fig. 6.18**). Specific membrane proteins regulate the movement of such substances into and out of the cells. To cross a lipid bilayer, a polar substance must shed some or all of the hydration sphere and bind to a carrier protein form membrane translocation or pass through an aqueous protein channel. Transmembrane water movement is mostly accompanie by ion transport. In contrast, non-polar substances diffuse through the lipid bilayer down their concentration gradients.

Asymmetry

Protein asymmetry is a major feature of the unit membrane and fluid-mosaic models. Biological membranes are asymmetric: the lipid composition of each half of a bilayer is different (**Fig. 6.19**). For example, the human red blood cell membrane has substantially more phosphatidylcholine and sphingomyelin on its outer surface. The phospholipids of the outer monolayer are more saturated fatty acid chains. The inner monolayer of most of the membranes contains phosphatidylserine and phosphatidylethanolamine. The phospholipids of the inner monolayer contain terminal amino groups and less saturated fatty acid chains. As a result, the inner monolayer is more fluid than the outer lipid monolayer. Such asymmetry originates during membrane synthesis in SER, because phospholipid biosynthesis occurs on only one side of a membrane.

The protein components of the membranes also exhibit considerable asymmetry with distinctly different functional domains within the membrane and on the cytoplasmic and extracellular faces of membrane. **Spectrin** is the principal component of the protein meshwork that underlies the plasma membrane.

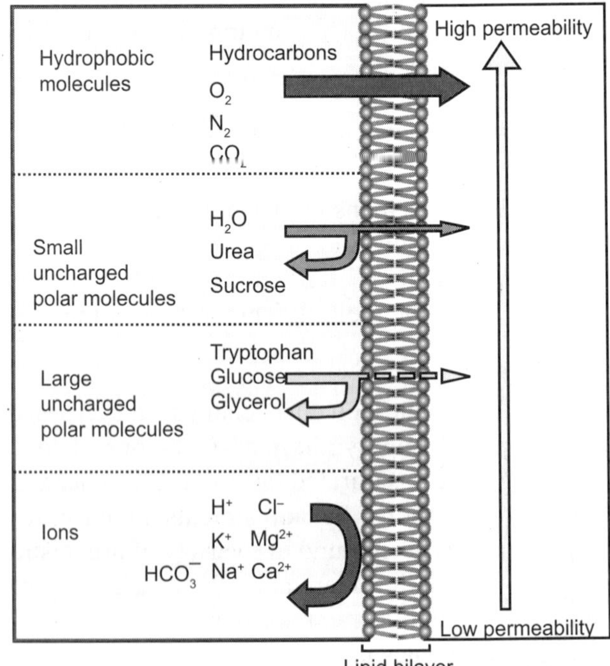

Fig. 6.18 Selective permeability of an artificial phospholipid bilayer to different classes of molecules. Some substances cross the bilayer more rapidly than others. The rate of transport depends on the size of the substance and its solubility in the hydrocarbon interior of the membrane. The phospholipid bilayer is not very permeable to ions or to polar molecules as large as glucose.

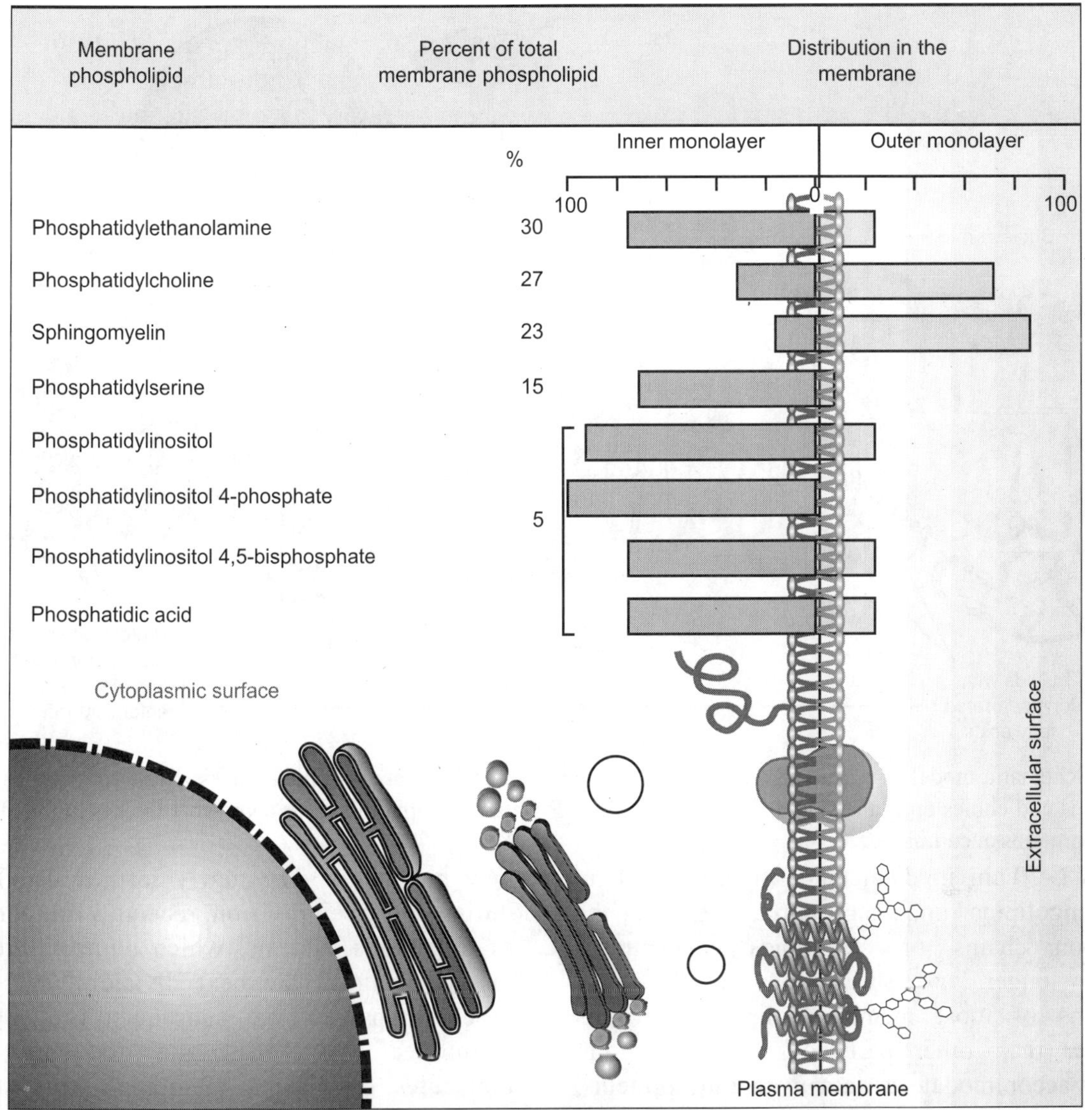

Fig. 6.19 The asymmetry of plasma membrane lipids. The membrane is bifacial, with distinct cytoplasmic and extracellular sides. The lipid composition of the two layers may differ, membrane proteins have specific orientations in the membrane and carbohydrates are restricted to the extracellular side of the membrane. Thus, there is an asymmetric distribution of phospholipids between the inner and outer monolayers of the erythrocyte plasma membrane.

It maintains the structural integrity and biconcave shape of the erythrocyte membrane.

The lipid bilayer can be thought to be composed of two more or less stable, independent monolayers with different physical and chemical properties. The biological role of lipid asymmetry is not well understood, except in a few cases.

Lipid rafts

Even though lipid molecules are uniformly distributed within each leaflet of the bilayer, it is only the proteins that make the membrane a mosaic. Actually, a large body of evidence suggests that the outer leaflet of the plasma membrane of many cells contain specialized regions, or **microdomains,** that possess a distinctive lipid composition. Cholesterol and sphingolipids tend to pack together tightly to form microdomains that are more gelated and highly ordered than the surrounding regions. As a result, these patches are referred to as **lipid rafts,** or platforms, for the attachment of proteins as membranes are moved around the cell and also during signal transduction (**Fig. 6.20**).

The sphingolipids, which tend to have long and saturated fatty hydrocarbon chains, have attractive forces which can be just strong enough to hold the adjacent molecules together transiently in small microdomains. The plasma membrane of animal cells is thought to contain many such tiny

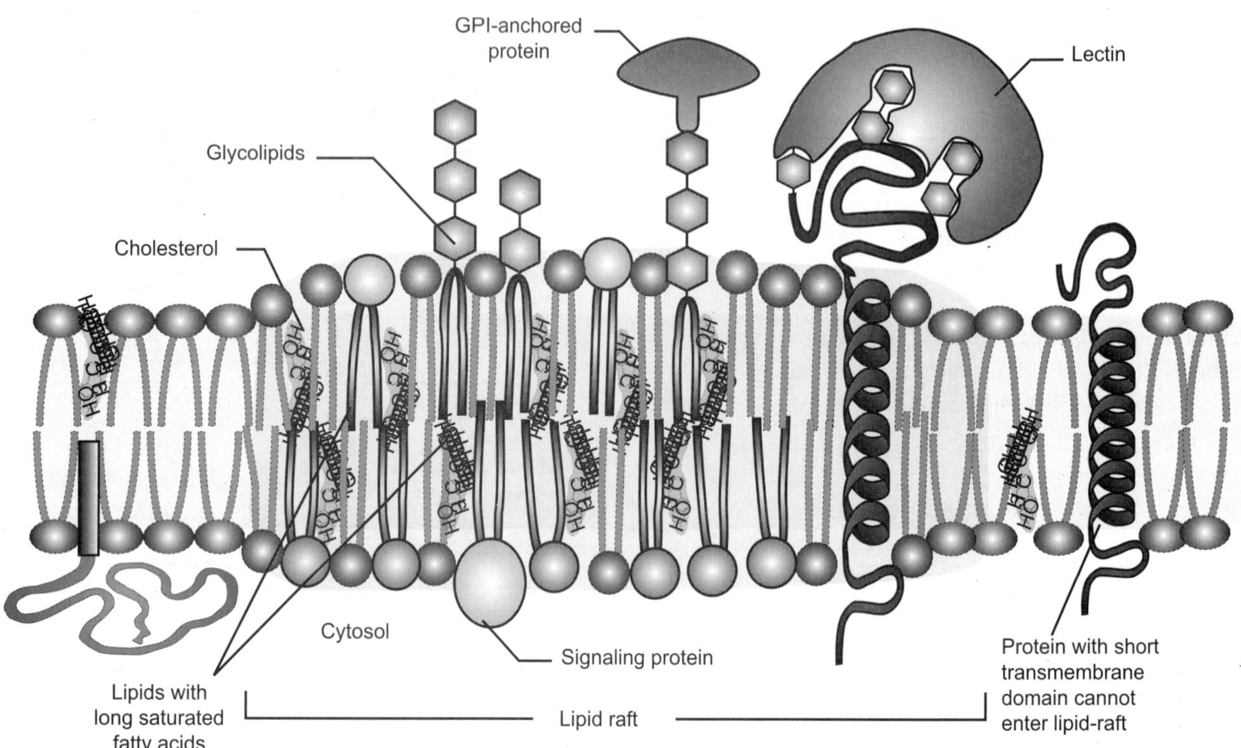

Fig. 6.20 Schematic model of a lipid raft. Lipid rafts are small, specialized areas in membranes where some lipids (primarily sphingolipids and cholesterol) and proteins are concentrated. Because the lipid bilayer is somewhat thicker in the rafts, certain membrane proteins accumulate.

lipid rafts (~70 nm in diameter), which are rich in both sphingolipids and cholesterol. Because the hydrocarbon chains of the lipids concentrated there are longer and straighter than the fatty acid chains of most membrane lipids, the rafts are thicker than other parts of the bilayer and can better accommodate certain membrane proteins, which therefore tend to accumulate there. Thus, lipid rafts help to organize such proteins either by concentrating them for transport in small vesicles or function in concert.

6.2.4 Biosynthesis of plasma membrane

The plasma membrance is the most important structure that maintains cellular integrity and dynamic molecular trafficking between the inner and the outer world. Injury to the cell membrane can be disastrous to a cell. The cell membranes need to be constantly replenished with newly synthesized membrane lipids without obstructing the cell functions (called **membrane turnover**). Also, for maintaining the characteristic membrane asymmetry, the membrane must be assembled with precisely the correct molecular topography.

During membrane biogenesis, the pre-existing membrane acts as a template for the addition of new precursors. The newly formed daughter cells during every cell division, receive a full complement of membrane systems, which undergo growth and maturation until the next division, and the same process is passed on to subsequent progeny. During membrane enlargement and maturation, protein molecules of plasma membrane are synthesized on both attached and free ribosomes. Proteins of plasma membrane synthesized on attached ribosomes of rough ER are inserted first into the membrane of RER and then transferred to the Golgi apparatus, where they are processed and ultimately dispatched to the plasma membrane via the secretory vesicles; but the synthesis of phospholipid molecules of the plasma membrane takes place by the smooth ER (SER). Like proteins, newly synthesized lipids are dispatched to the plasma membrane via small secretory vesicles.

6.2.5 Biosynthesis of important agricultural products on specialized membrane systems

Carbohydrates, proteins and fats (oil) are the three basic dietary products produced in plants and used by most life forms.

Oil bodies and dietary proteins formed by specialized ER domains

Two important agricultural products such as vegetable oils and dietary seed proteins are produced on the ER. The vegetable oils stored in the seeds (triacylglycerols) provide energy and membrane lipid building blocks during the early stages of embryo germination and growth. During their synthesis by ER enzymes, triacylglycerols partition into the interior or the ER bilayer and then accumulate at sites defined by molecules known as **oleosins**. Oleosins have a thumbtack-like architecture, with the 'shift' portion consisting of hydrophobic amino acids and the head exhibiting an amphipathic structure. As the newly synthesized triacylglyceride molecules accumulate between the bilayer leaflets at the oleosin sites, they give rise to spherical oil bodies that bud into the cytosol (**Fig. 6.21**). The surface layer of each oil body consists of a phospholipid monolayer, embedded oleosins and other proteins.

Seed storage proteins are an important dietary protein source for most organisms including humans. They are synthesized during seed development and serve as the principal source of amino acids for germination and seedling development. Cereals (e.g., rice, wheat) produce two classes of storage proteins: water-soluble **globulin** proteins and the more hydrophobic, alcohol-soluble **prolamins**.

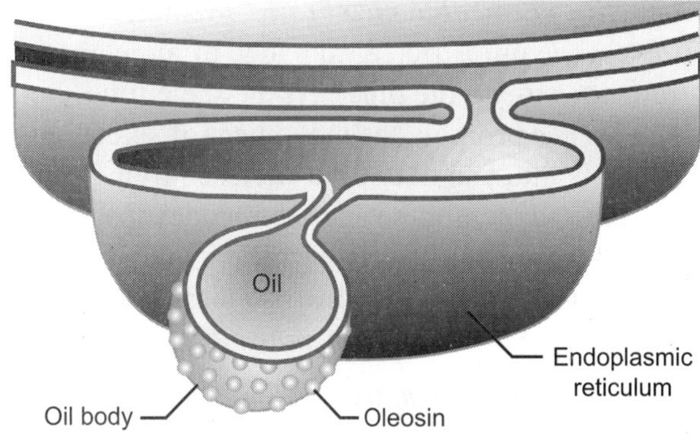

Fig. 6.21 Triacylglycerides accumulate between the two lipid monolayers of a smooth ER membrane and bud off as oil bodies at the sites defined by the oleosin molecules.

6.2.6 Membrane functions

Membranes are involved in a large variety of functions in living organisms. The important physiological functions are explained briefly in the following sections.

Compartmentalization: Membranes are continuous, unbroken sheets. They enclose and retain the entire cell contents (mechanical structure), whereas the nuclear and cytoplasmic membranes enclose diverse intracellular spaces. However, the various membrane-bound compartments of a cell possess selectively different contents. Membrane compartmentalization allows specific activities to take place inside without external interference.

Scaffold for biochemical activities: Membranes not only enclose cellular contents, but are also a distinct compartment themselves. They provide the cell with an extensive framework or scaffolding within which components can be ordered for effective interaction.

Membrane permeability: Permeability is fundamental to the functioning of the living cell or organelles and to the maintenance of intracellular physiological conditions. Because the contents of a cell are completely surrounded by the plasma membrane, all communication between the cell and the extracellular medium must be mediated through the membrane. Permeability determines which substances can enter the cell and it also regulates the outflow of excretory material and water from the cell. Among the most important of these are the transport of molecules and ions into and out of cells and organelles and the binding of hormones and other biomolecules. There are basically two means for the movement of substances through a membrane: *Passive transport* by *diffusion* or *active transport* by *energy-coupled transport* process.

The plasma membrane around the cytoplasm is very dynamic and allows the movement of small ions and molecules of various substances through it. This property of the plasma membrane is termed as *membrane permeability*. Based on the nature of permeability the membrane can be grouped into different types.

- ***Impermeable plasma membrane***: This type of membrane can be found in unfertilized eggs of certain fishes. It allows only gases to pass through.
- ***Semipermeable plasma membrane***: The selective membranes allow only water but no solutes to pass through it. Semipermeable membranes have not been found in living organisms. But synthetic membranes that function as semipermeable membranes have been used in research laboratories (e.g., dialysis membranes with different cut-off points).

- ***Selective permeable membrane***: Membranes prevent the unrestricted exchange of molecules from one side to the other. At the same time membranes promote the movement of select elements into and out of the enclosed living space. The plasma membrane and membranes inside the cytoplasm are selectively permeable membranes. The cellular membranes are also termed as *differentially permeable membranes*.
- ***Dialysis plasma membranes***: The plasma membranes of some cells have extraneous coats around them. The basement membranes of endothelial cells are the best examples of extraneous coats. Such membranes act as dialyzers. Water molecules and crystalloids are forced through them by the hydrostatic pressure forces.

Transport of solutes: The plasma membrane contains the structures for physically transporting substances from one side of the membrane to another—often from a region where the solute is present at lower concentration into a region where it is present at a much higher concentration. This facilitates cells to accumulate various chemical compounds (e.g., sugars, amino acids) that are essential for cellular metabolism. The plasma membrane is also able to transport specific ions by establishing ionic gradients.

Responding to external signals: The plasma membrane plays a critical role in the response of a cell to external stimuli through a process known as *signal transduction.* Membranes possess receptors that combine with specific molecules (or ligands) having a complementary structure. Different cell types have different membrane receptors and are, therefore, capable of recognizing and responding to different ligands in their environment. The interaction of a plasma membrane receptor with an external ligand may cause the membrane to generate a signal that stimulates or inhibits internal activities.

Intercellular interaction: The plasma membrane of multicellular organisms mediates the interactions between a cell and its neighbours. The plasma membrane allows cells to recognize and signal one another, to adhere when appropriate and to exchange materials and information.

Energy transduction: Membranes are intimately involved in the process by which one type of energy is converted into another type. The most fundamental energy transduction occurs during photosynthesis when energy from sunlight is absorbed by membrane-bound pigments converted into chemical energy and stored in carbohydrates. Membrances are also involved in the transfer of chemical energy from carbohydrates and fats into ATP. These energy conversion capabilities in eukaryotes are present within the membranes of chloroplasts and mitochondria.

6.3 TRANSPORT OF SUBSTANCES ACROSS CELL MEMBRANES

Both diffusion and active transport of solutes across the membrane lead to the net flux of a particular ion or compound. The term **net flux** indicates that the movement of the substance into the cell (**influx**) and out of the cell (**efflux**) is not balanced, but that one exceeds the other. The plasma membrane highly selective being maintains a constant internal environment (**homeostasis**).

Several different processes are known by which substances move across membranes.

- Simple diffusion through the lipid bilayer,
- Simple diffusion through aqueous protein-lined channels
- Facilitated diffusion by protein transporter
- Active transport, which requires an energy-driven protein 'pump' capable of moving substances against a concentration gradient.

The movement of substances across membranes is known as membrane transport and it involves two aspects: membrane permeability and transport mechanisms.

6.3.1 Energetics of solute movement

Classification of biological transport mechanisms is based on whether they require energy or not. The major types of biological transport are illustrated in **Fig. 6.22**.

In this diagram the relative sizes and bold letters indicate the directions of the concentration gradients.

1. Solute movement by different diffusion processes:
    - Simple diffusion through the membrane, which always proceeds from high to low concentration.
    - Simple diffusion through an aqueous channel formed within an integral membrane protein and the solute movement is always high to low concentration.
    - Facilitated diffusion in which solute molecules bind specifically to a membrane protein carrier and solute movement is always high to low concentration.

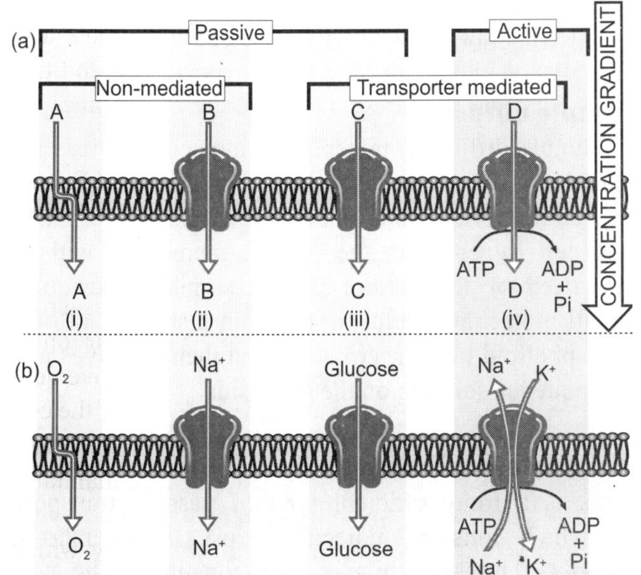

Fig. 6.22 Four basic mechanisms by which solutes move across the membrane.

- Active transport by means of a protein transporter with a specific binding site that undergoes a change in affinity driven with energy released by an exergonic process, such as ATP hydrolysis. Movement occurs against a concentration gradient.

2. Examples of each type of mechanism as it occurs in the membrane of a cell.

Diffusion is a spontaneous process in which a substance moves from a region of high concentration to a region of low concentration, eventually eliminating the concentration difference between the two regions (i.e., reaching the equilibrium). The general thermodynamic principles governing the transfer of substances across membranes or a substance from a region in which its concentration is C_1 to a place where its concentration is C_2 is given by

$$\Delta G = RT \ln[C_2]/[C_1]$$

$$\Delta G = 2.303\, RT \log_{10} [C_2]/[C_1]$$

where ΔG is the free-energy change, R is the gas constant, T is the absolute temperature and $[C_2]/[C_1]$ is the ratio of the concentration of the solute on the inside (2) and outside (1) surfaces of the membrane. According to this equation, if C_2 is less than C_1, ΔG is negative and the process is thermodynamically favourable. As more substance is transferred, C_1 decreases and C_2 increases, until $C_2 = C_1$. At this point $\Delta G = 0$ and the system is at equilibrium. Unless other factors are involved, this equilibrium is the ultimate state approached by transport across any membrane.

Diffusion of a substance depends on the random thermal motion of solutes and is an exergonic process driven by an increase in entropy. The free energy change when an uncharged solute (non-electrolyte) diffuses across a membrane depends on the magnitude of the concentration gradient, that is, the difference in concentration on each side of the membrane. If the solute is an electrolyte (a charged species), the overall charge difference between the two compartments must also be considered. As a result of the mutual repulsion of ions of like charges, it is thermodynamically unfavourable for an electrolyte to move across a membrane from one compartment into another with net charge of the same sign. Conversely, if the charge of the electrolyte is opposite in sign to the compartment into which it is moving, the process is thermodynamically favoured. The greater the difference in charge (voltage) between the two compartments, the greater the difference in the free energy. Thus the tendency of an electrolyte to diffuse between two compartments depends on two gradients: a chemical gradient and the electric potential gradient. Together these differences are combined to form an **electrochemical gradient**.

6.3.2 Diffusion of substances through membranes

For diffusion to occur two requirements must be met before a non-electrolyte can diffuse passively across a plasma membrane: the substance must be present at a higher concentration on one side of the membrane than the other and the membrane must be permeable to the substance. A membrane may be permeable to a given solute either (i) because that solute can pass directly through the lipid bilayer, or (ii) because that solute can traverse an aqueous pore that spans the membrane and prevents the solute from coming into contact with the lipid molecules of the bilayer.

Passive transport

Passive transport or passive diffusion, takes place basically by diffusion, in which an ion or molecule crossing a membrane moves down its electrochemical or concentration gradient. No external energy is required. The process is the same as the Brownian motion of molecules in any fluid, which is termed **molecular diffusion**. Passive transport ultimately results in the free concentration of the diffusing substance being the same on both sides of the membrane. The net rate of transport, J (in moles per square centimetre per second), is proportional to the concentration difference $(C_2 - C_1)$ across the membrane.

$$J = KD_1(C_2 - C_1)/l$$

where l is the thickness of the membrane, D_1 is the diffusion coefficient of the diffusing substance in the membrane and K is the partition coefficient for the diffusing material between membrane lipid and water. For ions and other hydrophilic substances, K is a very small number, with the result that diffusion of such substances through lipid membranes is extremely slow.

Osmosis

Water molecules move much more rapidly through a cell membrane than ions or small polar organic solutes. Because of this difference in the penetrability of water versus solutes, membranes are said to be semipermeable. The movement of water from a region of lower solute concentration to a region of higher solute concentration through a semipermeable membrane is called **osmosis** (Gr., *osmos* = pushing). The process by which the water molecules enter the cell is known as **endosmosis** and the exit of water molecules from the cell is known as **exosmosis**. In plant cells, during water stress, due to excessive exosmosis the cytoplasm along with the plasma membrane shrinks away from the cell wall. This process is known as **plasmolysis**.

When two compartments of different solute concentrations are separated by a semipermeable membrane, the compartment of higher solute concentration is said to be **hypertonic** (or **hyperosmotic**) relative to the compartment of lower solute concentration, which is described as being **hypotonic** (or **hypoosmotic**). When a cell is placed in a hypotonic solution, the cell rapidly gains water by osmosis and swells. Conversely, when a cell placed in a hypertonic solution, the cell rapidly loses water by osmosis and shrinks. If cells are placed in an **isotonic solution**, the cell volume is unchanged, because the intracellular and extracellular fluids contain the same concentration and no osmosis take place. These simple observations show that a cell's volume is controlled by the difference between the solute concentration inside the cell and that in the extracellular medium (**Fig. 6.23**). The hydrostatic pressure caused by the osmosis is known as **osmotic pressure**. The plasma membrane maintains a balance between the osmotic pressure of the intracellular and intercellular fluids. Thus, osmosis is an important factor in a multitude of bodily functions.

Simple diffusion

In simple diffusion, transport of molecules across the membrane takes place unaided, i.e., molecules of gases (O_2 and CO_2) and small molecules (e.g., ethanol) enter the cell by crossing the plasma membrane without the need of facilitation. During simple diffusion, a small molecule in aqueous solution dissolves into the phospholipid bilayer, crosses it and then dissolves into the aqueous solution on the other side.

Facilitated diffusion

This is a more efficient type of passive transport, in which ions or molecules cross the membrane rapidly. The lipid bilayer that constitutes the core of biological membranes is highly impermeable to charged substances (such as Na^+, K^+, Ca^{2+} and Cl^-). The rapid movement (**conductance**) of these ions across membranes plays a critical role in a multitude of cellular activities, including formation and propagation of a nerve impulse, secretion of substances into the extracellular space, muscle contraction, regulation of cell volume and the opening of stomatal pores on plant leaves.

Biologists have identified a variety of ion channels, each formed by integral membrane proteins that enclose a central aqueous pore. Most ion channels are highly selective and allow only one type of ion to pass through the pore. Often the diffusing

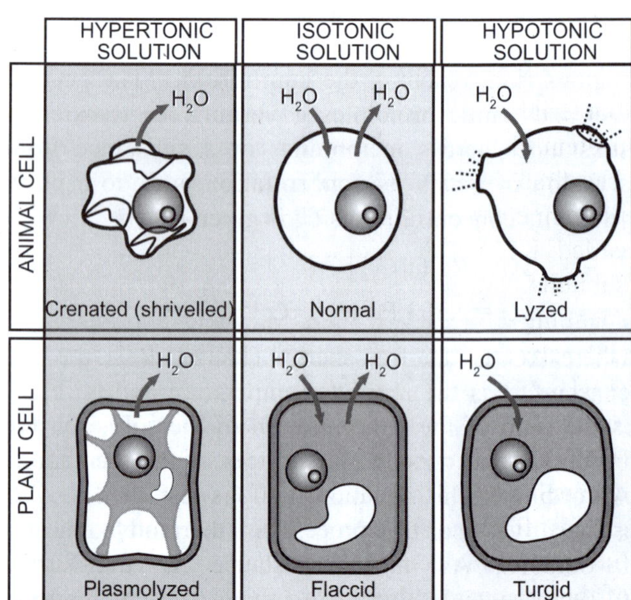

Fig. 6.23 Osmosis: Water balance in living cells.

substance first binds selectively to a membrane-spanning protein, called a **facilitative transporter** that facilitates the diffusion process. The binding of the solute to the facilitative transporter on one side of the membrane is thought to trigger a conformational change in the protein, exposing the solute to the other surface of the membrane, from where it can diffuse down its concentration gradient (**Fig. 6.24**). Because they operate passively, facilitated transporters can mediate the movement of solutes equally well in both directions. Facilitated diffusion is particularly important in mediating the entry and exit of polar solutes, such as sugars and amino acids that do not penetrate the lipid bilayer.

Channel proteins (pore-facilitated channels) and **carrier proteins** are the two major classes of membrane proteins.

6.3.3 Pore-facilitated transport

Ion channels can exist in both open and closed conformation, which are said to be **gated channels.** The opening and closing of gates are subject to complex physiologic regulation and can be induced by a variety of factors depending on the particular channel (**Table 6.2**). Two major categories of gated channels are:

1. ***Voltage-gated channels***: The Conformational state depends on the difference in the ionic charges on the two sides of the membrane.
2. ***Ligand-gated channels***: The Conformational state depends on the binding of a specific molecule (the ligand), which is usually not the solute that passes through the channel.

Bacterial membrane system contains many types of protein channels in the inner and outer membranes. The first atomic resolution image of an ion channel protein elucidated in bacteria was K^+ ion channel called KcsA. The KcsA channel consists of four subunits, and each is seen to contain two membrane-spanning helices (M1 and M2) and a pore region (P0 at the extracellular end of the channel) (**Fig. 6.25**). The lining of the selectivity filter contains a highly conserved pentapeptide-Gly-Tyr-Gly-Val-Thr (GYGVT). The X-ray crystal structure of the KcsA channel shows that the backbone carbonyl (C=O) groups from the conserved pentapeptide form the lining of the selectivity filter. The conserved residues of the selectivity filter create five successive rings of oxygen atoms. Each ring contains four oxygen atoms

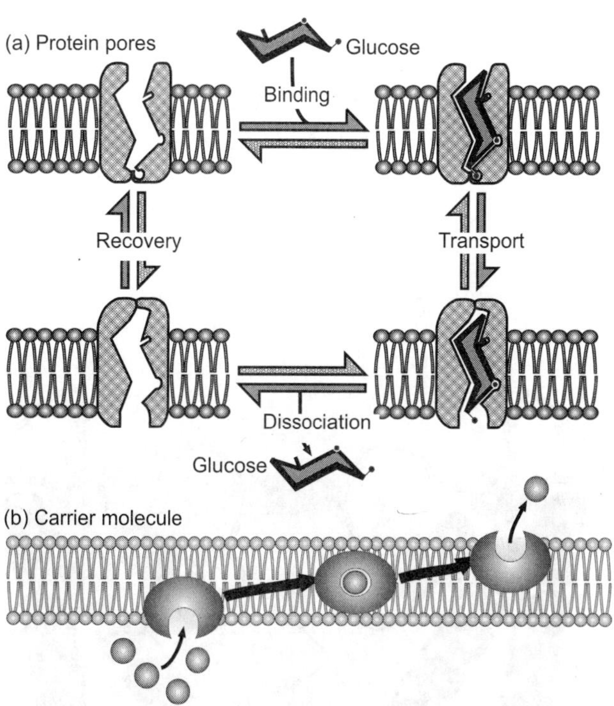

Fig. 6.24 Facilitated diffusion: (a) Protein pores: A schematic model for the facilitated diffusion of glucose depicts the alternating conformation of a carrier that exposes the glucose binding site to either the inside or outside of the membrane. (b) Carrier molecule.

Table 6.2 Types of gated ion channels

Gating effecter	Ion(s)	Example
Voltage	Na^+, K^+, Ca^{++}	Neuron
Mechanical	K^+	Auditory hairs
Odour	K^+	Olfactory epithelium
Taste	K^+, Na^+	Taste receptor
Neurotransmitter	K^+, Na^+	Acetylcholine receptor
Ion	K^+	Ca^{++}-activated K^+ channel
Cyclic nucleotide	Na^+	cGMP retinal rod channel
G protein	K^+	Muscarinic acetylcholine receptor

and has a diameter of approximately 3 Å, which is slightly larger than the 2.7 Å diameter of a K^+ ion that has lost its normal shell of hydration. Consequently, the electronegative O atoms that line the selectivity filter can substitute for the shell of water molecules that are displaced as each K^+ ion enters the pore.

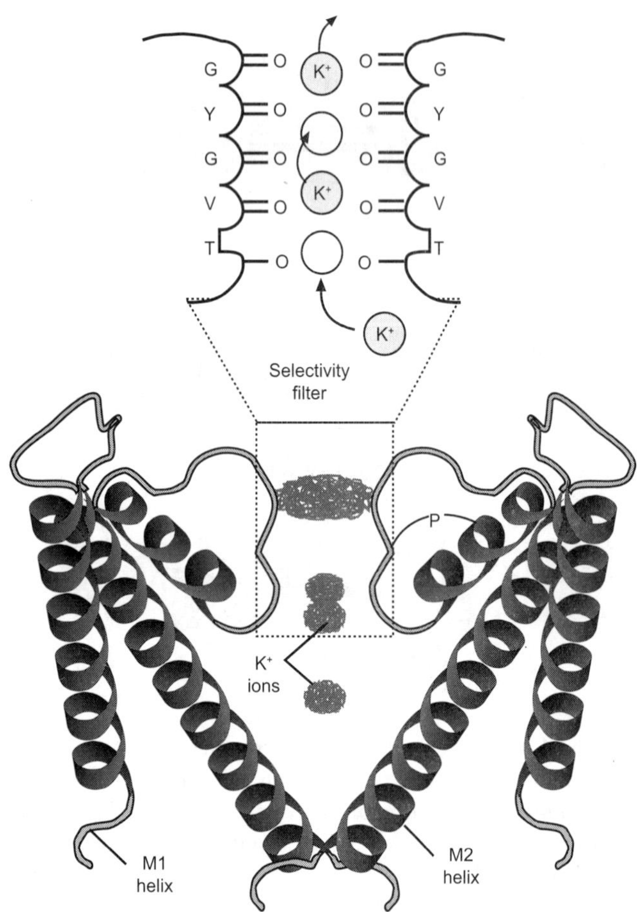

Fig. 6.25 Three-dimensional structure of the bacterial KcsA channel and the selection of K^+ ions (Source: Karp G, 2005).

Here the selectivity filter contains four potential K^+ ion-binding sites and is a precise fit for a dehydrated K^+ ion. Potassium ions move two at a time and at the entry a third K^+ ion into the selectivity filter creates an electrostatic repulsion that ejects the ion bound at the opposite end of the line. The movement of K^+ from one binding site to the next has no energy barrier, which accounts for the extremely rapid flow of ions across the membrane.

In facilitated diffusion, the transport of certain large or charged molecules occurs through tunnel-like transmembrane proteins. Each type is designated for the transport of a specific solute. Many channels are chemically or voltage-regulated. Chemically regulated channels, open or close in response to a specific chemical signals. For example, a chemically gated Na^+ channel in the nicotinic acetylcholine receptor complex opens when acetylcholine binds. Na^+ rushes into the cell and the membrane potential falls. Since the membrane potential is an electrical gradient across the membrane, a decrease in potential is membrane **depolarization**. Local depolarization caused by acetylcholine leads to the opening of nearby Na^+ channels (referred to as voltage-gated channels). **Repolarization**, the reestablishment of the membrane potential, begins with the diffusion of K^+ ions out of the cell through voltage-gated K^+ channels (become negative when K^+ goes out). Facilitated diffusion increases the rate at which certain solutes move down their concentration gradients. This process cannot cause a net increase in solute concentration on one side of the membrane.

6.3.4 Carrier-facilitated transport

Unlike pumps, carriers do not catalyze scalar reactions such as ATP hydrolysis. The transport process does not involve chemical modification of any of the compounds bound to the carrier. Rather, carriers catalyze solely vectorial reactions, i.e., the movement of inorganic ions and simple organic solutes across membranes. The array of ions and solutes translocated by carriers is vast. The principal inorganic nutrients, including NH_4^+, NO_3, Pi, K^+ and SO_4^{2-}, are all translocated into cells by plasma membrane carriers. Among the organic solutes translocated into cells by carriers are the sugars, amino acids and purine and pyrimidine bases. Carriers also play crucial roles at endomembranes. Tonoplast carriers catalyze sequestration of Na^+, Ca^{2+}, Mg^{2+} and NO_3^- as well as sucrose and amino acids. Many carriers at the vacuole and plasma membranes behave as H^+symporters or antiporters. As a result, the carrier substrate is translocated against its own electrochemical potential. Carrier proteins (permeases or transporters) bind the specific solute to be transported and undergo a series of conformational changes to transfer the bound solute across the lipid bilayer. When these pores are open, they allow specific solutes to pass through them and thereby cross the membrane. Antibiotic ionophores are the best known examples of carrier-facilitated transport.

6.3.5 Active transport: Transport against a concentration gradient

Concentration gradients can quickly reach equilibrium by the process of diffusion. But life cannot be sustained under equilibrium conditions. There will always be an imbalance of ions across the plasma membrane. The ability of a cell to generate such steep concentration gradients across its plasma membrane is generated by active transport. Unlike facilitated diffusion, the movement of a solute against a gradient requires the coupled input of energy. Consequently, the endergonic movement of ions or other solutes across the membrane against a concentration gradient is coupled to an exergonic process, such as the hydrolysis of ATP, the absorption of light, the transport of electrons, or the flow of other substances down their gradients (**Fig. 6.26**). Proteins that carry out active transport are often referred to as '**pumps**'. Some common protein pumps are related to Na^+,-K^+, Ca^{2+} and H^+.

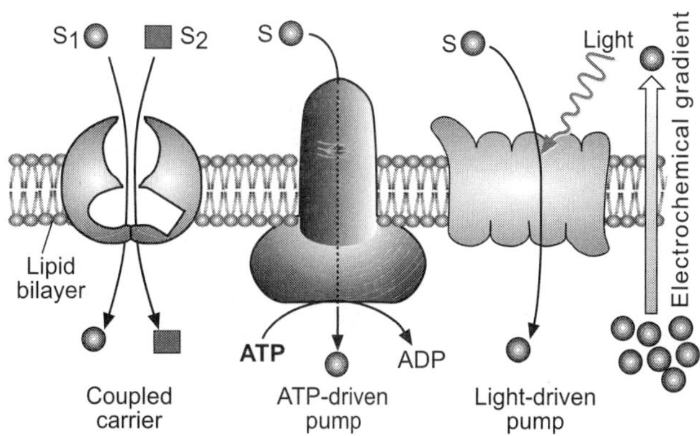

Fig. 6.26 Three ways of driving active transport. The actively transported molecule is round and the energy source is shown as a rectangle.

Ion gradient pumps

The two forms of active transport are primary and secondary. In **primary active transport**, the energy is derived from ATP. The Na^+-K^+ pump (also known as Na^+-K^+ ATPase) is a prominent example of a primary transporter. In **secondary active transport**, concentration gradients generated by primary active transport are harnessed to move substances across membranes. For example, the Na^+ gradient created by the N^+–K^+ ATPase pump is used in kidney tubule cells and intestinal cells to transport D-glucose.

The Na^+-K^+ pump

The sodium–potassium pump is found only in animal cells. Virtually, in all cells, the concentration of Na^+ is greater outside (150 mM) than inside (10 mM) the cell and the concentration of K^+ is higher inside (140 mM) than outside (5 mM). Na^+-K^+ ATPase is composed of two subunits. The larger of the subunits is a transmembrane protein, which crosses the bilayer seven times and acts as the catalytic centre and the other one is a smaller polypeptide that crosses only once and is important in some unknown way to the function of the pump.

The series of steps involved during the pumping cycle of the Na^+/K^+-ATPase is shown in **Fig. 6.27**. When the protein binds three Na^+ ions on the inside of the cell (step1) and becomes phosphorylated (step 2), it shifts from the E1 conformation to the E2 conformation (step 3). In doing so, the binding site becomes exposed to the extracellular compartment and the protein loses its affinity for Na^+ ions, which are then released outside the cell. Once the three sodium ions have been released, the protein picks two potassium ions (step 4), becomes dephosphorylated (step 5) and shifts back to the original E1 conformation (step 6). In this state, the binding site opens to the internal surface of the membrane and loses its affinity for K^+ ions, leading to the release of these ions into the cell. The cycle is then repeated.

The Na^+-K^+ ATPase has many important functions. It is largely responsible for the gradients of the two ions in excitable cells such as nerve and muscle. These provide the basis for the generation of action potentials and propagation of electric signal along the cell surface. In addition, the sodium pump helps to maintain cells in osmotic equilibrium with their environment and it is also essential in generating ion gradients for co-transport of metabolites.

Calcium pump

One of the best-studied pump proteins is the calcium transport protein from muscle cells. The **calcium pump** (also known as Ca^{2+}-ATPase) is present in the membranes of the endoplasmic reticulum, where it actively transports calcium ions out of the cytosol into the lumen of this organelle called the **sarcoplasmic reticulum**. Then in response to a nerve impulse, Ca^{2+} is rapidly released (through other channels) back into the cytosol to trigger muscle contraction. During contraction, Ca^{2+} around the myofilaments is about 10 μM. For relaxation to occur, this concentration must be reduced to 0.1 μM in less than a tenth of a second. This hundred-fold reduction is achieved by sequestering the Ca^{2+} inside the sarcoplasmic reticulum. Within the reticulum, the Ca^{2+} concentration reaches 10 mM. In this process, one molecule of ATP hydrolysis powers the movement of $2Ca^{2+}$ across the membrane into the reticulum. The protein responsible for calcium movement is termed the Ca^{2+} pump (**Fig. 6.28**). The Ca^{2+} pump is homologous to the Na^+–K^+ pump that maintains Na^+

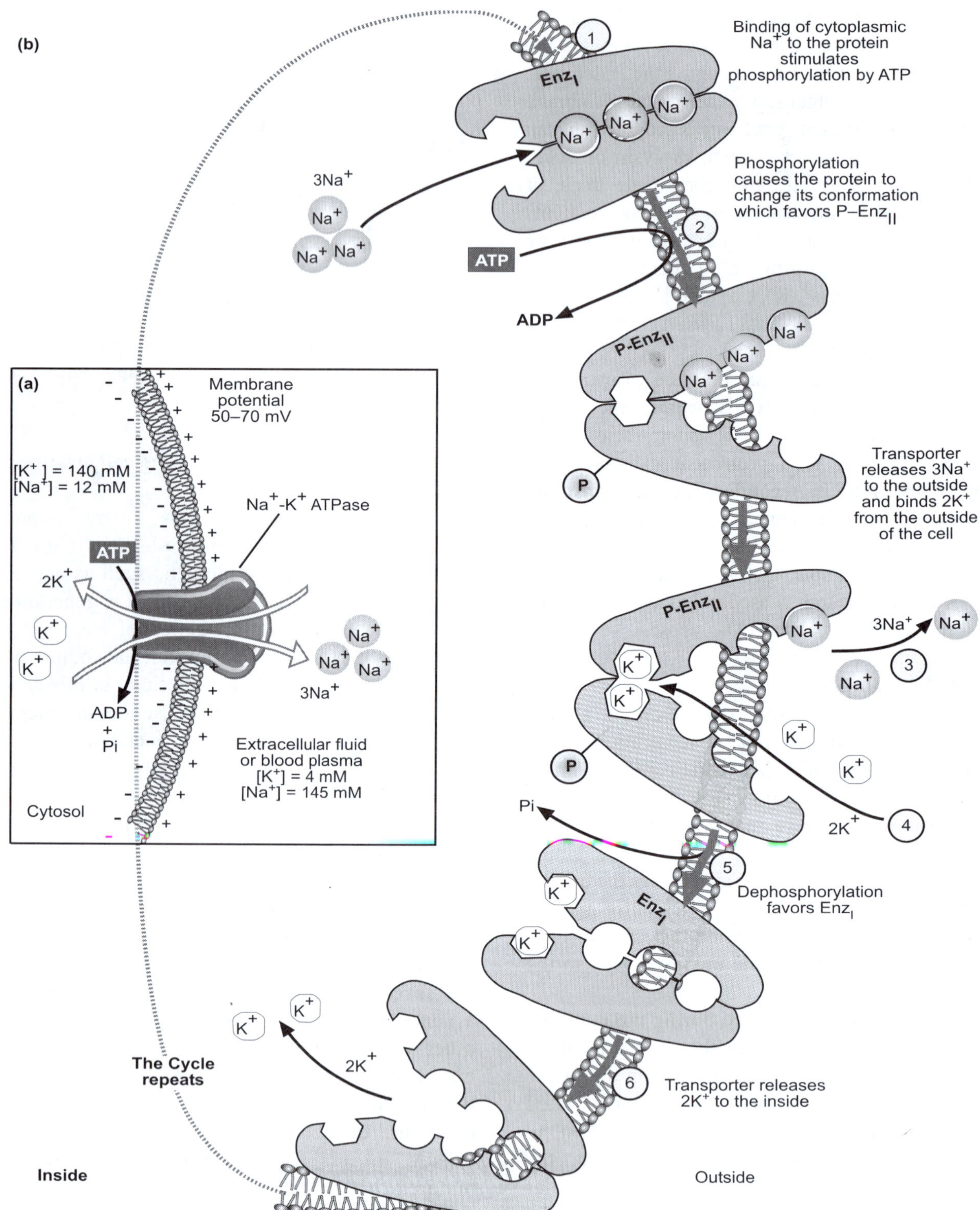

Fig. 6.27 Active co-transport of Na^+-K^+: (a) Na^+-K^+ ATPase and (b) Postulated mechanism of Na^+ and K^+ transport by the Na^+-K^+ ATPase. (1) Binding of three Na^+ to high-affinity sites on the large subunit of the transport protein on the inner surface of the membrane. This part of the large subunit also has the ATP-binding site. Phosphorylation of the transporter changes its conformation (2) and decreases its affinity for Na^+, leading to (3) Na^+ release on the outer surface. (4) K^+ on the outside now binds to high-affinity sites on the extracellular portion of the large subunit. (5) The enzyme is dephosphorylated, reducing its affinity for K^+. (6) K^+ is discharged on the inside of the cell. The transport protein is now ready for another cycle of Na^+ and K^+ pumping **(See Plate 3)**.

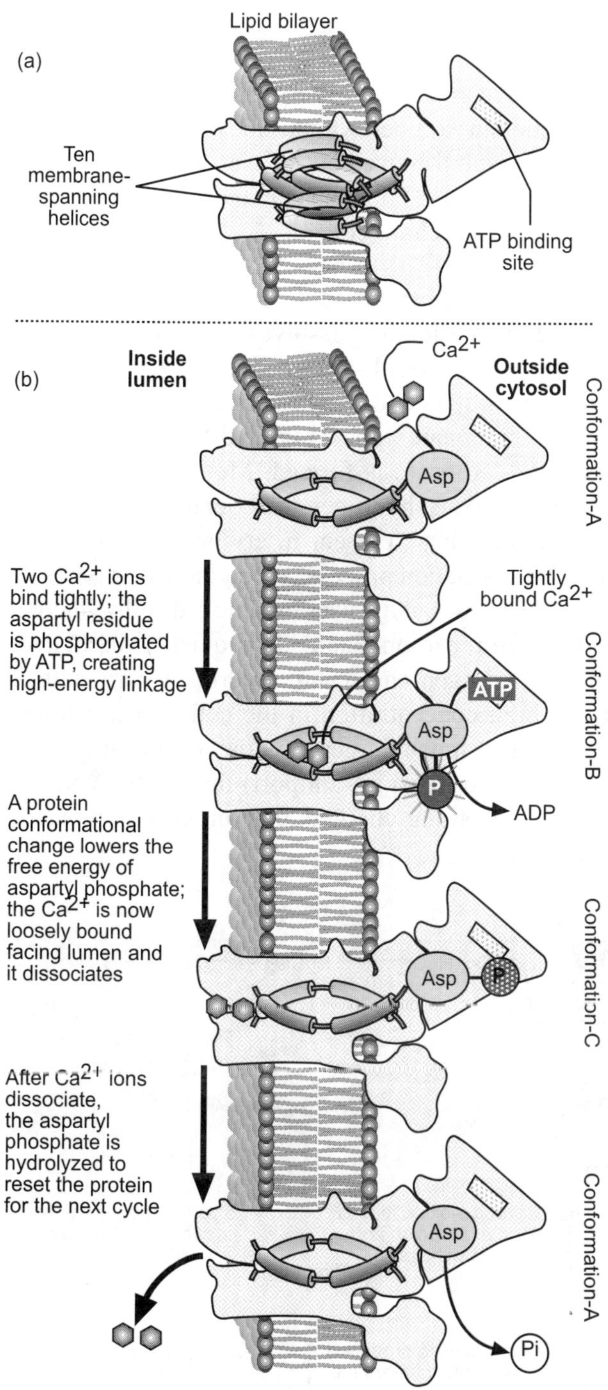

Fig. 6.28 The transport of calcium ions by the Ca^{2+} pump: (a) The structure of the calcium ion pump protein, formed from a single subunit of 994 amino acids; (b) A model for ion pumping. For simplicity only two of the four transmembrane α-helices that interact with calcium are shown.

and K^+ concentration differences across the plasma membrane, both being membranes of a family of P-type cation pumps (so named because these proteins are autophosphorylated during their reaction cycle). The transport of calcium ions by the Ca^{2+}-ATPase is accompanied by large conformational changes, which are thought to couple the hydrolysis of ATP to the changes in access and affinity of the ion binding sites.

Proton pump

A proton pump (or H^+-ATPase) is responsible for energy generation in chloroplasts, mitochondria and other membranes. Essentially, it does the pumping, using a proton gradient set up by asymmetric electron transport carriers in the membrane to make ATP. A second type of H^+-ATPase is found at the plasma membrane, where its function is to maintain the cytoplasmic H^+ concentration (pH) and, in the case of some plant cells, to acidify the cell wall during hormone-stimulated cell elongation. The third class of H^+ pumps occurs in the membranes of vacuoles, including the plant vacuole, lysosome and coated vesicle. Its role is to acidify the contents of the cellular organelle.

Co-transport

The establishment of concentration gradients, such as those of Na^+, K^+ and H^+, provides a means by which free energy can be stored in a cell. The potential energy stored in ionic gradients is utilized by a cell in various ways to perform work, including the transport of other solutes. The co-transport system can be well explained, for example, by studying physiologic activity of the intestinal epithelial cells (**Fig. 6.29**). Within the lumen of the intestine, enzymes hydrolyze high-molecular weight polysaccharides into simple sugars, which are absorbed by the epithelial cells that line the intestine. The movement of glucose across the apical plasma membrane of the epithelial cells, against a concentration gradient, occurs by co-transport with sodium ions. The Na^+ concentration is kept very low within the cells by the action of a primary active transport system located in the basal and lateral plasma membrane that pumps sodium ions out of the cell against a concentration gradient. The tendency for sodium ions to diffuse back across the apical plasma membrane down their concentration gradient is 'tapped' by the epithelial cells to drive the co-transport of glucose molecules into the cell against a concentration gradient. The glucose molecules are said to be driven by **secondary active transport (energy derived from a gradient)**. In this case, Na^+/glucose co-transporter, binds two sodium ions and one glucose molecule on the external surface of the apical plasma membrane. When the sodium ions are released on the inside of the cell into a solution of

Fig. 6.29 Transcellular transport of glucose. The transcellular transport of glucose across an intestinal epithelial cell depends on the non-uniform distribution of transport proteins in the cell's plasma membrane. Glucose is co-transported (symport) with Na^+ across the apical plasma membrane into the epithelial cell. It moves through the cell to the basal surface, where it passes into the blood via GluT2, a passive glucose transporter. The Na^+–K^+-ATPase continues to pump Na^+ outward to maintain the gradient of Na^+ that drives glucose uptake.

lesser sodium concentration, the conformation of the protein is changed so that it loses its affinity for the glucose molecule, which is then released into the cell. Once inside, the glucose molecules diffuse through the cell and are moved across the basal membrane by facilitated diffusion. The secondary active transport of glucose into the epithelial cells of the intestine or of sucrose into a cell, are examples of **symport**, in which the two transported species (Na^+ and glucose or H^+ and sucrose) move in the same direction. Numerous secondary transport proteins that engage in **antiport** have been isolated, in which the two transported species move in opposite directions. For example, cells often maintain a proper cytoplasmic pH by coupling the inward, downhill movement of Na^+ with the outward movement of H^+. Proteins that mediate antiport are usually called **exchangers**.

Carbon dioxide transport

When carbon dioxide dissolves in water, it reacts with it. Carbonic acid is formed and dissociates into hydrogen-carbonate ions. These can dissociate further to form carbonate ions (**Fig. 6.30**). At blood pH very little carbonate is formed. Most carbon dioxide is transported as hydrogen carbonate. The enzyme carbonic anhydrase present in erythrocyte speeds up carbonate formation process. When the blood picks up CO_2 in the tissues, it diffuses into the red cells, where it reacts to form carbonic acid. This quickly dissociates to form hydrogen carbonate. Hydrogen carbonate is now at high concentration in the cells, so it diffuses out into the much larger volume of plasma. It is exchanged (co-transport-antiport) for Cl^-, to avoid the red cells becoming charged. The process is reversed in the lungs.

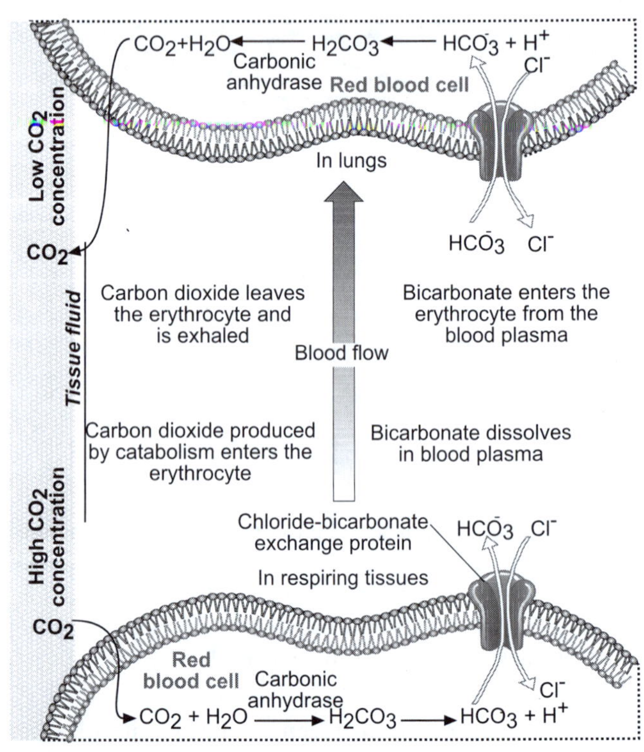

Fig. 6.30 Co-transport (antiport) of chloride and bicarbonate ions across the erythrocyte plasma membrane.

Action of bacterial lactose permease

In bacteria and yeasts, as well as in many membrane-enclosed organelles of animal cells, most active transport systems driven by ion gradients depend on H^+ rather than Na^+ gradients. The active-transport of many sugars and amino acids into bacterial cells, for example, is driven by the electrochemical H^+ gradient across the plasma membrane. One well studied H^+-driven symport is **lactose permease**, which transports lactose across the plasma membrane of *E. coli*.

ABC Transporters

Transport proteins, embedded in lipid membranes, facilitate the import of nutrients (metabolic products, lipids and sterols and drugs) into cells or the release of toxic products into the surrounding medium. The largest and arguably the most important family of membrane transporter proteins are the ABC transporters. ABC (**ATP-binding cassette**) transporters constitute the largest and most ancient family of membrane transport proteins. It is also known as the periplasmic binding protein-dependent transport system in Gram-negative bacteria and the binding lipoprotein-dependent transport system in Gram-positive bacteria. The ABC transporters form the largest group of paralogous genes in bacterial and archaeal genomes and the genes for the three components frequently form an operon.

The transporter shown is a common global organization with three types of molecular components (**Fig. 6.31**). Typically it consists of two integral membrane proteins each having six transmembrane segments, two peripheral membrane proteins that bind and hydrolyze ATP and a periplasmic (or lipoprotein) substrate-binding protein. The ATP-binding protein component is the most conserved, the membrane protein component is somewhat less conserved and the substrate-binding protein component is most divergent in terms of the sequence similarity. Importers and exporters represent the ABC transporters. ABC transporters include ATP-binding or nucleotide-binding domains (NBD1 and NBD2), transmembrane-spanning domains (MSD1 and MSD2) and solute-binding proteins (SBP1 and SBP2). In the case of exporters, the SBP domains are absent. Each MSD consists of α-helices which cross the phospholipid bilayer multiple times. The helices form between six to eleven membrane-spanning regions. The TDMs provide the specificity for the substrate and prevent unwanted molecules from using the transporter. In between the TMs is a ligand-binding domain, which is on the extracellular side of the protein for importers and on the cytoplasmic side for exporters. The ATP binding sites are located on the cytoplasmic side of the membrane. These domains are divided into parts or motifs, called Walker A and Walker B, which are separated by approximately 90–120 amino acids. In addition, there is a third short and highly conserved motif (called LSGGQ motif, C motif, or 'signature' motif) located after the Walker B motif.

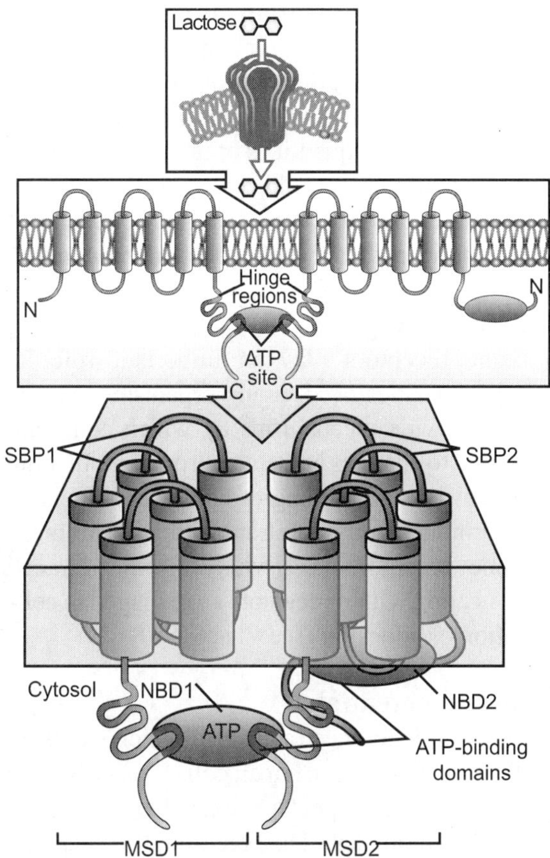

Fig. 6.31 A typical ABC transporter: ABC, ATP binding cassette. Many ABC transporters are dimers, with each polypeptide consisting of six transmembrane domains connected via a hinge region to an ATP-binding cassette. (ABC: ATP-binding cassette; MSD1 and MSD2: transmembrane-spanning domains; NBD1 and NBD2, nucleotide-binding domains; SBP1 and SBP2: substrate-binding domains)

When an ATP molecule binds to each cassette of an ABC transporter, it induces a conformational change in which the NBDs and MSDs shift to reveal a cavity called the translocation pathway. When the translocation pathway opens, molecules which were previously bound to the ligand-binding domains, are free to travel through the pathway. This action is dubbed the power stroke.

Transport by modification substrate

Molecules, on moving into a cell by passive or facilitated diffusion, often undergo chemical modification in such a way that can no longer return through the membrane. The net result is that quantities of the modified molecule steadily accumulate within the cell. Membranes are impermeable to the charged phosphorylated monosaccharides, thus these products remain in the cell.

Membrane receptors

Membrane receptors play an important role in the metabolic activities of all living organisms. Membrane receptors provide mechanisms by which cells monitor and respond to changes in their environment. In multicellular organisms the binding of chemical signals—such as the hormones and neurotransmitters of animals—to membrane receptors is a vital link in intracellular communication. Other receptors are engaged in cell–cell recognition or adhesion (see Chapter 10).

Membrane potentials

The potential difference (a charge) which exists across the cell surface membrane of all cells is usually negative inside the cell with respect to the outside; the membrane is said to be **polarized**. The potential difference across the membrane at rest is called the resting potential (–70 mV) and the resting potential is maintained by active transport and passive diffusion of ions. The gradients across the membrane are known as **electrochemical gradient**, which is used to conduct nerve impulses during the **action potential** (see Chapter 11).

6.4 BULK TRANSPORTATION

One of the important functions of the plasmalemma is to regulate the movement of various molecules into and out of the cytoplasm. The passage of macromolecules such as proteins through the cell membrane is limited because of their large size and ionic charge. Very few proteins that have hydrophobic signal sequence can pass through the membranes. But in most cases, large substances moving in and out of cells do not traverse the bilayer directly but do so indirectly via membrane-bound vesicles. Macromolecular trafficking across the cell membrane takes place by means of endocytosis and exocytosis, called **bulk movement**.

6.4.1 Endocytosis

Endocytosis is primarily a process by which the cell internalizes larger molecules, particles, organelles and cells from the surrounding medium through various mechanisms. The terminology 'endocytosis' was coined by Christian deDuve in 1963 to include both the ingestion of large particles (e.g., bacteria) and the uptake of fluids or macromolecules in small vesicles. Endocytosis can be divided broadly into two categories: **bulk-phase endocytosis** and **receptor-mediated endocytosis** (**Fig. 6.32**).

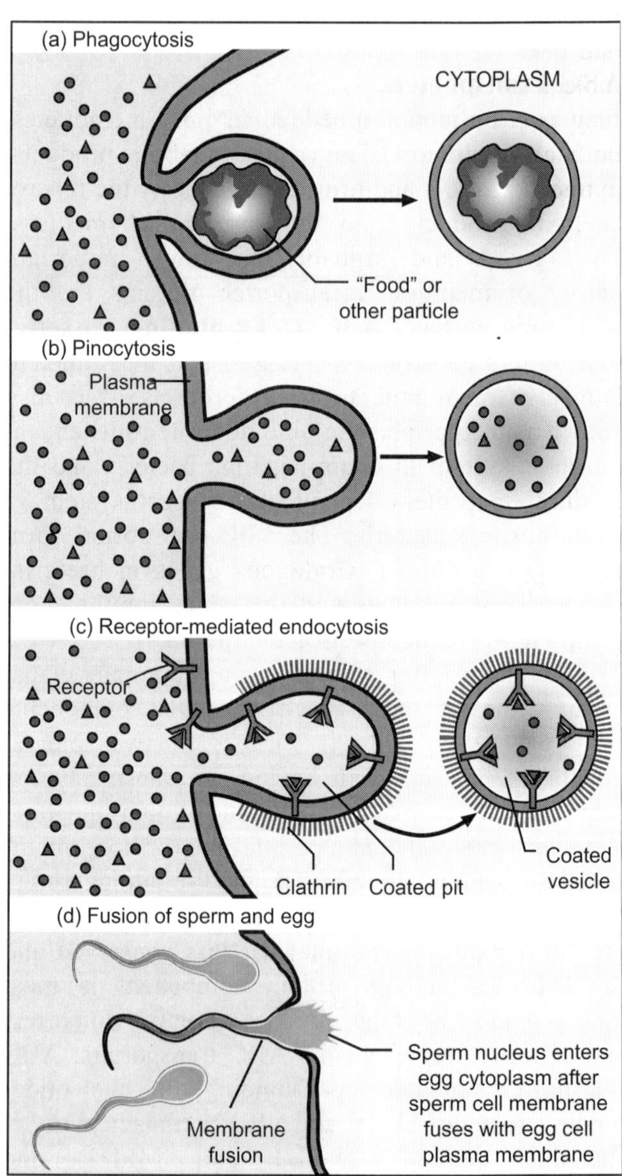

Fig. 6.32 Types of endocytosis (a) In phagocytosis, pseudopodia engulf a particle and package it in a vacuole. (b) In pinocytosis, droplets of extracellular fluid are incorporated into the cell in small vesicles. (c) In receptor-mediated endocytosis, coated pits form vesicles when specific molecules bind to receptors on the cell surface. Coated pits are reinforced on their cytoplasmic side by a fibrous protein named clathrin.

Bulk-phase endocytosis is non-specific and is of two types: **phagocytosis** (cell eating) and **pinocytosis** (cellular drinking). These processes are initiated by the adsorption of a solid particle or specific solute molecules on the surface of plasmalemma. The area of plasmalemma in contact with the solid/solute molecules begins to invaginate immediately. This invagination process continues until the invaginated part of plasma lemma is cut-off as a small vesicle to form endocytic vesicles containing material that had been outside the cell. Most of the endocytosed molecules and particles end up in lysosomes, where they are degraded and recycled. Endocytosis occurs both constitutively and as a triggered response to extracellular signals. Most of the plasma membrane components that are endocytosed are continually returned to the cell surface by exocytosis.

Phagocytosis

Phagocytosis (**Fig. 6.33**) is typically restricted to specialized mammalian cells for the uptake of relatively large particles (>0.5 μm diameter) from the surroundings. When extracellular particles greater than 0.2 μm in diameter are taken up by cell, the endocytic vesicle forms tightly around the particle, excluding most of the cextracellular fluid and this process is called phagocytosis. It may involve entire bacteria or even larger structures, as in the familiar single-celled protists, such as amoebas and ciliates, which make their livelihood by trapping

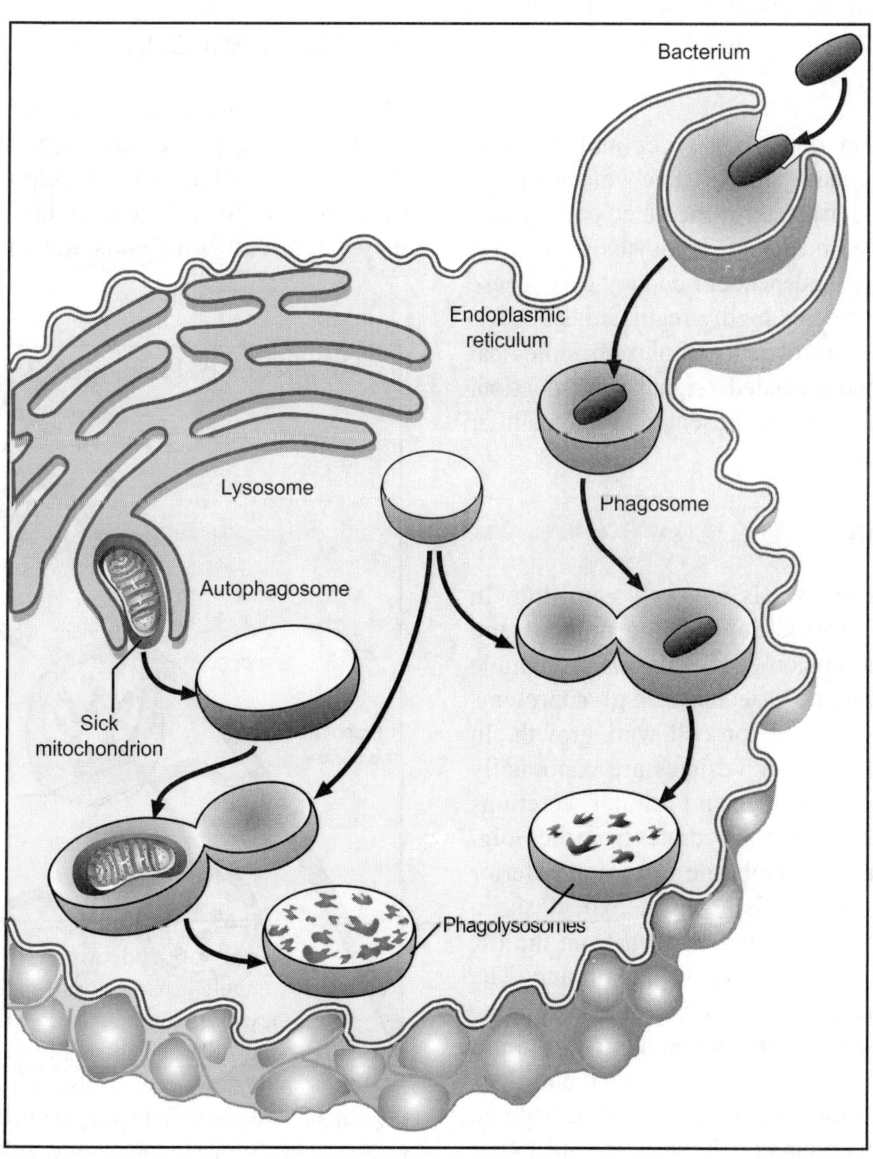

Fig. 6.33 Phagocytosis: Lysosomes in phagocytosis and autophagy.

food particles and small organisms and enclosing them within folds of the plasma membrane. The folds fuse to produce a vacuole (or *phagosome*) that pinches off inwardly from the plasma membrane. The phagosome then fuses with a lysosome and the material is digested within the resulting *phagolysosome*. However, in most animals, phagocytosis is a protective mechanism rather than a mode of feeding. Mammals possess a variety of 'professional' phagocytes, including macrophages and neutrophils.

Phagocytes in the liver line the blood sinusoids, clearing them of unwanted materials. In the lung, they line the airways and alveoli, helping to maintain sterility—often compromised by foreign material in the inhaled air. Polymorphonuclear leukocytes travel the bloodstream and are the major defence mechanism against infecting bacteria. Most of these phagocytic cells in animals are selective for the materials they digest.

Pinocytosis

Pinocytosis occurs in many cell types and in many organisms by at least four basic mechanisms: (i) macropinocytosis, (ii) clathrin-mediated endocytosis (CME), (iii) caveolae-mediated endocytosis, and (iv) clathrin- and caveolae-independent endocytosis. These mechanistically diverse and highly regulated endocytic pathways function to control such complex physiological process as hormone-mediated signal transduction, immune surveillance, antigen-presentation and cellular and organismal homeostasis.

6.4.2 Exocytosis

It is also called **emeiocytosis** and **cell vomiting**. In growing cells, this process delivers the proteins and lipids needed for expansion of the plasma membrane as well as the complex polysaccharides, glycoproteins and proteoglycans required for cell wall growth. In all eukaryotic cells secretory vesicles are continually carrying new plasma membrane and cellular secretions such as proteins, lipids and carbohydrates from the Golgi apparatus to the plasma membrane or to cell exterior by the process of exocytosis. During exocytosis, a vesicle carrying proteins usually budded from the ER or the Golgi migrates to the plasma membrane. The membrane of the vesicle and the plasma membrane come into close apposition, whereupon the lipid molecules of the two bilayers rearrange themselves (**Fig. 6.34**). The two membranes then fuse to become continuous and the contents of the vesicle spill to the outside of the cell (called secretion). Proteins can be secreted from cells by exocytosis in either a constitutive or a regulated fashion. In the regulated pathway, molecules are stored either in secretory vesicles or synaptic vesicles and release their contents by fusion with plasma membrane only when appropriate signal is received. In constitutive pathway, continual vesicular transport from the trans-Golgi network to the plasma membrane takes place. It operates in all eukaryotic cells. Many secretory cells use exocytosis to export their products (e.g., α-, ß- and acinus cells of pancreas).

This large-scale endocytic–exocytic bulk transportation is mediated largely by clathrin-coated vesicles, a process called receptor mediated endocytosis. Materials pass in or out by forming vesicles, after which the membrane spontaneously reassembles.

6.5 CELL RECOGNITION

The cell membrane separates the interior of the cell from its surrounding environment. As a boundary, it has an important role in determining with which other cells a given cell interacts. These interactions may be direct, with cells touching each other via

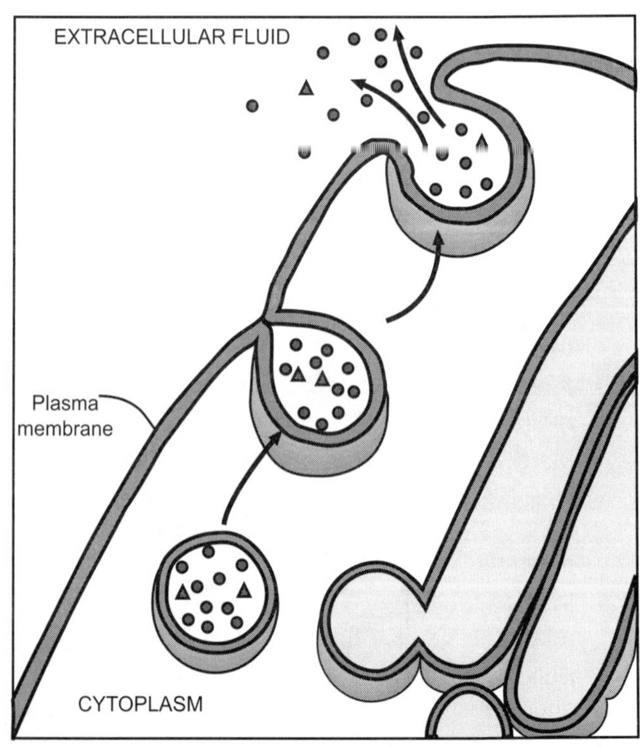

Fig. 6.34 Exocytosis. During exocytosis, vesicles fuse with the plasma membrane and dump their contents outside the cell.

glycoproteins and glycolipids of the cell surface or the interactions between cells may be indirect, i.e., when one cell sends out chemicals messengers that affect a distant cell. Cell–cell interactions will be discussed in detail in Chapter 13.

Summary

- All living cells are separated from the external world by a thin, fragile protective layer called the ***plasma membrane*** that defines its boundaries.
- Eukaryotic cells have developed an elaborate internal membrane system that includes ***endoplasmic reticulum***, ***Golgi apparatus***, ***mitochondria*** and other ***organellar membranes***.
- Cell membranes are not mechanically strong, in contrast to the rigid and tough cellulose walls of plants and bacteria.
- Singer and Nicolson proposed the concept of membrane structure based on the ***fluid-mosaic*** model. It describes the molecular organization of lipids and proteins in cellular membranes, and illustrates how a membrane's mechanical and physiological traits are defined by the physicochemical characteristics of its various molecular components.
- Membranes contain a wide variety of ***lipids***, all of which are ***amphipathic***; i.e., they contain both hydrophobic and hydrophilic regions. This dimorphic behaviour of lipids has a number of molecular consequences, including the spontaneous formation of surface monolayers, bilayers, micelles and vesicles, by lipids in contact with water.
- Most membrane lipids contain a phosphate group and are called ***phospholipids***.
- All the molecules in membranes are in constant motion and interactions are constantly being broken and re-formed.
- Another lipid component of certain membranes is ***cholesterol***. Cholesterol is absent from the plasma membrane of most plant and all bacterial cells.
- The plasma membranes of eukaryotic cells also contain ***carbohydrates*** of various types, which are covalently linked in both lipid and protein components.
- Several large families of membrane proteins contain an interior channel that provides an aqueous passageway through the lipid bilayer.
- Membranes are involved in a large variety of functions in living organisms. The important physiological functions are: ***compartmentalization***, ***scaffolding***, ***selective permeability***, ***transportation***, ***signal transduction*** and ***energy transduction***.
- Biological transport mechanisms are classified according to whether they require energy or not. ***Diffusion*** is a spontaneous process, two requirements must be met for it to occur: the substance must be present at higher concentration on one side of the membrane and the membrane must be permeable to the substance.
- ***Facilitated diffusion*** is a more efficient type of passive transport, in which ions or molecules cross the membrane rapidly.
- Many protein channels are chemically- or voltage-regulated.
- Unlike facilitated diffusion, movement of a solute against gradient requires the coupled input of energy, such as the hydrolysis of ATP, the absorbency of light, the transport of electrons, or the flow of other substances down their gradients.
- Proteins that carry out active transport are often referred to as '***pumps***'. Some common protein pumps are related to Na^+–K^+, Ca^{2+} and H^+.
- Bulk transportation across the membrane takes place by ***endocytosis*** and ***exocytosis***. Bulk-phase endocytosis is non-specific and of two types: ***phagocytosis*** (cell eating) and ***pinocytosis*** (cellular drinking).
- In all eukaryotic cells secretory vesicles continually carry new plasma membrane, proteins, lipids and carbohydrates from the Golgi apparatus to the plasma membrane or to cell exterior by the process of ***exocytosis***.

Exercises

1. List the different types and location of membranes in generalized prokaryotic and eukaryotic cells.
2. Explain with a diagram the fluid-mosaic model of plasma membrane.
3. Discuss the chemical composition of plasma membrane.
4. What are the biological features of plasma membranes?
5. List the different types of membrane functions.
7. Where and how does membrane synthesis take place in a cell?
7. Give an account of solute transportation across cell membranes
8. Differentiate between passive and active membrane transportation.
9. Explain briefly: uniport, symport and antiport systems.
10. Describe membrane receptors and explain their role.
11. Explain the process of endocytosis and exocytosis.

Further Reading

Books

Alberts B, Bray D, Johnson A, Lewis J, Raff M, Roberts K and Walter P (1998), *Essential Cell Biology: An Introduction to the Molecular Biology of the Cell,* Garland, New York.

Becker WM, Kleinsmith LJ and Hardin J (2000), *The World of the Cell*, 4th edition, Benzamin Cummings, New York.

Fahn A (1974), *Plant Anatomy*. Pergamon Press, Oxford, New York.

Hall ZW (1992), *An Introduction to Molecular Neurobiology*. Sinauer.

Hille B (2001), *Ionic Channels of Excitable Membranes*, 3rd edition, Sinauer.

Karp G (2005), *Cell and Molecular Biology*, 4th edition, John Wiley and Sons.

Lodish H, Baltimore D, Berk A, Zipursky SL, Matsudaira P and Darnell J (1995), *Molecular Cell Biology*, Scientific American Books, USA.

Mathews CK, Van Holde KE and Ahern KG (2005), *Biochemistry*, 3rd edition, Pearson Education, Inc.

Taylor DJ, Green NPO, Stout GW (1997), Soper R, (eds), *Biological Science*, Cambridge University Press.

References

Brown DA (2001), seeing is believing: visualization of rafts in normal membranes, *Proc. Nat'l. Acad. Sci.*, USA., 98:10517--18.

Caffrey M, (ed.), (2002), Membrane form and function in finer focus, *Curr. Opin. Struct. Biol.,* 12:471–530.

Channarayappa (1993), Resistance mechanism of *Lycopersicon* species to the vector, *Bemisia tabaci* and ultrastructural changes in leaf curl virus infected tomato leaves, PhD thesis submitted to the University of Agricultural Sciences, GKVK, Bangalore.

Channarayappa (1990), Cytogenetic analysis of clastogens and aneuploidogens by immunofluorescent and micronucleus methods, PhD thesis submitted to the West Virginia University, Morgantown, West Virginia, USA.

Kaplan JH (2002), Biochemistry of Na^+-K^+ATPase. *Annu. Rev. Biochem.*, 71:511–5.

Lai EC (2003), Lipid rafts make for slippery platforms, *J. Cell Biol.,* 162:365–370.

Maxfield FR (2002), Plasma membrane microdomains, *Curr. Opin. Cell Biol.*, 14:483–7.

Miller C (2003), A charged view of voltage-gated ion channels, *Nature Struct. Biol.,* 10:422–24.

Miller G (2003), The puzzling portrait of a pore, *Science*, 300:2020–22.

PopotJ-L and Engelman DM (2000), Helical membrane protein folding, stability and evolution, *Annu. Rev. Biochem.*, 69:881–922.

Sansom MSP and Law RJ (2001), Aquaporins—Channels without ions, *Curr. Biol.*, 11:R71–R73.

Stevens, J. and Samsó M (2003), Membrane proteins: The 'Wild West' of structural biology, *Trends Biochem. Sci.,* 28:137–144.

Von Heijne G (1994), Membrane proteins: from sequence to structure, *Annu. Rev. Biophys. Biomolec. Structure*, 23:167–92.

Wine JJ (2001), Cystic fibrosis: the "bicarbonate before chloride" hypothesis. *Curr. Biol.,* 11:R463–R467.

Yellen G (2002), The voltage-gated potassium channels and their relatives, *Nature,* 419:35–42.

7

Mitochondria: Powerhouse of the Cell

Learning objectives

After reading this chapter, you will be able to:

- Appreciate the mitochondrial structure in different organisms
- Understand the evolution and genetic basis of the structure of mitochondria
- List the various functions of mitochondria and the associated physiological processes

7.1 INTRODUCTION

Besides the nucleus, the mitochondria and chloroplasts are unique among cell organelles because they have genetic material of their own. Although a large portion of the DNA in most eukaryotes belongs to the nucleus, a significant amount of it is also present within the cell organelles, particularly in the mitochondria of animals, plants and fungi and the chloroplasts of plants. Both these organelles contain DNA of their own, which encode synthesis of several proteins within the organelle. Many proteins contain multiple subunits, some encoded by nuclear DNA and others by organellar genomes. The presence of DNA in mitochondria and chloroplasts reflects the ancient origin of these organelles as **endosymbionts**. Many lines of evidence indicate that mitochondria and chloroplasts evolved from bacteria that were endocytozed into ancestral cells containing a eukaryotic nucleus, forming endosymbionts.

Energy is essential for all living organisms for various functions and purposes. Biological systems harvest their energy either from solar radiation or by breaking of chemical bonds. Therefore, highly efficient methods of energy generation in biological systems evolved early in the history of life. This process was mainly based on membranes which enabled cells to acquire energy from a wide variety of sources. For example, the conversion of light energy into chemical bond energy in photosynthesis and aerobic respiration enables the use of oxygen to produce large amounts of ATP from food molecules. The most important molecule for capturing and transferring free energy in biological systems is **adenosine triphosphate (ATP)**.

This endergonic reaction, i.e., generation of the high-energy phosphoanhydride bond of ATP from ADP and inorganic phosphate (HPO_4^{2-}), requires an input of 7.3 kcal/mol. The energy required for this reaction is produced primarily by two main processes:

- ***Aerobic oxidation***, which occurs in nearly all cells
- ***Photosynthesis***, which occurs in leaf cells of plants and certain single-celled organisms.

In aerobic oxidation, fatty acids and sugars, principally glucose, are metabolized to carbon dioxide (CO_2) and water (H_2O) and the released energy is converted to the chemical energy of phosphoanhydride bonds in ATP. Prokaryotes use plasma membrane system for the production of ATP, but the eukaryotic cells employ specialized membranes inside the energy-converting organelles for the production of ATP. These membrane-enclosed organelles are **mitochondria**, which are present in all eukaryotic cells and **plastids**, which occur only in plants. The internal membrane provides the framework for an elaborate set of electron transport processes that produce most of the cell's ATP. The common pathway used by prokaryotes and cell organelles to harness energy for biological purposes operates by the process of **chemi-osmotic coupling**, reflecting a link between the chemical bond-forming reactions that generate ATP (**chemi-**) and membrane transport processes (**osmotic**). The electrochemical proton gradient is also used to drive other membrane-embedded protein machines.

7.2 CYTOPLASMIC INHERITANCE

Although there is definite evidence that most genes have specific loci on chromosomes and the behaviour of chromosomes explains Mendel's laws, there are exceptions to this theory. Not all the genes of a cell are a part of the chromosomes. For example, prokaryotic cells have plasmids, which carry one or a few genes. Genes in these extra-chromosomal DNA are inherited in distinctive ways. Plasmids replicate independently within the cell. When the bacterial cell divides, some plasmids may go to each daughter cell (or some daughter cells may not receive any plasmid at all). Eukaryotes have DNA in their mitochondria and plastids. Studies of mutants in yeasts and other single-celled organisms first indicated that mitochondria exhibit **cytoplasmic (extranuclear** or **maternal) inheritance**.

7.3 CYTOPLASMIC INHERITANCE IN MITOCHONDRIA AND CHLOROPLASTS

Mitochondrial and plastid DNA replicate and pass on their genes to daughter organelles, when the organelles divide. But, the organism as a whole inherits organelle genes in a peculiar way. Cytoplasmic genes were first observed in plants. In 1909, Karl Correns studied the inheritance of yellow or white patches on the leaves of an otherwise green ornamental plant. He found that colouration of the offspring was determined only by plants with flowers bearing the seeds and not by plants with flowers producing the pollen. Subsequent studies revealed that such variegated colouration of leaves is due to differences in pigment production by the plastids and is controlled by the genes located in the plastids.

Maternal inheritance is also the rule for mitochondrial genes in mammals. In mammals and most other multicellular organisms, the sperm contributes little (if any) cytoplasm to the zygote and virtually all the mitochondria in the embryo are derived from the egg and not the sperm. Only the mother's mitochondrial genes are passed on. Studies in mice have shown that 98.99% of mtDNA is maternally inherited, but a small part (0.01%), is inherited from the male parent. In higher plants, mtDNA is inherited exclusively in a uniparental fashion through the female parent (egg) and not the male (pollen). In each case, the simple rules of Mendelian genetics do not apply.

7.3.1 Maternal inheritance of mitochondria

Mitochondria are unique as they have their own DNA (mtDNA), which codes for some of the proteins in the respiratory chain while rest of the respiratory chain proteins and all those needed to repair mtDNA are coded by the nuclear genes (**Fig. 7.1**). This means that alterations in the mitochondria can be caused by either mitochondrial or nuclear DNA mutations.

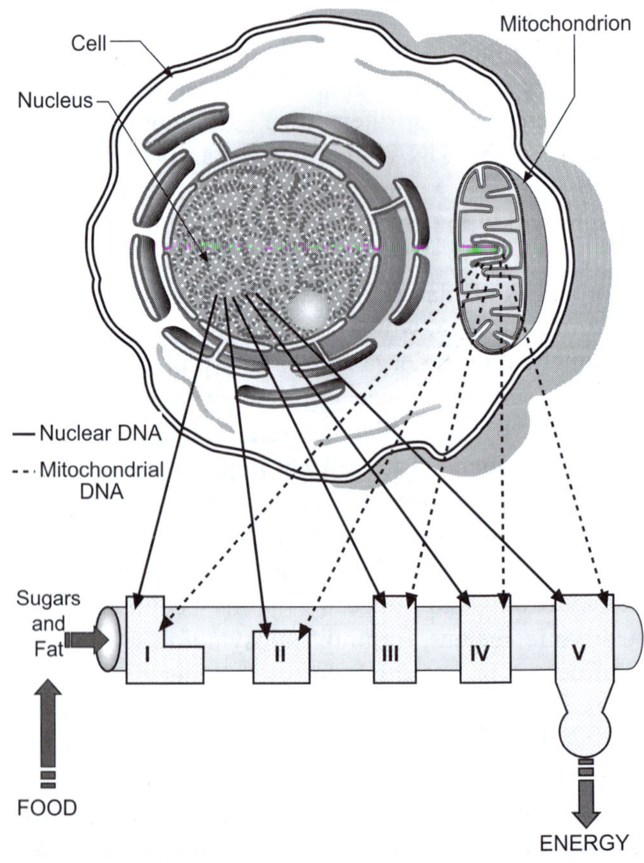

Fig. 7.1 Mitochondria contain their own DNA (mtDNA).

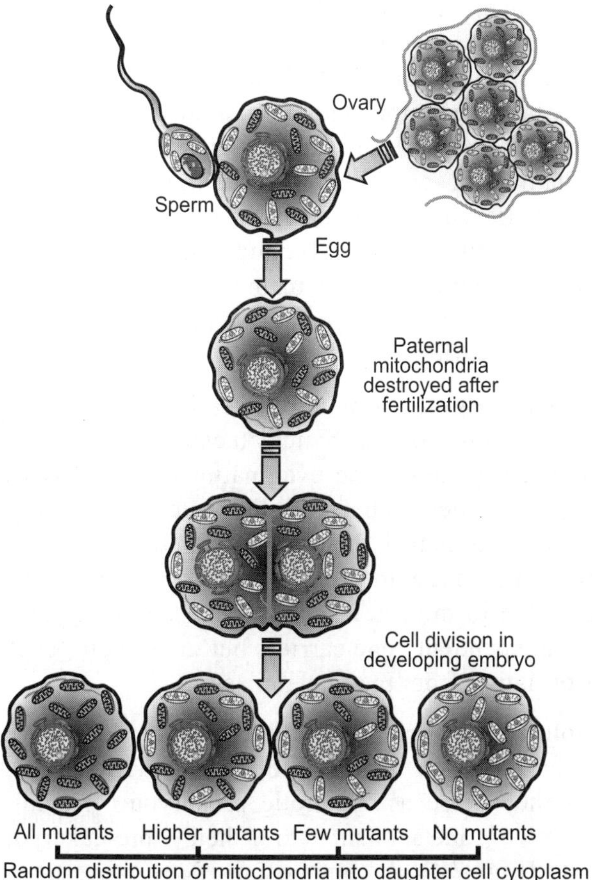

Fig. 7.2 Maternal inheritance of mitochondria.

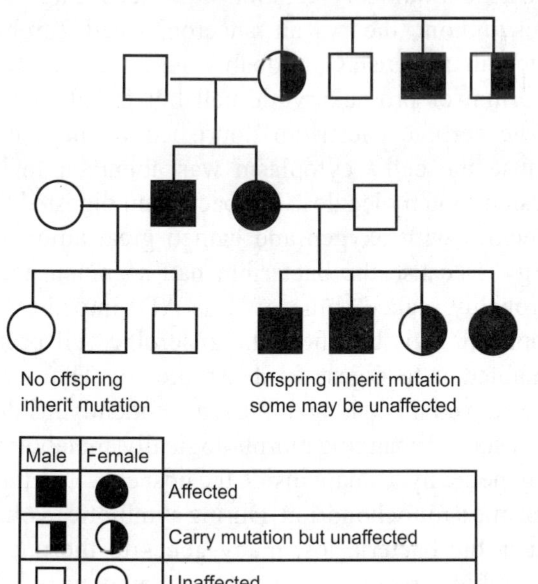

Fig. 7.3 Maternal inheritance of mutations. Mutation in mother's mitochondria is inherited only through the mother's egg. Sperm do not contribute mitochondria to the zygote, therefore, disease will not transmit through the male. The appearance of disease depends upon the number of affected mitochondria present in the egg cell.

In many higher organisms, including humans, the mitochondria are transmitted to the offspring only through the egg and not the sperm. Eggs are formed from precursor cells in the mother's ovaries. In a mother carrying a heteroplasmic mtDNA mutation, she would have normal and abnormal mitochondria, including in the precursor cells (**Fig. 7.2**). The proportion of mutant to normal mitochondria varies slightly between cells. For instance, if she has a 50:50 mix, only a small fraction of the mitochondria present in the precursor cell survives in the egg. If only ten survive and by chance the cell ends up with eight abnormal mitochondria and if that egg is fertilized, this high proportion of abnormal mitochondria will be passed on to the child. In this case, the child will be severely affected. On the other hand, if the egg cell contains only a few defective mitochondria, the child will be mildly affected.

Mitochondrial defects transmit to offspring maternally only. Although, both the egg and sperm contain mitochondria, following fertilization, the mitochondria from the sperm are destroyed and only the mother's mitochondria are passed on to the next generation. So even if a man has a mutation in his mitochondrial DNA, it will not be passed on to the zygote during fertilization. In contrast, a woman with a mutation can pass it on to all her children (**Fig. 7.3**). Moreover, although there is only one nucleus per cell containing one set of nuclear DNA, cells contain hundreds of mitochondria and each mitochondrion contains several copies of the mitochondrial DNA. Therefore, there are several factors to consider in the matter of inheritance.

How much the child is affected depends on the number of abnormal mitochondria in each tissue. As the fertilized egg divides, the mitochondria in the cells also divide and so the numbers of both normal and abnormal mtDNA molecules increase. As the embryo continues to develop, different cells are formed and go on to make the different tissues in the body. For example, if the cells that form the eye muscles have a particularly high proportion of abnormal mtDNA then those cells may not be able to make enough ATP, leading to symptoms. Either some of the patient's mtDNA is affected (**heteroplasmy**, containing more than two types), or in some cases all of it (**homoplasmy**, all have the same type); the severity depends primarily on how many abnormal mitochondria the egg contains.

7.4 ENDOSYMBIOTIC THEORY OF ORGANELLES

An endosymbiont is an organism that lives within the body or cells of another organism, i.e., forming an endosymbiosis (Greek, *endo* = inner and *biosis* = living). Many instances of endosymbiosis are obligate, that is, neither the endosymbiont nor the host can survive without the other. The endosymbiont theory attempts to explain the origins of organelles such as mitochondria and chloroplasts in eukaryotic cells engulfed through endophagocytosis. These cells and the bacteria trapped inside them enter a symbiotic relationship, i.e., a close association between different types of organisms over an extended time. There are several other examples for endosymbionts in different organisms.

7.4.1 Origin and evolution of mitochondria

There are at least four different views regarding the origin of mitochondrial multiplication.

Autonomous replication of chondriome: Chondriome is the term applied to all mitochondria of a cell, or to all hereditary determinants in a cell's mitochondria. This is because mitochondria are believed to be autonomous organelles with independent hereditary material and that they live in the cell in symbiosis. However, it is still not clear whether mitochondria are completely autonomous or only semi-autonomous.

Origin from promitocondrion: According to this theory, mitochondria arise from smaller particles. These particles are called promitochondria and arise from still smaller particles known as initials. These initials can be distinguished from other particles by the presence of a double membrane. These are generally found in meristematic cells. With the growth of the cell, these initials increase in volume and their membranes start forming folds at right angles to the surface. The initials follow the stage of promitochondria which grow vigorously in volume and the fold becomes differentiated into tubuli or cristae. It is also suggested that these initials really arise from the nuclear envelope. The outer membrane grows and forms the initials for plastids as well as mitochondria.

Division or budding: Like cell division in bacteria, mitochondria also undergo division. This division may take place by diaphragm-like foldings of the inner membrane; the outer membrane does not take part in this partition. Such a mitochondrion with septum will separate into two or more units, which then become mitochondria. It is also possible that a mitochondrion may divide into two due to the formation of constrictions. The mitochondria are also believed to originate by budding from pre-existing mitochondria. In budding, an initial will originate with a double membrane. This will then give rise to a pro-mitochondrion and subsequently to a mitochondrion.

Nuclear origin: As mentioned earlier, the formation of vesicles is brought about by means of evaginations from the nucleus. The evaginations would involve both the membranes of the nuclear envelope. The vesicles detach themselves from the nucleus and grow to give rise to proplastids or promitochondria. In such a form of development, these organelles are not independent gene carriers but draw their genetic information from the nucleoplasm.

Evolutionary origin of mitochondria: Endosymbiotic theory for the origin of mitochondria suggests that they arose as symbiotic prokaryotes providing oxidative metabolism to their pre-eukaryotic hosts (**Fig. 7.4**). A comparison of the properties of mitochondria and bacteria shows similarities, which indicate evolutionary relationship. According to the endosymbiont theory, an anaerobic cell (probably related to rickettsias), probably ingested an aerobic bacterium or pro-eukaryotic cell but failed to digest it. The aerobic bacterium flourished within the cell because the cell's cytoplasm was abundant in half-digested food molecules. The bacterium digested these molecules with oxygen and gained great amounts of energy. Because the bacterium had so much energy, it probably leaked some of it as ATP into the cell's cytoplasm. This benefited the anaerobic cell because it enabled it to digest food aerobically. The aerobic bacterium could no longer live independently inside the cell; it started changing morphologically, metabolically and genetically to adapt inside the host cell. Eventually, it became a mitochondrion. During symbiotic evolution the aerobic bacteria lost many genes to the nucleus; the symbiont retained only minimal amount of DNA, for its own propagation and functioning.

7.4.2 Endosymbiotic origin of plastids

Plastid genomes are similar to that of prokaryotes and this has lent great support to the endosymbiotic

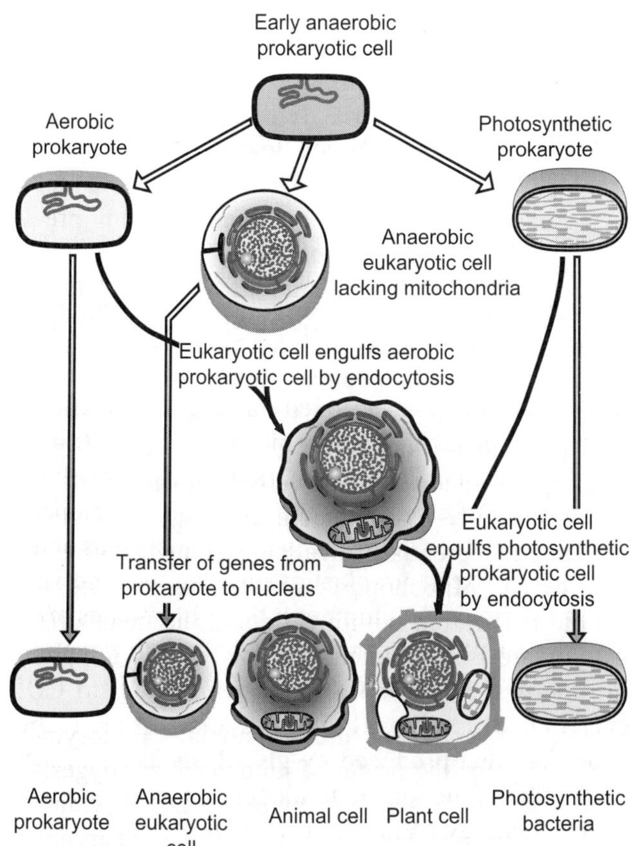

Fig. 7.4 Probable evolutionary pathway for the origin of mitochondria and chloroplast.

hypothesis of the origin of plastids in plant cells. This hypothesis was originally proposed by Schimper in the 1870s and was developed more fully by Mereshkovsky in 1905. These scientists proposed that the green organelles of photosynthetic plant cells arose from a symbiotic association between a photosynthetic bacterium and a non-photosynthetic eukaryotic organism. It is likely that the plastids in present-day plants have dual ancestry. The plastids of green algae and higher plants are probably derived from non-cyanobacterial, oxygenic bacteria exemplified by the present-day organism *Prochoron*. Red algae and cryptomonad plastids are probably derived from cyanobacteria-like organisms. It is suggested that an eukaryotic host cell, which would have contained a normal eukaryotic nucleus plus mitochondria, would have captured the endosymbiotic organisms by endocytosis. This would explain the double envelope membrane, which surrounds the plastids of green algae and higher plants. A proposed model for the endosymbiotic origin of plastids is shown in Fig. 7.4.

7.4.3 Evidence for the endosymbiont theory

For a long time, it was generally agreed that eukaryotic cells arose from prokaryotic cells. This can be verified by the following facts:

- Fossil records show that prokaryotic cells were present in rocks approximately 2.7 billion years old, i.e., one billion years before any evidence of eukaryotes arose.
- Both share many complex traits such as similar genetic codes, enzymes, metabolic pathways, plasma membranes, genome structure, gene expression mechanisms, antibiotic sensitivity and similarity between their gene and polypeptide sequences.

However, when compared to prokaryotes, in eukaryotic cells, other than the nucleus, cell organelles such as chloroplast and mitochondria, also contain genetic material of their own. It is thought that the chloroplast began as an ancestral cyanobacterium (a photosynthetic prokaryote), that became endocytoed by an ancestral eukaryotic cell and replicated within the cytoplasm. During evolution, some of the bacterial encoding genes moved into the nucleus and thus some of the chloroplast proteins got imported from nuclear coded genes into the organelle. Mitochondria probably originated by a similar process from a prokaryotic cell capable of oxidative phosphorylation; the plasma membrane of the prokaryote, the site of oxidative phosphorylation, became the inner mitochondrial membrane. The outer membranes of the mitochondrion and chloroplast may be evolutionary remnants of the membrane of the endocytic vesicle into which these prokaryotes were internalized. Over the evolutionary timescale, both chloroplasts and mitochondria have lost genes to the nucleus and have thus become dependent on their hosts. The existence of **promiscuous DNA** supports this. These DNA sequences may represent recent transposition events and although the transposition mechanism is unknown, it is clear that rare gene transfer from the cytoplasm to the nucleus does occur.

The evidence for mitochondria and chloroplasts being of symbiotic origin are:

- Both organelles arise only from pre-existing mitochondria and chloroplasts. They cannot be formed in cells lacking them.

- Both have their own genomes that resemble that of bacteria and not the nuclear genome.
- Both genomes consist of a single circular molecule of DNA.
- No histones are associated with the organellar DNA.
- Both have their own protein synthesizing machinery, which resembles that of bacteria. The first amino acid of the polypeptide is **fMet,** which is similar to bacterial protein synthesis in organelles.
- Antibiotics (such as streptomycin), that act by blocking protein synthesis in bacteria also block protein synthesis within mitochondria and chloroplasts.
- Conversely, inhibitors (example, diphtheria toxin), of protein synthesis by eukaryotic ribosomes do not have an effect on bacterial protein synthesis or on protein synthesis of organelles.
- The antibiotic, rifampicin, which inhibits the RNA polymerase of bacteria, also inhibits the RNA polymerase within mitochondria, but does not affect the RNA polymerase of eukaryotic nuclei.

The transfer of genes can be proved by finding one gene (subunit 9 of ATPase) in the mtDNA of some organisms but not in the nucleus in other organisms. Since the rate of production of base pair substitutions in organelles (e.g., mtDNA) is about 10 times that in nuclear DNA, it provides a highly sensitive tool for studying short-term divergences among related species.

7.4.4 Organelle mutants

Prior to the isolation and sequencing of mitochondrial DNA, studies of mutants in yeasts and other single-celled eukaryotes indicated that organelles exhibit cytoplasmic inheritance and thus must contain their own genetic system. Mutations affecting organelle genomes cause deficiencies in either photosynthesis or oxidative phosphorylation. Chloroplast mutants show clear morphological phenotypes (e.g., leaf variegation), which have been exploited in studies of maternal inheritance. Mitochondrial mutations have been extensively studied in microbial eukaryotes, where they generate a slow growth phenotype (termed 'petite' in yeast and 'poky' in *Neurospora crassa*).

7.5 MITOCHONDRIAL BIOLOGY AND GENETICS

Mitochondria (Gr. *mito*, thread + *chondrion*, granule), are granular or filamentous organelles present in the cytoplasm of all eukaryotic cells (protozoa, animal and plant cells) and contribute significantly in the evolution of complex animals. In the absence of mitochondria, present-day animal cells would be dependent on anaerobic glycolysis for all their ATP. When glucose is converted into pyruvate by glycolysis, only a very small fraction of the total free energy potentially available from the glucose is released. Mitochondria are the major site for aerobic respiration within cells and the metabolism of pyruvate is completed: the pyruvate is imported into the mitochondrion and oxidized by O_2 to CO_2 and H_2O. This allows 15 times more ATP to be made than that produced by glycolysis alone. Thus, the primary function of mitochondria is to provide chemical energy (ATP) for use by the cell.

In 1880, Kalliker for the first time recognized the structures now known as mitochondria, as granules in striated muscle. However, the credit for the discovery of mitochondria goes to W Flemming and R Altman. The term mitochondria, was coined by Benda (1897), who also developed the crystal violet staining technique. Michaelis studied mitochondria using vital stain Janus Green B. Meves (1934) discovered the presence of mitochondria in plant cell cytoplasm. Mitochondria were first isolated from liver by Bensley and Hoerr in 1934. Hogeboon et al (1948), demonstrated that mitochondria were indeed the site of cellular respiration. The mitochondria of a cell are collectively called **chondriome**.

7.5.1 Morphology

The cytological examination of mitochondria in living cells is difficult due to their low refractive index. They can be more easily examined in cultured cells rather than multicellular systems. They can be studied under the microscope by staining with Janus green, which stains the organelle with a greenish blue stain due to oxidation of the dye by cytochrome oxidase (maintains the vital dye in its oxidized form). The number of mitochondria per cell varies considerably and depends on the type of organism and nature of the cell (10–10,000 per cell).

Mitochondrial shape and size are also widely variable and are usually larger in active cells than in less active ones. They are usually depicted as stiff elongated cylinders with diameter 0.5–1 μm and length 1.5–10 μm, resembling *E. coli*. The morphology of mitochondria varies not only from one cell to another, but also within the same cell, depending upon certain functions. Mitochondria are remarkably mobile (mainly by cytoplasmic streaming) and plastic organelles, constantly changing their shape and even fusing with one another and then separating again. As they move about in the cytoplasm, they often seem to be associated with microtubules, which can determine the unique orientation and distribution of mitochondria in different types of cells. These organelles may be 1000–2000 per cell and occupy almost 20%–25% of the cell volume of an average mammalian liver cell. In some cells (mostly plant cells), mitochondria can move freely, carrying ATP where needed. The type of movement is both cytoplasmic streaming and bending and stretching cycles.

Mitochondria are the most sensitive indicators of the injury to the cell. Injury to mitochondria may cause irreversible changes if the alteration reaches a critical point. Consequently, there may be degenerative changes like fragmentation, intense swelling, or accumulation of material.

7.5.2 Mitochondrial membranes

Electron micrographs show that the mitochondrion is bounded by two membrane envelopes. They have a smooth outer membrane (6 nm thick) and a convoluted inner membrane (6 nm thick). The outer and inner membranes are separated by a space of 6–8 nm (**Fig.7.5**). The outer membrane is composed of approximately 50% lipid by weight and contains a mixture of enzymes involved in diverse activities. It also contains many copies of a transport protein called **porin**, which forms large aqueous channels through the lipid bilayer. The outer membrane thus resembles a sieve that is permeable by all molecules smaller than 5 kDa. Such molecules can enter the intermebrane space, but most of them cannot pass the impermeable inner membrane directly. When the porin channels are wide open, the outer membrane is freely permeable to molecules such as ATP, NAD and coenzyme A, which have key roles in energy metabolism within the mitochondrion.

Transport of proteins across the mitochondrial membranes

The inner membrane contains cardiolipin which renders it virtually impermeable. This requires transport mechanisms across the membranes that are more organized and regulated. Transport across the mitochondrial membranes requires concerted action by a number of translocation machineries. The machinery in the outer membrane is called the **Tom complex** (translocator outer membrane) and one in the inner membrane is called the **Tim complex** (translocator inner membrane). Proteins reaching the matrix have a $-NH_2$ cleavable signal sequence. Proteins in the cytoplasm must be uncoiled or stretched to go through the translocators. The translocation

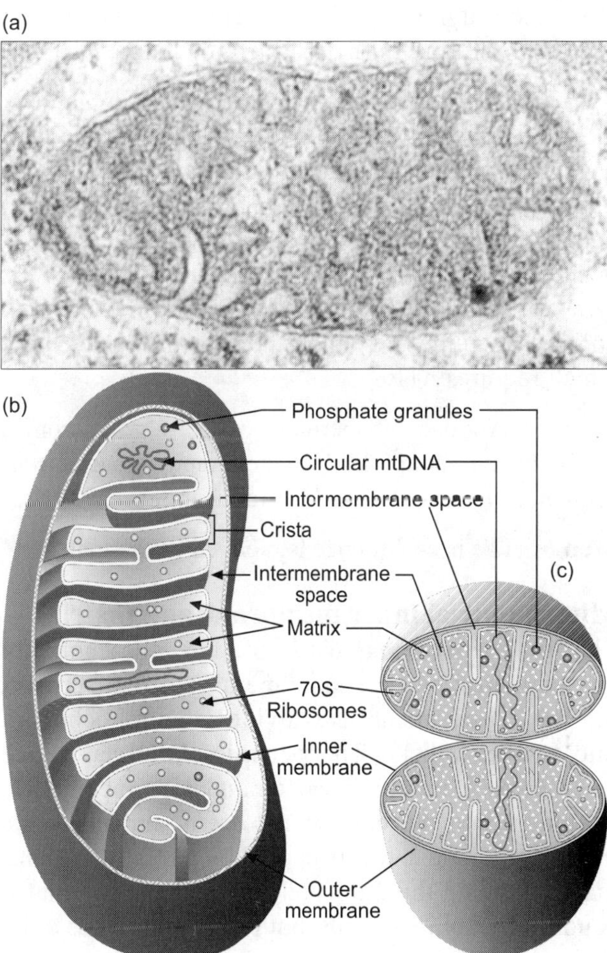

Fig. 7.5 Ultrastructure of mitochondria: (a) Electron micrograph of a cross section of a tomato leaf mitochondria (Channarayappa, 1992). Diagrammatic view of (b) longitudinal and (c) cross section of a mitochondrium. Inside of a mitochondrium shows extensive folding of the inner membrane called cristae.

of proteins from cytoplasm to matrix involves ATP binding and is monitored, stabilized and facilitated by various chaperone proteins. Thus, before the protein can go through the Tom complex, it must become 'translocation-competent'.

The entire sequence of events required to take a protein into the matrix is broadly divided into the following steps:

Step 1: The protein unfolds as it binds to hsp70 chaperone. The positively charged amino end of the protein indicates the targeting sequence (*signal sequence*). Chaperone binding is ATP-dependent.

Step 2: The signal sequence binds to receptor (usually Tom20).

Step 3: The receptor ushers the protein to the site of the translocator. Other Tom proteins are involved, but Tom40 is the core of the translocator channel.

Step 4: The protein is translocated and stimulated by the membrane potential. Electron transport complexes on the inner membrane pump H^+ across the intermembrane space, leaving the matrix more negative. This attracts the protein (the signal is positively charged). The protein moves through the Tim translocators. Tim 44 and hsp70 in the matrix continue to guide and pull the protein through the pore, which requires ATP.

Step 5: Another chaperone (called a chaperonin), hsp60, brings about folding of the protein into its tertiary sequence. This is also an ATP-requiring process.

Step 6: The pre-sequence is cleaved in the matrix.

Mitochondrial inner membrane and matrix

The major working parts of the mitochondrion are the matrix and the inner membrane that surround it. The inner membrane divides the mitochondria into two chambers. The **perimitochondrial space**, i.e., the space between the outer and inner membranes, continues into the core of the crests and inside the inner membrane and is occupied by a relatively dense material called the mitochondrial matrix. The matrix is normally homogeneous, but rarely appears as a fine filament or fibrous structures. In addition to various enzymes, the mitochondrial matrix also contains ribosomes and several molecules of DNA.

Mitochondrial cristae: As shown in **Fig. 7.5**, the inner membrane is convoluted, the folds consisting of double membranes called cristae (folds), in animals. In plants, these foldings appear tubular and are known as **tubuli** or **microvilli**. Generally these are incomplete septa or ridges that do not interrupt the continuity of the inner chamber. The shape and disposition of the cristae or crests vary in different cells and the numbers relate to the oxidative activity of the mitochondria. The number of crests per unit volume of a mitochondrion is also variable. Higher densities of cristae are found in the flight muscle mitochondria of insects. The inner membrane contains a high proportion of the 'double' phospholipid *cardiolipin*, which has four fatty acids rather than two and may help to make the membrane especially impermeable to ions. This membrane also contains a variety of transport proteins that make it selectively permeable to small molecules that are metabolized or required by the many mitochondrial enzymes concentrated in the matrix. The matrix enzymes include those that metabolize pyruvate and fatty acids to produce acetyl CoA and those that oxidize acetyl CoA in the citric acid cycle. The enzymes of the respiratory chain are embedded in the inner mitochondrial membrane and are essential to the process of oxidative phosphorylation, which generates most of the ATP in the animal cell.

Mitochondrial membrane structures are very dynamic. They can change shape depending on the tissue type, different developmental stages and physiological and pathological conditions. The cristae may be simple or branched forming a complex network, which provides access to the respiratory enzymes and also provide additional membrane surface. The inner membrane in the crest is found covered by 8.5-nm particles. These stalked particles are called elementary particles, inner membrane subunits or F1 particles (or coupling factor I, F1), placed regularly at intervals of 10 nm on the inner surface of the inner membrane. The asymmetric presence of F1 particles on the matrix side (*M side*), is related to the direction of the proton pump, i.e., the flow of H^+. Oxysomes are attached to the outer surface of these submitochondrial particles (***sonic particles***) and found to be responsible for respiratory chain phosphorylation. The oxysomes have been identified as molecules of ATPase enzyme (F1), responsible for catalyzing the terminal step of ATP synthesis. Each F1 particle is about 360 kDa and contains five different types of subunits.

7.5.3 Enzymes present in the mitochondria

Various enzymes with different functions are distributed throughout the mitochondria.

1. ***Outer membrane***: Choline phosphotransferase, monoamine oxidase, NADH–Cytochrome-C reductase, kynurenine hydroxylase and fatty acid CoA ligase.
2. ***Outer compartment/intermembrane space***: Adenylate kinase, creatine kinase, sulfite oxidase and nucleoside diphosphokinase.
3. ***Inner membrane***: Respiratory chain enzymes, ATP synthetase, succinate dehydrogenase, ß-hydroxybutyrate dehydrogenase and carnitine fatty acid acyl transferase, anion and cation translocases.
4. ***Matrix***: Malate and isocitrate dehydrogenase, fumarase and aconitase, citrate synthetase, α-keto acid dehydrogenases and β-oxidation enzymes, nucleic acids and protein synthesis systems.

7.5.4 Mitochondrial genome

Margit and Sylvan Nass (1963), demonstrated the presence of mitochondrial DNA (mtDNA) from a wide variety of cells. Study of the mtDNA organization is complicated by their large size and multipartite nature of genomes. The mtDNA is located in the mitochondrial matrix and may be attached to the matrix. They are extremely diverse and show striking characteristic differences between taxa in terms of size, structural organization, G+C content, the number and nature of the proteins they encode and even the genetic code itself; they are different in various organisms (**Table 7.1**). The mtDNA is usually found in cyclic, double-stranded, supercoiled molecules—the exceptions being the linear mitochondrial DNA molecules of *Tetrahymena* and *Paramoecium*. The mtDNA molecules are not packaged into nucleosomes. The genome exhibits a wide variation in size and form between related species.

Yeast mitochondrial DNA

Yeast mtDNA is five times as large (approximately 80 kbp in length), which is about the maximum size for fungal mtDNA. It contains some non-coding DNA punctuated with GC-rich regions containing origins of replication. Many yeast mitochondrial genes have large introns, which may contain open reading frames encoding proteins (~25 genes) that control intron splicing and transposition (e.g., homing introns). All three classes of self-splicing introns may be found in mitochondrial genes. The mtDNA of other fungi contain less repetitive DNA and the smallest genomes are approximately 20 kbp.

Table 7.1 Mitochondrial genes and their properties

Gene	Mammal	Yeast	Fungus (*Neurospora*)	Plant
Size (kbp)	14–18	78	19–108	200–2500
Length	Ca. 5	Ca. 25	–	Ca. 30
G + C content (%),	–	18	–	47
rRNAs				
Large subunit	16 S	21 S	21 S	26 S
Small subunit	12 S	15 S	15 S	18 S
5 S RNA	No	No	No	Yes
No. of tRNAs	22	23–25	23–25	≈30
Ribosomal protein (var-1),	No	Yes	Yes	?
Subunits 1, 2, 3 of Cyto C oxidase	Yes	Yes	Yes	Yes
Apocytochrome b subunit of $CoQH_2$—Cytochrome C reductase	Yes	Yes	Yes	Yes
Fo ATPase				
Subunit 6	Yes	Yes	Yes	Yes
Subunit 8	Yes	Yes	Yes	Yes
Subunit 9	No	Yes	No	Yes
Subunit α of F1 ATPase	No	No	No	Yes
No. of NADH–CoQ reductase subunits	7	0	6	6

When yeast cells are grown in the presence of mutagens such as ethidium bromide, 'petite' mutants are formed in which large sections of the mitochondrial genome are deleted with the remaining segments amplified so that the DNA molecules remain the same size. The mutants multiply readily in yeast cells but fail to synthesize mitochondrial proteins. Characterization of some petites has shown that mtDNA contain only adenine and thymine; yet they are able to replicate.

Mammalian mtDNA

Animal mitochondrial genomes are typically small and organized thriftily. There are no introns, little space between genes and a single region of non-coding DNA, which controls replication and gene expression. The space-saving mechanism of animal mtDNA is remarkable. The pangenomic transcription (full genome transcription), overlapping genes and missing termination codons, which are added post-transcriptionally by polyadenylation of the RNA are few features that explains the efficient use of mitochondrial genome.

Human mtDNA, a completely sequenced circular molecule of 16,569 bp, is among the smallest known (**Fig. 7.6**). It encodes two rRNAs found in mitochondrial ribosomes and 22 tRNAs used to translate mitochondrial mRNAs. Human mtDNA has 13 sequences (13 genes, most or all involved in electron transport) that begin with an ATG codon, end with a stop codon and are long enough to encode a polypeptide of more than 50 amino acids. Invertebrate mtDNA is about the same size as human mtDNA. Some of the protein-coding regions have only been identified as unassigned reading frames (URFs). None of the mammalian mitochondrial genes contain introns, but the only region of mtDNA which is non-coding is the D-loop region involved in the initiation of DNA replication. mtDNA can be separated into a heavy strand (**H-strand**) and a light strand (**L-strand**), due to differences in their density. In clockwise transcription, the H-strand is transcribed, while the L-strand is transcribed in the anticlockwise direction.

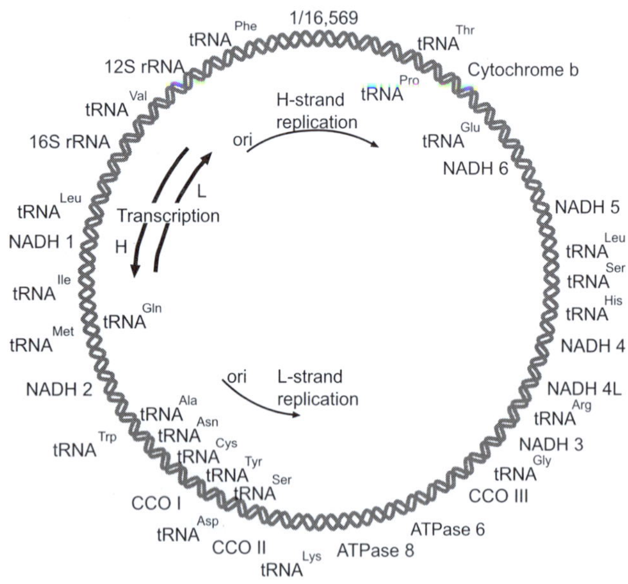

Fig. 7.6 Gene map of the human mitochondrial chromosome. Only nine loci are on L-strand (L); the rest are on heavy (H) strand. H-strand loci are labelled on the outside and L-strand loci are labelled inside the circle. CCO refers to Cytochrome -C oxidase (subunits I–III) and NADH refer to NADH dehydrogenase (subunits 1–4, 4L, 5 and 6).

Plant mtDNA

Plant mtDNA show an incredible diversity in size from 100 kbp to over 2.5 Mbp; the larger genomes contain a high proportion of repetitive DNA. In some species, the genome is of uniform size, but in many plants it has a complex organization involving linear and circular molecules of various sizes and structure generated by recombination. The full genome version is termed the **master circle**; smaller derivatives are **subgenomic circles**. Master circles contain repeats, which are sites of recombination—generally a profile of recombination products is seen for each genome. Abnormally rearranged mtDNAs, **sublimons**, may become amplified, perhaps by selfish replication like rho mutants in yeast.

Since several proteins are encoded by mtDNA in some species and by nuclear DNA in others, it is possible that all the genes moved from the mitochondrion to the nucleus, or vice versa, during evolution. Translocations of segments of mtDNA to nuclear DNA may still be occurring.

7.5.5 Mitochondrial gene expression

The mitochondrial RNA (mtRNA), differs from that of nuclear RNA and is resistant to the action of ribonuclease. All RNA transcripts of mtDNA and their translation products remain in the mitochondrion. There is export of these RNAs or proteins and all mtDNA-encoded proteins are synthesized on mitochondrial ribosomes. Mitochondria encode the rRNAs that form ribosomes, although all but one or two of the ribosomal proteins (depending on the species), are imported from the cytosol. In most organisms, such as plants, yeasts, fungi and mammals, all the tRNAs used for protein synthesis in the mitochondrion are encoded by mtDNAs. However, in wheat, parasitic protozoan *Trypanosoma brucei* and in ciliated protozoans all mitochondrial tRNAs are encoded by the nuclear DNA and imported into the mitochondrion.

Reflecting the bacterial ancestry of mitochondria, mitochondrial ribosomes resemble prokaryotic ribosomes and differ from cytoplasmic ribosomes in RNA and protein composition, size and sensitivity to certain antibiotics. For instance, chloramphenicol blocks protein synthesis by bacterial and mitochondrial ribosomes only, whereas cycloheximide inhibits protein synthesis by eukaryotic cytoplasmic ribosomes but has no effect on mitochondrial ribosomes.

Animal and fungal mitochondrial transcription is **polycistronic**, although there are many promoters in fungal mtDNA and only a single promoter for each strand of animal mtDNA, which is initially expressed as a **pangenomic** (full genome) transcript. In plants, most mitochondrial mRNAs are monocistronic, although they are transcribed from multiple initiation sites. All mitochondria use similar prokaryotic-like promoter sequences and encode RNA polymerases related to prokaryotic enzymes, including the use of accessory factors similar to the sigma-factor. Yeast and animal mtDNA transcription requires a transcription factor related to the HMG family of proteins.

7.5.6 Mitochondrial genetic codes are different from the standard nuclear code

The genetic code used in animal and fungal mitochondria is different from the standard code used in all prokaryotic and eukaryotic nuclear genes; the code even differs in mitochondria among species (**Table 7.2**). UGA, for example, is normally a stop codon, but is read as tryptophan by human and fungal mitochondrial translation systems; however, in plant mitochondria, UGA is still a stop codon. AGA and AGG are standard codons for arginine in fungal and plant mtDNA, but they are stop codons in mammalian mtDNA and serine codons in *Drosophila* mtDNA.

Table 7.2 Alterations in the standard genetic code in mitochondria

Codon	Standard nuclear code	Mitochondria				
		Mammals	*Drosophila*	*Neurospora*	Yeasts	Plants
UGA	Stop	Trp	Trp	Trp	Trp	Stop
AGA, AGG	Arg	Stop	Ser	Arg	Arg	Arg
AUA	Ile	Met	Met	Ile	MEt	Ile
AUU	Ile	Met	Met	Met	Met	Ile
CUU, CUC, CUA, CUG	Leu	Leu	Leu	Leu	Thr	Leu

The general principle is that plant mitochondria utilize the standard genetic code. As mentioned earlier, in certain organisms the genetic code is different even for nuclear encoded proteins. In mycoplasmas, UGA codes for tryptophan, not the normal 'stop' codon; in ciliated protozoans, such as *Tetrahymena*, the normal stop codons UAA and UAG code for glutamine in the mRNAs are synthesized from nuclear DNAs. The differences in codon usage between mitochondrial and cytosolic protein-synthesizing systems also make it difficult to explain how genes could have moved from mitochondrial to nuclear DNA.

7.5.7 Mitochondrial RNAs undergo extensive processing

RNA processing in mitochondria takes many forms. In plants, the rRNA transcripts are cleaved and matured, introns are spliced and some genes may be subjected to RNA editing. In animals, like their nuclear counterparts, mitochondrial rRNAs and mRNAs are generated by the enzymatic cleavage of longer precursors. The processing involves cleaving the polycistronic message into single gene fragments—a process possibly facilitated by the dispersed arrangement of tRNA genes—which could provide secondary structure motifs for endonuclease activity. Very often some of these cleavages occur on nascent chains. The cleavage sites must be precise to separate the mRNAs from the tRNAs, because the 5' end of each mRNA is immediately adjacent to the 3' end of a tRNA. In animal mitochondria, mRNA and rRNA may be polyadenylated, but contain no introns and very few untranslated sequences. RNA processing in fungal mitochondria involves a 12 nt long 3' end signal which controls maturation of the transcript.

7.5.8 Translation in mitochondria

Only a few proteins are encoded by mitochondrial DNA and synthesized on mitochondrial ribosomes. Most such proteins are subunits of integral proteins in the inner mitochondrial membrane. The vast majority of mitochondrial proteins are encoded by nuclear genes, synthesized on cytosolic ribosomes and imported into the mitochondrion. Some cytosol-synthesized proteins are transported to the intermembrane space, others to the outer membrane and still others to the inner membrane. The largest number are transported to the matrix, including the F1 ATPase subunits, ribosomal proteins, RNA polymerase and the enzymes of the citric acid cycle.

Plant mitochondria usually encode a complete set of tRNAs, although in certain species some are encoded by the nucleus and imported. Some algal and protozoan mitochondria possess a dramatically reduced set of tRNA genes or none at all. In these cases, many tRNAs are imported from the nucleus. Translation initiation involves prokaryotic-like ribosome binding sites and *N*-formylmethionine is the initiator amino acid. Translation is a predominant level of gene regulation in mitochondria.

7.5.9 Replication of mitochondrial DNA

Origins of replication have been identified in some plant and fungal mitochondria, but the replication mechanism is not yet understood fully. In some plant species, the system controlling replication must discriminate between the master circle and subgenomic circles. In animals, the leading strand (heavy strand or H-strand) replication is initiated by a large RNA primer synthesized by RNA polymerase, which is cleaved by the ribonucleoprotein endonuclease MRP. DNA synthesis is then initiated by DNA polymerase gamma, extending the primer and displacing the resident strand to generate a D-loop. The extension is often aborted at this point, probably due to motifs (termination associated sequences), within the D-loop. Read-through of these sites, or reinitiation of a terminated strand, results in successful replication. Replication of the complementary strand (light strand or L-strand), is initiated on the opposite side of the circle to the heavy strand and involves a specific mitochondrial primase. This unusual continuous replication mechanism means that mtDNA is single-stranded most of the time, which perhaps contributes to its relatively high mutation rate.

7.5.10 Kinetoplast DNA

The kinetoplast is a part of a highly specialized mitochondria found in certain groups of flagellated protozoa such as the trypanosomes. Like other mitochondria, it contains its own genome (kinetoplast DNA, kDNA). The kDNA consists of an interlinked series of many thousands of cyclic DNA molecules that vary in size from 0.6 kb to 2.4 kb depending on the type of trypanosome from which they are obtained. The organization of the kDNA genome is remarkable, with full genome-size **maxicircles** (about 30 kb in size), interlinked in a complex network with a multitude of shorter derivatives termed **minicircles**. Maxicircles appear to perform the conventional functions of mtDNA; minicircles are microheterogeneous in sequence and size and appear to play a role in providing a template for RNA editing. Both maxicircles and minicircles appear to replicate using a rolling circle mechanism. Primary transcripts from kDNA demonstrate the most prolific RNA editing, in extreme examples involving the deletion of more than 50% of the nucleotides. RNA editing is facilitated by guide RNAs, which appear to be the only product synthesized by the minicircles. Other unusual properties of protozoan kinetoplasts are the lack of organelle-encoded tRNAs and the synthesis of rRNA in short fragments. The rRNA genes are broken up and dispersed throughout the genome. Functional RNAs are presumed to arise by intermolecular base pairing.

7.5.11 Mitochondrial plasmids

In addition to their main chromosomes, a number of plant mitochondria contain small circular DNA species referred to as minicircular DNAs, or extra-chromosomal plasmids. Many plant mitochondria carry plasmids along with their genomes. They are diverse in structure, with genomes of linear or circular double-stranded DNA, or single- or double-stranded RNA.

These plasmids can take up three distinct topological states, i.e., supercoiled, relaxed circular and linear. The mitochondrial plasmids are invariably unrelated to the main mitochondrial chromosome with regard to their DNA sequences. This makes their origins and genetic functions uncertain and it is now believed that such plasmids make no genetic contribution either to the mitochondrion or to the plant as a whole.

In plants, many plasmids appear to be associated with the characteristic of **cytoplasmic male sterility** (cms), where cytoplasmic factors block the production of viable pollen. This promotes outcrossing and may therefore be of overall benefit to the plant. The cms phenotype can be suppressed by the expression of nuclear genes. Mitochondria of the cms T line, for instance, produce a novel, plasmid-encoded protein, which is responsible for the sterility phenotype in a poorly understood manner. The phenotype can be suppressed by increasing expression at two nuclear loci, Rf1 and Rf2 (for restoration of fertility), which act by preventing synthesis of the protein. Plasmids in fungal mitochondria may be cryptic or they may confer the characteristics of **longevity** or **senescence**. The latter trait is inherited in a cytoplasmic fashion and appears to involve rearrangement of the mitochondrial genome, induced by plasmid integration.

7.5.12 Functions of mitochondria

Mitochondria are often referred to as the 'powerhouse' of the cell, since they produce 95% of total ATP molecules in animal cells, but in plants, in addition to mitochondria, ATPs are also produced in chloroplasts. Cellular respiration (aerobic respiration), that is the oxidation of nutrients to generate energy in the form of adenosine 5'-triphosphate (ATP), takes place in the mitochondria. This energy is produced during the breakdown of food molecules including carbohydrates, fats and proteins, which involve:

- Embden–Meyerhof pathways (EMP or glycolysis)
- oxidative decarboxylation
- oxidative phosphorylation, which includes Kreb cycle or tricarboxylic acid (TCA) cycle and respiratory chain.

In the mitochondria, the metabolism of sugars is completed and the energy released is harnessed so efficiently that about 30 molecules of ATP are produced for each molecule of glucose (6 carbons) oxidized. By contrast, only two molecules of ATP are produced per glucose molecule by glycolysis alone. However, oxidation of an 18-carbon fatty acid can make 146 ATP molecules.

Glycolysis

The first stage of the energy releasing pathway is oxidation of glucose or glycogen (glycolysis), carried out in the cytosol by enzymes of glycolysis (**Fig. 7.7**). During the glycolytic pathway, a set of ten, water soluble cytosolic

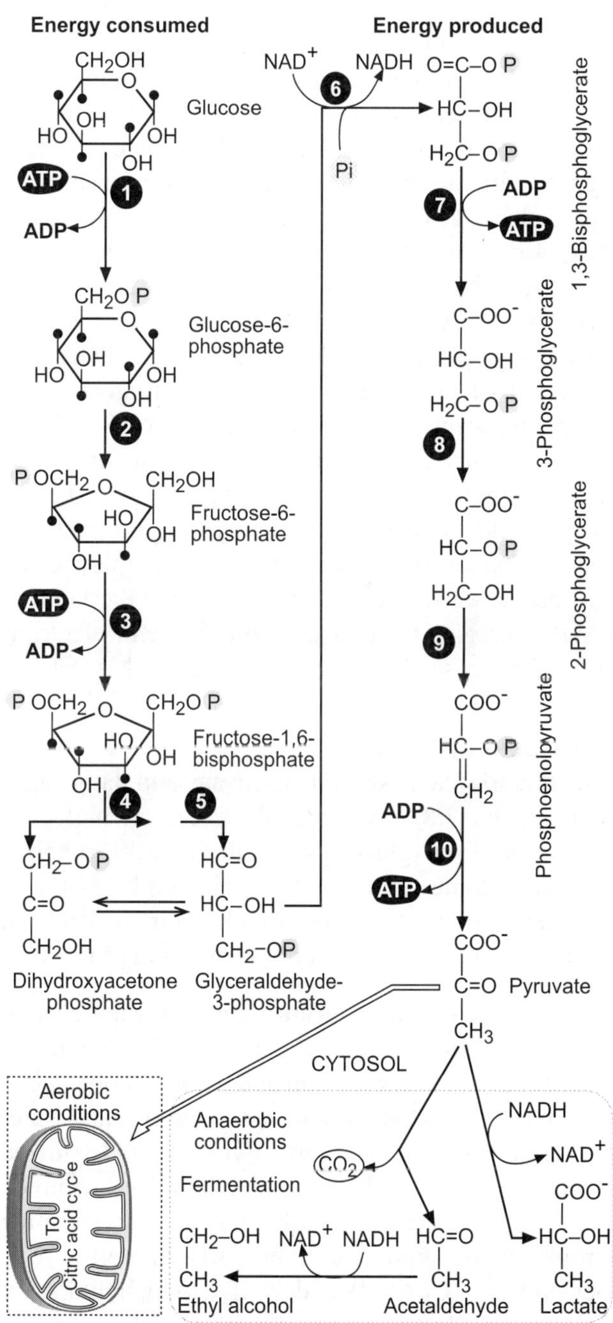

Fig. 7.7 The ten key reactions of glycolysis. Glucose is broken down to pyruvate, with the net formation of two molecules each of ATP and NADH.

enzymes catalyze the reactions, in which, one molecule of glucose is converted in to two molecules of pyruvate, two molecules of NADH and four molecules of ATP. Unlike ATP synthesis in mitochondria and chloroplasts, a proton motive force is not involved in substrate level phosphorylation. As energy carrying molecules, NADH contains a relatively small amount of energy stored in the two electrons that it picked up from the substrate, glyceroldehyde-3-phosphate. However, pyruvate still contains a large amount of chemical energy. The energy of these compounds must be transferred into a usable form (ATP). This occurs within the mitochondria.

The first steps of glycolysis are energy consuming. They proceed only when two ATP molecules each donate energy to the glucose backbone by transferring a phosphate group to it. Such transfers are called phosphorylations, which cause the backbone to split apart into two molecules of phosphoglyceraldehyde (PGAL). PGAL formation initiates the energy releasing process of glycolysis. Each PGAL is converted to unstable intermediate that gives up a phosphate group to ADPs, forming ATPs. Thus, a total of four ATP molecules have been formed by substrate level phosphorylation, resulting in the net energy yield of two ATPs. The hydrogen atoms and electrons released from each PGAL during this process are transferred to the coenzyme (an enzyme helper), NAD^+. The NAD^+ picks up hydrogen and electrons stripped from a substrate and so becomes NADH.

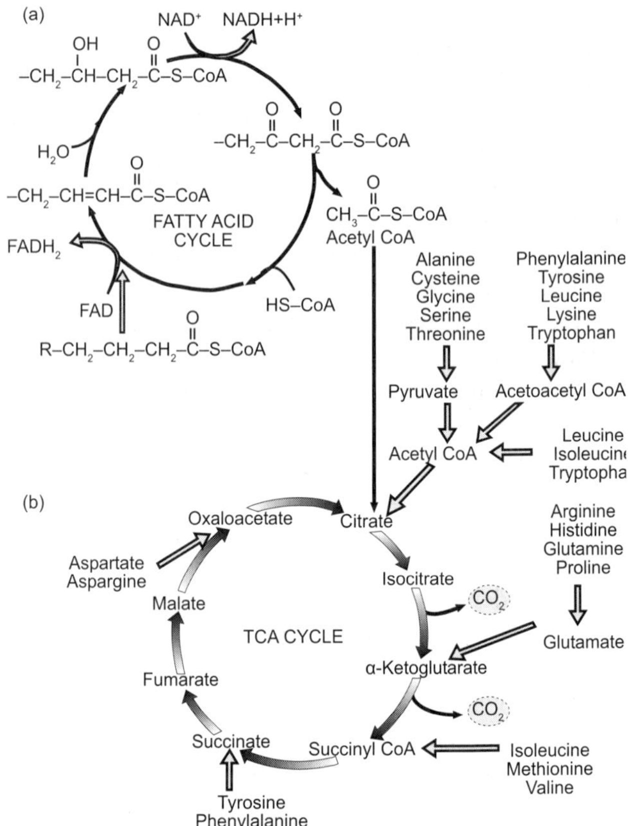

Fig. 7.8 Catabolic pathways generate compounds that are fed into the tricarboxylic acid cycle (TCA) cycle.

Citric acid cycle

Mitochondria can use both pyruvate and fatty acids as fuel (**Fig. 7.8**). Pyruvate (three carbon molecule, 3C) comes from glucose and other sugars, whereas fatty acids come from fats. Only the aerobic pathway continues inside this organelle and two more steps are required to complete the energy conversion process.

Oxidative decarboxylation: During the second stage, i.e., before entering the citric acid cycle, in the mitochondrial matrix, carbon and oxygen atoms move from pyruvate, in the form of carbon dioxide and water (*oxidative decarboxylation*). Pyruvate is oxidized by the pyruvate dehydrogenase enzyme complex (**Fig. 7.9**), which requires the bound cofactors thiamine pyrophosphate, lipoic acid and FAD as well as free coenzyme A (CoASH) and NAD^+, which links the citric acid cycle to glycolysis. The complex contains three separate enzymes: pyruvate dehydrogenase, dihydrolipoyl transacetylase and dihydrolipoyl dehydrogenase. The coenzymes accept hydrogen and

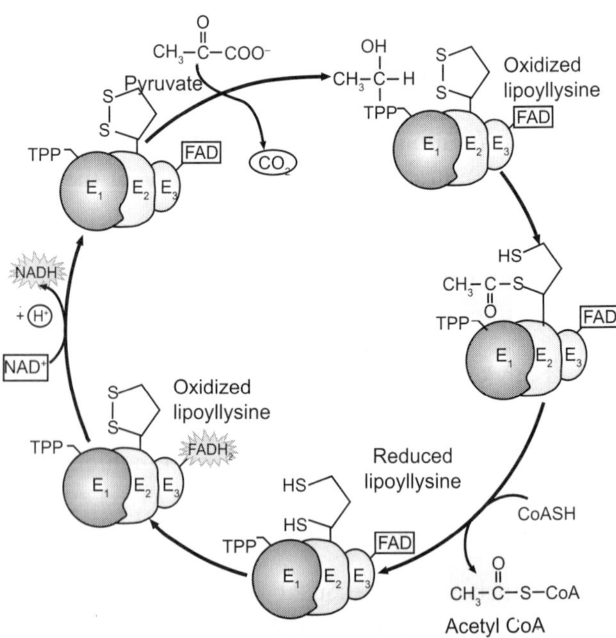

Fig. 7.9 Reaction sequence catalyzed by the pyruvate dehydrogenase complex.

electrons released during the reactions. The pyruvate dehydrogenase complex is subject to sophisticated regulatory mechanisms, including phosphorylation/ dephosphorylation by a protein kinase/phosphatase couple. ATP-dependent phosphorylation inhibits the dehydrogenase enzyme. In turn, various metabolites modulate the kinase activity. Pyruvate inhibits the protein kinase, ensuring that the dehydrogenase is active when the substrate is plentiful. Ammonia, on the other hand, stimulates the kinase and therefore inhibits pyruvate oxidation. The acetyl group (2C) gets picked up by a coenzyme, forming acetyl CoA which serves as a carrier of acyl groups in various metabolic reactions. During this process, one molecule of NAD^+ is reduced to NADH. Acetyl groups in acetyl CoA are then oxidized in the matrix via the citric acid cycle or *Kreb cycle* or *tricarboxylic acid cycle* (TCA cycle). The second stage reactions serve three functions:

- The hydrogen and electrons are transferred to NAD^+ and FAD, forming NADH and $FADH_2$.
- The substrate level photophosporylations produce two ATPs.
- Intermediates are rearranged into oxaloacetate.

Tricarboxylic acid cycle: The acetyl CoA formed by the above reaction enters the *citric acid cycle* (**Fig. 7.10**). The two-carbon (2C), acetyl group combines with oxaloacetate (4C), to yield citrate (6C).

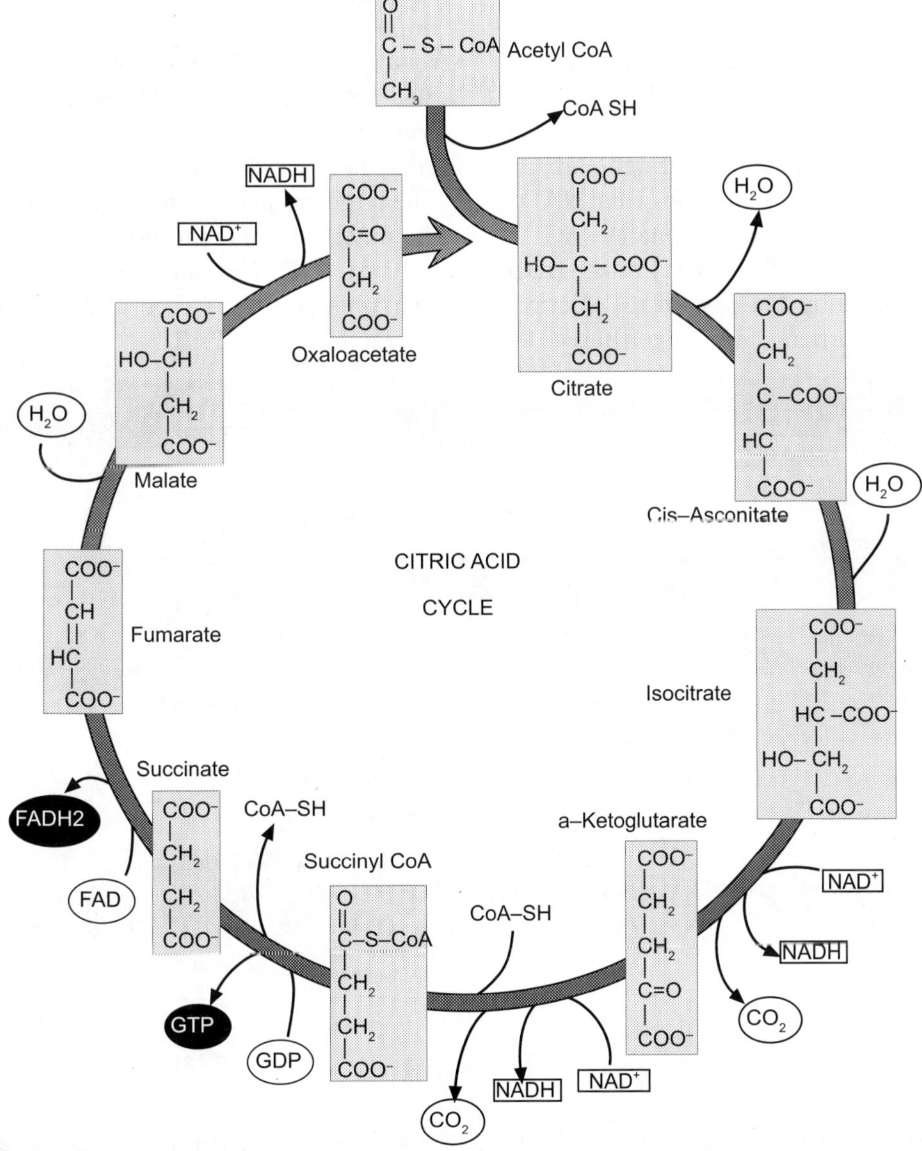

Fig. 7.10 The citric acid cycle. A two-carbon acetyl group is transferred from acetyl CoA to oxaloacetate, forming citrate, which gets into the citric acid cycle. Each turn of the cycle generates one molecule of GTP, three of NADH and one of $FADH_2$.

Further, through eight series of reactions, two carbons of citrate are completely oxidized to CO_2 and oxaloacetate is regenerated. During this process, one high-energy phosphate bond is formed in GTP, which is used directly to drive the synthesis of one ATP molecule. In addition, each turn of the cycle generates three molecules of NADH and one molecule of reduced *flavin adenine dinucleotide* ($FADH_2$). The citric acid cycle completes the oxidation of glucose to six molecules each of CO_2 and water. Four molecules of ATP are obtained directly from each glucose molecule (two from glycolysis and two from citric acid cycle, i.e., one for each molecule of pyruvate). In addition, ten molecules of NADH (two from glycolysis, two from the conversion of pyruvate to acetyl CoA and six from the citric acid cycle) and two molecules of $FADH_2$ are also formed. The remaining energy derived from the breakdown of glucose comes from the reoxidation of NADH and $FADH_2$, with their electrons being transferred through the electron transport chain to reduce O_2 to H_2O (**Fig.7.11**). During oxidative phosphorylation, the transfer of electrons from NADH to O_2 releases a large amount of free energy ($\Delta G°' = -53$ kcal/mol for each pair of electrons transferred). The process takes place by the passage of electrons through a series of carriers, which constitute the *electron transport chain* (*ETC*). The components of the ETC are located in the inner mitochondrial membrane of eukaryotic cells. In aerobic bacteria, ETC is located in the plasma membrane. Electrons from $FADH_2$ enter the ETC at a lower energy level, so their transfer to O_2 yields less usable free energy, only two ATPs.

Conversion of oxidation energy into ATP by chemi-osmotic process

The citric acid cycle, even though a part of the aerobic metabolism does not use oxygen. Only in the last catabolic reactions that take place on the inner mitochondrial membrane, molecular oxygen (O_2), is directly utilized. Nearly all the energy available from burning carbohydrates, fats and other foodstuffs in the earlier stages of their oxidation is initially saved in the form of high-energy electrons removed from substrates by NAD^+ and FAD. These electrons, carried by NADH and $FADH_2$, are then combined with O_2 by means of the respiratory chain embedded in the inner mitochondrial membrane. The large amount of energy released is harnessed by the inner membrane to drive the conversion of ADP + Pi to ATP. For this reason, the term **oxidative phosphorylation** is used to describe this last series of reactions (**Fig. 7.12**).

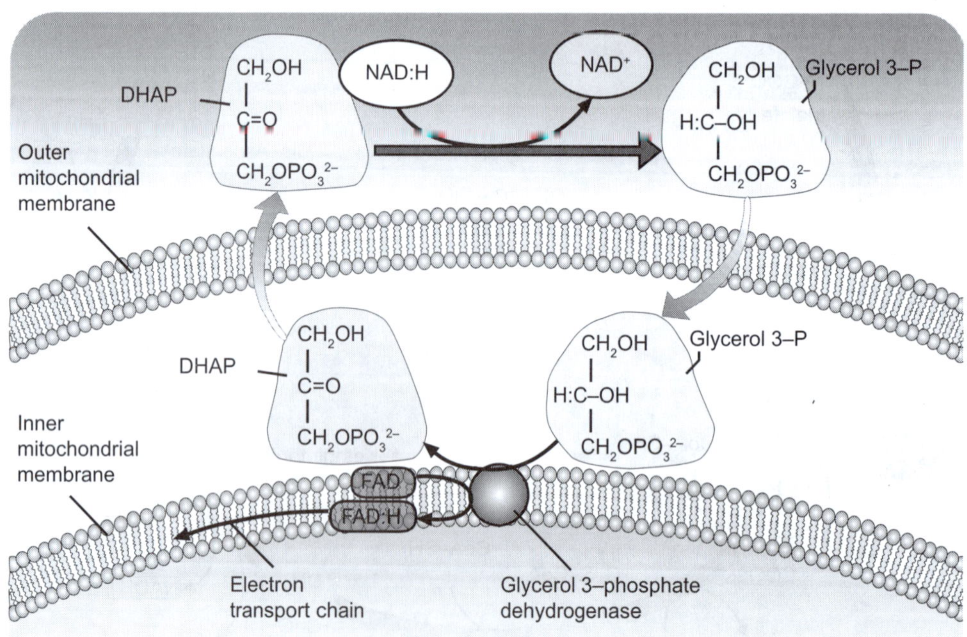

Fig. 7.11 The glycerol phosphate shuttle. During this process electrons are transferred from NADH to dihydroxy acetone phosphate (DHAP) to form glycerol 3-phosphate, which shuttles them into the mitochondrion. The electrons then reduce FAD at the inner membrane, forming $FADH_2$, which can transfer the electrons to a carrier of the electron transport chain.

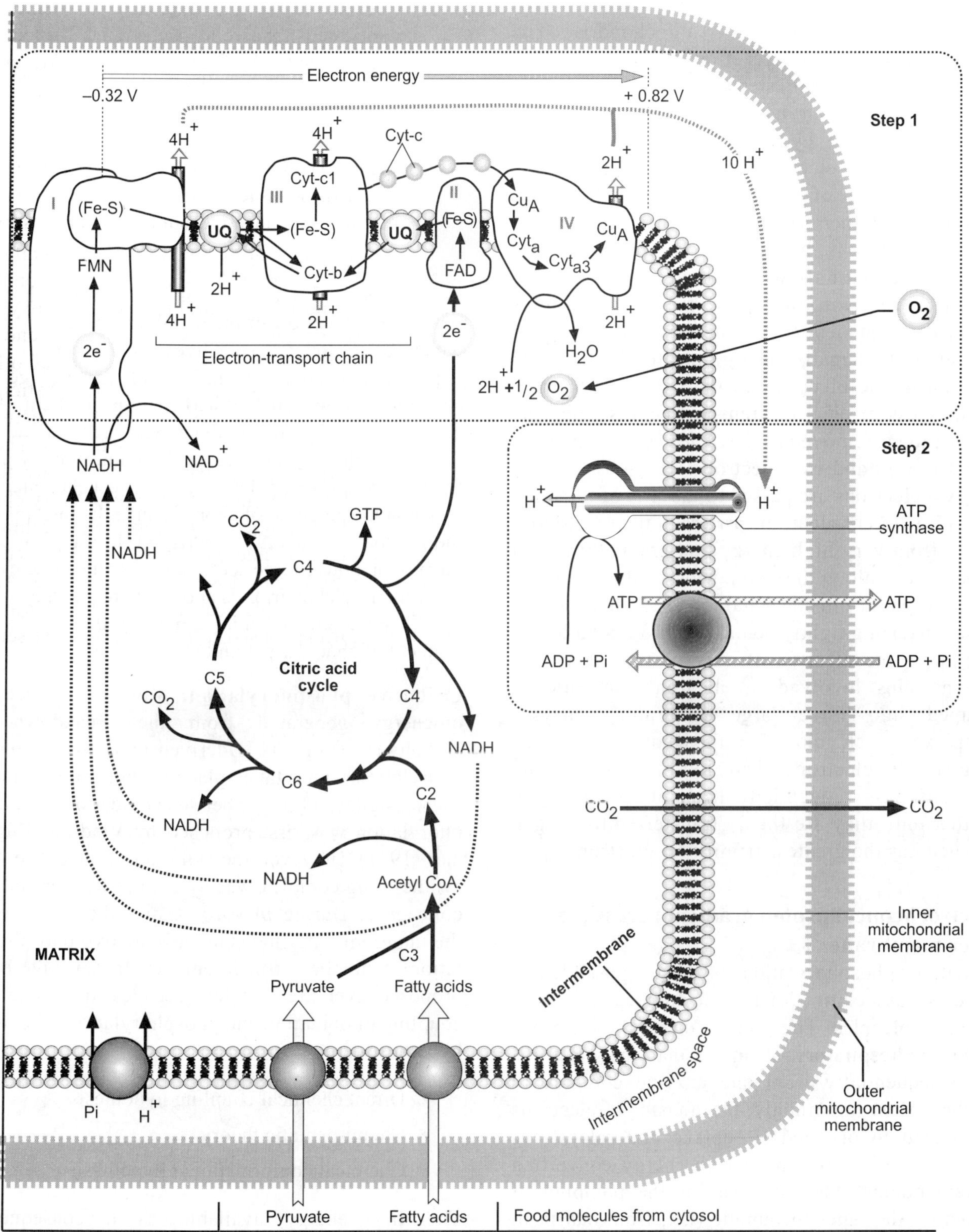

Fig. 7.12 A schematic diagram of energy generating metabolism in mitochondria. The protons move down the electrochemical gradient through an ATP synthesizing complex. The energy stored in the gradient is used to synthesize ATP. The two essential steps (Steps 1 and 2) of oxidative phosphorylation form the basis of the chemi-osmotic mechanism.

Electrons are transferred from NADH to oxygen through three large respiratory enzyme complexes on the inner membrane of mitochondria. The energetically favoured reaction $H_2 + {}^1/_2O_2 \rightarrow H_2O$ occurs in many small steps, so that most of the energy released can be stored instead of being lost as heat. The hydrogen atoms first separate into protons and electrons; then the electrons pass through a series of electron carriers in the inner mitochondrial membrane. At several steps along the way, protons and electrons are transiently recombined. But only when the electrons reach the end of the electron transport chain are the protons returned permanently, when they are used to neutralize the negative charge created by the final addition of the electrons to the oxygen molecule. The process of electron transport begins when the hydride ion is removed from NADH and is converted into a proton and two electrons ($H^- \rightarrow H^+ + 2e^-$). The two electrons are passed to the first of the more than 15 different electron carriers in the respiratory chain, from very high energy and gradually lose it as they move along the chain. Most of the times, the electrons pass from one metal ion to another, each of these ions being tightly bound to a protein molecule that alters the electron affinity of the metal ion. The proteins involved in electron transport are grouped into three large respiratory enzyme complexes. Each complex in the chain has a greater affinity for electrons than its predecessor and electrons pass sequentially from one complex to another until they are finally transferred to oxygen (oxygen has the greatest affinity for electrons).

Electrochemical proton gradient across the inner membrane

Oxidative phosphorylation is made possible by the close association of the electron carriers with protein molecules. The protons direct the electrons along the respiratory chain so that the electrons move sequentially from one enzyme complex to another. Most importantly, the transfer of electrons is coupled to oriented H^+ uptake and release, as well as to allosteric changes in energy converting protein pumps. The net result is the pumping of H^+ across the inner membrane (i.e., from matrix to intermembrane space), driven by the energetically favourable flow of electrons. Because protons are electrically charged particles, the potential energy stored in the proton gradient is electric as well as chemical in nature. Thus, movement of H^+ has two major consequences:

1. It generates a pH gradient across the inner mitochondrial membrane, with the pH higher in the matrix than in the cytosol, where the pH is generally close to 7.
2. It generates a voltage gradient (membrane potential) across the inner mitochondrial membrane, with the inside negative and the outside positive.

The pH gradient (ΔpH), drives H^+ back into the matrix and OH^- out of the matrix, thereby reinforcing the effect of the membrane potential (ΔV), which acts to attract any positive ion into the matrix and to push any negative ion out. The ΔpH and ΔV together constitute an **electrochemical proton gradient**. The electrochemical proton gradient exerts a proton motive force, which can be measured in millivolts (mV). Since each ΔpH of 1 pH unit has an effect equivalent to a membrane potential of about 60 mV, the total proton motive force equals $\Delta V - 60(\Delta pH)$. In a typical cell, the proton motive force across the inner membrane of a respiring mitochondrion is about 200 mV and is made up of a membrane potential of about 140 mV and a pH gradient of about –1 pH unit.

Oxidative phosphorylation: Most of the usable bioenergy generated from the breakdown of carbohydrates or fats is derived by oxidative phosphorylation which takes place within mitochondria. The existence of linkage between oxidation and phosphorylation was first proposed by Vladimir Engelhart (1931). However, the isolation of the entire ATP synthesizing complex was by Efraim Racker and his colleagues. During *in vitro* studies they uncovered the underlying mechanisms of oxidative phosphorylation. A number of different hypotheses have been proposed over the past few decades to explain the coupling of oxidation and phosphorylation. The three hypotheses are:

- Direct chemical coupling hypothesis
- Chemi-osmotic theory
- Chemical conformational hypothesis

The free energy available as a consequence of transferring two electrons from NADH or succinate to molecular oxygen is –57 kcal/mol and –36 kcal/mol, respectively. Oxidative phosphorylation traps this energy as the high-energy phosphate of ATP.

For oxidative phosphorylation to proceed, two principal conditions must be met. These are:

1. The inner mitochondrial membrane must be physically intact so that proteins can only re-enter the mitochondrion by a process coupled to ATP synthesis.
2. A high concentration of protons must be developed on the outside of the inner membrane.

The energy of the proton gradient is known as the chemi-osmotic potential, or proton motive force (PMF). This potential is the sum of the concentration difference of protons across the membrane and the difference in electrical charge across the membrane. The two electrons from NADH generate a six-proton gradient. The energy of the gradient is used to drive ATP synthesis as the protons are transported back down their thermodynamic gradient into the mitochondrion.

The mechanism of coupling electron transport to ATP generation, **chemi-osmotic coupling**, is a good example of the linkage between structure and function in cell biology. Chemi-osmotic coupling was first proposed in 1961 by Peter Mitchell, who suggested that ATP is generated by the use of energy stored in the form of proton gradients across biological membranes, rather than by direct chemical transfer of high-energy groups. Now it is recognized as a general mechanism of ATP generation, where ATP is generated via a proton gradient across the plasma membrane.

The ATP synthase: The process of **oxidative phosphorylation** in the inner mitochondrial membrane drives ATP synthesis by the membrane-bound enzyme *ATP synthase* (*Complex V* or *H*$^+$-translocating ATP synthase). This complex encompasses 16 polypeptides, two of which (ATPase6 and ATPase8) are encoded by the mtDNA. This enzyme creates a hydrophilic pathway across the inner mitochondrial membrane that allows protons to flow down their electrochemical gradient. The electrochemical gradient ($\Delta\rho = \Delta\psi + \Delta pH$), is generated by complexes I, III and IV of mitochondrial electron transport chain. This enzyme uses a rotary catalytic mechanism that is close to 100% efficient. As these ions thread their way through the ATP synthase, they are used to drive the energetically unfavourable reaction between ADP and Pi that make ATP (**Fig. 7.13**).

The ATP synthase (also called the F_0F_1 complex) is a multisubunit protein, the F_0 component contains three types of integral membrane proteins, designated a, b and c. In bacteria and in yeast mitochondria, the most common subunit composition is $a_1b_2c_{10}$, but animal mitochondria have 12c subunits and chloroplasts have 14c. In all cases, the c subunits form a doughnut-shaped ring in the plane of the membrane. The F_1 is a complex, shaped like a *lollipop head* composed of a ring of five distinct polypeptides with the composition of $\alpha_3\beta_3\gamma\ \delta\ \varepsilon$, projects on the matrix side of the inner mitochondrial membrane. The lower part of the $F_1\gamma$ subunit is a coil that fits into the centre of the c-subunit ring of F_0 and appears rigidly attached to it. The $F_1\varepsilon$ subunit is rigidly attached to γ and also forms rigid contacts with several of the c-subunits of F_0.

Based on the concept of rotary catalysis, complex V has been divided into two functional components: 'rotor' and 'stator' (**Fig. 7.13**). The rotor consists of a hydrophobic wheel of 12 c-(ATP9) subunits (c-ring), lying in the plane of the inner membrane that rotates with its axle extending into the matrix. The axle is composed of the ε-subunit at the base and the γ-subunit projecting up into the F_1 barrel. The stator is composed of a base made by the a-subunit (ATP6) embedded in the membrane, adjacent to the subunit wheel. The a-subunit anchors a pair of b-subunits that project out of the membrane

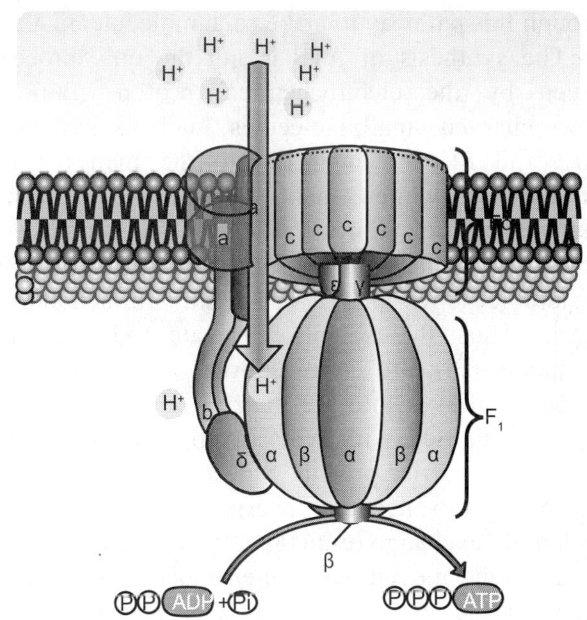

Fig. 7.13 Schematic diagram of the bacterial ATP synthase. The enzyme consists of two major portions, called F_1 and F_0. The F_1 head consists of five different subunits in the ratio 3α:3β:1δ:1γ:1ε. The F_0 base, which is embedded in the membrane, consists of three different subunits in the apparent ratio 1a:2b:10–14c.

and bind to the δ-subunit of F_1, which is attached to an α-subunit of the α-β barrel. Thus, F_0 of a-subunit, is linked to the 3α:3β barrel in a static structure through the b- and δ-subunits.

The stator F_0 a-subunit-a and rotor c-subunit interface mediate the proton flux from the intermembrane space to the matrix. As protons pass through a narrow channel formed at the stator–rotor contact, their movement causes the rotor ring to spin. Thus driving the rotor's rotation and causing the γ-axle to spin within past the a-subunit stator, the rotor's axle, ending in the γ-subunit, rotates inside the hollow α- and β-subunit (lollipop head), thus mediating the two components of the catalytic cycle. As a result, the energy of proton flow down a gradient is converted into the mechanical energy of two sets of protons rubbing against each other: rotating stalk proteins pushing against a stationary ring of head proteins. Three of the six subunits in the head contain binding sites for ADP and inorganic phosphate. These are driven to form ATP as mechanical energy is converted into chemical bond energy through the repeated changes in protein bond formation that the rotating stalk creates. In this manner, the ATP synthase is able to produce more than 100 molecules of ATP per second. Three or four protons need to pass through this pathway to make each molecule of ATP.

The synthesis of ATP is not the only process driven by the electrochemical proton gradient; many charged small molecules, such as pyruvate, ADP and Pi, are pumped into the matrix from the cytosol, while others, such as ATP move into the cytosol. Carrier proteins that bind these molecules couple their transport to the energetically favourable flow of H^+ into the mitochondrial matrix. Thus, for example, pyruvate and inorganic phosphate (Pi), are co-transported inward with H^+ as the H^+ moves into the matrix. In contrast ADP is co-transported with ATP in opposite directions by a single carrier protein. Since an ATP molecule has one more negative charge than ADP, each nucleotide exchange results in a total of one negative charge being moved out of the mitochondrion. This ADP–ATP co-transport is thereby driven by the voltage difference across the membrane (**Fig. 7.14**). In eukaryotic cells, the proton gradient is thus used to drive both the formation of ATP and the transport of certain metabolites across the inner mitochondrial membrane. In bacteria, the proton gradient across the bacterial plasma membrane is harnessed for both types of functions. And in the plasma membrane of

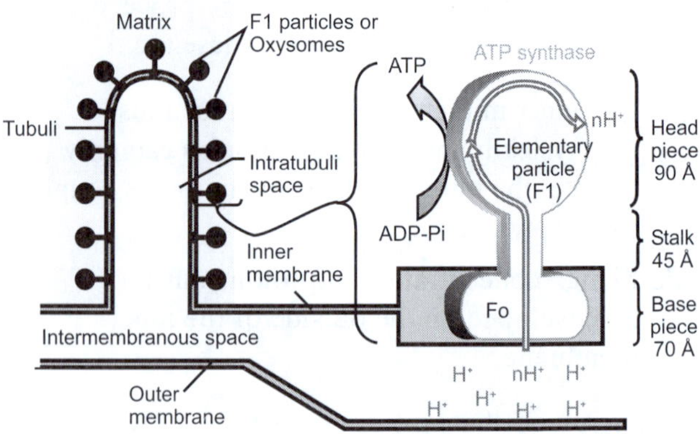

Fig. 7.14 The position and orientation of the F1 particles of mitochondria on cristae (left) and the general morphology of the F1 particle (ATP synthase).

motile bacteria, the gradient also drives the rapid rotation of the bacterial flagellum, which propels the bacterium along.

The carrier proteins in the inner mitochondrial membrane facilitate rapid exchange of ATP for ADP and the ADP molecules produced by ATP hydrolysis in the cytosol rapidly enter the mitochondria for recharging. The ATP molecules formed in the mitochondrial matrix by oxidative phosphorylation are rapidly pumped into the cytosol, where they are needed.

ATP synthase can also work in reverse: It can use the energy of ATP hydrolysis to pump H^+ across the inner mitochondrial membrane. Thus, ATP synthase acts as a reversible coupling device, interconverting electrochemical proton gradient and chemical bond energies. Although the exact number of protons needed to make each ATP molecule is not known with certainty, it can be assumed that one molecule of ATP is made by the ATP synthase for every 3 protons driven through it. Whether the ATP synthase works in its ATP-synthesizing or its ATP-hydrolyzing direction at any instant depends, in this case, on the exact balance between the favourable free energy changes for moving the three protons across the membrane into the matrix.

7.5.13 Electron transport chains and proton pumps

It is very important to understand the mechanisms that underlie the membrane-based energy conversion process, since similar chemi-osmotic mechanisms are used by mitochondria, chloroplasts, *Archea* and bacteria. These mechanisms underlie the function of nearly all living organisms, including anaerobes that derive all their energy from electron transfers between two inorganic molecules.

Protons are unique positive ions and by far the most abundant in living organisms. They are plentiful not only in carbon-containing biological molecules, but also in water molecules. The protons in water are highly mobile, flickering through the hydrogen-bonded network of water molecules by rapidly dissociating from a water molecule to associate with its neighbour. Protons are thought to move across a protein pump embedded in a lipid bilayer in a similar way; they transfer from one amino acid side chain to another, following a special channel through the protein. Protons are also special with respect to electron transport. Whenever, a molecule is reduced by acquiring an electron, the electron (e^-) adds a negative charge to it. Often this charge is rapidly neutralized by the addition of a proton (H^+) from water, so that the net effect of the reduction is to transfer an entire hydrogen atom, $H^+ + e^-$. Similarly, when a molecule is oxidized, a hydrogen atom removed from it can be readily dissociated into an electron and a proton, allowing the electron to be transferred separately to a molecule that accepts it, while the proton is passed to water. Therefore, in a membrane in which electrons are passed along an electron transport chain (ETC), pumping protons from one side of the membrane to another can be relatively simple. The electron carrier merely needs to be arranged in the membrane to pick up a proton from one side of the membrane and to release the proton on the other side of the membrane, as the electron is transferred to the next carrier molecule in the chain.

In biological systems, during biochemical reactions, electrons removed from a molecule are always passed on to another, so that whenever one molecule is oxidized, another is reduced. Like any other chemical reaction, the tendency of such oxidation–reduction reactions (redox–reactions), to proceed spontaneously depends on the free energy change (ΔG), for the electron transfer, which in turn depends on the relative affinities of the two molecules for electrons. Electron transfer provides most of the energy for living things.

NADH is a strong electron donor, because its electrons are held in a high-energy linkage and thefree energy change for passing its electrons to many other molecules is favourable. Therefore its redox partner, NAD^+, is of necessity a weak electron acceptor. The voltage difference is defined as the **redox potential (Fig. 7.15)**; electrons move spontaneously from a redox pair like $NADH/NAD^+$ with a low redox potential (a low affinity for electrons), to a redox pair like O_2/H_2), with a high redox potential (a high affinity for a electrons). Thus, NADH is a good molecule for donating electrons to the respiratory chain, while O_2 is well suited to act as the sink for electrons at the end of the pathway. The difference in redox potential, ΔE_0', is a direct measure of the standard free energy change ($\Delta G°$), for the transfer of an electron from one molecule to another. Those pairs of compounds that have the most negative redox potentials have the weakest affinity for electrons and therefore contain carriers with the strongest tendency to donate electrons. Conversely, those pairs that have the most positive redox potentials have the strongest affinity for electrons and therefore contain carriers with the strongest tendency to accept electrons.

In living systems NADH donates electrons to O_2 gradually via many electron carriers in the ETC. Since each successive carrier in the chain hold electrons more tightly, the highly energetically favourable reaction $2H^+ + 2e^- + {}^1/_2O_2 \rightarrow H_2O$ occurs in many small steps. This enables nearly half the released energy to be stored, instead of being lost to the environment as heat.

7.5.14 Respiratory chain complexes

Electron carriers: Many of the electron carriers in the respiratory chain absorb visible light and

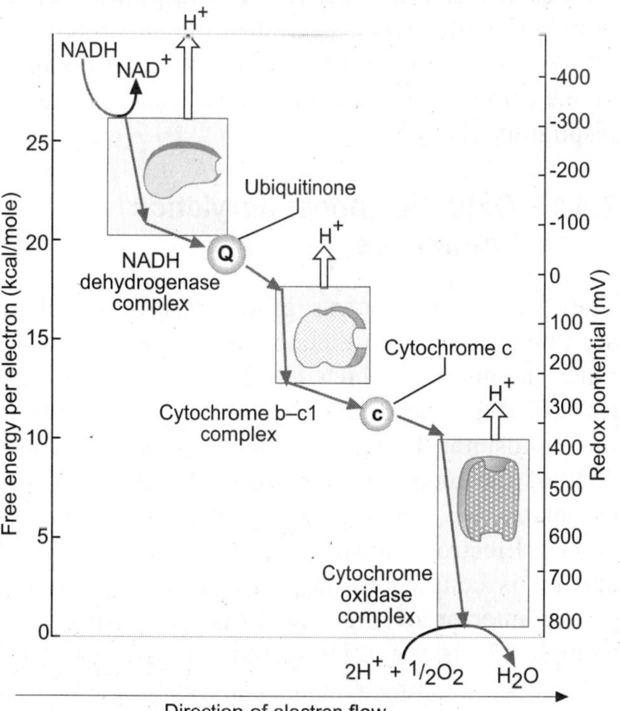

Fig. 7.15 Redox potential changes along the mitochondrial electron transport chain.

change colour when they are oxidized or reduced. It was therefore possible to identify them. Thus, the cytochromes were discovered as compounds that undergo rapid oxidation and reduction in living organisms such as bacteria, yeasts and insects. Based on their absorption spectra, three types of cytochromes were identified and designated cytochromes a, b and c. The cytochromes constitute a family of coloured proteins that are related by the presence of a bound **heme group**, whose iron atom changes from the ferric oxidation state (Fe^{3+}), to the ferrous oxidation state (Fe^{2+}), whenever it accepts an electron. The heme group consists of a porphyrin ring with a tightly bound iron atom held by four nitrogen atoms at the coroners of a square (**Fig. 7.16**). Iron–sulphur proteins are a second major family of electron carriers. In these proteins, either two or four iron atoms are bound to an equal numbers of sulphur atoms and to cysteine side chains, forming an iron–sulphur centre on the protein. Like the cytochromes these centres carry one electron at a time. The simplest of the electron carriers in the respiratory chain and not part of a protein is a small hydrophobic molecule that is freely mobile in the lipid bilayer known as ubiquinone, or coenzyme Q. A quinone (Q), can pick up or donate either one or two electrons; upon reduction, it picks up a proton from the medium along with each electron it carries. The respiratory chain proteins involve more than 60 different proteins in all. As one would expect, the electron carriers have higher affinities for electrons as they move along the respiratory chain.

7.5.15 Oxidative phosphorylation complexes

The various components of the respiratory chain are integrated within the molecular structure of the inner membrane, which has high protein content (60%–70%). The respiratory chain, together with the phosphorylating system represents about 35% of the membrane proteins. David Green and associates have recognized four main complexes (I–IV). Electron transport through complexes I, III and IV is coupled to the transport of protons out of the interior of the mitochondrion (**Fig. 7.17**). Complex V is the ATP synthase (ATPsyn). Thus the energy yielding reactions of electron transport are coupled to the transfer of protons from the matrix to the intermebrane space, which establishes a proton gradient across the inner membrane.

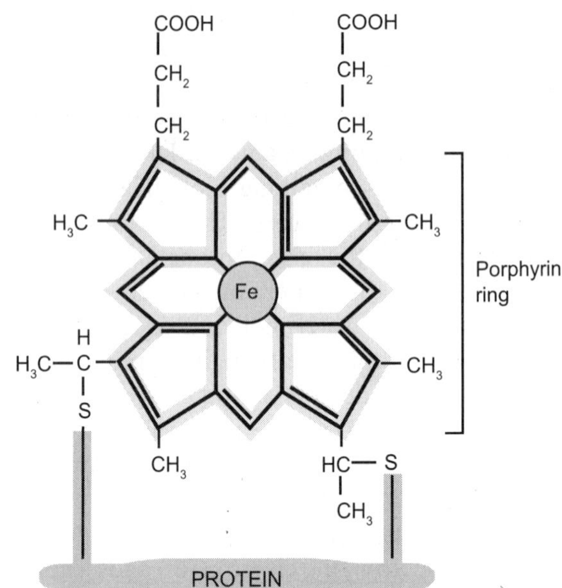

Fig. 7.16 Structure of the heme group attached covalently to cytochrome c.

The four respiratory chain enzyme complexes embedded in the inner membrane and involve in translocation of protons across the membrane are NADH dehydrogenase, succinate dehydrogenase, cytochrome b and c and cytochrome oxidase.

Complex I

The NADH dehydrogenase complex (complex I) is the largest of the respiratory enzyme complexes, containing more than 46 polypeptide chains (500kDa), seven (MTND1, 2, 3, 4, 4L, 5 and 6) of which are encoded by the mtDNA. These encompass 3 flavoproteins, 7 iron–sulphur proteins (ISPs) and over 24 hydrophobic proteins. It accepts electrons from NADH and passes them through a flavin and at least nine iron–sulphur centres (includes N-1a, N-1b, N-2, N-3, N-4 and N-5), to ubiquinone. Ubiquinone then transfers its electrons to a second respiratory enzyme complex, the cytochrome bc_1 complex. The NADH reaction site is at the M-side. Contact between this protein and CoQ is made in the middle of the membrane, in the hydrophobic area. Electrons are transferred from M-side to C-side and use the energy released to pump protons across the inner-membrane. The redox differential across complex I is form –320 mV for NADH to + 60 mV for CoQ_{10}.

Complex II

Complex II, the smallest of the four electron transfer complexes, corresponds to **succinate-Q-reductase** (i.e., succinate dehydrogenase), which can

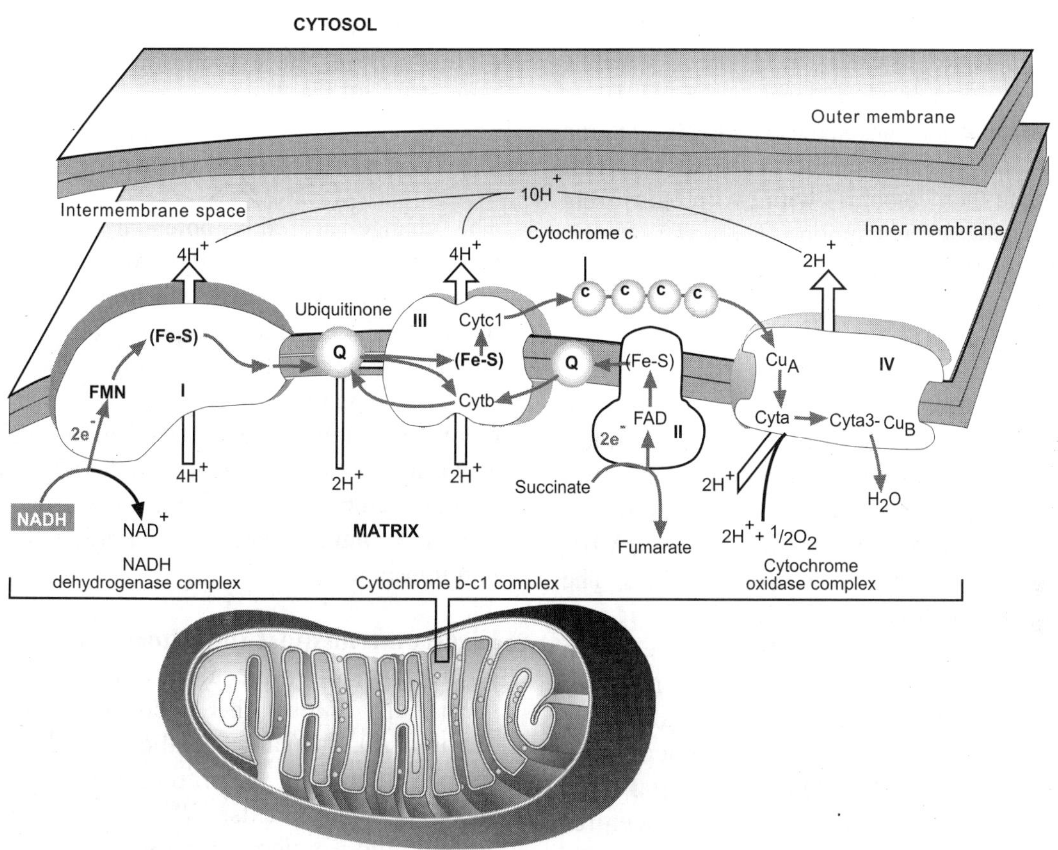

Fig. 7.17 The path of electrons through the three respiratory enzyme complexes. I–IV are protein complexes in the inner membrane. Complex V involved in ATP synthesis is not shown in this diagram.

transfer electrons directly from succinate to CoQ_{10}. The enzyme is composed of four subunits: a 70-kDa flavoprotein, a 27-kDa ISP, a 15-kDa membrane polypeptide and a 5–7 kDa membrane polypeptide. All four complex II subunits are nuclear encoded. The 70-kDa flavoprotein contains the succinate binding site and a covalently bound FAD moiety. The 27-kDa ISP contains three iron–sulphur clusters that transport electrons to CoQ_{10}: centre 1 $[2Fe\text{-}2S]^{2+,1+}$, centre: 2 $[4Fe\text{-}4S]^{2+,1+}$ and centre 3 $[3Fe\text{-}4S]^{1+,0}$ CoQ_{10} is bound to the two membrane intrinsic subunits, CII-3 and CII-4. A b_{560}-type heme is also associated with CII-3(SDHD) and CII-4 (SDHC), but its function is unclear. The succinate binding site appears to be on the M-side (matrix side). This complex cannot translocate protons across the membrane.

Complex III (cytochrome bc_1 complex),

The complex III, also known as cytochrome bc_1, complex contains at least 11 different polypeptide chains, one (cytochrome b, MTCYB), encoded by the mtDNA and cytochrome c1 and iron–sulphur protein. It functions as a dimer, each monomer (240 kDa), contains three hemes bound to cytochromes and an iron–sulphur protein. The complex accepts electrons from ubiquinone (oxidizes ubiquinol, $CoQ_{10}H_2$) and passes them on to reduce ferricytochrome c to ferrocytochromec and uses the energy released to pump protons across the mitochondrial inner membrane via the Q cycle. The polypeptides of complex III include two core proteins (subunit I and subunit II), MTCYB (subunit III), cytochrome c_1 (subunit IV), the 'Rieske' ISP (subunit V) and seven smaller polypeptides, subunits 6–11. The cytochrome b, cytochrome c_1 and the Rieske ISP provide the major redox centres of the enzyme.

Proton translocation of complex III is linked to electron transport through the intermembranes of CoQ_{10} oxidation–reduction in the proton motive Q cycle. Complex III has two CoQ_{10} binding sites, one on the cytoplasmic side (Q_o or Q_p site) and one on the matrix side (Q_i or Q_n site), of the inner membrane. $CoQ_{10}H_2$ binds to the Q_o site and transfers one electron to the Rieske ISP, which passes it on to cyto c_1. The resulting ubisemiquinone donates the other electron to the adjacent B_L heme located on the cytoplasmic side of the membrane. The transfer of the two electrons from $CoQ_{10}H_2$ releases two ubiquinol protons to the outside of the mitochondrial membrane. The electron at the

B_L heme is then transported to the more matrix B_H heme, where it is transferred to CoQ_{10} at Q_i site to generate ubisemiquinone. A second molecule of $CoQ_{10}H_2$ is oxidized in the same manner and the resulting B_H electron is donated to the ubisemiquinone at the Q site. This reduced quinone then combines with two protons from the matrix to generate ubiquinone ($CoQ_{10}H_2$). Hence, two protons are transported through the membrane for each pair of electrons that exits complex III.

Complex IV

Complex IV (cytochrome c oxidase), is a large complex consisting of 13 different polypeptides, three (COI, COII and COIII) encoded by mtDNA and the remaining nuclear encoded (designated as IV, Va, Vb, VIa, VIb, VIc, VIIa, VIIb, VIIc and VIII) include two cytochromes and two copper atoms. The crystal structure of complex IV reveal that it is a functional dimer. The complex accepts one electron at a time from reduced cyto c (ferrocytochrome c) → CuA → cyto a → the bimetallic cyto a_3/CuB active site → finally to oxygen to give water, concurrently pumping four protons across the mitochondrial inner membrane. The cytochromes–iron sulphur centres and copper atoms can carry only one electron at a time. Yet each NADH donates two electrons and each O_2 molecule must receive four electrons to produce water. There are several electron-collecting and electron-dispersing points along the electron transport chain where these changes in electron number are accommodated. The most obvious of these is cytochrome oxidase.

Cytochrome oxidase

An iron–copper centre in the cytochrome oxidase catalyzes efficient O_2 reduction. Because oxygen has a high affinity for electrons, it releases a large amount of free energy when it is reduced to form water (**Fig. 7.18**). The O_2 in the air we breathe is slow in picking up its first electron. This fact allows its initial reaction in cells to be controlled closely by enzymatic catalysis. But once a molecule of O_2 has picked up one electron to form a superoxide radical (O_2^-), it becomes highly reactive and rapidly takes up an additional three electrons wherever it can find them. The cell can use O_2 for respiration only because cytochrome oxidase holds onto oxygen at a special bimetallic centre, where it remains clamped between a heme-linked iron atom and a copper atom until it has picked up a total of four electrons. Only then can the two oxygen atoms of the oxygen molecules be safely released as two molecules of water. Because of this reason, the cytochrome oxidase complex is crucial for all aerobic life.

7.5.16 Energy for H⁺ pumping

The change in redox potential between any two electron carriers is directly proportional to the free energy released when an electron transfers between them. Each enzyme complex acts as an energy conversion device by harnessing some of this free energy change to pump H^+ across the inner membrane, thereby creating an electrochemical proton gradient as electrons pass through that complex. This conversion can be demonstrated by purifying each respiratory enzyme complex and incorporating it separately into liposomes.

Electron transport inhibitors

A number of molecules have been found to specifically inhibit the electron transport process (**Table 7.3**). In many metabolic studies, inhibitors are invaluable in determining the correct order of ETC components. When electron transport is inhibited, oxygen consumption is reduced or eliminated. Oxidized ETC components accumulate on the oxygen side of the site of inhibition. Reduced ETC components accumulate on the non-oxygen side of the site of inhibition. For example, antimycin A inhibits cyt b; rotenone and amytal inhibit NADH dehydrogenase (complex I); carbon monoxide (CO), azide (N_3^-) and cyanide (CN^-), inhibit cytochrome oxidase.

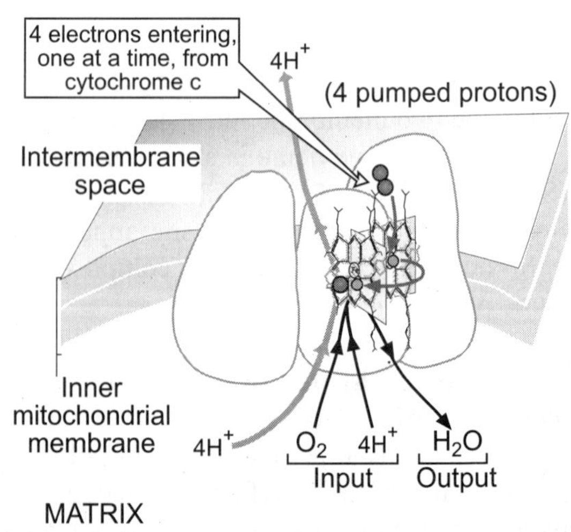

Fig. 7.18 Reaction of O_2 with electrons in cytochrome oxidase.

Table 7.3 Inhibitors of oxidative phosphorylation

Name	Function	Site of Action
Rotenone	e^- transport inhibitor	Complex I
Amytal	e^- transport inhibitor	Complex I
Antimycin A	e^- transport inhibitor	Complex III
Cyanide	e^- transport inhibitor	Complex IV
Carbon monoxide	e^- transport inhibitor	Complex IV
Azide	e^- transport inhibitor	Complex IV
2,4-Dinitrophenol	Uncoupling agent	Transmembrane H^+ carrier
Pentachlorophenol	Uncoupling agent	Transmembrane H^+ carrier
Oligomycin	Inhibits ATP synthase	OSCP fraction of ATP synthase

7.5.17 H^+ ionophores

Several substances, such as 2,4-dinitrophenol, have been known to act as uncoupling agents, uncoupling electron transport from ATP synthesis. The addition of these organic compounds to cells stops ATP synthesis by mitochondria without blocking their uptake of oxygen. The uncoupling agent decreases the H^+ gradient by providing the pathway for the flow of H^+ across the inner mitochondrial membrane that bypasses the ATP synthesis. As a result, the proton motive force is dissipated completely and ATP can no longer be made.

Some natural uncouplers convert the mitochondria in brown fat into heat generating machines. In these cells, mitochondrial respiration is normally uncoupled from ATP synthesis. Most of the energy of oxidation is dissipated as heat rather than by conversion into ATP. The inner membranes of the large mitochondria in these cells contain a special transport protein that allows protons to move down their electrochemical gradient, bypassing ATP synthase. As a result, the cells oxidize their fat stores rapidly and produce more heat than ATP. Tissues

Table 7.4 Human mitochondrial diseases

Disease	Mitochondrial gene mutated
Acyl CoA dehydrogenase deficiency	Reduced fatty acid oxidation
Carnitine deficiency	No transport of fatty acyl CoA
Kearns–Sayre syndrome	Cytochrome oxidase deficiency
Leber's hereditary optic neuropathy	A mutation in one of several mitochondrial genes that encode respiratory proteins, including ND1, ND2, CO1, ND4, ND5, ND6 or cytb
Menk's disease	Cytochrome aa_3 deficiency
Neurogenic muscle weakness	A mutation in ATPase6 gene that encodes a subunit of the mitochondrial ATP synthetase required for ATP synthesis
Mitochondrial myopathy	Pyruvate dehydrogenase deficiency
Mitochondrial myopathy	Altered F_1-ATPase
Mitochondrial myopathy	Cytochrome b deficiency
Myoclonic epilepsy and ragged-red muscle fibers	Mutation in a gene that encodes tRNA for lysine
Porphyria	Uncoupling of oxidative phosphorylation
Zelleweger syndrome	No oxidation of succinate

containing brown fat serve as 'heating pads' helping to revive hibernating animals and protect sensitive areas of newborn human babies from the cold.

7.5.18 Mitochondrial diseases of humans

The human mitochondrial genome is maternally inherited, since it is transmitted from mother to offspring via the cytoplasm of the egg. Therefore, the transmission of human mitochondrial diseases follows a strict maternal inheritance pattern (**Table 7.4**). Diseases involving mutations in mitochondrial genes are usually chronic degenerative disorders that affect the brain, heart, muscle, kidney and endocrine glands. For example, Leber's hereditary optic neuropathy (LHON) is a disease that affects the optic nerve. It may lead to progressive loss of vision in one or both the eyes. LHON can be caused by defective mutation in one of the several mitochondrial genes.

Summary

- Eukaryotes have DNA in their mitochondria and in plastids, which exhibit ***cytoplasmic inheritance.***
- ***Mitochondria*** are the major sites of aerobic respiration within cells and the metabolism of pyruvate is completed: the pyruvate is imported into the mitochondrion and oxidized by O_2 to CO_2 and H_2O. This allows 15 times more ATP to be produced than that by glycolysis alone.
- Mitochondrial shape and size are tremendously variable and are usually larger in active cells than in less active ones.
- Electron micrographs show that the mitochondrion is bounded by two membrane envelopes. They have a smooth outer membrane and a convoluted inner membrane (***cristae***).
- The cristae may be simple or branched, forming a complex network which provides access to the respiratory enzymes and also provides additional membrane surface. The inner membrane in the cristae is covered by stalked particles placed regularly at intervals of 10 nm on the inner surface of the inner membrane.
- The mtDNA is located in the mitochondrial matrix. Only a few proteins are encoded by mitochondrial DNA and synthesized on mitochondrial ribosomes. The vast majority of mitochondrial proteins are encoded by nuclear genes, synthesized on cytosolic ribosomes and imported into the mitochondrion.
- The genetic code used in mitochondria is different from the standard code used in all prokaryotic and eukaryotic nuclear genes
- Mitochondria are often referred to as the '***powerhouse***' of the cell, since they produce 95% of the ATP molecules in animal cells.
- ***Cellular respiration*** the oxidation of nutrients to generate energy in the form of adenosine 5'-triphosphate (ATP), occurs in the mitochondria.
- ***Oxidative phosphorylation***, is made possible by close association of electron carriers with protein molecules. The protons direct the electrons along the respiratory chain such that the electrons move sequentially from one enzyme complex to another. The transfer of electrons is coupled to oriented H^+ uptake and release.
- Oxidative phosphorylation in the inner mitochondrial membrane drives ATP synthesis by the membrane-bound enzyme ***ATP synthase***, a multisubunit protein.
- The various components of the respiratory chain are integrated within the molecular structure of the inner membrane, which has high protein content.
- Electron transport through complexes I III and IV is coupled to the transport of protons out of the interior of the mitochondrion. Complex V is the ATP synthase (ATPsyn). Thus, the energy-yielding reactions of electron transport are coupled to the transfer of protons from the matrix to the intermembrane space, which establishes ***a proton gradient*** across the inner membrane.
- ***The human mitochondrial genome is maternally inherited***, since it is transmitted from mother to offspring via the cytoplasm of the egg. Therefore, the transmission of human mitochondrial diseases follows a strict ***maternal inheritance pattern***.

Exercises

1. Explain cytoplasmic inheritance with suitable examples.
2. Describe the steps involved in the endo symbiotic theory of organellar evolution.
3. Explain with a diagram the ultrastructure of mitochondria.
4. Describe the steps involved in oxidative phosphorylation.
5. Explain in detail the electron transport chain in the inner membrane of the mitochondria.
6. Give a brief description of oxidative phosphorylation complexes.
7. Briefly explain the cytochrome oxidases.
8. Explain electron transport inhibitors.

Further Reading

Books

Fahn A (1974), *Plant Anatomy*, Pergamon Press, Oxford, New York.

Gillham NW (1978), *Organelle Heredity*, Raven Press, New York.

References

Bereiter–Hahn J (1985), Behaviour of mitochondria in the living cell, *Int. Rev. Cytol.*, 122:1.

Birky CW (2001), The inheritance of genes in mitochondria and chloroplasts: Laws, mechanisms and models, *Annu. Rev. Genet.* 35:125–48.

Brandt U (1997), Proton–translocation by membrane-bound NADH: ubiquinone–oxidoreductase (complex I), through redox-gated ligand conduction. *Biochem. Biophys. Acta*,1318:79–91

Breyton C (2000), The cytochrome b_6f complex: Structural studies and comparison with the bc_1 complex. *Biochem. Biophys. Acta*, 1459:467–74.

Capaldi R and Aggeler R (2002), Mechanism of the F_0F_1-type ATP synthase—a biological rotary motor. *Trends Biochem. Sci.*, 27:154–60.

Channarayappa (1993), Resistance mechanism of *Lycopersicon* species to the vector *Bemisia tabaci* and Ultrastructural changes in leaf curl virus infected tomato leaves, PhD thesis submitted to the University of Agricultural Sciences, GKVK, Bangalore, India.

Clayton DA (2000), Transcription and replication of mitochondrial DNA. *Hum. Reprod.*, 2 (Suppl.):11–17.

Ernster L and Schatz G (1981), Mitochondria: A historical review, *J. Cell Biol.*, 91:227.

Forgac M (1999), Structure and properties of the vacuolar (H^+), ATPases, *J. Biol. Chem.*, 274:12951–54.

Frey TG and Mannella CA (2000), The internal structure of mitochondria, *Trends Biochem. Sci.*, 25:319–24.

Gray MW (1999), Evolution of organellar genomes, *Curr. Opin. Genet. Dev.*, 9:678–87.

Gray MW, Burger G and Lang BF (1999), Mitochondrial evolution, *Science,* 283:1476–81.

Grivell L (1983), Mitochondrial DNA, *Sci. Am.*, 248:78–88.

Hatefi Y (1985), The mitochondrial electron transport and oxidative phosphorylation system, *Annu. Rev. Biochem.*, 54:1015–70.

Kurland CG (1992), Evolution of mitochondrial genomes and the genetic code, *BioEssays*, 14:709–14.

Mattson MP (1997), Mother's legacy: Mitochondrial DNA mutations and Alzheimer's disease, *Trends Neurosci.*, 20:373–75.

Miesfeld RL (1999), *Applied Molecular Genetics*, Wiley–Liss, New York.

Nicholls DG and Ferguson SJ (2002), *Bioenergetics*, 3rd edition, Academic Press, San Diego.

Palmer JD, et al. (2000), Dynamic evolution of plant mitochondrial genomes: mobile genes and introns and highly variable mutation rates, *Proc. Nat'l. Acad. Sci.*, USA 97:6960–66.

Sadava DE (1993), *Cell Biology–Organelle Structure and Function*, Jones and Bartlett, Boston.

Schultz BE and Chan SI (2001), Structures and proton–pumping strategies of mitochondrial respiratory enzymes. *Annu. Rev. Biophys. Biomol. Struct.*, 30:23–65.

Shadel GS and Claton DNA (1997), Mitochondrial DNA maintenance in vertebrates, *Annu. Rev. Biochem.*, 66:409–35.

Starr C and Taggart R (1995), *Biology, The Unity and Diversity of Life*, Wadsworth Publishing Company, a division of International Thomson Publishing Inc.

Wallace DC (1999), Mitochondrial diseases in man and mouse. *Science,* 283:1482–88.

8 Plastids: Food Factory of the Cell

Learning objectives

After reading this chapter, you will be able to:

- Identify the different components of plastids and their respective roles
- Trace the various steps in photosynthesis: light reaction and dark reaction

8.1 INTRODUCTION

Every animal and most microorganisms on Earth, rely continuously on the uptake of large amounts of organic compounds from their environment. These compounds are used to build biomolecules of various structures and to obtain the metabolic energy that drives a cellular process. The earliest life forms obtained their raw materials and energy from simple organic molecules dissolved in their aqueous environment. These organic molecules formed **abiotically**, that is, as a result of non-biological chemical reactions. Organisms that depend on an external source of organic compounds are called **heterotrophs**. The early heterotrophs were severely restricted, because spontaneous production of organic molecules occurs very slowly. Only those life forms that could employ a new metabolic strategy had better survival fitness. Unlike their predecessors, these organisms could manufacture their own organic nutrients from the simplest types of inorganic molecules, such as carbon dioxide (CO_2) and hydrogen sulphide (H_2S). Organisms capable of surviving on CO_2 as their principal carbon source are called **autotrophs**.

The manufacture of complex organic molecules from CO_2 requires the input of large amounts of energy. Over the course of evolution, two main types of autotrophs have evolved and can be distinguished by their source of energy. **Chemo-autotrophs** utilize the energy stored in inorganic molecules (such as ammonia, hydrogen sulphide, or nitrites) to convert CO_2 into organic compounds, while **photo-autotrophs** utilize the radiant energy of the sun to accomplish this task. Chemo-autotrophs are prokaryotes and their relative contribution to the formation of biomass on earth is very small. Photo-autotrophs (plants, eukaryotic algae, various flagellated protists and members of several groups of photosynthetic bacteria) on the other hand, are responsible for capturing the energy that fuels the activities of most organisms on earth. All these organisms carry out **photosynthesis**, a process by which energy from sunlight is transformed into chemical energy and stored as carbohydrates and other organic molecules.

8.2 PHOTOSYNTHESIS: BIOLOGICAL OXIDATION–REDUCTION PROCESS

During photosynthesis, relatively low-energy electrons are removed from a donor compound and converted into high-energy electrons using energy absorbed from light. These high-energy electrons are then employed in the syntheses of reduced biological molecules, such as starch and oils. Photosynthesis is a biological oxidation–reduction (redox) process in which electrons are driven uphill away from H_2O to a weaker electron acceptor CO_2. Oxidation reactions (respiration, combustion or oxidation of gasoline or carbohydrates in wood to CO_2 and H_2O) are spontaneous processes that release energy. These relationships are summarized in **Fig. 8.1**.

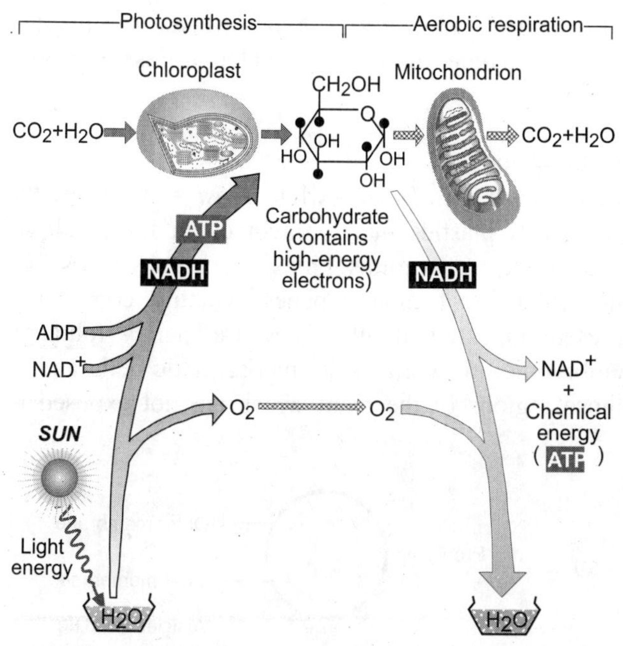

Fig. 8.1 A comparative overview of the energetics of photosynthesis and aerobic respiration.

Photosynthetic organisms convert light energy into stable chemical products. CO_2 is the electron acceptor and H_2A is any reduced compound that can serve as the electron donor. (CH_2O) represents the carbohydrate generated by the reduction and A represents the product formed by oxidation of H_2A:

$$CO_2 + 2H_2A \xrightarrow{\text{light}} (CH_2O) + 2A + H_2O$$

It is likely that the first groups of photo-autotrophs, which dominated primitive earth, may have performed **anoxygenic** photosynthetic reactions and using hydrogen sulphide as their source of electrons for photosynthesis, carrying out the overall reactions:

$$CO_2 + 2H_2S \xrightarrow{\text{light}} (CH_2O) + H_2O + 2S$$

There are numerous bacteria that carry out this type of photosynthesis. These bacteria retain the capacity to reduce CO_2 to carbohydrate and in some cases the mechanism of the carbon fixation reactions is almost identical to those found in oxygenic organisms. However, hydrogen sulphide is not abundant on earth and therefore, a new type of photosynthetic prokaryote appeared on earth (about 2.7 billion years ago) that was able to utilize a much more abundant source of electrons, namely **water**. This new form not only allowed the use of water but also enabled these organisms to exploit a much more diverse array of habitats on earth. It also produced a by-product of enormous consequence for all forms of life—molecular oxygen (O_2), which is formed, in the overall reaction:

$$CO_2 + H_2O \xrightarrow{\text{light}} (CH_2O) + O_2.$$

The switch from H_2S to H_2O as the substrate for photosynthesis was very difficult. The redox potential of the S–H_2S couple is –0.25 V as compared to +0.816 V for the O_2–H_2O couples. In other words, the sulphur atom in a molecule of H_2S has lesser affinity for its electrons (and is thus easier to oxidize) than does the oxygen atom in a molecule of H_2O. Thus, if an organism is going to carry out oxygenic (oxygen-releasing) photosynthesis, it has to generate a very strong oxidizing agent as part of its photosynthetic metabolism in order to pull out the tightly held electrons from water.

8.2.1 Site of photosynthesis

At some point of time, one of these ancient oxygen-producing cyanobacteria took up residence inside a mitochondria-containing, non-photosynthetic pro-eukaryotic cell. Over time, the symbiotic cyanobacterium was transformed from a separate organism living within a host cell into a cytoplasmic organelle, the **chloroplast** (see Chapter 7). Since the early days of primitive earth, most of the organic compounds required by the living cells, have been provided by the photosynthetic organisms, including photosynthetic bacteria. The most advanced photosynthetic bacteria (existing since early days) are the cyanobacteria, which have minimal nutrient requirements. They derive electrons from water and energy from sunlight, when they convert atmospheric CO_2 into organic compounds by a process called **carbon fixation**.

At present, plants and algae are the major source of photosynthetic activities. The chloroplast—in which photosynthesis occurs in a unique and highly specialized intracellular organelle—developed much later. Chloroplasts perform photosynthesis in the presence of sunlight (daylight hours). The primary products of photosynthesis are NADPH and ATP, which are used by the photosynthetic cells to produce many organic compounds. In plant cells, the next immediate products include a low-molecular weight sugar (usually sucrose) that is exported to meet the metabolic needs of the many non-photosynthetic cells of the organism.

8.3 PLASTIDS

Plastids are major organelles found only in plant and algal cells. The term 'plastid' (Gr., *plastikas* = formed as moulded) includes the cell organelles which are primarily involved in the formation and storage of both soluble and insoluble carbohydrates. Plastids belong to a special category of cell organelles and all plastids within a given plant contain identical genomes. The morphology of the plastid in a given plant tissue, however, can be quite different to that of the plastids in another tissue such as leaves, petals, stem, roots or tubers. This indicates that the morphology of plastids is controlled from outside and is not dictated by the plastid genome. Various types of plastids can be distinguished. They differ from each other in structure and function, but develop from similar primordial organelles.

8.3.1 Differentiation of plastids

The chloroplast is one of several types of organelles termed plastids, all of which are formed from the same precursor **proplastid** (**Fig. 8.2**), and all of which contain the same chloroplast DNA. Proplastids are small (0.5–1 μm) undifferentiated organelles present in rapidly dividing cells of plant roots, shoots and embryonic tissues, which then develop into various types of mature plastids according to the needs of differentiated cells. Proplastids are composed only of an inner- and outer-membrane and a small stromal space that contains the chloroplast DNA (ctDNA). Plastid development is controlled both by environmental signals and by intrinsic programs of cell differentiation. During the process of chloroplast development, the thylakoid membrane formed by vesicles bud out from the inner membrane of the plastid envelope and arrange themselves in stacks; these vesicles incorporate essential proteins and chlorophyll and are transformed into mature thylakoid vesicles. Chloroplasts develop only in the presence of light. Beside photosynthesis, plastids are involved in nitrogen assimilation, synthesis of some amino acids, many fatty acids, lipids and even some plant hormones.

Depending on the plant tissue and environmental cues, proplastids differentiate into several types of plastids (**Fig. 8.3**): **chromoplasts** (pigmented organelles other than green) and **leucoplasts** (without pigments).

Leucoplasts

Leucoplasts (Gr., *leuco* = white, *plast* = living) are the colourless plastids mainly involved in the synthesis and storage of various kinds of carbohydrates and in synthesis of monoterpenes—volatile compounds present in essential oils. They are mainly found in embryonic and germ cells, meristematic cells and in those regions of the plant which are not exposed to

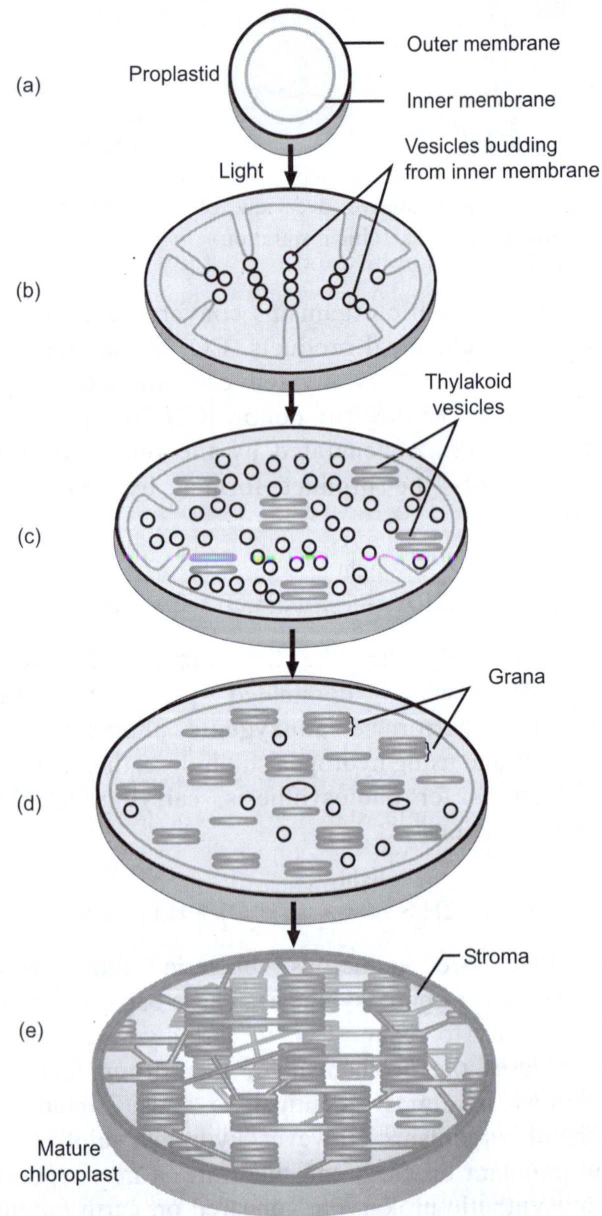

Fig. 8.2 Steps in light-induced differentiation of a proplastid into a chloroplast.

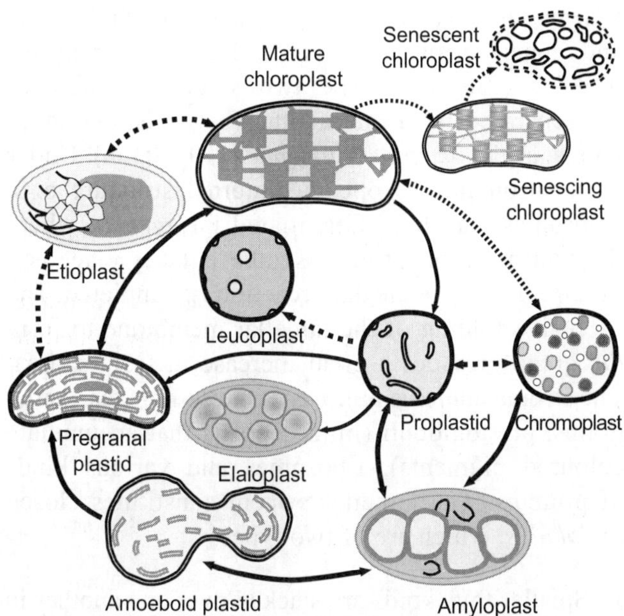

Fig. 8.3 The proplastid and its differentiated progeny organelles.

light. **True leukoplasts** occur in fully differentiated cells such as epidermal and internal plant tissues. These will not become green again, since they do not contain thylakoids and ribosomes. Depending on the compounds they synthesize and store, the leucoplasts are called amyloplasts, elaioplasts, proteinoplasts and aleuroplasts.

Amyloplasts (L., *amylo* = starch; Gr., plast = living) are unpigmented, starch-containing plastids found in tissues such as potato tubers. The outer membrane of the amyloplast encloses the stroma and contains 1 to 8 starch granules. Their name is derived from amylose, a linear, water-soluble polysaccharide composed of $\alpha(1\rightarrow4)$-glucopyranosyl units, which is a major component of starch. Starch grains, which can be quite massive, occur free in the stroma. In potato tubers, for example, starch grains commonly fill the entire plastid, except for a thin layer of stroma lining the inner membrane of the envelope.

Elaioplasts store droplets of lipids (oils), and are generally found in seeds. They also include sterol-rich **sterinochloroplast**. **Proteinoplasts** or **aleuroplasts** are involved primarily in the storage and synthesis of protein granules. They occur mostly in seeds and contain a few thylakoids. Thus, these plastids in non-photosynthetic tissues store essential energy sources, such as starch, lipids and proteins. What controls the differentiation of these various plastids from a common progenitor plastid is not known. **Eoplasts** contain small fragmented thylakoids. They are often formed in meristems and seeds. It has been suggested that they represent the progenitor of all other plastid types. **Etioplasts** are found in leaves where light is deprived. They are usually smaller than chloroplasts, and are spherical in shape. Their most conspicuous feature is a three-dimensional lattice of tubules, the prolamellar body.

Chromoplasts

Chromoplasts (Gr., *chroma* = colour; *plast* = living) are the coloured plastids containing various kinds of pigments. They impart colour (e.g., yellow, orange and red) to certain portions of plants such as the petals, fruits and some roots. Their structure is quite diverse; they may be round, ellipsoidal, needle–shaped or crystalline. For example, **carotenoids** are present in green algae and in higher plants; **pheoplasts** (Gr., *phaeo* = dark) contain the pigment fucoxanthin, which absorbs light, e.g., Brown algae, diatoms, dinoflagellates; **rhodoplasts** (Gr., *rhode* = red) contain the pigment phycoerythrin, e.g., red algae; **blue green chromoplasts**, contain phycocyanin, phycoerythrin, chlorophyll a and carotenoids, e.g., blue green algae. Chromoplasts are responsible for the colour of flowers, ripening fruits (e.g., tomato contains red pigment **lycopene**), autumn leave and some roots. They can arise either from chloroplasts or rarely from leucoplasts. In general, chromoplasts have a reduced chlorophyll content and are photosynthetically less-active.

Chloroplasts

The most important plastids are **chloroplasts** (Gr., *chloro* = green), a prominent member of the plastid family, which carry out photosynthesis by converting light energy into chemical energy.

Shape, size and number: The shape, size and number of chloroplasts vary in different species and also in different cells within a species. In leaves of higher plants, each cell contains a large number of spheroid, ovoid or discoid chloroplasts. In algae, chloroplasts are found in various shapes such as ribbon-like, reticulate, cup, spiral band, stellate, sheet and disc-like. In higher plants, the chloroplasts are usually lens-shaped (**Fig. 8.4**) and are about 3–10 μm (average 5 μm) in diameter. The size of the chloroplasts varies from species to species. Chloroplasts generally measure 2–3 μm in thickness and 5–10 μm in diameter. Differences in size may sometimes be genetically controlled or may depend on genetic and environmental conditions.

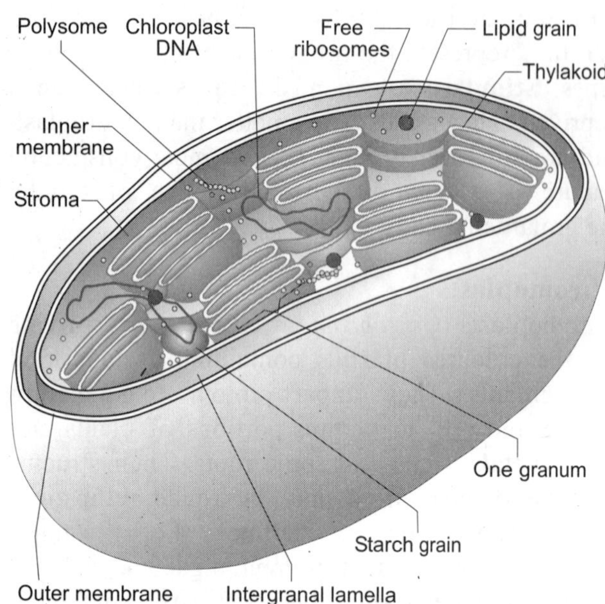

Fig. 8.4 Structure of a chloroplast (**See Plate 4**).

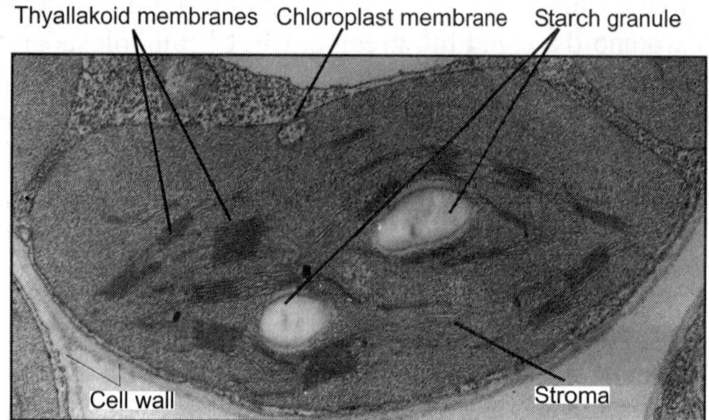

Fig. 8.5 TEM micrograph of cross section of a tomato leaf chloroplast.

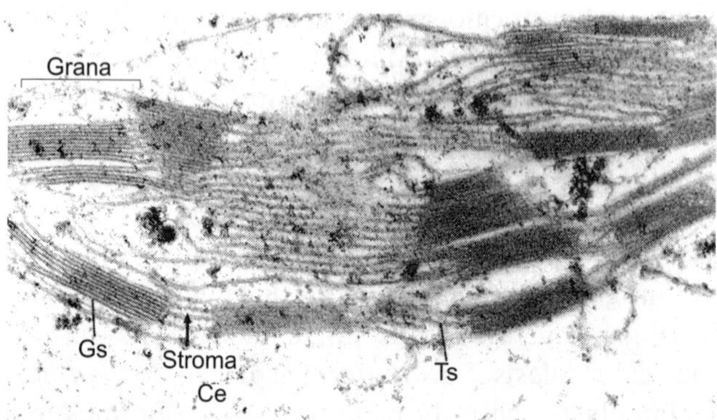

Fig. 8.6 Thin section of tomato thylakoid membrane showing stacked grana (Gs) and unstacked stroma (Ts) membrane regions (Channarayappa, 1992).

Ultrastructure: Chloroplasts have permeable outer and inner membranes embedded with transport proteins (**Fig. 8.5**). The two smooth membranes are separated from one another by a distinct space known as the *periplastidial space* (10–30 nm). Under an electron microscope the internal structure of a chloroplast can be differentiated into *grana*, where the light reaction of photosynthesis takes place and the *stroma*, where the dark reaction is completed. The extensive folding of the lamellar membrane to form the grana provides a great increase in surface area. The stroma contains multi-enzyme complexes, starch grains, plastoglobuli (lipid deposits that accumulate coloured pigments), ribosomes and various kinds of proteins. The lamellar system constitutes closed *thylakoids*, which are of two types:

1. Smaller thylakoids are stacked over one another in order to give rise to *grana lamellae*, so that each granum represents a column of flattened oppressed discs, piled one on top of the other (**Fig. 8.6**).
2. Larger thylakoids connect the grana and are known as *intergranal lamellae* or *stroma lamellae*.

The number of thylakoids per granum varies depending on the organism. The grana lamellae contain light absorbing pigments called *quantasomes*. These quantasomes are the smallest units capable of carrying out the photochemical reaction and each of them contains about 200 chlorophyll molecules. The quantasomes may also be present on stroma lamellae with lower number.

Photosynthetic pigments are compounds that appear coloured because they only absorb light of particular wavelengths within the visible spectrum. The main component of chloroplast is *chlorophyll*, a magnesium-containing molecule with hydrophilic porphyrin rings. Chloroplasts in the leaves are green, because they contain large quantities of chlorophyll, which absorbs most strongly in the blue and red wavelengths, leaving the intermediate green wavelengths to be reflected. The chloroplast can be found either as chlorophyll a or chlorophyll b, the latter having an aldehyde (-CHO) group in place of $-CH_2$ group of chlorophyll a. Chlorophyll can be synthesized by the action of light from a precursor substance, *protochlorophyll*. Chlorophyll pigments consist of a porphyrin ring attached to a long hydrocarbon side chain (**Fig. 8.7**). The chlorophyll molecule at the centre contains a Mg^{2+} ion and has additional five member rings. The alternating singles and double bonds along the edge of the porphyrin ring delocalize electrons, which form

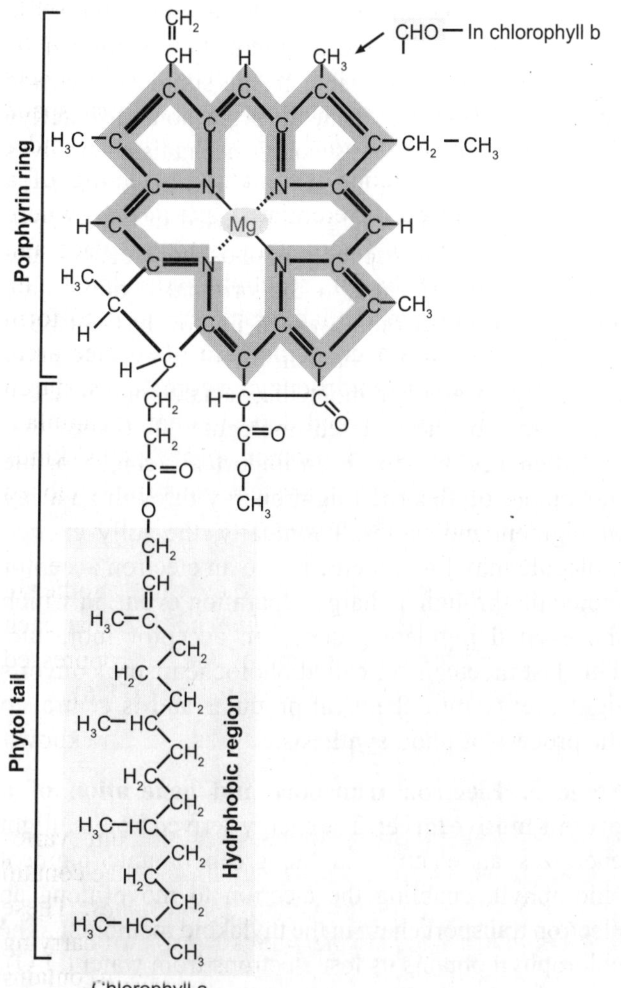

Fig. 8.7 The structure of chlorophyll a. A magnesium atom is held in a porphyrin ring. Electrons are delocalized over the alternate single and double bonds.

a cloud around the ring. Systems of this type are said to be conjugated and are strong absorbers of visible light. The absorbed energy causes a redistribution of the electron density of the molecule, which in turn favours loss of an electron to a suitable acceptor. The energy of the absorbed light is used to remove electrons from water, forming oxygen and hydrogen ions. Several classes of chlorophyll, differing from one another in the side groups attached to the porphyrin ring, occur among photosynthetic organisms. In addition to chlorophyll, terrestrial plants also contain orange and red accessory pigments called **carotenoids**. These absorb light primarily in the blue and green region of the spectrum, while reflecting those of the yellow, orange and red regions.

Functions of the chloroplast: The most significant function of the chloroplast is to carry out the process of photosynthesis. In most common oxygen-evolving (oxygenic) photosynthesis, water serves as the reductant. Water is oxidized and the electrons released are energized and ultimately transferred to CO_2, yielding oxygen and carbohydrate. This light-driven, endergonic (energy-requiring) reaction has a free energy change (ΔG°) of +2840 kJ per mole of hexose formed. Plants, algae and cyanobacteria utilize the following water-cleaving photosynthetic reaction:

$$CO_2 + 2H_2O \xrightarrow{\text{light}} (CH_2O) + O_2 + H_2O$$

8.4 PHASES OF PHOTOSYNTHESIS: LIGHT AND DARK REACTIONS

The many reactions that occur during photosynthesis can be grouped into two broad categories: light reaction and dark reaction. The two phases of photosynthesis occur in different regions of the chloroplast: the light reaction occurs on thylakoid membranes and dark reaction occurs in the stroma.

8.4.1 Light reaction

Light reaction is also called **photosynthetic electron transfer reaction**, **photochemical reaction** or **light-dependent reaction** (or **Hill reaction**). In the late 1930s, Robert Hill found that various redox-active compounds, including non-physiological iron-containing compounds, could serve as electron acceptors and could support the evolution of oxygen by chloroplasts in the light, even in the absence of CO_2. During the 1950s and1960s, Daniel I Arnon and colleagues showed that isolated chloroplasts can convert CO_2 to carbohydrate in light. R Hill and F Bendall proposed the Z scheme for electron flow in the light reaction. It describes the flow of electrons from water, through the two photosystems, to NADP.

The light reaction is further divided into three stages:

- absorption of light,
- electron transport leading to formation of O_2 from H_2O, reduction of $NADP^+$ to NADPH and generation of a proton-motive force
- synthesis of ATP.

All these, stages are catalyzed by the proteins present in the thylakoid membrane.

Stage 1. Light absorption and energy conversion: The initial step in photosynthesis is the absorption of light by chlorophylls attached to the proteins in the thylakoid membranes. Light has properties of both particles and waves. Light travels in packets (or quanta) of energy called photons. The energy content of a photon depends on the wavelength of the light, according to the equation ($E = hc\lambda$). Where, h = Planck's constant (1.58×10^{-34} cal s), c is the speed of light in vacuum and λ is the wavelength of light. One mole (6.02×10^{23}) of photons (or an Einstein) of 680 nm wavelength, contains approximately 42 kcal of energy, which is equivalent to a change in redox potential of approximately 1.8 V. Relating this data to the +2840 kJ/mol required for CO_2 fixation give a some idea of the theoretical minimum number of quanta required to convert six CO_2 molecules to one molecule of hexose sugar. When a molecule absorbs photon, an electron becomes sufficiently energized to be pushed from an inner to an outer orbital. The molecule is said to have shifted from the ground state to an excited state. The excited state of a molecule is unstable and can be expected to last only about 10^{-9} s. There are several consequences that can befall an excited electron, depending on the circumstances (**Fig. 8.8**). If the electron of the excited molecule drops back to a lower orbital, the energy it is absorbed must be released. In the simplest case, the energy is released as heat during a non-**radioactive** decay (relaxation). A second mechanism involves the emission of a photon in a process known as fluorescence. If the energy is released in the form of heat or light (fluorescence or phosphorescence), the molecule has returned to the original ground state and the energy of the absorbed photon has not been utilized. A third mechanism involves the transfer of energy to another molecule, usually one in close proximity to the excited molecule. This process, called energy transfer, is an important vehicle for the movement of absorbed light energy through an array of pigment molecules. Eventually, the fully excited molecule may lose an electron to an electron acceptor molecule through a charge separation event, in which the excited pigment reduces an acceptor molecule. This last mechanism, called photochemistry, converts light energy into chemical products and is central to the process of photosynthesis.

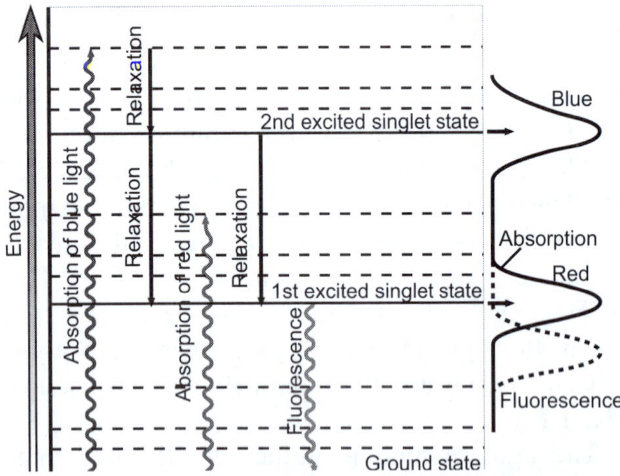

Fig. 8.8 Energy levels in the chlorophyll molecule. Absorption of blue or red light causes chlorophyll molecules to transform into an excited state, with blue light absorption resulting in a higher excited state relative to red light. Internal conversions or relaxations result in the conversion of higher excited states to the lowest, regardless of whether excitation was from red or blue light, with a concomitant loss of energy as heat. Light may be re-emitted from the lowest excited state through the process of fluorescence. The spectra for fluorescence and absorption are shown on the right of the graph.

Stage 2. Electron transport and generation of a proton-motive force: The energy derived from sunlight energizes an electron in the green organic pigment chlorophyll, enabling the electron to move along an electron transport chain in the thylakoid membrane. The chlorophyll obtains its lost electrons from water (H_2O), producing O_2 as a by-product. Electrons move from the quinone primary electron acceptor through a series of electron carriers until they reach the ultimate electron acceptor, usually the oxidized form of nicotinamide adenine dinucleotide phosphate ($NADP^+$), reducing it to NADPH. The transport of electrons in the thylakoid membrane is coupled to the movement of protons (H^+) from the stroma to the thylakoid lumen (pumped across the thylakoid membrane), forming a proton (pH) gradient across the membrane.

Stage 3. Synthesis of ATP: The resulting electrochemical proton gradient drives the synthesis of ATP in the stroma. Protons move down their concentration gradient from the thylakoid lumen to the stroma through the F_0F_1 complex (ATP synthase), which couples proton movement to the synthesis of ATP from ADP and P_i.

Photochemistry of chlorophyll molecules

Photosynthesis is unique not only for carbon fixation, but also for driving the photosynthetic electron transfer reactions in the chloroplast that generate the ATP and the NADPH needed for production of carbohydrates from CO_2 and H_2O. The energy required for this process is derived from sunlight absorbed by different chlorophyll molecules (**Fig. 8.9**). The process of energy conversion begins

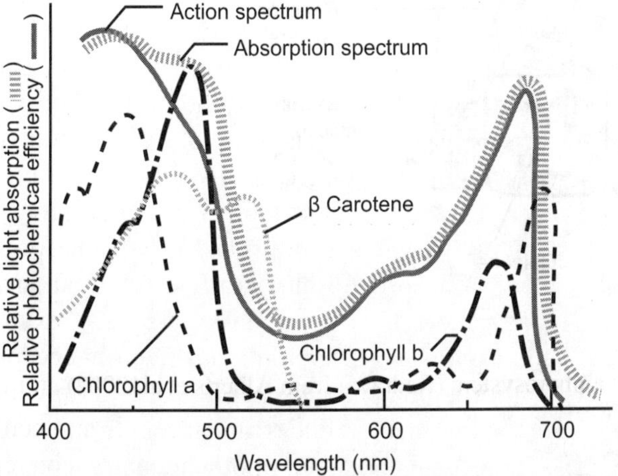

Fig. 8.9 Absorption and action spectrum for photosynthesis. The action spectrum indicates the relative efficiency with which light of various wavelengths is able to promote photosynthesis in the leaves of a plant. An action spectrum can be generated by measuring the O_2 produced by the leaves following exposure to various wavelengths. The hatched lines indicate the absorption spectra of each of the major photosynthetic pigments. The thick hatched line shows the combined absorption spectrum of all pigments.

when a chlorophyll molecule is excited by a quantum of light (a photon) and an electron is moved from one molecular orbital to another of higher energy. Such an excited molecule is unstable and tends to return to its original, unexcited (ground) state in one of three ways as illustrated in **Fig. 8.10**. The last two mechanisms are exploited in the process of photosynthesis.

A photosystem (multiprotein complexes) catalyzes the conversion of the light energy captured in excited chlorophyll molecules to useful forms. It consists of two closely linked components: an **antenna complex** consisting of a large set of pigment molecules and a **photochemical reaction centre**, consisting of a complex of proteins and chlorophyll molecules (**Fig. 8.11**). The antenna complex is important for capturing light. In chloroplasts it consists of a number of distinct membrane protein complexes (known as light harvesting complexes); together, these proteins bind several hundred chlorophyll molecules per reaction centre, orienting them precisely in the thylakoid membrane. Depending on the plant, different accessory pigments called **carotenoids**, which protect the chlorophyll from oxidation and can help collect light of other wavelengths, are also located in each complex. When chlorophyll molecules in the antenna complex are excited, the energy is rapidly transferred from one molecule to another by resonance energy transfer until it reaches a special pair of chlorophyll molecules in the photochemical reaction centre.

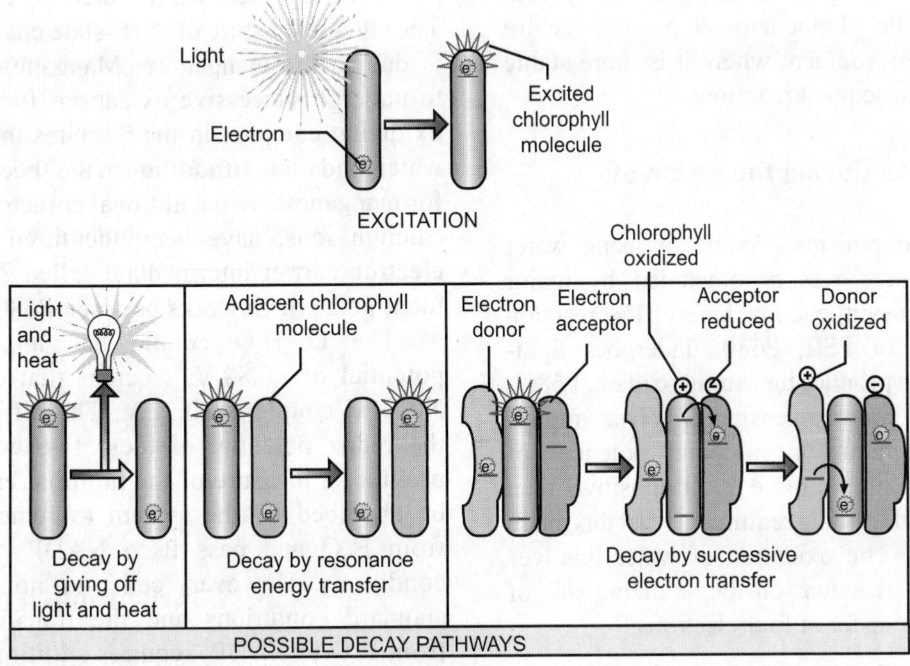

Fig. 8.10 Three ways by which an excited chlorophyll molecule can return to its original unexcited state.

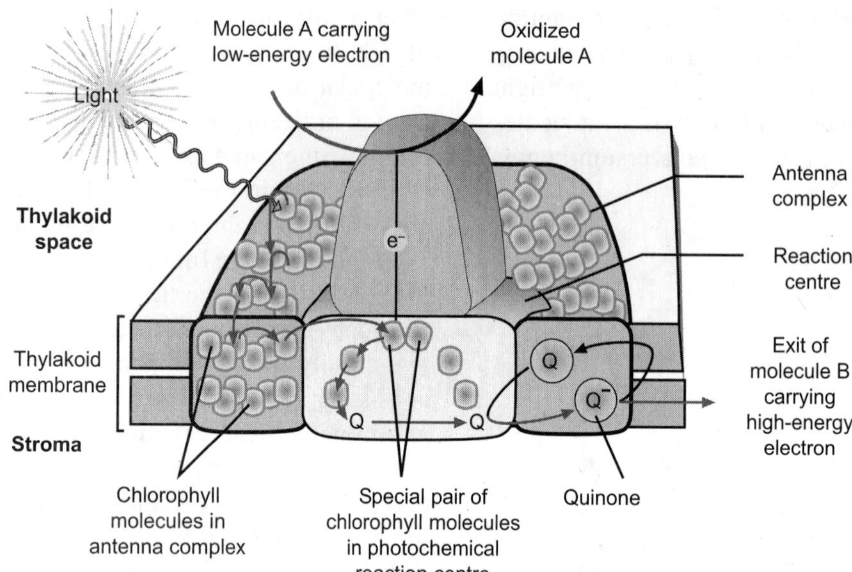

Fig. 8.11 The antenna complex and photochemical reaction centre in a photosystem (Redrawn from Alberts et al, 2002).

Each antenna complex thereby acts as a funnel, collecting light energy and directing it to a specific site where it can be used effectively. The photochemical reaction centre is a transmembrane protein–pigment complex that lies at the heart of photosynthesis. The special pair of chlorophyll molecules in the reaction centre acts as an irreversible trap for excitation quanta, because its excited electron is immediately passed to a chain of electron acceptors that are precisely and closely positioned in the same protein complex. By moving the high-energy electron rapidly away from the chlorophylls, the photochemical reaction centre transfers it to an environment where it is more stable and available for subsequent reactions.

Oxygen generation during the action of photosystem II

The evolution of organisms capable of using water as an electron source was accompanied by major changes in the photosynthetic machinery. The reaction centre chlorophyll of PSII, P680, undergoes light-induced oxidation to produce the strong oxidant, $P680^+$, which can oxidize water, releasing O_2. This implies that the oxidation–reduction midpoint potential of the $P680/P680^+$ couple is more electropositive than +820 mV at pH 7 and splitting requires at least this much oxidizing potential. The oxidation of water involves a complex series of reactions on the oxidizing side of PSII involving the transfer of four electrons:

$$2H_2O \longrightarrow O_2 + 4H^+ + 4e^-$$

The mechanism of O_2 evolution was explained through the S-state model. This model postulates a light-driven charge-accumulation mechanism by which the oxygen-evolving machinery of PSII progresses through five successive states (S_0–S_4) of increasing oxidation, with S_4 being a strong oxidant capable of oxidizing water. Each charge separation event in the P680 reaction centre yields $P680^+$, which ultimately oxidizes the charge accumulator, advancing it to the next S-state and increasing its charge by +1. The only state from which O_2 is evolved is S_4. The chemical nature of the S-state charge accumulator is due to the manganese. Manganese atoms appear to undergo successive oxidations to yield a strongly oxidizing complex in the S_4 states that is capable of water oxidation. In addition to the specific requirement for manganese, two additional cofactors, chloride and calcium ions, have been identified. An additional electron carrier intermediate called Z, has been also identified, which works between P680 and S-state.

The O_2–H_2O couple has a standard redox potential of +0.82 V, whereas that of the $NADP^+$–NADPH couple is –0.32 V. The difference between the redox potential of these two couples (1.14 V) provides a measure of the minimal energy that must be absorbed by the system to remove an electron from H_2O and pass it to $NADP^+$ under standard conditions. However, cells do not operate under standard conditions and the transfer of electrons from H_2O to $NADP^+$ requires additional energy. It is estimated that during operations in the chloroplast, more than 2 V of energy are utilized to carry out this oxidation–reduction reaction. As earlier noted one mole of photons of 680 nm wavelength (red light) is equivalent to a change in redox potential of 1.8 V. Theoretically it is possible for one photon of red light to boost an electron to an energy level required for reducing $NADP^+$ under standard conditions. However, the process is accomplished in the cell through the combined action of two different light-absorbing reactions.

Photosystem II

The first of the two photosystems, photosystem II (PSII), has the unique ability to withdraw electrons from water (i.e., oxidation of water molecules). Photosystem II (**Fig. 8.12**) is a large membrane-spanning protein–pigment complex containing at least 23 components. The most prominent of these is the reaction centre, a protein-pigment complex composed of two polypeptide subunits known as D_1 and D_2, cytochrome b_{559} and a special pair of chlorophyll a molecules (**P680**) that absorb light at 680 nm. The oxygen-evolving components of PSII consists of manganese stabilizing protein (MSP), a membrane-associated complex that contains a manganese cluster and a tyrosine residue (tyr^{161}), often referred to as Y_z, located on D_1. The D_1/D_2 dimer is also associated with several electron acceptors such as pheophytin-a, a chlorophyll-like pigment that accepts an electron

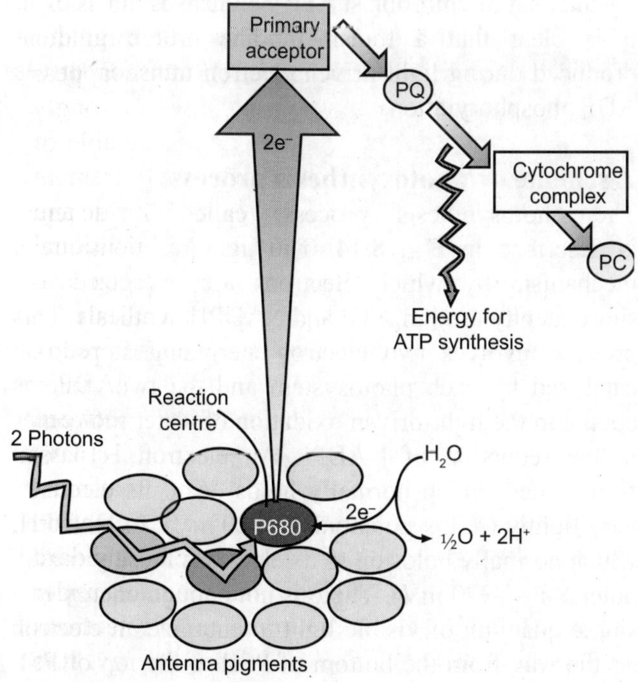

Fig. 8.12 Photosystem II.

from P680. This electron is donated to two forms of plastoquinone (PQ). Several hundred antenna pigment molecules are also associated with the reaction centre. The oxygen of two water molecules binds to a cluster of manganese atoms in a poorly understood water-splitting enzyme. This enzyme enables electrons to be removed one at a time from the water, as required to fill the electron-deficient holes created by light in chlorophyll molecules in the reaction centre. As soon as four electrons have been removed from the two water molecules (requiring four quanta of light), O_2 is released. Photosystem II thus catalyzes the reaction:

$$2H_2O + 4 \text{ photons} \longrightarrow 4H^+ + 4e^- + O_2.$$

The quinones pass their electrons to H^+ pump called the **cytochrome b_6f** complex, which resembles the cytochrome bc_1 complex, pumps H^+ into the thylakoid space across the thylakoid membrane and the resulting electrochemical gradient drives the synthesis of ATP by an ATP synthase.

Excitation energy is passed from the outer-antenna pigments of LHCII to a small number of inner antenna chlorophyll molecules situated within the core of PSII. From there, the energy is ultimately passed to the PSII reaction centre. When LHCII absorbs a photon, its energy is transferred to P680 PSII. An energized electron is then donated to pheophytin- a. Reduced pheophytin- a passes this electron to Q_A. When a second electron is transferred from P680 and 2 protons are transferred from the stroma, reduced plastoquinone (Q_AH_2) (also referred as plastoquinol) is formed. One at a time, Q_AH_2 donates its 2 electrons to Q_B. Two additional stroma proteins are used in the reduction of Q_B. Reduced Q_B (Q_BH_2) donates its electrons to the cytochrome b_6f complex. Finally, the cytochrome b_6f complex donates its electrons to plastocyanin (PC), a mobile peripheral membrane protein. PC, a single-electron carrier, then transfers these electrons to P700 in PSI. The **oxygen-evolving complex**, composed in part of the manganese cluster in MSP and the tyrosine residue located on D_1, is responsible for the transfer of electrons from water to P680. The evolution of one O_2 requires splitting two H_2O, which releases 4 protons and 4 electrons. During this reaction water is converted to O_2 by a mechanism referred to as the water-oxidizing clock. The O_2-evolving complex has five oxidation states: S_0, S_1, S_2, S_3 and S_4. S_0 is the most reduced state and S_4 is the most oxidized state of the complex. It is now believed that Mn cluster, a group of four manganese atoms bound to MSP near the PSII reaction centre, is mostly responsible for these translations. As the oxygen-evolving complex abstracts electrons and protons from water, it cycles through the five oxidation states. The electrons are transferred one at a time to a tyrosine residue on the D_1 polypeptide and then to the P680 reaction centre. The protons released in the process remain in the thylakoid lumen, where they contribute to the pH gradient that drives ATP synthesis.

Photosystem I

The electron acceptor in this electron-transport chain is the second photosystem, photosystem I (PSI), which accepts an electron into the electron-deficient hole created by light in the chlorophyll molecules in its reaction centre. Each electron that enters PSI is finally boosted to a very high-energy level that allows it to be passed to the iron-sulphur centre in ferredoxin and then to NAD^+ to generate NADPH (**Fig. 8.13**). $NADP^+$ is a large membrane-spanning protein–pigment complex composed of several polypeptides. The largest of these are two nearly identical subunits designated A and B. These molecules, referred as a special pair, are located in the core complex of PSI, the AB dimer. Because they absorb light at 700 nm, the special pair within PSI is P700. In addition, the AB dimer contains a series of single electron carriers: A_0, A_1 and A_x. A_0 is a specific chlorophyll *a* molecule that accepts an energized electron from

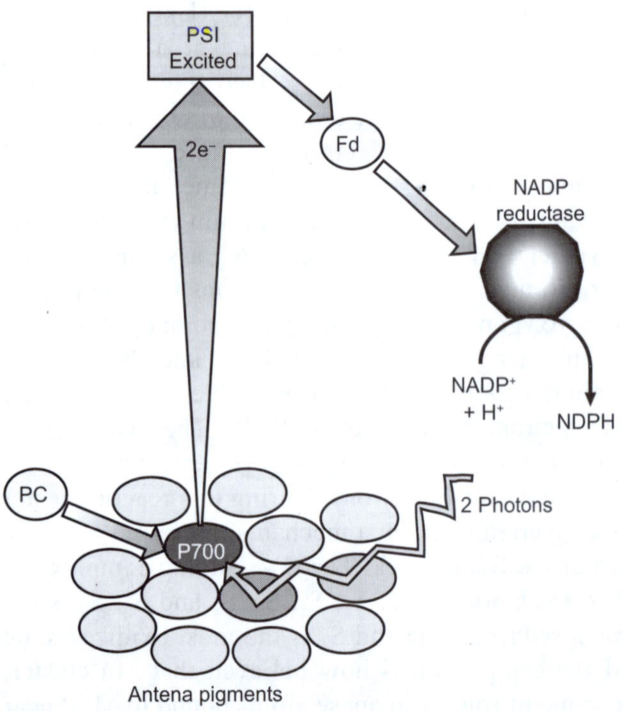

Fig. 8.13 Photosystem I.

P700 and transfers it to A_1. A_1, which has been identified as phylloquinone (vitamin K_1), is similar in structure to ubiquinone (Q). The electron is then transferred from A_1 to F_x, a 4Fe–4S centre. Chlorophyll a molecules other than the special pair, as well as small amounts of chlorophyll b and carotenoids that occur in PSI, act as antenna pigments. Antenna pigments absorb light energy and transfer it to the reaction centre.

Cytochrome b_6f complex: This cytochrome complex contains one cytochrome f, two cytochromes of b_{563}, one FeS centre and a polypeptide. Cytochrome b_6f complex, uniformly distributed throughout the thylakoid membrane, is similar in structure and function to the cytochrome bc_1 complex in mitochondrial inner membrane. The cytochrome b_6f complex plays a critical role in the transfer of electrons from PSII to PSI. An iron–sulphur site on the complex accepts electrons from the membrane-soluble electron carrier plastoquinone and donates them to a small water-soluble copper-containing protein called plastocyanin.

ATP synthase: The chloroplast ATP synthase, also referred to as CF_0CF_1ATP synthase, is structurally similar to the mitochondrial ATP synthase. The CF_0 component is a membrane-spanning protein complex that contains a proton-conducting channel. The CF_1 head-piece, which projects into the stroma, possesses an ATP-synthesizing activity. Although the actual mechanism of chloroplast ATP synthesis is not known, it is clear that a transmembrane proton gradient produced during light-driven electron transport drives ADP phosphorylation.

Z scheme or photosynthesis process

The photosynthesis process, called Z scheme, is described in **Fig. 8.14**. The light reactions are a mechanism by which electrons are energized and subsequently used in ATP and NADPH synthesis. This process involves two electron-energizing steps, one catalyzed by each photosystem and the two systems couple in the light-driven oxidation of water molecules to the reduction of $NADP^+$. An electron is passed from water, which normally holds on to its electrons very tightly (redox potential = +820 mV), to NADPH, which normally holds on to its electrons loosely (redox potential = –320 mV). There is not enough energy in a single quantum of visible light to energize an electron all the way from the bottom of PSII to the top of PSI, which is presumably the energy change required to pass

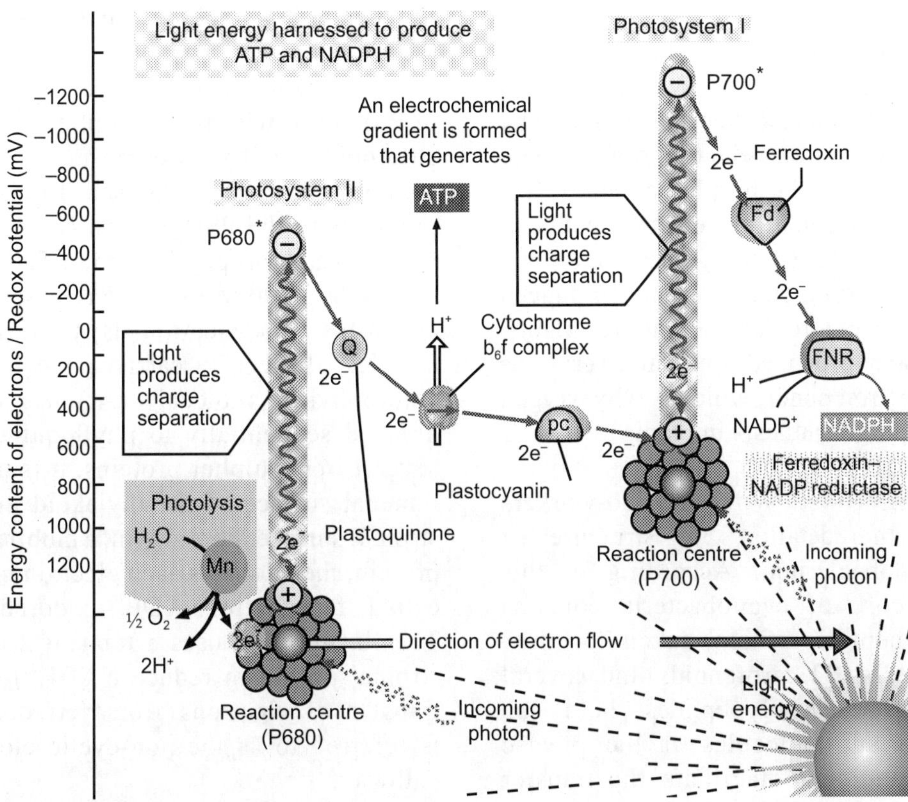

Fig. 8.14 Changes in redox potential (energy content of electrons in volts) during photosynthesis.

an electron efficiently from water to $NADP^+$. The use of two separate photosystems in series means that the energy from two quanta of light is available for this purpose. In addition, there is enough energy left over to enable the electron-transport chain that links the two photosystems to pump H^+ across the thylakoid membrane, so that the ATP synthase can harness some of the light-derived energy for ATP production.

ATP and NADPH production by photo-phosphorylation: In the sequence of electron transfers that take place in the reaction centre (**Fig. 8. 15**), light causes a net electron transfer

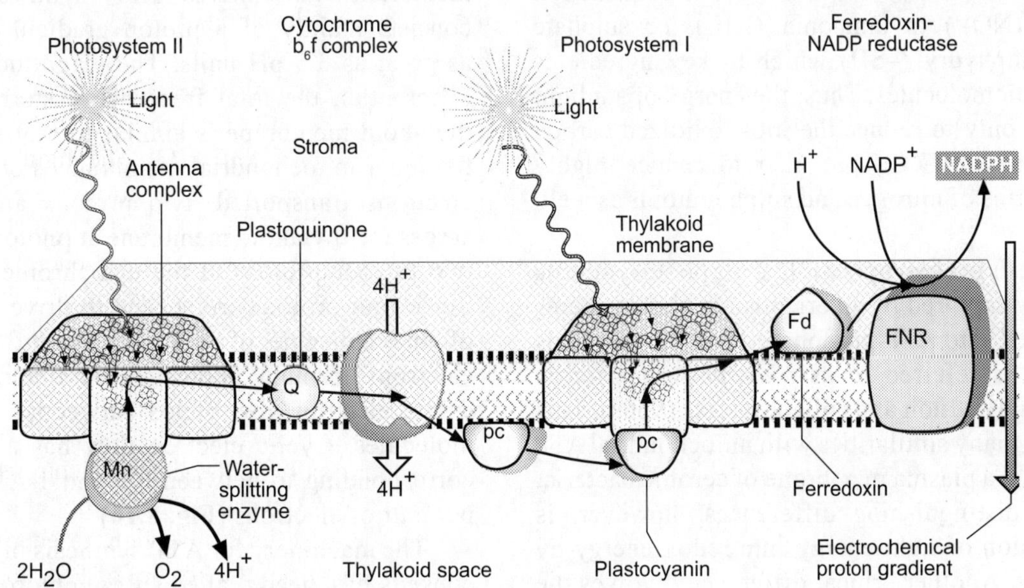

Fig. 8.15 Electron flow during photosynthesis in the thylakoid membrane.

from a weak electron donor to a molecule that is a strong electron donor in its reduced form. The excitation energy in chlorophyll that would normally be released as fluorescence or heat is thereby used instead to create a strong electron donor where none had been before. In the purple bacterium, the weak electron donor used to fill the electron-deficient hole created by light-induced charge separation is a cytochrome and the strong electron donor produced is a quinone. In the chloroplasts of higher plants, a quinone is similarly produced, but water serves as the initial weak electron donor, which is why oxygen gas is released by photosynthesis in plants.

NADPH synthesis: Horst Witt and co-workers (2001) reported the detailed 3D structure of PSI from a cyanobacterium. According to this model, the PSI core of cyanobacteria contains 12 different polypeptides which, taken together, bind 96 chlorophylls, 22 carotenoids and several other elements. These polypeptides hold the functionally important molecules at the precise distance and orientation to facilitate the transfer of excitation energy and electrons through the photosystem.

The photochemical events that take place in PSI begin with the absorption of photons by the chlorophyll and ß-carotene molecules of the core complex. Energy is passed from these antenna pigments to the PSI reaction centre pigment P700, which is a chlorophyll a dimer. The absorption of a photon by P700 leads to the release of an energized electron. This electron is then passed through a series of electron carriers, the first of which is chlorophyll a molecule (A_0). As the electron is donated sequentially to phylloquinone (Q) and to several iron–sulphur proteins, it is moved from the lumenal surface of the thylakoid membrane to its stromal surface. Ferredoxin, a mobile, water-soluble protein, then donates each electron to a flavoprotein called ferredoxin–NADP oxidoreductase (FNR). The flavoprotein uses a total of 2 electrons and a stromal proton to reduce $NADP^+$ to NADPH. The transfer of electrons from ferredoxin to $NADP^+$ is referred to as the non-cyclic electron transport pathway.

$$2\ \text{Ferredoxin}_{red} + H^+ + NADP^+ \xrightarrow{\text{Ferredoxin–NADP}^+\ \text{Reductase}} 2\ \text{Ferredoxin}_{ox} + NADPH$$

Not all electrons passed to ferredoxin in PSI are used for NADPH synthesis; alternate roots can be taken depending on the particular organism and conditions. For example, electrons from PSI can be used to reduce various inorganic acceptors. These paths for electrons can lead to the eventual reduction of nitrate (NO_3^-) to ammonia (NH_3), or sulphate (SO_4^{2-}) to sulphydryl (–SH), which are key ingredients of various biomolecules. Thus, the energy of sunlight is used not only to reduce the most oxidized carbon atoms (those in CO_2), but also to reduce highly oxidized forms of nitrogen and sulphur atoms as well.

ATP synthesis in chloroplast: During photosynthesis light energy captured by an organism's photosystems is transduced into ATP phosphate bond energy. This conversion is referred to as *photophosphorylation* (discovered by Arnon and collaborators, 1950s). This process has many similarities with mitochondrial ATP synthesis and a plasma membrane of aerobic bacteria. The most distinguishing differences, however, is the conversion of light energy into redox energy by chloroplasts. Another critical difference involves the permeability characteristics of mitochondrial inner membrane. In contrast, the thylakoid membrane is permeable to Mg^{2+} and Cl^-. Therefore, Mg^{2+} and Cl^- move across the thylakoid membrane as electrons and protons are transported during the light reaction. The electrochemical gradient across the thylakoid membrane that drives ATP synthesis therefore consists mainly of a proton gradient that may be as great as 3.5 pH units. The magnitude of this pH differential, the total free energy stored across the thylakoid membrane is similar to that stored across the inner mitochondrial membrane. For each pair of electrons transported, two protons are transferred across the thylakoid membrane at photosystem II and two to four protons at the cytochrome *bf* complex. Since four protons are needed to drive the synthesis of one molecule of ATP, passage of each pair of electrons through photosystems I and II by non-cyclic electron flow yields between 1 and 1.5 ATP molecules. Cyclic electron flow has a lower yield, corresponding to between 0.5 and 1 ATP molecules per pair of electrons (**Fig. 8.16**).

The machinery for ATP synthesis in a chloroplast consists of a head (called CF_1 in chloroplast), which contains the catalytic site of the enzyme and a base

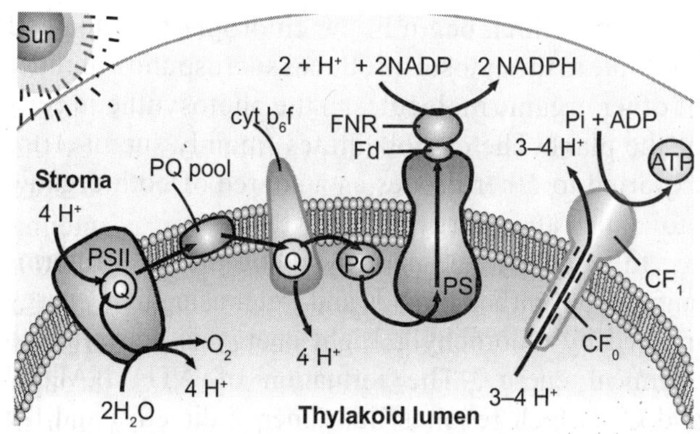

Fig. 8.16 Light-dependent reactions.

(CF_0), which spans the membrane and mediates proton movement. The thylakoid ATP synthase complex contains numerous subunits (**Table 8.1**). The two parts are connected by rotary stalk. The CF_1 heads project outward into the stroma in keeping with the orientation of the proton gradient, which has its higher concentration within the thylakoid lumen. Thus, protons move from higher concentration in the lumen through the CF_0 base of the ATP synthase and into the stroma, thereby driving phosphorylation of ADP. Measurements made during periods of maximal ATP synthesis suggest that differences in H^+ concentrations of 1000- to 2000-fold exist across the thylakoid membranes, corresponding to a pH gradient of more than 3 units.

Non-cyclic photophosphorylation: The pigments and proteins responsible for the light-dependent reactions of photosynthesis are found within the thylakoid membrane. Most of these molecules are organized into the working units of photosynthesis. Photosynthesis in plants and cyanobacteria produces both ATP and NADPH directly by a *non-cyclic photophosphorylation*. During this process, two photosystems, called photosystem II and I are used in series to energize an electron, the electron can be transferred all the way from water to NADPH. As the high-energy electrons pass through the coupled photosystems to generate NADPH, some of their energy is siphoned off for ATP synthesis.

Cyclic photophosphorylation: In the non-cyclic photophosphorylation, high-energy electrons leaving PSII are harnessed to generate ATP and are passed on to PSI to drive the production of NADPH. This produces slightly more than one molecule of ATP for every pair of electrons that passes from water to $NADP^+$ to generate a molecule of NADPH. But 1.5 molecules of ATP per NADPH are needed for carbon fixation. To produce extra ATP, the chloroplasts in some species of plants can switch PSI into a cyclic mode so that it produces ATP instead of NADPH. In this process, the high-energy electrons from PSI are transferred to the cytochrome b_6f complex (or *bf* complex) rather than being passed on to $NADP^+$. From the bf complex, the electrons are passed back to PSI at a low energy state, completing a cycle of electron transport. The only net result, besides the conversion of some light energy to heat, is that H^+ is pumped across the thylakoid membrane by the bf complex as electrons pass through it, thereby increasing the electrochemical proton gradient that drives the ATP synthase. Electron transfer from photosystem I can thus generate either ATP or NADPH (**Fig. 8.17**), depending on the metabolic needs of the cell.

8.4.2 Dark reaction

Chloroplasts perform diverse metabolic reactions in green leaves. In addition to CO_2 fixation, the syntheses of almost all amino acids, fatty acids and carotenes, pyramidines and purines occur in chloroplasts. However, the synthesis of sugars from CO_2 is the most extensively studied biosynthetic pathway

Table 8.1 Polypeptide subunits of the ATP synthase complex

Protein	Gene	Location of gene	Mol. mass (kDa)	Functions
CF1	*A*			
α-subunit	*atp*	C	55	Catalytic
β-subunit	*atpB*	C	54	Catalytic
γ-subunit	*atpC*	N	36	Proton gating
δ-subunit	*atpD*	N	20	Binding of CF_1 to CF_0 regulation
ε-subunit	*atpE*	C	15	ATPase inhibition
CF_o				
I	*atpF*	C	17	Binding of CF_0 to CF_1
II	*atpG*	N	16	Binding of CF_0 to CF_1
III	*atpH*	C	8	Proton translocation
IV	*atpI*	C	27	Binding of CF_0 to CF_1

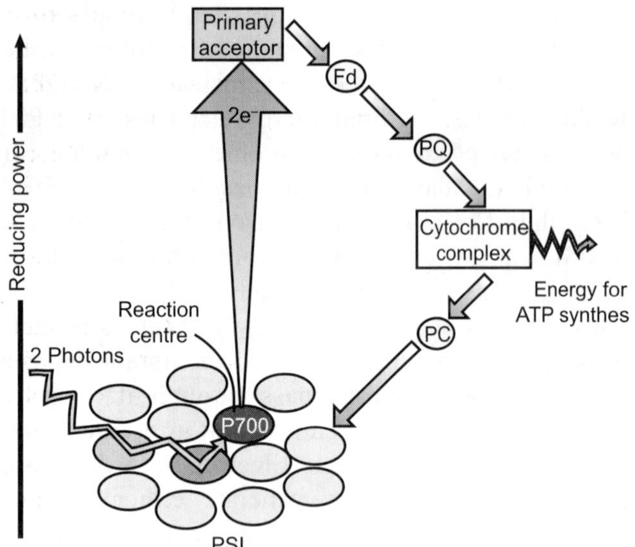

Fig. 8.17 Changes in reducing power during cyclic photophosphorylation.

in plant cells. Melvin Calvin, along with colleagues, made a decade-long study of the enzymatic reactions by which carbon dioxide is assimilated into the organic molecules of the cell.

Dark reaction is also called as the **carbon fixation reaction** or **light-independent reaction**. Dark reaction does not require light, but can occur in light as well. It takes place in stroma and involves the use of the reducing power of NADPH to change CO_2 into carbohydrates (**Fig. 8.18**). The carbon fixation reactions, which begin in the chloroplast stroma and continue in the cytosol, produces sucrose and a number of other organic molecules in the photosynthetic cells of the plant. The carbohydrates (mainly sucrose) are exported to other tissues as a source of both organic molecules and energy for growth.

In summary, the photosynthetic process produces sugars from carbon dioxide and water using light energy trapped by chlorophyll. Light energy is converted to chemical energy. The formation of ATP, NADPH and O_2 (which requires light energy directly) and the conversion of CO_2 to carbohydrate (which requires light energy indirectly) are separate reactions.

Carbon fixation

During carbon fixation, an atom of inorganic carbon is converted to organic carbon: CO_2 from the atmosphere combines with the five-carbon compound ribulose 1,5-bisphosphate plus water to yield two molecules of the three-carbon compound 3-phosphoglycerate. This 'carbon fixing' reaction is catalyzed in the chloroplast stroma by a large enzyme called ribulose 1,5-bisphospate carboxylase (also called *rubisco*). Ribulose bisphosphate carboxylase is a large enzyme (~500 kDa) composed of eight identical large subunits and eight identical small subunits. One subunit is encoded in chloroplast DNA and the other is by nuclear DNA. It constitutes more than 50% of the total chloroplast protein, and is hence the most abundant enzyme.

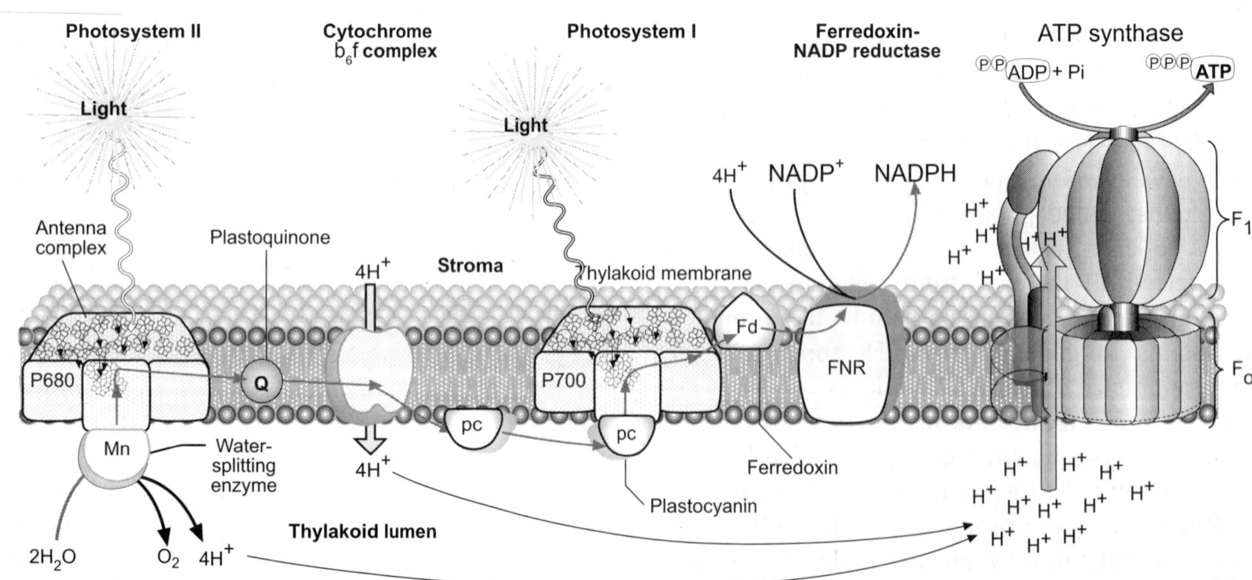

Fig. 8.18 Major steps in light reaction of photosynthesis at thylakoid membrane in chloroplast (**See Plate 4**).

The CO_2 fixation reaction is energetically favourable because of the reactivity of the energy-rich compound ribulose 1,5-bisphosphate, to which each molecule of CO_2 is added. The complete metabolic pathway that produces ribulose 1,5-bisphosphate requires both NADPH and ATP (**Fig. 8.19**). The carbon fixation cycle (also called the **Calvin cycle**) starts when 3 molecules of CO_2 are fixed by rubisco to produce six molecules of 3-phosphoglycerate. The six molecules of 3-phosphoglycerate then undergo a cycle of reactions that regenerates the 3 molecules of ribulose 1,5-bisphosphate used in the initial carbon-fixation step. This leaves one molecule of glyceraldehyde 3-phosphate (3 carbon atoms) as the net gain. At least nine enzymes are required to regenerate ribulose 1,5-bisphosphate from 3-phosphoglycerate. A total of 3 molecules of ATP and 2 molecules of NADPH are consumed for each CO_2 molecule converted into carbohydrate. The net equation is:

$$3CO_2 + 9ATP + 6NADPH + H_2O \xrightarrow{\text{rubisco}} \text{glyceraldehyde 3-phosphate} + 8Pi + 9ADP + 6NADP^+$$

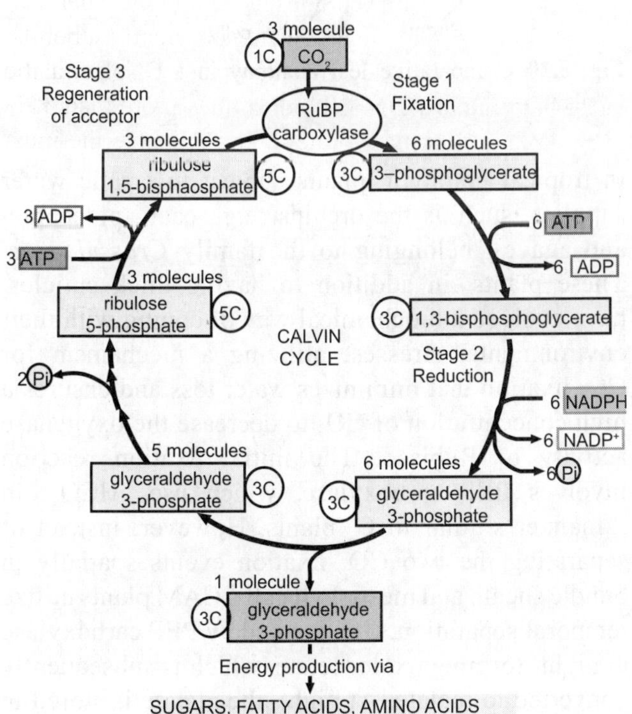

Fig. 8.19 The carbon fixation cycle, which forms organic molecules from CO_2 and H_2O. 3 molecules of CO_2 fixed give a net yield of 1 molecule of glyceraldehyde 3-phosphate at a net cost of 9 molecules of ATP and 6 molecules of NADPH. The number of carbon atoms in each type of molecule is indicated in the white circle.

Synthesis of starch in chloroplast and sucrose in cytosol: The glyceraldehyde 3-phosphate produced in chloroplasts by the carbon fixation cycle is a three-carbon sugar that also serves as a central intermediate in glycolysis. Most of the glyceraldehyde 3-phosphate that remains in the chloroplast is converted to starch in the stroma. Like glycogen in animal cells, starch is a large polymer of glucose that serves as a carbohydrate reserve.

Generally, most glyceraldehyde 3-phosphate molecules are transported into the cytosol in exchange for phosphate. The final steps of sucrose synthesis occur in the cytosol of leaf cells. In these reactions, one molecule of glyceraldehyde 3-phosphate is isomerized to dihydroxyacetone phosphate. This compound condenses with a second molecule of glyceraldehyde 3-phosphate to form fructose 1,6-bisphosphate, which is the reverse of the aldose reaction in glycolysis. Later, fructose 1,6-bisphosphate is converted primarily to sucrose.

Efficient carbon fixation at low CO_2: Rubisco preferentially fix CO_2 to ribulose 1,5-bisphosphate, however, at lower CO_2 concentrations, it can use O_2 as a substrate in place of CO_2. This process is called *photorespiration* and its ultimate effect is to use up O_2 and liberate CO_2 without the production of useful energy stores. In many plants, about one-third of the CO_2 fixed is lost again as CO_2 because of photorespiration. Photorespiration is a serious problem in hot, dry conditions, which cause them to close their stomata to avoid excessive water loss. Consequently, the CO_2 levels in the leaf decreases precipitously falling below the K_m of rubisco for CO_2. Under these conditions, the rate of photosynthesis is slowed and photorespiration is greatly favoured.

In many plants (e.g., corn, sugarcane) this problem has been overcome by a special adaptation. In hot and dry environments, in these plants, the carbon fixation cycle occurs only in the chloroplasts of specialized bundle sheath cells, which contain all the rubisco. The pathway has been named as the C_4 pathway because $[^{14}C]$ CO_2 labelling showed that the first radioactive molecules formed during photosynthesis in this pathway are four carbon compounds, such as oxaloacetate and malate, rather than the three carbon molecules that begin the Calvin cycle (C_3 pathway).

The bundle sheath cells are protected from the air and are surrounded by a specialized layer of mesophyll cells that use the energy harvested by their chloroplasts to 'pump' CO_2 into the bundle sheath cells. This supplies the rubisco with a high concentration of CO_2, thereby greatly reducing photorespiration.

In C_4 plants, the process of CO_2 harvesting is produced by a reaction cycle that begins in the cytosol of the mesophyll cells. A CO_2-fixation step in mesophyll cells is catalyzed by an enzyme *phosphoenolpyruvate carboxylase* (or *PEP carboxylase*), which is found exclusively in C_4 plants, it is insensitive to O_2 and has higher affinity for CO_2. PEP carboxylase binds CO_2 (as bicarbonate) and combines it with an activated three-carbon molecule to produce a four-carbon molecule, oxaloacetate. The oxaloacetate is later reduced to malate (4C) in mesophyll cells, and then transported by a special transporter to the bundle sheath cells, where it is broken down to release CO_2 and generate a molecule with three carbons (*pyruvate*). The pumping cycle is completed when pyruvate is returned to the mesophyll cells and converted back to its original activated form *phosphoenolpyruvate (PEP)* (**Fig. 8.20**). Because of the transport of CO_2 from mesophyll cells, the CO_2 concentration in the bundle sheath cells of C_4 plants is much higher than it is in the atmosphere. Bundle sheath cells are also unusual in that they lack PSII and carry out only cyclic electron flow catalyzed by PSI, so no O_2 is evolved. The high CO_2 and reduced O_2 concentrations in the bundle sheath cells favour the fixation of CO_2 by rubisco to form 3-phosphoglycerate and inhibit the utilization of ribulose 1,5-bisphosphate in photorespiration.

The transport of CO_2 into the bundle sheath cells in C_4 plants, however, costs extra energy. Since, one ATP is utilized in the cyclic C_4 process to generate PEP from pyruvate, the overall efficiency of the photosynthetic production of sugars from NADPH and ATP is lower that it is in C_3 plants, which use only the Calvin cycle for CO_2 fixation. However, in hot and dry environments, this cost can be much less than the energy lost by photorespiration in C_3 plants. Therefore, C_4 plants have a potential advantage by fixing about twice as much net carbon as C_3 plants per unit of water lost.

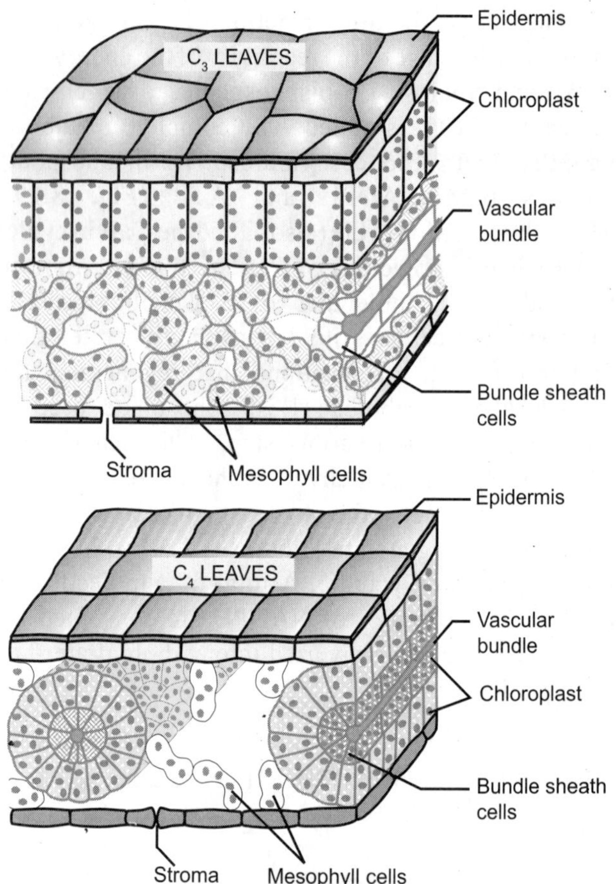

Fig. 8.20 Comparative leaf anatomy in a C_3 plant and a C_4 plant.

Crassulacean acid metabolism (CAM) plants

CAM photosynthesis is commonly associated with plants that inhabit extremely arid environment and in tropical epiphytic plants subject to erratic water supplies, such as the orchids (e.g., cacti, pineapple, and agave), belonging to the family *Crassulaceae*. These plants, in addition to having thick cuticles, have evolved a biochemical way of coping with their environmental stresses, utilizing a mechanism for CO_2 fixation that minimizes water loss and ensures a high concentration of CO_2 to decrease the oxygenase activity of Rubisco. The initial fixation reaction involves PEP carboxylase, which uses HCO_3^- in a manner similar to C_4 plants. However, instead of separating the two CO_2 fixation events spatially in bundle sheath and mesophyll cells, CAM plants utilize temporal separation. CO_2 is fixed by PEP carboxylase at night, forming oxaloacetate, which is subsequently converted to malate. At night, the malate is stored in vacuoles, where it can reach very high concentrations. As the light period commences, malate is transported out of the vacuole and decarboxylated to produce CO_2 and pyruvate. The CO_2 released is then refixed by Rubisco and the conventional Calvin cycle converts this into carbohydrate. Other notable feature of CAM plants is their regulation of stomatal opening and closing. Stomata open when the temperature is cool and close when it is hot.

8.5 REGULATION OF PHOTOSYNTHESIS

Plants adapt to a wide variety of environmental conditions and therefore, the mechanism of regulation of photosynthesis is complex. Although all the regulatory mechanisms of the photosynthetic process are not known, some of them, have been well-characterized.

Light control of photosynthesis

Since, the photosynthetic rate depends on temperature, cellular CO_2 concentration and light, investigation becomes complicated. Light is the most important factor that regulates photosynthesis. Most of the enzymes involved in the Calvin cycle for catalyzing CO_2 fixation are rapidly inactivated in the dark, thereby conserving ATP that is synthesized in the dark for other reactions, such as lipid and amino acid biosynthesis. Light assists in activation of certain photosynthetic enzymes and deactivation of several enzymes in degradative pathways. Among the light-activated enzymes are ribulose-1,5-bisphosphate carboxylase, $NADP^+$-glyceraldehyde-3-phosphate dehydrogenase, fructose-1,6-bisphosphatase, sedoheptulose-1,7-bisphosphatase and phosphoribulokinase. Examples of light-inactivated enzymes include phosphofructokinase and glucose-6-phosphate dehydrogenase. Light affects enzymes by indirect mechanisms. Among the best-studied are pH, Mg^{2+}, the ferredoxin–thioredoxin system and phytochrome.

Control of ribulose-1,5-bisphosphate carboxylase

Rubisco is found within the chloroplast (the L subunit) and the nucleus (the S subunit). The activation of these genes is mediated by an increase in light intensity (illumination). Once the S subunit is transported from the cytoplasm to the chloroplast, both subunits assemble to form the L_8S_8 holoenzyme. The activity of rubisco is modified by a number of metabolic signals. Rubisco is spontaneously activated in the presence of high CO_2 and Mg^{2+} concentrations.

In order to activate rubisco, the active site of the L subunit must be carbamoylated at a specific lysine residue-191, forming a carbamate group that then binds a Mg^{2+} ion. The rate of carbamoylation is dependent on the CO_2 concentration and an alkaline pH. Under normal conditions, however, with ambient levels of CO_2, the reaction requires catalysis by *rubisco activase*, an enzyme that simultaneously hydrolyzes ATP and uses the energy to attach a CO_2 to the lysine. The level of activation is cooperative and increases as more of the eight subunits are modified. This regulatory mechanism ensures that CO_2 fixation only occurs at an appreciable rate when the concentration of CO_2 and available energy are high.

8.6 BIOGENESIS OF CHLOROPLAST AND OTHER PLASTID TYPES

The biogenesis of a chloroplast is similar in many respects to that of a mitochondrion, although many key steps are not understood in molecular detail. Some chloroplast proteins are encoded by chloroplast DNA and are translated by chloroplast ribosomes in the organelle using the 'standard' genetic code. Others are fabricated on cytosolic ribosomes and are incorporated into the organelle after translation; these imported proteins are synthesized with *N*-terminal uptake-targeting sequences that direct each protein to its correct sub-compartment and are subsequently cleaved. Chloroplasts, like mitochondria, grow by expansion and then fission. This process can be observed easily in unicellular algae, such as *Chlamydomonas*, that contain a single large chloroplast.

8.6.1 Chloroplast genome

Chloroplast DNAs (cpDNA) are circular molecules of 120–160 kb present in multiple copies per organelle. They contain approximately 150 genes (**Fig. 8.21**). Each chloroplast may consist more than one circular DNA molecule and the number can vary between 30 and 200 copies of genomes per chloroplast. The amount of DNA/plastid varies with the stage of leaf and chloroplast development.

Of the ≈120 genes in cpDNA, about 60 are involved in RNA transcription and translation, including genes for rRNAs, tRNAs, RNA polymerase subunits and ribosomal proteins. About 20 genes encode subunits of the chloroplast photosynthetic electron transport complexes and the F_0F_1-ATPase complex; also encoded is the larger of the two subunits of ribulose 1,5-bisphosphate carboxylase. Chloroplasts have their own protein-synthesizing machinery and are also partly genetically independent of the nucleus. Transcription here resembles transcription

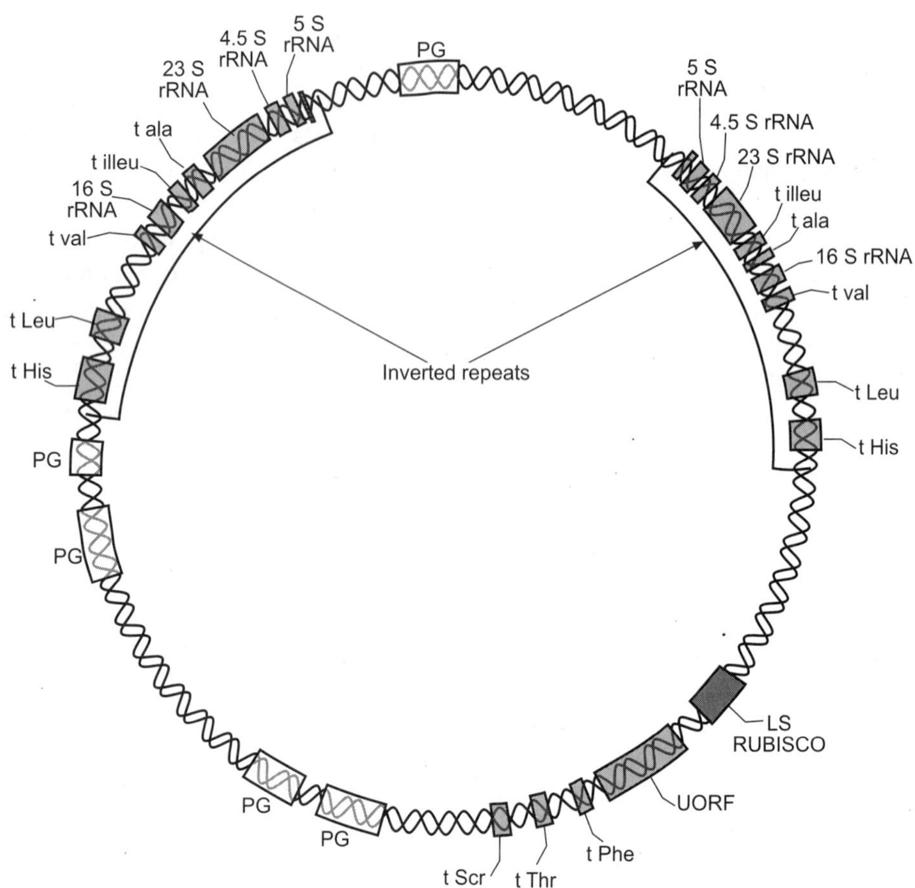

Fig. 8.21 A genetic map of the circular maize chloroplast DNA showing the location of several tRNA genes as well as photogenes (PG), the large subunit gene of RUBISCO anf the large unidentified open reading frame (UORF).

in prokaryotes in many respects. Chloroplasts contain 70 S ribosomes resembling those of bacteria. The nuclear encoded polypeptides are imported and the degree of nuclear/cytoplasmic integration is high.

The chloroplast genome in some plant species is an almost totally unique sequence; others, however, are organized as two unique regions separated by inverted repeats, the repeats containing several genes including the rRNA genes. Recombination occurs frequently between the repeats, generating inversion isomers and concatemers. The major part of the plastid genome encodes its own genetic apparatus or proteins, which are involved in the photosynthetic apparatus in plant cells. Therefore, a vast majority of the plastid genome tends only to be expressed in photosynthetically active cells. Plastid gene expression is light-regulated and is associated with the development of non-differentiated plastids into photosynthetically competent chloroplasts. A large proportion of chloroplast ribosomes are bound to thylakoid membranes during protein synthesis.

Reflecting the endosymbiotic origin of chloroplasts, plastid genomes show a number of prokaryotic features. Some regions of ctDNA are strikingly similar to those of the DNA of present-day bacteria. These include the clustering and co-transcription of functionally related genes. The order of such genes is often similar to that in prokaryotes such as *E. coli*. Another prokaryotic feature is the presence of a number of overlapping genes. Normally, the two gene products of the overlapping genes are found to be part of the same protein complex, as in prokaryotes. Yet another non-prokaryotic feature found in the plastid genome is the presence of introns. For example, more than 16 plastid genes have now been found to contain introns (suggest the presence of RNA splicing mechanism). An interesting and unexpected finding is that all the plastid genomes so far sequenced contain DNA sequences, which are predicted to encode respiratory chain NADH dehydrogenases, similar to those found in human mitochondria. These genes are highly expressed in tobacco plastids, which may suggest the existence of a respiratory chain in the chloroplasts of higher plants.

8.6.2 Gene expression in chloroplast

Chloroplast genes are controlled by *cis*-acting elements similar in structure to bacterial promoters and are able to function in *E. coli*. Transcription is often complex, generating polycistronic mRNAs of differing sizes, which may be cleaved at specific processing sites. Gene regulation occurs primarily at the levels of RNA stability and protein synthesis and a number of host-encoded proteins with roles in posttranslational regulation of chloroplast genes have been isolated. Many chloroplast genes contain **self-splicing introns** of the class II type and demonstrate other intriguing RNA processing reactions such as *trans*-**splicing** and **RNA editing**.

8.7 PEROXISOMES

Peroxisomes are single membrane-enclosed small cytoplasmic organelles (**Fig. 8.22**), They contain membrane proteins critical for various functions, such as importing proteins into the organelles and aiding proliferation. Peroxisomes are ubiquitous organelles in eukaryotes that function to get rid of toxic substances. They contain enzymes involved in a variety of metabolic reactions related to energy metabolism. Although peroxisomes do not have their own genomes, all their proteins, **peroxins** (Pex 1, Pex 2), are synthesized from the nuclear genome. But they are similar to mitochondria and chloroplast in that they replicate by division (self-replication). Most human cells contain about 500 peroxisomes. Peroxisomes were called **microbodies** when first observed with the electron microscope in animal cells in the 1950s and in plant cells in the 1960s. In 1965 Christian de Duve isolated similar organelles from liver cells by centrifugation, he called them 'peroxisomes' because they both generated and destroyed hydrogen peroxide.

Functions of peroxisomes

Peroxisomes have diverse functions and contain more than 50 different enzymes, which are involved in a variety of biochemical pathways in different types of cells. Peroxisomes in general carry out oxidation reactions by oxidative enzymes such as catalase, D-amino acid oxidase and uric acid oxidase, which lead to the production of hydrogen peroxide. By using molecular oxygen, hydrogen atoms are removed from specific organic substrates (labelled as R), in an oxidative reaction, producing hydrogen peroxide (H_2O_2):

$$RH_2 + O_2 \longrightarrow R + H_2O_2$$

However, hydrogen peroxide is harmful to the cell and therefore peroxisomes contain the detoxifying enzyme **catalase**, which decomposes hydrogen peroxide either by converting it to water or by using it to oxidize another organic compound. Catalase uses H_2O_2 generated by other enzymes in the peroxisome to oxidize other substances, including phenols, formic acid, formaldehyde and alcohol, by means of the peroxidation reaction:

$$H_2O_2 + R'H_2 \longrightarrow R' + 2H_2O$$

This reaction is important in liver and kidney cells where the peroxisomes detoxify various toxic substances that enter the blood. About 25% of the ethanol we ingest is oxidized to acetaldehyde in this way. In addition, when excess H_2O_2 accumulates in the cell, catalase converts it to H_2O through the catalytic reaction:

$$2H_2O_2 \longrightarrow 2H_2O + O_2$$

A variety of other substrates broken down by such oxidative reactions in peroxisomes include uric acid, amino acids, purines, methanol and fatty acids.

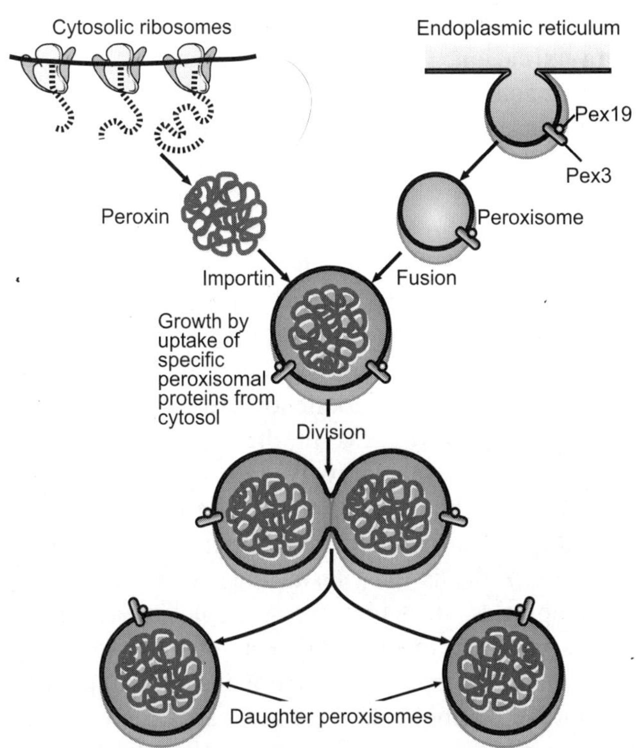

Fig. 8.22 A model for assembly of new peroxisomes.

The oxidation of fatty acids, by a process called **beta-oxidation**, is very important, since it provides a major source of metabolic energy. In this process, the fatty acids are broken down two carbons at a time, converted to acetyl CoA, and then transported back to the cytosol for further use. In animal cells fatty acids are oxidized in both peroxisomes and mitochondria, but in yeasts and plants fatty acid oxidation is confined to peroxisomes only.

Peroxisomes are also involved in biosynthesis of lipids and the amino acid, lysine. In animal cells, cholesterol and dolichol are synthesized in peroxisomes as well as in the ER. In the liver, peroxisomes are also involved in the synthesis of bile acids, which are derived from cholesterol. In addition, peroxisomes contain enzymes required for the synthesis of **plasminogens**—a family of phospholipids in which one of the hydrocarbon chains is joined to glycerol by an ether bond rather than an ester bond. Plasmalogens are important membrane components in some tissues, particularly heart and brain, although they are absent in others. Peroxisomes carry out different biochemical reactions in different tissues.

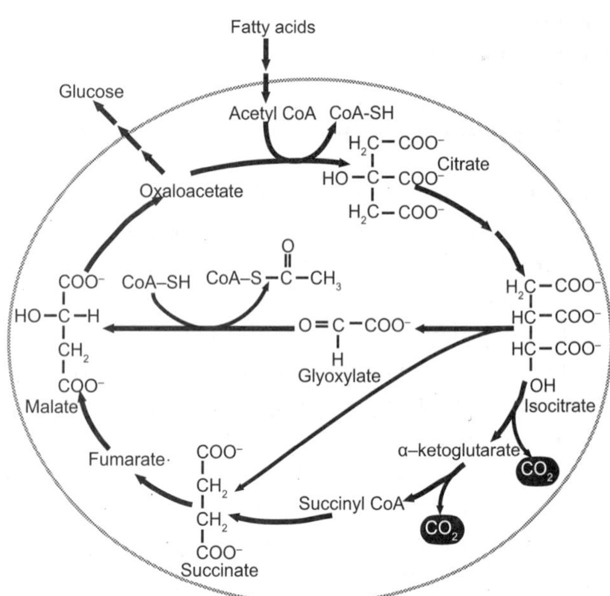

Fig. 8.23 The glyoxylate cycle. Plants are capable of synthesizing carbohydrates from fatty acids via the glyoxylate cycle, which is a variant of the citric acid cycle.

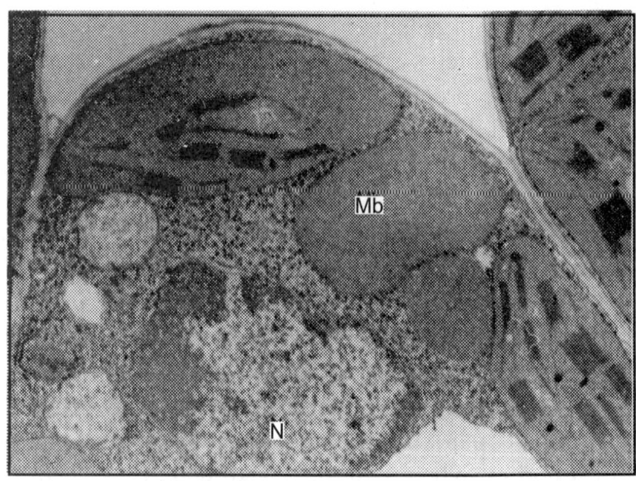

Fig. 8.24 Transmission electron micrograph of tomato leaf tissue showing the presence of peroxisome (Mb) in between two chloroplasts (Channarayappa, 1992).

Glyoxylate cycle

Studies on the peroxisomes of leaf cells have revealed a striking example of interdependence among different organelles. Peroxisomes play two important roles in plants. Peroxisomes in seeds are responsible for the conversion of stored fatty acids to carbohydrates, which is critical to providing energy and raw material for growth of the germinating plant. This occurs via a series of reactions termed the glyoxylate cycle, which is a variant of the citric acid cycle (**Fig. 8.23**). The peroxisomes in which this takes place are sometimes called **glyoxysomes**. The electron micrograph (**Fig. 8.24**) shows the peroxisome of a leaf cell with proximity the surface of adjacent chloroplasts and to nearby mitochondria. This arrangement reflects an underlying biochemical relationship whereby the products of one organelle serve as substrates in another organelle.

Peroxisomes assembly

Recent experiments indicate that assembly of peroxisome assembly begins on the RER, where two peroxins, Pex3 and Pex19, initially localize. Pex3 is an integral transmembrane protein while Pex19 is a farnesylated protein found largely in the cytosol. Pex3 recruits Fex19 to the ER membrane, where their interaction causes Pex3/Pex19–containing vesicles to bud off the ER. These vesicles may then fuse either with pre-existing peroxisomes or with one another to form new peroxisomes. Pex3, Pex19 and other peroxisomal membrane proteins then act as receptors for import of the other peroxins, which are translated on free cytosolic ribosomes and then transported into peroxisomes as complete and folded polypeptides. They are targeted to the interior of peroxisomes by at least two pathways, which are conserved from yeasts to humans.

Photorespiration in plants

Peroxisomes in leaves are involved in photorespiration, a second important role, which serves to metabolize a side product formed during photosynthesis (**Fig. 8.25**). CO_2 is converted to carbohydrates during photosynthesis *via* a series of reactions called the Calvin cycle. The first step is the addition of CO_2 to the five-carbon sugar ribulose-1,5-bisphosphate, yielding two molecules of 3-phosphoglycerate (3C each). In this reaction, rubisco functions as a **carboxylase**, adding atmospheric CO_2 to ribulose-1,5-bisphosphate. However, the enzyme rubisco sometimes function as **oxygenase**, catalyzes the addition of O_2 instead of CO_2, producing one mol-

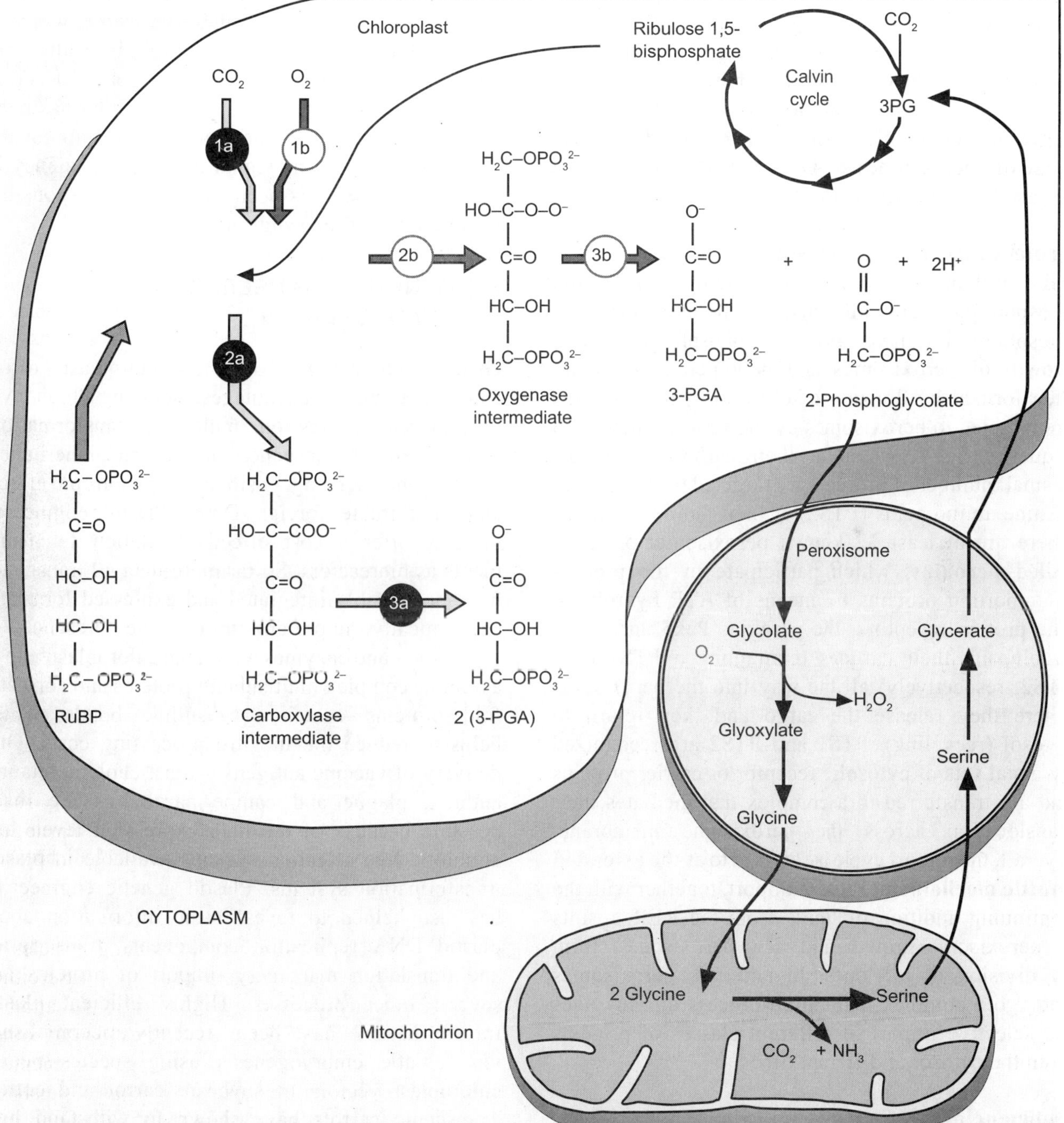

Fig. 8.25 The photorespiration reaction. Ribulose-1,5-bisphosphate can be diverted from the Calvin cycle when the concentration of CO_2 is low. Rubisco can catalyze two different reactions with RuBP as a substrate (shown in the enediol state within the plane). If the RuBP reacts with O_2 (step 1b), the reaction produces an oxygenase intermediate (step 2b) that breaks down into 3-PGA and 2-phosphoglycolate (step 3b).

ecule of 3-phosphoglycerate and one molecule of phosphoglycolate (2C). This is a side reaction and phosphoglycolate is not a useful metabolite. It is first converted to glycolate and then transferred to peroxisomes, where it is oxidized and converted to glycine. Glycine is then transferred to the mitochondria, where two molecules of glycine are converted to one molecule of serine, with the loss of CO_2 and NH_3. Serine is then returned to peroxisomes, where it is converted to glycerate. Finally, the glycerate is transferred back to the chloroplasts, where it re-enters the Calvin cycle. Photorespiration does not appear to be beneficial for the plant, since it is essentially the reverse of photosynthesis—O_2 is consumed and CO_2 is released without any gain of ATP.

Protein import in peroxisomes

All proteins destined for peroxisomes and phospholipids are imported to peroxisomes, via phospholipid transfer proteins, phospholipids help growth of peroxisomes and new peroxisomes are then formed by division of old ones. Most proteins are targeted to peroxisomes by the simple amino acid sequence Ser–Lys–Leu at their carboxy terminus. A small number of proteins are targeted by a sequence of nine amino acids (PTS2) at their amino terminus. There are at least 32 known peroxisomal proteins, called **peroxins**, which participate in the process of importing proteins by means of ATP hydrolysis. The protein receptors, the peroxins Pex5 and Pex7, accompany their cargoes (containing a PTS1 or a PTS2, respectively) all the way into the peroxisome where they release the cargo and then return to cytosol (recycling). PTS1 and PTS2 are recognized by distal sets of cytosolic receptor or carrier proteins and are transferred to a complex that mediates their translocation across the peroxisome membrane. Overall, the import cycle is referred to as the **extended shuttle mechanism**. Protein import, together with the continuing addition of lipids from the RER, results in peroxisome growth and new peroxisomes form by division of old ones. In addition, peroxisomes undergo a complex maturation process that involves the selective import of different classes of proteins from the cytosol at different times.

Deficiencies

The components of the peroxisomes import pathways have been identified not only as mutants but also as mutations associated with serious human diseases involving disorders of peroxisomes. In some diseases single peroxisomal enzyme is deficient, but in other diseases (defects in peroxisome function) multiple peroxisomal enzymes fail to be imported into peroxisomes. The second group of diseases results from deficiencies in the PST1 or PST2 pathways responsible for peroxisomal protein import. For example, a deficiency in the protein import can lead to empty peroxisomes, leading to abnormalities in the brain, called **Zelleweger syndrome**, which is lethal within the first ten years of life. It results from mutations in at least ten different genes affecting peroxisomal protein import, one of which has been identified as the gene encoding the receptor for the peroxisomes targeting signal PTS1. A deficiency in the peroxin Pex2 has been shown to be responsible for one form of the syndrome.

8.8 GENETIC ENGINEERING OF CHLOROPLASTS

Genetic engineering of the chloroplast offers a number of unique advantages, including a high level of transgene expression, multigene transformation, containment of transgenes via maternal inheritance, lack of gene silencing, position and pleiotropic effects and undesirable foreign DNA. Plastid engineered systems offer a cost effective solution for using plants as bioreactors. So far more than 40 transgenes have been stably integrated and expressed for many economically important traits. The presence of chaperones and enzymes within the chloroplast help to assemble complex multisubunit proteins and correctly fold proteins containing disulfide bonds, which helps to reduce the in vitro processing costs. Oral delivery of vaccine antigens against cholera, tetanus, anthrax, plague and canine parvovirus are made possible because of the high expression levels and antibiotic-free selection systems available in plastid transformation systems. Plastid genetic engineering has also helped to reveal basic information about plastid DNA replication components, transcription and translation machinery, import of proteins and several other processes. Highly efficient plastid transformation has been recently accomplished via somatic embryogenesis using species-specific chloroplast vectors in soybean, carrot and cotton. Transgenic carrots have shown to withstand high salt concentrations. Recently, plastid engineering has been efficiently employed for the production of therapeutic proteins, vaccines and biomaterials using an environmentally friendly approach.

Summary

- The manufacture of complex organic molecules from CO_2 requires the input of large amounts of energy.
- Over the course of evolution, two main types of autotrophs: ***Chemo-autotrophs*** and ***photo-autotrophs***.
- These organisms carry out ***photosynthesis***, a process in which energy from sunlight is transformed into chemical energy that is stored in carbohydrates and other organic molecules.
- Plastids are major organelles found only in plant and algal cells. Various types of plastids can be distinguished.
- The chloroplast is one of several types of organelles termed ***plastids***, all of which are formed from the same precursor ***proplastid*** and all of which contain the same chloroplast DNA.
- Depending on the plant tissue and on environmental cues, proplastids differentiate into different types of plastids: ***Chromoplasts*** and ***Leucoplasts***. The most important plastids are chloroplasts, which carry out photosynthesis.
- Chloroplasts have permeable outer and inner membranes embedded with transport proteins. The internal structure of a chloroplast can be differentiated into ***grana***, where the light reaction of photosynthesis takes place and the ***stroma***, where the dark reaction is completed.
- The most significant function of chloroplasts is to carry out the process of photosynthesis. In most common oxygen-evolving (***oxygenic***) ***photosynthesis***, water serves as the reductant. Water is oxidized and the electrons released are energized and ultimately transferred to CO_2, yielding oxygen and carbohydrate.
- The photosynthetic process is divided into the light reaction and the dark reaction.
- In the light reaction, a ***photosystem*** catalyzes the conversion of the light energy captured in excited chlorophyll molecules to useful forms. It consists of two closely linked components: an ***antenna complex*** and a ***photochemical reaction centre***.
- The first of the two photosystems, ***photosystem II*** (PSII), has the unique ability to withdraw electrons from water. Photosystem II is a large membrane-spanning protein–pigment complex that has at least 23 components. The most prominent of these is the reaction centre, a protein–pigment complex composed of two polypeptide subunits known as D_1 and D_2, cytochrome b_{559} and a special pair of chlorophyll a molecules (***P680***) that absorb light at 680nm. The electron acceptor in this electron-transport chain is the second photosystem, ***photosystem I*** (PSI), which accepts an electron into the electron-deficient hole created by light in the chlorophyll molecules in its reaction centre. Each electron that enters PSI is finally boosted to a very high-energy level that allows it to be passed to the iron–sulphur centre in ferredoxin and then to NAD^+ to generate NADPH.
- The dark reaction, also called ***carbon fixation reaction***, takes place in the stroma and involves the use of the reducing power of NADPH to change CO_2 into carbohydrates.
- Carbon fixation reactions— which begin in the chloroplast stroma and continue in the cytosol— produce sucrose and a number of other organic molecules in the photosynthetic cells of the plant.
- The carbohydrates are exported to other tissues as a source of both organic molecules and energy for growth.
- Plants adapt to a wide variety of environmental conditions; therefore, the mechanism of regulation of photosynthesis is complex.
- Peroxisomes in leaves are involved in ***photorespiration***, a second important role, which serves to metabolize a side product formed during photosynthesis.
- The enzyme rubisco sometimes functions as *oxygenease*, catalyzes the addition of O_2 instead of CO_2, producing one molecule of 3-phosphoglycerate and one molecule of phosphoglycolate (2C).

Exercises

1. List the different types of plastids and explain their biological role.
2. Describe with a diagram the mechanism of light reaction of photosynthesis.
3. Explain the different steps involved in photophosphorylation.
4. What are the main differences between cyclic and non-cyclic photophosphorylation?
5. Discuss the different steps involved in the dark reaction.
6. How does a C_3 plant differ from the C_4 plant?
7. How are peroxisomes formed and what are their main functions in cells?
8. Describe the role of cell organelles other than mitochondria and chloroplast.

Further Reading

Books

Albert B, Johnson A, Lewis J, Raff M, Roberts K and Walker P (2002), *Molecular Biology of the Cell*, 5th edition, Garland Scienses.

Barber J, (ed.), (1976), *The Intact Chloroplast*, Elsevier Scientific Publishing Company, Amsterdam.

Blankenship RE (2002), *Molecular Mechanisms of Photosynthesis*, Blackwell Science Limited.

Clayton RK (1980), *Photosynthesis: Physical Mechanisms and Chemical Patterns*, Cambridge University Press.

Fahn A (1974), *Plant Anatomy*, Pergamon Press, Oxford, New York.

Gillham NW (1978), *Organelle Heredity*, Raven Press, New York.

References

Allen JF (2002), Photosynthesis of ATP-electrons, proton pumps, rotors and poise, *Cell*, 110:273–76.

Barber J (2002), Photosystem II: a multisubunit membrane protein that oxidizes water, *Curr. Opin. Struct. Biol.*, 12:523–30.

Barber J and Anderson B (1994), Revealing the blueprint of photosynthesis, *Nature*, 370:31–4.

Bendall DS and Manasse RS (1995), Cyclic photophosphorylation and electron transport, *Biochem. Biophys, Acta*, 1229:23–38.

Bogorad L (1981), Chloroplasts, *J. Cell Biol.*, 91:256.

Channarayappa (1993), Resistance mechanism of *Lycopersicon* species to the vector *Bemisia tabaci* and ultrastructural changes in leaf curl virus infected tomato leaves, PhD Thesis Submitted to the University of Agricultural Sciences, GKVK, Bangalore, India.

Chitnis PR (2001), Photosystem I: Function and physiology, *Annu. Rev. Plant Physiol. Mol. Biol.*, 52:593–626.

Daniell H, Khan MS and Allison L (2002), Milestones in chloroplast genetic engineering: An environmentally friendly era in biotechnology, *Trends Plant Sci.*, 7:84–91.

Des Marais DJ (2000), When did photosynthesis emerge on Earth? *Science*, 289:1703–5.

Douce R and Joyard J (1990), Biochemistry and function of the plastid envelope, *Annu. Rev. Cell Biol.*, 6:173–216.

Douglas SE (1998), Plastid evolution: Origins, diversity, trends, *Curr. Opin. Genet. Dev.*, 8:655–61.

Gray MW (1999), Evolution of organellar genomes, *Curr. Opin. Genet. Dev.*, 9:678–87.

Hatch MD (1987), C4 photosynthesis: a unique blend of modified biochemistry, anatomy and ultrastructure, *Biochem. Biophys. Acta*, 895:81–106.

Heathcote P, Fyfe P and Jones M (2002), Reaction centres: the structure and evolution of biological solar power, *Trends Biochem. Sci.*, 27:79–87.

Huber R (1990), a Structural Basis of Light Energy and Electron Transfer in Biology, *Eur. J. Biochem.*, 187:283–305.

Nugent J (Ed), (2001), Photosynthetic water oxidation, *Biochim. Biophys. Acta*, 1503, #1.

Palmer JD (1985), Comparative organization of chloroplast genomes, *Ann. Rev. Genet.*, 19:325–54.

Prince RC (1996), Photosynthesis: The Z–Scheme revised, *Trends Biochem. Sci.*, 21(4),:121–22.

Rao K and Hall DO (1999), *Photosynthesis*, Cambridge University Press.

Rogner M, Boekema EJ and Barber J (1996), How Does Photosystem II Split Water? The structural Basis of Efficient Energy Conversion, *Trends Biochem. Sci.*, 21(2),:44–9.

Sadava DE (1993), *Cell Biology–Organelle Structure and Function*, Jones and Bartlett Pub., Boston.

Sage R and Colemana J (2001), Effects of low atmospheric CO_2 on plants: more than a thing of the past, *Trends Plant Sci.*, 6:18–24.

Sager R (1985), Chloroplast genetics, *BioEssays*, 3:180–84.

Spreitzer RJ and Salvucci ME (2002), Rubisco: structure, regulatory interactions and possibilities for a better enzyme, *Annu. Rev. Plant Physiol. Mol. Biol.*, 53:449–75.

Stern DB, Higgs DC and Yang JJ (1997), Transcription and translation in chloroplasts, *Trends Plant Sci.*, 2:308–15.

Youvan DC and Marrs BL (1987), Molecular mechanism of photosynthesis, *Sci. Am.*, 256(6):42–8.

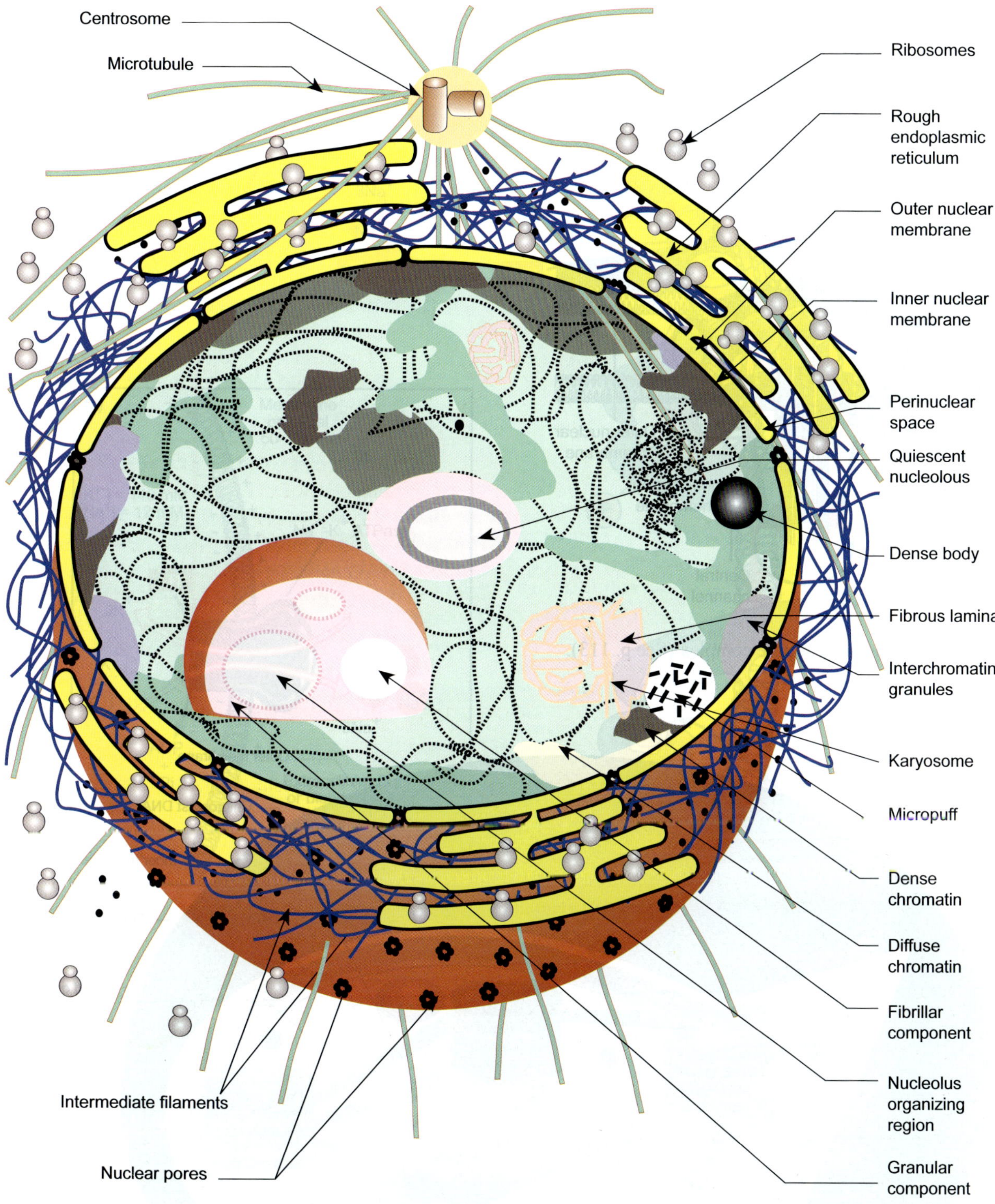

Fig. 5.14 Cross-sectional view of a typical nucleus (See p. 108).

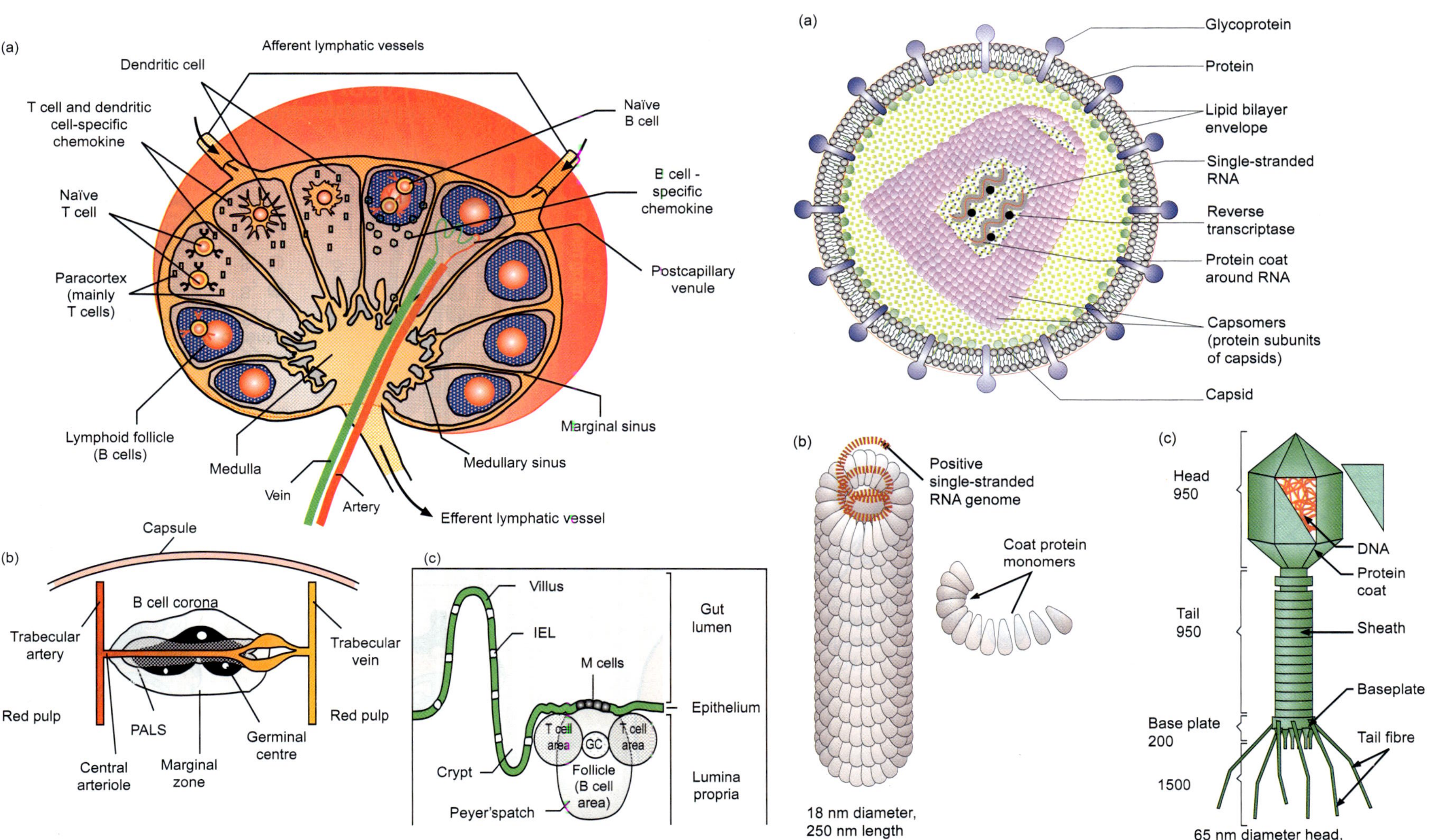

Fig. 18.16 Peripheral lymphoid organs: (a) Lymph node structure and lymphocyte activation; (b) Structure of white pulp in the spleen (PALS: periarteriolar lymphoid sheath); (c) Organization of gut-associated lymphoid tissue (GALT), including Peyer's patch, lamina propria and intra-epithelial lymphocytes (IEL) GC: germinal centre (See p. 522).

Fig. 19.6 Body plans of viruses: (a) Structure of HIV virus, (b) Morphology of tobacco mosaic virus (TMV), and (c) Complex viruses, such as the T-even bacteriophages (not drawn to scale) (See p. 554).

9 Cell Division: Propagation of Genetic Information

Learning objectives

After reading this chapter, you will be able to:

- Understand the features of the steps in cell division—mitosis and meiosis
- Identify key features of difference between mitosis and meosis

9.1 INTRODUCTION

In multicellular organisms (animals and plants), growth occurs by cell elongation (increase in volume) and cell division (increase in number). However, organisms cannot grow purely by cell elongation; as the surface of a spherical organism increases by the square of its radius (*r*), its volume increases more than proportionately by the cube of this number. The inner constituents of such an expanding organism have proportionately less surface area from which to obtain food, oxygen and various metabolites. In the absence of cell division, death would quickly ensue. Some form of cell division must, therefore, have been a primary necessity for the maintenance of life. Every cell was formed by cell division. In this chapter the mechanics of cell division will be discussed.

Cellular reproduction means producing a new generation of cells or multicelled individuals. It begins with the division of single cells. Parent cells provide their blueprint with hereditary instructions and enough cytoplasmic machinery for the daughter cells to start up their own operation. Unless the daughter cell receives DNA, proteins, carbohydrates, lipids, necessary instructions and other building blocks of cells (including cell organelles) for making cellular components, it cannot grow or function properly. The cells of plants, animals and other eukaryotic organisms divide the DNA by **mitosis** or meiosis. Both mechanisms sort out and package DNA molecules into new nuclei for forthcoming daughter cells. Multicellular organisms grow by **mitosis** and cytoplasmic division of body cells, which are called somatic cells. Besides this, many protistants, fungi, plants and even some animals reproduce asexually by mitotic cell division. In contrast, **meiosis** occurs only in germ cells and proceeds to the formation of gametes, such as sperm and eggs. However, prokaryotic cells (bacteria) reproduce asexually by way of a different mechanism called **binary fission**.

Before a cell starts preparing for mitosis, its DNA molecules are stretched out like a thread, with many proteins attached to them. Each DNA molecule, along with its attached proteins is a **chromosome**. Chromosomes undergo duplication while they are in this thread-like form. Each duplicated chromosome consists of two DNA molecules (called **sister chromatids**), which stay together until late in mitosis. Chromosomes have a region of attachment sites—the **centromere**—for microtubules that help to move the chromosome during nuclear division.

Mitotic cell division is actually duplication or multiplication rather than just division. In unicellular animals, however, cell duplication equates with reproduction, since two new individuals are formed from the original parent. In higher organisms, growth occurs through mitotic cell division with subsequent enlargement and differentiation of the cells produced. Each human being, for example, begins as a single fertilized cell or zygote, which develops eventually into a person with billions of cells. The complex body of a multicellular organism eventually contains a variety of specialized cells.

9.2 EUKARYOTIC CELL CYCLE

In higher organisms most of the growth occurs by an increase in cell mass, duplication of the genetic material and division so that each progeny cell receives an equal complement of the genetic material to ensure perpetuation of the cell line. All these steps occur in an ordered progression during the life cycle of the cell. Mitosis is only one phase of the **cell cycle,** starting with new cells being produced and ending when cell division is completed. The **interphase** (between divisions) is usually the longest part of a cell cycle. During this period, a cell increases its mass, roughly doubling the number of its cytoplasmic components. Chromatin is dispersed throughout the nucleus and there are no chromosomal features. Finally, it duplicates its chromosomes. The standard eukaryotic cell cycle is divided into discrete non-overlapping phases (**Fig. 9.1**). These phases of the cell cycle include:

1. ***Mitosis (M)***: Nuclear division, commonly followed by cytoplasmic division;
2. ***Interphase***: Between the four phases.

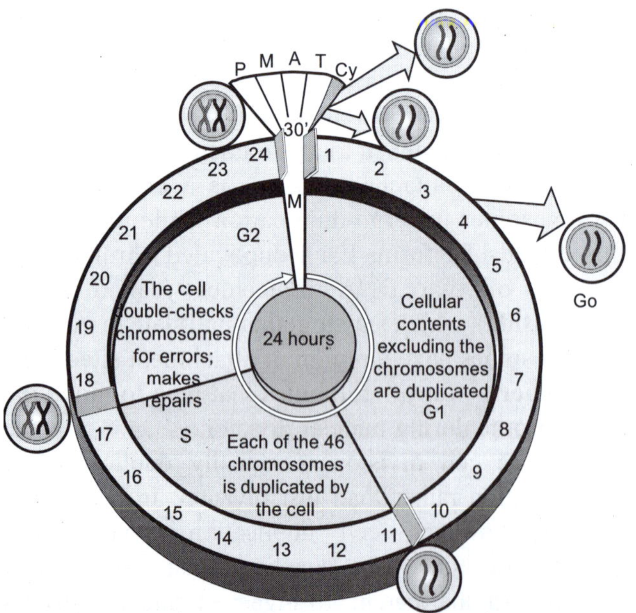

Fig. 9.1 Stages in eukaryotic cell division. The amount of time required for human fibroblast cell to complete one cell division is approximately 24 hours. The relative amount of time taken for each stage is shown in a 24-hour clock.

The process of passing from one phase of the cell cycle to the next is a **cell cycle transition.** As we have discussed in Chapter 8, the events of the cell cycle are regulated by the cell-cycle control system. The core of the control system consists of various **cyclin-dependent kinases (Cdks)**, which are activated in sequence to trigger various steps of the cycle. The **M-phase Cdk (M-Cdk)** triggers a cascade of protein phosphorylation that initiates the M phase. These phosphorylations are responsible for the many morphological changes that occur during cell division. Triggered protein degraded by the **anaphase-promoting complex (APC)** plays an equally important regulatory role in mitosis. It initiates the separation and segregation of the replicated chromosomes and it inactivates M-Cdk at the end of mitosis.

9.3 MITOSIS

The word 'mitosis' comes from the Greek word *mitos*, meaning 'thread' and was coined by Walther Fleming in 1882 to describe thread-like chromosomes that appeared in animal cells before they divided into two. Mitosis is a process of nuclear division in which the replicated DNA molecules of each chromosome are faithfully partitioned into two nuclei. Mitotic cell division takes place in localized regions called **meristems** (in plants) or **stem cells** (in animals). The development of an individual from the zygote to adult stage takes place through mitotic cell division. Both plant and animal cells, for example, are essentially alike: they have genes arranged in linear arrays: they have of chromosomes. Chromosomes are involved in the division cycles of individual cells as well as in the basic reproduction processes of the entire organism. ***Mitotic cell division is the process by which single cells reproduce and multicellular organisms grow.*** All normal members of the same species have the same total number of chromosomes in their somatic cells. Any cell having two of each type of chromosome is a **diploid cell**. Such cells exist in humans, gorillas, pea plants and many other organisms. With mitosis, a diploid parent cell produces two diploid daughter cells. The chromosome number indicates how many of each type of chromosome are present in a cell. Diploid cells with two of each type are $2n$, where the n stands for chromosome number of gametes. With mitosis, the chromosome number remains constant, division after division, from one cell generation to the next. Thus, if a parent cell is diploid ($2n$), its two daughter cells will also be diploid ($2n$).

The M phase

The main objective of cell division in the M phase is to accurately separate and distribute (segregate) its chromosomes that are replicated in the preceding S phase, so that each new daughter cell receives an identical copy of the genetic material. Most eukaryotic cells have developed a similar mechanism to achieve this feat: they assemble specialized cytoskeletal network—first to pull the duplicated chromosomes to the poles and then to split the cytoplasm into equal parts.

Chromosomal disjunction

During cell division, chromosomes duplicate in the S phase of the interphase. The two copies of each replicated chromosome remain tightly bound together as identical **sister chromatids**. The sister chromatids are glued together by multisubunit protein complexes called **cohesins**, which are deposited along the length of each sister chromatid as the DNA is replicated. This cohesion between sister chromatids is very important for chromosome segregation and is broken only late in mitosis to facilitate chromosomal disjunction. The first visible sign of a cell that enters into mitosis is the progressive compaction of the replicated chromosomes, which become visible as thread-like structures (a process called **chromosome condensation**) and the condensation process is carried out by proteins called **condensins**. Activated M-Cdk phosphorylates some of the condensin subunits, triggering the assembly of condensin complexes on DNA and thereby, the progressive condensation of the chromosomes. The cohesins and condensins are structurally related and work together to configure the replicated chromosomes (**Fig. 9.2**).

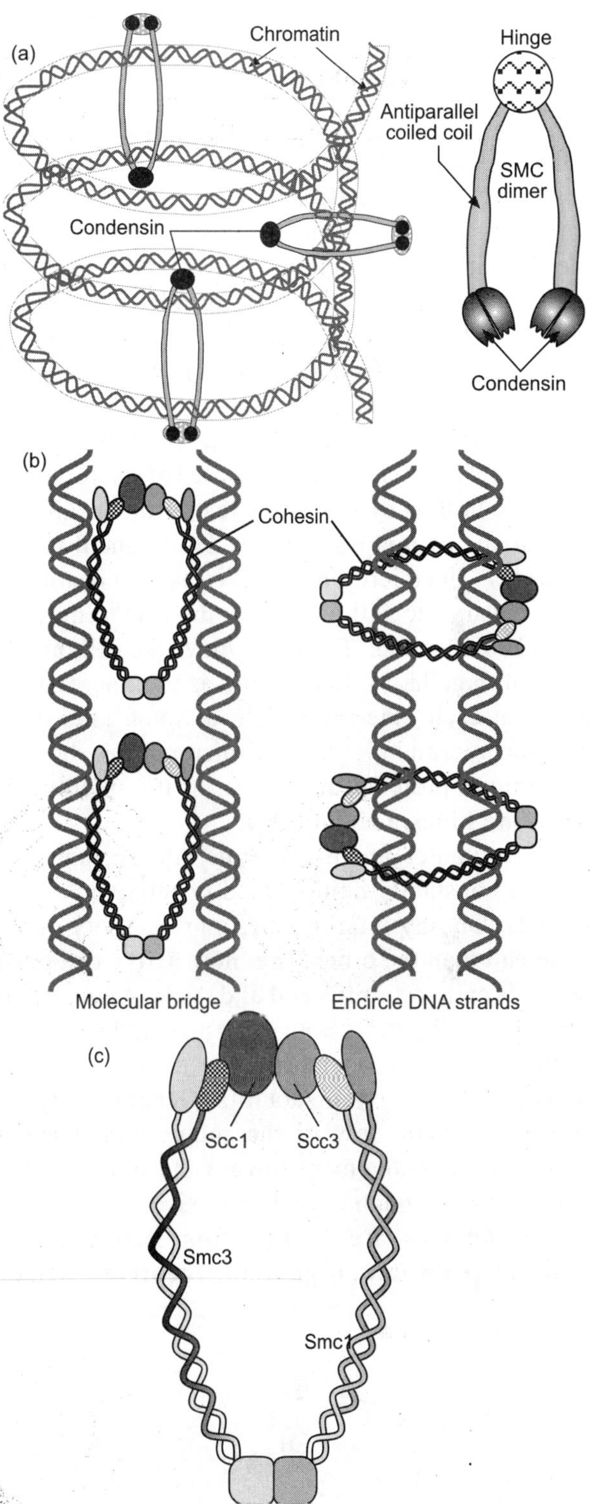

Fig. 9.2 Models for the role of condensin and cohesin in the formation of mitotic chromosomes: (a) Condensin brings about chromosome compaction by forming a ring around supercoiled loops of DNA within the chromatin. (b) Cohesin rings function either as a molecular bridge or encircle a pair of DNA molecules. (c) The subunit organization of a single cohesin complex.

Cohesins and sister chromatids

Cohesins are located at various sites along a pair of sister chromatids. At metaphase, each sister chromatid pair is in equilibrium on the spindle, being connected to both poles by microtubules. When the connecting glue dissolves, the sister chromatids are released from one another and pulled towards the opposite poles. This marks the transition from metaphase to anaphase.

The mitotic spindle

Following chromosome condensation, a distinct cytoskeletal machine is assembled to perform the mechanical process of separating the replicated chromosomes to each daughter cell. In all eukaryotic cells this task is performed during mitosis by a bipolar mitotic spindle, which is composed of microtubules and various proteins that interact with them. Mitotic spindles in plant and animal cells differ in spindle pole structure. In animal cells, the spindle pole is usually broad. The focussed pole of an animal mitotic spindle is usually built around a conspicuous structure called a **centrosome**. Plant cells have no centrosomes. Instead, the plant spindle has diffuse poles, containing numerous microtubules (**Fig. 9.3**). Lacking centrosomes, the poles of mitotic spindle in plant cells have been compared to the poles of the meiotic spindles of oocytes, which in many animal species also lack centrosomes.

The spindle apparatus plays a very important role in mitosis. This structure is composed of microtubules, which are hollow cylinders made of protein subunits; each subunit (**dimer**) is composed of α- and ß-tubulin molecules. The microtubules give shape and structure to the eukaryotic cell and allow the cell to move its internal components. They also move the cell through cilia and flagella. Motion occurs as the microtubules slide past each other; a vesicle of some kind slides along the microtubules and the microtubules shortens. Two unique proteins that make up the microtubule motors are **kinesin** and **dynein**.

Microtubules are in dynamic equilibrium, with subunits constantly being added or removed at both the ends. On any microtubule, more activity occurs at one end than the other. The more active end of the tubules is called the **plus end** and the less active is the **minus end** (**Fig. 9.3**). Both ends may add or remove subunits, or the plus end may be adding while the minus end is removing subunits. Generally, dynein causes movement toward the minus end, whereas kinesin causes movement toward the plus end of a microtubule although exceptions exist.

Microtubules are formed from active centres called **microtubule organizing centres** (**MOC**). **Centrioles**, composed of two cylinders, which contain microtubules, are the MOC for cilia and flagella; in this case the centrioles are referred to as **basal bodies**. The centrioles were also originally believed to organize spindles. However, for most organisms, the MOC is called the centrosome. In some organisms, such as fungi, a different cell organelle, the **spindle pole body** serves this function. In most animals, the centrosome contains a centriole, but in higher plants the centriole is absent. Many experiments demonstrated that the centriole is not necessary for spindle formation. The centriole when present, replicates during the S and G2 phases. When mitosis begins, the centrosome divides and moves to opposite poles of the cell around the nucleus. The centrosomes trail microtubules, forming the spindle that at this point begins at each centrosome and overlaps in the middle of the cell. These are called **interpolar microtubules**.

The spindle microtubules in mitotic cells can be grouped into three classes (**Fig. 9.4**). **Astral microtubules** radiate in all directions from the centrosomes and contribute to the forces that separate the poles. They also act as 'handles' for orienting and positioning the spindle in the cell. **Kinetochore microtubules** attach end-on to the kinetochore which forms at the centromere of each duplicated chromosome. They serve to attach the chromosomes to the spindle. **Overlap microtubules** interdigitate at the equator of the spindle and are responsible for the symmetrical, bipolar shape of the spindle. All the three classes of microtubules have plus ends projecting away from their centrosomes. The behaviour of each class is thought to be different because of the different protein complexes that are associated with the plus and minus ends.

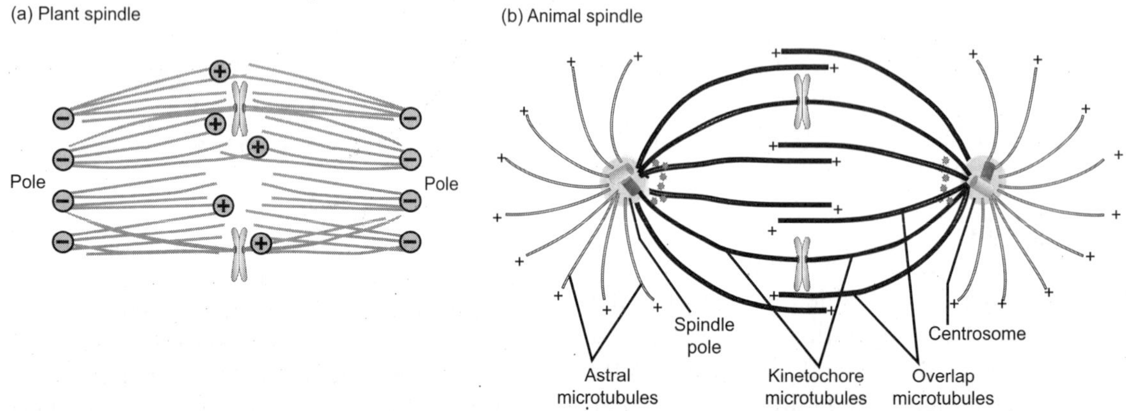

Fig. 9.3 Schematic comparison of the organization of the mitotic spindle in the cells of animals and plants.

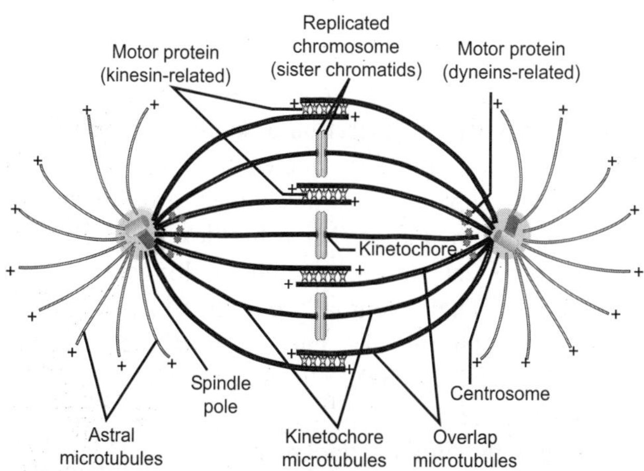

Fig. 9.4 The three classes of microtubules of the fully formed mitotic spindle in an animal cell.

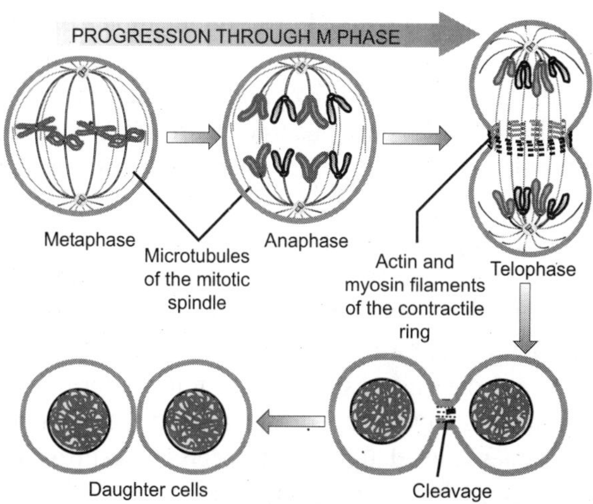

Fig. 9.5 Two cytoskeletal machines that operate in the M phase.

Cytoskeletal proteins and cytokinesis

The final process in cell division is cytokinesis, which is facilitated by different cytoskeletal structures. In animal cells and many unicellular eukaryotes, the **contractile ring** is responsible, whereas in most plant cells, it is the **phragmoplast**. The contractile ring contains both actin and myosin filaments and forms around the equator of the cell, just under the plasma membrane. As the ring contracts, it pulls the membrane inward, thereby dividing the cell into two (**Fig. 9.5**). Plant cells, which have a cell wall, divide their cytoplasm by a different mechanism. The phragmoplast constructs a new cell wall from within the cell, between the two sets of replicated chromosomes (**Fig. 9.6**).

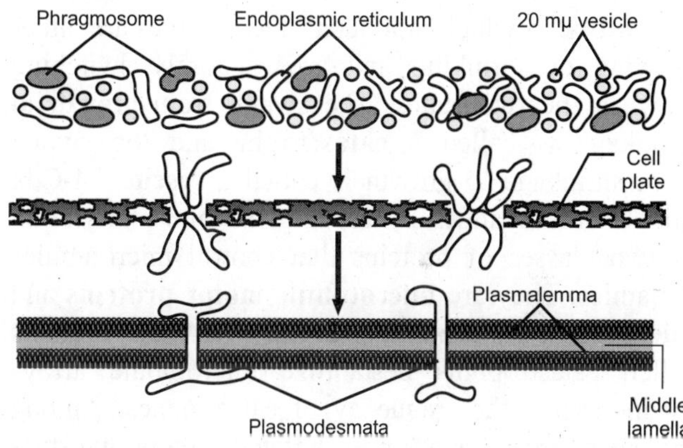

Fig. 9.6 Assembly of cell plate and middle lamella by the aggregation and fusion of vesicles and cisterns of endoplasmic reticulum (ER). At several points, the ER (plasmodesmata) runs across the middle lamella and connects the cytoplasm of neighbouring cells.

Centrosome duplication

At the end of interphase, DNA replication and centrosome duplication must be completed to progress cell division into the M phase. The centrosome is the principal **MOC** in animal cells. It consists of a cloud of amorphous material (called the **centrosome matrix** or **pericentriolar material**) that surrounds a pair of centrioles. During interphase, the centrosome matrix nucleates a cytoplasmic array of microtubules, with fast-growing plus ends projecting outward toward the cell perimeter and the minus ends buried within the centrosome. The matrix contains a variety of proteins, including microtubule-dependent motor proteins, coiled coil proteins that link the motors to the centrosome, structural proteins, components of the cell cycle control system and **r-tubulin ring complex,** which is mainly responsible for nucleating microtubules.

The process of centrosome duplication and separation is known as the **centrosome cycle** (**Fig. 9.7**).

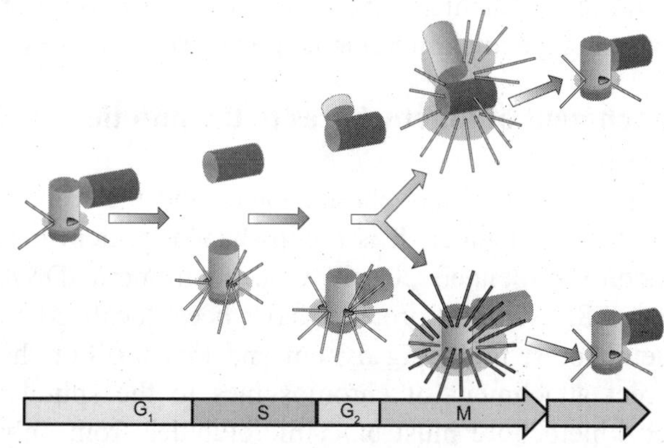

Fig. 9.7 Centriole replication.

The mechanism involved in duplication of the centrosome is not yet understood fully As mitosis proceeds, the centrosome complex splits into two and each centriole pair becomes part of a separate microtubule organizing centre that nucleates a radial array of microtubules called an **aster**. The two asters move to the opposite sides of the nucleus to initiate the formation of the two poles of the mitotic spindle. When the nuclear envelope disintegrates, the plus ends of the spindle capture the chromosomes and separate them toward the end of mitosis. As mitosis terminates and the nuclear envelope re-forms around the separated chromosomes, each daughter cell receives a centrosome in association with its chromosomes.

Microtubule assembly and disassembly

Animal cells in interphase contain a cytoplasmic array of microtubules radiating from the single centrosome. The interphase array will be in a state of **dynamic instability**, in which individual microtubules are either growing or shrinking and are randomly switching between the two states. The switch from growth to shrinkage is called a **catastrophe** and the switch from shrinkage to growth is called a **rescue**. M-Cdk initiates the changes by causing the phosphorylation of two classes of proteins that control microtubules dynamics. They are **microtubule motor proteins** and **microtubule-associated proteins (MAPs)**. Proteins called catastrophins destabilize microtubule arrays by increasing the frequency of catastrophes. Among the catastrophins is a kinesin-related protein that does not function as a motor. In general, MAPs have the rescue effect, stabilizing microtubules in a number of ways: they can increase the frequency of rescues by switching microtubules from shrinkage to growth, or increase the growth rate, or decrease the shrinkage rate of microtubules. Thus, in principle, changes in catastrophins and MAPs can make microtubules much more dynamic in the M phase by increasing or decreasing the total microtubule polymerization rates.

Attachment of kinetochores to the mitotic spindle

The kinetochore links chromosomes and the spindle in a flexible manner. It is a cup-shaped, protein-rich structure associated closely with centromeric DNA (**Fig. 9.8**). Various proteins have been localized to kinetochores, including dynein and kinesin. For the correct attachment of chromosomes to the spindle, each kinetochore must bind microtubules from only one spindle pole and sister kinetochores must bind microtubules from opposite poles. Each kinetochore typically binds dozens to hundreds of microtubules, forming a fibre. The attachment and movement seems to involve a 'search and capture' mechanism, in which microtubules nucleated from each of the rapidly separating centrosomes grow outward from the chromosomes. Microtubules attached to a chromosome get stabilized, and no longer undergo catastrophes. They eventually end up attached end-on at the kinetochore. The end-on attachment to the kinetochore is through the plus end of the microtubule, which is now called a kinetochore microtubule.

During the initial attachment capture (**Fig. 9.9**), it appears that the kinetochore first binds to the side of the microtubule, then slides rapidly along the microtubule towards one of the centrosomes. The lateral attachment to the chromosome is rapidly converted to an end-on attachment. At the same time, microtubules growing from the opposite spindle pole attach to the kinetochore on the opposite side of the chromosome, forming a bipolar attachment (**Fig. 9.10**). Further, the chromosomes are tugged back and forth, eventually assuming a position equidistant from the two spindle poles, a position called the **metaphase plate**.

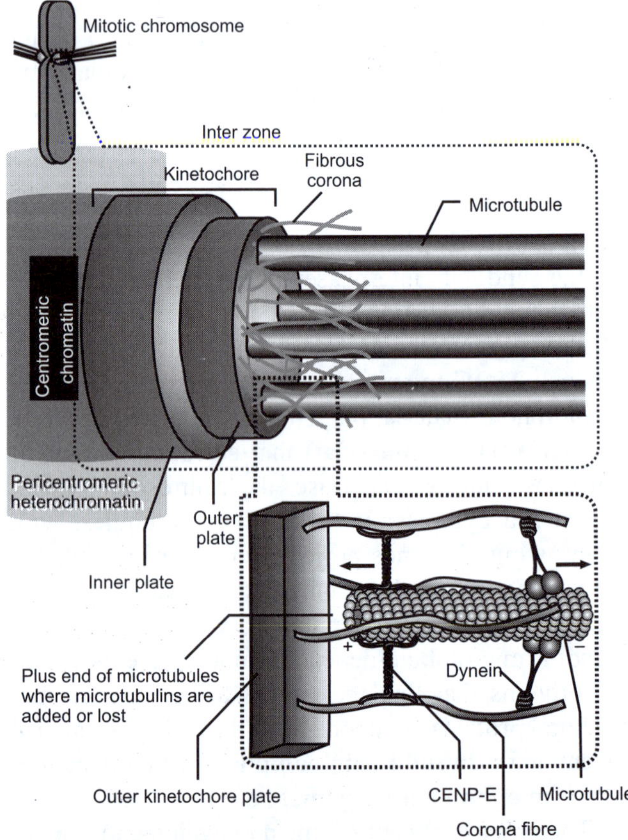

Fig. 9.8 Schematic representation of the kinetochore.

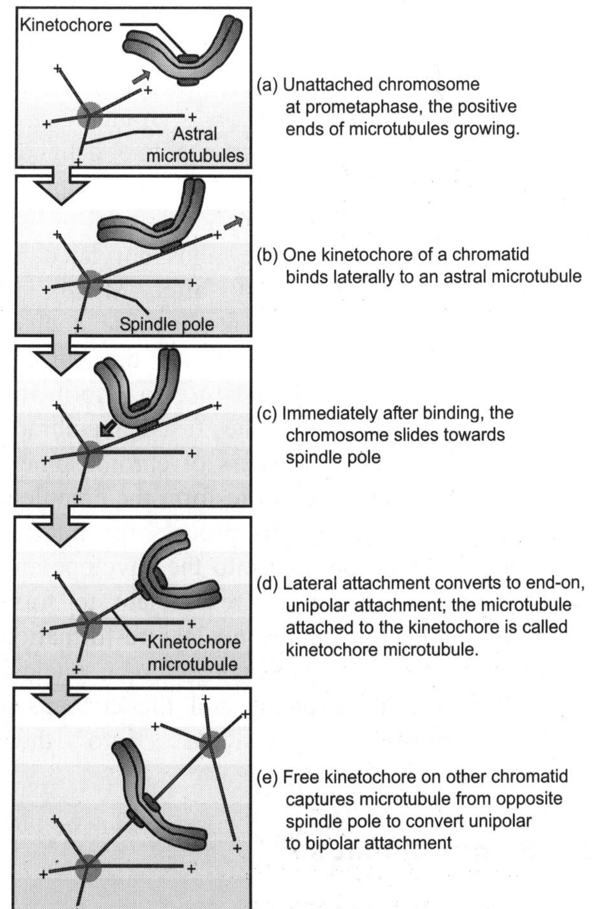

Fig. 9.9 Capture of microtubules by kinetochores.

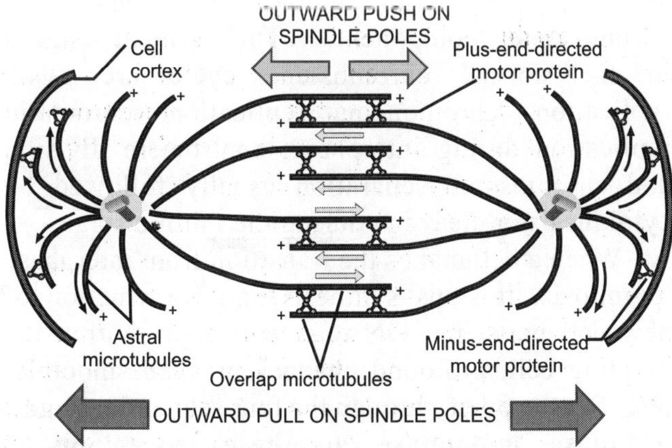

Fig. 9.10 A model for how motor proteins may act in anaphase B.

Kinetochores show plate-like organization and are associated with both plus-end-directed and minus-end-directed microtubule motor proteins. But what is not very clear yet is how the plus end microtubules attach themselves to the kinetochores.

Chromosomal movements at mitosis

Chromosomes move mainly in two phases of cell division: prometaphase and anaphase. At prometaphase chromosomes 'congress', moving from the random positions they held when first captured by the mitotic spindle to gather at the metaphase plate. The kinetochore and overlap microtubules exhibit a behaviour called poleward flux, with a net addition of tubulin subunits at their plus end, balancing a net loss at their minus ends, near the spindle poles. In vertebrate cells, continuous oscillatory movement of the chromosomes takes place at the metaphase plate. These movements switch between two states—a poleward (P) state which is a minus-end-directed pulling movement and an away-from-the-pole (AP) state, which is a plus-end-directed movement. Kinetochores pull the chromosomes toward the poles, while an **astral ejection** force is thought to push the chromosomes away from the poles, toward the spindle equator. The spindle attachment checkpoint monitors the attachment of the chromosomes to the mitotic spindle. The unattached kinetochores emit a signal that delays anaphase until they are all properly attached to the spindle.

Sister chromatids at anaphase

At anaphase, the replicated sister chromatids, previously held together at the centromeres, uncouple and move apart to pole. At the start of anaphase, the sister chromatids uncouple synchronously, an activity which requires a topo-isomerase. Chromosome movement begins abruptly with the release of the cohesin linkage that holds the sister chromatids together at the metaphase plate. This metaphase-to-anaphase transition is triggered by the activation of the **anaphase-promoting complex (APC)**. Once the proteolytic complex APC is activated, it takes part in two crucial functions: cleaves and inactivates the M phase cyclin (M-cyclin), thereby inactivating M-Cdk; and cleaves the inhibitory protein (**securin**), thereby activating a protease called **separase**. Separase then cleaves the subunit in the cohesin complex to unglue the sister chromatids (**Fig. 9.11**). The sister chromatids immediately separate (now called daughter chromosomes) and move to the opposite poles. The chromosomes move by independent and overlapping processes. The first process referred to as **anaphase A**, is an initial poleward movement of the chromosomes. It is accompanied by shortening of the kinetochore microtubules at their attachment to the chromosome and to a lesser extent by the

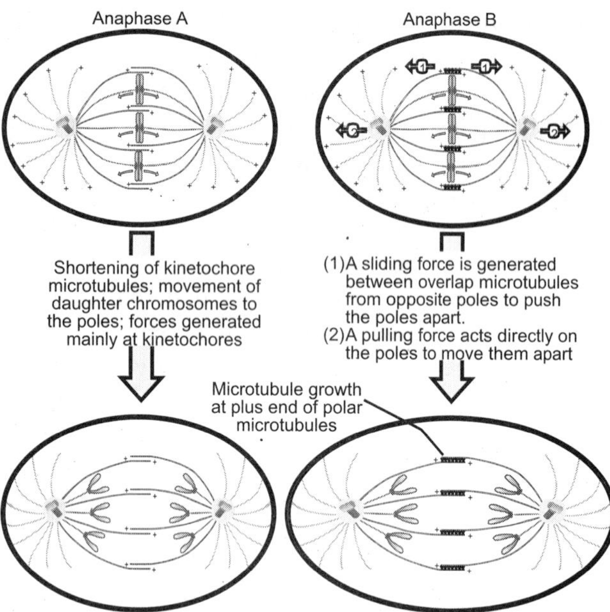

Fig. 9.11 Forces that separate daughter chromosomes at anaphase in mammalian cells.

depolymerization of spindle microtubules at the two spindle poles. The second process, referred to as **anaphase B**, is the separation of the poles themselves, which begins after the sister chromatids have separated and the daughter chromosomes have moved some distance apart. Anaphase A depends on motor proteins at the kinetochore, whereas anaphase B depends on motor proteins at the poles that pull the poles apart, as well as on motor proteins at the central spindle that push the poles apart.

Kinetochore microtubules disassemble at both ends during anaphase A. Each daughter chromosome moves poleward, its kinetochore microtubules depolymerize to the extent that microtubules completely disappear at telophase. The poleward flux with the continuous loss of tubulin subunits from both the overlap and kinetochore microtubules at the poles continues through anaphase. In anaphase B, the overlap microtubules actually elongate to push the spindle poles apart. Anaphase B is driven by two distinct forces. The first depends on plus end-directed microtubule motor proteins in the central spindle that form a bridge between the overlapping microtubules of opposite polarities. The translocation of these motors towards the microtubule plus ends slides the microtubules past one another, pushing the poles apart. The second force contributing to anaphase B depends on minus end-directed motor proteins that interact with both the astral microtubules and the cell cortex and pull the two poles of the spindle apart (**Fig. 9.11**).

In telophase, the final stage of mitosis, the nuclear envelope re-assembles around each group of chromosomes to form the two daughter interphase nuclei. The transition from metaphase to anaphase initiates the dephosphorylation of the many proteins that were phosphorylated at prophase. This happens only after M-Cdk is switched off. Shortly thereafter, at telophase, nuclear membrane fragments associate with the surface of individual chromosomes and fuse to reform the nuclear envelope. Initially, the fused membrane fragments partly enclose clusters of chromosomes: the fragments then coalesce to re-form the complete nuclear envelope. During this process the nuclear pore complex is incorporated into the envelope and the dephosphorylated lamins reassociate to form the nuclear lamina. Once the nuclear re-formation is complete, the pore complexes pump in nuclear proteins, the nucleus expands and the condensed mitotic chromosomes decondense into their interphase state.

9.3.1 Stages of mitosis

Many of the important events of the chromosome cycle cannot be distinguished under the light microscope as they occur in the interphase nucleus. Some can be inferred using techniques such as autoradiography with radioactive precursors of chromosome components. The critical stages in the normal chromosome cycle are DNA replication, chromonema replication, chromatid replication during interphase, centromere division and chromosome separation usually followed by cytokinesis or cleavage during the mitotic stages.

When a cell makes the transition from interphase to mitosis, it stops synthesis and construction of new cell parts, but DNA is replicated. Within the dividing cell, profound changes proceed smoothly, one after the other, through the four sequential stages: **prophase**, **metaphase**, **anaphase** and **telophase**. Once mitosis begins, the cell division process is usually rapid, although the timing of different mitotic stages may vary (**Table 9.1**). A bipolar spindle, composed of two sets of microtubules, positions the chromosomes and moves them to specific locations during these stages. **Figure 9.12** shows these stages for a dividing animal cell and **Fig. 9.13** for a plant cell.

Table 9.1 Duration of mitotic stages

	Temperature °C	Minutes				
		P	M	A	T	Total
Plants						
Pea (root tip)	20	78	14.4	4,2	13.2	110
Pea (endosperm)	20	40	20	12	110	182
Onion (root tip)	20	71	6.5	2.4	3.8	83.7
Broad bean (root tip)	19	90	31	34	34	155
Oatgrass (stigma)	19	36–45	7–10	15–20	20–35	78–110
Spiderwort (stamen hair)	20	181	14	15	130	340
Animals						
Sea urchin embryo	12	19	17	12	18	66
Mouse (spleen)	38	21	13	5	20	59
Grasshopper (neuroblasts)	38	102	13	9	57	181
Frog (fibroblasts in culture)	20–24	32	20–29	–	6–11	–
Chicken (mesenchyme cells in culture)	39	30–60	2–10	3–7	2–10	37–87
Salamander (embryo kidney)	20	59	55	6	75	195

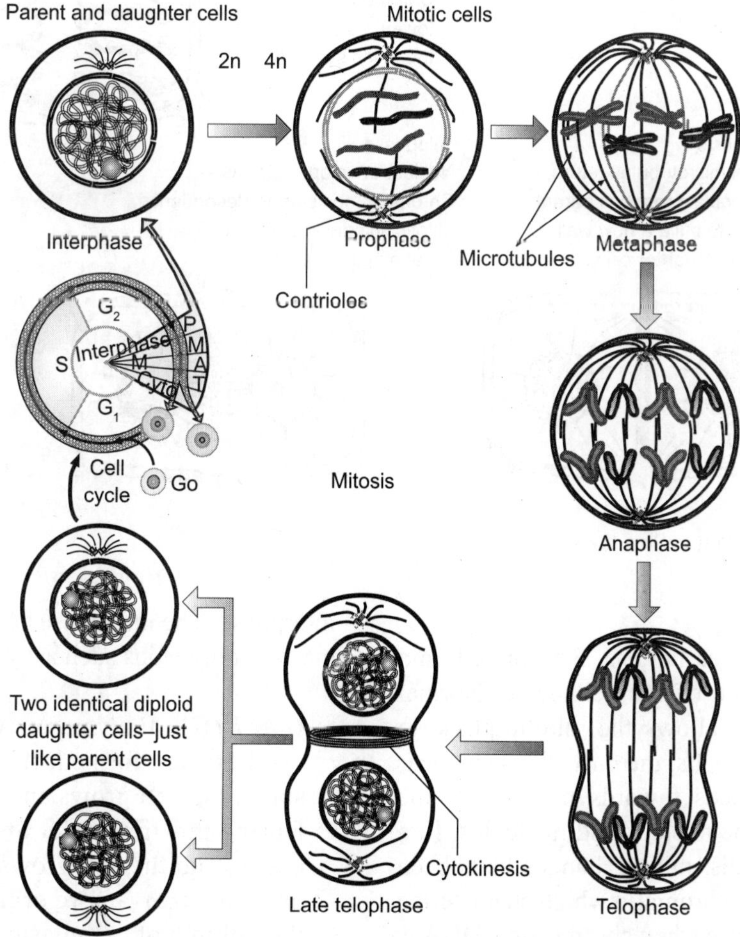

Fig. 9.12 An overview of mitosis. Mitosis perpetuates the chromosome constitution of the cell. Two pairs of homologous chromosomes is shown. Mitosis proceeds through four stages: prophase, metaphase, anaphase and telophase.

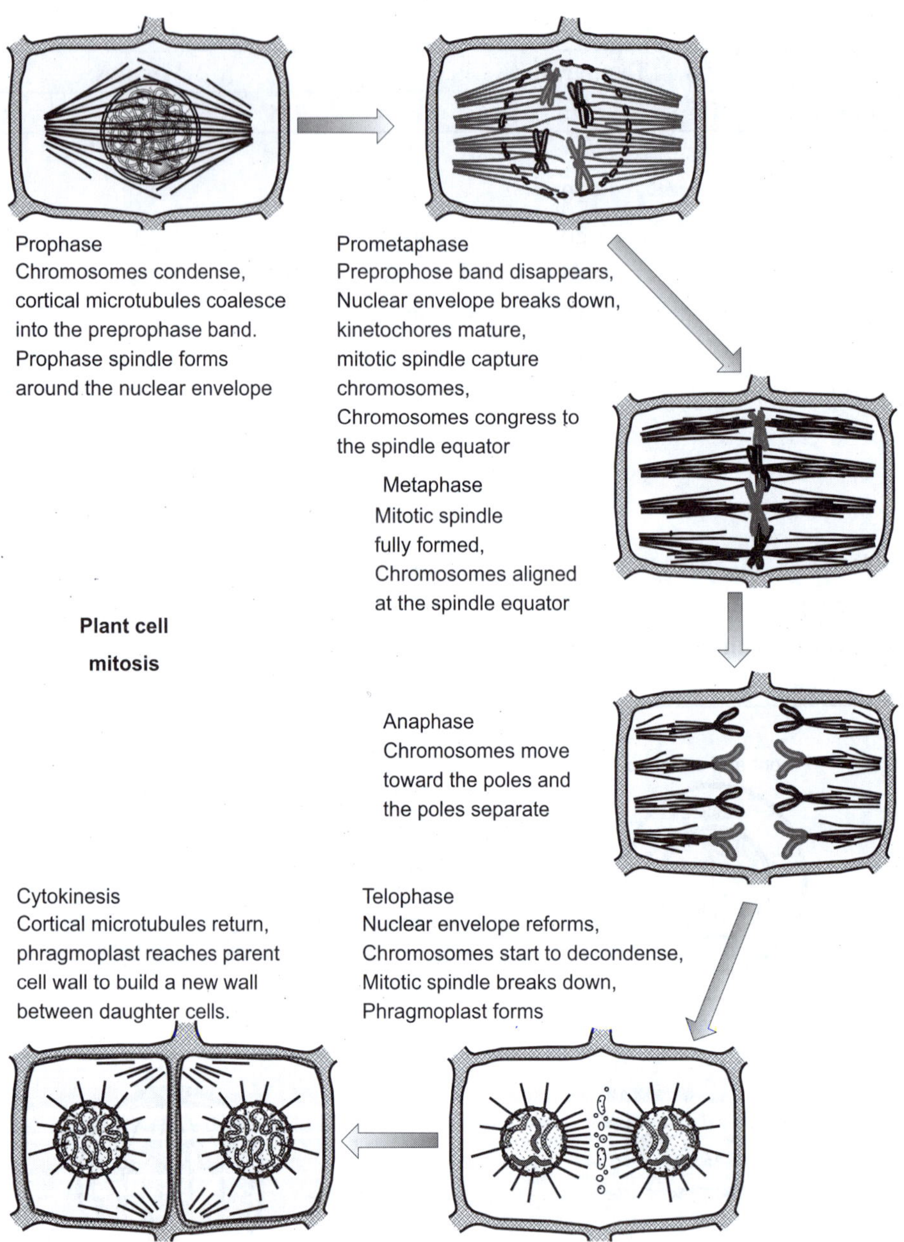

Fig. 9.13 An overview of plant cell mitosis.

Interphase

Interphase as described earlier, is the longest and metabolically active stage of the cell cycle. Normal completion of interphase allows the mitotic phase to begin. In most interphase cells, the only visible nuclear changes as a cell proceeds towards mitosis are the increasing volumes of the nucleus and nucleolus. In a few plant and animal cells, minor changes have been seen in the distribution of chromatin which characterize the S period of the cell cycle when chromosome DNA is synthesized and the subsequent interval before mitosis begins. The S period usually occupies about half of interphase. The interphase has been further divided into many subdivisions such as:

- ***Gap 1 (G_1)***: Enzymes essential for DNA synthesis are formed; this takes 7 hours. In this stage, the cell grows and functions normally. During this time, a lot of protein synthesis occurs and the cell grows (to about double its original size)—more organelles are produced and the volume of cytoplasm increases. If the cell is not to divide again, it will remain in this phase.

- ***Synthesis (S)***: Synthesis of DNA (Duplication of genes); takes 6 hours. In this stage, the cell duplicates its DNA (via semi-conservative replication).
- ***Gap 2 (G_2)***: Preparation of the cell for division; takes 5 hours. In this stage, the cell resumes its growth in preparation for mitosis.
- ***Gap 0 (G_0)***: In the cell cycle interphase is proceeded by telophase and cytokinesis of the M phase. Alternatively, interphase is sometimes interrupted by G_0 phase, which, in some circumstances, may then end and be followed by the remaining stages of interphase (S and G 2). After the successful completion of the G2 stage, the cell proceeds to prophase or in plants to preprophase, which is the first stage of mitosis. G_0 phase is viewed as either an extended G_1 phase where the cell is neither dividing nor preparing to divide and as a distinct quiescent stage which occurs outside of the cell cycle.

The time spent in interphase and in each stage of interphase is variable and depends on both the type of cell and the species of organism it belongs to. Some human cells divide every 24 hours; in this case interphase takes about 22 hours. The cell enters into prophase to divide either by mitosis (for somatic cells and spermatogonia or oogonia) or meiosis (for gametes: Meiosis I and Meiosis II). In programmed cell death, interphase is followed or pre-empted by apoptosis.

Prophase

The beginning of prophase (Gr., *pro* = before; *phasis* = appearance) starts by appearance of thin thread-like condensing chromosomes (**Fig. 9.14**) and marks the first phase of mitosis.

Early prophase: In most cells entering mitosis, a band of microtubules (*centrioles* in animal cells) encircles the nucleus and is called the preprophase band which lies in the cell cortex near the plasma membrane. As mitosis proceeds, microtubules begin to appear at the surface of the nucleus. The DNA and its associated proteins begin to condense and coil within the nuclear envelope. The chromosomes become visible in the light microscope as thread-like forms. There is a decrease in length and increase in thickness due to coiling of the chromosomes. In early prophase, the number of coils in each chromosome is very large but the diameter of the each coil is relatively small. There is a progressive increase in the diameter of the coils so that their number per chromosome becomes correspondingly smaller, causing chromosomes to become thicker. The overall change in the length of the chromosome is due to *condensins*. These introduce positive supercoils in the DNA requiring ATP hydrolysis and the presence of topoisomerase I. The two sister chromatids of each chromosome are coiled in relation to each other and referred to as *relational coiling*. *Plectonimic* and *paranimic* are the two types of relational coiling. In *plectonimic* coiling sister chromatids are twisted round each other as two strands of a rope. As a result, the two sister chromatids of a chromosome cannot separate from each other without rotating. Plectonimic coiling is found between sister chromatids during the mitotic prophase. In *paranimic* coiling, however, the sister chromatids are not twisted round each other; the coils of one of the sister chromatids are simply slipped over the other. As a result, paranimically coiled sister chromatids (PCSC) are easily separated without rotating the chromosome. This type of coiling is found during meiotic prophase. Every chromosome that duplicated

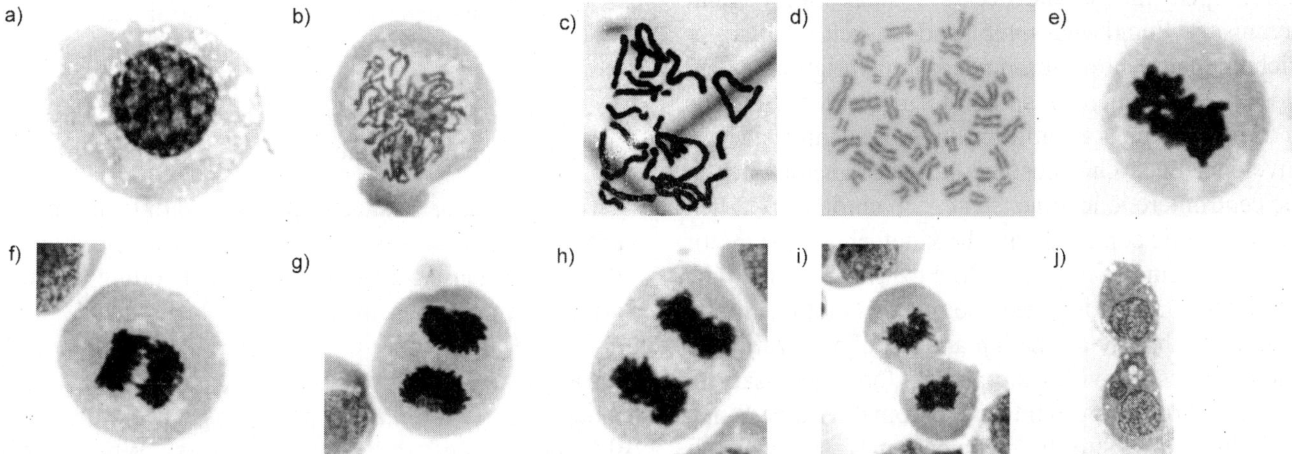

Fig. 9.14 Light microscopic photos showing different stages (a–j) of mitotic cell division in animal cells (Channarayappa, 1990).

in an earlier phase has the centromere, which serves as an attachment site for microtubules of the spindle.

Late prophase: Chromosomes continue to condense. First, the combined helical DNA and protein structure fold back on themselves to form a secondary coil, in a process that is regulated by the concentration of free calcium. New microtubules are assembled and they move one of two centriole pairs towards the opposite end of the cell. When a nucleus is about to divide, most of the cytoskeletal microtubules disassemble, so that the unbound subunits reassemble as new microtubules. The nuclear envelope starts to break up and the nucleolus disappears.

Transition to metaphase: Many changes takes place during transition from prophase to metaphase (*transitional metaphase* or *prometaphase*). The nuclear envelope breaks up completely and microtubules penetrate the nuclear region. Together they form a spindle apparatus. They become attached to the sister chromatids of each chromosome. If both sister half-chromosomes attach to the same pole, this is soon corrected so that the two half-centromeres of each chromosome are attached to opposite poles. This is known as *amphitelic* centromere attachment. Other fibres do not link with the chromosomes but provide the spindle framework. Once a chromosome has both its kinetochores harnessed, microtubules from both poles start pulling on it. The chromosomes move at a rate of about 1 μm per minute towards the equator of the spindle, from where they leave after the nuclear envelope dispersed. All the push-pull forces are balanced when the chromosomes reach the midpoint of the spindle. The formation of the bipolar spindle is now complete.

Prometaphase: The nuclear envelope breaks down and microtubules invade the nuclear space. This is called open mitosis and occurs in most multicellular organisms. Fungi and some protists such as algae or trichomonads, however, undergo a variation called closed mitosis where the spindle forms inside the nucleus or its microtubules are able to penetrate an intact nuclear envelope. Each chromosome has two kinetochores at the centromere, one attached at each chromatid. When a microtubule connects with the kinetochore, the motor activates, using energy from ATP to 'crawl' up the tube toward the originating centrosome. This motor activity, coupled with polymerization and depolymerization of microtubules, provides the pulling force necessary to separate chromosomes into two chromatids later. When the spindle grows to the sufficient length, kinetochore microtubules begin searching for kinetochores to attach to. A number of non-kinetochore microtubules find and interact with corresponding non-kinetochore microtubules from the opposite centrosome to form the mitotic spindle.

Metaphase

The metaphase stage is the most conspicuous stage of the mitotic process. Most cytological analyses are made during this stage only. Metaphase comes from the Greek word *meta* meaning 'after'. As microtubules find and attach to kinetochores in prometaphase, the centromeres of the chromosomes convene along the metaphase plate or equatorial plane—an imaginary line that is equidistant from the two centrosome poles. This even alignment is due to the counterbalance of the pulling powers generated by the opposing kinetochores. In certain types of cells, chromosomes do not line up at the metaphase plate and instead move back and forth between the poles randomly, lining up only roughly along the midline. Because proper chromosome separation requires that every kinetochore be attached to a bundle of microtubules, it is thought that unattached kinetochores generate a signal to prevent premature progression to anaphase without all the chromosomes being aligned. The signal creates the mitotic spindle checkpoint. During the metaphase, four major cytological events take place:

- Disappearance of the nucleolus,
- Breakdown of the nuclear envelope and distribution of its components into the endoplasmic reticulum,
- Appearance of spindle apparatus,
- Arrangement of the chromosomes on a single plane called the equatorial plate. It lies in the middle of the two poles of the spindle apparatus, but need not necessarily be the central plane of the cell. It is an imaginary plane and it does not represent any structural components. The movement of chromosomes to and their orientation on the equatorial plate is termed as ***metakinesis.***

The other major changes observed during this phase are: the chromosomes are at their shortest and thickest (highly condensed state) and relational coiling between sister chromoatids is absent.

Anaphase

Anaphase (Gr., *ana* = up, against, back or re) is the shortest of the mitotic phases. When every kinetochore is attached to a cluster of microtubules and the chromosomes have lined up along the metaphase plate, the cell proceeds to anaphase. Two events then occur:

- The proteins that bind sister chromatids together are cleaved, allowing them to separate. These sister chromatids, which have now become distinct sister chromosomes, are pulled apart by shortening kinetochore microtubules and move toward the respective centrosomes to which they are attached. They are separated from each other and move to opposite poles, at 0.2–4 μm per minute dragging the arms behind them.
- The non-kinetochore microtubules elongate, pushing the centrosomes (and the set of chromosomes to which they are attached) apart to opposite ends of the cell. The force that causes the centrosomes to move towards the ends of the cell is still unknown, although one theory suggests that the rapid assembly and breakdown of microtubules may cause this movement. These two stages are sometimes called *early* and *late anaphase*. Early anaphase is usually defined as the separation of the sister chromatids, while late anaphase is the elongation of the microtubules and the microtubules being pulled farther apart. At the end of anaphase, the cell has succeeded in separating identical copies of the genetic material into two distinct populations.

Telophase

Telophase (Gr., *telo* = end) is a reversal of prophase and prometaphase events. It begins once the two daughter chromosomes arrive at the opposite spindle poles. At telophase, the non-kinetochore microtubules continue to lengthen, elongating the cell even further. Corresponding sister chromosomes attach at opposite ends of the cell. Chromosomes decondense. They are no longer harnessed to kinetochore microtubules and they return to the thread-like form. A new nuclear envelope, using fragments of the parent cell's nuclear membrane, forms around each set of separated sister chromosomes. Both sets of chromosomes, now surrounded by new nuclei, unfold back into chromatin. The nucleolus reappears in a particular place on a chromosome where the ribosomal RNA genes are located. This location corresponds with a structural feature called a **secondary constriction** in some chromosomes. At the end of telophase, two daughter nuclei are formed. Each will be diploid, with two of each type of chromosome—just like the parent nucleus. Mitosis is complete, but cell division is not yet complete.

Post-mitotic interphase

Once the nuclear membrane is complete, the nuclei expand to become approximately spherical and nucleoli reach their full size. It is not possible to distinguish interphase nuclei from post-mitotic nuclei of differentiating cells. Except in some young embryos, interphase is the longest stage, often occupying 90% of the cycle.

9.3.2 Cytokinesis

The cell division cycle culminates in the division of the cytoplasm through cytokinesis. The cytoplasm usually divides at some time between late anaphase and the end of telophase. Cytokinesis proceeds by different mechanisms in plant and animal cells. The cells of most land plants have fairly rigid walls and cannot be merely pinched into two. Plant cell cytoplasm divides by a mechanism that involves **cell plate formation**. At this location, cellulose deposits accumulate and form a cross-wall that divides the parent cell into daughter cells. Early steps in cytokinesis include formation of vesicles at the mid-plane of the mitotic apparatus and the coalescence of these vesicles. Starting at the centre of the spindle, a cell plate is formed and later converted to a **middle lamella** with new cross-walls between the daughter cells deposited on each side of the middle lamella. The separation of the cytoplasm begins in the centre of the cell and gradually extends outwards on either side in a plane perpendicular to the axis of the spindle.

By contrast, animal cells do not have a cell wall and most of them divide by a mechanism called **cleavage** (cell furrow). The first visible change of cytokinesis in the animal cell is the sudden appearance of a cleavage furrow on the cell surface. The furrow rapidly deepens and spreads around the cell until it completely divides the cell in to two. A layer of deposits forms around microtubules at the midsection of most animal cells. The structure that accomplishes cytokinesis is the contractile ring—a dynamic assembly composed of actin filaments, myosin II filaments and many structural and regulatory proteins. The contractile ring assembles just beneath the plasma membrane and constrict the cell into two. Microfilaments are arranged to slide past one another, as they do, they also pull the plasma membrane inward and cut the cell into two. At the same time, a new membrane is inserted into the plasma membrane adjacent to the contractile ring by the fusion of intracellular vesicles. Thus, cytokinesis can be considered to occur in four stages—initiation, contraction, membrane insertion and completion. The essential event in cytokinesis, however, is to ensure that it occurs at the right time and in the right place.

Even though nuclear and cytoplasmic phases are distinct, they are highly coordinated processes. The entire process may require two hours to several days, depending on the type of organism and environmental conditions. At the completion of mitosis, the process ensures that each daughter cell receives a full complement of chromosomes. In addition, each daughter cell must also inherit all the other essential cell components, including the membrane-enclosed organelles. Since organelles (mitochondria and chloroplasts) are usually present in large enough numbers, they are safely inherited. The ER in interphase cells is continuous with the nuclear membrane and is organized by the microtubule cytoskeleton. When the cell enters the M phase, the reorganization of the microtubules releases the ER, which fragments and re-emerges at telophase.

Cytokinesis in plants

Higher plant cells are enclosed by a semi-rigid cell wall; hence, cytokinesis differs from animal cells. In plant cells the partitioning takes place from the inside out by the construction of a new cell wall, called the cell plate, between the two daughter nuclei. The orientation of the cell plate determines the positions of the two daughter cells relative to neighbouring cells. The first visible sign that a plant cell has committed to divide in a particular plane can be seen as early in G2, when the cortical array of microtubules disappears in preparation for mitosis. A circumferential band of microtubules and actin filaments forms a ring around the entire cell just beneath the plasma membrane, called the **prophase band**. The band becomes thinner as the cell progresses to prophase and it disappears before onset of metaphase, but the division plane is well established. When the new cell plate forms later during cytokinesis, it grows outward to fuse with the parental wall precisely at the zone formerly occupied by the prophase band. The assembly of the cell plate begins in late anaphase and is guided by a structure called the **phragmoplast**, contains overlap microtubules of the mitotic spindle that interdigitate at their growing plus ends. Small vesicles, derived from the Golgi apparatus and filled with polysacharides and glycoproteins required for synthesis of the new cell wall matrix and plasma membrane are transported along the microtubule-dependent motor proteins. The vesicles fuse to form a disc-like, membrane-enclosed structure called the **early cell plate**. The cell plate expands outward by further vesicle fusion until it reaches the plasma membrane and the original cell wall and divides the cell into two. Later, cellulose microfibrils are laid down within the matrix of the cell plate to complete the construction of the new cell wall (**Fig. 9.15**).

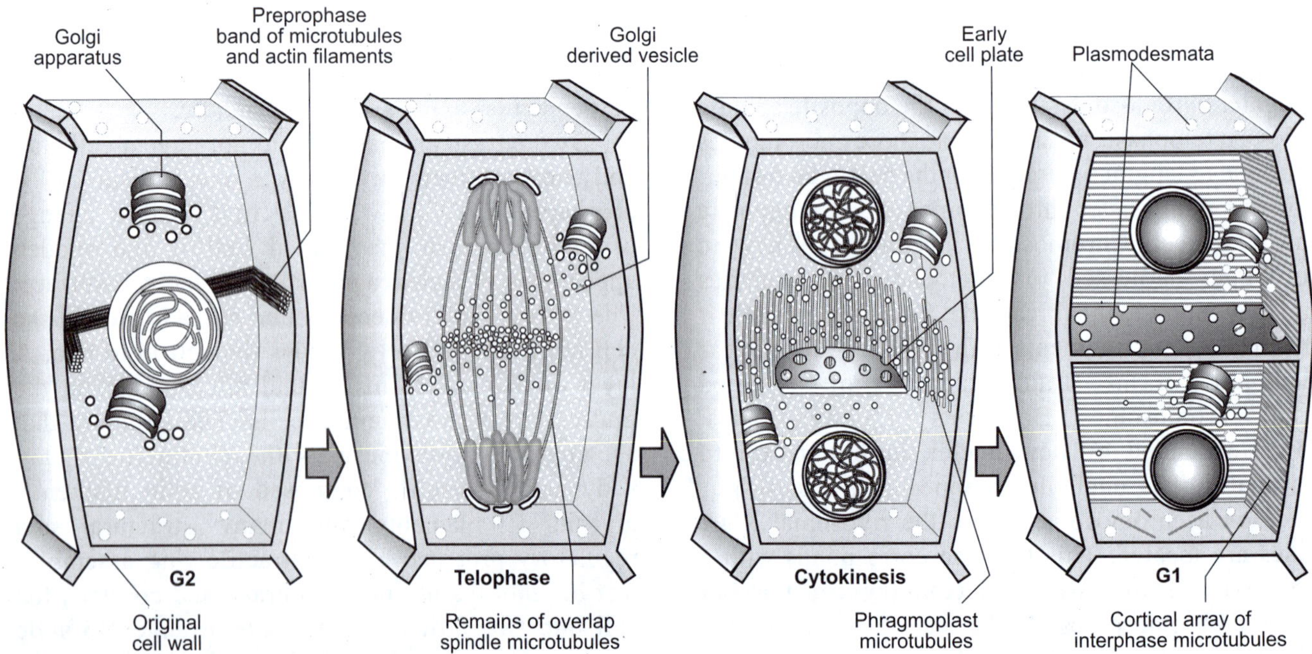

Fig. 9.15 General features of cytokinesis in a higher plant cell.

9.3.3 Regulation of cell division

One of the important events regulating cell division is the mechanism that delays initiation of cell division until sister chromatids properly segregate from each other. In the yeast, *Schizosaccharomyces pombe* onset of cytokinesis is controlled by a signalling cascade—the **septum initiation network (SIN)**. All eukaryotic cells have to coordinate the segregation of their genetic material (karyokinesis) with cell division (cytokinesis) in order to maintain ploidy level and prevent chromosome mis-segregation. In mitosis, two signalling cascades have been identified for regulating the metaphase–anaphase transition and the exit of mitosis/onset of cytokinesis, respectively. The first signalling cascade is the spindle checkpoint that monitors kinetochore attachment to microtubules and delays loss of cohesion between sister chromatids until all chromosomes are correctly attached to the mitotic spindle. In budding yeast, several proteins (Mad1p, Mad2p and Mps1p) assemble on unattached kinetochores and act in a concerted fashion to block chromosome segregation. The cell cycle can thus be arrested in a prometaphase-like state.

In late mitosis, a second network of proteins regulates septum initiation in fission yeast (SIN) and the exit from mitosis in budding yeast (**mitotic exit network, MEN**), respectively. The components of this network localize mostly to the spindle pole body (**SPB**)—the yeast functional equivalent of the centromere of higher eukaryotes. The mechanism of regulation of late anaphase in higher eukaryotes is a SIN-like pathway which is likely to be a conserved mechanism to prevent premature onset of cytokinesis.

9.3.4 Variations in mitosis

Plant cells

Most of the higher plant cells do not contain visible centrioles. An analogous region of the cell acts as a microtubule-organizing centre, from which the spindle microtubules radiate. The shape of the plant cell does not vary greatly in mitosis, since it has a strong cell wall. During telophase, the new cell membrane and cell wall are formed from membrane vesicles that fuse together in a plane that is perpendicular to a line separating the two nuclei.

Modifications of mitosis

In a highly differentiated multicellular organism, not all the dividing cells are the same in behaviour or appearance. As some cell differentiation is compatible with continued mitotic activity, dividing cells differ in respect of size, shape, cytoplasmic content, position and orientation of the mitotic spindle and timing of the cycle. There are a number of other controlled changes affecting the karyotype during development. These changes result in differences between cells and tissues in chromosome size, in the number of replicates of the genome or in the number of individual chromosomes. Most of these changes occur during the development of somatic tissues. For those that occur in the germ line, the cell lineage from zygote to zygote is usually reversed at some other stage of the life cycle in order to preserve a stable karyotype from one generation to the next. In *Drosophila*, early development begins in an unusual way: a series of nuclear divisions without cell division creates a **syncytium (Fig. 9.16)**.

Many changes in plant expression at nematode feeding sites (NFSs) have been reported. At NFS the giant cells and syncytia are activated to undergo multiple rounds of shortened cell cycles leading to genome amplification, but in different ways. The giant cells go through repeated (acytokinetic) mitosis, whereas nuclear divisions have never been observed inside syncytia, which go through repeated S phase (**endo-reduplication**), apparently shunting mitosis (**Fig. 9.17**). The cyst nematodes activate the

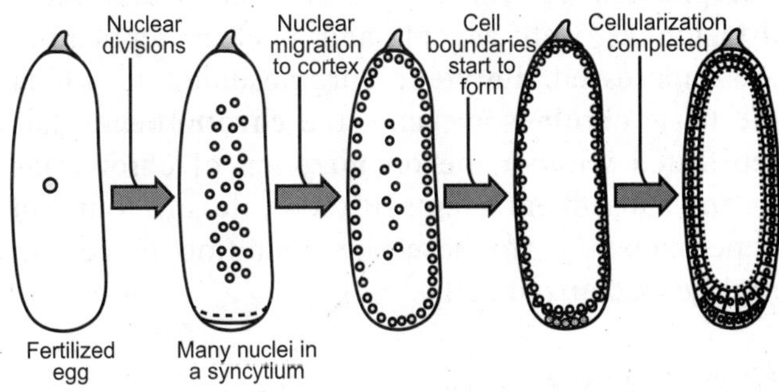

Fig. 9.16 Mitosis without cytokinesis in the *Drosophila* embryo. The first 13 nuclear divisions occur synchronously and without cytoplasmic division to create a large syncytium. Most of the nuclei then migrate to the cortex and the plasma membrane extends inward and pinches off to surround each nucleus to form individual cells in a process called cellularization.

cell cycle in the feeding cells without mitotic events. The cells repeatedly go through the S phase, also through G_1 and G_2 but shunting mitotic phase, called **acytokinetic mitosis**.

Changes in chromosome size

Differences in chromosome size, with no differences in DNA content, occur between different tissues of the same plant and between cells at different stages of development of the same tissue. Chromosome size appears to be correlated with metabolic activity and increases with nutrient availability.

Changes in ploidy level

These accompany the normal alteration of generations by fusion in the zygote and endosperm of the meiotically reduced complements found in the spores and gametes. Similar process in purely somatic cells is rare. **Binucleate** cells and after subsequent cycles, multinucleate cenocytes are formed when mitosis is suppressed (**Fig. 9.18**). Polyploidy with double the number of chromosomes in the nucleus results when both cell and nuclear replication are omitted. This is the most frequent cause of enlarged nuclei in differentiated somatic cells of many animal and plant tissues. The increase usually occurs without chromosome condensation within an intact nuclear envelope by a process referred to as **endomitosis**, producing **endo-polyploid** nuclei. Occasionally the reversion to a full mitotic cycle occurs naturally and polyploid and diploid mitoses are found together in a meristem—a condition known as **polysomaty**. **Diplochromosomes**, chromosomes with four or more chromatids attached by one unreplicated centromere, result when centromere division as well as cell and nuclear replication are suppressed. **Polytene** chromosomes, in which the basic chromosome unit, the **chromonema**, has replicated without the organization of chromatids or any subsequent events of the cycle, are found in a permanently prophase-like condition in certain highly specialized cells.

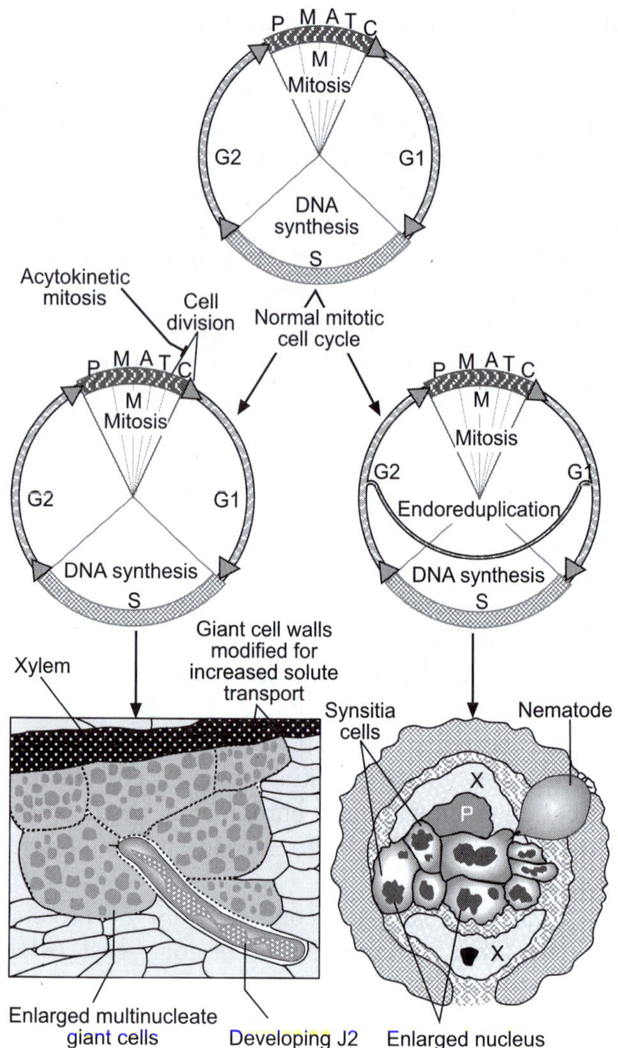

Fig. 9.17 The cell cycle.

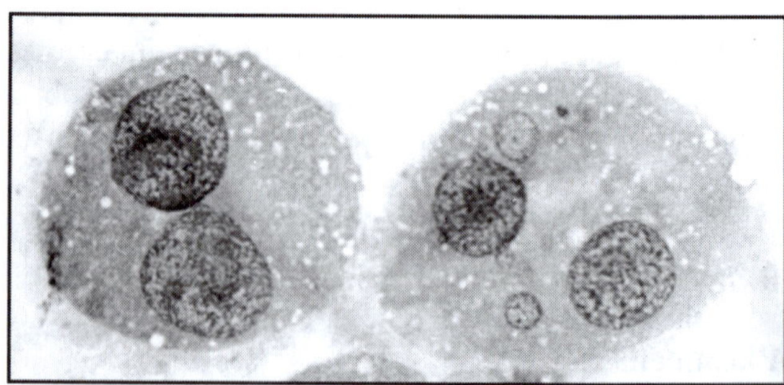

Fig. 9.18 Two binucleated animal cells. Cytokinesis of mitosis division is inhibited by treating cells with cytochalazin-B, consequently, two daughter nuclei remained in the same cell. This resulted in normal karyokinesis but no cytokinesis (Channarayappa, 1990).

Karyotype segregation

While in most organisms there are changes in the number of genomes in the nuclei, in certain tissues, each genome usually remains intact during development. However, in certain plants and animals, controlled non-disjunction at predetermined stages of development results in the gain or loss of specific chromosomes and the production of cell lines with an altered karyotype. The classic example of this somatic segregation is in *Parascaris equorum*, a roundworm, where the two large polycentric chromosomes found in the zygote are maintained in the germline but fragment and lose their acentric terminal segments during embryonic mitoses, thus giving rise to somatic tissues. A related situation occurs in some plants with supernumerary B chromosomes. In a variety of sexually reproducing organisms, similar mechanisms exist in the germline to alter the number of autosomes, sex chromosomes or even whole genomes.

9.3.5 Significance of mitosis

The process of mitosis and the subsequent replication of chromosomal material in the succeeding interphase have the inevitable result of creating from one cell, two new cells that are genetically identical. Mitosis produces two nuclei which have the same number of chromosomes as the parent cell. Since these chromosomes were derived from parental chromosomes by the exact replication of their DNA, they are genetically identical to their parents. This results in **genetic stability** within populations of cells derived from the same parental cells. The number of cells within an organism increases by mitosis and this is the basis of **growth** in multicellular organisms. Some animals are able to **regenerate** whole parts of the body, such as legs in *Crustacea* and arms in starfish. Replacement of cells and tissues also involve mitosis. Cells are constantly dying and being replaced with new cells (e.g., skin, bone marrow) through mitosis.

9.3.6 Consequences of mitotic errors

Although errors in mitosis are rare, they still happen, especially during early cellular divisions in the zygote. Mitotic errors can be especially dangerous to the organism because future offspring from this parent cell will carry the same disorder. A common mitotic error may be non-disjunction; a chromosome may fail to separate during anaphase. For example, one daughter cell will receive both sister chromosomes and the other will receive none. This results in the former cell having three chromosomes (two sisters and a homologue), a condition known as **trisomy** and the latter cell having only one chromosome (the homologous chromosome), a condition known as **monosomy**. These cells are considered aneuploidic cells and these abnormal cells can cause cancer.

Mitosis is a traumatic process and the cell may undergo dramatic changes in ultrastructure. Chromosomes are jostled constantly by probing microtubules and occasionally chromosomes may become damaged. For instance, an arm of the chromosome may be broken and the fragment lost, causing deletion. The fragment may incorrectly reattach to another,, non-homologous chromosome, causing translocation. It may re-attach to the original chromosome, but in reverse orientation, causing inversion. Or, it may be treated erroneously as a separate chromosome, causing chromosomal duplication. The effect of these genetic abnormalities depends on the specific nature of the error. It may range from no noticeable effect to cancer induction or organism death.

Division of cell organelles

The growth and division of mitochondria and chloroplasts is not coupled to nuclear division and the incorporation of proteins and lipids into these organelles occurs continuously during the interphase period of the cell cycle. As these organelles increase in size, one or more daughters 'pinch off' in a manner similar to the way in which bacteria grow and divide. At mitosis, each daughter cell receives approximately the same quantities of organelles and organelle DNA, but there is no system for apportioning exactly equal amounts of organelle DNA to the daughter cells.

Partition and cytokinesis

During the bacterial cell division, the partition of the replicated chromosome marks the culmination of the chromosome cycle and is followed by cell division. A septum forms at the midpoint of the parental cell, which is identified by a periseptal annulus, a region of modified cell envelope where the inner and outer membranes are joined together around the circumference of the cell. Additional annuli form by duplication and migrate to positions equivalent to one-quarter and three-quarter cell lengths and these are the sites of **septation** in daughter cells during the next round of cell division. Once the septum has formed, the cell undergoes cytokinesis and divides by binary fission (see Chapter 4). The partition of bacterial chromosomes involves the association of the chromosome with the cell membrane and may be regulated by replication: the origin and terminus of replication, as well as active replication forks are membrane-associated.

Viral multiplication cycles

Viruses multiply in a variety of ways depending on the viral type and the host cell in which they replicate. Nearly all multiplication cycles include five basic types:

- The virus attaches to a host cell.
- The whole virus or its genetic material alone penetrates the cell's cytoplasm.
- In an act of molecular piracy, the viral DNA or RNA directs the host cell into producing many copies of viral nucleic acids and proteins, including enzymes.
- The viral nucleic acids and proteins are put together to form new virus particles.
- Newly formed virus particles are released from the infected cell.

Figure 9.19 shows the multiplication of bacteriophage. Two pathways are common among bacteriophages. In a **lytic pathway**, steps 1 through 4 proceed rapidly, with new particles released by lysis. In **lysogenic pathways**, an infection enters a latent period; the host cell is not killed outright. Genetic recombination may take place during latency. A viral enzyme cuts the host chromosome and integrates viral genes into it. The recombinant DNA is replicated and passed on to the infected cell's descendants. Later, if viral genes move out of the chromosome, the multiplication cycle will resume.

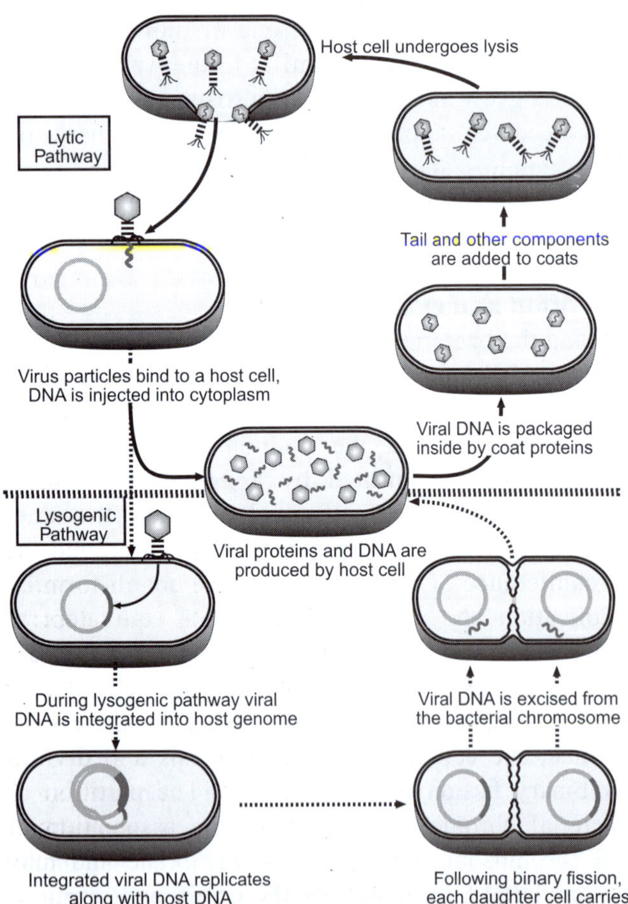

Fig. 9.19 Multiplication cycle for some bacteriophages.

9.4 MEIOSIS

Meiosis was discovered and described in sea urchin eggs in 1876, by noted German biologist Oscar Hertwig and later in 1883 by Edouard Van Beneden in *Ascaris* eggs. The significance of meiosis for reproduction and inheritance was described only in 1890 by German biologist August Weismann, who noted that two cell divisions were necessary to transform one diploid cell into four haploid cells if the number of chromosomes had to be maintained. In 1911 American geneticist TH Morgan observed crossover in *Drosophila melanogaster* meiosis and provided the first genetic evidence that genes are transmitted on chromosomes.

Meiosis (Gr., *meio* = to reduce) is a critical, highly conserved stage occurring during the sexual reproduction of all eukaryotes. It is a unique event in the sexual life cycle of plants and animals and differs from mitosis in several complex ways, but still has many fundamental features in common with somatic cell division. It is likely that meiosis arose from mitosis as sexual reproduction evolved as a means of generating more genetic variation. Adaptation, speciation and long-term evolution depend on the generation of genotypic, and thus phenotypic, variation. The doubling of the chromosome number at fertilization is compensated by an equivalent reduction in chromosome number at a stage prior to the formation of the gametes. In a haploid asexual life cycle, there will usually be a rapid elimination of a new mutant, while in a diploid asexual life cycle the mutation will usually not be expressed in the heterozygote produced. Only in a sexual life cycle can the products of mutation be fully exploited. Fertilization (**syngamy**) between mutant individuals can bring these mutations together within one organism of a later generation. Meiosis then replaces new allelic combinations by recombining not only alleles on the same chromosome but also those on other chromosomes. Random fertilization of recombinant gametes will then produce individuals with new combinations of phenotypic characters.

Meiosis is concerned with sexual transmission of chromosomes where the genes from two different parent cells eventually reside in a single daughter cell. In meiosis, haploid gametes with half (n) amount of DNA such as pollen (sperm) and ovules (egg) are formed from diploid cell ($2n$) in the germinal tissue of higher organisms. Fertilization then makes the event in which two haploid nuclei join to reform a diploid cell. Meiosis begins with the interphase containing G_1 and S phases and a relatively short G_2 period when meiosis becomes irreversible. During meiosis, a single round of DNA replication is followed by two cycles of cell division, termed **meiosis I** and **meiosis II**.

9.4.1 Variations in eukaryotic meiosis

Extensive studies made in eukaryotic meiosis revealed the existence of marked differences with respect to the stage within the life cycle at which meiosis occurs and the duration of the haploid phase. Based on these differences, meiosis is divided into the following three general groups:

Gametic or **terminal meiosis:** This group includes all multicellular animals and many protists. In male vertebrates, for example, meiosis occurs just prior to the differentiation of the spermatozoa. The spermatogonia become primary spermatocytes, which then undergo the two divisions of meiosis to produce four relatively undifferentiated spermatids. Each spermatid undergoes a complex differentiation to become the highly specialized sperm cell. In female vertebrates, oogonia become primary oocytes, which then enter a greatly extended meiotic prophase. During this stage the primary oocyte grows and becomes filled with yolk and other materials; it then undergoes meiotic division. Vertebrate eggs are typically fertilized at a stage before the completion of meiosis and meiosis is completed after fertilization, while the sperm resides in the egg cytoplasm.

Zygotic or **initial meiosis:** This group includes only protists and fungi; the meiotic divisions occur just after fertilization to produce haploid adult generation. Consequently, the diploid stage of the life cycle is restricted to a brief period after fertilization when the individual is still a zygote.

Sporic or **intermediate meiosis:** Plants and some algae belong to this group; the meiotic divisions take place at a stage unrelated to either gamete formation or fertilization. In this case the life cycle begins with the union of a male (pollen) and female (egg) gametes; the diploid zygote undergoes mitosis and develops into a diploid *sporophyte.* At some stage in the development of the sporophyte sporogenesis occurs, producing spores that germinate directly into a haploid *gametophyte.* The gametophyte can be either an independent stage or a tiny structure retained within the ovules. In either case, the gametes are produced from the haploid gametophyte by mitosis.

9.4.2 Stages of meiosis

Because meiosis is a 'one-way' process, it cannot be said to engage in a cell cycle as mitosis does. However, the preparatory steps that lead up to meiosis are identical in pattern and name to the interphase of the mitotic cell cycle. To compare the events of mitosis and meiosis, we need to examine the fate of the chormatids. In contrast to mitosis, meiosis is spread over a considerably longer duration. As a rule the meiotic divisions follow a standard scheme during which two successive divisions of the chromosomes occur (**Fig. 9.20**).

The first division represents a reduction division in which members of homologous pairs of chromosomes are separated into daughter cells without duplication, but their numbers are reduced to half. Although variations have been discovered, one of the significant and common features of meiosis is the initial pairing and subsequent separation of homologous chromosomes.

Premeiotic interphase

The interphase stage just prior to the entry of a cell into meiosis is known as premeiotic interphase. In many cases, cells enter meiosis with a degree of synchrony. This has frequently developed over one or more preceding mitotic cycles so that synchronous premeiotic mitoses may also be seen. Nuclei entering meiotic prophase are usually considerably larger than other meristematic nuclei and in the case of oocytes and embryo sac mother cells, the cells are also unusually large. Most of the nuclear DNA is replicated at the end of premeiotic interphase, shortly before division begins (about 0.3% does not replicate during the S phase). The S phase is longer than that of mitosis. The long G_2 period between DNA synthesis and the onset of prophase typical of most mitoses is absent from the meiotic cycle. In plants, the meiotic

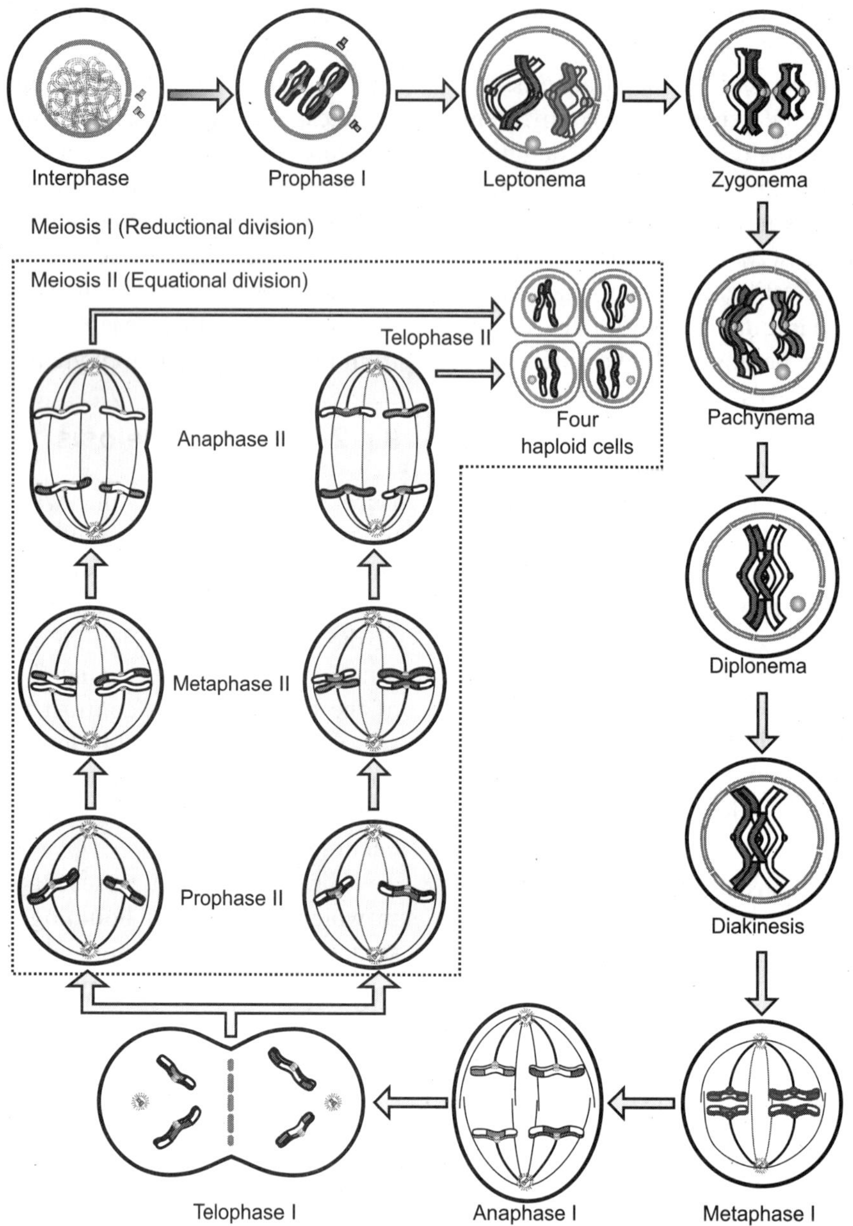

Fig. 9.20 The nuclear division mechanism by which the parental number (*2n*) of chromosomes is reduced by half for forthcoming gametes (*n*).

synchrony may extend throughout an anther and even to other anthers of the same bud.

During the S phase of the premeiotic interphase, a special type of histone specific to cells preparing for meiosis is synthesized. This histone is not found in cells undergoing mitosis and it may be related specifically to the entry of cells into meiosis. The entry of cells into meiosis also depends on some substances synthesized during G_2 substage of premeiotic interphase; the identity of all these substances is not yet definitely known. During meiosis, although the nucleus divides twice. The S phase occurs only once during the premeiotic interphase. Thus, during meiosis, chromosomes replicate only once, while the nuclei divide twice.

Crossover and meiosis

As discussed in the previous section, sister chromatids replicated during the S phase initially link by **cohesin complexes** at multiple positions throughout the chromosome (**Fig. 9.21**). As chromosomes condense, cohesin complexes become localized to the centromeric region only. At metaphase the sister chromatids composing each (replicated) chromosome

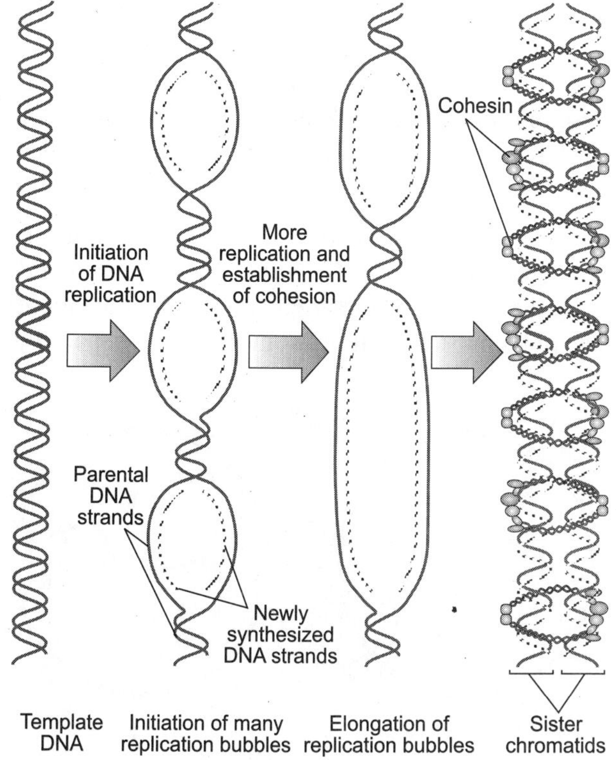

Fig. 9.21 The events of S phase.

associate with microtubules emanating from opposite spindle poles. Later, separase-catalyzed cleavage of the Scc1 cohesin subunit permits movement of the chromatids toward the spindle poles, signalling the onset of anaphase. In metaphase of meiosis I, both sister chromatids in one (replicated) chromosome associate with microtubules emanating from the same spindle pole, rather than from opposite poles as they do in mitosis. Two physical links between homologous chromosomes are thought to resist the pulling force of the spindle until anaphase: (i) crossing over between chromatids, one from each pair of homologous chromosomes and (ii) cohesin crosslinks between chromatids distal to the crossover point.

During meiosis I, the cohesin crosslinks between chromosome arms are cleaved by separase, allowing the homologous chromosomes to separate, but cohesin complexes at the centromere remain linked.

Normal segregation of chromosomes to opposite spindle poles occurs only when homologous chromosome pairs associate. This process is called **synapsis** and can be visualized microscopically in eukaryotes with large chromosomes. In higher eukaryotes, processes in addition to recombination and chromatid linking through cohesin complexes contribute to synapsis in meiosis I.

Meiosis I

The most significant events of meiotic division I are homologous chromosome pairing, crossing over, separation and their migration to the opposite poles. As a result, the two daughter nuclei produced by this division receive only half of the chromosomes present in somatic cells. Thus, the first meiotic division is called **reduction division**.

Prophase I: This stage of meiotic division is the longest in duration. In female animals, the first prophase is especially long and may last for several days to several years. In the human female, for example, oocytes initiate prophase I of meiosis prior to birth and then enter a period of prolonged arrest. When human females are born, all the oocytes present in their ovaries are in first prophase; they stay in this stage for 12–60 years. Oocytes resume meiosis just prior to the time they are ovulated, which occurs every 28 days or so after an individual reaches puberty. Very important events take place during the first meiotic prophase, homologous chromosomes pair and there is recombination between the chromosomes. Each chromosome starts to twist and fold into more condensed forms. It pairs up with its homologue and the two typically swap segments. The first meiotic prophase thus divides into several stages that are similar in all sexually reproducing eukaryotes.

Leptotene or **leptonema:** The first stage of prophase I is leptotene, during which the chromosomes become visible in the light microscope. Once the cells enter the leptonema phase, they are committed to meiosis. It represents the end of interphase, where chromosome replication has occurred. The chromosomes appear as long threads and in fact consist of a pair of sister chromatids, which are identical and held together by a centromere. In addition, a complex of RNA and proteins associated with DNA called a *lateral component* or lateral element is deposited along the chromatids during leptonema. Lateral element synthesis is conducted by both sister chromatids. In addition to chromosome condensation and lateral element synthesis, the telomeres of the chromosomes begin to associate with a region of the nuclear envelope called the attachment site, which is devoid of nuclear pores and slightly thickened. This site may align homologous chromosomes.

Zygotene or **zygonema:** In this phase, the homologous chromosomes become paired longitudinally; this process of chromosome pairing is called synapsis. Chromosome compaction and synapsis during zygotene simply make this chromosomal arrangement visible under the microscope. In most organisms, the individual chromosomes occupy discrete regions within nuclei rather than being randomly dispersed throughout the nuclear space. The telomeres of leptotene chromosomes are distributed throughout the nucleus. Then, near the end of leptotene, there is a dramatic reorganization of chromosomes in many species so that the telomeres become localized at the inner surface of the nuclear envelope at one side of the nucleus. Clustering of telomeres at one end of the nuclear envelope occurs in a wide variety of eukaryotic cells and causes the chromosomes to resemble a bouquet of flowers.

Electron photomicrographs indicate that chromosome synapsis is accompanied by the formation of a complex structure called the *synaptonemal complex*. The synaptonemal complex (**Fig. 9.22**) is a ladder-like structure. This complex comprises two lateral elements and a central region, which is approximately 100 nm wide and is bisected by a narrow band called the central component, which like the lateral elements is composed primarily of RNA and proteins. The lateral elements are composed primarily of cohesin, which presumably binds together the chromatin of the sister chromatids. A possible hypothesis to account for the function of the synaptonemal complex in meiotic genetic exchange (recombination or crossing over) is that the lateral components may serve as scaffold complexes through which important DNA sequences may be carefully threaded during leptonema in preparation for the zygonema phase where the two homologous chromatids must align precisely. When synapsis is initiated, the number of visible structures is half that in leptonema and the visible bodies in the nucleus are now homologous chromosome pairs of bivalents or a tetrad rather than single chromosomes.

Pachytene or **pachynema:** The synapsed homologous chromosomes (or bivalents) continue to become shortened and thickened during this stage. Simultaneously, the chromomeres become fewer, larger and more conspicuous. Synaptonemal complex (SC) formation and subsequent pairing of the chromosomes is complete and recombination of genetic material is also conducted during this stage. The exchange event takes place first and appears visually as a *chiasma* (crossing over). The synaptonemal complex gradually disappears towards the end of the pachynema phase. Under the electron microscope, a number of electron-dense bodies (100 nm) are seen within the centre of the SC. These structures have been named *recombination nodules* because they correspond to the sites where crossing over takes place.

Diplotene or **diplonema:** There appears to be separation between the homologous chromatids during diplonema. There are two types of attachments that prevent the chromosomes from separating

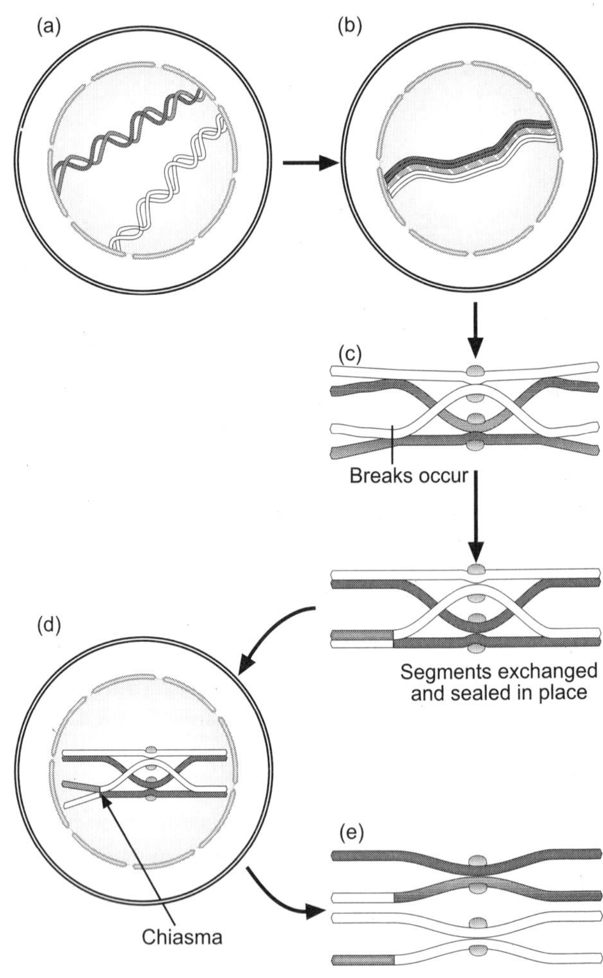

Fig. 9.22 Key events during prophase I, the first stage of meiosis: (a) chromosomes duplicate before meiosis begins; (b) the two chromosomes become zippered together, so that all four chromatids are intimately aligned; (c) one or more crossovers occur at intervals along the chromosomes; (d) as prophase I ends, the chromosomes continue to condense, becoming thicker, rod-like forms. They detach from the nuclear envelope and from each other—except at chiasmata and (e) Crossing over breaks up old combinations of alleles and puts new ones together in pairs of homologous chromosomes.

completely. First, the sister chromatids are joined by the centromeres and second, the non-sister chromatids are usually held together at one of more sites that are referred to as chiasmata, a cross-like conformation. Chromatids continue to shorten and thicken and the four chromatids in a group, now called a tetrad, become move apart. Paired chromatids appear to repel each other, causing the strands to separate longitudinally in some areas and to form loops. The synaptonemal complex degrades and homologous chromosomes separate from one another a little. The chromosomes themselves uncoil a bit, allowing some transcription of DNA. However, the homologous chromosomes of each bivalent remain tightly bound at chiasmata, the regions where crossing-over occurred. The chiasmata remain on the chromosomes until they are severed in Anaphase I. Later the chiasmata may move towards the end of the chromosome arm. This is known as *terminalization* and is a mechanical process whereby the chiasma, but of course not the point of genetic exchange, moves away from the centromere.

In human foetal oogensis all developing oocytes develop to this stage and stop before birth. This suspended state is referred to as the *dictyotene stage* and remains so until puberty. In males, only spermatogonia exist until meiosis begins at puberty.

Diakinesis: In diakinesis, the chromosomes are totally condensed and connected to each other through the chiasmata. The bivalents then move to the periphery of the nucleus and lie, well separated, on the inside of the nuclear membrane. This is the first point in meiosis where the four parts of the tetrads are actually visible. The sites of crossing over entangle together, effectively overlapping, making chiasmata clearly visible. Later, the nucleolus and the nuclear envelope either disappear or detach from its associated chromosome and the meiotic spindle begins to form.

Prometaphase I: As the nuclear membrane disintegrates, a spindle forms; is similar to the mitotic spindle, but there is an important difference in the way the centromeres attach. In regular diploid meiosis, the bivalent is the structural unit and each centromere attaches to one pole by means of microtubules. The two chromosomes of a bivalent are held together by attraction of chromatids distal to the chiasmata. The attachment of the two centromeres to opposite poles brings the whole bivalent to an equilibrium position on the spindle chiasmata on the equator and the centromeres equidistant on opposite sides. There is always a random assortment of maternal and paternal centromeres on each side of the equator, which in heterozygous genotypes accounts for the recombination of alleles at different loci on non-homologous chromosomes. As in mitosis, movement to the equator is known as *congression* and a bivalent with localized centromeres shows *co-orientation* as a result of the *syntelic* attachment of each of the two centromeres to opposite poles.

Metaphase I: The bivalents then align on the spindle equator; the nuclear envelope and usually the nucleolus disappear and coiling is at a maximum and often clearly visible. The major coil in the two chromatids can later separate without complications. The homologous chromosomes migrate towards the poles and remain attached to each other through the chiasmata and not through centromeres as in mitosis. In the first division of meiosis, apparently due to the different attachment of the kinetochores, the homologous chromosomes do not separate so that both chromatids migrate to the pole. It is the sister chromatid attraction distal to interstitial chiasmata or the end-to-end attachment at fully terminalized chiasmata, which prevents the separation of homologues.

Anaphase I: The chromatids continue to be joined through chiasmata until the anaphase I stage. A sudden lap of attraction of chromatids distal to chiasmata, or separation of chromosome end when the chiasmata have fully terminalized, allows whole chromosomes to separate and be drawn by their undivided centromeres towards the nearest spindle pole. Chromosomes at this stage have four distinct arms, if the centromere is median or submedian, or two distinct arms in telocentric and acrocentric chromosomes. Disjunction is a synchronous event throughout the cell and results in two groups of chromosomes at opposite spindle poles. Each group has one chromosome from each bivalent and therefore has half the zygotic chromosome number. Since all the chromosomes are replicated, the DNA content remains the same as at mitotic anaphase in the sporophyte. The chromosomes in each pair look similar but are not genetically identical as a result of exchanges at chiasma formation with one or both of the chromatids of the homologue, now at the other pole. The centromeres and proximal segments at each pole will usually be a random mixture of those inherited from the parents, as a result of random assortment of centromeres on the equator. Eventually, separation of the chromosomal pairs is completed, the chiasmata will be destroyed and each pair of chromosomes migrates to each pole.

Telophase I: Events between the completion of chromosome separation and the onset of the second division of meiosis differ between species, but in no case is there a replication of DNA between divisions, which makes this period almost unique among division cycles. The chromosomes unfold and this occurs concomitantly with the reappearance of the nuclear membrane and the nucleoli. The cytoplasm divides and two haploid ($2n$) cells result. Each cell has one chromosome of each type, although these are still in the duplicated state.

Interphase: This phase separates the two meiosis divisions in higher plants. There is no replication of chromosomal DNA in contrast to interphase preceding mitosis. In some, such as *Trillium* species, the chromosomes enter the second division almost immediately with little change in coiling. However, in other cases the from groups chromosomes but no cytokinesis occurs (e.g., lily). In lily and other species, the cytokinesis or cleavage occurs after the first meiotic division to produce two distinct cells, called *dyads*, each with a single interphase nucleus. Where such an interface occurs, the chromosomes enter the second division like mitotic chromosomes.

Meiosis II: The events in the second meiotic division parallel those observed in mitosis. It is an equational division and although it reduces the amount of genetic material per cell by half, it does not further reduce the chromosome number per cell. Meiosis begins with a diploid chromosome number and produces four haploid cells.

Prophase II: During the transition to prophase II, the two centrioles move apart and a new spindle will assemble in each newly formed cell. At prophase II, chromosomes become attached to microtubules and start moving toward the spindle equator.

Metaphase II: Following disintegration of the nuclear envelopes and contraction of the chromosomes (if interphase nuclei reformed between divisions), each daughter group from meiosis I forms on a separate spindle, often in a distinct dyad cell. The chromosomes behave in the same way mechanically as mitotic chromosomes, with auto-orientation of amphitelic centromeres. All the duplicated chromosomes are now positioned midway between the two poles of the spindle.

The orientation of second division spindles varies. In some, particularly embryo sacs and fungal asci, the tetrad is linear, with all 3 spindles parallel to each other and in line. Pollen tetrads are usually tetrahedral, isobilateral (where the second division spindles are parallel to each other, but perpendicular to the first division spindle), or decussate (where second divisions are perpendicular to each other and to the first division).

Anaphase II: The attachment between the two chromatids of each chromosome breaks. The former 'sister chromatids' are now called sister chromosomes in their own right and move to the opposite poles of the spindle.

Telophase II: Telophase II is similar to telophase I and is marked by uncoiling and lengthening of the chromosomes and the disappearance of the spindle. Nuclear envelopes re-form and cleavage or cell wall formation eventually produces a total of four daughter cells, each with a haploid set of chromosomes. Meiosis is now complete and ends up with four new daughter nuclei (tetrads). When the cytoplasm divides, each new cell has a haploid number of chromosomes, all in the unduplicated state. And each cell will return to interphase state.

Cytokinesis: At the end of meiotic telophase II, in some species, cytoplasm of each of the two cells of a dyad divides into two parts. As a result, one parent cell produces four haploid daughter cells after completion of the two meiotic divisions. The four daughter cells produced from a parent cell remain together and are known as tetrad. However, in some species, dyads are not produced and at the end of the second telophase four haploid nuclei are present in the same cell. In such species, cytokinesis occurs at the end of telophase II dividing each cell into four daughter cells, each of which contains a single haploid nucleus. These final products of meiosis may differentiate into gametes (animals) or spores (plants). The spores ordinarily give rise to the gametophytic generation (which produced gametes through mitosis) that may be independent (in lower plants) or totally dependent (in higher plants) on the diploid sporophytic stage (the generation, which gives rise to spores through meiosis). In plants, gametes are generally produced through mitosis in the gametophyte.

9.4.3 Modification of meiosis

Some organisms (e.g., many insects) have either polycentric chromosomes or diffuse centromeres. In such organisms, the meiotic cell division is so modified that the first division of meiosis is equational, while the second division is reductional. For example, in Coccid *Puto*, chromosomes pair normally during zygotene, but chiasma terminalization is so complete that the two homologous chromosomes forming each bivalent separate completely and during metaphase I all the $2n$ chromosomes are present as univalents. During anaphase I, sister chromatids of each chromosome separate and move to the opposite poles so that each of the two daughter nuclei receives $2n$ number of chromosomes (chromatids). During the second prophase of meiosis, the homologous chromosomes pair again (**secondary pairing**) and during anaphase II they separate and move to the opposite poles so that each daughter nucleus receives only n number of chromosomes. Several variations to this basic scheme occur in different species. But in each of the cases homologous chromosomes separate only during the second anaphase of meiosis.

9.4.4 Key events during meiosis

Meiosis generates four haploid cells containing genetically different chromosomes. Another event that contributes enormously to the key adaptive advantage of sexual reproduction during prophase I and metaphase I of meiosis is, the production of offspring with new combinations of alleles (hence variation in traits). The genetic variability is attributed to **chromosome migration** and **recombination** of genetic material during prophase I of meiosis.

Prophase I of meiosis is a time of major gene shuffling. During prophase I in germ cell, all chromosomes will condense to thread-like form. Each condensed chromosome is drawn close to its homologue by a process called synapsis. The intimate parallel array favours **crossing over**, a type of molecular interaction between the chromatids of a pair of homologous chromosomes. The non-sister chromatids break at the same places along their length and exchange corresponding segments (i.e., genes) at break points.

Major shuffling of whole chromosomes begins during the transition from prophase I to metaphase I, the second stage of meiosis. During the chromosomal orientation at metaphase, the maternal and the paternal ones attach to the spindle microtubules randomly. The human germ cell has 23 pairs of homologous chromosomes, so $(2)^{23}$ or 8,388,608 combinations of maternal and paternal chromosomes are possible every time a germ cell gives rise to sperm or eggs! And in each sperm or egg, many hundreds of alleles inherited from the mother might differ from their counterparts from the father. The gametes that form after meiosis are not all the same.

More gene shuffling takes place at fertilization. The parental chromosome number is restored at fertilization, when the nuclei of two haploid gametes fuse. Fertilization contributes to further variation in offspring.

9.4.5 Significance of meiosis

Meiosis occurs in all organisms that reproduce sexually. Cytologically, the basic significance of meiosis is the formation of four monoploid nuclei from a single diploid nucleus in two successive divisions, thus balancing off the doubling or chromosome numbers that result from syngamy. During fertilization the nuclei of two gametic cells fuse (**Fig. 9.23**). Each gamete has one set of chromosomes; the product of fusion is zygote which has two sets of chromosomes. This fusion of different gametes can create a new individual.

Meiosis provides opportunities for new combinations of genes to occur in the gametes. If genes are located on chromosomes, the process of meiosis generates genetic variability in two important ways:

- ***Random assortment*** of paternal and maternal chromosomes
- ***Crossing over***: Arrangement of each pair of synapsed homologs on the metaphase I spindle is random, that is, the paternal member of each pair has an equal chance of being oriented towards either pole, as does the maternal member.

The second division will simply increase the number of each gametic gene arrangement from one to two. The number of possible gamete types occurring after the second meiotic division is 2n (n pairs of chromosomes). For organisms with many chromosomes the number of possible combinations becomes very large. In humans, for example, with 23 pairs of homologous chromosomes, the probability that any particular gamete will have a specific combination of chromosomes (excluding crossing–over) is (½)23, or about 1 in 8 million.

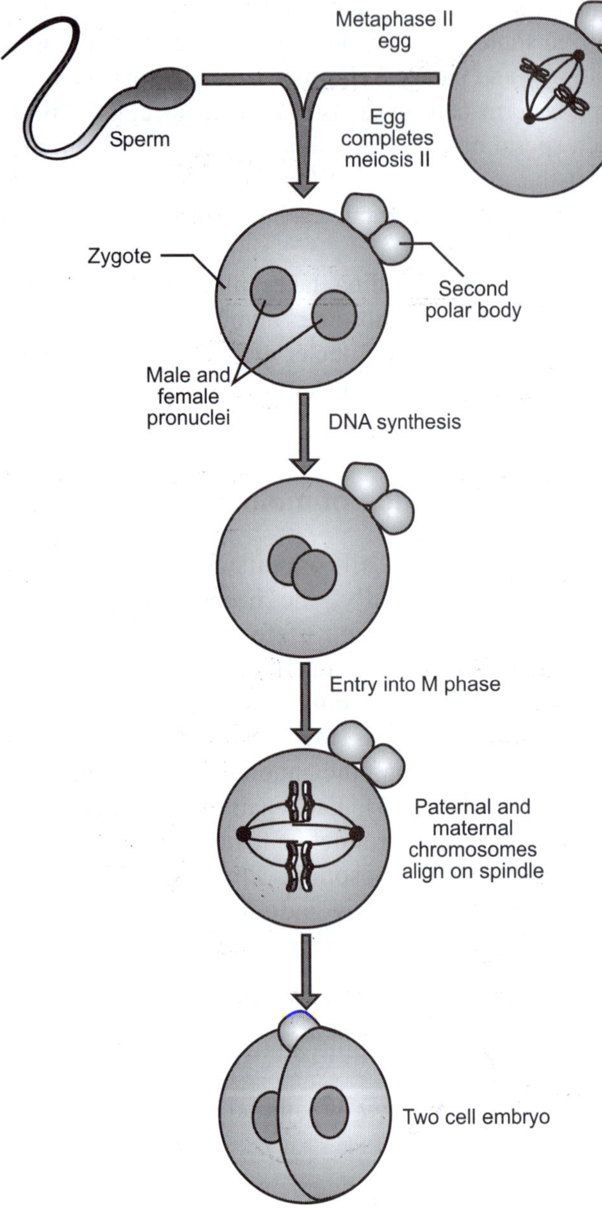

Fig. 9.23 Fertilization and completion of meiosis. Fertilization induces transition from metaphase II to anaphase II, leading to completion of oocyte meiosis. The sperm nucleus decondenses, so the zygote contains two haploid nuclei.

Due to crossing over, *more allelic combinations* are possible. The process of creating new arrangements, either by crossing over or by independent segregation of homologous pairs of chromosomes, is called recombination. Possible zygotic genotypes resulting from random fusion of these gametes rise to about 64 trillions. This leads to *genetic variation* in the offspring produced by fusion of the gametes. Independent assortment of chromosomes and crossing over are two ways in which genetic variation occurs during meiosis.

Role of chiasmata in chromosome segregation

In addition to re-assorting genes, chromosomal crossing-over is crucial in most organisms for the correct segregation of the two duplicated homologs to separate daughter nuclei. This is because the chiasmata created by crossover events has a crucial role to play in holding the maternal and paternal homologs together until the spindle separates them at anaphase I. Mutant organisms that have a reduced frequency of meiotic crossing over fail to segregate normally and many of the resulting gametes contain too many or too few chromosomes. The duplicated homologs are held at chiasmata only because the arms of sister chromatids are glued together along their length by **cohesin** proteins. The arms of sister chromatids suddenly become unglued at the start of anaphase I, when the cohesins holding the arms together are degraded, allowing the duplicated homologs to separate and be pulled to opposite poles of the spindle. However the sister chromatids of each duplicated homolog remain attached at the centromere by meiosis-specific cohesins, which are degraded at anaphase of meiotic division II (anaphase II), only then can the sister chromatids separate.

In meiotic division II, the kinetochores on each sister chromatid have attached kinetochore microtubules pointing in opposite directions, so that the chromatids are drawn into different daughter cells at anaphase. In meiotic division I, by contrast, the kinetochores on both sister chromatids behave as a single functional unit, as their attached kinetochore microtubules all point in the same direction so that the sister chromatids stay together when the duplicated homologs separate. In budding yeast, a meiosis-specific protein located at the kinetochores of meiosis I chromosomes has been shown to be required for this special behaviour.

Recombination nodules: Sites of genetic recombination

The crossover events that take place during prophase I can occur randomly along a chromosome. Although they are random, there are recombination 'hot spots', where DNA breaks seem to be preferentially induced by the meiotic endonuclease called *Spo10*. Moreover, both genetic and cytological experiments indicate that the occurrence of one crossover event decreases the probability of a second occurring at a nearby chromosomal site. This interference seems to ensure that the limited number of crossovers are spread out, so that even small chromosomes get at least one, as required for the homologs to segregate normally.

There is strong indirect evidence that the general genetic recombination events in meiosis are catalyzed by recombination nodules. These are very large protein complexes that sit at intervals on the synaptonemal complex (**Fig. 9.24**). These nodules contain Rad51, which mediates general recombination.

Two main types of recombination nodules have been identified.

- ***Early nodules***, which are present before pachytene and are thought to mark the sites of the initial DNA strand-exchange events of the recombination processes.
- ***Late nodules***, which are less numerous, present during pachytene and thought to mark the sites where the initial strand-exchange events are being resolved as stable crossovers.

9.4.6 Spore formation in plants

Gamete formation in plants, like that in animals requires a reduction in chromosome number from $2n$ to n. The process of meiosis is similar to that in animals, but the life cycle of plants is somewhat more complicated. Gamete formation usually does not follow meiosis directly. Sporogenesis in plants involves the formation of spores rather than sex cells. Spores produce gametophytes and gametes come from gametophytes in which the chromosome number is already reduced. Diploid plants, called **sporophytes**, undergo meiosis to produce spores with reduced chromosome number. Spores develop into haploid **gametophytes**, which ultimately produce gametes capable of fertilization. Zygotes resulting from fertilization develop into sporophytes, completing the cycle. Sporophyte and gametophyte phases vary in length and importance in different plants.

9.4.7 Meiosis and Mendel's principles

Mendelian segregation and independent assortment are simple corollaries of meiosis. The physical bases of these two principles are inherent in the chromosome mechanism of meiosis. By observing microscopically distinguishable pairs of chromosomes during a reduction division, the maternal and paternal chromosomes can be followed. In **Fig. 9.25**, the three distinguishable pairs of maternal and paternal chromosomes are symbolized as AA'BB'CC' and the various observed arrangements of the division spindles are illustrated. The independent alignment

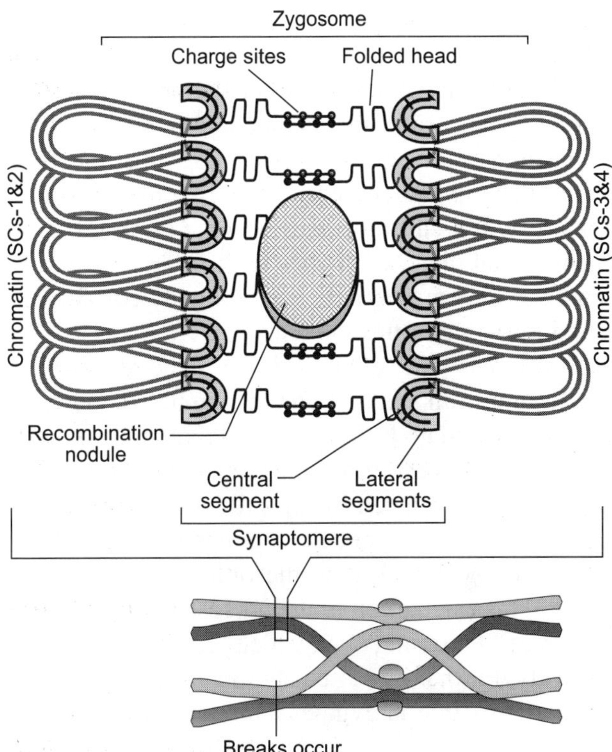

Fig. 9.24 A mature synaptonemal complex.

(a)

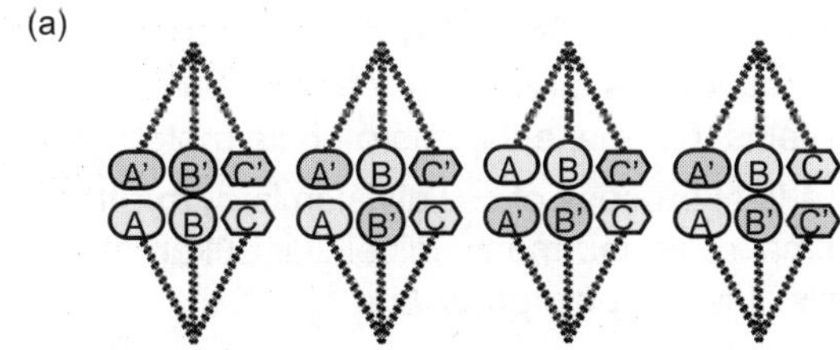

(b)

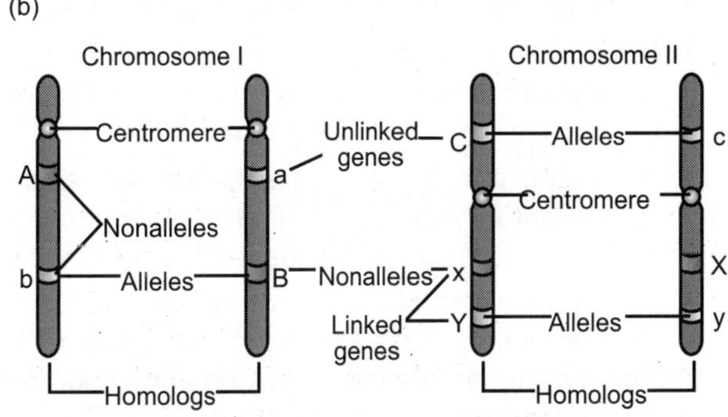

Fig. 9.25 (a) Arrangements of maternal and paternal chromosomes that can be identified by microscopic studies. (b) Alleles and non-alleles and linked and unlinked genes.

of pairs of homologous chromosomes during the reduction division of meiosis explains Mendel's second principle. A chance distribution of maternal and paternal chromosomes occurs in such a way that each has an equal chance of facing one pole or the other on the equatorial plate of the reduction division. When chromosomes become established on the equatorial plate, each goes to the pole it faces and eventually becomes a part of the progeny cell formed around that pole of the spindle.

In animals and plants with a large number of chromosomes, an almost infinite number of possible combinations may be expected. In organisms with 23 pairs of chromosomes, for example, the probability that a gamete produced by an individual in the population will have any specific combination of chromosomes is $(\frac{1}{2})^{23}$, which is in the order of one in eight million. This calculation is an underestimate due to the occurrence of crossing over, which is another source of variability. Further increased numbers of gene combinations are possible in zygotes that result from random fertilization.

9.4.8 Non-disjunction

The normal separation of chromosomes in meiosis I or sister chromatids in meiosis II is termed disjunction. However, when the separation is not normal, it is called non-disjunction. This results in the production of gametes which have either too many or too few of a particular chromosome and is a common mechanism for formation of aneuploidy. Non-disjunction can occur in meiosis I or meiosis II, phases of cellular reproduction, or during mitosis. Non-disjunction causes several medical conditions in humans:

Down syndrome: trisomy of chromosome 21

Patau syndrome: trisomy of chromosome 13

Edward syndrome: trisomy of chromosome 18

Klinefelter syndrome: extra X chromosomes in males, i.e., XXY, XXXY, XXXXY
Turner syndrome: lack of one X chromosome in females, i.e., XO
Triple X syndrome: an extra X chromosome in females
XYY syndrome: an extra Y chromosome in males.

Summary

- ***Mitotic cell division*** is actually duplication of entire cell into two. The first visible sign of a cell that enters into ***mitosis*** is the progressive compaction of the replicated chromosomes, which become visible as thread-like structures.
- Following the *chromosome condensation*, a distinct cytoskeletal machine is assembled to perform the mechanical process of separation of its replicated chromosomes to each daughter cell.
- Chromosomes move as the microtubules slide past each other. ***Kinesin*** and ***dyneins*** are two unique proteins that make the microtubule motors move.
- The final process of cell division is ***cytokinesis***, which is facilitated by different cytoskeletal structures. In animal cells and many unicellular eukaryotes, the ***contractile ring*** is responsible; whereas, in most plant cells, it is the ***phragmoplast***.
- The sequential stages of mitosis are ***prophase, metaphase, anaphase*** and ***telophase***.
- The process of mitosis and the subsequent replication of chromosomal material in the succeeding interphase have the inevitable result of creating from one cell two new ones that are genetically identical.
- Mitosis produces two nuclei which have the same number of chromosomes as the parent cell. Since these chromosomes were derived from parental chromosomes by the exact replication of their DNA, they are genetically identical to their parents.
- ***Meiosis*** is concerned with sexual transmission of chromosomes where the genes from two different parent cells eventually reside in a single daughter cell. In meiosis, haploid gametes with half (n) amount of DNA such as pollen (sperm) and ovules (egg) are formed from diploid cell ($2n$) in the germinal tissue of higher organisms.
- The first division of meiosis represents a reduction division in which members of homologous pairs of chromosomes are separated into daughter cells without duplication; but their numbers are reduced to half. The prophase I of meiosis is long and subdivides into ***leptotene, zygotene, pachytene, diakinesis*** and ***diplotene***.
- The second meiotic division is parallel to those observed in mitosis and is an equational division. Meiosis begins with a diploid chromosome number and produces four haploid cells.
- Defects in chromosome distribution can lead to different kinds of ***aneuploidy,*** *e.g.,**trisomies*** .

Exercises

1. Explain the different stages of mitosis.
2. What makes chromosomes align at metaphase plate and later move to the poles?
3. Which proteins involve themselves in chromosome condensation and holding of sister chromatids together?
4. Discuss the process of cytokinesis in both animals and plants.
5. How is mitotic cell division regulated?
6. Briefly explain the product of mitosis and its biological significance.
7. Give the different steps involved in bacterial cell division.
8. Draw and explain the life cycle of a virus.
9. Discuss briefly the different steps involved in meiosis.
10. What are the main differences between mitotic and meiotic cell division?
11. Explain the biological significance of meiosis.
12. How do you relate the meiotic events with Mendelian laws?

Further Reading

Books

Alberts B, Johnson A, Lewis J, Raff M, Roberts K and Walter P (2002), *Molecular Biology of the Cell*, 4th edition, Garland Science.

Lewin B, Cassimeris L, Lingappa VR and Plopper G (2007), *Cells,* Jones and Bartlett Publishers.

Starr C and Taggart R (1995), *Biology: The Unity and Diversity of Life*, Wadsworth Publishing Company.

Swanson CP and Webster P (1977), *The Cell*, Prentice-Hall, Englewood Clifs, NJ.

Sybenga J (1975), *Meiotic Configurations*. Springer-Verlag, Berlin and New York.

Yeoman MM, (ed.), (1976), *Cell Division in Higher Plants*, Academic Press, London and New York.

References

Ben–Yehuda S, Rudner DZ and Losick R (2003), RacA, a bacterial protein that anchors chromosomes to the cell poles, *Science,* 299:532–36.

Bridges CB (1916), Non-disjunction as proof of the chromosome theory of heredity, *Genetics*, 1:1–52, 107–63.

Brinkhous KM and JB Graham (1950), Hemophilia in the female dog, *Science* 111:723–24.

Buonomo SB, Clyne RK, Fuchs J, Loidl J, Uhlmann F and Nasmyth K (2000), Disjunction of homologous chromosomes in meiosis I depends on proteolytic cleavage of the meiotic cohesion Rec8 by separin, *Cell*, 103:387–98.

Channarayappa (1993), Resistance mechanism of *Lycopersicon* species to the vector *Bemisia tabaci* and ultrastructural changes in leaf curl virus infected tomato leaves, PhD thesis submitted to the University of Agricultural Sciences, GKVK, Bangalore, India.

Channarayappa (1990), Cytogenetic analysis of clastogens and aneuploidogens by immunofluorescent and micronucleus methods, PhD thesis submitted to the West Virginia University, Morgantown, West Virginia, USA.

Gheysen G and Fenoll C (2002), Gene expression in nematode feeding sites, *Annu. Rev. Phytopath.,* 40:191–219.

Gruneberg U and Nigg EA (2003), *Regulation of cell division: Stop the SIN!, Trends Cell Biol*., 13:159163.

Guacci V, Hogan E and Koshland D (1994), Chromosome condensation and sister chromatid pairing in budding yeast, *J. Cell Biol*., 125:517–30.

Hirano T (1999), SMC-mediated chromosome mechanics: A conserved scheme from bacteria to vertebrates? *Genes Dev.*, 13:11–19.

Hirano T (2000), Chromosome cohesion, condensation and separation, *Ann. Rev. Biochem*., 69:115–44.

Keeney S (2001), Mechanism and control of meiotic recombination initiation, *Curr. Top. Dev. Biol*., 52:1–53.

Murray AW and Szostak JW (1985), Chromosome segregation in mitosis and meiosis, *Annu. Rev. Cell Biol*., 1:289–315.

Nysmyth K (2001), Disseminating the genome: Joining, resolving and separating sister chromatids during mitosis and meiosis, *Annu. Rev. Genet.,* 35:673–745.

Nysmyth K (2002), Segregating sister genomes: The molecular biology of chromosome separation, *Science,* 297:559–65.

Page SL and Hawley RS (2003), Chromosome choreography: The meiotic ballet, *Science,* 301:785–89.

Skibbens RV, Corson LB and Koshland D (1999), Ctf7p is essential for sister chromatid cohesion and links mitotic chromosome structure to the DNA replication machinery, *Genes Dev.*, 13:307–19.

Uhlmann F, Wernic D, Poupart MA, Koonin EV and Nasmyth K (2000), Cleavage of cohesin by the CD clan protease separin triggers anaphase in yeast, *Cell,* 103:375–86.

Westergaard M and von Wettstein D (1972), The Synaptonemal Complex, *Annu. Rev. Genet.*, 6:71–110.

10 CELL SIGNALLING

Learning objectives

After reading this chapter, you will be able to:

- Identify key features of cellular communication
- Understand the role of transmitters and receptors in cell signalling
- List the features of selected receptors

10.1 INTRODUCTION

All living cells receive and respond to various kinds of signals from their surroundings. Even the simplest unicellular bacteria can sense and either move towards or away from the stimuli (e.g., host, nutrients, toxins). However, the eukaryotic organisms, particularly multicellular organisms, have elaborate cell communication mechanisms. They can communicate with one another and with the surrounding environment in complex ways for the benefit of all the cells in the organism and for communal living. These communication mechanisms depend heavily on extracellular signal molecules, which are produced by cells to signal their neighbours or cells further away (surroundings). Multicellular organisms communicate by means of chemical and neural signals. Chemicals released by one cell are detected by other cells or vice versa. This cellular communication mainly depends on elaborate systems of proteins and enables the cell to respond to a particular subset of these signals in a cell-specific way. These proteins include **cell surface receptor proteins**, which bind the signal molecules (e.g., hormones) and a variety of intracellular signalling proteins and chemicals (**second messengers**) that distribute the signal to appropriate parts of the cell.

Cell signalling brings together several major biochemical concepts. It allows the organism to function as a coherent system. Thus, cell signalling affects virtually every aspect of cell structure and function, namely conformational changes in proteins, protein kinases, transcription factors, cell cycle regulation and gene control in eukaryotes and so on. Eukaryotic cells respond to signals released by other cells. The binding of most signalling molecules to their receptors initiates a series of intracellular reactions that regulate virtually all aspects of cell behaviour, including metabolism, movement, proliferation, survival and differentiation. No cell lives in isolation and the activities of individual cells are controlled such that they are an integrated organism as a whole. Cells within small groups often communicate by direct cell–cell contact. In addition, the establishment of specific cell–cell interactions between different types of cells is necessary in the development of many tissues. In some cases, a type of protein on one cell binds to a receptor on the surface of an adjacent target cell, triggering its differentiation.

A large number of different signals are used to coordinate cellular activities. Cells communicate principally by means of **extracellular signalling/ messenger molecules**. Cells continually exchange chemical messengers called **transmitters**. These specific substances are synthesized and released by signalling cells and produce a specific response only in target cells that have receptors for the signalling molecules. An enormous variety of chemicals, including small molecules, peptides and proteins, are used in this type of cell-to-cell communication. The extracellular products synthesized by signalling cells can diffuse away or be transported in the blood, thus providing a means for cells to communicate over longer distances than is possible by chains of direct cell–cell contacts. These responses usually include modulation of the expression of specific genes as well as more direct effects on metabolism, not involving gene control.

10.2 CELLULAR COMMUNICATION

Signals used by cells to communicate with one another are often small proteins, growth factors, neuropeptides, neurotransmitters, cytokines and other small molecules such as amino acids and steroids. External signals such as odourants and tastes carried in the atmosphere or in fluids and stretch, pressure and other mechanical effects as well as heat, pain and light can also act as signals. Out of the huge variety of signals to which a cell is exposed, the ones to which it responds depend on the signals that are received by the receptors. A cell is sensitive only to those signals for which receptors are present.

10.2.1 Signalling molecules and their receptors

Signal transduction refers to any process by which a cell converts one kind of signal or stimulus into another. This involves an ordered sequence of biochemical reactions inside the cell, which are carried out by enzymes and linked through second messengers resulting in what is known as a 'second messenger pathway'. In many signal transduction processes, the number of proteins and other molecules participating in these events increases as the process emanates from the initial stimulus, resulting in a 'signal cascade' and often results in a relatively small stimulus eliciting a large response.

Signalling molecules are simply molecules of the right shape which can bind through non-covalent bonds to their receptors with great specificity. Nothing happens to them in the course of signalling, though they are subsequently destroyed to terminate the signal. Many different kinds of molecules transmit information between cells of multicellular organisms. Although all these molecules act as ligands that bind to receptors expressed by their target cell, there is considerable variation in structure and function. Structurally, the signalling molecules range in complexity from simple gases to proteins. Some of these molecules carry signals over long distances, whereas others act locally to convey information between neighbouring cells. They also they differ in their mode of action on their target cells: some are able to cross the membrane and bind to intracellular receptors in the cytoplasm or nucleus, whereas most bind to receptors expressed on the surface of the target cell. Signalling molecules in animals can be classified as:

1. Neurotransmitters
2. Endocrine hormones
3. Growth factors and cytokines
4. Vitamin derivatives.

This classification is important in terms of nomenclature and physiology, but they are all signalling molecules, which bind to cellular receptors and elicit responses.

10.2.2 Molecular mechanism of signal transduction

Signal transducers are very specific and highly sensitive. Specificity is achieved by precise molecular complementarities between the signal and receptor molecules (**Fig. 10.1a**), mediated by the same kind of non-covalent forces that occur in enzyme–substrate and antigen–antibody interactions. Specificity is also determined by the presence of specific receptors in specific cell types. Three factors contribute to the high specificity of signal transducers: the high affinity of receptors for signal molecules, cooperation in the ligand–receptor interaction and amplification of the signal by enzyme cascades. **Amplification** by enzyme cascades results when an enzyme associated with a signal receptor is activated and, in turn, catalyzes the activation of many molecules of a second enzyme, each of which activates many molecules of a third enzyme and so on (**Fig. 10.1b**). Following activation, the sensitivity of receptor systems is subject to modification. When a signal is present continuously, there is **desensitization** of the receptor system (**Fig. 10.1c**); when the stimulus falls below a certain threshold, the system can again become sensitive. When the system receives multiple signals, it produces a unified response (**integration**) appropriate to the needs of the cell or organism (**Fig. 10.1d**). Different signal pathways converse with each other at several levels, generating a wealth of interactions that maintain homeostasis in the cell and the organism. However, all signal transductions are able to alter the activity of specific proteins, either at

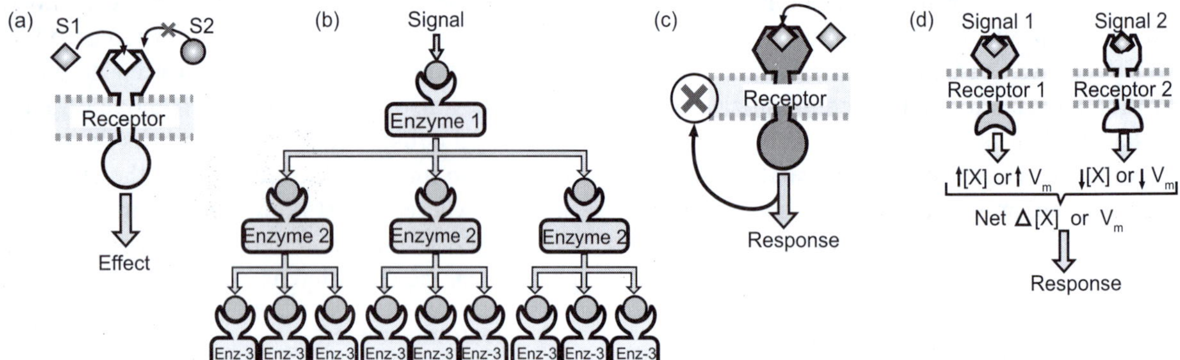

Fig. 10.1 Four features of signal transducing systems: (a) Specificity: Signal molecule fits binding site on its complementary receptor; other signals do not fit. (b) Amplification: When enzymes activate enzymes, the number of affected molecules increases geometrically in an enzyme cascade. (c) Desensitization/Adaptation: Receptor activation triggers a feedback circuit that shuts off the receptor or removes it from the cell surface. (d) Integration: If two signals act oppositely on a metabolic characteristic or the membrane potential V_m, the regulatory outcome results from the integrated input from both receptors. (Source: Nelson and Cox, 2000).

the cell surface or intracellularly, which then change the cell's biological behaviour in some way.

10.2.3 Extracellular communication

The various kinds of signalling molecules are: proteins, small peptides, amino acids, nucleotides, steroids, retinoids, fatty acid derivatives and even dissolved gases such as nitric oxide and carbon monoxide and many other physical signals (**Table 10.1**). Most of these molecules are secreted from the signalling cell into the extracellular space (exocytosis) and others are released by diffusion through the plasma membrane and some are exposed to the extracellular space while remaining tightly bound to the surface of the signalling cell. Regardless of the nature of the signal, the target cell responds by means of a specific protein called a **receptor**, which specifically binds the signal molecule and then initiates a response in the target cell. The signal is transmitted through an **intracellular signalling pathway**. Signalling pathways are the information superhighways of the cell. Overall communication by extracellular signals usually involve following steps:

- Synthesis and release of signalling molecules by the signalling cell,
- Transport of the signal to the target cell,
- Detection of the signal by a specific receptor protein(s),
- A change in cellular metabolism or gene expression triggered by the receptor–signalling molecule complex.

Removal of the signal often results in termination of the cellular response.

In many microorganisms (e.g., yeast, molds and protozoans), secreted molecules coordinate the aggregation of free-living cells for sexual mating or differentiation under certain environmental conditions. Chemicals released by one organism that can alter the behaviour or gene expression of other organisms of the same species are called **pheromones**. They are released by many organisms and are usually dispersed into air or water, to attract members of the opposite sex. In plants (**Fig. 10.2**) and animals extracellular signalling molecules control metabolic processes within the cell, such as, the growth of tissues, synthesis and secretion of proteins and composition of intracellular and extracellular fluids. When compared to animals,

Table 10.1 Some major extracellular signals

Chemical signals	Physical signals
Antigens	Light
Cell surface glycoproteins/ oligosaccharides	Mechanical pressure
Developmental signals	Sound
Extracellular matrix components	Temperature
Growth factors	Humidity
Hormones	Wind
Neurotransmitters	Radiation
Odourants	Gravity
Pheromones	
Tastants	

to date, only a few receptors for plant signalling molecules have been identified. Thus, most information on the understanding of signalling is more advanced in animals and microorganisms than in plants.

A large variety of molecules can function as extracellular carriers of information. Some examples are: amino acids and their derivatives, glutamate, glycine, acetylcholine, epinephrine, dopamine, thyroid hormone and so on. Steroids are derived from cholesterol. They regulate sexual differentiation, pregnancy, carbohydrate metabolism and excretion of sodium and potassium ions. Eicosanoids are non-polar molecules containing 20 carbons. That are derived from a fatty acid named arachidonic acid. They regulate a number of processes including pain, inflammation, blood pressure and blood clotting. A wide variety of polypeptides and proteins play an important role in signalling. For example, transmembrane proteins on the surface of an interacting cell and a large number of proteins excreted into the extracellular environment are involved in regulating processes such as cell division, differentiation, immune response, or cell death and cell survival (**Fig. 10.3**).

Classification of transmitters and receptors

There are three types of receptors: **ionotropic**, **metabotropic cell surface** and **intracellular receptors**. Further, a transmitter that is rapidly broken down or taken up into cells acts only near its release site, while one that is broken down slowly can diffuse and act on cells located for away.

Ionotropic cell surface receptors: Ionotropic cell surface receptors are channel proteins that open when a specific stimulus activates the channel protein. Many ion channel receptors are located in plasma membrane, but some are also located on the internal membranes as well. For example, chemical signal molecules bind to the extracellular face of the channel protein and activate the channel protein to

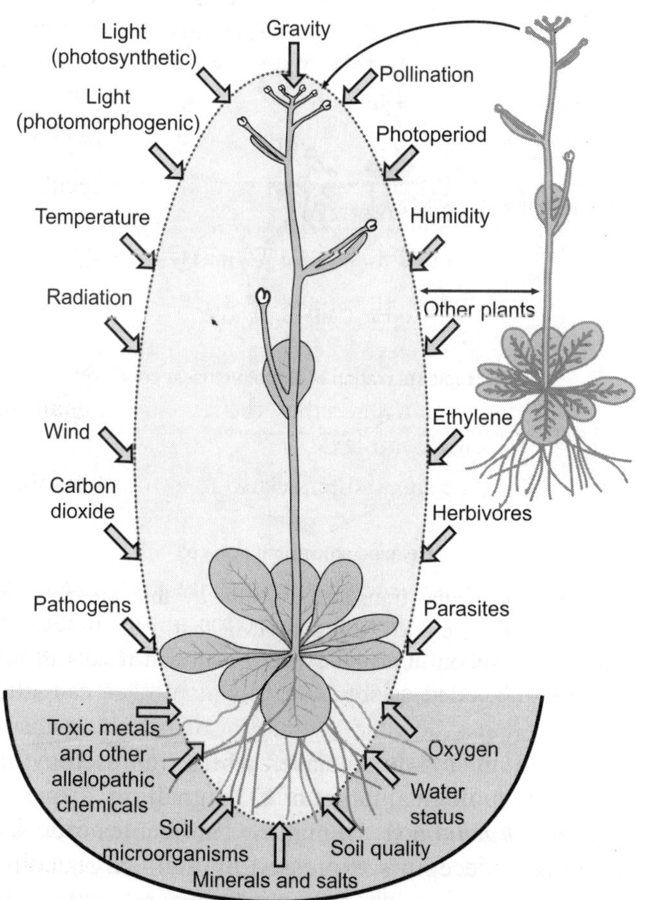

Fig. 10.2 External signals that affect plant growth and development include many aspects of the plant's physical, chemical and biological environments.

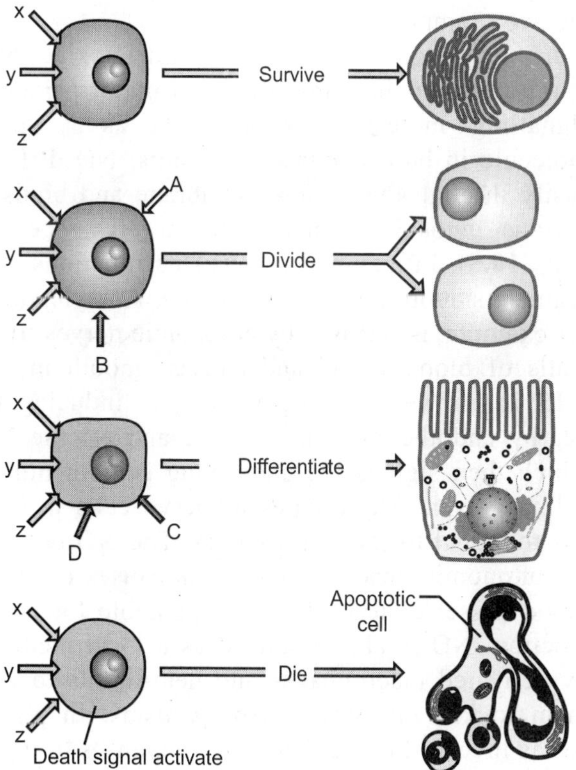

Fig. 10.3 Dependence of an animal cell on multiple extracellular signals. Each cell has a specific set of receptors that enables it to respond to a corresponding set of signal molecules produced by other cells. A cell requires multiple signals to survive (x, y, z) and additional signals to divide (A and B) or differentiate (C and D). If deprived of necessary factors, a cell may undergo suicide (apoptosis).

open. For the acetylcholine receptor, in the absence of a transmitter (acetylcholine), the channel is closed. When acetylcholine binds, the channel opens and allows sodium and potassium ions to pass through (**Fig. 10.4**). The cell organelles (mitochondria, chloroplast, ER, lysosomes, sarcoplasm, peroxisomes) contain diverse channel receptors that regulate various kinds of ion flux in response to specific stimulus.

Metabotropic cell surface receptors: Metabotropic cell surface receptors are linked to enzymes. Most enzyme-linked receptors are protein kinases. For example, when the ADP receptor binds to extracellular ADP, it activates G_q and hence the phospholipase Cß.

Intracellular receptors: Not all the extracellular signal molecules that bind to receptors on the cell surface are big and hydrophilic; some are hydrophobic and small enough to readily cross the target cell plasma membrane. Once inside, it directly regulates the activity of a specific intracellular protein. Receptors for nitric oxide and steroid hormones are the two examples.

***Nitric oxide* gas**: The simple gas nitric oxide (NO) is a transmitter in many tissues (a major paracrine signalling molecule), which acts as a single molecule in both animals and plants. NO diffuses easily through the plasma membrane and binds to various cytosolic proteins, called NO receptors (**Fig. 10.5**). In mammals, one of its functions is to regulate smooth muscle contraction. Acetylcholine, for example, is released by autonomic nerves in the walls of blood vessel and causes smooth muscle cells in the vessel wall to relax by inducing the nearby endothelial cells to make and release NO, which then signals the underlying smooth muscle cells to relax. Many types of nerve cells use NO gas to signal to their neighbours. The NO released by autonomic nerves in the penis causes the local blood vessel dilation that is responsible for penile erection. NO is also produced as a local mediator by activated macrophages and neutrophils to help them to kill invading microorganisms. In plants, NO is involved in the defensive responses to injury or infection.

Another simple gas, carbon monoxide (CO), also functions as a signalling molecule in the nervous system. CO is closely related to NO and acts similarly as a neurotransmitter and mediator of blood vessel dilation. The synthesis of CO in the brain cells is stimulated by neurotransmitters and it can also stimulate guanylate cyclase, which represents the major physiological target of CO signalling.

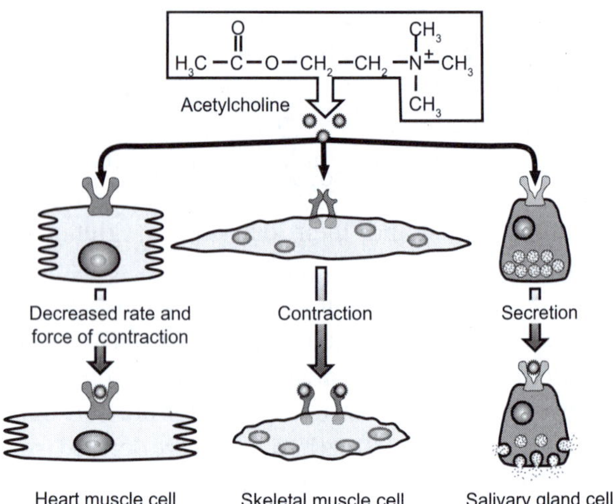

Fig. 10.4 Various responses induced by the neurotransmitter acetylcholine on different cell types.

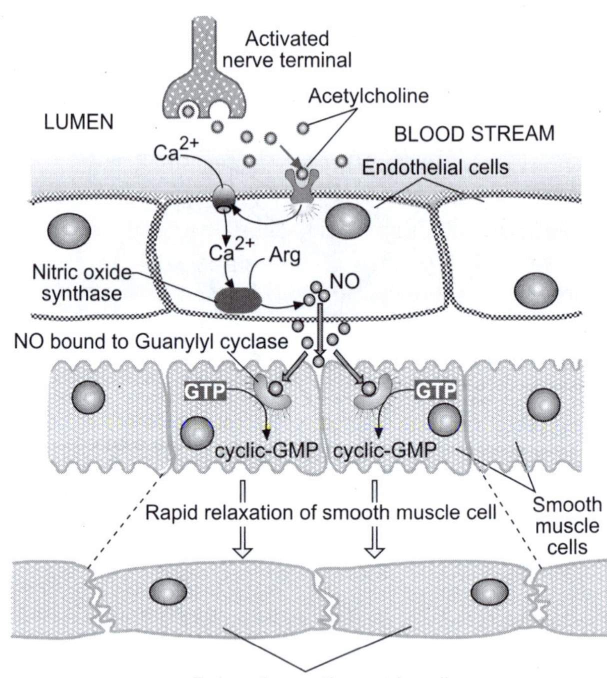

Fig. 10.5 A signal transduction pathway that operates by means of nitric oxide (NO) and GMP that leads to smooth muscle relaxation in a blood vessel wall and results in the dilation of blood vessels.

Steroid hormones: Signalling molecules act by binding to receptors expressed by their target cells; in most cases, on the target cell surface. But some receptors are intracellular located in the cytosol or in the nucleus. Steroid hormones have intracellular receptors such as the glucocorticoid receptor

Steroid hormones

Steroid hormones have diverse functions and origin. For example, testosterone, estrogen, progesterone and ecdysone are synthesized from cholesterol. Testosterone, estrogen and progesterone are the sex steroids which are produced by the gonads. The corticosteroids are produced by the adrenal gland. They include the glucocorticoids, which act on a variety of cells to stimulate production of glucose and the mineralocorticoids, which act on the kidneys to regulate salt and water balance. Ecdysone is an insect hormone that plays a key role in development by triggering the metamorphosis of larvae to adults. The brassinosteroids are plant-specific steroid hormones that control a number of developmental processes, including cell growth and differentiation.

Although, thyroid hormone, vitamin D_3 and retinoic acid are structurally and functionally distinct from the steroids, they share a common mechanism of action in their target cells. For example, *thyroid hormone* is synthesized from tyrosine in the thyroid gland; it plays an important role in development and regulation of metabolism. *Vitamin* D_3 regulates Ca^{2+} metabolism and bone growth. *Retinoic acid* and related compounds (*retinoids*) are essential for vertebrate development. Since the steroid hormones and related molecules are hydrophobic in nature, they can enter cells by diffusing across the plasma membrane. In the cytoplasm, they bind to intracellular receptors and regulate their function as both activators and repressors of their target genes.

(see box). In the cytosol, this enzyme remains in the inactive form because it is bound to an inhibitor protein. Upon binding of glucocorticoid hormone to its receptor, the inhibitor protein is displaced. The complex, glucocorticoids receptor and hormone move into the nucleus and bind to a 15-bp sequence known as the hormone response element (HRE), which is a transcriptional enhancer sequence that lies upstream of the TATA box.

10.2.4 Intercellular (cell–cell signalling) communication

Extracellular signal molecules can act over either short or long distances. Many signalling molecules remain bound to the surface of the signalling cell and influence only cells that come into contact with them. Such **contact-dependent signalling** is especially important during development and in immune responses. In most cases, however, signal molecules are secreted and may be carried far afield to act on distant targets, or they may act as **local mediators**, affecting only cells in the immediate environment of the signalling cell. In animals, signalling by extracellular, secreted molecules can be classified into different types, based on the distance over which the signal acts (e.g., autocrine, paracrine or endocrine). In addition, certain membrane-bound proteins on one cell can directly signal an adjacent cell (**Fig. 10.6**).

Autocrine signalling

In autocrine signalling, cells respond to substances released by themselves. During development, for example, once a cell has been directed along a

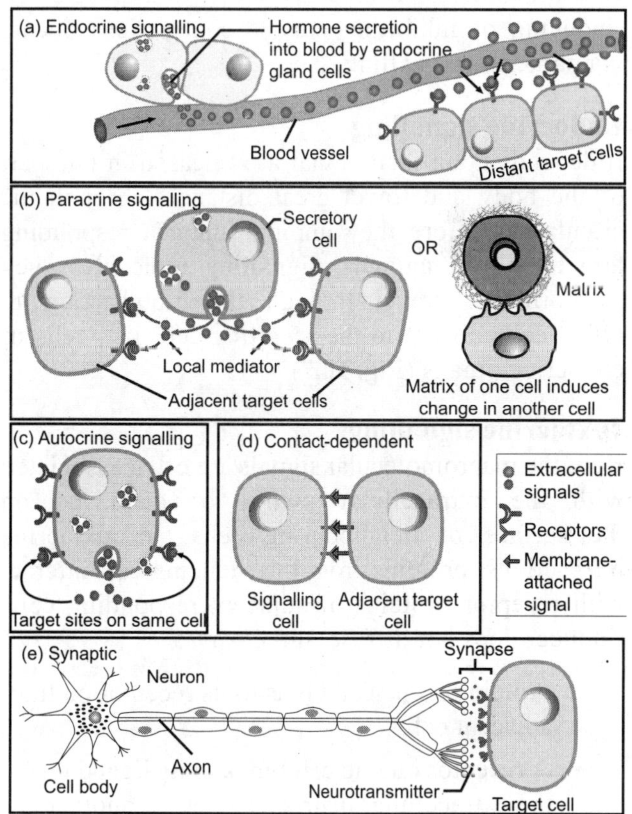

Fig. 10.6 General scheme of intercellular signalling in animals. (a–c) Cell-to-cell signalling by extracellular chemicals; (d) Proteins attached to the plasma membrane can interact directly with receptors on an adjacent cell; (e) Endocrine signalling is carried out by neurons that transmit signals electrically along their axons and release.

particular pathway of differentiation, it may begin to secrete autocrine signals to itself that reinforce this developmental decision. Autocrine signalling is an important mechanism underlying the 'community effect' that is observed in early development, during which a group of identical cells can respond to a differentiation-inducing signal. Many growth factors act in this fashion and cultured cells often secrete growth factors that stimulate their own growth and proliferation. Tumour cells commonly overproduce and release growth factors that stimulate inappropriate, unregulated proliferation of the tumour cells, as well as adjacent non-tumour cells; this process may lead to the formation of tumour mass.

Paracrine signalling

Here, the signalling molecules released by a cell affect only target cells in close proximity to it. The conduction of an electric impulse from one nerve cell to another or from a nerve cell to a muscle cell (inducing or inhibiting muscle contraction) occurs via paracrine signalling.

Endocrine signalling

It depends upon factors that are released in one part of the body and travel great distances through the circulation before they impinge upon a responding target cell. In animals, signalling molecules such as hormones travel through the bloodstream to affect cells distant to the secreting cell, i.e., cells of endocrine organs (**Fig. 10.7**).

Juxtacrine signalling

Here the macromolecular signals are often associated with the extracellular matrix or displayed on the surface of neighbouring cells. In juxtacrine interactions, proteins from the inducing cell interact with receptor proteins of adjacent responding cells (inducer does not diffuse) three ways:

- Protein on one cell binds to its receptor on the adjacent cell.
- A receptor on one cell binds to its ligand on the extracellular matrix secreted by another cell.
- The signal is transmitted directly from the cytoplasm of one cell through small conduits into the cytoplasm of an adjacent cell.

Some protein hormones, such as epidermal growth factor (EGF), are synthesized as the exoplasmic part of plasma-membrane protein. Membrane-bound EGF can bind to and signal an adjacent cell by direct contact. Cleavage by a protease releases secreted EGF, which acts as an endocrine signal on distant cells.

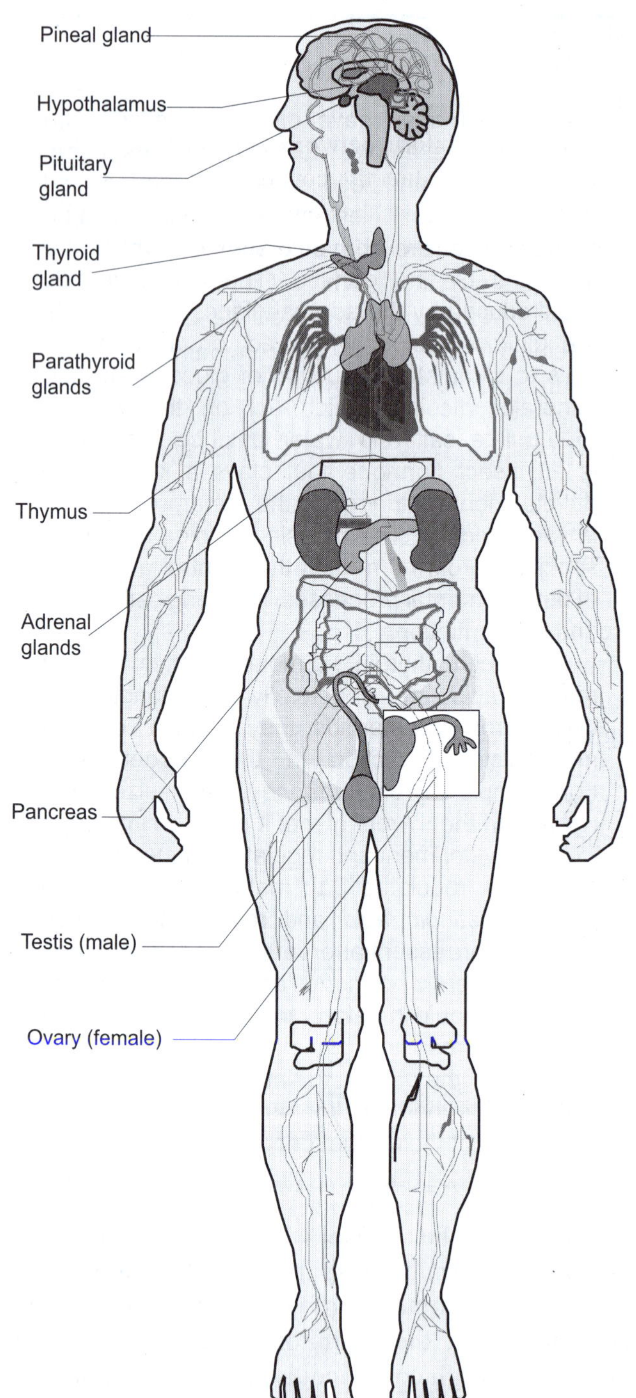

Fig. 10.7 The major human endocrine glands and their central nervous system control centres. Some other tissues (e.g., lining of the gastrointestinal tract) also produce hormones.

Signalling gap junctions

Coordinated activities of neighbouring cells take place through gap junctions. They are specialized cell–cell junctions that can form between closely apposed plasma membranes, and directly connect the cytoplasms of the joined cells via narrow water-filled channels. The channels allow exchange of small intracellular signalling molecules, such as Ca^{2+} and cyclic AMP, but not of proteins or nucleic acids. Thus, cells connected by gap junctions can communicate with each other directly.

Peptide hormones and growth factors

A large number of peptides act as signalling molecules in animals. **Table 10.2** lists some peptide hormones, neuropeptides and polypeptide growth factors. Hormone synthesis release and degradation are highly regulated. Due to their potent effects, hormones and neurotrasmitters must be carefully regulated. The release and degradation of some signalling compounds are regulated to produce rapid, short-term effects; others to produce slower-acting but longer lasting effects. In some cases, complex regulatory networks coordinate the levels of hormones whose effects are interconnected.

Feedback control of hormone levels: The synthesis release of many hormones are subject to feedback regulation. This type of regulation is particularly important in coordinating the action of multiple hormones on various cell types during growth and differentiation. Often, the levels of several hormones are interconnected by **feedback circuits**, in which changes in the level of one hormone affect the levels of other hormones. One example involves the regulation of **estrogen** and **progesterone**—steroid hormones that stimulate the growth and differentiation of cells in the **endometrium**, the tissue lining the interior of the uterus. Changes in the endometrium prepare the organ to receive and nourish an embryo. The levels of both hormones are regulated by complex feedback circuits involving several other hormones (**Fig. 10.8**).

Eicosanoids

Various kinds of lipids serve as signalling molecules by binding to cell surface receptors. The most important are the eicosanoids, which include **prostaglandins**, **prostacyclin**, **thromboxanes** and **leukotrienes**. The eicosanoids break down rapidly and act locally in autocrine or paracrine signalling pathways.

Table 10.2 Representative peptide hormones, neuropeptides and growth factors

Signalling molecule	Number of amino acids	Important activities
Peptide hormones		
Insulin	A = 21, B = 30	Regulation of glucose uptake; stimulation of cell proliferation
Glucagon	29	Stimulation of glucose synthesis
Growth hormone	191	General stimulation of growth
Follicle-stimulating hormone (FSH)	198	Stimulation of milk production
Neuropeptides and neurohormones		
Substance P	11	Sensory synaptic transmission
Oxytocin	9	Stimulation of smooth muscle contraction
Vasopressin	9	Stimulation of water re-absorption in the kidney
Enkephalin	5	Analgesic
ß-endorphin	31	Analgesic
Growth factors		
Nerve growth factor (NGF)	118	Differentiation and survival of neurons
Epidermal growth factor (EGF)	53	Proliferation of many types of cells
Platelet-derived growth factor (PDGF)	A=125, B=109	Proliferation of fibroblasts and other cell types
IL-2	133	Proliferation of T lymphocytes
Erythropoietin	166	Development of red blood cells
TGF-α	-	Important for normal wound healing
TGF-β	-	Anti-inflammatory (suppresses cytokine and class II MHC expression), promotes wound healing, inhibits macrophage and lymphocyte proliferation
IGF-1	-	Promotes proliferation of many cell types
IGF-2	-	Promotes proliferation of many cell types primarily of foetal origin

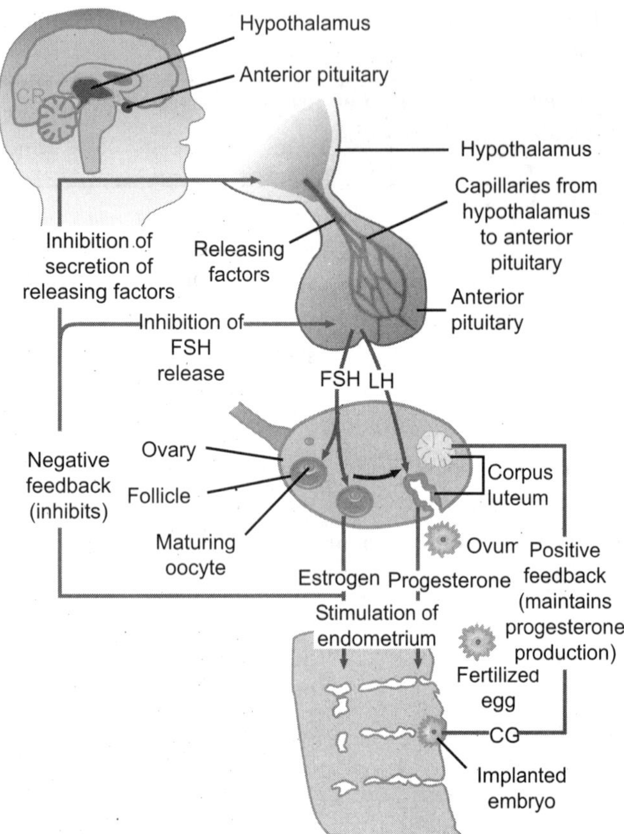

Fig. 10.8 Feedback control of estrogen and progesterone levels in the blood of female mammals. FSH: follicle-stimulating hormone, LH: lutenizing hormone, CG: chorionic gonadotropin. (Source: Lodish et al, 1995)

They stimulate a variety of responses in their target cells, including blood platelet aggregation, inflammation and smooth muscle contraction.

10.2.5 Response to specific combinations of extracellular signals

A typical cell in a multicellular organism is exposed to hundreds of different signals in its environment. The cell has to respond to these combinations of signals selectively, based on its own specific character acquired through progressive cell specialization. Most of the cells in a complex animal are programmed to depend on a specific combination of signals simply to survive. Deprivation of these signals, activates a suicide program and the cell dies by a process called **programmed cell death** (PCD) or **apoptosis**. Therefore, different types of cells require different combinations of survival signals; each cell type is restricted to a specific environment.

10.3 RECEPTORS AND SIGNALLING PATHWAYS

Recognition and coupling domains on receptors

Most receptors have at least two functional domains. A **recognition domain** binds the ligand and a second region generates a **signal** that couples ligand recognition to some intracellular function. Coupling (**signal transduction**) occurs in two or more general ways. The dual functions of binding and coupling ultimately define a receptor and it is the coupling of ligand binding to signal transduction—so called **receptor–effector coupling**—that acts as the first step in amplification of the hormonal response. This dual purpose also distinguishes the target cell receptor from the plasma carrier proteins that bind ligands but do not generate a signal.

Cell surface receptors and their ligands

Most of the signalling in an organism occurs via membrane receptors. The cellular response to a particular extracellular signalling molecule depends on its binding to a specific **receptor protein** located on the surface of a target cell or in its nucleus or cytosol. The signalling molecule (a hormone, pheromone, or neurotransmitter) referred to as the **ligand**, binds to, or 'fits' a site on the receptor. Binding of a ligand to its receptor causes a conformational change in the receptor that initiates a sequence of reactions leading to a change in cellular function.

The response of a cell or tissue to specific hormones is dictated by the particular hormone receptors it possesses and by the intracellular reactions initiated by the binding of any one hormone to its receptor. Therefore, a receptor protein is characterized by **binding specificity** for a particular ligand and the resulting hormone–ligand complexes exhibit **effector specificity** (i.e., mediate a specific cellular response). In most receptor–ligand systems, the ligand appears to have no function except to bind to the receptor. The ligand is not metabolized to useful products, is not an intermediate in any cellular activity and has no enzymatic properties. The only function of the ligand appears to be to change the properties of the receptor, which then signals to the cell that a specific product is present in the environment. Target cells often modify or degrade the ligand and, in so doing can modify or terminate their response or the response of neighbouring cells to the signal. Hormones fall into three broad categories:

1. Small lipophilic molecules (group I hormones) that diffuse across the plasma membrane and interact with intracellular receptors,

2. Hydrophilic molecules,
3. Lipophilic molecules that bind to cell surface receptors.

10.3.1 Types of cell surface receptors

The different types of cell surface receptors that interact with water-soluble ligands are schematically represented in **Fig. 10.9–10.10**. Binding of ligand to some of these receptors induces second messenger formation, whereas ligand binding to others does not. As mentioned earlier, all water-soluble signal molecules bind to specific receptor proteins on the surface of the target cells that they influence. These cell surface receptor proteins act as signal transducers. Most cell surface receptor proteins belong to one of the following classes.

G protein-coupled receptors (GPCR)

A group of cell surface receptors, which transmits signals to intracellular targets via the intermediary action of guanine nucleotide-binding proteins, is called G proteins (**Fig. 10.9a**). Ligand binding to many different mammalian cell surface receptors activates a signal-transducing G protein that, in turn, activates an effector enzyme to generate an intracellular second messenger. The G protein-coupled receptors (GPCRs) constitute a very large superfamily of receptors and examples of the receptor type are found universally in organisms from yeast and insects to humans. More than a thousand such G protein-coupled receptors have been identified, including the receptors for eicosanoids, many neurotransmitters, neuropeptides and peptide hormones, and at least 130 G protein-linked receptor genes have been cloned from various mammals.

G protein receptor: The GPCRs *N*-terminal region is present on the outside of the cell and is often the ligand-binding domain, although some ligands (e.g., adrenaline) bind within the membrane domain. The internal *C*-terminal region is associated with a heterotrimeric

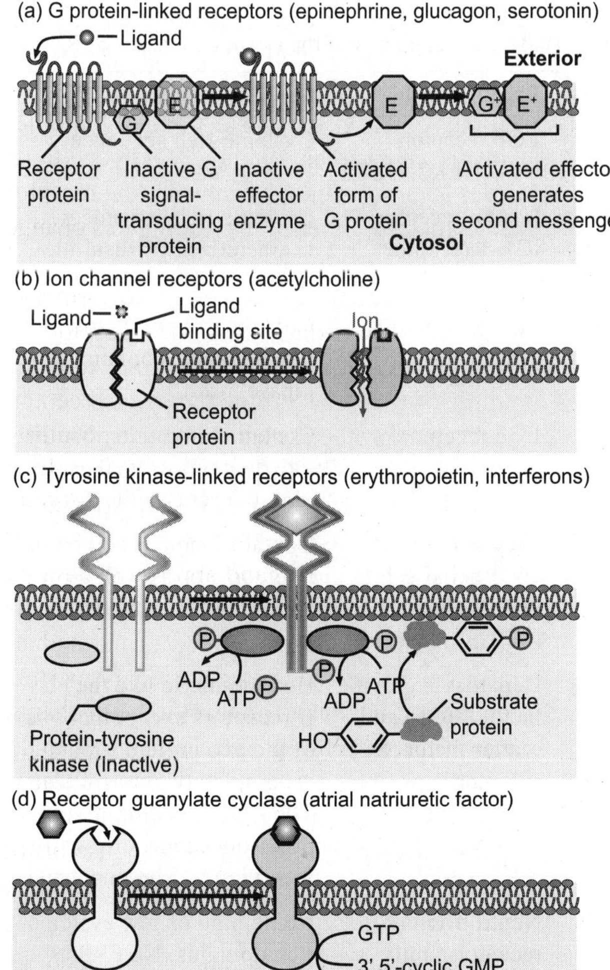

Fig. 10.9 Ligand triggered cell surface receptors.

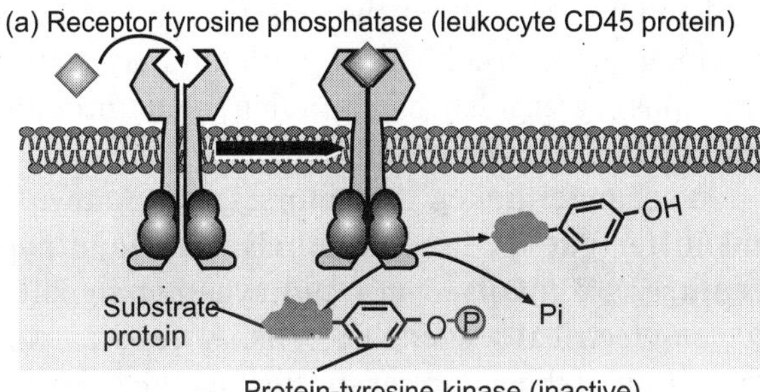

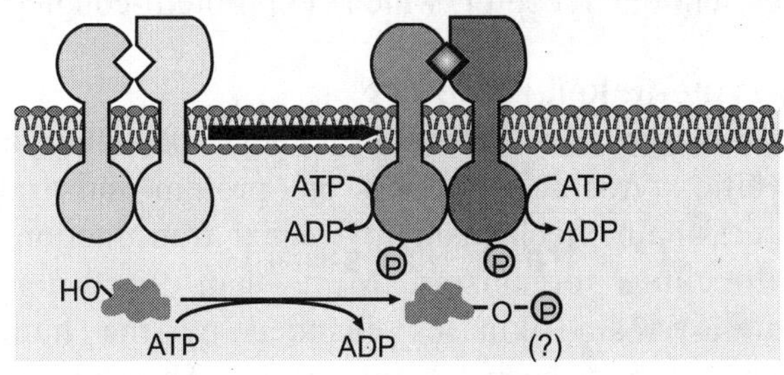

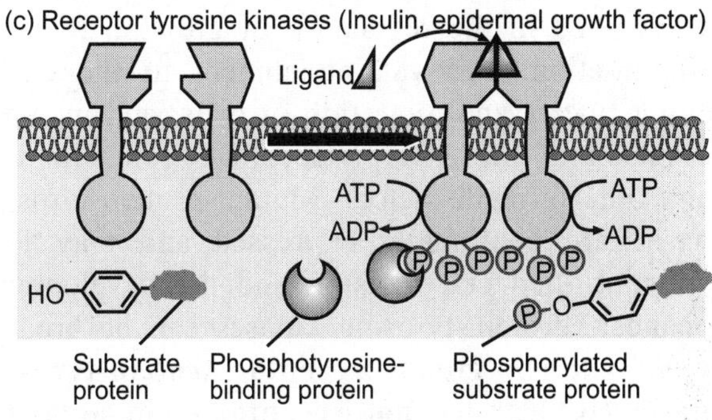

Fig. 10.10 Ligand triggered cell surface receptors.

guanine nucleotide binding protein, or G protein present on the cytoplasmic side of the cell. There are three loops present on the outside of the cell. Together they form the ligand binding site. The cytoplasmic side of the receptor is also associated with a G protein.

Ion channel-linked receptors

Ion channels are basically water-soluble pores in the cell membranes whose activity is controlled by opening and closing for allowing or preventing the movement of ions and other small molecules in or out of the cell (**Fig. 10.9b**). Ligand binding changes the conformation of the receptor so that specific ions flow through it. Some are ligand gated, i.e., they open in response to the binding of a particular ligand. Ion channels found in neurons are voltage-gated, i.e., they open in response to the electrical changes associated with an action potential. The third class, are **second messenger-gated ion channels**, which respond to second messengers in the cell (e.g., calcium ions, cyclic nucleotides and lipids). Ion channels may allow the passive movement of ions along an electrochemical gradient, may actively pump ions against such a gradient and through a process which requires ATP.

Ion channel-linked receptors, also known as **transmitter-gated ion channels** or **ionotropic receptors**, are involved in rapid synaptic signalling between electrically excitable cells. A good example is the action of the neurotransmitter, acetylcholine, on the heart muscle. Heart muscle cells have a different acetylcholine receptor, which is G protein-coupled.

Enzyme-linked receptors

Protein-tyrosine kinases: They phosphorylate specific tyrosine residues on protein substrates as the main mechanism for signal transduction in multicellular organisms. More than 90 different protein-tyrosine kinases encoded by the human genome have been characterized. These kinases are involved in the regulation of growth, division, differentiation, survival, attachment to the extra-cellular matrix and migration of cells, and includes the receptors for EGF, NGF, PDGF, insulin and many other growth factors. Mutant protein-tyrosine kinases are continually expressed and may lead to uncontrolled cell division and the development of cancer. Protein-tyrosine kinases can be broadly divided into two groups: **receptor protein-tyrosine kinases** (**RTKs**) and **non-receptor** or **cytoplasmic protein-tyrosine kinases**.

Receptor tyrosine kinases (RTKs): The RTKs signal transduction pathway was one of the pathways to unite various areas of developmental biology. RTKs form a large and important class of cell surface receptors. Their ligands are soluble or membrane-bound peptide/protein hormones such as insulin and epidermal growth factor (**Fig. 10.10c**). The human genome encodes about 58 RTKs, including the receptors for EGF, NGF, PDGF, insulin and many other growth factors (**Table 10.4**). The proteins encoding RTKs contain four major domains:

- An extracellular ligand binding domain
- An intracellular tyrosine kinase domain
- An intracellular regulatory domain
- A transmembrane domain.

All RTKs comprise an *N*-terminal extracellular domain containing a ligand binding site, a single hydrophobic transmembrane α-helix and cytosolic C-terminal

Table 10.4 Characteristics of the common classes of RTKs

Class	Examples	Structural features
I	EGF receptor, NEU/HER2, HER3	Cysteine-rich sequences
II	Insulin receptor, IGF-1 receptor	Cysteine-rich sequences, characterized by disulfide-linked heterodimers
III	PDGF receptors, c-Kit	Contain 5 immunoglobulin-like domains; contain the kinase insert
IV	FGF receptors	Contain 3 immunoglobulin-like domains as well as the kinase insert, acidic domain
V	Vascular endothelial cell growth factor (VEGF) receptor	Contain 7 immunoglobulin-like domains as well as the kinase insert domain
VI	Hepatocyte growth factor (HGF) and scatter factor (SC) receptors	Heterodimeric like the class II receptors except that one of the two protein subunits is completely extracellular. The HGF receptor is a proto-oncogene, originally identified as Met oncogene
VII	Neutrotrophin receptor family (trkA, trkB, trkC) and NGF receptor	Contain no or few cysteine-rich domains; NGFR has leucine-rich domain

domain with protein-tyrosine kinase activity. Binding of a ligand to this type of receptor stimulates the receptor's protein tyrosine kinase activity, which subsequently stimulates a signal transduction cascade leading to changes in cellular physiology patterns of gene expression. RTK signalling pathways have a wide spectrum of functions including regulation of cell proliferation and differentiation, promotion of cell survival and adjustments in cellular metabolism. Binding of ligand causes most RTKs to dimerize (**Fig. 10.11**), the protein kinase of each receptor monomer then phosphorylates a distinct set of tyrosine residues in the cytosolic domain of its dimer partner and the process termed **autophosphorylation**. The phosphotyrosine residues of activated RTKs play a crucial role in transducing hormone signals to intracellular signalling molecules.

Some RTKs have been identified in studies on human cancers associated with mutant forms of growth-factor receptors (**Table 10.5**), which send a proliferative signal to cells even in the absence of growth factor. One such mutant receptor, enclosed at the *neu* locus, contributes to the uncontrolled proliferation of certain human breast cancers. Other RTKs that leads to blocks in differentiation of certain cell types have been uncovered during analysis of

Table 10.5 Fibroblast growth factor receptor (FGGR) mutants and associated syndromes

Affected receptor	Syndrome	Phenotypes
FGFR1	Pfeiffer	Broad first digits, hypertelorism
FGFR2	Apert	midface hypoplasis, fusion of digits
FGFR2	Beare–Stevenson	midface hypoplassia, corrugated skin
FGFR2	Crouzon	midface hypoplassia, ocular proptosis
FGFR2	Jackson–Weiss	midface hypoplassia, foot anomalies
FGFR2	Pfeiffer	Same as for FGFR1 mutations
FGFR3	Crouzon	midface hypoplassia, acanthosis nigrans, ocular proptosis
FGFR3	Non-syndromatic	
craniosynostosis	Digit defects, hearing loss	

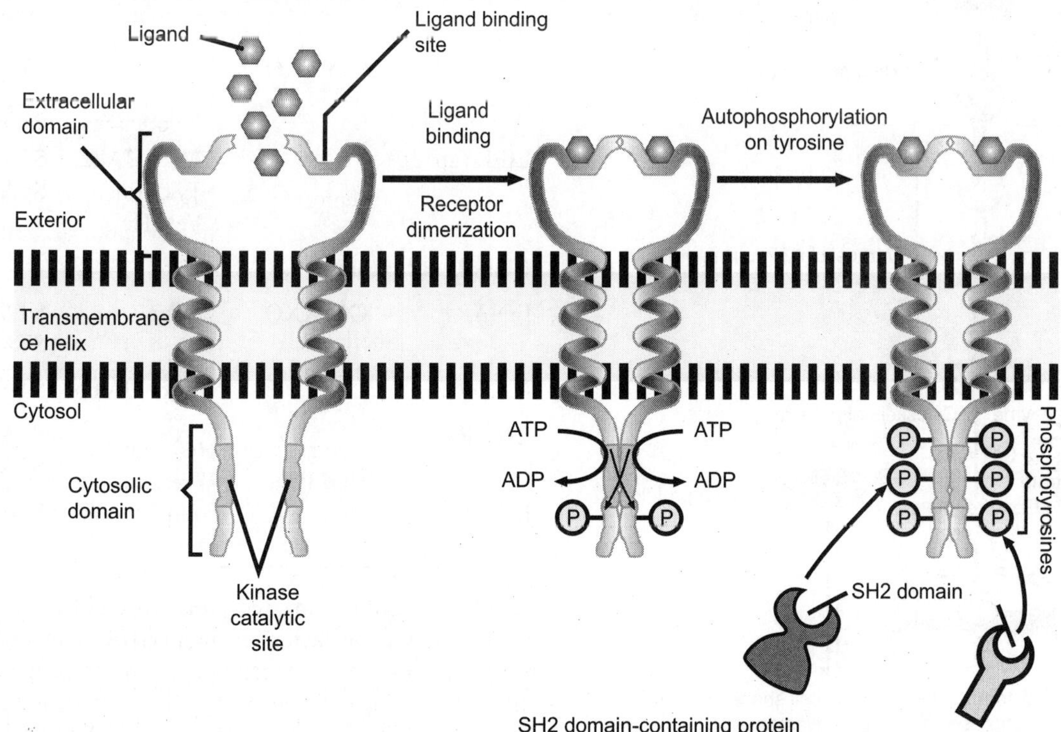

Fig. 10.11 Dimerization and activation of most receptor tyrosine kinases (RTKs) is induced by ligand binding. (Source: Lodish et al, 1995).

developmental mutations of certain human breast cancers and analysis of developmental mutations in *C. elegans*, *Drosophila* and the mouse.

Non-receptor protein tyrosine kinases(PTKs): There are numerous intracellular PTKs that are responsible for phosphorylating a variety of intracellular proteins on tyrosine residues following activation of cellular growth signals (**Fig. 10.12**). Now two distinct families of non-receptor PTKs have been recognized. The archetypal PTK family is related to the **Src protein** and the **Janus kinase (JAK)**.

The intracellular domain of each receptor is constitutively associated with a protein-tyrosine kinase of the Janus family (**Janus kinase or JAKs**; **Table 10.6**). Members of the JAK family appear to be universally required for signalling from cytokine receptors, indicating that JAK family kinases play a critical role in coupling these receptors to the tyrosine phosphorylation of intracellular targets. Phosphorylation of the STATs induces dimerization and translocation to the nucleus, where they act as transcriptional regulators.

Additional non-receptor protein-tyrosine kinases belong to the Src family, which consists of Src and eight closely related proteins. It was initially identified as the oncogenic protein of Rous sarcoma virus and was the first protein shown to possess protein-tyrosine kinase activity. Members of the Src family play key roles in signalling downstream of receptor protein-tyrosine kinases, from antigen receptors on

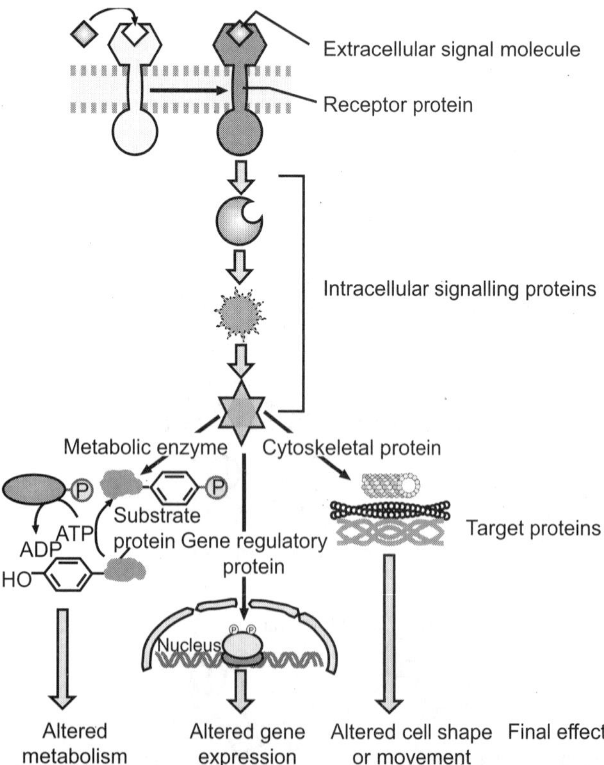

Fig. 10.12 A general intracellular signalling pathway activated by an extracellular signal molecule.

Table 10.6 Signal transduction components of the class I and class II cytokine receptors

Receptor complex	Recruitment motif	JAKs	STATs
Class I receptors			
EPOR	XXYLVL/V	JAk2	STAT5
G-CSFR		JAk1, JAk2	STAT3
GHR		JAk2	STAT1
GM-CSFR (ßc)		JAk1, JAk2	STAT5
IL-2Rb (γc)	DAYLSL, DAYCTF	JAk1, JAk2 (also Lck, Syk)	STAT5
IL-3R (ßc)		JAk1, JAk2	STAT5
IL-4R (γc)	XGYKPFG, GYKAFS	JAk1, JAk3	STAT6
IL-5R (ßc)		JAk1, JAk2	STAT5
IL-6R (gp130)	XXYXPQX	JAk1, JAk2, Tyk2	STAT1, STAT3
	XXYXXQ		STAT3
IL-7R (γc)		JAk1, JAk3	STAT5
LIFR (gp130)		JAk1, JAk2, Tyk1 (also Yes, Hck)	STAT3, STAT1
Class II receptors			
IFNAR1	XXYXXQ	Tyk2	STAT1, STAT2
IL-10R		JAk1, JAk2	STAT1
INFAR2		JAk1	STAT1, STAT3
INFGR		JAk1, JAk2	STAT1, STAT3

EPOR: erythropoietin receptor; G-CSFR: granulocyte colony-stimulating factor receptor; GHR: growth hormone receptor; GM-CSFR: granulocyte macrophage colony-stimulating factor receptor; IL: interleukin; LIFR: leukemia inhibitory factor receptor; IFNA(GR): interferon a(g) receptor; JAK: Janus kinase; STAT: signal transducers and activators of transcription. (Source: Twyman. 1998).

B and T cells, and from integrins at sites of cell attachment to the extracellular matrix.

Another example of receptor signalling through protein interaction is the insulin receptor (IR). This receptor has intrinsic tyrosine kinase activity but does not directly interact, following autophosphorylation, with enzymatically active proteins containing SH2 domains. Signal transduction by RTKs is usually terminated by internalization of the receptor, although, the exact mechanism is not fully known.

Receptors serine/threonine kinases (RSTKs): Although several types of enzyme-linked receptors stimulate protein-tyrosine phosphorylation, some receptors are associated with other enzymatic activities. Binding of ligand to some receptors activates guanylate cyclase activity or protein phosphatase activity in the cytosolic domain (**Fig. 10.13**). The receptors for insulin and many growth factors are ligand-triggered protein kinases; in most cases, ligand binding leads to dimerization and activation of either serine/threonine kinase or tyrosine kinase activity (**Fig. 10.14**). These receptors are often referred to as receptor serine/threonine kinases (RSTKs). The receptors for the TGF-β superfamily have intrinsic serine/threonine kinase activity. A more complete description of the TGF-β signalling cascade can be found in the signalling by Wnts and the TGFs-β/BMP families.

Receptors for members of the transforming growth factor ß (TGF-ß) superfamily of signalling molecules are classified according to size and ligand-binding specificity (**Table 10.7**). A number of type I and type II receptors are known to possess intrinsic serine/threonine kinase activity. Type I receptors typically possess a smaller cysteine-rich extracellular domain and a smaller *C*-terminal tail than the type II receptors and also have a unique G_S domain (glycine and serine rich) adjacent to the kinase domain. Signal transduction involving receptor serine/threonine kinases may require receptor oligomerization. TGF-ß initially bind to the class II TGF-1 ß receptor, which is a constitutively active kinase. Binding recruits a class I TGF-ß receptor, which is activated by phosphorylation on serine residues and can itself phosphorylate downstream targets. Other signalling molecules in the superfamily bind directly to type of receptors, notably the bone morphogenetic proteins.

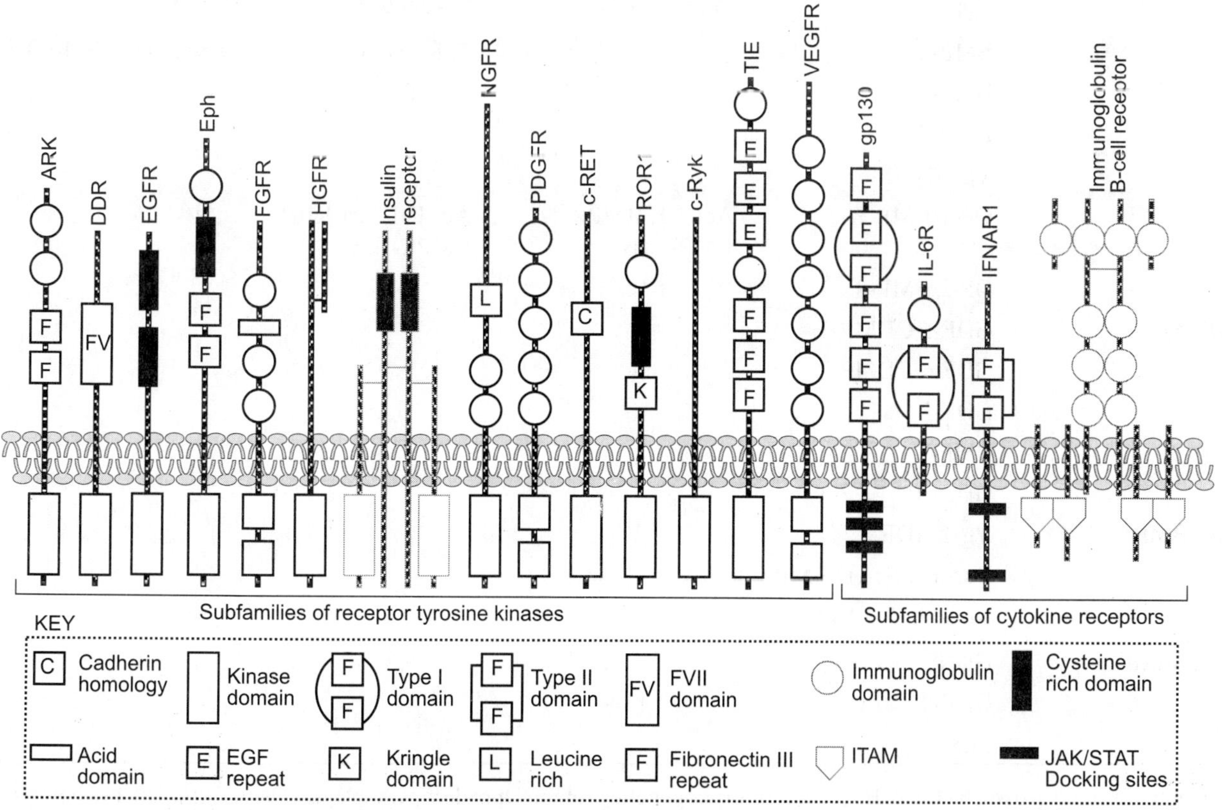

Fig. 10.13 Structure of receptors with intrinsic or associated tyrosine kinase activity.

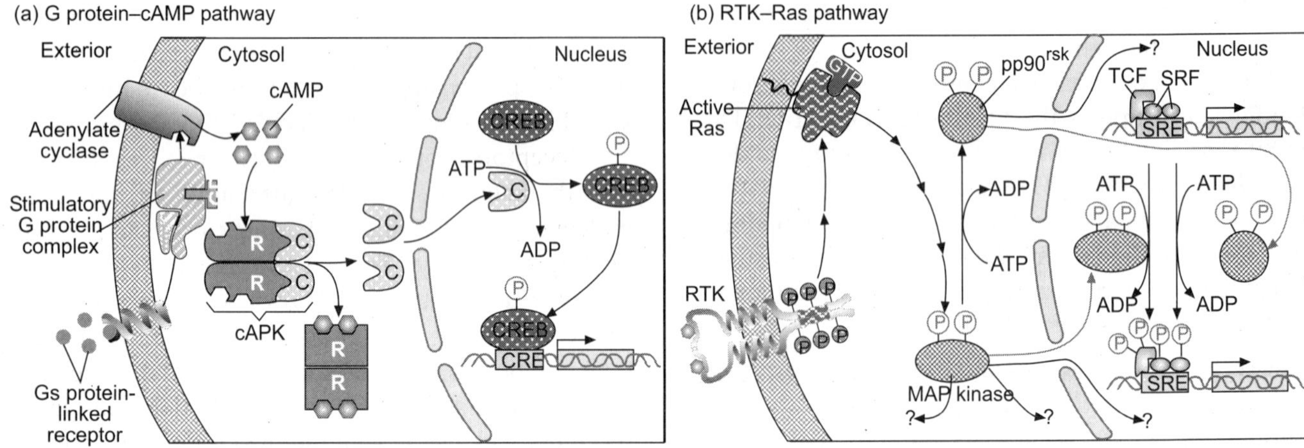

Fig. 10.14 Signalling pathways leading to activation of transcription factors and modulation of gene expression following (a) ligand binding to certain G_s protein-linked receptors and (b) receptor tyrosine kinase. [Source: Lodish et al, 1995]

Table 10.7 The TGF-ß superfamily of growth factors, their receptors and downstream targets

Family	Ligands	Type I receptor	Type II receptor	Targets where known
TGF-ß	TGF-β1/5			
	TGF-2	TβR-I (Alk-5)	TβR-II	SMAD 2 (Smad 3)
	TGF-3			
Activin/inhibin	Activin	ActR-I, Act-IB, Atr-1	Act-RII, ActRIIB	
	Several inhibins		Punt (Atr-II)	
BMP (DPP)	Dpp/	Punt (Atr-II)	Saxophone,	MAD
	BMP-2		Thick veins	SMAD 1 (SMAD 5)
	BMP-4	BMPR1A, BMPR1B	BMPR-II	
(OP-1)	BMP-5			
	Vg-1/BMP-6			
	OP-1/BMP-7	ActR-1. BMPR-1A. BMPR-1B	Act-RII, Act-RIIB, BMPR-II, C14	
	OP-2/BMP-8			
(GDF-5)	GDF-5/CDMP-1			
	GDF-6/CDMP-2			
	GDF-7			
(BMP-3)	BMP-3			
	GDF-10			
Other BMPs	Vg-1/GDF-1			
	Dorsalin/Screw/BMP-9			
	NodaGDF-3/Vgr-21			
Unclassified	GDNF			
	GDF-9			
	MIS		C14	

TGF-ß: transforming growth factor-ß; BMP: bone morphogenetic protein; Dpp: decapentaplegic (*Drosophila* BMP-2 homologue); OP: osteogenic protein; CDMP: cartilage-derived morphogenetic protein; GDF: growth/differentiation factor; GDNF: glial-derived neurotrophic factor; MIS: Miillerian inhibiting substance.

Several downstream targets for TGF-ß signalling are cyclin-dependent kinase inhibitors, which, by inactivating cyclin-dependent kinases, prevent phosphorylation of the retinoblastoma protein and block progression into the S-phase. Activated type I receptors also phosphorylate proteins of the SMAD family on the *C*-terminal consensus motif SSV/MS. SMADs then form heterodimers and translocate to the nucleus, where they interact with DNA-binding proteins and influence transcription, or bind to DNA directly.

Non-receptors serine/threonine kinases: There are several serine/threonine kinases that function in signal transduction pathways. The two more commonly known are cAMP-dependent protein kinase (PKA) and protein kianse C (PKC). Additional serine/threonine kinases important for signal transduction are the mitogen-activated protein kinases (MAP kianses).

Protein kinase C was originally identified as a serine/threonine kinase that was maximally active in the presence of diacylglycerols (DAG) and calcium ion. It is now known that there are at least ten proteins of the PKC family. Each of these enzymes exhibits specific patterns of tissue expression and activation by lipid and calcium. PKCs are involved in the signal transduction pathways initiated by certain hormones, growth factors and neurotransmitters. The phosphorylation of various proteins by PKC, can lead to either increased or decreased activity. Of particular importance is the phosphorylation of the EGF receptor by PKC, which down regulates the tyrosine kinase activity of the receptor.

MAP kinases were identified by virtue of their activation in response to growth factor stimulation of cells in culture; hence the name mitogen-activated protein kinases. MAP kinases are also called ERKs for extracellular signal-regulated kinases.

Other receptors: Besides the above types of receptors the atrial natriuretic peptides, for instance, bind receptors with intrinsic guanyl cyclase activity. These increase the intracellular levels of cGMP and activate protein kinase G (PKG). A number of receptors with protein tyrosine phosphatase activity have also been described, including the leukocyte common antigen CD45 receptor. The tumour necrosis factor receptor family signal uses death domains, which bind to cytoplasmic proteins with similar domains (apoptosis).

Cytoplasmic and nuclear receptors

Most receptors for extracellular signals are on the cell surface because the signalling molecules are hydrophilic and unable to penetrate the plasma membrane. Other molecules can cross the plasma membrane directly and interact with cytoplasmic or nuclear receptors and these often initiate simple signal transduction pathways (e.g., steroid and thyroid hormones and retinoic acid and the gas nitric oxide). The steroid hormone receptor superfamily includes receptors for steroid and thyroid hormones and vitamins A and D and their derivatives, including the important developmental molecule retinoic acid. Interaction with the ligand causes a conformational change (transformation) which stimulates DNA binding activity. Some receptors are located in the cytoplasm (e.g., glucocorticoid receptor) and transformation allows nuclear translocation. Others are already present in the nucleus. The transformed receptors are transcription factors that interact with DNA through highly conserved zinc-binding domain. They can be divided into several families based on the architecture of the recognition site. The activity of some receptors can be regulated by phosphorylation as well as ligand binding (e.g., progesterone and estrogen receptors), allowing integration with other signalling pathways.

Nuclear receptors: The nuclear receptor superfamily consists of a diverse set of transcription factors that were discovered because of a sequence similarity in their DNA-binding domains. These nuclear receptors have several common structural features. All have a centrally located **DNA-binding domain (DBD)** that allows the receptor to bind with high affinity to a response element. The DBD contains two zinc finger-binding motifs that direct binding either as homodimers, heterodimers, or monomers. A multifunctional **ligand binding domain (LBD)** is located in the carboxyl terminal half of the receptor. A highly variable **hinge region** separates the DBD from the LBD. There is also a highly variable amino terminal region that contains another transactivation domain referred to as **AF-1**, which may provide for distinct physiological functions through the binding of different coregulator proteins.

Nuclear receptor proteins diffuse directly across the plasma membrane of target cells and bind to intracellular receptor proteins. Examples include steroid hormones, thyroid hormones, retinoids and vitamin D.

10.3.2 Intracellular signal transduction

Most cell surface receptors activate intracellular target enzymes either directly linked or indirectly coupled to G protein receptors. These intracellular enzymes serve as downstream signalling elements that propagate and amplify the signal initiated by ligand binding. Generally, a chain of reactions transmits signals from the cell surface to a variety of intracellular targets: a process called intracellular signal transduction. The targets of such signalling pathways frequently include transcription factors that function to regulate gene expression. Intracellular signalling pathways thus connect the cell surface to the nucleus, leading to changes in gene expression in response to extracellular stimuli.

The activation of receptor tyrosine kinases or receptor-associated tyrosine kinases transduce a signal across the cell membrane without transferring the ligand into the cell. Signal transduction is often mediated by a cascade of sequential enzyme activation: one signalling component phosphorylates and activates a second component, which performs the same process on a third and so on until the ultimate target is reached (**Fig. 10.15**). The signalling process may also be inhibitory.

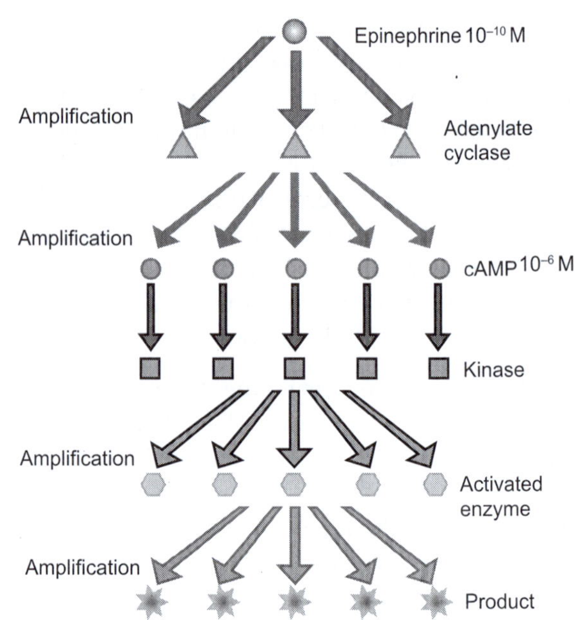

Fig. 10.15 Intracellular transduction of an extracellular signal via a cascade of sequential reactions produces a large amplification of the signal. In this example, binding of a single epinephrine molecule to one receptor molecule induces synthesis of a large number of cAMP molecules and these in turn activate multiple enzyme molecules.

MAP kinase signalling pathways

MAP kinase (mitogen-activated protein kinase, also known as **ERK**, extracellular-signal regulated kinase) is a serine/threonine kinase activated by many growth factors. The pathway to MAP kinase activation involves Ras and Raf phosphorylating a kinase upstream of MAP kinase, termed MAP kinase (MAPKK) or MEK (MAPK/Erk kinase). MAP kinase requires phosphorylation of tyrosine and threonine residues to become activated and MEK is an example of a **dual specificity kinase**, with the combined activities of both a tyrosine kinase and a serine/threonine kinase. The MAP kinase pathway channels mitogenic signals from the cell surface to the nucleus and many components of the pathway are therefore oncogenic when inappropriately activated. The activation of MAP kinase causes it to be translocated to the nucleus, where it phosphorylates and activates a number of transcriptional regulators including Elk-1, C/EPBß and cMyc. MAP kinase also phosphorylates a further kinase called Rsk, which translocates to the nucleus and may activate other transcription regulators, such as the serum response factor (SRF) (**Fig. 10.16**).

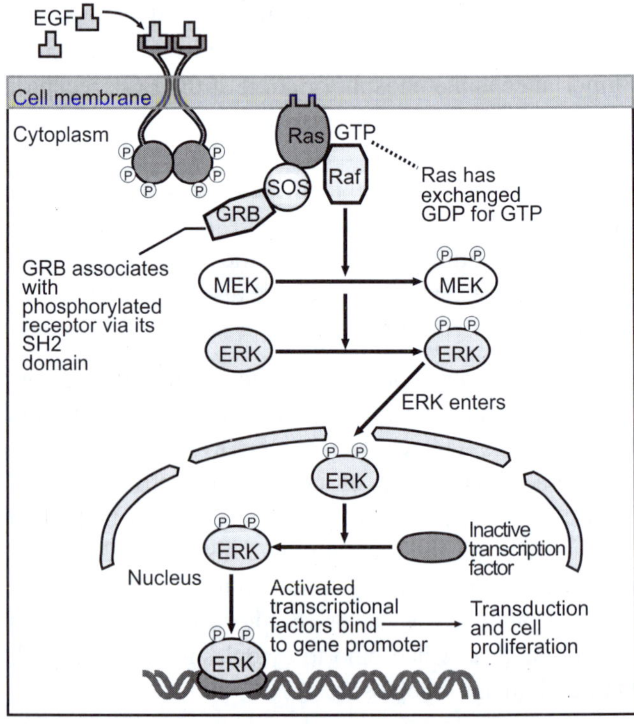

Fig. 10.16 The Ras pathway of signal transduction. Raf, MEK and ERK are protein kinases, each of which phosphorylates the next component of the pathway.

The JAK/STAT pathways

This pathway connects protein-tyrosine kinases and transcription factors. In this pathway, protein-tyrosine phosphorylation directly affects transcription factor localization and function. The key factors in this pathway are the **STAT proteins** (signal transducer and activator of transcription). These proteins are a family of transcription factors that contain SH2 domains and are inactive in unstimulated cell cytoplasm.

Interferons, the protective proteins released by cells infected by viruses, were the first to be associated with this type of pathway, but many cytokines are known to use it. Ligand binding causes dimerization of receptors and association of a **JAK kinase** (Janus the Roman god with two faces) with the cytoplasmic domain of each receptor in the dimer (**Fig. 10.17**).

Stress-activated kinases

Stress-activated kinases (**SAP kinases**) are serine/threonine kinases, related to MAP kinase, which activate the transcription factor c-1 Jun in response to stresses such as UV irradiation and inflammatory cytokines, but only poorly in response to growth

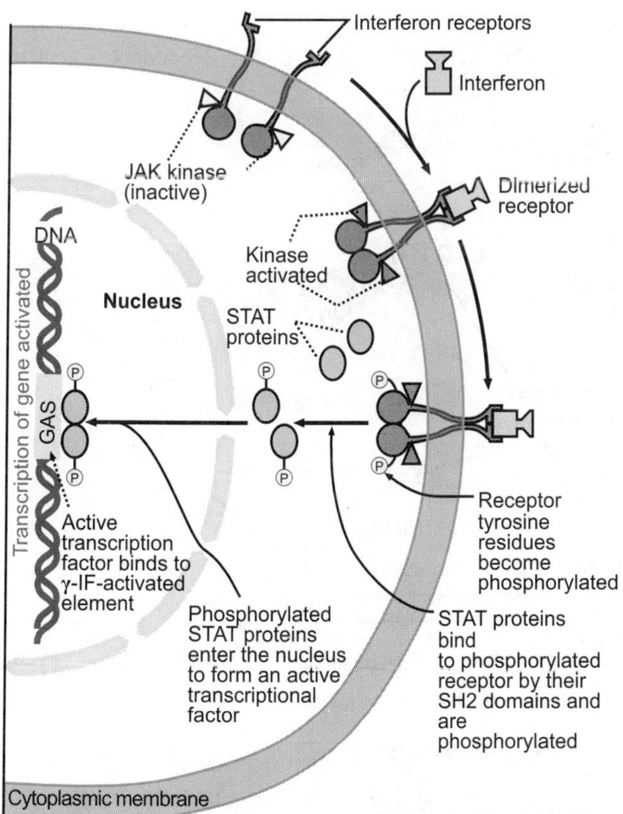

Fig. 10.17 The signalling pathway by which γ-interferon activates specific gene transcription. STAT: signal transducer and activator of transcription; GAS: γ-interferon-activated sequence element.

stimulation. The SAP kinase pathway is also active during T cell development and can induce proliferation in some cells. There are at least eight SAP kinase proteins, derived from three genes by alternative splicing. The individual SAP kianses are named Jnk-1, Jnk-2 and so on. They phosphorylate the *N*-terminal Ser63 and Ser73 residues of Jun as well as other transcription factors such as Elk-1 and ATF-2. Like the MAP kinases, SAP kinases require tyrosine and threonine phosphorylation for activation.

10.3.3 Second messengers

The binding of ligands to many cell surface receptors leads to a diverse range of signals and requires a huge repertoire of receptors. The range of responses is much smaller, since many of the signals arriving at the cell surface, for instance, cause the cell to either divide or withdraw from the cell cycle. Others induce the expression of characteristic groups of genes, which protect the cell from stress. Therefore, the early signalling pathways converge into a small number of intracellular signalling networks and this allows the cell to convert the complex information arriving at the cell surface into simple biochemical signals in the cytoplasm. The molecules involved in this process are termed second messengers.

Cyclic nucleotide second messengers

The first second messengers to be discovered were the cyclic nucleotides (e.g., cAMP and cGMP). The concentrations of these nucleotides in the cells are controlled by the opposing activities of nucleotidylate cyclases, which catalyze the reaction NMP → cNMP and cyclic nucleotide phosphodiesterases (PDEs), which catalyze the reverse reaction. G protein linked receptors of the G_s and G_i families stimulate and inhibit adenylate cyclase, respectively, whereas transducin (G_t) stimulates cGMP PDE activity. A small number of cell surface receptors have intrinsic guanidylate cyclase activity. Receptor **guanidylate cyclases** bind nitric oxide via a heme group. Downstream of the cyclic nucleotides are effectors dependent on cyclic nucleotides for their activity: protein kinase A (PKA) is cAMP-dependent and protein kinase G (PKG) is cGMP dependent. Effectors are serine/threonine kinases with a variety of substrates. There are also cyclic nucleotide-gated ion channels and feedback pathways involving cyclic nucleotide dependent PCEs. Other PDEs are regulated by phosphorylation or by calmodulin. See **Figs. 10.18, 10.19** and **10.20**.

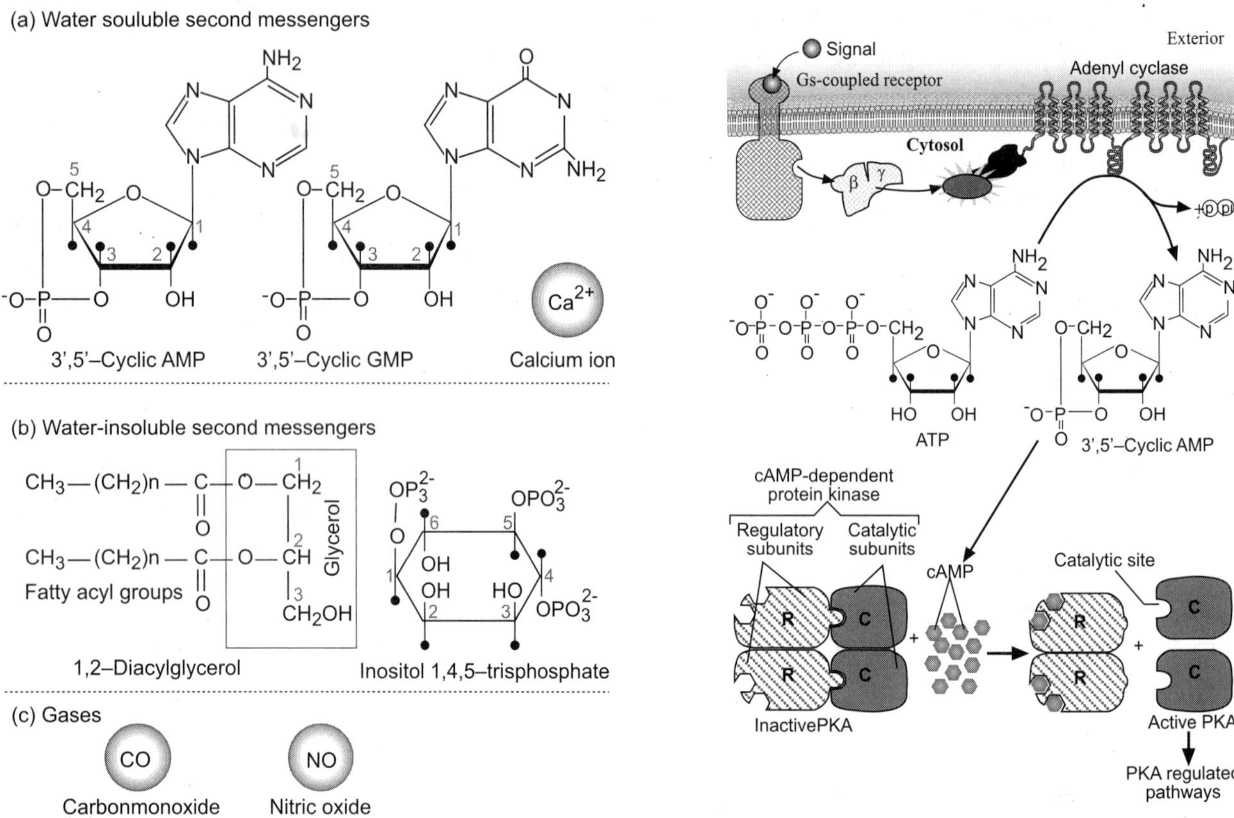

Fig. 10.18 Structures of five common intracellular second messengers.

Fig. 10.20 Activation of cAMP-dependent protein kinase by cAMP.

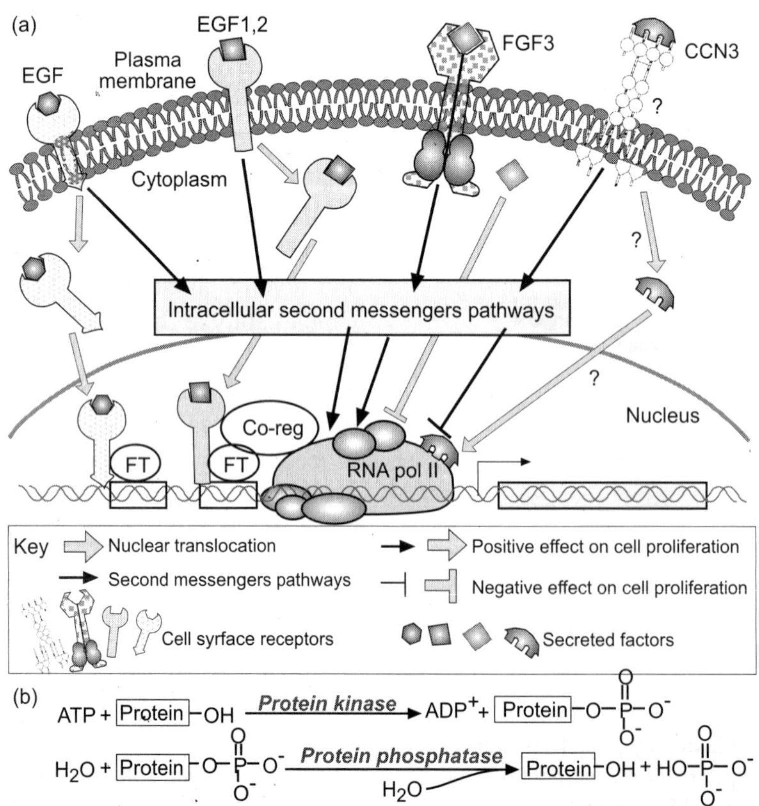

Fig. 10.19 Activation of cAMP–dependent protein kinase by cAMP.

Ca^{2+} as second messenger

The level of calcium ions in the cell controls diverse processes including metabolic processing, proliferation and specialized functions such as membrane excitability and muscle contraction. The Most intracellular Ca^{2+} ions are sequestered in the mitochondria and endoplasmic reticulum (ER) or other cytoplasmic vesicles (**Fig. 10.21**). The concentration of Ca^{2+} ions free in the cytosol is usually kept below 0.2 μM. Ca^{2+} ATPase pumps cytosolic Ca^{2+} ions across the plasma membrane to the cell exterior or into the lumens of the endoplasmic reticulum or other intracellular vesicles that store Ca^{2+}. Ca^{2+} concentrations can actually be detected in specific regions of the cytosol. Two major pathways stimulating calcium release are the G protein–linked receptors associated with G_q family proteins and the RTKs. Both these receptors stimulate phospholipases which increase levels of Ins(1,4,5)P_3, leading to the opening of Ins(1,4,5)P_3 -gated Ca^{2+} channels in the ER membrane. Small increases in the level of cytosolic Ca^{2+}, which are often mediated by a rise in IP_3 trigger many cellular response. In secretory cells, such as the insulin-producing ß cells in the pancreatic isles, a rise in Ca^{2+} triggers the exocytosis of secretory vesicles and the release of insulin. In smooth or striated muscle cells, a rise in Ca^{2+} triggers contractions; in both liver and muscle cells, an increase in Ca^{2+} activates the degradation of glycogen to glucose 1-phosphate.

A small cytosolic protein called *calmodulin*, which is ubiquitous in eukaryotic cells, mediates many cellular effects of Ca^{2+} ions. Each calmodulin molecule binds four Ca^{2+} ions (**Fig. 10.22a**) in a cooperative fashion (i.e., binding of one Ca^{2+} ion facilitates binding of additional Ca^{2+} ions). Binding of Ca^{2+} causes calmodulin to undergo a conformational change that enables the Ca^{2+}-calmodulin complex to bind to and activate many enzymes (**Fig. 10.22b**). Because binding of Ca^{2+} is cooperative, a small change in the level of cytosolic Ca^{2+} leads to a large change in the level of active calmodulin.

Ca^{2+} ions and cAMP: Stimulation of muscle cells by nerve impulses cause the release of Ca^{2+} ions from the sarcoplasmic reticulum and an increase in the cytosolic Ca^{2+} concentration. This rise not only triggers muscle contraction but also stimulates degradation of glycogen to glucose 1-phosphate, which fuels prolonged contraction. Glycogen degradation and synthesis are also modulated by ß-adrenergic activation of the cAMP cascade. However, the key regulatory enzyme

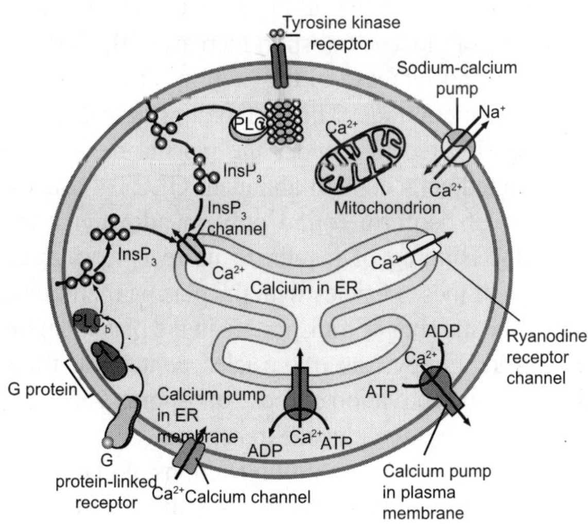

Fig. 10.21 Overview of calcium regulation in cells. Cytosolic calcium concentration is lowered by the actions of the ER calcium pumps, the plasma membrane calcium pump, sodium-calcium exchangers and the mitochondria. Calcium concentration increases in the cytosol because of the opening of calcium channels in the plasma membrane and the release of calcium through the InsP3 or ryanodine receptor channels in the ER membrane.

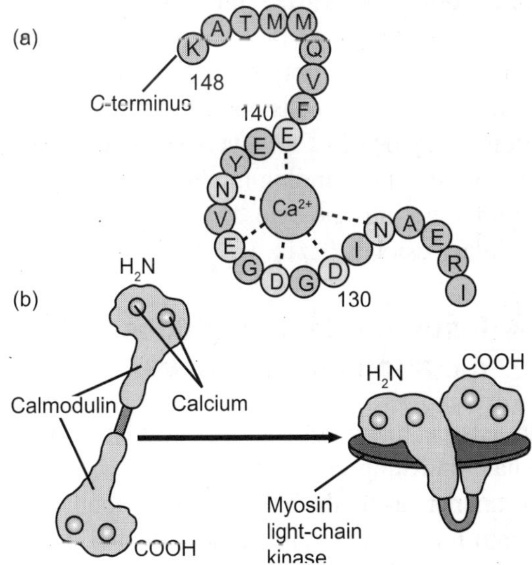

Fig. 10.22 Structure of calmodulin, a cytoplasmic protein of 148 amino acids that binds Ca^{2+} ions: (a) Amino sequence of one Ca^{2+}-bonding site in calmodulin. Each site contains aspartate, glutamate and aspargine residues (hatched circles) whose side chains establish ionic bonds with a Ca^{2+} ion, forming a loop in the backbone. (b) On binding Ca^{2+}, calmodulin undergoes a major conformational change, enabling it to bind to a target protein. [Redrawn from Lodish et al, 1995].

in glycogenolysis, glycogen phosphorylase kinase (GPK), is activated by a rise in the level of cytosolic Ca^{2+}, as well as by phosphorylation by cAMP-dependent protein kinase **(Fig. 10.23)**.

When nerve stimulation of muscle cells ceases, Ca^{2+} ions are pumped back into the sarcoplasmic reticulum, the level of cytosolic Ca^{2+} declines, the activity of glycogen phosphorylase kinase is reduced and less glycogen is converted to glucose phosphate. In several other cells, a rise in cytosolic Ca^{2+} increases the rate of pumping Ca^{2+} out of cells and causes a more rapid drop in cytosolic Ca^{2+} to the resting state, terminating all Ca^{2+}-induced signals unless additional Ca^{2+} is released. This rapid decline is due to Ca^{2+}-calmodulin activation of the plasma-membrane Ca^{2+} ATPase.

Lipids as second messengers

A well-characterized second-messenger system involves products of the hydrolysis of a minor phospholipid component of the inner cell membrane, phosphaticylinositol-4,5-bisphosphate [PtdIns(4,5) P_2]. Activation of G protein-linked receptors associated with G_q family proteins (G_q, G_{11}, G_{12}) stimulates the activity of phospholipase [Ins(1,4,5)P_3] and 1,2-diacylglycerol (DAG). Another PLC isoform, PLC–γ), is activated by RTK signalling by recruitment to the membrane through its SH2 domain. Both Ins(1,4,5)P_3 and DAG are important second messengers.

Binding of many hormones to their cell surface receptors on liver, fat cells and other cells induces an elevation in cytosolic Ca^{2+} even when Ca^{2+} ions are absent from the surrounding medium. In this situation, Ca^{2+} is released into the cytosol from the ER and other intracellular vesicles.

10.3.4 Insulin: Effect on glucose metabolism and growth

Insulin acts on many cells in the body to produce both immediate and long-term effects. Its immediate effects include an increase in the rate of glucose uptake from the blood into muscle cells and adipocytes and modulation of the activity of various enzymes involved in glucose metabolism. These effects occur within minutes, do not require new protein synthesis and occur at insulin concentrations of 10^{-9} to 10^{-10} M. Continued exposure to insulin produces longer-lasting effects, including increased expression of liver enzymes that synthesize glycogen and adipocyte enzymes that synthesize triacylglycerols; insulin also functions as a growth factor for many cells

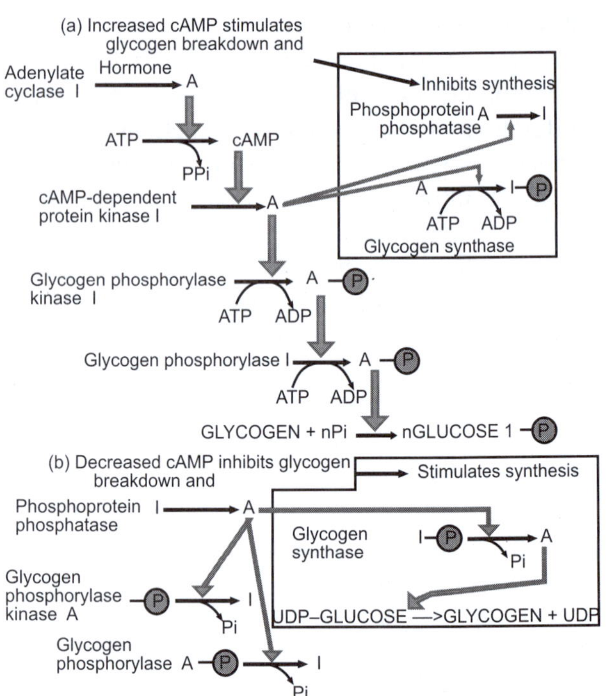

Fig. 10.23 Regulation of glycogen breakdown and synthesis by cAMP in liver and muscle cells.

(e.g., fibroblasts). These effects are manifested in hours and require continuous exposure to $\approx 10^{-8}$ M insulin.

Within 15 min of the addition of insulin to adipocytes, the rate of glucose transport increases 10–20 times, mainly due to an increase in the number of transporters in the plasma membrane. The membranes of endosome-like vesicles located just under the plasma membranes of unstimulated cells contain abundant GLUT-4 glucose transporters, an isoform found only in adipocytes and muscle cells. Binding of insulin to its receptor induces rapid fusion of these vesicles with the plasma membrane, increasing the number of transporters in the plasma membrane tenfold. Exocytosis of vesicles containing transporters and some activation of these transporters accounts for the large increase in glucose transport.

Many of the growth-promoting actions of insulin, for example, induction of DNA synthesis, may be initiated by its binding to the receptor for insulin-like growth factor 1 (IGF-1), rather than to the insulin receptor. IGF-1, which is similar in structure and sequence to insulin, is produced mainly in the liver in response to growth hormone secreted by the pituitary gland and appears to be a primary regulator of growth in an organism.

Insulin receptor substrate

Biochemical studies have identified a 130 kDa polypeptide, called **insulin receptor substrate 1** (IRS1), containing tyrosine residues that were rapidly phosphorylated following insulin stimulation of cultured cells. Most other RTKs do not phosphorylate IRS1, but the related IGF-1 receptor does. Although IRS1 does not bind to the insulin receptor, several lines of evidence suggest that insulin action depend on phosphorylation of IRS1 (**Fig. 10.24**).

Insulin and glucagon act together

Glucagon is released from G cells in the islets of the pancreas in response to low levels of blood glucose and low levels of insulin. The glucagon receptor, found primarily on liver cells, is a seven-spanning G_s protein-linked receptor, like the epinephrine receptor. It is coupled to adenylate cyclase and the cAMP cascade

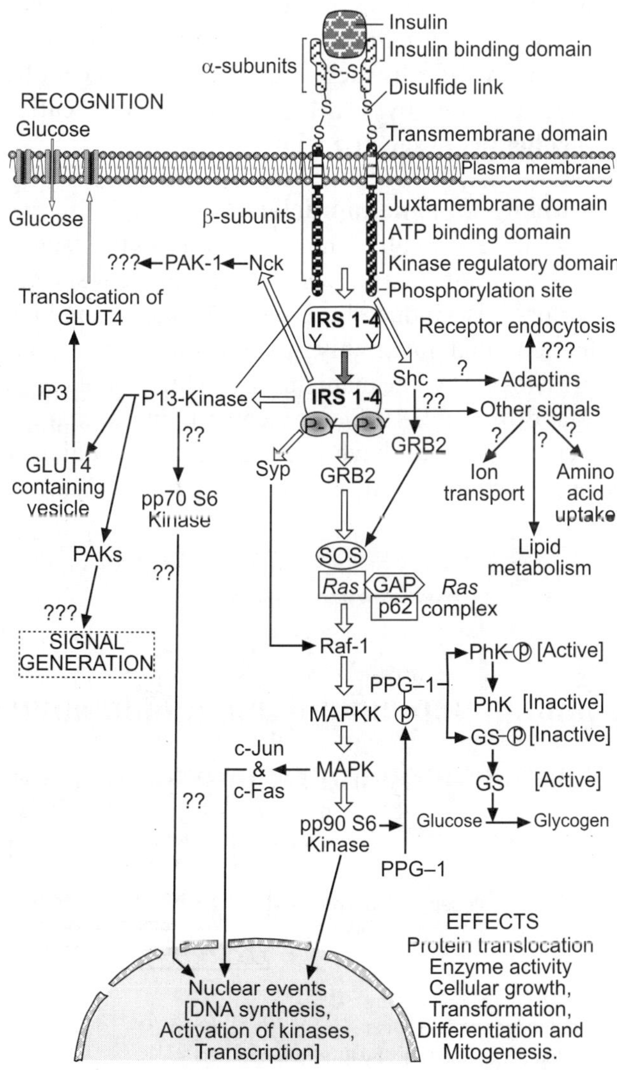

Fig. 10.24 Schematic representation of mechanism of insulin action. Formation of insulin-receptor complex seems to transmit signals via two signalling proteins, IRS-1 and Shc. [Redrawn from: Gupta BBP (1997), *Curr. Science,* 73:993–1003.]

that leads to glycogenolysis and inhibition of glycogen synthesis. The availability of glucose for cellular metabolism is regulated during periods of abundance (following a meal) or scarcity (following fasting) by the adjustment of insulin and glucagon concentrations in the blood (**Fig. 10.25**). When, after a meal, blood glucose rises above its normal level of 80–90 mg/100 ml, the pancreatic ß- cells of the islets respond to the rise in glucose or amino acids by releasing insulin into the blood, which transports the hormone throughout the body. By binding to muscle and adipocyte cell-surface receptors, insulin causes glucose to be removed from the blood and stored in muscle cells as glycogen. Insulin also affects hepatocytes, primarily by inhibiting glucose synthesis from smaller molecules, such as lactate and acetate and by enhancing glycogen synthesis from glucose. If the blood glucose level falls below ≈80 mg/100 ml, the α-cells of the islets start secreting glucagon. Glucagon binds to glucagon receptors on liver cells, triggering degradation of glycogen and the release of glucose into the blood.

10.4 REGULATION OF CELL-SURFACE RECEPTORS

The number and activity of functional hormone receptors on the surface of cells is not constant. The

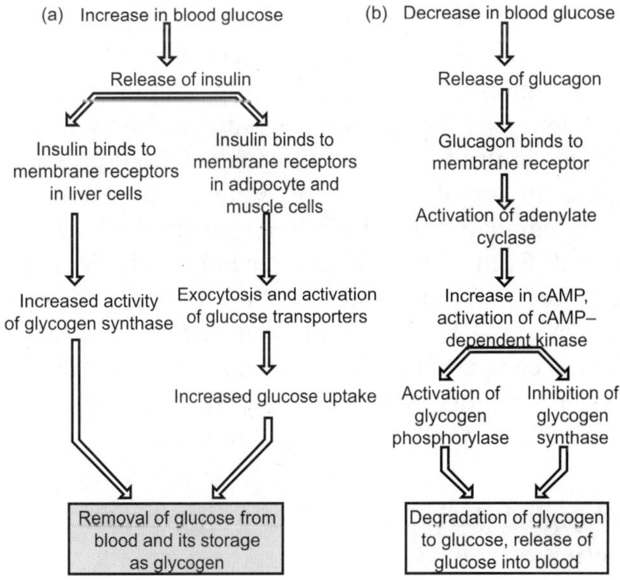

Fig. 10.25 Regulation of blood glucose level by the opposing effects of insulin and glucagon: (a) Insulin causes an increase in glucose uptake, mainly in muscle cells and adipocytes and stimulates storage of glucose as glycogen, mainly in liver cells; (b) Glucagon acts mainly on liver cells to stimulate glycagon degradation. This effect is mediated by the second messenger cAMP.

level of receptors for a particular hormone can be increase (up-regulation) or decrease (down-regulation), permitting cells to respond optimally to small changes in the hormone level. Prolonged exposure of a cell to a high concentration of a hormone usually results in a reduction in the number of its functional receptors, thereby desensitizing the cell to that hormone. Up to a point, an increase in hormone concentration causes typical hormone-induced responses, rather than excessive ones.

Down-regulation of cell surface hormone receptors can occur in several ways. Receptors can be internalized by endocytosis, thus decreasing the number on the cell surface and then either destroyed or storing in intracellular vesicles. Alternatively, the number of receptors on the cell surface is not decreased but their activity is modified, so that they either cannot bind ligand, or bind the ligand but form a receptor–ligand complex that does not induce the normal cellular response. Receptors for many hormones are regulated by two or more of these mechanisms.

Receptors for peptide hormones down-regulated by endocytosis

Endocytosis is the principal mechanism for down-regulating the receptors for many peptide hormones including insulin, glucagon, EGF and PDGF. The receptor–hormone complex is brought into the cell by receptor-mediated endocytosis. The internalized hormone is subsequently degraded in lysosomes—a fate similar to that of other endocytosed proteins, such as low-density lipoproteins. Internalization and degradation most likely terminate the hormone signal. Even if the hormone receptors recycle to the cell surface by exocytosis, as they do in many cases, a substantial function will be in the internal membrane compartments at any one time. Therefore, fewer receptors will be on the cell surface and thus available to bind extracellular hormone.

Phosphorylation and cell surface receptor activity

Cultured cells that are exposed to epinephrine for several hours can bind epinephrine, but adenylate cyclase is not activated and the cAMP level does not rise. This receptor desensitization can be reversed by removing epinephrine from the medium; within several hours, the activity of the ß-adrenergic receptor is restored to normal. This desensitization of the ß–adrenergic receptor results from phosphorylation of several serine and threonine residues in the cytosolic domain (**Fig. 10.26**). Phosphorylation of four residues is catalyzed by a cAMP-dependent protein kinase (cAPK). The activity of this kinase is enhanced by the high cAMP level induced by epinephrine, explaining the receptor desensitization observed after prolonged exposure to epinephrine. If the phosphorylated receptor is purified and incorporated into cells, it can bind ligands normally but cannot activate G_s proteins or stimulate cAMP synthesis in response to epinephrine. Subsequent enzymatic removal of these phosphate groups reactivates the ability of the receptor to stimulate adenylate cyclase. cAMP-dependent protein kinase phosphorylates all G protein-linked cell surface receptors, not just the ß-adrenergic receptor. Since the activity of all such receptors is reduced by phosphorylation by cAMP, the process is called **heterologous desensitization**.

Plasma membrane to nucleus

Many water soluble hormones, growth factors and neurotransmitters, which bind to cell surface receptors, also induce long-term changes in cellular behaviour that invariably involve changes in gene expression. The mechanisms by which binding of a ligand to a cell-surface receptor induces changes in gene expression have been thoroughly studied in several systems. In all cases, binding of a ligand to

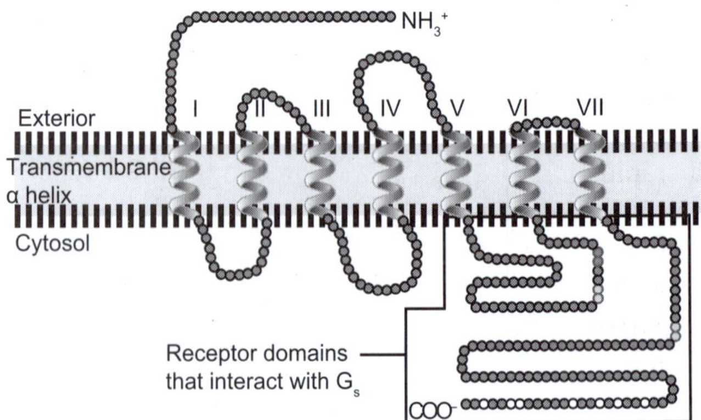

Fig. 10.26 The activity of ß-adrenergic receptors is reduced by phosphorylation of certain serine/threonine residues in two cytosolic domains.

its receptor stimulates protein kinases that directly or indirectly phosphorylate serine, threonine, or tyrosine residues on specific transcription factors.

10.5 ACTIVATION OF TRANSCRIPTION FACTORS

Studies on modulation of gene expression by water soluble hormones and growth factors have identified the CRE-binding (CREB) protein and serum response factor (SRF) as transcription factors that stimulate expression of many genes.

Activation of CREB protein

In mammalian cells an elevation in the cytosolic cAMP level induces the expression of many genes, including the somatostatin gene in certain endocrine cells and the liver genes encoding several enzymes involved in converting three-carbon compounds to glucose (gluconeogenesis). All such genes contain a *cis*-acting DNA sequence, called cAMP-response element (CRE), that is essential for induction by cAMP. The phosphorylated form of the transcription factor called CREB protein binds to this sequence. Binding of neurotransmitters and hormones to G_S protein-linked receptors activates adenylate cyclase, leading to an increase in cAMP and subsequent activation of the catalytic subunit of cAMP-dependent protein kinase. The catalytic subunit then translocates to the nucleus where it phosphorylates serine-133 on CREB protein; phosphorylated CREB protein binds to CRE-containing target genes, stimulating their transcription (**Fig. 10.27**).

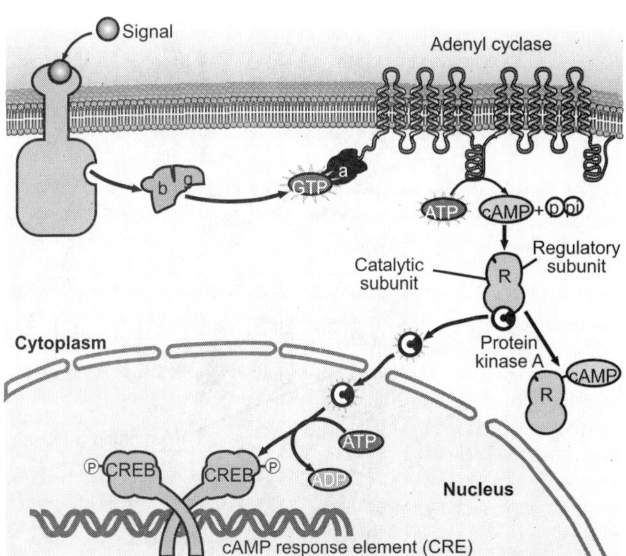

Fig. 10.27 The interaction of heterotrimeric G proteins, cAMP and the regulation of gene expression in animal cells. After synthesis by a plasma membrane-bound adenyl cyclase, cAMP can regulate gene expression through phosphorylation of CREB by an activated protein kinase A (**See Plate 4**).

Activation of SRF

Addition of a growth factor (e.g., EGF or PDGF) to quiescent cultured mammalian cells in G_0 causes a rapid increase in the expression of as many as 100 different genes. These are called early-response genes because they are induced well before cells enter the S phase and replicate their DNA. One important early response gene encodes the transcription factor c-Fos; together with other proteins, c-Fos induces expression of many genes encoding proteins necessary for cells to progress through the cell cycle. Induction of the c-fos gene itself by various growth factors is mediated via several intracellular signalling pathways that lead to activation of different protein kinases: those involving cAMP (cAMP-dependent protein kinase), Ca^{2+}/DAG (protein kinase C) and Ras protein (MAP kinase). The regulatory sequence of the c-fos gene contains a serum response element (**SRE**), so named because it is activated by many growth factors in serum. Mutational analysis of the SRE in the c-fos gene led to the surprising finding that different signalling pathways act through different sequences in this response element. For example, cells carrying certain mutations in SRE cannot respond to protein kinase C activation (induced by TPA) but can respond to growth factors.

Binding of serum response factor and associated proteins to SREs activates gene expression. The RTK–Ras signalling pathway activates the MAP and kinase translocates it to the nucleus where it phosphorylates specific sites in the *C*-terminal domain of a protein called **ternary complex factor (TCF)**. Phosphorylated TCF associates with two molecules of SRF, forming an active trimeric DNA-binding factor. SRF and TCF may remain bound to promoters in vivo. Mutational studies have shown that if TCF lacks the serine residues phosphorylated by MAP kinase, it does not activate gene expression in transfection experiments. Growth factor induced phosphorylation of SRF on serine-103 may also stimulate transcription from SREs. Serine-103 of SRF is rapidly phosphorylated in response to growth factors; this modification increases the rate and affinity of SRF binding to the SRE in vitro. A MAP kinase-activated kinase called pp90 has been shown to phosphorylate this serine in SRF in vitro.

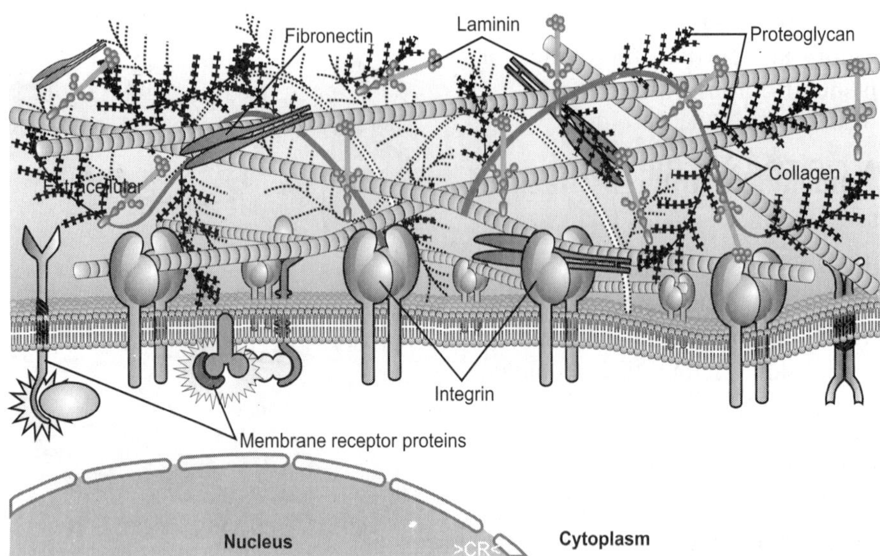

Fig. 10.28 An overview of the macromolecular organization of the extracellular matrix. The proteins depicted (fibronectin, collagen, laminin and other membrane proteins) contain binding sites for one another, as well as binding sites for receptors (integrins) that are located at the cell surface. The proteoglycans are huge protein–polysaccharide complexes that occupy much of the volume of the extracellular space (**See Plate 5**).

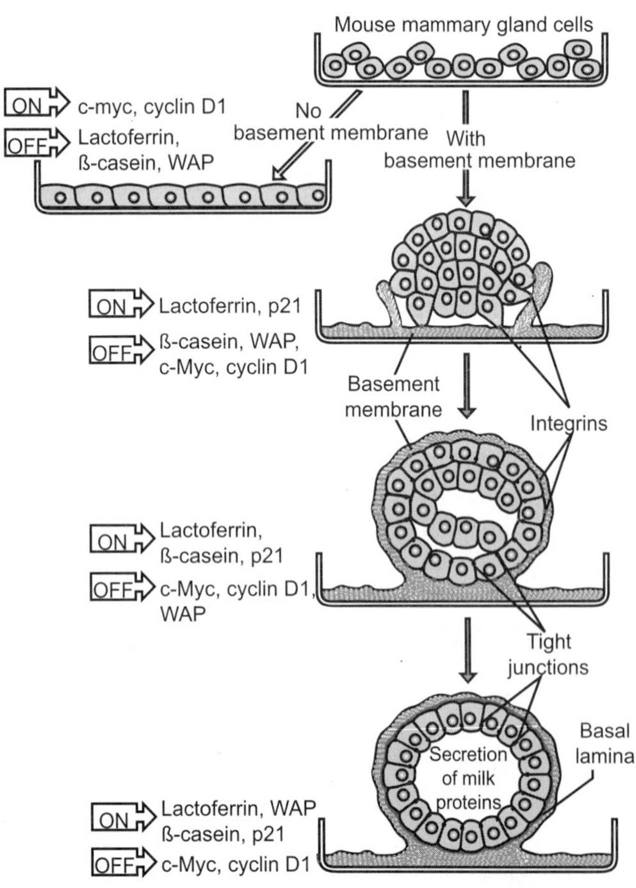

Fig. 10.29 Basement membrane-directed gene expression in mammary gland tissue.

10.6 EXTRACELLULAR MATRIX: A SOURCE OF DEVELOPMENTAL SIGNALS

The extracellular matrix (ECM), a proteinaceous cytoskeletal structure, consists of macromolecules secreted by cells into their immediate environment (**Fig. 10.28**). These macromolecules form a region of non-cellular material in the interstices between the cells. The ECM is a critical region for animal development. Cellular functions such as adhesion, cell migration and the formation of epithelial sheets and tubes all depend on the ability of cells to attach ECM. For example, in case of epithelia, these attachments have to be extremely strong and in other cases it merely serves as a permissive substrate to which cells can adhere or migrate to (see Chapter 13). The ECM may also provide the directions for cell movement or the signal for a developmental event (see **Fig. 10.29**).

10.7 SIGNAL TRANSDUCTION AND THE CYTOSKELETON

Other than the signalling pathways that regulated changes in metabolism, gene expression and cell behaviour in response to different stimuli, now it is clear that the functions of most cells are affected directly by cell adhesion and the organization of the cytoskeleton. The receptors responsible for cell adhesion thus act to initiate intracellular signalling pathways that regulate cell behaviour, including gene expression. The components of the cytoskeleton thus act as both receptors and targets in cell–cell signalling pathways, integrating cell shape and movement with other cellular response.

Signalling in development and differentiation

Understanding the molecular mechanisms that govern animal development is one of the major challenges of contemporary cell and molecular biology. Starting from only a single cell, the fertilized egg, all the diverse cell types of the body are produced and organized into tissues and organs. Both cell differentiation and development of body structures are regulated by intricate pathways of cell–cell signalling that coordinate the activities of individual cells and ultimately give rise to organisms.

Convergence, divergence and crosstalk among signalling pathways

The signalling pathways are the major channels of information flow. They are connected to one another.

1. Signals from a variety of unrelated receptors, each binding to its own ligand, can ***converge*** (**Fig. 10.30**) to activate a common effector (e.g., Ras or Raf).
2. Signals from the same ligand can ***diverge*** (**Fig. 10.31**) to activate a variety of different effectors, leading to diverse cellular responses (e.g., EGF or insulin).
3. Signals can be passed back and forth between different pathways, a phenomenon known as ***crosstalk*** (e.g., cAMP).

Crosstalk can be seen in numerous tissues, wherein two or more signalling pathways reinforce each other. In some cell types, initiation of gene expression requires two signals. Lymphocyte differentiation (**Fig. 10.32**), for example, requires two signalling genes, each one producing one of the two peptides of a transcription factor needed for

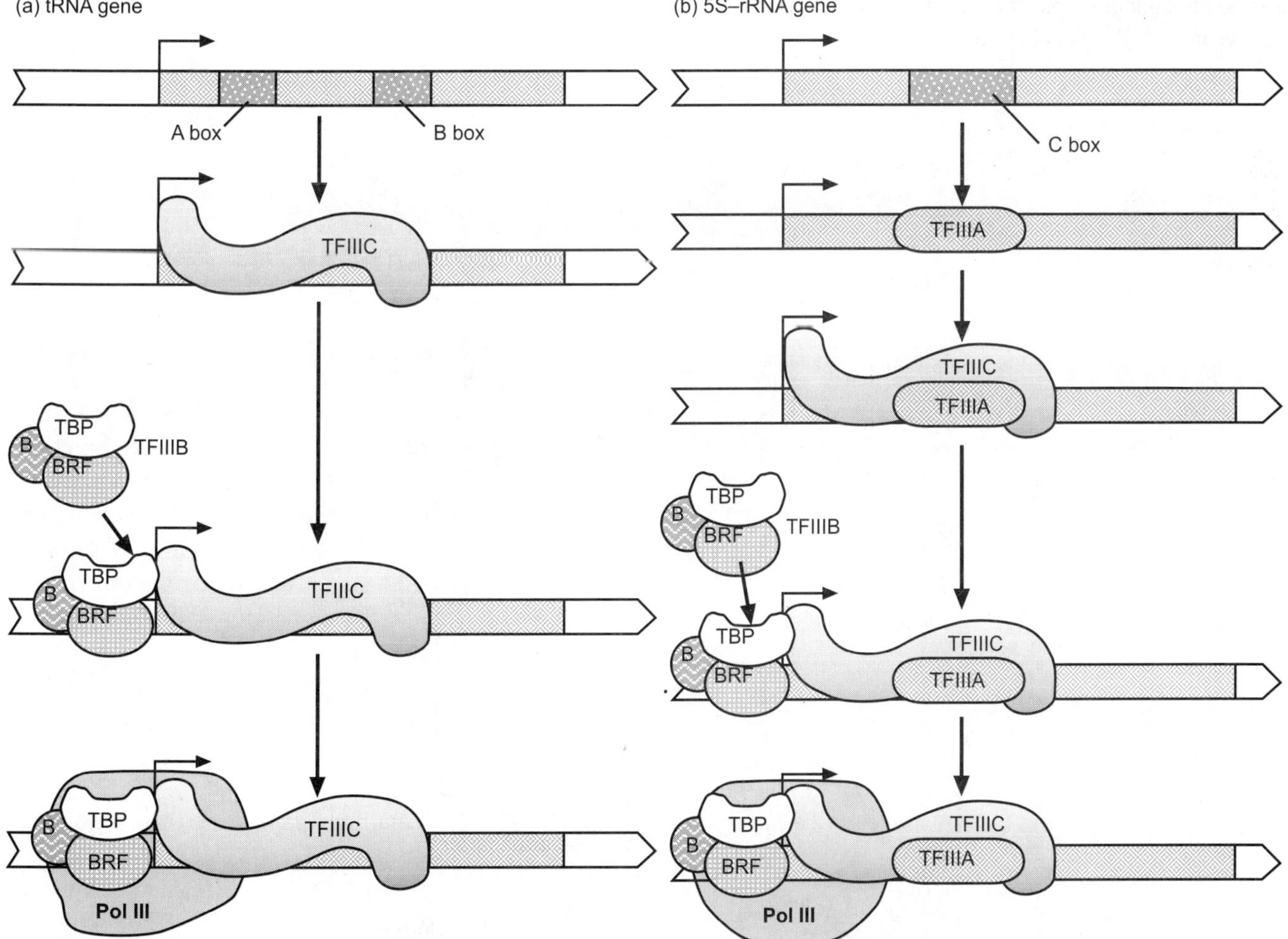

Fig. 10.30 Signals transmitted from a G protein-coupled receptor, an integrin and a receptor tyrosine kinase all converge on Ras and are then transmitted along the MAP kinase cascade.

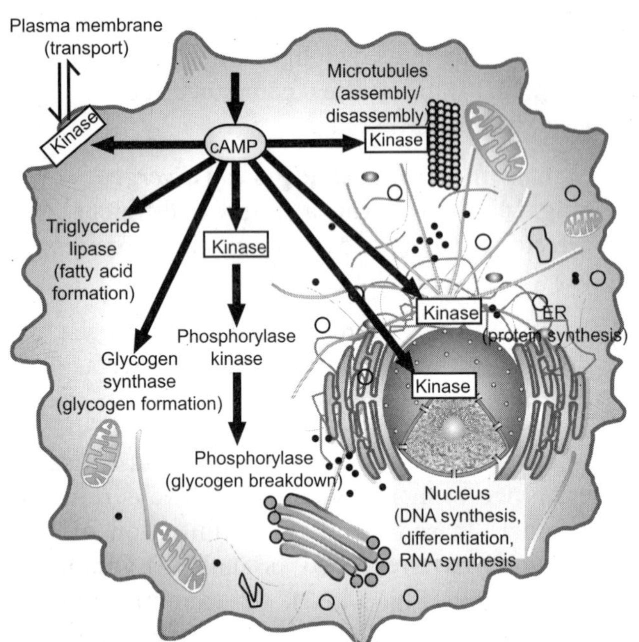

Fig. 10.31 Schematic illustration of the variety of processes (divergent signalling pathways) that can be affected by changes in cAMP concentration.

the production of interleukin 2 (*IL-2*). One peptide, *c-Fos*, is required by the binding of the T cell receptor to an antigen. This signal activates the *Ras* pathways, creating a transcription factor, *Elk-1*, that activates the *c-fos* gene to synthesize *c-Fos*. The second signal comes from the B7 glycoprotein on the surface of the cell presenting the antigen. This signal activates a second cascade of kinases, eventually producing c-Jun. The two peptides, c-Fos and c-Jun, can join to make the AP-1 protein, a transcription factor that binds to the IL-2 enhancer and activates its expression.

There are many examples to show that one receptor can activate several different pathways. The fibroblast growth factors, for instance, can activate the RTK pathways, the STAT pathway, or even a third pathway that involves lipid turnover and increases the levels of calcium ions in the cell.

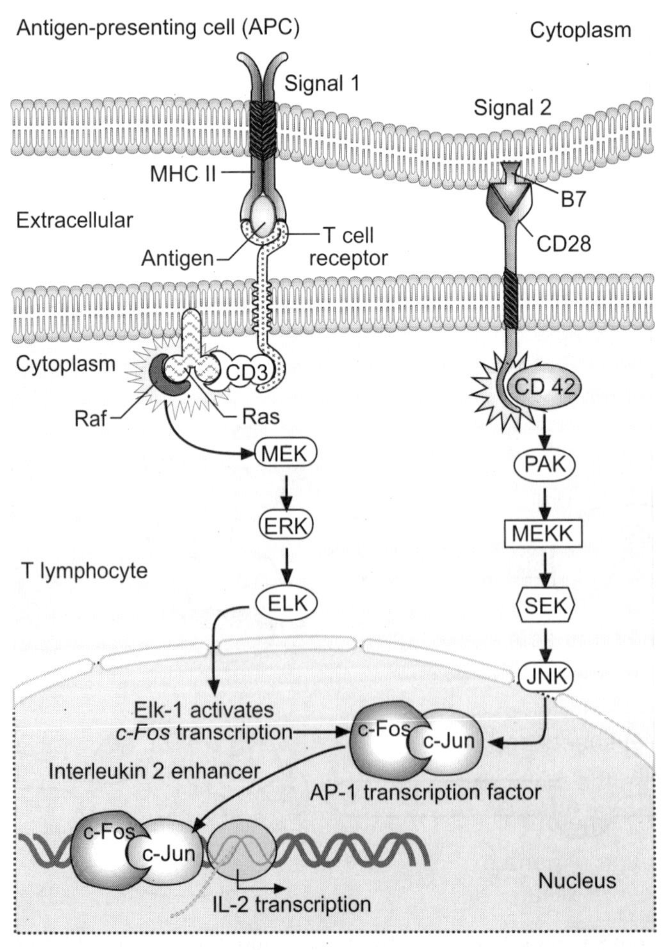

Fig. 10.32 Involvement of two signals to effect the differentiation of the T lymphocyte.

Summary

- All living cells receive and respond to various kinds of signals from their surroundings. These communication mechanisms depend heavily on ***signal molecules***, which are produced by cells to signal to their neighbours or to cells further away.
- ***Cell signalling*** brings together several major biochemical concepts. It allows the organism to function as a coherent system.
- ***Signal transduction*** involves an ordered sequence of biochemical reactions inside the cell, which are carried out by enzymes and linked through ***second messengers*** as a 'second-messenger pathway'.
- Signalling molecules include proteins, small peptides, amino acids, nucleotides, steroids, retinoids, fatty acid derivatives and even dissolved gases such as nitric oxide and carbon monoxide and many other physical signals.
- Extracellular signal molecules can act over either short or long distances in different ways: ***autocrine, paracrine, endocrine*** and ***juxtacrine*** interactions.
- Coordinative activities of neighbouring cells also take place through ***gap junctions***. The cellular response to a particular signalling molecule depends on its binding to a specific receptor protein located on a target cell or in its nucleus or cytosol.
- A group of cell surface receptors, which transmit signals to intracellular targets via the intermediary action of guanine nucleotide-binding proteins, is called ***G protein***.
- The ***G protein-coupled receptors*** (***GPCRs***) constitute a very large superfamily of receptors and examples of the receptor type are found universally in most organisms.
- ***Ion channels*** are basically water-soluble pores in the cell membranes. They regulate the movement of ions and other small molecules in the cell.
- Other important membrane receptors are the ***protein-tyrosine kinases, MAP kinase (mitogen-activated protein kinase)***—a serine/threonine kinase activated by many growth factors and the *pathway JAK/STAT*, that connects protein-tyrosine kinases and transcription factors.
- The binding of ligands to many cell surface receptors leads to a diverse range of signals and requires a huge repertoire of ***receptors***.
- Early signalling pathways often converge into a small number of intracellular signalling networks that allow the cell to convert the complex information arriving at the cell surface into simple biochemical signals in the cytoplasm.
- The level of receptors for a particular hormone can be increased (up-regulation) or decreased (downregulation), permitting cells to respond optimally to small changes in the hormone level.
- ***Endocytosis*** is the principal mechanism for down-regulating the receptors for many peptide hormones including insulin, glucagon, EGF and PDGF.
- The receptor-hormone complex is brought into the cell by ***receptor-mediated endocytosis***. The internalized hormone is subsequently degraded in ***lysosomes***. Even if the hormone receptors recycle to the cell surface by exocytosis, as they do in many cases a substantial function will be in the internal membrane compartments.
- The ***extracellular matrix*** (***ECM***), a proteinaceous cytoskeletal structure, consists of macromolecules secreted by cells into their immediate environment. These macromolecules form a region of non-cellular material in the interstices between the cells.
- The ECM is a critical region for animal development. Cellular functions such as adhesion, cell migration and the formation of epithelial sheets and tubes all depend on the ability of cells to attach ECM.
- ***Signalling pathways*** are the major channels of information flow. They are connected to one another.
- Crosstalk takes place in numerous tissues, wherein two or more signalling pathways reinforce each other.
- In some cell types, initiation of gene expression requires two signals. There many examples where one receptor can activate several different pathways.

Exercises

1. Briefly explain cellular communication by chemical signals.
2. Describe briefly the different types of metabotropic cell surface receptors.
3. Describe briefly the role of transmitters and receptors in cellular signalling.
4. What are the different modes of cell-to-cell communication?
5. Discuss briefly the peptide hormones and growth factors.
6. Write the mechanism of signal generation.
7. Give a brief account of cell surface receptors.
8. Explain with examples the G protein-coupled receptors.
9. Draw and explain the ion channel-linked receptors.
10. Explain the different types of intracellular signal transduction.
11. How does cAMP act as a second messenger in regulation of cellular metabolism?
12. Why is insulin so important in sugar metabolism?
13. Explain how extracellular matrix play a role in signal transduction.
14. In what way do cytoskeletal filaments play a role in signal transduction in development?

Further Reading

Books

Alberts B, Bray D, Lewis J, Raff M, Roberts K and Watson JD (2002), *Molecular Biology of the Cell*, 4th edition, Garland Science, New York.

Elliott WH and Elliott DC (2001), *Biochemistry and Molecular Biology*, Oxford University Press.

Gilbert SF (2000), *Developmental Biology*, 6th edition. Sinauer Associates, Inc.

Nelson DL and Cox MM (2000), *Lehninger Principles of Biochemistry,* 3rd edition, WH Freeman.

Lodish H, Baltimore D and Berr A (1995), *Molecular Cell Biology*, WH Freeman.

Twyman RM (1998), *Advanced Molecular Biology*, Bios Scientific Publishers, Oxford; Viva Books Private Limited. New Delhi.

References

Aaronson DS and Horvath CM (2002), A road map for those who don't know JAK–STAT, *Science,* 296:1653–55.

Berridge M (1993), Inositol triphosphate and calcium signalling, *Nature*, 361:315.

Berridge MA, Bootman MD and Lipp P (1998), Calcium—a life and death signal, *Nature*, 395:645–47.

Bootman MD and Berridge MJ (1995), The elemental principles of calcium signalling, *Cell*, 83:675–78.

Burgering BMT and Bos JL (1995), Regulation of *Ras*-mediated signalling: More than one way to skin a cat, *Trends Biochem. Sci.,* 22:18–22.

Cantley LC (2002), The phosphoinositide 3-kinase pathway, *Science,* 296:1655–57.

Clapham DE (1995), Calcium signalling, *Cell,* 80:259–268.

Darnell JE Jr, Kerr IM and Stark GR (1994), Jak–STAT pathways a transcriptional activation in response to IFNs and other extracellular signalling proteins, *Science*, 264:1415.

Elion EA (1998), Routing MAP kinase cascades, *Science*, 281:1625–26.

Fantl WJ, Johnson DE and Williams LT (1993), Signalling by receptor tyrosine kinases, *Annu. Rev. Biochem*., 62:453.

Heldin CH (1995), Dimerization of cell surface receptors in signal transduction, *Cell*, 80:213–23.

Hoey T and Schindler U (1998), STAT structure and function in signalling, *Curr. Opin. Genetics Dev*., 8:582–87.

Hunter T (2000), Signalling—2000 and beyond, *Cell*, 100:113–27.

Hynes RO (2002), Integrins: bi-directional, alosteric signalling machines, *Cell*, 110:673–87.

Ihle JN (1996), STATs and MAPKs: Obligate or opportunistic partners in signalling, *BioEssays*, 18:95–8.

Johnson GL and Lapadat R (2002), Mitogen-activated protein kinase pathways mediated by ERK, JNK and p38 proein kinases, *Science,* 298:1911–12.

Karin M and Hunter T (1995), Transcriptional control by protein phosphorylation: Signal transmission from the cell surface to the nucleus, *Curr. Biol*., 5:747–57.

McCormick F (1999), Signalling networks that cause cancer, *Trends Cell Biol*., 9:M53–M56.

Miranti CK and Brugge JS (2002), Sensing the environment: a historical perspective on integrin signal transduction, *Nature Cell Biol*., 4:E83–E90.

Monteith GR and Roufogalis BD (1995), The plasma membrane calcium pump—a physiological perspective on its regulation, *Cell Calcium*, 18:459–70.

Neer EJ (1997), Intracellular signalling: Turning down G protein signals, *Curr. Biol*., 7:R31–3.

Neves SR, Ram P and Iyengar R (2002), G protein pathways, *Science*, 296:1636–39.

Nishizuka Y (1994), Protein kinase C and lipid signalling for sustained cellular responses, *FASEB J*., 9:484–96.

Pawson T (1994), Look at a tyrosine kinase, *Nature*, 372:726–27.

Pitcher JA, Freedman NJ and Lefkowitz RJ (1998), G protein-coupled receptor kinases, *Annu. Rev. Biochem.*, 67:653–92.

Scott JD and Pawson T (2000), Cell communication, The inside story, *Sci. Am.*, 282(6):7279.

Sheen J (2002), Phosphorelay and transcriptional control in cytokinin signal transduction, *Science,* 296:1650–52.

Stark GR Kerr IM, Williams BRG, Silverman RH and Schrieber RD (1998), How cells respond to interferons, *Annu. Rev. Biochem.*, 67:227–64.

Struhl K (1998), Fundamentally different logic of gene regulation in eukaryotes and prokaryotes, *Cell,* 98:1–4.

Wedel B and Garbers D (2001), The guanylyl cyclase family at Y2K, *Annu. Rev. Physiol.*, 63:215–33.

Whitmarsh AJ and Davis RJ (1998), Structural organization of MAP kinase signalling modules by scaffold proteins in yeast and mammals, *Trends Biochem. Sci.,* 23:481–85.

Krugman S and Welch H (1998), PI 3-kinase, *Curr. Biol.*, 8:R828.

Williams JG (1999), Serpentine receptors and STAT activation: More than one way to twin a STAT, *Trends Biochem. Sci.,* 24:333–34.

11 SENSORY SIGNALLING

Learning objectives

After reading this chapter, you will be able to:

- Appreciate the key features of the nervous system
- Understand the process of sensory signalling in the various organs
- Appreciate various aspects of memory functions

11.1 INTRODUCTION

The most powerful computers available today are mere playthings compared to the complexity, efficiency and information processing capacity of the human brain. Underlying the brain's far superior design are the billions of connections between the brain cells called **synapses** that form vast neural networks in which brain cells, or **neurons**, are each connected to thousands of other neurons. These networks and their ability to be shaped by experience enable us to receive, process, store and retrieve all manner of information about the world. This activity goes on continuously until the end of life. However, the main obstade in understanding the neural intricacy fully is the extremely tiny size of the synapses and the limitations of conventional experimental techniques that have hampered detailed studies of these essential structures. Neural connections are so tiny and dense, that more than a trillion synaptic compartments, or **dendritic spines**, could fit into a thimble.

The **peripheral nervous system** (PNS) consists of all parts of the nervous system, except the brain and spinal cord, which are the components of the **central nervous system** (CNS). The PNS connects the CNS to the rest of the body and is the conduit through which neural signals are transmitted to and from the central nervous system. Within the PNS, sensory neurons transmit impulses to the CNS from sensory receptors. The motor neurons are an oppositely oriented system, that transmit neural signals from the CNS to effectors (glands, organs and muscles).

Neurons are highly differentiated cells of the nervous system. They communicate between sense organs and the brain, as well as between the brain and muscles and among the neurons themselves. Thus, the nervous system does much more than transfer information: it can also process information and execute action.

11.2 NERVOUS SYSTEM

11.2.1 Neurons

All organisms respond to external stimulation. Nerve cells (or neurons) are specialized for collection, conduction and transmission of information which is coded in the form of fast-moving electrical impulses. The basic parts of a typical neuron are illustrated in **Fig. 11.1**. The nucleus of the neuron is located within an expanded region called the **cell body**, which is the metabolic centre of the cell and the site where most of the metabolites are synthesized. Extending from the cell bodies of most neurons are a number of fine extensions, called **dendrites**, which receive incoming information from external sources, typically from other neurons. Also emerging from the cell body is a single, more prominent extension, the **axon**, which conducts outgoing impulses away from the cell body and toward the target cell(s). While some axons are only a few micrometres long,

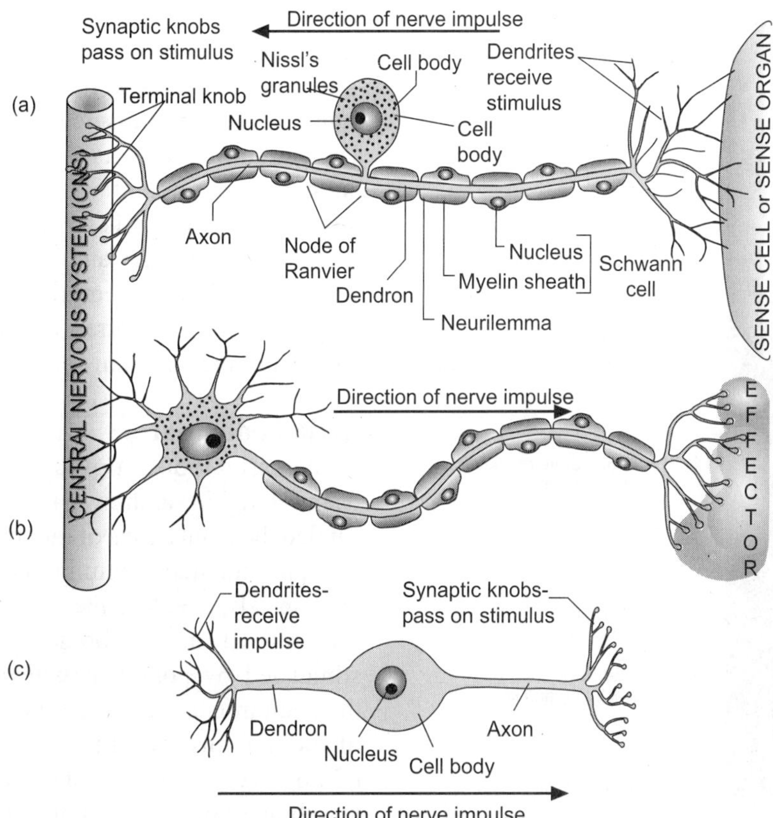

Fig. 11.1 (a) Sensory neuron, (b) motor neuron, (c) interneuron (**See Plate 5**).

others extend to several metres as in large vertebrates, such as giraffes or whales. Most axons split near their ends into smaller processes, each ending in a terminal knob—a specialized site where impulses are transmitted from neuron to the target cell. Some neurons in the brain may end in thousands of terminal knobs, allowing these brain cells to communicate with thousands of potential targets. Most neurons in vertebrates are wrapped in a lipid-rich **myelin sheath**.

Nerves are bundles of axons. Cell bodies are found only in clusters called ganglia or within the CNS of vertebrates (**Fig. 11.2**). Many neurons are found fully within the CNS. Others hang out of the CNS into the body, their axons forming parts of nerves. Most contain a mixture of cells, which bring information from receptor cells to the CNS (sensory neurons) and motor neurons, which carry signals out to effector organs, notably muscles.

Nerve impulses

Neurons carry information much faster and more precisely than hormones. The high amount of energy and materials needed to build and maintain neurons justify that rapid cell communication has great survival benefits. Information travels along neurons as a stream of pulses. Each pulse consists of a cycle of changes in the axon membrane. These changes involve the opening and closing of ion channels and the movement of sodium and potassium ions. The electrical gradient across the membrane breaks down for a few milliseconds before being restored. This depolarization triggers depolarization of the next piece of membrane and so the impulse moves down the axon.

Resting potential

An electric potential difference arises between two points, such as the inside and outside of the plasma membrane, when there is an excess of positive ions at one point and an excess of negative ions at the other point. Voltage across plasma membranes can be measured by inserting a fine micro-electrode into the cytoplasm of a cell, placing another electrode in the extracellular fluid outside the cell and connecting the electrodes to a voltmeter (**Fig. 11.3**). When this experiment was first carried out on a giant axon of the squid, a potential difference of approximately 70 mV was recorded, the inside being negative with respect to the outside, indicated by a negative (–70 mV).

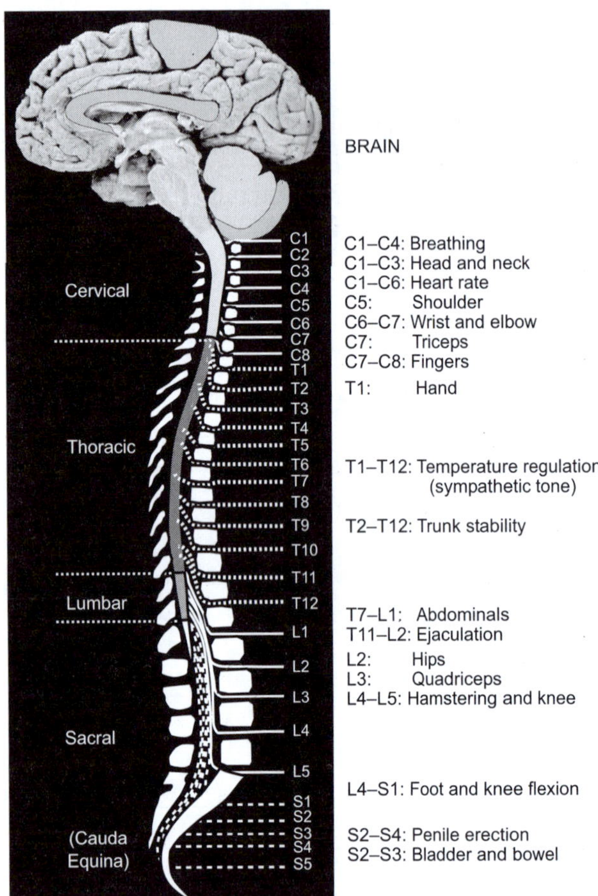

Fig. 11.2 Human central nervous system (CNS).

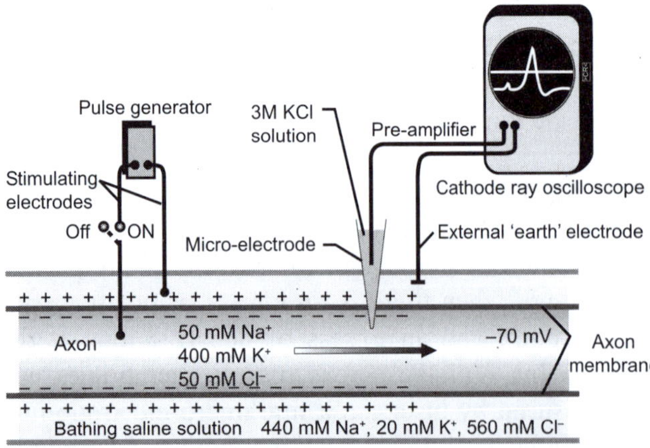

Fig. 11.3 The basic apparatus used to record electrical activity in the axon of an isolated neurone. The stimulator produces a current which generates an action potential in the axon, which is detected and recorded using a micro-electrode inserted into the axon. The signal is sent to a dual-beam cathode ray oscilloscope.

The membrane potential is present in all types of cells, the magnitude varying between –15 mV and –100 mV. For non-excitable cells, that is, those other than neurons and muscle cells, this voltage is termed as the **membrane potential**. In a nerve or muscle cell, the membrane potential is referred to as the **resting potential** because it is subject to dramatic change.

The magnitude and direction of the voltage across the plasma membrane are determined by the differences in concentrations of ions on either side of the membrane and their relative permeabilities.

Action potential

In the early 1950s, a group of physiologists (notably Alan Hodgkin andrew Huxley and Bernard Katz) studied the giant axons of the squid. These axons, which are approximately 1 mm in diameter, carry impulses at high speeds, enabling the squid to escape rapidly from predators. If the membrane of a resting squid axon is stimulated by poking it with a fine needle, some of the sodium channels open, allowing a limited number of sodium ions to diffuse into the cell. The movement of positively charged ions into the cell reduces the membrane potential, making it less negative. Because the positive change in membrane voltage causes a decrease in the polarity between the two sides of the membrane, it is called **depolarization**.

If the stimulus causes the membrane to depolarize by only a few mV, say from –70 mV to –60 mV, the membrane rapidly returns to its resting potential as soon as the stimulus has ceased. If however, the stimulus depolarizes the membrane beyond a certain point, called the **threshold**, which occurs at about –50 mV, then a new series of events is launched. The change in voltage causes the voltage-gated sodium channels to open. As a result, sodium ions diffuse freely into the cell down both their concentration and electric gradients. The increased permeability of the membrane to Na^+ ions and the corresponding movement of positive charge into the cell causes the membrane to reverse potential briefly, becoming positive at about +40 mV, which approaches the equilibrium potential for Na^+.

After approximately 1 ms, the sodium channels spontaneously deactivate, blocking further influx of Na^+ ions. Meanwhile, the change in membrane potential caused by Na^+ influx triggers the opening of the voltage-gated potassium channels. As a result, potassium ions diffuse freely out of the cell down their steep concentration gradient. The decreased permeability of the membrane to Na^+ and increased permeability to K^+ causes the membrane potential to swing back to a negative value of about –80 mV, approaching that of the K^+ equilibrium potential. The large negative membrane potential makes the voltage-gated potassium channels to close, which returns the membrane to its resting state. Collectively, these changes in the membrane potential are called action potential. The entire series of changes during an action potential takes only about 5 ms in the squid axon and less than 1 ms in myelinated mammalian nerve cell. Because the Na^+ channels cannot reopen for several milliseconds after their deactivation, the membrane enters a brief refractory period following an action potential in which it cannot be re-stimulated.

Although the action potential changes membrane voltage dramatically, only a minute percentage of the ions on the two sides of the membrane are involved in any given action potential. The striking changes in membrane potential are not caused by changes in Na^+ and K^+ ion concentrations on the two sides of the membrane (which are insignificant); they are caused by the movement of charge in one direction or the other that result from the fleeting changes in permeability to these ions. Those Na^+ and K^+ ions that do change places across the membrane during an action potential are eventually pumped back by the N^+/K^+ATPase.

Once the membrane of a neuron is depolarized to the threshold value, a full-blown action potential is triggered without further stimulation. This feature of nerve cell function is known as the **all-or-none law**, but there is no in-between. An action potential is not an energy requiring process, but it results from the flow of ions down their respective electrochemical gradients. The movement of ions across the plasma membrane of nerve cells is the basis for neuronal communication. Certain local anesthetics, such as procaine and novocaine, act by closing ion channels in the membranes of sensory cells and neurons. As long as these ion channels remain closed, the affected cells are unable to generate action potentials and thus unable to inform the brain of events occurring at the skin or teeth.

Propagation of action potential as an impulse

Once an action potential has been initiated, it does not remain localized at a particular site but is propagated as a **nerve impulse** down the length of the cell to the nerve terminals. Nerve impulses are propagated along a membrane because an action potential at one site has an effect on the adjacent site. The large depolarization that accompanies an action potential creates a difference in charge along the inner and outer surfaces of the plasma membrane (**Fig. 11.4**). This local flow of current causes the membrane in the region just ahead of the action potential to become depolarized. Because the depolarization accompanying the action potential is very large, the membrane in the adjacent region is readily depolarized to a level greater than the threshold value, which opens the sodium channels in this adjacent region, generating another action potential. Thus, once triggered, a succession of action potentials passes down the entire length of the neuron without any loss of intensity, arriving at its target cell with the same strength it had at its point of origin.

Because all impulses travelling along a neuron exhibit the same strength, stronger stimuli cannot produce 'bigger' impulses than weaker stimuli. Yet, we are clearly capable of detecting differences in the strength of a stimulus. The ability to make

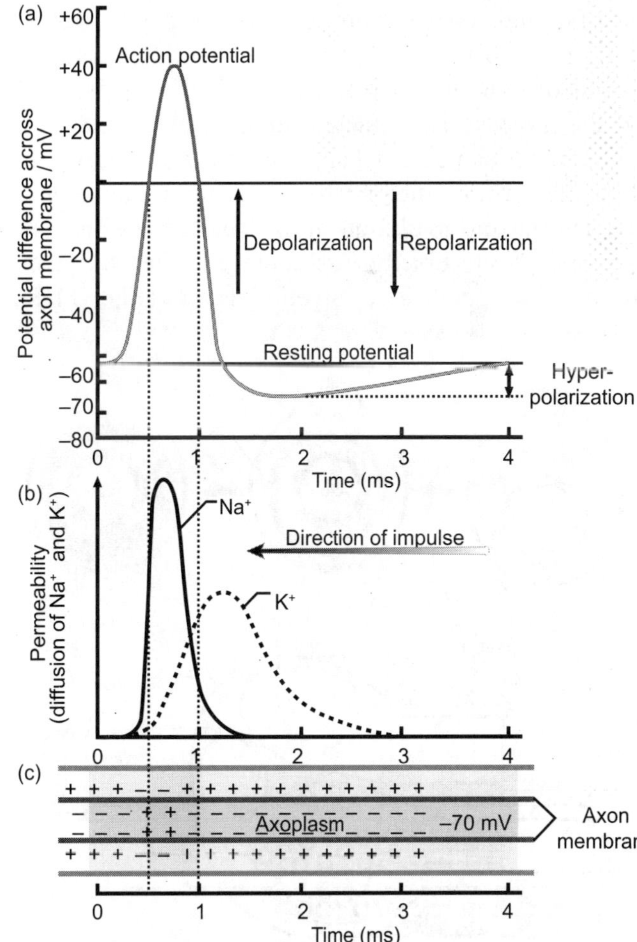

Fig. 11.4 Typical action potential in the axon and changes in the axon during the propagation of a nerve impulse: (a) the membrane potential, showing the electrical events associated with the nerve impulse; (b) change in permeability to ions; (c) the net charge across the axon membrane during production of an action potential.

sensory discriminations depends on several factors. For example, a stronger stimulus activates more nerve cells than does a weaker stimulus. It also activates 'high-threshold' neurons that would remain at rest if the stimulus were weaker. Stimulus strength is also encoded in the pattern and frequency by which action potentials are launched down a particular neuron. In most cases, the stronger the stimulus, the greater the number of impulses generated.

Speed of neuronal transmission

The properties of the axon have a major effect on how rapidly it can pass along a single pulse. The greater the diameter of an axon, lesser the resistance to local current flow and more rapidly an action potential at one site can activate adjacent regions of the membrane. Some invertebrates, such as squid and tubeworms, have evolved giant axons that facilitate escape from danger. There is, however, a limit to this evolutionary approach. Because the speed of conduction increases with the square root of the increase in diameter, an axon that is 480 μm in diameter can conduct an action potential only four times faster than the one that is 30 μm in diameter.

During the evolution of vertebrates, an increase in conduction velocity was achieved when the axon became wrapped in a myelin sheath (**Fig. 11.5**).

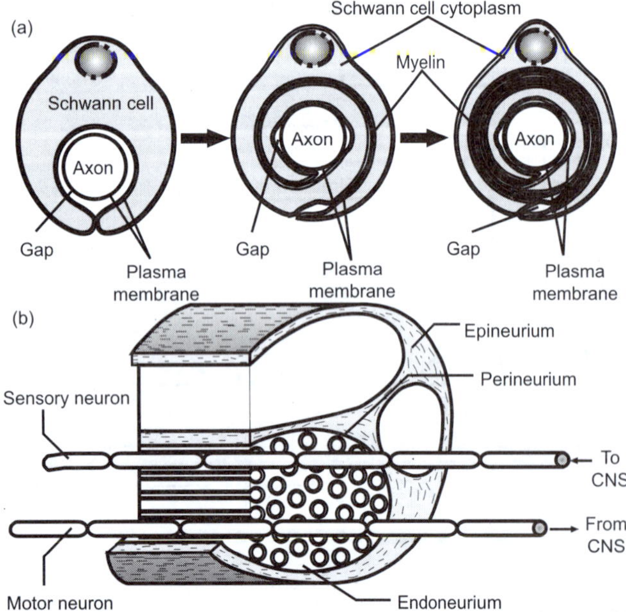

Fig. 11.5 (a) TS myelinated nerve fibre. The different stages in myelination of an axon is shown. The myelin membranes are usually low in protein content and correspondingly high in lipids. Thus, they make excellent insulators. (b) A section of the myelinated nerve.

Because it is composed of many layers of lipid-containing membranes, the myelin sheath is ideally suited to prevent the passage of ions across the plasma membrane. In addition, nearly all the Na^+ ion channels of a myelinated neuron reside in the unwrapped gaps, or **nodes of Ranvier**, between adjacent Schwann cells or oligodendrocytes that make up the sheath. Consequently, the nodes of Ranvier are the only sites where action potentials can be generated. An action potential at one node triggers an action potential at the next node, causing the impulse to jump from node to node without having to activate the intervening membrane. Propagation of an impulse by this mechanism is called **saltatory conduction** (**Fig. 11.6**). Impulses are conducted along a myelinated axon at speeds up to 120 metres per second, which is more than 20 times faster than the speed of impulses that travel in an unmeylinated neuron of the same diameter.

The importance of myelination is dramatically illustrated by **multiple sclerosis** (MS), a disease associated with deterioration of the myelin sheath that surrounds axons in various parts of the nervous system. Consequently, conduction of action potentials is slowed, with an accompanied loss of motor coordination. Manifestations of the disease usually begin in young adulthood; patients experience weakness in their hands, difficulty in walking and problems with their vision.

Neurotransmission

Neurons are linked with their target cells at specialized junctions called synapses. It is a junction

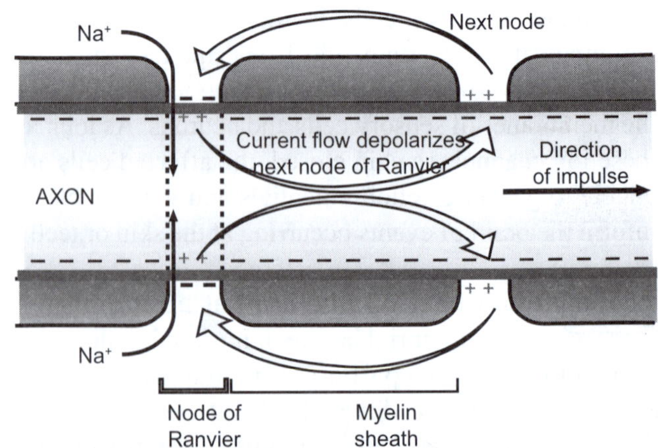

Fig. 11.6 Saltatory conduction. During this type of conduction, only the membrane in the nodal region of the axon becomes depolarized and capable of forming an action potential. This is accomplished as current flows directly from an activated node to the next resting node along the axon.

between an axon of one neuron and the cell membrane of another. Careful examination of a synapse reveals that the two cells do not make direct contact but are separated by a narrow gap of about 20–50 nm. This gap is called the **synaptic cleft**. A **pre-synaptic cell** (a sensory cell or a neuron) conducts impulses toward a synapse and a **post-synaptic cell** (a neuron, muscle or gland cell) always lies on the receiving side of a synapse. **Figure 11.7** shows a number of synapses between the terminal branches of an axon and a skeletal muscle cell; synapses of this type are called **neuromuscular junctions**.

How does an impulse in a pre-synaptic neuron jump across the synaptic cleft and affect the post-synaptic cell? Studies indicate that a chemical substance is involved in the transmission of an impulse from one cell to another. In the electron microscope image, the tips (terminal knobs) of the branches of an axon appear to contain large numbers of **synaptic vesicles** (**Fig. 11.8**) that serve as storage sites for the chemical which transmits (**neurotransmitters**) impulses to the body's skeletal and cardiac muscles. **Neurotransmitters** are chemical substances, which travel across the space, or synaptic cleft, from the pre-synaptic neuron and are received by receptors on the opposite cell membrane, or post-synaptic membrane of a target cell.

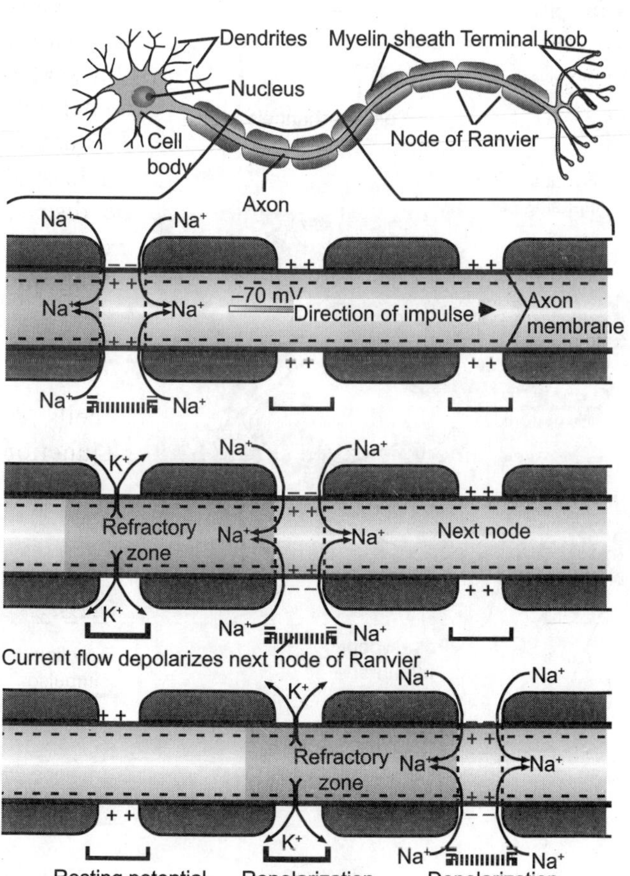

Fig. 11.7 Propagation of an impulse results from the local flow of ions. An action potential at one site on the membrane depolarizes an adjacent region of the membrane, triggering an action potential at the second site. The action potential flows only in the forward direction since the portion of the membrane that has just experienced an action potential remains in a refractory period.

During synaptic transmission (**Fig. 11.9**), when an impulse reaches a terminal knob (step 1), the accompanying depolarization induces the opening of a number of voltage-gated Ca^{2+} channels in the plasma membrane of this part of the pre-synaptic nerve cell (step 2). Calcium ions are normally present at very low concentration within the neuron (about 100 nM), as in all cells. When the gates open, calcium ions diffuse from the extracellular fluid into the terminal knob of the neuron, causing the Ca^{2+} to rise more than a thousand-fold within localized microdomains near the channels. The elevated Ca^{2+} triggers the rapid fusion of one or a few nearby synaptic vesicles with the plasma membrane, causing the release of neurotransmitter molecules into the synaptic cleft (step 3). The neurotransmitter molecules diffuse across the narrow gap and bind selectively to receptor molecules that are concentrated directly across the cleft in the post-synaptic plasma membrane (step 4). A neurotransmitter molecule can have one of the two opposite effects depending on the type of receptor on the target cell membrane to which it binds:

1. The bound transmitter can trigger the opening of cation-selective channels in the membrane, leading primarily to an influx of sodium ions and a less negative (more positive) membrane potential. This depolarization of the post-synaptic membrane excites the cell, making it more likely to respond to this or subsequent stimuli by generating an action potential of its own (steps 5a and 6).
2. The bound transmitter can trigger the opening of anion-selective channels, leading mainly to an influx of chloride ions and a more negative (hyperpolarized) membrane potential. Hyperpolarization of the post-synaptic membrane makes it less likely that the cell will generate an action potential because greater sodium influx is subsequently required to reach the membrane's threshold (step 5b).

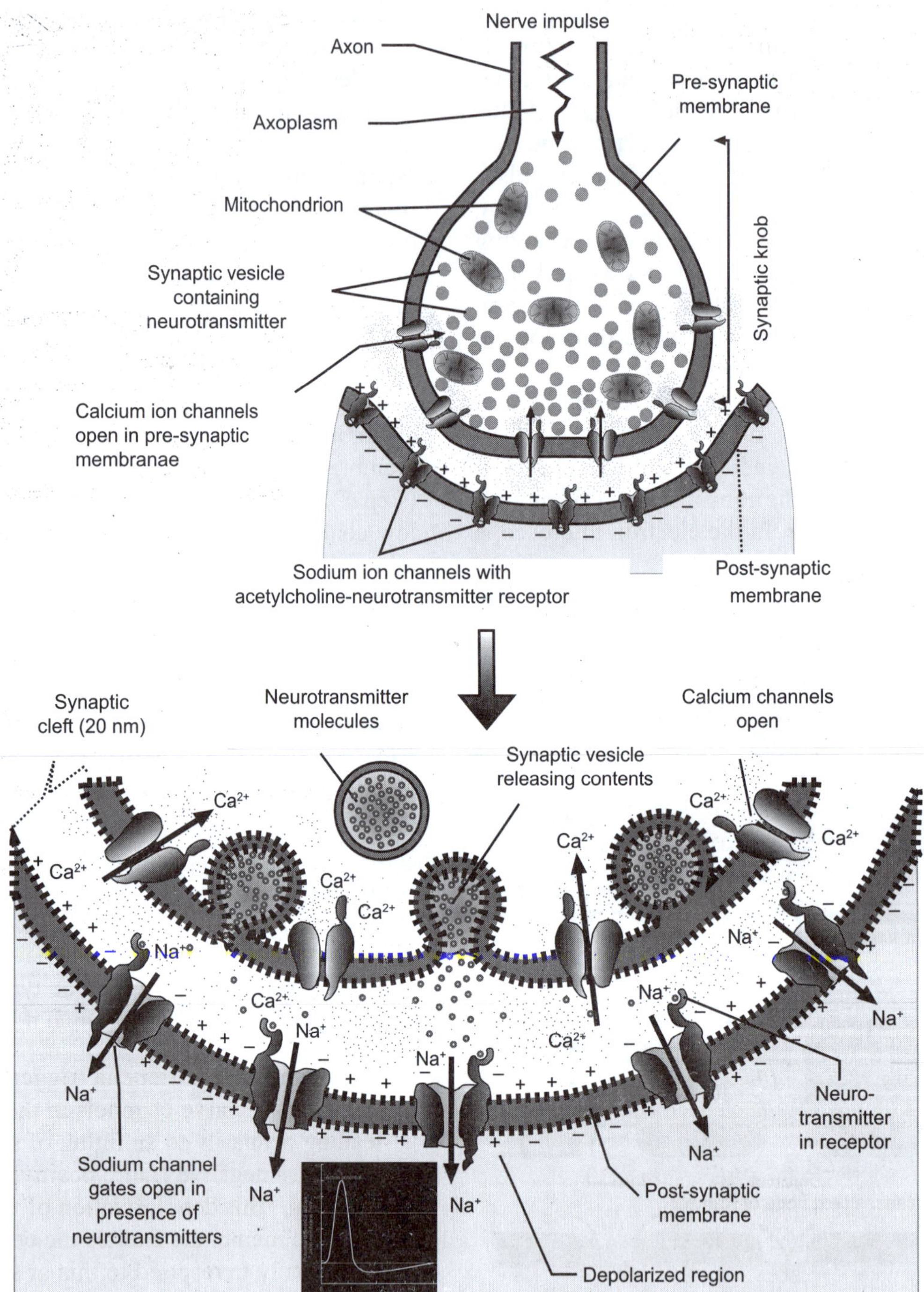

Fig. 11.8 Structure of a synapse.

Most nerve cells in the brain receive both excitatory and inhibitory signals from many different pre-synaptic neurons. It is the summation of these opposing influences that determine whether or not an impulse will be generated in the post-synaptic neuron. All terminal knobs of a given neurotransmitter may have a stimulatory effect on another. Acetylcholine (ACh), for example, inhibits contractility of the heart but stimulates contractility of skeletal muscle. Within the brain, glutamate serves as the primary excitatory neurotransmitter and gamma-aminobutyric acid (GABA) as the primary inhibitory neurotransmitter. A number of general anesthetics, as well as valium and its derivatives, act by binding to the GABA receptor and enhancing the activity of the brain's primary 'off' switch.

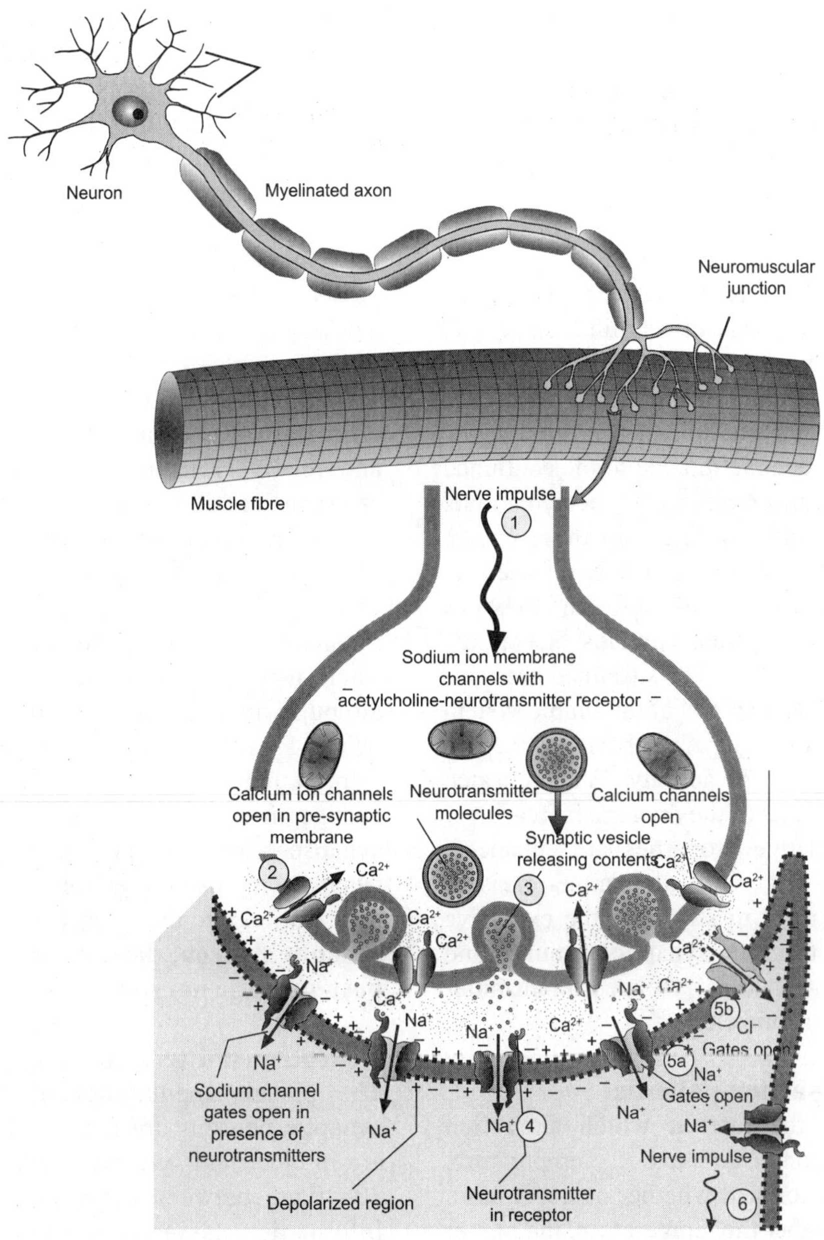

Fig. 11.9 Structure of the vertebrate neuromuscular junction. Communication between nerve and muscle cell is by release of the neurotransmitter acetylcholine from synaptic vesicles into the synaptic cleft. Acetylcholine diffuses across the cleft and binds to acetylcholine receptors on the muscle cell membrane (**See Plate 6**).

Action of drugs on synapses

A number of extremely toxic substances have their effect by blocking the action of the specific ion gates necessary for development of the action potential. These substances are often called **neurotoxins. Tetradotoxins** are found in some organs of the puffer fish. They bind specifically to the Na^+ channel, blocking all ion movement. The same effect is produced by **saxitoxin**, contained in the marine dinoflagellates responsible for 'red tide'. Another very poisonous substance, **veratridine**, is found in the seeds of a plant of the lily family, *Schoenocaulon officinale*. This toxin also binds to the Na^+ channels but blocks them in the 'open' configuration. These toxins have proved to be useful in studies of axonal structure and conduction.

It is important that a neurotransmitter has only a short half-life following its release from a pre-synaptic neuron; otherwise the effect of the neurotransmitter would be extended and the post-synaptic neuron would not recover. A neurotransmitter can be eliminated from the synapse in two ways:

by enzymes that destroy neurotransmitter molecules in the synaptic cleft and by proteins that transport neurotransmitter molecules back to the pre-synaptic terminals—a process known as **re-uptake**. Because of the destruction or re-uptake of neurotransmitter molecules, the effect of each impulse lasts no more than a few milliseconds.

Interfering with the destruction or re-uptake of neurotransmitters can have dramatic physiological and behavioural effects. Acetylcholinesterase, an enzyme localized within the synaptic cleft, hydrolyzes acetylcholine. If this enzyme is inhibited by exposure to the nerve gas DFP, for example, the muscles of the body undergo violent contraction due to the continued presence of high concentration of acetylcholine. Cocaine, on the other hand, interferes with the re-uptake of the neurotransmitter dopamine that is released by certain nerve cells in a portion of the brain known as the limbic system. This system contains the brain's 'pleasure' or 'reward' centres. The sustained presence of dopamine in the synaptic clefts of the limbic system produces a short lived feeling of euphoria, as well as a strong desire to repeat the activity. With repeated use, the pleasurable effects of the drug are increasingly reduced, but its addictive properties are enhanced. Amphetamines also act on dopamine-releasing neurons; they are thought to stimulate the excessive release of dopamine from pre-synaptic terminals and interfere with the re-uptake of the neurotransmitter molecules from the synaptic cleft.

Factors influencing action potential

Three things affect the rate at which an action potential moves along an axon: temperature, the thickness of the axon and whether or not it has a myelin sheath. The higher the temperature, the higher the speed of conduction, as most chemical processes respond to temperature. The thicker the axon, the faster the impulse travels, which is clearly seen in many invertebrates including insects and the squid, which have a few giant axons up to 1 mm in diameter—as much as 100 times thicker than other axons. These carry information fast: impulses travel around 30 m/s. In vertebrates, even without giant axons, a myelin sheath greatly speeds up conduction. Myelinated axons carry signals as quickly as invertebrate giant axons, but more efficiently.

Synapses and information processing

Each neuron has synapses with other neurons. In vertebrates, most are in the brain and spinal cord. These cells have an important role in information processing. Impulses move down the axon of a neuron. Their frequency and so their meaning, does not change as they travel. Individual neurons have to process incoming information and 'decide' at what frequency impulses should be sent off down their own axons. Information enters the neuron through its synapses. Incoming information is summed and passed to the axon hillock, a region which acts as the gateway to the axon and fires off action potentials on the basis of this information.

There are two quite different types of synapses: electrical and chemical. The former are quick, simple and relatively uncommon. The two neurons meet at a gap junction. The closeness of the membranes and the pore complexes, which allow free diffusion of ions mean that action potentials pass across unchanged. Electrical synapses are found in small numbers in both vertebrates and invertebrates. Their general significance is not well understood although their role in the 'tail flick' escape response of fish is well documented. Chemical synapses pass information slowly and allow the possibility of some change in the signal. Also, unlike many electrical synapses, information can only cross a chemical synapse in one direction. The neuron carrying information into the synapse is referred to as the **pre-synaptic neuron**; the one carrying it onwards is the **post-synaptic neuron**.

Neurotransmitters and synaptic signalling

The plasma membranes of the pre- and post-synaptic neurons are not physically linked, as they are in electrical synapses. Instead, there is a gap of 20–30 nm between them called the synaptic cleft. Information passes across the cleft in the form of diffusing molecules: neurotransmitters (Fig. 11.9). They are stored in vesicles close to the pre-synaptic membrane. They fuse with the membrane when action potentials arrive, releasing the neurotransmitter into the cleft. It is bound by receptors in the post-synaptic membrane, which leads to a response. There are not many possible responses, found in many parts of the CNS as well as where motor neurons meet skeletal muscle. Noradrenaline, used widely in the CNS as well as in the sympathetic nervous system; and glutamate, probably the most common excitatory neurotransmitter in the brain. Others usually inhibit depolarization. These include GABA and dopamine, both restricted to the CNS. It is not the transmitter itself that is inhibitory, but the combination of transmitter and receptor. So less commonly, noradrenaline may be inhibitory and dopamine excitatory.

What triggers the release of a neurotransmitter? The arrival of an action potential triggers a flow of Ca^{2+} into the end of the pre-synaptic neurone, due to voltage-gated calcium channels in the plasma membrane (**Fig. 11.10**). This acts as a second messenger, leading ultimately to vesicle fusion. Neurotransmitters work by binding to and opening ion channels in the post-synaptic membrane. Excitatory transmitters open sodium channels, leading to depolarization. Inhibitory transmitters open other channels which allow chloride ions to diffuse in, making the cytosol even more negative than before. Neurotransmitters need to be removed from the synaptic cleft. Otherwise, the post-synaptic membrane will keep responding indefinitely. Some neurotransmitters are destroyed by enzymes in the synaptic cleft, for example, acetylcholine by acetylcholinesterase. Others, like noradrenaline, are actively re-absorbed into the pre-synaptic terminal for re-use.

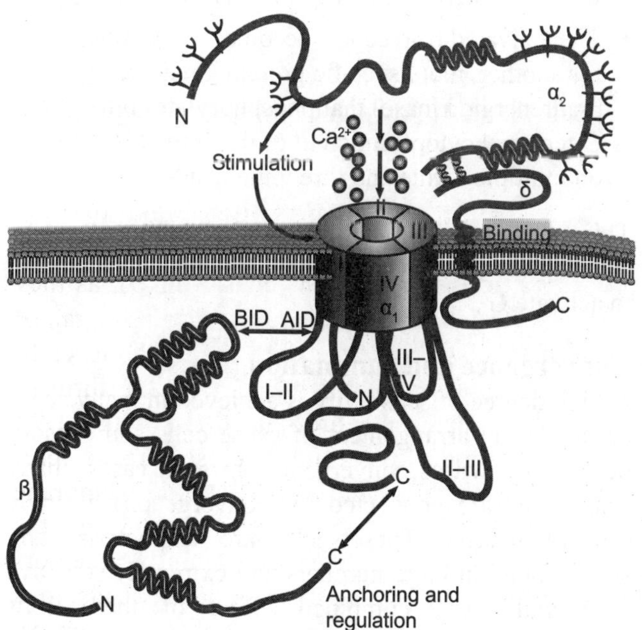

Fig. 11.10 Structural organization of animal voltage-gated calcium channel. The basic channel is composed of four different poroteins (α_1, α_2, β and δ), of which the sites of interaction are indicated by double-headed arrows. The major membrane-spanning subunit of α_1 consists of four homologous domains (I–IV). The α_2 protein is linked by a disulfide bridge to δ, which in turn interacts with α_1. BID: β-subunit interacting domain; AID: α-subunit interacting domain.

11.2.2 Sensory receptors

The coordinated activity of an organism relies upon a continuous input called **stimulus** from the internal and external environments and this information leads to a change in activity or behaviour of the animal. The specialized region of the body detecting the stimulus is known as a **sensory receptor**. The simplest receptor is the **Pacinian corpuscle**, a single sensory neuron, which detects the stimulus and passes it on to the CNS. More complex receptors, known as sense cells, consist of modified epithelial cells able to detect stimuli (e.g., taste buds). The most complex receptors are **sense organs** such as the eyes and ears. Structures transforming stimulus energy into electrical responses (nerve impulses) in axons are known as **transducers.** Here receptors act as **biological transducers**. All receptors trigger nerve impulses when they receive a particular environmental stimulus. Thus, receptors encode a variety of stimuli into nerve impulses which pass into the CNS where they are decoded and used to produce the required responses. In the following sections, we discuss some of the important mammalian receptors.

Transduction is the process by which a receptor converts a stimulus into a nerve impulse. The stimulus may be light, electricity or touch. All sensory cells are excitable and they share with nerve cells and muscles the ability to respond to an appropriate stimulus by producing a rapid change in their electrical properties. When not stimulated, sensory cells maintain a resting potential.

Properties of receptors

Sensations and the perceptions they evoke in the mind, begin with excitation of sensory receptors. They transmit information about changes in the external and internal environment of the animal. Receptors are generally modified neurons, which occur singly or in groups within sensory organs such as the eyes and ears. Receptors are specialized to respond to various stimuli (heat, light, pressure and chemicals) and all these stimuli represent forms of energy. The general function of receptor cells is to convert the energy of stimuli into the electrochemical energy of action potentials and carry those action potentials into the nervous system. There are various ways in which the effectiveness of receptors can be increased.

Thresholds of sensory cells

Some sense organs, such as stretch receptors in muscles are composed of many sense cells, which have a range of thresholds. If the cell has a low threshold it responds to a weak stimulus. As the strength of the stimulus increases, the cell can respond by producing an increasing number of impulses in the sensory neuron leaving the cell. However, at a given point saturation occurs and the frequency of impulses in the neuron cannot be increased. A further increase in the intensity of the stimulus will then excite sense cells with higher thresholds and these too produce a frequency of nerve impulses, proportional to the intensity of the stimulus. In this manner, the range of the receptor is extended.

Adaptation of receptors

Cells and organisms typically can detect the same percentage change in a signal over a wide range of stimulus intensities. At the cellular level, this requires that target cells undergo a process of adaptation or desensitization, when they are exposed to a stimulus for a prolonged period. In this way, a cell reversibly adjusts its sensitivity to the stimulus. In the case of chemical signalling, adaptation enables cells to respond to changes in the concentration of a signalling ligand over a wide range of absolute concentrations. Most receptors initially respond to a strong constant stimulus by producing a high frequency of impulses in the sensory neuron. The frequency of these impulses gradually declines and this reduction in response, with time, is called **adaptation**. For example, on entering a room you may immediately notice the clock ticking but after a while become unaware of its presence. The rate and extent of adaptation in a receptor cell is related to its function. Adaptation to chemical signals can occur in various ways. In some cases, it results from a gradual decrease in the number of specific cell surface receptor proteins and in others it results from a rapid de-activation of such receptors, which can occur in minutes. In some other cases it is due to change in the proteins involved in transducing the signal following receptor activation, which usually occurs with an intermediate time course.

Slowly adapting receptors

Slowly adapting receptors, or tonic receptors, register a constant stimulus with slowly decreasing frequency of impulses. After a protein hormone or growth factor binds to its receptor on the surface of a target cell, it is usually ingested by receptor-mediated endocytosis and delivered to early endosomes. Most receptors discharge their ligands in the acidic environment of early endosome and recycle back to the plasma membrane for reuse, while the ligand is delivered to lysosomes and is degraded. This process, therefore, represents a major pathway for the breakdown of many signalling proteins.

Rapidly adapting receptors

Rapid adapting receptors, or phasic receptors, respond to changes in stimulus level by producing a high frequency of impulses at the moments when the stimulus is switched 'on' or 'off'. The best understood example is the β_2-adrenergic receptor, which activates adenylyl cyclase via the stimulatory G protein G_s. When cells are exposed to a high concentration of adrenaline, they can desensitize within minutes by two pathways that depend on β_2-adrenergic receptor phosphorylation.

- The rise in cAMP caused by adrenaline binding activates A-kinase, which phosphorylates the β_2-receptor on a serine residue, thereby interfering with the receptor's ability to activate G_s.
- The activated β_2-receptor becomes a substrate for another, more specific protein kinase (called β-adrenergic kinase) that phosphorylates the *C*-terminal cytoplasmic tail of the activated receptor on multiple serine and threonine residues.

This phosphoryalted tail binds an inhibitory protein called **β-arrestin**, which blocks the receptor's ability to activate G_s.

Convergence and summation

A high degree of sensitivity is achieved in many sense organs by an arrangement of sense cells and sensory neurons known as convergence. In such cases, many sense cells are connected to (converge on) a single sensory neuron. These cells are characteristically small, found in large numbers and extremely sensitive to stimuli. The combined effect of these cells simultaneously producing stimulation in the sensory neuron is cumulative and is known as summation. Rod cells of the retina in the eye provide a good example for convergence and summation.

Spontaneous activity

Some receptors produce nerve impulses in sensory neurons in the absence of stimulation. This system has two important advantages.

- It increases the sensitivity of the receptor by enabling it to respond to a stimulus that would normally be too small to produce a response in the sensory neuron. A slight change in the intensity of the stimulus will now add to the existing potential in the receptor and produce a change in the frequency of impulses along the sensory neuron.
- The direction of the change in stimulus can be registered by this system as an increase or decrease in the frequency of the response in the sensory neuron.

Feedback control of receptors

The threshold of some sense organs can be raised or lowered by efferent (outward) impulses from the CNS. This 'resets' the sensitivity of the receptor to respond to different ranges of stimulus intensities. In many cases, the mechanism of control involves feedback from the receptors, which produces changes in accessory structures enabling the receptor cells to function over a new range.

11.3 SENSORY SYSTEM

A sensory system is a part of the nervous system responsible for processing sensory information. It consists of sensory receptors, neural pathways and parts of the brain involved in the sensory perception. The commonly recognized sensory systems are those for visual (vision), auditory (hearing), somatosensory (sensation of touch, temperature, pain, posture and movement), gustatory (taste) and olfaction (smell) activity (**Fig. 11.11**). The respective sensory field is the specific part of the world to which a receptor organ and receptor cells respond.

11.3.1 Sense organs: Transducers of the environment

Most animals have a central nervous system and sensory organs. Sense organs are the structures that convert information about the outside world and the inside environment into a form suitable for processing by the central nervous system. The most usual form into which this information is transformed is a series of propagated nerve impulses. Sense organs consist of appropriate sensory cells and subsidiary physical structures that present aspects of the environment in the form of a stimulus. Some sense organs are under control of the central nervous system, either as a relay of feedback information or in an independent manner.

The five human senses

Sensory systems code for four aspects of a stimulus: type (modality), intensity, location and duration. For example, arrival time of a sound pulse and the phase differences of continuous sound are used for localization of sound sources. Receptors send impulses in certain patterns to indicate the intensity of a stimulus. The location of the receptor that is stimulated gives the brain information about the location of the stimulus. The duration of the stimulus is conveyed by firing patterns of receptors. The human sensory system consists of the following subsystems:

Visual (Sight): The eye is the organ of vision. It has a complex structure consisting of a transparent lens that focuses light on the retina. The retina is covered with two basic types of light-sensitive cells: rods and cones.

- The ***cone cells*** are sensitive to colour and are located in the part of the retina called the *fovea*, where the light is focused by the lens.
- The ***rod cells*** are not sensitive to colour, but have greater sensitivity to light than the cone cells.

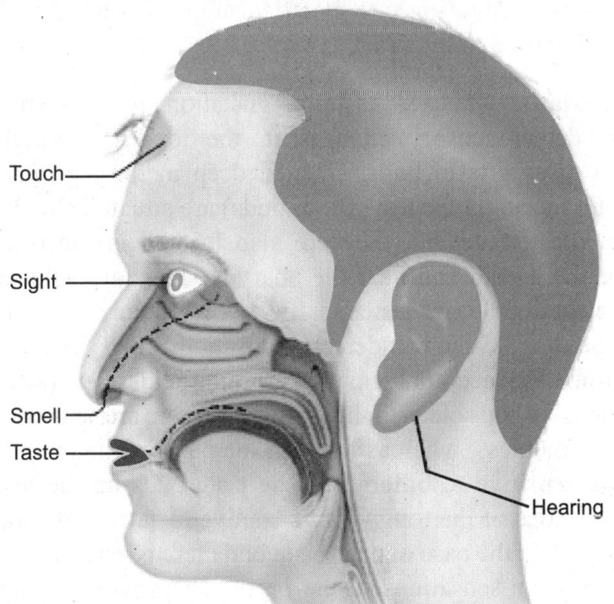

Fig. 11.11 Five human senses that communicate between the external world and internal functional centres: Visual (sight), auditory (hearing), somatosensory (touch, pain, temperature, posture and movement), gustatory (taste) and olfaction (smell)

These cells are located around the fovea and are responsible for peripheral vision and night vision. The eye is connected to the brain through the optic nerve. The point of this connection is called the 'blind spot' because it is insensitive to light. The back of the brain maps the visual input from the two eyes and combines it into a single three-dimensional image. In addition, even though the image on the retina is upside-down because of the focusing action of the lens, the brain compensates and provides the right-side-up perception.

Auditory (hearing): The ear is the organ of hearing. The outer ear protrudes away from the head and is shaped like a cup to direct sounds towards the tympanic membrane, which transmits vibrations to the inner ear through a series of small bones. The inner ear, or *cochlea*, is a spiral-shaped chamber covered internally by nerve fibres that react to vibrations and transmit impulses to the brain via the auditory nerve. The brain combines the input of the two ears to determine the direction and distance of sounds. Once the sound is received by the ear, the *pinna* leads the sound inside through the external auditory meatus. After the sound passes through the *meatus*, it goes to the eardrum, or *tympanus*, then vibrates its way through the tiny *ossicles*, the hammer (malleus), anvil (incus) and stirrup (stapes), then to the cochlea. The cochlea converts vibration into electrical impulses, which are transmitted to the brain.

Gustatory (taste): Taste, or gustation, is the ability to detect sensory changes in the tongue, through the use of taste buds, situated deep in the *papillae*. The receptors for taste (taste buds) are situated chiefly in the tongue, but they are also located in the roof of the mouth and near the pharynx. They are able to detect four basic tastes: salty, sweet, bitter and sour. The tongue also can detect a sensation called 'umami' from taste receptors sensitive to amino acids. Generally, the taste buds close to the tip of the tongue are sensitive to sweetness, whereas those in the back of the tongue are sensitive to bitterness. The taste buds on the top and sides of the tongue are sensitive to salty and sour tastes. At the base of each taste bud there is a nerve that sends the sensations to the brain. The sense of taste functions in coordination with the sense of smell.

Olfactory (smell): The nose is the organ responsible for the sense of smell or olfaction. The cavity of the nose is lined with mucous membranes that have smell receptors connected to the olfactory nerve. Smell is received by the olfactory bulb and is connected to the brain by the olfactory nerve. The smells consist of vapours of various substances. The smell receptors interact with the molecules of these vapours and transmit the sensations to the brain. The nose also has a structure called the *vomeronasal organ* whose function has not been determined, but which is believed to be sensitive to pheromones that influence the reproductive cycle. The smell receptors are sensitive to seven types of sensations that can be characterized as *camphor, musk, flower, mint, ether, acrid* or *putrid*. The sense of smell is sometimes temporarily lost when a person has a cold. Dogs have a sense of smell much more sensitive than humans.

Somatosensory (touch): The somatosensory system consists of receptors and transmitters (pathways) that experience the sensations labelled as touch or pressure, temperature, pain and the sensation of muscle movement and joint position including posture, movement and facial expression (collectively called *proprioception*). The sense of touch (cutaneous sense) is distributed throughout the body. Nerve endings in the skin and other parts of the body transmit sensations to the brain. Four kinds of touch sensations can be identified: *cold, heat, contact* and *pain*. Hairs on the skin magnify the sensitivity and act as an early warning system for the body. Some parts of the body have a larger number of nerve endings and, therefore, are more sensitive. The fingertips and the sexual organs have the greatest concentration of nerve endings.

Other sensory systems: The entire sensory system, including both physical sensation and interpretation (or cognition) of information from the senses, is referred to as *sensorium*. In addition to the five senses mentioned above, humans have also evolved other senses such as awareness of balance, pressure, temperature, pain and motion, all of which may involve the coordinated use of multiple sensory organs. The sense of balance is maintained by a complex interaction of visual inputs, the proprioceptive sensors (which are affected by gravity and stretch sensors found in muscles, skin and joints), the inner ear vestibular system and the central nervous system.

Kinesthetic sense: It is the sense of posture and movement and also referred to as proprioception. It is the precise awareness of muscle and joint movement that allows us to coordinate our muscles when we walk, talk and use our hands. It is the sense of kinesthesia that enables us to touch the tip of our nose with our eyes closed or to know which part of the body we should scratch when we itch.

Synesthesia: Some people experience a phenomenon called synesthesia in which one type of stimulation evokes the sensation of another. For example, the hearing of a sound may result in the sensation of the visualization of a colour, or a shape may be sensed as a smell. Synesthesia is hereditary and it is estimated that it occurs in 1 out of 1000 individuals with variations of type and intensity. The most common forms of synesthesia link numbers or letters with colours.

Vestibular sense: It is the sense of balance and gives awareness of the position and movement of the head. The receptors are *maculae* and *cristae ampullaris*. Maculae are located in utricle and saccule. Cristae ampullaris is found in semicircular canals.

Organic sense: It refers only to sensation from the internal organs, or viscera, but can be expanded to include certain physiological processes, such as hunger, thirst, drowsiness and shortness of birth. It is also referred to as *interoception*.

Types of sensory receptors

The various types of sensory receptors can be divided on the basis of location into two broad groups:

- ***Exteroreceptors*** that receive information from the outside world (light, sound, touch, smell)
- ***Interoreceptors*** that provide information about the body's internal environment

Alternatively, receptors can be grouped in terms of the energy stimulus to which they respond. Based on this criterion the different types of receptors are: mechanoreceptors, electromagnetic receptors, photoreceptors, thermoreceptors, chemoreceptors and nociceptors.

Mechanoreceptors: These are considered the most primitive type of receptors and are stimulated by their physical deformation caused by mechanical stimuli such as touch, pressure, motion, vibration (e.g., cochlea receptor, **Fig. 11.12**), stretch and all forms of mechanical energy. Bending or stretching of the plasma membrane of a mechanoreceptor increases its permeability to both Na^+ and K^+ ions, resulting in depolarization (receptor potential).

Electromagnetic receptors: Electromagnetic radiation consists of visible light, electricity and magnetism. Snakes have extremely sensitive infrared receptors that detect the body heat of prey standing out against a colder background. Some fishes discharge electric currents and use special electroreceptors to locate objects, prey, that disturb the electric currents. There is also evidence that some animals that home or migrate use the magnetic field lines of the Earth to help orient themselves. Photoreceptors, which detect the radiation (visible light) are often organized into eyes.

Photoreceptors: Photoreceptor cells of the retina are covered by cell bodies and axons that link them to the brain. There is a photoreceptor layer (outermost layer) containing photosensitive cells—*rods* (**Fig. 11.13**) and *cones*, partially embedded in the pigmented epithelial cells of the choroid. The intermediate layer, containing bipolar neurons with synapses, connects the photoreceptor layer to the cells of the third layer. The third layer is the internal surface layer containing *ganglion cells* with dendrites in contact with bipolar neurons and axons of the optic nerve.

In the photoreceptor layer, the retina of the mammalian eye contains light receptors called rod cells. These respond to light intensity. Mammals with colour vision also have cone cells mixed with the rods. Each cone is linked to its own neuron, whilst any one of a group of rods can fire a single neuron. The result is that cones are less effective for vision in dim light, but allow finer resolution. Rods contain the light-sensitive pigment *rhodopsin*, which is attached to the outer surface of the vesicles. Rhodopsin is a molecule formed by the combination of a protein called *retinene (retinal)*, which exists in

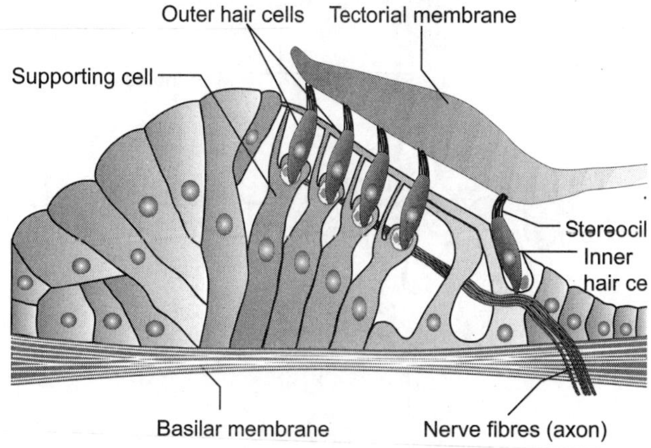

Fig. 11.12 Auditory hair cells. Cross section of the auditory apparatus in the inner ear of a mammal.

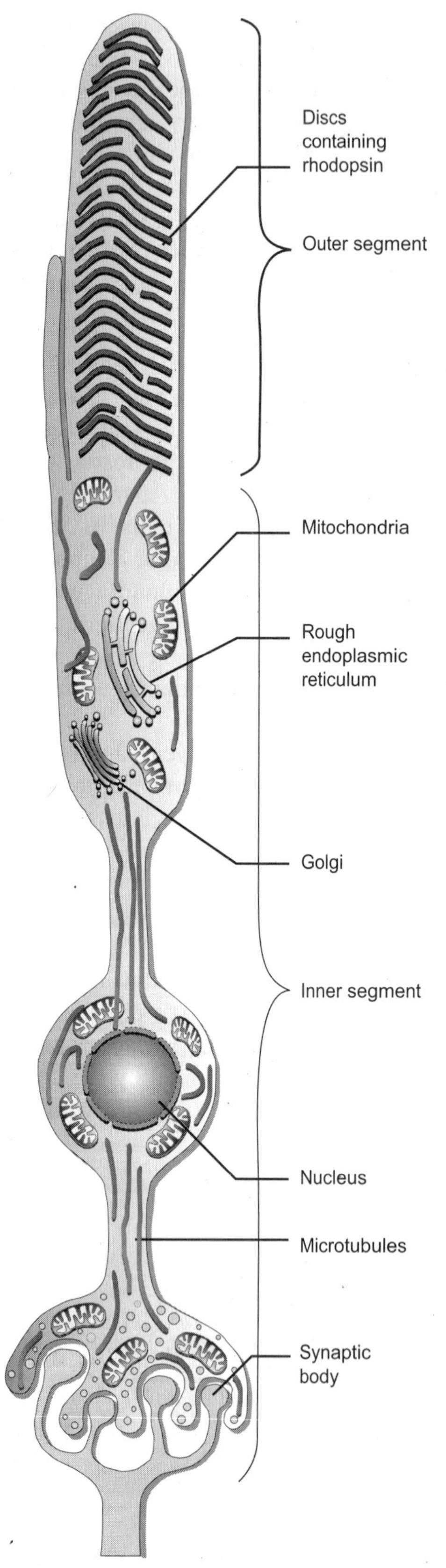

Fig. 11.13 Schematic diagram of a human rod cell (**See Plate 6**).

two isomeric forms according to the light conditions shown in **Fig. 11.14**. When a molecule of rhodopsin is exposed to light, one photon of light produces the isomeric change. The rhodopsin breaks down into retinene and scotopsin, a process known as *bleaching*. Rhodopsin is re-formed immediately in the absence of further light stimulation. *Trans*-retinene is first converted into *cis*-retinene and then recombined with scotopsin. This process is called *dark adaptation*. The change in rhodopsin leads to the creation an action potential. This involves changes in the outer and inner segment membranes of the rod (**Fig. 11.15**). The Na^+ pump, in the inner segment of the rod continuously pumps out Na^+. In darkness the outer segment membrane is leaky and allows the sodium back in again by diffusion, reducing the negative charge inside the cell (–40 mV instead of the normal –70 mV). In the light, the permeability of the outer segment membrane to Na^+ decreases whilst the inner segment continues to pump out Na^+, thus making the inside of the rod more negative. This causes hyperpolarization of the rod. This situation is exactly opposite to the effect normally found in sensory receptors where the stimulus produces a depolarization and not a hyperpolarization. The hyperpolarization reduces the rate of release of excitatory transmitter substance from the rod, which is released at its maximum rate during darkness. The bipolar neuron linked by synapses to the rod cell also responds by producing hyperpolarization, but the ganglion cells of the optic nerve supplied by the bipolar neuron respond to this by producing an action potential.

Chemoreceptors: They include both general receptors that transmit information about the total solute concentration in a solution and specific receptors that respond to individual kinds of molecules. Osmoregulators in the mammalian brain, for example, are general receptors that detect changes in the total solute concentration of the blood. Most animals also have receptors specific to important molecules, including glucose, oxygen, carbon dioxide and amino acids. The stimulus molecule binds to a specific site on the membrane of the receptor cell and initiates changes in membrane permeability.

Two other groups of chemoreceptors show intermediate specificity. Gustatory (taste) and olfactory (smell) receptors respond to categories of related chemicals. Humans have an ability to distinguish more than 10,000 distinct smells. Olfactory receptor cells line the upper portion of

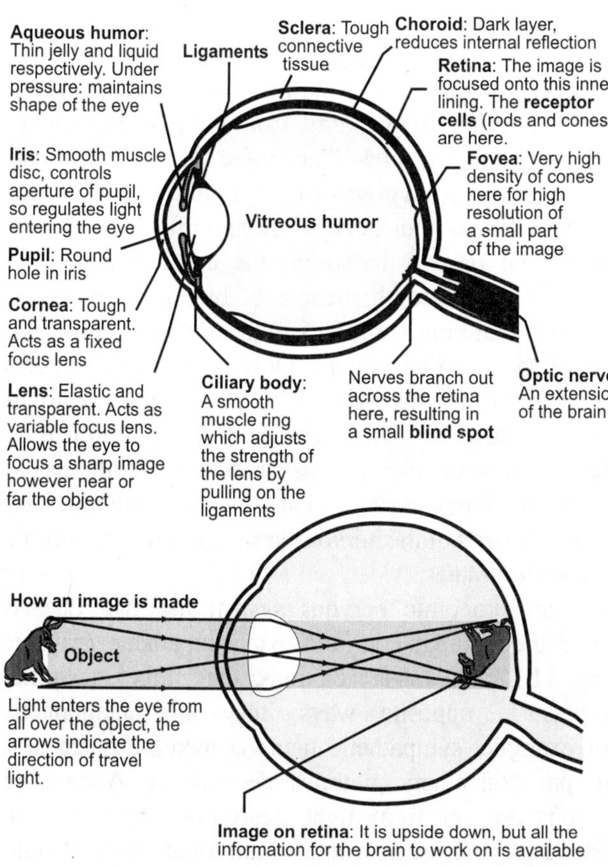

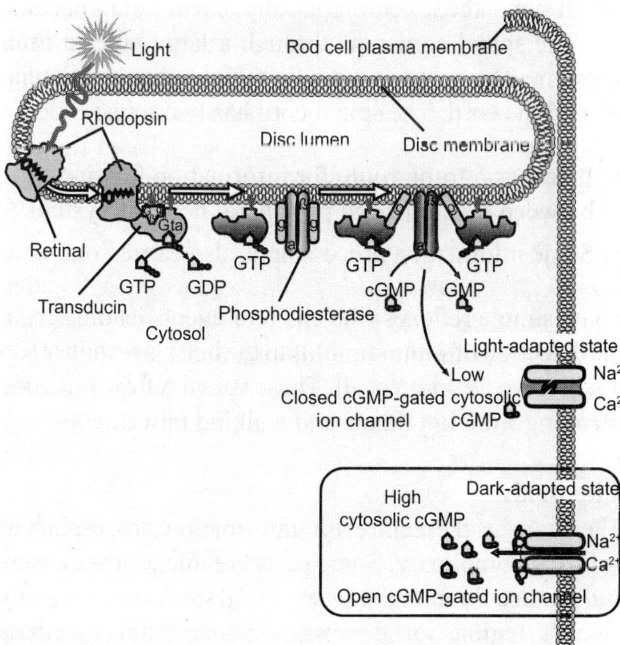

Fig. 11.14 Functional model for rhodopsin-induced closing of cation channels in rod cells. Absorption of light by retinal activates the G protein-coupled receptor rhodopsin. The subunit of transducin then stimulates cGMP phosphodiesterase, leading to a decrease in intracellular levels of cGMP.

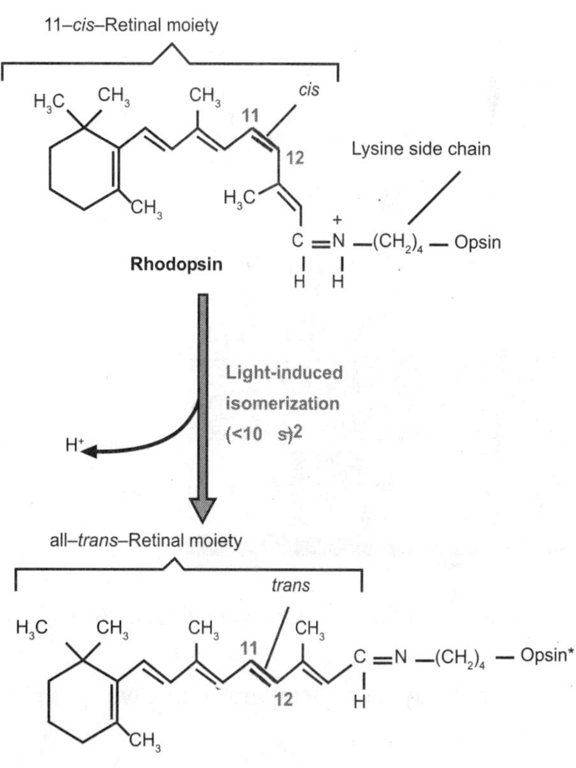

Fig. 11.15 The light-triggered step in vision that causes rapid photo-isomerization of *cis*-retinal to the all *trans*-isomer.

the nasal cavity and send impulses along their axons directly to the olfactory bulb of the brain. The receptive ends of the cells contain cilia that extend into the layer of mucus coating the nasal cavity. When an odorous substance diffuses into this region, it binds to specific receptor molecules in the olfactory hairs and depolarizes the olfactory receptor cell. These specialized olfactory receptor neurons in the lining of the nose recognize odours by means of specific G protein-linked olfactory receptors, which are displayed on the surface of the modified cilia that extend from each cell (**Fig. 11.16**). The receptors act as olfactory-specific G protein (known as G_{olf}), which in turn activates adenylyl cyclase. The resulting increase in cAMP opens cAMP-gated cation channels, thereby allowing an influx of Na^+, which depolarizes the olfactory receptor neuron and initiates a nerve impulse that travels along its axon to the brain.

Thermoreceptors: The free nerve endings in the skin are the major structures detecting temperature. There are separate hot and cold receptors that help regulate body temperature by signalling both

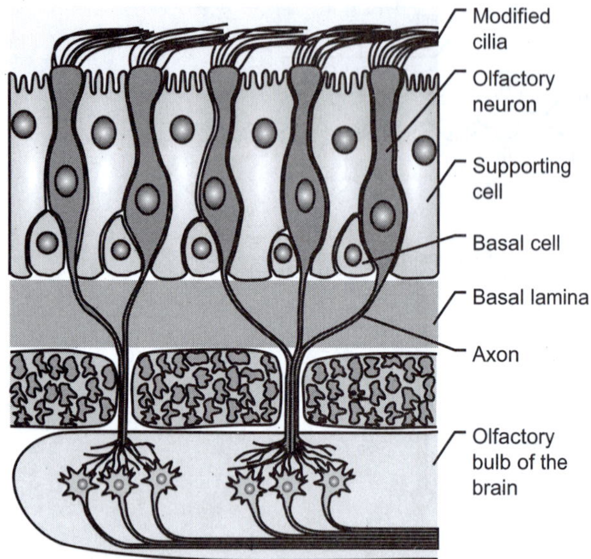

Fig. 11.16 Section of an olfactory epithelium in the nose.

surface and body core temperature. Two possible receptors are:

- ***Ruffini's end organs,*** which may be heat receptors
- ***End-bulbs of Krause,*** which may be cold receptors.

Nociceptors: Pain is associated with stimulation of the receptors. It is one of the most important senses because the stimulus becomes translated into a negative reaction, such as withdrawal from danger. Pain is detected by a class of naked dendrites called nociceptors. Different groups of pain receptors respond to excess heat, pressure, or specific classes of chemicals released from damaged or inflamed tissues. Some of the chemicals that trigger pain include histamine and acids. Prostaglandins increase pain by sensitizing the receptors, that is, lowering their threshold (aspirin reduces pain by inhibiting prostaglandin synthesis). Nociceptive neurons carry impulses to the spinal cord, synapsing with neural pathways leading to the brain.

11.3.2 Nervous system

The nervous system can be divided into the central nervous system (CNS) and the peripheral nervous system. The CNS comprises the brain and spinal cord. The peripheral nervous system has three main elements: sensory neurons, motor neurons and autonomic nervous system. Sensory neurons carry signals from receptor cells to the CNS. They may be linked directly to receptor cells, or there may be other local neurons in between, carrying out low-level information processing within the receptor organ. Motor neurons carry signals from the CNS to skeletal muscles—the muscles that move bones. They are largely involved in controlling voluntary actions.

The autonomous nervous system affects the action of internal organs, including the cardiac muscle of the heart and smooth muscle of blood vessel walls, sphincters and gut wall. It is rather independent of the rest of the nervous system. Only a few distinct parts of the brain are involved in it. Heart rate, breathing cycle, control of body temperature by controlling blood flow to the skin and hair erection, are all largely under autonomic control. They can be influenced by voluntary parts of the nervous system but the autonomic system dominates.

The autonomic nervous system has two distinct parts, the sympathetic and parasympathetic systems (**Fig. 11.17**). In most situations, they affect the same process in opposite ways: they are antagonistic. For example, sympathetic neurons increase heart rate, but parasympathetic neurons decrease it. Autonomic neurons do not form tight neuromuscular junctions. Instead, they release neurotransmitters more diffusely into the tissue. Parasympathetic neurons release acetylcholine, whilst sympathetic neurons usually release noradrenaline.

The spinal cord runs through a large hole in each vertebra. The spine as a whole forms a protective tube around the cord. The spinal cord has two roles:

- It acts as a trunk route for information flowing between the brain and peripheral nervous system.
- Some information processing is also carried out here.

Some simple reflexes (the chain of events leading from the detection of some stimulus to the body's response) do not involve the brain at all. These spinal reflexes include recoiling from hot things and walking movements.

The brain

The brain is the centre for information processing of amazing complexity. Some parts are thicker than others and it is deeply folded. **Figure 11.18** shows some visibly distinct regions of the brain. Some brain functions seem to be located in specific parts of the lower brain, for example, the cerebellum controls many reflex actions which need to be learned, including the control of the body's posture, running and the mechanics of speech. The autonomic nervous system is largely controlled by the hypothalamus and medulla.

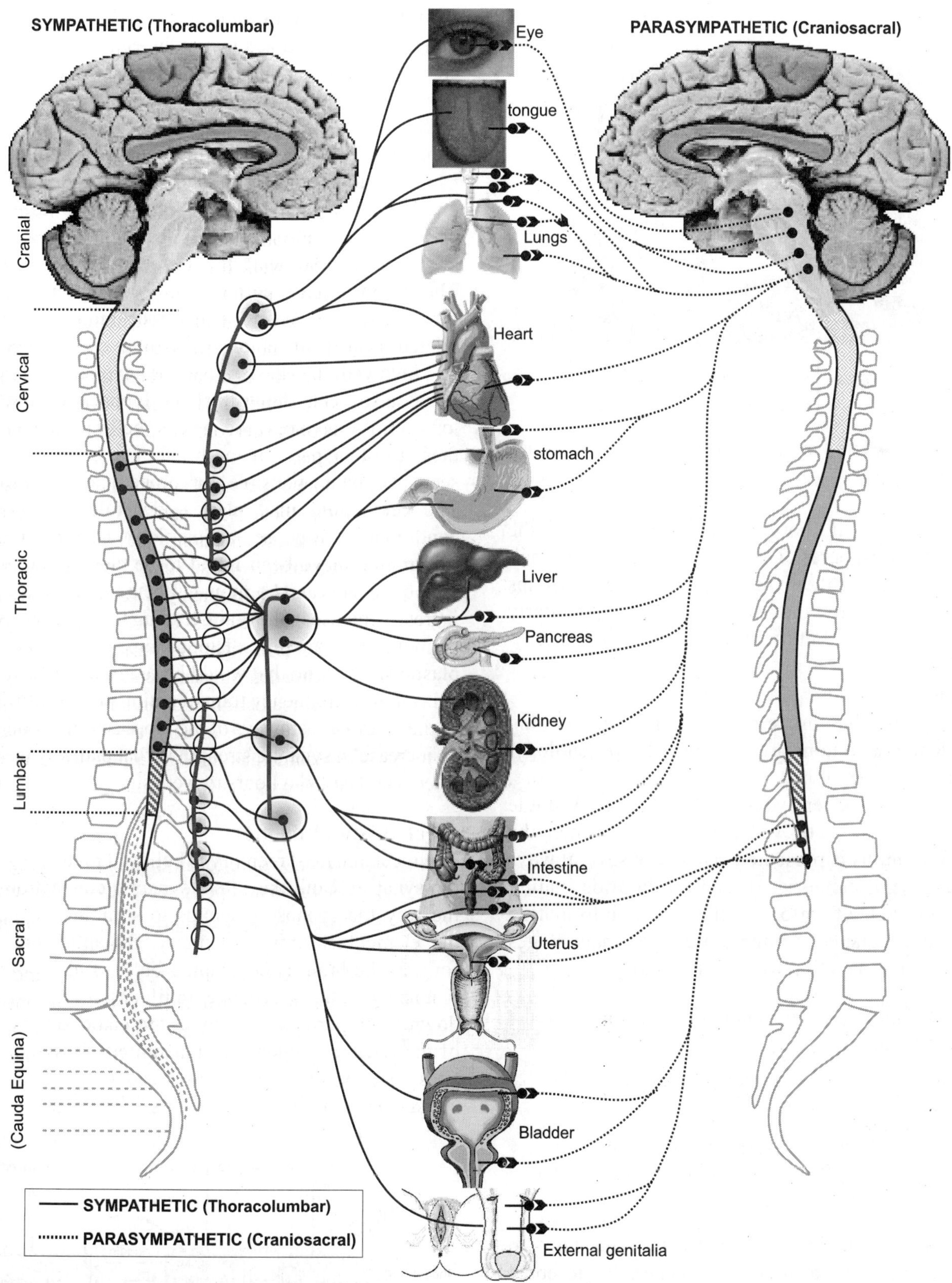

Fig. 11.17 The autonomic nervous system: Sensory neurons and motor neurons run between the central nervous system and various internal organs (heart, viscera, and glands) (**See Plate 7**).

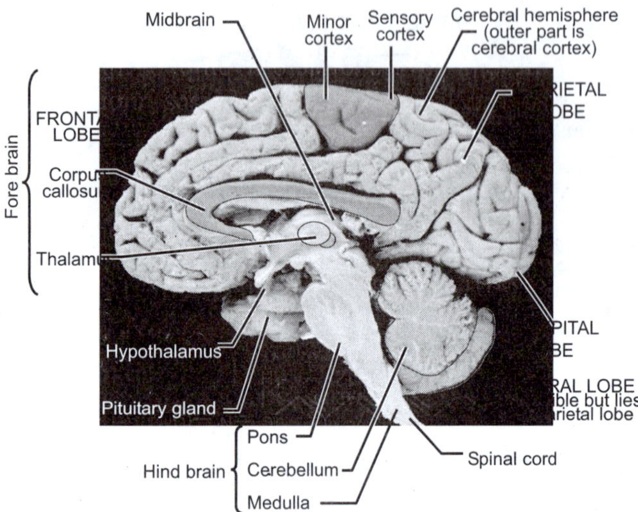

Fig. 11.18 External view of human brain. A section through the brain is shown.

The cerebral cortex has two halves: the cerebral hemispheres. These communicate via a belt of neurons below them: the corpus callosum. Groups of neurons within the cortex have extracellularly precise functions, but they are not visibly distinct.

Memory

Memory is the process by which a learning experience is retained over time. In psychology, memory is the ability of an organism to store, retain and subsequently recall information. A single memory can be retrieved several times when the proper stimulus is presented. There are several ways of classifying memories, based on duration, nature and retrieval of information. From information processing perspective, there are three main stages in the formation and retrieval of memory:

- ***Encoding:*** Processing and combining of received information,
- ***Storage:*** Creation of a permanent record of the encoded information,
- ***Retrieval/recall:*** Calling back the stored information in response to some cue for use in a process or activity.

Even though, there is no consensus on how to classify memory, a basic and generally accepted classification of memory is based on the duration of memory retention and identifies three distinct types of memory (**Fig. 11.19**): sensory memory, short-term memory (STM) and long-term memory (LTM).

Functioning of memory

Scientists have realized that at the end of the memory chain there are neurons that respond to specific items. It also appears that memories are distributed within the various subsections of the brain and in complex situations it may be distributed throughout many sections involving vision, sound, touch and so on. Memories are composed of elements from any or all of the sensory modalities and may therefore be truly diverse as far as storage is concerned. The hippocampus, part of the forebrain, is crucial for long-term memory. The hippocampus and some related nearby areas of the cortex are thought to be involved in declarative or explicit memory. It has also been shown that the hippocampus stores long-term memories for weeks before transferring them to other parts of the brain. The basal ganglion seems to be responsible for procedural knowledge. The cerebellum may play a key role in classical conditioned response acquisition. Certain neurotransmitters have been found to be closely related to memory events and hormonal effects are also known to increase glucose in the brain, leading to improved event memory at certain times. Hippocampus has remarkable plasticity, the kind that would be required for learning. A brief high-frequency train of action potentials in one of the neuronal pathways of the hippocampus produces an increase in synaptic strength in that pathway and the effect can last from hours to weeks.

Mechanism of memory

The mechanism of memory is believed to take place at the synapse. Both the actual synaptic connection and the way the synapse responds to stimulus, represent the elementary unit of memory. Synaptic junctions form as the brain is developing but it is also possible that new junctions form later in life. Memories are laid down when permanent changes are made to the effect of the release of chemical transmitters. Post-synaptic sites may become more sensitized to the effect of chemicals or additional receptor sites may grow.

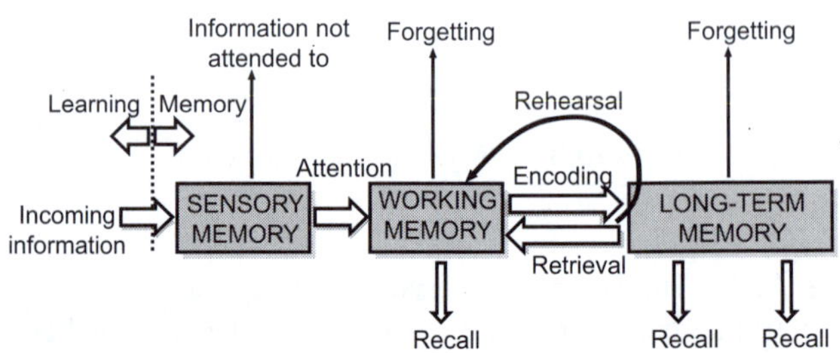

Fig. 11.19 The three types of memory.

Brain plasticity

Brain plasticity, or neuroplasticity, is the lifelong ability of the brain to reorganize neural pathways based on new experiences, newly learnt or acquired knowledge and skills through instruction or experience. In order to learn or memorize a fact or skill, there must be persistent functional changes in the brain that represent the new knowledge. The ability of the brain to change learning is what is known as neuroplasticity. Neuroplasticity includes several different processes that take place throughout a lifetime. It occurs during normal brain development and as an adaptive mechanism to compensate for lost function to maximize remaining functions in the event of brain injury.

Synaptic plasticity

During the early years of life, the brain grows rapidly. As each neuron matures, it sends out multiple branches (axons and dendrites) increasing the number of synaptic contacts and laying the specific connections from neuron to neuron. At birth, each neuron in the cerebral cortex has approximately 2,500 synapses. By the time an infant is two or three years' old, the number of synapses is approximately 15,000 synapses per neuron. As we age, old connections are deleted through a process called **synaptic pruning (Fig. 11.20)**. This eliminates weaker synaptic contacts while stronger connections are strengthened. Experience determines which connections will be strengthened and which will be pruned. The connections that have been activated most frequently are preserved. Neurons must have a purpose to survive. Without a purpose, neurons die through a process called **apoptosis** in which neurons that do not receive or transmit information become damaged and die. Ineffective or weak connections are 'pruned'. It is synaptic plasticity that enables the process of developing and pruning connections, allowing the brain to adapt itself to its environment.

Synaptic plasticity is the ability of the connection, or synapse, between two neurons to change in strength. There are several underlying mechanisms that cooperate to achieve synaptic plasticity, including changes in the amount of neurotransmitter released into a synapse and changes in how effectively cells respond to those neurotransimitters. Since memories are postulated to be stored in synapses of the brain, synaptic plasticity is one of the important neurochemical foundations of learning and memory. Two known mechanisms exist for synaptic plasticity:

- involves modification of existing synaptic proteins (typically protein kinase) resulting in altered synaptic function;
- depends on the second messenger neuro-transmitters regulating gene transcription and changes in the levels of key proteins at synapses.

The strength of the synapse also depends on the number of ion channels it has. Several facts suggest that neurons change the density of receptors on their post-synaptic membranes as a mechanism for changing their own excitability in response to stimuli.

11.3.3 Endocrine system

Multicellular animals have both nervous and chemical communication between cells. Although, the nervous and endocrine systems seem unrelated, they depend on each other and have many chemical similarities. The hypothalamus–pituitary system clearly links the two. Neurons manufacture, transport and control the release of hormones. Neurotransmitters are another link.

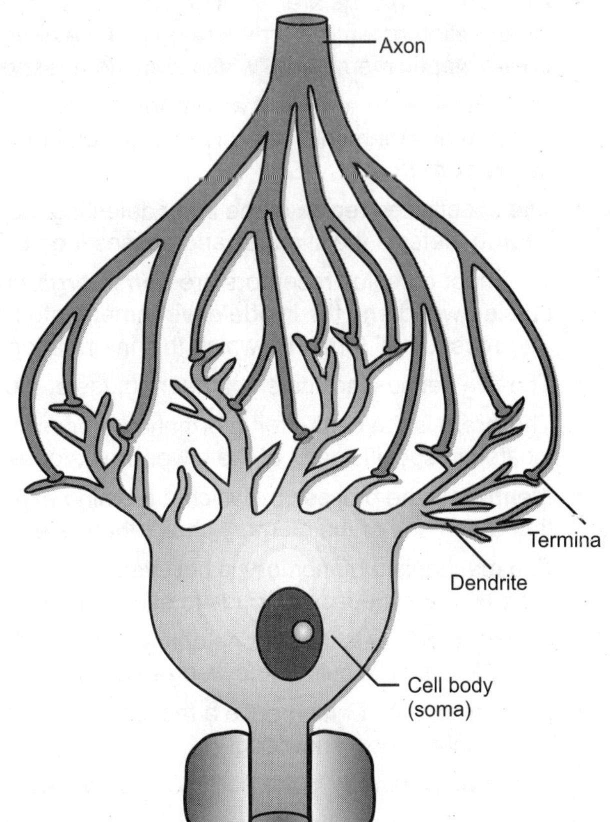

Fig. 11.20 Schematic diagram of synaptic contacts.

In their release, movement between cells and binding to receptors, they are like local hormones or paracrines. In the adrenal medulla, neurotransmitters actually become hormones. The sympathetic neurons, which end in the medulla, convert most of their noradrenaline to adrenaline before releasing both into the blood. They act on the same types of target cells, but throughout the body instead of locally and for longer because they are removed from the blood more slowly.

The cell signalling pathways involved in responding to hormones are shared by most types of cells, including neurons. The role of calcium as a second messenger in neurotransmitter release is well known and it is also involved in the NMDA receptor memory mechanism. Neuropeptides are short peptides made in the cell bodies of some neurons, by chopping up larger proteins. They travel down axons in vesicles and are released at synapses. Instead of binding to ion channels, they bind to G protein- or enzyme-linked receptors. They bring about changes by gene switching or enzyme activation. The encephalins are neuropeptides, which reduce pain perception. Pain is an early warning system for problems in the body and several types of neurons are concerned with it. Encephalins are released onto some types of pain neurons at synapses and are strongly inhibitory.

Summary

- The ***brain*** is the most powerful information processing organ, which can outperform even the most advanced computers available today.
- The complete neural system is divided into the ***peripheral nervous system*** and the ***central nervous system*** (CNS).
- Nerve cells (or ***neurons***) are specialized for the collection, conduction and transmission of information, which is coded in the form of fast-moving electrical impulses. Information travels along neurons as a stream of pulses.
- Each pulse consists of a cycle of changes in the axon membrane. These changes involve the opening and closing of ion channels and the movement of sodium and potassium ions.
- Neurons are linked with their target cells at specialized junctions called ***synapses***.
- ***Neurotransmitters*** are stored in vesicles close to the presynaptic membrane and fuse with the membrane when action potentials arrive, releasing the neurotransmitter into the cleft. It is bound by receptors in the post-synaptic membrane, which leads to a response.
- The coordinated activity of an organism relies upon a continuous input of information from the internal and external environments and this information leads to a change in activity or behaviour of the animal, called ***stimulus.***
- The specialized region of the body detecting the stimulus is known as a ***sensory receptor***. A single sensory neuron, detects the stimulus and passes it on to the CNS.
- The most complex receptors are ***sense organs*** such as eyes and ears. They convert information about the outside world and the inside environment into a form suitable for processing by the central nervous system. The most usual form into which this information is transformed is a series of propagated nerve impulses.
- The five senses include sight, hearing, taste, touch and smell and some other evolved senses in human.
- The ***brain*** is the centre for information processing of amazing complexity. Some brain functions seem to be located in specific parts of the lower brain; for example, the *cerebellum* controls many reflex actions.
- ***Memory*** is the process by which a learning experience is retained over time. Studies have demonstrated that at the end of the memory chain, there are neurons that respond to specific items.
- The mechanism of memory is believed to take place at the synapse. Both the actual synaptic connection and the way the synapse responds to stimulus, represent the elementary unit of memory.
- ***Neuroplasticity*** is the lifelong ability of the brain to reorganize neural pathways based on new experiences, newly learnt or acquired knowledge and skills through instruction or experience.
- In order to learn or memorize a fact or skill, there must be persistent functional changes in the brain that represent the new knowledge.
- ***Synaptic plasticity*** is the ability of the connection, or synapse, between two neurons to change in strength.
- In most situations, the neurons and endocrine systems work in coordination.

Exercises

1. Draw and explain the three neurons involved in cellular communication.
2. Describe nerve impulses.
3. What are synapses?
4. Give a brief account of neurotransmitters with one or two examples.
5. Explain briefly the different types of sensory receptors.
6. How do the target cell receptors undergo adaptation?
7. What is signal convergence and summation?
8. Draw and explain the structure and functions of different sensory receptors.
9. Define memory. Briefly discuss the different types of memory.
10. Explain brain plasticity.

Further Reading

Books

Gilbert SF (2000), *Developmental Biology*, 6th edition, Sinauer Associates Inc.

Goldman C (2000), *Textbook of Medicine*, 21st edition, WB Saunders Co., New York.

Guyton AC and Hall JE (2000), *Textbook of Medical Physiology*, 10th edition, WB Saunders Co., New York.

Keynes RD and Aidley DJ (2001), *Nerve and Muscle*, 3rd edition, Cambridge University Press.

Siegel GJ, Agranoff BW, Alberts RW, Fisher SK and Uhler MD (1999), *Basic Neurochemistry*, Lippincott–Raven, Philadelphia.

Sutton S (1998), *Biology*, Macmillan Foundations.

Tortora GJ and SR Grabowski (2000), *Principles of Anatomy and Physiology*, 9th edition, John Wiley and Sons Inc., New York.

References

Berridge MA, Bootman MD and Lipp P (1998), Calcium—a life and death signal, *Nature*, 395:645–47.

Jessell TM, Kandel ER et al. (1993), Synaptic transmission: Nine reviews, *Cell*, 72:Suppl. 1–149.

Karin M and Hunter T (1995), Transcriptional control by protein phosphorylation: Signal transmission from the cell surface to the nucleus, *Curr. Biol.*, 5:747–57.

McCormick F (1999), Signalling networks that cause cancer, *Trends Cell Biol.* 9:M53–M56.

Miranti CK and Brugge JS (2002), Sensing the environment: A historical perspective on integrin signal transduction, *Nature Cell Biol.*, 4:E83–E90.

Mombaerts P (1999), Seven transmembrane proteins as odorant and chemosensory receptors, *Science*, 286:707–11.

Monteith GR and Roufogalis BD (1995), The plasma membrane calcium pump—a physiological perspective on its regulation, *Cell Calcium*, 18:459–70.

Neer EJ (1997), Intracellular signalling: turning down G–protein signals, *Curr. Biol.*, 7:R31–3.

Scott JD and Pawson T (2000), Cell communication: The inside story, *Sci. Am.*, 282(6):7279.

Struhl K (1998), Fundamentally different logic of gene regulation in eukaryotes and prokaryotes, *Cell*, 98:1–4.

12 DIFFERENTIATION AND DEVELOPMENT

Learning objectives

After reading this chapter, you will be able to:

- Describe the processes of cell differentiation and their molecular pathways

12.1 INTRODUCTION

How does a single cell, the zygote, get transformed into a uniquely complex organism made up of so many distinctive cells and tissues? How do cells that are derived from the same fertilized egg establish different programs of gene expression? How does the embryo mark and measure space and time, so that organs and tissues develop on schedule and in the right locations? How are the three dimensions of space and one of time encoded in the one-dimensional DNA molecule? And how does each species develop into a highly organized body pattern and shape?

These and many other questions about the nature of development have fascinated scientists for many years. Some insights came after the 1970s, when the development of high-powered molecular techniques allowed the isolation and analysis of mutations affecting early developmental stages. Studying development in mammals is difficult, because the embryo develops inside the uterus; it takes weeks or months and genetic analysis is slow and complex. Therefore, most of the developmental studies came from a few representative animals including fruit flies, nematodes, zebra fish, frog and mice. From these and other subjects, scientists have learnt that, many similar genes operate in a vast array of evolutionally dissimilar organisms. Among the many mechanisms that involve in building multicellular organisms, cell division and differentiation play an important role in the developmental process.

Development is the process by which a multicellular organism arises, from a single cell. Development is progressive in a very orderly fashion; this strategy is also known as **epigenesis**, in contrast to the early theory of **pre-formation**. **Embryology** is a study of the developmental stages, between fertilization and birth of an organism. To develop into an embryo, a fertilized cell (or embryo) needs a number of life-supporting processes (e.g., respiration without lungs, nutrients without digestive system, formation of bones and neurons from a pulpy tissue). The embryo mediates between genotype (inherited genes) and phenotype (an adult organism). **Developmental biology** is the study of the process by which organisms grow and develop. Modern developmental biology studies the genetic control of cell growth, differentiation and morphogenesis—the process that gives rise to tissues, organs and anatomy. The development of a single zygote to multicellular adult and the reproduction of an organism involves:

- growth
- cell division
- differentiation
- pattern formation
- morphogenesis
- reproduction
- evolution
- interaction with the environment.

Differentiation is the term used to describe the developmental processes that cause structural or functional distinction (specialization) between parts of an organism. In a single-cell organism it refers to parts within the cell. However, in multicellular organisms it refers to groups of cells (e.g., tissues, organs) which differ morphologically, structurally and functionally. Multicellular organisms begin their lives with a fairly simple organization (viz., fertilized egg) and then proceed step-by-step towards a more complex arrangement (adult). As this process proceeds, cells divide, migrate and change their characteristics, as they become highly specialized units within a multicellular individual. Based on the type and structural differences of proteins, cells can be characterized as differentiated cells. In the early stages of differentiation, differences between cells are not easily visible due to expression of a few genes. However, as the process continues the cells become **determined** or **committed**, with respect to their developmental potential. In animals, for example, muscle cells allow an organism to move, while intestinal cells facilitate the absorption of nutrients. This division of labour among the various cells and organs of the individual collectively promotes its survival and competitive fitness. ***Developmental genetics is concerned with the role genes play in orchestrating the changes that occur during development.***

Cell differentiation is controlled by a variety of external signals, ranging from cell surface proteins to secreted polypeptide cytokines and molecules of the extracellular matrix. However, the external signals that stimulate differentiation are often referred to as being instructive and are very selective. There are rarely options available at any given time; these options are set by the cell's internal state, which in turn reflects its developmental history. Some self-renewing stem cells can give rise to a wide range of cell types under the influence of different external signals. Differentiation is a gradual process with each generation becoming progressively more differentiated. In many cases, however, there is a conflict between cell division and cell differentiation, with cell cycle arrest being necessary for cell differentiation. Terminally differentiated cells rarely divide (red blood cells, skeletal muscle and nerve cells). For cells that divide after differentiation, the differentiated state, like the determined state, is passed on through all subsequent cell divisions to the progeny. Thus, the pattern of gene activity is the key feature in cell differentiation.

For developmental studies, scientists chose a few experimental organisms, such as the fruit fly, nematode, frog, mouse and *Arabidopsis* and worked towards the differentiation and characterization of the genes required for running their developmental programs. In the case of the fruit fly, most of the genes that play a critical role in the early stages of development have been identified. Scientists are now trying to understand the exact role of these genes and their expression in differentiation.

12.2 MOLECULAR MECHANISMS OF DEVELOPMENT

Although the genomes of all the cells are identical, the cells differ from each other, not because they contain different genetic information, but because they express different sets of genes. During the second half of the 20th century, the types of molecules in embryonic development were identified. Transcription factors are the key regulators that determine which genes are expressed in cells. Transcriptional control in the various differentiated cell types allows each type of cell to express different amounts of the possible proteins. The transcription factors are regulated by signal transduction pathways, which relay signals from outside the cells to the nucleus. Signal transduction pathways often involve receptors, receptor ligands and enzymes such as protein kinases. Cell adhesion proteins are among the key regulators of morphogenesis. Selective gene expression controls the four essential processes (**Fig. 12.1**) by which the embryo is constructed:

- proliferation
- specialization
- interaction
- movement.

During embryogenesis, all these processes happen in different ways in different parts of the organism.

During differentiation, cells become specialized to perform particular functions and this reflects the synthesis of characteristic sets of proteins. With a few exceptions, the genome of all cells in a developing organism remains the same, indicating that the basis of differentiation is selective gene expression. The patterns of gene activity in differentiated cells vary enormously from one cell type to another. The three strategies by which cells are instructed to express specific sets of genes during development are:

- mRNA localization
- cell-to-cell contact
- signalling through the diffusion of a secreted signalling molecule

In order to understand the molecular basis of differential gene expression during cell differentiation and development, we need to first determine how a gene can be expressed in a cell-specific manner.

Gene expression can be regulated at many levels and some examples of developmental gene regulation are given in **Table 12.1**. Molecular analysis of genomic DNA shows that genome structure is typically the same in all cells regardless of which genes are expressed.

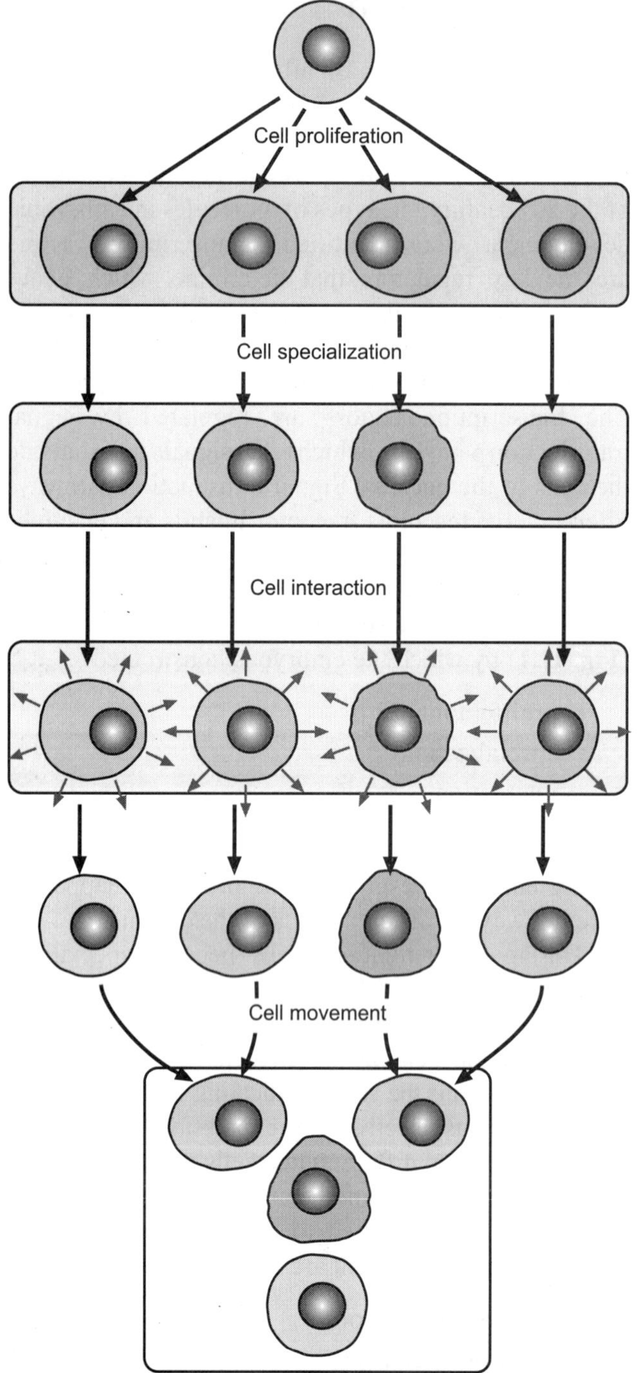

Fig. 12.1 The four essential processes by which a multicellular organism is made.

Functional experiments show that differentiated cells are, under certain circumstances, able to **dedifferentiate** and follow alternative pathways of development, a phenomenon termed **transdifferentiation** or **metaplasia**. This implies that genes, which are not utilized in a particular cell type are latent and can be re-activated. The ultimate functional test is to regenerate an entire organism from a single differentiated cell. This is quite possible in plants, where differentiated cells of several species can be routinely dedifferentiated in culture and which will then proliferate to form a clone of disorganized undifferentiated cells or **callus**. These calluses can be

Table 12.1 Regulation of differentiation at different levels of gene expression

Level of gene expression	Developmental system
Transcription	Differentiation of muscle cells is controlled by synthesis of transcription factors such as MyoD1, which activates muscle-specific genes.
RNA processing	Differentiation of sexes in *Drosophila* is controlled by alternative splicing of primary transcripts of the sex lethal, transformer and double sex genes. Precocious translation of a number of mRNAs in oocytes of the moth *Manduca sexta* is prevented by delayed capping.
RNA turnover	Limb bud development in the chick is controlled in part by the expression of FGF-2. Antisense RNA complementary to FGF-2 mRNA is expressed in the chick limb bud and targets the message for degradation.
Protein synthesis	Establishment of anterior and posterior structures in early *Drosophila* development is controlled in part by repression of translation. Bicoid protein represses translation of caudal mRNA in the anterior and Nanos protein represses translation of hunchback mRNA in the posterior. Synthesis of the *Xenopus* fibroblast growth factor receptor is blocked in oocytes by proteins bound to the Xfgfr1 mRNA; the repressor is lifted in mature eggs.
Protein function	Establishment of dorso-ventral polarity in the *Drosophila* embryo depends on the transfer of the dorsal protein to the nucleus, which is brought about by phosphorylating (thus inactivating) the inhibitory protein, cactus.

exposed to plant hormones and then regenerate to form a complete plant. It has been impossible to obtain the same results using differentiated animal cells, but it has been possible to produce viable embryos only by transferring the nucleus of a differentiated amphibian cell into an enucleated egg. Although embryos do not develop into adults, these experiments indicate that, unlike plant cells differentiated animal cells are unable to fully reverse their developmental commitment. Performing the same technique using nuclei from cells of an amphibian embryo, however, allows development to the adult stage and nuclei from pre-implantation embryos of certain mammals also support full development. These results reveal that the animal nucleus progressively loses its potency as development proceeds. Despite this restriction, a sheep has been cloned (Dolly) recently by transferring the nucleus of a fully differentiated mammary gland cell into an enucleated egg, demonstrating for the first time that the differentiated animal cell nucleus retains all the information required to produce a fully functional organism.

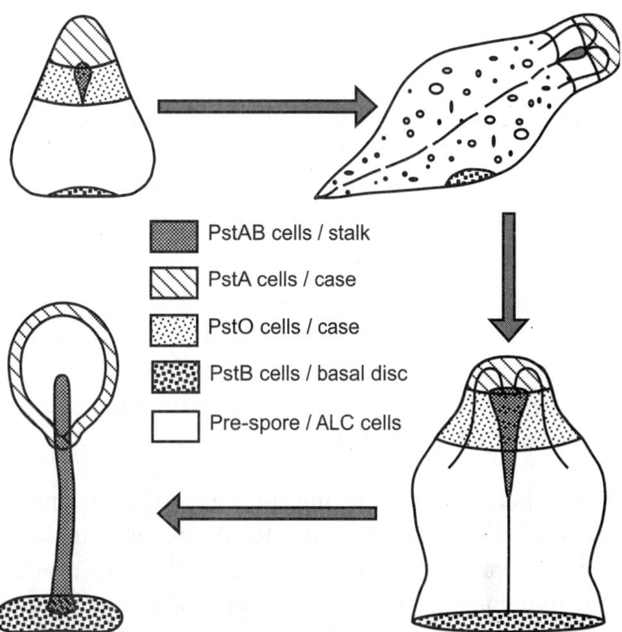

Fig. 12.2 Distribution and relative movement of different cell types during the multicellular phase of the *Dictyostelium* life cycle.

12.3 SIMPLE DEVELOPMENTAL MODELS

Although the mechanism of development of a number of organisms has been studied, a handful of organisms provide most of our knowledge about developmental mechanisms. The development of a multicellular organism is a complex process, because it involves not only the diversification of cell types, but also their precise spatial organization. Unicellular organisms may also differentiate in response to their environment and they provide simple model systems involving the same principles seen in multicellular development—differential gene expression controlled by intrinsic and extrinsic factors—but in the absence of pattern formation. The simplest developmental model is probably the organization of viruses and a little more organized model is **sporulation** seen in *Bacillus subtilis*, where two cell types, a spore and a mother cell, differentiate upon nutrient depletion. Yeast cells demonstrate a simple form of **trans differentiation** in their ability to spontaneously switch from one mating type to another, a process, which involves programmed non-reciprocal recombination. The slime mould *Dictyostelium discoideum* alternates between unicellular existence and a simple multicellular organism and is perhaps the most informative model for development in complex organisms (**Fig. 12.2**). The aggregate of cells demonstrates many of the properties of animal embryos, including morphogenesis, simple mechanisms of regional specification and the ability to regulate for missing parts. Sea urchins and frogs were the main animals used for the first experimental investigations, because their developing embryos are easy to obtain. Among vertebrates, the frog *Xenopus*, the mouse, the chick and more recently the zebra fish are the main model systems studied. Among invertebrates, the fruit fly *Drosophila* and the nematode *Caenorhabditis elegans* are the most studied organisms. In the following sections, a brief mention is made of a few organisms and the developmental process of *Drosophila* has been discussed in more detail.

12.3.1 Differentiation in viruses

Viruses have the simplest organization among all life forms and so have been well studied. Development at the level of viruses involves, as in higher organisms, the production of proteins that provide specific phenotypes to their carriers. In general, the developmental stages of viruses involve genome replication and formation of virion (assembly of head and tail structures into a mature virus). The assembly of virion takes place in specific sequence; for each step of the sequence products of specific genes are required. This indicates that different genes become activated at different points during phage development (i.e., expression in a pre-programmed sequence). Although mechanisms of pre-programming are unique to each viral type, all the viruses share some common features.

- The genes of different viruses can be grouped into 2–4 sets, depending on the time of gene expression starting from the time of infection. The members of different sets are active at different points of time, but always simultaneously.
- The gene products of one or more genes of a set have the tendency to repress the members of that set, while activating the set that follows. This differential gene expression has been regulated primarily at transcription level.

The rod-shaped tobacco mosaic virus (TMV) consists of only two components: RNA genetic material and coat protein. During this self-assembly process approximately 2100 identical protein subunits interact with each other independently, beginning at the 5' end of a single TMV RNA and coat protein components; no other information appears necessary to erect this simple structure. However, the morphology of many viruses is far more intricate. In general, complex assembly systems prevail which depend upon the timed arrival of different kinds of proteins as well as other molecules and their sequential incorporation into a relatively elaborate morphological structure.

In most cases, the sequential gene expression is regulated by modification of specificity of RNA polymerases. The polymerase can be either of viral or bacterial origin. The simplest system of regulation operates in phage T4 of *E. coli*. When this phage infects an *E. coli* cell, the bacterial RNA polymerase transcribes its 'early' set of genes. One gene of the 'early' set codes for a T4 RNA polymerase, which transcribes only those genes that belong to the 'late' set; it does not transcribe the 'early' genes.

12.3.2 Differentiation in bacteria

The major morphological changes that occur during spore formation have received considerable attention because some of the biochemical and genetic steps can be analyzed in detail. Differentiation in bacteria can be clearly explained in spore formation (**Fig. 12.3**). Bacterial spores are dormant structures without any metabolic activity (do not replicate). They are resistant to many environmental stresses (e.g., heat, UV irradiation, desiccation, organic solvents and enzymes) and survive for hundreds of years under adverse conditions. When favourable conditions are present, they germinate to yield normal vegetative bacterial cells. Spore formation (sporulation) is generally triggered by a scarcity of nutrients, especially carbon and nitrogen. Prespore production can be observed in some *Bacillus* and *Clostridium* species also. Various experiments have suggested that sporulation in bacteria is associated with the replacement of the sigma subunit in DNA-directed RNA polymerase. The entire sporulation process can be divided into seven morphologically distinct stages, each controlled by a particular set of genes.

12.3.3 Differentiation in yeast

Yeast fission is a unique model of development and differentiation. It is similar to eukaryotic species, which reproduce sexually through differentiation of two alternate sexes; yeasts have also evolved unique types of sexual differentiation mechanisms. The best understood example is that of the commonly known baker's yeast, *Saccharomyces cerevisiae*. This yeast is referred to as a homothallic yeast as it can switch within the same population between two alternate mating types, namely, *a* and α (alpha). Thus, a cell with mating type *a* can switch to α and vice versa, in a highly regulated manner,

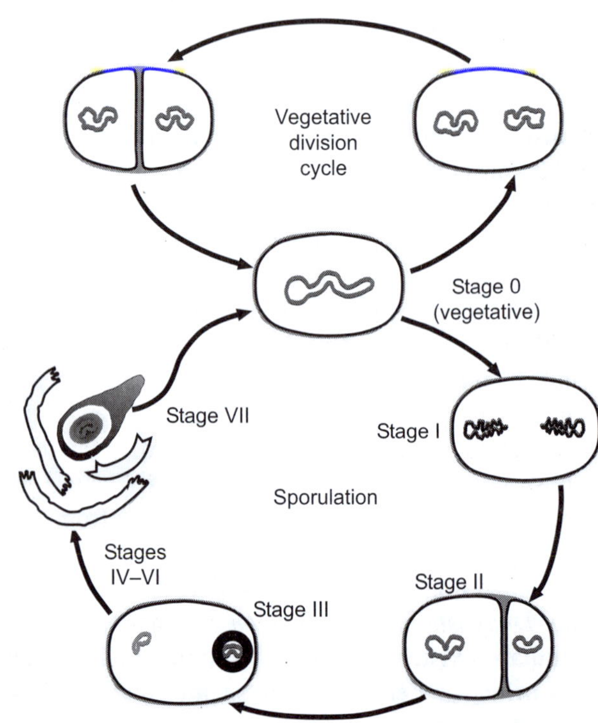

Fig. 12.3 Life cycle of *Bacillus subtilis*, showing the vegetative division cycle and sporulation.

such that two out of four granddaughters, which are sister cells, switch to the opposite mating types (**Fig. 12.4**). A haploid yeast cell divides by budding and replicated chromosomes are distributed between two asymmetric cells—the larger progenitor cell (or mother cell) and a smaller bud (or daughter cell). A mother cell and its daughter cells can exhibit different mating types. This difference arises by a process called **mating type switching**. It is regulated by the product of the HO gene, a sequence-specific endonuclease. HO triggers gene conversion within the mating type locus (MAT locus) by creating a double strand break at one of the two silent mating type cassettes. The HO is activated in mother cell only, but it is kept silent in the daughter cell due to the selective expression of a repressor called ash1 and this is why the daughter cell does not switch mating type. The ash1 gene is transcribed in the mother cell prior to budding, but the encoded RNA becomes localized within the daughter cell. During budding, the ash1 mRNA attaches to the growing ends of microtubules. Several proteins also function as 'adapters' that bind the 3'UTR of the ash1 mRNA and also to the microtubules.

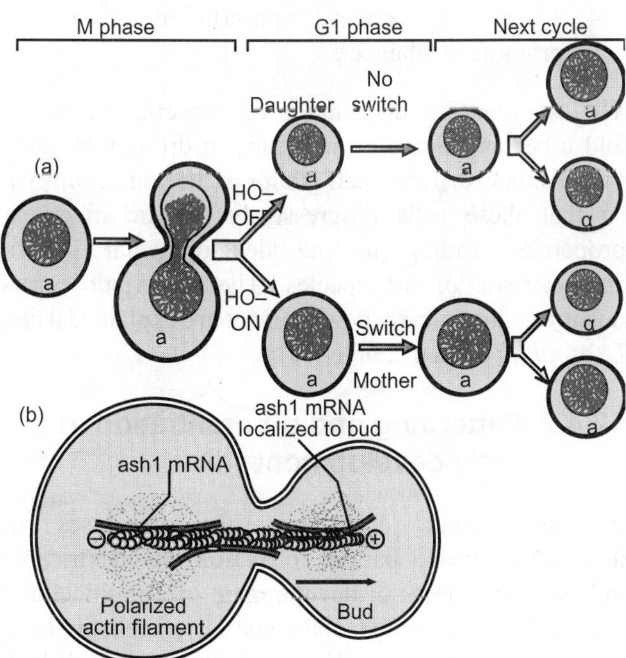

Fig. 12.4 A haploid yeast cell of mating type a undergoes budding to produce a mother cell and smaller daughter cells: (a) Initially, both cells are mating type *a*, but sometimes the mother cell can undergo switching to the α type. (b) Localization of ash1 mRNA during budding (Source: Watson et al, 2004).

The microtubules extend from the nucleus of the mother cell to the site of budding and in this way the ash1 mRNA is transported to the daughter cell. Within the daughter cell, the ash1 mRNA is translated into a repressor protein that inhibits the transcription of the HO gene. This silencing of HO expression in the daughter cell prevents that cell from undergoing mating type switching.

12.4 EVOLUTION OF CELL DIFFERENTIATION

One of the biological wonders was the creation of multicellular organisms. In these organisms, the vast variety of cell types was generated through division and differentiation from the descendants of a single cell, the zygote (**Fig. 12.5**). The different cell types have highly divergent morphology and are specialized for carrying out a unique metabolic function.

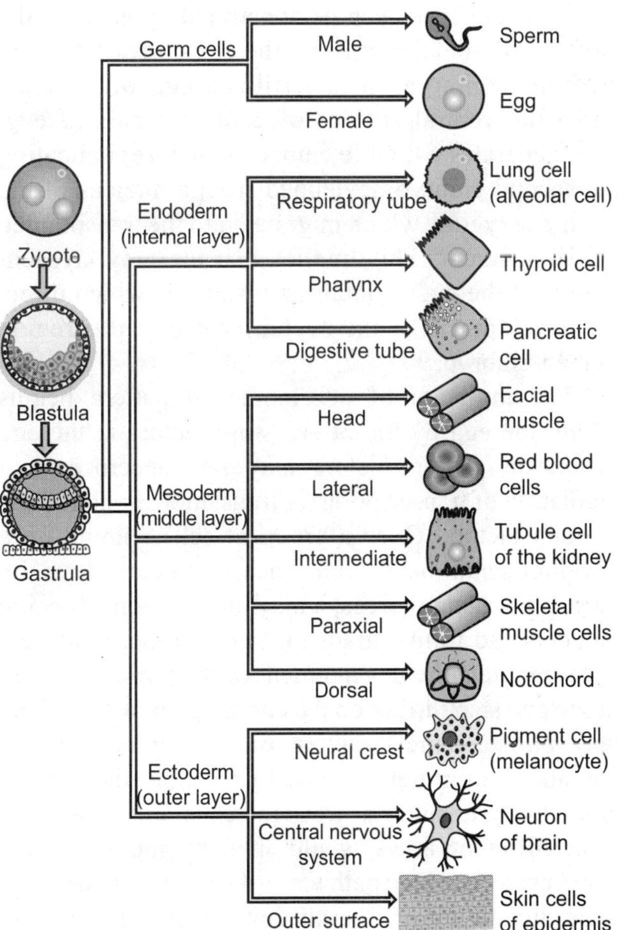

Fig. 12.5 Schematic illustration of some representative differentiated cell types of the vertebrate body.

12.4.1 Generalized scheme for differentiation

Egg fertilization initiates intense protein synthesis, rapid DNA replication and cell division. Until the stage of gastrula, protein synthesis in embryos depends on the mRNA molecules already present in the egg; new mRNA synthesis begins only after the embryos reach gastrula stage. There is an intense RNA synthesis during meiosis in oocytes; in vertebrates, this synthesis occurs on lamp brush chromosomes. Huge amounts of rRNA, mRNA and other components synthesized during meiosis in oocytes support translation until the gastrula stage. The distribution of mRNAs within the egg cell is uneven and specific mRNAs become localized to pre-determined areas in the egg cytoplasm. Thus, in most vertebrates, the genetic programs controlling early development (up to the blastula stage) are established during oogenesis. These programs are present in the egg in the form of localization of specific mRNAs. This has been demonstrated by grafting the enucleated dorsal section of the egg cortex (the gray crescent region) from a fertilized egg of *Xenopus* on to the ventral surface of another fertilized egg. This generates a double embryo structure, indicating the second embryo developed from a nucleus of the recipient zygote, which migrates into the transplanted dorsal segment. This implies that the gray crescent region of the eggs contains substances, which direct the development of the nuclei located in this region into an embryo.

The initiation of developmental pattern begins within the egg by the pre-existing factors in the egg cytoplasm. These factors may be concerned with regulation of transcription, or translation, or activation of such factors. Once a group of cells commit on a developmental path (cell lineage commitment) by these pre-existing forces, it may produce some specific inducers and form a gradient. Some of these inducers may stimulate the development of the pre-determined tissue one step further on the chosen path, while others may initiate a new pattern of development in the neighbouring group of cells. In addition, the inducer gradients present in a group of cells and emanating from different areas would interact and determine the developmental pathway of the other targeted cells. In this manner, each developmental step may generate the stimulus for initiation of the next step leading to the complex developmental patterns of various tissues and organs. The inducer gradients and their interactions would enable the individual cells to 'know' their relative position in the developing embryo. Thus specific groups of genes would be activated in cells due to these inducer gradients and their interactions depending upon the position of cells in the embryo. Since these gradients occur in a pre-programmed manner, the sequences and the end product of developmental events generated by them in different individuals of the same species will be quite uniform.

The asymmetrical distribution of factors responsible for early development is typified by *Drosophila*. But in case of mammals, the egg itself is homogeneous and the asymmetry is acquired during the initial division cycles. In all these cases, development involves differential gene expression mediated by transcriptional factors, which include a variety of proteins. Some important ones are:

1. Chromatin-remodelling—those that change the chromatin structure at a promoter region
2. Transcription initiation factors—those that are needed to initiate transcription at a particular active promoter
3. Those that regulate the activity of an enhancer
4. Those that activate transcription at specific promoters/enhancers

The asymmetrical distribution of transcription factors and other associated events leads to differential gene expression in different cell groups of the young embryo, so that these cells progressively acquire divergent properties leading to the developmental pattern characteristic of the species. The important events occurring during early development are explained using *Drosophila* as an example.

12.4.2 Patterning and differentiation in early development

In multicellular organisms, the processes of differentiation and pattern formation are inextricably linked. The process of development of a multicellular organism is a complex orchestration of cell division, migration, death, growth and differentiation, which leads to the final genetically determined architecture of the body. The early stages of development involve the segregation of different lineages (e.g., three germ layers: ectoderm, mesoderm and endoderm in vertebrate embryos) and the establishment of the first crude positional cues, in the form of the major body axes.

As mentioned earlier, the developmental program is controlled by both intrinsic and extrinsic information. In most animals and plants, the course of development is initiated by intrinsic biochemical asymmetry in the egg. Particular molecules are segregated into different cells as the egg cleaves and cells arising at different positions in the embryo are therefore non-equivalent. Such molecules are termed cytoplasmic **determinants** and are products of the maternal genome deliberately placed in the egg at certain positions. Hence, it is the maternal genotype, which determines the early developmental phenotype of the embryo—a phenomenon referred to as the **maternal effect**.

Further developmental cues may arise through external stimuli. In all vertebrate embryos, for instance, the dorso-ventral axis is specified by an external physical stimulus in the environment. Gastrulation then begins on the dorsal surface of the embryo and in doing so establishes the antero–posterior axis. In frogs, the dorso-ventral axis of the embryo is specified by the act of fertilization—the region opposite the site of sperm entry becomes the dorsal side via the mechanism of cortical rotation. Gravity and rotation are also important in the polarity of avian and mammalian embryos. In each case, homologous molecules are expressed in the dorsal organizing centres.

12.4.3 Mosaic and regulative development

The relative predominance of cytoplasmic determinants and cell–cell interactions in early development differs across the animal kingdom. Generally, cytoplasmic determinants appear to be most important in invertebrate embryos and cell–cell interactions in vertebrate development. These two sources of information define two extreme mechanisms of development, termed mosaic and regulative (**Fig. 12.6**). A **fate map** of an egg or an embryo shows the fate (normal developmental outcome) of each region. In mosaic development, the fate of an individual cell is determined entirely by its intrinsic characteristics (i.e., its cytoplasmic determinants), whilst in regulative development, the fate of each

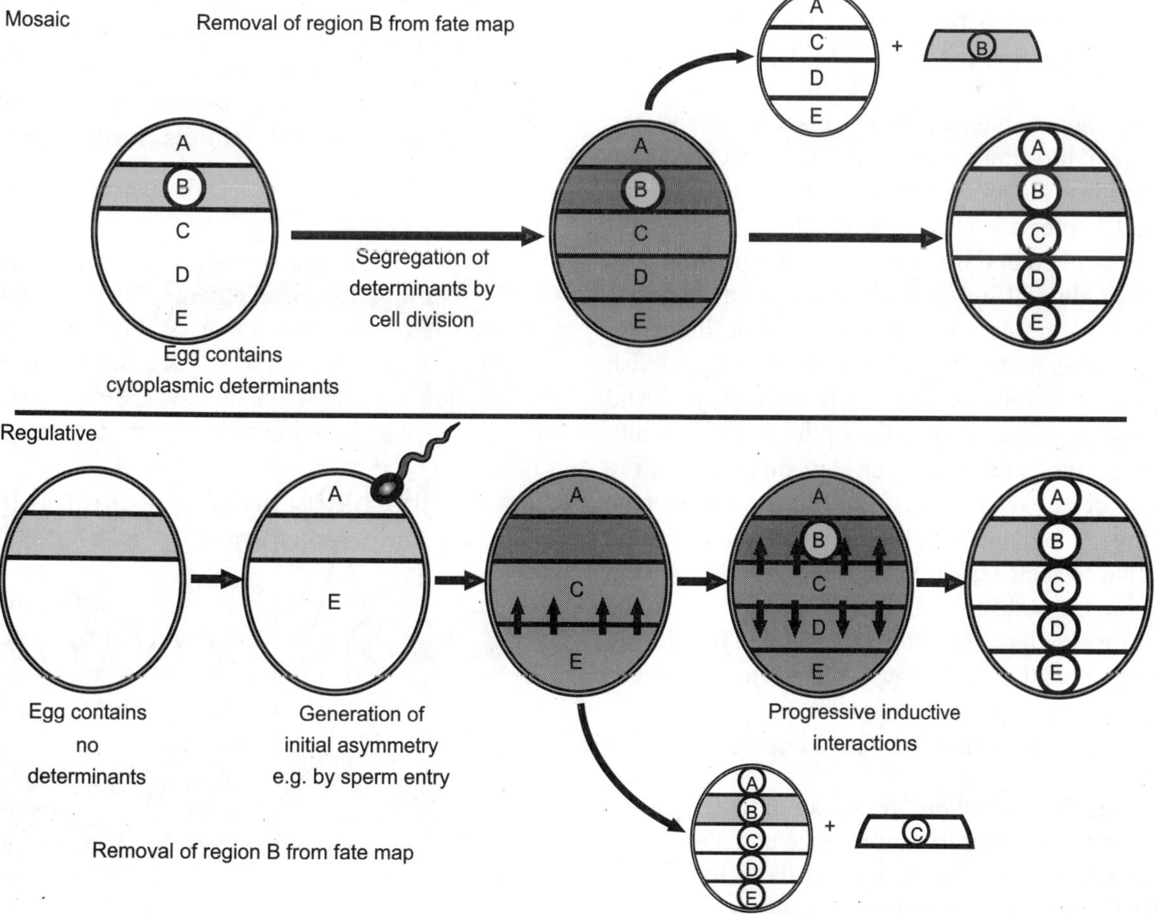

Fig. 12.6 Behaviour of embryos and explants during mosaic and regulative development. (Source: Twyman RM, 2003)

cell is determined by its interactions with other cells. In mosaic development, therefore, each cell undergoes **autonomous speciation**, i.e., in principle, if removed from the embryo it will develop according to its intrinsic instructions. An isolated cell, thus differentiates into the appropriate part of the embryo and the remainder of the embryo will lack that part (**community effect**). The fate map of the embryo can be reconstructed from the fate of individual cells differentiating in isolation. In regulative development, each cell undergoes **conditional specification**, i.e., if it is removed from the embryo it will fail to fulfil its fate because it lacks the appropriate interactions with other cells. The remainder of the embryo, however, can regulate itself and replace the missing parts. This is because the appropriate interactions have yet to occur and therefore the fates of the remaining cells are not determined. The region of the embryo driving such interactions is termed an **organizer**. It has the special property of being able to induce the formation of new structures when transplanted to another region of the embryo, because it is the source of the primary inductive signals.

12.4.4 Cell–cell interactions in development

Cells responds to signals from other cells and the environment by altering patterns of gene expression and activities of protein molecules already in the cell (see Chapter 10 Signal transduction). Both these responses are potential routes to differentiation and where a single cell influences the developmental fate of another, the process is termed **induction**. Inductive signalling takes many forms. First, one can distinguish between interactions involving different cell types and those involving equivalent cells. In the former, locally secreted active substances may mediate signalling between cells over short distances (**paracrine signalling**) or by signals released by distant cells and transported to their target through the vascular system (**endocrine signalling**). Cell–cell contact may be required for signal transduction (also termed **juxtacrine signalling**), or the cell may respond to molecules secreted into the extracellular matrix.

There are two kinds of inductive interactions.

- In ***instructive induction***, the responding cells adopt different fates in the presence and absence of the inducer, i.e., the inducing cells instruct the responding cells to follow a particular developmental pathway.
- In ***permissive induction***, the responding cells are already committed to a certain fate but need a particular signal to permit them to continue, i.e., there is no choice involved.

Responses to induction can also be diverse. The classical response is uniform, i.e., the inductive signal causes the responding cell to adopt a single, characteristic fate. This often occurs when two layers of tissue are brought together (appositional induction) or where induction is considered at a single cell level. A **morphogen**, however, is a diffusible signal, which establishes a concentration gradient over a given population of cells (a morphogenic field) and elicits different responses at different concentration thresholds. **Homogenetic induction** is a propagated response where the responding tissue is induced to differentiate into the inducing tissue and then induces its neighbouring cells to follow the same fate. These different types of induction are summarized in **Fig. 12.7**.

The induction mechanisms all concern interactions between non-equivalent cells. Differences can also arise within populations of equivalent cells by a process termed **lateral inhibition**. In this model (Fig. 12.7), equivalent, undifferentiated cells secrete a signal which suppresses differentiation in surrounding cells and inhibits further signalling.

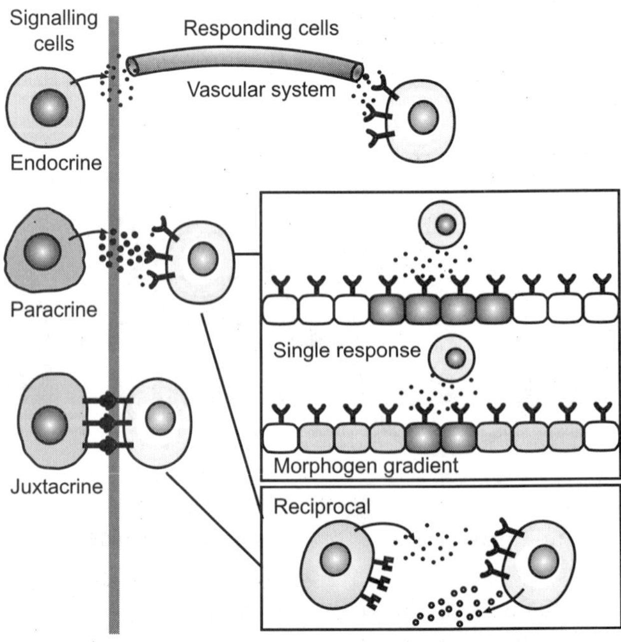

Fig. 12.7 Ranges and types of induction.

An equilibrium is thus established where all cells signal their neighbours in low amounts. A slight imbalance arising by chance can disrupt this signalling process and become amplified by feed-back. A random burst of signal released by a central cell represses signalling in the surrounding cells and enables the central cell to differentiate. However, the surrounding cells not only cease to inhibit differentiation, but also cease to inhibit signalling from the central cell. The central cell thus establishes dominance by constitutive high-level signalling, whilst the surrounding cells are able to neither signal nor differentiate. This mechanism is thought to underlie the differentiation of neural precursor cells in both the *Drosophila* ectoderm and the vertebrate brain and the differentiation of the anchor cell in the *Caenorhabditis elegans* uterus.

The converse of lateral inhibition is the **community effect**, where cells behave differently in isolation compared with a population of equivalent cells. This effect is brought about by **autocrine signalling** (the ability of a cell to respond to a signal secreted by itself). Here, equivalent cells secrete a signal, which is necessary for their own differentiation and that of their neighbours.

12.4.5 Commitment and maintenance of the differentiated state

The diversification of the embryo by progressive cell–cell interactions allows an initially small number of cell types to be elaborated into the many distinct cell types in the mature animal or plant. This occurs as a hierarchy, as more cell types arise; more types of inductive interactions are possible. A prominent feature of animal cells, discussed briefly at the beginning of this chapter, is their tendency to become committed to particular developmental pathways. Once the early vertebrate embryo has diversified into ectoderm, mesoderm and endoderm, the cells of those lineages are restricted in their fate and can only differentiate into the spectrum of the cell type characteristic of their own lineage and not those of other lineages. Animal development thus proceeds through a series of irreversible decisions, where cells become progressively restricted to narrower and narrower fates until they are terminally differentiated. At this point, the cells either cease to proliferate or produce only the same cell type as themselves. ***The major exceptions are stem cells and germ cells.***

Commitment occurs in several stages. A cell may be specified to adopt a certain fate by its intrinsic properties and differentiates this way in isolation. However, when exposed to a range of different environments, its fate may be altered by interactions with other cells. A cell becomes **determined** when its commitment is irreversible, i.e., it will follow the same developmental pathway regardless of its environment. The timing of determination can be demonstrated by transplant experiments. Ectoderm in *Xenopus* embryos is specified to become the epidermis but can be induced by underlying dorsal mesoderm to form the neural plate.

12.5 PATTERN FORMATION

Pattern formation is the process by which cells adopt precise spatial organization, a phenomenon, which ensures that all members of a species are morphologically similar. The components of pattern formation are regional specification and morphogenesis. **Regional specification** is the process by which cells acquire **positional values**, i.e., molecular information causing them to behave in a manner appropriate for their position in the organism as a whole. Morphogenesis describes the process by which cells form structures, by interpreting the positional information and behaving accordingly.

12.5.1 Regional specification

Two forms of regional specification can be distinguished. One controls the differentiation and causes particular differentiated cell types to arise in particular positions and another controls morphogenesis in cells which are equivalent with respect to their state of differentiation, i.e., it makes them behave differently so that they give rise to regionally specific structures. Examples of both processes can be found in the vertebrate limb. The limb comprises a number of different cell types which form the bone, muscle, skin and so on and these become organized into a roughly concentric pattern so that the bone is on the inside, surrounded by muscle and with skin on the outside. Once this ground plan has been laid, structural differences arise between identical cell types in different locations, e.g., between the forelimb and the hindlimb and between cells at various points along the three principal limb axes. Such differences reflect the individual behaviour of the cells in each position, e.g., differential growth or changes in shape.

12.5.2 Positional information

For an organism, for example, the fruit fly, to develop into a segmented pattern with unique morphological and cellular features, each cell of the body must somehow know its position relative to the other cells. An embryonic cell may respond to positional information in various ways. For example, positional information may stimulate a cell to divide into two daughter cells, or it may cause a cell to differentiate into a particular cell type. Positional information may also cause a cell or group of cells to migrate in a particular direction from one region of the embryo to another, or it may tell a cell that it is supposed to die (**apoptosis**).

Cells acquire their positional values through positional information conferred by patterns of gene expression. The source of positional information was first determined in *Drosophila* through the analysis of mutants whose body pattern was fundamentally disrupted. Two sets of genes control positional values along the antero-posterior axis. The **segment polarity genes** specify the positions of specific cell types in each segment and the **homeotic selector genes** instruct those cells to generate the appropriate regional structures. The combination of genes expressed in a particular cell thus gives it an 'address' and assigns it to a particular region of the embryo. Mutations in the segment polarity genes cause the loss of particular regions of each segment and their replacement by a mirror image of the remaining cells (that is, cell types are specified incorrectly). Whereas mutations in the homeotic selector genes generate spectacular and sometimes bizarre phenotypes where one body part is replaced by the kinases of another (that is, the correct cell types are present but they behave aberrantly).

The animal homeotic genes are not found in the plant kingdom: the fundamentally different body plans and developmental mechanisms suggest entirely different regulatory strategies. However, a family of transcription factors related by their possession of a MADS box (a motif also found in the mammalian serum response factor) has been identified in *Arabidopsis thaliana* and its loss of function causes homeotic transformations in the reiterated structures of the flower (petals, sepals, stamens and carpels).

12.6 MORPHOGENESIS IN DEVELOPMENT

The study of morphogenesis involves an attempt to understand the processes that control the organized spatial distribution of cells that arises during the embryonic development of an organism and which give rise to the characteristic forms of tissues, organs and overall body anatomy. **Morphogenesis** (Greek, *morphe* = shape and *genesis* = creation) is one of the three fundamental aspects of developmental biology along with the control of **cell growth** and **cellular differentiation**. It is the creation of form and structure and can be regarded as the 'third' component of development, taking cells which have been differentiated and endowed with positional values and translating the instructions they have been given into particular movements and behaviour which create precise arrangements in space. However, while morphogenesis may indeed act as the effector of the developmental program, it also plays a very active role in its control. From the earliest stages of development, the manner in which cells behave, grow and divide, adhere to each other or change shape is controled by cellular interactions in the embryo. Thus the framework within which differentiation and regional specification takes place is well established.

12.6.1 Mechanisms of morphogenesis

The basis of morphogenesis is cell behaviour. It may be considered as an isolated process or as the cell interacting with its environment. Intrinsic processes which drive morphogenesis include the rate of cell growth and proliferation, the nature of cell division, changes in cell shape and cell death. For some species, these processes are tightly regulated. In the nematode worm, for instance, every cell division and cell death is written into an invariant developmental program. The predictable movements mean that all cell–cell inductive interactions are also invariant and each individual contains precisely the same number of cells in precisely the same positions. Several genes have been identified in *C. elegans*, which regulate this cell behaviour. An example is lin-14, which encodes a transcription factor required in the first-stage larva. Modulations in the level of lin-14 protein disrupt several cell lineages by causing cells to behave in a **heterochronic** fashion. That is, appropriate for a previous or subsequent stage of development. Overexpression of lin-14 causes many cells to behave immaturely and repeat division characteristic of an earlier stage of development. Conversely, it may cause cells to skip certain divisions and behave in a manner more appropriate for later development. In larger organisms, cell growth and cell division tend to be controlled en masse by growth factors and hormones rather than by an intrinsic program. There is, therefore, a greater degree of variability in cell number and a statistical likelihood, rather than a programmed certainly, that a given cell will divide. Such stochastic influences probably play a major role in the phenomenon of developmental noise, a source of variation in isogenic populations, which are maintained in a constant environment.

Extrinsic processes which drive morphogenesis include cell–cell adhesion and cell–matrix interactions (that is, adherence to extracellular matrix, **ECM**). Cells can sort out in different layers due to the differential adhesion or thermodynamic model. This model states that cells sort themselves based upon differences in adhesion between the cells. Cells that move to the centre of a mixed aggregate of cells have the strongest adhesion. Cells can change the repertoire of **cell adhesion molecules (CAMs)** they express and thus alter their relationship with the cells around them. Two families of cell adhesion molecules are recognized: the Ca^{2+}-dependent cadherins and the Ca^{2+}-independent CAMs of the immunoglobulin superfamily (for example, N-CAM).

12.7 ENVIRONMENTAL REGULATION OF DEVELOPMENT

Much of the phenotypic variation observed in populations arises through the influence of the environment during development, as well as the stochastic influences behind developmental noise. Most organisms use environmental cues in the course of their normal development and at the same time many exogenous compounds found in the environment can divert development from its usual path and cause congenital abnormalities.

12.7.1 Environmental cues and normal development

The inclusion of environmental cues into normal development occurs in most organisms developing under natural conditions. In mosquitoes (*Aedes aegypti*) egg production is triggered by a blood meal. Without blood meal, females do not make vitellogenin yolk protein. The digested products of the blood meal stimulate the brain to secrete the egg development neuro-secretory hormone (EDNH). This hormone stimulates the ovary to make ecdysteroids, which instruct the fat body cells to make vitellogenin for the oocytes. Vitellogenin is critical for egg production.

In some organisms, development of an individual is possible only by the presence of another individual of a different species. It is called **developmental symbiosis**. For example, in case of leafhopper (*Euscelis incisus*), the symbiosis occurs within the egg. Symbiotic bacteria are found within the egg cytoplasm and are transferred through generations—like mitochondria. These bacteria multiply only in the cytoplasm of leafhopper cells; in turn the host completes embryogenesis only in the presence of bacteria. The bacteria are also found to be essential for formation of the embryonic gut.

12.7.2 Predictable environmental differences provide cues for development

If the environment contains predictable components (such as gravity, seasons) these can become part of the development of the organism. The use of temperature and daylight length, for example, is used by numerous species to adjust their development to a changing environment.

For example, many species of insects have developed a survival strategy called **diapause**. It is a suspension of development that can occur at the embryonic, larval, pupal or adult stage, depending on the species. In some species of aphids, diapause is facultative and is often seen in the temperate zone, where it is induced by changes in the photoperiod (the relative lengths of day and night). The day length at which 50% of the population has entered diapause is called the critical day length and is genetically determined property.

Gravity and pressure also provide the developmental cues in many organisms. For example, in case of chicks, the formation of several bones depends on stresses occasioned by the movement of the embryo. Recently, it has been experimentally proved that several bones in the chicken do not form if embryonic movement is suppressed in the egg. In case of humans, such stresses have been known to be responsible for the formation of the human patella after birth.

12.7.3 Polyphenism and reaction norms

The ability of an individual to express different phenotypes under different environmental conditions is called **phenotypic plasticity**. There are two types of phenotypic plasticity: **polyphenism** and

reaction norms. Polyphenism refers to discontinuous phenotypes elicited by the environment. For example, migratory locusts exist in two mutually exclusive forms: a short-winged, uniformly coloured solitary phase and a long-winged, brightly coloured gregarious phase. Cues in the environment (mainly population density) determine which morphology a young locust will take. In other instances, the genome encodes a range of potential phenotypes and the environment selects the phenotype that is usually the most adaptive. For example, the microhabitat of a young salamander can cause its colour to change, but within a genetically defined limit. This continuous range of phenotypes expressed by a single genotype across a range of environmental conditions is called the **reaction norm**. It is thus a property of the genome and can also be selected.

The two phenotypes of the butterflies, *Araschnia levana* and moth *Nemoria arizonaria* depend on the diet. Not all polyphenisms are controlled by seasons. In bees, the size of the female larva at its metamorphic moult determines whether the individual would be a worker or a queen. A larva fed nutrient-rich 'royal jelly' retains the activity of her corpora allata during her last instar stage. The juvenile hormone secreted by these organs delays pupation, allowing the resulting bee to emerge larger and more specialized in her anatomy. The JH level of larvae destined to become queen is 25 times more than the larvae destined to become worker. This type of differentiation is called **nutritional polyphenism**.

12.7.4 Environment-dependent sex determination

Like many developmental processes, sex determination involves both genetic and environmental factors, but this process is unusual in the extent to which these factors can be separated and their relative predominance in different organisms. There are many species in which the environment determines whether an individual is to be male or female. The temperature dependence of sex determination in fishes and reptiles has provided the best-studied case.

12.8 LIFE CYCLES AND THE EVOLUTION OF DEVELOPMENTAL PATTERNS

The life cycle of multicellular organisms starts fertilization and ends in death. The environment has enormous influence on the evolution of an organisms life cycle. For example, in the animal kingdom, an incredibly large variety of embryonic types exist, but most patterns of embryogenesis are variations of five themes:

Cleavage: Soon after fertilization, cleavage (a series of rapid mitotic divisions) occurs. These cells, called *blastomeres* at the end of cleavage, form a spherical structure known as *blastula*.

Gastrula: At the end of blastula the mitotic cell division slows down and blastomeres undergo movements wherein they change their positions relative to one another. This series of extensive cell rearrangements is known as *gastrulation* and the embryo is said to be in *gastrula* stage (**Fig. 12.8**). As a result of gastrulation, the embryo contains three germ layers: ectoderm, mesoderm and the endoderm.

Organogenesis: Once the three germ layers are established, the cells interact with one another and re-arrange themselves to produce tissues and organs and this process is called organogenesis. However, many organs contain cells originated from more than one germ layer. For example, the outer layer of skin comes from the ectoderm while the inner layer (dermis)

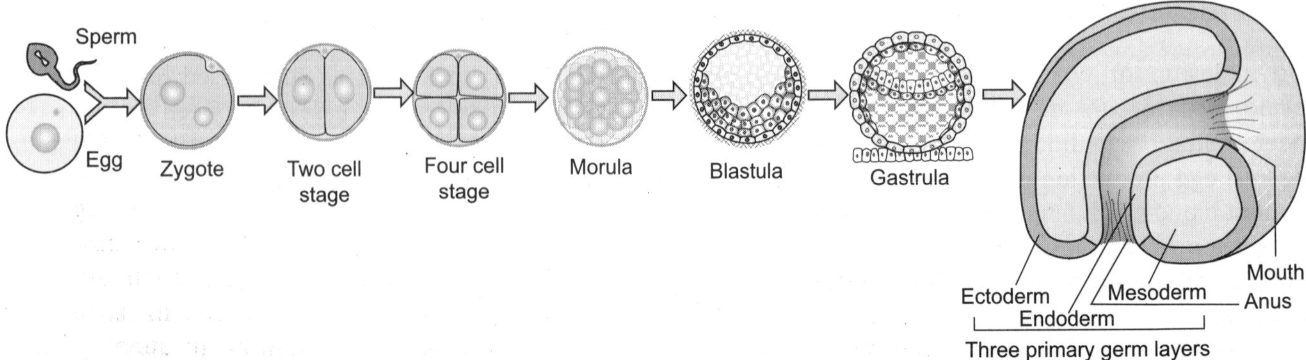

Fig. 12.8 Early embryonic developmental stages of vertebrates. The basic animal body plan, has a sheet of ectoderm on the outside, a tube of endoderm on the inside and endoderm sandwiched between them.

comes from the mesoderm. During organogenesis, certain cells undergo migrations from their place of origin to their target locations. The precursor blood cells, lymph cells, pigment cells and gametes are some examples of migrating cells.

Somatic and germ cells: In most multicellular organisms, a specialized portion of zygote cytoplasm generates cells that are precursors to the *gametes* (sperm and egg) and collectively these cells are called *germ cells* (meant for reproduction). All the remaining cells of the body are called *somatic cells*. The separation of early embryonic cells into germ cells and somatic cells is often said to be the first differentiations to occur. The germ cells eventually migrate to the gonads, where they differentiate into gametes (called *gametogenesis*).

Larva/young organism: In many species, the organism hatches from the egg but is sexually immature and morphologically very distinct from an adult. Larvae mostly constitute the stage of life that is used for feeding or dispersal.

12.9 THE EGG

In most multicellular organisms, the fertilized egg is the beginning of differentiation and development. Through successive cleavages, it produces the many stages of the embryo. Observations in sea urchin eggs revealed that unfertilized eggs contain protein-forming mRNA. These findings indicate that the protein synthesis that occurs upon fertilization utilizes mRNA that has been stored in the egg during its maternal maturation. It has been suggested that pre-fertilization ribosomes, prevented from functioning due to masking proteins, can bind mRNA. Masked mRNA remains inactive until it is released and becomes attached to functioning ribosomes. Differentiation in early stages is therefore probably caused by maternally produced differences in the egg cytoplasm or egg cortex. Biochemical evidence has shown that both RNA and DNA synthesis is firmly controlled by the egg cytoplasm.

12.9.1 Overview of normal human embryonic development

Fertilization of the ovum usually takes place in the fallopian tube. Like every cell in the body, the zygote contains all the genetic information unique to an individual. Half the genetic information comes from the mother's egg and the other half from a single sperm. The zygote spends a few days travelling down the fallopian tube, while dividing mitotically. Embryogenesis begins and development of the foetus is completed in about 40 weeks, undergoing various stages of embryonic development (**Table 12.2**).

A brief summary of the different stages of development of the human foetus from 1 to 40 weeks given in **Table 12.3**.

12.10 MODEL ORGANISM: INVERTEBRATE

Although the development of invertebrates is very diverse, certain features common to the development of most invertebrates are also seen in vertebrate development. These include cleavage, formation of a blastula or blastoderm and gastrulation. However, in invertebrates, *Drosophila melanogaster* has been pivotal in our understanding of developmental genetics. The molecular details of how development is regulated are better understood in this system than in any other animal embryo.

12.10.1 The fruit fly *Drosophila melanogaster*

Drosophila has emerged as an excellent model organism for the study of genetics and development.

Table 12.2 Organ derivatives of the three germ layers

Germ layer	Derivatives
Ectoderm	Epidermis of the skin Central nervous system (brain and spinal cord) All peripheral nerve tissues (neural crest) Some bones of the face Retina, pituitary gland, adrenal medulla Sensory epithelia of the eye, ear and nose
Endoderm	Epithelial parts of Thyroid and parathyroid glands Respiratory tract Gastrointestinal tract, liver, pancreas Terminal part of urogenital systems
Mesoderm	Connective tissue Cartilage and bone Blood cells (red and white) All muscle types (cardiac, skeletal, smooth) Lining of body cavities (pericardial, pleural, peritoneal) Cardiovascular system Testes, ovaries Kidneys, adrenal cortex, spleen

Table 12.3 Development of the human foetus

Weeks (Period of implantation)	Foetal development
1–2	**Fertilization**
A sperm fertilizing an ovum 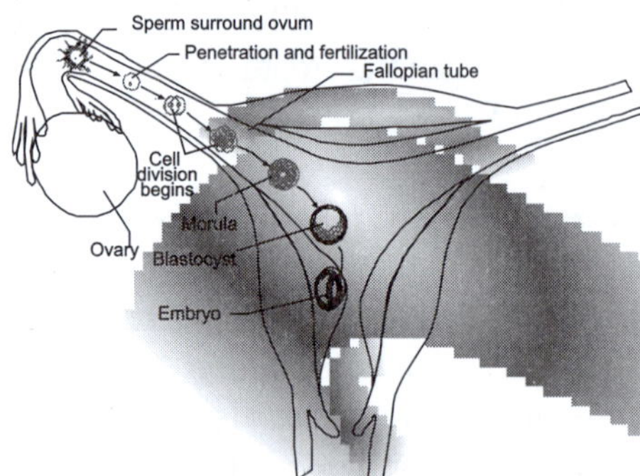	Fertilization of ovum to form zygote; when semen is deposited in the vagina, the spermatozoa travel through cervix → uterus → fallopian tubes. Fertilization of the ovum (egg cell) usually takes place in the fallopian tube. Many sperm battle to penetrate the thick protective shell-like barrier that surrounds the ovum. The first sperm that penetrates fully into the egg (successful one) donates its genetic material (DNA). The egg then quickly polarizes, repelling any additional sperm. The resulting combination is called a **zygote**. The zygote spends the next few days travelling down the Fallopian tube. Meanwhile it divides several times to form a ball of cells called a **morula**. Further cellular division is accompanied by the formation of a small cavity between the cells. This stage is called a **blastocyst**. Up to this point there is no growth in the overall size of the embryo, so each division produces successively smaller cells. Fast mitotic cell division, cleavage and a hollow cavity forms marking the blastocyst 4–5 days after fertilization (more than 100 cells). The blastocyst hatches from its protein shell (zona pellucida) and implants onto the endometrial lining of the mother's uterus. Implantation 6–9 days after fertilization. Trophoblast cells surrounding the embryonic cells proliferate and invade deeper into the uterine lining. They will eventually form the placenta and embryonic membranes. The three basic germ layers of the embryo develop, namely ectoderm, mesoderm and endoderm. Formation of yolk sac.
	Embryonic period
The embryonic period begins at fertilization (12–24 h after ovulation, generally between the 2nd and 3rd week of gestational age) and continues until the end of the 10th week of gestation (8th week by embryonic age).	
3 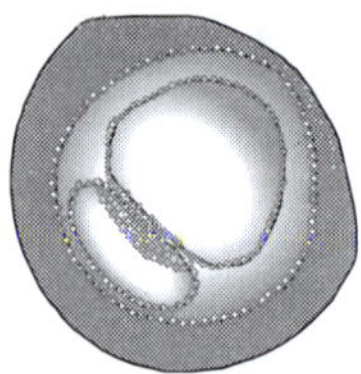	***Block of menstruation:*** This may be the first sign of pregnancy in a female. Beginnings of the backbone. A notochord forms in the centre of the embryonic disk. Neural tube develops over the notochord, the beginning of the brain and spinal cord (first organs). Embryo about 2 mm long.
4 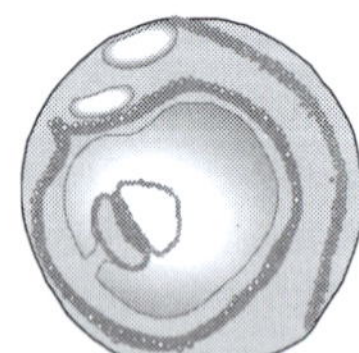	Somites, the divisions of the future vertebra form. Heart tubes begin to fuse. Heart, blood vessels, blood and gut start forming. Umbilical cord developing. Embryo about 5 mm long.
5–6 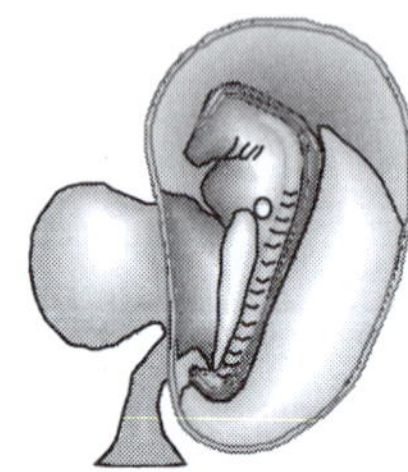	***Brain developing:*** 'Limb buds', small swellings which are the beginnings of the arms and legs. Heart is a large tube and starts to beat, pumping blood. Bronchial arches, grooves which will form structures of the face and neck form. Neural tube closes. The ears begin to form as otic pits. Arm buds and a tail are visible. Embryo about 8 mm long. Lens pits and optic cups form the start of the developing eye. A primitive mouth and nasal pits form. The brain divides into 5 vesicles. Leg buds form and hands form as flat paddles on the arms. Rudimentary blood moves through primitive vessels connecting to the yolk sac and chorionic membranes. The embryo measures 13 mm.
7–8 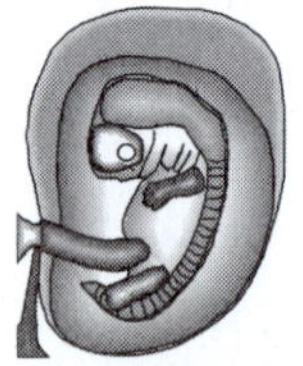	All major internal organs developing. Face forming. Eyes have some colour. Mouth and tongue. Beginnings of hands and feet. Nipples and hair follicles begin to form. Location of the elbows and toes are visible. Foetus is 17 mm long.

Table 12.3 Continued...

Weeks (Period of implantation)		Foetal development
		Foetal period
	The foetal period begins at the end of the 10th week of gestation (8th week of development). Since the precursors of all the major organs are created by this time.	
12	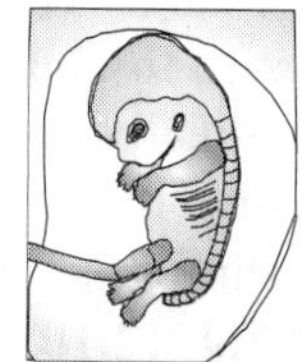	Foetus fully formed, with all organs, muscles, bones, toes and fingers. The head comprises nearly half the size of foetus. The baby teeth appear. Foetus can make a fist with its fingers. Sex organs well developed. Red blood cells are produced in liver. Foetus is moving. For the rest of the gestation period, it is mainly growing in size. Foetus is 56 mm long from head to bottom. Pregnancy may be beginning to show.
16	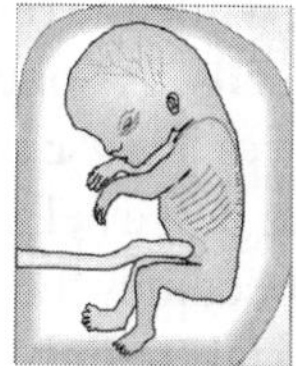	The foetus reaches a length of about 15 cm (6 inches). Fine hair called lanugo develops on the head. Fetal skin is almost transparent. More muscle tissue and bones develop, and the bones become harder. The foetus makes active movements. Sucking motions are made with the mouth. Meconium is made in the intestinal tract. The liver and pancreas produce fluid secretions.
20		The lanugo covers the entire body. Hair begins to grow, including eyebrows and eyelashes. Fingerprints developed. Fingernails and toenails growing. Firm handgrip. Between 16–20 weeks the baby usually moves for the first time. Foetus is 160 mm long from head to bottom.
24	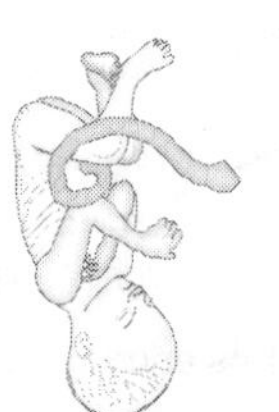	Eyelids open. Foetus has a hand and startle reflex. Alveoli (air sacs) are forming in lungs.
26	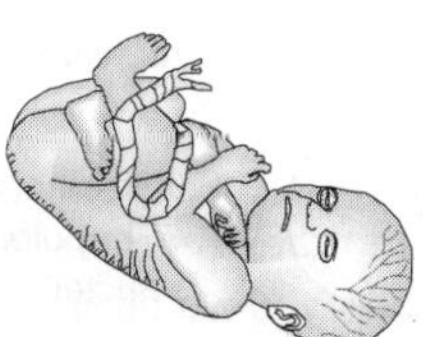	The brain develops rapidly: The nervous system develops enough to control some body functions. The respiratory system has developed to a point where gas exchange is possible. The foetus has a good chance of survival if born prematurely.
28	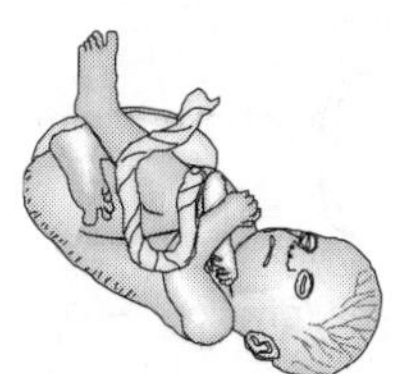	Foetus moving vigorously. Responds to touch and loud noises. Swallowing amniotic fluid and urinating. The amount of body fat rapidly increases. Thalamic brain connections, which mediate sensory input, form.
30	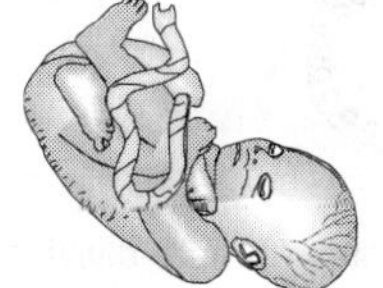	Lanugo begins to disappear. Usually lying head down ready for birth. Baby is 240 mm from head to bottom.
38-40	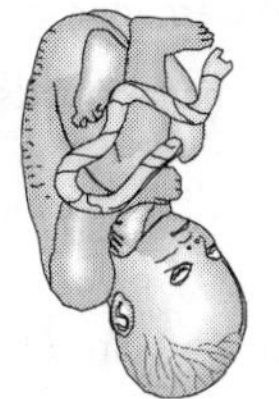	The foetus reaches a length of about 40–48 cm (16-19 inches) and weighs about 2.5 to 3 kg (5 lb 12 oz to 6 lb 12 oz). Lanugo begins to disappear; body fat increases and fingernails reach the end of the fingertips. A baby born at 36 weeks has a high chance of survival, but may require medical interventions. The foetus is considered full-term at the 37th week of pregnancy. Small breast buds are present on both sexes. Hair on head is now coarse and thicker. (9 months) Birth.

The advantages of the fruit flies as experimental material are many. Other than being of small size and with short life cycle (10–14 days), accurate genetic maps available. The life cycle of *Drosophila* consists of embryogenesis, three larval stages, a pupal stage and the adult stage. A number of mutant strains with altered developmental pathways have been identified. At larval stage, *Drosophila* is large enough to conduct transplantation experiments, yet small enough to determine where particular genes are expressed at critical stages of development. Recently, the entire genome has been sequenced and the presence of around 13,600 genes has been reported. The complete life cycle of *Drosophila* is shown in **Fig. 12.9**. The making of the fly occurs as the mother begins to build the egg. The egg includes storage material, proteinecious yolk, lipid vesicles, mitochondria and ribosomes. Some kind of positional information is needed to make the pattern of the fly and to organize the cells that they make in the pattern of the fly, i.e., to so organize the cells that they make all the body parts in their proper places. This information must interact with the genetic information carried in the nuclei of the egg and the sperm. This is the process that determines the pattern formation.

12.10.2 Early stages of embryonic development

The progressive growth of a fertilized egg into an adult organism involves four types of cellular events: cell division, movement, differentiation and death.

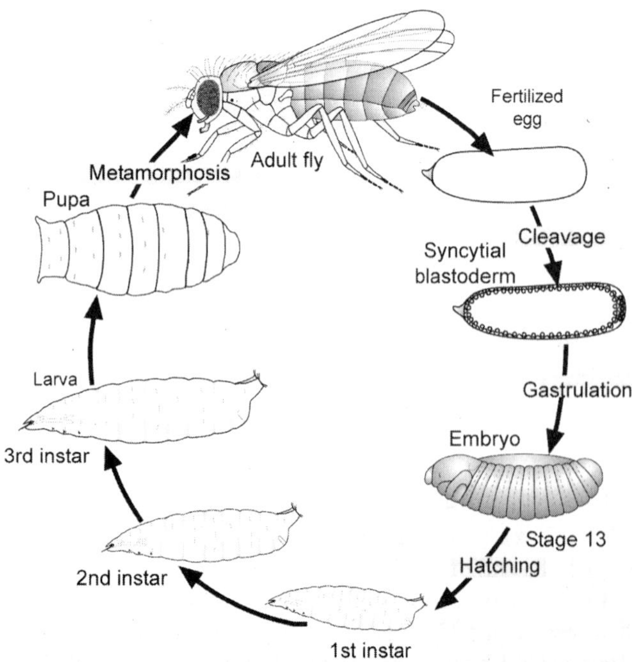

Fig. 12.9 Life cycle of *Drosophila melanogaster*.

It is the coordination of these four events that leads to the formation of a body with a particular pattern. *Drosophila* development begins within a follicle that contains the oocyte surrounded by follicle and nurse cells. The asymmetry in molecular distribution is established during oogenesis. A single cell of the ovary undergoes meiosis to give rise to one haploid cell, which divides four times mitotically to produce 16 cells, all of which are enclosed by a layer of somatic cells, called **follicle cells**; this entire structure constitutes a follicle in the ovary (**Fig. 12.10**). One of the 16 cells develops as an oocyte, while the remaining 15 haploid cells serve as **nurse cells**. At one end of the oocyte, **cytoplasmic bridges** or **ring canals** are formed between the oocyte and the nurse cells that are adjacent to it. This end of oocyte becomes the anterior end of the egg. Cytoplasmic material (mainly, protein and RNA) is transported from nurse cells to the oocyte and accounts for a considerable volume of the egg.

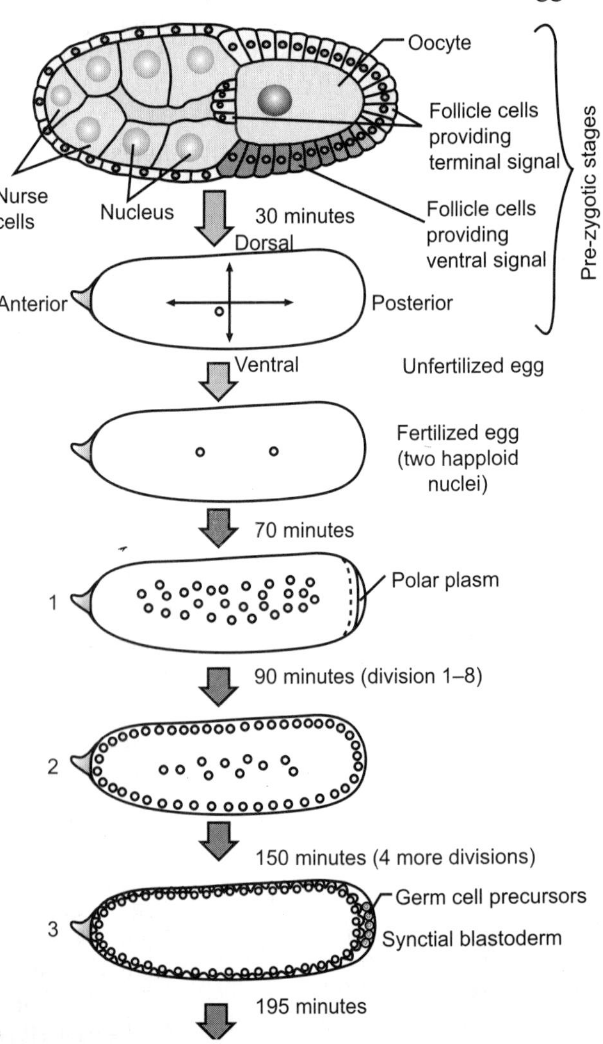

Fig. 12.10 Early development: Stages 1–3 in *Drosophila melanogaster*.

The *Drosophila* egg is sausage-shaped and the future anterior end is recognizable by the micropyle, a nipple-shaped structure in the tough external coat surrounding the egg. The sperm enters the anterior end of the egg through the micropyle. After fertilization, the egg has two parental haploid nuclei and its posterior region is distinct from the rest of the egg, having a polar plasm. The two haploid nuclei fuse to produce a diploid nucleus, which divides in the common egg cytoplasm. The diploid nuclei undergo a series of rapid mitotic divisions—one about every 9 minutes, or a total of 13 times in the space of about 3 1/2 hour—forming a **syncitium**, or a group of nuclei without cell membrane present in a common cytoplasm. The embryo remains single cell during its early development. At division 7, some nuclei migrate towards the cortex or periphery of the egg and divide (another three divisions) to form a layer at the surface of the egg (called **syncytial blastoderm**). The nuclei divide four more times (formation of a monolayer of about 6,000 nuclei), after which membranes surround them to form somatic cells. This gives rise to cellular **blastoderm (Fig. 12.11)**.

Before the formation of cell membranes, the nuclei are **totipotent** or uncommitted. Just after cellularization, however, nuclei become irreversibly '**determined**' to differentiate into specific tissues in the adult fly. At this point in development, particular regions of the egg are determined to become particular types of adult structures. Determination begins with the migration of the nuclei to the egg surface to generate syncytial blastoderm and progresses gradually over the next few cell divisions. The fate of individual regions of the egg becomes increasingly restricted. The type of tissue or organ that will develop from a given cells depends on its precise location relative to the anterior–posterior and dorso-ventral gradients in the blastoderm.

About 15 or so end up at the posterior end of the embryo and develop into **pole cells**, which later give rise to germ cells. Because of the formation of a syncytium, even large molecules such as proteins can diffuse between nuclei during the first 3 hours of development. All the future tissues are derived from the single epithelial layer of the cellular blastoderm. For example, the prospective mesoderm is located in the most ventral region, while the future midgut derives from two regions of prospective endoderm, one at the anterior and the other at the posterior end of the embryo. Endodermal and mesodermal tissues move to their future positions inside the embryo during gastrulation, leaving the ectoderm as the outer layer. Gastrulation starts at about 3 hours after fertilization when the future mesoderm in the ventral region invaginates to form a furrow along the ventral midline. The mesodermal cells are initially internalized by formation of a mesodermal tube (similar to neural tube formation in vertebrates). The mesoderm cells then separate from the surface layer of the tube and migrate under the ectoderm to

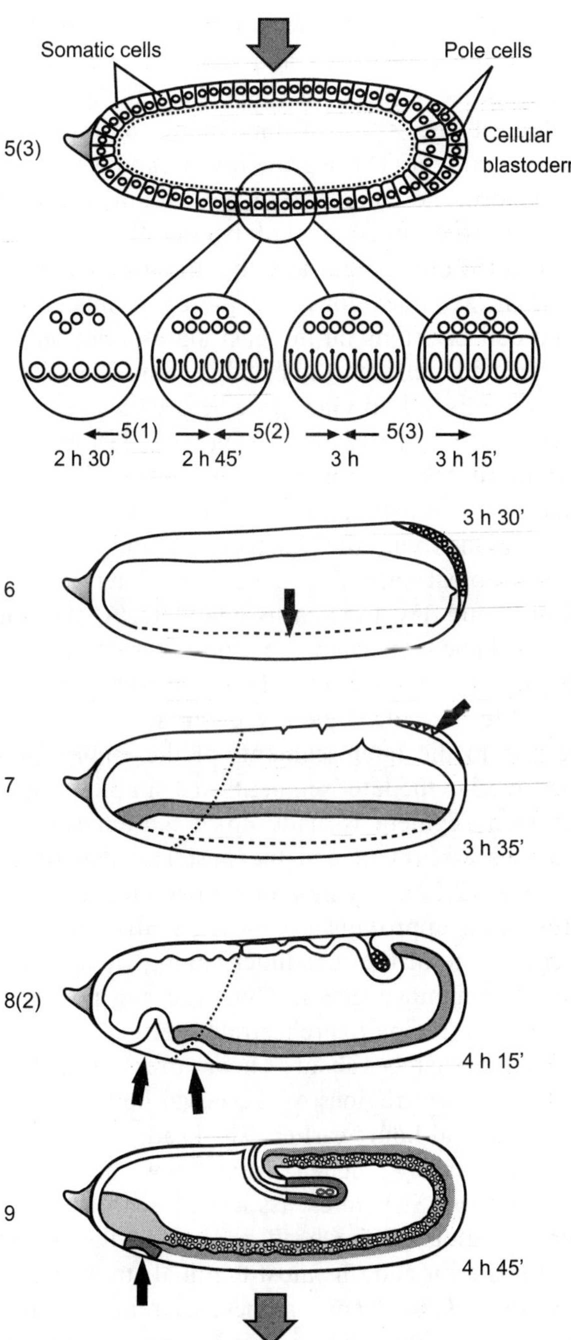

Fig. 12.11 Early development: Stages 5–9. The arrows indicate features that help staging embryos.

internal locations, where they later give rise to muscle and other connective tissue.

Development then proceeds through gastrulation, in which cells grow inward, forming the basic germ layers of the embryo. From these layers, various adult structures will arise. In insects the main nerve cord lies ventrally, rather than dorsally as in vertebrates. Shortly after the mesoderm has invaginated, ectodermal cells of the ventral region that will give rise to the nervous system leave the surface individually and form a layer of neuroblasts between the mesoderm and the outer ectoderm. At the same time, two tube-like invaginations develop at the sites of the future anterior and posterior midgut. These grow inward and eventually fuse to form the endoderm of the midgut, while the ectoderm is dragged inward behind them at each end to form the foregut and the hindgut. The outer ectoderm layer develops into the epidermis. There are no cell divisions during gastrulation, but once it is completed, cells start to divide again. The cells of the epidermis only divide twice before they secrete a cuticle. During gastrulation, the ventral blastoderm or **germ band**, which comprises the main trunk region, undergoes extension, which drives the posterior trunk regions round the posterior end and onto what was the dorsal side. The germ band later retracts, as embryonic development is complete. At the time of germ band extension the first external signs of **segmentation** can be seen. The early segments that are visible are called **para-segments**. They do not give rise to the later segments of the embryo, but rather overlap the later segments in a simple fashion: Each later segment is made up of the anterior end of one para-segment and the posterior end of the next (**Fig. 12.12**). The mother deposits RNA species representing approximately 80% of all genes into the egg; some of the remainder belongs to a special class of regulatory genes. Thus, the egg possesses the complete recipe to orchestrate life.

The larva hatches about 24 hours after fertilization, but the different regions of the larval body are well defined several hours earlier. The head is a complex structure, largely hidden from view before the larva hatches. Structures associated with the most anterior region of the head are called the **acron**. At the posterior end, the most terminal structures are called the **telson**. Between these extremities, three thoracic segments and eight abdominal segments can be distinguished by specializations in the cuticle secreted by the epidermis. On the ventral side of

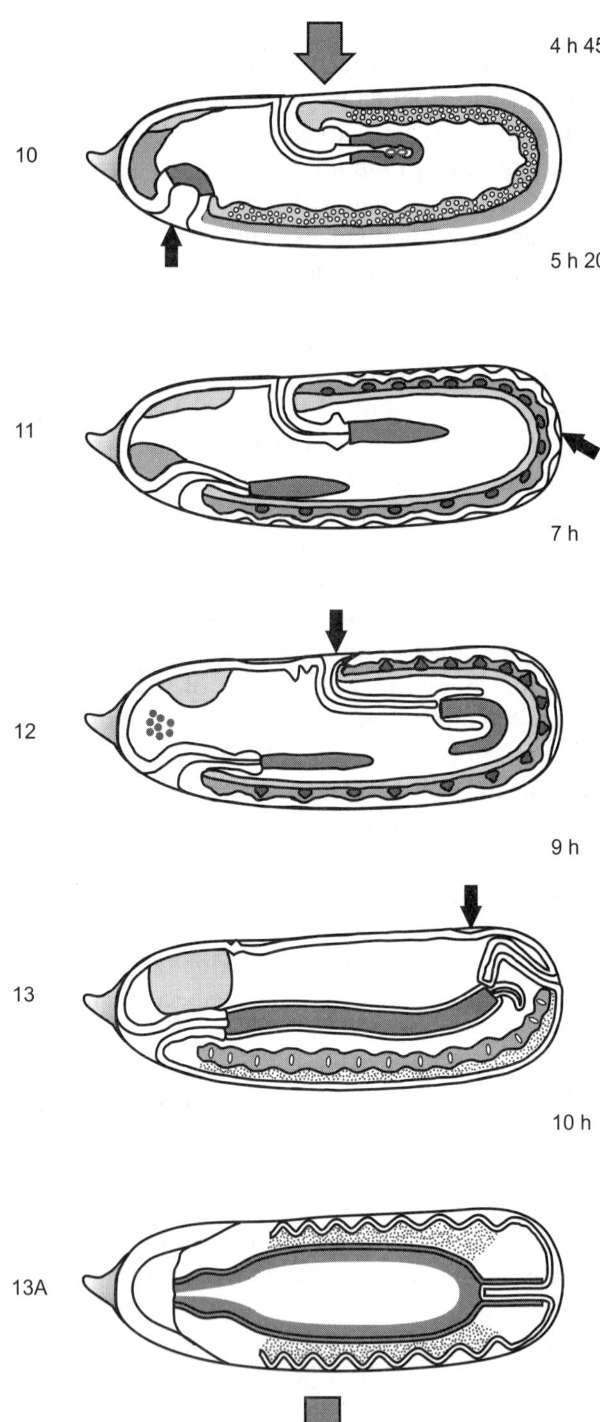

Fig. 12.12 Early development: Stages 10–13A. The arrows indicate features that help staging embryos.

each segment are dentricle belts and other cuticular structures characteristic of each segment. The fates of these segments have been determined by treating them with harmless dyes. A projection of adult structures on embryonic tissue is called a fate map. As the larva feeds and grows, it moults, shedding its cuticle, which occurs twice, each stage being called **instar**. The larva persists for about 4 days, before forming the pupa.

Larva undergo hormone-induced metamorphosis within the pupa after the third instar to develop adult parts. Most of the larval tissues, except the brain and a few others, are destroyed and used to fuel the construction of new adult organs. These structures are, however, already present in the larva as **imaginal discs**—small sheets of prospective epidermal cells derived from the cellular blastoderm and usually containing about 40 cells each. These discs grow throughout larval life and form folded sacs of epithelia to accommodate their increase in size. There are imaginal discs for each of the six legs, two wings and the two halteres and for the genital apparatus, eyes, antennate and other adult head structures (**Fig. 12.13**). In the segments of the abdomen there are groups of about ten blast cells. These cells do not divide but will contribute to the epidermis of the adult at metamorphosis.

12.10.3 Establishment of body axes

In multicellular organisms, the spatial arrangement of their tissues and organs is the most unique feature.

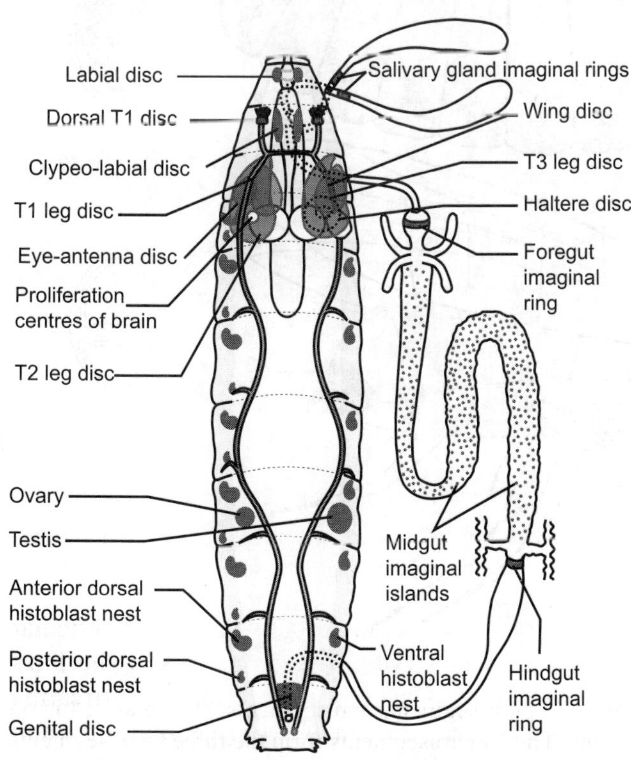

Fig. 12.13 The origin of the adult cells.

Each body part is not only different, but occupies localized positions that are fairly uniform in all individuals of the species. Therefore, it is evident that for such precise and repeatable patterns to occur, they must be preceded by a precisely ordered pattern of development. Generally, initiation of pattern begins with the egg. Uneven deposition of maternal substances in the egg cytoplasm or by external factors such as gravity, undoubtedly helps to initiate localized development. **Fig. 12.14** shows how some cells or tissues can obtain information regarding their positions because they are affected by such gradients. This positional information in turn can influence subsequent activity so that different groups of cells follow different developmental patterns, such as moving to new locations and/or causing the production

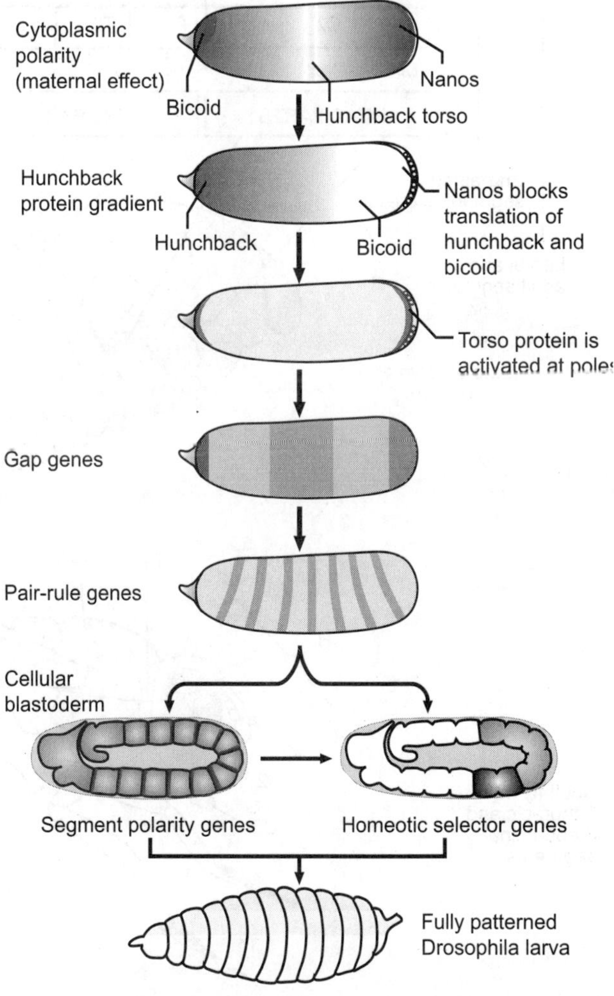

Fig. 12.14 Generalized diagram of anterior–posterior patterning of the egg is determined by three independent systems. The early pattern is established by maternal effect genes that form gradients and regions of morphogenetic proteins.

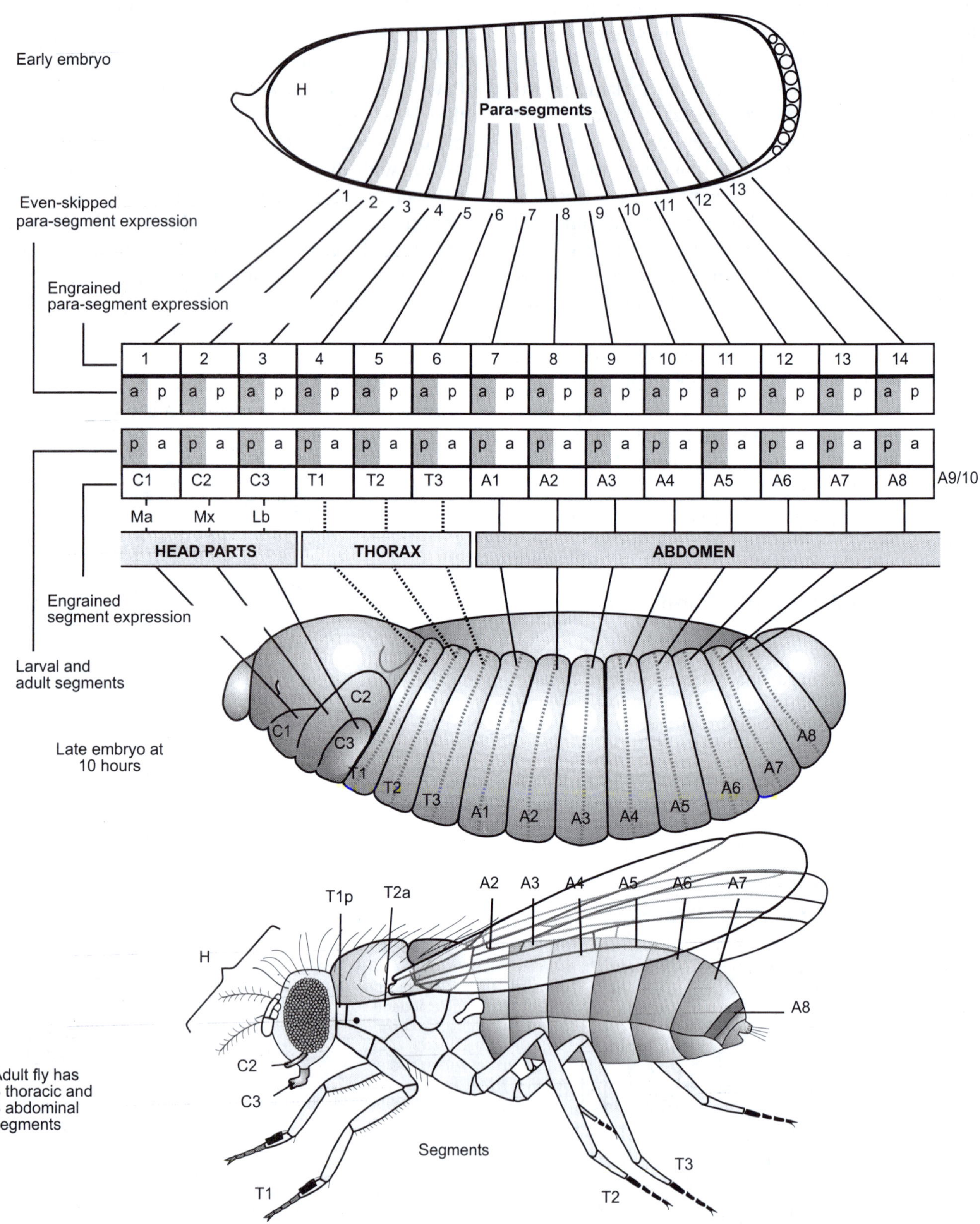

Fig. 12.15 *Drosophila* development proceeds through the registration of para-segments and segments. There are 14 main parasegments in the trunk of the embryo, plus head (H) and tail regions. The 14 parasegments form the three head segments C1–C3 (or Ma, Mx, Lb), the three thoracic segments (mandibula, maxillary and labial), T1–T3 and the eight abdominal segments A1–A8. a: anterior p: posterior.

of inducer substances that set up developmental gradients of their own. It has been suggested each new set of positional coordinates stimulates special 'selector' genes that, in turn, cause cells to commit themselves to a particular developmental direction. Subsequent exposure to further positional influence stimulates a different battery of selector genes leading to further commitment.

In general, discrete regions in the embryo (larva) determine the specific parts of the adult body. These regions are superficially apparent in the distribution pattern of **denticles** (small hairs) on the cuticular surface of larva. The pattern of denticles shows features that vary along both anterior-posterior and dorso-ventral axes. Along the anterior-posterior axis, denticles form 11 discrete bands that correspond on 1:1 basis with the 11 segments of the adult. Along the dorso-ventral axis, the denticles become closer as we move from dorsal to the ventral side of larva. Surprisingly, the pattern of denticle distribution is diagnostic of the overall body organization in both axes. Therefore, much of the analysis of development in mutants has been based on the alterations in their denticle distribution pattern.

12.10.4 Segments and parasegments

The adult *Drosophila* consists of a series of segments, viz., three thoracic (T1–T3) and eight abdominal (A1–A8) segments. The pattern of segmental units is established by the blastoderm stage when the main mass of the embryo is divided into a series of alternating anterior (a) and posterior (p) compartments. A pair of anterior and posterior compartments constitutes a segment, e.g., segment T3 consists of the compartment pair T3A and T3P. In fact, the p and a compartments are not separated morphologically at the blastoderm stage. They are, however, discernible in terms of molecular differentiation. But in the 5- to 6-hour-old gastrula stage, the embryo's shallow grooves separate the adjacent regions called para-segments, which are the earliest morphological units of the embryo. Each para-segment consists of the p compartment of one segment and the a compartment of the following segment. For example, para-segment 9 consists of p compartment of segment A3 and a compartment of segment A4. When segments first form in nine-hour-old embryol, the location of grooves moves to the extent that the boundary of each segment now represents the centre of a para-segment. Thus the anterior part of a segment is derived from one para-segment and its posterior part is derived from the next (**Fig. 12.15**).

12.10.5 Hormone action and determination

Insect genomes contain different developmental programs for larval, pupal and adult stages of their life cycle. Metamorphoses from larva to pupa and to adult stages require sudden changes in the programs of gene expression. Although the detailed mechanisms of insect metamorphosis differ among species, the general pattern of hormone action is very similar. The metamorphosis of insects is regulated by effector hormones, which are controlled by neurohormones in the brain. Insect moulting and metamorphosis are controlled by two effector hormones: the steroid **20-hydroxyecdysone** and the lipid **juvenile hormone (JH)** (**Fig. 12.16**).

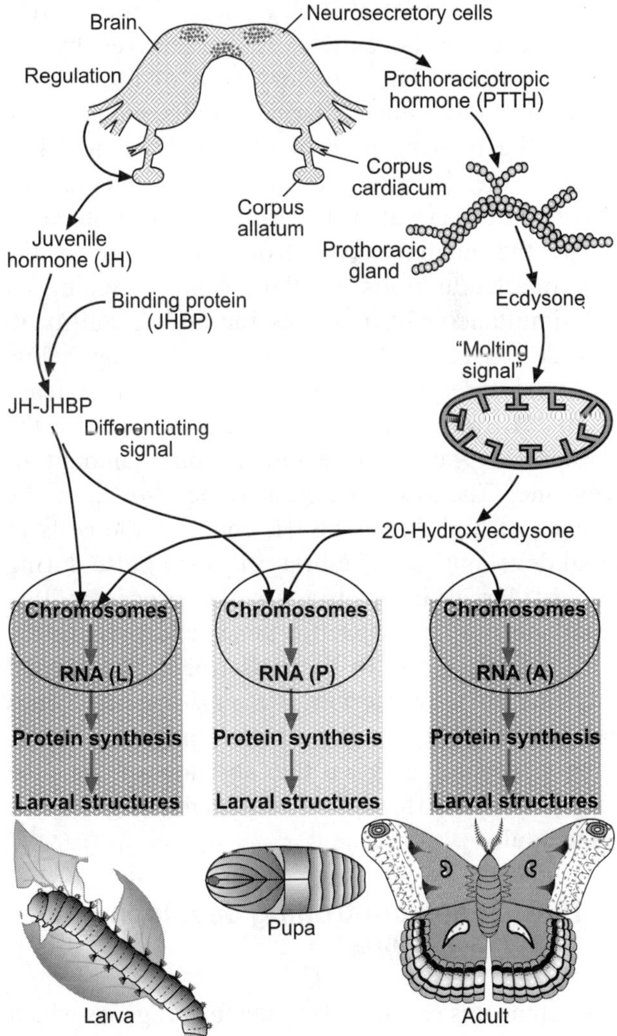

Fig. 12.16 General pathway of regulation of insect metamorphosis.

The steroid 20-hydroxyecdysone initiates and coordinates each moult and regulates the changes in gene expression that occur during metamorphosis. On the other hand, JH prevents the ecdysone-induced gene expression that is necessary for metamorphosis. The presence of JH during a moult ensures that the result of that moult produces another instar and not a pupa or an adult. The moulting process is initiated in the brain, where neurosecretory cells release **prothoracicotropic hormone (PTTH)** in response to neural, hormonal or environmental signals. PTTH stimulates the production of ecdysone by the **prothoracic gland**. This ecdysone is modified in peripheral tissues to become the active moulting hormone 20-hydroxyecdysone. Each moult is initiated by one or more pulses of 20-hydroxyecdysone and these pulses commit and stimulate the epidermal cells to synthesize enzymes that degrade and recycle the components of the cuticle. JH is secreted by the **corpora allata**. The secretory cells of the corpora allata are active during larval moults but inactive during the metamorphic moult. As long as JH is present, the hydroxyecdysone-stimulated moults end in a new larval instar. Only in the last larval instar, the medial nerve from the brain to the corpora allata inhibits the gland from producing JH and simultaneously increases the body's ability to degrade existing JH. When JH levels drop below a critical threshold value, it triggers the release of PTTH from the brain. PTTH stimulates the prothoracic glands to secrete a small amount of ecdysone. The resulting hydroxyecdysone, in the absence of high levels of JH, commits the cells to pupal development. The first ecdysone pulse during the last larval instar triggers the processes that inactivate the larva-specific genes and prepare the pupa-specific genes to be transcribed. The second ecdysone pulse transcribes the pupa-specific genes and initiates the moult. At the imaginal moult, when ecdysone acts in the absence of JH, the imaginal discs differentiate and the moult gives rise to the adult.

12.10.6 Genes controlling development in *Drosophila*

Development is regulated by a number of genes which normally act in a predetermined and precise order. Further, groups of genes appear concerned with a specific aspect of the development. Different groups of genes often seem to act independent of each other. The genes governing *Drosophila* development are classified into two groups.

Maternal effect genes: These are genes expressed in maternal cells and whose products are transported into the egg. They express during oogenesis and establish anterior, posterior and terminal organizing centres in the syncytial embryo (formation of gradients in the egg). These genes encode regulatory factors, which set up opposing morphogen gradients and activate downstream zygotic genes.

Zygotic genes: These genes are expressed in the embryo during development, rather than in the surrounding maternal cells in a concentration-dependent manner. Progressive hierarchical activation of different classes of zygotic genes divides the syncytium into stripes defined by domains of gene expression. These form the basis of the segments, which become patterned by later acting genes. These genes are further subgrouped into:

- ***Segmentation genes* are** expressed after fertilization and determine the number and polarity of segments.
- **Homeotic genes** govern the identity of a segment but do not affect the number, polarity or size of segments.

12.10.7 Establishment of gradients in the egg

A **gradient** comprises a differential distribution of a molecule, either a protein or RNA, in the cytoplasm. For example, the RNA and protein products of Nanos gene are localized at the posterior end of *Drosophila* egg/embryo; this constitutes a gradient. The RNA however, has to be translated into a protein to generate its effect on development. A protein whose local concentration or activity causes the surrounding region to take up a particular structure or fate is called a **morphogen**. A morphogen may be a regulator of transcription or an activator of a transcription factor in a localized region. At the start of development, the egg has two well-established gradients: anterior-**posterior** and **dorso-ventral**. The anterior end of the egg gives rise to the head; the posterior end forms the tail; the dorsal side develops in the back of larva or adult; and the ventral side gives rise to the lower side of larva/adult.

The anterior-posterior radients is generated by differential distribution of molecules along the anterior-posterior axis of the egg; this gradient is established after fertilization. It controls the positional information along the larva, e.g., location of head, thorax, abdomen. The dorso-ventral gradient system, on the other hand, comprises differential activation of molecules leading to differential distribution of some other molecules; it is generated somewhat later than the anterior-posterior gradient. This system regulates tissue differentiation, i.e., specification of distinct embryonic tissues, including mesoderm, neuro-ectoderm and dorsal ectoderm. The experimental demonstration for the existence of gradients in egg cytoplasm is shown in **Fig. 12.17**.

12.10.8 Anterior and posterior polarity

Anterior pathway

The anterior system of pattern development is responsible for the development of head and thorax. This system generates gradients opposite to that produced by the posterior pathway. But both anterior and posterior systems control the development of body segments so that a defect in either system affects segment development. Bicoid protein functions as a transcription regulator of the gene hunchback and probably of other segmentation and homeotic genes. Thus biocoid is instructive as it activates the expression of genes needed for anterior development. The bicoid RNA is transcribed in nurse cells and the mRNA is transported into the oocyte, where it is localized at the anterior end of the egg. Bicoid is a homeodomain-containing protein, which binds to both DNA and RNA. It activates transcription of the hunchback gene in the zygotic nuclei and represses translation of the maternal caudal mRNA in the anterior of the egg.

Translation of bicoid mRNA begins soon after fertilization and bicoid protein establishes a characteristic stable until after the blastoderm stage. Hunchback is transcribed only when bicoid concentration is above a certain threshold level. In this way, the quantitative differences in the concentration of a morphogen are converted into qualitative differences in cell function and, ultimately, structures during embryonic development. As a result, hunchback product forms a band in the anterior part of the embryo (**Fig. 12.18**). Products of gene swa

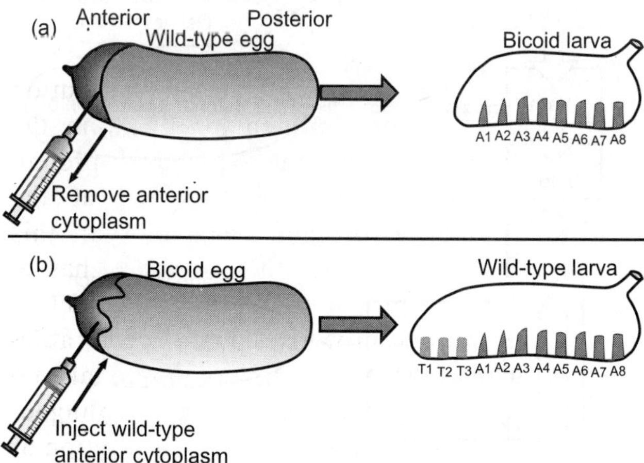

Fig. 12.17 Experiments to demonstrate that a cytoplasmic localization at the anterior end of the fruit fly egg determines anterior structures. (a) Cytoplasm from anterior end of egg belonging to a wild-type was removed; consequently larva appears lacking in anterior structures, similar to a bicoid mutant. (b) Anterior cytoplasm from a wild-type egg was injected into the anterior of the bicoid mutant egg, resulting in a larva similar to the wild-type.

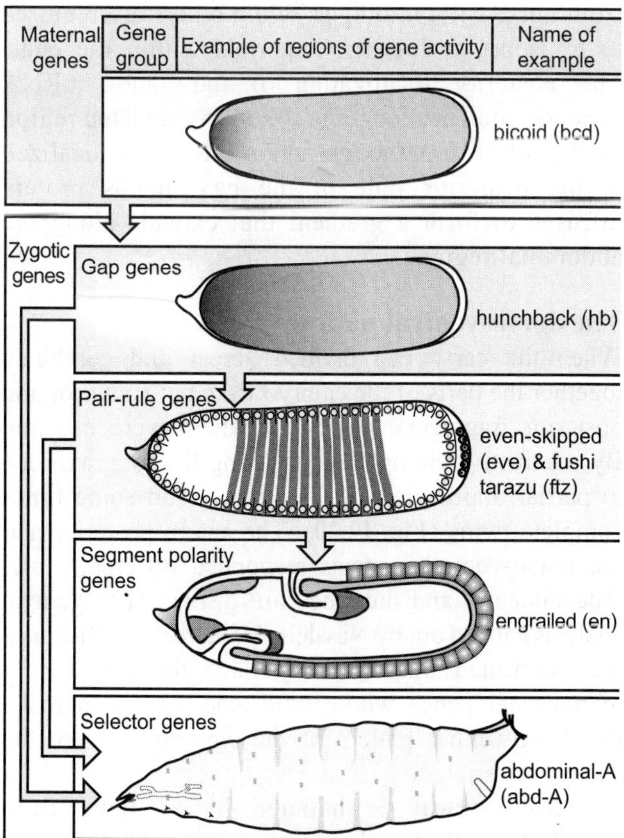

Fig. 12.18 The sequential expression of different sets of genes establishes the body plan along the antero–posterior axis.

(swallow) and exu (exuperantia) are also needed for anterior development. But these gene products are required either for transport of bicoid mRNA into the oocyte or for its localization in the anterior part of the egg. The 3' untranslated region (UTR) of the bicoid mRNA is attached to the anterior cytoskeleton by the products of the exu and swa genes. Translation of biocoid mRNA is initially prevented by its very short polyadenylate tail, but after fertilization, the products of the cortex, grauzone and staufen genes facilitate further polyadenylation and allow synthesis of Bicoid protein.

The posterior system

This system is responsible for the development of the segment of abdomen. A large number of events cause the localization of nanos mRNA at the posterior end of the egg, i.e., at the pole the nanos protein functions as the morphogen (**Fig. 12.19**). Nanos protein leads to a localized repression of the translation of hunchback mRNA, thereby enforcing the anterior localization of hunchback protein. Some of the genes concerned with the posterior development pathway are involved in the export of materials, including nanos mRNA, from nurse cells into eggs, while others are required to transport or localize materials within the eggs. The posterior localization of the nanos mRNA depends upon sequences in the 3' untranslated region of the mRNA. Although nanos mRNA is localized at the posterior pole of the egg, nanos protein diffuses to form a gradient that extends along the abdominal region.

The dorso-ventral pathway

When the early egg divided across and combined together the parts of the embryo from both anterior and posterior fragments, pieces of pattern were missing. By contrast, dividing the egg along the long axis led to partial duplication of the embryo and some times complete twins (**Fig. 12.20**). The antero-posterior and the dorso-ventral systems responded differently and independently and thus they are distinct. The genetic analysis carried out by Nusslein-Volhard and colleagues has confirmed this, for they have found a family of maternal genes whose wild-type role is to make the dorso-ventral pattern of the embryo, but not the antero-posterior.

This pathway is initiated by a signal from a follicle cell on the ventral side of the egg. The signal is perceived by the transmembrane (probably vitelline membrane) receptor encoded by toll, whose activation leads to a differential localization of dorsal gene product, which is a transcription factor. Toll is necessary for dorso-ventral polarity and in its absence the embryo is strongly dorsalized. Interestingly, toll is distributed uniformly in the egg. A total of 11 maternal genes are needed to establish the dorso-ventral axis between the time of fertilization and the cellular blastoderm. The dorso-ventral development depends on a complex interaction between the oocyte and follicle cells.

Maternal genes of terminal pathway

The terminal system is responsible for the development of the egg, viz., acron at head and the telson at the tail. The initial events that created asymmetry occur in the follicle cells. The key players in this group are the torso and torso-like genes. These events lead to a localized production of an extracellular ligand for the activation of a uniformly distributed transmembrane (plasma membrane of egg) receptor encoded by the gene, torso. Activated

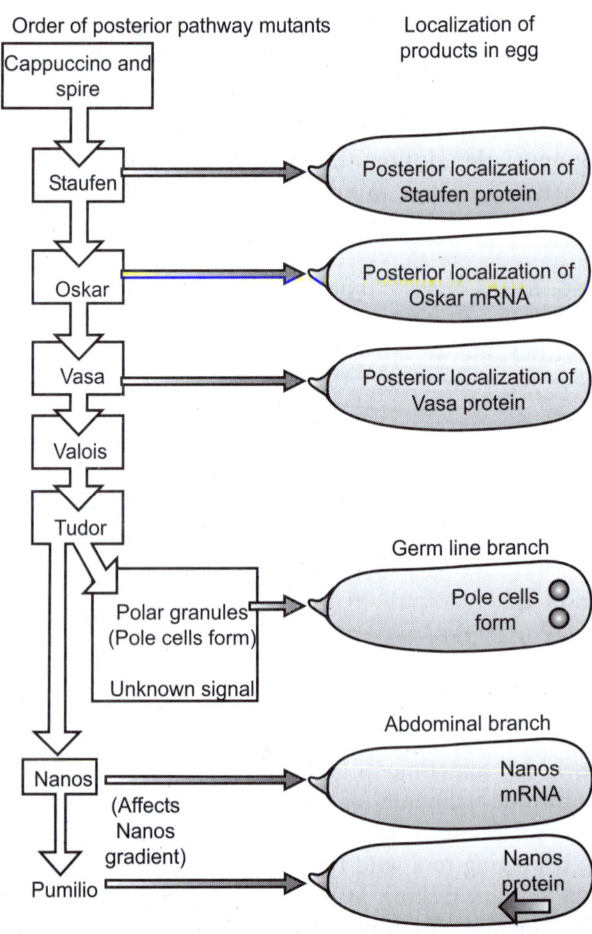

Fig. 12.19 Genes involved in the posterior pathway determine abdominal development and germ cell formation.

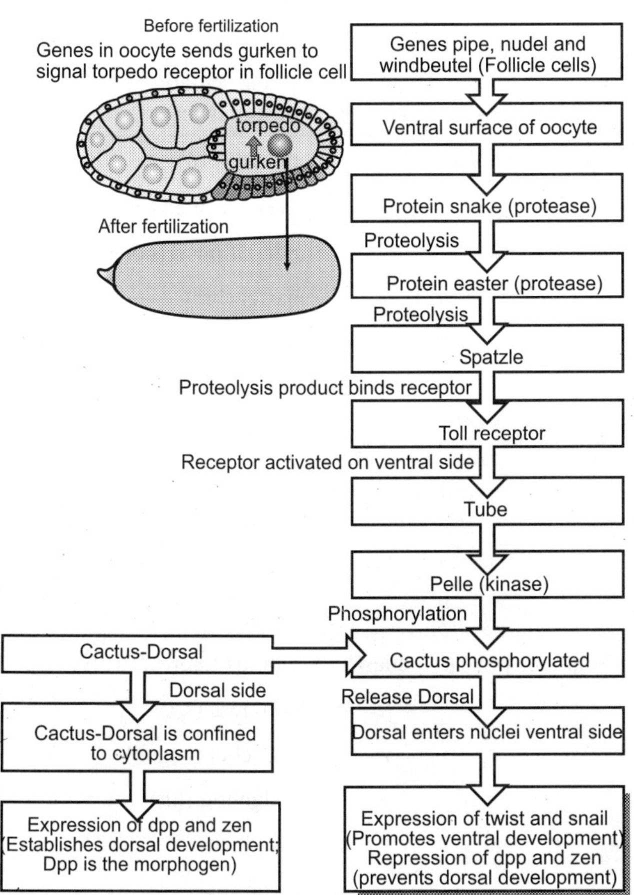

Fig. 12.20 Genes involved in the dorso-ventral system of embryo development in *Drosophila*. Involves interactions between follicle and oocyte.

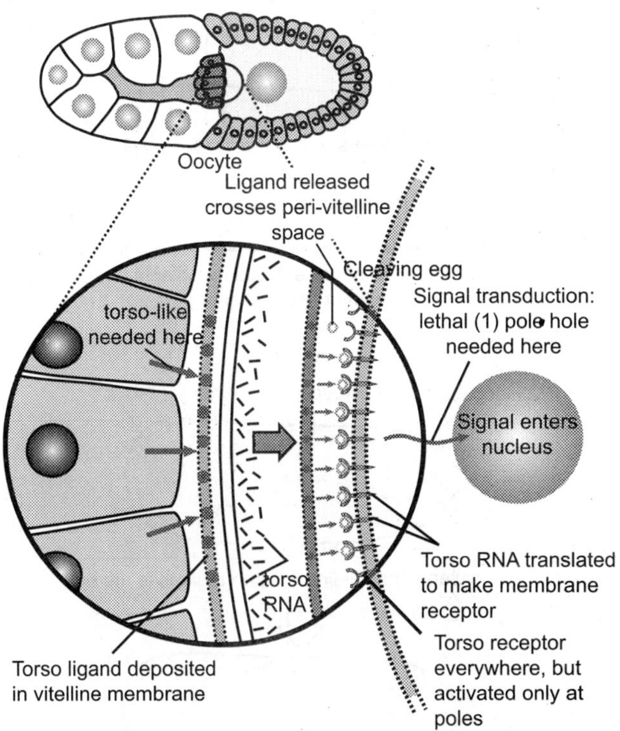

Fig. 12.21 A model of the terminal system. Local gene expression in the follicle cells results in localized activation of the torso receptor in the egg. Only some of the genes in the pathway are known.

torso has a kinase activity; this activity initiates a cascade (cascade involving Ras, Raf and MAP kinase) that results in the localized expression of the genes tailless and huckebein. These two genes, encode for transcription regulates, which activate genes specifying terminal structures (**Fig. 12.21**). Constitutively active torso mutants are epistatic to the anterior and posterior group genes—the phenotype is '**hyperterminal**', with large acron and telson but no body in between. The various genes of the terminal system are collected and ordered in a pathway of action by epistasis studies. Thus, the patterns of the four pathways of early embryo development in *Drosophila* are established by the action of maternal genes that establish the initial asymmetry in the egg and early embryo.

Segmentation genes

The most significant development of a *Drosophila* larva is the regular segmentation of the larval cuticle along the antero-posterior axis, each segment carrying cuticular structures that define it as, for example, thorax or abdomen. This pattern of segmentation is carried over into the adult, in which each segment has its own identity. The maternal genes—bicoid, hunchback and caudal—encode transcription factors which control the expression of zygotic genes involved in the formation of the segmental body plan. Once the general body plan of the fly is set, development continues in the formation of para-segments and segments. The various organs of the fly's body are produced from these segments. The formation of discrete regions that are determined to give rise to different body parts depends on a differential expression of the genes of the embryo per se and these genes are called **zygotic genes**. The zygotic genes are identified by segmentation mutants (**Fig. 12.22**); they are also known as **segmentation genes**. They are expressed after fertilization and their products form transverse bands that differentiate individual regions on the anterior-posterior axis. Approximately 30 different loci are involved in segment formation. A mutant showing aberrant segment is called **segmentation mutant**. These genes fall into three general groups: gap genes, pair-rule genes and segment polarity genes

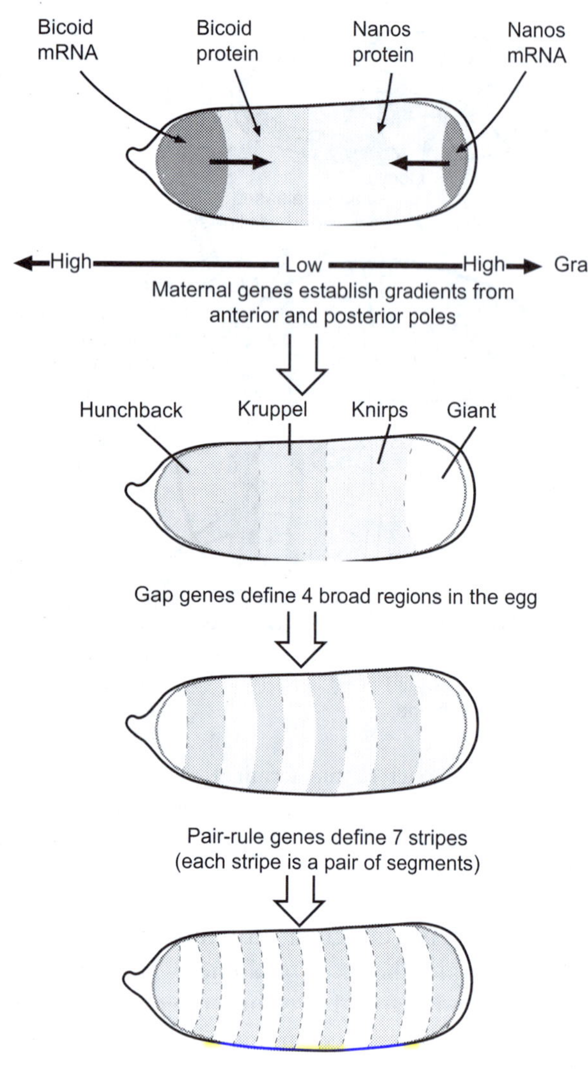

Fig. 12.22 Maternal and segmentation genes divide *Drosophila* embryos into successively smaller segments.

(**Table 12.4**). These genes are activated sequentially (Maternal genes → Gap genes → Pair rule genes → Segment polarity genes); each by the genes activated before it; each group controls a smaller and more focussed domain of the fly's development.

The maternal effect genes of the anterior-posterior system have created bicoid and nanos gradients, which are discussed in the previous sections. The segmentation genes increment, narrow and focus these gradient signals until 14 distinct bands form, corresponding to the 14 parasegments that develop, creating compartments that the tissues of the fly arise from.

Table 12.4 Segmentation genes in *Drosophila*

Class	Locus (allelic designation)
Pair rule genes	paired (prd)
	even-skipped (eve)
	odd-skipped (odd)
	barrel (brr)
	runt (run)
	engrailed (en)
Gap genes	kruppel (Kr)
	knirps (kni)
	hunchback (hb)
	giant
	tailless
Segment-polarity genes	cubitus interruptus (ci)
	wingless (wg)
	gooseberry (gsb)
	hedgehog (hh)
	fused (fu)
	patch (pat)

Segment polarity genes

The expression of the pair rule genes defines the anterior boundaries of all 14 para-segments; like the gap genes, their activity is temporary. However, the final segment boundaries of the larval epidermis are established by the segment **polarity genes**. Segment polarity genes are activated in response to pair rule gene expression. They are each expressed in 14 transverse stripes, one stripe corresponding to each para-segment. Mutations in these genes most often lead to the loss of a part of the p compartment of each segment, but some mutations lead to a loss of a compartment or the middle of segments. When P compartment is affected, it is replaced by a mirror image duplication of a compartment of the same segment. So far, 16 polarity genes are known, viz., *engrailed*, *wingless*, *gooseberry*, *patched*, *hedgehog*, *coatal2*, *fused* and so on. The expression of these genes occurs at nuclear division 13 and is regulated by the products of pair rule genes; these genes are expressed in 14 stripes in every p compartment (but not in a compartment). When engrailed stripes first appear, they are only one-cell-wide, but later they spread to 3–4 cells for each stripe. The pattern of engrailed stripes becomes established over a period of 30 min moving from the anterior to the posterior region. Initially, one stripe is apparent, then every other segment is striped and in the end every segment has a 3- to 4-cell-wide stripe.

The expression of engrailed stripes defines the anterior boundary of every p compartment of adult segments. The initial one-cell-wide stripes appear at the anterior boundaries of both ftz and eve stripes, i.e., within the border region of eve and ftz expression. The pattern of engrailed (and other genes) expression is explained on the basis of the following two models.

Combinatorial model: According to this model, each pair rule gene is expressed in a pattern of stripes and the stripes of different pair rule proteins overlap because they are out of phase with one another. As a result, different cells of the cellular blastoderm express different combinations of pair rule genes. Each compartment is defined by the particular combination of the genes that are expressed in them and these combinations determine the expression of different genes of the next set of genes.

Boundary model: According to this model, a compartment is defined by the striped pattern of gene expression. Subdivisions of these compartments arise due to interactions, involving cell-to-cell communication at the boundaries of the compartments. Thus in case of engrailed, juxtaposition of cells having ftz or eve protein with those that do not triggers some unique event that induces engrailed expression.

The functions of other segment polarity genes are not clear, but their products include secreted proteins, transmembrane proteins, kinases, cytoskeletal proteins as well as transcription factors. Obviously, at this stage, cell-to-cell interactions become important.

Homeotic genes

Homeotic genes (also called **memory genes**) act during embryogenesis and control the identity of a segment, but do not change the number, polarity or size of segments. Homeotic genes ensure the unique differentiation of each segment. Mutations in these genes cause cells of one compartment to develop the phenotype of a different compartment. Some homeotic mutations 'transform' part of a segment or an entire segment into another type of segment. Most homeotic genes are expressed in a spatially restricted manner that corresponds to para-segments. The expression of these genes depends on the signals established by segmentation genes. Many homeotic genes code for transcription factors that act upon other homeotic genes. The segment polarity genes are active at the same time as the peak expression of homeotic genes.

Homeotic mutations cause one segment of the abdomen to develop as another, legs to develop in place of antennae or wings to develop in place of eyes. These genes do not create new patterns; rather change cell fates by switching the set of genes that functions in a particular compartment. Two major homeotic gene complexes are known in *Drosophila*: antennapedia complex (**ANT-C**) and bithoracic complex (**BX-C**); these two complex loci together called **Hom-C**, regulate the development of all the segmented units of adult insect body and each complex locus contains several homeotic genes (**Fig. 12.23**). Complex loci have the same general organization as typical split genes; their unusual properties arise because they have an array of regulatory elements that control their expression on different parts of the embryo. The genes are extremely large and usually there are two or more genes within a complex locus.

The homeobox

The regulatory transcription factors encoded by different genes affecting development have diverse DNA-binding domains. For example, several gap proteins, e.g., hunchback, kruppel and knirps, have zinc finger motifs, while giant (another gap protein) has a basic zipper motif. The transcription factors encoded by homeotic genes activate the next category of genes, collectively known as **realizator genes**. These genes produce morphological characteristics of each segment. Many segmentation and homeotic genes contain an opa box, which encodes a stretch of polyglutamic acid; the functions of opa domain are, however, not known. Approximately, 40 *Drosophila* genes contain a homeobox, a well conserved motif of 180 bp located near the 3' end of the transcription unit of several segmentation and homeotic loci. Homeobox encodes 60 amino acids called the **homeodomain**, which is constant in size and fairly basic in sequence. Almost all the genes having homeobox are known to be involved in the development. The homeodomains encoded by *Drosophila* genes fall in several groups of which antennapedia group is the major group.

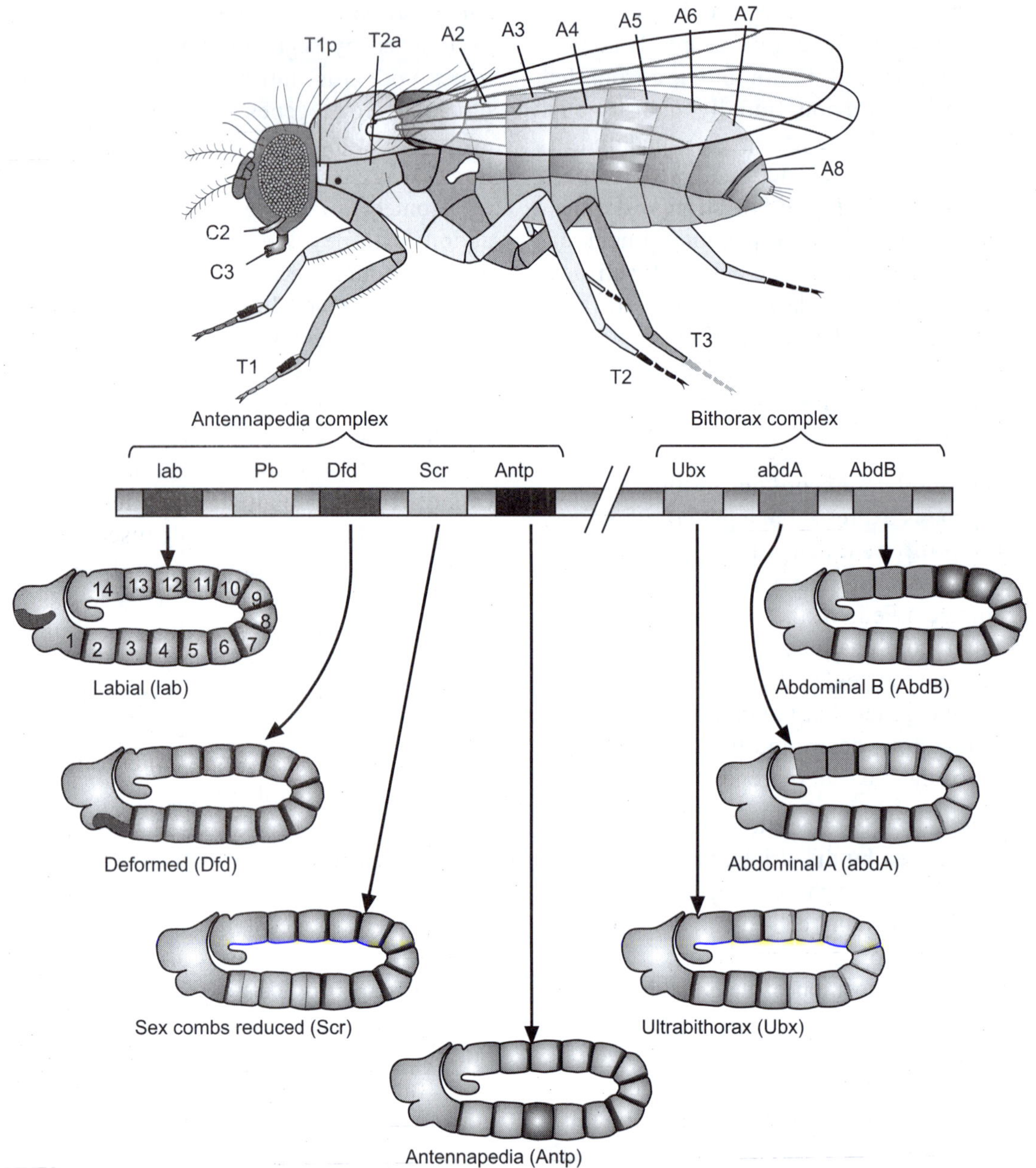

Fig. 12.23 Homeotic gene expression in *Drosophila* (**See Plate 8**).

Genes containing homeoboxes have been identified in frog, mouse and human genomes, which are expressed in early embryogenesis. The mammalian genes related to HOM-C group, are organized in clusters and are called **Hox genes**. A Hox genes cluster may be 20 kb to 100 kb long and contain up to 10 genes; four Hox clusters, viz., A, B, C, D, have been characterized in mouse and man. On the basis of homeobox sequences and, sometimes, sequences of other short regions, Hox genes can be placed into groups that correspond to HOM-C genes of *Drosophila*. For example, HoxA4 and HoxB4 are best related to dfd gene of *Drosophila*. Such genes, that is, corresponding genes of different loci (gene clusters), are often called **paralogues**. The mammalian genes are divided into groups I (most anterior acting) to 13 (most posterior acting); of these, groups 1–9 are related to the fly HOM-C genes. Groups 10–13 appear to have arisen by tandem duplication and divergence of group 9 genes (homologous of AbdB gene of fly). The Hox genes are located in chromosomes in the same sequence as that of the region of the embryo in which they act.

12.11 TISSUE-SPECIFIC DNA METHYLATION AND DIFFERENTIATION

In most plants and animals, cytosine residues are methylated (to yield 5-methyl cytosine) after DNA replication. The extent of methylation varies from species to species (about 2%–7% of the cytosine residues in mammals). The methyl group of 5-methyl cytosine lies exposed in the major groove of DNA; thus it has the potential to be involved in the interaction of DNA with the various proteins. This conclusion is strongly supported by the fact that the addition or removal of a single methyl ($-CH_3$) group from the *lac* (operon) operat or DNA sharply changes its affinity for the *lac* repressor.

Such direct evidence for the involvement of methylation in gene regulation is lacking in eukaryotes, but there are three indirect evidence:

1. The degree of DNA methylation has been found to be negatively correlated with gene expression, that is, the higher the degree of methylation the lower the extent of gene expression.
2. In some cases methylation patterns are tissue-specific, i.e., different tissues of the same organism show different patterns of methylation.
3. The base analogue 5-azacytidine, which is also used as a drug, is incorporated into DNA in place of cytosine but it cannot be methylated. The incorporation of 5-azacytidine in DNA (following a treatment with the drug) has been shown to result in the expression of genes in those tissues where they normally are not expressed, e.g., foetal haemoglobin (*c* haemoglobin) and, to a lesser extent the embryonic haemoglobin (*h*-haemoglobin) genes in anemic adult baboons and adult humans with severe ß-thalassemia (inability to synthesize the ß-chain of haemoglobin) and sickle cell anemia. These genes are normally not expressed in the RBCs of adults. In one of these studies, DNA in the regions of γ–haemoglobin and ε–haemoglobin genes was shown to be hypomethylated, i.e., contained fewer methyl groups than normal.

In most eukaryotes, more than 90% of the methylation occurs on cytosine residues present in CG dinucleotide sequence. Further, the CG dinucleotide sequences are symmetrically methylated. The semi-conservative replication of such a symmetrically methylated sequence would yield two progeny DNA molecules; in each molecule, the CG dinucleotide sequence of one of the two strands will be methylated, i.e., the CG sequences will be half-methyalted or hemi-methylated. A DNA methylase specific for such half-methylated sequences would methylate the unmethylated CG dinucleotide sequences of the newly synthesized DNA strand. Some bacterial methylases are known to be specific for such half-methylated sequences. Consequently, once such a methylation pattern is generated in a cell, is transmitted to all the daughter cells produced through mitosis, i.e., such methylation patterns are clonally heritable. It seems likely that tissue-specific DNA methylation may be involved in gene regulation during differentiation and development. But a crucial question remains to be answered. How are the tissue-specific methylation patterns generated? According to an attractive model, these patterns are produced by tissue-specific demethylases; these enzymes remove methyl groups from the critical sites of those genes that are to be expressed in a specific cell type or tissue. Thus, according to this model, all the genes are uniformly methylated in the zygote and in all the cells of undifferentiated embryo. The various genes are differentially activated in the different cell types due to the action of tissue-specific demethylases. But thus far no such enzyme has been identified, which remains a serious limitation of this model.

Summary

- ***Development*** is the process by which a multicellular organism arises, initially from a single cell.
- ***Differentiation*** is the term used to describe the developmental processes that cause structural or functional distinction between parts of an organism.
- Cell differentiation is controlled by a wide variety of external signals, ranging from cell surface proteins to secreted polypeptide cytokines and molecules of the extracellular matrix.
- ***Fertilization*** initiates intense protein synthesis, rapid DNA replication and cell division.
- Until the stage of ***gastrula***, protein synthesis in embryos depends on the mRNA molecules already present in the egg; new mRNA synthesis begins only after the embryos reach gastrula stage.
- Once a group of cells commit on a developmental path by these pre-existing forces, it may produce some specific inducers and form a ***gradient***. Some of these inducers may stimulate the development of the predetermined tissue one step further on the chosen path, while others may initiate a new pattern of development in the neighbouring group of cells.
- Further developmental cues may arise through external stimuli. In all vertebrate embryos, for instance, the dorso-ventral axis is specified by an external physical stimulus in the environment. ***Gastrulation*** then begins on the dorsal surface of the embryo and in doing so establishes the antero–posterior axis.
- Cells respond to signals from other cells and from the environment by altering patterns of gene expression and the activities of protein molecules already in the cell.
- The diversification of the embryo by progressive cell–cell interactions allows an initially small number of cell types to be elaborated into the many distinct cell types in the mature animal or plant.
- ***Commitment*** occurs in several stages. A cell may be specified to adopt a certain fate by its intrinsic properties and will differentiate this way in isolation. However, when exposed to a range of different environments, its fate may be altered by interactions with other cells.
- ***Pattern formation*** is the process causing cells to adopt a precise spatial organization, a phenomenon, which ensures that all members of a species are morphologically similar. The components of pattern formation are regional specification and morphogenesis.
- ***Morphogenesis*** describes the process by which cells form structures by interpreting the positional information and behaving accordingly.
- Two forms of ***regional specification*** can be distinguished. One controls the differentiation and causes particular differentiated cell types to arise in particular positions and another controls morphogenesis in cells which are equivalent with respect to their state of differentiation.
- Much of the phenotypic variation observed in populations arises through the influence of the environment during development as well as the stochastic influences behind developmental noise.
- The inclusion of environmental cues into normal development occurs in most organisms developing in natural conditions. In some organisms, the development of one individual is possible only by the presence of another individual of a different species, called as ***developmental symbiosis***.
- Like many developmental processes, sex determination involves both genetic and environmental factors.
- Although the development of invertebrates is very diverse, certain features common to the development of most invertebrates are also seen in vertebrate development.
- In invertebrates, ***Drosophila melanogaster*** has been pivotal in our understanding of developmental genetics. The molecular details of how development is regulated are better understood in this system than in any other animal embryo.
- Tissue-specific DNA methylation may be involved in differentiation. In most plants and animals, cytosine residues are methylated after DNA replication. The methyl group of 5-methyl cytosine lies exposed in the major groove of DNA; thus it has the potential to be involved in the interaction of DNA with the various proteins.

Exercises

1. Differentiation is regulated at different levels of gene expression. Explain the mechanism.
2. Discuss at least two simple developmental model organisms.
3. How does the switching take place from one form of mating type to another in yeast?
4. Explain the mechanism of differentiation.
5. How does pattern formation determine development?
6. What is morphogenesis?
7. Explain the regulation of morphogenesis by environmental factors.
8. Write the different steps involved in normal human embryogenesis.
9. The pattern of structures in the adult organism determines the early stages of embryonic development. Explain.
10. Draw and explain the formation of the body plan during early embryonic development.
11. What are the fate maps of development?
12. Discuss genes involved in regulation of development in *Drosophila*.
13. How do maternal genes play a role in early development of the egg?
14. How do gradients determine early development in *Drosophila*?
15. Explain the different groups of segmentation genes.
16. What is the homeobox?

Further Reading

Books

Gilbert SF (1997), *Developmental Biology*, 5th edition, Sinauer, Sunderland.

Gurdon JB (1974), *The Control of Gene Expression in Animal Development*, Clarendon Press, Oxford.

Lawrence PA (1992), *The Making of a Fly: The Genetics of Animal Design*, Blackwell Scientific Publications.

Slack JMW (1991), From egg to embryo: Regional specification in early development, In Jonathan BL Bard, Barlow PW, David L Kirk, (eds), *Developmental and Cell Biology Series*, Vol. 26, 2nd edition, Cambridge University Press, Cambridge.

Slack JMW (2005), *Essential Developmental Biology*, 2nd edition, Blackwell Publishing.

Twyman RM (2001), *Developmental Biology*, Viva Book Private Limited, New Delhi.

Wolpert L (2002), *Principles of Development*, 2nd edition, Oxford University Press, Oxford.

References

Bryant PJ (1978), Pattern formation in imaginal discs. In Ashburner M and Wright TRF, (eds), *The Genetics and Biology of Drosophila*, Vol. 2c, Academic Press, London, 229–335.

Chan LN and Gehring W (1971), Determination of blastoderm cells in *Drosophila melanogaster, Proc. Nat. Acad. Sci.*, 68:2217–21.

Doe CQ and Bowerman B (2001), Asymmetric cell division: Fly neuroblast meets worm zygote, *Curr. Opi. Cell Biol.*, 13:68–75.

Duncan I and EB Lewis (1982), Genetic control of body segment differentiation in *Drosophila*, In Subtelny S, (ed.), *Developmental Order: Its Origin and Regulation*, Alan-Liss, New York.

Freeman M and Gurdon JB (2002), Regulatory principles of developmental signalling, *Annu. Rev. Cell. Dev. Biol.*, 18:515–39.

Garcia–Bellido A and P Ripoll (1978), Cell lineages and differentiation in *Drosophila*. In Gehring WJ, (ed.), *Genetic Mosaics and Cell Differentiation*, Springer–Verlag, Berlin.

Gaul U and Jackle J (1990), Role of gap genes in early *Drosophila* development. *Adv. Gen.*, 27:239–75.

Godfrey SS and M Susman (1982), The genetics of development in *Dictyostelium discoideum*, *Annu. Rev. Genet.*, 16:358–404.

Gurdon JB and Bourillot PY (2001), Morphogen gradient interpretation, *Nature,* 413:797–803.

Ingham PW (1988), The molecular genetics of embryonic pattern formation in *Drosophila*, *Nature*, 335:25–34.

Johnstone O and Lasko P (2001), Translational regulation and RNA localization in *Drosophila* oocytes and embryos, *Annu. Rev. Genet.*, 35:365–406.

Krumlauf R (1994), Hox genes in vertebrate development, *Cell,* 78:191–201.

Kwon S and Schnapp BJ (2001), RNA localization: Shedding light on the RNA–motor linkage, *Curr. Biol.*, 11:R166–68.

Lawrence PA and Struhl G (1996), Morphogens, compartments and pattern: Lessons from *Drosophila*? *Cell,* 85:951–61.

MacOnochie M, Nonchev S, Morrision A and Krumlauf R (1996), Paralogous Hox genes: Function and regulation, *Annu. Rev. Genet.*, 30:529–56.

Mannervik M, Nibu Y, Zhang H and Levine M (1999), Transcriptional coregulators in development, *Science*, 284:606–9.

Morisato D and Anderson KV (1995), Signalling pathways that establish the dorso-ventral pattern of the *Drosophila* embryo, *Annu. Rev. Genet.*, 29:371–99.

Nusslein-Volhard C (1996), Gradients that organize embryo development, *Sci. Am.*, 275:54–55; 58–61.

Nusslein–Volhard C and Wieschaus E (1980), Mutations affecting segment number and polarity in *Drosophila. Nature*, 287:795–801.

Pankratz MJ and Jackle H (1990), Making stripes in the *Drosophila* embryo, *Trends Genet.*, 6:287–92.

Peifer M, Karch F and Bender W (1987), The bithorax complex: Control of segmental identity, *Genes and Devel.*, 1:891–98.

Pick L (1998), Segmentation: Painting stripes from flies to vertebrates, *Dev. Genet.*, 23:1–10.

Riechmann V and Ephrussi A (2001), Axis formation during *Drosophila* oogenesis, *Curr. Opin. Genet. Dev.*, 11:374–83.

Riverapomar R and Jackle H (1996), From gradients to stripes in *Drosophila* embryogenesis—filling in the gaps, *Trends Genet.*, 12:478–83.

Scott MP and Carroll (1987), The segmentation and homeotic gene network in early *Drosophila* development, *Cell*, 51:689–98.

Scott MP and O'Farrell PH (1986), Spatial programming of gene expression in early *Drosophila* embryogenesis, *Annu. Rev. Cell Biol.*, 2:49–80.

Singh J (1999), Mating type switching in fission yeast, a unique model system of development and differentiation, *Curr. Sci.*, 77:1262–72.

Sprenger F and Nusslein–Volhard C (1992), Cellular terminal regions of the *Drosophila* egg, *Cell,* 71:987–1001.

St. Johnston D and Nusslein–Volhard C (1992), The origin of pattern and polarity in the *Drosophila* embryos, *Cell*, 68:202–19.

Stragier P and Losick R (1996), Molecular genetics of sporulation in *Bacillus subtilis*, *Annu. Rev. Genet.*, 30:297–341.

Strigini M and Cohen SM (1999), Formation of morphogen gradients in the *Drosophila* wing, *Semin. Cell. Dev. Biol.*, 10:335–44.

Struhl G (1989), Morphgen gradients and the control of body pattern in insect embryos, *Ciba Found. Symp.*,144:65–86, 86–9, 92–8.

Struhl G, Johnston P and Lawrence PA (1992), Control of *Drosophila* body pattern by the hunchback morphogen gradient, *Cell,* 69:237–49.

Twyman RM (2003), *Developmental Biology*, Bios Scientific Publisher Ltd.

Wilson JE and Macdonald PM (1993), Formation of germ cells in *Drosophila*, *Curr. Opin. Genet. Dev.*, 3:562–65.

Wright TRF (1970), The genetics of embryogenesis in *Drosophila*, *Adv. Genet.*, 15:261–395.

13 BUILDING MULTICELLULAR ORGANISMS

Learning objectives

After reading this chapter, you will be able to:

- Appreciate the various theories of evolution of multicellular organisms
- Understand the basic components of cellular organization in different organisms

13.1 INTRODUCTION

Development is the orderly sequence of changes that produce increasing complexity during the growth of an organism. It is controlled by the differential expression of genes. The central problem of development is explaining **genomic equivalence**, that is, how a fertilized egg cell develops into a complex multicellular organism with identical genetic material in different cell types. The transition to multicellularity is unique.

A multicellular organism the activities of the individual cells are coordinated and the cells themselves are either in contact or close enough to interact strongly. During the development of a multicellular organism, progenitor cells differentiate into distinct structural, compositional and functional forms. The coordinated action of many types of cells within tissues and multiple specialized tissues in an organ permits the organism as a whole to move, metabolize, reproduce and carry out other essential activities.

The assembly of cells in a tissue and organization of distinct tissues into organs are determined by molecular interactions at the cellular level. This organization is possible with the temporally and spatially regulated expression of a wide array of adhesive molecules. Cells in tissues can adhere directly to one another (**cell–cell adhesion**) through specialized integral membrane proteins called **cell-adhesion molecules** (**CAMs**). Cells in animal tissues also adhere indirectly (**cell-matrix adhesion**) through the binding of **adhesion receptors** in the plasma membrane to components of the surrounding extracellular matrix (ECM)—a complex interconnecting meshwork of proteins and polysaccharides secreted by cells into the spaces between them (**Fig. 13.1**). These two basic types of interactions not only allow cells to aggregate into distinct tissues but also provide for the bidirectional transfer of information between the exterior and interior of cells.

Over the past decade, developmental biologists have made remarkable progress in understanding genomic equivalence using molecular methods, including mutagenesis and recombinant DNA technology. Until recently scientists believed that development might take place through permanent changes in chromosomes during development. However, the cloning of a mammal, such as the sheep Dolly, from the cell of an adult nucleus can give rise to the whole organism and all its cellular types, indicating that the chromosomes are intact. Also, many general approaches were applied to study the developmental decisions involved in multicellular organisms, such as the fruit fly *Drosophila* and the soil worm *Caenorhabditis elegans*. These studies have revealed that many developmentally important genes encode transcription factors, suggesting that regulation of transcription is a primary strategy for controlling development in all organisms.

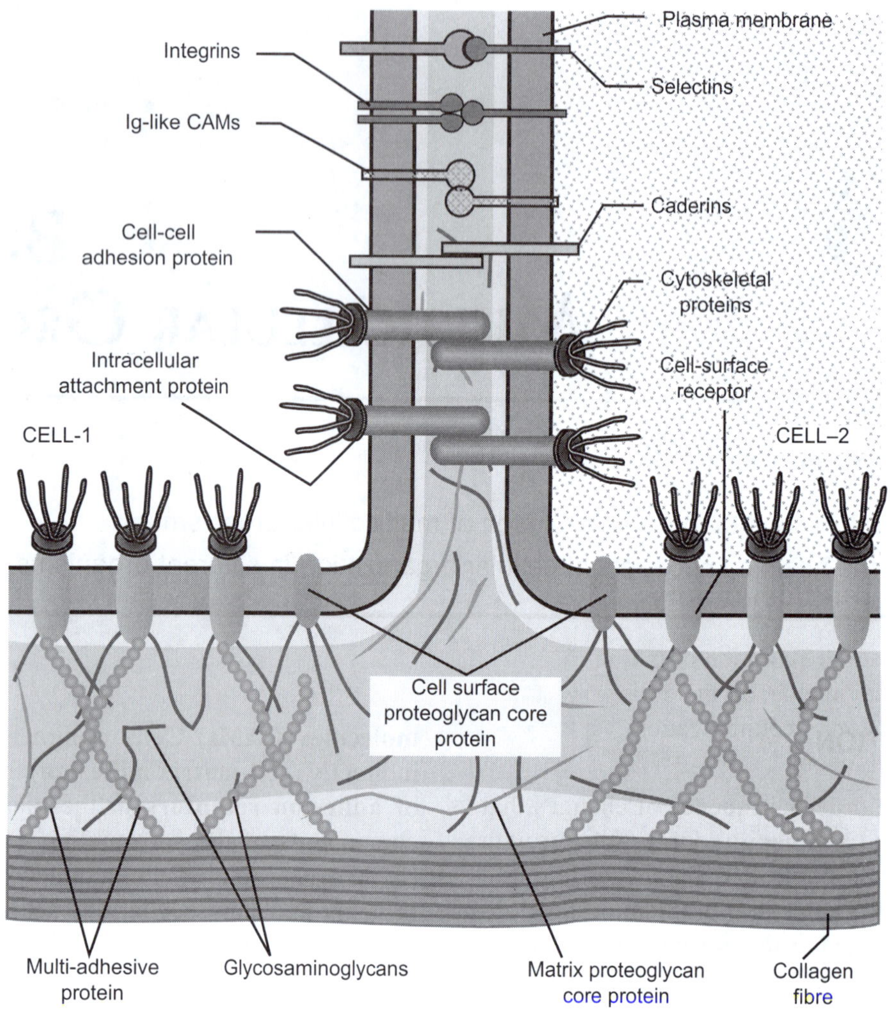

Fig. 13.1 Cell-adhesion molecules bind cells together and via intracellular attachment proteins, form anchors for cytoskeletal proteins. Other plasma membrane receptor proteins form connections with components of the extracellular matrix and also, via attachment proteins, with cytoskeletal proteins. Multi-adhesive proteins bind to other matrix components. Proteoglycans also participate in adhesion of cells to each other and to the protein components of the matrix (**See Plate 9**).

13.2 FROM UNICELLULAR TO MULTICELLULAR LIFE

The first organisms that existed—bacteria and protozoa—were unicellular; they have been successful in adapting to diverse environments on planet Earth. Unicellular organisms comprise more than half the total biomass on earth. They can live in extreme conditions, can synthesize all the substances they need from a few simple nutrients and can divide more than once every few hours. Then, what were the selective advantages that led to the evolution of multicellular organisms? The attempt to become multicellular seems to have happened by chance. For example, a pair of cells failed to separate during cell division, or two cells stuck together accidentally. If natural selection favoured this new form, it would survive and propagate. One theory suggests that since there is always an open ecological niche for large-size organisms, evolution favours bigger size to escape competition with the smaller ones. Further collaboration and division of labour allows organisms to exploit resources that no single cell could. Multicellularity, for instance, enables a plant to become physically large, have roots, where one set of cells can take up water and nutrients and leaves and another set of cells can efficiently capture radiant energy from the sun. Specialized cells in the stem of the plant form channels for transporting water and nutrients between the roots and the leaves. Yet another set of specialized cells forms a layer of epidermis to prevent water loss and to provide protection to all the internal cells. In a changed environment, further evolution occurred and various other organisms with different lifestyles evolved.

Multicellular organisms have arisen independently a number of times starting from about 1.5 billion years ago. From the fossil record and from extensive comparative studies of living organisms, it is certain that eukaryotes arose from prokaryotes. Because of the similarities in the genetic systems, metabolic pathways and molecular features of all life forms, we must infer that eukaryotes arose from prokaryotic ancestry rather than by independent pathways. Multicellular life began between 0.5 and 1.0 billion years ago. It is clear that multicellular plants, animals and fungi arose independently and at various times from various lineages of unicellular eukaryotic protists.

13.2.1 Origin of multicellularity

It is hypothesized that the evolution of multicellularity could have happened in one of the following three ways: symbiotic theory, cellularization (syncytial) theory or the colonial theory.

Symbiotic theory

This theory suggests that the first multicellular organisms might have occurred from symbiotic (cooperation) process of different species of single-celled organisms, each with different tasks (see Chapters 7 and 8). Over time, these organisms became so dependent on each other that they could not survive independently, eventually leading to their genomes being incorporated into one multicellular organism with each organism becoming a differentiated cell. However, the problem with this theory is that it is still not known how each organism's DNA could be incorporated into one genome, therefore allowing replication of the whole organism. Although such symbiosis is known to have occurred in mitochondrial and chloroplast endosymbiosis, it has only happened extremely rarely and did not directly lead to multicellularity.

Cellularization (syncytial) theory

According to this theory, a unicellular organism could have developed internal membrane partitions around each of its nuclei. For example, many protists such as the ciliates or slime moulds have several nuclei, confirming to this theory. However, mere presence of multiple nuclei is not enough to support the theory. Also, multiple nuclei of ciliates are dissimilar and have clear differentiated functions: the macronucleus serves the organism's needs while the micronucleus is used for sexual-like reproduction with exchange of genetic material. Slime moulds and syncytia form from individual amoeboid cells, such as syncytial tissues of some multicellular organisms, but not the other way round. This theory needs further evidence about the mechanism of generation of a multicellular organism from a pre-existing syncytium.

Colonial theory

The third and most convincing explanation of multicellularization is the colonial theory, which was proposed by Haeckel in 1874. Multicellularity probably began when a number of single-celled eukaryotes aggregated into a colony. Although each cell was capable of surviving on its own if necessary, living in a colony offered advantages over living alone. The adaptive advantages of multicellular construction include:

- Replenishment of individual cells and a longer life expectancy due to cellular repair and replacement;
- More reliable modes of reproduction through specialized cells set apart for this purpose;
- Larger body size, which leads to greater internal stability and to a variety of body plans and lifestyles;
- Cellular differentiation whereby functional efficiency increases for the organism as a whole. It also provides increased protection from predators and improved ability to locate food. Gradually, cells in the colony became specialized for specific functions. With increased specialization, life could be conducted more efficiently.

As different animals and plants appeared, they changed the environment in which further evolution occurred. Many other changes in talents, movements, sensory detection, communication and social organization all enabled eukaryotic organisms to compete, propagate and survive in ever more complex ways.

13.2.2 Colonization of cellular organizms

An early step in the evolution of multicellular organisms was probably the association of unicellular organisms to form cellular clusters or colonies. The simplest way to achieve this is for the daughter cells to remain together following cell division. The presence of some kind of social behaviour in many single-cellular prokaryotes has been reported. Slime molds are a living example of the attempts to become multicellular organism; they transform from single cells to one multicellular organism under adverse conditions. Similarly, *Myxobacteria* live in the soil and feed on insoluble organic molecules that they break down by secreting degradative enzymes. These bacteria stay together in loose colonies in which the digestive enzymes secreted by individual cells are pooled, thus increasing the efficiency of feeding (**wolf pack effect**). When food supplies are exhausted, the myxobacterial cells aggregate tightly together and form a multicellular fruiting body, within which the bacteria differentiate into spores that can survive even in extremely hostile conditions. When conditions are more favourable, the spores in a fruiting body germinate to produce a new swarm of bacteria.

Green algae are eukaryotes, which can exist as unicellular, colonial or multicellular forms. Different species of green algae can be grouped in order of complexity, indicating the progression towards the evolution of higher plants and animals. Unicellular green algae, such as ***Chlamydomonas***, resemble flagellated protozoa, except that they possess chloroplasts, which enable them to carry out photosynthesis. In closely related genera, groups of flagellated cells live in colonies held together by a matrix of extracellular molecules secreted by the cells themselves. The simplest species (genus *Gonium*) have a conclave of 4, 8, 16 or 32 cells, in which each cell is equivalent and divides to form individual colonies. Their flagella beat independently, but in the same direction, to propel the colony through the water. Larger colonies are found in other genera, the most interesting one being ***Volvox***, wherein some species have as many as 50,000 or more cells linked together to form a hollow sphere. In a *Volvox* colony, the individual cells are connected by fine cytoplasmic bridges so that the beating of their flagella is coordinated to propel the entire colony along like a rolling ball. Within the colony, there is some division of labour among cells, with a small number of cells being specialized for reproduction and serving as precursors of new colonies and the others that cannot live in isolation.

13.2.3 Specialization and cooperation of cells

The evolution of multicellular organisms probably began by colonization of single cellular organisms. *Volvox*, for example, is more like a multicellular organism than a simple colony. In an organized colony, all the flagella beat in synchrony as it spins through the water; the colony is structurally and functionally polarized and can swim towards a distant source of light. The reproductive cells are usually confined to one end of the colony, where they divide to form new miniature colonies, which are initially sheltered inside the parent sphere. Thus, in a primitive way it displays two multicellular features: the cells become **specialized** and they **cooperate**. By specialization and cooperation, the cells combine to form a coordinated single organism with more capabilities than its individual component parts.

Organized patterns of cell differentiation occur in many other organisms. For example, many kinds of cyanobacteria remain together after cell division, forming filamentous chains running to several metres. Within this cellular chain, some cells specialize to become heterocysts, in which fixation of atmospheric nitrogen takes place.

13.2.4 Cohesion between cells

To form a multicellular organism, the cells must remain together and eukaryotes have evolved a number of different strategies for this purpose. In *Volvox*, for example, the cells do not separate entirely at cell division but remain connected by cytoplasmic bridges. In higher plants, the cells not only remain connected by cytoplasmic bridges (plasmodesmata), they are also imprisoned in a rigid honeycomb of chambers with cellulose that the cells themselves have secreted (cell walls).

The cells of animals have developed different mechanisms. They are bound together by a relatively loose mesh of large extracellular organic molecules (extracellular matrix) and by adhesions between their plasma membranes. Often side-to-side attachments between the cells hold them together to form a multicellular sheet, or epithelium.

13.2.5 Anastomosis in fungi

Fungi, like animals and plants, have evolved complex multicellular development. They differ from most other eukaryotes as their vegetative body shows indeterminate growth. Fungal mycelia continue to grow and invade new substrates for as long as new substrates remain available and growth conditions remain satisfactory. In the most highly adapted behaviour, in model basidiomycetes, a basidiospore germinates to produce homokaryotic mycelium with uninucleate cells, called monokaryon. When two monokaryons meet, hyphal anastosomes occur, forming a heterokaryon, in which the hyphal compartments contain two sexually compatible haploid nuclei. Hyphal anastomosis is the fusion between hyphae or hyphal branches. In saprotrophic basidiomycetes, hyphal anastomoses occur readily as part of the maturation process of the mycelium. The process involves breakdown of two hyphal walls and union between two separate plasma membranes to bring the cytoplasms of the fusing hyphae into continuity with each other. Once they are in continuity, they can exchange nuclei and other organelles. The important point is that anastomosis is not limited to sexual reproduction; rather, hyphal fusions are essential to the efficient functioning of the mycelium of filamentous ascomycete and basidiomycete fungi because they convert the initially radiating system of hyphae into the fully interconnected three-dimensional network. The interconnections they establish enable transport of nutrient and signalling molecules within the colony.

Basidiomycetes create the most dramatic and prominent fruit bodies produced by fungi, including puffballs, different types of mushrooms and the bracket fungi often seen on trees, fallen logs and in organically rich soil under hard surfaces (**Fig. 13.2**). The process of anastomosis is the key to the form and structure in fungi and a plausible mechanism for the control of fungal morphogenesis. The morphogenesis in fungi is compartmentalized into a collection of distinct developmental processes called subroutines. These separate subroutines can be recognized at the levels of organs (e.g., cap, stem, veil), tissues (e.g., hymenophore, context, pileipellis), cells (e.g., basidium, paraphysis, cystidium) and cellular components (e.g., uniform wall growth, growth in girth, length and wall thickness). They are distinct genetically and physiologically and may run in parallel or in sequence. There are many mechanisms that control fungal morphogenesis. Homologues and analogues of all the mechanisms known in animals and plants can be found in fungi.

13.2.6 Cell–cell communication

Cells in multicellular organisms are not just bound together mechanically and connected by junctions that seal off the interior from the exterior environment. They also communicate with each other along the body of an organism. For example, in *Hydra*, if one end of a body is cut off, the remaining cells react to the absence of the amputated part by adjusting their characteristics and rearranging themselves so as to regenerate a complete animal, clearly indicating that signals pass from one part of the organism to the other. Moreover, these signals are independent of the nervous system.

Fig. 13.2 Left: Diagram of an umbrella mushroom. Right: Photograph of a wild mushroom lifting the hard crest of tar in the road, displaying the strength of stalk formed by the process of anastomosis (Channarayappa).

The more complex higher organisms have evolved from simpler ancestors resembling coelenterates and they owe their complexity to more sophisticated exploitation of the body, creating sheltered compartments and controlled internal environments in which specialized functions are performed by differentiated cells. Specialized cells interact and communicate with each other, setting up signals to govern the characteristics of each cell according to its place in the structure as a whole. To show how it is possible to generate multicellular organisms of such size, precision and complexity as a plant, insect, or mammal, however, it is necessary to consider more closely the sequence of events in **development**.

13.2.7 Cell memory

Cells of every multicellular organism are generated by repeated division from a single precursor cell and they constitute a **clone**. As the proliferation continues and the clonal mass grows, some of the cells within the clonal mass become differentiated from others, adopting a different structure and function, in response to cues received from their neighbours. Remarkably, the progeny cells usually persist in their differently specialized states even after the influences that originally directed their differentiation have disappeared, suggesting these cells have some form of **memory**. Consequently, their final character is determined by the entire sequence of influences to which the cells have been exposed in the course of development. Thus, as the body continues to develop, the finer details of adult body pattern become specified, creating an organism of gradually increasing complexity, the ultimate form being the expression of a long developmental history.

13.2.8 Conservation of basic developmental programs in evolution

Evolution of multicellular organisms was gradual and subject to stringent selection to reach coordination among various specialized cellular structures. During the long history of multicellular organisms, many developmental devices that evolved have been conserved as basic principles for the construction of their more complex descendants. For example, organization of cells into epithelia or formation of nerve cells, are found throughout the animal kingdom. Studies have also revealed an astonishing number of developmental resemblances at a fundamental genetic level, even between species as remotely related as mammals and insects. Many basic steps appear to have been 'frozen' into developmental processes, just as the DNA replication, genetic code or protein synthesis mechanisms have become universal.

13.3 MULTICELLULAR ORGANISMS

Diversity

The wealth of diversity found among the cells of multicellular organisms is fascinating. In a vertebrate, more than 200 distinct cell types are distinguishable and many of these types of cells include, under a single name, a large number of more subtly different varieties. Much of our current knowledge of the general properties of the eukaryotic cells depend on the study of such specialized types of cells; because they demonstrate particular features on which all cells depend on in some measure—for example, the neuromuscular junction, where just three types of cells are involved: a muscle cell, a nerve cell and a Schwann cell, each with different role.

Genes can be switched on and off

Multicellular organisms contain numerous distinct cell types. Although cell types are often distinguished morphologically, the fundamental property characterizing a cell type is its particular pattern of gene expression. Certain genes are expressed in all cells; others in just a small population of cells; yet others are expressed only under certain physiological conditions. It seems paradoxical, that all the cells in a multicellular organism combrise the same kind of DNA and are descendent from same precursor cell (e.g., fertilized egg). Common lineage implies a similar set of genes, but the question is how did the differences arise? Specialisation purely depends on changes in **gene expression**, not on the loss or acquisition of genes. Eukaryotic cells have evolved far more sophisticated mechanisms for controlling gene expression than bacteria and these affect entire systems of interacting gene products. Groups of genes are activated or repressed in response to both external and internal signals. Membrane composition, cytoskeleton, secretory products, even metabolism and other features must change in a coordinated manner when cells become differentiated.

Building multicellular organisms

The transition to multicellularity is unusual among the major transitions in evolution because it has occurred

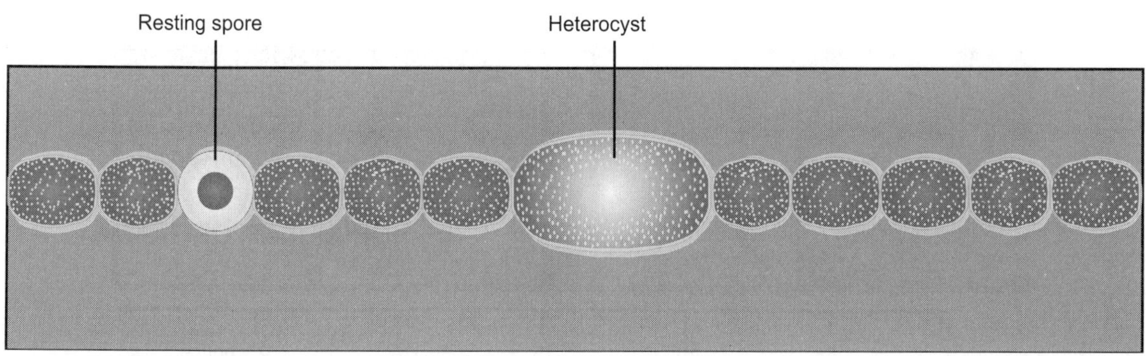

Fig. 13.3 Cyanobacterium—a premier photo-autotroph. Resting spores form when conditions do not favour growth. Heterocyst is a unique differentiated cell involved in nitrogen-fixing function.

numerous times. Novel multicellular forms are still being discovered. In a multicellular organism the activities of individual cells are highly coordinated and the cells themselves are either in contact or close enough to interact strongly. The multicellular condition has independently arisen in each of the organic kingdoms and in some phyla several times in the course of their evolution. For example, sponges are believed to have arisen from chromoflagellates separately from all other animals. The seed plants, the fungi, the brown algae and the red algae all gained their multicellularity in separate events. Microorganisms might have also evolved in similar ways. After cyanobacteria, cellular slime molds, myxomycetes and mycobacteria independently adapted multicellularity. These transitions can be viewed as different trials of an experiment conducted by Nature to test the fitness and the advantages of multicellular life.

Life's unicellular origin implies that the fundamental genetics and epigenetic systems of all animals and plants were originally devised for these single cells. If so, how were those systems enhanced to manage a multicellular life cycle? What are the regulatory circuits of single and multicellular forms and the new nodes of epigenetic control available in a few multicellular organisms such as cyanobacterium (Nostock), *Caenorhabditis elegans*, *Drosophila melanogaster*, *Homo sapiens* and other organisms?

13.3.1 Cyanobacteria

Filamentous cyanobacteria are the earth's oldest known multicellular organisms. Fossilized remains discovered in the 3465-Myr sedimentary rocks of northwestern Australia record unsheathed filaments resembling modern species. Assuming that life began with single cells, the antiquity of these multicellular filaments, as well as their morphological diversity, imply that multicellularity was advantageous on the Archaen earth. The unique feature of cyanobacteria is that in addition to photosynthesis, many species fix atmospheric dinitrogen. Since nitrogenase, the fixation catalyst, is very oxygen-sensitive, cyanobacteria arrange to separate nitrogen fixation from photosynthesis (nitrogen fixation occurs only in heterocysts). **Figure 13.3** shows a long chain of Nostoc vegetative cells in which several heterocysts are visible as larger darker cells. The intercellular exchange of metabolites illustrates the metabolic interdependence of vegetative cells and heterocysts. These cells are also developmentally interdependent as the heterocysts differentiate at fairly regular spatial intervals.

13.3.2 Myxobacteria

Some organisms become multicellular by aggregation rather than by growth and cell division. Myxobacterial cells aggregate to build quite elaborate fruiting bodies within which they sporulate. The life cycles of these prokaryotes appear similar to development of the (eukaryotic) cellular slime molds, such as *Dictyostelium*. All myxobacteria form fruiting bodies and lie on the boundary between unicellular and multicellular organisms. Although they grow and divide as proper Gram-negative bacteria, they constitute a primitive multicellular organism whose cells feed socially as multicellular swarm units. They eat particulate organic matter in the soil using hydrolytic enzymes. They feed like packs of microbial wolves. They also build multicellular fruiting bodies. When they have exhausted their food, 100,000 cells build these compact and symmetrical structures, often with stalks, branches and multiple cysts. A selective advantage for their multicellular state is their cooperative feeding.

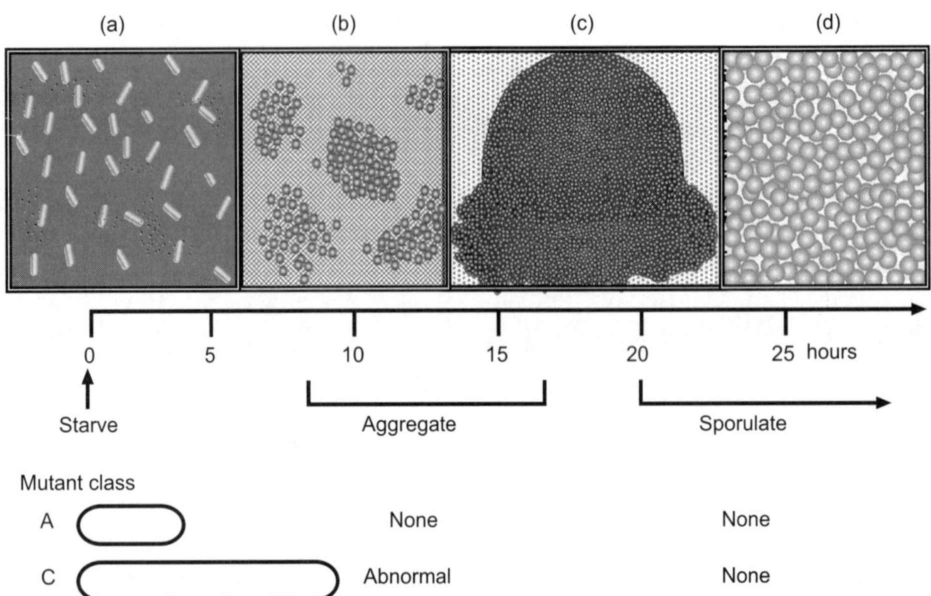

Fig. 13.4 Development of fruiting body in *Myxococcus xanthus* depends on the A and C signals. Fruiting body development viewed at (a) the start, (b) 4 hours, (c) 24 hours and (d) after cracking the fruiting body open. Below the time scale are shown the aggregation and sporulation processes. A signalling mutants express a few early developmentally regulated genes but no aggregation or sporulation. C signalling mutants express more developmentally regulated genes, aggregation but no sporulation.

The selective advantage for multicellular fruiting bodies is that after the spores germinate from the fruiting body they can cooperate for efficient feeding; and the second advantage is a 0.2 mm fruiting body can adhere to an animal and get carried to distance. Sporulation is a great strategy for survival when food runs out.

How spores become localized is explained by cell-to-cell signalling. About 1000 cells of irregular aggregates from disorganized biofilm appear in 4 hours (**Fig. 13.4**). Then many cells stream into these foci from all directions so that in 24 hours a hemispherical mound of 10^5 cells has been created. At least extracellular signals are necessary for making spores. Signalling to coordinate sporulation in myxobacteria may be due to a number of genes. A fixed amount of A-factor is (a set of six particular amino acids: Trp, Pro, Phe, Tyr, Leu and Ile) like a **quorum sensor** produced by each cell. By contrast to A-factor, C-factor is an approximately 20 kDa protein, has a hydrophobic *N*-terminal that keeps itself associated with the cell surface. C-factor signalling requires end-to-end cell contact for signal transmission.

The fruiting body is dense because the cells within are closely packed. The surrounding rod-shaped cells are organized into raft-like clusters and will not differentiate into spherical spores; only the central cells differentiate into spherical spores. Two forces appear to have combined in selecting the myxobacterial grade of multicellularity for more efficient feeding and dispersal.

13.4 MULTICELLULARITY IN HIGHER ORGANISMS

The evolution of multicellular higher organisms permitted specialized cells and tissues to form. For example, a typical flowering plant has at least 15 types of cells and in a vertebrate, hundreds. The specialized cells carry out a particular task, for example, a phloem and xylem in plants, a liver, a muscle, or a heart. Different types of cells in a tissue are precisely arranged in patterns often exhibiting staggering complexity. In the human brain, for instance, there are hundreds of different types of neurons interconnected to one another through a network of some 10^{15} synaptic connections! The coordinated functioning of many types of cells within tissues and multiple specialized tissues, permits the organism as a whole to move, metabolize, reproduce and carry out other essential activities.

Cells in multicellular organisms are separated from one another by **intercellular** or **extracellular space**, which is modified in various ways. A key step in the evolution of multicellularity must have been the ability of cells to contact tightly and interact specifically with other cells. **Cell surface proteins** enable many animal cells to adhere tightly and specifically with cells of a similar type. These interactions allow populations of cells to segregate into distinct tissues. Following aggregation, cells elaborate specialized junctions that stabilize these interactions and promote local communication between adjacent cells. Animal cells also secrete a complex network of secreted proteins and carbohydrates—the **extracellular matrix**—that fills the spaces between cells. The matrix helps bind the cells in tissues together and is a reservoir for many hormones controlling cell growth and differentiation. It also provides a lattice through which cells can move, particularly during the early stages of differentiation. However, in case of plants, the **cell walls** that surround plant cells are thicker, more impermeable and more rigid than the extracellular matrix around animal cells. Most importantly, plant cells can grow larger but do not move in relation to their neighbours. The plant cell wall is an elaborate extracellular matrix that encloses each cell in a plant (**Fig. 13.5**).

Local interactions between cells often lead to changes in cell shape and movement and to changes in gene expression leading to cellular differentiation. During development, local interactions between cells and their environment induce the formation of different cell types and their organization in tissues; they also guide the migration of cells and direct the formation of precise patterns of synaptic connections among developing nerve cells.

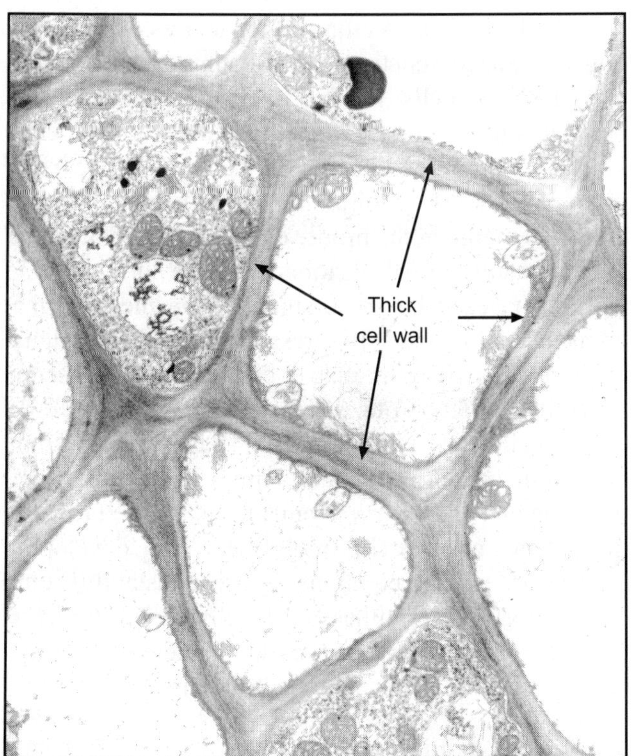

Fig. 13.5 Electron photomicrograph illustrating the plant cells surrounded with a thick cell wall, which may be an equivalent of extracellular matrix of animal tissues (Channarayappa, 1992).

13.5 THE EXTRACELLULAR MATRIX

Somewhat similar to cell walls surrounded by the bacterial and plant cells, many of the animal cells in tissues are embedded in an extracellular matrix (ECM) consisting of secreted proteins and polysaccharides. The term 'matrix' implies a structure made up of a network of interacting components, which is aptly applicable to ECM which contains a number of proteins and other molecules that interact with one another in highly specific ways. The ECM is often referred to as the connective tissue. It plays a key regulatory role in determining the shape and many cellular activities. In animals, it facilitates organization of cells into tissues, and coordinates their cellular functions by activating intracellular signalling pathways that regulate cell growth, proliferation and gene expression. Animals contain many types of extracellular matrices, each specialized for a particular function such as strength (in a tendon), filtration (in the kidney) or adhesion. The ECM is composed of three major classes of biomolecules, which are widely distributed in the tissues:

- Collagen fibres
- Ground substance: Proteoglycans
- Multi-adhesive matrix proteins.

Many functions of the matrix require transmembrane adhesion receptors that bind directly to ECM components and also interact, with the cytoskeleton through adapter proteins. The extracellular matrix principally consists of different combinations of **collagen** proteins, which form either fibres or porous sheets. The long polysaccharide **hyaluronan** and **proteoglycans**, are macromolecules consisting of a core protein covalently attached to highly charged polysaccharides called **glycosaminoglycans**. **Multi-adhesive proteins** are also important matrix constituents. They bind to receptor proteins on the surfaces of cells and to other matrix components, imparting strength and rigidity to the matrix. In the epidermis of the skin, for example, there is little extracellular space between the cells (**Fig. 13.6**). Most epithelia and other organisation groups of cells (such as muscle) rest upon and are tightly bound to, a thin matrix called the **basal lamina**, which is lined

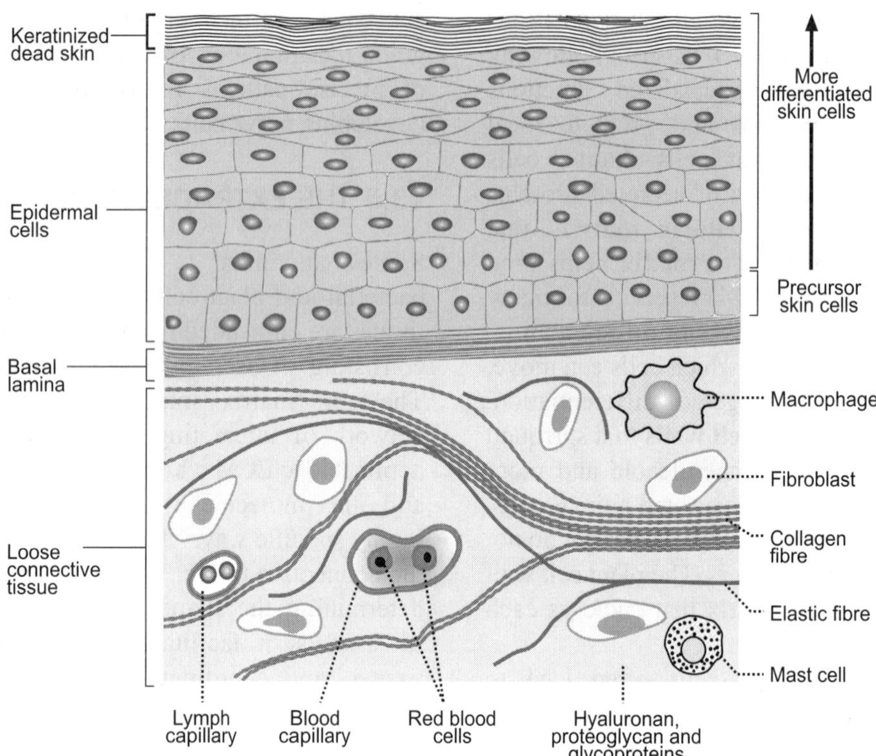

Fig. 13.6 A section through the outer layer of the skin of a pig.

to the plasma membrane by specific receptor proteins. In the glomerulus of the kidney, the basal lamina forms a porous filter that allows water, ion and small molecules in the blood to cross into the urinary space while retaining proteins and cells in the blood.

Animals contain abundant connective tissue in which much of the tissue volume consists of space between cells. **Loose connective tissue** forms the bed on which most small glands and epithelia lie and connect to the basal lamina around the cells. Proteins such as fibrous collagens and the rubber-like flexible protein **elastin** give tissues their shape, rigidity and flexibility. Most of the space is filled with the highly hydrated hyaluronan and proteoglycans, giving tissues a gel-like consistency. Loose connective tissue is highly cellular and contains numerous fibroblasts, which synthesize much of the extracellular matrix and also lymphocytes. **Dense connective tissue**, including bone, cartilage and tendon, is a major component of organs in which strength or flexibility or both are essential attributes (e.g., skeletal system). It consists entirely of fibrous extracellular materials, very few cells, densely packed collagens and other fibrous proteins surrounded by glycoproteins, proteoglycans and other substances (e.g., elastin) that are produced and secreted by the relatively few cells present.

The ECM is also required for many cells to carry out specialized functions. For example, hepatocytes, the principal cells in the liver, must contact the basal lamina or other extracellular matrix in order to synthesize proteins specific to cell-type. Like most other cells, it also requires contact with an appropriate matrix to maintain the proper cytoskeletal architecture. It binds many growth factors and other hormones and can either sequester these signals from the cells that contact it or, conversely, present the hormonal signal to these cells. Specific matrix components are needed for cells to differentiate, that is, to acquire the functions of a specific cell type. The later stage of embryonic development, during which form is achieved by cell movements and rearrangement, is also critically dependent on matrix molecules. In developing organisms, matrix components are constantly being remodelled and resynthesized locally. Even in adults, for example, in areas of wounding, degradation and resynthesis of matrix components occurs.

13.5.1 Collagen

Collagen is the major class of insoluble fibrous protein in the ECM and in connective tissue. It is the single most abundant protein in the animal kingdom (25% of all human body proteins) and is noted for its high tensile strength. Collagen fibres provide mechanical strength and resilience. Collagen is responsible for the integrity of the fibrous tissue of bone, teeth, cartilage, skin and tendons. The chemistry of collagen is quite distinctive in several respects:

- Large portions of collagen contain two rare amino acids: hydroxylysine and hydroxyproline. These two amino acids account for up to 25% of the total amino acids and are essential for the structure and modification of the protein.
- Glycine accounts for about one-third of the amino acids, occupying almost every third position of the sequence (-X-Y-Gly-), with X often proline and Y hydroxyproline.
- Short oligosaccharide chains composed only of glucose and galactose are attached to the protein.
- Most importantly, collagen fibres can fold and cross-link into fibrils or networks.

Collagen protein-types

Collagens comprise a family of fibrous glycoproteins that are present only in extracellular matrices (**Table 13.1**). Collagen is produced primarily by fibroblasts and also by smooth muscle cells and epithelial cells. More than 20 distinct types of collagens

Table 13.1 Representative members of the collagen family

Collagen class	Types	Characteristics and tissue distribution
Fibril-forming	I	Low carbohydrate. Most connective tissues: bone, skin, tendons, ligaments, cornea, internal organs
	II	>10 hydroxylysines per chain: cartilage, invertebral disc, notochord, vitreous humor of eye
	III	Has cytosine, low hydroxylysine; extensible connective tissues (e.g., skin, blood vessels and lung)
	V	Elevated hydroxylysine, low alanine: Tissues containing collagen I
	VI	Cartilage
	XI	Cartilage, invertebral disc, notochord, vitreous humor of eye
	XXIV	Bonc and cornca
	XXVII	Eye, ear and lung
Fibril-associated	IX	Cartilage
	XII	Tissues containing collagen I, tendons, ligaments
	XIV	Tissues containing collagen I
	XVI	Widespread
	XIX	Widespread
	XX	Cornea
	XXI	Many tissues
	XXII	Cell junctions
	XXVI	Testis and ovary
Network-forming	IV	Extensive lysine and hydroxylation. Basal laminac
	VIII	Many tissues
	X	Cartilage
Anchoring fibrils	VII	Attachments of basal laminate to underlying connective tissue
Transmembrane	XVII	Skin hemidesmosomes
	XXV	Nerve cells
Others	XVIII	Basal lamina around blood vessels

have been identified. Types I, II and III are the most abundant; these collagen molecules pack together to form long, thin fibrils of similar structure. Type IV is non-fibrillar, forms a two-dimensional reticulum and is the principal component of the basal lamina. The structural features of each type of collagen make it suitable for a particular function. Although there are many differences among the members of the collagen family, all share at least two important structural features:

- All collagen molecules are trimers consisting of three polypeptide chains, called A chains.
- Along at least part of their length, the three polypeptide chains of a collagen molecule are wound around each other to form a unique, rod-like triple helix.

Fibrous type I collagen structural unit is a long (300 nm), thin (1.5 nm) protein that consists of three coiled subunits [two α_1(I) chains and one α_2(I)]. Each chain contains precisely 1050 amino acids wound around each other in a characteristic right-handed triple helix (**Figs. 13.7a** and **13.7b**). All collagens were eventually shown to contain three-stranded helical segments of similar structure. The unique properties of each type of collagen are due mainly to segments that interrupt the triple helix and that fold into other kinds of three-dimensional structures. There are three amino acids per turn of the collagen helix and every third amino acid is glycine, the smallest amino acid. The rigid five-member pyrolidone rings in these amino acids are on the outside of the triple helix; the fixed angle of the C–N peptidyl–proline or peptidyl-hydroxyproline bond enables each polypeptide chain to fold into a helix with a geometry such that three polypeptide chains can twist together to form a three-stranded helix. Many regions of collagen chains are composed of the repeating motif of Gly-Pro-X, where X can be any amino acid. Hydrogen bonds linking the peptide bond NH of a glycine residue with a peptide carbonyl (C=O) group in an adjacent polypeptide help hold the three chains together.

Given their abundance and widespread distribution, undoubtedly abnormalities in fibrillar collagen formation can lead to serious disorders. Burns or traumatic injuries to internal organs can lead to an accumulation of scar tissue, which consists largely of fibrillar collagen. Mutations in genes encoding type I collagen can produce **osteogenesis**

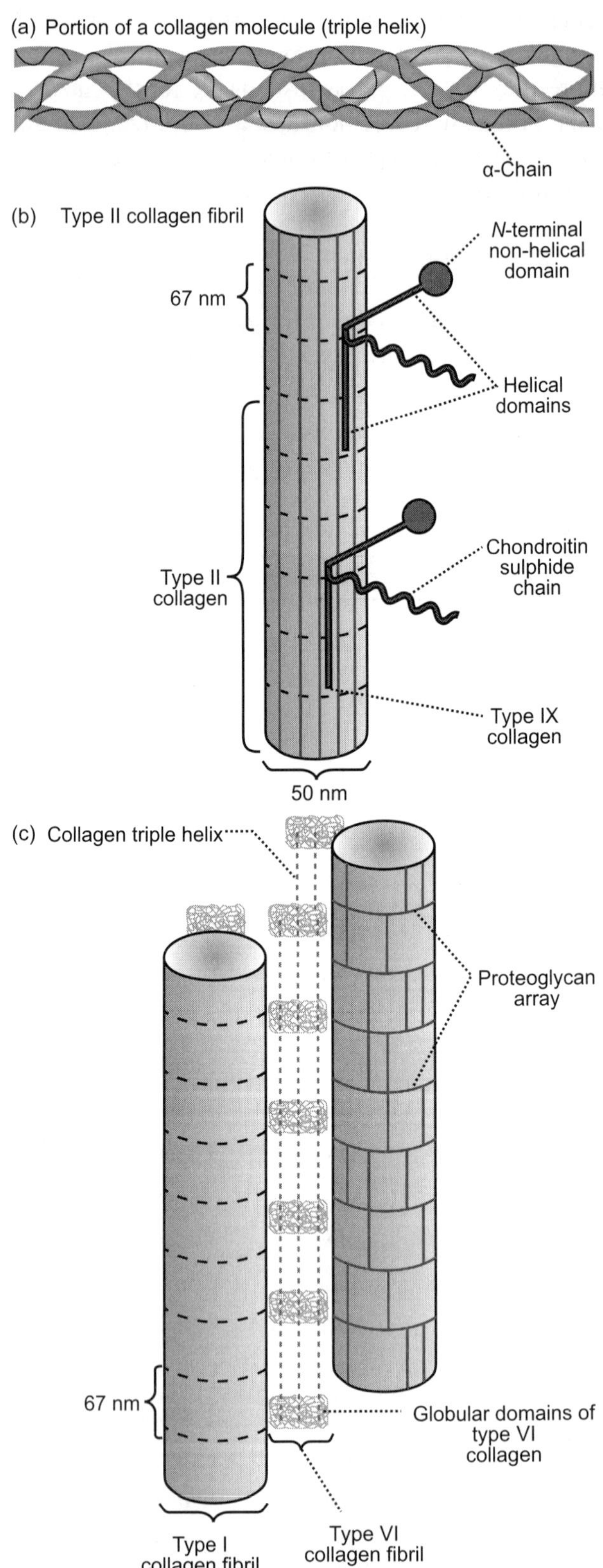

Fig. 13.7 Collagen structure and interaction of fibrous and non-fibrous collagens: (a) The basic structural unit is a triple-stranded helical molecule. (b) Types II and IX in a cartilage matrix. (c) Organization of the major fibrous components in the extracellular matrix of tendons.

imperfecta, a potentially lethal condition characterized by extremely fragile bones, thin skin and weak tendons. Mutations in genes encoding type II collagen alter the properties of cartilage tissue, causing dwarfism and skeletal deformities. Mutations in a number of other collagen genes can lead to a variety of distinct but related defects in collagen matrix structure.

Clinical significance of collagen disorders

Collagens are the most abundant proteins in the body. Alterations in collagen structure resulting from abnormal genes or abnormal processing of collagen proteins results in numerous diseases (**Table 13.2**). Ehlers–Danlos syndrome is actually the name associated with at least ten distinct disorders that are biochemically and clinically distinct, yet manifest as structural weakness in connective tissue as a result of defects in the structure of collagens. Osteogenesis imperfecta also causes more than one disorder.

Table 13.2 Human disorders due to the defects in collagen proteins

Disorder	Collagen defect	Symptomology
Ehlers–Danlos IV	Decrease in type III	Arterial, intestinal and uterine rupture, thin, easily bruised skin
Ehlers–Danlos V	Decreased cross-linking	Skin and joint hyperextensibility
Ehlers–Danlos VI	Decreased hydroxylysine	Poor wound healing, musculoskeletal deformities, skin and joint hyperextensibility
Ehlers–Danlos VII	*N*-terminal propeptide not removed	Easily bruised skin, hip dislocations, hyperextensibility
Oseteogenesis imperfecta	Decrease in type I	Blue sclerae, bone deformities
Scurvy	Decreased hydroxyproline	Poor wound healing, deficient growth, capillary weakness

13.5.2 Hyaluronan

Hyaluronan (HA) is a major component of the extracellular matrix that surrounds migrating and proliferating cells, particularly in embryonic tissues. It is also a major structural component of the complex proteoglycons that are found in many extracellular matrices, particularly cartilage. Due to its remarkable physical properties, HA imparts stiffness and resilience as well as lubricating quality to many types of connective tissue such as joints. HA is the only extracellular oligosaccharide that is not covalently linked to a protein. Each hyaluronan molecule consist of as many as 50,000 repeats of the simple disaccharide glucuronic acid β(1→3) *N*-acetylglucosamine β(1→ 4).

Hyaluronan is bound to the surface of many migrating cells by a 34 kDa receptor protein termed CD44, or by a homologous protein in the CD44 family (**Fig. 13.8**).

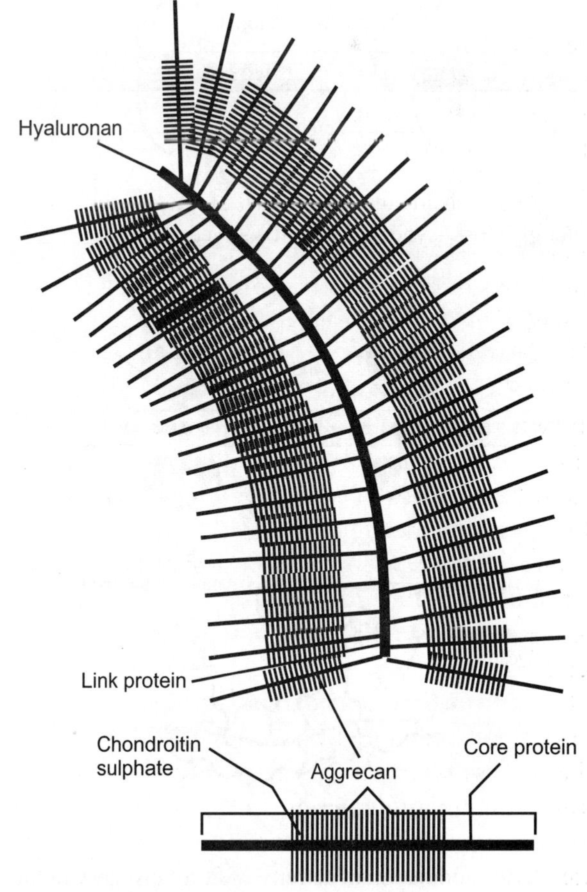

Fig. 13.8 Complex of aggrecan and hyaluronan.

13.5.3 Ground substance: proteoglycans

Proteoglycans are a group of highly viscous glycoprteins that cushion cells and bind a variety of extracellular molecules. Proteoglycans are found in all connective tissues and extracellular matrices and on the surface of many cells. They consist of a core protein covalently attached to one or more polysaccharides called **glycosaminoglycans (GAGs)**. Proteoglycans are remarkable for their diversity. A given matrix may contain several different types of core proteins and each may contain several different numbers of GAGs with differing lengths and compositions. Each GAG chain is composed of a repeating disaccharide; that is, it has the structure –A–B–A–B–A–, where A and B represent two different sugars. One of the most important extracellular proteoglycans is **aggrecan**, the predominant proteoglycan in cartilage. Aggrecan forms very large aggregates, termed proteoglycan aggregates. A single aggregate, one of the largest macromolecules known, can be more than 4 μm long and have a volume larger than that of a bacterial cell! These aggregates give cartilage its unique gel-like properties and its resistance to deformation, essential for distributing the load in weight-bearing joints. The central component of the cartilage proteoglycan aggregate is a long molecule of hyaluronan. Bound to it, tightly but non-covalently, at 40 nm intervals, are aggrecan molecules. Aggrecan core protein has one *N*-terminal globular domain that binds with high affinity to a decasaccharide sequence in hyaluronan; this binding is facilitated by a **link protein** that binds to the aggrecan core protein and hyaluronan. Covalently attached to each aggrecan core protein, via the trisaccharide linker, are multiple chains of chondroitin sulphate and keratan sulphate.

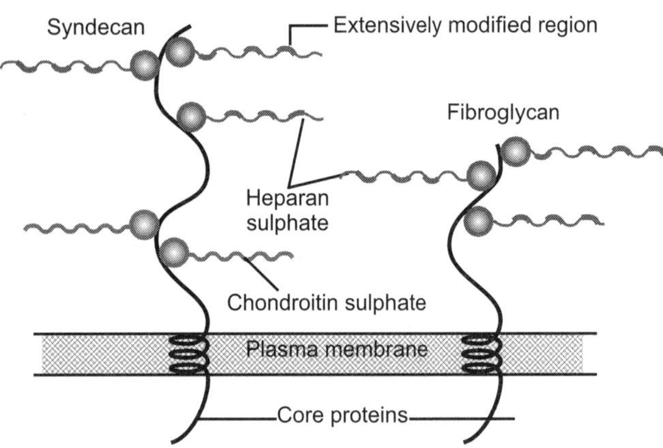

Fig. 13.9 Schematic diagrams of two cell surface proteoglycans—syndecan and fibroglycan.

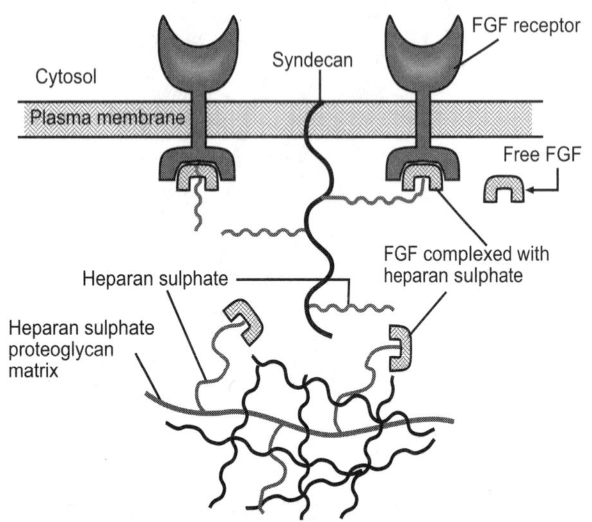

Fig. 13.10 Modulation of activity of fibroblast growth factor (FGF) by heparan sulphate proteoglycans.

Cell surface proteoglycans: These are attached to the surface of many types of cells, particularly epithelial cells. Two examples are **syndecan** and **fibroglycan;** the general structures are illustrated in **Fig. 13.9**. The core protein of cell surface proteoglycans spans the plasma membrane and contains a short cytosolic domain, as well as a long external domain to which a small number of heparan sulphate chains are attached; some cell surface proteoglycans also contain chondroitin sulphate. The attached heparan sulphate chains bind to fibrous collagens and to the large glycoprotein fibronectin, which are present in the interstitial matrix surrounding the basal lamina. In this manner, cell surface proteoglycans anchor cells to matrix fibres.

Besides acting as structural components of the extracellular matrix and anchoring cells to the matrix, both extracellular and cell surface proteoglycans, particularly those containing haparan sulphates, also bind many proteins (fibroblasts) growth factor (FGF). These stimulate proliferation of many types of cells other than fibroblasts as well. FGF binds tightly to the heparan sulphate chains in extracellular proteoglycans. Since the bound growth factor is resistant to degradation by extracellular proteases, it serves as a reservoir of matrix-bound FGF (**Fig. 13.10**).

13.5.4 Multi-adhesive matrix proteins

Collagens and proteoglycans are the two important components of all extracellular matrices. Another major group of proteins bind not only to collagens and proteoglycans but also to specific cell surface receptors. These multi-adhesive proteins are important for organising the other components of the matrix and also for regulating cell attachment to the matrix, cell migration and cell shape.

Basal lamina

In animals, epithelia and other organized groups of cells are surrounded by the basal lamina, a sheet-like mesh of ECM components usually around 60–120 mm thick. All basal laminae contain a common set of proteins and glycosaminoglycans (**Fig. 13.11**). **Type IV collagen** is a trimeric molecule with both rod-like and globular domains that form a two-dimensional network. **Perlecan** (heparan sulphate proteoglycans) is a large multidomain proteoglycan that binds to and cross-links many ECM components and cell surface molecules. **Entactin** (also called **nidogen**) is a rod-shaped molecule with three globular domains; it binds tightly to laminin and is often considered a laminin subunit. Nidogen also binds to type IV collagen and to proteoglycans and thus could mediate the assembly of all the components of the basal lamina into a network. **Laminins** are a family of multi-adhesive proteins of accessory molecules almost 70 nm long. In adult animals, laminin contains three polypeptides and has high-affinity binding sites for other components of the basal lamina.

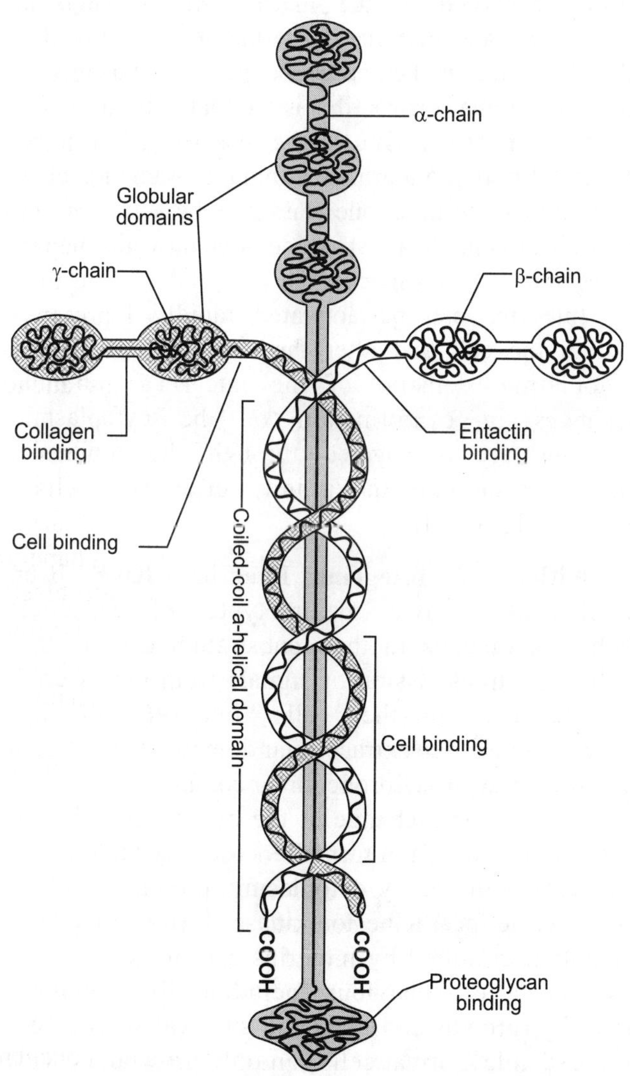

Fig. 13.11 Structure of lamini. They consist of three polypeptides (a, b and g) that are disulphide bonded into an asymmetric cross-like structure. Some of the binding sites for entactin, type IV collagen, proteoglycans and cell surface receptors are indicated.

The basal lamina is structured differently in different tissues. Different types of cells that interact with a basal lamina, such as epithelial cells and smooth and striated muscle may use different cell surface receptors that bind to matrix components. In columnar and other epithelia it is a foundation on which only one surface of the cells rests. In muscle or fat, the basal lamina surrounds each cell. Basal laminae play an important role in regeneration after tissue damage and in embryonic development. Extracellular laminins can greatly influence cell migration, growth and differentiation. For example, laminins facilitate the migration of primordial germ cells. These cells arise in the yolk sac, which is located outside the embryo and then migrate through the bloodstream and embryonic tissues to the developing gonad, where they eventually give rise to sperm or eggs. During their migration the primordial germ cells traverse surfaces that are particularly rich in laminin.

Integrins

Integrins are a family of integral membrane proteins found only in animals and bind to specific substances in the extracellular environment. They are a large class of cell surface receptors that bind different components of the extracellular matrix; some of these transmembrane proteins also mediate cell–cell interactions. Most integrins are expressed on a variety of cells and most cells express several integrins, enabling them to bind to several matrix molecules. Individual integrins often bind more than one ligand and individual matrix molecules are often bound by more than one integrin, because they play multiple roles in differentiation and in adult organisms.

Integrin structure and diversity: Integrins are composed of two membrane-spanning polypeptides (heterodimers of α- and β-subunits), that are non-covalently linked. The ligand-binding site is composed of parts of both chains (**Fig. 13.12**). In mammals, at least 29 integrin heterodimers, comprised of 18 different α-subunits and 8 different β-subunits, are known. A single β-chain can interact with multiple α-chains, forming integrins that bind different

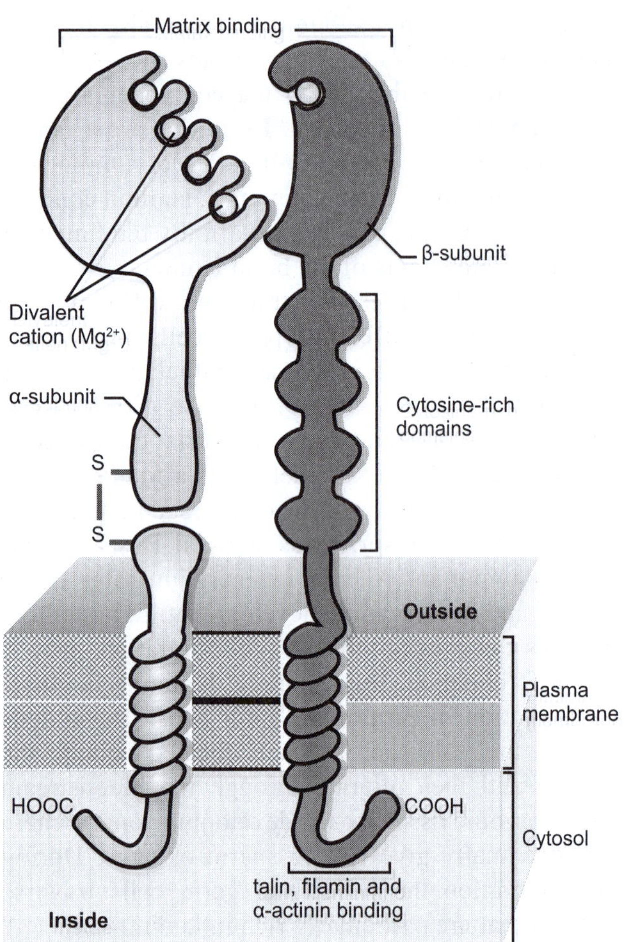

Fig. 13.12 The subunit structure of an integrin cell surface matrix receptor. The integrins are heterodimers of two transmembrane subunits, designated as α and β.

ligands. Electron microscopic observations of integrin molecules suggested that the two subunits are oriented to form a globular extracellular head connected to the membrane by a pair of elongated legs. The legs of each subunit extend through the lipid bilayer as a single transmembrane helix and end in a small cytoplasmic domain of about 20–70 amino acids. In general, integrins containing the β_1-subunit bind components of the extracellular subunit. For example, $\alpha_1\beta_1$- and $\alpha_2\beta_1$-integrins both bind a segment within the *C*-terminal domain of type IV collagen; these integrins, as well as the widely expressed $\alpha_2\beta_1$, also bind at least two different regions of laminin; and $\alpha_5\beta_1$ binds fibronectin. In contrast, the β_2-subunit is expressed exclusively on leukocytes along with three leukocyte-specific α-subunits. These integrins mediate cell–cell interactions, rather than cell–matrix interactions. Several integrin subunits in mammals have alternatively spliced cytosolic domains. This adds a further level of complexity, since these integrins probably bind different intracellular attachment proteins and thus interact differently with components of the cytoskeleton.

Regulation of integrin activity: Even though integrins are present on the cell surface, they may require activation in order to bind their ligand and anchor the cell to the extracellular matrix, or to another cell. Integrins typically exhibit relatively low affinities for their ligands, compared with the high affinities of typical cell surface hormone receptors. However, the multiple weak interactions generated by binding hundreds or thousands of integrin molecules to extracellular matrix proteins allows a cell to remain firmly anchored to the matrix. Alternatively, in situations where cells are migrating, it is essential that they be able to make and break specific contacts with the extracellular matrix; this is facilitated if individual contacts are weak. One of the reasons cells express several different integrins that bind the same ligand is that, at any one time, one class may be activated and thus able to bind to a matrix component, while another class of integrins is inactivated.

Integrins can be activated rapidly by events within the cell that alter the conformation of the cytoplasmic domains of the integrin's subunits. Changes in conformation of the cytoplasmic domains are propagated through the molecule, increasing the integrin's affinity for an extracellular ligand (**Fig. 13.13**).

Activities of integrins: Integrins have been implicated in two major types of activities: adhesion of cells to their substratum (or to other cells) and transmission of signals from the external environment to the cells interior(outside-in). The binding of the extracellular domain of an integrin to a ligand, such as fibronectin or collagen, can induce a conformational change at the cytoplasmic end of the integrin, which in turn alters the way the integrin interacts with nearby cytoplasmic proteins, such as the enzyme focal adhesion kinase (FAK). Outside-in signals transmitted by integrins can influence many aspects of cell behaviour, including differentiation, motility, growth and even the survival of the cell. For example, normal cells can only grow and divide if they are cultured on a solid substratum; if they are placed in suspension cultures, they die. In suspension, integrins of normal cells are not able to interact with extracellular substrates as a result; they are not able to transmit life-saving signals to the interior of the

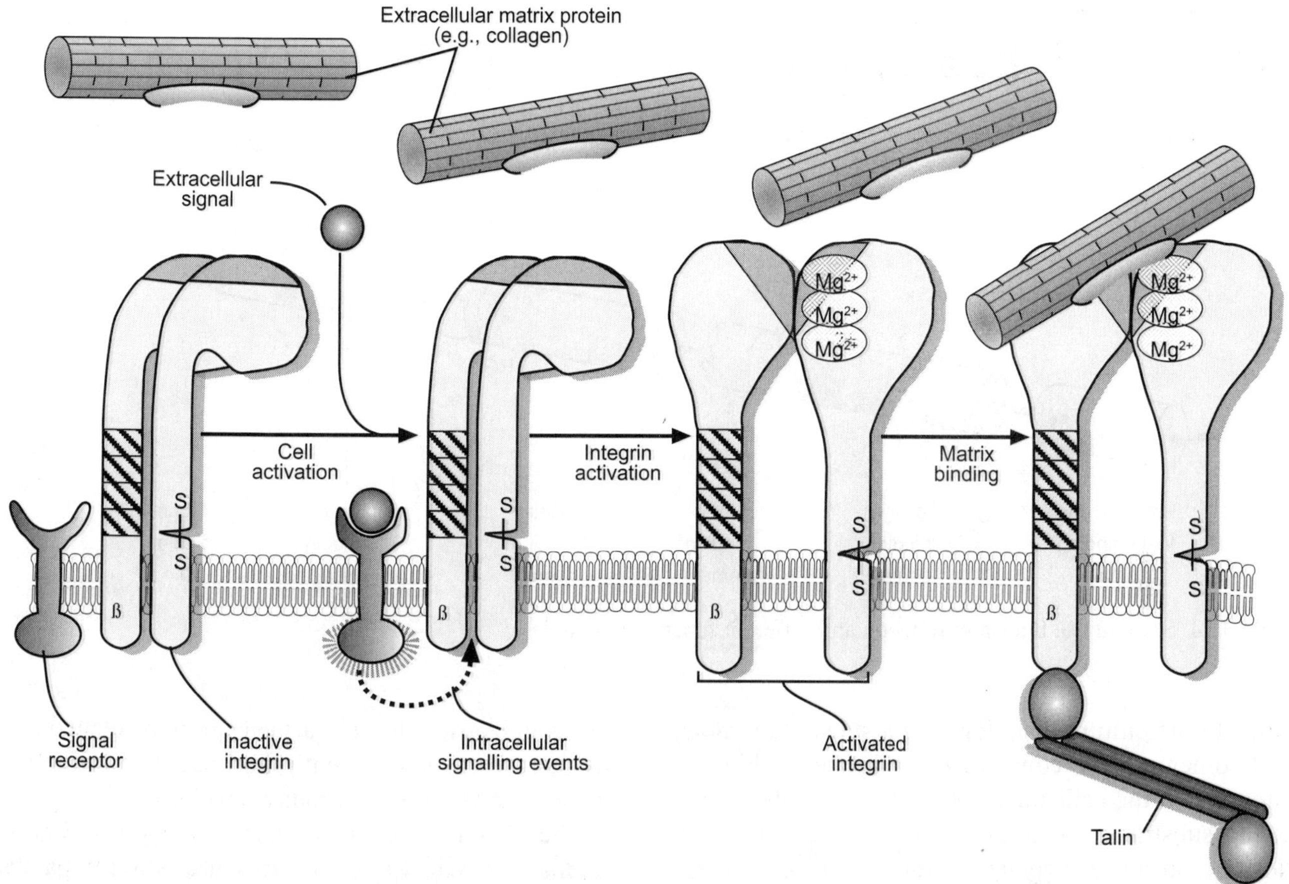

Fig. 13.13 General structure of the integrin family of cell surface receptors. Integrins are heterodimers of two transmembrane subunits, designated α and β. The α-subunit binds divalent cations (Mg^{2+}). The matrix binding region consists of portions of both subunits.

cell. In contrast, most malignant cells can grow in suspension cultures, because their survival no longer depends on integrin binding.

Fibronectins

Fibronectins are another important class of multi-adhesive matrix proteins. Their primary role is attaching cells to a variety of extracellular matrices, especially those that contain the fibrous collagens (types I, II, III and V). The presence of fibronectin on the surface of non-transformed cells first led to the identification of fibronectin as an alternative protein. By their attachments, fibronectins regulate the shape of cells and the organization of the cytoskeleton; they are essential for migration and cellular differentiation of many cell types during embryogenesis. Fibronectins are also important for wound healing, because they facilitate migration of macrophages and other immune cells into the affected area.

Fibronectins are dimers of two similar polypeptides linked at their *C*-terminii by two disulphide bonds. Each fibronectin polypeptide is constructed from a sequence of approximately 30 independently folding Fn modules.

At least 20 different fibronectin chains have been identified, all of which are generated by alternative splicing of the RNA transcript of a single fibronectin gene. Analysis of digests of fibronectin with low amounts of proteases shows that the polypeptides consist of six tightly folded domains. Each domain, contains small repeated sequences. The similarities in amino acid sequence allows them to be classified into three types (**Fig. 13.14**). Different domains of fibronectin possess specific high-affinity binding sites for cell surface integrins, denatured forms of collagen, fibrin and sulphated proteoglycans.

Cell adhesion to the substratum: Eight vertebrate integrins bind fibronectin; most of these bind specifically to the RGD sequence found in fibronectin and certain other extracellular matrix proteins. Many cultured animal cells secrete fibronectin, which

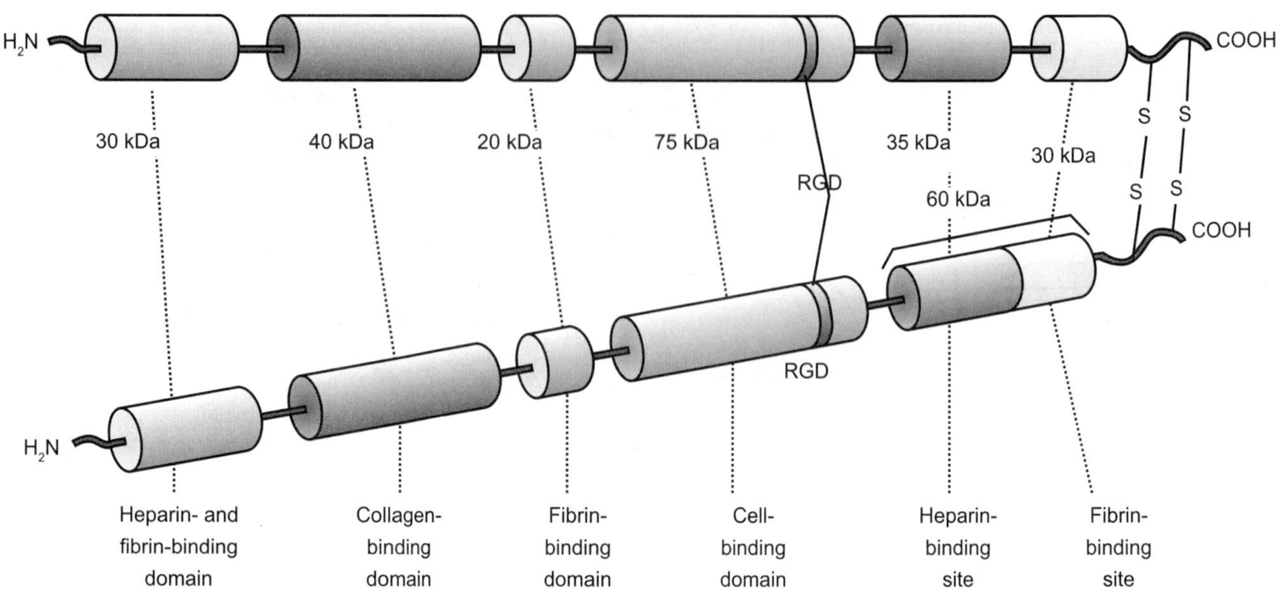

Fig. 13.14 Structure of human fibronectin and different functional modules.

binds to the culture dish, forming a substratum along with other matrix components. Fibronectin-binding integrins on the cells then interact with the fibronectin in the substratum, causing the cells to adhere to it. Non-migrating cells contain abundant actin-rich stress fibres that maintain the cell shape and attach to the plasma membrane where it adheres to substratum. Integrin $\alpha_5\beta_1$, a fibronectin receptor, is localized to these focal contacts and may anchor both fibres to the opposite sides of the plasma membrane. Many tumour cells growing in culture do not adhere to plastic culture dishes. Such cells synthesize little fibronectin and contain few stress fibres. Regulation of the number of integrin molecules on cells can modulate cell attachment to the substratum. An important example is the $\alpha_4\beta_1$-integrin, a fibronectin receptor, found on many hematopoietic cells. For these cells to proliferate and differentiate, they must be attached to extracellular matrix fibronectin synthesized by supportive (stromal) cells in the bone marrow. The hematopoietic cells directly contact the stromal cells as well as attach to the matrix. A decrease in the number of $\alpha_4\beta_1$-integrin molecules present on hematopoietic cells at a later stage in their differentiation is thought to allow mature blood cells to detach from the matrix and stromal cells in the bone marrow and subsequently enter the circulation.

Cell migration: Fibronectins are abundant constituents of several types of extracellular matrices, along with fibrous collagens and specific proteoglycans. The six-armed protein tenascin is frequently associated with fibronectin, particularly in connective tissues in many developing organs and in the mesenchyme that underlies many epithelia. In many cases, these matrices are loosely packed with several fibrillar components secreted by various interstitial cells and provide trails along which other cells can migrate. The importance of fibronectin and other extracellular proteins is particularly evident when tissues undergo dynamic activities, such as during embryonic development. Development is characterized by waves of cell migration during which different cells follow different routes from one part of the embryo to another. For example, the cells of the neural crest, which migrate out of the developing nervous system into virtually all parts of the embryo, traverse pathways rich in fibrils composed of fibronectin. Many other organs of the body, including the salivary gland, kidneys and lungs are formed by a process of branching, when the epithelial layer gets divided by a series of clefts. In adult organisms, cells involved in wound healing (fibroblasts, macrophages and other immune system cells) migrate the affected area and move across blood clots, which are composed primarily of the fibrous protein, fibrin.

13.6 CELL–CELL INTERACTIONS

Detailed cytological studies of thin section through a major organ of an animal reveals a complex architecture countaining a variety of cells. However, little is known about the mechanisms responsible for generating the complex 3D cellular arrangements found within developing organs. Studies also indicate that cells can specifically recognize the surface of other cells, ignoring non-target cells. Interactions between cells, as well as between cells and the extracellular matrix, are critical to the development and function of multicellular organisms. Some cell interactions are transient (e.g., cells of immune system) and in other cases, stable cell–cell junctions play a key role in the organisation of cells in tissues. In multicellular organisms, the cells adhere to one another and to various types of extracellular surface, with the exception of certain cells such as leukocytes and erythrocytes. Adhesion of like cells is a primary feature of the architecture of many tissues. A sheet of absorptive epithelial cells, for instance, forms the lining of the small intestine and sheets of hepatocytes that are two cells thick make up much of the liver. A number of cell surface adhesive molecules are now known (**Table 13.3**); a single cell type may use more than one such molecule to mediate such homophilic adhesion.

13.6.1 Cell adhesive proteins

Cell–cell adhesion is a selective process: cells adhere only to other cells of specific types. Various types of embryonic tissues dissociate into single cells and reassociate in the presence of other cell types. Invariably, it was found that cells of a particular type, such as liver, kidneys or retinal re-associate preferentially with cells of similar origin to form small aggregates. If a single large mass is formed, one cell type may localize to the centre, surrounded by cells of the other type. Adhesion of similar cells extend across species boundaries. Cell adhesive proteins are now known to fall into three principal classes (**Fig. 13.15**): selectins, cadherins and immunoglobulin (Ig) superfamily.

Selectins

Selectins are a family of integral membrane glycoproteins that recognize and bind to a particular arrangement of sugars in the oligosaccharides that project from the surfaces of other cells. They (**Fig. 13.16**) possess a small cytoplasmic domain, a single membrane-spanning domain and a large extracellular segment that consists of a number of separate modules, including an outermost domain that acts as the lectin—a compound that binds to specific carbohydrate groups. There are at least three known selectins: E selectin, present on endothelial cells; P selectin, present on platelets and endothelial cells; and L selectin, present on leukocytes. All selectins recognize a particular grouping of sugars that is found at the ends of carbohydrate chains of certain complex glycoproteins.

Cadherins

Cadherins are a large family of glycoproteins that mediate Ca^{2+}-dependent cell–cell adhesion and transmit signals from the ECM to the cytoplasm. They are important for tissue differentiation and structure. Cadherins join cells of similar type to one another and do so predominantly by binding the same

Table 13.3 Cell adhesion molecules

Family	Ligands recognized	Stable cell junctions
Selectins	Carbohydrates	No
Integrins	Extracellular matrix	Focal adhesion and hemidesmosomes
	Members of Ig superfamily	No
Ig superfamily	Integrins	No
	Homophilic interactions	No
Cadherins	Homophilic interactions	Adherens junctions and desmosomes

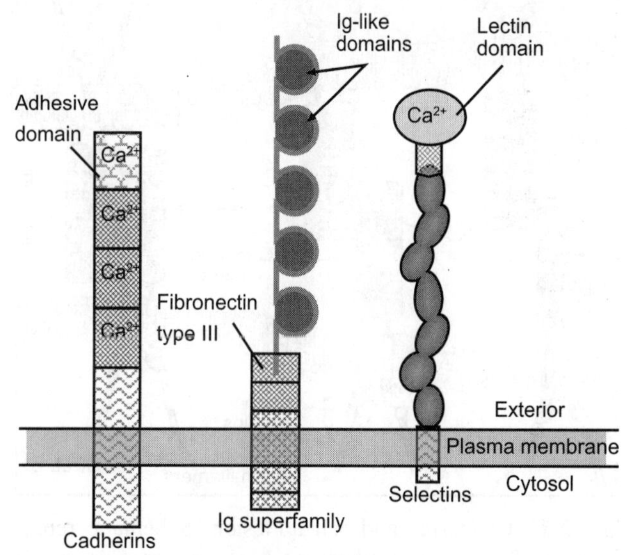

Fig. 13.15 Major families of cell adhesion proteins, built with multiple domains.

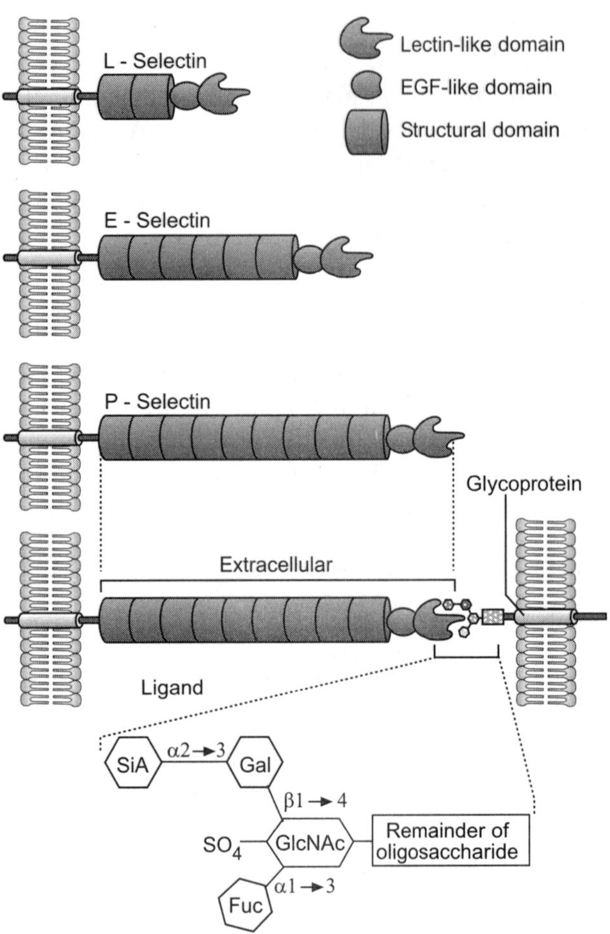

Fig. 13.16 Schematic drawing showing the three types of known selectins and a common carbohydrate ligand at the ends of oligosaccharide chain on glycoproteins.

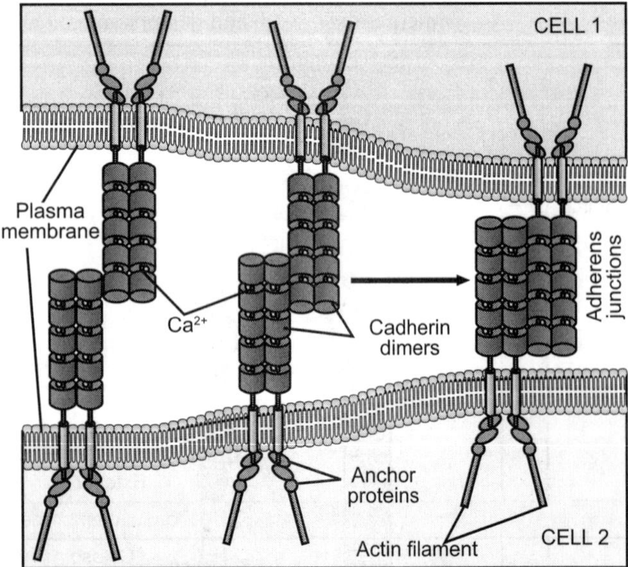

Fig. 13.17 Cadherins and cell adhesion. Schematic representation of two cells adhering to one another as the result of interactions between similar types of cadherins projecting from the plasma membrane of each cell (**See Plate 9**).

cadherin present on the surface of the neighbouring cell (**Fig. 13.17**). Cadherins are found on the surfaces of many different cell types in animals, with each member of the cadherin family having a specific distribution within the body. The most widely expressed, particularly during early differentiation, are the E cadherin (epithelial), P-cadherin (placental) and N cadherin (neural). Over 30 different cadherins are known. The brain expresses the largest number of cadherins, presumably due to the necessity of forming very specific cell–cell contacts. Each cadherin is an integral membrane glycoprotein of 720–750 amino acids with a relatively large extracellular segment consisting of five tandem domains of similar size and structure, a single transmembrane segment and a small cytoplasmic domain. The cytoplasmic domain is often associated with members of the **catenin** family of cytosolic proteins, which have a dual role: they tether the cadherins to the cytoskeleton and transmit signals to the cytoplasm. On an average, 50%–60% of the sequence is identical among different cadherins. Importantly, each cadherin has a characteristic tissue distribution.

Cadherins play a key role in cell differentiation and tissue morphogenesis. Embryonic development is characterized by changes in gene expression, cell shape, motility, cell adhesion and so on. After fertilization and fusion of mouse sperm and egg pronuclei, the egg divides several times into a mass of cells called a morula. At this stage, the major cell surface adhesive protein expressed is E cadherin. At a much later stage in differentiation, the embryo consists of two layers of epithelial cells, both of whìch express E caderin. The outermost embryonic cell layer called the **ectoderm**, gives rise to the skin, nervous system and other tissues; and the inner **endoderm** differentiates into the gastrointestinal tract and other tissues. First, some of the ectoderm cells lose their E cadherin, releasing them from the epithelium; these cells move to the middle of the embryo and form the mesoderm. After a number of additional cell divisions, a region of the ectoderm along the middle of the embryo folds inward to form the neural tube that runs the length of the embryo. As these ectodermal cells invaginate, they lose E cadherin and synthesize a different but related cadherin, N cadherin. During further differentiation of the central nervous system from the neural tube, N cadherin becomes a major adhesive protein; P cadherin is also expressed by several types of cells, allowing for selective cell–cell interactions.

Table 13.4 Some cell adhesion molecules (CAMs) of vertebrates

Molecule	Ca^{2+} dependence	Tissue distribution
Liver CAM	Yes	Epithelia, others
P Cadherin	Yes	Epithelia, others
N Cadherin	Yes	Epithelia, others
E Cadherin	Yes	Epithelia
Uvomorulin	Yes	Blastomeres
Neural CAM	No	Neurons, muscles, others
Axonal CAM	No	Axons
MAG	No	Neuron-glia
ICAM	No	WBCs, endothelium
LEC–CAM	No	Leukocytes, endothelium

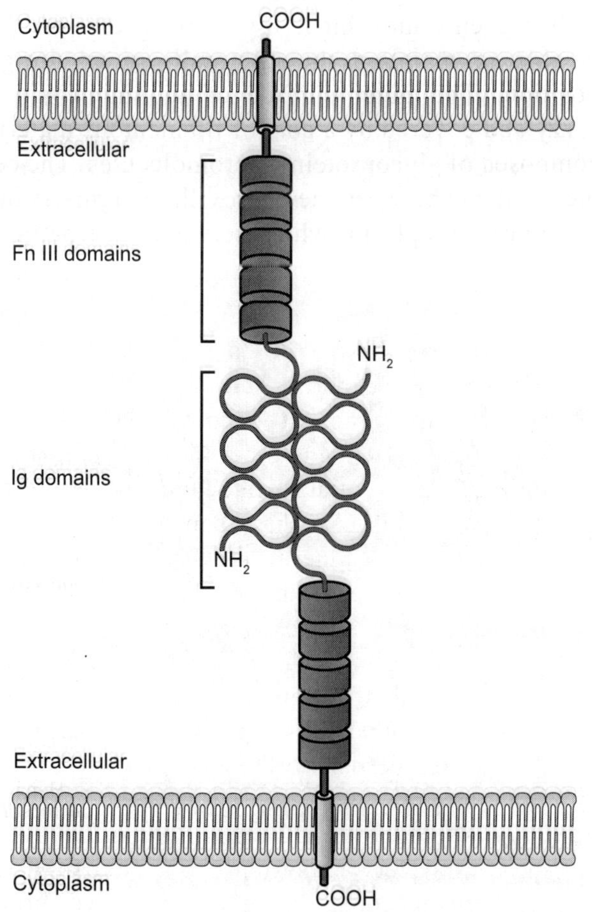

Fig. 13.18 A model of L1, a cell adhesion molecule of the immunoglobulin (Ig) superfamily.

Immunoglobulin-like domains

The immunoglobulins (or Igs) were found to consist of polypeptide chains composed of a number of similar domains. Each of these Ig domains consists of 70–110 amino acids organized into a tightly folded structure. Later, it became apparent that Ig-type domains were present in a wide variety of proteins, which together constitute the immunoglobulin superfamily (or IgSF). Most members of the IgSF are involved in various aspects of immune function, but some of these proteins mediate calcium-independent cell–cell adhesion (**Table 13.4**). In fact, the discovery of Ig-like domains in cell-adhesion receptors in invertebrates—animals that lack a classic immune system—suggests that Ig-like proteins originally evolved as cell adhesion mediators and only secondarily took on their functions as effectors of the vertebrate immune system.

Most IgSF cell adhesion molecules mediate specific interactions of lymphocytes with cells required for an immune response. However, some IgSF members, such as vascular cell adhesion molecule (VCAM), neural cell adhesion molecule (NCAM) and L1, mediate adhesion between non-immune cells. NCAM and L1 (**Fig. 13.18**), for example, play important roles in nerve outgrowth, synapse formation and other events, during the development of the nervous system.

Neural cell adhesion molecules (NCAMs), a group of Ca^{2+}-dependent cell adhesion proteins in vertebrates, belong to the Ig superfamily. Like adherins, NCAMs primarily mediate homophilic interactions, binding together cells that express similar NCAM molecules. Unlike cadherins, NCAMs are encoded by a single gene. Their diversity is generated by alternative mRNA splicing and by differences in glycosylation (**Fig. 13.19**).

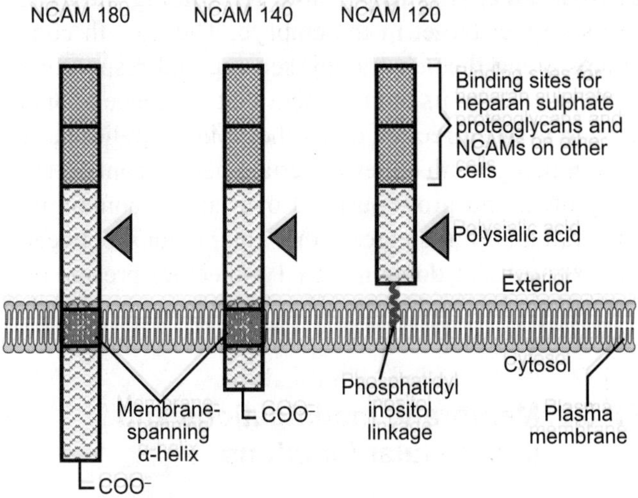

Fig. 13.19 Three forms of NCAM produced by alternative splicing.

NCAMs appear during morphogenesis as the neural tube. They are uniformly distributed along the neural tube and maintain neuron–neuron adhesion. NCAMs disappear from the neural crest cells during migration but are expressed again once the cells stop migrating and form ganglia. Differentiating muscle, glial and nerve cells all express NCAMs. The adhesive properties of NCAMs are modulated by long chains of sialic acid—a negatively charged sugar. NCAMs that are heavily sialylated form weaker homophilic interactions than less sialylated forms, possibly due to repulsion between the negatively charged sialic acid residues. In embryonic tissues such as the brain, polysialic acid constitutes as much as 25% of the mass of NCAMs. In contrast, NCAMs from adult tissues contain only one-third as much sialic acid. The lower adhesive properties of embryonic NCAMs enable cell–cell contacts to be made and then broken—property necessary for specific cell contacts to form in the developing nervous system. The higher adhesive properties of the adult forms of NCAM stabilize these contacts. Thus, the strength of cell–cell adhesions is modified during differentiation by differential glycosylation of the NCAMs. Additionally, binding heparan sulphate proteoglycans to a site near the *N*-terminus of NCAMs can also modify cell–cell adhesion during neuronal differentiation.

In humans, mutations in the L1 gene can have devastating consequences (hydrocephalus, 'water in the brain'). Autopsies on patients who died of L1 deficiency disease reveal two nerve tracts missing, one that runs between the brain and the spinal cord, which suggests that L1 is involved in the growth of axons within the embryonic nervous system. The tip of a growing axon is a highly motile portion called a **growth cone**. It is responsible for guiding the axon to its correct target in the embryo. The growth cone creeps along the substratum, sensing and responding to substances in its environment that influence axonal growth in one direction or another. Many studies have shown that growth cones of certain nerves contain the L1 protein and grow outward over an L1-containing substratum. This suggests that interactions between the extracellular domains of L1 molecules present on the cell surface and in the substratum generate signals within the growth cone that promote its extension.

13.6.2 Membrane modifications and intercellular junctions

In multicellular organisms, when cells form a tissue, the membranes of adjacent cells come quite close to one another and components of the membranes allow the cells to interact in specific ways. Plasma membrane modifies in various ways, which are adapted for functions such as absorption, secretion, fluid transport, mechanical support and interactions with neighbouring cells. One of the most important membrane modifications that takes place for absorption and secretion is the formation of microvilli.

Microvilli

Microvilli are the cytoplasmic processes covered by the plasma membrane and are 0.6–0.8 μm long and about 0.1μm diameter. The core of the microvilli contains fine microfilaments made of actin (**Fig. 13.20**), which are attached to the tip of the microvilli by α-actinin. The F actin undergoes binding and cross-linking by other proteins such as fimbrin and villin. The actin molecules produce contraction of the microvilli. At the base of the microvilli, the microfilaments form a terminal web, which in addition to actin contain myosin, α-actinin, tropomyosin and spectrin. The terminal web prevents movement by exerting tension that is able to maintain the microvilli in an upright position. The outer surface of the microvilli is fuzzy and covered by a coat of filamentous material composed of glycoprotein macromolecules. The cell membrane in between the microvilli invaginates into the apical cytoplasm, which serves as the pathway

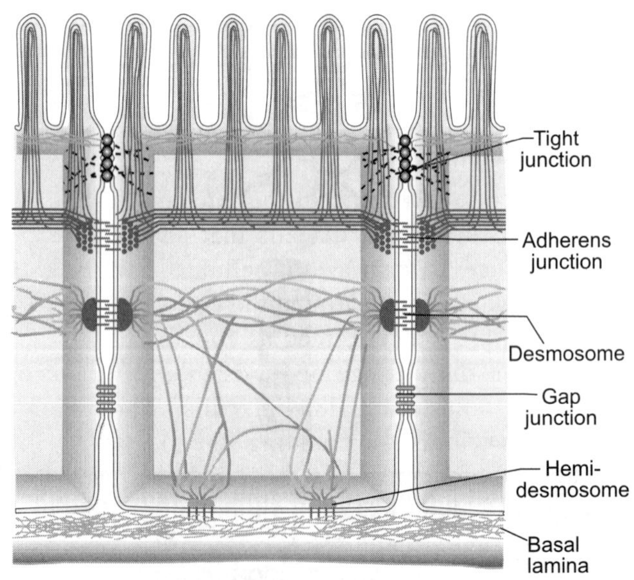

Fig. 13.20 Various cell junctions found in a vertebrate epithelial cell of the small intestine.

for pinocytosis. Microvilli are found in a variety of cells such as intestinal, mesothelial, epithelial cells of the gall bladder, uterus and yolk sac, in hepatic cells and so on. Each intestinal cell possesses as many as 3000 microvilli and present 2×10^8 microvilli in a square millimetre. These increase the surface of absorption.

Cell junctions

Adhesive macromolecules merely bring cells into contact with each other. The formation of intercellular junctions 'cements' the relationship. For cells in tissues to function in an integrated manner, specialized junctions are essential. The major intercellular junctions are listed below:

Occluding junctions

Tight junctions (vertebrates only)
Separate junctions (invertebrates mainly)

Anchoring junctions

Cell–cell junctions (adherens junctions)
Cell–matrix junctions (focal adhesions)
Cell–cell junctions (desmosomes)
Cell–matrix junctions (hemidesmosomes)

Communicating junctions

Gap junctions
Chemical synapses
Plasmodesmata (only in plants)

Occluding junctions: These seal cells together in an epithelium in such a manner that even small molecules leaking from one side of the sheet to the other. The occluded types include tight junctions and separate junctions.

Tight junctions are areas where the plasma membranes of two cells come in closest contact (i.e., sealing the extracellular space). A simple epithelium, like the lining of the intestine or lungs, is comprised of a layer of cells that adhere tightly to one another to form a thin cellular sheet. Tight junctions occur as belts around the circumference of the cell and are hence termed as *zonula occludens*. They prevent the passage of fluids to and from the lumen. For polarized epithelial cells to function, extracellular fluids surrounding their apical and basolateral membranes must be kept separate. This is accomplished by tight junctions, which connect adjacent epithelial cells. These specialized regions of the plasma membrane form a barrier that seals off body cavities such as the intestinal lumen and pancreatic ductules from the blood. They also prevent diffusion of membrane proteins and glycolipids between the apical and basolateral regions of the plasma membrane, ensuring that these regions contain different membrane components. Tight junctions are composed of thin bands of plasma membrane proteins that completely encircle a polarized cell and are in contact with similar thin bands on adjacent cells. In intestinal epithelial cells, tight junctions are usually located just below the apical microvillar surface (**Fig. 13.21a**). When thin sections of cells are viewed in an electron microscope, the plasma membranes of adjacent cells appear to touch each other at intervals and even to fuse. Because tight junctions prevent diffusion of small molecules directly from the intestinal lumen into the blood, intestinal epithelial cells must transport nutrients through the cells.

The microvillar tight junction appears to comprise an interlocking network of ridges in the cytosolic face of the plasma membrane. High magnification reveals that these ridges are made up of particles, believed to be proteins. The tight junction is formed by a double row of these particles, one row donated by each cell (**Fig. 13.21b**). Protein particles on the two cells probably form extremely tight links with each other, essentially fusing the two plasma membranes and creating an impenetrable seal. Tight junctions also serve as 'fences' to maintain the polarized character of epithelial cells. Tight junction strands are mainly composed **occludin** and **claudin** proteins. At least 24 different claudin proteins have been identified and differences in the distribution of these proteins may explain differences in tight junction permeability. Another important function of tight junctions that has recently come to light is the prevention of water loss through epidermis. Tight junctions keep the epidermis watertight. They are also present between the endothelial cells that line the walls of capillaries. These junctions are particularly evident in the brain where they help to form the blood–brain barrier, which prevents substances from passing from the bloodstream into the brain.

In order to function in an integrated manner, the aggregates of cells that compose organized tissues must adhere to each other and to the surrounding extracellular matrix; they also control the movement of ions and small molecules between them. Two principal types of junctions, the desmosome and the gap junction, perform these functions; like the tight junction, these are specialized regions of the plasma membrane.

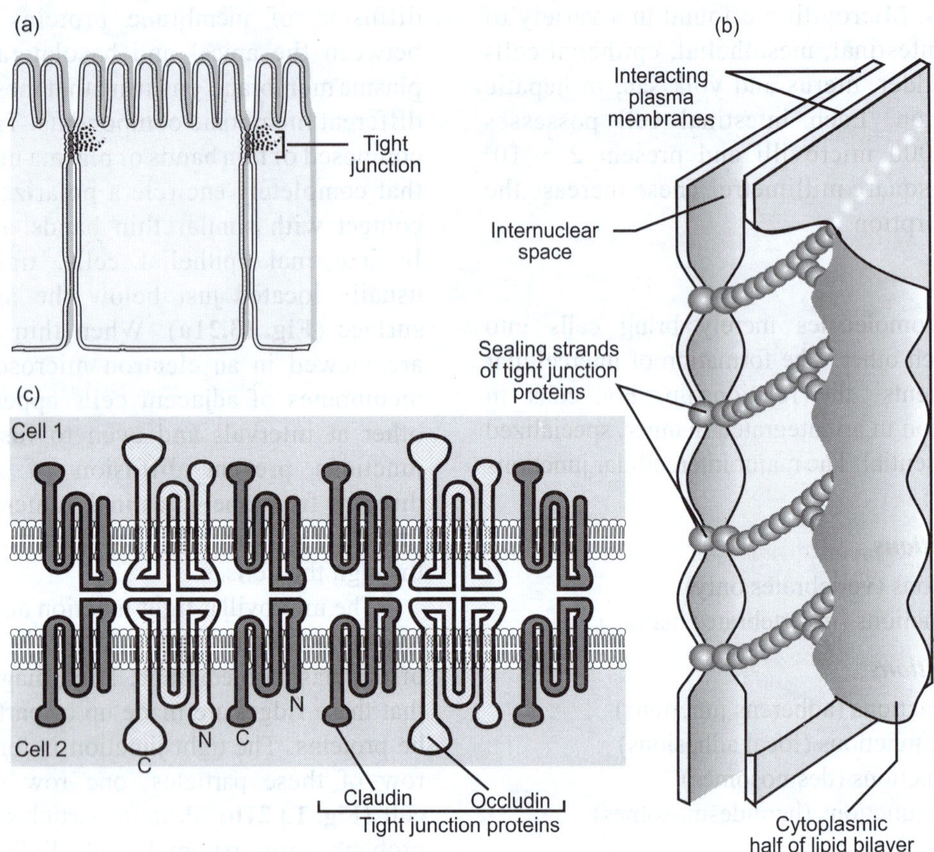

Fig. 13.21 Structure of a tight junction between epithelial cells of the small intestines: (a) The junctions are shown schematically. (b) Current model of a tight junction. (c) Transmembrane claudin and occludin proteins in a tight junction.

Anchoring junctions: These mechanically attach cells to their neighbours or to the extracellular matrix. The cells of certain tissues (e.g., epithelia and cardiac muscle) are notoriously difficult to separate from one another because they are held together tightly by specialized calcium-dependent adhesive junctions. There are two main types of adhesive junctions: cell–cell (adherens junctions and desmosome) and cell–matrix (focal adhesion and hemidesmosome).

Adherens junctions are found in a variety of sites within the body. They are particularly common in epithelia, such as the lining of the intestine, where they occur as a belt (or *zonulae adherens*) that encircles each of the cells near its apical surface, binding that cell to its neighbours. The cells of an adherens junction are held together by calcium-dependent linkages formed between the extracellular domains of cadherin molecules that bridge the 30-nm gap between neighboring cells (**Fig. 13.22**). The cytoplasmic domain of these cadherins is linked by α- and β-catenins to a variety of cytoplasmic proteins, including actin filaments of the cytoskeleton. Thus, the cadherin clusters of an adherens junction connect the external environment to the actin cytoskeleton and provide a potential pathway for signals to be transmitted from the cell exterior to the cytoplasm. An example is the adherens junctions situated between endothelial cells that line the walls of the blood vessels and transmit and are ensure the survival of the cells.

Desmosomes (*desmos* = bond) or macular adherens are thickened regions (~1 μm in diameter) of the plasma membrane where cells are tightly attached to their neighbours or to the extracellular matrix (e.g., cardiac muscle and epithelial layers of the skin and uterine cervix). They increase the rigidity of tissues, such as epithelia by holding cells securely together and act as an anchor for intracellular fibres. These junctions have only recently been purified and several of the major protein components have been identified. The main types of desmosomes recognized are *spot desmosomes* found in all epithelial cells and many other tissues, such as smooth muscle. They are button-like points of contact between cells, often thought of as a 'spot-weld' between adjacent plasma membranes. *Hemidesmosomes* anchor the plasma membrane to regions of the extracellular matrix. Bundles of intermediate filaments course through the cell and interconnect spot desmosomes and

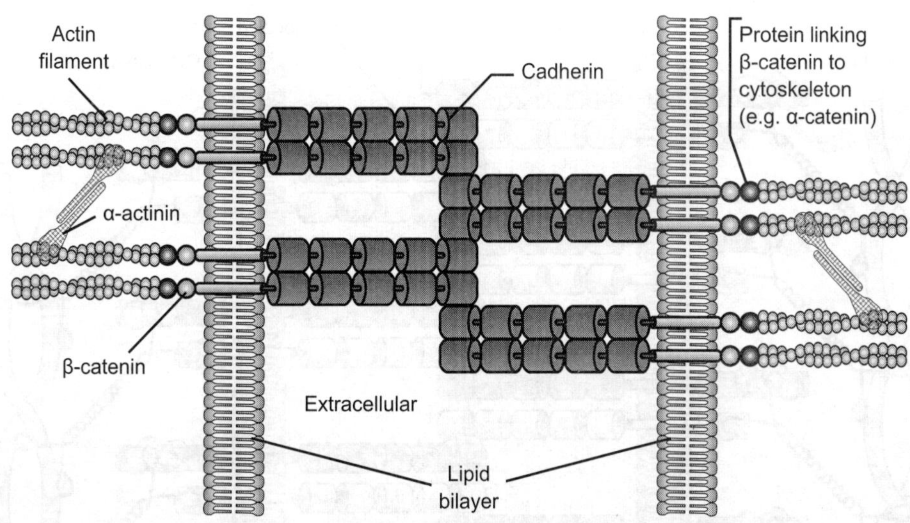

Fig. 13.22 The structure of adherens junction.

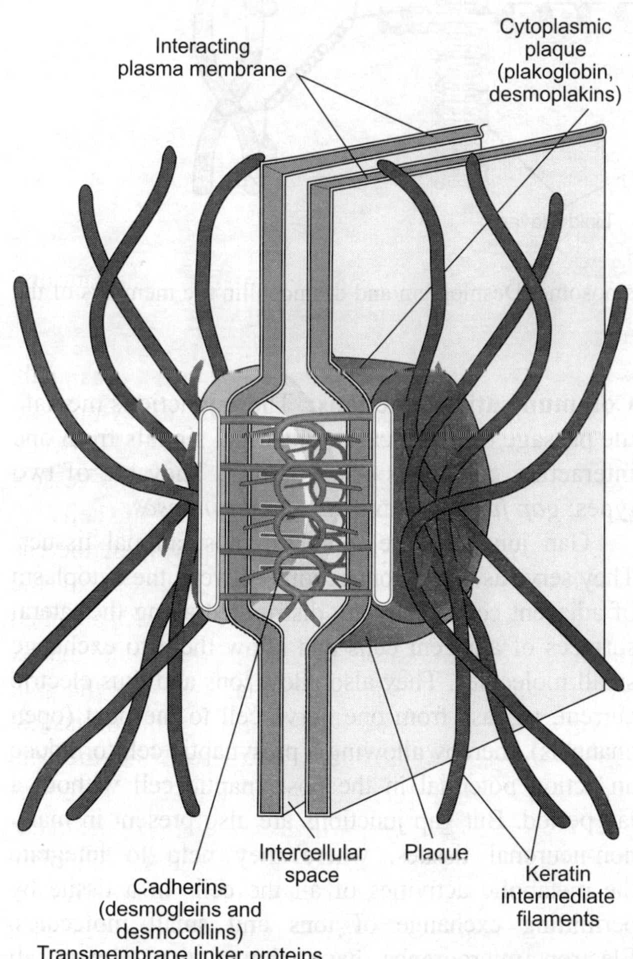

Fig. 13.23 Components of a spot desmosome between epithelial cells and attachments to the sides of keratin intermediate filaments which criss-cross the interior of cells.

hemidesmosomes; likewise, bundles of actin filaments interconnect belt desmosomes and adhesion plaques.

Spot desmosome consists of proteinaceous cytoplasmic adhesion plaques (0.5 µm in diameter) attached to the cytosolic face of the plasma membranes of adjacent cells and connected by transmembrane linker proteins (**Fig. 13.23**). Inside the plasma membrane there is a discoidal intracellular plaque, toward which numerous filaments called *tonofilaments* converge. The tonofilaments loop back into the cytoplasm and are made of the protein keratin, which represent a type of intermediate filament (10 nm diameter). The thinner filaments that arise from the dense plaques and traverse the cell membrane into the intracellular space forming the so-called transmembrane linkers. *Plakoglobin* is a major constituent of the plaques; it is closely related in sequence to vinculin, a protein that underlies the plasma membrane at adherens junctions and interconnects transmembrane proteins to the actin cytoskeleton. Fibrous transmembrane linker proteins, *desmoglein* and *desmocollin* (**Fig. 13.24**), bind to plakoglobin and other proteins in the plaques and extend into the intercellular space where fibres from the two cells form an interlocking network that binds the two together. Desmoglein is related in sequence and structure to the cadherins, in particular to E cadherin, a transmembrane protein found in adherens junctions. In the epithelial cells, keratin intermediate filaments near the cytoplasmic plaques of spot desmosomes are linked to them by the *desmoplakin proteins*. Some of these filaments run parallel to the cell surface and others penetrate and traverse the cytoplasm. Spot desmosomes also transmit shearing forces from one region of a cell

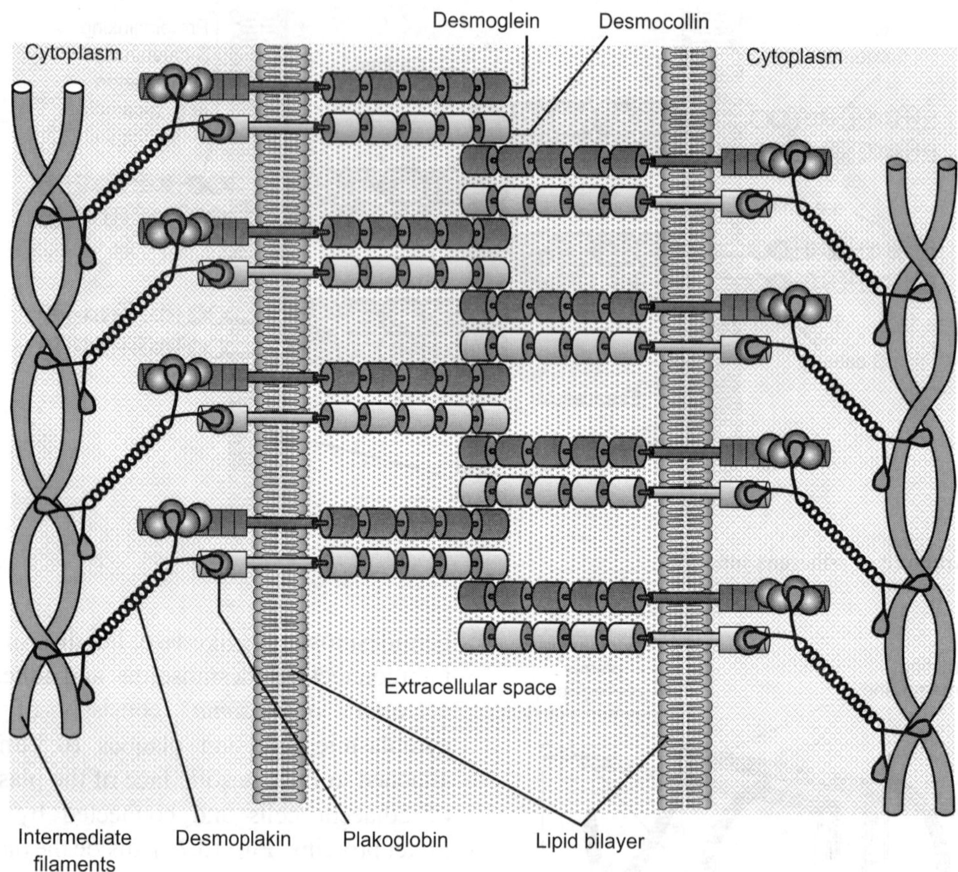

Fig. 13.24 Schematic model of the molecular architecture of a desmosome. Desmoglein and desmocollin are members of the cadherin family of adhesion proteins.

layer to the epithelium as a whole; they thus provide strength and rigidity to the entire epithelial cell layer. Cardiac (heart) muscle cells are also interconnected by spot desmosomes. Desmin intermediate filaments interconnect these desmosomes, allowing the stress and strain of the contractile force of one muscle cell to the transmitted to be other. In smooth muscles, desmin filaments are anchored to the dense bodies, regions of the plasma membrane that anchor together adjacent cells.

Hemidesomosomes are found along the basal surface of epithelial cells where they form tight points of contact with the basal lamina. By interconnecting the intermediate filaments of the cytoskeleton with the fibres of the basal lamina, they increase the overall rigidity of epithelial tissues. Hemidesmosomes consist of two proteinaceous plaques on the cytosolic face of the plasma membrane, which are attached to the ends of intermediate filaments (**Fig. 13.25**). $\alpha_6\beta_4$-integrin is localized to hemidesmosomes and is thought to bind to proteins within the plaques and to the extracellular matrix protein to laminin. In this way, the cell is anchored to the extracellular network.

Communicating junctions: These junctions mediate the passage of chemical or electrical signals from one interacting cell to its other partner. They are of two types: *gap junctions* and *chemical synapses.*

Gap junctions are found in most animal tissues. They serve as direct connections between the cytoplasm of adjacent cells. They are distributed along the lateral surfaces of adjacent cells and allow them to exchange small molecules. They also allow ions and thus electric current, to pass from one nerve cell to the next (open channels), thereby allowing a presynaptic cell to induce an action potential in the postsynaptic cell without a lag period. But gap junctions are also present in many non-neuronal tissues, where they help to integrate the metabolic activities of all the cells in a tissue by permitting exchange of ions and small molecules. Electron micrographs have shown that almost all animal cells that come in contact with each other have regions of junctional specialisation characterized by an intercellular gap, which is filled by a well-defined set of cylindrical particles (molecular pipelines). Morphologists named these regions gap junctions,

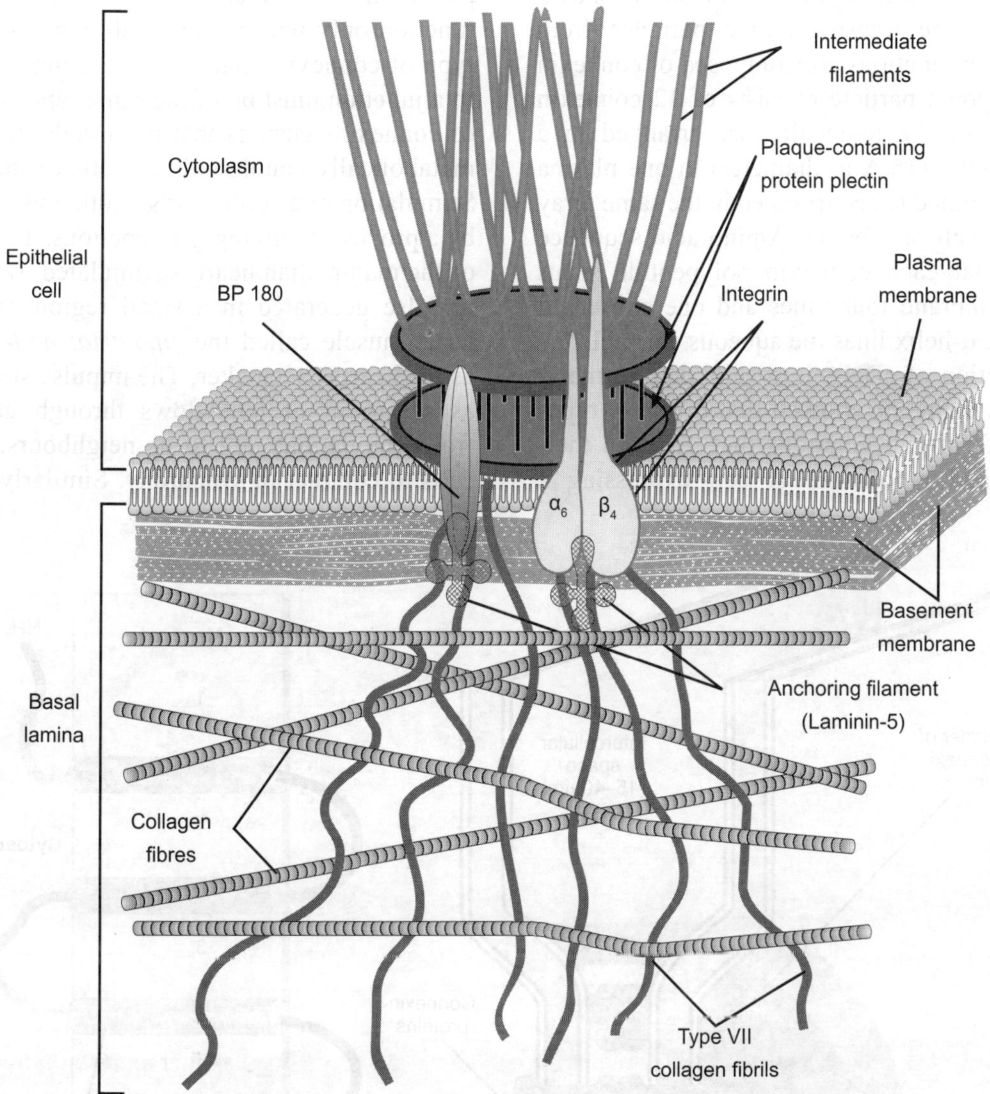

Fig. 13.25 Schematic model of a hemidesmosome connecting an epithelial cell to the basal lamina (**See Plate 10**).

but in retrospect the gap is not their most important feature. The cylindrical particles, which are tiny water-filled channels, are the key to the function of gap junctions. These channels directly link the cytosol of one cell to an adjacent cell, providing a passage for movement of very small molecules and ions between the cells.

Gap junctions between mammalian cells permit the passage of molecules as large as 1.2 nm in diameter (range 2–4 nm). Thus, ions and many low-molecular weight building blocks of cellular macromolecules, such as amino acids and nucleoside phosphates, can pass from cell to cell. An example of this cell–cell transfer is the phenomenon of *metabolic coupling*, or *metabolic cooperation*, in which a cell can transfer molecules to a neighbouring cell that is unable to synthesize them. An important aspect of gap junction physiology is that the channels close in the presence of very high concentrations of Ca^{2+} in the cytosol. The membrane of one cell closes the channels that connect the cell with its neighbours and thus prevents leakage of the cytoplasm of all epithelium cells. Even a slight increase in cytosolic pH can decrease the permeability of gap junctions. Thus cells may modulate the degree of coupling with their neighbours. Precisely why and how they accomplish this is a matter of investigation.

In liver and other tissues, large numbers of individual gap junctions cluster together in an area of about 0.3 μm in diameter. Electron micrographs of stained, isolated gap junctions reveal a lattice of hexagonal particles with hollow cores as intercellular channels. A model of the structure of the gap junction, based on these and other studies, is shown in **Fig. 13.26a**. The transmembrane particles from purified liver gap junctions are composed of connexin 32. Each hexagonal particle consists of 12 connexin molecules: six of the molecules are arranged in a hexagonal cylinder (16 Å in diameter) in one plasma membrane and joined to six arranged in the same array in the adjacent cell membrane. Amino acid sequence data suggest that each connexin polypeptide spans the plasma membrane four times and one conserved transmembrane α-helix lines the aqueous channel.

Microinjection of cDNA into cells that normally do not express connexin showed that different types of connexins form gap junctions that differ in their channel size and regulation. A cell expressing a particular type of connexin generally forms a gap junction only with another cell expressing the same type of connexin, that is, all 12 connection subunits in a junction must be of the same type. This property of connexins ensures that most cells in a tissue will metabolically couple only to cells of the same type. Stimulation of a cardiac or smooth muscle cell occurs by a process involving gap junctions. The contraction of the mammalian heart is stimulated by an electrical impulse generated in a small region of specialized heart muscle called the *sinoatrial node*, which acts as the hearts pacemaker. The impulse spreads rapidly as a current of ions flows through gap junctions from one cardiac cell to its neighbours, causing the cells to contract in synchrony. Similarly, the flow of

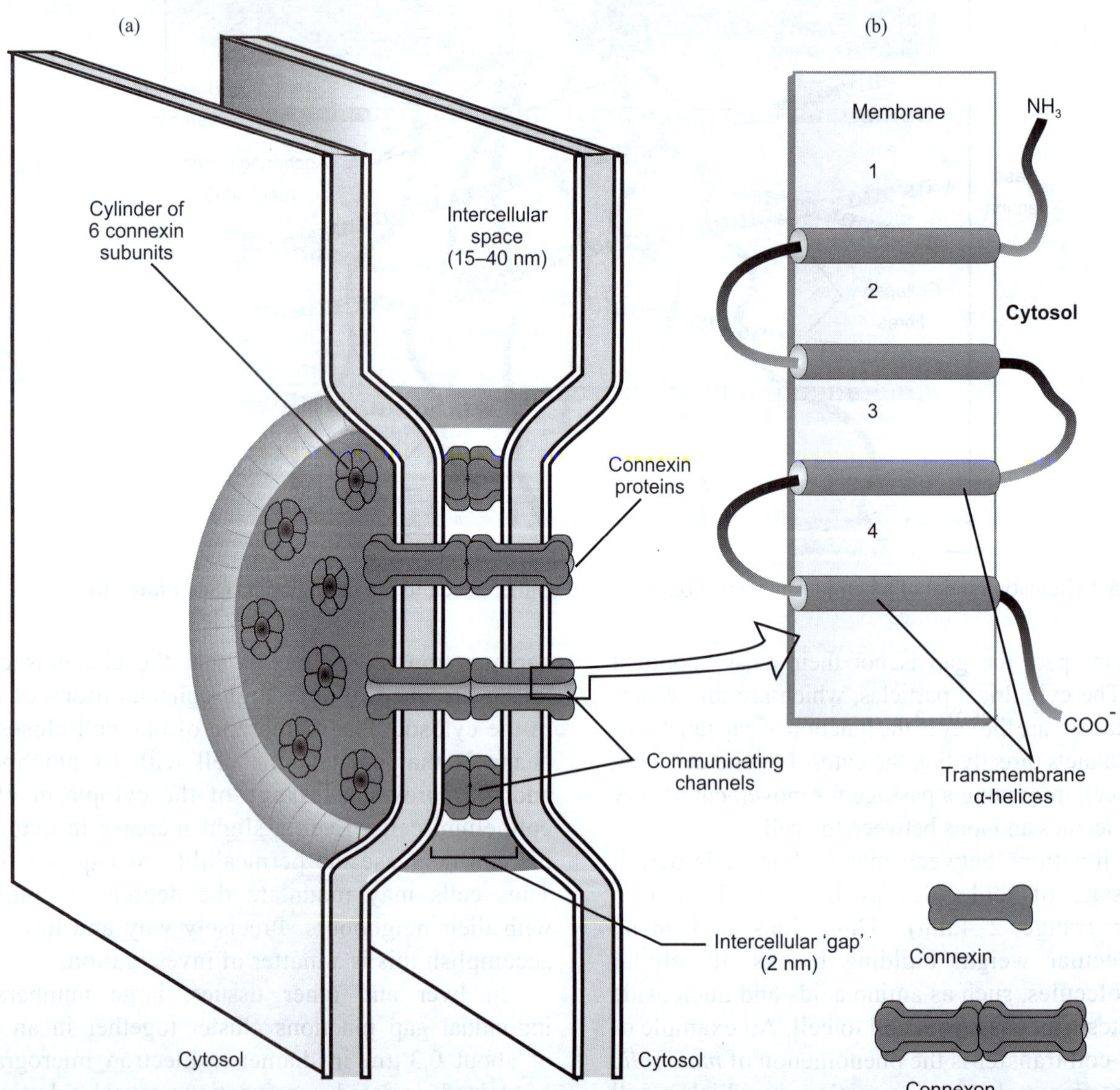

Fig. 13.26 Gap junction: (a) Model of a gap junction, based on electron microscopic and X-ray diffraction analysis. (b) Membrane spanning domains of connexin proteins.

ions through gap junctions that interconnect smooth muscle cells in the wall of the oesophagus or intestine leads to the generation of coordinated peristaltic waves that move down the length of the wall. Gap junctions can bring a large number of cells of a tissue into intimate cytoplasmic contact. This has important physiologic consequences, because a number of highly active regulatory substances, such as cAMP and inositol phosphates can easily pass through the gap junction channels. As a result, gap junctions have the potential to integrate the activities of individual cells of a tissue into a functional unit. Gap junctions also allow cells to cooperate metabolically by sharing key metabolites (ATP, sugar phosphates, amino acids and many coenzymes) by passing through these intercellular channels.

Chemical synapses: In many multicellular organisms, cell–cell junction allows signals to pass from a nerve cell to another. In a chemical synapse, the signal is carried by a diffusible neurotransmitter and in an electrical synapse a direct connection is made between the cytoplasms of the two cells via gap junctions.

Plant cell adhesion and plasmodesmata: Unlike animals, whose cells make intimate contact with one another, plant cells are separated from each another by a substantial barrier—the cell wall. Cell–cell adhesion in plant cells is mediated by their cell walls rather than by transmembrane proteins. In particular, specialized pectin-rich region of the cell wall called the middle lamella acts as glue to hold adjacent cells together. These cells communicate with each other through cytoplasmic connections called *plasmodesmata* which functions analogous to animal cell gap junctions. Plasmodesmata allow the passage of molecules with a molecular size of less than about 800 nm, which is similar to the molecular weight cut-off for gap junctions. In each plasmodesmata the plasma membrane of one cell is continuous with that of its neighbour, creating an open channel between the two cytosols. An extension of the smooth endoplasmic reticulum passes through the pore (desmotubule), leaving a ring of surrounding cytoplasm through which ions and small molecules are able to pass freely between the cells. Plasmodesmata are dynamic structures, can expand in response to appropriate stimuli, permitting the regulated passage of macromolecules between adjacent cells. In addition, there is evidence of specific protein targeting to plasmodesmata (**Fig. 13.27**). Plasmodesmata may thus play a key role in plant development by controlling the trafficking of regulatory molecules—such as transcription factors or RNAs—between cells.

13.6.3 Cellular interactions during embryonic development

Dorso-ventral patterning during embryogenesis

Local interactions between cells and the extracellular matrix play a prominent and widespread role in regulating the development of multicellular organisms. In early vertebrate embryogenesis, for example, local cellular interactions determine regionalization along the major axes of the embryo. Local interactions from different cell populations, however, is the primary mechanism regulating the formation of internal organs such as the kidney, lungs and pancreas. The vast number of highly specialized cells and their stereotyped arrangement in different tissues is also a consequence of locally acting signals. Within the developing nervous system, local interactions guide the developing neuronal processes to their targets.

As discussed earlier, during early development the embryo is divided progressively into domains

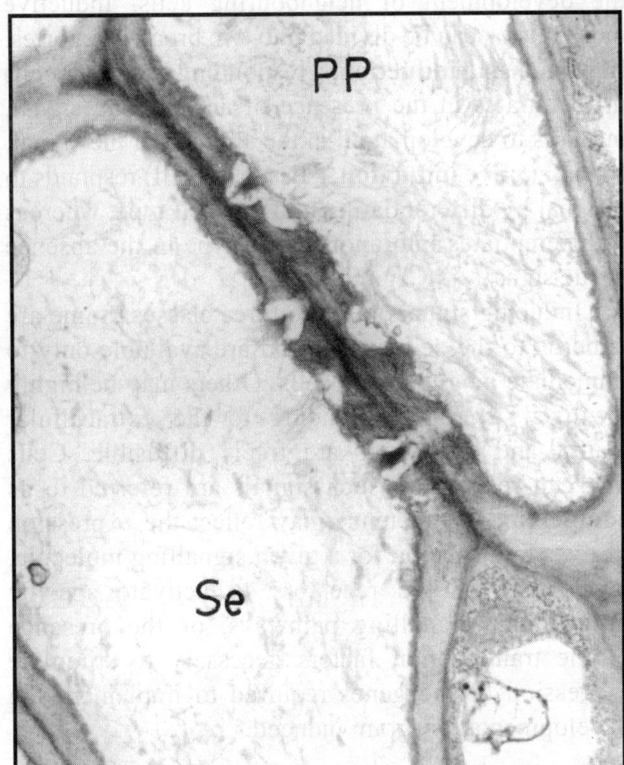

Fig. 13.27 TEM micrograph of plasmodesmatas involved in macromolecular trafficking between tomato leaf cells (Channarayappa, 1992).

through the action of patterning genes. The early *Drosophila* embryo is a syncitium and many of the important patterning events are controlled by the spatial distribution of transcription factors. Two mechanisms operate to subdivide the embryo along the two axes:

- Gradients of transcription factors, acting as morphogens, activate different sets of genes at different threshold concentrations.
- The action of one patterning gene may induce expression of another patterning gene, which in turn functions to control the expression of other genes.
- During later stages of *Drosophila* embryogenesis and in early vertebrate embryos, the cells of which are separated by membranes, extracellular molecules act as morphogens and induce expression of other extracellular signals.

Induction

In embryology, the term induction refers to any mechanism whereby one cell population influences the development of neighbouring cells. Inductive mechanisms can be divided into two broad categories. In **permissive induction**, a responding tissue (or cell) differentiates in the presence of an inducing signal, but fails to develop at all in the absence of the signal. In **instructive induction**, a tissue (or cell) responds to a signal by differentiating into one cell type, whereas it differentiates into another cell type in the absence of the signal.

Inducing signals fall into three classes. Some are tethered to the cell surface and are available only to immediate neighbouring cells. Others may be highly localized by tightly binding to the extracellular matrix and yet others are freely diffusible. Cells that can respond to such signals are referred to as **competent**. Competence may reflect the expression of receptors specific for a given signalling molecule, the ability of the receptors to activate specific intracellular signalling pathways, or the presence of the transcription factors necessary to stimulate expression of the genes required to implement the developmental program induced.

Transforming growth factor β (TGF-β)

A number of related extracellular signalling molecules that play an important role in regulating development in both invertebrates and vertebrates constitute the **TGF-β superfamily**. Some TGF-β superfamily members have anti-proliferative effects on certain tissues and proliferative effects on others. These growth factors also promote the production of cell-adhesion molecules (**Fig. 13.28**).

Sequential induction events

The unfertilized *Xenopus* egg has an inherent asymmetry with a pigmented animal hemisphere (pole or cap) and an unpigmented vegetal hemisphere. Entry of sperm into the animal pole induces the actin-rich cortical region to rotate 30 degrees relative to the yolk cytoplasm of the vegetal pole. Cortical rotation results in a second asymmetry that leads to the future dorso-ventral axis with the point of sperm entry defining the ventral side of the embryo. A stereotyped pattern of rapid synchronous cell divisions then ensues. As these divisions occur, a cavity called the **blastocoel**, forms; the developing embryo at this stage is referred to as a **blastula**. After the first 12 cell divisions, the rate of cell division slows and becomes asynchronous as zygotic transcription begins. During this stage, the three germ layers get arranged in a simple layered pattern along the animal/vegetal axis. The animal pole will give rise to the ectoderm (e.g., skin and neural tissue), the vegetal pole to endoderm (e.g., gut), and the marginal zone between the two poles to mesoderm (e.g., notochord, muscle, blood). During gastrulation, the mesodermal and endodermal cells move inside the embryo. The movement of cells into the embryo is initiated at a structure on

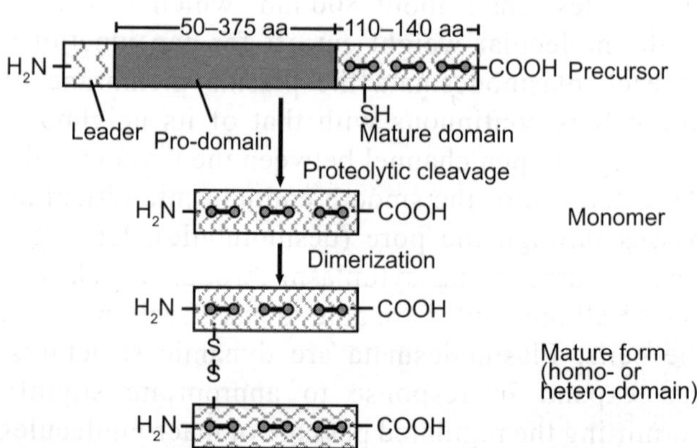

Fig. 13.28 Schematic diagram of formation of mature dimeric TGF-β proteins from secreted monomeric TGF-β precursors (*Source*: Lodish et al, 1995).

the dorsal side of the embryo called the **dorsal lip of the blastopore**. At the completion of gastrulation, virtually the entire surface of the embryo is covered by the ectoderm. The endoderm is found towards the centre of the embryo and a layer of mesodermal cells separates the endoderm and ectoderm.

Formation of internal organs

The formation of internal organs such as the kidneys, gut, pancreas and lungs is regulated by interactions between epithelial and mesenchymal cells. Epithelial cells form continuous sheets of polarized cells with the apical and basal regions separated by tight junctions. In contrast, the mesenchyme comprises loosely associated non-polarized cells. Epithelial cells in different organs are derived from one of the three germ layers: the ectoderm, mesoderm or endoderm. Mesenchyme is derived from either the mesoderm or the ectoderm. Extracellular matrix material separates the epithelial and mesenchymal cell populations. Underlying the epithelium is a basal lamina containing laminin, type IV collagen, nidogen and certain proteoglycans. Several integrins on the surface of epithelial cells bind to components of the basal lamina, anchoring them to the extracellular matrix. The basal lamina is linked to the mesenchyme through a reticular network of extracellular components containing collagen (type I and III) and fibronectin. Two examples (formation of the salivary gland and kidney) for epithelial–mesenchymal interactions that control organ development are discussed below.

Salivary gland formation: In the developing salivary gland, groups of epithelial cells, derived from the endoderm, secrete proteins and proteoglycans that form a basal lamina, grow and form clusters of cells that become the secreting cells of the mature gland. Mesodermal cells are attached under the basal lamina. These groups of epithelial cells together with the basal lamina can be separated from the mesodermal cells by microdissection of embryos. **Figure 13.29** demonstrates how two cell types—epithelial and mesodermal—and basal lamina interact and are necessary for the endodermal cells to differentiate into the mature secreting cells of the salivary gland.

Regulating kidney induction: Both the epithelium called the **ureteric bud** and the mesenchyme that gives rise to the kidneys are mesodermal derivatives. As embryonic development proceeds, the ureteric bud grows and branches, eventually differentiating into the collecting ducts of the kidney. The mesenchyme is transformed into a distinct epithelium that will form the proximal and distal tubules and the glomeruli. Both the mesenchyme and the uretric bud send and receive inductive signals, a phenomenon referred to as **reciprocal induction**. Many *in vitro* and *in vivo* experiments involving disruption of kidney development have demonstrated that the epithelium is necessary for conversion of the mesenchyme into tubules. However, kidney mesenchyme can also be induced to form tubules by many different tissues.

Little is known about the signalling molecules that mediate specific epithelial–mesenchymal interactions, but in vitro experiments suggest direct cell–cell contact is necessary for kidney induction. Moreover, recent genetic studies have shown that c-ret, a proto-oncogene, is critical for kidney morphogenesis. This proto-oncogene encodes a transmembrane receptor tyrosine kinase, c-ret. Developmental studies have shown that c-ret is a receptor for a cell surface signalling molecule produced by mesenchyme that induces budding of the kidney epithelium.

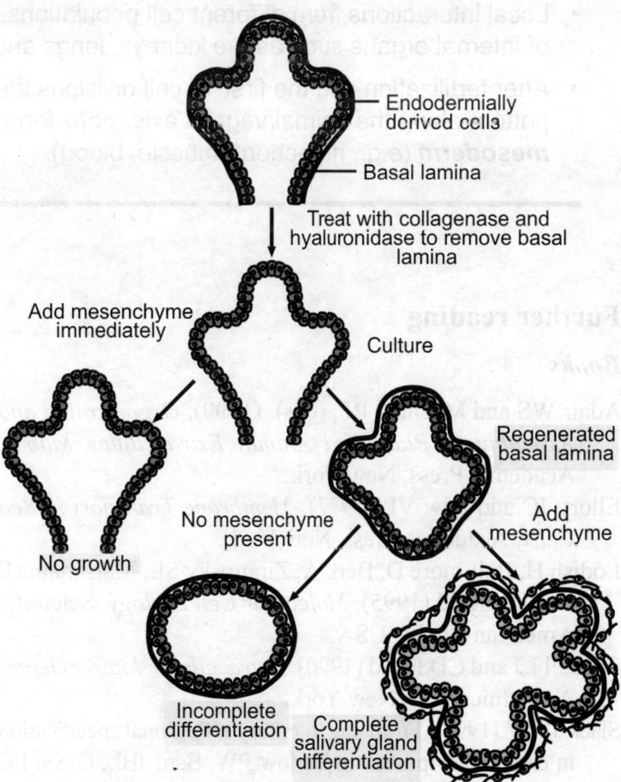

Fig. 13.29 Elements required for salivary gland differentiation.

Summary

- The general belief is that life evolved first as single cellular organisms and later developed into ***multicellular organisms*** (plants, animals and colonial microorganisms).
- Some organisms become multicellular by aggregation rather than by growth and cell division.
- The evolution of higher multicellular organisms permitted specialized cells and tissues to form.
- Similar to cell walls surrounded by the bacterial and plant cells, many of the animal cells in tissues are embedded in an ***extracellular matrix*** (ECM) consisting of secreted proteins and polysaccharides.
- The ECM in animals facilitate cells to organize themselves into tissues and coordinates their cellular functions by activating ***intracellular signalling*** *pathways* that regulate cell growth, proliferation and gene expression.
- The ECM is composed of three major classes of biomolecules, which are widely distributed in the ECM of all tissues: ***collagen fibres***, ***proteoglycans*** and ***multi-adhesive matrix proteins***.
- ***Cell–cell adhesion*** is a selective process, such that cells adhere only to other cells of specific types. If a single large mass formed, one cell type may localize to the centre, surrounded by cells of the other type.
- Cell adhesive proteins are now known to fall into three principal classes: the ***cadherins***, ***selectins*** and ***immunoglobulin (Ig) superfamily***.
- Adhesive macromolecules merely bring cells into contact with each other. The formation of intercellular junctions 'cements' the relationship.
- For cells in tissues to function in an integrated manner, specialized junctions are essential. The major kinds of intercellular junctions are: (i) ***Occluding junctions***: Tight junctions and separate junctions, (ii) ***Anchoring junctions***: Cell–cell junctions, cell–matrix junctions and (iii) ***Communicating junctions***: Gap junctions, chemical synapses and plasmodesmata.
- Local interactions between cells and the ECM play a prominent and widespread role in regulating the development of multicellular organisms. In early vertebrate embryogenesis, for example, local cellular interactions determine regionalization along the major axes of the embryo.
- Local interactions from different cell populations, however, is the primary mechanism regulating the formation of internal organs such as the kidneys, lungs and pancreas.
- After fertilization and the first 12 cell divisions the three germ layers become arranged in a simple layered pattern along the animal/vegetal axis: ***ectoderm*** (e.g., skin and neural tissue), ***endoderm*** (e.g., gut) and ***mesoderm*** (e.g., notochord, muscle, blood).

Further reading

Books

Adair WS and Mecham RP, (eds), (1990), *Organization and Assembly of Plant and Animal Extracellular Matrix*, Academic Press, New York.

Ellory JC and Lew VL (1977), *Membrane Transport in Red Cells*, Academic Press, New York.

Lodish H, Baltimore D, Berk A, Zipurosky SL, Matsudaina P and Darnell J (1995), *Molecular Cell Biology*, Scientific American Books, USA.

Sandell LJ and CD Boyd (1990), *Extracellular Matrix Genes*, Academic Press, New York.

Slack JMW (1991), From egg to embryo: Regional specification in early development. In Barlow PW, Bard JBL, Green PB and Kirk DL, (eds), *Developmental and Cell Biology Series*, 2nd edition, Vol. 26, Cambridge University Press, Cambridge.

Zern MA and L Reid, (eds), (1993), *Extracellular Matrix: Chemistry, Biology and Pathology with Emphasis on the Liver*, Marcell Dekker.

References

Bartsch U (2003), Neural CAMs and their role in the development and organization of myelin sheaths, *Front. Biosci.*, 8:D477–D490.

Bell GI (1978), Models for the specific adhesion of cell to cell, *Science*, 200:618.

Catterall WA (1995), Structure and function of voltage-gated ion channels, *Annu. Rev. Biochem.*, 64:493–531.

Channarayappa (1990), Cytogenetic analysis of clastogens and aneuploidogens by immunofluorescent and micronucleus methods, PhD thesis submitted to the West Virginia University (WVU), Morgantown, West Virginia, USA.

Exercises

1. Explain the major events involved during evolution of multicellular organisms from unicellular ones.
2. Discuss the symbiotic theory of cell organellar evolution.
3. How did development of multicellular organisms probably begin?
4. What is the role of extracellular matrix in multicellular evolution?
5. Explain briefly the role of multi-adhesive matrix proteins and the cell surface receptors.
6. What are the different cell adhesive proteins involved in the cell-cell interaction?
7. What is the role of cell junctions in cellular interactions?
8. Describe with diagrams different anchoring proteins.
9. How do gap junctions play a role in cellular communication?
10. What are the different interactive mechanisms involved during development?
11. Explain briefly the different sequential induction events that regulate early *Xenopus* development.

Channarayappa, (1992), Resistance mechanism of *Lycopersicon* species to the vector *Bemisia tabaci* and ultrastructural changes in leaf curl virus infected tomato plants, PhD thesis submitted to the University of Agricultural Sciences (UAS), Bangalore, India.

Clandinin TR and Zipursky SL (2002), Making connections in the fly visual system, *Neuron,* 35:827–41.

Cosgrove DJ (2001), Plant cell walls: Wall-associated kinases and cell expansion, *Curr. Biol.* 11:R558–R559.

Eyre DR, MA Paz and PM Gallop (1984), Cross-linking in collagen and elastin, *Annu. Rev. Biochem.,* 53:717–48.

Fuchs E and Raghavan S (2002), Getting under the skin of epidermal morphogenesis, *Nature Rev. Genet.,* 3(3):199–209.

Green KJ and Gaudry CA (2000), Are desmosomes more than tethers for intermediate filaments? *Nature Revs.,* MCB 1:208–16.

Gumbiner BM (1996), Cell adhesion: Molecular basis of tissue architecture and morphogenesis, *Cell,* 84:345–57.

Heinlein M (2002), Plasmodesmata: Dynamic regulation and role in macromolecular cell–cell signalling, *Curr. Opin. Plant Biol.,* 5:543–52.

Hohenester E and Engel J (2002), Domain structure and organization in extracellular matrix proteins, *Matrix Biol.,* 21(2):115–28.

Hynes RO (2002), Integrins: bidirectional, allosteric signalling machines, *Cell,* 110:673–87.

Jamora C and Fuchs E (2002), Intercellular adhesion, signalling and the cytoskeleton, *Nature Cell Biol,.* 4:E101–E108.

Juliano RL (2002), Signal transduction by cell adhesion receptors and the cytoskeleton: Functions of integrins, cadherins, selectins and immunoglobulin superfamily members. *Annu. Rev. Pharmacol. Toxicol.,* 42:283–314.

Knust E and Bossinger O (2002), Composition and formation of intercellular junctions in epithelial cells, *Science,* 298:1955–59.

Miranti CK and JS Brugge (2002), Sensing the environment: A historic perspective on integrin signal transduction, *Nat. Cell Biol.,* 4:E83–E90.

McDonald JA (1988), Extracellular matrix assembly, *Annu. Rev. Cell Biol.,* 4:183–207.

Paulsson M (1992), Basement membrane proteins: structure, assembly and cellular interactions, *Crit. Rev. Biochem. Mol. Biol.,* 27:93–127.

Shaw LM and BR Olsen (1991), FACIT collagens: Diverse molecular bridges in extracellular matrices, *Trends Biochem. Sci.,* 16:191–94.

Stevenson BR and Keon BH (1998), The tight junction: Morphology to molecules, *Annu. Rev. Cell Biol.,* 14:89–109.

Takei K and Haucke V (2001), Clathrinmediated endocytosis: membrane factors pull the trigger, *Trends Cell Biol.,* 11:385–91.

Tsukita S and Furuse M (2002), Claudin-based barrier in simple and stratified cellular sheets, *Curr. Opin. Cell Biol.,* 14:531–36.

Unwin N (1993), Neurotransmitter action: Opening of ligand-gated ion channels. *Cell 72/Neuron,* 10 (Suppl):31–41.

Yamada KM (1991), Adhesive recognition sequences, *J. Biol. Chem.,* 266:12809–12.

Yurchenco PD and JC Schittiny (1990), Molecular architecture of basement membranes. *FASEB J.,* 4:1577–90.

Zambryski P and Crawford K (2000), Plasmodesmata, *Annu. Rev. Cell Dev. Biol.,* 16:393–421.

Zamir E and Geiger B (2001), Molecular complexity and dynamics of cell-matrix interactions, *J. Cell Sci.* 114:3583–91.

14

Cytoskeleton and Cell Motility

Learning objectives

After reading this chapter, you will be able to:

- Identify the various structures that help cell movement
- Outline the role of protein transport in cell motility

14.1 INTRODUCTION

The cytoplasm in eukaryotic cells not only contains extensive membrane systems and various types of organelles, but also an intricate supportive network of proteinaceous fibres and filaments called the **cytoskeleton** (meaning, skeleton of a cell), also called as **cytomusculature**. The cytoskeleton provides a structural framework (architecture) for the cell, serving a scaffold that determines cell shape and the general organization of the cytoplasm. The existence of organized fibrous arrays (or cytoskeleton) in the protoplasm was postulated by Koltzoff (1928). Cytoplasm is highly organized, with most soluble proteins either bound to filaments or localized in specific regions. In animal cells, which lack a rigid cell wall, it is the cytoskeleton that determines cell shape. Several types of protein filaments visible with the electron microscope (EM) criss-cross the eukaryotic cell, forming a three-dimensional meshwork of the cytoskeleton. It extends from the plasma membrane and nucleus of eukaryotic cells, as an interconnected system of bundled fibres, slender threads and lattices (**Fig. 14.1**). This system gives eukaryotic cells their internal organization, overall shape, capacity to move and ability for various other activities such as endocytosis and exocytosis. Some parts of the cytoskeleton reinforce the plasma membrane and hold its proteins in place; other parts do the same for the nuclear envelope. The main proteins that are present in the cytoskeleton are microtubules, actin and intermediate filaments.

The ability of cells to adopt a variety of shapes and to carry out coordinated and directed movements is an important event of the evolutionary process. In prokaryotes, some bacterial cells move by beating flagella powered by a rotary motor in the cell membrane. In multicellular organisms, the migration of single cells and groups of cells from one part of an embryo to another is critical to the development of the organism. In adult animals, a specific cell type migrates to find out foreign invaders as part of a host's defences against infection. Motile sperm cells can use the flagella to move long distances. Other structures such as cilia are commonly used for locomotion in many protozoa.

Cells have evolved two basic mechanisms for generating movement. One entails the assembly and disassembly of cytoskeletal filaments. The other requires a special class of enzymes called **motor proteins**, which use energy from ATP to move or slide along a microfilament or microtubule and ferry organelles and vesicles with them. In general, cytoskeletal filaments are involved in major functions such as:

- establishing cell shape
- providing mechanical strength
- cellular locomotion
- chromosome separation during mitosis and meiosis
- intracellular transport of organelles and other structures.

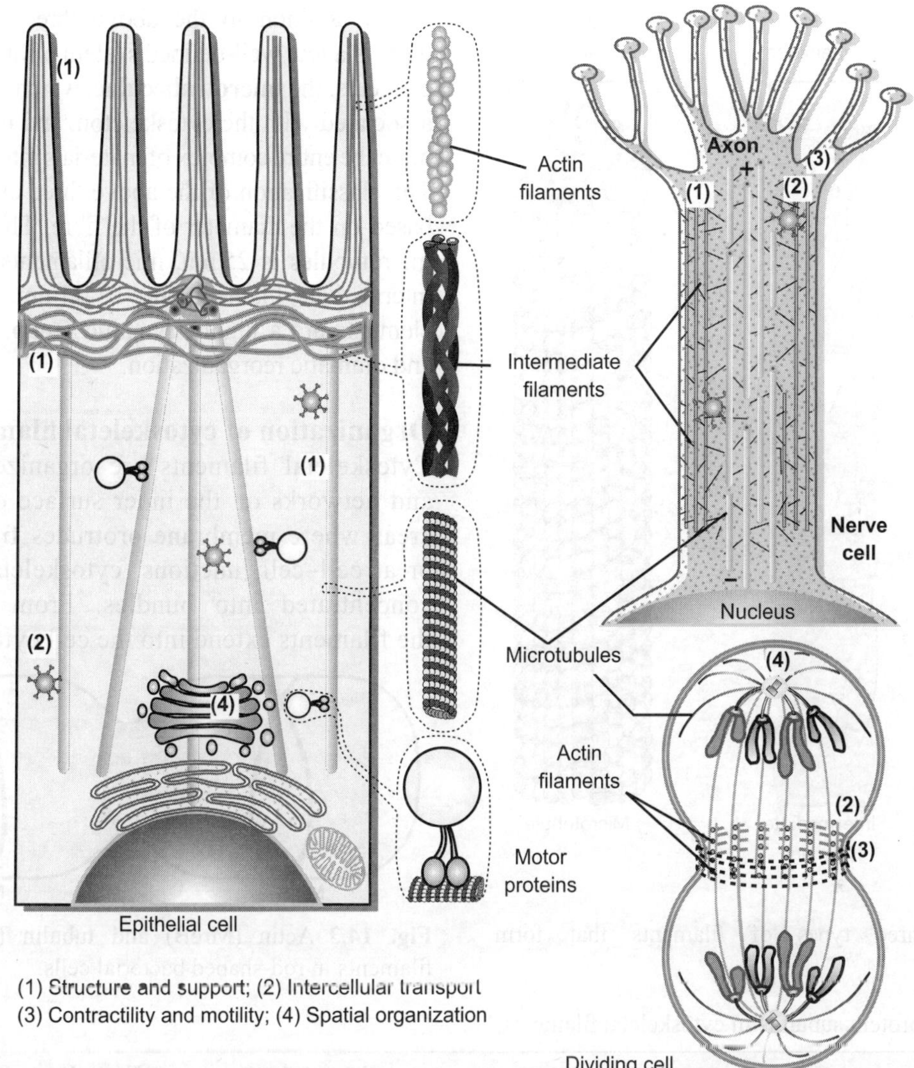

Fig. 14.1 An overview of the structure and functions of the cytoskeletal filaments and their location in different cell types (**See Plate 10**).

14.2 CYTOSKELETAL FILAMENTS

The cytoskeleton is a cytoplasmic system of fibres. It is composed of three well-defined fibrous elements, which can be distinguished based on their diameter (**Fig. 14.2**).

- Actin filaments (microfilaments)
- Intermediate filaments (IF)
- Microtubules

The cytoskeletal filaments are linear, unbranched polymer of protein subunits that may extend several microns through the cytosol. They are highly dynamic, altering their organization in response to the needs of the cell. Most eukaryotic cells contain all the three types of cytoskeletal filaments, often concentrated in distinct locations. The subunits that make up the three filament systems have been highly conserved throughout evolution (**Table 14.1**). A mutation in a cytoskeleton protein subunit could disrupt the assembly of filaments and their binding to other proteins.

The absence of IF-like proteins in bacteria and unicellular eukaryotes implies that IF developed later in evolution. The first IF protein to arise was most likely a nuclear lamin from which cytosolic IF proteins later evolved. The simple bacterial cytoskeleton controls cell length, width and the site of cell division. The **FtsZ protein**, a bacterial homologue of tubulin, is localized around the neck of dividing bacterial cells, suggesting that FtsZ participates in cell division. The prokaryotic version FtsZ protein appears to be the ancestor of all plant and animal tubulins. Another bacterial protein, **MreB**, has been found to be similar to actin in an atomic structure and filament structure, suggesting that actin evolved from MreB (**Fig. 14.3**). Thus, MreB may be involved mainly in the regulation of width of rod-shaped bacteria.

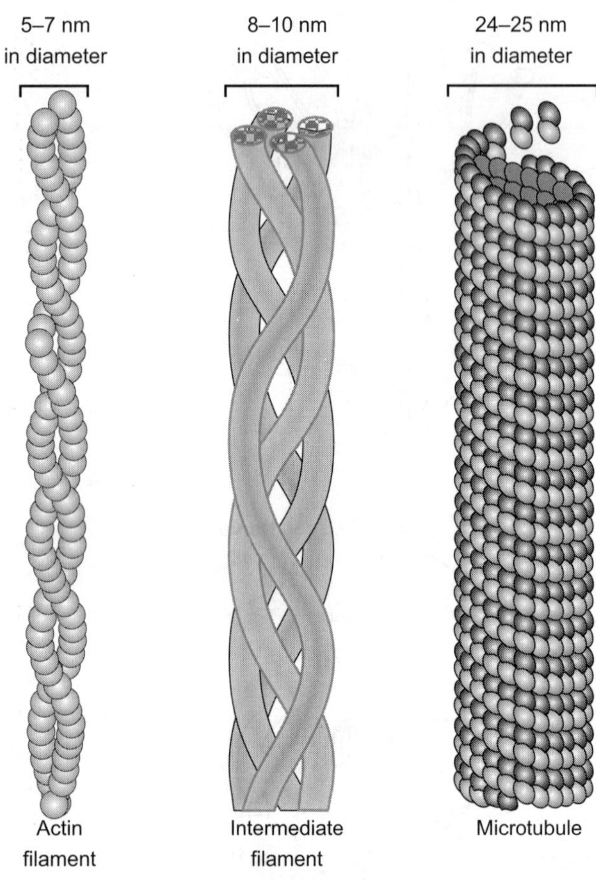

Fig. 14.2 The three types of filaments that form the cytoskeleton.

In addition to the above three fibrous elements, there is a less well-defined system of fibrous cytoplasmic elements, the **microtrabeculae**, which appears to be well associated with the cytoskeleton, and may sever to integrate the entire complex of materials into a functional unit. The classification of the above three elements is mainly based on the diameter of the fibre. The diameter of the microtubules is 25 nm, microfilaments are 5–7 nm and intermediate filaments are 10–11 nm. The cytoskeletal elements are a highly dynamic group, capable of rapid and dramatic reorganization.

Organization of cytoskeletal filaments

Cytoskeletal filaments are organized into bundles and networks on the inner surface of a cell. At the areas where membrane protrudes from cell surface or at cell–cell junctions, cytoskeletal filaments are concentrated into bundles. From these bundles, the filaments extend into the cell cytosol, where they

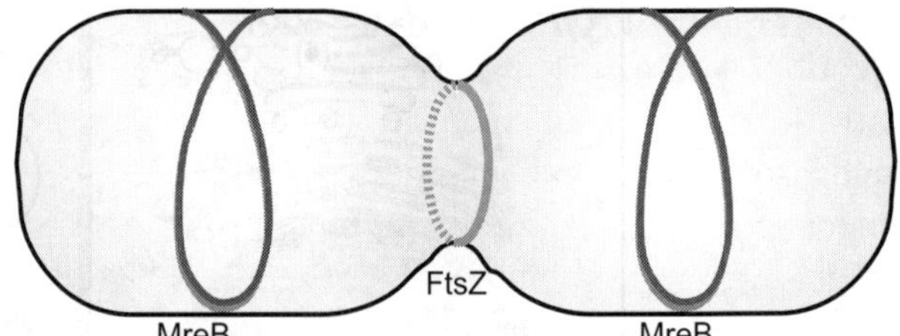

Fig. 14.3 Actin (MreB) and tubulin (FtsZ) homologue filaments in rod-shaped bacterial cells.

Table 14.1 Major protein subunits in cytoskeletal filaments

Protein subunits	Structure	Expression	Functions
Microfilaments			
Actin	42 kDa, diameter 7 nm, two intertwined strands of actin	Fungal, plant, animal	Structural support, motility, membrane extension
MreB	36 kDa	Rod-shaped bacteria	Width control
Microtubules			
Tubulin (and ß)	58 kDa, diameter 25 nm, hollow tubes; wall consists of 13 columns of tubulin proteins	Fungal, plant, animal	Structural support, motility, cell polarity
FtsZ	58 kDa	Bacteria	Cell division
Intermediate filaments			
Lamins	Diameter 8–10 nm, hollow tubes; various	Plant, animal	Support for nuclear membrane, maintains nuclear structure
Desmin, keratin, Vimentin, others	Various	Animal	Cell adhesion
Other			
MSP	50 kDa	Nematode sperm	Motility

fan out and become part of a cytoskeletal network of filaments. These two structures—bundles and networks—are the most common arrangements of cytoskeletal filaments in the interior of a cell (**Fig. 14.4**). In bundles, the filaments are closely packed in parallel arrays, but in a network, the filaments criss-cross and are loosely packed. Networks are either associated with the nuclear and plasma membrane, which is planar (is 2D) or present within the cell, giving the cytosol gel-like properties (3D). All the filaments in these structures are held together by various cross-linking proteins.

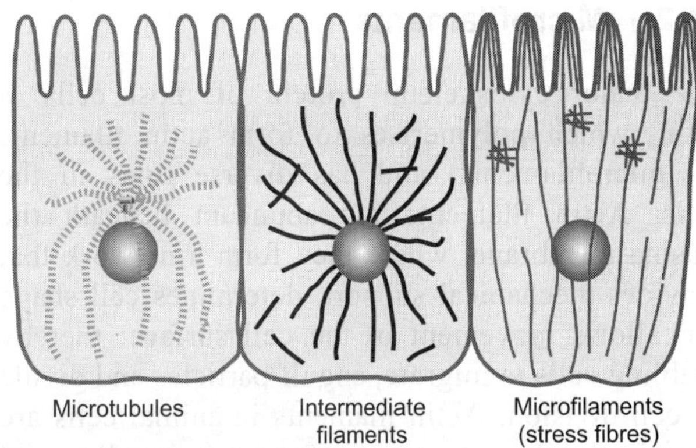

Fig. 14.4 Distribution of cytoskeletal filaments. Typical spatial organization of three major cytoskeletal filaments in eukaryotic cells.

Any particular shape of a cell depends on the organization of actin filaments and proteins that connect microfilaments to the membrane. These are called **membrane–microfilament binding proteins**, and act as spot welds that tack the actin cytoskeleton framework to the overlying membrane. When attached to a bundle of filaments, the membrane acquires the finger-like shape of a microvillus or similar projection. The membrane–cytoskeleton connections entail the binding of integral membrane proteins directly to actin filaments. Many complex linkages connect actin filaments to integral membrane proteins through peripheral membrane proteins that function as adapter proteins. The richest are a of actin filaments in many cells lies in the cortex, a narrow zone just beneath the plasma membrane. In the cortex region, actin networks exclude most organelles from the cortical cytoplasm.

The simplest cytoskeleton is the network of actin filaments adjacent to the erythrocyte plasma membrane. Since a red blood cell must squeeze through narrow blood capillaries without rupturing its membrane, the strength and flexibility of the erythrocyte plasma membrane depend on a dense cytoskeletal network that underlies the entire membrane and is attached to it at many points. The primary component of the erythrocyte cytoskeleton is **spectrin**, a 200-nm long fibrous protein. The cytoskeleton is organized in a spoke-and-hub network as shown in **Fig. 14.5**.

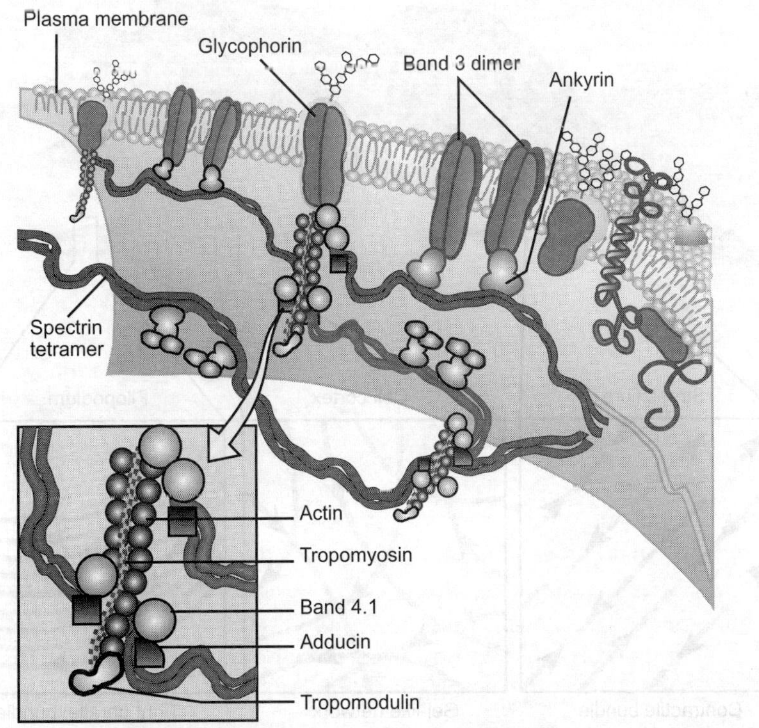

Fig. 14.5 Cortical cytoskeleton supporting the plasma membrane in red blood cells.

14.2.1 Microfilaments

The major cytoskeletal protein of most cells is actin, which polymerises to form actin filaments (or microfilaments) and has diverse roles in the cells. Actin filaments are abundant beneath the plasma membrane, where they form a network that provides mechanical support, determines cell shape and allows movement of the cell surface; thereby enabling cells to migrate, engulf particles and divide by cell division. Actin filaments in animal cells are organized into two types of arrays: bundles and web-like (gel-like) networks (**Fig. 14.6**). Actin filament cross-linking proteins are two types: bundling proteins and gel-forming proteins. Bundling proteins cross-link actin filaments into a parallel array, while gel-forming proteins hold two actin filaments together at a large angle to each other, thereby creating a looser mesh. Both types of cross-linking proteins generally have two similar actin-binding sites. The spacing and arrangement of these two filament-binding domains determines the type of actin structure that a given cross-linking protein forms.

- ***Fimbrin*** is a small cross-linker with two actin-binding domains close together in a single polypeptide chain. It is enriched in the parallel actin filament bundles in filopodia at the leading edge of cells and it is responsible for the tight association of these actin filaments (**Fig. 14.6**).
- ***α-Actinin*** contains two actin-binding domains that are further apart and is concentrated in stress fibres, where it is responsible for the relatively loose cross-linking of actin filaments in these contractile bundles.
- ***Villin*** is another bundling protein that, like fimbrin, has two actin filament-binding sites very close together in a single polypeptide chain. Villin helps to cross-link the 20–30 tightly bundled actin filaments found in microvilli, the finger-like extensions of the plasma membrane on the surface of many epithelial cells (**Fig. 14.7**). A single absorptive epithelial cell in the human small intestine has several thousand microvilli on its apical surface.

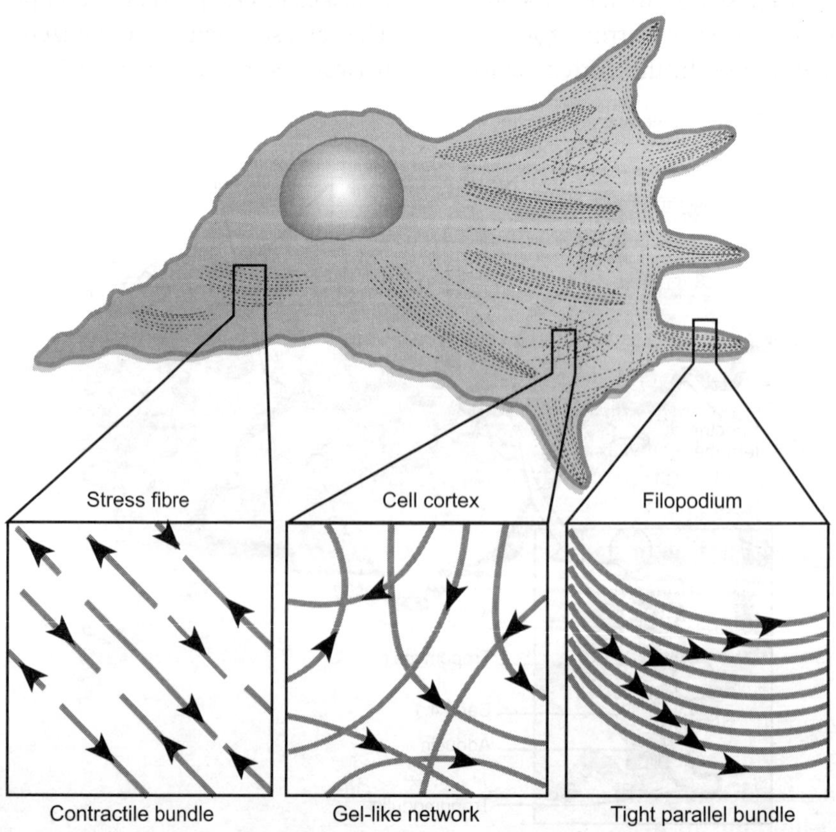

Fig. 14.6 Actin arrays in a crawling cell.

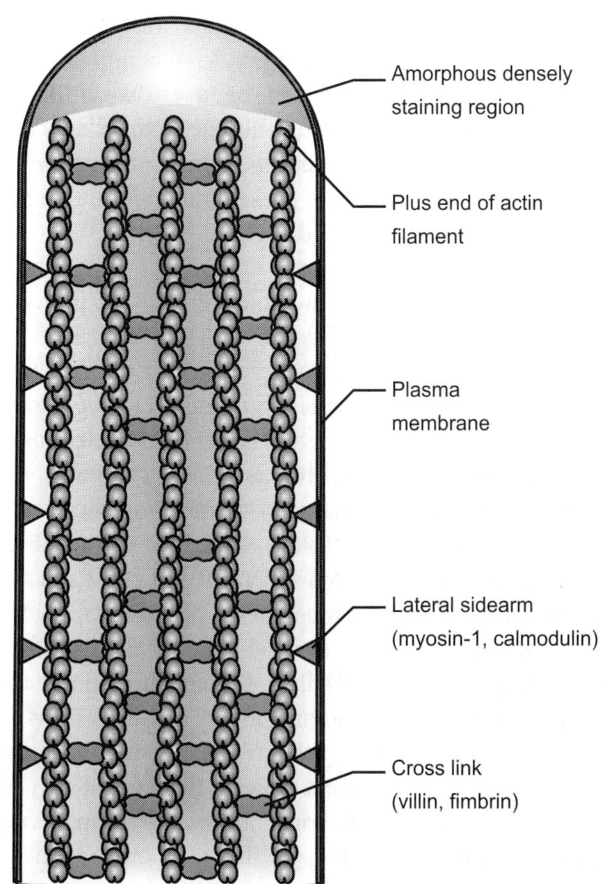

Fig. 14.7 Microvillus. A bundle of parallel actin filaments cross-linked by the actin-bundling proteins villin and fimbrin forms the core of a microvillus. Lateral sidearms connect the sides of the actin filament bundle to the overlying plasma membrane. All the plus ends of actin filaments orient toward tip of the microvillus, where they are embedded in an amorphous, densely staining substance of unknown composition.

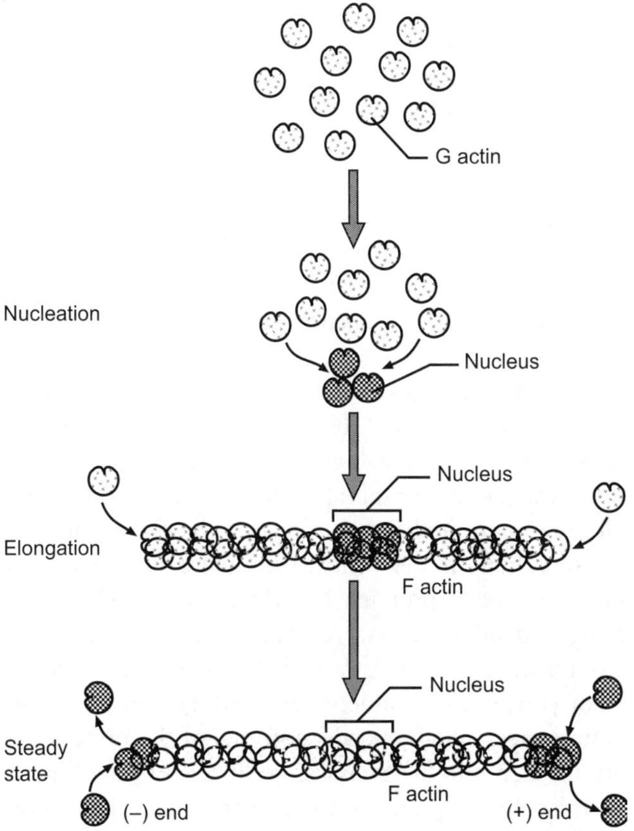

Fig. 14.8 In vitro polymerization of G actin occurs in three phases.

Structure

The microfilament is a fibrous element with a diameter of approximately 7 nm and length of several micrometres. Actin filaments were first isolated from muscle cells, in which it constitutes approximately 20% of total cell protein. It is composed of protein subunits known as globular actin (G actin), thus the monomer is termed as G actin, while the filament that forms it is referred to as **F actin**. Each actin filament is composed of two chains of actin monomers twisted around one another like two strands of beads (**Fig. 14.8**). The filaments are cross-linked into imperfect bundles and networks. The filament lengths are highly variable. This organizational flexibility of the actin cytoskeleton permits a cell to assume many shapes. The equilibrium between G and F actin is affected by many factors including **actin-binding proteins** (**Fig. 14.9**). The growth of the filament is prevented by **profilins**,

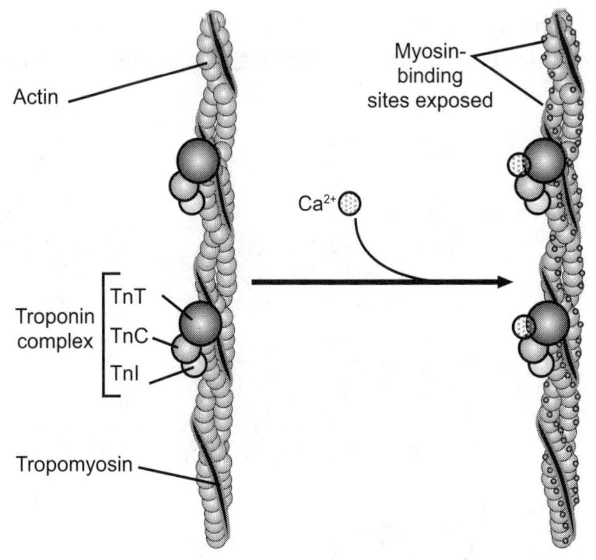

Fig. 14.9 F actin and its associated proteins, tropomyosin and troponin. The proteins present in the thin filaments of striated muscle—F actin, tropomyosin and troponins (TnI, TnC and TnT)—are shown.

which binds G actin filament ends. In contrast, filament nucleation proteins such as actin-related protein (Arp) act as a base on which new filaments can form. Cross-linking proteins stick to two existing filaments, forming a mechanically strong lattice. Of these **villin**, found in microvilli, generates parallel bundles, while related proteins bind criss-crossing filaments together to form a viscous, three-dimensional cytoplasmic gel. As with tubulin, actin is encoded by a family of closely related genes whose polypeptide products are highly conserved among all eukaryotes.

Actin is the most abundant intracellular protein in most eukaryotic cells (10% by weight in muscle cells). It is one of the most highly conserved proteins, with 80.2% sequence conservation at the gene level between *Homo sapiens* and *Saccharomyces cerevisiae* and 95% conservation of the primary structure of the protein product. Actin proteins (42 kDa) are encoded by a large and highly conserved gene family, which arose from a bacterial ancestor and then evolved further as eukaryotic cells became specialized. Humans have six actin genes, which encode isoforms of the protein and are divided into three classes (α, β and γ) according to their isoelectric point. However, some plants have more than 60 actin genes, but most of them are pseudogenes. Although differences between these isoforms are minor, each has different functions: α-actin is associated with contractile structures; γ-actin filaments involve in stress fibres and β-actin is at the front, or leading edge, of moving cells where actin filaments polymerize.

The monomeric and polymeric states exist in a dynamic equilibrium. Each actin filament monomer rotated by 166° with respect to neighbouring subunits, thus, every 36 nanometers or every 13 subunits eclipse each other at what appears to be a crossover of a double-stranded helix. Because all the actin filaments are oriented in the same direction, actin filaments have a distinct polarity: plus and minus ends (also referred to as **pointed** and **barbed**). Actin filaments grow ten times faster at plus (+) end than at minus (–) end, and the filament assembly is promoted by cytosolic protein **profiling** (15 kDa). The first step in actin polymerization is the formation of a small aggregate consisting of three actin monomers. The process of assembly of actin filaments into a double stranded in vivo and in vitro is termed as **nucleation**. Actin filament nucleation occurs most frequently at the plasma membrane. The actin filaments in the layer underlying the plasma membrane, called the **cell cortex**, determine the shape and movement of the cell surface. The nucleation of actin filaments at the plasma membrane is frequently regulated by external signals, allowing the cell to change its shape and stiffness rapidly in response to changes in its external environment.

As mentioned earlier, the two ends of an actin filament grow at different rates; this result in a difference in the critical concentration of monomers needed for polymerization at the two ends. When the polymerization rate at the plus end equals the depolymerization rate at the minus end, the result is a **treadmilling** effect, where the filament moves, in vitro or in vivo, without changing its overall length, which illustrates the dynamic behaviour of actin filaments (**Fig. 14.10**). For the system to be at an overall steady state, the concentration of free actin monomers must

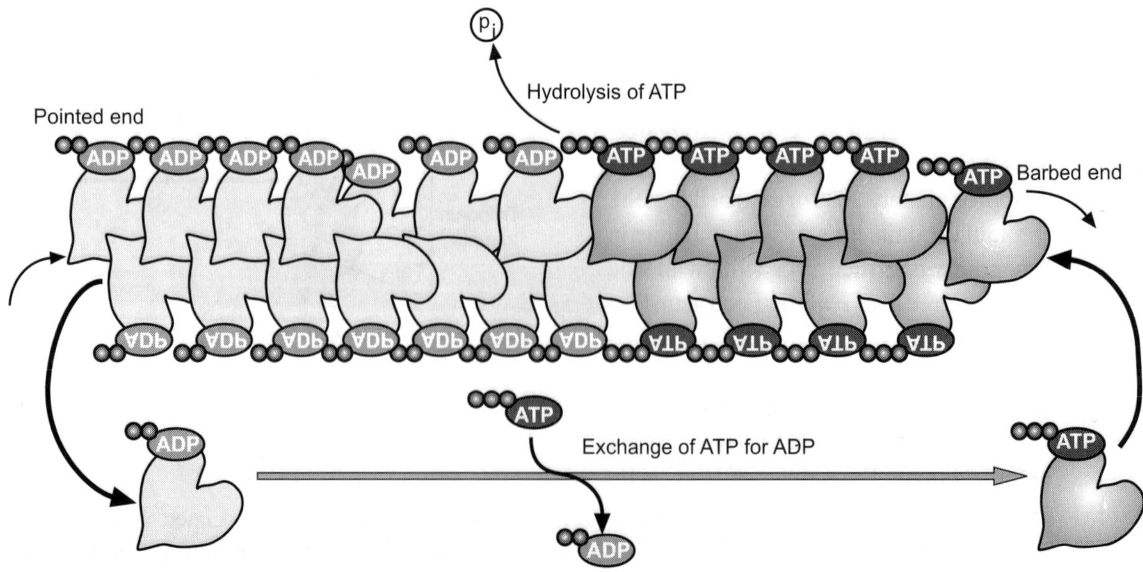

Fig. 14.10 Treadmilling and the role of ATP in microfilament polymerization.

be intermediate between the critical concentrations required for polymerization at the plus and minus ends of the actin filaments. Treadmilling requires ATP, with ATP–actin polymerizing at the plus end of filaments and ADP–actin dissociating from the minus end.

Actin-binding proteins

In addition to actin the cytoplasm of eukaryotic cells also contains some other proteins, which are complex in nature. They are designated as actin-binding proteins (**Fig. 14.11**). These proteins can be classified into three functional groups:

- Those that promote cross-linking and produce gelatin of microfilaments (e.g., fimbrin, spectrin, alpha actinin, filamin),
- Those able to serve the long actin filaments by capping to short filaments, a process that, results in solation, i.e., diminished viscosity (e.g., gelsolin),
- Those proteins that tend to stabilize actin in the monomeric form (G actin), also causing solation (e.g., profilin). G actin polymerize to produce F actin (a filamentous form).

Inhibition of actin polymerization: Several drugs useful in cell biology act by binding to actin and affecting its polymerization. For example, the cytochalasins bind to plus ends of actin filaments and block their elongation. This results in changes in cell shape as well as inhibition of some types of cell movements, indicating that actin polymerization is required for these processes. Another drug, phalloidin, binds tightly to actin filaments and prevents their dissociation into individual actin molecules. Phalloidin labelled with a fluorescent dye is used to visualize actin filaments by fluorescence microscopy.

Assembly and dis-assembly of actin filaments

Within the cell, actin-binding proteins regulate both the assembly and disassembly of actin filaments. The key to understanding this process came form the discovery of a complex of proteins, the **actin-related**

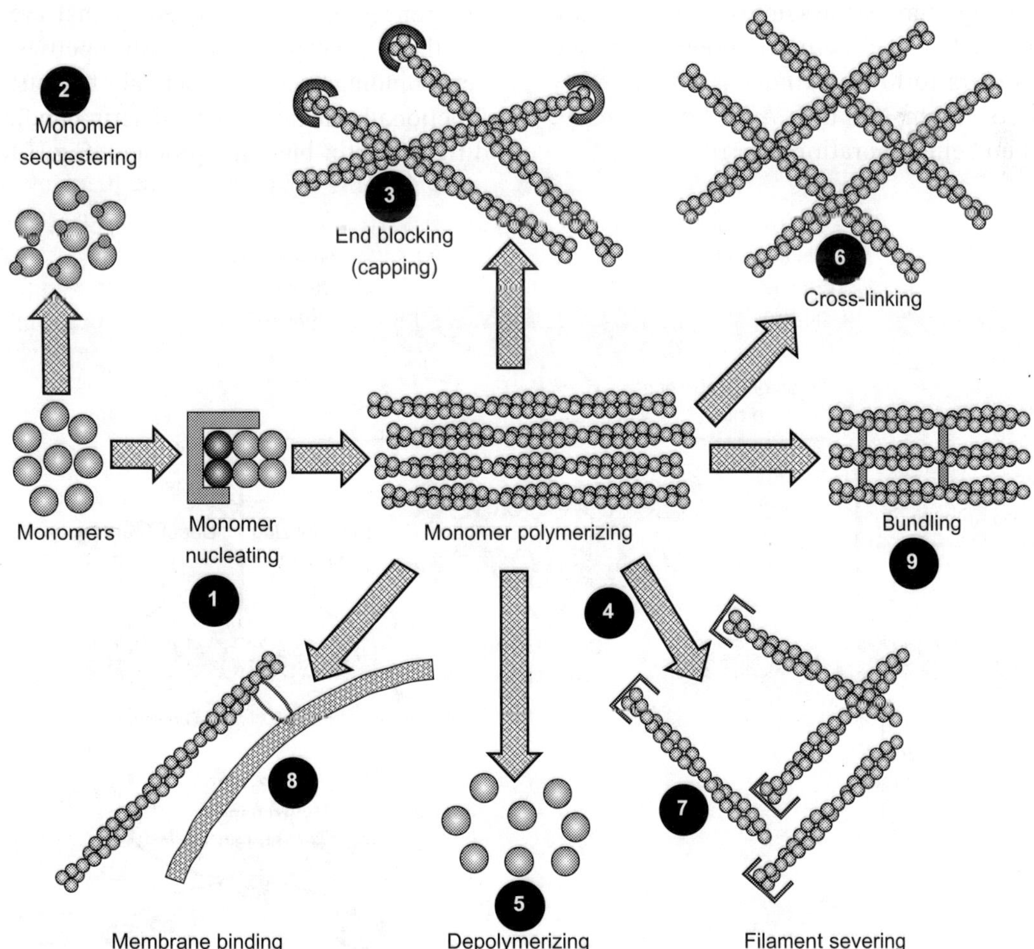

Fig. 14.11 The different roles of actin-binding proteins.

protein complex (**Arp2/3**), which binds to the actin/ATP at the plus ends of microfilaments (**Fig. 14.12**). The Arp2/3 is the major factor that drives actin nucleation and actin filament branching and initiates filopodial/lamellipodial membrane protrusion. The Arp2/3 complex contains two actin-related proteins and five other subunits (p40, p34, p21, p20 and p16). The complex on its own has very low activity but is activated by several proteins that bind. The activation of the Arp2/3 complex is regulated by the *C*-terminal VCA (verprolin homology, cofilin homology, acidic) domains of WASP proteins in concert with their activators Rac/Cdc42 and PIP2. Other proteins play key roles in the dynamic properties of actin filaments by coupling them to the cell surface (Ena/VASP family), capping their ends (capping proteins, severing them (gelsolin, ADF/cofilin), promoting dissociation from the pointed ends (ADF/cofilin), regulating association at the barbed end (profilin) and cross-linking them (α-actinin, filamin). ADF/cofilin can also serve actin filaments generating more plus ends, further enhancing filament assembly. ADF/ cofilin preferentially bind to ADP-actin, so it remains bound to actin monomers following filament dis-assembly and sequesters them in the ADP-bound form, preventing their reincorporation into filaments.

Organization of actin filaments

Individual actin filaments are assembled into two general types of structures—actin bundles and actin networks—which play different roles in the cell (**Fig. 14.13**). In bundles, actin filaments are cross-linked into closely packed parallel arrays. In networks, the actin filaments are cross-linked in orthogonal arrays that form 3D meshworks with the properties of semisolid gels. The formulation of these structures is governed by a variety of actin-binding proteins, which cross-link actin filaments in distinct patterns. All the actin-binding proteins involved in cross-linking contain at least two domains that bind actin. The proteins that cross-link actin filaments into bundles (called **actin-bundling proteins**) are usually small rigid proteins that force the filaments to align closely with one another. In contrast, the proteins that organize actin filaments into networks, tend to be large flexible proteins that can cross-link perpendicular filaments. Actin cross-linking proteins appear to be modular proteins and are separated by spacer sequences that vary in length and flexibility. It is these differences in spacer sequences that are responsible for the distinct cross-linking properties of different actin-binding proteins. There are two structurally and functionally distinct types of actin bundles, involving different actin-bundling proteins (**Fig. 14.14**).

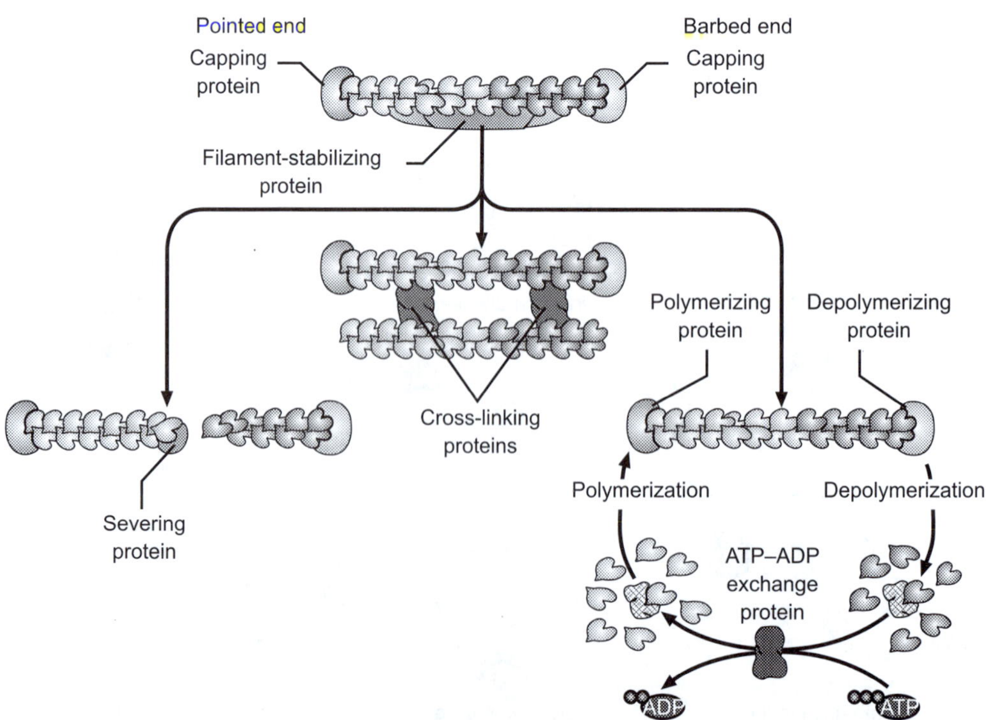

Fig. 14.12 Regulation of actin filament organization by actin-binding proteins.

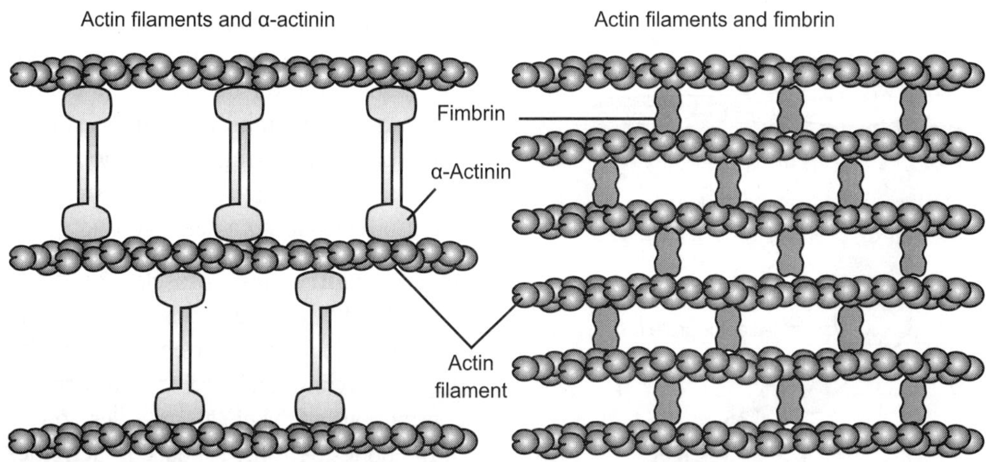

Fig. 14.13 Cross-linking proteins with distinct properties organize assemblies of actin filaments differently. Two types of actin filament bundles form. α-Actinin is a homodimer; it cross-links actin filaments into loose bundles, which allow the motor protein myosin II to participate in the assembly. Fimbrin cross-links actin filaments into tight bundles, which exclude myosin.

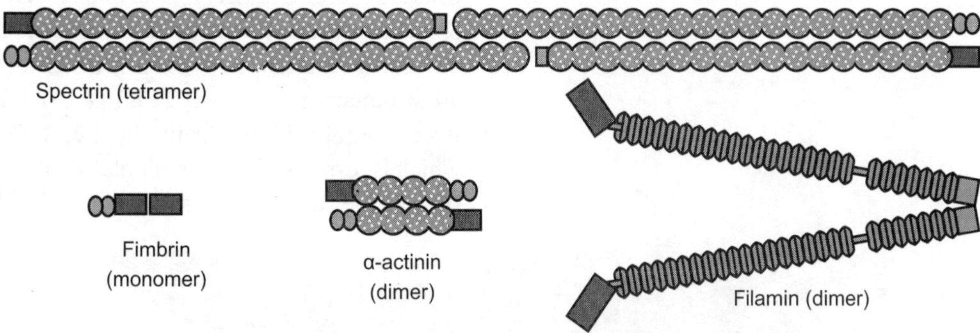

Fig. 14.14 The modular structures of four actin cross-linking proteins.

14.2.2 Functions of actin filaments

Microfilaments serve a number of cellular functions. They form a band just beneath the cell membrane that:

- provides mechanical strength to the cell;
- links transmembrane proteins (e.g., cell surface receptors) to cytoplasmic proteins;
- anchors the centrosomes at opposite poles, of the cell during mitosis;
- forms major component of a cell contractile machinery, which facilitates cell cleavage during cytokinesis phase of cell division;
- generates locomotion in cells such as some leukocytes (white blood cells) and the amoebae
- plays a major role in muscle contraction, during which they interact with myosin ('thick') filaments in skeletal muscle fibres to provide the force of muscular contraction (see following sections)
- takes part in nerve outgrowth, tubular gland formation, movement of intestinal microvilli, generation of cytoplasmic streaming in some cells, gastrulation and so on.

In most cases, the evidence of the involvement of actin containing filaments is indirect, but substantial.

Association of actin filaments with the plasma membrane: Actin filaments are highly concentrated at the periphery of the cell, where they form a 3D network beneath the plasma membrane. This network of actin filaments and associated actin-binding proteins, called the *cell cortex*, determines cell shape and is involved in a variety of cell surface activities, including movement. The association of the actin cytoskeleton with the plasma membrane is thus central to cell structure and function.

Anchorage of cells to extracellular matrix: Cells are anchored to the extracellular matrix through transmembrane proteins such as *integrins.* These are dimeric proteins, that have an extracellular domain that binds to collagen and other extracellular matrix proteins and an intracellular domain that attaches to actin microfilaments on the inside (**Fig. 14.15**).

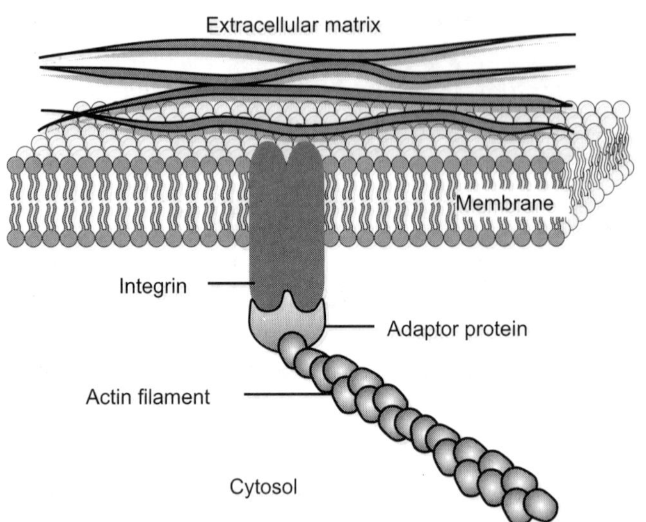

Fig. 14.15 Integrins anchor the actin cytoskeleton to the extracellular matrix.

Cell cleaving during cytokinesis: During cytokinesis of cell division in animals, the microfilaments are involved in the formation of the contractile rings just beneath the membrane. These rings shrink as the rings shorten during contraction of microfilaments. As a result of this, the membrane starts pinching in, eventually dividing a cell into two.

Muscle contraction: Muscle contraction is the most studied and the best understood form of movement in animals. In vertebrates, walking, running, playing, skiing and flying, all depend on the rapid contraction of skeletal muscles on the scaffolding of the bones. All these muscular activities depend on the ATP-driven sliding of highly organized arrays of actin filaments, against arrays of myosin II filaments. Muscle cells are highly specialized for a single task: contraction. There are three distinct types of muscle cells in vertebrates. *Striated muscle* (most of it is skeletal muscles, about 40% of the human body mass), which is responsible for all voluntary movements (**Fig. 14.16**); *Cardiac muscle*, which pumps blood from the heart; *Smooth muscle*, which is responsible for involuntary movements of organs such as the stomach, intestine, uterus and blood vessels.

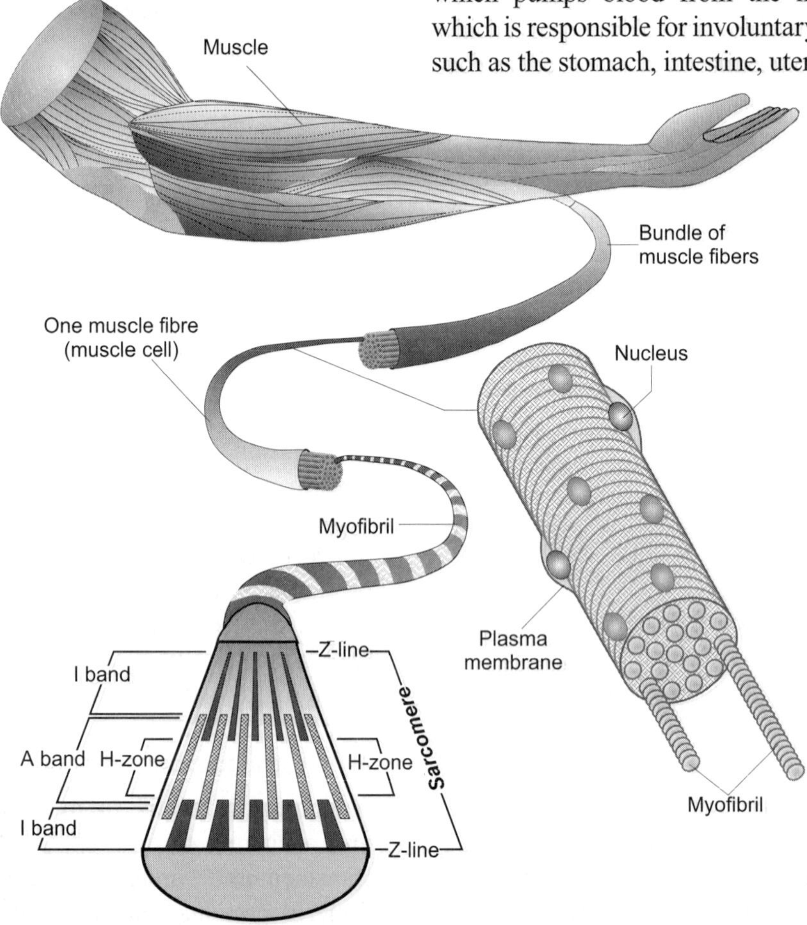

Fig. 14.16 Levels of organization in striated skeletal muscle. These huge multinucleated cells form by the fusion of many muscle cell precursors, called myoblasts. In an adult human, a muscle cell is typically 50 μm in diameter and can be up to several centimetres long (**See Plate 11**).

Steps involved in the action of muscle contraction: The events leading to contraction are initiated in the central nervous system, either as voluntary activity from the brain or as reflex activity from the cord. A motor neuron in the ventral horn of the spinal cord is activated and an action potential transmitted outward in a ventral root of the cord. The axon branches to supply a number of muscle fibres called a motor unit and the action potential is conveyed to a motor end plate on each muscle fibre. At the motor end plate, the action potential causes the release of the neurotransmitter, acetylcholine into the synaptic clefts on the surface of the muscle fibre. Acetylcholine generates the electrical **resting potential** under the motor end plate and initiates an action potential (Na+ flux into fibre), which passes in both directions along the surface of the muscle fibre. At the opening of each transverse tubule (T tubule) onto the muscle fibre surface, the action potential spreads inside the muscle fibre. At each point where a T tubule touches a part of the sarcoplasmic reticulum (SR) a Ca2+ ion is released from the SR terminal cisternae. The calcium ions bind troponin. Myosin ATPase is activated and ATP is hydrolyzed. This causes a tropomyosin shift from actin binding site and actin–myosin cross-bridge forms. The repeated formation and breakage of cross-bridges result in sliding of filaments and sarcomere shortening (**Fig. 14.17**). The movement of troponin and tropomyosin on their thin filaments enables the myosin molecule heads to 'grab and swivel' their way along the thin filament and the length of the sarcomere changes. This is the driving force of muscle contraction.

Sliding filament model: The basis for understanding muscle contraction is the observation that during muscle contraction, the sarcomere, the I and H bands become narrower, while the A band does not, coupled with the observation that thick and thin filaments do not shorten (at very short lengths the thin filaments may either push through the Z line or fold up like an accordion), suggested the sliding filament model (**Fig. 14.18**). According to this model, the thick and thin filaments simply slide past each other. There is general agreement that the sliding filament model is an accurate description of what happens during muscle contraction. Basically, it is an interaction between the actin and myosin filaments that generate movements relative to one another. Actin filaments are capable of interacting with another, in a highly specific manner. Heavy meromyosin (HMM), one of the subfragments of myosin attaches to the actin filaments to form a complex, which is identifiable in the EM. Actin filaments are present in muscle cells along with myosin, troponin and tropomyosin. These protein molecules constitute the contractile machinery of the muscle cells. Striated muscle found attached to bones or the heart gains that appearance because passing along the cell it encounters in turn regions of parallel microfilaments and regions of thick filaments of myosin. The complete repeating unit is called a *sarcomere* and is delineated by the *Z disc*, which holds

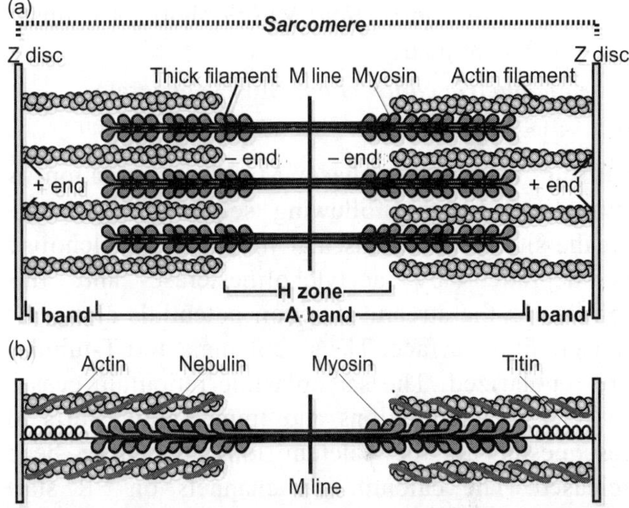

Fig. 14.17 Structure of the sarcomere: (a) Organization of actin (thin) and myosin (thick) filaments in the indicated regions. (b) Titin and nebulin (**See Plate 11**).

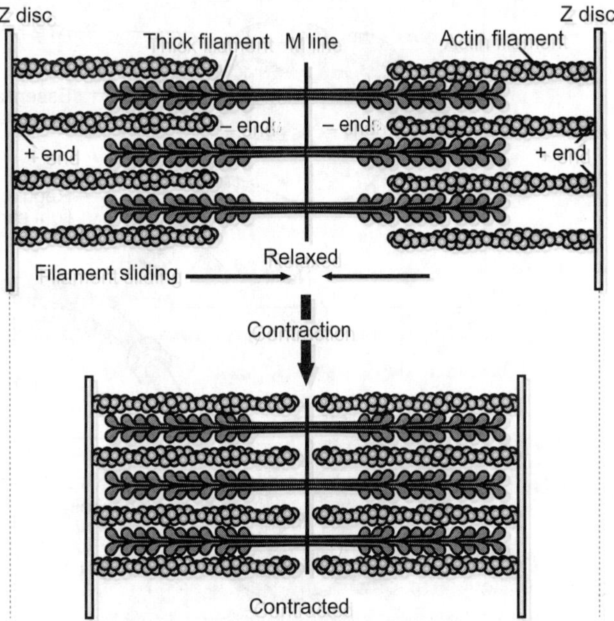

Fig. 14.18 Sliding filament model of muscle contraction.

the microfilaments in a regular pattern, so that striated muscle has a crystalline appearance in transverse section. The myosin molecule has a distinctive structure, consisting of a tail and two globular heads. The thick filament is formed from a large number of myosin molecules arranged in a tail-to-tail manner. This design means that the thick filament is a bipolar structure with myosin heads at both ends. In a resting striated muscle, the myosin heads are unable to operate on the actin microfilaments.

There are two forms of contraction or it could be a blend of both:

- ***Isotonic contraction:*** Muscle shortens against constant load.
- ***Isometric contraction:*** Muscle generates increased tension in an attempt to shorten.

Muscle relaxation phase: Muscle contraction is turned off by the following sequence of events. At the site of neuromuscular junction, acetylcholine is degraded by acetylcholinesterase and this terminates the stream of action potentials along the muscle fibre surface. The sarcolemma and T-tubules are repolarized. The sarcoplasmic reticulum ceases to release calcium ions and immediately starts to resequester all the calcium ions that have been released. The calcium ion channels on SR start pumping calcium back to the SR terminal cisternae by active ion transporters. In the absence of calcium ions, a change in the configuration of troponin and tropomyosin occur by terminating the actin–myosin cross-bridges and then blocks the action of the myosin molecule heads and eventually contraction ceases. Tropomyosin returns to the actin-binding site. Mg^{2+} complex is formed with ATP. Filaments slide passively and sarcomeres return to resting state. In the living animal, an external stretching force, such as gravity or an antagonistic muscle, pulls the muscle back to its original length.

Actin and myosin in non-muscle cells

Myosins are a large super family of mechanochemical motor proteins and exist in eight myosin gene families. Three family members, myosin I, II and V, are present in nearly all eukaryotic cells and are the best understood motor proteins. All myosins consist of one or two heavy chains and several light chains (**Fig. 14.19**), which generally have a regulatory function. A characteristic head, neck and tail domain organization is found in all myosin heavy chains. All myosin head domains have ATPase activity and in conjugation with the neck domain couple ATP hydrolysis to movement of a myosin molecule along an actin filament via a common mechanism involving cyclical binding and hydrolysis of ATP and attachment/detachment of myosin and actin. The role of particular myosin in vivo is related to its tail domain. The number and type of light

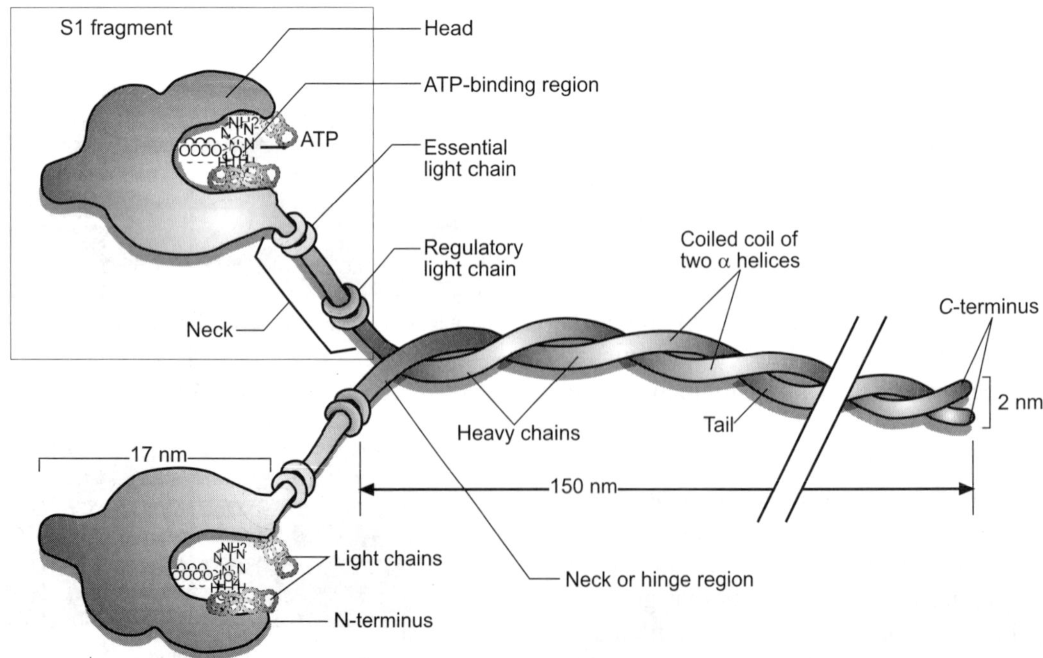

Fig. 14.19 Structure of a myosin II molecule. It is composed of two heavy chains (2000 amino acids long) and four light chains.

chains bound in the neck region vary among the different myosins.

In non-muscle cells, actin may appear in a variety of structural forms. Contractile assemblies of actin and myosin resembling small-scale versions of muscle fibres are also present in non-muscle cells. In non-muscle cells, the actin filaments in these contractile assemblies are interdigitated with bipolar filaments of myosin II, which produce contraction by sliding the actin filaments relative to one another (**Fig. 14.20**). Myosin II is similar to muscle myosin but does not assemble into filaments to the same extent, probably because the levels of force required within the non-muscle cells is relatively small. The most dramatic example of actin–myosin contraction in non-muscle cells, however, is provided by cytokinesis—a phase that leads to the division of a cell into two cells following mitosis (**Fig. 14.21**). Towards the end of mitosis in yeast and animal cells, a **contractile ring** consisting of actin filaments and myosin II assembles just underneath the plasma membrane. The contraction pulls the plasma membrane progressively inward, constricting the centre of the cell and pinching it in two.

One family of non-muscle myosins are membranes of the **myosin I** family (**Fig. 14.22**). The myosin I proteins contain a globular head group that acts as a molecular motor, like that of myosin II. The tails of myosin I bind to other structures, such as membrane vesicles or organelles. The movement of myosin I along an actin filament can then transport its attached cargo. **Myosin V** is also two-headed and is responsible for carrying cargo along actin filaments (**Fig. 14.23**). Unlike the microtubule-associated motors, dynein and kinesin, which can make long journeys, myosin V can only make short excursions along an actin filament before falling off. Other structural forms are:

- Actin filaments exhibit well-organized regular polarization in the intestinal microvilli,
- Bundles of actin filaments constitute stress fibres in spreading moving cells in culture,
- Cross-linked network in pseudopodia, ruffling membranes and in the bulk of the cytoplasm, hhese are disposed at random and form a cross-linked network.

Cell locomotion

The movement of cells across a surface represents a basic form of cell locomotion, employed by a

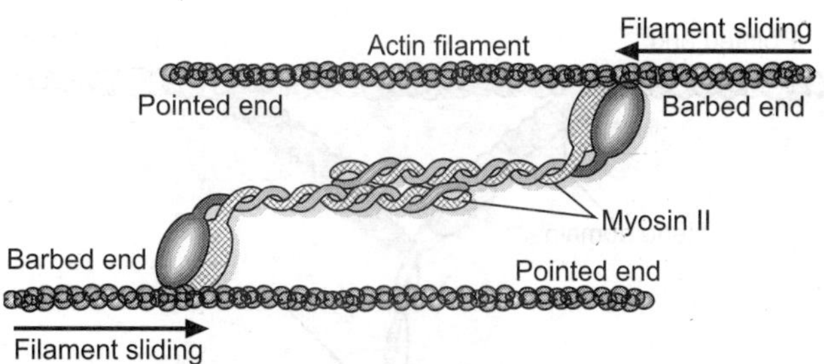

Fig. 14.20 Contractile assemblies of myosin II produce contraction by sliding actin filaments in opposite directions.

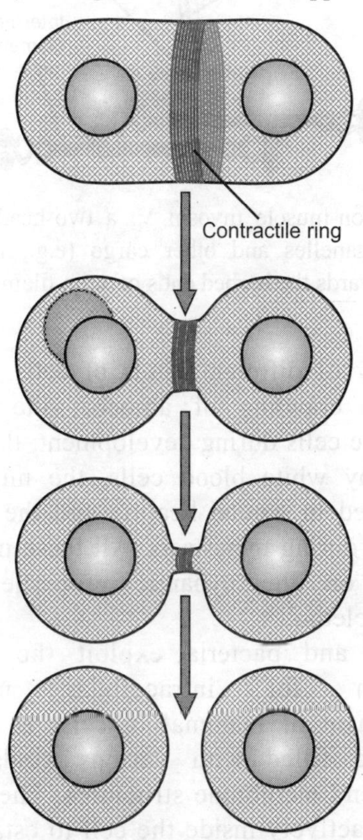

Fig. 14.21 Cytokinesis. During the later stage of mitosis a contractile ring consisting of actin filaments and myosin II divides the cell in two.

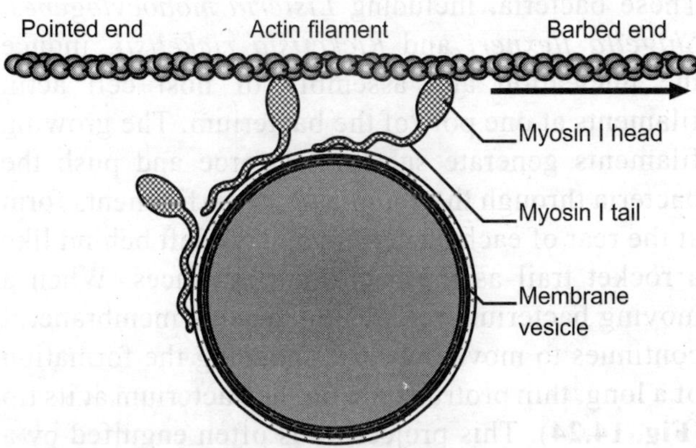

Fig. 14.22 Myosin I contains a head group and a short tail. It helps to move along actin filaments carrying a variety of cargoes.

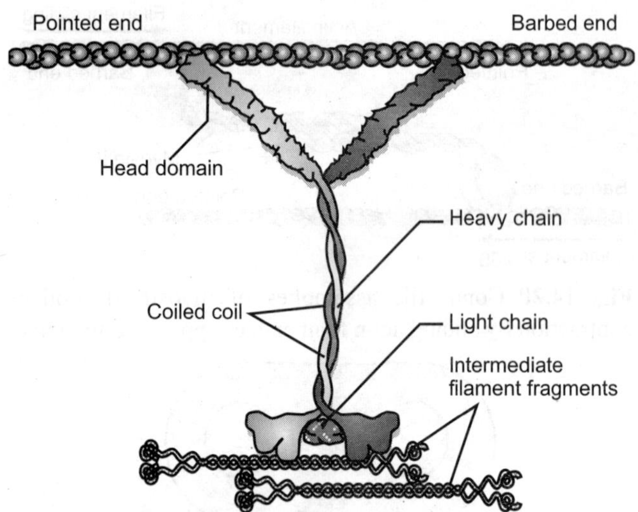

Fig. 14.23 Non-muscle myosin V, a two-headed myosin, transports organelles and other cargo (e.g., intermediate filaments) towards the barbed ends of actin filaments.

wide variety of different kinds of cells. Examples include the crawling of amoeba, the migration of embryonic cells during development, the invasion of tissues by white blood cells, the migration of cells involved in wound healing and the spread of cancer cells during metastasis. All these movements are based on the dynamic properties of the actin cytoskeleton.

Viruses and bacteria exploit the host cell cytoskeleton for intracellular movement. The cytoplasm of mammalian cells is extremely viscous and filled with various kinds of cell organelles and membrane structures. The pathogen must move actively inside the cell to establish and spread the disease. Several bacteria that replicate in the host cell cytosol have adapted a mechanism for movement, which depends on actin polymerization. These bacteria, including *Listeria monocytogenes*, *Shigella flexneri* and *Rickettsia rickettsii*, induce the nucleation and assembly of host cell actin filaments at one pole of the bacterium. The growing filaments generate substantial force and push the bacteria through the cytoplasm. New filaments form at the rear of each bacterium and are left behind like a rocket trail as the bacterium advances. When a moving bacterium reaches the plasma membrane, it continues to move outward, inducing the formation of a long, thin protrusion with the bacterium at its tip (**Fig. 14.24**). This projection is often engulfed by a neighbouring cell allowing the bacterium to enter the neighbour's cytoplasm.

Cell locomotion results from the coordination of motions generated by different parts of a cell. All moving cells exhibit polarity, i.e., certain structures always form at the front of the cell, whereas others are found at the rear. For unicellular organisms such as amoebae to crawl across a surface it must generate traction. Less than 1% of the ventral surface of a crawling cell is in contact with the surface over which it is moving, so that the cell does not slither along its belly but rather 'tiptoes'

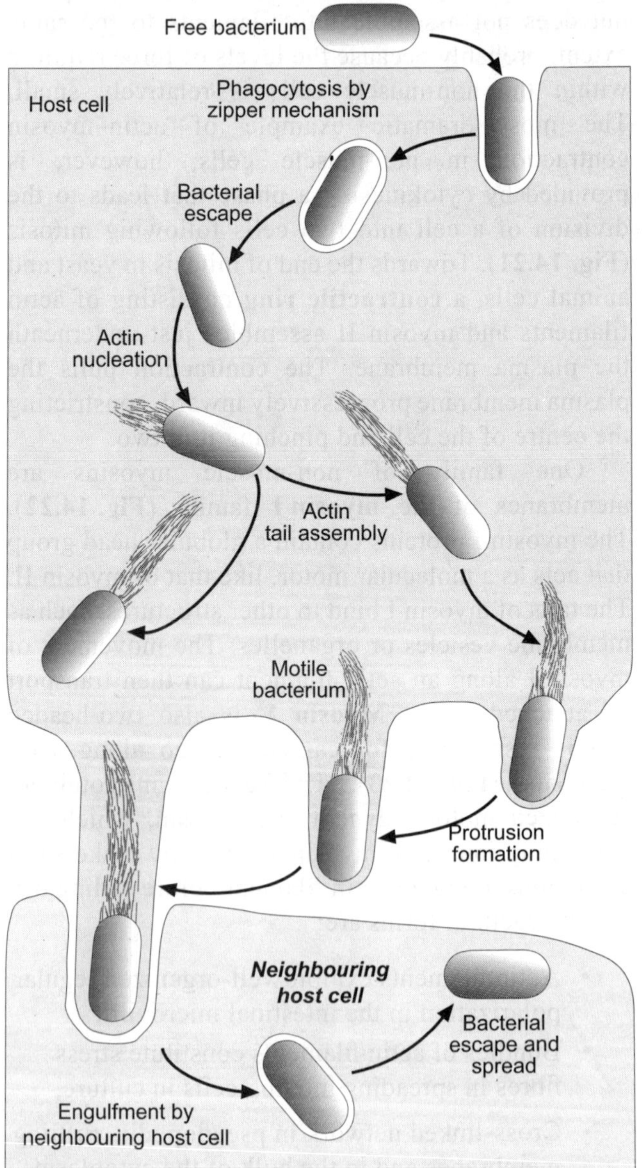

Fig. 14.24 The actin-based movement of *Listeria monocytogenes* within and between host cells. These motile bacteria activate the assembly of actin rich tails in the host cell cytoplasm, which enable them to spread from cell to cell by forming membrane-enclosed protrusions that are engulfed by neighboring cells. (Source: Alberts et al, 2002)

delicately across the surface. The transient points of attachment with the surface are called **focal contacts**. At the leading edge actin polymerization causes the cell to extend projections or **pseudopodia** (or **lamellipodium** in vertebrate cells) in the direction of progress. In some cases, slender, finger-like membrane projections, called **filopodia**, also extend from the leading edge. The overall increase in actin polymerization and cross-linking in this region causes the cytoplasm to form a viscous ectoplasm (**Fig. 14.25**). In contrast, raised cytosolic calcium concentrations in the tail activate **actin-severing proteins** such as **gelsolin**. Calcium also activates myosin II, the resulting contraction driving the now fluid endoplasm forward and displacing the existing actin cytoskeleton backward. Since actin is connected via the focal contacts to the surface, the backward movement of the ectoplasm drives the cell forward, opening rather like the tracks of a military tank.

External signals such as growth factors and chemoattractants induce the assembly and organization of the cytoskeleton and the establishment of an internal gradient of trimeric G proteins and calcium. The resulting polarization of the cell leads to locomotion. Extracellular factors trigger Ras-linked signal transduction pathways that activate actin polymerization at the leading-edge membrane as an early event, and the formation of focal adhesions as a later event. An interesting example is the killing of specific target cells by cytotoxic T lymphocytes (**Fig. 14.26**).

Cytoplasmic streaming

Even though the cells themselves do not move, the cytoplasm of many plant cells also shows active

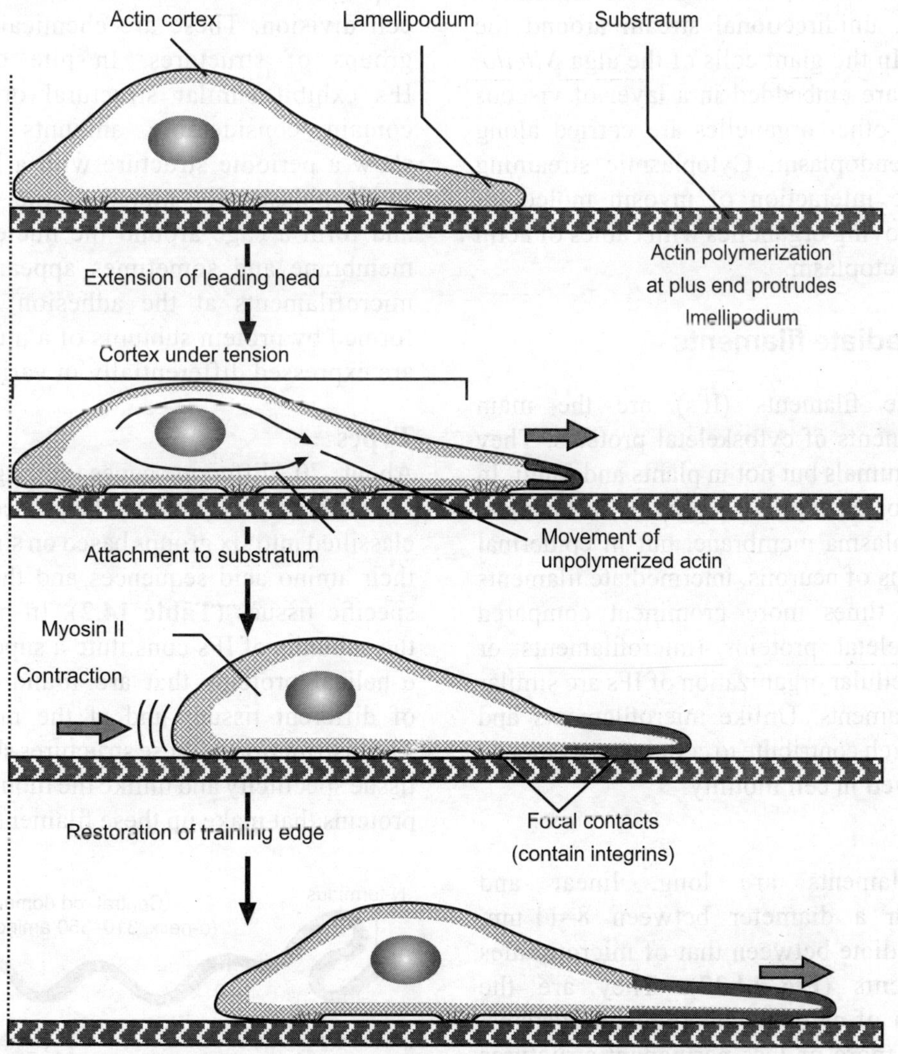

Fig. 14.25 A model of how forces generated in the actin-rich cortex move a cell forward.

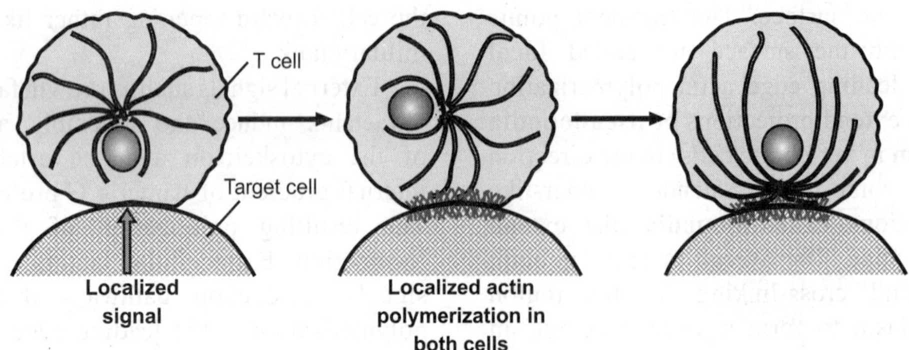

Fig. 14.26 The polarization of a cytotoxic T cell (Tc) after target-cell recognition. Once the cytotoxic T cell made contact with a target cell, cytoskeleton of a Tc cell changes. The initial recognition event results in signals that cause actin polymerization in both cells at the site of contact, from which microtubules emanate and cause reorientation of centrosome. This also causes the associated Golgi apparatus to be oriented directly apposed to the target cell.

cytoplasmic streaming. In the epidermal cells of the leaves of the water plant, *Elodea*, a belt of moving cytosol carries the chloroplasts and other organelles in a continuous, unidirectional stream around the central vacuole. In the giant cells of the alga *Nitella*, the chloroplasts are embedded in a layer of viscous ectoplasm while other organelles are carried along in the flowing endoplasm. Cytoplasmic streaming is driven by the interaction of myosin molecules attached to the moving organelles with cables of actin filaments in the ectoplasm.

14.2.3 Intermediate filaments

The intermediate filaments (IFs) are the main structural components of cytoskeletal proteins. They are found in all animals but not in plants and fungi. In the epithelium, for example, IFs provide mechanical support for the plasma membrane, but in epidermal cells and the axons of neurons, intermediate filaments are at least 10 times more prominent compared to other cytoskeletal proteins (microfilaments or microtubules). Cellular organization of IFs are similar to actin microfilaments. Unlike microfilaments and microtubules, which contribute to cellular movements, IFs are not involved in cell motility.

Structure

Intermediate filaments are long, linear and unbranched with a diameter between 8–11 nm, which is intermediate between that of microtubules and microfilaments **(Fig. 14.27)**. They are the most stable form of cytoplasmic filament systems. They tend to be more or less permanent structures in tissues such as skin and hair; in fact, in these non-living tissues, they are the only proteins. In other cell types, IFs are phosphorylated when required to be dis-assembled, for example, during cell division. These are chemically heterogeneous groups of structures. In spite of heterogeneity, IFs exhibit similar structural organization. They contain considerable amounts of α-helix and show a periodic structure with a length of 21 mm. In fibroblasts, IFs attach to the nuclear envelope and form a cage around the nucleus at the plasma membrane and sometimes appear associated with microfilaments at the adhesion plaques. IFs are formed by protein subunits of a multigene class that are expressed differentially in various cell types.

Types

About 70 different genes coding for more than 50 different IF proteins have been identified and classified into six groups based on similarities between their amino acid sequences and their distribution in specific tissues (**Table 14.2**). In higher vertebrates, the subunits of IFs constitute a superfamily of highly α-helical proteins that are found in the cytoplasm of different tissues and at the nuclear membrane. The distribution of these structures shows considerable tissue specificity and unlike the tubulins and actins, the proteins that make up these filaments reflect the tissue

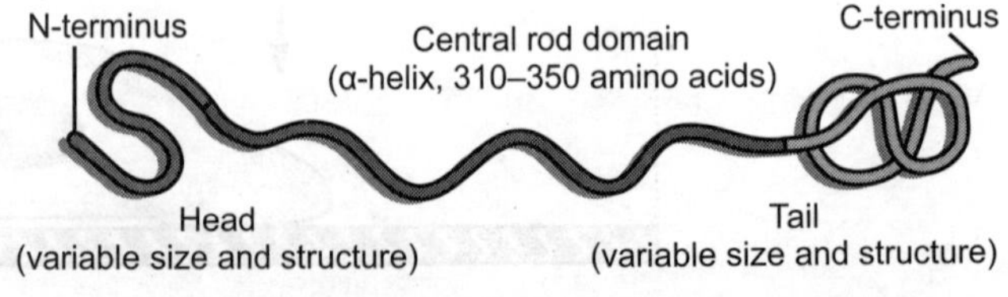

Fig. 14.27 Structure of intermediate filament proteins.

type and state of differentiation of the cells in which they are synthesized. The different types of IFs can be divided into four groups based on the similarities in sequence and their patterns of expression in cells. For example, the nucleus contains **lamins**, nerve cell axons contain **neurofilaments**, muscle cells contain **desmin** filaments (structural components of the sarcomeres). **Vimentin** filaments—the most widely distributed of all IF proteins—can be found in fibroblasts, leukocytes and blood vessel endothelial cells. They support the cellular membranes and keep some organelles in a fixed place within the cytoplasm. Epithelial cells contain filaments composed of **epithelial keratins** (about 20) and **trichocytic keratins** (about 13) (hair keratins) which make up hair, nails, horns and reptilian scales. **Peripherin** found in peripheral neurons and GFAP (glial fibrillary acidic protein) is found in astrocytes and other glia.

Table 14.2 Intermediate filament proteins

Type	Protein	Size (kDa)	Site of expression
I	Acidic keratins (~15 proteins)	40–60	Epithelial cells
II	Neutral or basic keratins (~15 proteins)	50–70	Epithelial cells
III	Vimentin	54	Fibroblasts, white blood cells and other cell types
	Desmin	53	Muscle cells
	Glial fibrillary acidic protein	51	Glial cells
	Peripherin	57	Peripheral neurons
IV	Neurofilament proteins		
	NF-L	67	Neurons
	NF-M	150	Neurons
	NF-H	200	Neurons
	α-internexin	66	Neurons
	Synemin	-	-
	Syncoilin	-	-
V	Nuclear lamins	60–75	Nuclear lamina of all cell types
VI	Nestin	200	Stem cells of the nervous system

Assembly of intermediate filaments

The beginning of filament assembly is the formation of dimers in which the central rod domains of two polypeptide chains are wound around each other in a coiled-coil structure, similar to that formed by myosin II heavy chains. The dimers then associate in a staggered anti-parallel fashion to form tetramers, which can assemble end to end to form proto-filaments. The final intermediate filament contains approximately eight proto-filaments wound around each other in a rope-like structure (**Fig. 14.28**). Consequently, in contrast to actin filaments and microtubules, intermediate filaments are apolar; they do not have distinct plus and minus ends. Many intermediate filaments further bundle themselves by self-association; for example, the neurofilament proteins NF-H and NF-H contain a *C*-terminal domain that extends outward from the surface of the assembled intermediate filament and binds to a neighbouring filament. Thus, a group of neurofilaments form robust parallel arrays that are held together by multiple lateral contacts, giving strength and stability to the long cell processes of neurons.

Other types of IF bundles are held together by accessory proteins, such as filaggrin, which bundles keratin filaments in differentiating cells of the epidermis to give the outermost layers of the skin

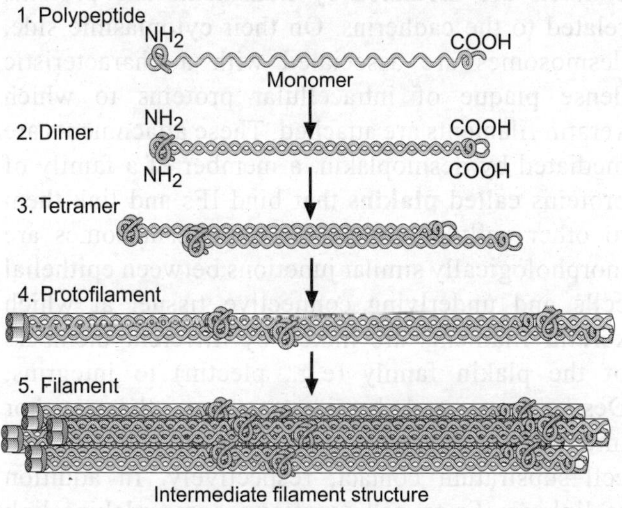

Fig. 14.28 A model of internediate filament assembly and architecture (steps 1–5).

their special toughness. Plectin, which makes bundles of vimentin, is a particularly interesting cross-linking protein. Besides bundling IFs, it also links them to microtubules, actin filament bundles and filaments of the motor protein myosin II, as well as helping to attach IF bundles to adhesive structures at the plasma membrane. Mutations in the gene for plectin cause a devastating human disease that combines **epidermolysis bullosa** (disruption of skin keratin filaments), muscular dystrophy (disruption of desmin filaments) and neuro-degeneration (disruption of neuro-filaments). Mice lacking a functional plectin gene die within a few days of birth, with blistered skin and abnormal skeletal and heart muscles.

Intracellular organization of intermediate filaments

Intermediate filaments form an elaborate network in the cytoplasm of most cells, extending from a ring surrounding the nucleus to the plasma membrane. Both keratin and vimentin filaments attach to the nuclear envelope, apparently serving to position and anchor the nucleus within the cell. In addition, IFs can associate not only with the plasma membrane but also with the other elements of the cytoskeleton, actin filaments and microtubules. IFs thus provide a scaffold that integrates the components of the cytoskeleton and organizes the internal structure of the cell. The keratin filaments of epithelial cells are tightly anchored to the plasma membrane at two areas of specialized cell contacts, **desmosomes** and **hemidesmosomes** (**Fig. 14.29**). Desmosomes are junctions between adjacent cells, at which cell–cell contacts are mediated by transmembrane proteins related to the cadherins. On their cytoplasmic side, desmosomes are associated with a characteristic dense plaque of intracellular proteins to which keratin filaments are attached. These attachments are mediated by desmoplakin, a member of a family of proteins called **plakins** that bind IFs and link them to other cellular structures. Hemidesmosomes are morphologically similar junctions between epithelial cells and underlying connective tissue, at which keratin filaments are linked by different members of the plakin family (e.g., plectin) to integrins. Desmosomes and hemidesmosomes thus anchor intermediate filaments to regions of cell–cell and cell–substratum contact, rèspectively. In addition to linking IFs to cell junctions, some plakins link IFs to other elements of the cytoskeleton. Plectin, for example, binds actin filaments and microtubules in addition to IFs, so it can provide bridges between these cytoskeletal components.

Bacterial cytoskeletal proteins

It is expected that any cellular system must have some kind of cytoplasmic skeletal filaments that maintain the cell shape and structure, provide mechanical strength and reinforce cell motility. Recently, the presence of various kinds of proteins that have functions similar to cytoskeletal proteins of eukaryotes have also been discovered in bacteria.

Crescentin is a protein, which is a bacterial relative of the IF found in eukaryotic cells. Crescentin was recently discovered in the prokaryote *Caulobacter crescentus*, an aquatic bacterium, which uses its crescent-shaped cells for enhanced motility. It is located on the concave face of these cells and appears to be necessary for their shape and the mutants lacking the protein form rod-shaped cells. Cresentin organizes into filaments and is present in a helical structure in the cell. This protein has 430 residues and the sequence consists of a pattern of seven repeated residues, which form a coiled-coil structure. Crescentin has a central rod made up of four coiled segments, but lacks certain consensus sequence elements at the ends of the rod domain, which are conserved in animal and keratin proteins. Crescentin protein can assemble into about 10-nm wide filaments and later into protein bundles, indicating an evolutionary linkage with IF.

FtsZ was the first protein of the prokaryotic cytoskeleton to be identified. FtsZ is protein encoded by the *ftsZ* gene that assembles into a ring at the future site of the septum of bacterial cell division. FtsZ, named after filamenting temperature-sensitive mutant Z, is a prokaryotic homologue to the eukaryotic protein tubulin. During cell division, FtsZ is the first protein to move to the division site and is essential for recruiting other proteins that produce a new cell wall between the dividing cells. FtsZ's role in cell division is analogous to that of actin in eukaryotic cell division, but will not provide the mechanical force that physically divides the cell. The formation of FtsZ ring in dividing chloroplasts and some mitochondria, further establishes their prokaryotic origin and ancestry.

MreB is a protein found in bacteria that has been identified as a homologue of actin, as indicated by similarities in tertiary structure and conservation of active site peptide sequence. Recent studies have found that MreB proteins polymerize to form

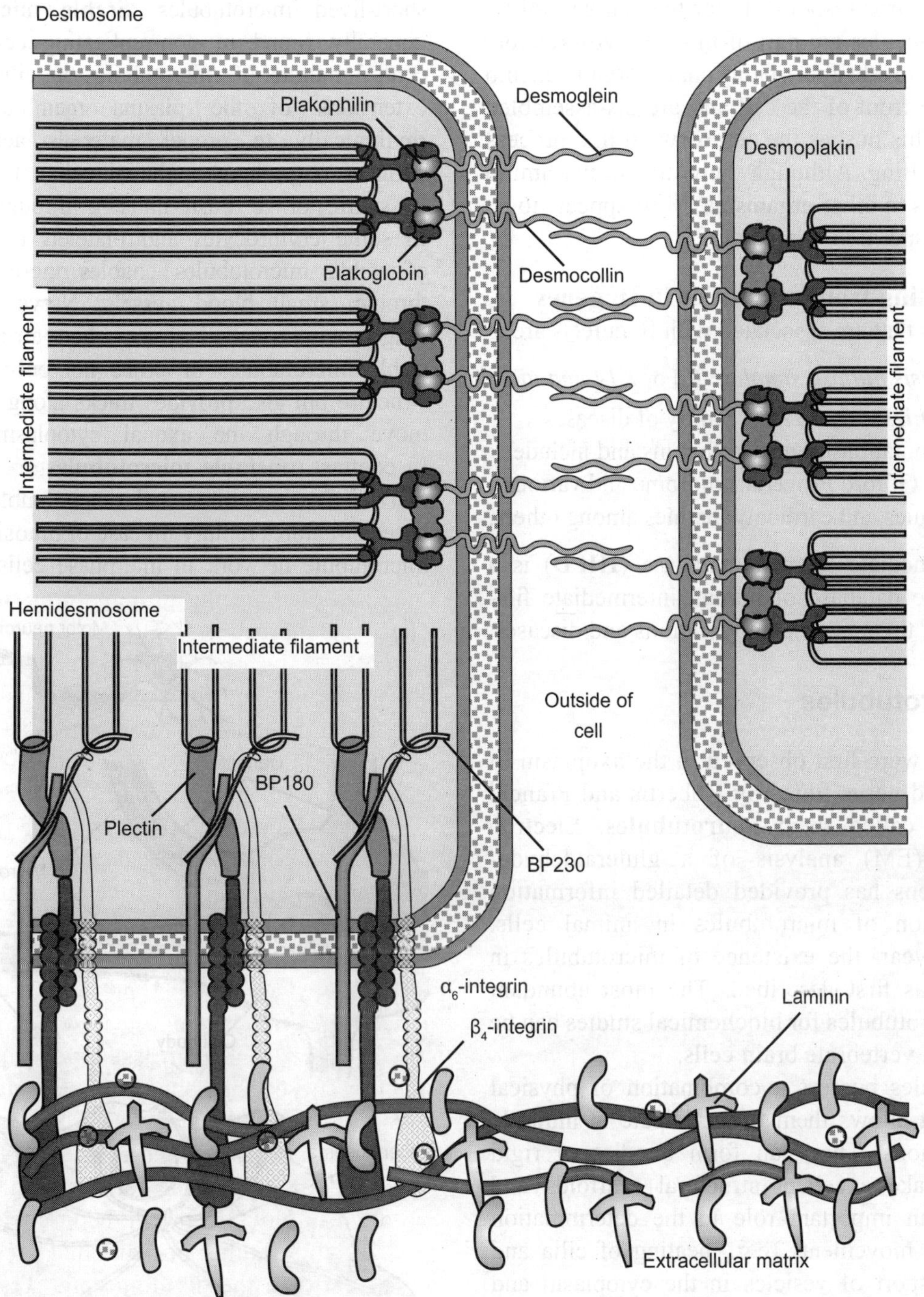

Fig. 14.29 Attachment of intermediate filaments to desmosomes and hemidesmosomes.

filaments that are similar to actin microfilaments. MreB controls the width of rod-shaped bacteria, such as *E. coli*. A mutant of *E. coli* that creates defective MreB proteins is spherical instead of rod-like. Also, bacteria that are naturally spherical do not have the gene encoding MreB. Prokaryotes carrying the *mreB* gene can also be helical in shape. MreB filaments sit right beneath the bacterial cell membrane and give bacteria a strong, rigid structure, as if the bacterium were being stretched out from the inside.

Proteins with cytoskeletal functions

The **major sperm protein,** commonly abbreviated to MSP, is the most abundant protein in nematode sperm, making up about 15% of the total protein in the sperm cell. It is responsible for the cell's motility. The MSP molecules are part of the cell cytoskeleton. They are built up into a set of long chains organized into bundles at the front of the cell and are dis-assembled at the back. This pushes the cell forward in a process called treadmilling. Although there are similar amino acid sequences in other organisms, MSP appears to be unique to the sperm of nematodes.

Diseases arising from mutations in IF genes

Some diseases that are associated with IF defects are:

- ***Epidermolysis bullosa simplex:*** *K5* or *K14 mutation,*
- ***Laminopathies:*** These are a family of diseases caused by mutations in nuclear lamins and include Hutchinson Gilford Progeria syndrome and various lipodystrophies and cardiomyopathies among others.

Human intermediate filament database (**HIFD**) is a comprehensive database of human intermediate filament proteins, their associated variations and diseases.

14.2.4 Microtubules

Microtubules were first observed in the axoplasm of the myelinated nerve fibres by Robertis and Franchi (1953), who called them **neurotubules**. Electron microscopic (EM) analysis of a gluteraldehyde-fixed specimens has provided detailed information on organization of microtubules in animal cells. In the same year, the existence of microtubules in plant cells was first described. The most abundant source of microtubules for biochemical studies can be obtained from vertebrate brain cells.

Microtubules possess a combination of physical properties that allows them to participate in multiple cellular functions. They can form bundles of rigid fibres that make excellent structural scaffolds and hence serve an important role in the determination of cell shape, movements (e.g., beating of cilia and flagella), transport of vesicles in the cytoplasm and other activities. These movements happen from the polymerisation and depolymerisation of microtubules or the actions of **motor proteins**. In addition to contributing to cell motility, microtubules play a major role in organizing the cell through a special structure called the **microtubule-organising centre** (**MTOC**). MTOC is located near the nucleus and directs the assembly and orientation of microtubules, direction of vesicular trafficking and the orientation of organelles.

Cells normally contain two populations of microtubules: stable, long-lived microtubules and unstable, short-lived microtubules. **Stable microtubules** are generally found in non-replicating cells, such as a central bundle of microtubules in cilia and flagella, extensions of the plasma membrane that beat rhythmically to propel materials across epithelial surfaces (e.g., lung epithelial cells), to enable sperm to swim, or to push an egg through the oviduct. In some erythrocytes and platelets a marginal band of stable microtubules enables these cells to pass through small blood vessels. Nerve cells maintain long processes called axons, where an internal core of stable microtubules in axons not only supports their structure but also provides tracks along which vesicles move through the axonal cytoplasm (**Fig. 14.30**). In contrast, **unstable microtubules** are found in cells that need to assemble and dis-assemble microtubule-based structures rapidly. In case of mitosis, the cytosolic microtubule network in interphase cells dis-assembles

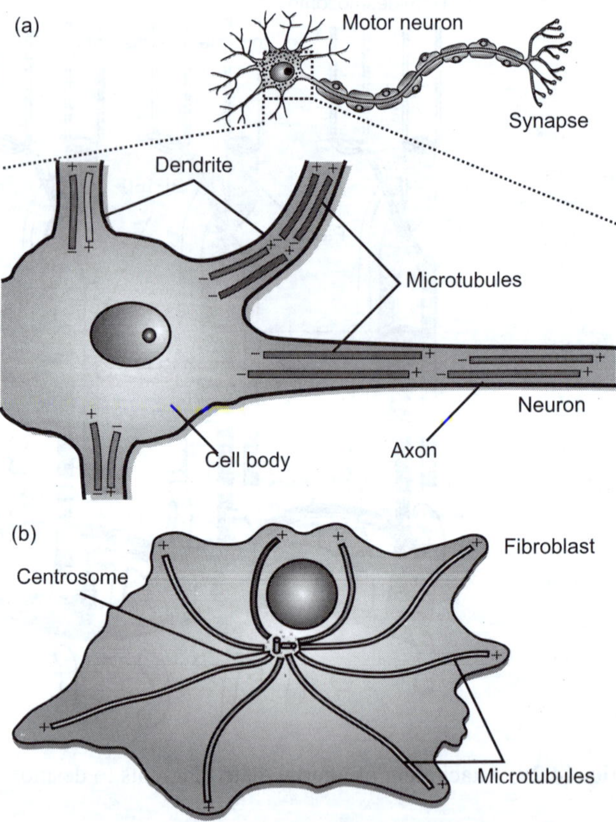

Fig. 14.30 Organization of microtubules in nerve and fibroblast cells: (a) In dendrites microtubules are oriented in both directions, with their plus ends pointing both towards and away from the cell body. In contrast, all the axon microtubules are oriented with their plus ends pointing toward the tip of the axon. (b) In a fibroblast, microtubules emanate outward from the centrosome in the middle of the cell.

and in metaphase cells it polymerizes to form spindle-shaped apparatus that partitions chromosomes equally to the daughter cells.

Microtubule structure

The microtubules have been observed in all the eukaryotic cells that have been studied and like other cytoskeletal filaments, microtubules are not randomly distributed in cells. They are small, hollow cylindrical structures called tubules. Typically a tubule has an outer diameter of approximately 25 nm, a wall diameter of approximately 5 nm leaving the lumen diameter of 15 nm. The wall of a microtubule is a polymer composed of globular subunits. In a cross section the wall nearly always reveals 13 subunits making up the complete circumference of the wall. The subunits are arranged in longitudinal rows, termed proto-filaments that are aligned parallel to the long axis of the tubule (**Fig. 14.31**). Each microtubule is surrounded by a of low electron-dense regions in which ribosomes or other particles are absent.

Chemically, the microtubules are composed of **tubulin subunits**. These are the building blocks of microtubules and exist as a heterodimer of closely related α- and β-tubulins. Each monomer contains about 450 amino acids and is of 55 kDa, found in all eukaryotes. These two proteins are found to be highly conserved across the species, for example, sea urchin flagella and chick brain cell β-tubulins differ only by one amino acid. A third tubulin, γ-tubulin, is not part of the tubulin subunit, but it may probably nucleate the polymerization of subunits to form αβ-microtubules. Each tubulin subunit binds two molecules of GTP. One GTP-binding site, located in α-tubulin, binds GTP irreversibly and does not hydrolyze it. The second site located on β-tubulin binds GTP reversibly and hydrolyzes to GDP. Tubulin is a GTPase like bacterial FtsZ protein. Longitudinal contacts between the ends of adjacent subunits link the subunit head to tail into a linear proto-filament. Through lateral interactions, proto-filaments associate side by side into a sheet or cylinder—α-microtubule. Thus, a polymer made of asymmetric subunits, all oriented in the same direction along the axis, is itself a polarized structure with one end distinguishable from the other. The two ends of a microtubule designated the (+) and (–) ends, differ in their rates of assembly and critical concentrations. The (+) end corresponds to the β-tubulin end of a microtubule. The assembly and stability of microtubules are temperature - dependent. At 4°C, microtubules depolymerize into αβ-tubulin dimers; at 37°C in the presence of GTP, the tubulin dimers polymerize into microtubules. This structural polarity enables these organelles to participate in directed mechanical movements. Some tissues may possess multiple subtypes of both the α- and β-species, thus exhibiting molecular heterogeneity.

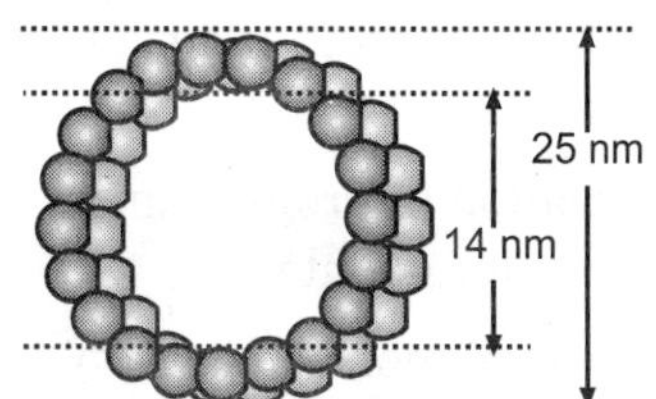

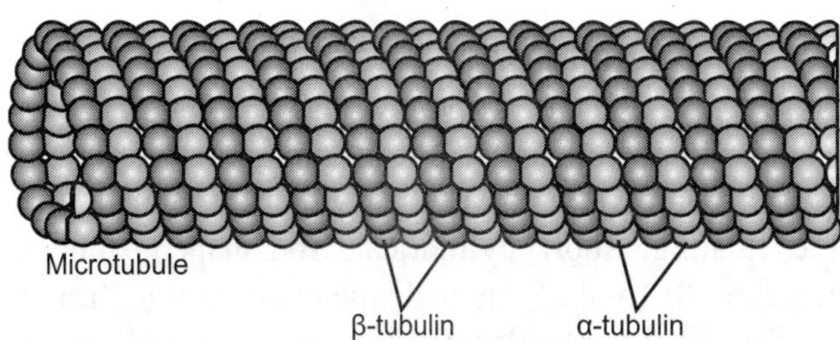

Fig. 14.31 Structure of microtubules.

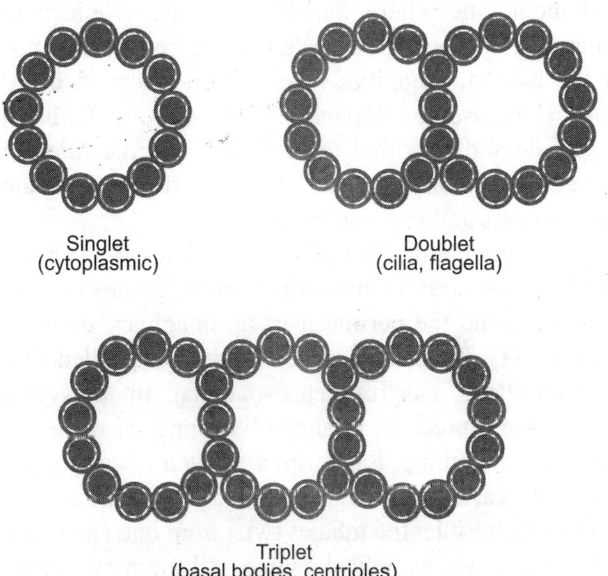

Fig. 14.32 Arrangement of proto-filaments in singlet, doublet and triplet microtubules.

A singlet microtubule is built from 13 proto-filaments and can also exist in **doublet** or **triplet microtubules** in specialized structures such as cilia and flagella and centrioles and basal bodies. Each doublet or triplet microtubule contains one complete 13-proto-filament microtubule (A tubule) and one or two additional tubules (B and C) consisting 10 proto-filaments (**Fig. 14.32**). Microtubule assembly comprises three steps:

- assembly of proto-filaments from αβ-tubulin subunits,
- association of proto-filaments to form the wall of the microtubule,
- addition of more subunits to the ends of the proto-filaments which elongates the microtubule.

Centrioles and basal bodies

Close to the nucleus, in the cytoplasm of animal cells is a dense region called the **centrosome** (described by Theodor Boveri, 1888), which contains two free floating, short cylindrical, rod-shaped, tubular structures (0.15–0.25 μm in diameter and 0.3–0.7 μm in length) called **centrioles**. During mitosis, microtubules extend from centrioles within a centrosome to the cell membrane where flagella and cilia originate in ciliated and flagellated cells. When centrioles exist at the base of flagellum or cilium, it is called **basal body** (kinetosome, blepharoplast, basal granule, basal corpuscle and proximal centrioles). Centrioles are not present in prokaryotes, red algae, yeast, conifers and angiosperms and in some protozoans.

The common feature of centrioles and basal bodies is the array of nine triplet microtubules equally spaced around the perimeter of an imaginary cylinder (**Fig. 14.33**). The central space of centrioles is filled with an amorphous, electron-dense material. In transverse section the triplets are arranged like vanes or blades of pinwheel or turbine. Each triplet is tilted inward to the central axis at an angle of about 450 to the circumference. Within each triplet the tubules twist from one end to the other, described as helical course. All the nine triplets that make up the wall are basically similar in centrioles and basal bodies. The three subunit microtubules that make triplet have been designated A, B and C, with the innermost tubule being A. The individual tubules are 200–260Å in diameter. Only the A tubule is round and the others are incomplete, C-shaped and share their wall with the preceding tubule. The A tubule of each triplet is linked with C tubule of neighbouring triplet by protein linkers at intervals along their entire length. These linkers hold the cylindrical array of the microtubules and maintain the typical radial tilt of the triplets.

In centrioles, central microtubules or special arms are absent. However, often faint protein spokes radiate out to each triplet from a central core, forming a pattern like a **cartwheel**. Such cartwheel structures, determines the proximal end of a centriole and thus provides a structural and functional **polarity** to it. **Satellites** or **perpendicular bodies** are electron-dense structures lying near the centriole, which may be nucleating sites for the microtubules. Microtubules of centrioles and basal bodies contain tubulin along with lipid molecules and also contain high amounts of ATPase enzymes. The origin of basal bodies is well characterized in ciliated paramoecium and tetrahymena. In ciliated vertebrate cells, centrioles of the precursor cell give rise to the many basal bodies required to nucleate the cilia in the mature cell. During differentiation of the ciliated epithelial cells that line the oviduct and the trachea, the centriole pair migrates from its normal location near the nucleus to the apical region of the cell where the cilia will form. There, each centriole in the pair forms numerous electron-dense fibrogranular satellites. Many basal bodies then arise from these satellites and migrate to the membrane to initiate the formation of cilia.

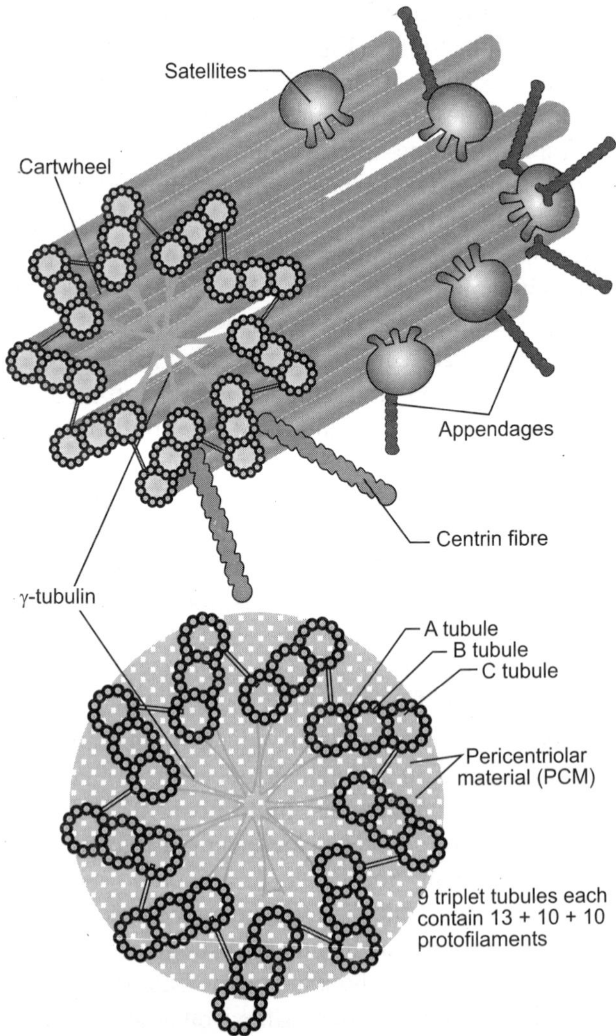

Fig. 14.33 (a) Structure of centriole (TS). (b) Schematic diagram of a centrosome (CS) showing the surrounding pericentriolar material (PCM).

But there are certain cases where centrioles seem to arise de novo. Unfertilized eggs of many animals lack functional centrioles and use the sperm centriole for the first mitotic division (for cleavage). Like the centrioles, the basal bodies are found to possess some capacity for self-assembly and may appear suddenly, for example, in *Naegleria*, as it changes from its amoeboid form to a typical ciliate.

The functions of centrioles and basal bodies are very diverse. Basal bodies initiate the cilia. Centrioles in most animal cells are the core unit of the centrosome. The centrosome (cell centre) organises the array of cytoplasmic microtubules during interface and duplicates at mitosis to nucleate the two poles of the **mitotic spindle**. Often centriole functions can keep changing, for instance, prior to each division in *Chlamydomonas*, the two flagella resorb and the basal bodies leave their position to act as mitotic poles. In spermatozoon, one centriole gives rise to the tail fibre of flagellum. Centrioles and basal bodies are also found to be involved in ciliary and flagellar beat. Centriole and basal bodies have major roles in the reception of optical, acoustic and olfactory signals.

Proteins regulating microtubule dynamics

A large number of other proteins, termed **microtubule-associated proteins** (**MAPs**) have recently been identified at the surface of the microtubule (**Table 14.3**) and may be present as regularly spaced projections (**Fig. 14.34**). These proteins greatly influence the assembly and stability of microtubules and their association with other cell structures. MAFs constitute about 15%–20% of the proteins. The MAPs of mammalian brain cells fall into two major groups:

- a group of ***high-molecular weight MAPs*** (HMW proteins), which have molecular weight > 200–300 kDa (e.g., MAP 1 and MAP 2).
- a group of low-molecular weight MAPs (40–60 kDa) termed as ***tau proteins***.

The first group stabilizes microtubules and the structure of a stabilizing MAP consists of two domains: a basic microtubule-binding domain and an acidic projection domain. The arm can bind to membranes, intermediate filaments or other microtubules and its length controls how far apart microtubules are spaced. When stabilizing MAPs (MAP 1A and MAP 1B) coat the outer wall of a microtubule; tubulin subunits are unable to dissociate from the ends of that microtubule. In most cases, control of length is accomplished by

Table 14.3 Some microtubule-associated proteins (MAP)

Protein	Major source	Mol. wt. (kDa)
MAP 1	Brain: white matter; cell bodies, axons	350
MAP 2	Brain: grey matter; mostly dendrites	270
MAP 3	Brain: glia; myelinated axons	180
MAP 4	Brain: glia; cultured cells	200
Tau	Brain: axons	60
Kinesi	Squid axon	110
Dynein	Cilia, flagella	> 400
Dynein (cytoplasmic)	Brain: axons	> 350
Dynamin	Brain	100

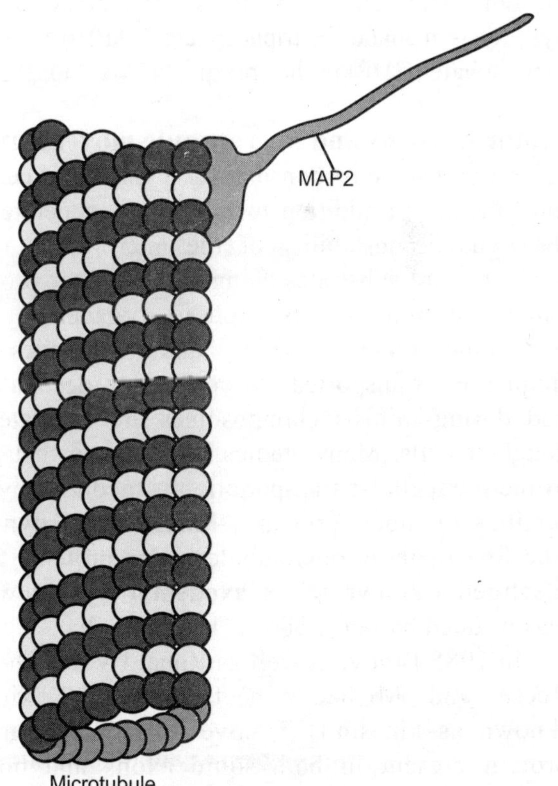

Fig. 14.34 Microtubule-associated proteins (MAPs).

the reversible phosphorylation of the MAP projection domain.

In the cell, the microtubules are in equilibrium with free tubulin. Phosphorylated MAPs are unable to bind to microtubules; thus they promote microtubule disassembly. **MAP kinase**, a key enzyme for phosphorylating MAPs is involved in many signal transduction pathways and MAPs-are thus targets of many extracellular signals. MAPs, especially MAP 4, are also phosphorylated by a cyclin-dependent kinase (CDK) that plays a major role in controlling the activities of various proteins in the course of the cell cycle. A second group of MAPs directly destabilizes microtubules in many cell types. A group called katanin, severs intact cytosolic microtubules by an ATP-dependent process. Internal bonds between tubulin subunits in the microtubule wall are broken, causing microtubules to fragment. This activity may release microtubules at the MTOC.

The assembly and dis-assembly of tubulin constitutes a polarized phenomenon. The centrioles and basal bodies of cilia are microtubule orientation centres from which the polymerization is directed. In a microtubule, the assembly of tubulin dimers takes place at one end while dis assembly is prevalent at the other end. The assembly is accompanied by the hydrolysis of guanosine triphosphate (GTP) to guanosine diphosphate (GDP) in the presence of Mg^{2+} ions.

Motor proteins and microtubule movement

Microtubules are responsible for a variety of cellular movements. In addition to movement generated by the dynamic instability of the microtubule itself, the fibres are substrates along which motor proteins can move. Within cells, proteins, organelles, other membrane-limited vesicles and organelles are frequently transported to different destinations, and during mitosis chromosomes are separated to daughter cells. Many studies have shown that most of the intracellular transportation is mediated by two families of motor proteins—**kinesins** and **dyneins**. The first of these microtubule motor proteins to be identified was dynein, as **axonemal dynein** which was isolated by Ian Gibbons (1965) in cilia.

In 1985 Brady, as well as Ronald Vale, Thomas Reese and Michael Sheetz identified kinesin (known as kinesin I), a novel microtubule motor protein present in both squid axons and bovine brain. Eighteen different kinesis are encoded in the genome of *C. elegans* and 45 kinesins are known in humans. Most of these, like kinesin I, move cargo along microtubules toward the plus (+) end (anterograde transport). Other kinesins move in the opposite direction towards the minus end and three kinesins do not move along microtubules at all but use their tubulin-binding properties to anchor microtubules to other structures. Plus end-directed kinesins have *N*-terminal motor domains, whereas, minus end-directed kinesins have *C*-terminal motor domains and others have the middle of the heavy chain. For example, the plus ends of microtubules in axons are all oriented away from the cell body: vesicles and organelles are transported away from the cell body towards the tip of the axon. Kinesin I is a molecule of approximately 380 kDa consisting of two heavy chains (120 kDa each) and two light chains (64 kDa each). The heavy chains have long α-helical regions that wind around each other in a coiled-coil structure (**Fig. 14.35**). The amino terminal globular head domains of the heavy chains are the motor domains of the molecule. They bind to both microtubules and ATP, the hydrolysis of which provides the energy required for movement. The tail portion of the kinesin molecule consists of light chains in association with the carboxy terminal domains of the heavy chains. This portion of kinesin responsible for binding to other cell components is transported along microtubules by the action of kinesin motors. Most kinesins walk towards the plus end of a microtubule which, in most cells, entails transporting cargo from the centre of the cell towards the periphery, which is known as **anterograde transport**. Two mechanisms have been proposed to explain how this movement occurs (**Fig. 14.36**).

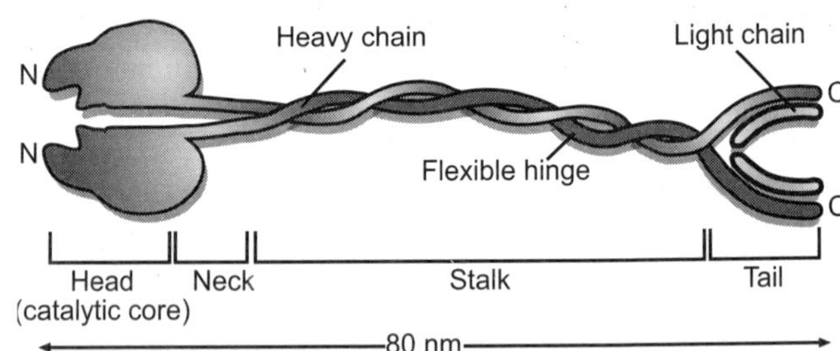

Fig. 14.35 Structure of a conventional kinesin molecule.

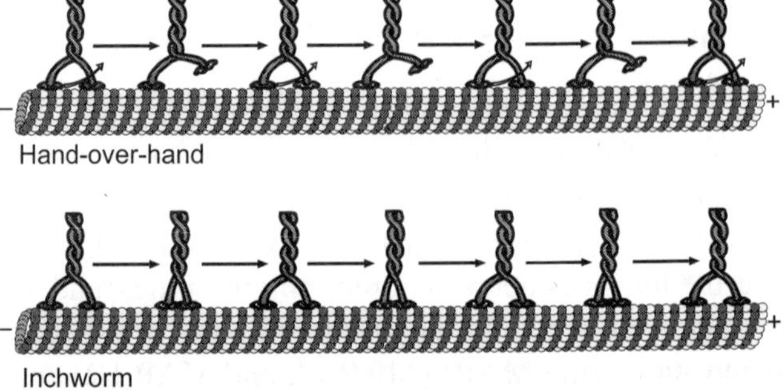

Fig. 14.36 A kinesin molecule moving along a microtubular track. Two alternate models (hand-over-hand and inchworm) are shown.

In the **hand-over-hand** mechanism, the kinesin heads step over one another, alternating the lead position. In the **inchworm** mechanism, one kinesin head always leads, moving forward a step before the trailing head catches up. Despite some controversy, a majority of the studies favour the hand-over-hand mechanism.

The other microtubule-associated motor protein, cytoplasmic dynein, is an extremely large molecule (up to 2000 kDa), which consists of two or three heavy chains (each about 500 kDa) complexed with a variable number of light and intermediate chains, which range from 14 to 120 kDa(**Fig. 14.37**). The heavy chains form globular ATP-binding motor domains that are responsible for movement along microtubules. The basal portion of the molecule, including the light and intermediate chains, is thought to bind to other subcellular structures, such as organelles and vesicles. In many situations, cytoplasmic dynein acts along with a protein called **dynactin** to move cargo over long distances along the microtubules. There are at least two types of cytoplasmic dyneins and several types of axonemal dyneins. In contrast to kinesins, dyneins transport cargo towards the (–) end (**retrograde transport**).

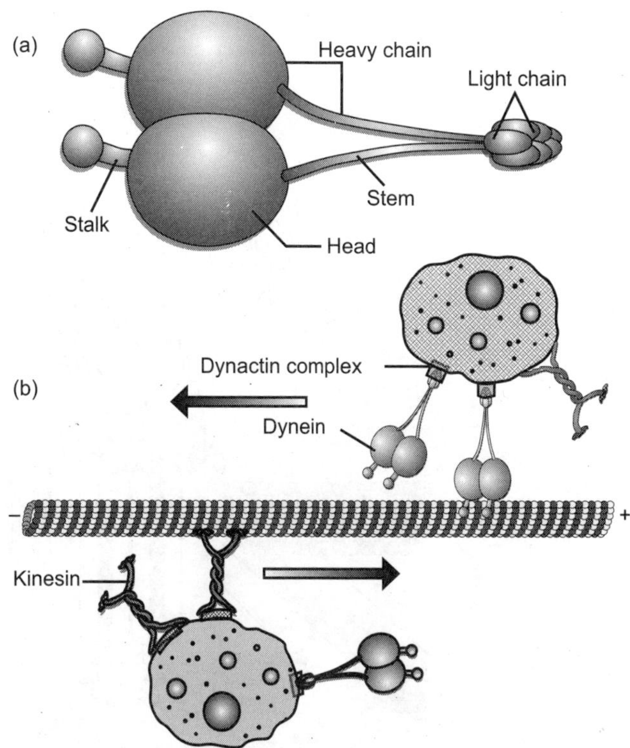

Fig. 14.37 Dynein: (a) Structure of a conventional dynein molecule; (b) Schematic diagram of two vesicles moving in opposite directions along the same microtubule, one powered by kinesin moving towards the plus end of the track and the other by cytoplasmic dynein moving towards the minus end of the track.

Functions of microtubules

A number of functions have been attributed to microtubules. One of the major roles of microtubules is to transport macromolecules, membrane vesicles and organelles through the cytoplasm of eukaryotic cells.

Shaping of the cell and mechanical support: The orientation and distribution of the microtubules in a cell are responsible for shaping some processes arising from the cell. They also provide the framework that determines the shape of the cell. Microtubules are commonly seen to be lined up parallel with and close to cell surfaces and in general conform to whatever particular shape a cell may exhibit. This arrangement is particularly evident in flattened cells, such as the erythrocytes of lower vertebrates, or in cells with highly elongated processes such as neurons, or the heliozoan protozoans.

Morphogenesis: Microtubules play an important role in shaping of the cell during differentiation. For example, the elongation of the nucleus of a spermatid during spermiogenesis is accompanied by the production of orderly array of microtubules, which wrap around the nucleus in double helical manner. In plant cells, cytoskeletons play an important role in defining the direction of cell growth. The direction in which a growing plant cell extends is basically determined by the direction in which the cellulose microfilaments run. However, these are laid down by cellulose synthase, a multiprotein transmembrane enzyme complex that spans the cell membrane and synthesizes cellulose from UDP-glucose in the cytosol. The growing cellulose strand remains bound to the enzyme as it is synthesized and is translocated across the plasma membrane to the exterior of the cell through a pore created by multiple enzyme subunits. In expanding cells, the newly synthesized cellulose microfibrils are deposited at right angles to the direction of cell elongation—an orientation that is thought to play an important role in determining the direction of further cell expansion (**Fig. 14.38**). The cortical microtubules underlying the plasma membrane define the orientation of newly synthesized cellulose microfibrils, possibly by determining the direction of movement of the cellulose synthase complexes in the membrane. The cortical microtubules thus define the direction of cell wall growth, which

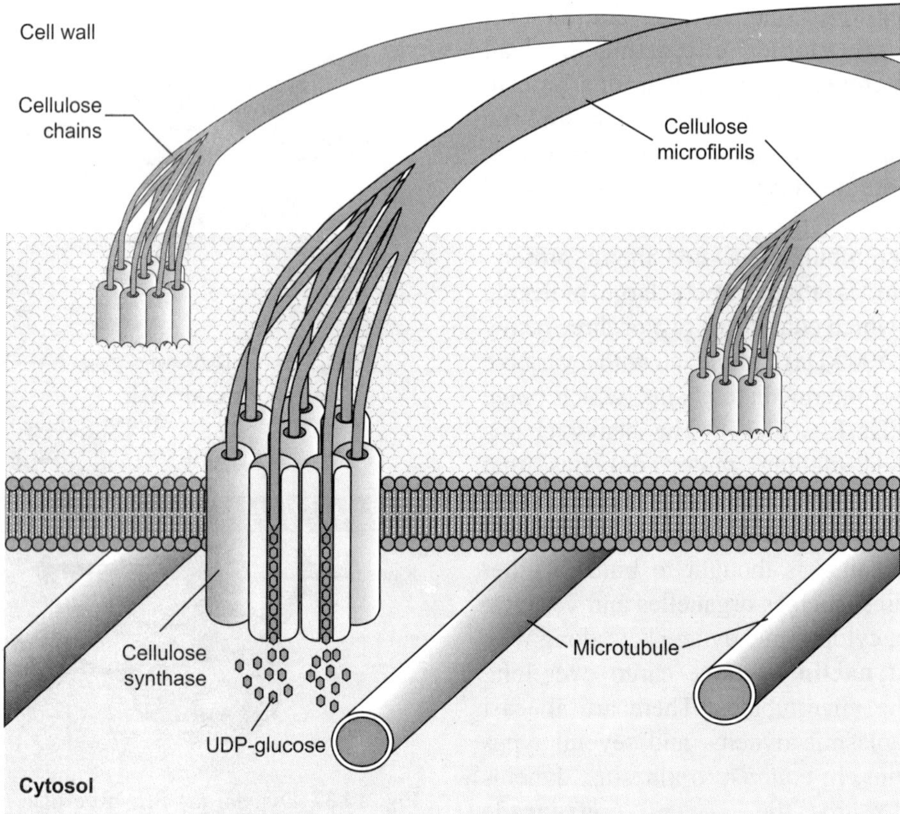

Fig. 14.38 Cellulose synthesis during cell elongation.

in turn determines the direction of cell expansion and ultimately the shape of the entire plant.

Cellular motility

Microtubules have also been implicated in the movement of cells. Cells move for a variety of reasons. Millions of human spermatozoa swim towards an ovum. The soil amoeba (*Acanthamoeba*) crawls over and between soil particles, engulfing bacteria and small organic particles. The invasive properties of some cancer cells is due to their reverting to this highly motile embryonic state. The microtubules are components of more complex structures such as cilia and flagella. Another interesting example is chromosomal movement during cell division. The microtubular network of the interphase cell disassembles and then re-assembles into mitotic spindles, which help in chromosomal movement toward their respective poles.

Cilia and flagella: Cilia and flagella are microtubule-based projections of the plasma membrane that are responsible for the movement of a variety of eukaryotic cells. Eukaryotic cilia and flagella are very similar structures (**Fig. 14.39**), each with a diameter of approximately 0.25 μm. They are composed of three major parts: a **central axoneme** or shaft, the surrounding **plasma membrane** (or **ciliary membrane**) and the interposed cytoplasmic **matrix**. Though the ciliary membrane is physically continuous with the plasma membrane of a cell, it contains a very low amount of proteins and some proteins which are very specific to it, act as a barrier to loss of ATP and certain essential ions. This membrane also has unusual features, like multiple strands (2–11) of particles, called **ciliary necklace** at the base of the organelle. The ciliary necklace is found at a region in the cilium where microtubules and basal bodies make contact with the membrane. This structure may help to position the underlying basal body from which the cilium originates and also in the differentiation of ciliary membrane. The bounded space of the cilium contains a fluid substance known as **matrix**, in which 11 microtubules of axoneme and other interconnecting proteins are embedded. The basic microtubular structure of cilia and flagella is called **axoneme**. It is about 0.2–10 μm in diameter and may range from a few microns to 1–2 mm in length.

Other associated structures of the doublets are **dynein arms**, **nexin links** and **radial spokes**. From each a subfibre, two dynein arms, an outer arm and

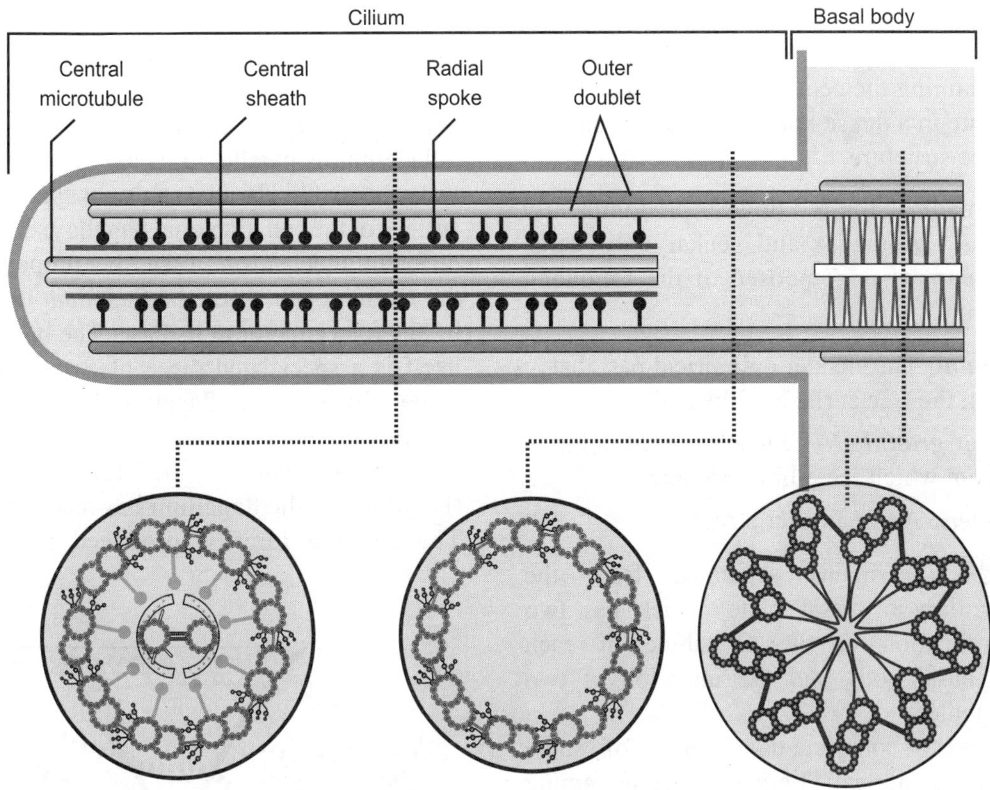

Fig. 14.39 Schematic diagram of the structural relationship between the microtubules of the basal body and the axoneme of a cilium or flagellum.

an inner arm extend in the same direction. This orientation is clockwise when the axoneme is viewed from the base end. Dynein is a large protein complex (about $2x10^6$ Da) composed of 9–10 polypeptide chains, the largest of which are about 450 kDa. Dynein is Mg^{2+}- and Ca^{2+}-activated ATPase enzyme which after solubilization can recombine at the same position on the A microtubule. Dynein contains two or three elliptical or globular **heads** linked to a common root, foot or base by the thin flexible strands or stalks (**Fig. 14.40**). Thus, the base of the dynein molecule attaches only to A subfibre, leaving the heads free to make contact with the adjacent B tubules of the neighbouring doublet. Since the B tubule has a different structure, the base of dynein cannot attach to it. The resulting asymmetry is required to prevent a fruitless tug-of-war between the neighbouring microtubules. Adjacent doublets are linked by peripheral, **interdoublet** or **Nexin links**. The nexin links have a periodicity of 86 nm and extend from A tubule of one doublet to B tubule of the adjacent doublet. Nexin proteins (150–160 kDa) are highly elastic and stretch from 30 nm to 250 nm without breaking. They maintain integrity of the axoneme

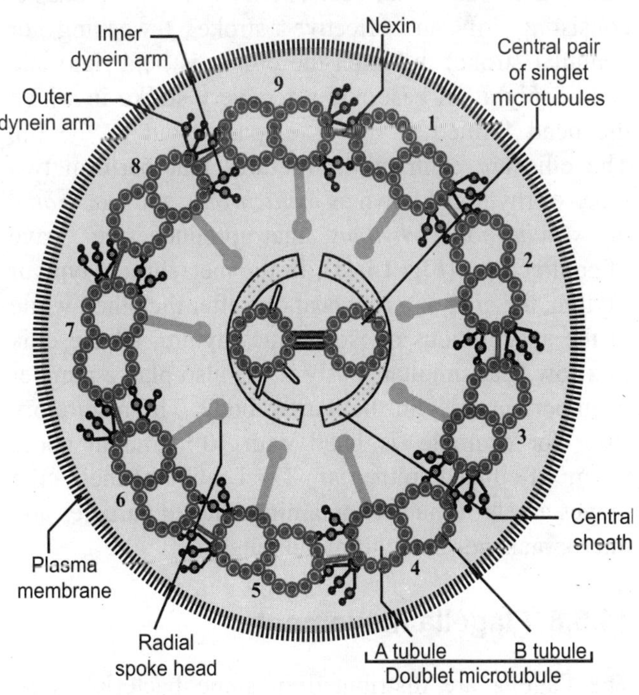

Fig. 14.40 Cross section of an axoneme of a typical cilium or flagellum.

during the sliding motion. There are 36 nm long **radial bridges** or **spokes** between the A subfibre and the sheath containing the central microtubules. These spokes terminate in a dense knob or head, which may have a fork-like structure.

Ciliary movement: Cilia are present predominantly in prokaryotes, protozoa and eukaryotic cells. The ciliary apparatus is composed of the following basic components:

- ***Shaft or cilium:*** The slender cylindrical part that projects from the free surface of the cell.
- ***Basal body* or *granule:*** A centriole-like cellular organelle from which the cilium originates.
- ***Ciliary rootlets:*** A fine fibrilar structure.

The basal body remains separated from the cilium by a ciliary or basal plate, which has two functions: termination of the C tubule of each triplet of the basal body and the initiation of two central microtubules.

Cilia are very conspicuous features of some protists. Heavy they are called ciliates. The swimming actum of a paramoecium, for example, is generated by the coordinated motion of several thousand cilia on the cell surface. Cilia are distributed uniformly throughout the cell surface. The movement is biphasic, consisting of an effective stroke (sweeping or pendular stroke) in which the cilium is held rigid and bends only at its base and a recovery stroke in which the bend formed at the base, passes out to the tip. The cilia are contractile structures that perform two types of rhythms known as *metachronic* and *isochronic* or *synchronous rhythms* that produce the wave of contractions (**Fig. 14.41**). In the metachronic type of rhythm, the cilia of a row beat one after the other, while in the synchronous or isochronic rhythm, all the cilia of a row beat simultaneously. Cilia also play a number of important roles in the human body. The respiratory tract, for example, is lined with 10^{12} cilia in about 0.5 m^2 of ciliated epithelium. The beating of these cilia moves a belt of mucus-containing inhaled particles and microorganisms away from the lungs.

14.6.8 Flagella movement

The flagella are distributed in some bacterial cells, protozoa of the class *Flagellata*, choanocyte cells of the sponges, spermatozoa of the *Metazoa* and among plants in the algae and gamete cells. Flagella are less (1–2) in number and wriggle like eels (**Fig. 14.42**) and are generally found at one end of the cell. They usually beat independently and generate waves that pass along their length, usually from the base to the tip at constant amplitude. Thus, the movement of water by a flagellum is parallel to its axis while a cilium moves water perpendicular to its axis, i.e., perpendicular to the surface of the cell. Bacterial flagella use a fundamentally different mechanism. Like the propeller of a boat, the motion of the bacterial flagellum is entirely driven by the rotary motor at its base. The bacterial flagellum itself is a specialized piece of extracellular cell wall, made of one protein (flagellin) that has no similarity to tubulin or dynein.

The mechanism of force and movement (bending) by the flagellum has recently been studied. The bending force is produced by the sliding of

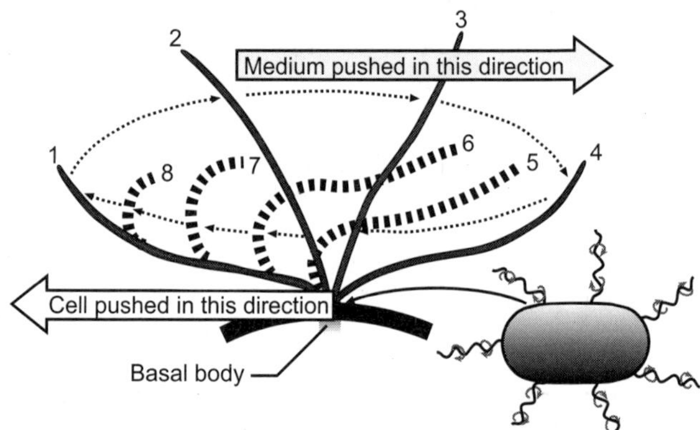

Fig. 14.41 The movement of cilium: Effect of direction of cilium rotation (arrows).

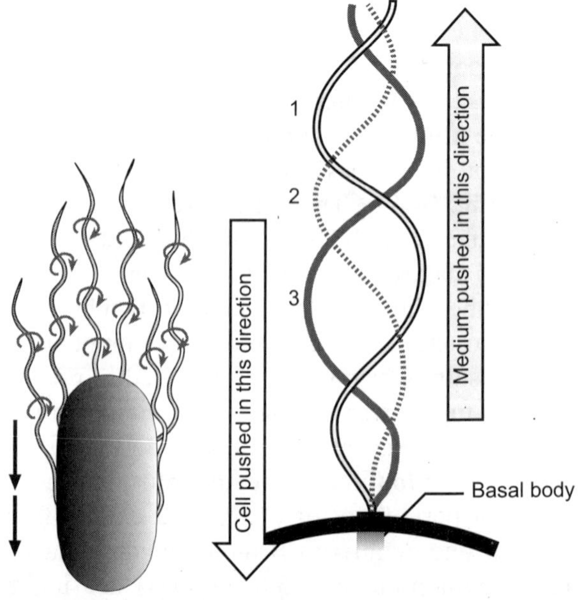

Fig. 14.42 The movement of flagellum: Effect of direction of flagellar rotation.

microtubles. The adjacent peripheral doublets can actively slide against each other once they have been freed of their lateral cross-links (such as nexin). In intact structure this sliding movement is converted to bending. Further, since the adjacent outer doublets actively slide against each other, a force must be generated between them. This force is apparently generated by dynein arms, which 'walk' along the doublets.

Intracellular transport

Other than the beating of cilia and flagella, the microtubules are also associated with the transport of macromolecules from one part of the cell to another (**Fig. 14.43**). The motility is a general property of microtubules within cells. The pseudopodia of *Actinosphaerium* (a protozoan) contain about 500 microtubules disposed in a helical configuration. The cytoplasmic particles migrate back and forth in these pseudopodia. This kind of movement is also seen in some specialized cell types such as the pigment cells called **chromatophores** in the skin of fish and amphibia. The inward and outward movement of pigment granules along radial arrays of microtubules underlies the remarkable colour changes such animals are able to display. The pigment granules found in the erythrophores of fish scales move at a speed of 25–30 μm per second between the microtubules.

The intracellular transport assisted by microtubules can be better explained in the case of a nerve cell axon. The axons extend up to 1 m from the cell body, in which organelles, small vesicles and even mRNA are transported in both directions in a phenomenon referred to as **axonal transport** (**Fig. 14.44**). The outward movement is called **anterograde** transport and the inward movement is called **retrograde** transport. Both are dependent on microtubules that

Fig. 14.43 Diagrammatic illustration of kinesin- and dynein-mediated transport of various components in a non-polarized cell. ER: Endoplasmic reticulum.

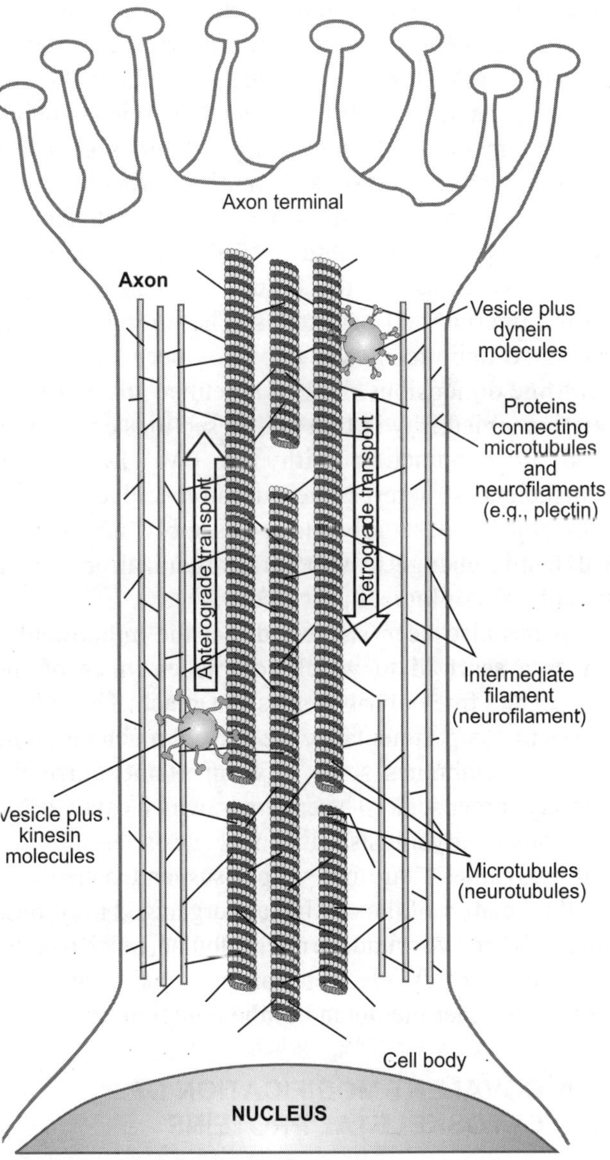

Fig. 14.44 Schematic drawing of anterograde and retrograde axonal transport.

are abundant in nerve cells. The axonal transport are dependent upon molecular motors (**dynein** and **kinesin**) associated with the microtubules. **Dynein** moves vesicles and organelles in the retrograde direction (**Fig. 14.45**) while **kinesin** movement along microtubules is in the anterograde direction (towards the minus end). But the dynein moves along the microtubule in a plus to minus direction. The specificity of transporting a specific cargo is imparted by having multiple dyneins and kinesins in a single cell.

14.3 CYTOPLASMIC MICROTRABECULAR SYSTEM (LATTICE)

Keith Porter (1976) reported the presence of a new cytoskeletal network, termed as **microtrabecular lattice**. This network contains microtrabeculae ranging in diameter 3–10 nm, extending throughout the cell and associated with all the organelles, microtubules, microfilaments and intermediate filaments. These filaments branch, fuse and ramify throughout the cytoplasm. In association with other cytoskeletal elements these constitute a 'form-fitting support', or cytoskeleton, for the outer cell membrane and all the internal cell organelles. The microtrabeculae remained unnoticed over the years because of their very fine dimensions, lack of a defined structure and relative invisibility in thin plastic sections commonly used in conjunction with the EM. Like other cytoskeletal structures, the microtrabeculae are also subject to dis-assembly and re-assembly, undergoing a dynamic change in structural organization during specific physiological processes.

It has also been suggested that the microtubulae act as a scaffold to which soluble enzymes of the cytosol and free polyribosomes can attach. This lattice serves to compartmentalize the cytosol much the same way as membranes serve to compartamentalize the cell on a larger scale. The enzymes, which catalyze the reactions of glycolysis, gluconeogenesis, amino acid and biosynthesis, are not simply suspended freely in the fluid matrix of the cell but are organized into some type of 3D array around the microtubulae, much like the components of the electron transport chain organized within the inner membrane of the mitochondrion.

14.4 COVALENT MODIFICATION OF CYTOSKELETAL PROTEINS

A large number of cytoskeletal proteins undergo covalent modification after translocation.

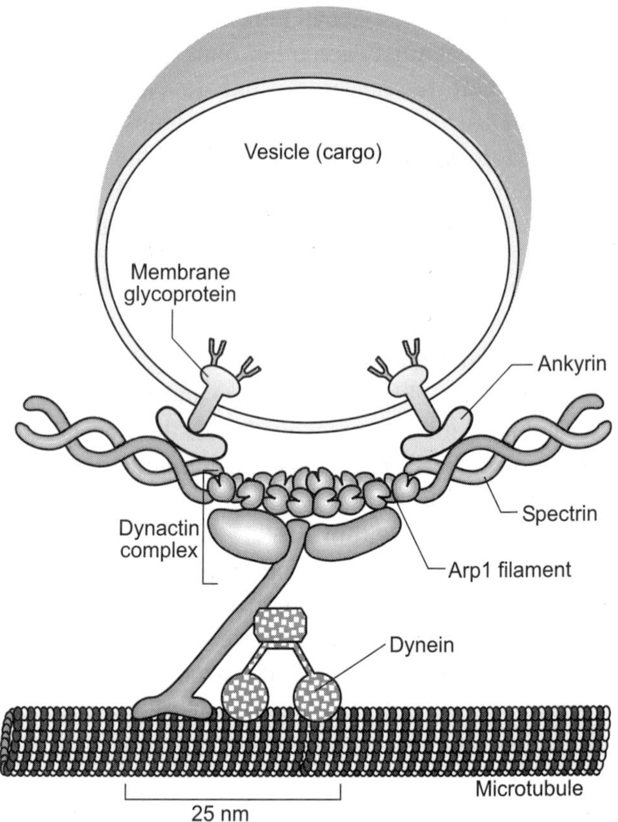

Fig. 14.45 Model for the attachment of dynein to a membrane-enclosed organelle.

Phosphorylation is one of the most common modifications. Proteins undergoing phosphorylation may be heavy and light chains of myosin, filamin, fibronectin, vinculin, β-tubulin, vinebtin, desmin, α- and β-tropomyosine and spectrin. The majority of these proteins are phosphorylated on serine and to a lesser extent on threonine. Vinculin, filamin and vimentin are the only cytoskeletal proteins showing phosphorylation on tyrosine. Some of the functions of phosphorylation are:

- Non-muscle myosin light chains are activated by phosphorylation.
- Myosin heavy chains dis-assemble with phosphorylation.
- Many microtubule-associated proteins are more extensively associated with microtubules, when phosphorylated.

In addition to phosphorylation, a number of other covalent modifications have been detected, which include:

- ***Tubulin***—post-translationally tyrosinlated and may by glycosylated.
- ***Actin***— shows two types of covalent modifications
  - contains *N*-methyl histidine and *N*-terminal acetyl group. The methyl histidine is shared with myosin.
  - contains trimethyllysine groups, e.g., calmodulin and *Acanthamoeba*.
- ***Acyl group***—which may be a 16-carbon side chain; present in vinculin to varying extents.

14.5 CYTOSKELETAL ARCHITECTURE

The fundamental design of the cytoskeleton is very complex and finds expression in forms specific to species and tissues. However, the core structural proteins in general are highly conserved, but specific modifications of the accessory proteins have permitted finetuning of the cytoskeleton for the functions of each cell type. Thus, most cells are characterized by a particular spectrum of cytoskeletal proteins present in a particular stoichiometry and arrangement.

The cytoskeleton confers upon the cell:

- shape
- ability to attach to substrates or to other cells
- freedom of movement
- capacity to move materials within the cell
- capacity to export materials from the cell.

14.5.1 Red blood cells

Mammalian red blood cells exhibit the simplest pattern of cytoskeletal architecture. It consists of a sub-membranous mesh of actin and spectrin. The actin is predominantly in the *F*-form since it contains ADP rather than ATP. The spectrin contains multiple binding sites. The head to tail tetramers of spectrin can cross-link the oligomers of actin by binding to the sides of the actin proto-filaments. The spectrin molecules also contain on the β-spectrin subunit a binding site for ankyrin. Ankyrin binds to both spectrin and integral membrane protein, Band 3, to form a bridge between the actin–spectrin mesh and the membrane. The actin spectrin associations are stabilized by yet another protein, Band 4.1, which binds near the end of the spectrin molecule occupied by the side of the actin filament.

Non-mammalian RBCs contain a peripheral band of microtubules called the marginal band. This bundle appears to be cross-linked and is organized around a pair of centrioles.

14.5.2 Platelet activation

Platelets are tiny cells without a nucleus that circulate in the blood and help to form clots at the sites of injury. Upon activation the platelets can change shape and attach to the surfaces. The resting platelet is a discoidal, symmetrical cellular fragment and it contains short actin filaments capped by CapZ, surrounded by a large pool of actin monomer bound to profilin. Activation causes numerous filopodia to extend from surface followed by 'spread' form, which spreads on the surface. The discoid form contains a marginal band of microtubules cross-linked by microtubule-associated proteins. In addition, actin and numerous actin-associating proteins are also found, which undergo coordinated rearrangements during filopodial extension and spreading. Microtubules are found in the marginal band, which undergoes distortion and rearrangements during activation.

The process of platelet activation shows how a cell can regulate its actin accessory proteins that mediate severing, capping and cross-linking to generate rapid and dramatic morphological changes. When the platelet is activated by physical contact with the edge of a damaged blood vessel or chemical clotting signal (e.g., thrombin), a rapid intracellular signal transduction cascade results in a massive influx of Ca^{2+} into the platelet cytosol. The Ca^{2+} activates gelsolin, which cleaves the capped filaments into tiny fragments, each now capped by gelsolin. A rise in PIP2 levels, inactivates both gelsolin and CapZ, removing them from the plus ends. The free plus ends are then rapidly elongated by the monomeric actin pool, forming many long filaments. Some of them are cross-linked into a gel by filamin, while others are bundled by α-actinin and fimbrin. This causes the activated platelet to extend lamellipodia and filopodia and to spread itself across the clot, attaching to the clot by transmembrane adhesion proteins called integrins. Once the PIP2 signal subsides, the CapZ returns to the ends of the filaments, rendering them stable against depolymerization and locking the platelet into its spread form. Finally, myosin II uses ATP hydrolysis to slide the long actin filaments relative to one another, causing a contraction of the platelet that pulls the edges of the wound together. The distribution of a number of proteins during various stages is depicted in **Table 14.4**.

Table 14.4 Platelet cytoskeletons

	Discoidal		Filopodial			Spread		
Region	1	2	1	2	3	1	2	3
Tubulin	++	–	–	–	+	±	–	Tangles
Actin	Granular	Granular	++	++	±	+	++	±
Myosin	Granular	Granular	–	+	Granular	±		–
α-Actinin	Granular	Granular	–	+	±	++		±
Filamin	Granular	Granular	+	+	±	±		+
Tropomyosin	Diffuse	Diffuse	++	++	–	±		+

14.5.3 Fibroblasts

Fibroblasts perform many functions. They:

- secrete extracellular matrix and contribute to tissue compaction,
- help in wound healing,
- undergo numerous movements during embryogenesis and give rise to mesenchymal derivatives.

Because of varied roles, the fibroblasts display complex cytoskeletal organisation. When the fibroblast responds to contact with a suitable substrate, numerous filopodia extend. If fibroblasts make contact with a non-adherent particle, frequently the filopodium will attach to the particle and retract with it. As soon as a sufficient number of contacts have been made, the cell begins to ruffle around its periphery; the ruffling and filopodial extensions may convert into each other. At this stage, the actin is found extensively associated with the filopodial ruffles and in a number of transnuclear cables. As the cell continues to spread, these transnuclear cables reorganize and begin to be distributed in the interior parts of the cell as a network of intersecting polygons. Over a period of hours these polygonal networks continue to be reorganized into stress fibres, until the characteristic interphase appearance of the cell is attained.

14.5.4 Epithelial cells

Epithelial cells cover the body surfaces, line the gut and form diverse glands within the body. When the epithelial cells form contacts with substratum, focal contacts are produced which contain α-actinin. Tropomyosin and α-actinin are more closely spaced than in fibroblasts; myosin is found periodically among the stress fibres. These cells contain extensive micronuclear networks, which frequently connect with subcellular organelles. Intermediate filament network is most diverse and highly specialized in epithelia. The intermediate filaments are composed of cytokeratin proteins, which are of at least seven major classes. The cytokeratin fibres bundle and anastamose in extended arrays throughout the cytoplasm and terminate at the desmosome.

14.6 EXTRACELLULAR SIGNALS

Other than the accessory proteins associated with cytoskeletal filament systems that can regulate filament length, location, organisation and dynamic behaviour, extracellular signals can also alter the activity of these accessory proteins, thereby changing cytoskeleton and cell behaviour. For the microfilaments, global structural rearrangements in response to external signals are triggered through diverse cell surface receptors. All these signals seem to converge inside the cell on a group of closely related monomeric GTPases that are members of the **Rho protein family**– *Cdc42*, *Rac* and *Rho*. Rho factors can act as molecular switches to control processes by cycling between an active, GTP-bound state and an inactive, GDP-bound state.

14.7 PLANTS

Plant cells contain both actin filaments and microtubules. But the evidence for intermediate filaments in plant cells remains inconclusive. The role of actin in plants is known to associate with cytoplasmic streaming, mediated through cables anchored in the plasma membrane. These cables of actin consist of bundles of microfilaments all with the same polity, i.e., with the barbed end anchored in the membrane.

Summary

- The cytoplasm in cells contains an intricate supportive network of proteinaceous fibres and filaments called the ***cytoskeleton***. It provides a structural framework for the cell, serving as a scaffold that determines cell shape and the general organization of the cytoplasm.
- The main proteins that are present in the eukaryotic cytoskeleton are ***microtubules***, ***actin*** and ***intermediate filaments***. In contrast, the simple bacterial cytoskeleton contains the ***FtsZ protein***, which participates in cell division and an ***MreB protein***, which is involved mainly in the regulation of the width of rod-shaped bacteria.
- Cytoskeletal filaments are organized into bundles and networks in the inner surface of a cell. The simplest cytoskeleton in eukaryotes is the network of actin filaments (microfilaments), which have a diameter of approximately 7 nm and up to several micrometers in length.
- The ***microfilaments*** also contain some other proteins, which are designated as ***actin-binding proteins***. Microfilaments serve a number of cellular functions: given mechanical strength, link transmembrane proteins to cytoplasmic proteins, anchor the centrosomes at poles during mitosis, form major component of contractile machinery, generate locomotion in cells and play a major role in muscle contraction. They are also involved in many other functions such as nerve outgrowth, tubular gland formation, movement of intestinal microvilli, generation of cytoplasmic streaming in some cells and gastrulation.
- The ***intermediate filaments*** (IFs) are the main structural components of cytoskeletal proteins. IFs are found in all animals but not in plants and fungi. In the epithelium, for example, IFs provide mechanical support for the plasma membrane, but in epidermal cells and the axons of neurons, intermediate filaments are at least 10 times more compared to other cytoskeletal proteins.
- IFs are long, linear and unbranched filaments having a diameter between 8–11 nm, which is intermediate between that of microtubules and microfilaments. IFs tend to be more or less permanent structures in tissues such as skin and hair; in fact, in these non-living tissues IF proteins are almost the only proteins. IFs are formed by protein subunits of a multigene class that are expressed differentially in various cell types.
- Recently, existence of various kinds of proteins that have functions similar to cytoskeletal proteins of eukaryotes have also been discovered in bacteria.
- Microtubules possess a combination of physical properties that allows them to participate in multiple cellular functions. They can form bundles of rigid fibres and serve an important role in the determination of cell shape, movements, transport of vesicles in the cytoplasm and other activities.
- Microtubules are small, hollow cylindrical structures called tubules. Typically, a tubule has an outer diameter of approximately 25 nm and lumen diameter of 15 nm. In addition to contributing to cell motility, microtubules play a major role in organizing the cell through a special structure called the microtubule-organising centre (MTOC).
- *The* ***centrosome*** is a dense region close to the nucleus of the cytoplasm of animal cells, in which two free floating, short, cylindrical and rod-shaped tubular structures called ***centrioles*** are present. The functions of centrioles and basal bodies are diverse.
- A large number of other proteins, termed ***microtubule-associated proteins*** (MAPs) have also recently been identified at the surface of the microtubule and present as regularly spaced projections.
- The major role of microtubules is to transport macromolecules, membrane vesicles and organelles through the cytoplasm of eukaryotic cells; they play an important role in shaping the cell during differentiation, in the movement of cells and in transportation.
- Another major role of microtubules is to transport macromolecules, membrane vesicles and organelles through the cytoplasm of eukaryotic cells. They are also important in shaping the cell during differentiation, and are implicated in the movement of cells (transportation).
- A cytoskeletal network, termed as ***microtrabecular lattice***, whose diameter ranges 3–10 nm, extends throughout the cell and is associated with all the organelles, microtubules, microfilaments and intermediate filaments.
- Large numbers of cytoskeletal proteins undergo covalent modification after translocation. ***Phosphorylation*** is one of the commonest modifications and a number of other covalent modifications have also been detected.

Exercises

1. What are the three cytoskeletal filaments present in the eukaryotic cells?
2. Describe the structure and composition of microfilaments.
3. Briefly explain the different roles of actin filaments in the cell.
4. Explain with diagrams the role of actin filaments in muscle contraction.
5. How do microfilaments act in cell locomotion?
6. Describe the structure and organization of intermediate filaments.
7. What are the different types of intermediate filaments present in the cell?
8. Explain the biological role of intermediate filaments.
9. Describe in detail the structure and organization of microtubules.
10. Describe the structure and role of the centriole in animal cells.
11. Explain with a diagram the organization of microtubules in cilia and flagella.
12. What are proteins and how do they function in cellular activity?
13. Describe the cytoskeletal architecture of red blood cells.
14. How are actin filaments organized in fibroblast cells?
15. Explain the role of different cytoskeletal filaments in plant cell functioning.

Further Reading

Books

Alberts B, Johnson A, Lewis J, Raff M, Roberts K and Walter P (2002), *Molecular Biology of the Cell*, 4th edition, Garland Science, Taylor and Francis Group.

Bray D (2001), *Cell movements*, Garland.

Darnell J, Lodish H and Baltimore D (1986), *Molecular Cell Biology*, Scientific American Books, New York.

Duston P (1978), *Microtubules*, Springer-Verlag, Berlin.

Howard J (2001), *The Mechanics of Motor Proteins and the Cytoskeleton*, Sinauer Associates Inc.

Inoue S and Stephens RE (1975), *Molecules and Cell Movement*, Raven Press, New York.

Kreis T and Vale R (1999), *Guidebook to the Cytoskeletal and Motor Proteins*, Oxford University Press.

Lloyd CW (1982), *The Plant Cytoskeleton*, Academic Press, New York.

Fulton AB (1984), *The Cytoskeleton: Cellular Architecture and Choreography*, Chapman and Hall, UK.

References

Berg JS, Powell BC and Cheney RE (2001), A millennial myosin census, *Mol. Biol. Cell*, 12:780–94.

Borisy GG and Svitkina TM (2000), Actin machinery: Pushing the envelope, *Curr. Opin. Cell Biol.*, 12:104–12.

Bresnick AR (1999), Molecular mechanisms of non-muscle myosin: II. Regulation, *Curr. Opin. Cell Biol.*, 11: 26–33.

Buss F, Luzio J and Kendrick-Jones J (2002), Myosin VI, an actin motor for membrane traffic and cell migration, *Traffic,* 3:851–58.

Channarayappa (1990), Cytogenetic analysis of clastogens and aneuploidogens by immunofluorescent and micronucleus methods. PhD thesis submitted to the West Virginia University, Morgantown, West Virginia, USA.

Cleveland DW (1982), Treadmilling of tubulin and actin, *Cell,* 28:689–91.

Cooper JA and Schafer DA (2000), Control of actin assembly and disassembly at filament ends, *Curr. Opin. Cell Biol.*, 12:97–103.

Condeelis J (2001) How is actin polymerization nucleated in vivo? *Trends Cell Biol.*, 11:288–93.

Furukawa R and Fechheimer R (1997), The structure, function and assembly of actin filament bundles, *Intl. Rev. Cytol.*,175:29–90.

Higgs HN and Pollard TD (2000), Regulation of actin filament network formation through ARP2/3 complex: activation by a diverse array of proteins, *Annu. Rev. Biochem.*, 70:619–76.

McGough A (1998), F actin-binding proteins, *Curr. Opin. Struc. Biol.*, 8:166–76.

Mermall V, Post PL and Mooseker MS (1998), Unconventional myodin in cell movement, membrane traffic and signal transduction, *Science*, 279:527–33.

Pollard TD, Blanchoin L and Mullins RD (2000), Molecular mechanisms controlling actin filament dynamics in nonmuscle cells, *Annu. Rev. Biophys. Biomol. Struc.*, 29:545–76.

Small JV, Strada T, Vignal E and Rottner K (2002), The lamellipodium: where motility begins, *Trends Cell Biol.*, 12:112–20.

Stossel TP et al. (2001), Filamins as integrators of cell mechanics and signalling, *Nature Rev. Cell Biol.*, 2(2):138–45.

Stuurman N, Heins S and Aebi U (1998), Nuclear lamins: their structure, assembly and interactions, *J. Struc. Biol.*, 122:42–66.

15 GROWTH, SEXUAL REPRODUCTION AND AGEING

Learning objectives

After reading this chapter, you will be able to:

- Classify the growth processes
- Understand the importance of and consequences of errors in cell growth
- Analyze the elements of sexual reproduction
- Appreciate the various theories of ageing, especially the most recent telomere length as evidence of ageing

15.1 INTRODUCTION

The questions of development, growth, sex and death arise in almost every area of biology. For example, humans start life as a single cell—the zygote; nearly eight months later, the child is born. Many astonishing changes take place in that time; it becomes much larger or **growth** has taken place. A single-celled zygote divides, by mitosis and this repeats many times. A zygote, which looks more or less spherical, **develops** obvious polarity and patterns of symmetry. Head (anterior) and tailposterior) ends are distinct, back (dorsal) and front (ventral) surfaces are clear and there is left–right (bilateral) symmetry. Many cell types form by the process of **differentiation**. They are organized into tissues and organs, fitting together in a tightly organized whole baby (development). After birth, there are fewer developmental changes, but growth continues for some years and consequently existing structures get bigger.

Every organism enters pubescence phase, wherein it becomes sexually active to reproduce. Most multicellular organisms have evolved a complex **sexual reproduction** mechanism that helps an organism to mix the genetic material from male and female and generate genetic diversity while maintaining the genomic integrity and phenotypic identity. Even though the process of ageing of an organism begins from the stage of zygote and lasts until death, ageing from late adolescence (often from the onset of sexual reproduction) is most significant (this stage is also called **senescence**). Senescence indicates decline in organismal metabolic activity and reaching the end of the life cycle, that is **death**. Death is the last stage of an organisms' somatic life cycle.

In this chapter, the process of growth, sexual reproduction and ageing are discussed briefly.

15.2 GROWTH

Growth is a fundamental characteristic of all living organisms. It is simply defined as an increase in the overall size of an organism—in terms of number, volume, weight or change in the function of cells. Growth is a quantitative character controlled by many genes. It may also include increase in the amount of **apoplasmatic substances,** such as the fibres and matrix of connective tissues. Thus, growth of an organism includes increase in both protoplasmic and apoplasmatic mass. The process of growth is an increase in body size, whereas development is the process by which a body takes on its mature form. However, the process of development is closely linked to growth.

The overall growth of an organism from an individual cell to a multicellular organism can be divided into three phases:

- ***Cell division:*** An increase in the number of cells as a result of mitotic cell division.
- ***Cell enlargement:*** An irreversible increase in cell size as a result of the uptake of water or the synthesis of living material.
- ***Cell differentiation:*** The specialization of cells.

These phases, however, occur at separate points in time and therefore it is difficult to define growth correctly. One widely accepted definition is that ***growth is an irreversible increase in dry mass of living material.*** All stages of growth involve biochemical activity, including DNA replication and RNA and protein synthesis. Enzymes regulate cellular activity and eventually result in a change in overall form and structure (called **morphogenesis**). Growth may be positive or negative. Positive growth occurs when synthesis of materials (**anabolism**) exceeds breakdown of materials (**catabolism**), whereas negative growth occurs when catabolism exceeds anabolism.

15.2.1 Size and proportion

Although **growth** is superficially familiar to all as simply an increase in size and height, it is one of the least understood aspects of development. By the time the general body plan is formed, the whole embryo will be about 1 mm long and each organ is formed as a small rudiment. In a large animal, such as the human, there will be around 10^9-fold increase in volume from this stage to that of the full grown adult. This enormous change is driven largely by cell division, but also by secretion of the extracellular matrix and by some increase in cell size. The intracellular events controlling cell division are now well understood. However, the biochemical mode of action of the growth factors that act on cells from their environment for the overall control of growth is still largely mysterious.

One problem relating to final size is why the animal stops growing at a particular point. The growth of cells in tissue culture is typically exponential in the presence of excess nutrients. However, exponential growth is rarely found in animals. The growth curve for an animal, or a part of an animal, will normally resemble **Fig. 15.1**, with a rate that is rapid at the beginning and tails off to zero as the final size is reached. Various equations have been proposed that give a mathematical description of overall growth; but none is entirely satisfactory because of the large number of individual processes that contribute in some way to the increase of size. A general explanation for the existence of the final size is that the need for nutrients will be limited by the number of terminal capillaries and the number of termini of a branched distribution system cannot increase as fast as the volume of a three-dimensional object. But this insight does not currently explain, for example, the 80-fold range of size of different breeds of domestic dogs.

Another issue is how coordination of growth between parts of the body is established. Differences in growth rate between body parts over a 10^9-fold expansion would lead to very large changes in relative proportions. This suggests that there must exist mechanisms that can sense the overall size of the organism and regulate the growth of individual parts accordingly. In fact, relative proportions do usually change systematically during growth. For example, the human head becomes smaller relative to the body.

Biochemistry of growth

Although cell division is fundamental to growth, it is not the only factor driving the increase in size. *Drosophila* wing discs have been used extenoively for the investigation of growth control because of the ease of overexpressing genes in clones by the FLP-out method, or removing them from clones by mitotic recombination. It is possible to drive a local increase in cell division rate by overexpressing the transcription factor E2F, which activates expression of a number of genes required for the S phase of the cell cycle. Conversely, cell division rate can be slowed

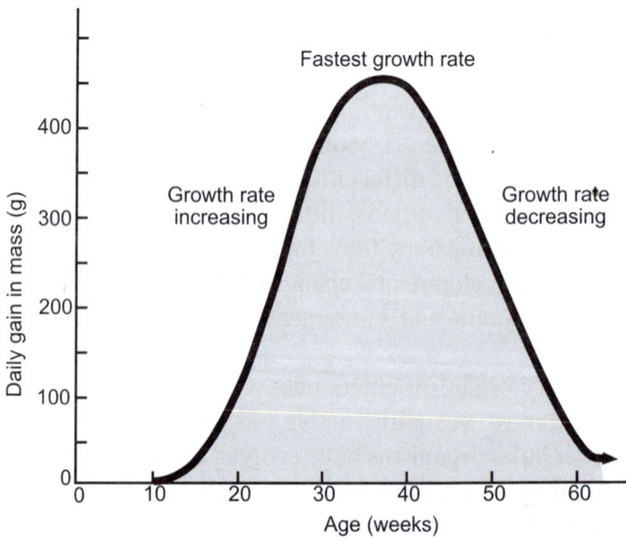

Fig. 15.1 Absolute growth rate curve obtained by plotting daily weight gained by an animal (growth rate) against age. Growth rate varies with age.

by overexpressing the RB protein, which inhibits the endogenous E2F. But neither of these manipulations have much effect on overall size and portion of the wing disc, because the rate of cell size increase does not change. In other words, the faster cell cycle generates a clone containing more of small cells and the slower cell cycle generates a clone containing fewer larg cells. In order to change the rate of growth it is necessary also to change the rate of cell size increase. In normal circumstances, but not necessarily in an experimental situation, the rate of cell division tends to be coupled to cell size such that a change in growth rate leads to altered volume of tissue composed of normal sized cells (**Fig. 15.2**).

Now it is confirmed that the key mediator of cell size is the insulin/IGF signalling system and the associated intracellular TOR (target of rapamycin) system. These are essentially similar in all animals, although the number of parallel components can vary.

The insulin/IGF signalling pathway is important for growth control in mammals. The two insulin-like growth factors, IGF1 and IGF2, are of central importance in growth control. They both operate through a receptor called the 'IGF1 receptor' which is similar to the generic insulin receptor. IGF2 is more important for supporting growth during the foetal period, largely through its effects on the placenta. IGF1 is more important in supporting postnatal growth. It is synthesized in a wide variety of tissues and is under the control of the growth hormone

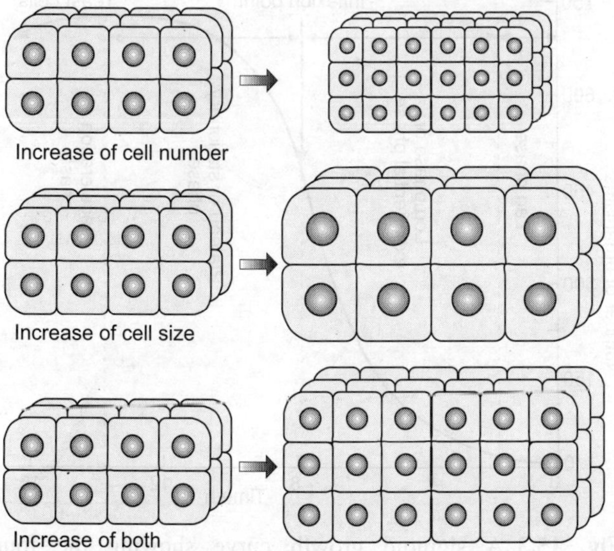

Fig. 15.2 Effect of increasing the cell division rate, the number and volume.

(GH or somatotropin) secreted by the somatotrophs of the anterior pituitary gland. Animals or humans lacking sufficient growth hormone grow and mature slowly and their final size is reduced. Conversely, individuals suffering from a pituitary tumour that secretes excessive growth hormone will grow faster during childhood and adolescence and reach a larger than normal final size (gigantism). Pituitary-deficient animals can be restored to near-normal growth rates by administration of either growth hormone or IGF1. There are normal hereditary differences in these components also. For example, there is a considerable difference of IGF1 levels in the serum of large and small breeds of dog. Mouse knockouts of IGF1 or IGF2 genes are about 60% normal size at birth and the double knockout is only 30% normal size and dies at birth.

Control of relative proportion

What are the factors that influence how different body parts expand in proportion to one another? Or, in cases where there are allometric deviations from proportional growth, how is this controlled? The data related to these questions are limited, although in general there is a tendency for graft to behave in an autonomous way, not adapting to the size of the host, nor affecting the relative growth of corresponding parts within the host. This is actually very surprising because it is hard to think of a mechanism to control relative proportions that does not involve a feedback inhibition of each body part on itself. For example, if each body part produces an inhibitor that is diluted into the total blood volume of the organism, then its concentration will measure the size of the part relative to the size of the whole. If the part gets a bit too big, the inhibitor will slow it down and if the part is too small the inhibitor level will be low, so its growth will accelerate. Such a hypothetical negative feedback signal is called **chalcone**.

Growth in stature

Growth in stature is determined by various factors. It results from the lengthening of bones via cellular divisions and is mainly regulated by human growth hormone (hGH) **somatotropin**, secreted by the anterior pituitary gland. Somatotropin also stimulates the release of another growth-inducing hormone, insulin-like growth factor (IGF1) mainly by the liver. Both hormones operate on most tissues of the body, have many other functions and continue to be secreted throughout life. Peak levels coincide with peak growth velocity and gradually decrease with age after adolescence. The bulk of secretion occurs in bursts (especially for adolescents) with the largest during sleep. Exercise promotes secretion and a positive net nutrition is also important, with proteins and various other nutrients.

The height of the human being, or overall status of a vertebrate animal, depends to a large extent on the length of the long bones in the limbs. These long bones, like most of the embryonic skeleton, are initially formed as a cartilage model, which are replaced by bone in later development. The majority of linear growth occurs as growth of cartilage at the epiphysis (ends) of the long bones which gradually ossify to from hard bone.

Measuring growth

The study of human growth is known as auxology. Growth and height have long been recognized as a measure of the health and wellness of individuals. For individuals, as indicators of health problems, growth trends are tracked for significant deviations and growth is also monitored for significant deficiency from genetic expectations. Genetics is a major factor in determining the height of individuals, though it is far less influential with regard to populations. Average height is increasingly used as a measure of the health and wellness of populations. Height is determined by the complex interactive combination of genetics and environment. Genetic potential plus nutrition minus stressors is a basic formula. Genetically speaking, the heights of mother and son and of father and daughter correlate, suggesting that a short mother will more likely bear a shorter son and tall fathers will have tall daughters. However, the precise relationship between genetics and environment and exact role of genetics itself is complex and uncertain since height is a multigenic trait.

Growth can be measured at various levels of biological organization, such as growth of a cell, organism or population. For example, the number of organisms in a population at different intervals of time can be counted and plotted against time to produce a population growth curve (**Fig. 15.3**). At the level of the organism a variety of parameters such as length, area, volume and mass are commonly used. In the case of growth in animals and plants, length and mass (fresh and dry mass) are two commonly measured parameters. In humans, changes in height and body mass are frequently used as indicators of growth. The process of growth can be explained by constructing a variety of growth curves produced by plotting different parameters such as length, height, mass, surface area, volume and number against time. The shape of these curves is described as **sigmoid**, meaning S-shaped and is typical of much growth. A sigmoid curve can be divided into four parts as shown in **Fig. 15.3**.

Growth abnormalities

Most intrapopulation variance of height is genetic. Generally, short stature and tall stature are usually not a health concern. If the degree of deviation from normal is significant, hereditary short stature is known as **familial short stature** and tall stature is known as **familial tall stature**. There are many diseases and disorders that cause growth abnormalities. Most notably, extreme height may be pathological, such as gigantism resulting from childhood hyperpituitarism and dwarfism which has various causes. Very short persons may be termed as having idiopathic short stature.

In general, height may not be causative of better health and longevity. However, being excessively tall can cause various medical problems, including cardiovascular issues, due to the increased load on the heart to supply the body with blood and issues resulting from the increased time taken for the brain to communicate with the extremities. Height extremes of either excessive tallness or shortness can cause social exclusion and discrimination for both men and women.

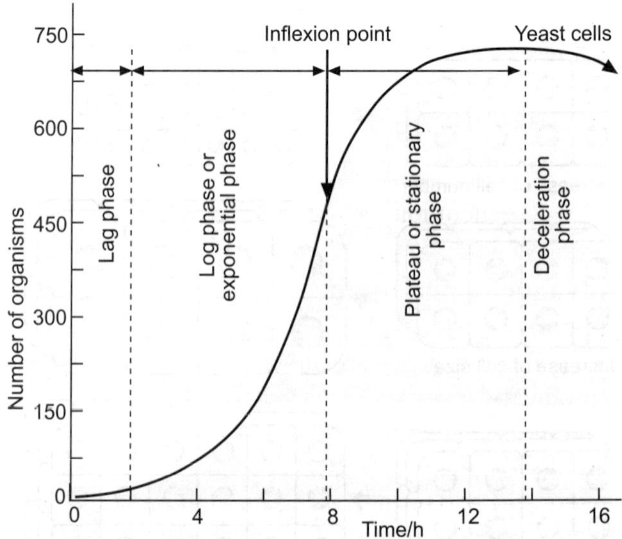

Fig. 15.3 A sigmoid growth curve showing the four characteristic growth phases and the inflexion point for the yeast. The inflexion point is the point of maximum growth rate where the curve is steepest.

15.3 REPRODUCTION

Reproduction is the inherent property of all living organisms to perpetuate and flourish their race. It is a process by which the living beings multiply or duplicate their own kind without any change or with little change. There are a number of different reproductive strategies, all with certain advantages and disadvantages. Basically, there are two principal modes of reproduction: **asexual** and **sexual**.

15.3.1 Asexual reproduction

The development of new individuals without the fusion of the male and female gametes (or genetic recombination) is known as asexual reproduction. The result would be a clone, a population of asexually produced, genetically identical organisms. Members of a clone only differ genetically as a result of random mutation. In case of plants and other lower organisms it is also called **vegetative propagation**. Vegetative propagation is an extension of the capacity of an organism for indeterminate growth. Asexual reproduction usually includes amitotic or mitotic division of the body (somatic cells); therefore, it is also known as **somatogenic** or **blastogenic reproduction**. Asexual reproduction is very common in lower plants and animals.

There are several types of asexual reproduction. Broadly, asexual reproduction may be grouped into the following types:

- Fission
- Budding
- Gemmule formation
- Regeneration.

Advantages and disadvantages of asexual reproduction

The main advantages are only one parent is required to reproduce; the offspring are genetically identical; there is rapid dispersal and spread and rapid multiplication. The disadvantage is that no genetic variation occurs among the offspring. Even though mutations can create some variations, the frequency is very low. If spores are produced, many will fail to find a suitable place and spreading of an organism asexually in the same area may result in overcrowding, exhaustion of nutrients and proneness to parasites and diseases.

15.3.2 Sexual reproduction

In case of single-celled organisms, reproduction is by simple mitotic division and plants can propagate vegetatively. Likewise, in the animal kingdom, a solitary multicellular *Hydra* can produce offspring by budding, and sea anemones can split into two halves; then each develops into a complete organism. Some species of lizards consist only of females, and reproduce without mating. In sexual reproduction, on the other hand, the genomes from two individuals are mixed to produce offspring that differ genetically from one another and from both their parents. The sexual mode of reproduction has many advantages over the asexual.

Significance of sexual reproduction

Sexual reproduction gives a comparative advantage to organisms in an unpredictably variable environment. It is universally credited with contributing to species fitness and generating new genotypes by bringing together the genes from two different individuals, male and the female, in the fertilized egg. The mechanism for generation of new diversity with the potential for evolutionary success generally involves differentiation into sexual forms in which reproduction occurs only as a result of union between sex cells (gametes) of different individuals. In most organisms (plants and many lower animals) male and female determining organs are combined within a single **hermaphroditic** or **monecious** individual, each capable of producing both types of sexual gametes. On the other hand, some plant species and most higher animals consist of separate male and female (**dioecious**) individuals that produce sperm or eggs but not both.

Determination of the sexual phenotype

Sexual reproductive strategies vary enormously between different organisms. For early single-cell eukaryotes reproducing sexually, two cells stick together and swap portions of their DNA in a process called conjugation. The fission of the participating cells then followes. The new products are genetically different from the parent cells prior to conjugation.

The more advanced single-cell eukaryotes like the ciliates have a more elaborate arrangement. It segregates the chromosomes into a micronucleus used exclusively for reproduction purposes and there is a macronucleus containing DNA used to direct the day-to-day operation of the cell. The haploid cells are made in the micronucleus through meiosis. After fertilization, the micronuclei of the two cells fuse, creating a diploid micronucleus, which is the foundation of a new generation.

Once eukaryotes became multicellular organisms, reproductive DNA remain in separate nuclei and are also sequestered in just a few special cells in the body called **germ cells**. Like micronuclei of protozoans, germ cells have only one function: the transmission of DNA from one generation to the next via sexual reproduction. The rest of the cells with identical sets of chromosomal DNA, carry out the body's daily non-reproductive functions. The somatic cells do not exchange or recombine DNA with one another—they do not undergo meiosis or have sexual activities. The only purpose of somatic cells is to optimize the survival and function of the true guardians of the DNA, the germ cells. While the germ cells can renew themselves through fertilization, the somatic cells would keep accumulating errors and mutations until they cannot function any more and die.

Male and female species in case of most plants cannot be differentiated from each other; but the structural characteristics of the flowers they produce help to distinguish between the sexes. In most animals, male and female individuals differ phenotypically for a number of characteristic features. Based on these morphological features, individuals can be grouped into primary and secondary sex characters.

- ***Primary sex*** characters are related to the gonads or the gameteproducing organs of the male and female individuals. Development of primary sex characters is genetically controlled and depends on genes present in the zygote.
- ***Secondary sex*** characters however, determine consistent differences between male and female individuals of a species. These include genital ducts, genitalia, many body characteristics such as, in case of humans, facial hair, mammary gland development, pitch of voice, degree of muscle development and subcutaneous fat deposition. Hormones produced in the gonads regulate development and function of secondary sex characters. The intensity of these characters can often be modified by external administration of human male (testosterone) and female (estrogen) hormones.

Differentiation of gonad: Primary sex determination is strictly chromosomal and is not usually influenced by the environment. Following fertilization, a zygote is destined to develop either as a male or a female. But in all the mammalian and avian species studied to date, gonads in the early stages of embryo development are undifferentiated (neither male nor female). In all vertebrate embryos, certain cells single out early in development as progenitors of the gametes. These *primordial germ cells* migrate to the developing gonads, which will form the *ovaries* in females and the *testes* in males. After a period of mitotic proliferation, the primordial germ cells undergo meiosis and differentiate into mature gametes (eggs or sperm). In most animals, when the egg is fertilized, it divides repeatedly to produce the cells of the early embryo. The cells that inherit specific molecules localized in a particular region of the egg cytoplasm become primordial germ cells. In mammals, by contrast, the egg is more symmetrical and the cells produced by the first few divisions of the fertilized egg are all totipotent. A small group of cells in the early mammalian embryo is induced to become primordial germ cells by signals produced by neighbouring cells. In the next few days, these primordial cells proliferate and sweep back into the embryo proper along with the invaginating hindgut (**Fig. 15.4**). They then actively migrate through the gut to their final destination in the developing gonads. As the primordial germ cells migrate through the embryo, they receive signals to survive, proliferate and migrate by various extracellular proteins produced by adjacent somatic cells. After primordial germ cells enter the developing gonad, which at this stage is called the genital ridge, the continue to proliferate for two or three more days. At this point, they commit to a developmental pathway that will lead them to either eggs or sperm. However, this depends not only on their own sex chromosome constitution, but also on whether the genital ridge has begun to develop into an ovary or a testis, respectively. The sex chromosomes in the somatic cells of the genital ridge determine which type of gonad the ridge becomes.

The developing gonads: The gonadal rudiments, have two normal options. On differentiation, they develop into either an ovary or a testis. But, before differentiation, the mammalian gonad first develops through *bipotential (indifferent) stage,* during which time it has neither female nor male characteristics. In humans, the gonadal rudiments appear in the intermediate mesoderm during week 4 and remain sexually indifferent until week 7. The gonadal rudiments are paired regions of the intermediate

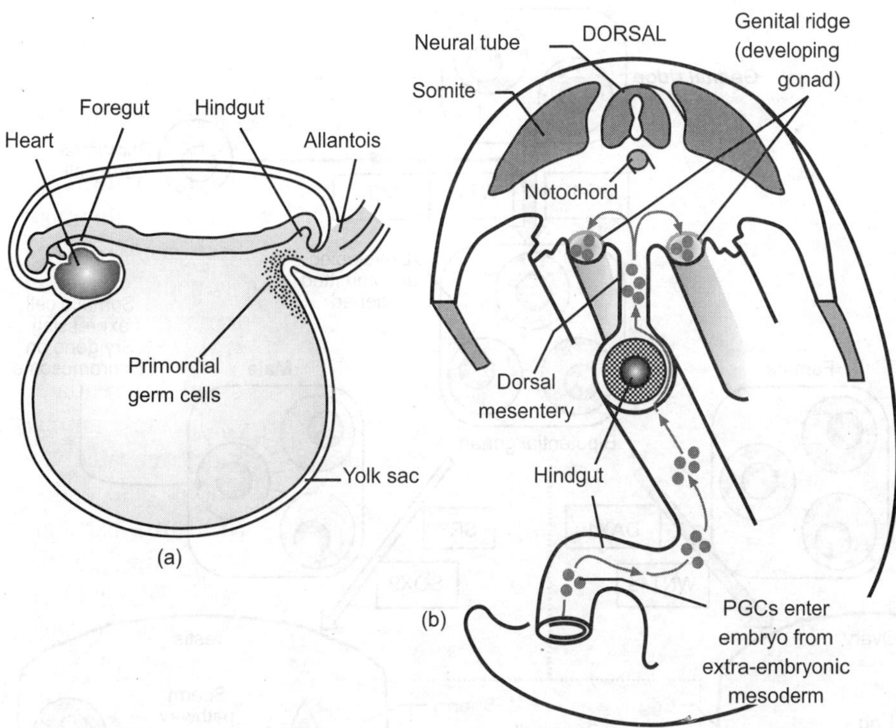

Fig. 15.4 Pathways for migration of mammalian primordial germ cells: (a) PGCs seen in the yolk sac near the junction of the hindgut and allantois. (b) The PGCs migrate through the gut and dorsally, up the dorsal mesentery and into the genital ridges PGC: Primordial germ cells.

mesoderm; they form adjacent to the developing kidney. The ventral portions of the gonadal rudiments are composed of the genital ridge epithelium. It proliferates into the loose connective mesenchymal tissue above it. These epithelial layers form the *sex cords*. The germ cells migrate into the gonad during week 6 and are surrounded by the sex cords. In both XY and XX gonads, the sex cords remain connected to the surface epithelium.

In the male foetus (XY), the sex cords continue to proliferate through the eighth week, extending deeply into the connective tissue. These cords fuse, forming a network of internal (medullary) sex cords and at its most distal end, the thinner *rete testis*. Eventually, the sex cords—now called *testis cords*—lose contact with the surface epithelium and become separated from it by a thick extracellular matrix, the *tunica albuginea*. During foetal life and childhood, the testis cords remain solid. At puberty, however, the cords will hollow out to differentiate into sperms. The cells of seminiferous tubules are called *Sertoli cells*. These cells of the testis cords nurture the sperm and secrete anti-Mullerian duct hormone. The sperms are transported from inside the testis through the rete testis, which joins the *efferent ducts*. These efferent tubules are the remnants of the mesonephritic kidney and they link the testis to the Wolffian duct. In males, the Wolffian duct differentiates to become the *epididymis* and the *vas deferens*—the tube through which the sperm pass into the urethra and out of the body. The interstitial mesnchyme cells of the testes differentiate into *Leydig cells*, which make testosterone.

In females, the germ cells reside near the outer surface of the gonad. The initial sex cords of the gonads degenerate. However, the epithelium soon produces a new set of sex cords, which do not penetrate deeply into the mesenchyme, but stay near the outer surface (cortex) of the organ. Thus, they are called *cortical sex cords*. These cords are split into clusters, with each cluster surrounding a germ cell. The germ cells become the ova and the surrounding cortical sex cords differentiate into the *granulosa cells*. The mesenchme cells of the ovary differentiate into the *thecal cells*. Together, the thecal and granulosa cells will form the *follicles* that envelop the germ cells and secrete steroid hormones. Each follicle will contain a single germ cell. In females, the Müllerian duct remains intact and it differentiates into the oviducts, uterus, cervix and upper vagina.

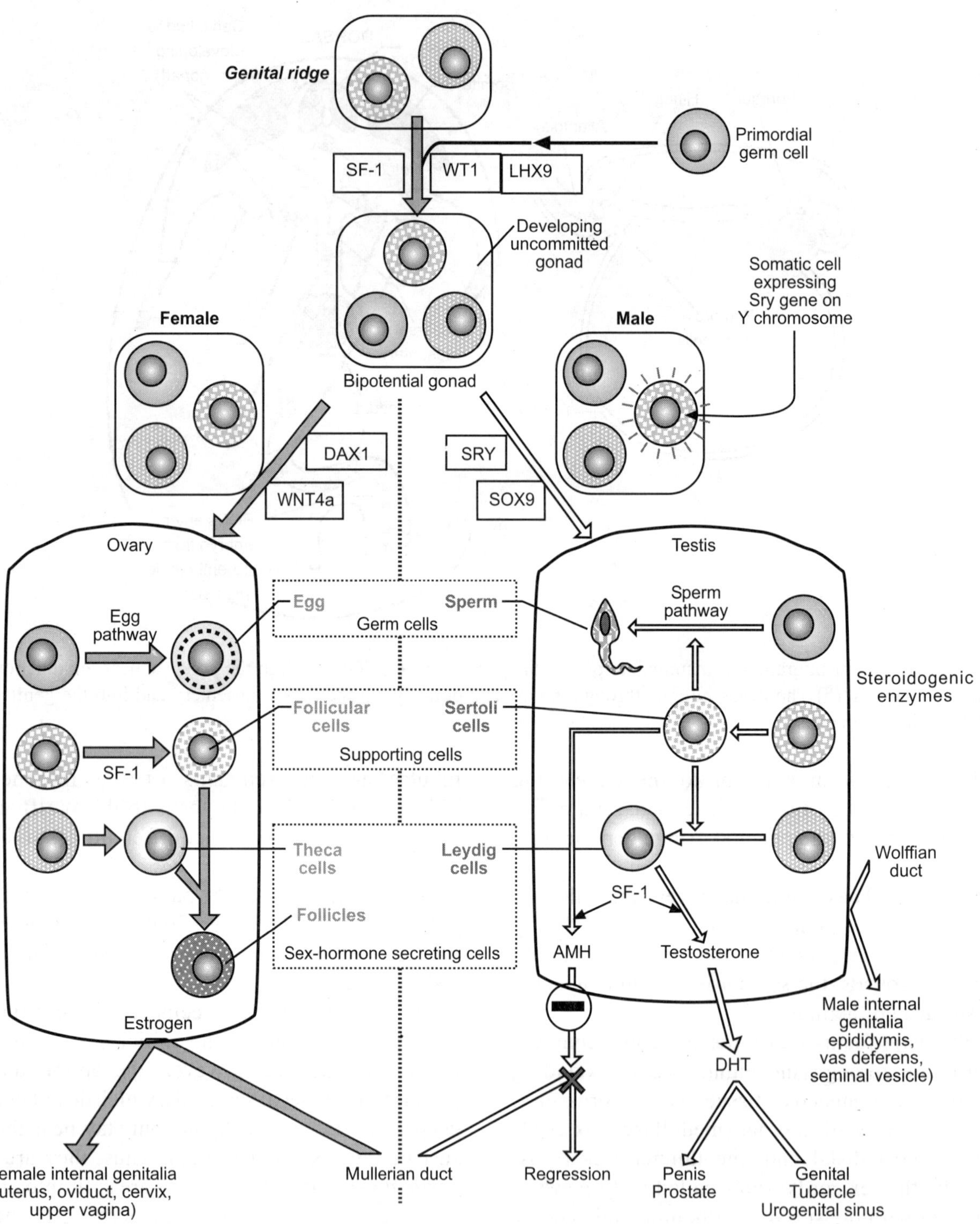

Fig. 15.5 Postulated cascades of molecular events in human sex determination. Development of the undifferentiated (bipotential) gonads from the genital ridge requires the products of many genes (SF-1, WT1, DAX1, WNT4a, SRY and SOX9 and many others).

The Wolffian duct, deprived of testosterone, degenerates (**Fig. 15.5**).

Mechanisms of primary sex determination

The gonads in mammals develop in close association with the *mesonephros*, an embryonic kidney that contributes to both the male and female reproductive organs. Associated with the mesonephros on each side of the body are the *Wolffian ducts*, which run down the body to the cloaca, an undifferentiated opening. Another pair of ducts, the *Mullerian ducts*, run parallel to the Wolffian ducts and also open into the cloaca.

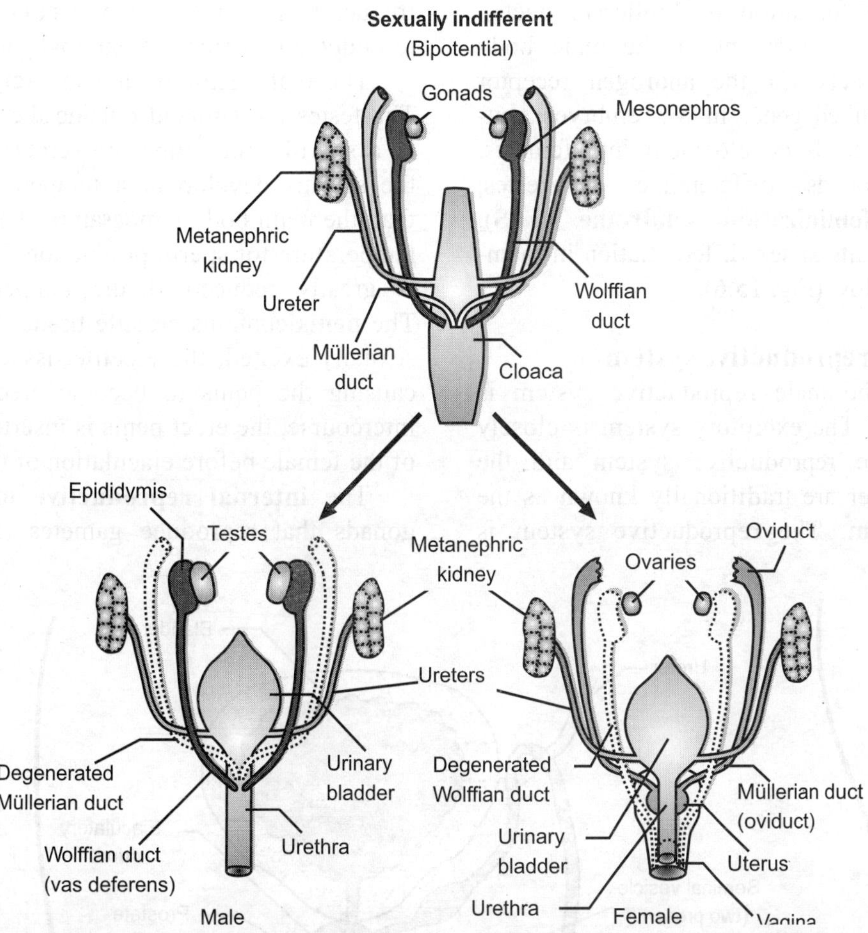

Fig. 15.6 The gonads and their ducts in mammals.

In early mammalian development, before gonadal differentiation, both sets of ducts are present **(Fig. 15.6)**. In females, in the absence of the testes, the Müllerian ducts develop into the *oviducts* (Fallopian tubes), which transport eggs from the ovaries to the uterus, while the Wolffian ducts degenerate. In males, **Sry (sex determining of Y)** encodes a gene-regulatory protein that activates the transcription of other gene-regulatory proteins required for Sertoli cell development, including the Sry-related protein Sox9. In the absence of either Sry or Sox9, the genital ridge develops into an ovary. The supporting cells become **follicle cells** instead of Sertoli cells. Other somatic cells become **theca cells** instead of Leydig cells and, beginning at puberty secrete the female sex hormone estrogen instead of testosterone. The primordial germ cells develop into eggs instead of sperm and the animal develops as a female. The expression of Sry results in the differentiation of Sertoli cells, which are the somatic cells of the testis and are essential for testis formation and for spermatogenesis as they retain the germ cells that migrate into the gonad. The development of sertoli cell involves upregulation of the gene Sox9 and the secretion of Müllerian-inhibiting substance, which induces regression of the Müllerian duct, largely by apoptosis. The interstitial cell lineage in the testis then differentiates into Leydig cells, which produce testosterone. This causes the Wolffian duct to become the vas deferens, the duct that carries sperm to the penis. There is evidence that the extracellular signalling molecule Wnt4 represses testosterone production in the undifferentiated gonad and that one of functions of the Sry gene is to cause the down regulation of Wnt4 expression, which occurs when Sertoli cells differentiate. FGF9 is required for Sertoli cell differentiation and zygotes lacking it develop into females.

Once testes and ovaries are differentiated, their somatic cells secrete male or female hormones to direct the development of male or female reproductive ducts, genitalia and other sex characters. Testes secrete testosterone (which induces the differentiation of Wolffian ducts) and **anti-Müllerian duct hormone** (AMH; also known as Mullerian-inhibiting substance, **MIS)** which prevents the formation of Mullerian ducts), which ensures the development of the male body type. Testosterone acts via the androgen receptor encoded by an X-linked gene. In XY embryos, mutation of this gene leads development into females, although their gonads differentiate as testes, called **testicular feminization syndrome** (**TFS**). The sequence of events in sex differentiation in mammals is discussed below (Fig. 15.6).

The human male reproductive system

The structure of the male reproductive system is shown in **Fig. 15.7**. The excretory system is closely associated with the reproductive system and the two systems together are traditionally known as the **urinogenital system**. The reproductive system is frequently described as two sets of organs, the internal reproductive organs and external genitalia (Fig. 15.7).

The male genitalia are the **scrotum** and **penis**. The testes are situated outside the abdominal cavity in a sac of skin called the scrotal sac. As a result, the sperms develop at a temperature 2–3°C lower than the main body temperature. This is the optimum temperature for sperm production. The life of sperms is greatly reduced if the temperature is higher. The **penis** contains erectile tissue. When the male is sexually excited, the erectile tissue fills with blood, causing the penis to become erect. During sexual intercourse, the erect penis is inserted into the vagina of the female before ejaculation of the semen.

The **internal reproductive organs** consist of gonads that reproduce gametes (sperm cells) and

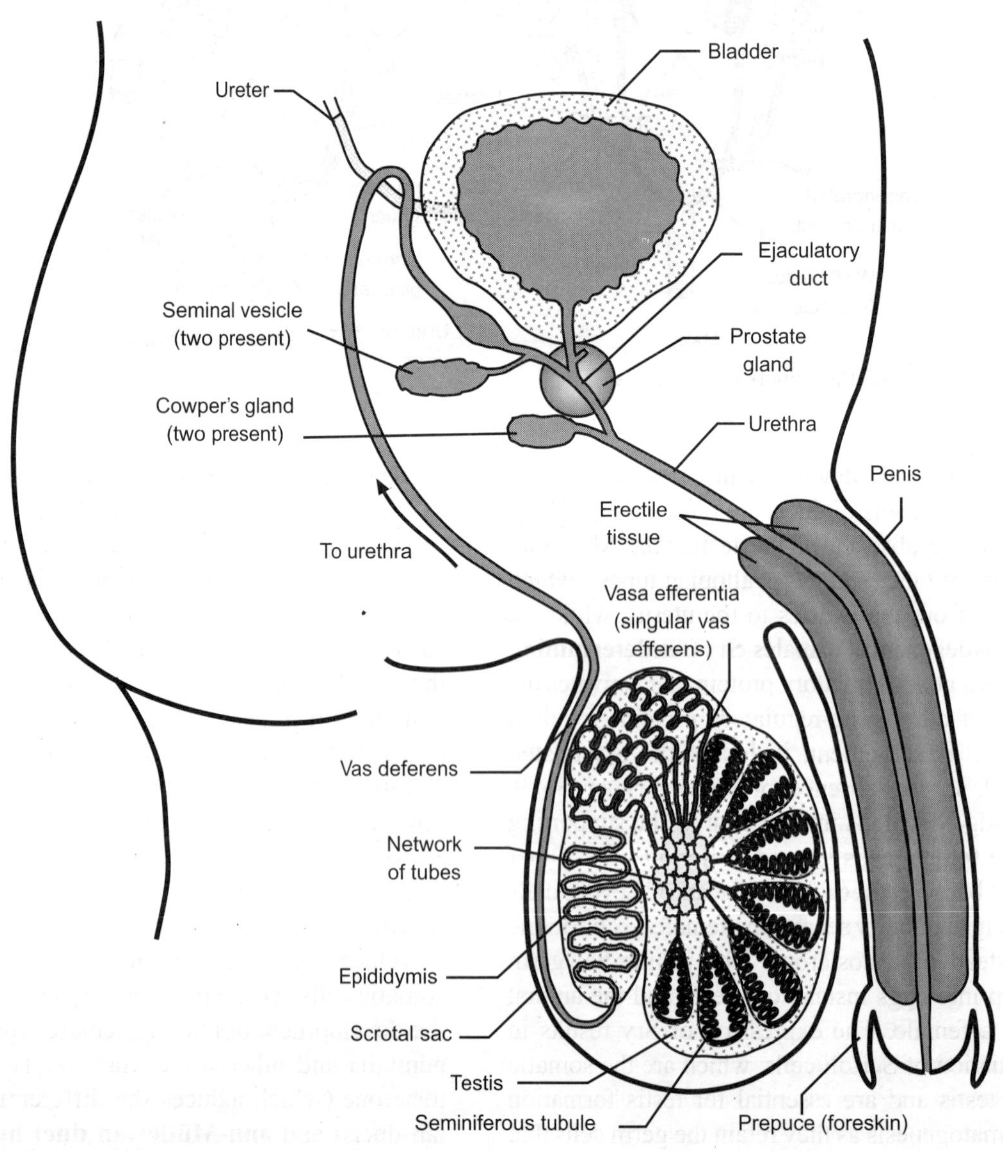

Fig. 15.7 Longitudinal view of the human male reproductive system.

hormones, accessory glands that secrete products essential for sperm movement and a set of ducts that carry the sperm and glandular secretions. The male gonads, **testes** (singular: testis), are a pair of tightly coiled tubes surrounded by several layers of connective tissue. These tubes are the **seminiferous tubules**, where sperm form. Inside each testis are about 1000 coiled seminiferous tubules. Each tubule is about 50 cm long and 200 μm in diameter. The wall of the tubules produce the sperm (spermatozoa), through a process called **spermatogenesis**. **Leydig cells**, also known as **interstitial cells**, between the tubules produce the male sex hormone **testosterone**. From the seminiferous tubules, the sperms pass into the coiled tubules of the **epididymis**, which stores the sperm and is the site of their maturation. It is a much coiled tube, about 6 m long, pressed against the testis. Sperms take several days to pass through this tube. Sperms are concentrated here to about 5000 millions per cm^3 by the re-absorption of fluid secreted originally by the seminiferous tubule. They also develop the ability to swim and pass to the base of the tube for a short period of storage before entering the vas deferens. During ejaculation, the sperm are propelled from the epididymis through the muscular **vas deferens**. These ducts run from the scrotum around and behind the urinary bladder where they join to form a short **ejaculatory duct**. It opens into the **urethra**, the tube draining both the excretory system and reproductive system.

In addition to the testes and ducts, the male reproductive system contains three sets of glands that add their secretions to the **semen**, the fluid that is ejaculated. The **seminal vesicles** contribute about 60% of the total volume of the semen. This pair of glands lies below and behind the bladder and empties into the ejaculatory duct. The seminal fluid is thick and clear, containing mucous, amino acids and fructose, which provide energy for the sperm. The seminal vesicles also secrete **prostaglandins**, a hormone-like compound. These hormone-like compounds, once in the female reproductive tract, stimulate contractions of the uterine muscles that help movement of the sperm up the uterus. The prostate gland also secretes mucous and a slightly alkaline fluid, which is released during ejaculation and helps to neutralize the acidity of the vagina, making the sperm more active. The **bulbourethral glands** are a pair of small glands along the urethra below the prostate. Their function is still unclear; the fluid probably helps to lubricate the penis and the vagina. **Cowper's glands** secrete mucous and an alkaline fluid into the urethra. The alkaline fluid neutralizes the acidity of any remaining urine.

The human female reproductive system

The females contains a pair of gonads, a system of ducts and chambers to conduct the gametes as well as to house the embryo and the foetus and external genitalia that facilitate reproduction (**Fig. 15.8**). The female gonads, the **ovaries**, lie in the abdominal cavity below the digestive system. Each ovary is enclosed in a tough protective capsule and contains many **follicles**. A follicle consists of one egg cell surrounded by one or more layers of follicle cells, which nourish and protect the developing egg cell. All the 400,000 follicles a woman will ever have are formed at birth. Of these, only several hundred will be released during the woman's reproductive years. After puberty, one (rarely two or more) follicle matures and releases its egg during each menstrual cycle. The cells of the follicle also produce the primary female sex hormones, the estrogens. When ovulation occurs, the egg is expelled from the follicle and the remaining follicular tissue grows within the ovary to form a solid mass called the **corpus luteum**. The corpus luteum secretes progesterone, the hormone of pregnancy and additional estrogen. If the egg is not fertilized, the corpus luteum degenerates and a new follicle matures during the next cycle.

The female reproductive system is not completely closed and the egg cell is expelled into the abdominal cavity near the opening of the **oviduct**, or Fallopian tube. The oviduct has a funnel-like opening and cilia on the inner epithelium lining the duct help collect the egg cell by drawing fluid from the body cavity into the duct. The cilia also convey the egg cell down the duct to the uterus, commonly called the **womb**. The uterus is a thick, muscular organ shaped much like an upside down pear. It is remarkably small and the uterus of a woman who has never been pregnant is about 7 cm long and 4–5 cm wide at its widest point. The unique arrangements of muscles that make up the bulk of the uterine wall allow it to expand to accommodate a 4-kg foetus. The inner lining of the uterus, the **endometrium**, is richly supplied with blood vessels.

The narrow neck of the uterus is the cervix, which opens into the vagina. The **vagina** is a thin-walled chamber that forms the birth canal through which the baby is expelled and it is also the repository for sperms during copulation. Although it is much thinner

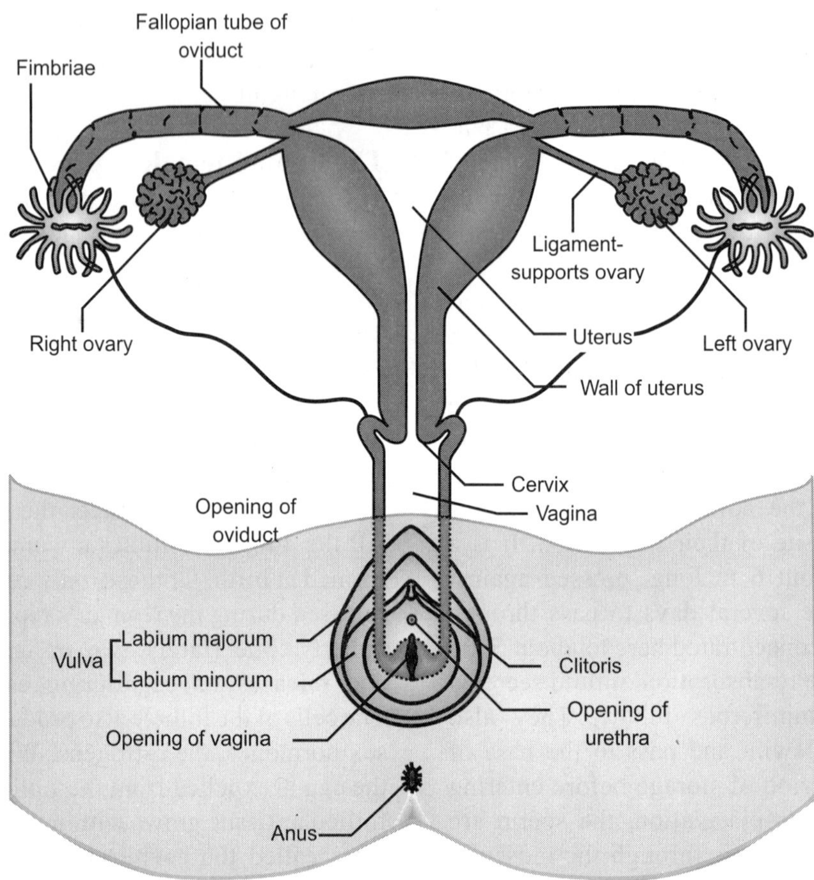

Fig. 15.8 The human female reproductive system. The uterus and vagina are shown in section. The external genitalia and the openings of the urethra and anus are shown.

than the uterus, the walls of the vagina are sufficiently muscular so that they can contract to hold any penis tightly or expand to allow passage of a baby's head.

The vagina is the terminal portion of the female reproductive system, but it is covered externally by two pairs of skin-folds that form a **vestibule** containing the vaginal orifice and the opening of the urethra. At birth and until intercourse or vigorous physical activity takes place, the human vaginal orifice is covered by a thin membrane called the **hymen**, which has no known function. The vestibule is bordered by the slender **labia minora**, which in turn are protected by a pair of thick, fatty ridges called the **labia majora**. Like the vagina, the labia minora are composed of erectile tissue and enlarge during arousal and intercourse. At the top of the vestibule a small bulb of erectile tissue peeks out from a hood of skin. This **glans clitoris**, or clitoris, is the female equivalent of the glans of the penis. Like that organ, it is composed of erectile tissue and is one of the most sensitive points of stimulation in sexual response.

The **mammary gland**, or breast, is another structure important to mammalian reproduction, although it is not part of the reproductive tract itself **(Fig. 15.9)**. The secretory apparatus consists of a series of alveoli—small sacs of epithelial tissue that secrete milk. The alveoli drain into a series of ducts that open at the nipple. Deposits of fatty tissue form the main mass of the mammary gland of a non-lactating mammal. The lack of estrogen in males prevents the development of both the secretory apparatus and the fat deposits, so the breasts remain small and the nipple is not connected to the ducts.

The external genitalia of both sexes arise from common **primordia**, or undifferentiated embryonic structures. Sex chromosomes determine the balance of male and female sex hormones in the embryo and in turn the hormonal balance determines whether the primordial genitalia develop into female or male structures. If androgens are present, male structures form. Female structures form in the absence of androgens.

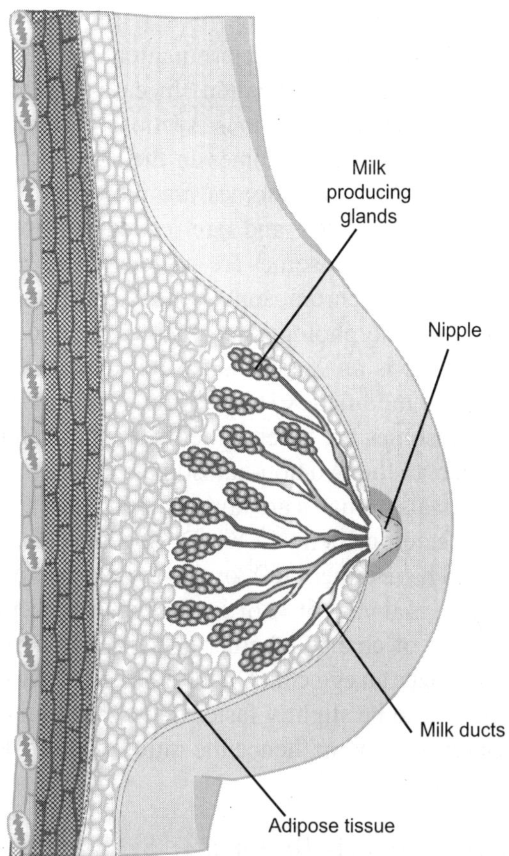

Fig. 15.9 Breast of a woman (**See Plate 12**).

Factors affecting sex determination

Sex differentiation is obviously affected by environmental, chromosomal and genetic factors and often these factors interact with one another in determining the sex of an individual. In all organisms, beyond the sets of genes required to develop both types of gonads and sexual organs, additional selector or master genes determine which of the two sets of genes is actually being used in developing an individual's sex. Generally, one decisive master gene is present but allotted to only one-half of the offspring. This is referred to as **genotypic sex determination**. It has been suggested that these factors or genes act as a switch and turn on a set of (hypothetical) sex determining gene S1, S2, S3 and so on concerned with sex differentiation (**Fig. 15.10**). Little is known about the identity of the genes S1 and d1. However, in case of humans, a master locus Sry regulating testis differentiation has been identified. Alternatively, master genes are activated differentially by environmental influences. Sex determination in most organisms is essentially similar and can be grouped into two main categories: environmental and chromosomal and genic mechanisms.

Temperature: We now know that the sex of mammals is determined by its sex chromosomes and not the environment. However, in animals, zygotes do not differ genetically for sex differentiation. In such cases, sexual differentiation is determined by environmental factors. In contrast to what seem to be purely genetic mechanisms, more environmentally dependent sex-determining mechanisms are known in many animals. In environmental sex determination, the decision is made only after fertilization and external cues, such as temperature or presence of a sexual partner, make the choice, presumably by influencing the activity of master genes. For instance, in the polychaete annelid *Ophryotrocha puerilis,* two individuals that encounter each other by chance influence each other by means of chemical signals (pheromones) and eventually one individual plays the role of the male and the other the female. In some animals, e.g., turtles, alligators, crocodiles and some lizards , the temperature at which eggs are incubated have a decisive effect on the sex of the animals that hatch from them.

Location: In the echiuroid sea worm, *Bonnelia*, the sex of the worm depends on where a larva settles. If the free-swimming larva settles on the ocean floor, it develops into a 10 cm-long female. If the larva is attracted to a female's proboscis, it travels along the tube until it enters the female's body. There it differentiates into a minute (1–3mm long) male that is essentially a sperm-producing symbiont of the female. These males lack digestive organs and exist in parasitic fashion in the genital ducts of the female. Males are very small and only their reproductive organs are developed; they ultimately migrate to the uterus of females where they live as parasites. It is suggested that some hormone produced by the proboscis of female *Bonelia* induce the larvae to differentiate into males.

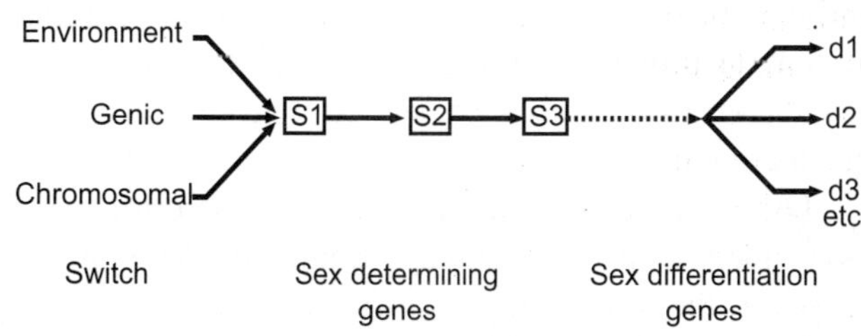

Fig. 15.10 Functional relationships among various factors in sex differentiation.

Size of the egg: In the case of another sea worm, *Dinophilus*, the size of the eggs appears to determine the sex of animals developing from them. Animals from eggs of relatively larger size are females, while those obtained from comparatively smaller eggs are males.

Visual stimulation: In some coral reef fishes, sex changes seem to occur as a result of visual stimulation: females become males when the surrounding fish in the group are relatively small.

Social behaviour: In certain fish, social behaviour can influence sex. In some sea fishes, e.g., Labroides, the social order of an individual in the colony of which it is a member, determines its sex. A colony has a single male member, while the remaining members of the colony are females. The male of a colony is physically strong and the body size of the females of the colony are according to their social status in the colony. Therefore, the most dominant female is the largest among the females, while the fish that is the last in social order is the smallest. The gonads of female fish also contain the elements of male gonads in an undeveloped state. When the male of a colony dies, the largest or the number one female of the colony begins to differentiate into a male. This transformation may take 14–18 days. Thus, females of the Labroides are, in fact, hermophrodite anatomically, but are functionally female. However, they can revert to maleness once they are elevated to the number one position in the colony. Thus, the social order of such a fish appears to regulate its sexual differentiation.

Sex determination in plants: Sex in various plants is also affected by environmental conditions. The horsetail plant, *Equisetum*, shows female characteristics when raised under good growth conditions and male characteristics when raised under poor conditions. In some plants, considerable changes in sex expression can occur through changes in day length and temperature, which affect various hormonal and chemical processes. For example, the appearance of female flowers in the cucumber and musk melon has shown to be highly correlated with production of the chemical, ethylene.

Other factors that affect sex determination have been reported in many organisms. In general, however, the most common forms of sex-determining mechanisms in diploids seem to be chromosomal, associated with differences between two types of sex chromosomes: X and Y.

Genic sex determination

In genotypic or genic sex determination, the decision is left to chance in fertilization. In case of humans, as explained in the previous sections, individuals carry $2 \times 23 = 46$ chromosomes in their diploid cells. Of these, 2×22 chromosomes are indistinguishable in males and females and are called **autosomes**. In females, the 44 autosomes are complemented by two more homologous chromosomes, the X chromosomes. In males, two morphologically distinct chromosomes are found: one is an X (similar to the female X) and the other is a modified X, termed the Y chromosome. XY chromosomes are often called sex chromosomes. With respect to the two non-autosomes, in humans and other mammals, females are homogametic (XX), while males are heterogametic (XY). During meiosis, each gamete will have either the X or the Y sex chromosome and the ratio is always 50% of each type. In conception, it is a matter of chance whether a sperm cell carrying X or Y fertilizes an egg cell. On the average, Y-carrying sperm appear to be slightly faster in the race to the egg than X-carrying sperm, hence the ratio of 105 males to 100 females.

Sex determination in *Drosophila*: Sex determination in *Drosophila* appears superficially similar to that in mammals because females possess two X chromosomes and males possess a single X chromosome plus a Y chromosome. However, the mechanism used in flies, differ from mammalian sex determination in two important respects. First, sex in *Drosophila* is determined by the number of X chromosomes. The ratios between the number of X chromosomes and the two sets of autosomes (AA) in each diploid cell, is decisive in determining the sex of the individual cell. For example, XX/AA (a ratio of 1:1) is female and X/AA (a ratio of 1:2) is male. The Y chromosome in *Drosophila* is of significance in spermatogenesis in the adult male but not in early sex development (**Fig. 15.11**). Secondly, sex characters in *Drosophila* (genitals, size, wing length, body markings and the presence of sex combs in males) are determined genetically in a cell-autonomous fashion (the secondary characters in mammals are determined by hormones).

Sex determination in bees

Haplodiploidy in insects: Many species of ants, bees and wasps *(order Hymenoptera)* are highly evolved in social hierarchy and have very distinct forms of sex determination. In these organisms males are haploid whilst females are diploid (**Fig. 15.12**). Individuals

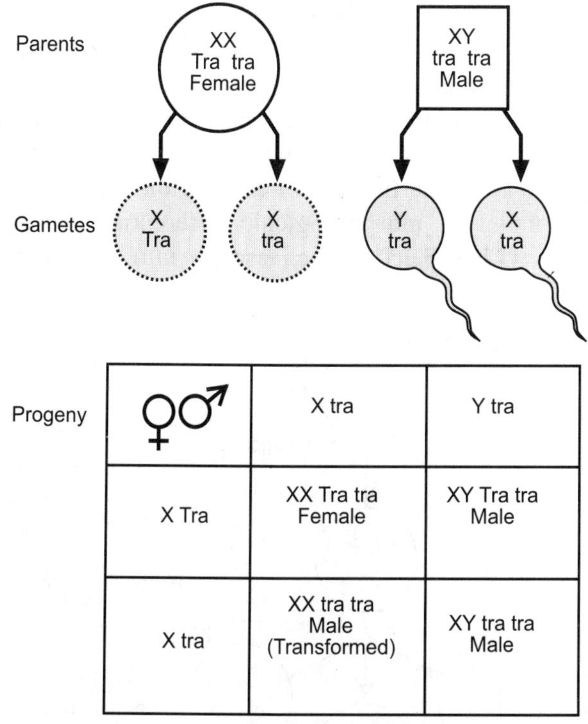

Fig. 15.11 Effect of an autosomal recessive gene tra on the ratio of males and females recovered in the progeny of *Drosophila*.

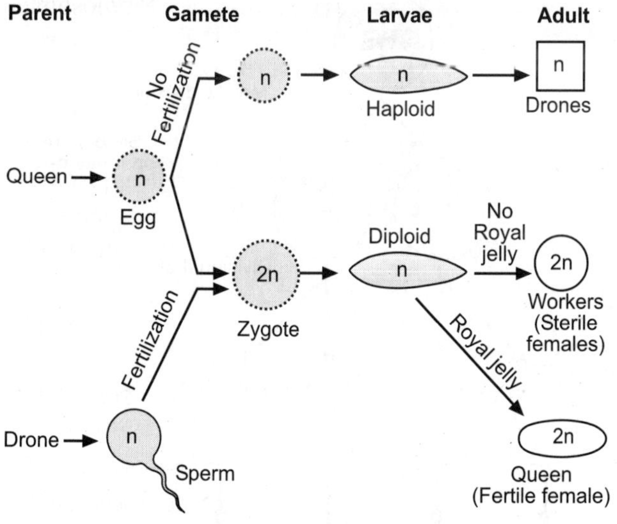

Fig. 15.12 Development of haploid males and diploid females in honeybees from unfertilized and fertilized eggs.

come in several distinct forms (castes) specialized for different roles. Females are diploid, but make haploid eggs by meiosis. Males are haploid and produce haploid sperms by mitosis. If an egg is fertilized, it develops into a female, whereas unfertilized eggs develop into males. In sexual diploids, parent and offspring have a relatedness of 0.5 (full sisters). However, in haplodiploid insects, full sisters have a relatedness of 0.75.

Gametogenesis

Gametogenesis is the process of forming gametes (by definition haploid, *n*) from diploid cells of the germ line. Most species produce just two types of gametes and they are radically different. The **egg** is among the largest cells in an organism, while the **sperm** (spermatozoon) is often the smallest. The egg and the sperm are optimized in opposite ways for the propagation of the genes they carry. The egg is non-motile and aids the survival of the maternal genes by providing large stocks of raw materials for growth and development, together with an effective protective wrapping. The sperm by contrast, is optimized to propagate the paternal genes by utilizing resources in the egg. The sperm is usually highly motile and shaped for speed and efficiency in the task of fertilization. During spermatogenesis, all four meiotic products develop into gametes; oogenesis places most of the cytoplasm into the large egg. The other cells, the polar bodies, do not develop. All the cytoplasm and organelles go into the egg. Human males produce 200,000,000 sperms per day, while the female produces one egg (usually) in each menstrual cycle.

Spermatogenesis: Spermatogenesis is the process of forming sperm cells by meiosis in specialized organs known as gonads (in males). After division, testes the cells undergo differentiation to become sperm cells. Sperm production begins at puberty and continues throughout life, with several hundred million sperms being produced each day. Once sperms form, they move into the epididymis, where they mature and are stored.

A sperm is a 'stripped-down' cell, equipped with a strong flagellum to propel it through an aqueous medium unencumbered by cytoplasmic organelles; it contains many mitochondria that can power the flagellum. Each sperm consists of two morphologically and functionally distinct regions enclosed by a single plasma membrane: the *tail*, which propels the sperm to the egg and helps it to burrow through the egg coat and the *head*, which contains a condensed haploid nucleus (**Fig. 15.13**). The DNA in the sperm nucleus is packed extremely tightly, so that its volume is minimized for transport and transcription is shut down. The chromosomes of sperms are free of histones

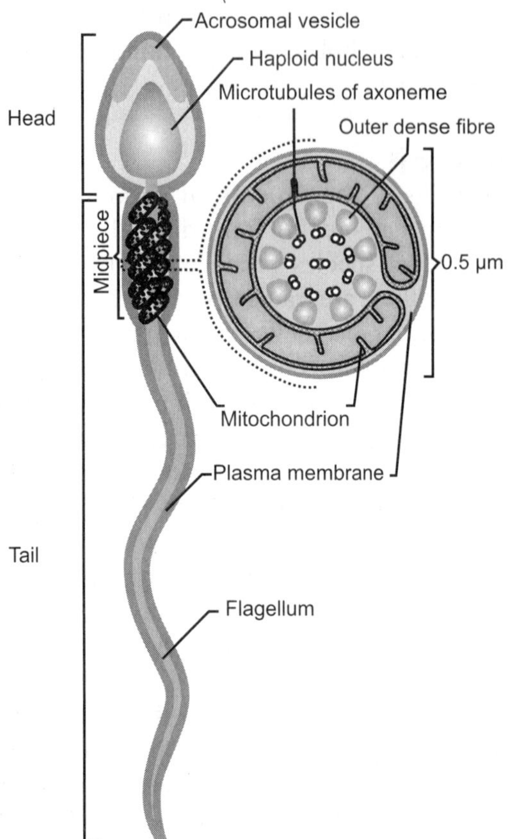

Fig. 15.13 Longitudinal section of a human sperm. Midpiece of a mammalian sperm as seen in cross section in an electron microscope.

of somatic cells and are packed instead with simple, highly positively charged proteins called *protamines*. *Acrosomal vesicle* is a specalized secretory vesicle. It is closely apposed to the anterior end of the nuclear envelope. This vesicle contains hydrolytic enzymes that may help the sperm to penetrate the egg's outer coat. When a sperm contacts an egg, the contents of the vesicle are released by exocytosis in the so-called *acrosomal reaction*. The motile tail is a long flagellum, whose central axoneme emanates from a basal body situated just posterior to the nucleus.

In males, meiosis and spermatogenesis begin in the testes only after puberty and then go on continuously in the epithelial lining of very long, tightly coiled tubes, called seminferous tubules. Immature germ cells, called spermatogonia, are located around the outer edge of these tubes next to the basal lamina, where they proliferate continuously by mitosis (**Fig. 15.14**). Some of the cells stop proliferating and differentiate into primary spermatocytes. These cells enter the first meiotic prophase, in which their paired homologue chromosomes participate in crossing over and then precede with division I of meiosis to produce two secondary spermatocytes, each containing haploid chromosomes. The two secondary spermatocytes derived from each primary spermatocyte proceed through meiotic division II to produce four spermatids. These haploid spermatids then undergo morphological differentiation into sperms (**Fig. 15.15**), which escape into the lumen

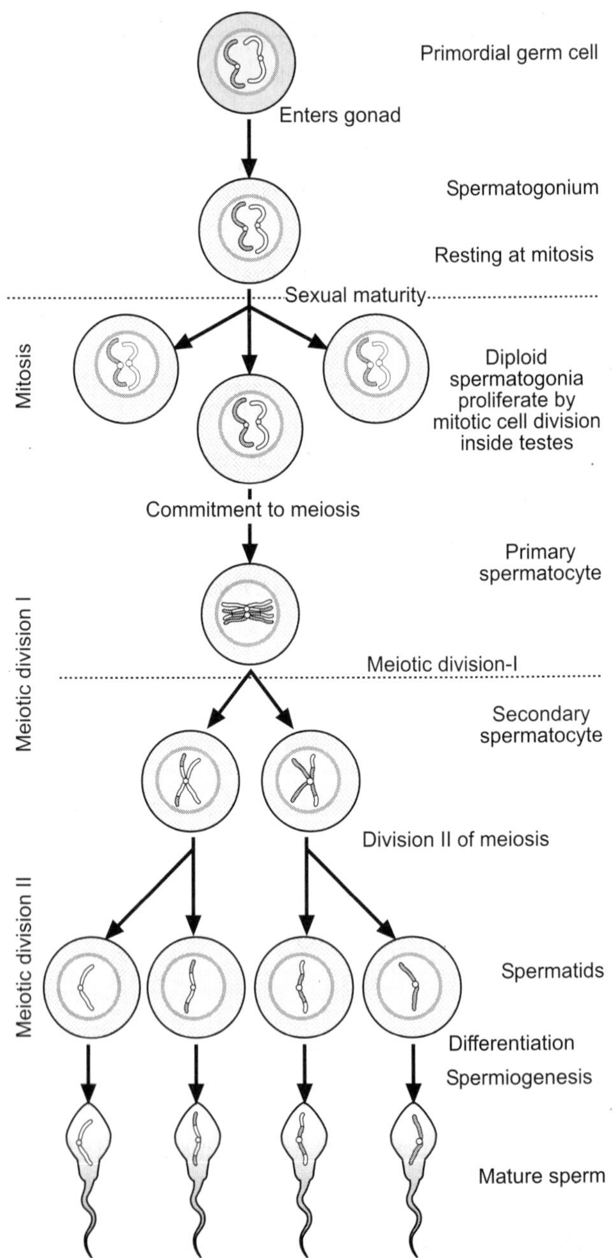

Fig. 15.14 The stages of spermatogenesis. Primordial germ cells of the embryonic testes differentiate into spermatogonia; the diploid cells that are the precursors of the sperm are located near the outer wall of the seminiferous tubules.

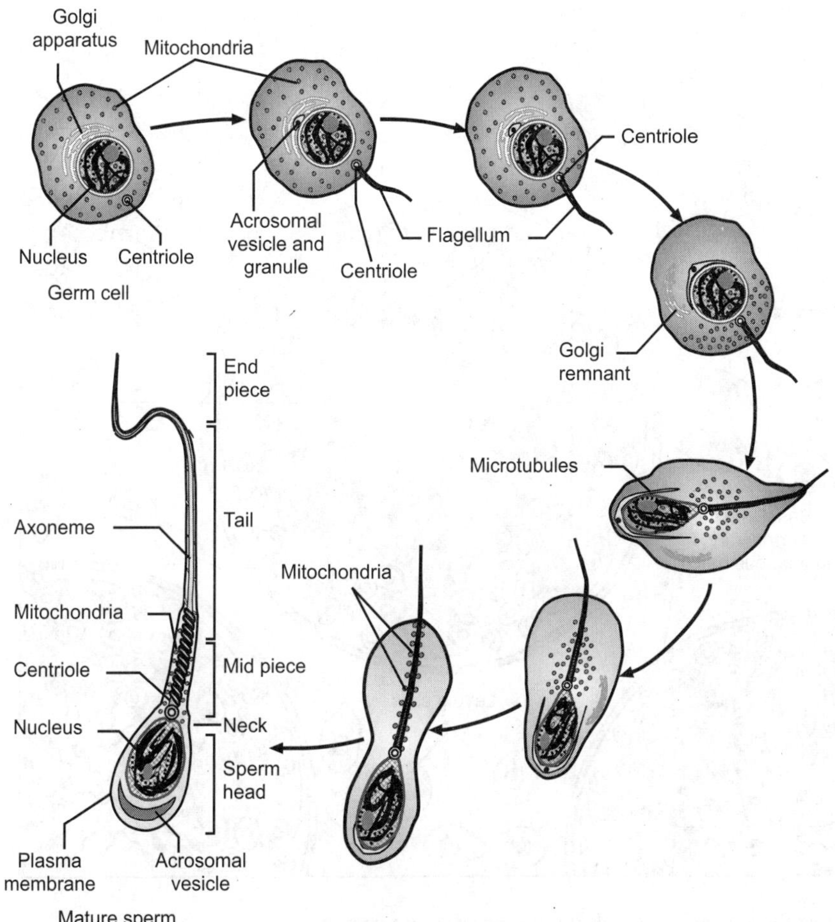

Fig. 15.15 The modification of a germ cell to form a mammalian sperm.

of the seminiferous tubule (**Fig. 15.16**) The sperms subsequently pass into the epididymis, a coiled tube overlying the testis, where they undergo further maturation and are stored.

During spermatogenesis the developing male germ cells fail to complete cytoplasmic division (cytokinesis) during mitosis and meiosis. Consequently, large clones of differentiating daughter cells that have descended from one maturing spermatogonium remain connected by cytoplasmic bridges, forming a syncytium (**Fig. 15.17**). The cytoplasmic bridges persist until the very end of sperm differentiation, when individual sperms are released into the tubule lumen. Sperms undergo most of their differentiation after their nuclei have completed meiosis to become haploid. The presence of cytoplasmic bridges between them, however, means that each developing haploid sperm shares a common cytoplasm with its neighbours.

Oogenesis: It is the process of forming an ovum (egg) by meiosis in specialized gonads known as ovaries. The ovary contains many follicles composed of a developing egg surrounded by an outer layer of follicle cells. Each egg begins oogenesis as a primary oocyte (**Fig. 15.18**). At birth, each female carries a lifetime supply of developing oocytes, each of which is in Prophase 1. A developing egg (secondary oocyte) is released each month from puberty until menopause, approximently 400–500 eggs.

Plant gametogenesis

The process of gamete formation in plants mirrors that seen in animals, however many features are similar to animal gametogenesis. Although there is conservation of process and substantial information can be obtained by the comparative analysis of distantly related organisms there remain distinct differences, which are specific to the organisms, concerned. These differences include:

- ***Germline specification:*** Plants have the ability to produce gametes at a late developmental stage that has not been predetermined, whereas gamete formation in animals is confined to distinct regions that are established at an early embryonic stage.

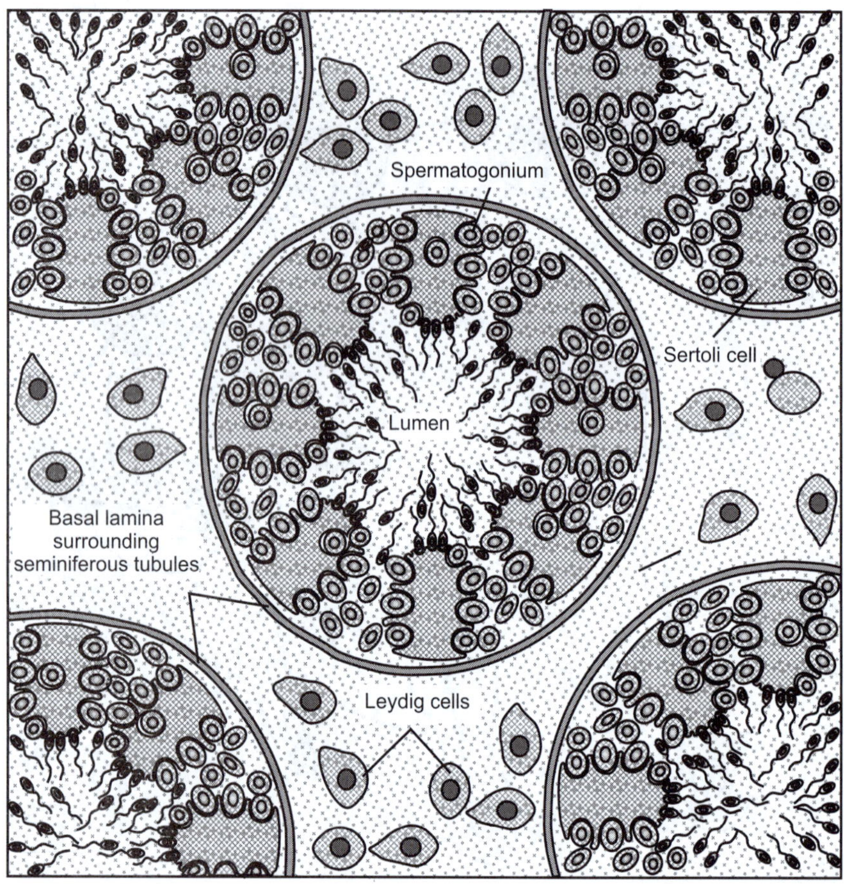

Fig. 15.16 Cross section of a seminiferous tubule in mammalian testis.

- ***Fertility may be frequently compromised by the mutation of genes associated with gamete formation:*** Defects frequently result in reduced fertility in animals, but they are rarely lethal in plants. Such defects can often be maintained since most plants produce male and female gametes and often the formation of one type of gamete remains unaffected by mutation. Plants can also reproduce asexually or via non-fertilized means (e.g., apomixis; see Chapter 17).

Gamete recognition and fertilization: Fertiliation is the fusion of two gametes (haploid genomes) to form the diploid genome of the zygote. Two major types of fertilisation have evolved: external and internal fertilization. The process of fertilization plays an important part in sexual reproduction. In *external fertilization*, the eggs are shed by the female and fertilized by the male in the environment. In this case, the gametes of each sex may move about on random currents (e.g., in water) and there is higher probability for the gametes of many species to mix together. In contrast, *internal fertilization* occurs when the sperm are deposited in (or nearby) the female reproductive tract and egg and sperm fuse within the female body. In this case, there is a definitive pathway to the egg and mixing of gametes from other species is avoided.

Egg attracts the sperm: Since the sperm is not deposited directly on the egg, the gametes possess a mechanism to attract each other at a distance. It has been shown that eggs release potent chemo-attractant signals. Sea urchin eggs, for example, release substances that have profound effect on sperm behaviour even at very low concentrations. It appears that sperm can swim up a concentration gradient towards the egg. With internal fertilization, the female reproductive tract often plays an important role in the early part of the sperms' journey. In mammals, for example, muscular contractions of the uterus are primarily responsible for transporting the sperms from the neck of the cervix to the oviduct. The cervix is normally blocked by thick mucous. This mucous becomes thinner in the first part of the menstrual cycle, allowing the penetration of sperms to the uterus. Progesterone causes it to become thicker in the second half of the cycle.

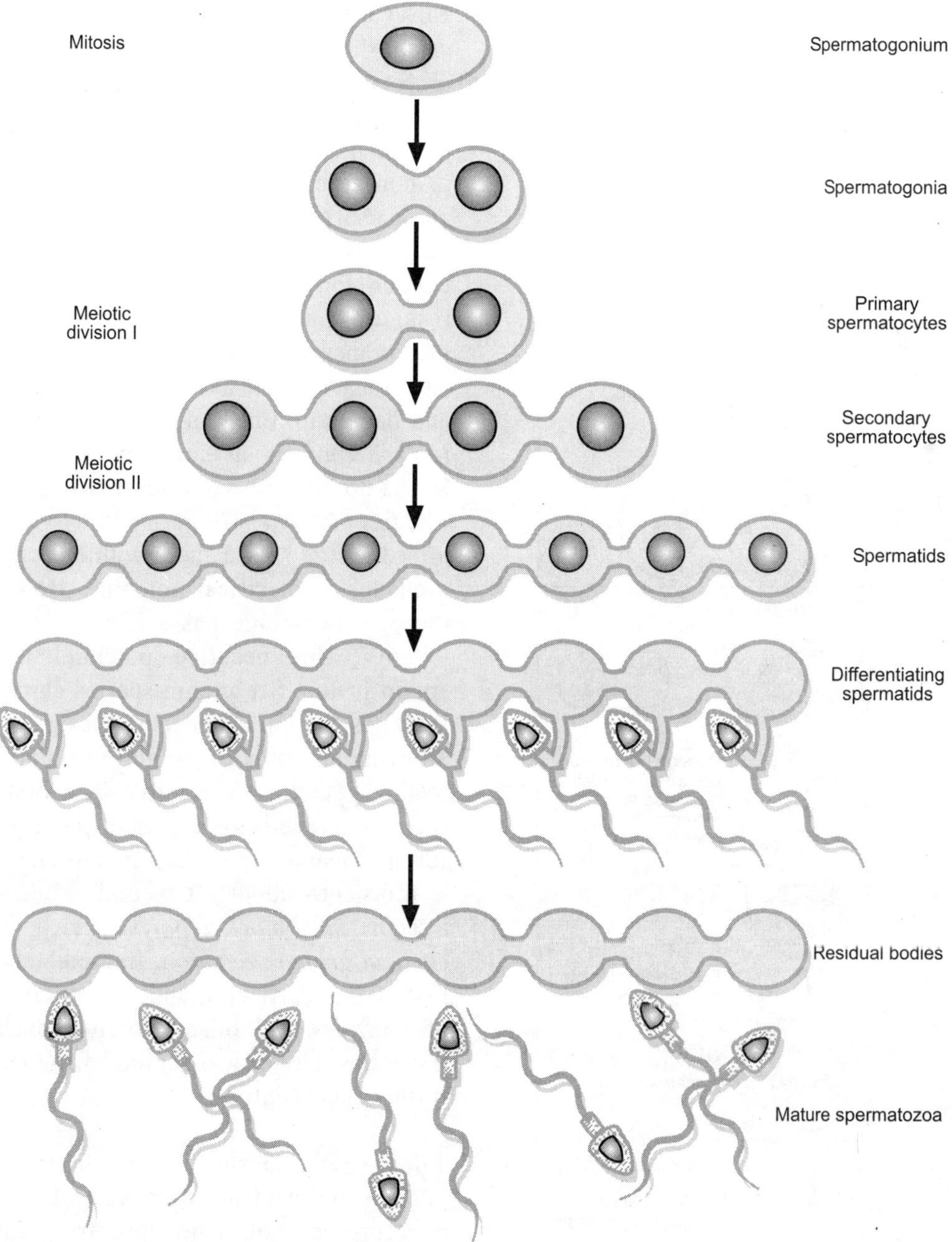

Fig. 15.17 Cytoplasmic bridges in developing sperm cells and their precursors.

Passage of sperm to egg: Investigations have shown that some sperms pass from the vagina through the uterus and to the top of the oviducts within five minutes. In the oviduct, sperms become motile and respond to signals, either from the egg itself or from the ovarian follicle, which increase motility and provides a directional cue. The contractions of the uterus and oviducts could be caused by chemicals in the semen (prostaglandins) and hormones such as oxytocin released by the female during sexual intercourse. It is more likely, however, to be due to the action of cilia which line the uterus and oviducts. It takes about 4–8 hours for most sperms to reach the oviducts and 1–3days to survive in the female.

Penetration of sperm into the egg: Once the sperm reaches the egg, it must penetrate the egg. This is achieved by the *acrosome reaction.* In all animals, sperm maturation involves the formation of an *acrosomal* **vesicle** from the Golgi apparatus of the spermatocyte. The acrosomal vesicle contains enzymes that digest the protective layers surrounding the egg, allowing the sperm access to the egg membrane.

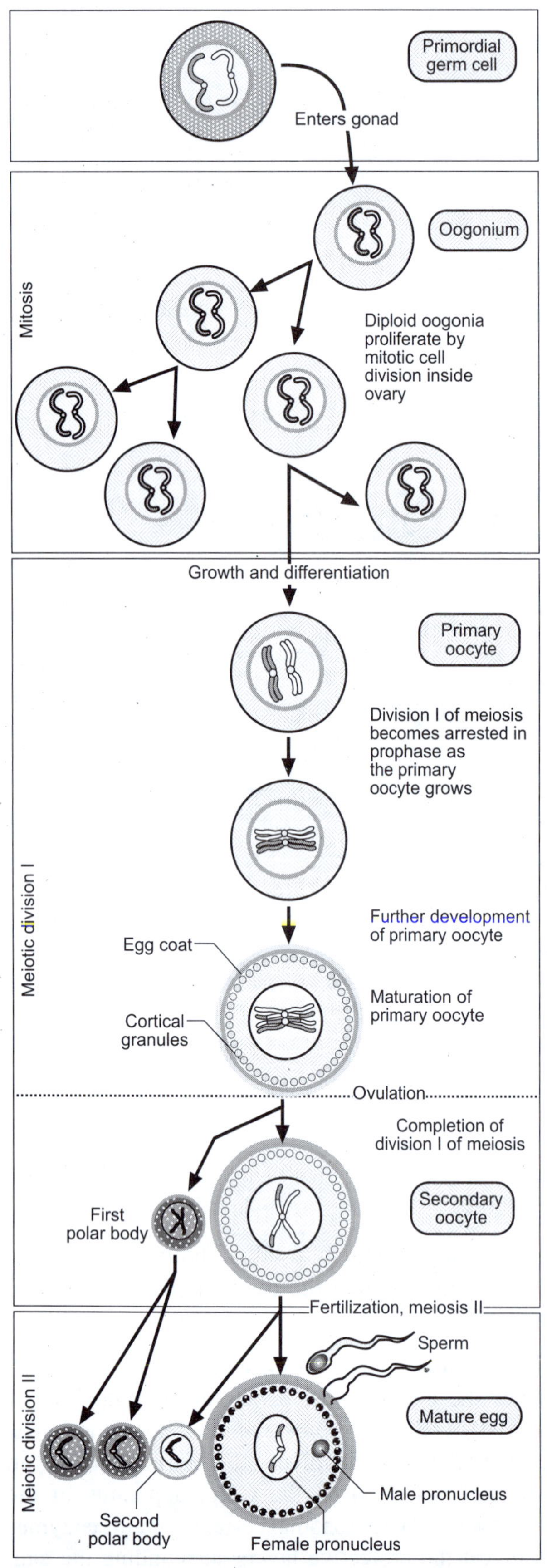

Fig. 15.18 The stages of gametogenesis in vertebrates.

Entry of sperm into the egg: Once contact has been established between the gametes, acrosomal enzymes digest part of the vitelline envelope or zona pellucida, allowing direct contact between the sperm and egg plasma membranes (**Fig. 15.19**). In mammals, fusion is induced by specific *fusogenic proteins* on the inner acrosomal membrane of the sperm, such as the *fertilins*. Fusogens recognize receptors on the egg surface and contain long stretches of hydrophobic amino acid residues that facilitate membrane fusion. Once the membranes fuse, the sperm nucleus enters into the egg along with the associated centriole, mitochondria and flagellum. Entry of multiple sperms into the egg *(polyspermy)* is avoided by various mechanisms. In sea urchin embryo at least two mechanisms exist. The *fast block to polyspermy* is activated immediately after the first sperm fuses with the egg and involves a change in electrical potential. Before fertilization the egg membrane has a resting potential of about –70 mV; this negative potential is required for sperm fusion. Sea urchin sperms carry a protein that activates sodium channels in the egg membrane, thus allowing an influx of sodium ions and increasing the resting potential to +20 mV. The first sperm fusion therefore, reverses membrane polarity and prevents further fusion. Since the membrane depolarization is transient—about 20 seconds after sperm entry—a second *slow block to polyspermy* begins, termed the *cortical granule reaction*. Immediately after gametic fusion the cortical granules, containing a cocktail of enzymes and other proteins, become active in the actin-rich cortical region of the egg and prevent further sperm entry.

The zygote: During fertilization only the sperm nucleus and centriole survive (a few sperm nucleus may survive). Following the entry, the nucleus decondenses as the sperm-specific histones *protamines* are replaced by histones from the egg cytoplasm and become known as the male *pronucleus*. The egg nucleus is the *female pronucleus*. The sperm centriole activates pronuclear fusion and later cleavage. The male and female pronuclei are brought together by the formation of a microtubule aster using the sperm centriole as an organizing centre and egg-derived tubulin subunits. In sea urchins, the egg is fertilized after meiosis and the two nuclei fuse to form a diploid zygotic nucleus in the single-celled zygote. In mammals, the female pronucleus completes meiosis before fusion and ejects the second polar body. The male and female pronuclei undergo S phase

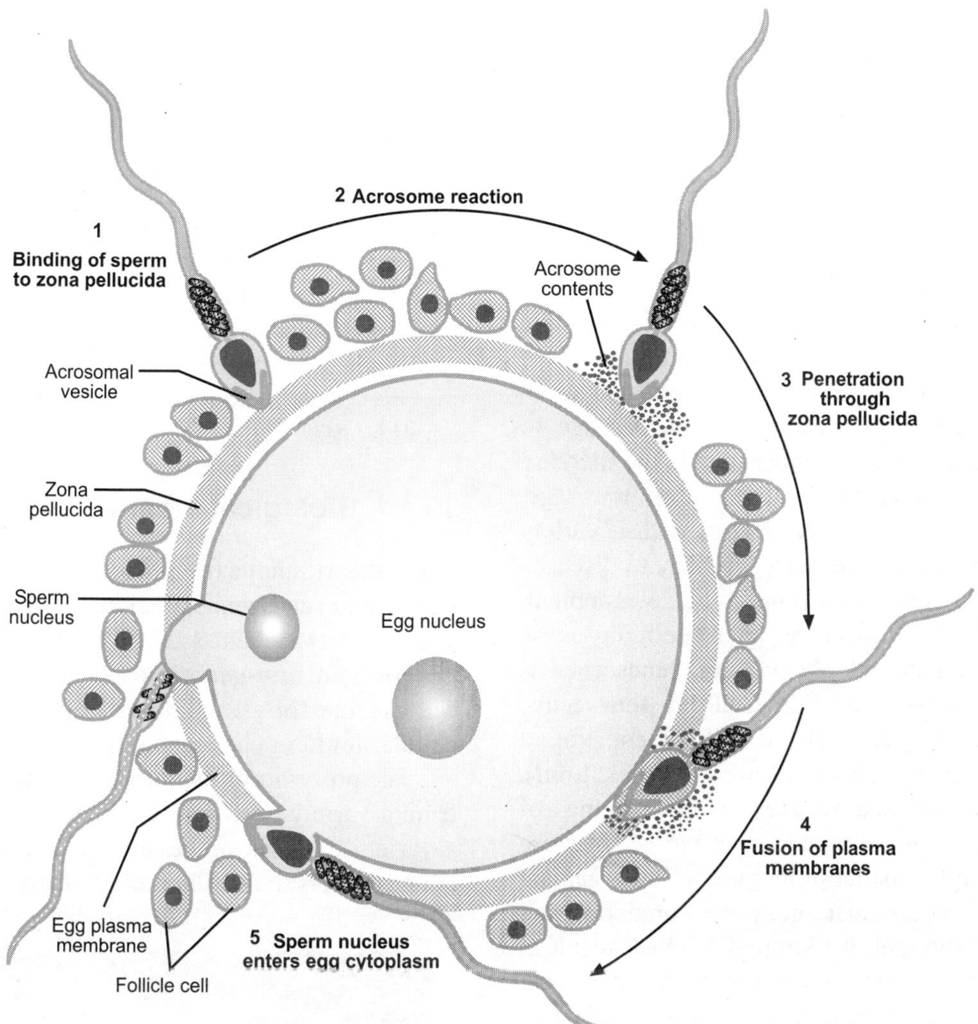

Fig. 15.19 The acrosomal reaction of sperm with mammalian egg during fertilization.

as they move towards each other and then undergo mitosis using the same spindle apparatus. The first diploid nuclei form after the first cleavage division.

As the zygote passes down the oviduct, it divides by successive nuclear and cell divisions into a small ball of cells by a process called **cleavage**.

15.4 AGEING

Organisms continue to change their structural and functional features after completing their embryonic development. Some noticeable changes, as previously discussed, are growth and maturation, metamorphosis from larval to adult forms in insects, wound healing and regeneration of lost body parts. Similar to other developmental processes, ageing may be programmed. Death is the ultimate consequence of ageing and differences in maximum lifespan between species correspond to different 'rates of ageing'. For example, inherited differences in the rate of ageing make a mouse elderly at 3 years and a human elderly at 90 years. Some researchers in gerontology regard ageing itself as a 'disease, that may be curable. Genetic and environmental interventions are known to affect the lifespan of model organisms. This gives hope that human ageing can be slowed, halted or reversed.

What is ageing?

Ageing is the reverse process of development. No organism is immortal, even if it escapes from diseases or accidents. The developed adult organism maintains its identity for some time; later deterioration surfaces in the rate of maintenance and the organism ages. Therefore, *ageing can be defined as* ***the time-related deterioration of the physiological functions necessary for survival and fertility***. Put simply, ageing is the process of becoming older. Ageing and a limited lifespan apparently are characteristic of all animals that stop growing after reaching a standard mature size. The process of ageing is the progressive decline in all the vital organs of an organism that terminate in death, and goes on at different rates in animals of different species and in different individuals of the same species. A number of mechanisms are involved in the ageing of the body. Many evolutionary biologists consider that ageing is the default state after an animal has fulfilled the requirements of natural selection. The biology of ageing is treated in detail in senescence. With ageing—the passage of time—senescence occurs. **Senescence** is the increase of impairment of physiological functions with age, resulting in decreased ability to deal with a variety of stresses and an increased susceptibility to disease.

Ageing is a continuous process of development, but senescence may reflect the later declining phase of an organism's growth. In biology, senescence is the state or process of ageing. **Cellular senescence** is a phenomenon where isolated cells demonstrate a limited ability to divide in culture (Hayflick limit), while **organismal senescence** is the ageing of organisms. Ageing is believed to have evolved because of the increasingly smaller probability of an organism still being alive at an older age, due to predation and accidents. It is thought that strategies which result in higher reproductive rate at younger age, but shorter overall lifespan, result in a higher lifetime reproductive success and are therefore favoured by natural selection. Ageing is therefore the result of investing resources in reproduction, rather than maintenance of the body (**disposable soma theory**).

Bodily changes associated with ageing

Ageing people's bodies normally change in appearance and show changes in height, weight, composition and facial expressions (**Fig. 15.20**).

Change in **height** varies among different people. Generally, our height increases until our late forties and then decreases about two inches by age 80. The reasons for height loss may be:

- changes in posture
- changes in the growth of a vertebrae
- forward bending of the spine
- compression of the discs between the vertebrae
- increased curvature of the hips and knees
- decreased joint space in the trunk and extremities
- joint changes in the feet
- flattening of the arches.

Change in **weight** is seen. In men, weight increases until their midfifties and then decreases, with weight being lost faster in the late sixties and seventies. In women, body weight increases until the late sixties and then decreases at a rate slower than that of men.

The change in **body composition** varies depending on the individual lifestyle. The proportion of body fat doubles between the ages 25 and 75. Exercise and diet may prevent or reverse much of it, which is essential for a healthy life.

15.4.1 Biological theories of ageing

Many theories have been proposed to explain cellular ageing and senescence. Each theory has its advocates and limitations as well. These theories include **genetic** and **non-genetic** aspects. No single theory can account for all the observations of ageing at the cellular level.

The process of senescence is complex may derive from a variety of different mechanisms and exist for a variety of different reasons. However, senescence is not universal and scientific evidence suggests that

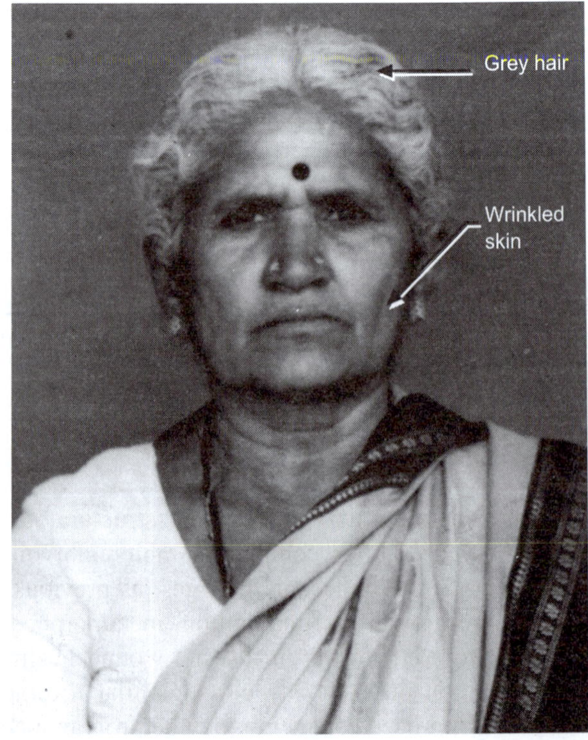

Fig. 15.20 Ageing person. Hair turns grey and wrinkles appear on the face.

cellular senescence evolved in certain species as a mechanism to prevent the onset of cancer. In some species, senescence is negligible and they do not have 'post-mitotic' cells. There has long been a debate on whether ageing is really a developmental process. The argument in favour is that senescence and death arise from preprogrammed events, which are under developmental control. The argument against is that such events are simply due to progressive accumulation of mutational and other genetic damages, which have an inherently random character.

Genetic control of ageing

A number of genetic components of ageing have been identified using model organisms, ranging from the simple budding yeast *Saccharomyces cerevisiae* to worms such as *Caenorhabditis elegans* and fruit flies (*Drosophila melanogaster*). Genes program senescence (ageing) and death of selected parts of organisms throughout development. The genetic control of ageing can be understood in terms of the **disposable soma theory**. It explains that natural selection tunes the life history of the organism so that sufficient resources are invested in maintaining the repair mechanisms that prevent ageing at least until the organism has reproduced and cared for its young. For example, mice which start reproducing when just a few months old, need to maintain their repair mechanisms for much less time than do elephants, which only start to reproduce when around 13 years old. In most animals in the wild, few individuals live long enough to show obvious signs of senescence, and senescence need only be delayed until reproduction is complete. The maximum recorded lifespan of animals show dramatic differences in their longevity. Humans can live as long as 120 years, the Indian elephant 70 years, chimpanzees for about 45 years, some owls 68 years, Xenopus 15 years, mice 31/2 years and the nematode about 25 days.

Mutations in genes that affect lifespan have been identified in both *C. elegans* and humans and may give clues to the mechanisms involved. A *C. elegans* worm that hatches as a first larva in an uncrowded environment with ample food, grows to adulthood and can survive for 25 days. However, in crowded conditions and when food is short, the worm enters a larval state known as the **dauer** larval state, where it does not eat grow or reproduce until food becomes available again. This state can last for 60 days. An insulin/IGF-1 pathway in *C. elegans* plays an important role in controlling lifespan. Mutations that lower the level of DAF2 expression, which encodes a receptor in this pathway, can cause the animal to remain youthful and live twice as long as normal. This pathway also regulates entry into the **dauer state**. However, the exact mechanism is not known; there probably may be a lowering of metabolic rate. Surprisingly, restoring the DAF2 pathway to neurons alone was sufficient to specify wild-type lifespan, suggesting the key role of the nervous system in regulation of longevity.

Other genes regulate ageing in yeast by increasing the resistance to oxidative stress. Superoxide dismutase, a protein that protects against the effects of mitochondrial free radicals, can extend yeast lifespan in stationary phase when overexpressed.

Program restriction theory: It is now well established that genetic information depends on base sequences of DNA. Only a small proportion of information in DNA is utilized by cells and the sequences appropriate for a particular tissue only are transcribed and translated. The remaining DNA is 'turned off'. The concept of program restriction states that ageing may be due to increased 'turn off' of DNA leading to impaired ability of the cell to transcribe necessary RNAs including mRNAs. It is also possible that transcription of messages may be normal but the ability to translate these messages into functional proteins is restricted in the cells of older individuals. Though DNA segments may not be physically lost, chromatin structure seems to be modulated with age in a way that favours restriction of DNA transcription. This might lead to a functional loss of portions of the gene library and impair overall cellular functions.

Cellular senescence: As mentioned above, senescence is not universal and is not observed in single-celled organisms that reproduce through the process of cellular mitosis. Moreover, cellular senescence is not observed in many organisms, including sponges, corals and lobsters. In these species, where cellular senescence is observed, cells eventually become *post-mitotic* when they can no longer replicate themselves through the process of cellular mitosis and cells experience replicative senescence. Even though how and why some cells become post-mitotic in some species is not understood, it is widely believed that cellular senescence evolved as a way to prevent the onset and spread of cancer. Somatic cells that divide many times will accumulate DNA mutations and would therefore be in danger of becoming cancerous if cell division continue.

The cells isolated from an animal, even if they are cultured in the presence of adequate medium, growth factors and optimal environment, will not grow continuously. For example, mammalian fibroblasts (connective tissue cells) will only grow through a limited number of cell doublings in culture and then stop dividing.

Telomere lengths in ageing: Recently the role of telomeres in cellular senescence has aroused general interest. One of the features shared by senescent cells in culture and *in vivo* is shortening of the telomeres. These are the repetitive DNA sequences at the ends of the chromosomes that preserve chromosome integrity and ensure that chromosomes replicate themselves completely without loss of the DNA containing information at the ends. In the somatic cells of mammals, the telomere repeats have been proposed to provide each cell with a counting mechanism that prevents the unlimited proliferation of wayward cells in adult tissues. Somatic cells develop with a full complement of telomeric repeats, but later the telomerase enzyme is turned off in a tissue, so that each time a cell divides, it loses 50–100 nt from each of its telomeres. After many cell generations, the descendent cells will inherit defective chromosomes and will consequently withdraw permanently from the cell cycle and cease dividing—a process called *replicative cell senescence*. The idea that telomere length acts as a 'measuring stick' to count cell divisions and thereby regulate the cell's lifetime has been tested in several ways. The length of the telomeres is reduced in older human cells progressively during DNA replication. This suggests that the telomere ends are not completely replicated, which leads to senescence. If the enzyme telomerase, which maintains telomere length and is normally absent from cells in culture, is expressed in those cells, senescence in culture does not occur. The scientists who made significant contributions in this ava have been honoured with a Nobel Prize (See box).

Neuroendocrine theory: All the body activities are directly dependent on neurological and endocrine systems. During ageing the loss of neurons and endocrine cells takes place. Decrease in brain weight results from cell loss and losses of fluids and ground substance. Changes in endocrine function result from cell loss and changes in receptors for hormone action. This modulation in gene expression may be due to oxidative DNA damage at promoter regions in the genome. Genes that are *downregulated* over the age of 40 include: GluRI AMPA receptor subunit, NMDA R2A receptor subunit (involved in learning), subunits of the GABA-A receptor, genes involved in long-term potentiation (e.g., calmodulin 1 and CAM kinase II α), calcium signalling genes, synaptic plasticity genes, synaptic vesicle release and recycling genes. Genes that are *upregulated* include genes associated with stress response and DNA repair and antioxidant defence.

Immune theory: The immune system in mammals plays a critical role in protecting against various pathogens, infectious agents, toxins and aberrant cells and cellular fragments. Defect or decline in the immune system can be fatal to an organism. The immune theory of ageing is based on two facts:

- Age-dependent decline in the ability of immune system, especially thymus-derived immunity (T cell or cellular immunity),
- Autoimmunity: Immune system starts treating autologous tissues as foreign.

These conditions may lead to vulnerability to diseases and abnormalities resulting from auto-immune response. Certain diseases of old age are characterized by some degree of immune system dysfunction. Some of these diseases include, cancer (due to failure of immune surveillance), maturity onset diabetes (auto-antibody activity against insulin receptors, insulin and Islet cells of pancreas), senile dementia (auto-antibody activity against neuronal surface and neuro-filaments), some vascular diseases and auto-immune effects bring about changes in proteins such as collagen and errors in transcription, translation or post-translational alterations.

Somatic mutation theory: Mutations are ubiquitous in all biological systems. They are the changes in the DNA of a cell. Somatic cells of the body of organisms develop spontaneous mutations. These mutations may be incorporated by errors in the replication of DNA during cell division. The impaired DNA may render the cell for deregulation in the gene expression and may have defects in the production of necessary enzymes resulting in cell ageing or even cell death. However, the somatic mutation theory fails to explain the following two points.

- The germ cell line is immortal and show no accumulation within a generation of deleterious mutations.

- Immortal but cancerous clones can emerge from senescent somatic cells by virus transformation.

Error and fidelity theory: This theory of ageing is one of the most extensively studied molecular theories of ageing. The errors result from a decline in the fidelity of all biological components. Here, fidelity refers to faithful reproduction of proper proteins through transcription and translation. The decrease in fidelity of transcription or translations is termed error and is distinct from DNA-based mutations. Errors in transcription or translation result in the insertion of improper amino acids into the sequence comprising of a particular protein. Murray and Holliday (1981) have demonstrated an increased error frequency of DNA polymerases derived from senescent human fibroblasts. There may be a point in error accumulation when it becomes overwhelming or catastrophic. Below this level, errors may be present

Telomeres

Three Americans, Elizabeth Blackburn, Jack Szostak and Carol Greider, were awarded the Nobel Prize for medicine in 2009 for revealing the existence and nature of telomerase, an enzyme which helps prevent the shortening of chromosomes that underlies ageing and cancer.

Elizabeth Blackburn mapped DNA sequences, when studying the chromosomes of *Tetrahymena*, a unicellular ciliate organism; she identified a DNA sequence that was repeated several times at the ends of the chromosomes. The function of this sequence, CCCCAA, was unclear at that time. At the same time, Jack Szostak had made the observation that a linear DNA molecule, a type of minichromosome, is rapidly degraded when introduced into yeast cells. In 1980, Jack Szostak's and Blackburn performed an experiment: from the DNA of Tetrahymena, Blackburn isolated the CCCCAA sequence and Szostak coupled it to the minichromosomes and put them back into yeast cells. They got striking results—the telomere DNA sequence protected the minichromosomes from degradation. As telomere DNA from one organism, Tetrahymena, protected chromosomes in an entirely different one, yeast, this demonstrated the existence of a previously unrecognized fundamental mechanism. Later, it became evident that telomere DNA, with its characteristic sequence, is present in most plants and animals—from amoeba to man.

Carol Greider (1984), then a graduate student under the supervision of Blackburn, discovered signs of enzymatic activity in a cell extract and they named the enzyme as telomerase, purified it and showed that it consists of RNA as well as protein. The RNA component turned out to contain the CCCCAA sequence. It serves as the template when the telomere is built, while the protein component is required for the construction work, i.e., the enzymatic activity. Telomerase extends telomere DNA, providing a platform that enables DNA polymerases to copy the entire length of the chromosome without missing the very end portion.

Scientists later began to investigate what roles the telomere might play in the cell. Szostak's group identified yeast cells with mutations that led to a gradual shortening of the telomeres. Such cells grew poorly and eventually stopped dividing. Blackburn and her co-workers made mutations in the RNA of the telomerase and observed similar effects in Tetrahymena. In both cases, this led to premature cellular ageing—senescence. In contrast, functional telomeres prevent chromosomal damage and delay cellular senescence. Greider's group showed that the senescence of human cells is also delayed by telomerase.

These discoveries had a major impact within the scientific community. Many scientists speculated that telomere shortening could be the reason for ageing, not only in the individual cells but also in the organism as a whole. But the ageing process has turned out to be complex and it is now thought to depend on several different factors, the telomere being one of them. Most normal cells do not divide frequently, therefore their chromosomes are not at risk of shortening and they do not require high telomerase activity. In contrast, cancer cells have the ability to divide infinitely and yet preserve their telomeres; the reason is that the cancer cells often have increased telomerase activity. It was therefore proposed that cancer might be treated by eradicating telomerase.

Some inherited diseases are now known to be caused by telomerase defects, including certain forms of congenital aplastic anaemia, in which insufficient cell divisions in the stem cells of the bone marrow lead to severe anaemia. Certain inherited diseases of the skin and the lungs are also caused by telomerase defects.

in non-critical levels and balanced by turnover and repair mechanisms.

Miscellaneous

Recently, early senescence has been alleged to be a possible unintended outcome of early cloning experiments. Most notably, the issue was raised in the case of Dolly the sheep, following her death from a contagious lung disease. The claim that Dolly's early death involved premature senescence has been vigorously debated.

But the fundamental concepts linking programmed death to the evolution of multicellularity were advanced as early as 1881 by August Weismann. He proposed that ageing and decay are not inherent to cellular life but are events that became integral to the development only in the course of evolution of multicellular organisms. Only multicellular organisms are inevitably doomed to death, through senescence. However, in a sense, single cell organisms like bacteria and protists are immortal. The life of bacteria and protists is terminated, either destroyed by parasites, or through environmental influences, but not through ageing. In single cell organisms, Weismann argued, there is no regular process of death to terminate individual life, because every individual cell is simultaneously a **generative cell** that secures the continuation of the species. By contrast, in multicellular organisms segregation of generative and somatic cells has evolved. **Somatic cells** can focus on a few functions of life and are finetuned to perform their particular task for the benefit of the entire cell community: some cells secrete enzymes and maximize the exploitation of food; other cells optimize sensory functions to exploit the environment and recognize dangers or opportunities. Death followed by senescence in metazoans is the regular fate of somatic cells. However, even in metazoans, there are, exceptions to this rule at cell level. Cells of metazoans (including humans) can be maintained in culture for long periods of time (immortal), just like protists. However, cells grown in petridishes have been subjected to selection and tend to change to other cell types or can become cancerous cells.

Multicellular organisms, after completing growth and reproductive phase, die. However, recent studies have indicated that there are genetic components to senescence and the genetically-determined lifespan characteristics of a species can be modulated by altering genes or diet. The maximum lifespan is the maximum number of years a member of species is known to survive and is a characteristic of a given species. For example, the maximum human lifespan is estimated to be 121 years. However, the average lifespan of humans is way below 100 years. The main signs of ageing are greying hair, sagging and wrinkling skin, joint stiffness, osteoporosis, loss of muscle fibres and muscular strength, memory loss, eyesight deterioration and the slowing of sexual responsiveness.

Non-genetic (error) theories

Error theories blame environmental impact on living organisms that induce cumulative damage at various levels as the cause of ageing.

Accumulation of chemical damage: One view of senescence is that it is the outcome of an accumulation of damage that eventually outstrips the ability of the body to repair itself and so leads to the loss of essential functions. Damage to longlived biopolymers, such as structural proteins or DNA, caused by ubiquitous chemical agents in the body such as oxygen and sugars, are in part responsible for ageing. The damage can include breakage of biopolymer chains, crosslinking of biopolymers, or chemical attachment of unnatural substituents (haptens) to biopolymers.

Glycation theory: It was proposed by Cerami (1985). According to this theory glucose acts as a mediator of ageing. The process involves non-enzymatic reactions that occur between reducing sugars and tissue proteins and nucleic acids. Sugars such as glucose and fructose can react with certain amino acids such as lysine and arginine and certain DNA bases such as guanine to produce sugar adducts, in a process called *glycation*. These adducts can further rearrange to form reactive species which can then cross-link the structural proteins or DNA to similar biopolymers or other biomolecules such as non-structural proteins. The end products of such reactions in case of proteins leads to dehydrated and rearranged irreversible structures referred to as advanced glycation end products. Similar end products are formed between DNA and reducing sugars. Glycated proteins exhibit loss of enzyme activity, inappropriate cross-linking and decreased degradation of abnormal proteins. Glycation of DNA, most usually uracil residues derived by de-amination of thymine, if not repaired by excision repair and uracil-DNA glycosylase, can induce mutations. Hemoglobin, collagen and lens proteins are among the proteins that undergo glycation with age. Glycation of collagen may lead to elasticity changes in blood vessels, tendons and much of the connective tissue. Glycation of nucleic acids is associated with diminished activity of uracil–DNA glycosylase in late passage cells in culture. Thus, glycation would seem to have profound cumulative effects through the lifespan, but it remains to be seen whether these glycation effects are the cause of ageing or a result of changes occurring in other fundamental processes.

Diabetic people, with elevated blood sugar, develop senescence-associated disorders much earlier than the general population. There is evidence that sugar damage is linked to oxidant damage in a process termed *glycoxidation*. Free radicals can damage proteins, lipids or DNA. Glycation mainly damages proteins. Damaged proteins and lipids accumulate in lysosomes as lipofuscin. Chemical damage to structural proteins can lead to loss of function; for example, damage to collagen of blood vessel walls can lead to vessel-wall stiffness and thus hypertension and vessel wall thickening and reactive tissue formation (atherosclerosis); similar processes in the kidneys can lead to renal failure. Damage to enzymes reduces cellular functionality. Lipid peroxidation of the inner mitochondrial membrane reduces the electric potential and the ability to generate energy. It is known that nearly all the so-called 'accelerated ageing diseases' are due to defective DNA repair enzymes. These changes bring impairment of the cellular functions.

Free radicals and anti-oxidants

Free radicals are charged molecules, which are produced transiently during normal metabolism. They are highly reactive and powerful oxidants, which can destroy cell components. Mammalian tissues contain an enzyme, superoxide dismutase (SOD), which acts to remove free radicals. Long-lived species contain high levels of SOD. Vitamin E also acts as a free radical scavenger. Some antioxidants have been reported to increase the life expectancy or average lifespan of mice.

15.3.3 The process of ageing

As mentioned earlier, ageing and organismal death is a complex process and there is no single to describe it. Ageing can occur at molecular, cellular and organismal levels.

Limits set at molecular level

Ageing of proteins: Cell survival and functioning mainly depend on active metabolism and cellular integrity. The metabolic activities of a cell are driven by a number of enzymes, which perform both anabolic and catabolic processes. Enzymes and other proteins are not in the energetically lowest conformational state when they are biologically active; they are in a metastable state, created during their synthesis, sometimes with the assistance of chaperones. Many factors such as temperature, changing ionic strength and the pH of the solute or collision with radicals, may throw a protein out of its metastable condition into a deeper energetic level (denatured). The probability of spontaneous renaturation is minimal and denatured proteins are useless and must be replaced by newly synthesized ones. A terminally differentiated cell will have difficulties replacing all its proteins. For example, a heart muscle cell cannot replace its contractile apparatus without interrupting its rhythmic beating. A nerve cell in a brain cannot repair its numerous dendritic and axonal fibres and hundreds of synaptic connections while staying prepared to react.

Stability of genetic information: Similar to proteins and other biomolecules in a cell, the DNA also suffers from physical, chemical and biological damage. Though a cell has multiple mechanisms of DNA repair, the mechanisms function as long as one of the two DNA strands remains intact and can serve as a template of correct base sequence. In mammals, a correlation has been found between the capacity of the DNA repair system and the average lifespan of the species (**Fig. 15.21**).

Accumulation of damage at the cellular and organismal levels

At cellular level, at least three ageing processes are involved.

Decline in functional efficiency: It is the possible decline in the functional efficiency of non-dividing highly specialized cells, such as nerve cells, muscle cells and so on.

Progressive stiffening of structural proteins: The structural protein collagen constitutes more than one-third of all the body proteins and serves as the general binding substance of the skin, muscular and vascular systems. It is also the substance of cartilage and tendons that fills up the spaces between muscle fibres and between the cells of many organs, serving as

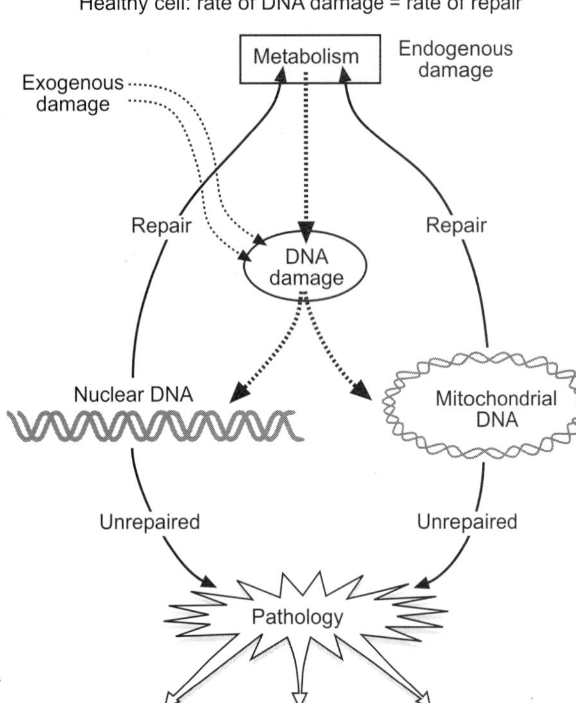

Fig. 15.21 Pathological effects of poor DNA repair

the stabilising fibre of connective tissue. In the skin a thick matting of collagen fibres forms the *Corium* that gives the skin its toughness and plasticity. During the ageing process, collagen accumulates changes in the molecular structure that affects its integrity and function. The ageing of collagen may provide an objective index for determining the 'biological age'. Once the fibres of collagen are laid down in the body they cannot be renewed. A collagen molecule consists of three strands wound around one another in helices. During collagen ageing cross-links increase in number between the different strands and different molecules making it less flexible. This rigidity can cause strain on muscular movements associated with different vital organs and affect organ function.

Limitations of cell division: Hayflick (1968) has experimentally demonstrated that as the population of human fibroblasts approaches the end of their lifetime (**Fig. 15.22**), aberrations often crop up in the chromosomes. Such aberrations and cell division peculiarities related to age have also been observed in human leukocytes and in liver. In humans, several organs lose weight after middle age, which can be directly attributed to cell loss. The human brain weighs considerably less in old age than it does in the middle years. The kidneys also show a large reduction in nephrons accompanying cell loss and the number of taste buds per papilla of the tongue drops from an average of 245 in young adults to 88 in the aged.

As age progresses, the DNA of dividing cells may become clouded with an accumulation of copying errors. Normally, the coding and decoding

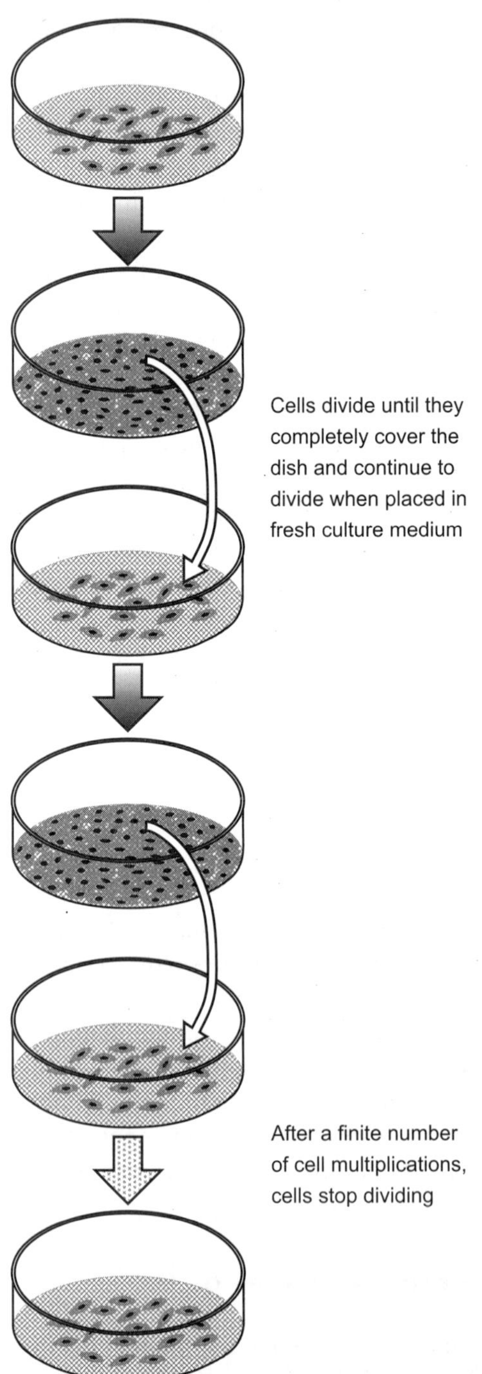

Fig. 15.22 Vertebrate fibroblasts can only take part in a limited number of divisions in culture.

systems that govern the replication of DNA operates with a high degree of accuracy, but the accuracy need not be absolute. Certain enzymes involved in the transcription of information from DNA for the synthesis of proteins may deteriorate with age. At all cellular events, the ability of cells to divide or to function is controlled by the inherited information in the molecules, but it appears that degeneration of these molecules may hold the key to the ageing and eventual death of cells.

Cellular changes during ageing

Ageing causes functional changes in cells at various levels, which includes morphological, physiological and sub-cellular changes.

Morphological changes: The most obvious morphological changes are decrease in overall cell volume accumulation of peroxidation by-products, condensation of nuclear material (shrunken; pyknosis) and increase in lipid vacuoles.

Physiological changes: Some of the major physiological changes are collagen stiffening, accumulation of Ca^{2+} and loss of ribosomal DNA (rDNA). Losses of rDNA might result in lowered amounts of protein synthesis leading to the presence of fewer ribosomes in the cytoplasm.

DNA repair

It has been reported that the ability to repair DNA damage induced by UV light correlates well with the lifespan. Cell culture experiments have demonstrated the ability of connective tissue from seven mammalian species of widely differing lifespans to repair UV damage to DNA in culture. It has revealed that animals with longer lifespan (man and elephant) have better ability to repair DNA damage than those with lower lifespan.

RNA synthesis and cellular ageing

Decline in transcription may result from alterations in chromosome organisation, specifically at the level of chromosomal proteins that complex with DNA to form chromatin. Chromatin modifications include an increase in cross-linking between DNA and chromosomal proteins, a condition that limits transcription in vitro. Histone acylation undergoes an age-related decrease, which also leads to a structure of chromatin not conducive to transcription. But the quantity and kind of histones associated with DNA in chromatin appears unaffected with age.

Protein synthesis and cellular ageing

The rate of protein synthesis universally declines with age in all eukaryotic organisms. All cellular components involved in protein synthesis have been implicated at one or another time. Ageing also causes alterations in tRNA, including changes in acetylation of tRNAs (required for amino acid binding). Elongation factors (specifically, EFI), which are necessary for stabilization and continued elongation of the polypeptide chain from the polyribosomes, decrease with age. Post-translational modifications also occur during ageing. The three-dimensional confirmation of proteins may be altered with age. De-amination of eye lens proteins has been reported with age, which might be related with the capacity of lens. Racemization (conversion of protein amino acids from L to D isomer) has been observed in tooth enamel.

Sub-cellular changes associated with ageing

A number of sub-cellur structures of eukaryotic cells have been found to be affected by ageing, mainly decrease or loss of functions.

- ***Plasma membrane and ageing:*** Overall functional capacity of plasma membrane gets perturbed during ageing. Its permeability changes, Ca^{2+} accumulates in the membrane and the total lipids of membrane decreases with age. The cholesterol and phospholipid ratio in cellular membranes increases with age, resulting in diminished fluidity of the membranes.
- ***Endoplasmic reticulum and ageing:*** The amount of granular ER decreases in the cytoplasm of old cells. In the nerves of older animals, Nissl granules (rRNA) are found to decrease.
- ***Mitochondria and ageing:*** Carbohydrate metabolism (Krebs cycle) in aged people is low and also affects glycolysis. Studies have shown that mitochondria are involved in the ageing process of the animals; in older tissues the mitochondria degenerate. The rate of respiration in multicellular organisms also becomes slow and somewhat dehydrated.
- ***Nucleus and ageing:*** Change in nucleocytoplasmic ratio acts as an important index for natural senescence and death. Incidences of loss of one X in females, Y in males and 21 chromosomes in humans have been reported in ageing cells.

Biochemical changes associated with ageing

As age advances, a variety of biochemical changes can be observed. Some of the known changes that have been detected are:

- Increase in cholesterol and triglyceride levels
- Increase in blood globulin
- Decrease in alkaline and acid phosphatases
- Decrease in cellular respiration
- Increase in serum creatine.

Ageing and organ systems

As people get older, a number of changes occur in various organ systems of the body. The rate of senescence is slow and progressive. It progresses at different rates in different systems and in different individuals, with increasing probability of organ system failures and death. Some major changes in various organ systems are discussed below.

Nervous system: Noticeable structural as well as functional changes can be observed in the nervous system of ageing organisms.

- Atrophy of the brain and increase in size of brain ventricles
- Loss of weight (1400 grams at age 20 to 1334 at age 60)
- Loss of white matter
- Shrinkage of dendrites
- Loss of nerve cells
- Accumulation of age pigment, lipofuscin
- Increase in cerebral amyloid
- Decrease in synaptic density in cerebellum
- Loss of cerebellar Purkinje cells
- Decline in cerebellar integrative functions
- Decline in shortterm memory and recall from long-term storage
- Decline in learning.

Sense organs: The effects of age on human body manifest prominently by impairing the function and structure of various organs. Some of the changes commonly observed as people get older are:

- Decrease in the number of taste buds
- Decline in olfactory sensitivity
- Reduction in sensory cells in nasal passages
- Occurrence of dropping eye lids (senile ptosis)
- Decline in lacrymal secretion
- Development of astigmatism, cataract and presbyopia
- Decrease in the aperture of the iris (senile miosis)
- Retinal detachment
- Diminished fidelity of colour vision
- Physiological degeneration of auditory system (presbycusis) resulting in bilateral loss of hearing
- Degeneration of vestibular system
- Increase in touch threshold.

Immune system: As previously discussed in this chapter, the immune system is strongly affected by the ageing process in the dollowing ways:

- Implant of both cellular and humoral immunity
- Involution or regression of the thymus gland
- Decline in T cell function
- Increase in susceptibility to infection
- Increase in auto-immune diseases such as inflammation of thyroid, joints (arthritis), nerve endings (multiple sclerosis), testes (sterility).

Endocrine system: The endocrine system that regulates various metabolic and behavioural responses is also affected during ageing. A few common measurable changes are:

- Decline in TSH (thyroid stimulating hormone secretion from the pituitary gland)
- Decrease in the synthesis of sex hormones from the gonads
- Decrease in renin and aldosterone levels.

Gastrointestinal systems: The gastrointestinal tract is the main route of nutrient intake. It functions throughout the lifespan of an organism. A large number of changes can be observed in the gastrointestinal system during ageing. They are:

- Atrophy of oral mucosa and degeneration of oral connective tissue
- Hardening of tooth enamel
- Calcification of tooth pulp leading to tooth loosening
- Reduction in salivary secretion
- Reduction in peristaltic activity
- Atrophy of gastric mucosal lining and diminished gastric digestion

- Impairment of intestinal digestive and absorptive processes
- Loss of liver weight and decrease in the number of hepatocytes per unit volume of liver
- Decreased motor functions in the colon leading to frequent constipation
- Increase in colorectal cancer.

Integument: Integuments, which are located at bone joints, play a vital role in movement. In addition to wear-and-tear, succumb to the ageing process, mainly cross the linking of collagen fibres of connective tissue reducing the elasticity of the skin.

Blood vascular system: With age the circulatory system shows many changes, such as:

- Diminution of heart muscle mass and increased fibrous tissue
- Increase in atrial volume
- Accumulation of lipofuscin pigment in heart muscles
- Degenerative calcification of aortic valves
- Increase in arterial intimal layer, due to accumulation of smooth muscle cells
- Development of atherosclerosis
- Increase in systole, diastole and pulse pressure.

Muscles and bones: Very severe and highly noticeable changes can be observed in muscles and bones with ageing:

- Atrophy of skeletal muscles
- Decrease in the size and number of muscle fibres
- Decline in muscle tension
- Increase in osteoclast activity and decrease in osteoblast activity resulting in increased bone resorption
- Development of osteoporosis
- Loss of joint flexibility due to cross-linking of collagen molecules.

Respiratory system: The respiratory system, which involves an exchange of gases between the body and environment, also shows noticeable changes with ageing.

- Increase in volume of ducts leading to alveoli at the expense of alveolar volume, so that surface area for gaseous exchange decreases
- Increase in thickness of the bronchial mucous gland layer
- Decrease in vital capacity of the lungs
- Increase in dead space, or non-respiratory space
- Diminution of diffusion capacity in lungs
- Progressively difficult respiration
- Complementary increase in residual volume.

Excretory system: A number of age-related diseases are associated with excretory system. Some of them are:

- Decrease in weight and volume of the kidneys
- Loss of nephrons; replacement of scar tissue
- Decline in filtration rate
- Impairment of elimination of nitrogenous wastes
- Reduction in creatinine clearance
- Lower capacity and a higher volume of residual urine after voiding in the urinary bladder
- Frequent urination and incontinence
- Nocturnal micturition.

Reproductive system and sexual activity: Very common symptoms associated with ageing in reproductive organ functioning are:

- Menopause in female human beings; atrophy of ovaries
- Enlargement of prostate glands in males
- Atrophy of testes resulting in decreased size; loss of cells and decrease in size of cells in seminiferous tubules
- Increased sexual arousal and capacity for orgasm in post-menopausal females
- Decrease in penile erection.

15.4.2 Causes of ageing

Recent studies have shown that the causes of ageing can be attributed to molecular, cellular and organismal.

Oxidative damage: One theory convincingly explains that our metabolism itself causes no mutations. About 2%–3% of the oxygen atoms taken up by the mitochondria reduced insufficiently to *reactive oxygen species* (ROS). These ROS include the superoxide ion, the hydroxyl radical and hydrogen peroxide. ROS can oxidize and damage cell membranes, proteins and nucleic acids. Evidence for this theory includes the observation that *Drosophila* that overexpress enzymes that destroy ROS (catalase, which degrades peroxide and superoxide dismutase) live 30%–40% longer than do controls.

General wear-and-tear and genetic instability: Wear-and-tear and genetic instability have been considered as important factors for ageing. As one gets older traumas to the body build up. Point mutations increase in number and the efficiencies of the enzymes encoded by our genes decrease. Mutations in genes responsible for DNA replication, transcription and protein synthesis can produce a large number of faulty proteins. Likewise, DNA repair may be important in preventing senescence and species, whose member cells have more efficient DNA repair enzymes, live longer. Mutations in repair enzymes can cause premature ageing syndromes in humans.

Mitochondrial genome damage: Considerable attention is now being directed to changes in mitochondrial DNA (mtDNA) that might underlie age deficits in various mammalian tissues. It is estimated that the mutation rate in mitochondria is 10–20 times faster than nuclear DNA; therefore, it is thought mutations in mitochondria could

- lead to defects in energy production
- lead to production of ROS by faulty electron transport
- induce apoptosis.

Age-dependent decline in mitochondrial function are seen in many mammals, including humans. Recently, presence of hot spots for age-related mutations in the mitochondrial genome has been identified. Since, mutant mitochondrial DNAs have higher replication rate, these can not only make more ROS, mitochondrial DNA become more susceptible to ROS-related damages.

Genetic ageing programs: Large numbers of genes have also been shown to affect ageing. In case of humans, Hutchinson–Gilford progeria (hereditary diseases collectively called progeria) syndrome causes children to age rapidly and die as early as 12 years. It is caused by a dominant mutant gene and its symptoms include thin skin with age spots, resorbed bone mass, hair loss and arteriosclerosis.

Genetically released suicide program: Apoptosis or (programmed cell death), affects certain cells as early as embryogenesis and later in life; consequently many cells die a few days or weeks after their birth. However, the short-lived cells are replaced by substitutes generated from stem cells; but apoptosis in the longlived, irreplaceable cells sets the limit to the lifespan. Recent studies indicate that most cells survive only in social communities. Cells cut off from society commit suicide. The actual cause is the presence of survival factors, with which cells prevent each other from switching on their suicide program.

Two hypotheses on genetically programmed mechanisms of self-destruction have been revealed:

- programmed decay of the mitochondrial DNA
- the telomere theory of cell senescence.

Senescence of plants

Flowering and senescence are closely linked in many angiosperms. Individual flowers petals in some species senesce following pollination. Orchids, which stay fresh for long periods of time if they are not pollinated, are a good example. Fruit ripening (and ultimately over-ripening) is an example of organ senescence. Whole-plant senescence leads to the death of the entire sporophyte generation. Monocarpic plants flower once and then senesce. Polycarpic plants, such as the bristlecone pine, can live thousands of years (4900 years is the present record) and flower repeatedly. In polycarpic plants, death is by accident; in monocarpic plants, it appears to be genetically programmed. Flowers and fruits play a key role in the process and their removal can sometimes delay senescence. In some legumes, senescence can be delayed by removing the developing seed—in other words, the embryo may trigger senescence in the parent plant. During flowering and fruit development, nutrients are reallocated from other parts of the plant to support the development of the next generation. The reproductive structures become a nutrient sink and this can lead to whole-plant senescence.

15.5 DEATH

In multicellular organisms, death is an integral part of the life cycle and the death process has been fixed in the genome itself. But the biological meaning of death is organisms have to make room for their offspring. But the fundamental question is, why has potentially immortal life not emerged with offspring generated sparingly (for instance, through asexual budding, as in *Hydra*) in order to compensate for losses due to accidents and to generate colonies for new living spaces? Death makes way for new life.

Summary

- Development, growth, sex and death are linked to every area of biology. For example, humans start life as a single cell (zygote) and nearly eight months later, the child is born. Many astonishing changes take place in that time.
- The ***zygote*** as it divides by mitosis continues to divide a number of times and many cell types form by the process of differentiation. Differentiated cells organize themselves into tissues and organs, fitting together in a tightly organized whole baby (development).
- The overall growth of an organism includes ***cell division, cell enlargement and cell differentiation***.
- ***Reproduction*** is the inherent property of all living organisms to perpetuate and multiply their own kind with little or no change. Most multicellular organisms have evolved a complex sexual reproduction mechanism that helps an organism to mix the genetic material from males and females and generate genetic diversity while maintaining the genomic integrity and phenotypic identity.
- There are two principle modes of reproduction: ***asexual*** and ***sexual***. Sexual reproduction gives a comparative advantage to organisms in an unpredictably variable environment. It is universally credited with contributing to species fitness and generating new genotypes by bringing together the genes from male and female in the fertilized egg.
- Following fertilization, the zygote is destined to develop either as a male or a female. During the early stage of ***embryogenesis***, the primordial germ cells migrate to the developing gonads, which form the ovaries in females, and the testes in males.
- The primordial germ cells later undergo ***meiosis*** and differentiate into mature ***gametes*** (eggs or sperm). ***Gametogenesis*** is the process of forming gametes from diploid cells of the germ line.
- Most species produce two types of gametes: ***egg*** and ***sperm***. The egg is among the largest cells in an organism, while the sperm is often the smallest. The egg and the sperm are optimized in opposite ways for the propagation of the genes they carry.
- In males, ***meiosis*** and ***spermatogenesis*** begin in the testes only after puberty and then go on continuously in the epithelial lining of very long, tightly coiled tubes, called ***somniferous tubules.***
- ***Oogenesis*** is the process of forming an ***ovum*** (egg) by meiosis in specialised gonads known as ***ovaries***. The ovary contains many follicles composed of a developing egg surrounded by an outer layer of follicle cells.
- ***Fertilization*** is the fusion of two gametes (haploid genomes) to form the diploid genome of the zygote. The two major types of fertilization that have evolved are external and internal fertilization. The process of fertilization plays an important part in sexual reproduction.
- Organisms continue to change their structural and functional features after completing their embryonic development.
- Similar to other developmental process, ageing may be programmed.
- ***Death*** is the ultimate consequence of ageing and differences in maximum lifespan between species correspond to different 'rates of ageing' and it is the reverse process of development.
- ***Ageing*** and a limited life span are characteristic of all animals that stop growing after reaching a standard mature size. Ageing peoples bodies normally change in appearance as they get older, such as, changes in height, weight, composition and facial expressions.
- Many theories have been proposed to explain cellular ageing and senescence. Each theory has its advocates and limitations as well. These theories include genetic and non-genetic aspects.
- No single theory can account for all the observations of ageing at the cellular level.
- Even though the process of ageing of an organism begins from the stage of zygote and lasts until death, ageing from late adolescence is most significant.
- ***Senescence*** is the increase of impairment of physiological functions with age, resulting in decreased ability to deal with a variety of stresses and an increased susceptibility to disease.It indicates decline in organismal metabolic activity and reaching towards dead end of life cycle, that is, death.
- ***Death*** is the last stage of an organism's somatic life cycle.

Exercises

1. Define growth and explain the processes that contribute to growth.
2. Explain briefly a typical growth curve and its use in measuring growth of cell population
3. Explain with examples the mode of asexual reproduction.
4. What is sexual reproduction? Mention its biological significance.
5. Discuss briefly different stages of sex determination during embryogenesis of animals.
6. How does the environment play a role in sex determination?
7. Explain briefly the different steps involved in gametogenesis in animals.
8. How does plant gametogenesis differ from animal gametogenesis?
9. What are the different cellular recognition events that occur during fertilization?
10. What is ageing? Explain the different biological theories of ageing.
11. What are the different genetic and non-genetic errors that contribute to ageing?
12. Briefly discuss the different causes of ageing.
13. What major changes are generally observed in the organs of an ageing person?
14. What is senescence?
15. Discuss the genetic relevance of death of organisms.

Further Reading

Books

Alberts B, Johnson A, Lewis J, Raff M, Roberts K and Walter P (2002), *Molecular Biology of the Cell,* 4th edition, Garland Science.

Blumenthal HT (1983), *Handbook of Diseases of Aging*, Van Nostrand, New York.

Bonner J (1965), *The Molecular Biology of Development*, Oxford University Press, Oxford.

Brookbank JW (1990), *The Biology of Aging*, Harper and Row, Publishers, New York.

Finch CE and Schneider (1985), *Handbook of the Biology of Aging*, Van Nostrand Reinhold, New York.

Gilbert SF (2000), *Developmental Biology*, 6th edition, Sinauer Associates Inc.

Mittwoch U (1973) Genetics of Sex Differentiation. Academic Press, New York.

Muller WA (1997), *Developmental Biology*, Springer–Verlag, New York.

Strehler BL (1977), *Zime, Cells and Aging*, 2nd edition, Academic Press, New York.

Swanson CP and Webster P (1977), *The Cell*, Prentice-Hall, New Jersey.

Zwilling R and Balduini C (1992), *Biology of Aging,* Springer–Verlag, Heidelberg.

References

Bennett D (1980), The T complex in the mouse: An assessment after 50 years of study, *The Harvey Lectures*, 74:1–21.

Bull JJ an RC Vogt (1979), Temperature-dependent sex determination in turtles, *Science,* 206:1186–89.

Carter CS, Ramsey MM and Sonntag WE (2002), A critical analysis of the role of growth hormone and IGF-1 in aging and life span, *Trends in Genetics*, 18:295–301.

Cerami A (1905), A hypothesis: Glucose as a mediator of aging, *J. Am. Geriatr. Soc.*, 33:626.

Goodfellow PN and Lovell–Badge R (1993), SRY and sex determination in mammals, *Annu. Rev. Genet.*, 27:71–92.

Hayflick L (1985), Theories of biological aging, *Exp.Gerontol.*, 20:145.

Helfand SL and Rogina B (2003), Genetics of aging in the fruit fly *Drosophila melanogaster*, *Annu. Rev. Genet.*, 37:329–48.

Kenyon C (2001), A conserved regulatory system for aging, *Cell,* 105:165–68.

Kirkwood TBL and Austad SN (2000), Why do we age? *Nature,* 105:165–68.

Kozma SC and Thomas G (2003), Regulation of cell size in growth, development and human disease: PI3K, PKB and S6K, *Bioessays,* 24:65–71.

Morgan TH (1910), Sex limited inheritance in *Drosophila*, *Science,* 32:120–22.

Ohno S (1979), *Major Sex Determining Genes*, Springer–Verlag, Berlin.

Schafer AJ (1995), Sex determination and its pathology in man, *Adv. Genet.*, 33:275–329.

Shy JW, Wright WE (2001), When do telomeres matter? *Science,* 291:839–40.

Stock NW (1977), Biological theories of aging, In Birren J and Schaise W, (eds), *Handbook of the Psychology of Aging*, van Nostrand Reinhold, New York.

Swain A and Lovell–Badge R (1999), Mammalian sex determination: a molecular drama, *Genes Dev.*, 13:755–67.

Wallace DC (1992), Mitochondrial genetics: A paradigm for aging and degenerative diseases, *Science,* 256:628–32.

Wassarman PM, Jovine L and Litscher ES (2001), A profile of fertilization in mammals, *Nat. Cell Biol.*, 3:E59–E64.

Williams GC (1957), Pleotropy, natural selection and the evolution of senescence, *Evolution*, 11:398–411.

Wyle C (1999), Germ cells, *Cell,* 96:165–74.

16

Cell Death and Cell Renewal

Learning objectives

After reading this chapter, you will be able to:

- Outline the various factors that cause cell death
- Appreciate the recent developments in research in cancer and HIV progression as aberrations of cell death

16.1 INTRODUCTION

Death is an unavoidable process in all living organisms. It can happen naturally or accidentally by a number of death-causing agents. However, in multicellular organisms, in addition to the organismal death, (self-destructive) death also occurs at individual cell or tissue level. This kind of death is pervasive, organized, rational and very essential for the development and survival of multicellular organisms. Accidental cell death may be due to the result of mechanical trauma or exposure to some kind of toxic agent and often referred to as **necrosis**. In unicellular organisms, this is the only type of death possible. The other type of death may be deliberate and essential, due to a built-in suicide mechanism known as **apoptosis** (Greek: *apo* = from, *ptosis* = dropping off or falling off) or developmentally regulated cell death called **programmed cell death** (**PCD**).

The necrotic and PCD cell deaths are quite distinct in many respects. In cells that are damaged, ATP concentrations fall very low, consequently the Na^+/K^+ATPase can no longer operate and therefore ion concentrations are no longer controlled. This causes the cells to swell and then burst. The cell contents then leak out, causing the surrounding tissues to become inflamed. However, apoptosis is an orderly process that generally confers advantages during an organism's life cycle. For example, the differentiation of fingers and toes in a developing human embryo requires cells between the fingers to initiate apoptosis so that the digits can separate. The process involves an orchestrated series of biochemical events leading to a characteristic cell morphology and death. It is executed in such a way to safely dispose of cell corpses and fragments. The shrinking nucleus and the cell contents are packaged into small membrane-bound packets called blebs.

In multicellular organisms, PCD is very essential to achieve and maintain homeostasis. Cells undergo self-destruction when they are no longer needed or if they are damaged. The amount of PCD that occurs in developing and adult animal tissues is enormous. In the developing vertebrate nervous system, for example, more than half the nerve cells normally die soon after they are formed. In a healthy adult human, billions of cells die in the bone marrow and intestine every hour. This is accomplished by activation of genetically regulated cell suicide machinery that requires the active participation of the cell in a suicide process. PCD is now recognized as a common cellular default mechanism in that almost all cells appear to possess the machinery necessary to carry out the terminal steps of death.

In addition to its importance as a biological phenomenon, defective apoptotic processes have been implicated in a wide variety of diseases. Excessive apoptosis causes diseases, such as in ischemic damage, whereas an insufficient amount results in uncontrolled cell proliferation, such as cancer. An estimate suggests that about 50–70 billion cells die each day due to apoptosis in the average human adult. In a year, this amounts to the proliferation and subsequent destruction of a mass of cells equal to an individual's body weight.

16.2 EVOLUTIONARY ORIGIN OF PCD

Many molecular biology experiments show that mitochondria originated from bacteria that had incorporated as endosymbionts of larger, eukaryotic cells. This evolutionary step would have been more than risky for the primitive eukaryotic cells that began to engulf energy-producing bacteria and conversely, it must have been a perilous step for the ancestors of mitochondria that began to invade their proto-eukaryotic hosts. A similar kind of process exists today between human WBCs and bacteria. Mitochondriate cells live poised between life and death, because they still retain their collection of molecules that can trigger cell suicide. This process has now been evolved to happen only when programmed; given certain signals to cells, such as feedback from neighbours, stress or DNA damage, mitochondria release caspase activators that trigger the cell death-inducing biochemical cascade. As such, the cell suicide mechanism is now crucial to life.

16.2.1 Experimental evidence of PCD

In 1972, John Kerr andrew Wyllie and Alistair Currie put forward a concept of the apoptotic death of a cell. They suggested that death by apoptosis of selected individual cells within a tissue might be an active process mediated by the cell's own biochemical mechanisms and this might play an important role opposite to that of mitosis over a wide range of physiological and pathological circumstances. Programmed cell death (PCD), a term coined by Lockshin and Williams (1965), is a physiological process that leads to the selective elimination of unwanted cells in multicellular animals and plants. In other words, apoptosis is a distinct form of PCD. In fact, PCD is reported to play a major role in controlling cell number and turnover especially in the skin, gut and reproductive tract, tissue homeostasis and specialization, tissue sculpting and pattern formation, oncogenesis and other disease conditions, defence reactions and in maintaining overall integrity of the animal. Although PCD occurs in diverse processes and cell types, the resultant microscopic and ultrastructural changes are remarkably consistent, suggesting a fundamentally uniform mechanism.

The phenomenon of physiological cell death has been discovered independently several times over the past 150 years. The impact of this process on cell production and homeostasis was widely recognized, however, only after the discovery that cell death in multicellular organisms is subject to genetic control and that abnormalities in cell death regulation can cause diseases such as cancer, auto-immunity and possibly degenerative disorders.

More recent highlights in apoptosis research include the discovery of molecular nature of apoptosis, death receptors and their ligands, definition of the separate apoptosis signalling pathways and identification of the components of the common cell death effector machinery and their biochemical functions. This also led to the understanding of the process and manipulation of the events by molecular techniques.

16.3 MORPHOLOGICAL FEATURES OF APOPTOSIS

Although many pathways and signals lead to apoptosis, there is only one mechanism, which usually causes the death of the cell. After the appropriate stimulus is received by the cell and the necessary controls exerted, a cell undergoes the organized degradation of cellular organelles by activated proteolytic caspases. A cell undergoing apoptosis shows a characteristic morphology that can be observed with a microscope.

16.3.1 Cytological changes in apoptotic cells

Earlier, the morphological description of apoptosis, based on the microscopic and ultrastructural changes, was described in two discrete stages. The first comprised nuclear chromatin condensation, cytoplas miczhrinkage and vacuolization, plasma membrane blebbing and breaking up of the cell into a number of membrane-bound, ultrastructurally well-preserved fragments, called '**apoptotic bodies**'. In the second, apoptotic bodies were phagocytozed by surrounding cells and degraded within phagolysosomes. The dying cell modified its plasma membrane, signalling to macrophages, which respond by engulfing the blebs and the remaining cell fragments and by secreting cytokines that inhibit inflammation. The apoptosis-activated changes are the result of hydrolysis of cellular proteins by a family of proteases called caspases. They are present in all the cells in inactive form and become active by cleaving off of inhibitory domain by protease.

The advantage of this strategy is that no protein synthesis is required to activate the apoptotic pathway.

A number of studies also noted that, during the early stages of apoptosis, no visible changes occur in mitochondria, the endoplasmic reticulum or the Golgi apparatus. However, others have more recently reported swelling of the outer mitochondrial membrane and release of cytochrome C and apoptosis-inducing factor, an oxidoreductase-related flavoprotein, from the mitochondrial intermembrane space. The sequence of ultrastructural changes that characterize apoptosis occurring in a cell is shown diagrammatically in **Fig. 16.1**. The light microscopic micrographs describing various stages of the process occurring in vitro are shown in **Fig. 16.2**.

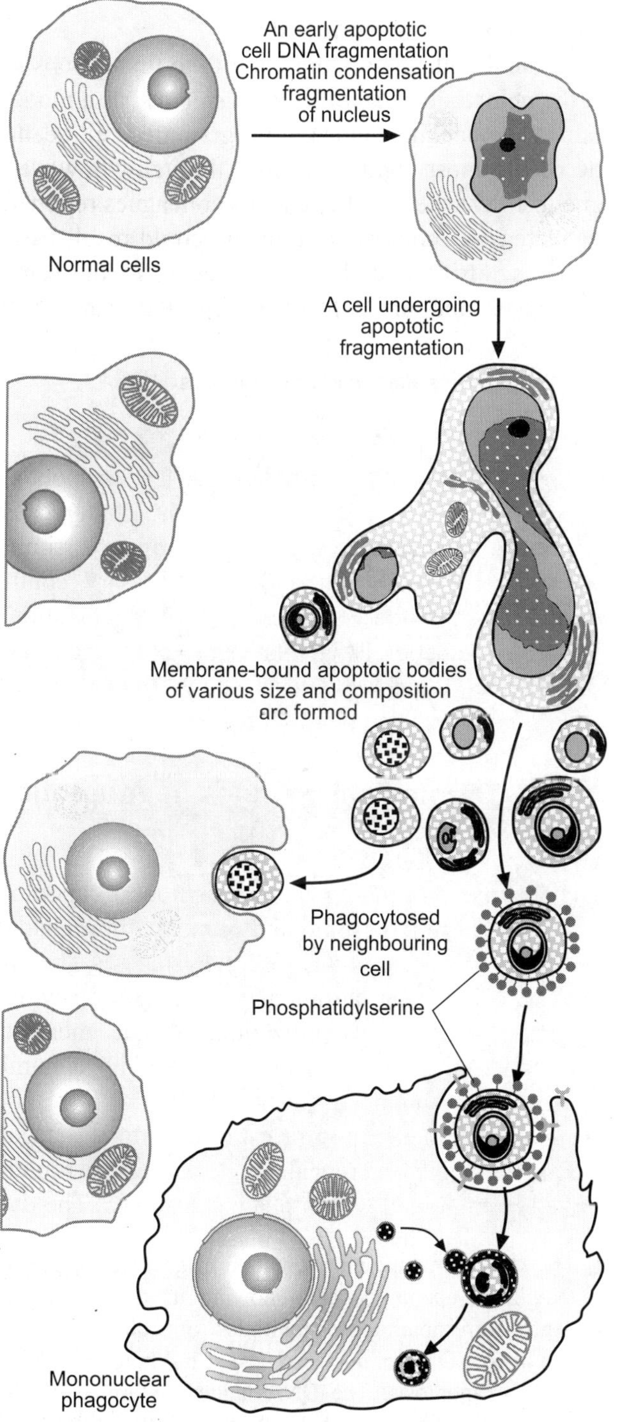

Fig. 16.1 Sequence of ultrastructural changes in cells undergoing apoptosis (**See Plate 12**).

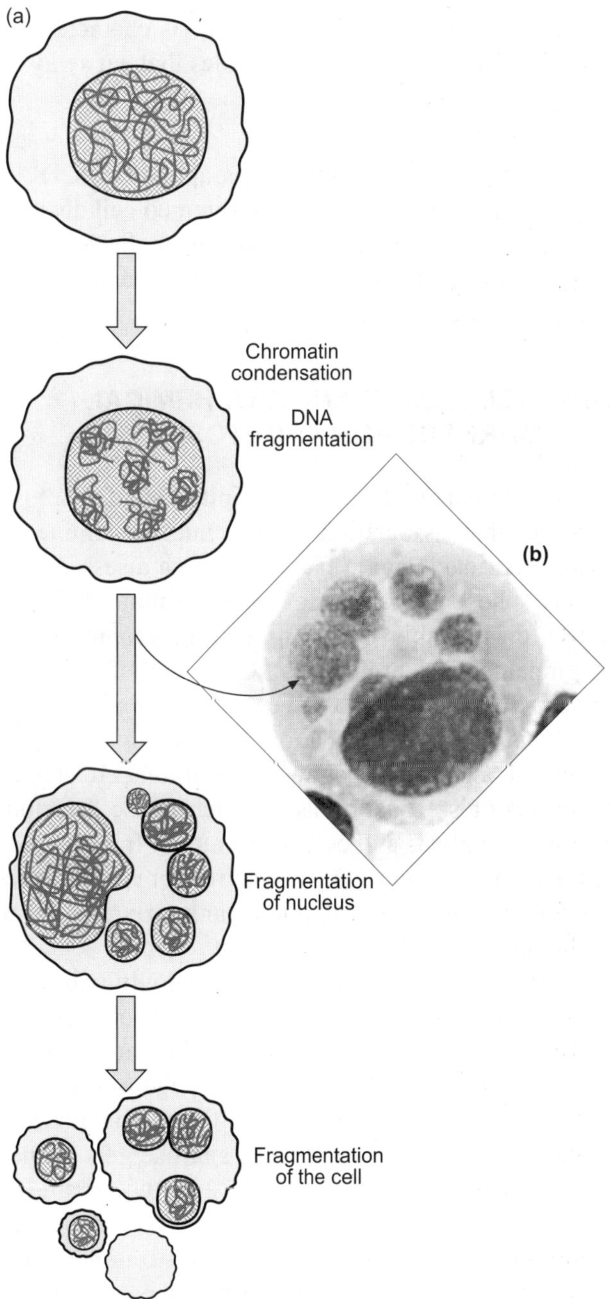

Fig. 16.2 (a) Diagrammatic representation of the events of apoptosis. (b) Light micrographs of apoptotic human peripheral lymphocytes treated with chemical clastogens, which illustrates condensation and nuclear fragmentation (Channarayappa, 1990).

16.3.2 Types of programmed cell death

Programmed cell death has been classified according to the pathways taken. The following classification is commonly adopted:

- ***Apoptosis*** (or ***Type I cell death***) is a particular form of programmed cell death and is described throughout this chapter.
- ***Autophagic*** (or ***Type II***) cell death is characterized by the formation of large vacuoles that eat away organelles in a specific sequence before the nucleus is destroyed.
- ***Other types*** of PCD called 'non-apoptotic PCD' (or 'caspase-independent programmed cell death' or necrosis-like PCD,) have also been discovered. These alternative routes to death are equally efficient mechanisms.

16.4 MOLECULAR AND BIOCHEMICAL MARKERS FOR APOPTOSIS

The movement of the pores at the initiation of apoptosis suggests that association with a nuclear lamina no longer restricts them; in fact, the lamina disassembles. While phosphorylation of the lamina subunits causes disassembly during mitosis in apoptosis it is accompanied by proteolysis. Cleavage of selective protein subsets is a key event in the execution of apoptosis. Protein degradation may play a role in the structural alteration that may result in cell suicide and it may also function as a switch in the decisions between apoptosis and cell proliferation. Investigation using highresolution 2D gel electrophoresis indicates the potential involvement of nuclear matrix proteins in the apoptotic process.

Thebiochemicalfeaturemostcommonlyassociated with apoptosis is the activation of endogenous endonuclease and the cleavage of DNA at the linker regions between nucleosomes into oligonucleosomal sized fragments. This results in the formation of a characteristic 'ladder' pattern in agarose gels stained with ethidium bromide (**Fig. 16.3**). Recent studies have shown that DNA fragmentation associated with apoptosis is mediated by a DNA fragmentation factor (DFF) which is activated by caspases, mainly caspase 3. The DNA ladder was the first biochemical marker of apoptosis Molecular changes also included the randomization of the distribution of phosphatidyl serine (PS) between the inner and outer leaflets of the plasma membrane. Techniques based on detecting some of these changes, such as terminal deoxynucleotidyltransferase-mediated incorporation of labelled nucleotides at DNA breaks, staining of cell surface-exposed PS with annexin V and detection of cells with apparent subdiploid DNA content by staining with DNA intercalating dyes, are the standard tools for demonstrating apoptosis.

Genetic control of apoptosis

Horvitz and colleagues were one of the first groups to use the nematode, *Caenorhabditis elegans*, as a model to study the genetic control of programmed cell death. One of the most significant breakthroughs, however, came in 1987 when Wyllie and his colleagues reported that increased apoptosis in tumours could result from a process intrinsic to the tumour cells. In the same year, Fesus and his colleagues reported that tissue

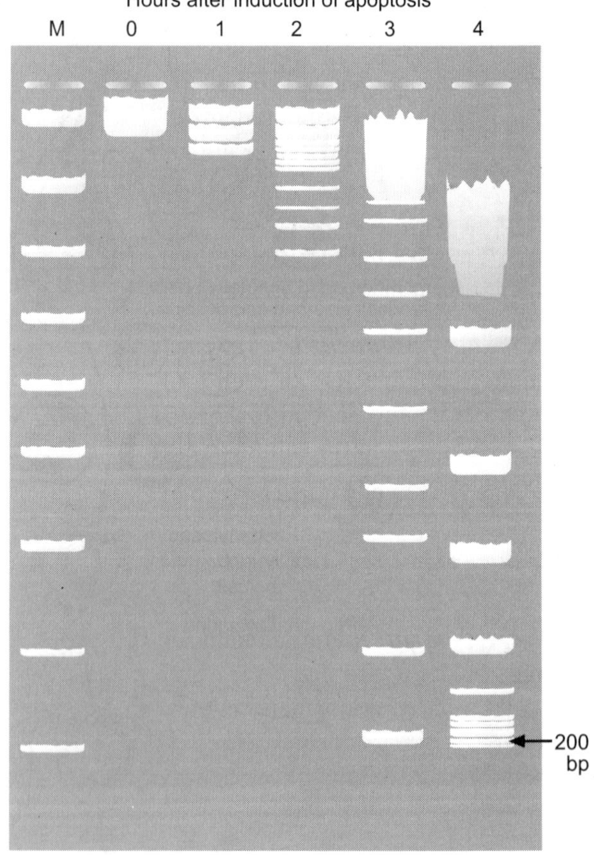

Fig. 16.3 Diagrammatic representation of gel electrophoresis of DNA from apoptotic cells, showing its degradation to fragments corresponding to multiples of 200 base pairs (the size of core DNA of nucleosomes) at 1–4 hours following induction of apoptosis. The DNA ladder is formed by the endonucleolytic cleavage of DNA during the process of apoptosis. Lane numbers 1–4 show DNA samples extracted from the cells exposed to drugs for different interval of times.

transglutaminase was induced and activated during the apoptotic process. Increase of cytosolic calcium in induction of apoptosis in some cases has also been reported. In separate studies, it was shown that apoptosis was responsible for the deletion of auto-reactive T cells in the thymus during the development of self-tolerance and for the selection of B cells in lymphoid germinal centres during the humoral immune response.

16.5 FUNCTIONS OF APOPTOSIS

The main function of apoptosis is to kill unwanted cells, for **development** and **homeostasis**, for **defence** and for **ageing** (senescence).

16.5.1 Cellular homeostasis

In the adult organism, the number of cells is kept relatively constant through cell death and division. Cells must be replaced when they become diseased or malfunctioning. But proliferation must be compensated by cell death. The balancing process is a part of the homeostasis required by living organisms to maintain their internal states within certain limits. Some scientists have suggested **homeodynamics** as a more accurate term. **Allostasis,** refers to a mor complex balance within the body. Homeostasis is maintained through cell death in the following manner:

- ***Cells that are produced in excess:*** In *C. elegans*, two male specific cells can each form a special part of the proctodeum, needed for mating. Only one of these cells lives and the others in excess die. The vertebrate neurons can be cited as examples of this category.
- ***Cells divide slower than they die:*** This results in a disorder of cell loss.
- ***Cell replacement:*** Mammalian blood contains various cell types that have very high turnover. The cells that are non-functional or damaged are removed by apoptosis and replaced by newly formed cells thus maintaining cellular homeostasis.

The organism must orchestrate a complex series of controls to keep homeostasis ongoing throughout life; therefore, it involves many different types of cell signalling. Impairment of any one of these controls can lead to a diseased state. For example, dysregulation of hedgehog signalling has been implicated in several forms of cancer. The hedgehog pathway, which conveys anti-apoptotic signals has been found to be activated in pancreatic adenocarcinoma tissues.

16.5.2 PCD in development

PCD is an integral part of both plant and animal tissue development. Development of an organ or tissue is often preceded by the extensive cell division and differentiation of a particular cell, with the resultant mass then 'pruned' into the correct form by apoptosis. Unlike cellular death caused by injury, apoptosis results in cell fragmentation into tightly packed cell bodies. This allows the cells to be efficiently phagocytozed and their components re-used without releasing potentially harmful intracellular substances into the surrounding tissue.

Cells removed during development

- ***Terminally differentiated cells:*** Examples for this category include the tadpole tail cells and those lining the intestinal and reproductive tracts in animals. In plants, they include root cap cells, cells undergoing senescence, vegetative cells of pollen, synergids, antipodals, anther tapetal and wall cells, transmitting cells of stigma, style and ovary wall and endosperm cells.
- ***Cells that are unwanted abinitio:*** A few examples in this category are the Müllerian duct cells that are required in females and stamen and carpel primordial cells in flowers.
- ***Cells that die during re-differentiation into specialized cell types:*** Keratinocytes at the skin surface of vertebrates, xylem tracheary elements, cells of certain trichomes, thorns and spines, sclerenchyma and cork cells are some examples.
- ***Cells that are present at wrong places:*** The cells present between the developing digits in the vertebrates may be mentioned as an example of this category. In rats, it has been shown that the neurons of the visual system that project in the wrong regions can be identified and subjected to cell death.

Signals for initiation of PCD

During development, apoptosis is tightly regulated and each tissue uses different signals for inducing apoptosis. In birds, bone morphogenetic proteins (BMP) signalling is used to induce apoptosis in the interdigital tissue. In *Drosophila*, steroid hormones regulate cell death. Developmental cues can also induce apoptosis, such as the sex-specific cell death or hermaphrodite-specific neurons in *C. elegans* males through low TRA-1 transcription factor activity (TRA-1 helps prevents cell death). During embryo development in vertebrates, structures called the notochord and the floor plate secrete a gradient of the signalling molecule Sonic hedgehog (Shh) and this gradient directs cells to form patterns in the embryonic neural tube. Cells that receive Shh in a receptor in their membranes called Patched (Ptc1) survive and proliferate; but in the absence of Shh, one of the ends of this same Ptc 1 receptor is cleaved by caspase 3, an action that exposes an apoptosis-producing domain.

16.5.3 Cell damage or infection

If cells are non-functional due to injury, ageing or infection by pathogens, those cells have to be removed to avoid ill-effects caused by them.

- ***Cells that are subjected to hypersensitive response:*** They are present both in animals and plants and die in response to pathogen infections.
- ***Cells that are damaged and unable to function further:*** Death of this category of cells removes the potentially harmful cells and prevents their multiplication and spread in the animals. Cells damaged beyond repair, infected with a virus or undergoing stress conditions such as starvation are examples. DNA damage from ionizing radiation or toxic chemicals can also induce apoptosis via the actions of the tumour-suppressing gene *p53*.
- ***Cells which through their death give rise to diseases:*** The death of T cells of the animal immune systems results in AIDS and cell death in neurons.
- ***Cancer cells:*** Apoptosis also plays a role in preventing cancer. If a cell is unable to undergo apoptosis, due to mutation or biochemical inhibition, it can continue dividing and develop into a tumour. Infection by papilloma viruses causes a viral gene to interfere with the cell's *p53* protein, an important member of the apoptotic pathway. This interference in the apoptotic capability of the cell plays a critical role in the development of cervical cancer.

16.5.4 Lymphocyte interaction

The growth of B lymphocytes and T lymphocytes in the human body is a complex process that effectively creates a large pool of diverse cells to begin with and then weeds out those potentially damaging to the body. Apoptosis is the mechanism by which the body removes both the ineffective and the potentially damaging immature cells and in T cells this is initiated by the withdrawal of survival signals. Cytotoxic T cells are able to directly induce apoptosis in cells by opening up pores in the target's membrane and releasing chemicals which bypass the normal apoptotic pathway. The pores are created by the action of secreted perforin and the granules contain granzyme b, a serine protease, which activates a variety of caspases by cleaving aspartate residues.

16.5.5 Stress-activated apoptosis in sick cells

Intracellular stimuli, such as irreparable genetic damage, high intracellular Ca^{2+} concentration, severe oxidative stress and lack of survival signals can trigger apoptosis by the intrinsic pathway (**Fig. 16.4**). In multicellular organisms, if a cell is

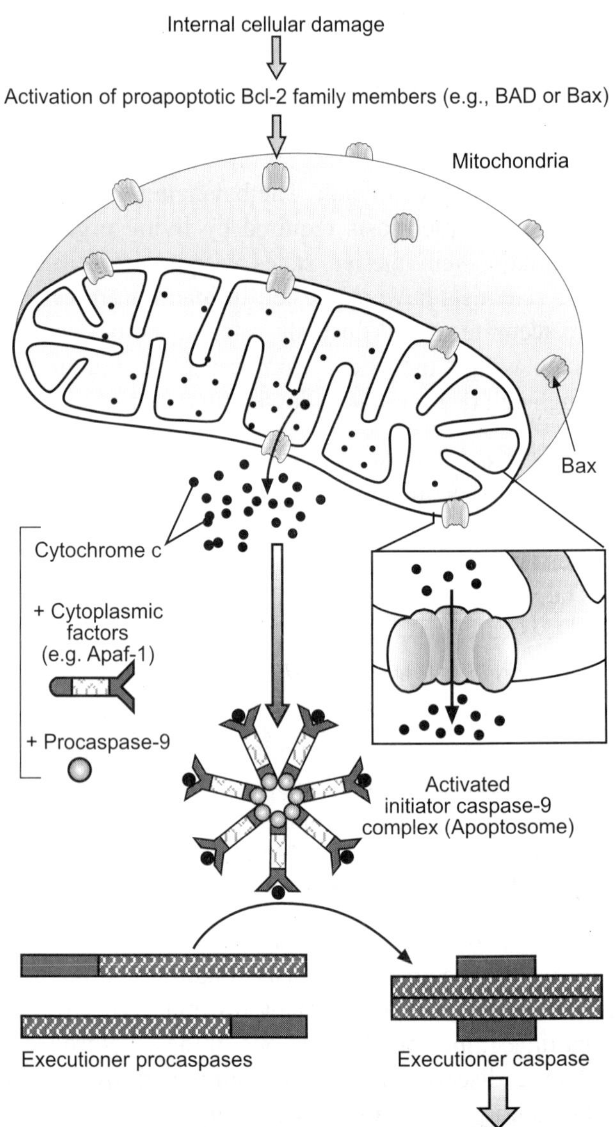

Fig. 16.4 The intrinsic (mitochondria-mediated) pathway of apoptosis.

stressed or damaged, the cells undergo quick suicide and a healthy neighbour replaces them through cell division .

Key trigger in PCD: Reactive oxygen intermediates

PCD as a response to stress is also a plant's response to infection. If a plant is able to recognize the pathogen, the host cells in the surrounding area of the infection site undergo rapid cell death. The death of the host tissue, thus, may kill the pathogen by exhausting the nutrients. Alternatively, the death of the host tissue may be a consequence of the mechanisms the plant uses to kill the pathogen. Reactive oxygen intermediates (ROI) are a key trigger in PCD accompanying the host response (HR) (**Fig. 16.5**). The production of ROI within hours of pathogen entry is known as oxidative burst. In both compatible and incompatible interactions, an oxidative burst occurs within the first hour after inoculation. However, incompatible interactions are characterized by a second, stronger and more prolonged oxidative burst and it is in these incompatible interactions that cell death occurs.

16.6 APOPTOSIS TRIGGERING MECHANISMS

There are a number of mechanisms that trigger apoptosis in response to diverse stimuli, which may originate either extracellularly (**extrinsic inducers**) or intracellularly (**intrinsic inducers**). Extracellular signals may include hormones, growth factors, nitric oxide or cytokines and therefore must either cross the plasma membrane or transduce to affect a response (**Fig. 16.6**). These signals can bind to nolecules and either induce (positive) or actively repress (negative) apoptosis.Intracellular signalling is a initiated by a cell in response to stress and

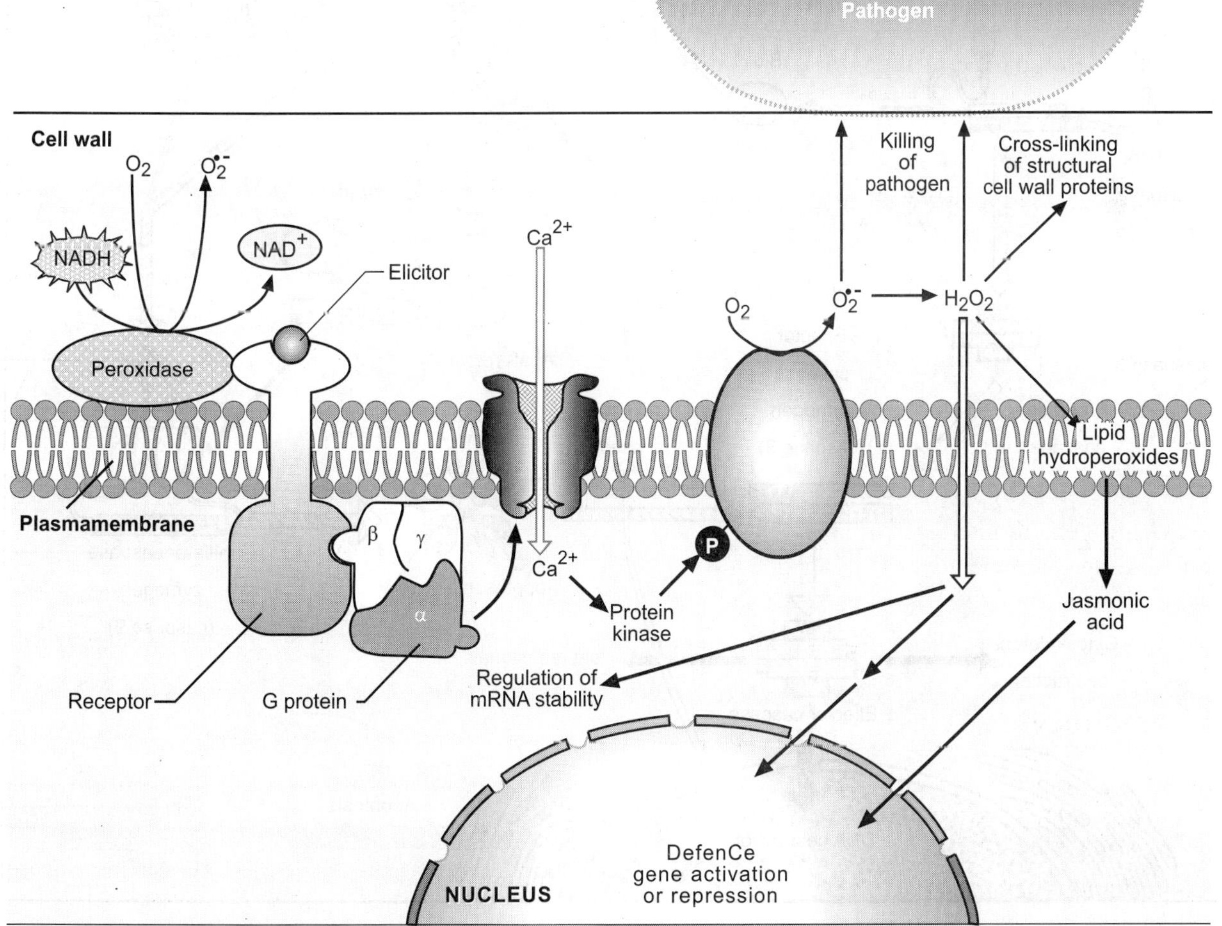

Fig. 16.5 A model of the relationship between pathogen attack on a host cell and the signal transduction pathway that generates reactive oxygen species (ROS).

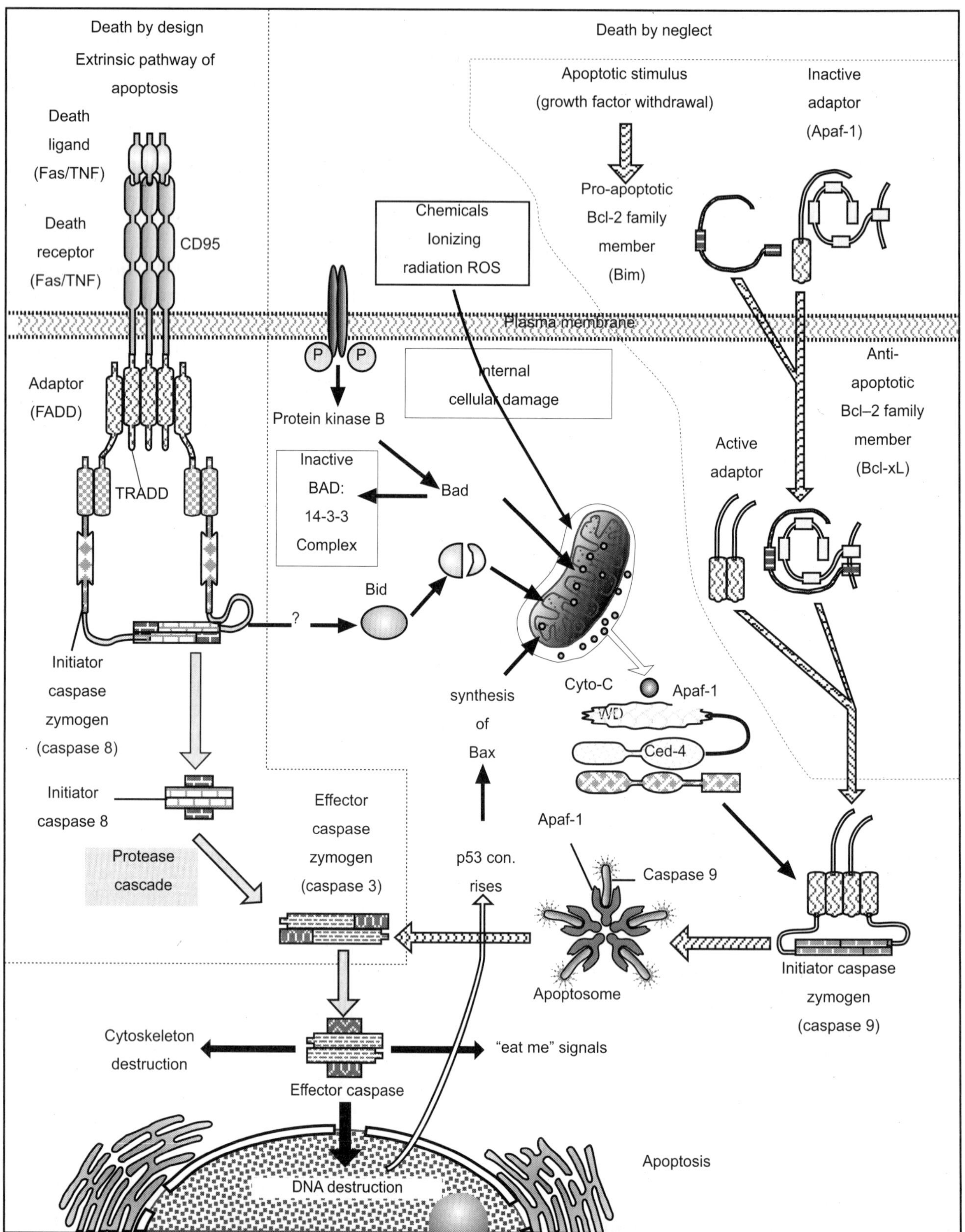

Fig. 16.6 Signalling through distinct pathways activates caspase. Apoptosis can be induced by stimulation of those members of the TNF receptor family, which have a death domain (e.g., CD95), or by a broad range of stimuli including growth factor withdrawal. Caspase 8-mediated activation of the pro-apoptotic BH3-only protein Bid may provide a link between these pathways in some cell types.

might ultimately result in cell suicide. The binding of nuclear receptors by glucocorticoids, heat, radiation, nutrient deprivation, viral infection and hypoxia are all factors which can lead to the release of intracellular apoptotic signals by a damaged cell. However, before the actual process of cell death is carried out by enzymes, apoptotic signals must be connected to the pathway by way of regulatory proteins.

The mechanisms that operate are:

- directly on mitochondria
- through the transcription factor p53
- through a protein kinase called p38.

Role of mitochondria in PCD

The mitochondria take part in the cytoplasmic control of PCD (**Fig. 16.7**). The basic role of mitochondria is oxidative phosphorylation, in cellular energy production and generation of reactive oxygen species (ROS). But the initiation of apoptosis has suggested a number of novel mechanisms for the mitochondrial pathology. Apoptotic proteins which target mitochondria affect them in different ways. They may cause mitochondrial swelling through the formation of membrane pores, or may increase the permeability of the mitochondrial membrane and cause apoptotic effectors to leak out.

Mitochondrial proteins such as SMACs (second mitochondrial-derived activator of caspases) are released into the cytosol following an increase in permeability. SMAC binds to inhibitor of apoptosis proteins (IAPs) and inhibits them, preventing the IAPs from arresting the apoptotic process and thus allowing apoptosis proteins to proceed. IAP also suppresses the activity of a group of cysteine proteases called caspases, which carry out the degradation of the cell. Therefore the actual degradation enzymes can be seen to be indirectly regulated by mitochondrial permeability.

Pro-apoptotic molecular Bax and the constitutive mitochondrial protein adenine nucleotide translocator (ANT) cooperate within the permeability transition pore complex (PTPC) to increase mitochondrial membrane permeability and to trigger cell death. Bid is an intracellular cross-talk agent that can amplify Fas/TNF apoptotic signal through the mitochondrial death pathway after caspase 8 cleavage induces conformational change of Bax. This is responsible for release of cytochrome C from mitochondria during apoptosis. Action of Bax on the outer mitochondrial membrane requires the presence of ATP both in vitro and in vivo and it is shown that ATP directly increases

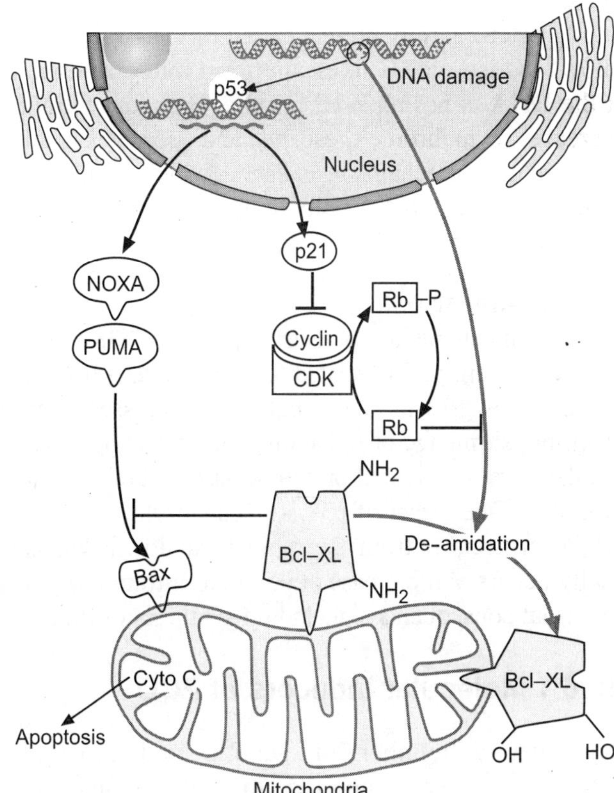

Fig. 16.7 A de-amidation death knell for tumour cells.

the amount of Bax inserted into the mitochondria. A cytosolic factor is required for mitochondrial cyto C efflux during apoptosis. It has been confirmed that mitochondrial permeability transition causes intracellular acidification.

A reduction in mitochondrial transmembrane potential occurring before nuclear changes in PCD is a constant observation in different types of PCD. Reduction of transmembrane potential correlates with a decrease in mitochondrial translation and a loss in mitochondrial gene transcription. Findings suggest that DNA damage by duocarmycin A induces production of hydrogen peroxide which causes loss of mitochondrial membrane potential and subsequently caspase 3 activation resulting in apoptosis.

In invertebrates, PCD frequently involves a relocalization of mitochondrial cyto C to the cytoplasm. This role in regulation of apoptosis is in addition to the primary function of cytoC in mitochondrial electron transport chain.

Transcription factor p53

A second mechanism operates through the transcription factor p53. The p53 concentrations increase when DNA is damaged. Consequently, DNA repair mechanisms are activated, but so is transcription of the gene for another Bcl-2 family protein called Bax. Like Bad, Bax allows cytochrome C to escape from mitochondria, so if the DNA is not repaired in time, Bax concentration increases, cytochrome C escapes and apoptosis results. Without this mechanism, cancers, which result from changes in the DNA of our body cells, would be far more abundant.

Protein kinase p38

A third mechanism operates through a p38 protein kinase. Protein p38 is related to MAP kinase. It is activated not by growth factors but by cell stresses such as swelling, shrinkage or radiation. The effects of p38 and another stress-activated protein kinase, called JNK, are complex. One effect of p38 is phosphorylation of p53, which protects it from the rapid breakdown that normally occurs. An increase of p53 concentration triggers the usual consequences, including synthesis of Bax.

16.6.1 Molecular inducers of PCD

To date, a number of genetic and molecular tools have been reported to induce apoptosis. A large number of genes have been identified that can be used to modulate apoptosis by transfection or ablation. The protein products of several of these genes, along with their known characteristics, are listed in **Table 16.1**. Most of the molecular inducers of apoptosis can be placed in one of the primary sites of action within the cell:

- the cell surface
- the cytosol
- the cytoskeleton
- the nucleus
- intracellular signalling pathways.

Table 16.1 categorizes several of these apoptosis-inducing reagents and lists their possible mechanisms. In general, apoptosis involves a syndrome of mechanisms with effectors, adaptors, regulators and signals as well as the activity of endonucleases.

Cell surface receptor ligands

So far, three factors tumour necrosis factor (TNF), Fas ligand (FasL) and TNF-related apoptosis inducing ligand (TRAIL) as well as four death factor receptors (Fas, TNFRI, DR3/Ws1-1 and CARI) have been identified. It is quite possible that many more death factor and receptor systems that regulate apoptosis in a

Table 16.1 Molecular agents for the manipulation of programmed cell death

Apoptosis-promoting gene product	Characteristics
Adenovirus E1A	Stabilizes p53 protein, which might in turn induce PCD
Bcl-2 family (Bax, Bcl-X_S)	Promote PCD in damaged and senescent cells
C-Myc	Induces PCD in growth factor deprived cells
Cytokines (Fas ligand, TNF)	Induce PCD in hemopoietic cells via receptor activation
ICE/Ced-3 family	Cysteine proteases involved in the execution of PCD
P53	Tumour suppressor associated with hypoxia and DNA damage -induced PCD
RB-1	Tumour suppressor that induces PCD when hypophosphorylated
Reaper	Universal activator of PCD in *Drosophila*
PCD-inhibiting gene product	**Characteristics**
Adenovirus E1B (19 kDa, 55 kDa)	Block E1A-induced PCD; thought to function by different mechanisms,
African swine fever virus, LMW5-HL	Homologyous to Bcl-2
Bcl-2 family (Bcl-2, Bcl-X_L Ced-9, Mcl-1)	Dimerize with and inhibit the function of pro-death Bcl-2 family members
Cowpox virus cram	Cowpox virus serpin that inhibits ICE/Ced-3 protease
IAP family	Inhibitor of apoptosis proteins having homology to baculoviral IAPs
Epstein–Barr virus LMP-1	Epstein–Barr viral protein that upregulates host cell Bcl-2 expression
Baculovirus p35	Protein inhibitor of ICE/Ced-3 proteases; distinct from IAP proteins

IAP: inhibitor of apoptosis proteins; ICE: interleukin-1ß-converting enzyme; PCD: programmed cell death; TNF: tumour necrosis factor (Source: Wertz and Hanley, 1996).

tissue-specific manner would be found. The apoptotic signals triggered by death factors are irreversible, that is, a protease cascade is activated by the death signal and the protease cleave various cellular components leading to morphological changes of the cell and nuclei.

TNF is a death-inducing cytokine, which is responsible for cell death associated with anti-tumour activity. TNF is produced mainly by activated macrophages and is the major extrinsic mediator of apoptosis. Although initially associated with necrotic cell death, TNF-induced cell killing is now recognized to proceed by PCD in many cell types. Cellular responses to TNF are mediated by receptors (TNF-R) and depending on the cell type and the other signals that the cell receives (**Table 16.1**). Recently, several components of the TNF-TNFR death response pathway have been identified (**Fig. 16.8**).

Effector molecules in apoptosis

The Bcl-2 family: The most potent inducers of cell death are a group of proteins that possess only a BH3 region and otherwise bear no resemblance to the

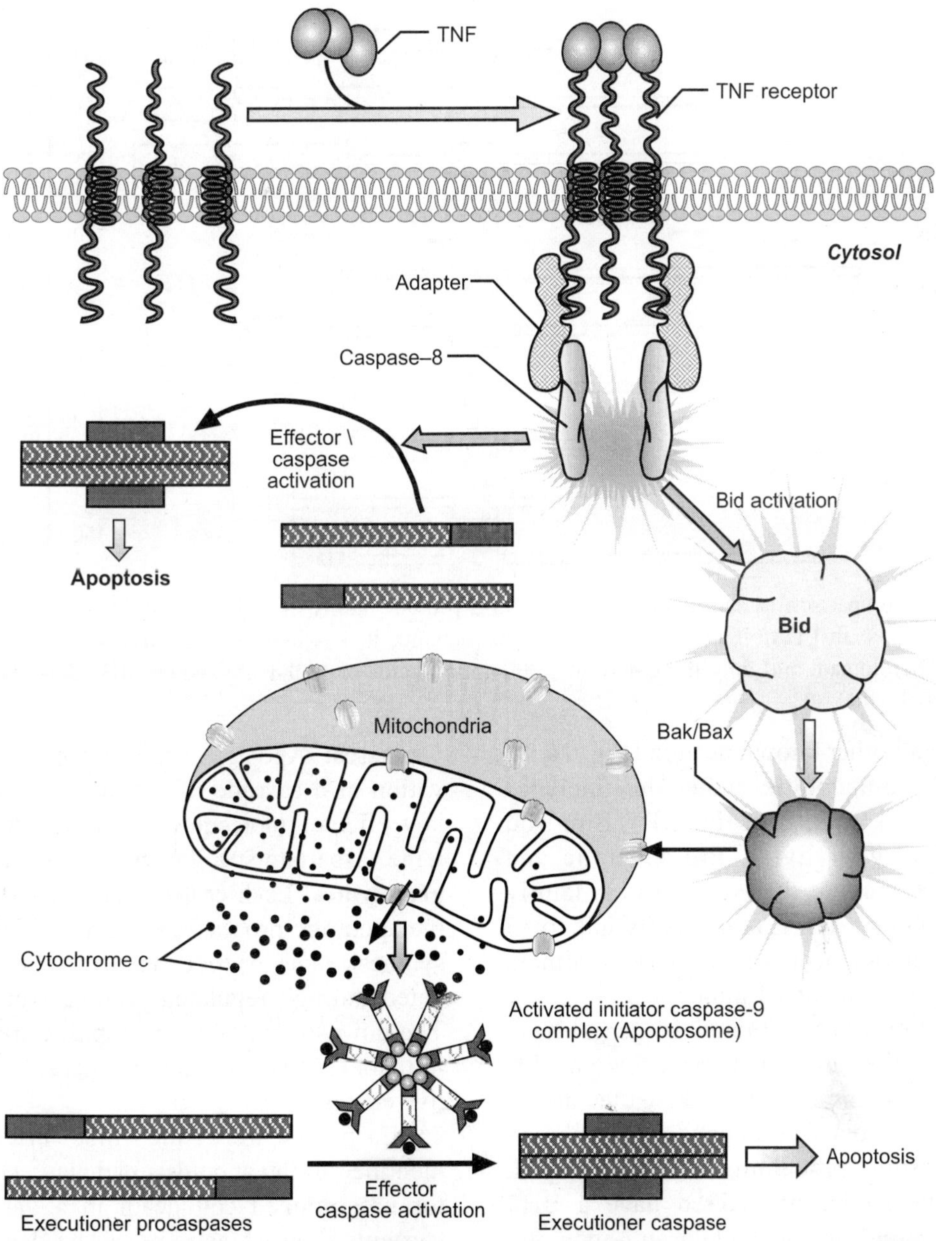

Fig. 16.8 TNF and other cell death receptors. TNF cell death receptor ligands consists of three polypeptide chains, so their binding to cell death receptors induces receptor trimerization. Caspase 8 is recruited to the receptor and activated via interaction with adaptor molecules. Activated caspase 8 can directly cleave and activate effector caspases (**See Plate 13**).

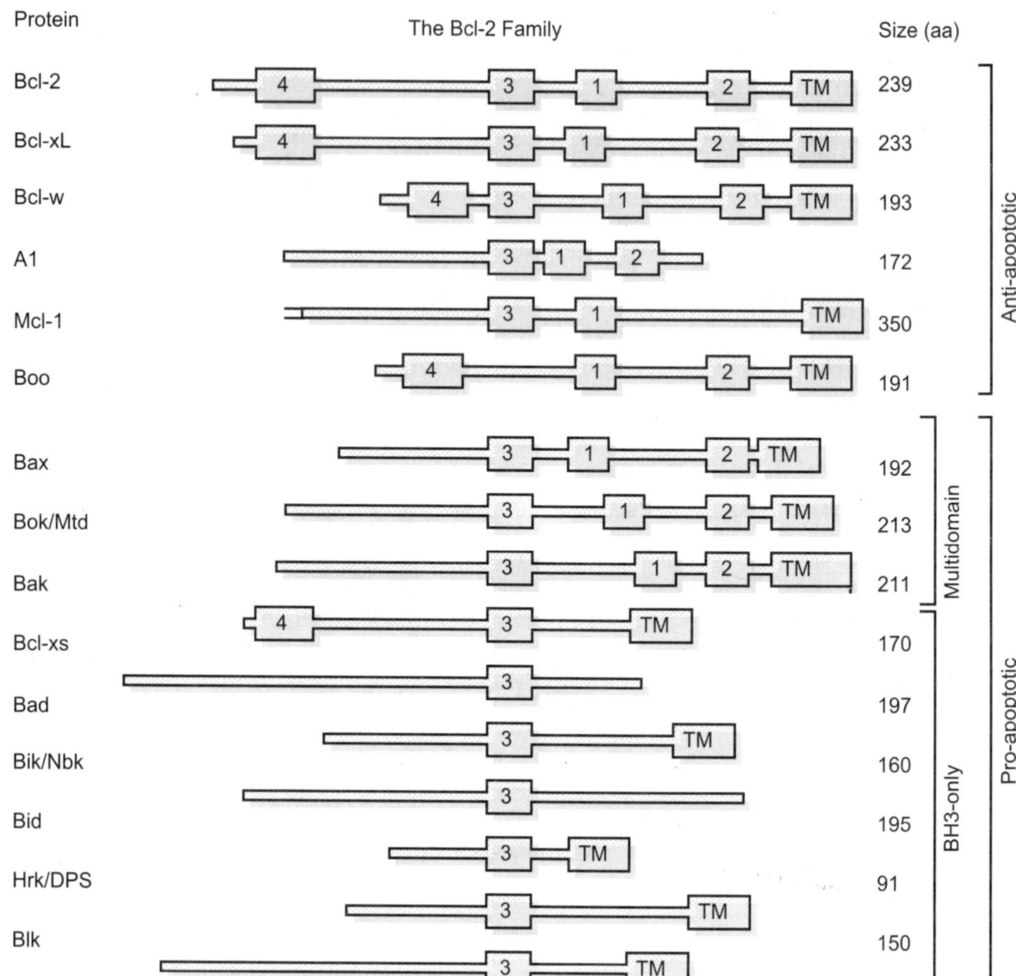

Fig. 16.9 Structural comparison of the members of the Bcl-2 protein family. The Bcl-2 family of proteins can be divided into two subgroups: those that inhibit apoptosis and those that enhance it. 1–4 indicates the BH1-4 Bcl-2 homology domains. Transmembrane (TM) domain and only the C-terminal 200 amino acids of Mcl-1 are shown. BH, Bcl-2 homology domains (Source: Strasser et al, 2000).

Bcl-2 family or any other known protein (**Fig. 16.9**). This group of pro-apoptotic molecules includes mammalian Bad, Bik/Nbk, Bid, Hrk/DP5, Bim/Bod and Blk and *C. elegans* Egl-1. Pro-apoptotic and anti-apoptotic members of the Bcl-2 protein family can physically interact, but it is presently unknown whether this interaction occurs in the form of dimers or as part of a conglomerate of molecules.

Members of Bcl-2 family (about 15 genes) have a conserved C terminal transmembrane region. This region localizes proteins to the outer leaflet of the nuclear envelope, the outer mitochondrial membrane and the endoplasmic reticulum. All of the pro-apoptotic proteins so far discovered have a BH3 (16 amino acid) region, but not all of the anti-apoptotic proteins have this region. Their common function is to bind and antagonize the anti-apoptotic effects of their ligands.

Caspases: Apoptotic cell death results from activation of a cascade of enzymes called caspases. The key effector molecules of animal PCD are the cysteine aspartate-specific proteases (**c-asp-ases**) and **granzymes**. Each caspase has a characteristic role in the apoptotic process. Upstream initiator caspases are synthesized as inactive zymogens that usually are activated through regulated protein–protein interactions. The downstream effector caspases are cleaved by the initiator caspase in order to generate a catalytically active enzyme.

Caspases were the first effector molecules identified in the apoptosis pathway. They are essential for programmed cell death in a variety of species. Cysteine protease activity can be detected in all cells undergoing apoptosis, regardless of their origin or the death stimulus. Mice lacking caspase 3 or caspase 9 have increased numbers of neurons in the brain and their lymphocytes are partially resistant to some apoptotic stimuli. Caspases are not merely responsible for degradation of cellular substrates during the end stage of apoptosis, but are critical regulators of cell death initiation.

At least 15 caspases that play a role in inflammation (group 1 caspases) and apoptosis (group 2 caspases) have been identified in mammals. Each caspase has a characteristic role in the initiation or execution of cell disassembly. Accordingly, they have been grouped into families known as **initiators** and **executioners**. A diverse set of signals from inside and outside the cell can activate initiator caspases (caspases 8, 9 and 10). Various kinds of cellular stresses induce cytochrome C release from the mitochondria can initiate a 'mitochondrial pathway' leading to caspase 9 activation. Executioner caspases (such as caspases 3 and 7) are usually activated by proteolytic cleavage between domains by a previously activated upstream caspase (**Fig. 16.10a**). Once activated, caspases cleave and thereby activate, other procaspases, resulting in an amplifying proteolytic cascade (**Fig. 16.10b**). Some of the activated caspases then cleave other key proteins in the cell.

Production and functioning of caspases during PCD depend on the sequential evocation and action of specific genes. These enzymes recognize tetrapeptide motifs and cleave their substrates on the carboxyl side of an aspartate residue. Individual caspases have distinct substrate specificities that are determined by the pattern of amino acids upstream of the cleavage site. Some caspases, particularly **effector caspases,** cleave and inactivate certain vital cellular proteins, such as DNA repair enzymes, lamins (**Fig. 16.11**), gelsolin, MDM2 and protein kinase c).

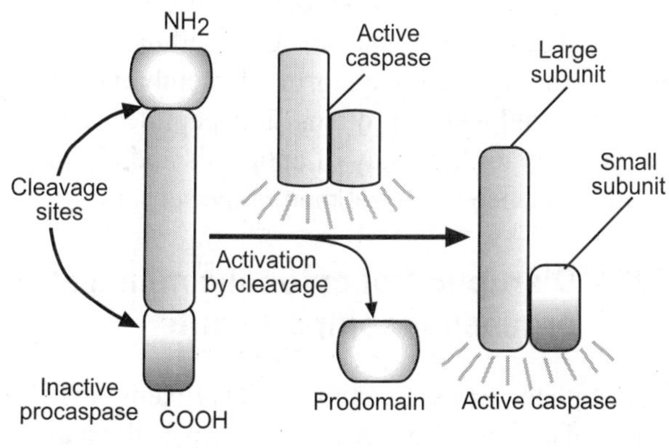

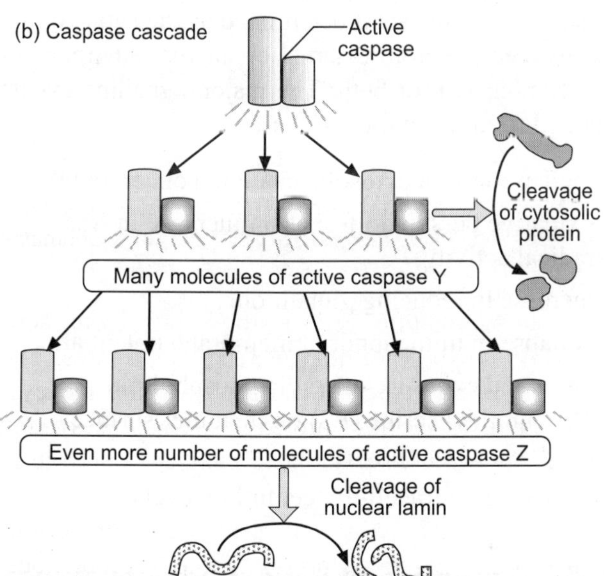

Fig. 16.10 The caspase cascade involved in apoptosis: (a) Conversion of procaspase into active caspase by proteolytic cleavage by another member of the caspase family; (b) Each activated caspase molecules can cleave many procaspase molecules, thereby activating them and these can in turn activate a small number of procaspases (called initiator caspases), which leads to amplifying chain reaction (a cascade). Some of the activated caspases then cleave a number of key proteins and nuclear lamins, leading to controlled cell death.

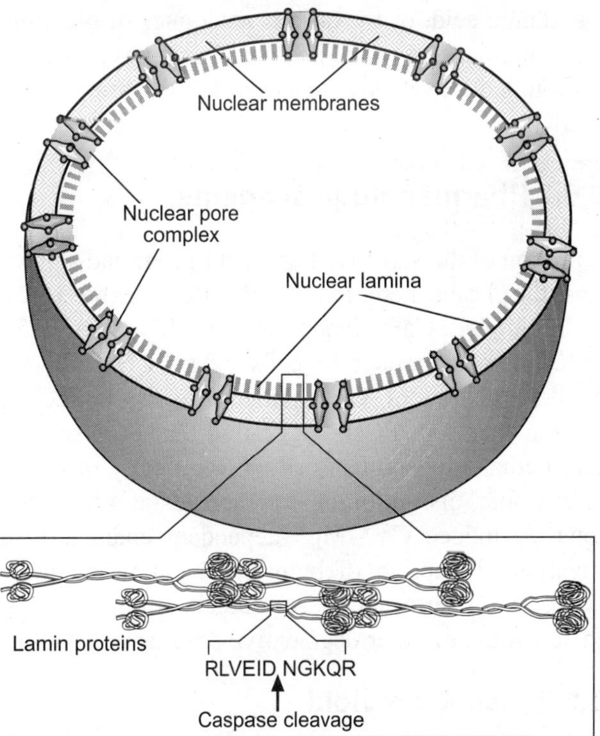

Fig. 16.11 Caspase cleavage. Nuclear lamins are one of the targets of caspase cleavage. Caspases cleave after aspartic acid (D) residues.

Adaptor proteins: A general principle that activates procaspases to initiate the caspase cascade is that the activation is triggered by **adaptor proteins** that bind to them via shared motifs and bring multiple copies of specific procaspases, known as **initiator procaspases**, together in a complex or aggregate. In some cases, the initiator procaspases have a small amount of protease activity and forcing them together into a complex causes them to cleave each other, triggering their mutual activation. In other cases, the aggregation is thought to cause a conformational change that activates the procaspase. Within moments, the activated caspase at the top of the cascade cleaves downstream procaspases to amplify the death signal and spread it throughout the cell.

Procaspase activation can be triggered from outside the cell by the activation of death receptors on the cell surface. Adaptor proteins are the links between the cell death effectors, caspases and cell death regulators, death receptors and Bcl-2 family members. They form bridges between caspases and upstream regulators of apoptosis.

Regulators: Examples of ligands that act at the cell surface include the neuronal excitotoxins. It has long been recognized that neurons, particularly those of the central nervous system, are sensitive to a range of lethal stimuli that do not affect other cell types. In particular, hyperstimulation of glutamate receptor subtypes either by glutamic acid, or by various analogues of glutamic acid, can induce neuronal necrosis, but it has also been shown to induce selective neuronal PCD, sparing axons and glia.

16.6.2 Pharmacological agents

Disruption of the selective transport of ions and solutes across the plasma membrane induces PCD (**Table 16.2**). For example, Ca^{2+} ionophores such as A23187 are lethal to a number of cells by a PCD pathway. The deregulation of Ca^{2+} homeostasis is, in fact, a common theme in PCD, suggesting that Ca^{2+} sensors exist, perhaps inherent in cellular regulatory proteins. For example, disruption of Ca^{2+} homeostasis has been shown to induce Ca^{2+}-/Mg^{2+}-dependent endonuclease activation, disruption of both actin and microtubule organisation, perturbation of mitochondrial function and activation of other degradative enzyme cascades.

16.6.3 Tissue amyloid

Extracellular deposition of 'amyloid' is characteristic of a number of tissue pathologies. In all cases, the formation of insoluble amyloid is correlated with the selective loss of specific cells. Increased evidence supports the view that associated pathologies are examples of inappropriate induction of PCD. For example, formation of aggregated β-amyloid protein has been implicated as initiating neuronal apoptosis in Alzheimer's disease. A number of mechanisms have been proposed, but it is still unclear how the aggregation of β-amyloid protein creates a PCD stimulus specific for discrete populations of neurons. Similarly, the pancreatic hormone, amylin, can induce pancreatic β-cell apoptosis when aggregated and is also highly toxic to neurons in this form. This suggests that certain cell populations are for unknown reasons, highly sensitive to extracellular amyloid, perhaps, through a novel form of membrane sensor. Thus, amyloidogenic proteins and peptides are cell-specific tools for the study of apoptosis and might also provide evidence for novel membrane sensors.

16.6.4 Disruption of cytosolic milieu and organellar compartments

The complete process of animal PCD is exemplified by a number of cell signalling events which may affect the cell death mechanism at every level. The decision of a cell to initiate the suicidal act is determined by the nature of the signals received from external or internal environment or a combination of both. The major signalling events reported so far in animal PCD are:

- increase in free cytosolic calcium concentration,
- oxidative burst manifest as an increase in free radical activity,
- increase in reducing glutathione,
- collapse in mitochondrial membrane potential,
- increased exposure of acidic phospholipid phosphotidyl serine (PS) on the outer surface of the plasma membrane,
- increase in intracellular ceramide levels.

16.6.5 Disruption of ionic gradients across intracellular membranes

Calcium ion gradients across the endoplasmic reticulum (ER) can be dissipated by thapsigargin, a potent and specific inhibitor. As a result, Ca^{2+} levels in the ER are depleted and there is a sustained elevation in cytosolic Ca^{2+}, which is now recognized to induce PCD in specific cell types. PCD induced by thapsigargin could result from improper protein folding within the ER, repression of translation initiation, disruption of

Table 16.2 Pharmacological tools for inducing programmed cell death

Site of perturbation	Possible mechanism	Example(s)
Cell surface	Activate cell membrane receptors	Excitotoxins: glutamate, NMDA, kinate
	Disrupt regulation of plasma membrane solute flux	Ionophores: A23187, ionomycin Transporter inhibitors: amiloride, ouabain Detergents: Triton X-100
	Membrane perturbation	Tissue amyloid
Cytosol and organellar compartments	Intracellular oxidants	Menadione
	Metal ion chelators	TPEN, desferrioxamine
	Alter protein synthesis	Inhibit transcription: actinomycin-D
		Inhibit translation: cycloheximide, puromycin
	Disrupt intracellular membrane ion gradients	Endoplasmic reticulum: thapsigargin
		Vacuoles: bafilomycin, concanamycin A
		Mitochondria: MPTP
Cytoskeleton	Alter microtubule organization	Taxol, colchicine, colcemid, vincristine
	Disrupt actin networks	Cytochalasins
Nucleus	Perturb DNA integrity	Alter structure; adriamycin, camptothecin, etoposide, teniposide
		Inhibit nucleotide synthesis: fluorinated pyrimidies, methotrexate
	Affect gene expression	Nuclear hormones: vitamin D, glucocorticoids, retinoids
Intracellular signalling	Alter protein phosphorylation	Enhancement: okadaic acid, PMA
		Inhibition: staurosporine, genistein
	Alter signal transduction	Enhancement: cAMP analogues, forskolin
		Inhibition: flutamide, tamoxifen
	Putative second messenger in TNF, Fas and radiation-induced PCD	Ceramide

MPTP = *N*-methyl-4-phenyl-1,2,3,6-tetrahydropyridine, NMDA = *N*-methyl-D-aspartate, PCD = programmed cell death, PMA = phorbol 12-myristate-13-acetate, TNF = tumour necrosis factor, TPEN = *N,N,N',N'*-tetrakis(2-pyridylmethyl)ethylenediamine (Source: Wertz and Hanley, 1996).

ER structure or direct activation of PCD-inducing enzymes by increased cytosolic Ca^{2+} concentrations. Thapsigargin, there-fore, points to several damage sensors, particularly those localized to the ER, which could monitor ER protein folding or Ca^{2+} levels.

16.6.6 Disturbances in the cytoskeleton

Microtubule perturbing agents such as taxol, a chemotherapeutic agent, have been attributed to their ability to enhance tubulin polymerization, thereby arresting cells in mitosis and also induce PCD. Cytochalasins are fungal metabolites that bind actin subunits to prevent polymerization and also have anti-neoplastic activity. PCD induced by cytochalasins is associated with the disruption of actin networks.

16.6.7 Alteration of signal transduction

Cyclic AMP is a second messenger central to several signal transduction cascades, including cAMP-dependent protein kinase (PKA)- and calcium-dependent protein kinase (PKC)-mediated cellular signalling. Treatment with cAMP analogues such as 8-bromo-cAMP (8-Br-cAMP) elevates intracellular cAMP and induces PCD. For example, differentiation and luteinization of rat primary granulosa cell cultures are associated with moderate cAMP elevation, but high levels of cAMP promote PCD. One of the major intracellular signalling pathways responsible for promoting cell survival is initiated by the enzyme **PI 3-kinase (Fig. 16.12)**, which is activated by either protein-tyrosine kinases or G protein-coupled receptors.

Alternatively, signal transduction can be inhibited using flutamide, which is an anti-androgen commonly used for prostate cancer therapy. Androgen-dependent cells require dihydrotestosterone in order to activate the androgen receptor-mediated transcription of essential gene products, such as replication enzymes. Flutamide inhibits this hormone-receptor interaction and induces PCD in androgen-dependent cells. Androgen ablation is also associated with enhanced expression of several genes, the products of which might promote PCD.

16.7 APOPTOSIS AND NORMAL PHYSIOLOGY

Extracellular survival factors suppress apoptosis: Most animal cells need signals from other cells not only to grow and proliferate, but also to survive. If deprived of such *survival factors*, cells activate their intracellular death program (suicidal) and die by apoptosis. This process ensures that cells survive only when and where they are needed. Neural cells for example, are initially produced in excess in the developing nervous system and then compete for

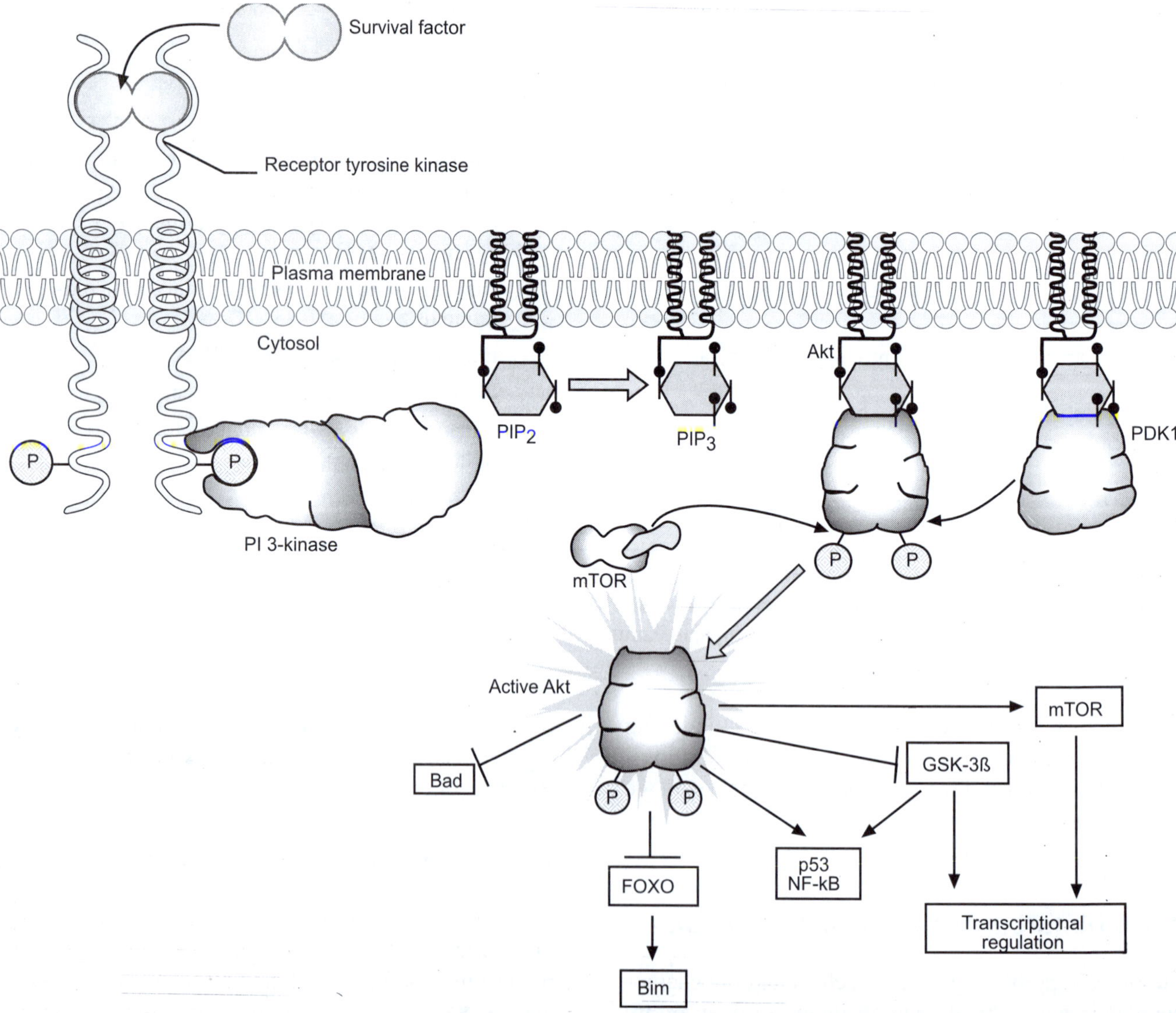

Fig. 16.12 The PI 3-kinase pathway and cell survival. A number of growth factors that signal cell survival activate receptor protein tyrosine kinases, leading to activation of PI 3-kinase, formation of PIP3 and activation of the protein kinase, Akt. Akt then phosphorylates a number of proteins that contribute to cell survival.

limited amounts of survival factors that are secreted by the target cells they contact (**Fig. 16.13**). Only those nerve cells alive receive enough survival factors, while the others die by apoptosis. A similar dependence on survival signals from neighbouring cells is thought to control cell numbers in other tissues, both during development and in adulthood. Survival factors, similar to mitogens and growth factors usually bind to cell surface receptors and activate signalling pathways that keep the death program suppressed, often by regulating members of the Bcl-2 family of proteins. Survival factors, either stimulate the increased production of apoptosis-suppressing members or act by inhibiting the function of apoptosis-promoting members of the family (**Fig. 16.14**).

Cell death is critical for animal development. In mammals, some developmental cell deaths are autonomous, meaning that pro-apoptotic signals from neighbouring cells are not needed for the demise of the such cells. In fact it is believed that, in many of these deaths, neighbouring cells provide survival signals and that cell death is initiated by withdrawal of growth factors and/or loss of cell attachment. These pathways to cell death are often referred to as *death by neglect* and can in many instances be blocked by anti-apoptotic members of the Bcl-2 protein family. Death receptors and their ligands are prime candidates for this function.

Gene knockout technology has been used to determine the essential functions of Bcl-2 family proteins. Mice deficient in Bcl-2 show increased cell death during nephrogenesis and premature demise of their mature lymphocytes. Mice deficient in Bcl-x die at embryonic day 13 and show abnormally increased apoptosis in the central nervous system and in erythroid cells in the foetal liver. It has been shown that erythropoietin can directly induce expression of Bcl-X_L via a signalling pathway. Mice lacking Bcl-w have defective spermatogenesis and mice lacking A1 have accelerated neutrophil apoptosis.

Bax, Bim and Bid are the only pro-apoptotic Bcl-2 family members for which studies with gene knockout analysis have been published. Cell production and cell

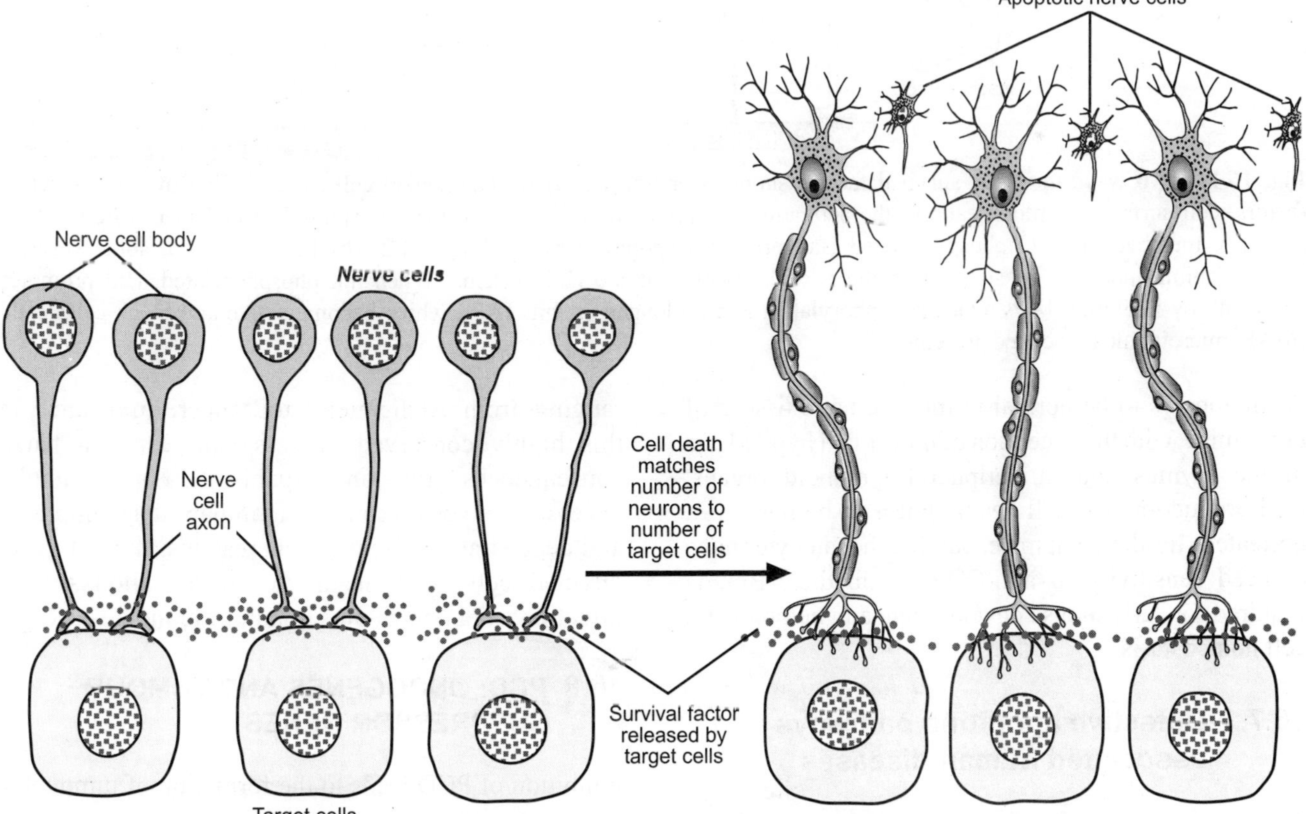

Fig. 16.13 The role of cell death in matching the number of developing nerve cells to the number of target cells they contact. Generally a greater number of neuronal cells are produced than can be supported by the limited amount of survival factors released by the target cells. But many cells die due to insufficient survival factors to keep their suicide program suppressed. This physiological culling ensures that all target cells are contacted by nerve cells and the extra nerve cells are automatically eliminated.

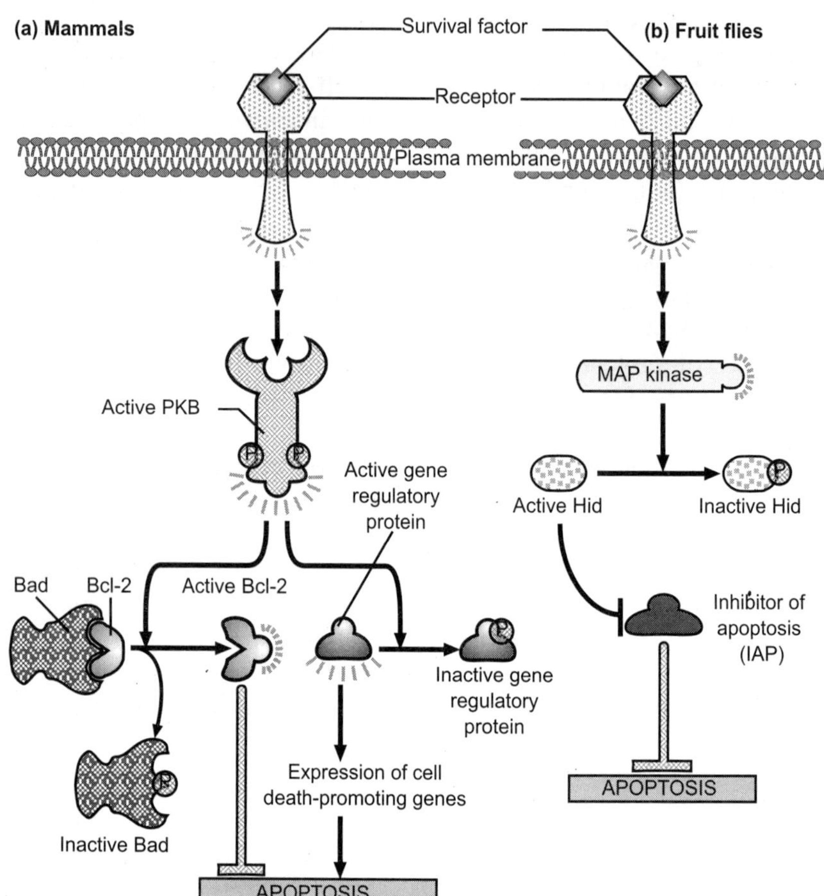

Fig. 16.14 Two ways in which survival factors suppress apoptosis: (a) In mammalian cells, the binding of some survival factors, cell surface receptors leads to the activation of various protein kinases, that phosphorylate and inactivate the Bcl-2 family member, Bad. Unphosphorylated Bad promotes apoptosis by binding Bcl-2. (b) In *Drosophila*, some survival factors inhibit apoptosis by stimulating the phosphorylation of the Hid protein. When not phosphorylated, Hid promotes cell death by inhibiting IAPs. Once phosphorylated, Hid no longer inhibits IAPs, which become active and block cell death. (MAP: microtubule associated proteins).

death appears to be normal in most somatic tissues of *bax*$^{-/-}$ mice, with the exception of mild T cell hyperplasia in the thymus and in peripheral lymphoid organs. Cell production and cell death appear to be normal in untreated Bid-deficient mice, but their hepatocytes have reduced sensitivity to anti-CD95 antibodies. Bim is required for early mouse development and hemopoietic cell homeostasis.

16.7.1 Defective apoptotic pathways and associated human diseases

Abnormalities associated with apoptosis can play a significant role in disease processes. Apoptosis plays a key role in both degenerative and proliferative diseases (**Table 16.3**). Detailed study of controlling genes and signalling pathways has provided evidence for an evolutionarily conserved link between diseases ranging from Alzheimer's to cancer. Imbalance in this highly conserved process can result in harsh consequences for an organism. For example, excessive cell death can lead to impaired development and degenerative diseases, whereas insufficient gene-directed cell death leads to cancer, allouxs viral infections and contributes to autoimmune diseases.

16.8 PCD: ONCOGENES AND TUMOUR SUPPRESSOR GENES

Inhibition of PCD leads to the formation of tumours in two ways:

- Accumulation of unwanted cells, which become tumorigenic on their own
- In the absence of cell death, cells live longer and become likely candidates to harbour oncogenic mutations resulting in tumour formation.

Many oncogenes and tumour suppressor genes have the ability to effect cell death.

p53 Dysregulation

In addition to death receptors, p53 (a tumour suppressor gene), normally a very shortlived protein, is a transcriptional regulator that can arrest growth and induce apoptosis in response to intracellular disruptions resulting from DNA damage, metabolite deprivation, heat shock, hypoxia and activated cellular or viral oncoproteins (**Fig. 16.15**). The dominant as well as the recessive mutations in the p53 can be

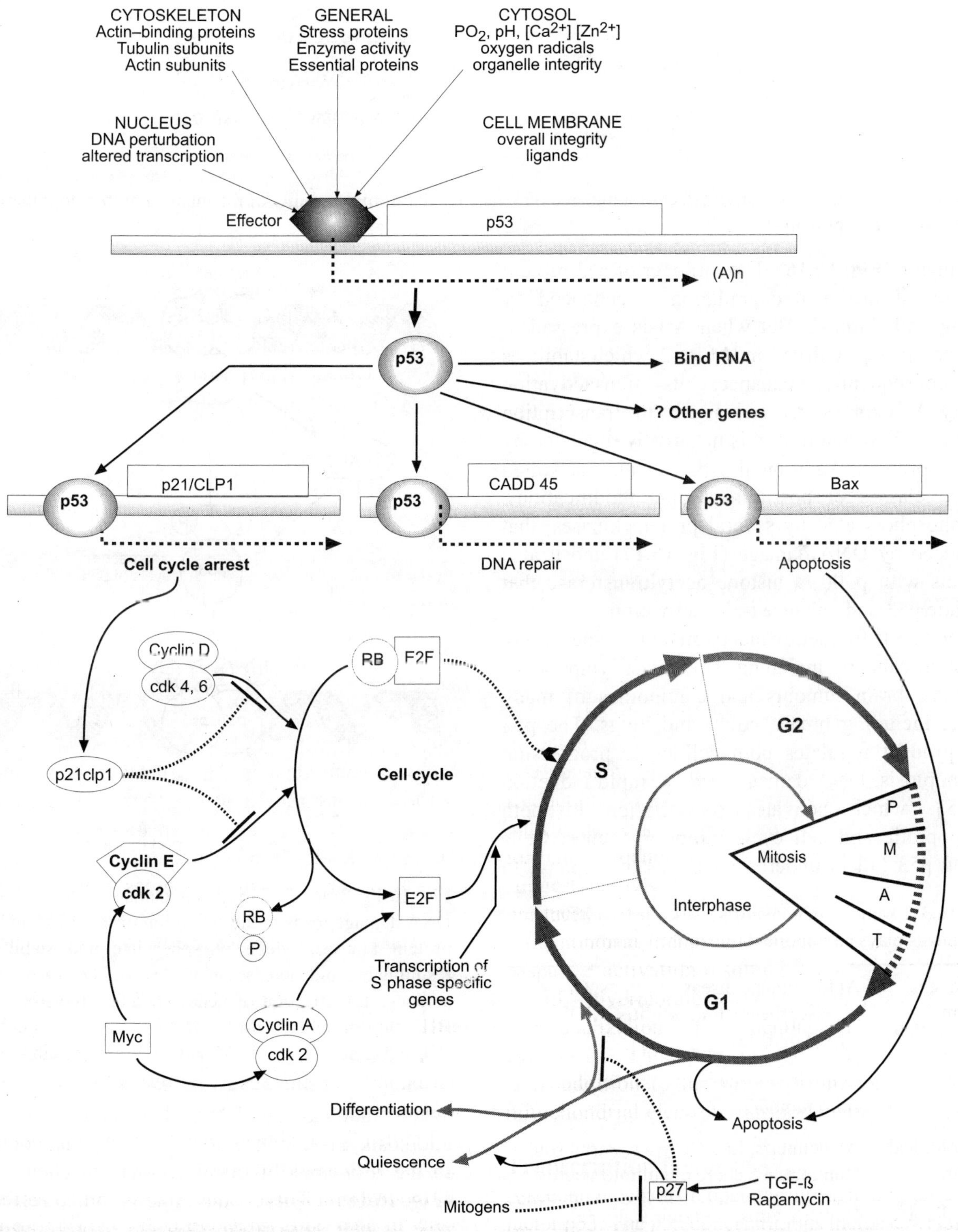

Fig. 16.15 Negative control of cell cycle by p53 and other cell cycle related proteins.

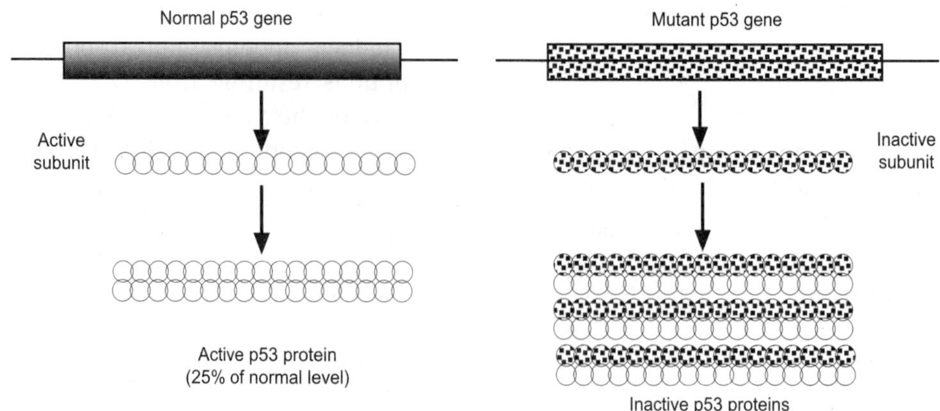

Fig. 16.16 The dominant–negative effect of a mutated p53 gene results from the ability of the mutated protein to dimerize and inactive the normal protein.

tumorigenic (**Fig. 16.16**). The turnover of p53 protein by ubiquitin-mediated degradation is enhanced by binding with Mdm-2. But when Arf is expressed, it forms complexes with p53 and Mdm-2, which stabilizes p53, consequently, enhances its transactivation activity. Arf expression is activated by the transcription factor E2F-1, which in turn is negatively regulated by the RB protein. Additionally, p53 is also activated by two classes of post-translational modifications. It is phosphorylated by several protein kinases, that is induced by DNA damage (**Fig. 16.17**) and it also interacts with p300, a histone acetyltransferase that acetylate p53 and enhance trans-activation.

The p53 is frequently inactivated in a wide variety of human cancers, including leukemias, lymphomas, sarcomas, brain tumours and carcinomas of many tissues, including breast, colon and lungs. The p53 gene product regulates both cell cycle progression and apoptosis. DNA damage leads to rapid induction of p53, which activates transcription of both pro-apoptotic and cell cycle inhibitory genes. Cells lacking p53 fail to undergo apoptosis in response to agents that damage DNA. This failure to undergo apoptosis in response to DNA damage contributes to the resistance of many tumours to chemotherapy (**Fig. 16.18**). These and many other effects of p53 on cell survival are thought to account for the high frequency of p53 mutations in human cancers.

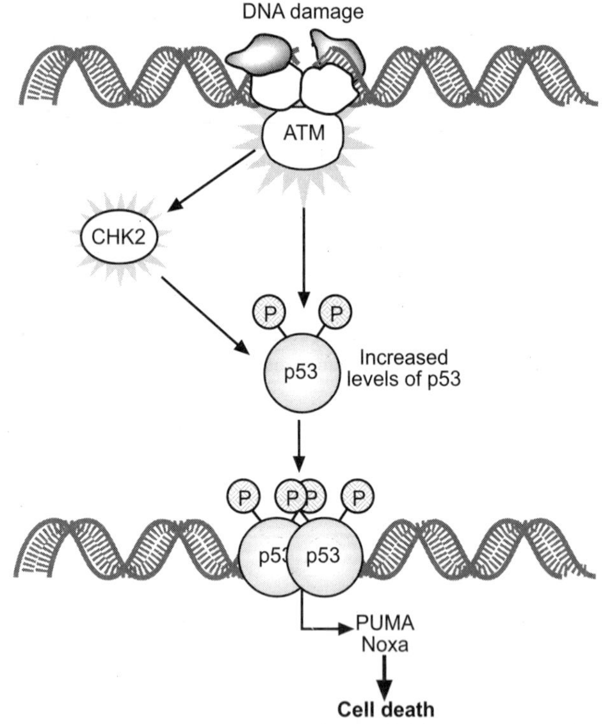

Fig. 16.17 Role of p53 in DNA damage-induced apoptosis. DNA damage leads to the activation of the ATM and CHK2 protein kinases, which phosphorylate and stabilize p53 resulting in rapid increases in p53 levels. The protein p53 then activates transcription of genes encoding the pro-apoptotic BH3-only proteins PUMA and Noxa, leading to cell death. BH, Bcl-2 homology; ATM and CHK are protein kinases.

Table 16.3 Examples of human diseases where dysregulation of apoptosis plays an important role in the syndrome

Immune system	Auto-immune diseases: Lupus, Graves disease, rheumatoid arthritis, AIDS, Type I diabetes
Cancer-related	Lymphomas, leukemia, colon cancer, solid tumours
Neurological system	Alzheimer's, Lou Gehrig's, Parkinson's, Huntington's chorea, multiple sclerosis
Viral diseases	Cowpox virus, adenovirus, Epstein–Barr virus

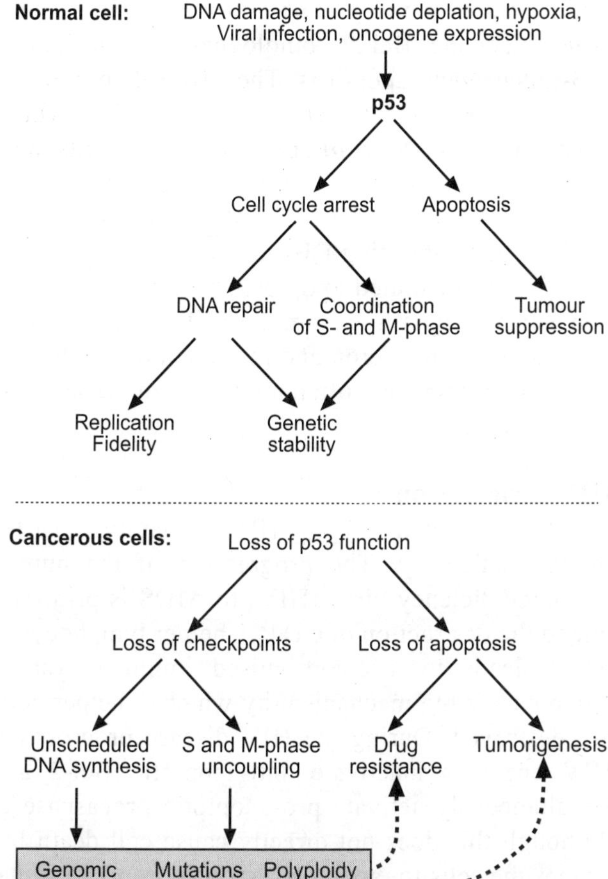

Fig. 16.18 p53-dependent activation of pathways leading to tumor suppression, genomic stability and the consequences of p53 loss or mutation in cancer cells

p21

The prototype cyclin-dependent kinase (CDK) inhibitor p21 (CIP1/WAF1/p21CAP20/SDI1MDA6) has become an intensely studied molecule. The p21 is a focal point that integrates many types of signals that impact processes of cell division and cell death. Besides its role in p53-dependent checkpoint control of cell cycle progression, p21 is involved in mediating a p53-independent growth arrest associated with terminal differentiation. Initial observations indicated that p53 regulates basal as well as DNA damage-inducible p21 expression. It was discovered as a senescent cell-derived inhibitor of DNA replication in human diploid fibroblasts. It is induced in a p53-dependent G1 arrest in normal human diploid fibroblasts resulting in the inhibition of CDK2-cyclin E kinase activity, thereby blocking its entry into the S phase.

Ras

Three variations of p21-ras proteins (Ha-ras, Ki-ras and N-ras) have been identified as key signal transducers. They promote cell proliferation and oncogenesis of proliferating cells and can also mimic the functions of NGF. Survival of neurons is NGF-dependent. It is believed that ras may be involved in the intracellular signalling pathway for survival effects of neurotrophic factors. A part of the amino acid of p21-ras has a homology with cystatins, an endogenous cysteine protease inhibitor. Probably ras-related proteins regulate Bcl-2 proteins and being cysteine protease inhibitors, prevent cell death.

C-myc

C-myc expression is induced by mitogenic stimulation. C-myc is downregulated and cells get arrested in G1 phase following removal of serum from culture media. Expression of endogenous c-myc is required for apoptosis induced by CD3/TCR stimulation. The same region of c-myc is required for transformation, transcription activation and induction of apoptosis which implies a close relationship between cell proliferation, transformation and apoptosis as a function of c-myc.

C-fos

Expression of the c-fos proto-oncogene has been associated with a variety of biochemical responses, including proliferation, differentiation and neural excitation. C-fos is very similar to c-myc and its continuous expression renders cells more sensitive to apoptotic signals and the signal transduction machinery of cell death may be shared with cell proliferation and differentiation. Both c-fos and c-myc can be induced by oxidative stress.

C-rel

The c-rel is a transcription factor and is a member of NF-kB family. The highest level of expression is detected in cells undergoing PCD. Overexpression of c-rel in primary avian fibroblast cells results in extended lifespans. Thus, it is possible that like any other cellular oncogenes c-rel can also regulate both cell proliferation and cell death.

16.9 INFECTION-INDUCED APOPTOSIS

In higher vertebrates, apoptosis is pivotal for the normal development and function of the immune system. Both self-reactive maturing immunocompetent cells and hyper-reactive mature cells are eliminated by apoptosis. Moreover it has been recognized as an important defence mechanism against viral, bacterial and parasitic pathogens during innate and adaptive immunity. Apoptosis of infected host cells facilitates the survival of the host by diminishing the production of pathogens. Also, apoptosis in specific cells contributes to the regulation of pathogen-induced immune responses. Viral, bacterial and protozoan pathogens have evolved different strategies to modulate apoptosis in the host cell.

16.9.1 Modulation of apoptosis by viruses

Viruses infect target cells, replicate large numbers of progeny virions and spread these progeny to initiate new rounds of infection. They encode highly efficient proteins to optimize such propagation events. In response to infection, the target organism contains both systemic and cell-based defences to limit viral spread, which include immune and inflammatory processes and the execution of suicide of infected cells. In spite of such powerful defence strategies, most viruses have evolved multiple mechanisms for modulating host cell apoptosis and produce proteins that are able to inhibit or delay protective actions. In addition, some viruses appear to encode products that actively induce apoptosis as a part of exit strategy to enhance virus spread. In these cases, a delicate balancing act between inhibition and induction of apoptosis is performed by a combination of viral products.

Some important virus families involved in apoptosis induction are: *Adenoviridae* (Adenovirus), *Papovaviridae* (SV40), *Parvoviridae* (MVM), *Circoviridae* (CAV), *Herpesviridae* (HHV6/7, EBV), *Baculoviridae* (AcNPV), *Hepadnaviridae* (Hepatitis B), *Retroviridae* (HIV), *Togaviridae* (Sindbis virus), *Flaviviridae* (Hepatis C), *Reoviridae* (Revirus, Rotavirus), and *Birnaviridae* (IBDV).

Alternatively, viruses have also developed a range of strategies to defend themselves against the host immune response and apoptosis. Several viruses have developed strategies to block p53-induced apoptosis. For example, the SV40 large T antigen binds directly to the p53 DNA binding domain, preventing its binding to and activation of cognate promoters. The ElB protein of adenovirus binds to the activation domain of p53 and blocks the activation p53-responsive genes. The E6 protein of papilloma viruses also bind p53, but marks it for degradation via the ubiquitin pathway. Many viruses encode Bcl 2h omologues, which inhibit p53-independent apoptosis. The EIB-19K protein of human adenovirus interacts with Bax and Bak, which are positive regulators of apoptosis and inhibits their action. The adenovirus E3 protein acts at multiple levels; it enhances clearance of Fas from the cell surface, thus inhibiting Fas-induced apoptosis and it blocks TNF-mediated apoptosis by interfering with downstream signalling from the TNF receptor. Some viruses target interferon and inflammatory pathways by using proteins that often mimic or counteract host cell functions.

HIV progression

People infected by HIV mainly suffer immunodeficiency. The progression of the human immunodeficiency virus (HIV) to AIDS is primarily due to the destruction of $CD4^+$ T helper lymphocytes, which leads to a compromised immune system. Apoptosis is the mechanism by which T helper cells are destroyed. During the HIV disease progression, HIV enzymes inactivate anti-apoptotic Bcl-2 and simultaneously activate pro-apoptotic procaspase 8. Although this does not directly cause cell death but primes the cells for apoptosis upon appropriate signal being received. HIV products may increase levels of cellular proteins, which have a promotive effect on Fas-mediated apoptosis. Viral proteins decrease the amount of CD4 glycoprotein marker present on the cell membrane of T helper cells. Released viral particles and proteins present in extracellular fluid are able to induce apoptosis in nearby T helper cells. HIV decreases the production of molecules involved in marking the infected cell for apoptosis, giving the virus time to replicate and continue releasing apoptotic agents and virions into the surrounding tissue.

16.10 APOPTOSIS IN PLANTS

Most events (structural and molecular changes) associated with PCD in animals are similar to that in plant cells. However, PCD in plants takes place in different categories of plant cells. A critical analysis of the results obtained so far pertaining to the processes associated with cell death in a few categories of cells are discussed below.

PCD in pollen prevents inbreeding

During pollination, plants enforce self-incompatibility (SI) as an important means to prevent self-fertilization. Research on the corn, poppy has revealed that proteins in the pistil on which the pollen lands interact with pollen and trigger PCD in incompatible (self) pollen.

PCD in plant development

Evidence of genetic controls and signalling molecules in plants undergoing PCD (analogous to animals) has emerged only recently. Examples of PCD in plant development include the timely death of petals after fertilization, during diploid parthenogenesis, the development of tracheary elements (**Fig. 16.19**) and senescence. Some category of plant cells (includes TEs, sclerenchyma fibres, sclerides and cork cells) undergo unique PCD. These cells stop functioning and consequently the cytological events accompanying cell death in these cells is oriented towards making them persist and function till the end of the life of the plant. They do not show cell shrinkage, but often enlarge and elongate during differentiation. The nuclei of the cells are not only metabolically very active but often undergo endo-duplication and polytenic changes during differentiation. They are transcriptionally very active and are responsible for the production of several chemicals that form part of the persisting specialized secondary cell wall. The presence of some types of proteases in these cells may clean up the cytoplasmic proteins. The same may be said of the endonucleases reported in animal PCD.

Apoptosis in plants also occurs during biotic and abiotic stress. Many studies have revealed that a population of cells subjected to abiotic/biotic stresses, undergoes PCD. The cells undergoing stress-induced PCD produce a special category of structural proteins rich in hydroxyproline, proline, glycine, arabino galactan and occasionally threonine, which are unique to plant cell PCD.

Cell death has also been assumed to play a role in the hypersensitive reaction (HR) that is characteristic of several incompatible plant–microbe interactions. HR-linked cell death requires active plant metabolism and depends on the activity of host transcriptional machinery. The resistance markers expressed in cells undergoing an HR cell death are the expression of so called pathogenesis-related proteins, generation of reactive oxygen intermediates, rapid influx of calcium, production of phyto-alexins and the cross-linking of components of the cell wall.

16.11 APOPTOSIS IN NEMATODE

Caenorhabditis elegans is a soil-dwelling nematode predominantly hermaphrodite, self-fertilizing nematode, which develops from egg to adult in about two days. Embryonic development is rapid, with hatching larvae produced in about 15 hours. The larvae undergo four moults to produce adult worms. Therefore, the phenomenon of apoptosis has been most extensively studied in *C. elegans* and it became a good model organisms for developmental studies, because of its invariant somatic cell lineage, producing hermaphrodite individual with exactly 959 somatic cells. Additionally, cells are transparent, which facilitate easy observations of cell death events and availability of detailed data on genetic and molecular analysis makes the dissection of the various aspects of cell death easy. Cell death in *C. elegans*

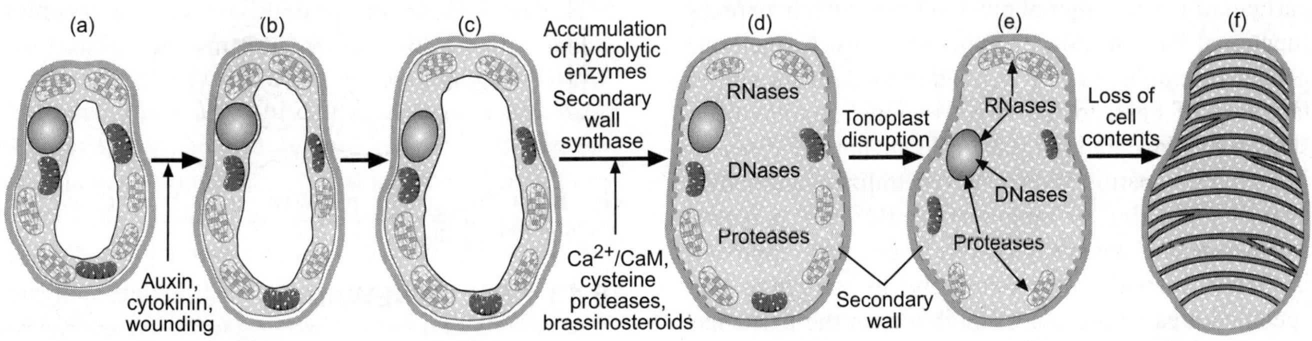

Fig. 16.19 The re-differentiation of cultured mesophyll cells into tracheary elements (TE) involves PCD. (a) A cultured mesophyll cell with its complement of cellular organelles; (b) After induction of modification of the hormone content of the culture medium, the cell differentiates. (c) Differentiate to form TE precursor cell; (d) An immature TE with characteristic secondary cell wall thickenings; (e) A mature TE cell; (f) A mature, dead, empty TE forms. In mature TE, lysis of the vacuole follows total degradation of the cellular contents.

can be divided into four stages (**Fig. 16.20**). During larval development, cells express pro-apoptotic factor, EGL-1, when they are destined to die. EGL-1 (egl = egg laying) is a BH3 domain-containing protein and is homologous to the vertebrates, BAD and BIK. EGL-1 binds to CED-9 (homologue of Bcl-2) and prevents CED-9 from binding to adaptor protein CED-4 (homologue of Apaf-1, cytochrome C). When released, CED-4 is capable of binding to CED-3 and activating the caspase CED-3. CED-3 is the key protein involved in executing the death of the cell.

16.12 APPLICATIONS OF APOPTOSIS

Apoptosis technology has great commercial as well as academic applications.

- To understand the developmental regulation of multicellular organisms and morphogenesis
- To study diseases (e.g., Alzheimer's and Parkinson's diseases), immunosuppression or an effect of trauma (e.g., stroke and heart attack); the pathological suppression of apoptosis is also important for cancer and viral infections
- To maintain several industrially useful cell lines (e.g., for production of antibodies and vaccines) which require costly growth factors. They could be maintained longer with less effective cell deaths by altering the genetic make up of these cells. Also, manipulation of such genes could be used to modify permanently the apoptotic tendency of these cell lines.

Cancer is a disease with excessive cell proliferation and insufficient cell death. The first objective would be to kill the tumour since some cells could be cancerous particularly, some lymphoid tumours, which express functional Fas. In order to avoid side-effects, methods need to be employed with local administration or proper targeting of FasL to the target regions. Fas has been shown to play an important role in human diseases like AIDS, hepatitis and so on. Neutralizing antibodies against them (by an inhibitor) would have potential as therapeutic agents. Research on this approach is being carried out extensively by targeting the specified area or the use of inhibitors in the particular cascade of apoptosis.

Apoptosis plays an important mechanism in hypersensitive reaction during many viral and microbial infections on crop plants. Understanding the molecular mechanism that triggers the apoptosis against invaders can help to develop plants resistant to the pathogens.

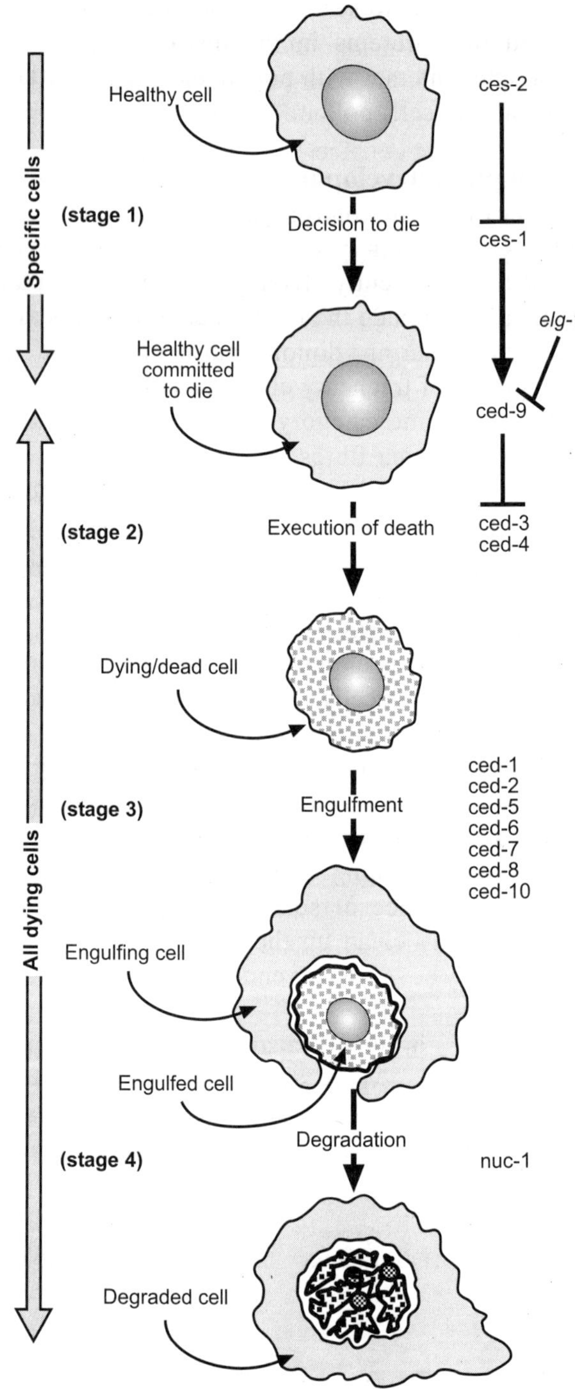

Fig. 16.20 The genetic pathway for PCD in *C. elegans*, showing four distinct stages.

16.13 CELL RENEWAL

During the early phase of development of multicellular organisms, the cellular mass increases by the rapid proliferation of embryonic cells, which then differentiate to form the specialized cells of adult tissues and organs. As cells differentiate, their rate of proliferation gradually decreases and most cells in adult animals are arrested in the G0 stage of the cell cycle. However, later during development, many cells are lost either due to injury or programmed cell death and this process continues throughout life. Thus, to maintain a constant number of cells in adult tissues and organs, lost cells must be balanced by cell proliferation. To maintain this cellular turnover, most tissues contain a group of cells that have ability to proliferate as required to replace cells that have died and renew themselves to maintain proliferative ability. Therefore, cell death and cell proliferation are carefully balanced to maintain optimal size and functioning of adult tissues and organs.

16.13.1 Multiplication of differentiated cells

Most mature cells in adult animals are terminally differentiated (no longer capable of cell division). If adult cells are lost, they will be replaced by the proliferation of less differentiated cells derived from self-renewing stem cells. The stem cells normally enter the G0 stage of the cell cycle but resume proliferation as needed. For example, G0-arrested fibroblast stem cells, which are dispersed in connective tissue rapidly proliferate if needed to repair damaged cells. A fully differentiated endothelial cell remains capable of proliferation (**Fig. 16.21**). Proliferation of endothelial cells allows them to form new blood vessels as needed for repair and re-growth of damaged tissue. Endothelial cell proliferation is triggered by a growth factor VEGF (vascular endothelial growth factor) produced by cells of the tissue, in which the new capillaries will invade. The secretion of VEGF is in turn triggered by a lack of oxygen. The result is a regulatory system, in which tissues that have a low oxygen supply resulting from insufficient circulation, stimulate endothelial cell proliferation and recruit new capillaries (also called **angiogenesis**).

Epithelial cells of some internal organs such as the liver and pancreas are also able to proliferate to replace damaged tissue. For example, liver cells, are normally arrested in the G0 phase of the cell cycle. Even if the liver loses a large number of cells (e.g., by surgical removal), the remaining cells can get stimulated to proliferate to replace the missing tissue (**Fig. 16.22**). The human liver is one of the few glands in the body that has the ability to regenerate from as little as 25% of its tissue. This is largely due to the unipotency of hepatocytes.

16.13.2 Stem cells

Stem cells are primal cells found in all multicellular organisms that retain the ability to renew themselves

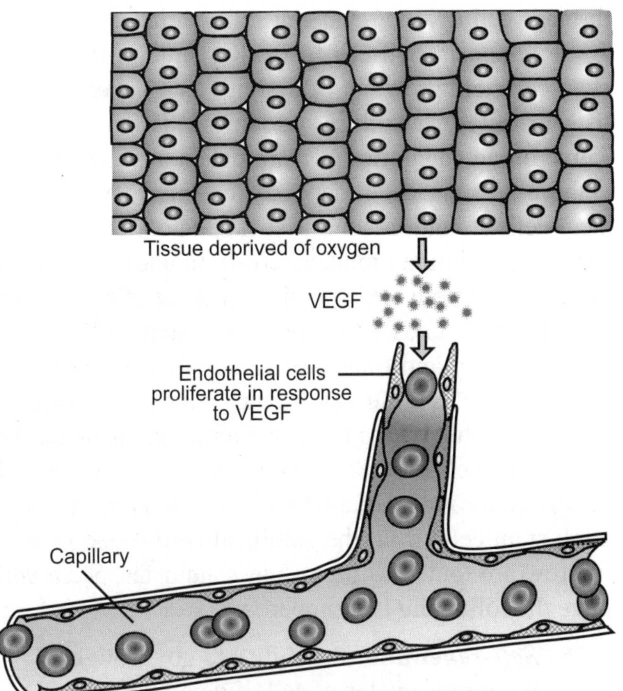

Fig. 16.21 Proliferation of endothelial cells.

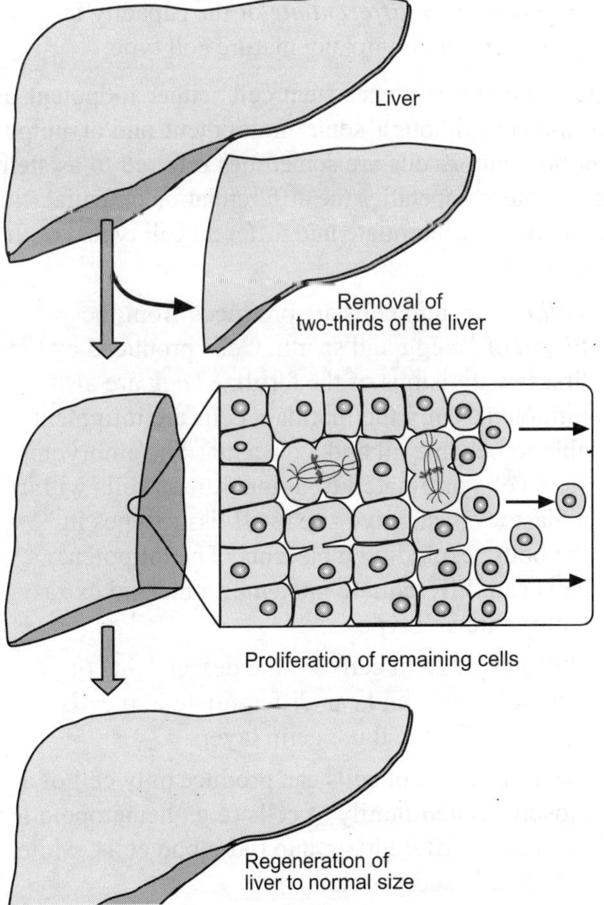

Fig. 16.22 Liver regeneration. Hepatocytes are normally arrested in G0 but resume proliferation to replace damaged tissue or surgically removed.

through cell division and can differentiate into a wide range of specialized cell types. Research in the human stem cell field grew out of findings by Canadian scientists, Ernest A McCulloch and James E Till in the 1960s. Three broad categories of mammalian stem cells exist: embryonic stem cells (derived from blastocysts), adult stem cells (found in adult tissues) and cord blood stem cells (found in the umbilical cord). In addition to these three normal stem cell types the existence of cancer stem cells has also been identified. Since stem cells can be readily grown and transformed into specialized tissues such as muscles or nerves through cell culture, their use in medical therapies has been proposed. In particular, embryonic cell lines, autologous embryonic stem cells generated through therapeutic cloning and highly plastic adult stem cells from the umbilical cord blood or bone marrow are touted as promising candidates. Stem cells have the following two propertis:

- ***Self-renewal*** or the ability to go through numerous cycles of cell division while maintaining the undifferentiated state,
- ***Potency to differentiate*** or the capacity to differentiate into any mature cell type.

These properties makes stem cells either totipotent or pluripotent, although some multipotent and/or unipotent progenitor cells are sometimes referred to as stem cells. Potency specifies the differentiation potential (the potential to differentiate into different cell types) of the stem cell.

- ***Totipotent*** stem cells are produced from the fusion of an egg and sperm. Cells produced by the first few divisions of the fertilized egg are also totipotent. Only the morula's cells are totipotent, able to become all and a placenta. The embryonic stem cells originated from inner mass cells within a blastocyst can give rise to all tissue types in the body, excluding a placenta. The totipotent cells can differentiate into embryonic and extra-embryonic cell types.
- ***Pluripotent*** stem cells are the descendents of totipotent cells and can differentiate into cells derived from the three germ layers.
- ***Multipotent*** stem cells can produce only cell of a closely related family of cells (e.g., hematopoietic stem cells differentiate into red blood cells, white blood cells and platelets).
- ***Unipotent*** cells can produce only one cell type, but have the property of self-renewal which distinguishes them from non-stem cells.

These properties can be illustrated in vitro, using methods such as clonogenic assays, where the progeny of single cell is characterized.

Embryonic stem cells

Adult stem cells have low or no proliferative ability and have short lifespan. Therefore, they are difficult to culture and not available for future use. When adult cells are removed from the body and maintained in culture, they generally maintain their original character and divide for a only few generations. Each differentiated cell type has a memory of its developmental events stabilized on its final specialized state. Therefore, it is relatively easy to isolate and propagate the stem cells, which may differentiate into one or more cell types, but restricted to adult types. However, it is possible through cell culture to derive early embryos or extraordinary class of stem cells called **embryonic stem cells**, or **ES cells**. Thede are cultures of cells derived from the epiblast tissue of the inner cell mass (ICM) of a blastocyst. A blastocyst is an early-stage embryo,approximately 4 to 5 days old in humans and consisting of 50–150 cells. ES cells are pluripotent and give rise during development to all derivatives of the three primary germ layers (ectoderm, endoderm and mesoderm). In other words, they can develop into each of the more than 200 cell types of the adult body when given sufficient and necessary stimulation for a specific cell type. When not given any stimuli for differentiation, ES cells will continue to divide in vitro and each daughter cell will remain pluripotent.

The pluripotency of ES cells has been rigorously demonstrated in vitro and in vivo and therefore they can be classified as stem cells. In appropriate culture conditions, these cells can be grown indefinitely as pure stem cell populations while maintaining the ability to give rise to all of the differentiated adult cell types. ES cells were first cultured from mouse embryos in 1981 (**Fig. 16.23**). Embryonic stem cells can be propagated indefinitely in vitro and all the cell types of the adult mouse can be obtained if re-introduced into the early embryos or induced into a variety of different types of cells in culture. Cells with properties similar to those of mouse ES cells can now be derived from early human embryos and from human foetal ovaries and testes, creating a potentially inexhaustible supply of cells that might be used for the replacement and repair of mature human tissues that are damaged. Thus, they find great potential for use, in regenerative medicine and tissue replacement after injury or disease.

Uses

Embryonic stem cells have been an important experimental tool in cell biology because they can be used to

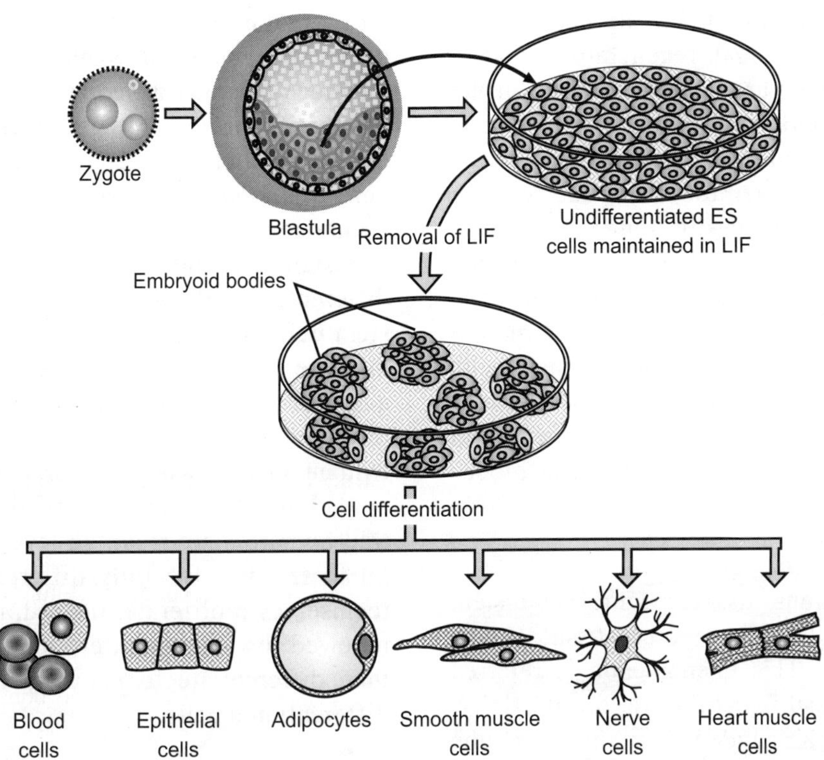

Fig. 16.23 Differentiation of embryonic stem cells. Mouse embryonic stem cells (ES cells) grow undifferentiated when cultured in the presence of LIF (leukemia inhibitory factor). In the absence of LIF in culture medium, the cells aggregate to form embryoid bodies and then differentiate into a variety of cell types.

introduce altered genes into an organism. Stem cells also provide an outstanding model for studying the molecular and cellular events associated with embryonic cell differentiation, so ES cells have long been of considerable interest to cell and developmental biologists. Recently, successful use of human embryonic stem cells in clinical transplantation therapies has been reported.

In mouse, embryonic stem cells are grown in the presence of a growth factor called LIF (leukemia inhibitory factor), which signals through the JAK/STAT pathway and is required to maintain these cells in their undifferentiated state (**Fig. 16.24**). If LIF is removed from the culture medium, the cells stop growing and the cells aggregate into structures that resemble embryos and then differentiate into a wide range of cell types, including neurons, adipocytes, blood cells, epithelial cells, vascular smooth muscle cells and even heart muscle cells.

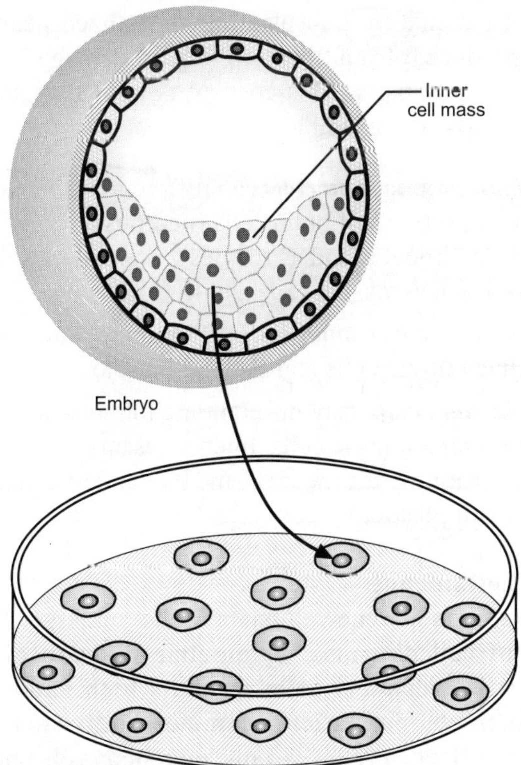

Fig. 16.24 Culture of mammalian embryonic stem cells. Embryonic stem cells are cultured from the inner cell mass of an early embryo (blastocyst).

Adult stem cells

Most terminally differentiated cells in adult animals are incapable of cell division. Nonetheless, they can be replaced by the proliferation of a subpopulation of less differentiated self-renewing cells called stem cells that are present in most adult tissues.

Adult stem cells found throughout the body divide to replenish dying cells and regenerate damaged tissues. Adult stem cells are generally found in the brain, bone marrow, peripheral blood, blood vessels, skeletal muscle, skin and liver. Adult stem cells may be classified as progenitor cells, due to their limited capacity for cellular differentiation. These cells typically generate the cell types of the tissue in which they reside. Since, they retain the self-renewal capacity and replace differentiated mature cells throughout the lifetime of an animal, stem cells play a critical role in the maintenance of most tissues and organs. Adult stem cells can be obtained from bone marrow, umbilical cord and peripheral blood-hematopoietic stem cells.

Stem cell plasticity

Under special conditions, tissue-specific adult stem ells can generate cell types of other tissues, even crossing germ layers. This phenomenon is referred to as stem cell **transdifferentiation** or **plasticity**. Stem cell plasticity can be induced, by modifying the growth medium and supplementing required growth factors. When cultured in vitro or transplanted in to an organ of the body different from the one they were originally isolated from, stem cells make it possible to treat patients by transplanting specialized healthy cells produced from them to repair damaged and diseased body parts. This concept is known as **stem cell therapy**. For example,

- ***Hematopoietic stem cells*** can differentiate into major types of brain cells (neurons, oligodendrocytes and astrocytes), skeletal muscle cells, cardiac muscle cells and liver cells.
- ***Bone marrow stromal cells*** differentiate into cardiac muscle cells and skeletal muscle cells.
- ***Brain stem cells*** may differentiate into blood cells and skeletal muscle cells. Intense research is on in to determining the mechanisms that underlie adult stem cell plasticity.

Stem cell lineage

To ensure self-renewal, stem cells undergo two types of cell divisions. **Symmetric division** gives rise to two identical daughter cells both endowed with stem cell properties. **Asymmetric division**, on the other hand, produces only one stem cell and a progenitor cell with limited self-renewal potential. Further progenitors can go through several rounds of cell division before terminally differentiating into a mature cell. It is possible that the molecular distinction between symmetric and asymmetric division lies in differential segregation of cell membrane proteins (e.g., receptors) between the daughter cells; however, the mechanism is not clear yet. Most stem cells have a unique property; they divide to produce one daughter cell that remains as a stem cell and the other one, which divides and differentiates. Because the stem cells produce new stem cells as well as differentiated daughter cells, they are self-renewing populations that can serve as a source for the production of differentiated cells throughout life. The importance of stem cells is more evident in the case of several tissues (blood cells, epithelial cells of the skin and lining of the digestive tract) where there is a high level of cell turnover. In all these sites, the fully differentiated cells do not themselves proliferate; instead, they are continually renewed by the proliferation of stem cells that then differentiate to maintain a stable number of differentiated cells.

Types of stem cells

Hematopoietic stem cells: Earnest McCulloch and James Till (1961) were first to identify stem cells in the hematopoietic (blood-forming) system in experiments showing that single cells derived from mouse bone marrow could proliferate and give rise to multiple differentiated types of blood cells. In the blood, there are several distinct types of blood cells with specialized functions. Most of these matured cells have limited lifespans ranging from less than a day to a few months and all are derived from the same population of hematopoietic stem cells. More than 100 billion cells are lost every day in humans and are replaced continually by the hematopoietic stem cells in the bone marrow. Descendents of the hematopoietic cell continue to proliferate and undergo several rounds of divisions. They become committed to specific differentiation pathways that are determined by growth factors that channel precursor cells along specific pathways of blood cell differentiation.

Epithelial stem cells: The intestine toward lumen side is lined by a layer of epithelial cells that are responsible for the digestion of food and absorption of nutrients. These cells are subjected to a harsh environment and have a lifetime of only a few days. After apoptosis they are shed into the digestive tract. Cell renewal of the intestinal epithelium is, therefore, a continual process through life. The stem cells at the bottom of intestinal crypts (**Fig. 16.25**) give rise to a population of transit-amplifying cells, which divide rapidly and occupy about two-thirds of the crypt. The transit-amplifying cells proliferate for 3–4 cell divisions and then differentiate into the three cell types of the colon surface epithelium.

Skin stem cells: Skin is another dynamic tissue where cell turnover is high. Stem cells responsible for continuous renewal of the skin and hair are also well-characterized. The skin and hair are also frequently exposed to harsh external environment. The differentiated cells are continuously renewed throughout life. In humans, the epidermis turns over every two weeks, with cells being sloughed from the surface. The epidermis is a multilayered epithelium, which is maintained by epidermal stem cells residing in a single basal layer (**Fig. 16.26**). The epidermal stem cells give rise to transit-amplifying cells, which undergo 3–6 divisions before differentiating and moving outward to the surface of the skin.

Adipose-derived adult stem cells (ASCs): They **(ASCs)** have been isolated from fat, usually by the method of liposuction in the case of humans. This cell population seems to be similar in many ways to mesenchymal stem cells (MSCs) derived from bone marrow. But it is possible to isolate more cells from adipose tissue and the harvest procedure is simpler. Human ASCs have been shown to differentiate in vitro into bone, cartilage, fat, muscle and might be able to differentiate into neurons, making them a possible source for future applications in the clinic.

Mammary stem cells: Mammary stem cells provide the source of cells for growth of the mammary gland during puberty and gestation and play an important role in carcinogenesis of the breast. Mammary stem cells have been isolated from human and mouse tissue as well as from cell lines derived from the mammary gland. A single such cell can give rise to both luminal and myo-epithelial cell types of the gland and has been shown to regenerate the entire organ in mice.

Neural stem cells: The existence of stem cells in the adult brain has been postulated following the discovery that the process of neurogenesis, birth of new neurons, continues into adulthood in rats. It has since been shown that new neurons are generated in adult mice, songbirds and primates, including humans. Normally adult neurogenesis is restricted to the subventricular zone, which lines the lateral ventricles of the brain and the dentate gyrus of the hippocampal formation. Under certain circumstances, such as following tissue damage in ischemia, neurogenesis can be induced in other brain regions, including the neocortex.

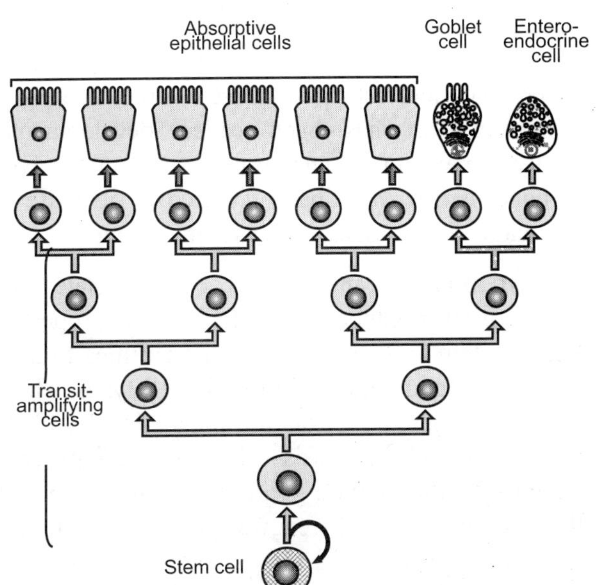

Fig. 16.25 Renewal of the intestinal epithelium.

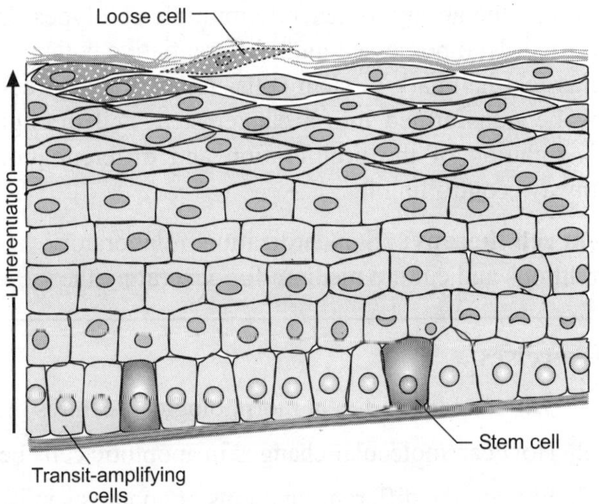

Fig. 16.26 Epidermal stem cells of the skin.

Olfactory adult stem cells: They have been successfully harvested from the human olfactory mucosa cells, the lining of the nose involved in the sense of smell. Adult stem cells isolated from the olfactory mucosa have the ability to develop into many different cell types if they are given the right chemical environment. These adult olfactory stem cells appear to have the same ability as embryonic stem cells in giving rise to many different cell types but have the advantage that they can be obtained from all individuals, even older people who might be most in need of stem cell therapies. Olfactory stem cells hold potential for therapeutic applications, since from their location, they can be harvested with ease without harm to the patient.

Stem cells have also been identified in many other adult tissues, including the retina, heart, lungs, kidneys, liver, mammary glands and pancreas. It has also been demonstrated that stem cells from one tissue might be able to differentiate into cells giving rise to other tissues—a phenomenon that has been called developmental plasticity. For example, hematopoietic stem cells derived from bone marrow have been reported to give rise to non-hematopoietic tissues, such as skin or lungs.

Cancer stem cells: Cancer stem cell theory states that tumours arise from cancer stem cells that have properties of normal stem cells, particularly the abilities to self-renew and differentiate into multiple cell types. These cells persist in tumours as a distinct population that likely cause disease and metastasis—they are the only cells capable of, by themselves, giving rise to new tumours. Cells with unlimited self-renewal potential must exist to sustain the growth of a tumour. Many tumours are heterogeneous and contain multiple cell types native to the host organ, suggesting that the cell that produced them had the ability to generate multiple cell types. In other words, it possessed multidifferentiative potential, a classical hallmark of stem cells. A normal stem cell may be transformed into a cancer stem cell through dysregulation of the proliferation and differentiation pathways controlling it.

Stem cell therapy: Standardization of laboratory conditions and culture medium for generating the required type and number of stem cells is on for adapting stem cells for treating various human diseases. Though stem cells have been found to have great potential for human cell therapy, it has not yet become the standard practice. Stem cells have been tried for a variety of human diseases:

- ***Cardiac***—for regeneration of cardiac tissue after myocardial infarct. So far, bone marrow is the main source for stem cells used in clinical cardiac repair.
- ***Central nervous system***—for brain diseases like Parkinson's, Huntington chorea and stroke.
- ***Spinal cord injury***—following successful animal experiments, a case report of improvement after stem cell therapy in spinal cord-injured patient has been reported.
- ***Burger's disease***–Korean scientists demonstrated symptomatic improvement in patients undergoing treatment.
- ***Diabetes***–restricted to anecdotal evidence of data from All India Institute of Medical Sciences, New Delhi, India. However, there is strong hope that diabetes can be cured one day by injecting β cells (insulin-producing cells) into pancreas of patients, whose β cells have been destroyed.

Exercises

1. What are the different morphological features that can be observed in apoptotic cells?
2. How can molecular changes in apoptotic cells be detected?
3. Explain the different functions of apoptosis in biological systems.
4. How is the apoptosis process involved in shaping organisms during their development?
5. What are the different cellular mechanisms that stimulate activation of apoptosis?
6. Explain briefly the intrinsic and extrinsic factors that stimulate apoptosis.
7. Discuss the role of effector molecules in initiation of apoptosis.
8. Briefly explain the different factors that disrupt cytosolic milieu that affects apoptosis.
9. How do cytoskeletal filaments influence initiation of apoptosis?
10. Explain the impact of defective apoptotic pathway in human health.
11. Explain the role of oncogenes and tumour suppressor genes in initiation of apoptosis.
12. Explain in detail the biological significance of apoptosis in plants.
13. What is cell renewal and why is it so important?
14. What are stem cells? Differentiate between embryonic stem cells and adult stem cells.
15. Give a brief account of the major types of stem cells in humans.
16. What is stem cell therapy?

Summary

- In multicellular organisms, in addition to organismal death, ***self-destructive death*** also occurs at the individual cell or tissue level. This kind of death is essential for the development and survival of multicellular organisms.
- Death may be deliberate and essential and results due to a built-in suicide mechanism known as ***apoptosis*** or developmentally regulated cell death called ***programmed cell death*** (PCD).
- The microscopic observations of ultrastructural changes in apoptotic cells described two discrete stages. In the first stage nuclear chromatin condensation, cytoplasmic shrinkage and vacuolization, plasma membrane blebbing and breaking up of the cell into a number of membrane-bound, ultrastructurally well-preserved fragments, called 'apoptotic bodies' takes place. In the second stage, the apoptotic bodies were ***phagocytozed*** by surrounding cells and degraded within ***phagolysosomes***.
- The main function of apoptosis is to kill unwanted cells for development and homeostasis, for defence and for ageing (***senescence***). PCD is an integral part of both plant and animal tissue development.
- If cells are non-functional due to injury, ageing or infection by pathogens, they have to be removed to avoid ill effects caused by them.
- There are a number of mechanisms that trigger apoptosis in response to diverse stimuli, which may originate either extracellularly (extrinsic inducers) or intracellularly (intrinsic inducers). Extracellular signals may include hormones, growth factors, nitric oxide or cytokines, whereas, intracellular apoptosis signals in response to stress may ultimately result in cell suicide.
- ***Caspases*** were the first effector molecules identified in the apoptosis pathway, which are essential for PCD in a variety of species. A general principle that activates ***procaspases*** to initiate the caspase cascade is that the activation is triggered by adaptor proteins that bind to them via shared motifs.
- Abnormalities associated with apoptosis can play a significant role in disease processes. Apoptosis plays a key role in both degenerative and proliferative diseases.
- Most events associated with PCD in animals are similar to that in plant cells. Examples of PCD in plant development include the timely death of petals after fertilization, during diploid parthenogenesis, the development of tracheary elements and senescence.
- Cell death has also been assumed to play a role in the hypersensitive reaction (HR) that is characteristic of several incompatible plant–microbe interactions.
- Apoptosis technology has great commercial as well as academic applications. It is useful for understanding developmental regulation of multicellular organisms and involvement in morphogenesis, to study the diseases where it is the cause for pathology.
- Another important aspect is the maintenance of several industrially useful cell lines which require costly growth factors. They could be maintained for longer times with less effective cell deaths by altering the genetic make up of the cells.
- Apoptosis plays an important role in hypersensitive reaction during many viral and microbial infections on crop plants.
- Most of the mature cells in adult animals are terminally differentiated. If adult cells are lost, they will be replaced by the proliferation of less differentiated cells derived from stem cells.
- Stem cells possess two properties: ***Self-renewal*** or the ability to go through numerous cycles of cell division while maintaining the undifferentiated state and ***potency*** to differentiate–the capacity to differentiate into any mature cell type. These properties makes stem cells either totipotent or pluripotent.
- Adult stem cells have low or no proliferative ability and have a short lifespan. Through cell culture, early embryos or embryonic stem cells can be generated.
- Embryonic stem cells have been an important experimental tool in cell biology because they can be used to introduce altered genes into an organism.
- Stem cells also provide an outstanding model system for studying the molecular and cellular events associated with embryonic cell differentiation.

Further Reading

Books

Alberts B, Johnson A, Lewis J, Raff M, Roberts K and Walter P (2002), *Molecular Biology of The Cell*, 4th edition, Garland Science.

Channarayappa (2006), *Molecular Biotechnology: Principles and Practices*, Universities Press, (India) Private Limited and CRC Press, USA.

Michael D, Jacobson MD and McCarthy N (2002), (eds), *Apoptosis: The Molecular Biology of Programmed Cell Death*, Oxford University Press, Oxford.

References

Barry M and McFadden G (1998), Apoptosis regulators from DNA viruses, *Curr. Opin. Immunol,*. 10:422–30.

Chernova OB, Chernov MV, Agarwal ML, Taylor WR and Stark GR (1995), The role of p53 in regulating genomic stability when DNA and RNA synthesis are inhibited. *TIBS,* 20:431–34.

Chang HY and Yang X (2000), Proteases for cell suicide: functions and regulation of caspases, *Microbial Mol. Biol. Rev.,* 64:821–46.

Channarayappa (1990), Cytogenetic analysis of clastogens and aneuploidogens by immunofluorescent and micronucleus methods, PhD thesis submitted to the West Virginia University, Morgantown, West Virginia, USA.

Channarayappa (1993), Resistance mechanism of *lycopersicon* species to the vector *Bemisia tabaci* and ultrastructural changes in leaf curl virus infected tomato leaves, PhD hesis submitted to the University of Agricultural Sciences, Bangalore.

Copelan EA (2006), Hematopoietic stem cell transplantation, *New Engl. J. Med.,* 354:1813–27.

Ellis RE, Yuan JY and Horvitz RA (1991), Mechanisms and functions of cell death, *Annu. Rev. Cell Biol.*, 7:663–98.

Enoch T and Norbury C (1995), Cellular response to DNA damage: Cell cycle checkpoints, apoptosis and the roles of p53 and ATM, *TIBS,* 20:426–30.

Fesq H, Bacher M, Nain M and Gemsa D (1994), Programmed cell death (apoptosis) in human monocytes infected by influenza virus, *Immunobiology*, 190:175–82.

Gilchrist DG (1998), Programmed cell death in plant diseases: the purpose and promise of cellular suicide, *Annu. Rev. Phytopathol.*, 36:393–414.

Hasnain SE et al. (2003), Host–pathogen interaction during apoptosis, *J. Biosci.,* 28:349–357.

Jarvelainen HA, Galmiche A and Zychlinsky A (2003), Caspase-1 activation by Salmonella. *Trends in Cell Biology*, 13:204–10.

Jacobson MD, Well M and Raff MC (1997), Programmed cell death in animal development, *Cell* 88:347–54.

Kerr JFR, Wyllie AH and Currie AR (1972), Apoptosis: a basic biological phenomenon with wide-ranging implications in tissue kinetics, *Br. J. Cance,r* 26:239–57.

Krishnamurthy KV, Krishnaraj R, Chozhavendan R and Christopher FS (2000), The programme of cell death in plants and animals—A comparison, *Current Science*, 79:1169–81.

Li H and Yuan J (1999), Deciphering the pathways of life and death, *Curr. Opin. Cell Biol.*, 11:261–66.

McCulloch EA and Till JE (2005), Perspectives on the properties of stem cells, *Nature Med.*, 11:1026–28.

Nagata S (1999), Fas ligand–induced apoptosis, *Annu. Rev. Genet.*, 33:29–55.

Roulston B, Marcellus RC and Branton PE (1999), Viruses and apoptosis, *Annu. Rev. Microbiol.*, 53:577–628.

Saran S (2000), Programmed cell death, *Curr. Sci.*,78:575–86.

Strasser A, O'Connor L and Dixit VM (2000), Apoptosis signalling, *Annu. Rev. Biochem.*, 69:217–45.

Vaux DL and Korsmeyer SJ (1999), Cell death in development, *Cell,* Jan 96:245–54.

Wertz IE and Hanley MR (1996), Diverse molecular provocation of programmed cell death, *TIBS*, 21:359–64.

17

Plant Growth and Development

Learning objectives

After reading this chapter, you will be able to:

- Identify the sequence of processes in plant growth—development, growth, differentiation, morphogenesis, phases of flowering, embryogenesis, germination and leaf development, development of root, shoot, leaf and flower, pollination
- Outline the various factors that influence these processes

17.1 INTRODUCTION

Plants and animals have evolved multicellular organization independently, but surprisingly, using the same initial set of genes inherited from their common eukaryotic ancestors. it is now clear that plants have evolved different developmental strategies as compared to animals. Land plants originated from freshwater green algae and the transition to land correlates with the evolution of an increasingly protected embryo. Two unique examples of embryo protection are the seed coat and the formation of fruit. In animal development also, a similar type of protective mechanism evolved. The major developmental differences between plants and animals are:

- ***Plants do not have a process similar to gastrulation***: Plant cells are trapped within a rigid cell wall; no cell migration takes place.
- ***Plants produce microspores not gametes***: During meiosis, plants produce spores.
- ***Germ cells***: No early separation of germ cells can be seen.
- ***Morphogenesis***: Plants have elaborate developmental changes that are mainly localized to meristems (similar to stem cells in animals) and take place throughout the life of the plant.
- ***Regeneration potential***: The whole plant can be regenerated from a single cell (called *totipotency*) and the lost part of a plant can be replaced from existing cells at the damaged sites.
- ***Genetic stability***: Plants can tolerate higher genetic alterations than animals, before the phenotype is affected. Many plants can exist in aneuploid and polyploid states.
- ***Lifecycle***: Most plants have evolved both haploid and diploid multicellular stages referred to as *alteration of generations*.
- ***Ability to adapt to varied environments***: Plants are highly plastic and influenced by environmental factors (e.g., light, temperature or altitude) and diverse morphologies can result from the same genotype.

Plants get their energy from sunlight. Their cells are encased in semi-rigid cell walls and are cemented together. The plan of the plant body is hence different from that of animals. Since plants cannot move, they cannot find their ideal environment by moving from place to place like animals. Instead, they adapt by altering the course of their development. Plants produce a specific type of organ (a leaf, a flower, or a root) from the fertilized egg by following different paths according to environmental cues. The mature plant is typically made of many copies of a small set of standardized modules (**Fig. 17.1**), repeated again and again. The positions and times at which those modules are generated are strongly influenced by the environment, causing the overall structure of the plant to vary.

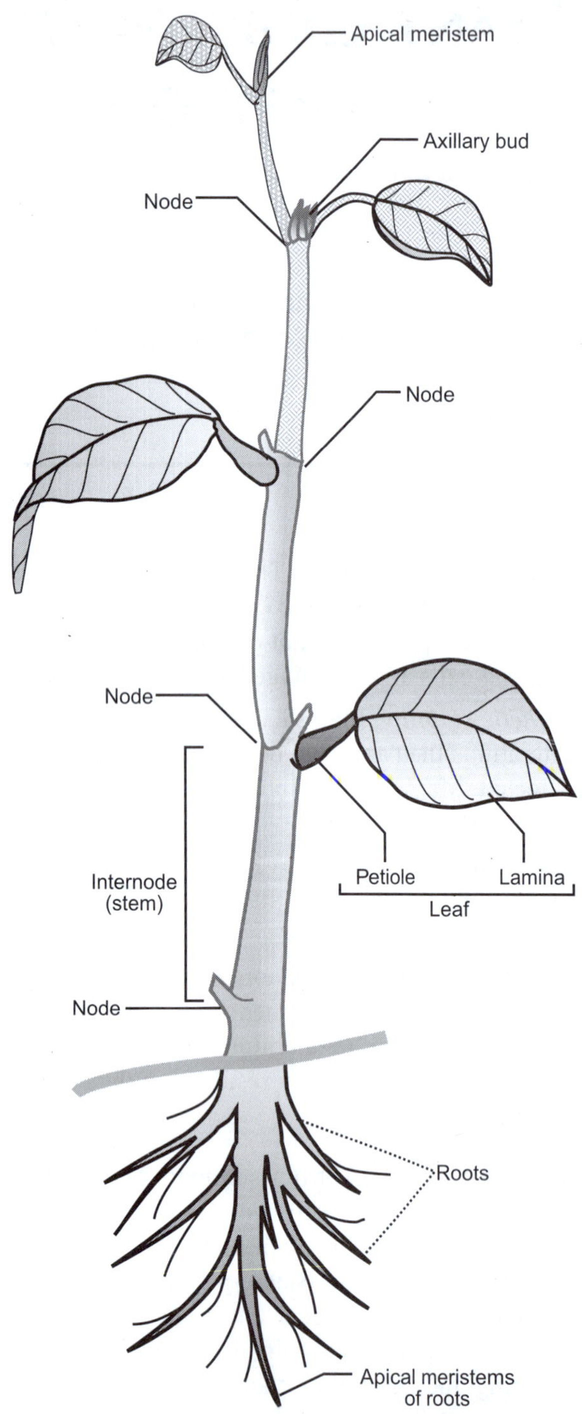

Fig. 17.1 Modular construction of a plant body. Each module consists of a stem, a leaf, bud and roots containing a potential growth centre or meristem.

17.2 PLANT DEVELOPMENT

It is very useful to separate the developmental events of an organism into several components: **growth**, **differentiation** and **morphogenesis**. Development is a familiar term but a very complex biological process to understand. ***The simplest definition of development is the complete morphological and functional changes of an individual during its life cycle***. Development thus, is not a one-way street but is cyclic following the reproductive cycles of the species and involves many intermediate stages (**Fig. 17.2**). Development in multicellular eukaryotes can be considered to commence at fertilization. The zygotic cell proliferates to form a multicellular embryo, within which polarity and axes of symmetry are established to provide

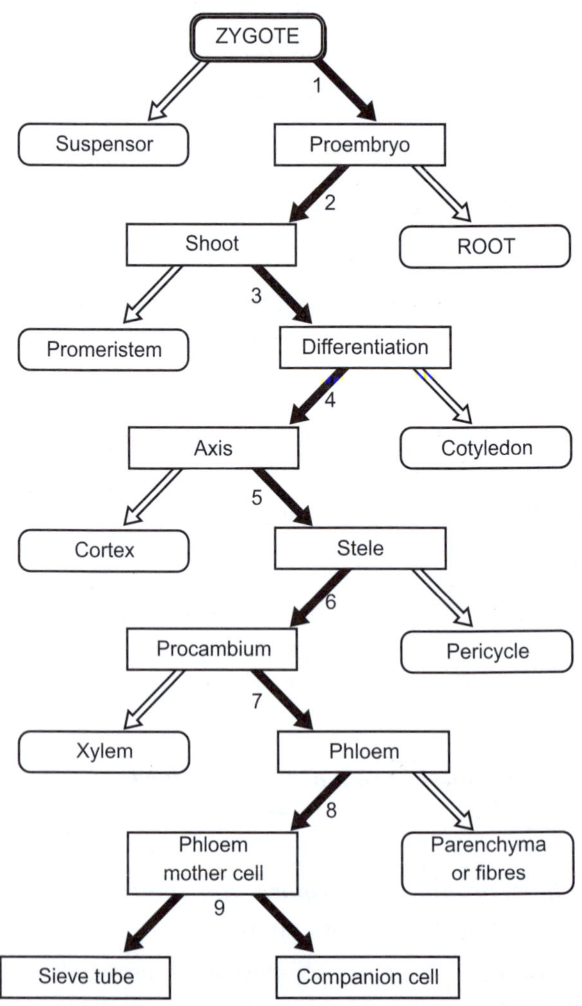

Fig. 17.2 Possible pathway of developmental choices. The dark arrows indicate one of the many possible paths through the developmental pathways. It requires only nine regulatory decisions to get from zygote to phloem sieve tube cell.

developmental framework for groups of cells to differentiate and develop into tissues and organs of the developing organism. Much of our understanding of the processes which control and coordinate developmental processes is derived from studies of animal development, using genetic and biochemical analyzes of early events in embryogenesis. These studies provide some but not complete insights into the analysis of plant development.

17.2.1 Plant growth

Growth is a fundamental characteristic of all living organisms. It is defined as a gain of volume and mass of an organism due to cell elongation and cell multiplication. Growth is the increase of size and weight, i.e., the quantitative changes during development. Growth of the whole organism can be easily identified by measuring external dimensions such as diameter, height, volume and weight. Every cell has its specific developmental cycle, the cell cycle (see Chapter). The duration of the cell cycle and the orientation of cell divisions have an obvious effect on developing organs and the final shape of the entire organism. This is reflected in the term formative growth, which contrasts with proliferative (gaining mass) growth. In plants, there are two different mechanisms of volume and mass growth: increase in **biomass** and **enlargement** by stretching. Plants gain biomass by mitosis. Stretching, however, does not increase organic substance. Volume increase of the whole cell is mostly due to the intake of water and this is accompanied by stretching and surface growth of the cell wall.

The process of growth is very closely linked with development, but development could be described as an increase in complexity. Starting from a single cell, growth of a multicellular organism can be divided into three phases:

- ***Cell division***: An increase in cell number as a result of mitosis and cell division.
- ***Cell enlargement***: An irreversible increase in cell size as a result of the uptake of water or the synthesis of living material.
- ***Cell differentiation***: The specialization of cells.

17.2.2 Differentiation

Differentiation describes alterations of the form and function of cells, resulting in the formation of new structures or changes in tissues and organs. It is a process of specialization of form and function and describes the qualitative changes. Ontogeny of most cell types and tissues results in morphological differences, which coincide with a specialized function. Therefore, differentiation involves **changes of morphology** and **function of cells, tissues and organs**. Differentiation is often equated with cell differentiation and limited to a single cell. It is visible externally because specific membrane structures and pigments are formed. Typically, differentiation leads from the meristematic cell to cells with special functions, which are terminally differentiated. Differentiation can also proceed in the opposite direction, i.e., re-embryonalization: closure of wounds with newly formed callus cells. Completely differentiated single cells can be stimulated to regenerate complete plants (totipotency) and this reflects the fact that single, completely differentiated plant cells can retain all necessary information for development into any specialized cell type whole organism.

Our knowledge of molecular mechanisms underpinning the establishment of plant embryos (embryogenesis) is much less than for animal development. Interestingly, plant cells retain a high degree of totipotency; thus developmental identity can be manipulated in tissue culture. Organized and apparently terminally differentiated tissues can be induced to dedifferentiate to generate disorganized callus tissue and this can in turn redifferentiate into recognizable organs or whole plants by simple alterations in the hormonal environment to which the cells are exposed. Such manipulations are now being exploited in order to identify genes and gene products whose expression is associated with developmental decisions made before the regulation of these genes.

Meristematic activity is also associated with cellular differentiation—the internal and external products of cambial cell division forming the xylem and phloem tissues, respectively. The shoot apical meristem is capable of being reprogrammed in the course of plant development. During vegetative growth of the plant, its principal activity is to contribute to stem elongation and to generate leaf primordia at subapical positions in a regular manner, giving a typically invariant pattern of leaves. However, due to an appropriate environmental or developmental signal, this pattern of development is disrupted and the vegetative apex undergoes terminal differentiation in function, producing the reproductive structures—the flowers. The molecular control of floral morphogenesis is one of the few developmental processes occurring in plants for which a molecular explanation is beginning to emerge.

Factors of cell differentiation

As discussed previously, cell differentiation can happen at many levels and ranges from changes in shape and internal structures, to biochemical changes. Biochemical differentiation involves a vast spectrum of metabolic products and provides many answers about cell differentiation. Other factors that contribute to differentiation include **chemical substances** (e.g., phytohormones) or **electrical** and **mechanical stimuli**. However, the role of these chemical and physical influences during differentiation is still to be investigated. The important role of genes and their effect on cell differentiation have been studied in many mutants that lack defined developmental traits. **Differential gene expressions**, i.e., from nucleotide sequence information to time- and space-dependent cell activities, are the primary determinants of differentiation. During the transition of a cell from differentiation state A to B, the activity of a single gene may not be influenced. But it is possible that its activity may be either higher or lower with regard to the total measurable activity of the genes. A group of genes, called **housekeeping genes**, express in all the cell types and are not influenced during transition from A to B. Although daughter cells are genetically identical, the cytoplasm is not. The maternal cytoplasm is not evenly distributed and the two nuclei in the daughter cells have different cytoplasmic environments. These dissimilar cell divisions are very important for the development of almost all groups of organisms. The uneven distribution of the cytoplasm is referred to as **cell polarity**, i.e., the composition and the features of the cytoplasm change from pole to pole along a polarity axis.

17.2.3 Morphogenesis

The changing shape (i.e., generation of precise structures) in a developing embryo, which is linked to changes in function, is called morphogenesis. Therefore, morphogenesis is a process of developing the form of the complete organism. Morphogenesis in plants entirely reflects the shape of their constituent cells and the orientation of their planes of division in space. During plant ontogenesis, we find all kinds of symmetries: the spherical early embryo has a **radial symmetry**; it later changes to **bilateral symmetry**. During the development of the flowers we often see **dorso-ventral symmetry**. **Metameric symmetry** is a third type, used to group events where elements along an axis are formed at the same distance and have similar orientation with respect to one another. Well-known examples are pinnate leaves and the scales that cover pine cones. Special forms develop when the axis is a spiral or a helix. Combinations of basic forms with two types of symmetry produce new **overlapping symmetries** that are generally complex. Morphogenesis explains the visible expression of a number of highly complex processes, which includes newly formed structures, changes in forms and functions (seedling development, formation of adult plant, flower development, seeds and fruit formation and senescence).

Regulation of cytoskeletal orientation appears to play an important role in determining the orientation of nuclear division and also in controlling the pattern of deposition of cellulose microfibrils within the cell wall. The orientation of these microfibrils constrains the direction in which turgor-driven cell expansion can occur and changes in the extent and orientation of cellulose deposition may ultimately act to limit the extent of cell growth. One of the most striking characteristics of plant development is the limitation of those cells able to divide and thus contribute to the growth and pattern of the plant. The continued cell division is restricted to meristematic zones (root and shoot apices, in lateral buds and secondary roots) that divide to produce additional cells, which subsequently expand longitudinally and contribute root and stem elongation.

Pathological or induced morphogenesis: Laboratory experiments can induce and spontaneous natural events may result in, morphogenic processes that are not a part of the normal development of the individual. These are, for example, the formation of tumours and galls. The study of these abnormal structures reveals the mechanism underlying morphological regulations.

17.2.4 Regulation of development

Differentiation and morphogenesis have temporal and spatial components. The changes that lead to differentiation and expression of a pattern follow a predetermined plan within certain limiting boundaries. **Pattern** is the non-random distribution of structures in individual organelles or in the organism. It describes the spatial organization of differentiating parts to a higher level of organization. In pattern formation, two separate phases are defined: an initial pattern specification and a final pattern realization. **Determination** is the commitment of cells along a particular and virtually irreversible path to differentiation. Plants have to coordinate developmental processes between distant organs. Therefore, long-distance communication between organs rely upon the **vascular system**. Long-distance transport of organic substances occurs mainly via the **sieve tubes**, which are connected with the companion cells by numerous plasmodesmata. Electrical signals may also function as transmitters from one organ to another in long-distance regulation. Interactions between plant organs may enhance or inhibit (correlative inhibition) developmental processes. For example, the promotion of cell division in the cambium of trees, where cell division starts during the spring, is influenced by developing buds. The signals are auxins and other phytohormones, which are transported from the buds downwards.

17.3 LIFE CYCLE OF FLOWERING PLANTS

Theoretically, there are two kinds of reproduction: **asexual** (vegetative) and **sexual** (generative). In the case of asexual reproduction, the propagative units originate from mitosis. Often the propagative units that result from mitosis are either single-celled or may contain very few cells (spores). Sexual propagation, however, requires that individuals of opposing sex are present in the same generation. Two cells with different sexes fuse (syngamy). During the following ontogeny, the zygote becomes a complete diploid organism. This produces, at a later date, more generative reproductive cells (gametes).

The lifecycle of a flowering plant is divided into three major phases, which are defined by physiological turning points. These phases are **embryogenesis** (covers fertilization to embryo maturity), **post-embryogenic vegetative development** (germination and seedling development) and **generative development** (development of reproductive parts and reproduction).

17.3.1 Embryogenesis in plants

In plants, the term embryogenesis covers development of embryo from the time of fertilization until dormancy occurs. The basic plan of the body of the sporophyte is established during embryogenesis, but is reinforced and expanded after dormancy breaks. The major changes of embryogenesis are:

- ***Establishment of body plan***: *Radial patterning* produces three tissue types and *axial patterning* establishes the apical–basal (shoot–root) axis.
- ***Formation of separate meristematic tissue***: Meristematic tissue will be set aside for post-embryonic development of the body structure (leaves, roots and flowers).
- ***Nutrient source***: It establishes an accessible food reserve for the germinating embryo until it becomes autotrophic.

Properties of plant embryos

The mature embryo of a flowering plant (e.g., *Arabidopsis*) comprises only two organ systems: the **embryonic axis** and the **cotyledons**. The embryonic axis contains the basic layout of the seedling and has a population of meristem cells at each end, which give rise to adult structures of the plant during post-embryonic development. The axis is therefore organized into regions representing the **shoot apical meristem (SAM)**, the future seedling shoot (**epicotyl** and **hypocotyl**), the embryonic root (**radicle**) and the **root apical meristem** (RAM) (**Fig. 17.3**). The cotyledons, which are attached to the hypocotyls, serve as storage organs that provide nourishment to the growing seedling. Bands are organ-forming regions. The embryo is also organized into radial bands representing the three fundamental cell layers common to all flowering plants: **L1**, **L2** and **L3**.

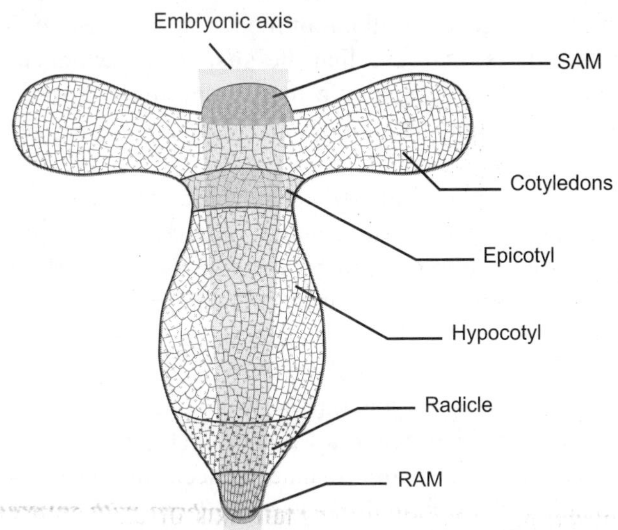

Fig. 17.3 Organ-forming regions of a dicot plant embryo.

17.3.2 Development of the embryo

The embryo develops in the ovule of the mother plant, which is connected via a stalk (the funiculus) to the wall of the ovary. At fertilization the ovary contains the female gametophyte—the embryo sac that is made of seven cells. Two of these cells, the haploid egg cell and the diploid central cell, are fertilized. In this way a diploid zygote and the triploid cell (forms endosperm later) will emerge. Embryogenesis in all flowering plants is more or less similar, even though the seedlings of monocotyledons and dicotyledonous plants differ in their outer appearance and construction. The major difference between the two is that in dicotyledonous plants two cotyledons flank the shoot primordium symmetrically, whereas in monocotyledonous plants only one cotyledon develops, which will overgrow the shoot meristem, pushing it to one side.

Immediately after fertilization, the zygote usually enters a dormant stage for a certain period. At the same time, the large vacuole, which is still present in the egg disappears and the cytoplasm becomes more homogenous. The zygote elongates to three times its original length along its future apical–basal axis of the embryo, which corresponds to the future shoot–root axis of the seedling. Usually, the elongated zygote begins to divide after the division of the endosperm nucleus. The first division is asymmetrical to establish the polarity of the future embryo. One product of this division is a small cell with dense cytoplasm (a **terminal/apical cell**), which will become the **embryo proper**. The other is a large vacuolated cell (a **basal cell**, closest to the micropyle) that divides further and generates the root meristem and the extra-embryonic structure called the **suspensor**, which in some ways is comparable to the umbilical cord in mammals. **Figure 17.4** shows how the signals trigger the next stage of development.

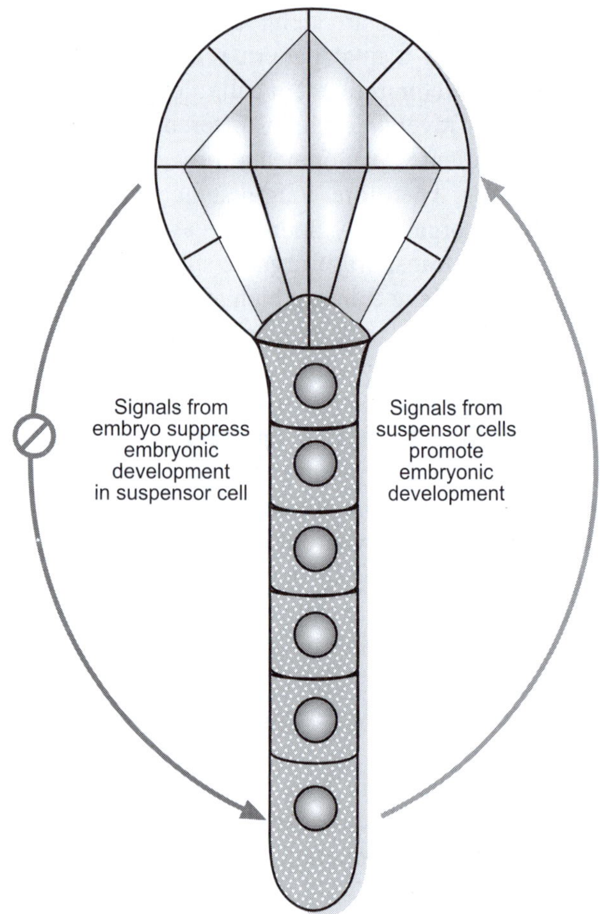

Fig. 17.4 Diagram showing how the embryo proper suppresses embryonic development in the suspensor and the suspensor provides feedback information to the embryo (redrawn from Gilbert, 2000).

The diploid embryo cell proliferates to form a globule of cells that have polarized structure. The radial and axial patterns develop as cell division and differentiation continues. This comprises two key groups of proliferating cells, one at the suspensor end of the embryo that collaborates with the uppermost suspensor cell to generate a root and one at the opposite end that generates a shoot. This root–shoot axis established in this manner reflects the head-to-tail axis of an animal. The cells of embryo proper (octant stage), divide in transverse and longitudinal planes to form a **globular stage** embryo with several tiers of cells. The emerging shape of the embryo depends on regulation of the planes of cell division and expansion. Cell division planes in the outer layer of cells become restricted and this layer is called the **protoderm**. Radial patterning emerges at the globular stage as the three tissue systems: the **dermal**, **ground** and **vascular** (**Fig. 17.5**). These three groups are comparable to the three germ layers of the animal embryo. The **dermal tissue** (epidermis), forms the protoderm and contributes to the outer protective layers of the plant. The **ground tissue** (**cortex** and **pith**) forms from the ground meristem, which is located just beneath the protoderm. The **procambium** that forms at the core of the embryo will give rise to the **vascular tissue** (xylem and phloem) involved in support and transportation.

As cotyledons (first leaves) give form to the globular mass shape of the embryo alters. Later, the rudiment of the shoot begins to produce the embryonic

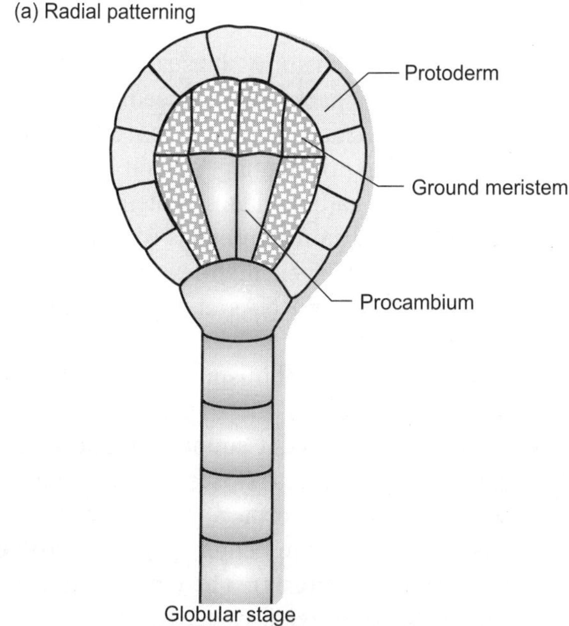

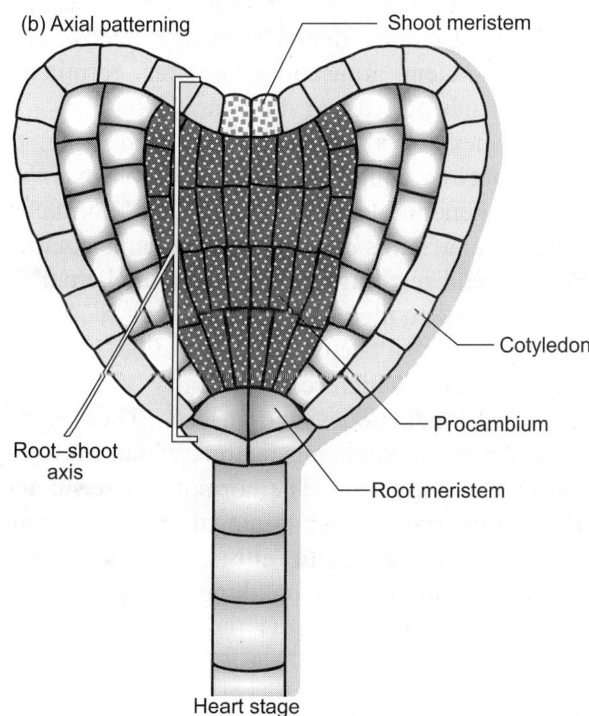

Fig. 17.5 Radial and axial patterning.

seed leaves, cotyledons one in the monocots and two in the dicots (this gives heart shape, called **heart stage** of development). Hormones (specifically, auxins) may mediate the transition from radial to bilateral symmetry. Cotyledons serve as the nutrient source for the germinating seedling. The shoot apical meristem and root apical meristem are clusters of stem cells that will persist in the post-embryonic plant and give rise to most of the sporophyte body.

Genetic studies indicate that the development of shoot and root meristems are independently regulated. The *dek23* maize mutant and the *shoot meristemless* (STM) mutant of *Arabidopsis* both form a root meristem but fail to develop shoot meristem. Mutations of the *hobbit* gene in *Arabidopsis*, for example, affect the hypophysis derivatives and eliminate root meristem function.

17.3.3 Genetic control

Genetic screens can be used in plants, just as in *Drosophila* or *C. elegans*, to identify the genes that regulate the developnent of the embryo. Genes involved in the formation of the seedling root, stem and apex with its cotyledons can be identified and grouped (**Fig. 17.6**). The second class of genes is required for the formation of three major tissue types viz., epidermis, ground tissue and vascular tissue (Fig. 17.6). Another class required is for the organized changes of cell shape that gives the embryo and seedling their elongated form. The first sign of apical–basal polarity in the *Arabidopsis* embryo is seen just after fertilization, when the zygote elongates along the future apical–basal axis

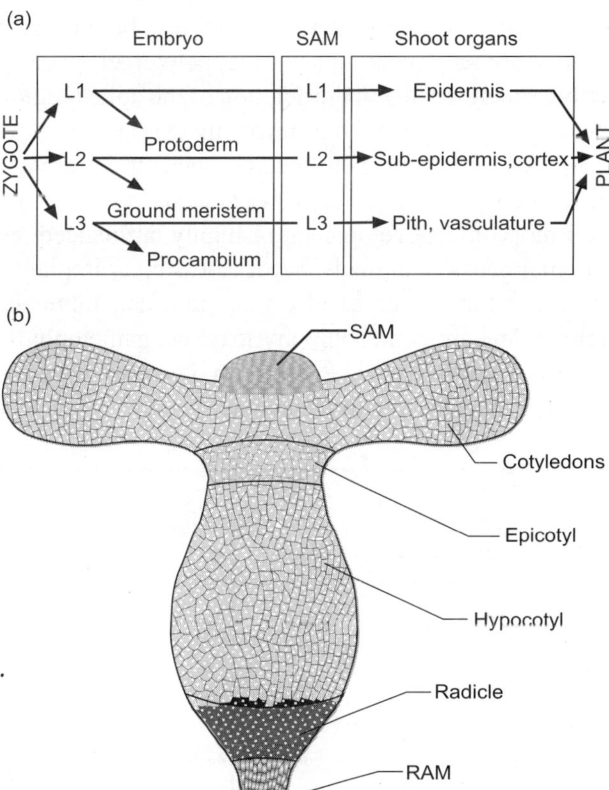

Fig. 17.6 (a) The fundamental cell layers of angiosperms and the tissues derived from them in the embryo and shoot. (b) Organ-forming regions of a dicot plant embryo.

and then divides asymmetrically to produce the small apical cell and long basal cell. The zygotic gene *gnom* is involved at this stage because in *gnom* mutants, the zygote divides more-or-less symmetrically producing apical and basal cells of similar sizes. At the heart stage, *gnom* embryos are more rounded than wild-type embryos and fail to specify the RAM.

Organ-forming regions of the *Arabidopsis* embryo are distributed along the apical–basal axis. A number of mutants have been identified in which particular organ-forming regions are never specified, or wrongly-specified. Such mutations are conveniently divided into three groups: **apical mutants**, **central mutants** and **basal mutants** (**Fig. 17.7**). As mentioned earlier, the lower tier of cells in the pro-embryo induces the most apical suspensor cell to form RAM. The *monopteros* gene is probably required to specify the lower tier cells, since in the (basal) *monopteros* mutant, the lower tier behaves like an additional upper tier and the hypocotyls, radicle and RAM are not specified. The apical mutant *gurke* shows the opposite phenotype, in which the cotyledons and SAM are deleted. The defect in this case appears to be in the control of cell division in the apical region of the embryo, during the transition from the globular embryo to the early heart stage. In weaker *gurke* alleles, rudimentary cotyledons may be produced, but no SAM form. Another apical mutant, doppelwurzel (means, double roots), results in the deletion of the apical region and its replacement with a mirror image basal region. The *fackel* gene appears to specify the central region as the hypocotyl is deleted in *fackel* mutants.

Embryonic development is highly influenced by maternal genes in animals, but it is less clear in plants. There are three sources of tissues that can influence embryo development: sporophytic tissue, gametophytic tissue and the polyploid endosperm, because these are in close association with the zygote.

17.3.4 Seed dormancy

As the embryo reaches maturity, it usually slows down to a halt and becomes packaged in a seed. During this process, the developmental plan shifts from constructing the basic body plan to creating a food reserve by accumulating storage carbohydrates, proteins and lipids. Genes coding for seed storage proteins were the first to be characterized due to their high-level expression of specific storage protein mRNAs at different times in embryonic development.

Eventually the metabolism slows and the connection of the seed to the ovary is severed by degeneration of the adjacent supporting sporophyte cells. The seed dessicates (loses water) and the integuments harden to form a tough seed coat. The seeds at this stage enter into **dormancy** (i.e., **ending embryogenesis**). The embryo in a seed is physiologically stabilized by dehydration and can remain dormant for weeks or years (e.g., lotus seed can survive for more than 100 years). Seeds found stored in ancient archaeological sites germinated after thousands of years of dormancy!

Dormancy is a precisely regulated program. The **viviparous** mutation in maize, for example, produces genetic lesions that block dormancy. The apical meristems of viviparous mutants continue to develop and seedlings emerge from the kernels on the ear attached to the parent plant. Recently, a group of plant genes have been identified that belong to the Polycomb group, which regulates early development in mammals, nematodes and insects. These genes encode chromatin-silencing factors, which may play a role in seed formation. The hormone **abscisic acid** (**ABA**) is important in maintaining dormancy in many species. In contrast, **gibberellins**, another class of hormones, are involved in breaking dormancy.

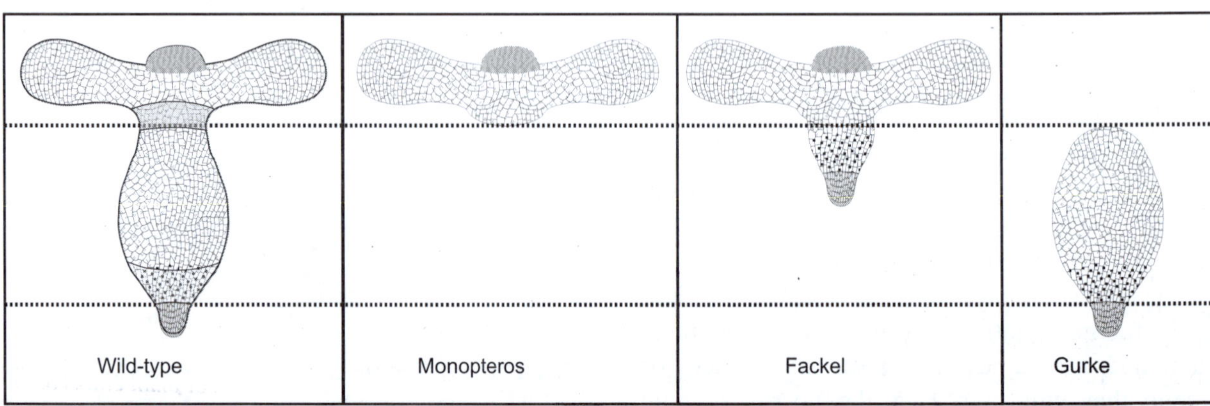

Fig. 17.7 Embryo axis patterning mutants in *Arabidopsis*.

17.4 GERMINATION AND VEGETATIVE DEVELOPMENT

The **post-embryonic vegetative phase** of plant development begins with **germination**. Apparently, the post-embryonic development commences with the germination of the seed, when the root of the seedling begins to grow. But, if the pattern of cell division is taken as a reference point, the transition from embryonic to post-embryonic development can be considered to be independent of germination. The intercalary growth that is characteristic of the embryo is, in many species, supplemented during late embryogenesis by localized (meristematic) growth, when the primary root meristem produces parts of the seedling's root.

17.4.1 Germination

Except in those seeds held in check by a specific dormancy mechanism, rehydration of seeds under conditions appropriate for growth lead to germination, seedling growth and ultimately the establishment of the new plant. Germination may be regarded as a resumption of growth, which has been previously stalled by desiccation. However, germinability of developing seeds changes during development and there are actually clear metabolic and physiological differences between a developing embryo and a germinating embryo. Some dormant seeds require a period of **after-ripening** during which low-level metabolic activities continue to prepare the embryo for germination.

When environmental conditions are favourable, seeds germinate and a new generation of plants begins to grow and develop (**Fig. 17.8**). The process by which the dormant embryo of the seed resumes active growth and forms a seedling is known as germination. During germination, the radicle of the embryo thrusts through the seed coat, forming a root that anchors the new plant. Favourable conditions for germination include cossect temperature, sufficient water, adequate oxygen and the presence or absence of light. Some seeds need more than these conditions to germinate. They must be exposed to freezing temperatures for exact periods. **Stratification** is the requirement for chilling (5°C) to break dormancy in some seeds. Germination ability of seeds of many species ranges from 25to30°C and extreme temperatures will not be suitable. Some seeds must travel through the digestive system of an animal, be battered and ground against rocks, or even be scorched by fire before they germinate. Of all the environmental factors, adequate moisture is often the most critical for seed germination and seedling development. Seeds are very dry, sometimes containing as little as 2% water and as a result, must be hydrated before germination can begin. As moisture enters the seed, tissues soak up water, a process called **imbibition**, the seed swells, metabolism speeds up and cellular division and growth begin in the embryo. However, the seed can also undergo dormancy. Inability on the part of viable seeds to germinate under suitable conditions is known as seed dormancy. The **radicle** (primary embryonic root) emerges from the seed first to facilitate water uptake.

During germination, the plant draws nutrients from the endosperm or cotyledons. Interactions between the embryo and endosperm in monocots use gibberellin as a signal to trigger the breakdown of starch into sugar. Digestion of food reserves proceeds mainly by hydrolysis: proteins by proteases, polysaccharides by carbohydrases (e.g., amylase, maltase) and lipids by lipases. The soluble products of digestion are

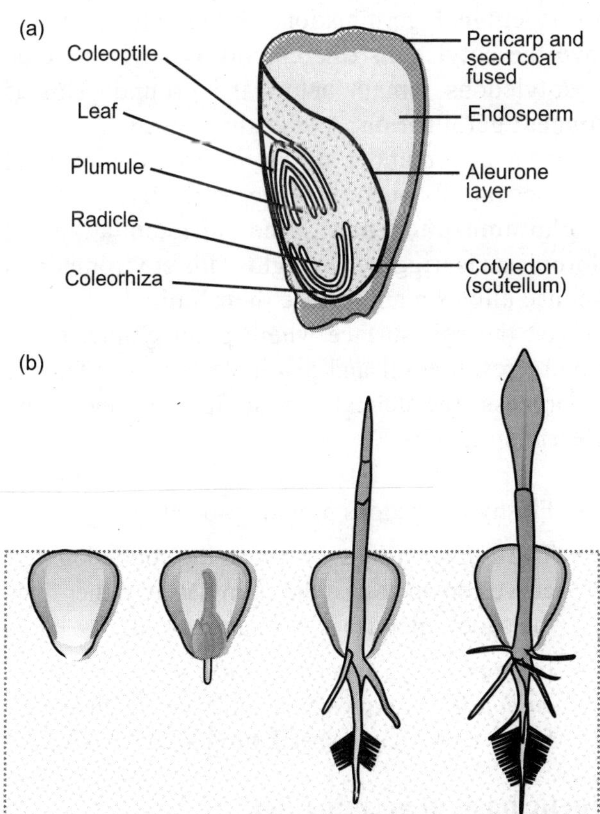

Fig. 17.8 Seed germination: (a) Structure of corn (*Zea mays*) with mature embryo and (b) Different stages of corn kernel germination.

then translocated to the growth regions of the embryo. The sugars, fatty acids and glycerol may be used to provide substrates for respiration in both the **storage** (food reserve) and **growth centres** (embryo). Both these centres derive energy from respiration. Rates of respiration in both storage tissues and embryos are high owing to the intense metabolic activity in both regions.

Within the embryo, growth occurs by cell division, enlargement and differentiation. The amount of proteins, cellulose, nucleic acids and so on increases steadily in the growing regions while dry mass of the food store decreases. The first visible sign of growth is the emergence of the embryo root, the radicle. This is positively geotropic and will grow down and anchor the seed. Subsequently, the embryo shoot, the **plumule**, emerges and being negatively geotropic (but positively phototropic) will grow upwards. There are two types of germination, according to whether or not the cotyledons grow above ground or remain below it. In dicotyledons, if that part of the shoot axis, or internode, just below the cotyledons (the hypocotyl) elongates, then the cotyledons are carried above the ground. This is **epigeal** germination, if the internode just above the cotyledons (the epicotyl) elongates, then the cotyledons remain below the ground. This is **hypogeal** germination. As the shoot growth reaches the surface, a number of phytochrome-controlled responses rapidly occur in leaves collectively known as **photomorphogeneis**. The differentiation of chloroplasts is triggered by light. This environmental response allows plants to use their limited resources to reach the soil surface where photosynthesis can be more resourceful and available lifelong. During this process, the delicate shoot tip is protected by different strategies:

- Fleshy-cotyledons provide protection.
- The ***epicotyl*** (the stem above the cotyledons) curves down so that the stem tissue, rather than the shoot tip, pushes through the soil.
- In monocots, a special leaf-like structure, the ***coleoptile***, provides mechanical strength and forms a protective sheath around the shoot tip.

Genetic basis of germination

The transition from embryogenesis to germination is controlled largely by two plant hormones, abscisic acid (ABA) and gibberellic acid (GA). Generally, GA promotes germination while ABA inhibits it. A number of *aba* mutants with viviparous phenotypes have been identified in many crops and these mutants show much lower levels of ABA than wild type embryos, suggesting a defect in ABA synthesis. Mutants representing genes in the GA biosynthesis pathway and the response to GA signalling (GA1, GA2, GA3) show delayed germination phenotypes. A number of other genes have also been identified that specifically affect the transition to germination when mutated, but are not involved in hormone synthesis; Instead they encode regulatory proteins that control downstream embryo-specific gene expression. Mutations in these genes cause appearance of trichomes' on the cotyledons and the accumulation of anthocyanins. This represents a partial transformation of cotyledons into leaves and genes are therefore grouped as the LEAFY COTYLEDON (LEC) class. The genes LEAFY COTYLEDON 1 and FUSCA 3, show viviparous mutant phenotypes as well as leaf-like cotyledon morphology and reduction in the accumulation of storage proteins. The ABA3, LEC1 and FUS3 genes are found to encode transcription factors with temporally overlapping expression patterns during embryogenesis. These genes appear to regulate each other and other genes involved in embryonic development, while suppressing genes involved in post-embryonic development.

Skotomorphogenesis and photomorphogenesis

The early stage of germination is called dark development or skotomorphogenesis and is characterized by the rapid extension of the shoot upward through the soil. The shoot is pale and spindly (called **etiolated**). The seedling emerging out from the soil is exposed to light, leading to photomorphogenesis. Consequently, the seedling turns green (as it begins to accumulate chlorophyll) photosynthesis begins, shoot-apical meristems are activated and true leaves begin to form.

17.5 GROWTH OF THE PRIMARY PLANT BODY

Plants do not have a preprogrammed body plan like animal embryonic cells. However, the plants retain areas where rapidly dividing, undifferentiated cells remain all through the life of the plant (called meristems). Meristematic tissue continues to rapidly divide producing undifferentiated cells which eventually differentiate to form tissue and cell types. The pattern of plant growth depends on the location of meristems.

17.5.1 Development of meristems

The plant embryo develops into an adult quite differently when compared to that of insect and animals. The parts of the adult plant are created sequentially by two groups of cells that proliferate to lay down additional structures at the plant's periphery. These groups of cells are called shoot and root meristems and contain a population of self-renewing stem cells. The shoot apical meristem (SAM) gives rise to all the aerial parts of the plant, while the root apical meristem (RAM) gives rise to the root network. The stem cells in a meristematic tissue divide to give a daughter cell that continues to be meristematic (slowly dividing central stem cells) and rapidly dividing peripheral initial cells that differentiate. There are three categories of meristems: **apical**, **lateral** and **intercalary**. As these cells divide, they leave behind a trail of progeny that become displaced from the meristem region, enlarge and finally differentiate. Other than the shoot and root apical meristems (which can generate all basic cell types needed to build leaves, roots and stems), many cells outside the apical meristems also have the capacity for further proliferation and differentiation. In this manner, trees and other perennial plants, for example, are able to increase the girth of their stems and roots as the years go by and can sprout new shoots from dormant regions.

Apical meristem: They occur at the growing shoot and root tips (**Fig. 17.9**). The *shoot apical* meristem can produce different cell types, which produce stems, leaves and reproductive structures. In addition to the shoot apical meristem, axillary-shoot apical meristems form in the axils. *Root apical* meristems produce the root cap, which consists of lubricated cells that are sloughed off as the meristem is pushed through the soil by cell division and elongation in more proximal cells. It also produces daughter cells that produce the three tissue systems of the root. New root apical meristems are initiated from tissue within the core of the root and emerge through the ground tissue and dermal tissue. Root meristems can also arise from the stem of the plant.

Lateral meristem: They are located near the periphery of the plant, usually in a cylindrical meristem found in shoots and roots that result in secondary growth (increase in stem and root girth by the production of vascular tissues). Lateral meristems are found in all woody and some herbaceous plants. Lateral meristems and secondary growth are found only in dicots.

Intercalary meristem: Monocot stems do not have lateral meristems they often have *intercalary* meristems inserted in the stems between mature tissues.

17.5.2 Morphogenesis of a developing plant

Since plant cells are completely confined within their cell walls, they cannot move as the plant develops. Therefore, the formation of different structures depends on orderly cell divisions followed by oriented cell expansions. However, most cells produced in the meristem, go through three distinct phases of development: division, elongation and differentiation. These three processes overlap in space and time

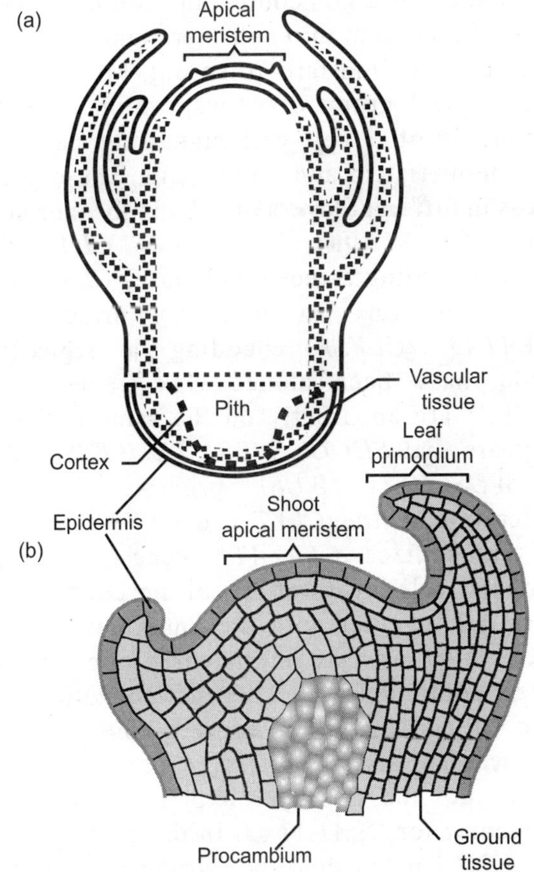

Fig. 17.9 (a) Longitudinal section of the shoot tip of a dicotyledonous angiosperm. (b) Shoot meristem develops from apical meristems; undifferentiated cells are clustered at their tips.

andgive rise to the characteristic architecture of a root tip or shoot tip meristems.

The apical meristems are self-perpetuating in nature. In perennial plants, they carry on with their functions indefinitely—as long as the plant survives—and are responsible for continuous growth and development. Apical meristems also give rise to a second type of outgrowth such as a leaf or flower, with a determined size and shape and a short lifespan. Thus, as a vegetative shoot elongates, its apical meristem lays down behind itself an orderly sequence of nodes, where leaves have grown and internodes form.

Genes required for meristem maintenance

As discussed earlier, the apical meristem maintains itself for years by continuously proliferating. It is just a cluster of cells and remains constant for days and years. Recently, genetic screens have identified a number of genes required for meristem maintenance. For example, mutations that disrupt the *WUSCHEL* gene, which codes for a homeodomain protein, convert the apical meristem into non-meristematic tissue, so that the seedling fails to sprout. Mutations in the *CLAVATA* group of genes, which code for components of a cell–cell signalling pathway, make the meristem abnormally big. These genes express in different layers of cells in the meristem region (**Fig. 17.10a**). The genes *CLAVATA1 (CLV1)*, encoding a receptor kinase, *CLAVATA2 (CLV2)*, encoding an accessory protein and *CLAVATA3 (CLV3)*, encoding a signalling peptide, have been cloned. The genes *CLV3* and *CLV1* are apparently on the same pathway. It appears that *SHOOT MERISTEMLESS (STM)* and *WUSCHEL (WUS)* promote meristem proliferation, while *CLV1* and *CLV3* inhibit it. *STM* antagonizes *CLAVATA* genes. The two superficial cell layers of apical meristem called L1 and L2 together with the uppermost part of the L3 layer are the meristem proper (which divides continuously). The meristematic cells of the L1 and L2 layers express *CLAVATA3*, a small secreted signal protein. A cluster of cells just beneath the L3 layer express *CLAVATA1*, the receptor for *CLAVATA3*. In the centre of this *CLAVATAL1* patch there is a group of cells that express the *WUSCHEL* gene regulatory proteins.

Figure 17.10b shows the negative feedback delivered by the *CLAVATA* signalling pathway restricting the size of the meristem.

17.5.3 Hormonal signals that regulate developmental events

In addition to apical growth, if a stem is to branch or flower, a new SAM must be produced. This process however, is well under the control of shoot apex. At each developing nodal region (between the leaf primordia and the stem), a bud is formed. This contains a group of cells derived from the apical meristem and retains meristematic character. These cells have the capacity to become the apical meristem of a new branch or the primordium of a structure for a flower, or alternatively remain quiescent as axillary buds.

The developmental direction of the axillary bud is dictated not only by its genes, but also by environmental conditions. Different parts at a distance experience different environments and react to it individually. The plant however, must continue to function as a whole, implying that the long-range signals play a role in such coordination. As it known, for example, by pinching off the tip of a branch one can stimulate side growth (quiescent axillary bud grows). This indicates the existence of a long-range signal from the apical meristem that regulates the axillary bud at a distance. The role of **auxin**, a member of the six known classes of plant growth regulators, is well known. The other five plant hormone classes (**Fig. 17.11**) are the **gibberellins**, **cytokines**, **abscisic acid**, **Sethylene** and **brassinosteroids**. These are small molecules (penetrate the cell wall) and are all synthesized by most plant cells and can either act locally or be transported to influence target cells at a distance. Each growth regulator has multiple effects and is modulated by the other growth regulators, as well as by environmental cues and nutritional status.

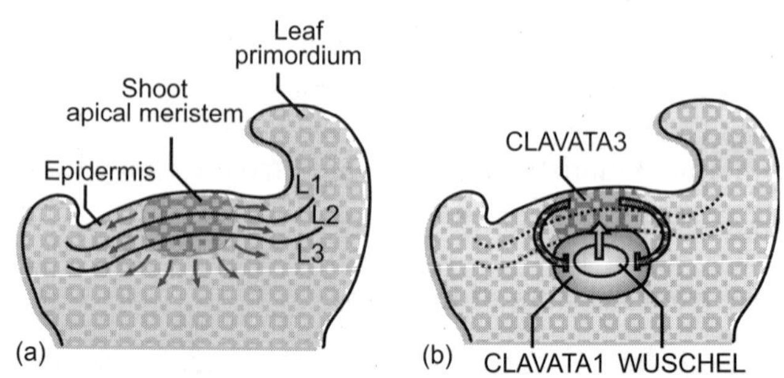

Fig. 17.10 The feedback loops that are assumed to maintain the shoot apical meristem: (a) The arrangement of cell layers constituting a shoot apical meristem. (b) The pattern of cell–cell communication that maintains the meristem. (Redrawn from Alberts et al, 2002).

Gibberellic acid (GA3), a gibberellin

Indole-3-acetic acid (IAA), an auxin

Zeatin, a cytokine

Abscisic acid (ABA)

Brassinolide, brassinosteroid

Ethylene

Fig. 17.11 Plant growth hormones that regulate different groups of developmental genes. A representative example of the five major classes of plant hormones is shown.

17.6 PRIMARY ROOT DEVELOPMENT

17.6.1 Primary root meristem

The most active part of a seedling during germination is the primary root meristem. Primary root meristem of the seedling is a regular structure (**Fig. 17.12**).

Root cap: It is a thimble-like covering, which protects the delicate apical meristem. The root cap is produced from the cells derived from the root apical meristem and secretes polysaccharide slime that lubricates the soil. It is constantly sloughed off and replaced.

Apical meristem: Just beneath the root cap is apical meristem. Apical meristem is a region of rapid cell division of undifferentiated cells and cell division is directed away from the root cap.

Quiescent centre: It is a population of cells in the apical meristem, which reproduce much more slowly than other meristematic cells. These cells are resistant to radiation and chemical damage. It is a reserve of cells which can be called into action if the apical meristem becomes damaged. Quiescent cells are surrounded by the initials, which are actively dividing stem cells. Towards the root cap, a layer of initials

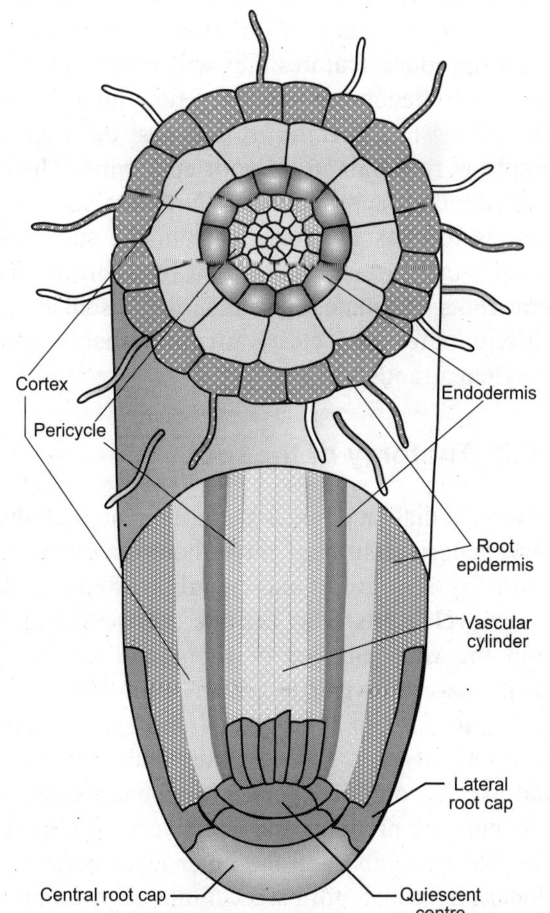

Fig. 17.12 Organization of the root meristem and the primary root (**See Plate 13**).

constantly produces new cells for the central part of the root cap, which is eroded away continuously as it spears its way through the soil.

Zone of cell division: The primary root grows mainly by cell division within its meristematic zone, called the zone of cell division. Three areas just above the apical meristem continue to divide for some time:

- *Protoderm*—outermost primary meristem that continues to divide for some time
- *Ground meristem*—central primary meristem that produces cells which become the dermal tissue
- *Procambium*—innermost primary meristem-produces cells which will become vascular tissue.

Zone of elongation: Newly produced cells elongate up to ten times their original length. This growth pushes the root further downward into the soil. The cells elongate along the longitudinal axis of the root and this also contributes to the growth.

Zone of maturation or differentiation: Finally the cells become the zone of differentiation, where they begin to produce features that will later characterize them. Cells become completely functional. In each layer of tissue, specific cell types develop; one example is root hair in the root epidermis. The cells of the endodermis produce the Casparian strip, which separates the vascular bundle from the surrounding cortical root tissue. The pericycle, from which lateral roots originate, surrounds the vascular tissue. It differentiates into clearly distinguishable phloem and xylem strands.

17.6.2 Anatomy of the root

In roots, radial and axial patterning started during embryogenesis continues throughout development as the primary root grows and lateral roots emerge from the **pericycle cells** (can become meristematic) deep within the root. Studies of root-radial organization mutants have revealed genes with layer-specific activity. In case of wild-type *Arabidopsis*, normally two layers of root ground tissues are present. The outer layer becomes the **cortex** (a ground tissue, stores photosynthetic products and is active in the uptake of water and minerals) and the inner layer becomes the endodermis, which forms a cellular tube around the vascular tissue core (xylem and phloem), contains the **Casparian strip**. The ***searecrow(scr)*** and ***short-root (shr)*** genes have mutant phenotypes with two layers (instead of one) of root ground tissue. The *scr* gene is necessary for an asymmetrical cell division in the initial layer of cells, yielding a smaller endodermal cell and a larger cortex cell (**Fig. 17.13**). The *scr* mutant expresses markers for both cortex and endodermal cells, indicating that the differentiation progresses in the absence of cell division, but cells in the *shr* mutant do not develop endodermal features and thus there is no Casparian strip.

The **root epidermis** that protects the root consist basically of two types of cells: hair cells and non-hair cells. The root hair primordial cells are initially shorter and make a local hump close to the basal cell wall and from there grow outwards as long tubes (the root hairs). The non-hair cells are longer, have no hump and are coated with deposits. The fate of the cells in epidermis (pattern formation) appears to be the ground state of differentiation in the root epidermis. Mutations in the constitutive triple response 1 *(ctr1)* gene increase the numbers of root hair cells by about 30%. Mutations in another gene, transparent testa glabra *(ttg)* also lead to the formation of root hairs from cells that otherwise would not differentiate this way. The ttg phenotype is suppressed by a constitutive expression of the *r* gene from maize, which itself prevents the formation of hair cells. The interaction of ttg with the *r* gene is similar to the process involved

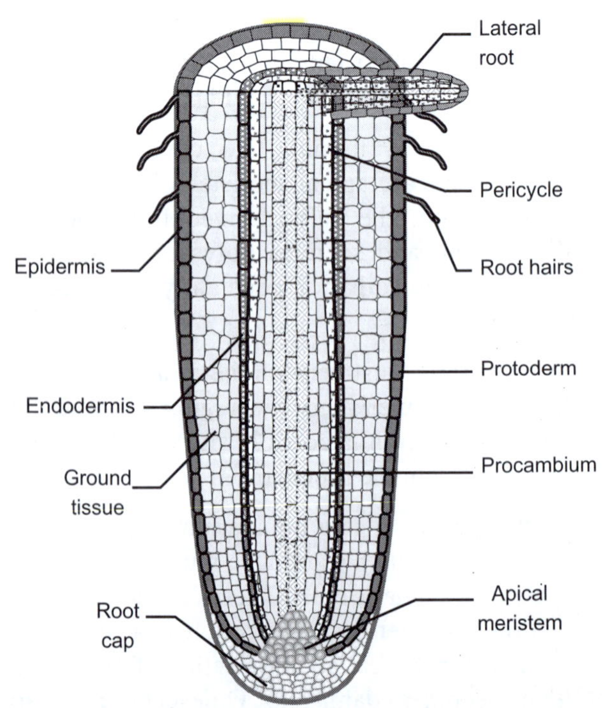

Fig. 17.13 Root meristem develops from apical meristems. Undifferentiated cells are clustered at their tips.

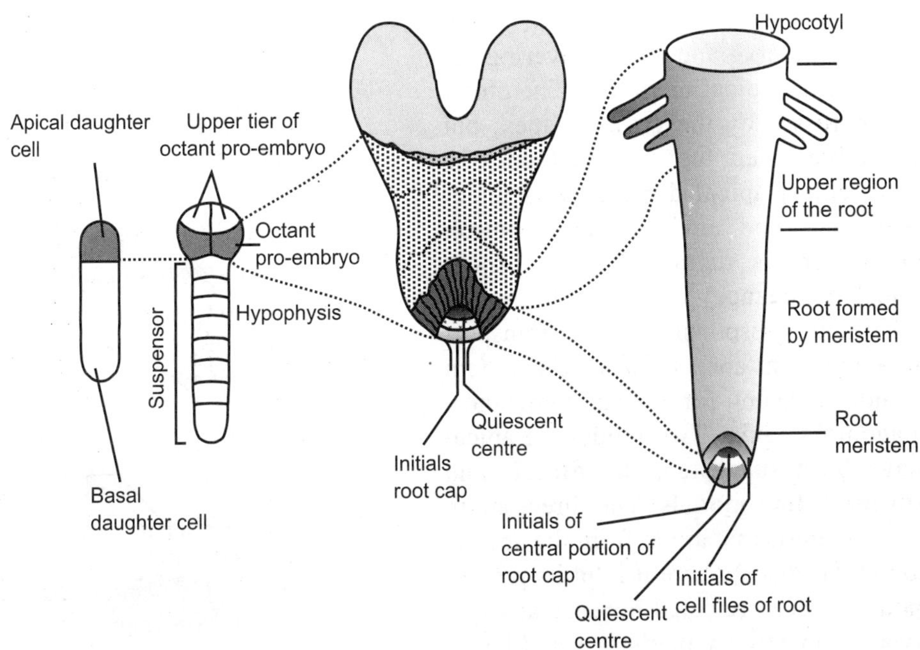

Fig. 17.14 Origin of the root and the root meristem.

in the production of **trichome cells** in leaves. Therefore, we can assume that local interactions between cells in the root epidermis determine whether a cell will be a hair cell or non-hair cell.

Primordial of lateral and adventitious roots may not arise from a defined cell group that set aside from the primary root meristem rather they emerge from the pericycle cells. The lateral roots however, emerge only from the pericycle cells that lie close to the xylem strands. The growing lateral root breaks through the adjacent cell layers and then forces its way through the epidermis. At the tip of the lateral root is a meristem, with an organization similar to that of the primary root meristem. Lateral roots differ from the primary root by their origin and the number of cell files per tissue layer (**Fig. 17.14**).

17.7 SHOOT DEVELOPMENT

During vegetative growth, the shoot apical meristem (SAM) gives rise to a series of primordia (doom-shaped mass of dividing cells) that form lateral organs (leaves and axillary buds). The SAM gives rise to three primary meristems: protoderm, ground meristem and procambium, just as root apical meristem. Leaves arise as leaf primordia on the flanks of apical meristem. **Leaf primordia** initiate at the periphery of the shoot meristem. The union of a leaf and the stem is called a **node** and stem tissue between nodes is called **internode**. The spatial arrangement of these primordial governs the phyllotaxy of the stem. **Phyllotaxy**, the positioning of leaves on the stem, involves communication among existing and newly forming leaf primordia. There is a delay between the initiation of successive primodia, known as a **plastochron**. The expression of the progress of plant development in terms of plastochrons allows normalization of the many external conditions that affect plant growth. Shoots are therefore composed of a series of modules called **phytomers**, each comprising a node (with an associated leaf and axillary bud) and an intervening internode (**Fig. 17.15**).

Apical dominance

The unique and well organized, above-the-ground structures of different plant species have their origins in shoot meristems and are affected by the amount of axillary bud outgrowth. Plants have many

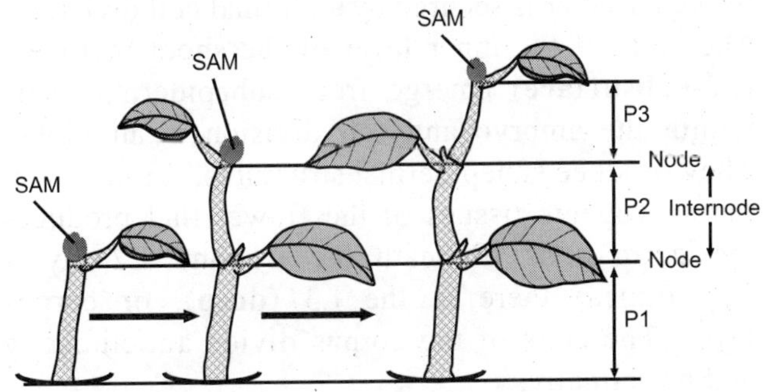

Fig. 17.15 Aerial growth of plants in phytomers. Each phytomer comprises a node and associated internode.

meristems, for instance, one at every leaf node. Most of these are in reserve and act as covering for potential damage to the apical meristem. Therefore, they must be suppressed by the growing apex; but they becomes active when the dominant apex is removed. The extent of apical dominance is under genetic, physiological and environmental control. The branching pattern is regulated by the shoot tip (called apical dominance) and plant hormones appear to be the factors responsible. Polar transport for taking the auxin from apex to base of the stem is necessary and sufficient for apical dominance. Cytokinin antagonizes auxin. Two models of apical dominance have been suggested: the **direct-** and **indirect-auxin inhibition models**. The direct auxin inhibition model suggests that auxin from the apex is transported to axillary buds where it inhibits growth (**Fig. 17.16**), but has been rejected because auxin is transported away from axillary meristems and levels increase after relief of apical dominance. The indirect model states that the development of vascular tissue directing photo-assimulate and nutrients and perhaps cytokinin to the growing apex, is favoured as long as most of the auxin comes from there and is maintained.

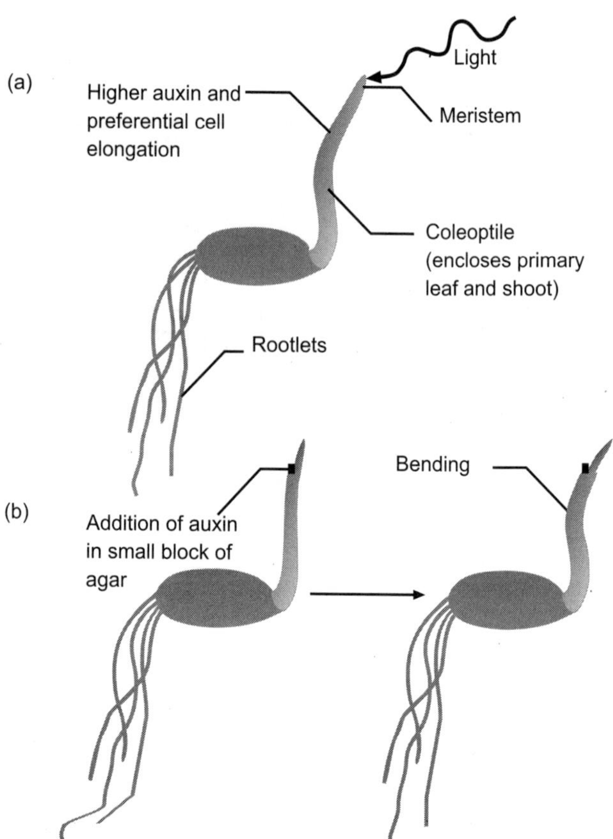

Fig. 17.16 Auxin-induced cell elongation causes phototropism in oat coleoptiles: (a) The release of auxin by the growing tip on the side of the coleoptile away from the light causes the cells on the far left side to elongate preferentially and bend the tip towards the light. (b) Phototropism can be demonstrated experimentally by removing the tips from coleoptiles grown in the dark, keeping the decapitated seedlings in the dark and applying an agar block containing auxin to one side. The coleoptiles subsequently bend away from the side at which auxin is applied.

17.7.1 Organization of primary shoot meristem

The primary shoot meristem of the seedling is organized into layers and zones. **Figure 17.17** shows the longitudinal section of the shoot tip of a dicotyledonous angiosperm. The apical meristem is surrounded by the developing leaf primordia. In most dicots and some monocots, the shoot apical meristem consists of three superimposed cell layers, which generate the whole shoot. The outer L1 (**superficial**) layer of three emerges from the epidermis of the embryo. Later, it forms the epidermal tissue of the shoot, the leaves and the flower organs. The integrity of the L1 layer is secured by anticlinal cell divisions. The cells of the inner layer of the shoot meristem (L2-**subsurface**) emerge from subepidermal cells within the embryo and cell division is anticlinal. They produce subepidermal structures, for example, the sporogenic tissues of the flower that produces the germ cells. Beneath the L1 and L2 layers (the tunica) there is the L3 (**deep**), or corpus layer. The cells of the corpus divide anticlinically and periclinally.

The shoot meristem is further subdivided into several zones, each fulfilling a different function.

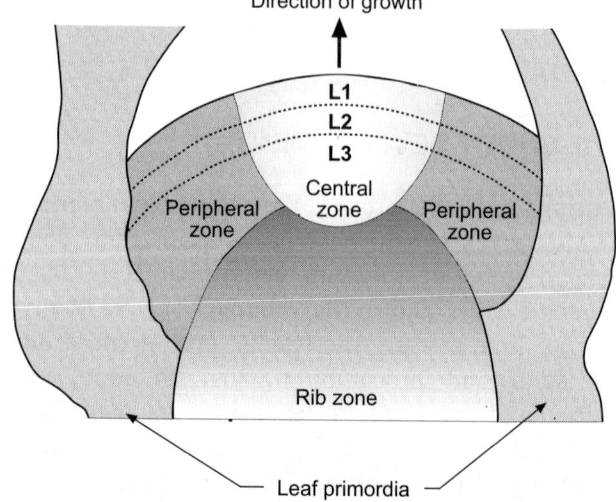

Fig. 17.17 Functional organization of the shoot meristem; shoot meristem is composed of layers (L1–L3).

***Central zone (CZ)*:** The central zone is at the centre of the tip of the shoot meristem (SM); it contains actively dividing cells (but at low rates), which continuously renew the meristem, producing daughter cells to the side and below. Cells remain undifferentiated, characteristically small and primarily consist of cytoplasm, with only small vacuoles. More importantly, undifferentiated cells retain the ability to initiate any organ type. The central zone is also important for the integrity of the shoot meristem. As the SM grows, cells from the CZ are 'left behind' to form the PZ. It is believed that positional information in the PZ triggers differentiation into primordia. Once differentiation begins the fate of the cell is determined. In other words, cells in the PZ might only be able to differentiate into leaf cells.

Peripheral zone (PZ): Cells of the peripheral zone surround the central zone. Here cells divide at a rapid rate and can differentiate. As cell division continues, the older PZ cells begin to develop into discrete primordia (e.g., leaf primordia) and are now referred to as OZ cells.

Organ zone (OZ): It is the site at which organ primordia (e.g., leaf, flower) become distinct and the leaves and axillary meristem form.

Rib meristem (RM): Rib meristem gives rise to vascular and interior stem structures. The subepidermal cells divide periclinally, so that the leaf precursor cells form leaf primordia. Beneath the central zone is the rib zone. Its cells contribute to the growth of the shoot.

During shoot apical meristem development, a few cells in each of the layers function as the initial cells of that layer. If one of the initial cells is marked by a phenotypically identifiable mutation (e.g., chlorophyll synthesis), a number of initial cells can be established. For example, mutant sectors occupying 1/3 of the circumference of the stem would indicate the three initial cells in each layer of the meristem. Interestingly, when excised shoot apical meristems are grown in sterile culture, the only plant hormone required for vegetative growth was IAA. These results suggest that the basic signal-generating system for vegetative programming of the meristem is located within the meristem itself, or in its environment.

Genetic regulation of shoot meristem organization

An essential gene for shoot meristem formation seems to be the *KNOTTED (KN)* gene from maize. This gene codes for a putative transcription factor with a DNA-binding homeodomain. Normally, the *KN* gene is expressed in the shoot meristem of maize but not in the founder cells of the leaf. Therefore, the *KN* protein seems to promote meristematic activity the cells. In *Arabidopsis*, recessive mutations in the *KN*-related gene *SHOOT MERISTEMLESS (STM)* lead to a defective shoot meristem. The STM gene has a similar homeodomain to that of *KN*. Another gene, *ZWILLE (ZLL)*, which is necessary for the formation of the shoot meristem in the embryo, has been identified in *Arabidopsis*, although it is not needed for the production of adventitiously formed shoot meristems. Mutations in the *WUSCHEL (WUS)* gene eliminate structures that would normally be formed by the central zones of the shoot and flower meristems. The *WUS* gene appears to be important for the formation of the central zone.

17.8 LEAF DEVELOPMENT

Leaves are determinate structures that arise from the lateral organ primorida of the shoot apical meristem. The leaves of dicot and monocot plants are distinct (**Fig. 17.18**). Dicot leaves comprise a **leaf blade** or **lamina** with a single central vein (**mid rib**) and subsidiary veins running off at oblique angles. The lamina is joined

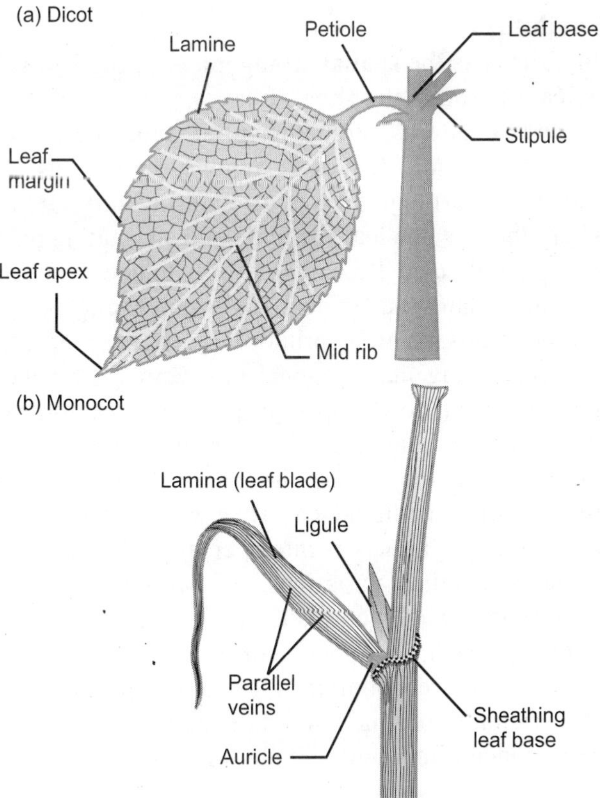

Fig. 17.18 Comparison of dicot and monocot leaves: (a) Parts of a typical dicot leaf and (b) Parts of a typical monocot leaf.

to the stem via a stalk (**petiole**) and **leaf base**. Dicots show greater diversity in leaf shape and size (e.g., ovoid, lobed or serrated). Many dicot plants have **simple leaves** but others have **compound leaves**. In contrast, monocot leaves are long and thin, with multiple parallel veins running along the long axis of the lamina. The lamina is attached to a **sheath** via a **ligule** and a pair of **auricles** and the sheath is attached to the stem.

Leaves are bilaterally symmetrical structures that are polarized along two axes (**Fig. 17.19**). The **apical–basal axis** runs proximal to distal, i.e., from stem to leaf tip. The **dorso-ventral axis** runs from the upper surface (relatively smooth and present few epidermal hairs called trichomes) of the leaf to the lower surface (protruding veins and regularly distributed gaseous exchange organs called **stomata**). The dorso-ventral axis is also called the **adaxial–abaxial axis** (**Fig. 17.20**); the dorsal surface is innermost, wrapped around the stem, while the ventral surface is furthest away from the stem. Leaf development begins with the emergence of a **leaf primordium** from the shoot apical meristem. The primordium expands laterally so that the leaf base partially encircles the stem and extends longitudinally to produce the petiole and midrib vein.

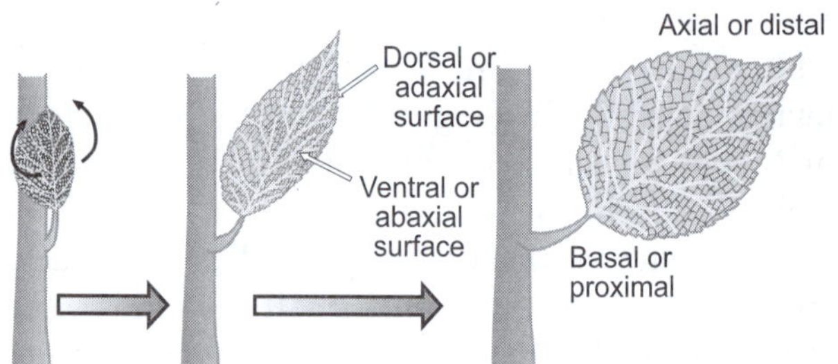

Fig. 17.19 Leaf axes. Since dicot leaves originate adjacent to the stem, the upper (dorsal) surface is also termed adaxial and the lower (ventral) surface abaxial.

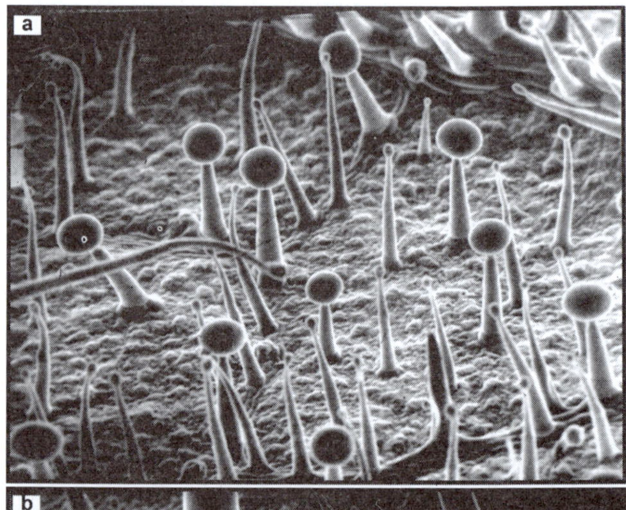

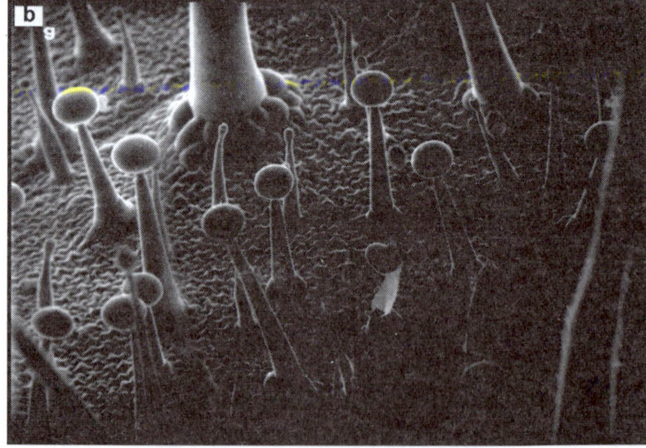

Fig. 17.20 Plant leaves are bilaterally symmetrical structures. Scanning electron microscopic photographs taken on both (a) upper (adaxial) surface and (b) lower (abaxial) surface of tomato *Lycopersicon hirsutum*. (Channarayappa, 1992)

Phyllotaxy

Phyllotaxy is the spatial arrangement of organs such as leaves around the shoot. Leaves may be arranged in various patterns, including spiral, pairs and whorls of three or more leaves at a node. In the case of maize, leaf arrangement is **alternating phyllotaxy**, where there is one leaf per node and alternating pairs are perpendicular. Remarkably, phyllotactic patterns of many plants have ideal **divergence angles** and the positions of new primordia can be predicted precisely using mathematical algorithms. However, it is not clear how a specific pattern of phyllotaxy gets started. Evidence of the molecular basis of phyllotaxy comes from the analysis of a maize mutant called terminal ear 1, in which the normal alternating phyllotaxy is interspersed with decussate, spiral and other variations. In wild-type plants, the terminal ear 1 gene is expressed in a horseshoe pattern in each phytomer, with the open end of the horseshoe corresponding to each leaf primordium. The orientation of the horseshoe alternates along the stem, concomitant with the phyllotaxy of the leaves.

Leaf determination

The primordium that gives rise to a leaf can also give rise to a branch of the stem, an inflorescence, a flower, or in some plants, a variant organ such as a thorn. However, the fate of the organ primordium is not specified until after it has emerged. In the presence of signals from the SAM, the primordium becomes determined as a leaf, while in the absence of those signals it may develop as a shoot. The timing of this commitment varies between the plants. The first process that distinguishes leaf development from lateral branch development is the appearance of dorso-ventral polarity. The phantastica mutant identified in both *Antirrhinum* and *Arabidopsis*, shows dorso-ventral polarity defects in emerging leaves. The upper leaves on mutant plants are ventrilized and form needle-like structures reminiscent of stems in their organization. The *phantastica* gene is, therefore, an important component of a dorsalizing signal early in leaf development.

The unique shapes of leaves results from regulation of cell division and cell expansion as the leaf blade develops. Leaves emerge from a bunch of founder cells that originate from the peripheral zone of the shoot meristem and contain all three layers of the meristem. The first sign that a leaf primordium is beginning to form is the disappearance of the knotted (*KN*) protein from a cell group still lying in the peripheral zone. The beginning of leaf formation becomes visible as a lateral hump, caused by periclinal cell division of the L2 layer. Developmental genetic approaches are being applied to leaf morphogenesis. A gene LEAFY, that is essential for the transition from vegetative to reproductive development also appears to play a role in compound leaf development. In some compound leaves, developmental decisions about leaf versus tendril formation are genetically regulated.

Development of the trichome

The leaf matures from the tip to the base: first, the cells at its tip stop dividing, then, cells at its base also stop. This gradient of development is mirrored in the development of the trichomes, specialized epidermal hairs. The genetics of how a specialized cell type can develop from a uniform cell layer can be studied by investigating trichome development. Trichomes are differentiated from single cells originating in the epidermis. They occur on stems and sepals, as well as on leaves. Trichomes develop from single epidermal cells in young leaf primordia and have remarkable cell morphogenesis (**Fig. 17.21**). First, the trichome cell enlarges and its DNA content increases until it has eight times more DNA than the surrounding cells. The enlarging cell grows outwardly from the leaf surface, so its nucleus is pushed into a subapical location. When the volume of the trichome leaf cell has increased ten-fold, it branches. The nucleus enlarges again by endo-re-duplication, while the two newly formed branches grow at their tips. The nucleus relocates to the centre between the two branch points. The mature trichome rests on a small podium formed by a circle of 8–12 epidermal cells. Trichome development is linked to leaf development.

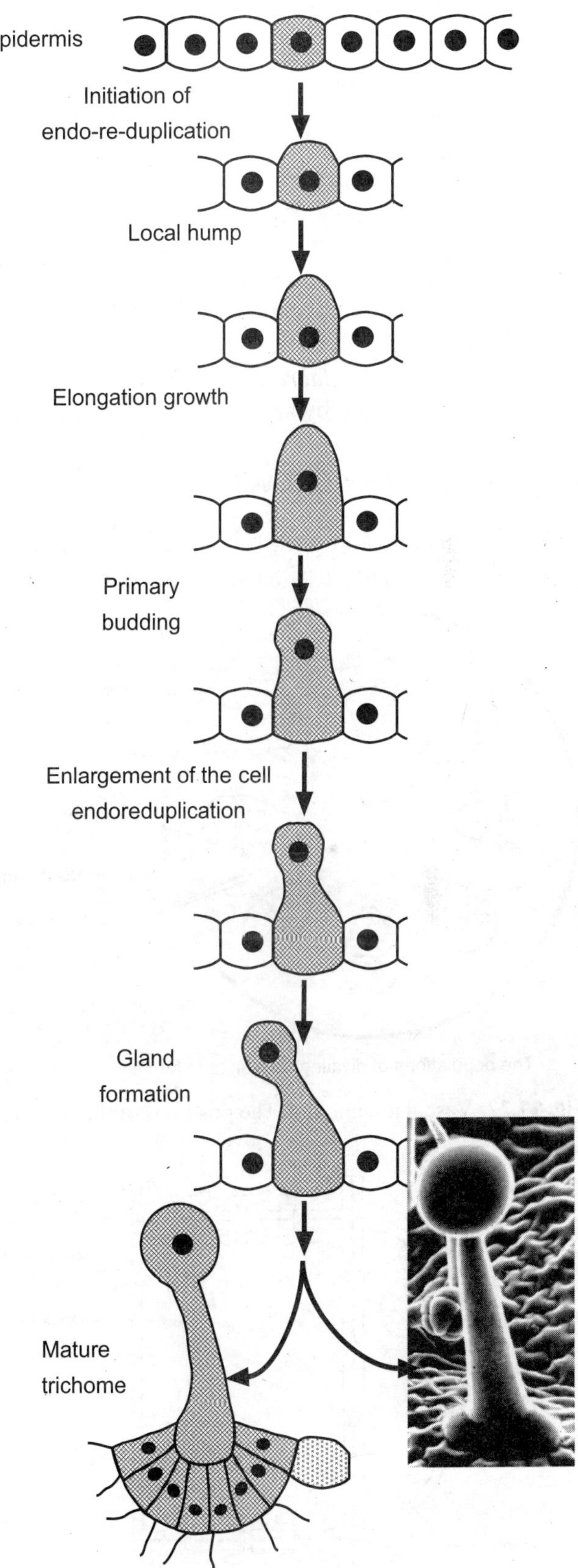

Fig. 17.21 Different stages in trichome development in plants. The insert is a SEM photomicrograph of tomato leaf (Channarayappa, 1992).

Two genes that are essential for trichome development, *glabra 1 (gl1)* and *transparent testa glabra (ttg)*, have been identical in *Arabidopsis*. These genes act early in the trichome development pathway and involve both to initiate trichome development and to control the spacing pattern. Both genes encode bHLH transcription factors and appear to encode inhibitors of trichome development that may antagonize each other. If either gene is deleted, the inhibitory effect of the other is unrestrained and no trichomes form at all. The outwardly growth of the cell from the epidermis is promoted by the *glabra 2 (gl2)* gene. Trichome branching is regulated by another gene *stichel (st1)*.

17.9 SECONDARY GROWTH

Lateral meristems add girth by producing secondary vascular tissue and periderm. Secondary growth begins with the initiation of the vascular cambium, a cylinder of meristematic tissue that produces additional xylic and phloeic tissues. The cells that eventually form the vascular cambium come from two sources: the **procambium** in the vascular bundles and the **interfascicular parenchyma** cells between vascular bundles. Close observation at the cells of the vascular cambium reveals two patterns of division: Initial cells can undergo multiplicative divisions or they can undergo additive divisions. Multiplicative divisions produce more initial cells and result in increased circumference of the vascular cambium (**Fig. 17.22**). Of the two cells produced by additive division, one is retained as an initial cell that will divide again and the other will become a phloem mother cell or a xylem mother cell. These mother cells will differentiate into their respective cell types (**Fig. 17.23**).

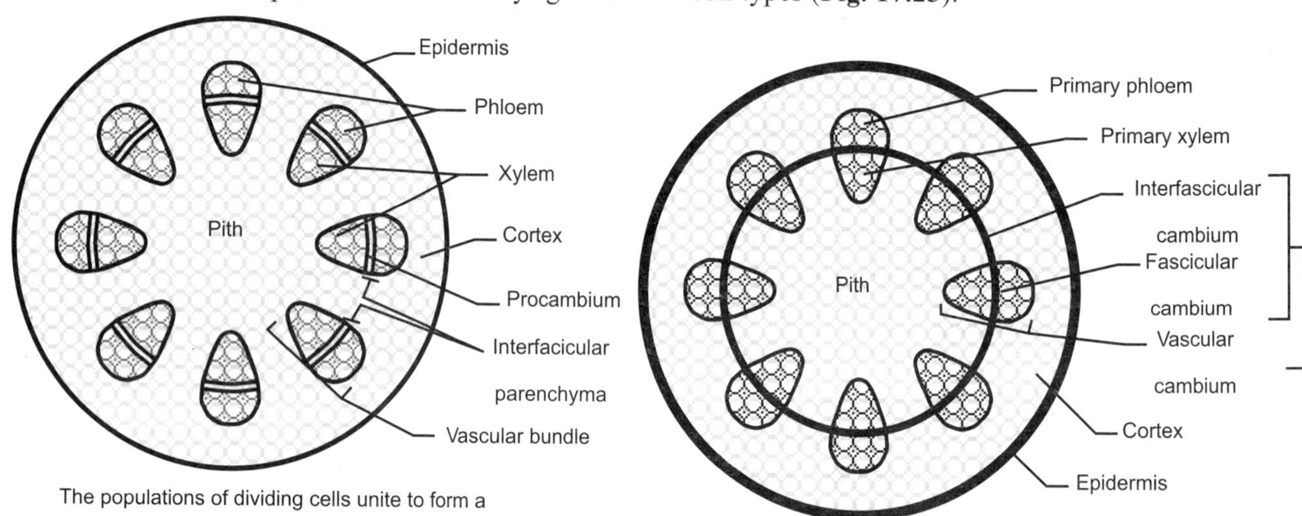

Fig. 17.22 Vascular cambium. The positions of the two populations of cells in a stem with only primary growth.

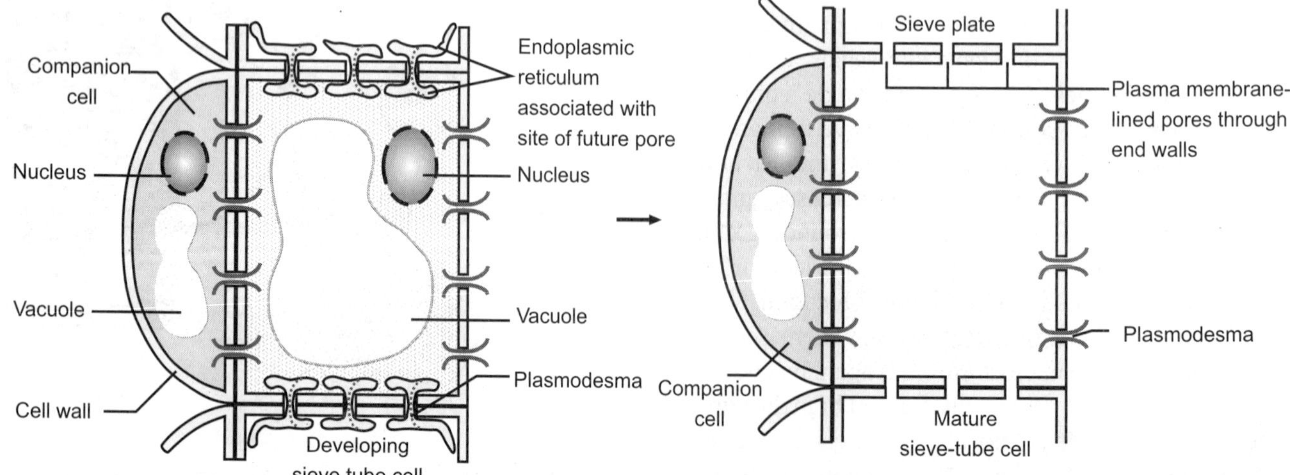

Fig. 17. 23 Modifications of the cell wall during formation of the phloem sieve tube cells. The developing sieve tube cell and its companion cell are derived from the same mother cell.

Both herbaceous and woody dicots exhibit secondary growth. In herbaceous dicots, secondary xylem and phloem are in a single ring of discrete bundles. But in woody dicots, the secondary xylem forms a continuous cylinder. During secondary growth, the epidermis produced by primary growth splits and falls off the stem. It is replaced by new protective tissues produced by the **cork cambium**. Cork cambium is a cylinder meristematic tissue that initially forms from the outer cortex of the stem. Cork cambium produces cork cells, which form exterior to the cork cambium. As cork cells mature, they secrete suberin (a waxy substance) in their cell walls and then die. Cork cells function as a barrier to protect the stem from physical damage and from pathogens. The cork cambium plus the cork are known as the **periderm**, where the bark of the tree consists of the periderm plus the phloem.

Monocot stems differ from dicot stems in that they lack secondary growth—no vascular cambium or cork cambium, stems usually uniform in diameter and scattered vascular bundles.

17.10 TRANSITION TO REPRODUCTIVE PHASE

The flowering plants (angiosperms) go through a phase of vegetative growth—producing more stems and leaves and a flowering phase where they produce the organs for sexual reproduction. The germ line is established only after the transition from vegetative to reproductive development, i.e., flowering (or generative phase). All the vegetative and reproductive structures of plants develop from shoot meristem formed during embryogenesis. During normal plant development, the vegetative shoot meristem produces leaves and lateral shoots continuously and then switches to producing generative organs such as flowers (**Fig. 17.24**). Flowers represent a fundamental change from vegetative to reproductive development, involving the specification of a determinate **floral meristem** that gives rise to a series of specialized organs found only in the flower (sepals, petals, stamens and carpels). This switch is not only endogenously controlled; there is also a clear interaction with environmental factors. The process of flower formation can be split into several discrete steps:

- ***Induction of flowering*** involves the switch from the vegetative to the generative phase.
- ***Evocation*** means transition from vegetative shoot meristem (VM) to inflorescence meristem (IM), with the associated generation of the primordia for flowers, lateral organs and the cauline leaves.
- ***Actual formation of the flower*** involves transition from inflorescence to flower meristem (FM), with associated formation of the primorida for flower organ precursors and their further differentiation.
- ***Functional phase*** when the flower organs gain their characteristic forms and function means the maturation of the reproductive organs, so that pollination and fertilization can take place.

17.10.1 Floral organs

The centre of sexual activity for all flowering plants is the flower. Flowers are reproductive structures whose evolutionary origins are unclear, but sometimes regarded as a collection of highly specialized leaves. It is the most advanced reproductive structure in the plant kingdom, coming in an array of sizes, shapes, colours, colour patterns and rearrangements. Generalized flowers are shown

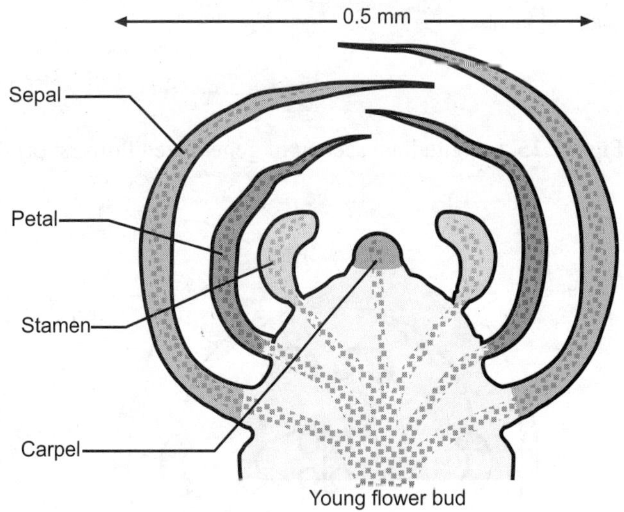

Fig. 17.24 Longitudinal section of a dicot flower showing the sequence of floral organ initiation in meristematic tissue. Flower structure can vary and is species specific, but generally consists of four concentrically arranged sets of structures that may each be regulated as modified leaves.

in **Fig. 17.25** and some of the terms used to describe the flower parts are explained below.

The arrangement of the floral organs on the receptacle may be spiral or whorled and both types of arrangement may occur in the same flower. In most flowers in which the arrangement is whorled, the organs of each whorl alternate with those of the neighbouring whorls. Floral organs may be free or fused. Fusion of organs of the same type is termed **cohesion** and that of different types of organs, **adnation**.

A flower containing all four whorls of modified leaves—sepals, petals, stamens and carpels (pistil)—is called a **complete flower**. Flowers lacking one or more whorls are **incomplete flowers**. Imperfect flowers contain either stamens or carpels, making them male or female flowers, respectively. The term **hermophrodite** (bisexual) is applied to whole plants and flowers, where male and female sex organs are borne on the same plant. **Dioecious** (unisexual) plants have male and female sex organs on separate plants. Such plants are either male or female. **Monoecious** plants are hermoprodite.

The flower provides a major sink for photosynthetic assimilates (carbon and nitrogen) in the plant. After the flowers have opened and been pollinated, senescence sets in and the sepals, petals and stamens wither. Fertilization sets in train a cascade of differentiation events in the pistil. The ovules produce the seeds. Fruit formation occurs with the initiation or redifferentiation of entire tissues.

Specification of floral organs: As mentioned earlier, flowers comprise four concentric whorls of floral organs and an important phase of flower development concerns the specification of organ identities (**Fig. 17.26**).

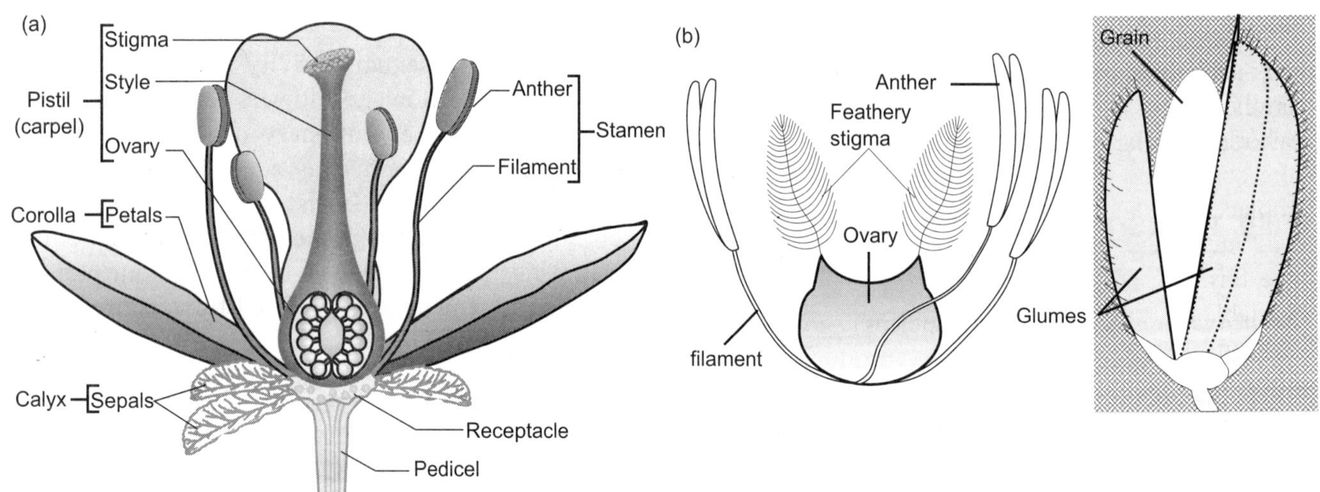

Fig. 17.25 Longitudinal section of generalized flowers: (a) Dicot flower and (b) Monocot flower (rice/paddy).

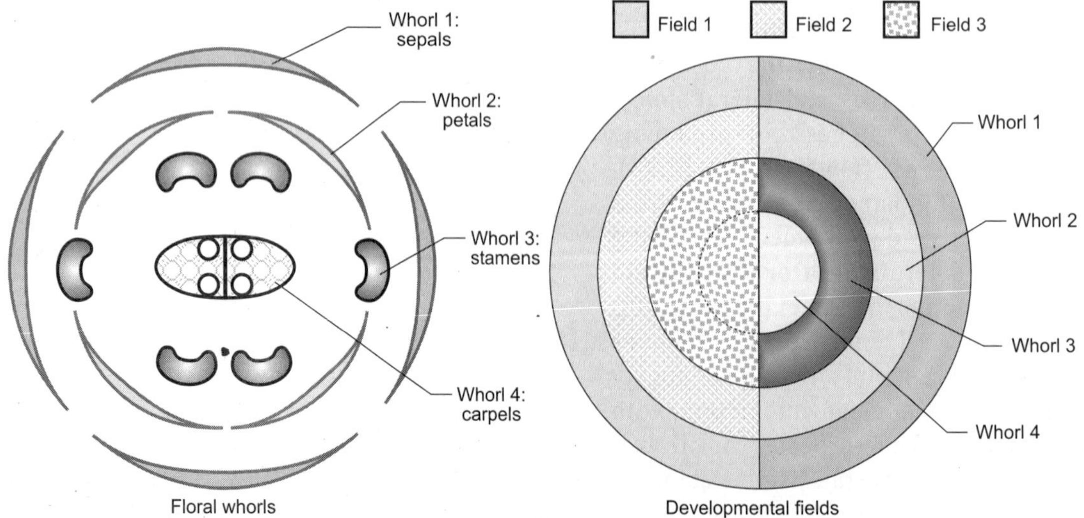

Fig. 17.26 The ABC model for floral organ specification in *Arabidopsis*. The flower consists of four nested whorls: Sepals (whorl 1), petals (whorl 2), stamens (whorl 3) and carpels (whorl 4).

Study of *homeotic mutants* has revealed where particular floral organs are transformed into the likeness of another. In *Arabidopsis*, at least five floral organ identity genes (*homeotic selector genes*) have been identified through the isolation of different types of homeotic mutant (**Table 17.1**). For example, in the mutant apetala2, the identities of whorls 1 and 2 are changed, so that carpels develop instead of sepals in whorl 1 and stamens develop in the place of petals in whorl 2. The mutant phenotypes are summarized in **Table 17.2**.

17.10.2 ABC model for specification of floral organs

Genes that affect the development of the floral organs can be identified by mutations that result in reduced, altered or misplaced floral organs.

Mutations in four genes, AP_2, *AGAMOUS (AG), PISTILLATA (PI)* and *APETALA3 (AP3)*, cause mis-specification of organ types (**Fig. 17.27**). Generally, it affects development in the first and second whorls and two of the effects of *ap2* mutations are the conversion of whorls 1 and 2, such that carpels replace sepals and stamens often replace petals. The *ap2* mutations also lead to loss of organs, a result of the formation of fewer organ primordia. This loss is variable but often all organs are missing from whorl 2 and only a few organs form in whorl 3. When organs form in both whorls, they are all stamens. Defective alleles of *AG* affect flower development in the third and fourth whorls: stamens are replaced by petals and the gynoecium is replaced by a second *ag* flower, so that the pattern sepals–petals–petals is repeated several times. The indeterminate development of *ag* flowers causes them to look like many-petalled cultivated roses.

Table 17.1 Floral organ identity genes grouped by class in *Arabidopsis* and *Antirrhinum*

Class (and expression)	*Arabidopsis thaliana*	*Antirrhinum majus*
Class A (whorls 1 and 2)	Apetala1 (Ap1)	
	Apetalai2 (Ap2)	Squamosa (Sqa)
Class B (whorls 2 and 3)	*Apetalai3 (Ap3)*	Deficiens (Def)
	Pistillata (Pi)	Globosa (Glo)
Class C (whorls 3 and 4)	*Agamous (Ag)*	Plena (Ple)

The mutant phenotypes can be explained by a simple **ABC model** in which floral organ identity is specified by a particular combination of homeotic selector genes. Genetic analysis of mutants—especially those found in the dicots, *Arabidopsis thaliana* and snapdragon (*Antirrhinum*) support the ABC model of flowering. This model postulates a group of genes that encode the transcription factors needed to run on the

Table 17.2 *Arabidopsis* floral homeotic mutant phenotypes in terms of ABC model of floral organ specification

Genotype	Phenotype				Expression domain pattern			
	Whorl 1	Whorl 2	Whorl 3	Whorl 4				
Wild type	sepal	petal	stamen	carpel	B	A + B	B + C	C
Loss of A function								
Apetalai1 or Apetalai2	carpel	stamen	stamen	carpel	C	B + C	B + C	C
Loss of B function								
Apetalai3 or Pistillata	sepal	sepal	carpel	carpel	A	A	C	C
Loss of C function								
Agamous	sepal	petal	petal	sepal	A	A + B	A + B	A
Loss of ABC function								
Apetalai2 + Apetalai3 + Agamous	leaf	leaf	leaf	leaf	0	0	00	0
Apetalai3, Pistillata	petal	stamen	stamen	petal	A + B	A + B	B + C	B + C

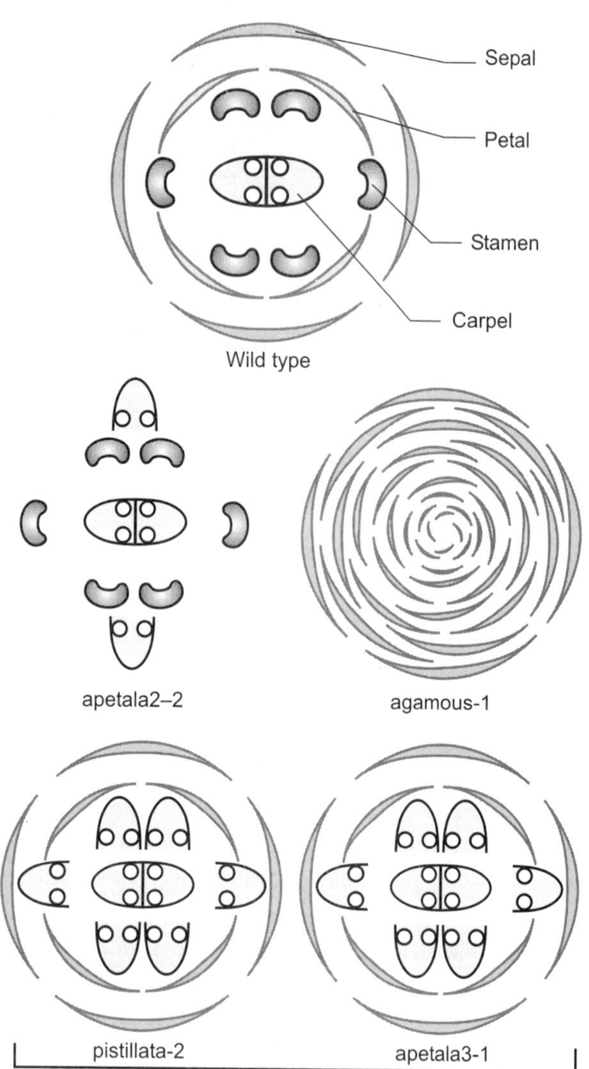

Fig. 17.27 The four original ABC mutants: *ap2, ap3, pi* and *ag*. Each mutant demonstrates a homeotic transformation (transformation in which one organ type replaces another).

genes for sepal, petal, stamens and carpel development. The "master switches" fall into 3 groups: **A**, **B** and **C** and these are the rules:

- Cells in which only **A genes** are expressed develop into **sepals**. This occurs at the lowest level of the floral meristem.
- Cells in which both **A** and **B genes** are expressed develop into **petals**. This occurs at the next level of the floral meristem.
- Expression of **B** and **C genes** turns on the developmental program to form **stamens**.
- Expression of **C genes** alone turns on the development of **carpels** in the innermost band of cells.

17.10.3 Induction of flower formation

Flowering is a crucial stage in plant development. It leads to various genetic recombinations during meiosis. These multiple trials allow the species to adapt themselves to changes in environmental conditions, as plants are immobile and therefore have adapted strategies for their survival. The process of sexual reproduction of an organism (studied in embryology) includes the study of processes such as structure and development of reproductive organs, gamete formation, fertilization, embryo development and formation of a new plant.

In plants, the germ line is established only after transition from vegetative to reproductive development, that is, flowering. Flowers are formed by a change in the specification of cell fate at shoot apical meristems, which switch from ordered production of leaf primordia in regular phyllotactic array to the induction of floral organs. The type, number and relative disposition of floral organs within the developing inflorescence is stringently genetically defined and regulated at numerous control points in different organs, resulting in a diversity of flowering times and reproductive architectures. The shape, size, colour and pattern of the inflorescence are diagnostic features of most plant species. A simplistic explanation of the flowering process is that a signal from the leaves moves to the shoot apex and induces flowering. The induction of floral development is either genetically preprogrammed or prompted by environmental cues. In many species the induction of flowering is triggered by the perception of light—the photoperiodic response.

Photoperiodism and stimulation of flowering

Photoperiodism defines the requirements of plants to undergo repeated cycles either of short days or long days before flowering is initiated. It turned out that it is not the day length (light period) that is decisive for stimulation of flowering but the dark period. A minimum light period is required for the production of enough assimilates. With the onset of light period, a physiological activity called the **photophilic stage** begins. After 9–12 hours of light the development of the plant is inhibited by further exposure to light; the plant enters its **skotophile** or darkness-loving phase. The next photophilic phase begins just a few hours later, long before dawn and is thus quite independent of the actual conditions of alternation of day and night. A reduced light program where the plant receives just a few hours of light every 72 hours is enough to synchronize the alternation of skotophile and photophilic phase. It should be noticed that the endogenous rhythm contributes decisively to the maintenance of an approximate 24-hour rhythm also called the **circadian rhythm**. Photoperiodism mainly controls the relative share the two mentioned phases have during a 24-hour period.

Flowering hormone

Phytochrome is believed to mediate a wide range of light-dependent phenomena in plants, in addition to photoperiodically related floral induction. It is a characteristic of these phenomena that they are activated by illumination with red light and this activation is reversed by a subsequent period of illumination with far-red light (**Fig. 17.28**). The *pfr* form of phytochrome is therefore regarded as the biologically active form of the molecule. In plants, which show a light-responsive pattern of floral induction, the perception of light, by phytochrome, occurs in the leaves. However, years of research have failed to reveal the nature of this flowering inducer, often termed 'florigen'. The general features of flowering inducer are described by grafting experiments as given below:

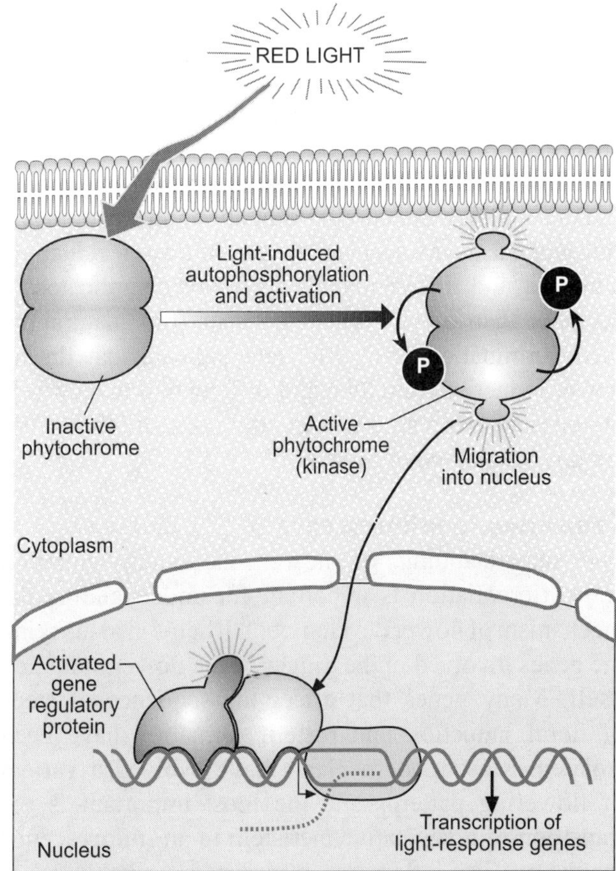

Fig. 17.28 A model of one way in which phytochromes mediate a light response in plant cells.

Florigen: In the leaves of plants a hormone-like substance (florigen) and a complex of substances are produced and conducted to the shoot meristems in order to stimulate them to pass from vegetative growth to flower formation.

Florigen is not species-specific: It can be transferred to members of the same species and to members of different genera. The transferability seems to be limited only by the acceptability of the graft. A good connection between the phloem of the two partners seems to be important. Grafting of monocots is very difficult and a florigen transfer by grafting has not been successful, but it could be shown with other techniques.

Florigen is physiologically not specific: It can be exchanged between short-day, long-day and day-neutral plants and is very likely identical in all plants. The main difference between short-day and long-day plants is that the florigen production occurs only under a certain (inducive) light program that differs in the two types.

Likewise a transferable substance exists in several long-day plants that is produced under short-day conditions and suppresses flower formation. It is called **antiflorigen**. The reaction is weak and cannot be detected in all combinations. Florigen and antiflorigen appear to be antagonists. The flower formation is mainly based on the ratio of the two substances.

17.10.4 Floral initiation events

The first stage to be controlled in floral induction is its timing, i.e., flower formation usually occurs at a specific time in the lifecycle of a plant affected by environmental cues (day length, temperature, e.g., in some species, chilling, can enhance flower initiation, called **vernalization**). Flower differentiation consists of three major phases. The first one is the **induction of flowering**. The induction phase is marked either by a shift or commitment to changes from a vegetative meristem to a floral meristem or by an induction of direct formation of floral meristems from differentiated cells without passing by a stage of vegetative meristem. It is followed by the **flower initiation phase** during which cell division within the vegetative meristem or within the sub-epidermal layer occurs at a particular rhythm and with a particular cell plate orientation. This leads to a pre-floral meristem characterized by a voluminous dome of meristematic cells. The third phase relates to **flower differentiation,** i.e., differentiation of new structures (floral organs such as sepal, petal, anther and ovary) and onset of new functions: pollination, fertilization and embryogenesis. Molecular aspects, of flower induction, initiation and differentiation are more complicated to decipher than other events (seeds or fruit maturation, seed germination, growth). Data indicate that during flower induction, initiation and differentiation, new sets of genes are expressed while certain sets of other genes are downregulated or turned off.

Transition to inflorescence

The understanding of flower induction, initiation and differentiation is important for understanding the mechanism of flower development. It helps in identifying the genes involved in the trigger of the flower induction itself. Many genes that affect the sequence of steps of floral induction and pattern formation have been isolated. Most modern plants have evolved a variety of flowering patterns and the most important is the transition of a vegetative meristem to an **inflorescence meristem**. The inflorescence meristem probably arises through the action of a gene that suppresses terminal flower formation. As for the flowering response to environmental factors, mechanisms of flower induction by photoperiodism (a response to the timing of light and darkness) are yet to be analyzed, although a single recessive mutation is known, which changes the requirement of photoperiodism. At least five genes that impart identity on floral meristem (**floral meristem identity genes**) are known. When mutated, these genes result in either shoots instead of flowers or abnormal flowers.

Formation of a flower organ

Even though the induction of flowering in many plant species is regulated by environmental factors (mainly photoperiodism), the endogenous species-specific developmental programs also take control of flower formation. The role and importance of essential genes and gene products have been well studied with the help of a wide variety of mutants. Two main groups can be distinguished:

- meristem mutants do not form flowers at all
- organ mutants have altered shapes and numbers of flower organs.

Floral development continues by the creation of organ primordia. The events that occur within the apex upon receipt of the flowering stimulus have been much more extensively characterized by the analysis of **homeotic mutations**. Upon floral induction, the vegetative meristem undergoes a transition to inflorescence meristem, which retains an indeterminate growth habit. However, the primordia, which arise on the flanks of the inflorescence meristem transform, not to leaves, but individual flowers. The inflorescence meristem thus gives rise to a series of floral meristems, each of which exhibits a determinate pattern of growth, i.e., it undergoes differentiation to produce the floral organs and when these reach maturity further growth ceases.

When meristematic re-programming occurs to initiate a flower, it is necessary that the cells within the meristem be able to interpret their position in order to develop the characteristic floral morphology. It is presumed that positional information is available to the cells within the meristem. Studies of mutations, which cause homeotic transformations, illustrate how such a definition of developmental fate may be achieved. These mutations occur in single genes, yet cause profound and specific changes in the identity and position of the floral organs. Two alternative roles may thus be envisaged for these genes. Either they are central in the establishment of positional information within the meristem, or they are primary regulators of families of organ specific genes, which respond to positional information.

Molecular analysis of homeotic genes DNA sequences corresponding to the *deficiens* and *floricaula* genes of *Antirrhinum* and the *agamous* gene of *Arabidopsis*, have been cloned. Sequence analysis of the protein-coding regions contained within the *deficiens* and *agamous* genes confirms their functional similarity, as they contain regions of striking homology within and also with other divergent species. Therefore, two of the genes establishing organ identity—through the specification of the 'b function' (*def*) and the 'c function' (*ag*)—can be predicted to act by controlling the expression of other genes at the level of transcription, an activity which may be further 'fine tuned' by interaction with protein kinases and phosphatases.

Factors that regulate flowering time

Plants in which flowering is promoted when they are grown under long-night/short-day conditions are classified as **short-day plants** and plants in which flowering is promoted in short-night/long-

Table 17.3 Photoperiod and flowering response in short-day and long-day plants

Light treatment	Dark treatment	**Developmental response**	
		Short-day plant	**Long-day plant**
		Flowering	Vegetative growth
12	**24**		
		Vegetative growth	Flowering
12	**24**		

day conditions are classified as **long-day plants** (**Table 17.3**). Flowering in a third class of plants, known as **day-neutral plants**, is not affected by photoperiod. Further variations include plants that flower in response to long days followed by short days and plants that flower in response to short days followed by long days.

Photoperiod is detected in the leaves and a chemical signal produced in the leaves (florigen) is transmitted to the apical meristems to start their conversion into floral meristems. The most recent evidence suggests that at least one compound is the mRNA encoded by the gene *FLOWERING LOCUS* T (FT). The FT protein translated from the mRNA of FT binds to and activates proteins in the meristem cells. These include a protein designated FD and the transcription factor LEAFY (LFY). These appear to be master switches that turn on expression of genes needed for flowering, e.g., *APETALA1*.

17.11 DEVELOPMENT OF REPRODUCTIVE CELLS

The process of reproduction in seed plants, is different from those in lower plants and animals. As with animals, the formation of gametes in plants begins with meiosis. Special diploid mother cells (sporocytes) inside the anthers and ovaries of flowers divide by meiosis to form haploid spores. Mother cells in the anthers are called **microspore mother cells** and the diploid mother cells inside ovaries are called **megaspore mother cells**. Microspores produced within anthers divide by mitosis to form pollen grains and in the ovary, megaspores divide by mitosis to form female gametophytes that produce eggs.

Plant genes express in a highly regulated and specific manner in response to specific environmental and hormonal stimuli. In recent years, it has become apparent that better understanding of regulation of reproductive tissue-specific genes presents possibilities for an efficient means of producing hybrid seeds and male sterile lines. Style-specific S (self-incompatibility) genes have cloned and characterized in both sporophytic and gametophytic self-incompatiblility systems. Growing number of pollen- and anther-specific cloned genes have been reported from many laboratories. Knowledge of the regulation of specific reproductive genes will undoubtedly throw light on the future direction of fundamental and applied research.

17.11.1 Development of male and female gametophytes

Although many similarities in the process of meiosis have been detected between animals and plants, little is known about plant signals that switch development from mitosis to meiosis. Once the functional flower has formed, there are no major visible changes that could be readily seen from outside. The most important event in this phase happening internally is **gametogenesis**. The most important step in this phase is the production of meiospores—the extremely reduced forms of micro- and mega-gametophytes. Pollination and fertilization that lead to the formation of the zygote and alter to embryogenesis are two other major steps in this phase of development.

The initiation of meiosis is a key stage in the developmental process of higher plants since it provides the trigger for splitting the diploid **sporophytic** cell and formation of haploid cells to generate a multicellular **gametophyte**. It is evident that meiosis and the subsequent development of the gametes are regulated in part by the developing gametophytes, but also by the surrounding sporophytic tissue. This is true for both male and female gametes. Male and female gamete production is similar in higher plants with many genes playing a role in the development of both types of gametes; however, the subsequent fate of the male and female meiotic product is distinct. In male gamete production, all the four meiotic products (microspores) develop to produce fully viable pollen grains. But in females, usually only one of the four megaspores will continue to develop, while the remaining meiotic products degenerate by a form of programmed cell death.

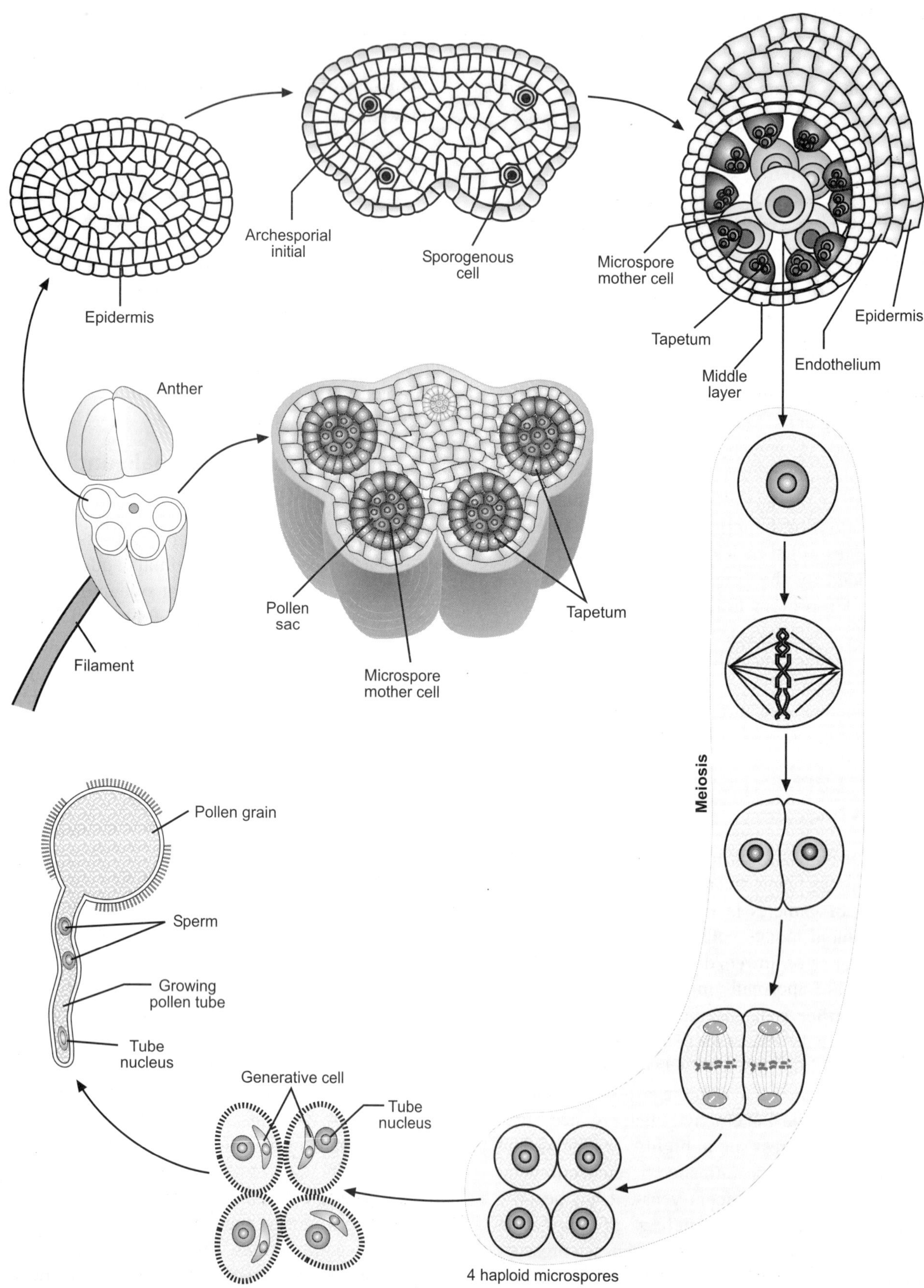

Fig. 17.29 Development of anther and microsporogenesis (**See Plate 14**).

17.11.2 Development of male gametes (microsporogenesis)

Anther

The formation of the male gametophyte occurs in the anther. In the stamens of whorl three, each stamen consists of an anther and a filament at their upper end. The anther contains a sterile central region (the connective) that supports two thecae. Each theca is composed of two pollen sacs, i.e., four pollen sacs in an anther. Dissection of the pollen sac shows that its wall is composed of four layers. These are, from the outermost to the innermost: the epidermis, the endothecium, the middle wall layer and the tapetum. They surround the pollen-forming archespore. Pollen grains are produced from the diploid sporogenous cells in the young anthers. The pollen mother cells of the archespore divide meiotically (microsporogenesis). Each produces four haploid pollen cells (or microspores). The filament contains a vascular bundle supplying food and water to the anther. **Figure 17.29** shows the internal structure of an anther, with its four pollen sacs containing pollen mother cells. The microsporocytes are produced by mitosis in the sporogenous tissue. These diploid cells undergo two meiotic divisions to yield four haploid microspores (or gametes).

Pollen

Immediately after meiosis, the young pollen grains form groups of four called **tetrads**. Each grain (or **pollen grain**) develops a thick wall, often with an elaborate sculpted pattern characteristic of the species or genus. The pollen grain is a very simple and multicellular structure. The protoplast of these cells is surrounded by a resistant two-layered wall called the **sporoderm**. The outer layer of the pollen grain, the **exine**, is made of a water-proof substance called **sporopollenin**, provided by both the tapetum and microspore. Sporopollenin is a polymer of phenolics and it is one of the most resistant and long lasting substances in nature and allows grain coats to survive unchanged (resistant to weathering) over long periods of time sometimes millions of years. The inner wall, the **intine** produced by the microspore, contains the typical components of plant cell walls. The development of the extremely reduced male gametophyte starts with an unequal mitotic division of a pollen cell with a single nucleus. This leads to a large vegetative cell and a smaller generative cell. The generative cell initially adheres to the wall of the vegetative cell. It later loosens, becomes spiral and is surrounded by the vegetative cell. By a further mitotic division, the generative cell produces two sperm cells whose cell nuclei are surrounded by only a thin layer of cytoplasm. A mature pollen grain thus consists of two cells (**tube cell** and **generative cell**), one within the other (**Fig. 17.30**). The tube cell nucleus guides pollen

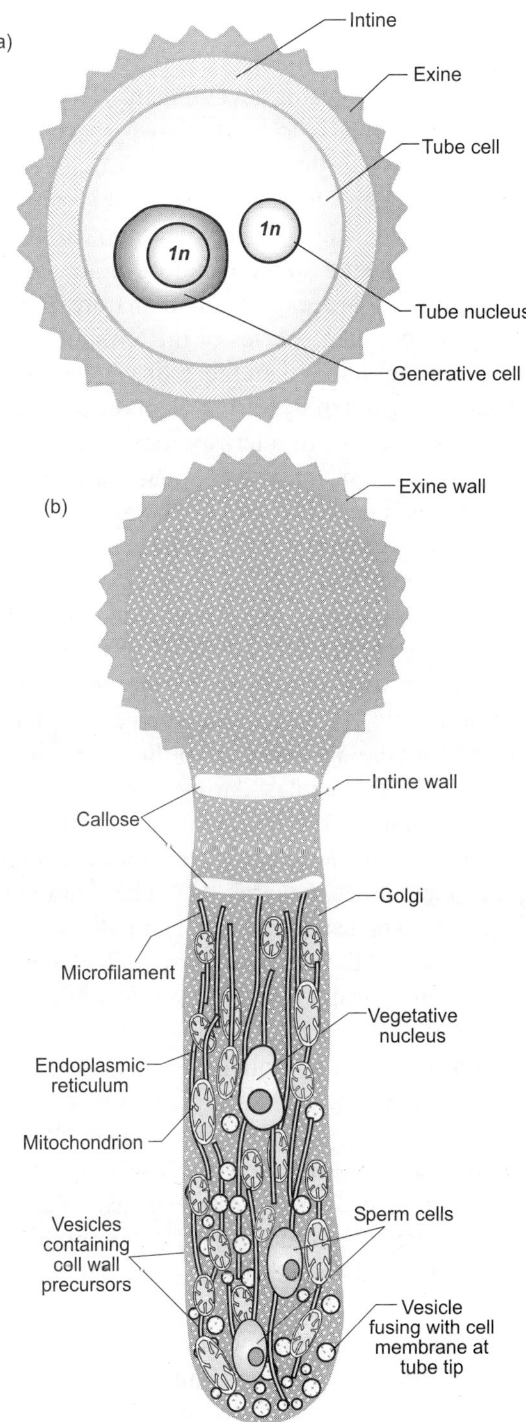

Fig. 17.30 (a) A pollen grain consisting of a cell within a cell. (b) Schematic illustration of a pollen tube growing by tip extension. Callose plugs limit the volume occupied by the cytoplasm to regions close to the growing tip.

germination and the growth of the pollen tube after the pollen lands on the stigma of a female gametophyte.

Gene expression in male gametophyte

The gametophyte of higher plants has many unique characteristics, which make it ideal for genetic analysis and manipulation. Basically pollen grains are haploid and available in a large population at synchronizing stages of development. The role of the male gametophyte is to produce two sperm cells and transport them via the pollen tube through the stylar tissue to the embryo sac where double fertilization occurs. Considering its many specialized functions, it is not surprising to find that as many as 20,000 different genes are expressed in mature pollen and the highly abundant species of mRNAs are present in higher copy numbers than those of vegetative shoots. The male program of flowering plants encompasses the events from meiosis of microsporocyte to fertilization. The processes of male haplophase are continuous and recently, as many as 26 character states from microspore formation until pollen tube production have been identified.

Estimates based on RNA/DNA hybridization studies suggest that 60%–90% of the genes that are expressed in the gametophyte are also expressed in sporophytic tissues. Much research has been focussed on the isolation of pollen- specific genes to determine their roles during pollen development and to understand the mechanisms of pollen-specific gene expression. With recent advances in molecular cloning techniques, several researchers have isolated 350 genes expressed exclusively in the *Arabidopsis* anther. Some of these genes are specifically expressed in the pollen grain. Others include pollen specific variants of proteins that are found in other parts of the plant. Five mature pollen cDNA clones (pLAT51, pLAT52, pLAT56, pLAT58, pLAT59) isolated from tomato and three pollen-specific clones (P1, P2 and P3) from *Oenothera* follow the transcription pattern for that of the late-expressed genes. The pollen-expressed gene from tomato, *LAT52*, contains several directly repeated sequences that may be involved in transcriptional regulation. The cDNA sequences of *LAT52* and *zml3*, pollen-specific clone from *Zea mays* share a number of common features.

17.11.13 Development of female gametes

Female gametogenesis occurs within the ovule and as in males, begins with meiosis. Although many variations are possible in the female cell division patterns that give rise to the embryo sac, more than 70% of flowering plant families demonstrate Polygonum-type embryo sac development, in which three of the products of meiosis undergo cell death and the fourth forms the megaspore.

The ovary

The fourth whorl of the flower forms the **carpel**, which gives rise to the female gametophyte (**Fig. 17.31**). The carpels of the fourth whorl fuse at their bases to form the ovary. The upper areas of the neighbouring carpels can fuse to varying extents, depending on the species. Often they are completely joined and the resulting structure is then called a pistil. Each carpel therefore consists of a **stigma** (where the pollen lands), the **style** and the **ovary**. Within the ovary one or more ovules (seed progenitors) develop. Each is attached to the ovary wall at a point called the **placenta** by a short stalk or **funicle** (contains vascular tissue) through which food and water pass to the developing ovule. Fully developed ovules are called **seeds**. The main body of the ovule is called

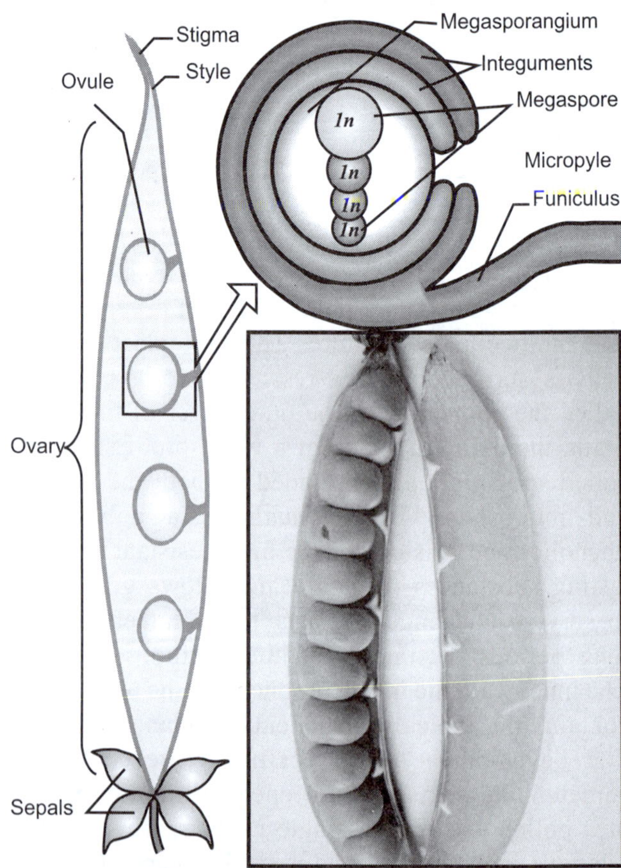

Fig. 17.31 The carpel contains one or more ovules. Insert: a pea pod containing many number ovules (Channarayappa, 2008).

the **nucellus** and is enclosed and protected by two sheaths or **integuments**, the inner of which is called the endothelium. They start at an area at the base of the ovule called the **chalaza**, which marks the border with the funiculus and surrounds the inner tissue area (nucellus) of the ovary. Only a small spot of tissue opposite the chalaza is not covered by the integuments, leaving an opening called the **micropyle** (through which the pollen tube grows).

Female gamete formation occurs within the ovule and more than 70% of flowering plants exhibit the Polygonum-type pattern of development, resulting in an eight- nucleate embryo sac. Each ovule comprises of a nucellus that encloses the embryo sac. Inside the nucellus, at the end nearest the micropyle, a single cytoplasm-rich cell enlarges—the **embryo sac mother cell** (or **mega-spore mother cell, MMC**). The MMC divides meiotically to form four haploid megaspores (microsporogenesis). The **megasporangium**, which contain a diploid sporophytic cell undergo meiosis and unequal cytokinesis, to produce four **megaspores** (**Fig. 17.32**). Of the four emerging haploid spores, only the one that lies in the direction of the chalaza will lead to the female gametophyte and the other three degenerate, via programmed cell death. This cell with its single nucleus (the megaspore) grows to form the embryo sac. Then, in typical cases, megagametogenesis begins with three consecutive

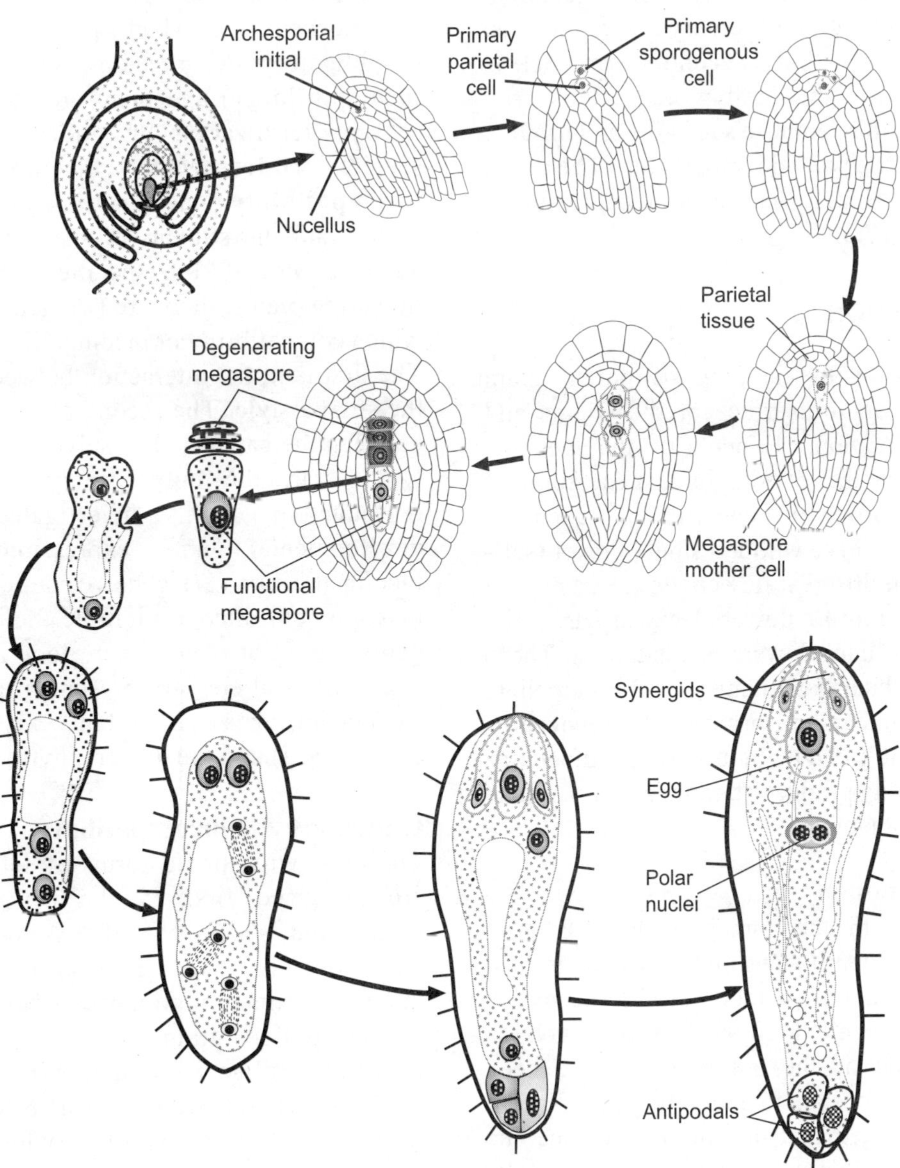

Fig. 17.32 Development of monosporic embryosac (female gametophyte) and successive stages of development into organized embryo sac.

nuclear divisions. The products of the first division move to opposite poles of the embryo sac; once there they divide twice again. There are now eight nuclei in total. Three of the four nuclei at each cell pole remain there and thin walls are formed which separate them (together with some cytoplasm) from the rest of the embryo sac and appear to serve many further functions (**Fig. 17.32**). The three upper cells nearest the micropyle form the egg apparatus, which consists of the large **egg cell** and the two smaller **synergids**. The three lower cells at the opposite end of the embryo sac are **antipodals**, which degenerate before or during embryonic development. The two remarking nuclei (polar nuclei) migrate to the centre of the embryo sac and once there fuse together to form the secondary diploid embryo sac nucleus, which will later fuse with the second sperm nucleus and develop into the polyploid endosperm. All together the mature embryo sac is the end product of female gametophyte development. The final appearance of the mature carpel at fertilization is illustrated in **Fig. 17.32**.

17.12 POLLINATION

Pollination refers to the landing of pollen grain on **papillar cells** on the surface of the stigma and subsequent germination. Hence, it involves an interaction between the gametophytic generation of the male and the sporophytic generation of the female. Pollination can take place within a single flower (self-fertilization), or on different flowers on the same or a different plant. **Staminate** flowers lack carpels, while **carpellate** (or pistillate) flowers lack stamens. Those plants, which produce both staminate and carpellate flowers on the same plant, are called **monoecious** (Greek: *mono*, 'one'; *oecos*, 'house'). The **dioecious** ('two houses') plants produce staminate and pistillate flowers on separate plants.

Germination of pollen

Pollen released from the anther is in a partially dehydrated state. When it lands on the stigma, it must hydrate and grow a pollen tube toward the ovule. In plants, with so-called 'wet' stigmas, such as tobacco, the lipids of the pollen coat have been thought to be less important to pollen function. Recent studies have suggested that the initial guidance cue for the emerging pollen tube is actually water and that the lipids in the pollen coat or stigma exudate establish a concentration gradient of water that directs growth of the pollen tube. Secondary metabolites such as flavonols can also facilitate pollen tube growth in some plants. Maize and petunia plants that are deficient in chalcone synthase—a key enzyme in the biosynthesis of falvonols—are self-sterile. Female tissue in the style can also provide flavonols to the flavonol-deficient pollen grains.

17.12.1 Fertilization incompatibility

The arrival of a viable pollen grain on a receptive stigma does not guarantee fertilization. The fertilization events do not usually come to completion if 'incorrect' or incompatible pollen lands on the stigmatic surface. In other words, there are mechanisms, which prevent self-fertilization in obligatory out breeding species (**intraspecific incompatibility**) and there are mechanisms, which prevent fertilization by pollen from a different species (**interspecific incompatibility**). **Self-incompatibility**—incompatibility between the pollen and the stigmas of the same individual—is an example of intraspecific incompatibility. In flowering plants, there are two general phases during which compatibility or incompatibility may manifest. The first is in the interaction between the pollen and the stigma style. The second phase is the interaction between the gametes themselves.

So far, the only system related to female reproductive organs investigated in terms of developmental gene expression is the self-incompatibility (*SI*) gene. Genetic studies indicate that *SI* is controlled by multiple alleles of the S nuclear gene present at one or more loci. *SI* operates when two similar alleles are present in pollen and pistil. Two major types of *SI* systems exist: the sporophytic system (SSI) and the gametophytic system (GSI).

Gametophytic incompatibility

The most widespread forms of self-incompatibility are governed first by the genotypes of pollen grains and second by the genotypes of diploid stigmatic or stylar tissues. These systems are known as gametophytic, because of the involvement of the pollen grain. The gametophytic control of compatibility is regulated by a single highly polymorphic genetic locus, the *S* locus. The *S* gene codes for an antigenic protein, which is glycosylated and is involved in cell recognition. Growth of a pollen tube carrying a given S allele is prevented, if, the same S allele is also carried by the stylar tissue

(**Fig. 17.33a**). If, the allele carried by the pollen is not carried by the stylar tissues, then successful pollen tube growth occurs. The incompatibility system may work at different possible points: either very early on the stigma surface or while growing through the stylar tissues.

Sporophytic incompatibility

In sporophytic self-incompatibility (SSI), the pollen grains can germinate and grow tubes only when they do not carry determinants that match the genotype of the female. In the sporophytic system, pollination fails if either of the S alleles carried by the female parent is matched by either of the alleles carried by the male parent (**Fig. 17.33b**). Recognition of incompatibility occurs very early, usually immediately after hydration of the pollen grain. The mutual recognition depends on substances present in the cell walls of both the stigmatic papillae and in the pollen grain. If the interaction is compatible, the pollen tube grows down in the stylar tissues, having dissolved the cuticle of the papillae by secretion of hydrolytic enzymes. In an incompatible reaction, pollen tube tip growth ceases quickly and is completely occluded by callose. The antigenic substances mentioned above are now known to be glycoproteins. As with the gametophytic system, we do not yet understand how a particular S glycoprotein recognizes another as being similar or different.

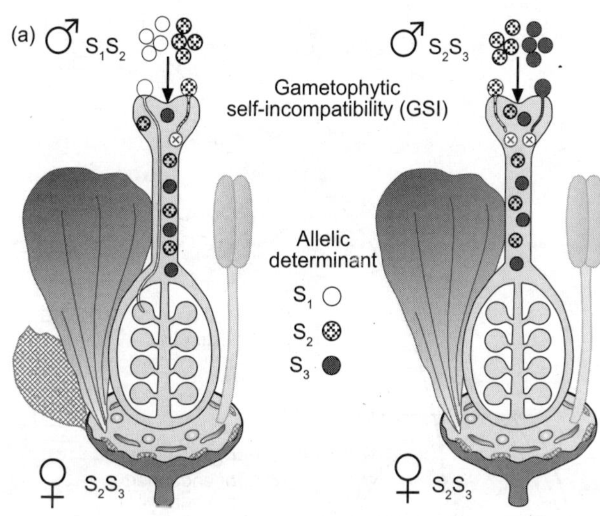

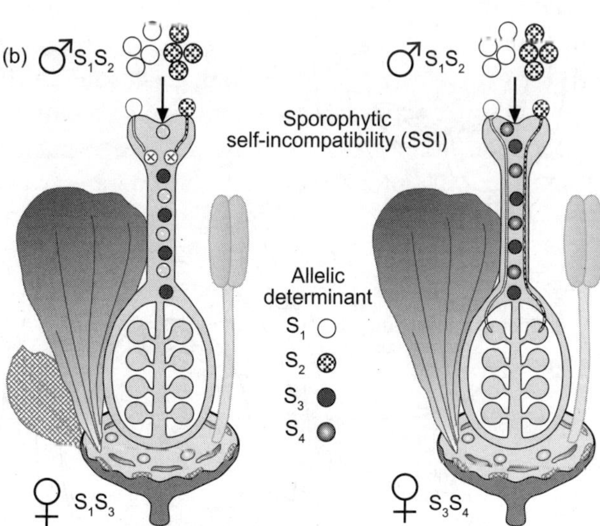

Fig. 17.33 Genetics of self-incompatibility: (a) In gametophytic self-incompatibility (GSI), pollen is successful only when the genotype of the pollen does not match the genotype of the female. (b) In sporophytic self-incompatibility (SSI), the pollen grains can germinate and grow tubes only when they do not carry determinant that matches the genotype of the female (**See Plate 15**).

Sporophytic self-incompatibility in the *Brassicaceae* is mediated by receptor-like protein kinases in the female. This receptor-like kinase that contains an extracellular domain—a transmembrane domain that anchors the protein in the plasma membrane—and a cytoplasmic domain that has serine/threonine kinase SRK activity. The extracellular domain of SRK frequently shares significant amino acid identity with the corresponding SLG protein. This sequence conservation is striking, considering that the amino acid sequences of SLCs (or SRKs) from different S alleles can vary substantially.

Barriers to interspecies pollination

Some closely related plant species are interfertile, but in the vast majority of instances, pollen from one species is incompatible with stigmatic or stylar tissues from another. Very often this incompatibility is governed by non-specific morphological or physiological barriers. The molecular basis of this particular interaction is not understood, particularly since there is evidence that the system can be 'fooled': provision of compatible pollen alongside the incompatible will often lead to the acceptance of the incompatible pollen.

17.13 FORMATION OF ENDOSPERM AND EMBRYO

The seed is the characteristic product of early development in flowering plants. It is a genetically chimeric structure, a consequence of the nature of the fertilization event in angiosperms.

17.13.1 Fertilization

The plant kingdom clearly exhibits the alternation of generations: the sporophyte generation bears 'spores', which germinate to give a gametophyte generation, which bears gametes. In some groups of plants, both generations are well developed (e.g., ferns). However, the independence of the gametophyte generation has been lost with evolutionary advancement and this is seen in flowering plants, where the gametophyte generation is reduced to an embryo sac containing eight cells in the female and two haploid nuclei carrying pollen grains in the male plants. In order to achieve pollination, the pollen grain must reach the receptive surface of the stigma; successful pollination later occurs in several steps.

The first step is the adhesion of pollen grain to the stigmatic surface (papillar cell walls). Once a pollen grain has landed on the stigma of a compatible species, it leads to hydration followed by germination of the vegetative cell to form the pollen tube. The pollen tubes grow through the papillar cell walls and via the middle lamellae of the pistil, reach the micropyle of the ovule. Sucrose secreted from the epidermal cells of the stigma stimulates germination (**Fig. 17.34**). A **pollen tube** emerges from one of the pores in the wall of the pollen grain and grows rapidly down the style to the ovary. Its growth is controlled by the tube nucleus of the pollen grain, which is found at the growing tip of the tube. Although this process requires active protein synthesis, there does not appear to be a need for synthesis of any RNA, suggesting that the complete protein synthetic machinery, including mRNA, is present in the ungerminated pollen grain, having been laid down during microsporogenesis. The penetration of the pollen tube and its subsequent growth in the style means that the gametophyte is essentially acting parasitically within female sporophyte tissues, but correct intercellular recognition is essential.

Pollen tube growth is stimulated by auxins produced by the gynoecium and the pollen tube is directed towards the ovary by a certain chemical, which is an example of chemotropism. Hydrolytic enzymes are secreted by the pollen tube to digest the female tissue as the pollen tube grows. During the growth of the pollen tube the **generative nucleus** of the pollen grain divides by mitosis to produce two male nuclei that represent the **male gametes**. A growing pollen tube therefore contains three separate nuclei. They cannot swim, unlike the sperm

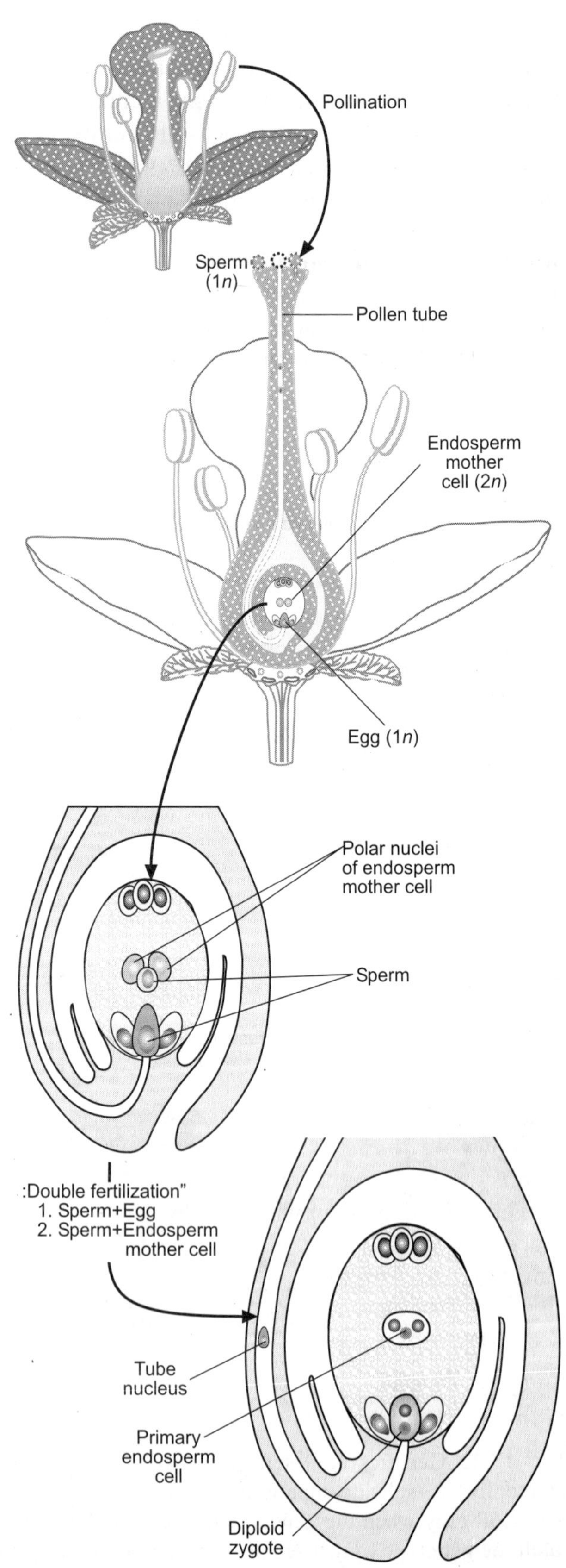

Fig. 17.34 Major events of pollination and fertilization in angiosperms.

of lower plants and depend on the pollen tube to reach the female gamete, which is located in the embryo sac of the ovule. The pollen tube does not directly reach the ovary in a straight line. It travels near the skin of the style and curls to the bottom of the ovary; near the receptacle, it breaks the ovule through the micropyle (an opening in the ovule wall). The tube nucleus degenerates and the tip of the pollen tube bursts, releasing two male gametes near the embryo sac, which they enter. This is the point when fertilization actually occurs. After being fertilized, the ovary starts to swell and become a fruit. With multiseeded fruits, multiple grains of pollen are necessary for syngamy with each ovule.

One nucleus fuses with the female gamete, forming a diploid **zygote** and the other fuses with the diploid nucleus forming a triploid nucleus known as the **primary endosperm nucleus**. This **double fertilization** is unique to flowering plants. It leads eventually to the two structures found in the seed, namely **embryo** and the **endosperm**. The zygote of the angiosperm produces only a single embryo, but the zygote of the gymnosperm produces two or more embryos after cell division begins, by a process known as cleavage embryogenesis. Recently, research has shown that in one primitive group of flowering plants, the water lilies (*Nuphar*), the endosperm is diploid, resulting from the fusion of a pollen nucleus with one, rather than two, maternal nuclei.

Apomixis

Fertilization is not an absolute prerequisite for many angiosperms to form embryos. Embryos can form within embryo sacs from haploid eggs and from cells that did not divide meiotically. This phenomenon is known as **apomixis** (Greek: without mixing) and results in viable seeds. Apomixis is a process of asexual reproduction in which no nuclear fusion takes place and occurs in place of sexual reproduction. Embryos can also develop from cultured sporophytic tissue and develop into no associated endosperm and seed coat. Maheswari (1950) distinguished four types of apomixis.

Non-recurrent apomixis: The megaspore mother cell undergoes the regular meiotic division to form the haploid embryo sac. The new embryo develops from the egg cell, i.e., haploid parthenogenesis, or from any other cell of the embryo sac, i.e., haploid apogamy. As plants thus formed contain only one set of chromosomes, they are usually sterile and so the process is not repeated in the next generation.

Recurrent apomixis: The embryo sac may develop from a sporogenous cell—generative apospory—or from other cells of the nucellus-somatic apospory. In this case, all the cells are diploid and the embryo may develop from the egg cell (diploid parthenogenesis) or from any other cell of the embryo sac (diploid apogamy).

Adventive embryony: The embryo does not develop from the cells of the embryo sac but from a cell of the nucelus or the integuments. This type of development is also known as sporophytic budding.

Vegetative reproduction: In this type of development the flowers, or parts of them, are replaced by bulbils or other vegetative propagules, which may germinate while still on the plant.

Polyembryony

It refers to the appearance of two or more embryos in a single seed. The process of apomixis is often accompanied by the formation of a few embryos from the same ovule. Sometimes a normal embryo may develop together with those produced by apomixis. In a single ovule of certain species of Citrus, for example, apomixial 3–12 adventitious embryos which develop from cells of the nucellus may develop alongside the normal embryo. Numerous embryos may develop as a result of the formation of growth on the multicellular body derived from the zygote and change, during the course of development, into independent embryos. Sometimes, additional embryos may be formed by the synergids or the antipodal cells that may be fertilized by male gametes brought into the embryo sac by additional pollen tubes, or even without fertilization. Pseudopolyembryony may also result when two or more nuclei, each of which gives rise to a normal embryo, fuse.

17.13.2 Development of the endosperm

In many angiosperms, the endosperm is a storage tissue that stores reserves temporarily until the developing embryo absorbs them into its own cotyledon storage tissues. In some seeds where the endosperm persists (e.g., lettuce, tomato) it may be only one to a few cells thick and may play a minor storage role compared with that of the embryo it surrounds. In other seeds, the endosperm may act as the major storage tissue, with the embryo containing few or different reserves. In castor bean, for example, the relatively massive endosperm consists of living cells that store oil and protein and produce the enzymes necessary for the mobilization of these reserves. In the endosperms of cereal and some endospermic legumes, most of the endosperm cells are nonliving at maturity. The storage reserves they contain are hydrolyzed by enzymes synthesized in and released from a surrounding thin layer of living aleurone. The endosperm develops from mitotic division of the **endosperm nucleus**. This division usually precedes that of the zygote. Further development of the endosperm differs in the various groups of plants and the following main types can be distinguished according to the manner of development: nuclear endosperm, cellular endosperm and helobial endosperm.

Deposition of stored reserves during seed development

There is no direct vascular connection between the parent plant and the embryo or endosperm. The vascular tissue ends in the seed coat or fruit coat of dicots or in the placenta–chalazal region of cereals. Assimilates are transported from the parental seed coat cells to the extracellular space (apoplast) by a passive, facilitated membrane-transport process and are subsequently taken up by the filial tissues of the endosperm or embryo. During early development, sucrose is generated by the parent plant, released from the phloem into the apoplast and hydrolyzed by invertases in the cell wall. The hexose products are loaded into the filial storage tissues. The amount of hexose made available to these tissues before synthesis of the reserves can influence embryogenesis. Seeds with high active seed coat invertase and the high concentrations of hexose promoters are associated with extended mitotic activity of the developing cotyledons. During the early stages of development, sucrose may be resynthesized from hexoses imported into the storage tissues by the sucrose–phosphate synthesis pathway. At later stages of seed development, during deposition of reserves, the ratio of sucrose to hexose entering the storage tissues increases. The major forms of nitrogen translocated into the seed in the phloem are asparagine and glutamine, although in some legumes (e.g., soybean and cowpea) about 10%–15% of nitrogen translocated into seeds is ureide-*N* (e.g., allantoin and allantoic acid). These compounds are converted to asparagine and glutamine by enzymes in the seed coat. The products of seed coat asparaginase and aspartate aminotransferase are imported into the embryo or endosperm and serve as substrates for syntehsis and interconversion of amino acids. Later, during development, aspargine is taken up directly into the cotyledons, where it is deaminated by a cotyledon-localized iso-enzyme of asparginase. Glutamine synthetase and glutamate synthase also become more active in cotyledons as development progresses.

The dry weight of mature seeds of most species is accounted for by two or three major storage reserves—carbohydrates, triacylglycerols and proteins (**Table 17.4**). The factors that determine which reserves are synthesized and stored have not been determined, although tissue-specific expression of genes that encodes biosynthetic enzymes certainly plays a key role. Within the seed, incoming assimilates must be allocated to various biosynthetic pathways to ensure that reserve compounds are produced in precise quantities and ratios.

Nectaries

Nectar, a sugar-containing solution, secreted by nectaries, most frequently occurs on insect- and bird-pollinated plants. Nectaries may consist of specialized tissue, which differs histologically from the

Table 17.4 Storage reserves of some important crop species

Crop species	Protein %	Oil %	Carbohydrate %	Storage
Cereals				
Barley	12	3	76	Endosperm
Maize	10	5	80	Endosperm
Oats	13	8	66	Endosperm
Wheat	12	2	75	Endosperm
Legumes				
Broad bean	23	1	56	Cotyledon
Garden pea	25	6	52	Cotyledon
Peanut	31	48	12	Cotyledon
Soybean	37	22	12	Cotyledon
Others				
Castor bean	18	64	Negligible	Endosperm
Oil palm	9	49	28	Endosperm
Pine	35	48	6	Megagame-tophyte
Rapeseed	21	48	19	Cotyledon

neighbouring tissues, i.e., structural nectaries, or non-structural nectaries. They may develop on all parts of the plant. Nectaries that are connected with the floral organs are termed floral nectaries and those developing on the vegetative parts of the plant as extrafloral nectaries. The following classification of the floral nectaries, according to their location, has been suggested.

- ***Perigonial nectaries*** develop on the perianth.
- ***Toral nectaries*** develop on the receptacle.
- ***Staminal nectaries*** are related to stamens.
- ***Ovarial nectaries*** develop on the ovary wall.
- ***Stylar nectaries*** develop at the base of the style.

Sucrose, glucose and fructose have been the most common constituents of nectar. Apart from these, mucilages, proteins and organic acids are also found. The cells constituting the nectar form the nectariferous tissue. The nectariferous tissue usually contains branches of vascular bundles. The phenomenon is consistent with the hypothesis that the secreted sugar is brought to the nectary by the vascular bundles. Nectar secretion (the process of release of sugar solution from protoplast to the apoplast) may occur in all the cells constituting the nectary, these are called secretory cells.

17.13.3 Development of seed

The embryo arises by mitosis from the fertilized zygote. Although the time taken from fertilization for the formation of a mature, dry seed may be between 40 and 60 days, the time taken for an embryo to become a fully differentiated, functionally mature structure is very short. Its development is the reverse of the reproductive phase, where reduction to a single cell is the theme. The acquisition of functional maturity can be recognized by the ability of an immature embryo, removed from the confines of the developing seed, to undergo premature or **precocious germination** when supplied with the most basic nutrients. For example, immature wheat embryos will germinate, in culture, between 10 and 14 days following fertilization, if removed from the grain and provided with a simple medium containing sucrose, amino acids, salts and vitamins. Such precocious germination does not occur while the grain is attached to the parent plant. The ability of the embryo to germinate within the grain is acquired only following dehydration of the grain. That the differentiation of the embryo and the acquisition of germinability is attained far in advance of the achievement of the maximum size and dry weight of the whole seed is due in large measure to the necessity to accumulate storage reserves (triglycerides, carbohydrates and storage proteins).

Dormancy in immature seeds may be imposed through the action of the plant growth regulator, abscisic acid (ABA). Demonstrating that ABA plays a central role in the maintenance of dormancy does not, however, indicate the regulatory pathway by which ABA acts. In pursuit of this end, experiments that identify gene products which are synthesized in isolated immature embryos, such as the mRNA and polypeptide products of genes, which express naturally towards the end of the period of embryonic dormancy has been extensively studied. These gene products accumulate to high levels in dry embryos and in consequence of this and their temporal appearance during the later phases of embryogenesis, have been designated 'late-embryogenesis-abundant' or lea gene products.

Immediately after fertilization, the ovule becomes known as the seed and the ovary, the fruit. The following changes take place. It will be useful to study the formation of seeds in order to understand these changes (**Fig. 17.35**).

- The zygote grows by mitotic divisions to become a multicellular embryo, which consists of a first shoot, the ***plumule***, a first root, the ***radicle*** and either one or two seed leaves called ***cotyledons***. These cotyledons are simpler in structure than the first true foliage leaves and may become swollen with food to act as storage tissue, as in the pea and broad bean. The plumule consists of a stem, the first pair of true foliage leaves and a terminal bud.
- The triploid primary endosperm nucleus undergoes repeated mitotic divisions to form the ***endosperm***, a mass of triploid nuclei, which are separated from one another by thin cell walls. In some seeds this remains the food store, as in cereals such as wheat and maize.
- If the cotyledons act as a food store they grow at the expense of the endosperm, which may disappear altogether. Some seeds store food in both endosperm and cotyledons. In legumes for example, the food is stored chiefly in the cotyledons and there is ***no endosperm*** (such seeds are called non-endospermic or ***exalbuminous***). In some dicotyledons and monocotyledons, the food is ***stored in the endosperm*** (such seeds are known as endospermic or ***albuminous***).
- As growth of the embryo continues, the surrounding nucellus breaks down and stops

supplying nutrients for growth. Further nutrients are supplied by the vascular bundle in the stalk of the ovule. Food materials stored in the seeds include carbohydrates, proteins and fats. Different types of seeds accumulate different food materials.

- The testa develops from the integument. It is a thin but tough protective layer.
- The micropyle remains a small pore in the testa through which oxygen and water enter when the seed germinates.
- The final stages in development of the seed involves a reduction in the water content of the seed from the normal levels in plant tissues of about 90% by mass to about 10%–15% by mass. This greatly reduces the potential for metabolic activity and is an essential step in ensuring seed dormancy.
- While the seeds develop, the ovary becomes a mature fruit and its wall is called as ***pericarp***. The changes that occur vary with species, but the fruit is adapted to protect the seeds and to help in their dispersal.
- The remaining flower parts wither and die or are lost.

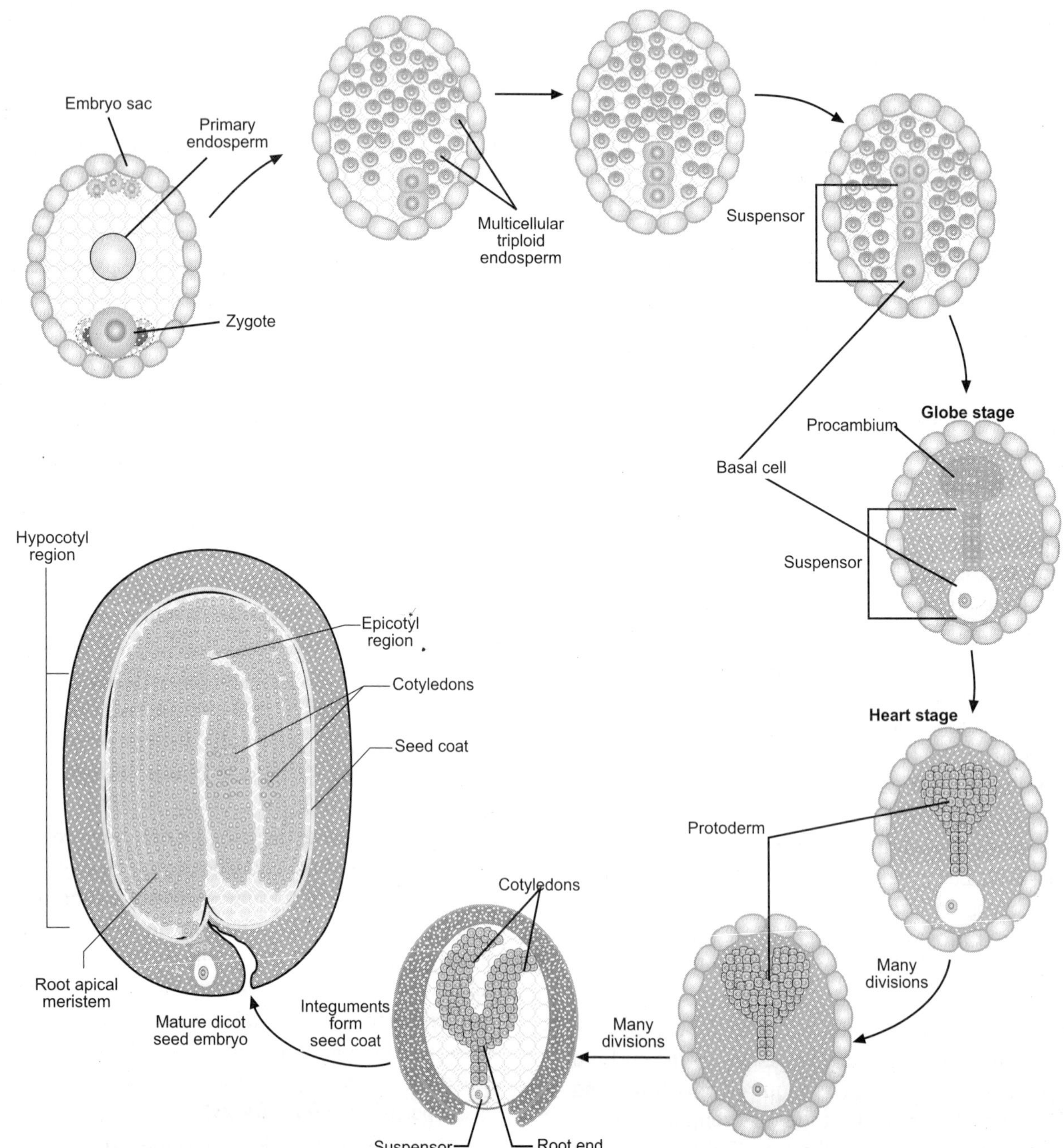

Fig. 17.35 Seed development in a typical dicot plant.

17.13.4 Development of fruit and ripening

Generally, the ovary undergoes two important changes immediately after fertilization. The ovules develop into seeds and the wall of the ovary thickens and ripens into pericarp (fruit wall). These changes of ovary together called **true fruit** (i.e., development from the gynoecium). However, in development of many fruits, other organs also participate. Such organs may be the tepals (Morus), the receptacle (Fragria), bracts (Ananas), the floral tube, which is formed by the floral organs together with the receptacle (*Pyrus malus*), or the enlarged axis of the inflorescence (Ficus). In cases where organs other than the gynoecium participate in the formation of the fruit, the fruit is termed as **false fruit** or **pseudocarp**.

The development of fruit in most species depends on successful fertilization and thus proceeds in step with the development of the enclosed seeds. For fleshy fruits and other fruit-like structures, an initial phase of cell division is followed by a phase of massive cell expansion, leading to the formation of very large vacuolate cells (up to 700 μm in length). This cell expansion phase naturally enough leads to fruit enlargement, for example, the volume of a mature watermelon may be as much as 30 litres.

There are different methods of fruit classification. The main criterion is the degree of hardness of the pericarp, i.e., the fruit wall—whether it is dry and hard or soft and fleshy or juicy. The second criteria is the ability of the fruit to dehisce or not when ripe.

Climacteric and non-climacteric fruit

Following cell expansion, fruit ripening occurs. For the cereal grain, this process parallels very closely the ripening of the enclosed seed, with the pericarp behaving like an extra seed coat and undergoing desiccation in parallel with the seed coat. For fleshy fruit, by contrast, the ripening of the fruit involves changes in colour, texture, flavour and aroma, these changes being associated with the attraction of suitable 'consumers' in order to facilitate the dispersal of the enclosed seeds. Such fruits have, on the basis particularly of their respiratory metabolism during ripening, been classified as either climacteric or non-climacteric. In the former, the rate of respiration falls as the expansion/maturation phase finishes and then rises significantly during ripening before finally declining again. There is also a massive increase in the rate of synthesis of the gaseous hormone, ethylene. In the latter, there is no clear 'burst' of respiration during ripening and ethylene production is very limited.

Exercises

1. Explain briefly the process of plant development and growth.
2. Explain with a diagram the mechanism of plant cell differentiation.
3. What is morphogenesis? How can it be induced?
4. Explain the different phases of the life cycle of a flowering plant.
5. Discuss the process of plant embryogenesis.
6. Explain the different stages of seed germination.
7. How is the development of plant meristem regulated?
8. Explain the formation of shoot and root meristem with a diagram.
9. What is apical dominancc?
10. Discuss in detail the transition of plants from vegetative to reproductive tissue.
11. Explain briefly: (a) Specification of floral organs and (b) Photoperiodism
12. What are the different factors that regulate flowering time under short- and long-day plants?
14. Discuss the process of development of male and female gametophytes.
15. What is the importance of pollination in plants?
16. How does seed development take place in plants?

Summary

- For a better understanding of the process of plant growth and development, it is very useful to separate the developmental events into several components: ***growth, differentiation*** and ***morphogenesis***.
- In plants, the term ***embryogenesis*** covers development of embryo from the time of fertilization until dormancy occurs. The basic body plan of the sporophyte is established during embryogenesis.
- The major changes of embryogenesis are: establishment of body plan, formation of separate meristematic tissues and nutrient source.
- Plants do not have a preprogrammed body plan like animal embryonic cells. However, the plants retain areas where rapidly dividing, undifferentiated cells remain all through the life of the plant (called meristems).
- The parts of the adult plant are created sequentially by two groups of cells called as ***shoot*** and ***root meristems***, which contain a population of self-renewing ***stem cells***.
- In roots, radial and axial patterning started during embryogenesis continues throughout development as the primary root grows and lateral roots emerge from the pericycle cells deep within the root.
- During vegetative growth, the ***shoot apical meristem*** (SAM) gives rise to a series of ***primordia*** that form lateral organs.
- The SAM gives rise to three primary meristems: ***protoderm***, ***ground meristem*** and ***procambium***. The apical meristem is surrounded by the developing leaf primordia.
- The leaf matures from the tip to the base: first, the cells at the tip stop dividing, then, the cells at the base also stop.
- Flowering plants go through a phase of vegetative growth—producing more stems and leaves—and a flowering phase where they produce the organs for sexual reproduction.
- The germ line is established only after transition from vegetative to reproductive development, i.e., ***flowering***.
- The flower provides a major ***sink for photosynthetic assimilates*** in the plant.
- Flowers comprise four concentric whorls of floral organs and an important phase of flower development concerns the specification of organ identities.
- The ***ABC model*** postulates a group of genes that encode the transcription factors needed to run on the genes for sepal, petal, stamens and carpel development. The 'master switches' fall into 3 groups: ***A, B*** and ***C***.
- The first stage to be controlled in floral induction is its timing. That is, flower formation usually occurs at a specific time in the life cycle of a plant affected by environmental cues.
- Flower differentiation consists of three major phases: ***induction of flowering, flower initiation phase*** and ***flower differentiation***.
- As with animals, the formation of gametes in plants begins with ***meiosis***.
- Pollination refers to the landing of pollen grain on ***papillar*** *cells* on the surface of the stigma and subsequent germination.
- There are mechanisms which prevent self-fertilization in obligatory out breeding species.
- ***Self-incompatibility*** is an example of intra-specific incompatibility.
- Generally, the ovary undergoes two important changes immediately after fertilization. The ovules develop into seeds and the wall of the ovary thickens and ripens into pericarp.
- The development of fruit in most species depends on successful fertilization and thus proceeds in step with the development of the enclosed seeds.

Further Reading

Books

Gilbert SF (2000), *Developmental Biology*, 6th edition, Sinauer Associates Inc., Publishers. USA.

Howell SH (1998), *Molecular Genetics of Plant Development*, Cambridge University Press, Cambridge, UK

Jordan BR, (ed.), (1993), *The Molecular Biology of Flowering*, CAB International, UK.

Peter Westhoff (1998), *Molecular Plant Development: From Gene to Plant*, Oxford University Press, Oxford, New York, Tokyo.

Slack JMW (1991), *From Egg to Embryo,* 2nd edition, Cambridge University Press, Cambridge, UK.

Starr C and Taggart R (1995), *Biology the Unity and Diversity of Life*, Wadsworth Publishing Company.

Taylor DJ, Green NPO and Stout GW (1997), In Soper S, (ed.), *Biological Science,* Cambridge University Press, Cambridge.

Twyman RM (2001), *Instant Notes Developmental Biology*, Viva Books Private Limited, India.

Westhoff P, Jeske H, Jurgens G, Kloppstech K and Link G (1998), *Molecular Plant Development from Gene to Plant*, Oxford University Press, Oxford.

Wolpert L (1998), *Principles of Development*, Oxford Univeristy Press, Oxford.

References

Barton MK and Poethig RS (1993), Formation of the shoot apical meristem in *Arabidopsis thaliana*: an analysis of development in the wild type and in the shoot meristemless mutant, *Development,* 119:823–31.

Bowman JL and Eshed Y (2000), Formation and maintenance of the shoot apical meristem, *Trends Plant Sci.*, 5(3):110–15.

Callos JD and Medford JI (1994), Organ positions and pattern formation in the shoot apex, *Plant Journal*, 6:1–8.

Caryl AP, Jones GH and Franklin FCH (2003), Dissecting plant meiosis using *Arabidopsis thaliana* mutants, *J.Exper.Bot.*, 54:25–38.

Channarayappa (1993), Resistance mechanism of *Lycopersicon* species to the vector *Bemisia tabaci* and Ultrastructural changes in leaf curl virus infected tomato leaves, PhD Thesis submitted to the University of Agricultural Sciences, Bangalore.

Channarayappa (2006), *Molecular Biotechnology: Principles and Practices*, Universities Press (India) Private Limited.

Clark SE (1997), Organ formation at the vegetative shoot meristem, *The Plant Cell*, 9:1067–76.

Coen ES (1991), The role of homoeotic genes in floral development and evolution, *Annu. Rev. Plant Physiol. Plant Mol. Biol.*, 42:241–79.

Coen ES and Meyerowitz EM (1991), The war of the whorls: Genetic interactions controlling flower development, *Nature,* 353:31–7.

Colasanti J and Sundaresan V (2000), 'Florigen' enters the molecular age: long-distance signals that cause plants to flower, *Trends Biochem. Sci.*, 25(5):236–40.

Dickinson HG (1990), Self-incompatibility in flowering plants, *Bio Essays* 12:155–61.

Goldberg RB, Beals TP and Sanders PM (1993), Anther development: basic principles and practical applications, *The Plant Cell*, 5:1217–29.

Jackson D (1996), Plant morphogenesis: Designing leaves, *Curr. Biol.*, 6:917–19.

Jurgens G, Torres Ruiz RA and Berleth T (1994), Embryonic pattern formation in flowering plants, *Annu. Rev. Genet.*, 28:351–71.

Kerstetter RA and Hake S (1997), Shoot meristem formation in vegetative development, *The Plant Cell*, 9:1001–10.

Larkin JC, Marks MD, Nadeau J and Sack F (1997), Epidermal cell fate and patterning in the leaves, *The Plant Cell*, 9:1109–1020.

Laux T and Jurgens G (1997), Embryogenesis—a new start in life. *The Plant Cell*, 9:989–1000.

Levy YY and Dean C (1998), The transition to flowering, *The Plant Cell*, 10:1973–89.

Ma H (1997), The on and off of floral regulatory genes, *Cell,* 89:821–24.

Malamy JE and Benfey PN (1997), Organization and cell differentiation in lateral roots of *Arabidopsis thaliana*, *Development,* 124:33–44.

McCormick S (2004), Control of male gametophyte development, *The Plant Cell*, 16:S142–S153.

Medford JI (1992), Vegetative apical meristems, *The Plant Cell*, 4:1029–39.

Nasrallah JB, Nishio T and Nasrallah ME (1991), The self-incompatibility genes of *Brassica*, *Annu. Rev. Plant Physiol. Plant Mol. Biol.*, 42:393–442.

Nonomura K, Miyoshi K, Eiguchi M, Suzuki T, Miyao A, Hirochika H and Kurata N (2003), The *mspi* gene is necessary to restrict the number of cells entering into male and female sporogenesis and to initiate anther wall formation in rice, *The Plant Cell*, 15:1728–39.

Poethig RS (1997), Leaf morphogenesis in flowering plants, *The Plant Cell*, 9:1077–87.

Ray A (1998), New paradigms in plant embryogenesis: Maternal control comes in different favours, *Trends Plant Sci.*, 3:325–18.

Reeves PH and Coupland G (2000), Response of plant development to environment: control of lwering by day length and temperature, *Curr. Opin. Plant Biol.*, 3(1):37–42.

Reiser L and Fischer RL (1993), The ovule and the embryo sac, *The Plant Cell,* 5:1291–1301.

Schoof H, Lenhard M, Haecker A et al. (2000), The stem cell population of *Arabidopsis* shoot meristems is maintained by a regulatory loop between the *clavata* and *wuschel* genes, *Cell,* 100:636–44.

Sinha N (1999), Leaf development in angiosperms, *Annu. Rev. Plant Physiol. Plant Mol. Biol.*, 50:419–446.

Taylor CB (1997), Plant vegetative development: From seed and embryo to shoot and root. *The Plant Cell,* 9:981–88.

Taylor LP and Hepler PK (1997), Pollen germination and tube growth, *Annu. Rev. Plant Physiol. Plant Mol. Biol.*, 48:461–91.

Yanofsky MF (1995), Floral meristems to floral organs: genes controlling early events in *Arabidopsis* flower development, *Annu. Rev. Plant Physiol. Plant Mol. Biol.*, 46:167–88.

18 IMMUNE RESPONSE

Learning objectives

After reading this chapter, you will be able to:

- Identify different methods of immune response
- List disorders of immunity
- Appreciate the latest developments in immune modulation, especially molecular and genetic engineering methods
- Understand the benefits of immunologic therapy in various diseases

18.1 INTRODUCTION

Most organisms live in groups, share a common habitat and compete for resources. Not all organisms are autotrophs (use CO_2 as their carbon source by the process of photosynthesis). Non-photosynthetic organisms are heterotrophs (use organic carbon as their carbon source). Therefore, it is not surprising that every organism has many parasites. Animals are subject to infection by viruses, bacteria, protists, fungi and animal parasites. During evolution, most organisms developed several mechanisms that allow them to recognize and destroy these infectious agents or inhibit their invasion. As a result, vertebrates are able to develop various kinds of resistance mechanisms against invading pathogens.

18.2 THE IMMUNE SYSTEM

Single-cell organisms protect themselves by a simple membrane barrier around them. They can distinguish themselves from the environment since the contents inside the membrane are own-self and outside are non-self. In contrast, for multicellular organisms, distinguishing self from non-self is but complicated, essential for their survival. Multicellular organisms consist of hundreds of distinct cell types that differ morphologically and functionally. They are very resourceful and provide a hospitable environment for parasitic invaders. Therefore, the evolution of a sophisticated mechanism that distinguishes self from non-self is required.

The agents that threaten the body of vertebrates have incredible diversity and generally include chemical and infectious agents (bacteria, viruses, parasites and against introduction of transfused cells, blood cells foreign cells greft tissues of different people or their own cancer cells). Astonishingly, vertebrates are often able to defend themselves against biotic and abiotic factors though they have not encountered them before. This implies that vertebrates have developed a complex but highly efficient defence system (immune system) to guard against a myriad agents. How does the immune system work? The key is to produce more than 10^8 distinct antibodies and 10^{12} T cell receptors, each of which presents a different surface for specifically binding a molecule from a foreign organism and initiating the destruction of the invader. However, the presence of this remarkable repertoire of defensive molecules poses one important challenge. How does this immune system avoid attacking its own molecules or cells? That is, distinguishing between self and non-self.

The coordinated reaction of these cells and molecules to infectious agents is the **immune response**. The study of the immune system and its response to invading agents is called **immunology**. The overall defence mechanism of an individual organism is termed **immunity**. Immunity was first, defined by Sir Macfarlane Burnet as "*the capacity to recognize the intrusion of material foreign to the body and to mobilize cells and cell products to help to remove that particular sort of foreign material with great speed and effectiveness.*" In short, it is a resistance against invaders. The immune system responds to macromolecules of several types and proteins are the most important class. The immune response itself is a complex interaction of various cell types, signalling pathways, and other components. The immune system of a mammal can destroy millions of different antigens without harming its own cells.

Key players in immune response

Three different types of white blood cells play a central role in immune response in vertebrates (**Fig. 18.1**). These cells are:

- B lymphocytes
- T lymphocytes
- Macrophages

The immune response may be **humoral** (antibody-mediated response) leading to antibody formation, or it may be **cellular** (cell-mediated immune response).

In this chapter, the main emphasis is on vertebrate defence systems, structures of proteins participating in the molecular recognition processes and on mechanisms for selecting cells that express molecules useful for protecting us from a specific pathogen.

18.3 THREE LINES OF DEFENCE

All vertebrate bodies have physical, chemical and cellular defences against invasion by viruses, bacteria and other agents of different diseases. The overall defence mechanism of vertebrates has been broadly divided into three lines of defence against pathogens.

18.3.1 Surface barriers to invasion

The outer surface of the body and the linings of its internal tracts provide an excellent barrier to penetration by viruses, bacteria, and parasites. If these surface barriers are breached, a series of immune responses are initiated that contain the invasion.

Intact skin: Most often, pathogens cannot get past the skin or linings of other surfaces. The outer layer of skin has low moisture, low pH, and a thick layer of dead cells. Generally, non-pathogenic bacteria are adapted to this habitat and provide protection against pathogenic bacteria by competing with them for space and food.

Mucous membranes at body surfaces: The gut's mucous lining and its resident bacteria keep pathogens in check. Barriers in the tubular airways leading to the lungs stop airborne pathogens. As air rushes past the turns and branches of the mucus-coated tube walls, it becomes turbulent and flings bacteria against the sticky sides. The mucous contains protective substances, for example, lysozyme, and an enzyme that digests the cell wall of bacteria and so contributes to their death.

Infection-fighting substances in tears, saliva and body fluids: Lysozymes and other enzymes in tears, saliva and fluid in the gut provide more protection. For example, tears provides eyes with sterile washing. Urine, with its low pH and flushing action, helps keep invaders from moving into the urinary tract.

18.3.2 Non-specific response

Inflammation: All animals react to tissue damage. An inflammatory response (redness, warmth, swelling and pain) develops when a tissue's cells are damaged or killed, as happens during an infection. Small blood vessels in the tissue dilate. As the cells in their walls pull apart slightly, the blood vessels become leaky to plasma proteins and white blood cells (**Fig. 18.1**). Neutrophils (phagocytize bacteria), eosinophils (secrete lytic enzymes), basophils (secrete histamine and other substances to induce inflammation) and macrophages (engulf and digest pathogens) leave the bloodstream. They defend and help repair a besieged tissue. Histamine, complement proteins, interleukins (communication signals), blood clotting proteins, and other chemicals take part in the inflammatory response, or in the mopping up and repair operations.

Organs with phagocytic functions: Peripheral lymphoid organs

The peripheral lymphoid organs, which consist of the lymph nodes, the spleen, and the mucosal and cutaneous immune systems, facilitate concentration of antigens, APCs and lymphocytes interact efficiently with each other and maximize to establish adaptive immune response. Since the T and B lymphocyte number is very low (about 1 in 1,000,000 cells), the anatomic organization of peripheral lymphoid organs enables lymphocytes in these organs to locate and respond to microbes. This organization is complemented by lymphocytes to circulate throughout the body in such a way that naïve

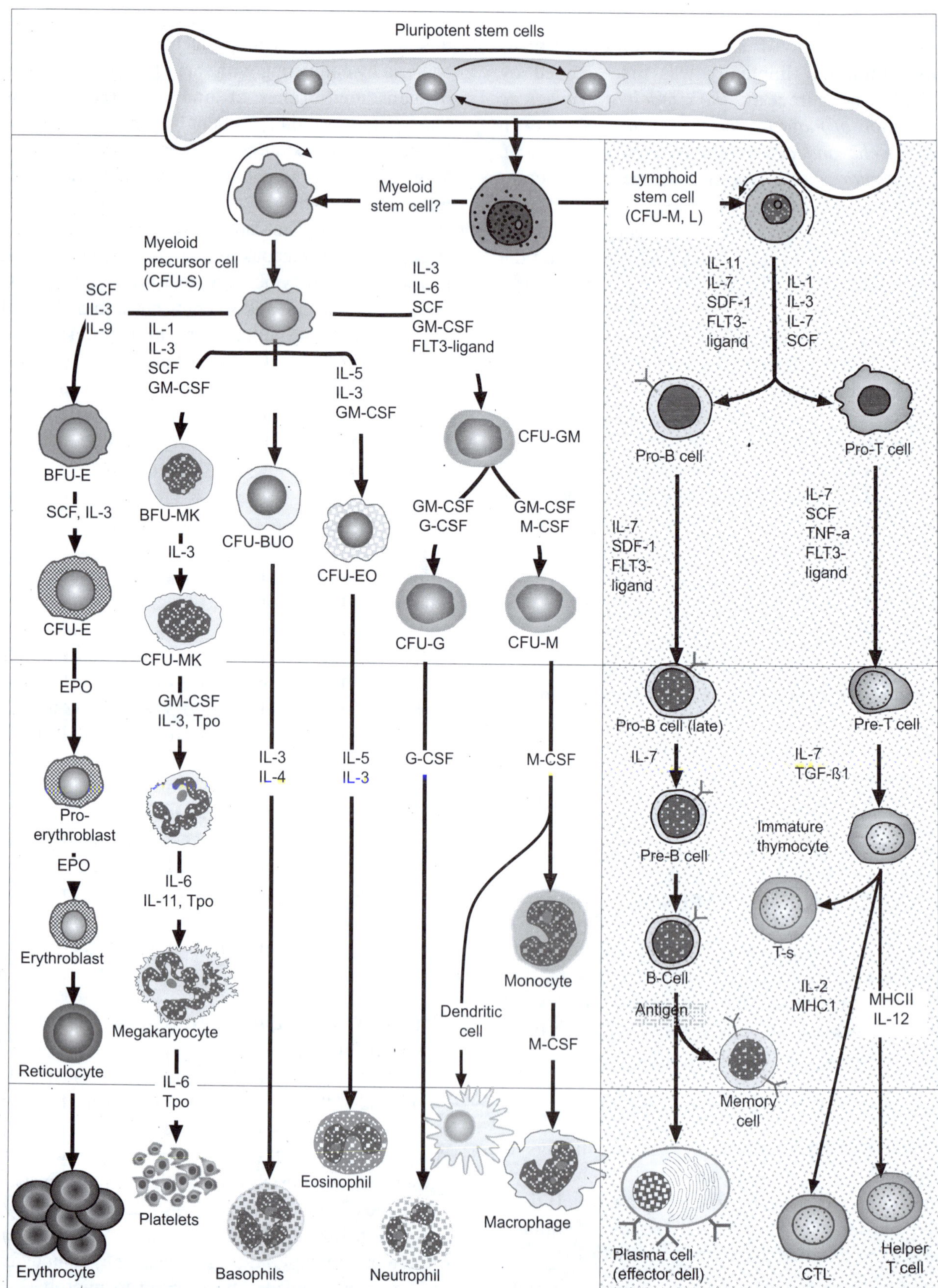

Fig. 18.1 A model for the origin of mammalian blood and lymphoid cells. All the cells in the bottom row are terminally differentiated blood cells.

Table 18.1 Major white blood cells and their roles in defence

Cell type	Main characteristics
Macrophage	Phagocyte: takes part in non-specific defence responses; presents antigen to T cells; cleans up and helps repair tissue damage
Neutrophil	Fast-acting phagocyte: takes part in inflammation but not in sustained responses
Eosinophil	Secretes enzymes that attack parasitic worms
Basophil and mast cell	Secrete histamines, prostaglandins, other substances that act on small blood vessels, producing inflammation; have a role in allergies
Lymphocytes	Immune response; following antigen recognition, all activated lymphocytes form clonal populations of effectors cells and memory cells.
B cell	Effectors secrete four classes of antibodies (IgA, IgE, IgG and IgM) that protect the host in specialized ways
Helper T cell	Effectors secrete interleukins that stimulate rapid division and differentiation of both B and T cells
Cytotoxic T cell	Effectors kill infected cells, tumour cells and foreign cells by way of a lethal hit
Natural killer (NK) cell	Cytotoxic cell of undetermined affiliation (may be a kind of lymphocyte); kills infected cells and tumour cells by way of a lethal hit.

lymphocytes preferentially to the specialized organs in which antigen is concentrated and effector cells go to sites of infection, from where microbes have to be eliminated. It is also essential that different types of lymphocytes often need to communicate to generate effective immune response. Thus, lymphoid organs play a very important role in bringing these rare cells together to have effective interactions.

Lymph nodes are nodular aggregates of lymphoid tissues located along the lymphatic channels throughout the body. Fluid from all epithelia and connective tissues and most parenchymal organs is drained by lymphatics and this fluid is called **lymph**, which will be transported from the tissues to the lymph nodes. As the lymph passes through the lymph nodes, lymphatic cells such as APCs in the nodes are able to sample the antigens of microbes that may enter through epithelia into tissues. Dendritic cells pick up antigens of microbes from epithelia and transport these antigens to the lymph nodes, where antigen will be presented to T cells.

The **spleen** is an abdominal organ. Its immune responses to blood-borne antigens role in is the same as that of lymph nodes in response to lymph-borne antigens. Blood entering the spleen flows through a network of channels (sinusods). In spleen, blood-borne antigens are trapped and concentrated by dendrites and macrophages. The densely resided phagocytes in the spleen ingest and destroy microbes that are circulated through the blood.

The **cutaneous** and **mucosal lymphoid** systems are located under the epithelia of the skin and the gastrointestinal and respiratory tracts, respectively. Pharyngeal tonsils and Peyer's patches of the intestine are two mucosal lymphoid tissues, in which immune response to antigens that breach epithelia take place.

Within the peripheral lymphoid organs, B and T cells are localized into different compartments. In lymph nodes, the B cells are concentrated in discrete structures, called follicles, located around the periphery, or cortex, of each node. On the other hand, T lymphocytes are concentrated outside, but adjacent to, the follicles, in the paracortex. The follicles contain a special dendritic cell called the follicular dendritic cells (FDCs) that are involved in the activation of B cells and paracortex dendritic cells present antigens to T lymphocytes. In the spleen, T lymphocytes are concentrated in periparteriolar lymphoid sheaths surrounding small arterioles, and B cells reside in the follicles. The anatomic organization of peripheral lymphoid organs creates a highly favourable environment for immune responses to develop.

Lymphocyte re-circulation

Lymphocytes constantly circulate lymphatic tissues in such a way that naive lymphocytes traverse the peripheral lymphoid organs, where encounter antigens and immune responses are initiated. The effector lymphocytes migrate to sites of infection, where infectious microbes are eliminated. Thus, lymphocytes at distinct stages of their lifecycles migrate to the different sites where they are needed for their functions. However, the sites of action vary to some extent between B and T cells. Since the effector T cells have to locate and eliminate microbes at any site of infection, they have to migrate out of the lymphoid system. In contrast, effector B lymphocytes remain in the lymphoid organs, and do not need to migrate to sites of infection. Upon activation, B cells secrete antibodies, and the antibodies enter the blood and find microbes and microbial toxins in the circulation or distance tissues.

18.3.3 Specific response (specific targets)

When physical barriers and inflammation are not enough to overwhelm an invader, an infection may become well established. The two major components of the immune system are the B and T lymphocytes. Hundreds and thousands of white blood cells (B and T lymphocytes) are formed and initiate war against the invaders. Lymphocytes are central to the body's third line of defence—the immune system. The operating principle for this immune system is, each kind of infectious agent (cell, virus or substance) bears unique molecular configurations that give it a unique identity. Individual lymphocytes are capable of distinguishing molecular 'self' from 'non-self' and then destroying those entities identified as non-self (foreign). In this manner, the immune system eliminates viruses, bacteria, and other pathogens and molecules that may pose a threat to the organism. It combats tumours and neoplastic cells and can reject transfused cells and transplanted organs from genetically non-identical animals.

18.4 NATURAL AND ACQUIRED IMMUNITY

Immune response can be classified as either natural (innate or native or inherited) immunity or adaptive (acquired or specific) immunity (**Fig. 18.2**) based on the ability of the body to distinguish between self and non-self.

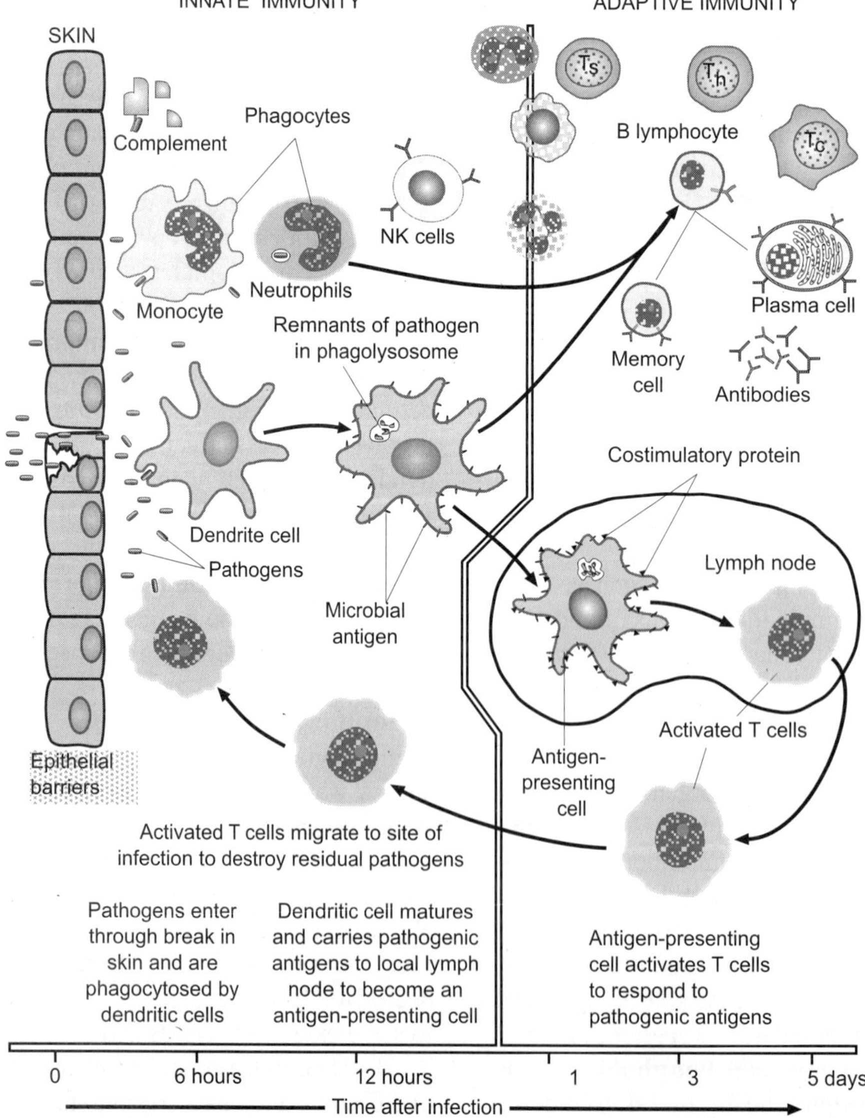

Fig. 18.2 Principle mechanisms of innate and acquired (adaptive) immunity. Innate immunity provides initial defence against infections. Innate immune response stimulates the adaptive immune system.

18.4.1 Natural immunity

All multicellular organisms, including, plants, invertebrates, and vertebrates possess various built-in intrinsic mechanisms for defending themselves against viral and microbial infections. Because these defence mechanisms are always present, and are ready to recognize and destroy the infectious agents, they are said to constitute innate or natural immunity (**Fig. 18.3**). The components of natural immunity make up the innate immune system. For many years it was thought that natural immunity is non-specific and weak. Now we know that, in fact, natural immunity specifically targets microbes and is a powerful early defence mechanism capable of mounting offense against the microbes and even eradicating infections before adaptive immunity gets activated. It not only provides early defence against infections but also instructs the adaptive immune system to respond to different microbes in a more effective manner. Conversely, the adaptive immune response often uses mechanisms of innate immunity to eradicate infections. Thus, there will always be a constant bi-directional crosstalk between natural and adaptive immunity.

Natural immunity of an animal does not require prior exposure to the foreign substance and a pre-formed defence system. The first line of defence in innate immunity is provided by specialized cells and natural antibiotics present in epithelia, which simply block the entry of invaders. If the invaders are successful in crossing the epithelial barrier and entering the tissues or circulation, they are attacked by cells of the monocytic lineage, polymorphonuclear leukocytes, and several plasma proteins—including proteins of the complement system. Natural immunity is relatively non-specific, although it normally distinguishes between self and non-self. It constitutes a potent, rapid-reacting, first-line defence against invasion and unwanted infection.

Recognition of microbes

The specificity of cells involved in natural immunity is very different in several respects from the specificity of cells of adaptive immunity. The components of innate immunity recognize structures that are shared by various classes of microbes and are not present on host cells. Each component of innate immunity can recognize multiple classes of pathogens (bacteria, viruses or fungi). The microbial molecules that are the targets of innate immunity are also called **molecular patterns** and the receptors of innate immunity that

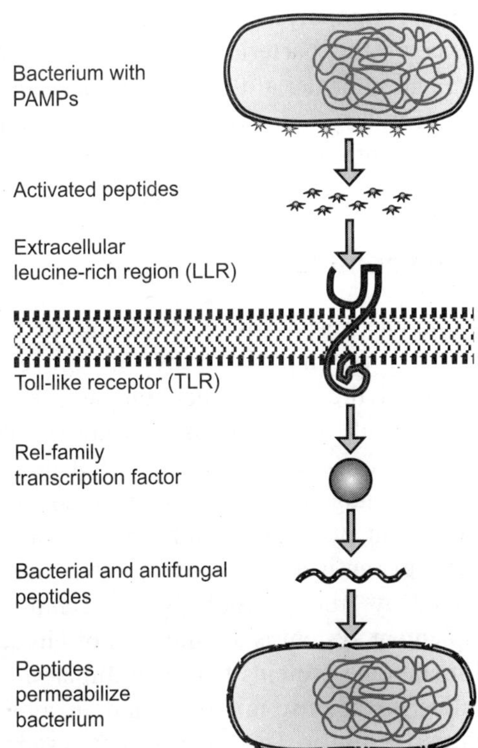

Fig. 18.3 Innate immunity is triggered by PAMPs.

recognize these shared structures are called **pattern recognition receptors**. Another characteristic of innate immunity is that its defence components have evolved to recognize structures of microbes that are often essential for their survival and infectivity. The receptors of the innate system are encoded in the germline and are not produced by somatic recombination of genes. However, the receptors are not diverse and cells are not clonally distributed to match all possible types of pathogens. Consequently many cells of innate immunity may recognize the same microbe.

When an invading pathogen makes its initial contact with the innate immune system, it will be greeted by the phagocytic cell (e.g., macrophage), whose function is to recognize foreign objects and launch an appropriate signal. Although innate immunity can effectively combat many infections, many pathogenic microbes have evolved to resist innate immunity. The innate immunity recognition depends mainly on the structures shared by classes of microbes and the phagocytic cells. These phagocytes possess receptor proteins on their surface that recognize conserved chemical motifs that are characteristic of pathogens. The best studied receptor proteins are the **Toll-like receptors (TLRs)**. At least 10 TLRs are expressed in humans; within this family are receptors that recognize the lipopolysaccharide or peptidoglycan components of the bacterial cell wall, the protein flagellin found in bacterial flagella, and unmethylated CpG dinucleotides. Activation of a TLR initiates a signal cascade within the cell that can lead to a variety of protective immune responses (Fig. 18.3).

18.4.2 The complement system

Natural immune responses to invading pathogens are typically accomplished by a process of inflammation at the site of infection (as previously explained). Blood also contains a group of soluble proteins called complement (or assist) that bind to extracellular pathogens, triggering their destruction. The involvement of serum factors that can augment the effects of antigen–antibody combination was first detected by Jules Bordet about a century ago. The complement system is a collection of circulating and cell membrane proteins that play important roles in host defence against microbes and in antibody-mediated tissue injury. The complement system may be activated by microbes in the absence of antibody, as part of the innate immune response to infection, and by antibodies attached to microbes, as part of adaptive immunity. Many complement proteins are proteolytic enzymes, and complement activation involves the sequential activation of these enzymes, also called **enzymatic cascade**, leading to the generation of effector molecules that participate in eliminating microbes in different ways. The complement system is tightly regulated by molecules present on normal host cells and prevents uncontrolled and potentially dangerous complement activation.

18.4.3 Pathways of complement activation

The complement cascade may be activated by one of three major pathways (**Fig. 18.4**), two initiated by microbes in the absence of antibodies, called the **alternative** and **lectin pathways** and the third initiated by certain isotypes of antibodies attached to antigens, called the **classical pathway**. Most of these involve the activation or fixation of the third complement component (C3).

The classical pathway is triggered after antibodies bind to antigens and is thus a component of the adaptive immunity. This pathway is triggered when IgM or IgG bind to antigens on microbes. As a result, the Fc regions of the antibodies become accessible to complement proteins and two or more Fc regions come close together. The C1 esterase complement protein then binds to two adjacent Fc regions. The attached C1 becomes enzymatically active, thus resulting in the binding and cleavage of two other proteins, C4 and C2. The resultant C4b2a complex gets covalently attached to the antibody and to the microbial surface where the antibody is bound. This complex functions as the 'classical pathway C3 convertase.' It breaks down C3, and the C3b that is generated again gets attached to the microbe. Some of the C3b binds to the C4b2a complex, and the resultant C4b2a3b complex functions as a C5 convertase.

The alternate pathway is triggered when a breakdown product of C3 hydrolysis, called C3b, is deposited on the surface of a microbe. The C3b forms stable covalent bonds with microbial proteins or polysaccharides and is thus protected from further degradation. The microbe-bound C3b becomes a substrate for the binding of another protein called factor B, which is broken down by a plasma protease to generate the Bb fragment. This fragment remains attached to C3b, and the C3bBb complex enzymatically breaks down more C3, functioning as the 'alternative pathway C3 convertase.' As a result of this convertase activity, many more C3b and C3bBb molecules are produced and become attached to the microbe. Some of the C3bBb molecules bind additional C3b and the C3bBb complex functions as a C5 and initiate the late steps of complement activation.

The lectin pathway is initiated in the absence of an antibody by the attachment of plasma mannose-binding lectin (MBL) to terminal mannose residues on the surface glycoproteins of microbes. MBL is structurally similar to a complement of C1 of the classical pathway and serves to activate C4. The subsequent steps are essentially the same as in the classical pathway.

18.4.4 Functions of the complement system

The complement system plays an important role in the elimination of microbes during both innate and adaptive immune responses. Activated complement proteins function as proteolytic enzymes to cleave other complement proteins. The central component of complement is a plasma protein called C3, which is cleaved by enzymes generated in the early steps. The major proteolytic fragment of C3, called C3b, becomes covalently attached to microbes and is able to activate downstream complement proteins on the

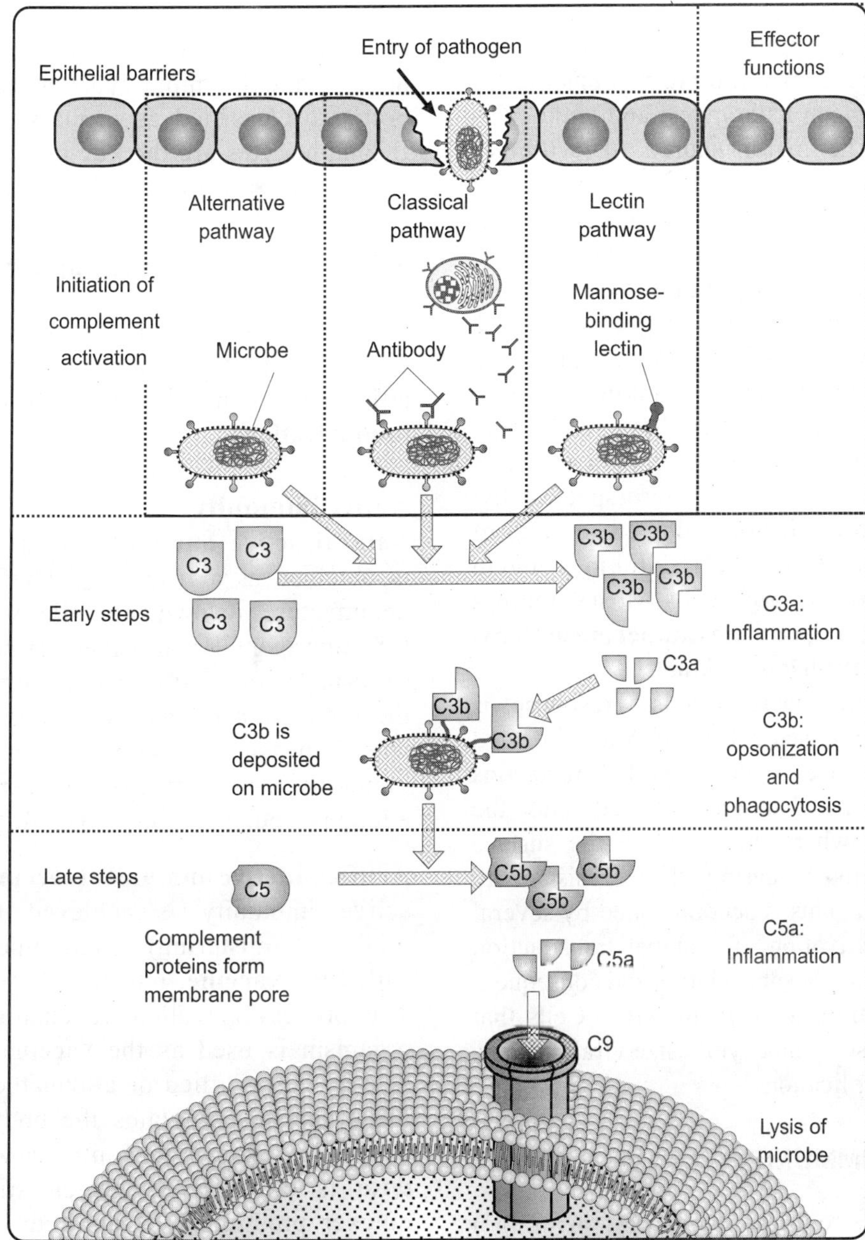

Fig. 18.4 Complement activation pathway. The complement activation may be initiated by three distinct pathways, all of which lead to the production of C3b. C3b initiates the late steps eventually polymerizing C9, which forms the membrane attack complex by making holes in the membrane.

microbial surface. Thc complement system serves three functions in host defence:

- C3b coats microbes and promotes the binding of receptors for C3b that are expressed on the phagocytes.
- Some breakdown products of complement proteins are chemo-attractants for neutrophils and monocytes and promote inflammation at the site of complement activation.
- Complement activation culminates in the formation of a polymeric protein complex that gets inserted into the microbial cell membrane, forming pores that lead to the influx of water and ions and death of the microbe.

18.4.5 Cytokines of innate immunity

In response to infections, macrophages and other cells secrete proteins called cytokines that mediate many of the cellular reactions of innate immunity. Cytokines are soluble proteins that mediate immune and inflammatory reactions and are responsible for communication between leukocytes, and leukocytes and other cells. The major source of cytokines is the macrophages activated by recognition of microbes. Cytokines are also produced in cell-mediated immunity. Cytokines that are secreted from lymphocytes are termed **lymphokines**, whereas those secreted by monocytes or macrophages are termed **monokines**. A large family of cytokines are produced by various cells of the body. Lymphokines are also known as interleukins (ILs), since they are not only secreted by leukocytes but are also able to affect cellular responses of leukocytes. More specifically, interleukins are growth factors targeted to cells of hematopoietic origin (**Table 18.2**). Cytokines bind to high-affinity receptors on target cells. Most cytokines act on the cells that produce them (**autocrine actions**) or adjacent cells (**paracrine actions**).

Another type of innate antiviral response is initiated within the infected cell itself. Virus-infected cells produce substances called type 1 **interferons** (interferon α and ß) that are secreted into the extracellular space, where they bind to the surface of non-infected cells rendering them resistant to subsequent infection. This is accomplished by several means, including activation of a signal transduction pathway that results in phosphorylation and consequent inactivation of the translation factor eIF2. Cells that undergo this response cannot synthesize viral proteins required for virus replication.

18.5 ADAPTIVE IMMUNITY

Adaptive immunity requires exposure ('priming') to the non-self material and also requires a lag period during which the immune system gears up for an attack against a foreign agent. Acquired immune response is highly specific and can discriminate between two very similar molecules; it also has a 'memory'. It is primarily mediated by lymphocytes and may be further divided into **cell-mediated** and **humoral** immune response. The immune response features a specialized array of cells and proteins. Blood cells undergo progressive cell division and differentiation in the presence of mitogenic and differentiation-inducing growth factors from stem cells, originating in the bone marrow (**Fig. 18.5**). The lymphoid cells include both B lymphocytes (B cells) and different types of T lymphocytes (T cells). The differentiated B lymphocytes mature in the bone marrow and T lymphocytes mature in the thymus; both are classified as primary lymphoid organs. B lymphocytes are the key cells in the production of antibodies following stimulation by an antigen.

18.5.1 Types of immune activities

Adaptive immunity can be described as active or passive. Both types may be acquired naturally or artificially. Artificially acquired immuity is called **immunization**.

Active immunity

Natural active immunity: This type of immunity is obtained as a result of an infection. The body manufactures its own antibodies when exposed to an infection agent. Because memory cells—produced on exposure to the first infection—are able to stimulate the production of massive quantities of antibodies when exposed to the same antigen again, this type of immunity is most effective and generally persists for a long period, sometimes even for life.

Artificial active immunity (vaccination): Artificial active immunity is achieved by injecting (or orally administering) small amounts of antigen, called the **vaccine**, into the body of an individual. The process is called vaccination. If the whole organism is used as the vaccine, it is first made safe by being killed or attenuated. The antigen on the organism stimulates the body to manufacture antibodies against the antigen. A second booster dose is often given, which stimulates a much quicker production of antibody and lasts longer; this protects the individual from the disease for a considerable time.

Passive immunity

Natural passive immunization: Passive immunization helps people who are already infected with pathogens, including the ones that cause diphtheria, tetanus, measles, and hepatitis B. A person receives injections of antibody molecules purified from other sources. The best source is another individual who has already produced a large amount of the required antibody molecules. They give immediate protection, unlike active immunity, which takes a few days or weeks to build up. However, the effects are not lasting, because the antibodies are not produced by the person's own

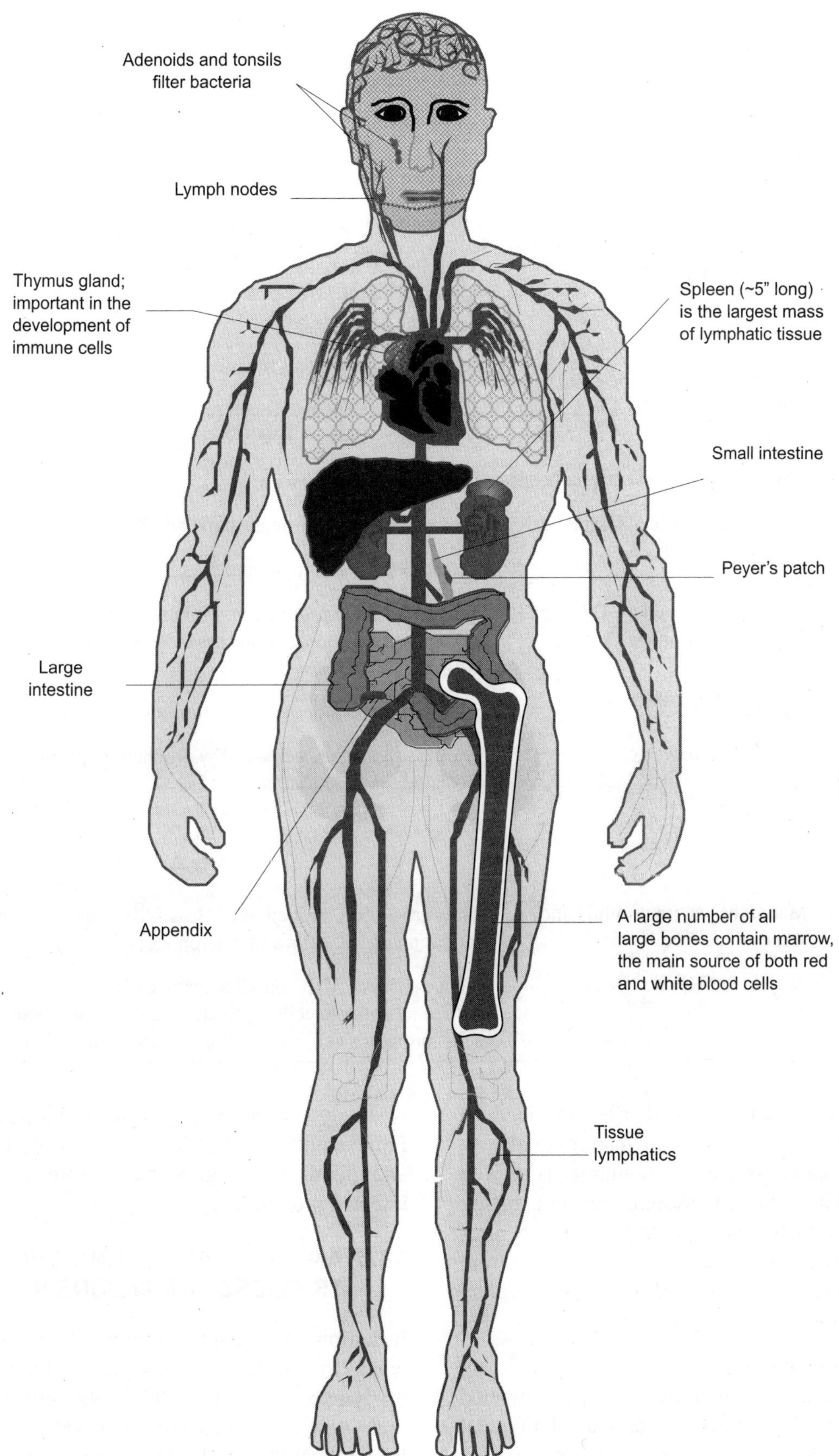

Fig. 18.5 Outline of lymphoid system in human beings.

Table 18.2 List of identified interleukins

Interleukins	Principal source	Primary activity
IL-1(α and β)	Macrophages and other APCs	Co-stimulation of APCs and T cells, inflammation and fever, acute phase response, haematopoiesis
IL-2	Activated TH_1 and NK cells	Proliferation of B cells and activated T cells, NK functions
IL-3	Activated T cells	Growth of hematopoietic progenitor cells
IL-4	TH_2 and mast cells	B cell proliferation, eosinophil and mast cell growth and function, IgE and Class II MHC expression on B cells, inhibition of monokine production
IL-5	TH_2 and mast cells	Eosinophil growth and function
IL-6	Activated TH_2 cells, APCs, other somatic cells	Acute phase response, B cell proliferation, thrombopoiesis, synergistic with IL-1 and TNF on T cells
IL-7	Thymic and marrow stromal cells	T and B lymphopoiesis
IL-8	Macrophages, other somatic cells	Chemo-attractant for neutrophils and T cells
IL-9	T cells	Hematopoietic and thymopoietic effects
IL-10	Activated TH_2 cells, $CD8^+$ T and B cells, macrophages	Inhibits cytokine production, promotes B cell proliferation and antibody production, suppresses cellular immunity, mast cell growth
IL-11	Stromal cells	Synergistic hematopoietic and thrombopoietic effects
IL-12	B cells, macrophages	Proliferation of NK cells, INF-γ production, promotes cell-mediated immune functions
IL-13	TH_2 cells	IL-4–like activities
Interferons		
INF-α and -β	Macrophages, neutrophils and some somatic cells	Antiviral effects, induction of class I MHC on all somatic cells, activation of NK cells and macrophages
INF-γ	Activated TH_1 and NK cells	Induces Class I MHC on all somatic cells and Class II MHC on APCs and somatic cells; activates macrophages, neutrophils, NK cells, promotes cell-mediated immunity, antiviral effects

B cells. Passive immunity can also be provided by *colostrums*, the first secretion of the mammary glands. An excellent example of passive immunity is seen in the newborn. Their immune systems are not mature enough to respond to many pathogens but are protected against infections by acquiring antibodies from their mothers through the placenta and milk. The baby absorbs the antibodies through its gut.

Artificial passive immunity: In this case, antibodies formed in one individual are extracted and injected into the blood of another individual of the same or different species. They can be used for immediate protection if a person has been, or is likely to be, exposed to a particular disease. For example, specific antibodies for combating tetanus and diphtheria used to be cultured in horses and injected into humans.

18.6 ANTIGEN-PRESENTING CELLS: TRIGGERS FOR IMMUNE RESPONSE

Initiation of adaptive immune response occurs when the antigen is recognized by the receptors of lymphocytes. B and T lymphocytes differ in their antigen recognition. The antigen receptors of B lymphocyte (e.g., IgM) can recognize a wide variety of macromolecules (proteins, polysaccharides, lipids

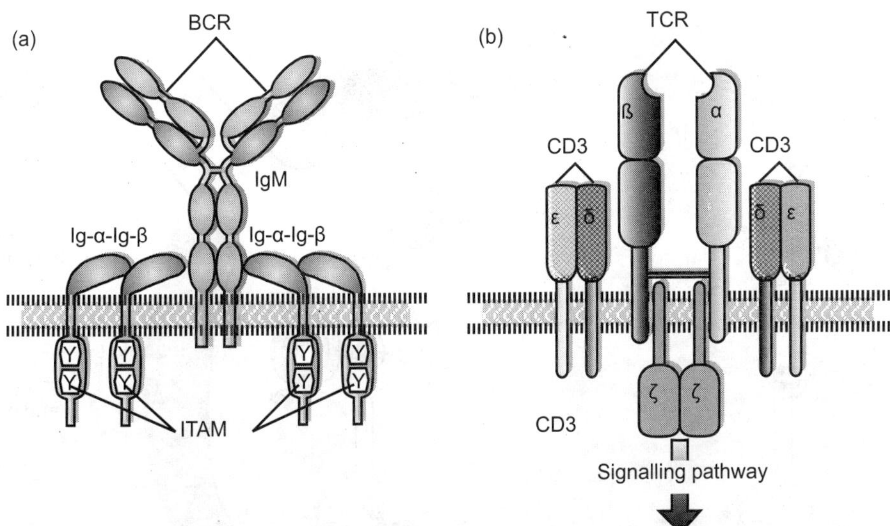

Fig. 18.6 Membrane-bound antigen receptor complexes. (a) The B cell receptor (BRC) contains membrane-bound IgM non-covalently bound to two Ig-α-Ig-β heterodimers. The intracellular domains of each of the Ig-α and Ig-β chains include an immunoreceptor tyrosine-based activation motif (ITAM). (b) In the T cell receptor (TCR), each subunit contains two Ig-like domains: one of the Ig-like domains of each subunit has variable amino acid sequence; the other has a constant amino acid sequence. The invariant polypeptided associated with BCRs and TCRs plays a key role in transmitting signals to the interior, leading to changes in activity of the B or T cell.

and nucleic acids) as well as small chemicals in soluble or cell surface-associated form. On the other hand, most T lymphocytes, have affinity for only peptide fragments of protein antigens, and only when they are presented by specialized peptide display molecules on host cells (**Fig. 18.6**).

The plasma membrane of every cell of an individual's body incorporates a variety of proteins. Among these are **major histocompatibility molecule** (MHC) markers. Some MHC markers are common to all the body cells. Others are unique to macrophages and lymphocytes. When bacteria enter the punctuated tissue, they multiply, inflammation follows, and then phagocytes complete the operation. Some of the phagocytes are macrophages that engulf the foreign cells and generate antigens for presentation. Their digestive enzymes fragment the antigen into bits, which get attached to MHC molecules. These antigen–MHC complexes are carted off in vesicles to the macrophage surface and presented for recognition by T lymphocytes. Dendritic cells are specialized cell types located in the epithelium; they primarily capture antigens and transport them to peripheral lymphoid tissues (lymph nodes), where they display portions of the antigens for recognition by T lymphocyte. In general, any cell that processes and displays an antigen with a suitable MHC molecule is known as an **antigen-presenting cell** (APCs). APCs have another important feature that gives them the ability to trigger T cell responses. These specialized cells respond to microbes by producing surface antigens and secreted proteins that activate naïve T lymphocytes, thus providing the 'second signals' for T cells and differentiation. These types of cells which display surface antigen and produce second signals are called '**professional**' APCs (e.g., dendritic cells).

Naive T lymphocytes need to encounter antigens presented by 'professional' APCs to initiate immune responses against protein antigens. Differentiated effector T cells again need to expose to antigens presented by various APCs, to activate the effector functions of the T cells in humoral and cell-mediated immune responses. Protein antigens of microbes that enter the body are captured by professional APCs and concentrated in the peripheral lymphoid organs where immune responses are initiated (**Fig. 18.7**). As mentioned earlier, different types of APCs serve distinct functions in T cell-dependent immune responses.

18.6.1 Processing of protein antigens

Extracellular proteins internalized by professional APCs into vesicles are processed and displayed by class II MHC molecules, whereas proteins in the

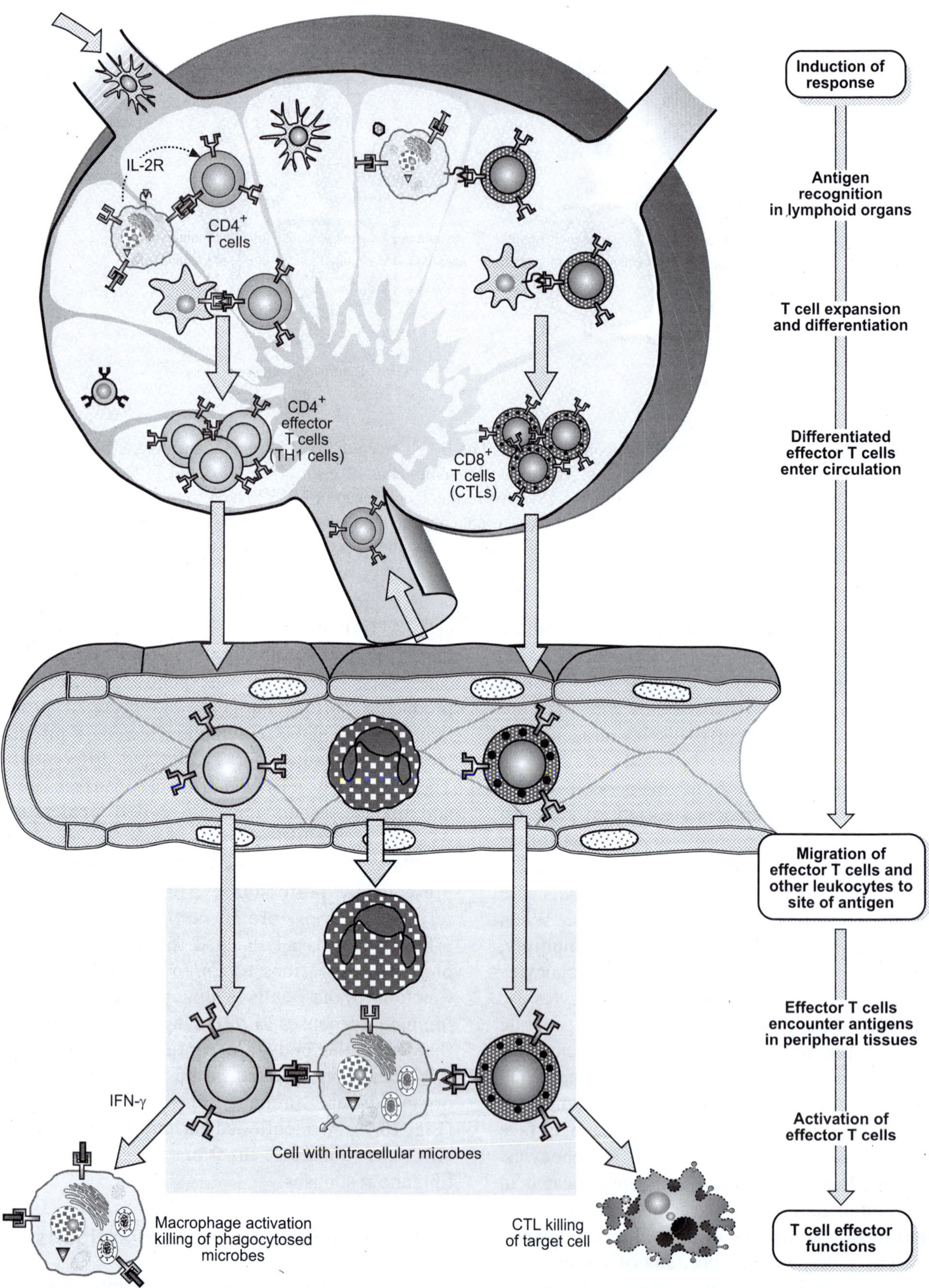

Fig. 18.7 The induction and effector phases of cell-mediated immunity (**See Plate 15**).

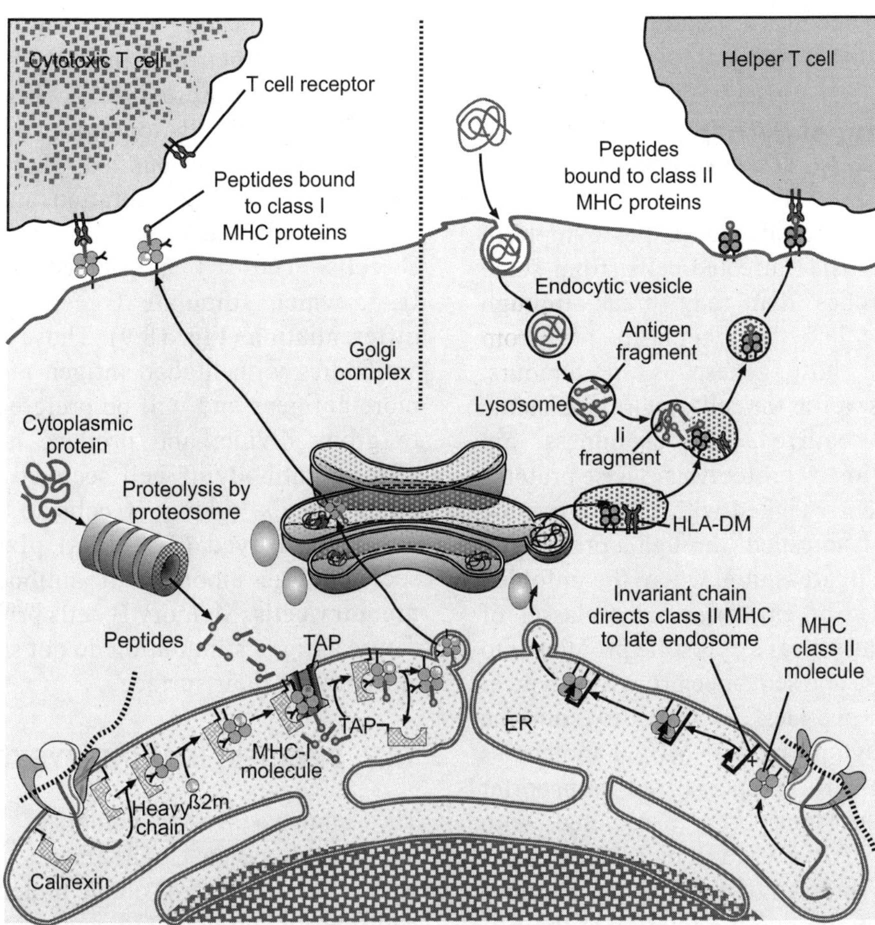

Fig. 18.8 Processing pathways for antigens associated with MHC class I (left of dotted line) and class II (right of dotted line) molecules.

cytosol of nucleated cells are processed and displayed by class I MHC molecules. These two pathways of antigen processing involve different cellular organelles and proteins (**Fig. 18.8**). The segregation of antigen processing pathways also ensures that different classes of T lymphocytes recognize antigens from different compartments.

18.6.2 Processing of internalized antigens for display by MHC II molecules

APCs internalize extracellular microbes or pathogenic agents by several mechanisms. Microbes bind to surface receptors specific for microbial products or to receptors that recognize antibodies or products of complement activation that are attached to the microbes. B cells internalize antigens that specifically bind to the cells' antigen receptors. Some APCs phagocytose microbes or pinocytose proteins without any specific recognition event. After internalization, proteins enter intracellular vesicles called **endosomes** or **phagosomes**. Within these vesicles, the proteins are broken down by proteolytic enzymes into fragments of varying lengths and sequences. APCs constantly synthesize class II MHC molecules in the endoplasmic reticulum (ER). Each newly synthesized class II molecule carries with it an attached protein called the **invariant chain**, which contains a sequence (called the class II invariant chain peptide, or CLIP) that binds tightly to the peptide-binding cleft of the class II molecule. This inaccessible class II molecule begins its transport to the cell surface in an exocytic vesicle, which then fuses with the endosomal vesicle containing broken down peptides derived from ingested extracellular proteins. The same endosomal vesicle contains a class II-like protein called DM, whose function is to remove CLIP from the class II MHC molecule. If the MHC II molecule either fails to bind to one of the peptides generated internally or finds outside the cell, it becomes unstable and gets degraded by proteases. Therefore, only the peptides from the intact antigen stimulate immune responses in that individual; such peptides are said to be **immuno-dominant epitopes** of the antigen.

18.6.3 Processing of cytosolic antigens for display by MHC I molecule

Antigenic proteins are produced in the cytoplasm from viruses living inside infected cells, from some phagocytosed microbes that may break through vesicles and escape into the cytoplasm, and from mutated or altered host genes, as in tumours. All these proteins, as well as the cell's own cytoplasmic proteins that have outlived their usefulness, are targeted for destruction by proteolysis. These proteins are unfolded, covalently tagged with a small peptide called ubiquitin, and 'threaded' through a proteolytic organelle called the **proteasome**, where the unfolded proteins are degraded by enzymes. Some classes of proteasomes efficiently cleave cytosolic proteins into peptides with the size and sequence properties typical of MHC I binding peptides. In order to combine peptides from the cytoplasm and MHC I molecules synthesized in the ER, a specialized transporter molecule called the transporter associated with antigen processing (TAP) is used. It picks up peptides from the cytoplasm and actively pumps the peptides across the ER membrane into the interior of the ER. Newly synthesized MHC I molecules are loosely attached to the interior face of the TAP molecule. If a MHC I molecule finds a peptide with the right fit, the complex will be stabilized and transported to the cell surface. If not, the molecule becomes unstable and is degraded by proteases. By segregating the MHC I and MHC II pathways of antigen processing the immune system is able to respond efficiently to extracellular and intracellular microbes, very and also combat them.

In addition to antigen display for recognition of T cells, APCs express 'second signals' for T cell activation. B lymphocytes use membrane-bound antibodies to recognize a wide variety of antigens. In response to antigen and other signals B cells differentiate into cells that secrete antibodies.

18.6.4 Lymphocyte activation

When an animal is exposed to an antigen, APCs (T and B cells) act in concert to stimulate production of antigen-specific **antibodies (Ab)**. Unless these cells interact successfully, the B cells fail to produce antigen-specific antibodies. A trimolecular complex forms between the T cell antigen receptor, the processed antigen and the MHC molecule on the APC as explained previously. This complex triggers the APCs and T cells to produce a variety of soluble factors such as interleukin-1 (IL-1) and IL-2, which stimulate T cell proliferation and differentiation. Properly activated T cells bind to antigen presenting B cells. These T cells then produce IL-4 and IL-5, which stimulate B cell multiplication and differentiation (**Fig. 18.9**). Those B cells producing antibodies with higher antigen affinity will attract more antigens and will be preferentially stimulated to grow, divide, and produce more of the high affinity antibody. T cell secreted IL-5, along with other signals, allows B cells to differentiate into either short-lived (3–4 days) **plasma cells** which secrete large amounts of antibody or long-lived **memory cells**. Memory B cells primed to respond to future antigen stimulation do not secrete but express antibody on their surface.

18.7 SPECIFIC IMMUNE SYSTEMS

The immune system comprises two parallel but interrelated systems. The production of antibodies occurs first. These are soluble proteins which combine with the foreign antigen resultingin protective events called antibody-mediated response (or humoral immunity). The second type of immunity is known as cell-mediated immunity. The cytotoxic T cells or killer T cells recognize abnormal cells in the body.

18.7.1 Humoral immunity

Humoral immunity is mediated by proteins called antibodies, which are produced by B lymphocytes. Antibodies are secreted into the circulation and mucosal fluids; they neutralize and eliminate foreign bodies. The most important properties of adaptive immunity are: fine specificity for structurally distinct antigens and memory of prior exposure to antigen. The specificity of immune response can be explained by the observation that prior exposure to an antigen results in elevated responses to subsequent challenge by that antigen but not by another. The immune system has enormous potential for distinguishing a billion different antigens or portions of antigens. The **memory lymphocytes**, which are long-lived cells induced during the primary immune response have rapid **secondary immune response** when subsequently exposed to the same antigen.

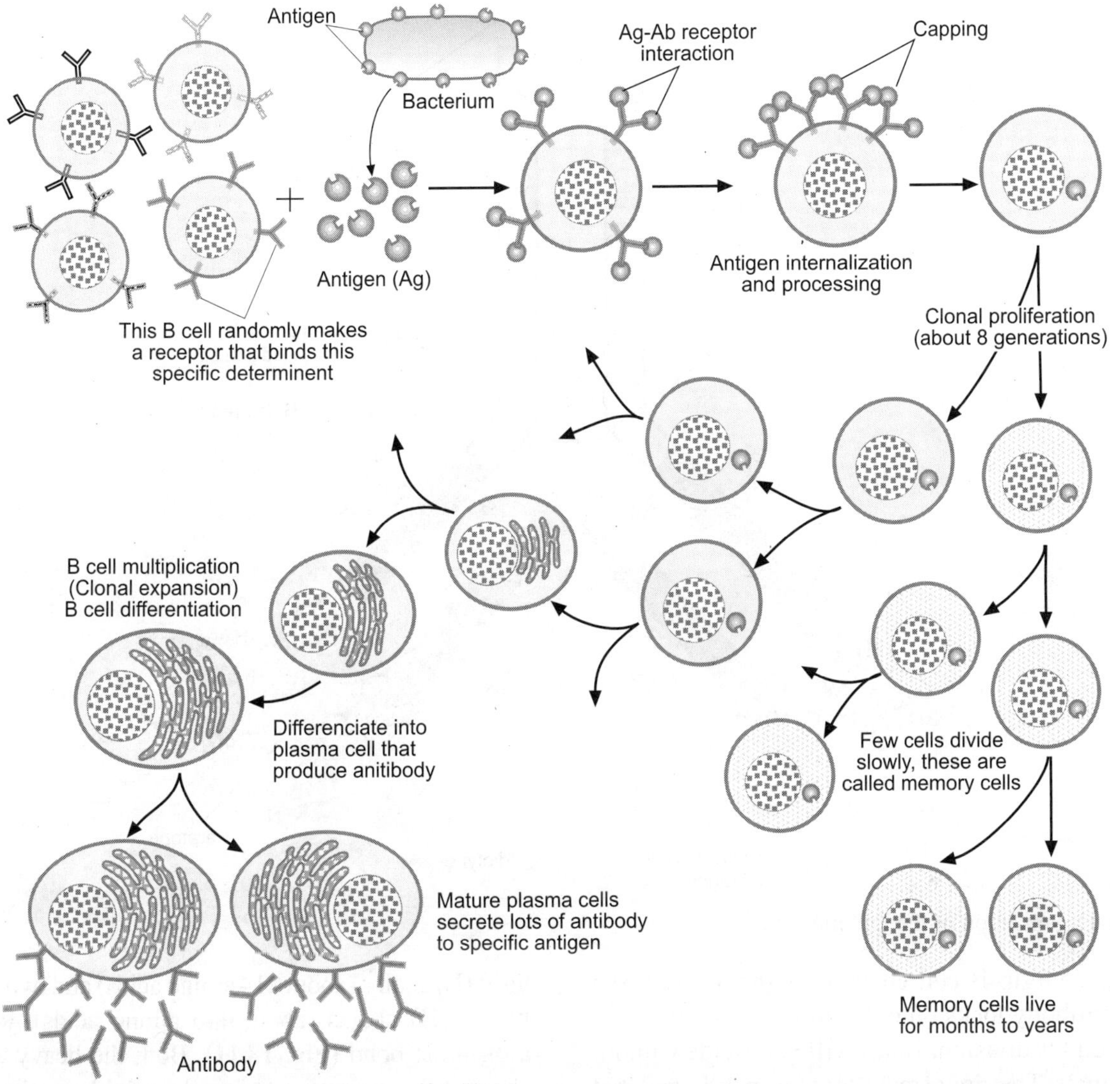

Antigen
Bacterium
Ag-Ab receptor interaction
Capping
Antigen (Ag)
Antigen internalization and processing
This B cell randomly makes a receptor that binds this specific determinent
Clonal proliferation (about 8 generations)
B cell multiplication (Clonal expansion) B cell differentiation
Differenciate into plasma cell that produce anitibody
Few cells divide slowly, these are called memory cells
Mature plasma cells secrete lots of antibody to specific antigen
Memory cells live for months to years
Antibody

Fig. 18.9 Antigen (Ag) internalization and processing of B cell.

18.7.2 B cells and targets of antibodies

The B cells go down a pathway that culminates in full differentiation. Along this pathway, B cells synthesize many copies of a single kind of antibody molecules. All antibodies are proteins, but each kind has binding sites that match up to only a single antigen. After synthesis, antibody molecules move to the plasma membrane of a maturing B cell (**Fig. 18.10**), where the tail (constant regions of heavy chains) of each one becomes embedded in the lipid bilayer, and the two arms stick out above it. The cell then enters the bloodstream as a virgin B cell with unique antibody receptor.

NH2
NH2
NH2
NH2
C C
Antibody molecule
Cell membrane
C C
Signalling molecule
Signal

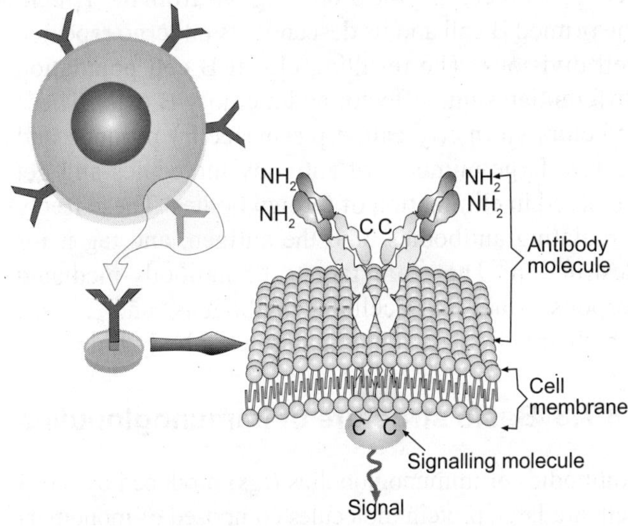

Fig. 18.10 An antibody anchored in the cell membrane of the B lymphocyte.

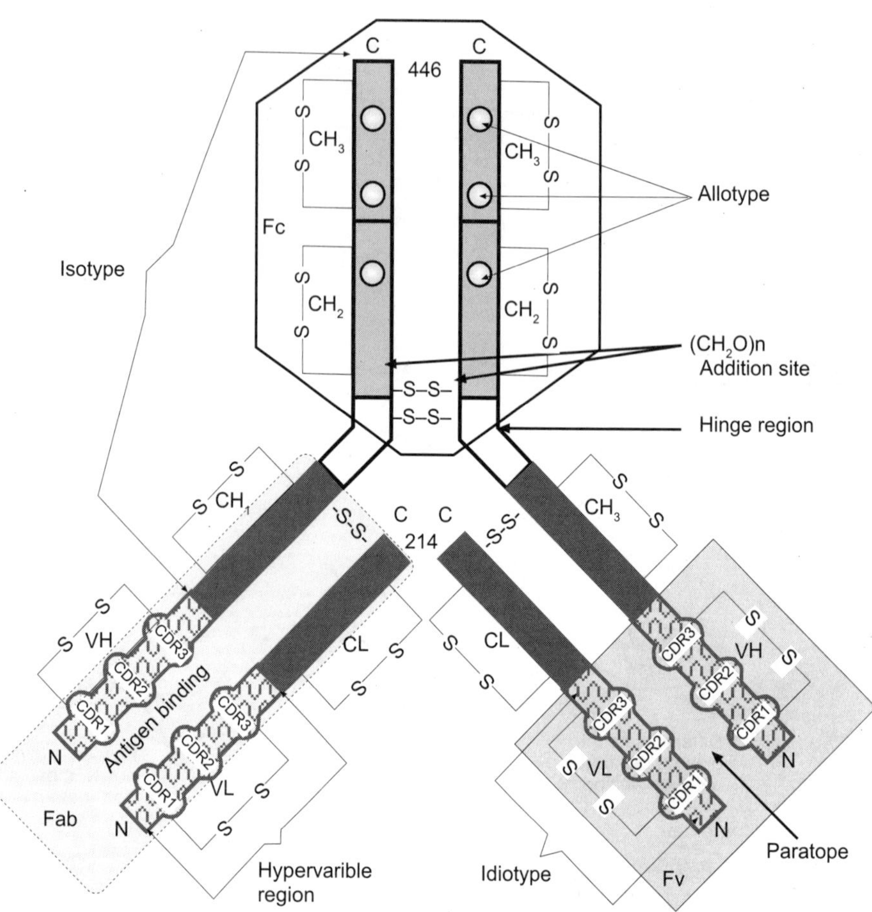

Fig. 18.11 Organization of the IgG molecule.

When a virgin B cell encounters the antigen, the receptor antibody locks onto the antigen. The B cell is now primed for division. But it will not divide without proper signals. The signal must come from a T_H cell that has already been activated by an antigen-presenting cell. After receiving the proper signals from the T_H cell, the primed B cell and its descendents undergo repeated cell divisions. The resulting clonal B cell population differentiates into effector and memory B cells. These effectors (formerly called plasma cells) produce and secrete large numbers of antibody molecules and get involved in elimination of foreign bodies. These freely circulating antibodies bind the antigen, and tag it for destruction. The main targets of antibody-mediated responses are extracellular pathogens and toxins, which are circulate freely in tissues or body fluids.

18.7.3 Basic structure of immunoglobulins

Antibodies or immunoglobulins (Igs) produced by the B cell, are large protein molecules composed of monomers or multimers of a basic antibody. Each basic antibody is a tetramer composed of four polypeptides, two identical **light** (**L**) chains (about 214 amino acids) and two identical **heavy** (**H**) chains (about 440 amino acids), joined by a disulfide bond (**Fig. 18.11**). Both the heavy and light chains have an amino-terminal **variable region**, within which the amino acid sequence varies among antibodies specific for different antigens, and a carboxy-terminal **constant region**, within which the amino acid sequence is the same for all Abs of a given Ig class, regardless of antigen-binding specificity. Each antibody has two antigen-binding sites or domains, each of which is formed by the variable regions of one light chain and one heavy chain. The other fragment, called Fc does not bind the antigen, but has other important functions, including the mediation of responses termed **effector functions** (**Fig. 18.12**). These functions include the initiation of the **complement cascade**, a process that leads to the lysis of target cells. The molecule adopts a conformation that resembles the letter Y, in which the stem portion corresponds to the Fc fragment comprising two carboxy-terminal domains of each H chain. The two arms of the Y correspond to the two Fab fragments formed by the two amino-terminal domains of each H chain and the two amino-terminal domains of L chains.

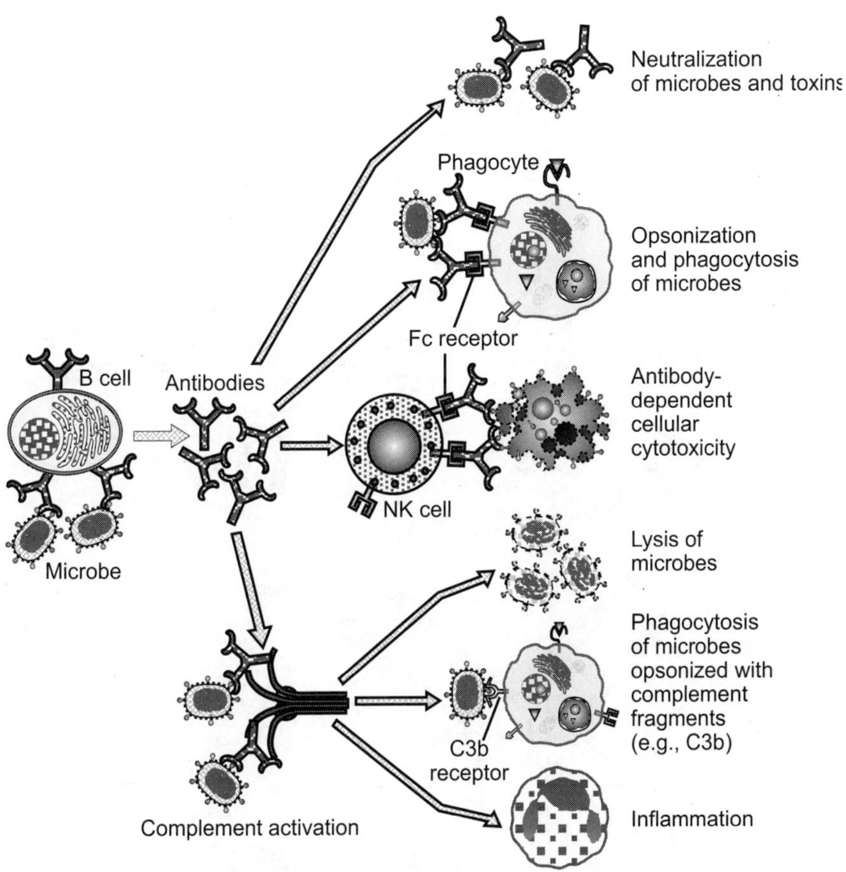

Fig. 18.12 Schematic of effector function of antibodies. Antibodies of different heavy chain classes (isotypes) perform different effector functions.

Although there are many types of Igs are now recognized, B cells produce five major types of antibodies during immune response. These five classes share the same type of combining site for the antigen. The five distinct classes of antibodies known are designated as IgG, IgM, IgA, IgD, IgE antibody molecules, the heavy chains of which are, γ, μ, α, δ and ε, respectively. These classes serve specialized functions in the immune response and exhibit small structural differences (**Table 18.3, Fig. 18.13**). IgG is the most abundant antibody class and has the simplest molecular structure. IgG, IgD, IgE and serum IgA antibody molecules each consist of a pair of heavy and light polypeptide chains. The chains are held together by disulfide bonds or, in the case of IgA2, by non-covalent linkages. IgM antibodies consist of five pairs of heavy and light chains, and are always the first class of antibody made by a developing B cell. The tail region of IgE molecules, which are

Table 18.3 Classes of human immunoglobulins and their general properties

Class	Heavy chain	Light chain	Molecular mass (kDa)	Properties
IgA	α	κ or λ	360–720	Present in tears, nasal mucus, breast milk, intestinal secretions
IgD	δ	κ or λ	160	Specifically present in B cell plasma membranes; function unknown
IgE	ε	κ or λ	190	Binds to mast cells, releasing histamine responsible for allergic reactions
IgG	γ	κ or λ	150	Primary blood-borne soluble antibodies; crosses placenta
IgM	μ	κ or λ	950	Present in B cell plasma membranes; mediates initial immune response; activates bacteria killing complement

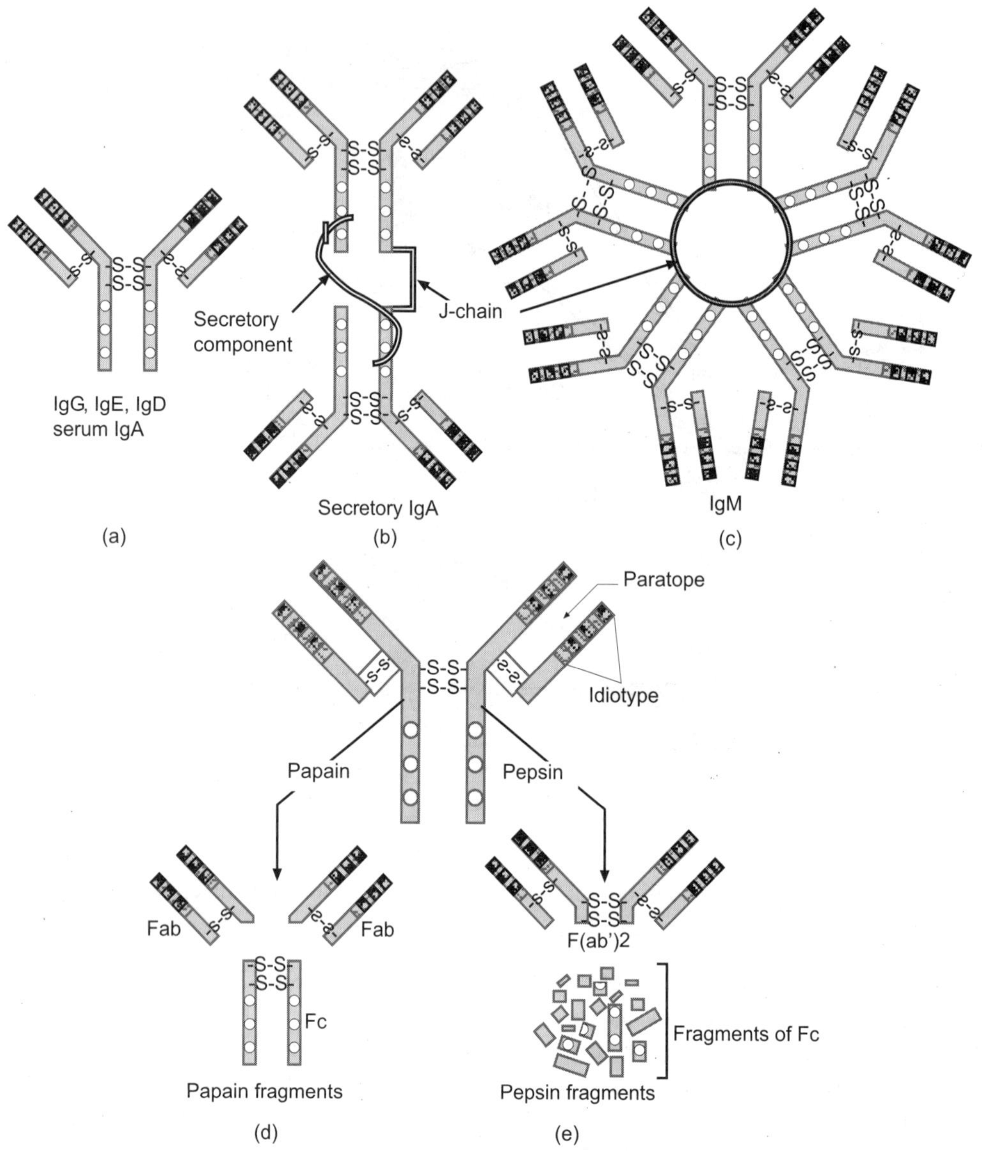

Fig. 18.13 Different forms of immunoglobulin molecules and creation of Fab, Fc and F(ab')2 fragments by specific proteolytic digestion of intact antibody molecules.

four-chain monomers, binds with unusually high affinity to yet another class of Fc receptors. These receptors are located on the surface of mast cells in tissues and basophils in blood. IgE molecules bound to them function as passively acquired receptors for antigen. Antigen binding triggers the mast cell or basophil to secrete a variety of cytokines and biologically active amines, especially **histamine (Fig. 18.14)**.

Antigen

An antigen is any agent that elicits an immune response in an animal circulatory system (either cellular or humoral). Most antigens consist of proteins or polysaccharides, but lipids and nucleic acids can also serve in this capacity. A small polypeptide can be made more antigenic by chemically fixing it to a larger molecule exhibiting a ligand capable of interacting with a T cell receptor. A substance may exhibit one or more antigenic sites (epitopes), which are capable of eliciting and interacting with the antibodies produced by an animal. Antibodies, which interact with a peptide exhibiting a specific conformation, define a conformational **epitope**. Antibodies to a conformational epitope will not react with the same peptide that unfolds (denatured). A sequence of amino acid residues constituting two or more groups of amino acids linked to one another is called continuous or contiguous epitopes. Antibodies which interact with a denatured molecule, but which do not interact with the corresponding native molecule recognize epitopes

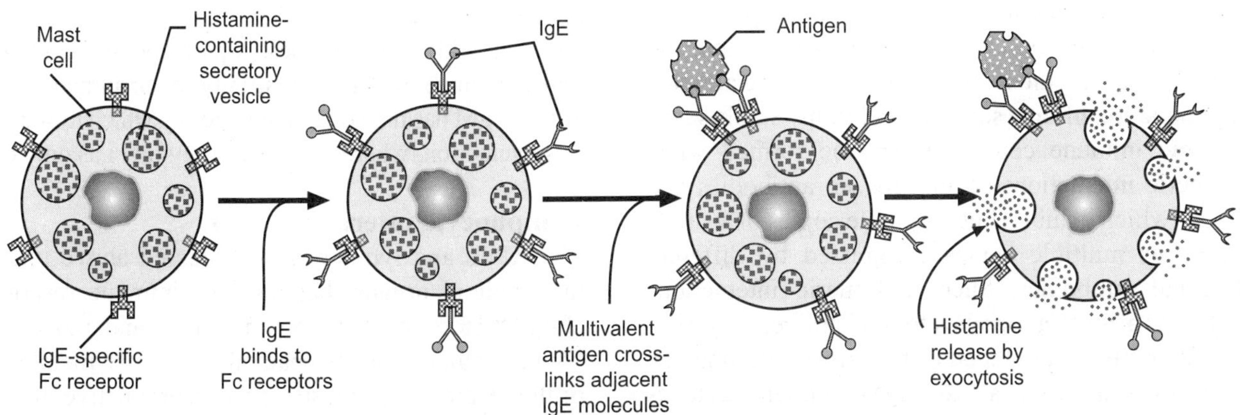

Fig. 18.14 The role of IgE in histamine secretion by mast cells.

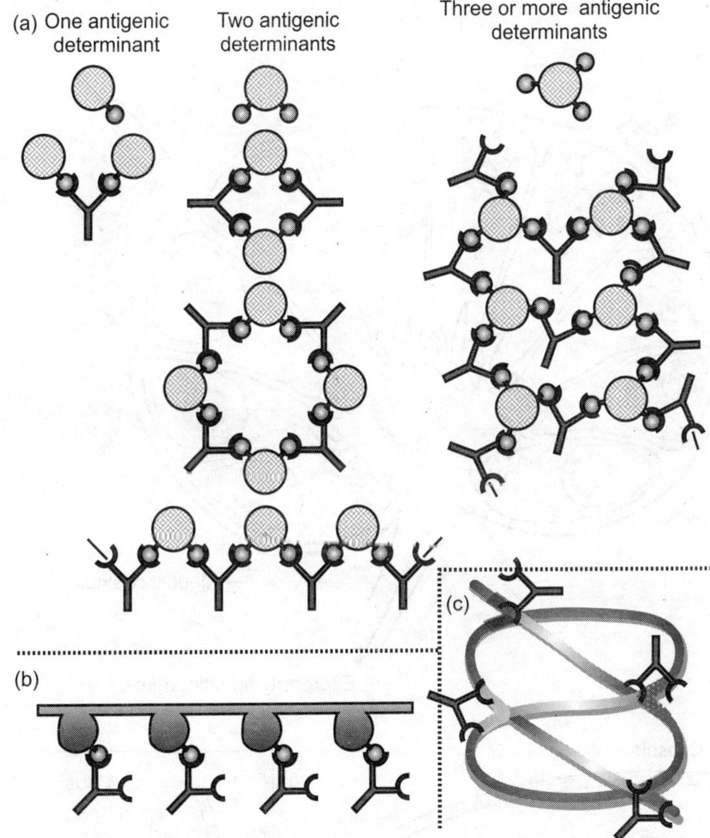

Fig. 18.15 Antibody–antigen interactions: (a) Antibodies have two identical antigen-binding sites, that can cross-link antigens. However, the types of antibody–antigen complexes that form depend on the number of antigenic determinants on the antigen. (b) Multiple identical antigenic determinants (a polyvalent antigen); (c) Multiple different antigenic determinants (a multivalent antigen).

known as **cryptotopes**. A **neotope** is an antigenic site on a monomeric subunit created by conformational changes brought about by the integration of the subunit with another molecule. Generally, the degree of antigenicity of a molecule subunit is a reflection of its location or prominence on the molecule or its physical or chemical makeup. Epitopes that are mobile and capable of flexing 1–2 Å usually interact more strongly with an antibody than rigid molecules.

Antigen–antibody interactions

A typical antibody has two identical antigen binding sites and is a Y-shaped molecule with two identical antigen binding sites (also called as **bivalent**), one at the tip of each arm of the Y. As long as an antigen has three or more antigenic determinants, bivalent antibody molecules can crosslink it into a large lattice (**Fig. 18.15**). An interaction between an antigen and an antibody involves hydrogen bonds, salt bridges, electrostatic charges and coulombic, hydrophobic or van der Waal's forces. This interaction is reversible and the strength of the interaction (affinity) can be affected by temperature, pH and solvent conditions. Stable semi-reversible immune complexes can be formed when multimeric interactions occur between antibodies and antigens which exhibit either multiple copies of a single epitope, or multiple copies recognized by different polyclonal antibodies. Such multimeric interactions enable low affinity antibodies to bind antigens tightly.

The ability of an antibody to form stable immune complexes is known as **avidity**. IgM with ten paratopes are more multimeric and have greater antigen avidity than IgG with two paratopes. A paratope of an antibody may exhibit a low degree of affinity for an antigen. A low affinity IgM has nine other paratopes by which to stabilize this interaction, whereas a low affinity IgG has only one.

Lymphocyte interaction sites

The APCs and lymphocytes discussed above interact in lymphoid organs that promote immune responses (**Fig. 18.16**). The location of tonsils and most other lymph nodules just beneath the mucous membranes of the respiratory, digestive, and reproductive systems

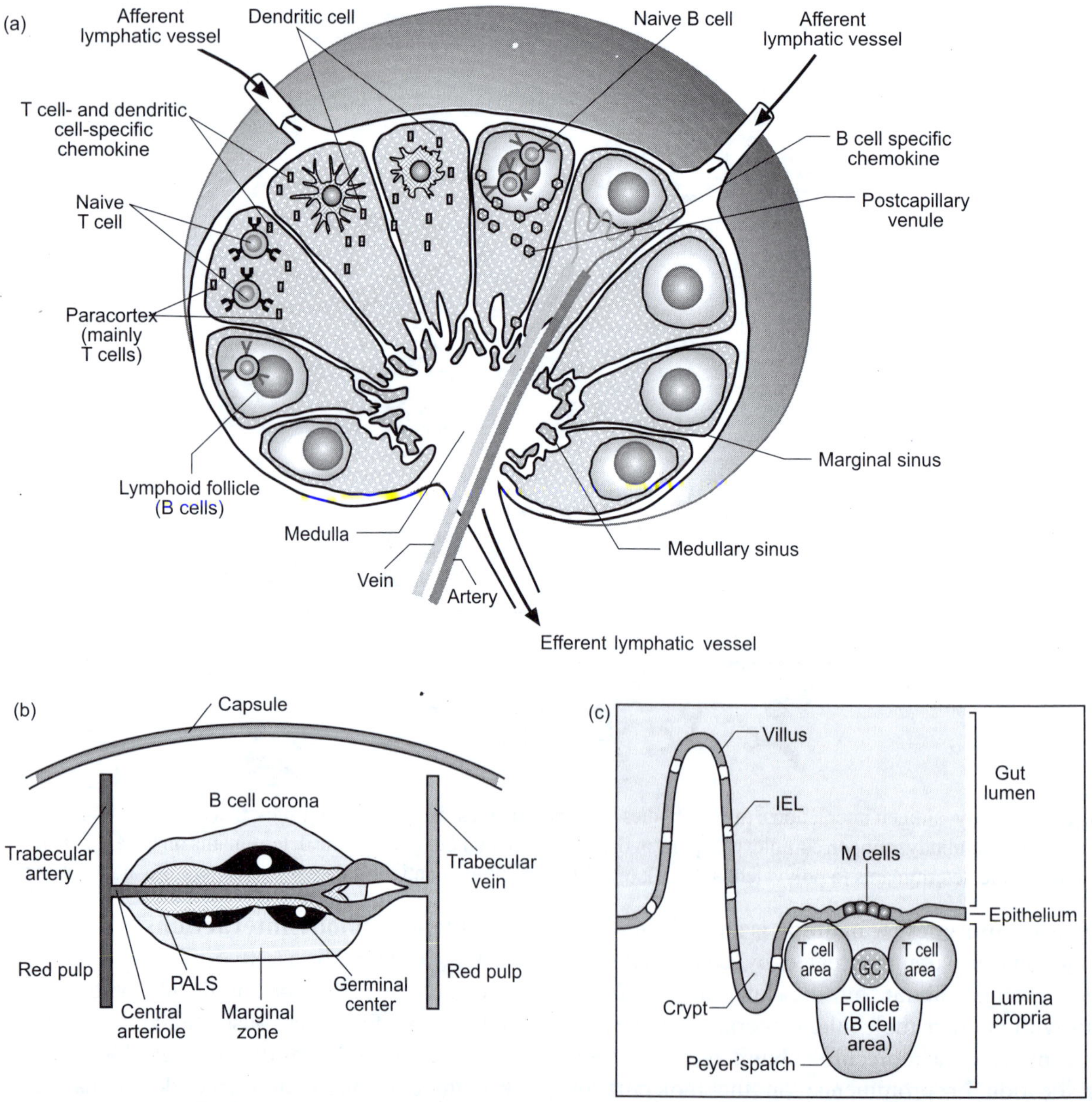

Fig. 18.16 Peripheral lymphoid organs: (a) Lymph node structure and lymphocyte activation, (b) Structure of white pulp in the spleen (PALS: periarteriolar lymphoid sheath); (c) Organization of gut-associated lymphoid tissue (GALT), including Peyer's patch, lamina propria and intraepithelial lymphocytes (IEL) GC: germinal centre (**See Plate 16**).

allow APCs and lymphocytes to intercept invaders just after they penetrate surface barriers. Cells are organized for maximum effect inside the lymphoid organs. Antigen-presenting cells make up the front line; antigen is engulfed and presented, and cell divisions produce populations of effector and memory lymphocytes.

Production of antibody polypeptide chains

The two main types of antibody polypeptide chains are: light chains and heavy chains. The light chains have been classified further depending on the number of V gene segments they possess. The two types are kappa(κ) and lambda(λ) light chains (**Fig. 18.17**). **Figure 18.18** shows the assembly of gene segments in heavy chain and **Fig. 18.19** shows an assembly of antibody in B lymphocyte.

18.8 GENERATION OF ANTIBODY DIVERSITY

How do lymphocytes make millions of different antigen-specific receptors required to detect pathogenic threats? A mammal can synthesize large amounts of specific antibody against virtually any foreign determinant, within a matter of days of being exposed to it. Antibody specificity is determined by the amino acid sequences of the variable regions of both light and heavy chains. The discovery of distinct variable and constant regions in the L and H chains raised the possibility that the genes that encode Ig have an unusual architecture that facilitates the generation of a diverse set of polypeptide products. All T and B cells in an individual carry the same genes. As the cell matures, a special recombination and mutation mechanism acts on the genes that specify antigen receptors. Different regions of these genes shuffled at random into one of millions of possible combinations. Genes for kappa (*k*), lambda (*λ*) and heavy chains are located on chromosomes 2, 22, and 14, respectively, in human beings. Each is a multigene complex.

The immune system generates a nearly limitless variety of specific proteins. Antibody proteins secreted by lymphocytes circulate in the blood. However,

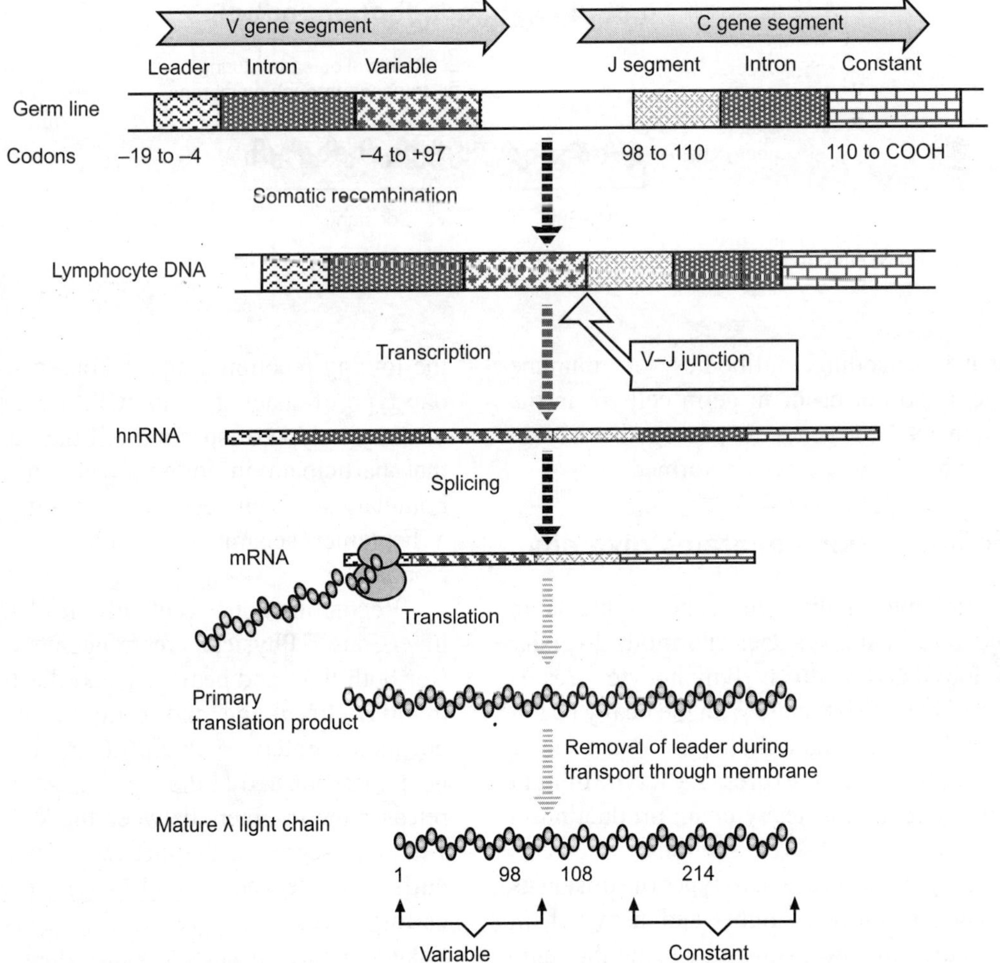

Fig. 18.17 Generation of functional λ light chain by somatic recombination and splicing.

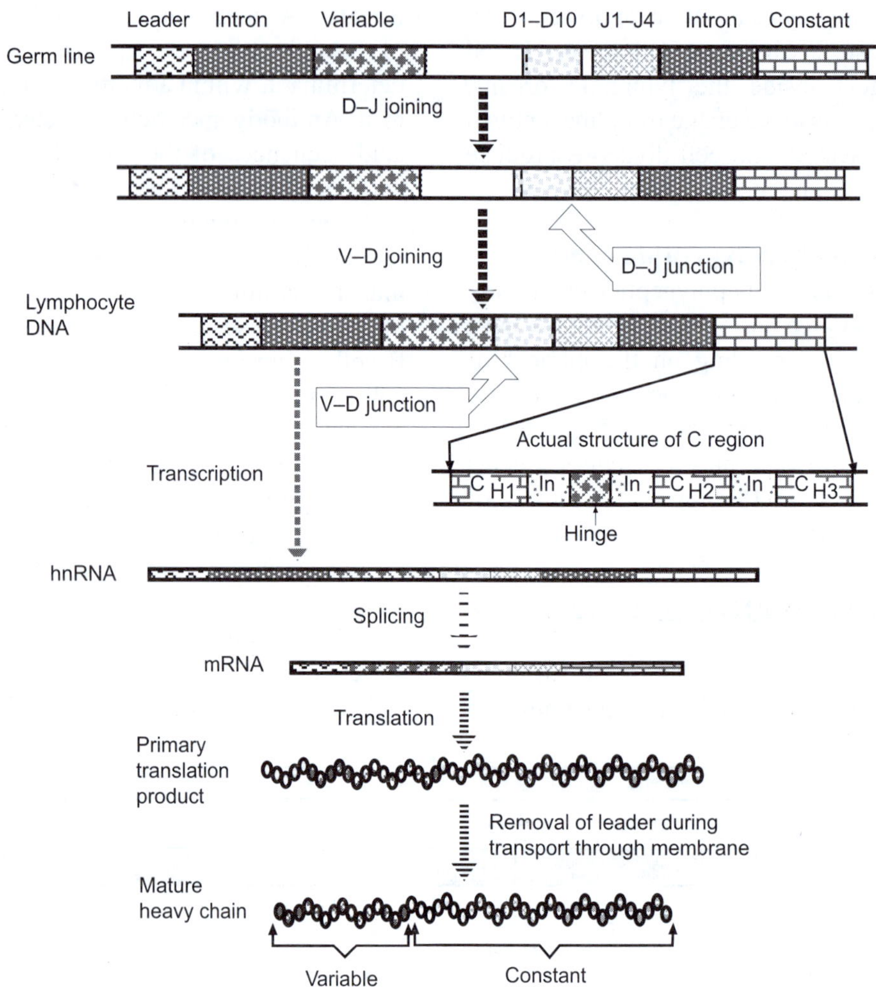

Fig. 18.18 Generation of functional heavy chain by sequential joining reactions.

complete genes encoding antibodies or immune receptor proteins do not occur in germ cells or in the early embryo. Instead, the genes are put together later in development, when lymphocytes are formed.

18.8.1 Recombination generates diversity

Recombination between different gene segments in both light and heavy chains creates enormous diversity in immunoglobulins. A given lymphocyte creates either a k or a λ chain to associate with the heavy chain. In humans, ~60% of the light chains are k and ~40% are λ. In mice, 95% of B cells express the k type of light chain. The single locus for heavy chain production in humans consists of many discrete sections.

Immune recombination uses two types of **consensus sequences**. The assembly of light- and heavy-chain genes involves the same mechanism by using the same consensus sequences. The common rule that governs the joining reaction is that a consensus sequence with one type of spacing, joins with a consensus sequence of another type of spacing. All the germline segments that participate in joining reactions have the same boundaries. Each consensus sequence consists of a heptamer separated by either 12 or 23 bp from a nonamer.

Recombination eventually leads to deletions or inversions. Physical rearrangement of sequences (for both light and heavy chains) during recombination involves **breakage** and **reunion**, but each reaction occurs separately. A double-strand break occurs at the heptamers located at the ends of the coding units. This releases the fragment between the V gene segment and J-C gene segment; the cleaved ends are called **signal ends**. The cleaved termini of this fragment are called **coding ends**. The two coding ends are covalently linked to form a coding joint which links the V and J segments.

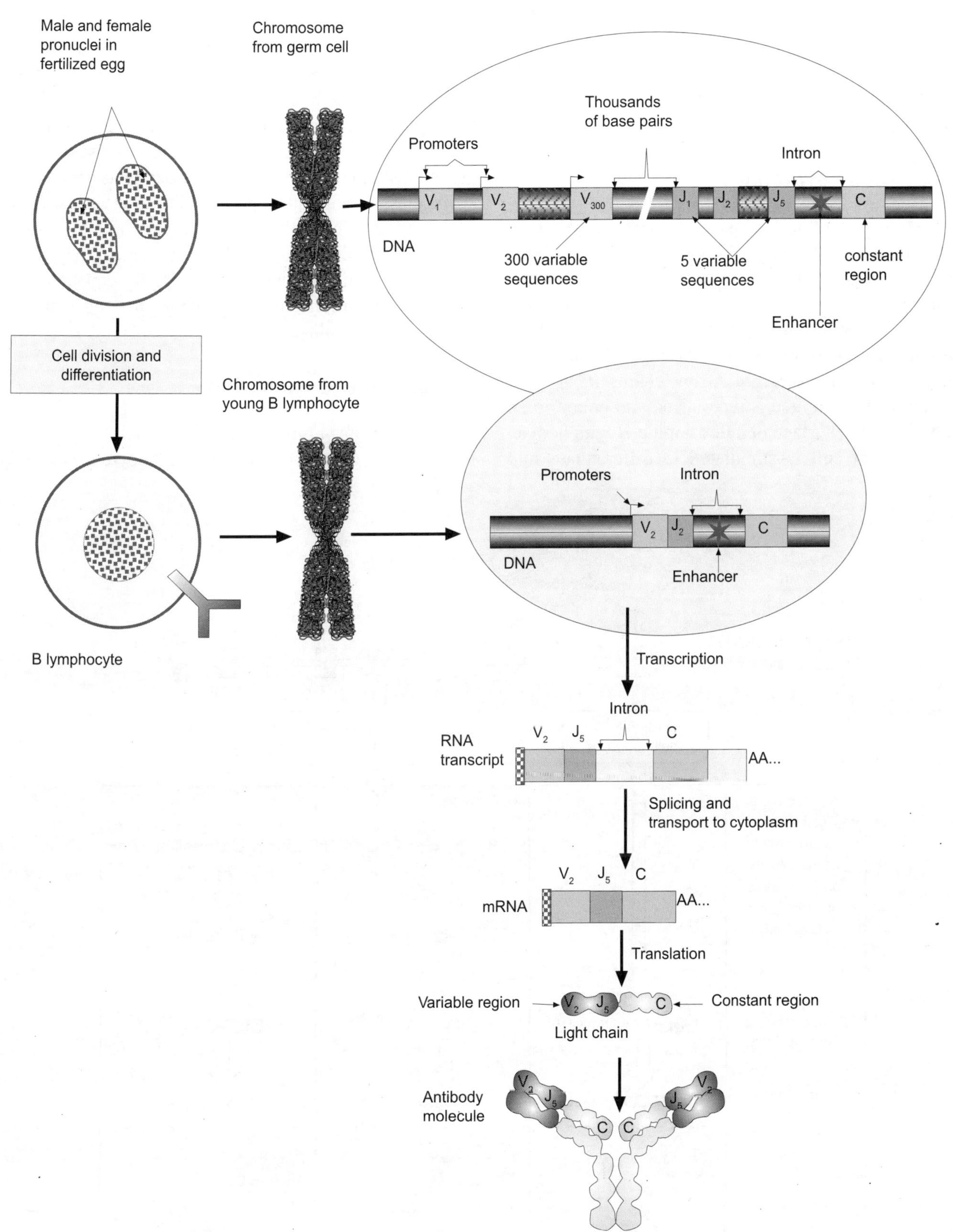

Fig. 18.19 Gene re-arrangement. Assembly of an active gene and its expression in B lymphocytes.

The proteins RAG1 and RAG2 catalyze (both cleavage and rejoining) of DNA for V(D)J recombination. Mice that lack these genes cannot recombine their immunoglobulins or T cell receptors, and consequently have immature B and T lymphocytes.

18.8.2 Class switching

A class of immunoglobulins is defined by, the type of C_H region it contains. At the time that antibody synthesis begins in the developing B lymphocyte, all the C_H gene segments are still present, separated from the newly formed L_H-V_HDJ_H gene segments by a short non-coding sequence. At this stage, all antibodies synthesized will have IgM heavy chains ($C_H\mu$ gene products). If an antigen is recognized and bound to an antibody on the surface of a developing B lymphocyte, however, that cell is stimulated to differentiate into a mature B lymphocyte. During this differentiation, some B lymphocytes switch from producing antibodies of class IgM to producing antibodies of another class (**Fig. 18.20**)—a phenomenon called **class switching**. It often involves further genome rearrangements during which the C_H gene segments closest to the previously joined L_H-V_HDJ_H gene segments are deleted. The class of antibodies produced after class switching is determined by which gene is brought into the closest proximity with the L_H-V_HDJ_H gene segment.

RNA splicing, appropriate signal sequences and somatic mutations at the joining sites also add to antibody variability.

18.8.3 Allelic exclusion

An important point about the genetic control of antibody synthesis is, each B lymphocyte makes only

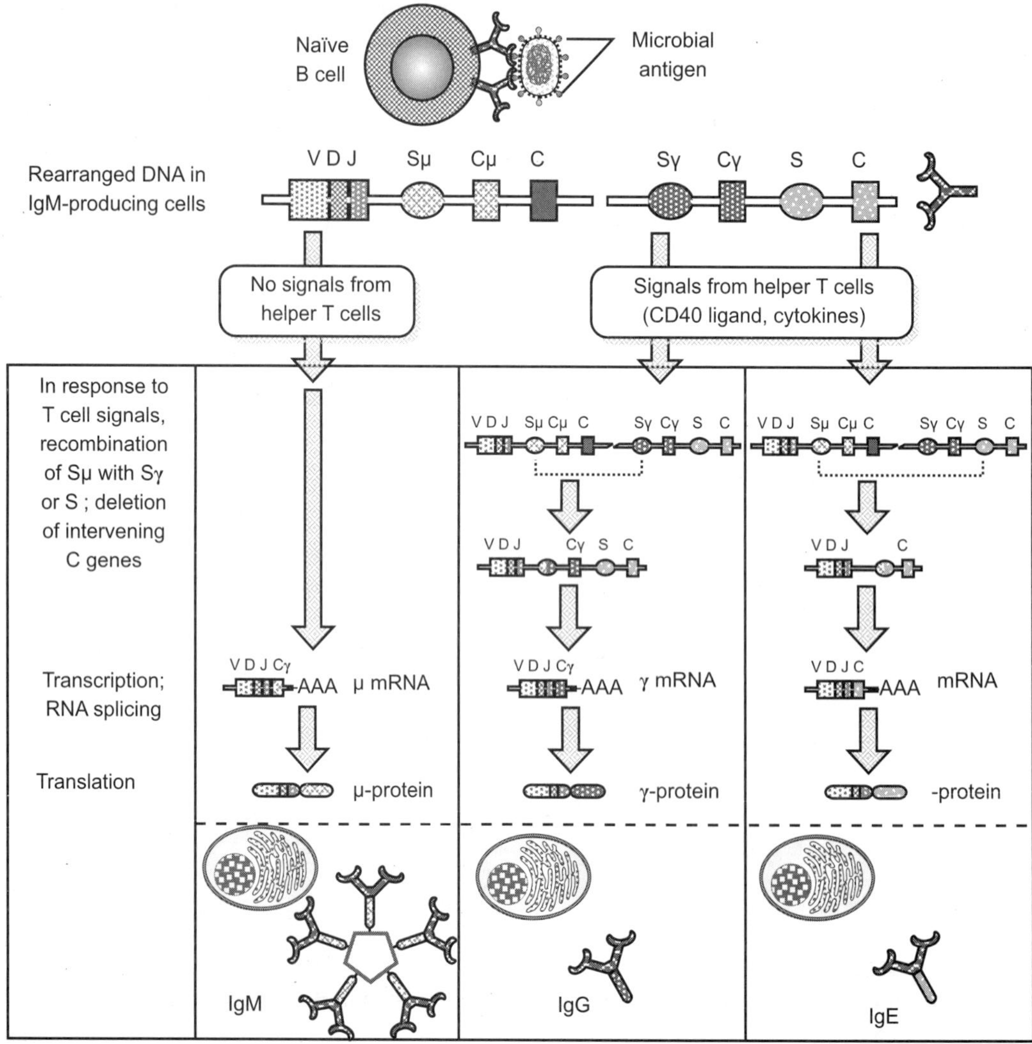

Fig. 18.20 Schematic illustraition of the mechanism of Ig heavy-chain class switching.

one type of antibody even though mammalian cells are diploid and carry two sets of genetic information coding for each of the antibody chains. But only one productive genome rearrangement of light chain coding sequences and one productive genome rearrangement of heavy- chain coding sequences occur in each B lymphocyte! This phenomenon is called **allelic exclusion,** because one of the 'alleles' is excluded from being expressed, i.e., each event involves the genes of only one of the homologous chromosomes, and alleles on other homologous chromosome are not expressed in the same cell.

18.9 ACTIVATION OF B CELLS TO PRODUCE ANTIBODIES

Activation of B cells involves an elaborate sequence of events.

18.9.1 Deletion of potentially self-reacting B cells in the bone marrow

The immature B cells formed in the bone marrow have the potential to produce antibodies to attack just about every macromolecular component in the body. Any B cell, whose antibody, if any, would work against self-component, must be eliminated. This happens during primary maturation in the bone marrow. As they mature, the B cells are exposed to the self-antigens that circulating B cells are likely to meet. Each immature B cell in the bone marrow produces about 10^5 molecules of its antibody and displays them on its surface (fixed into the membrane) inviting, as it were, recognition by an antigen. It does not release the antibody at this stage of its carrier. During primary maturation of a B cell in the bone marrow, if an antigen comes along and combines with its displayed antibody, it is assumed that it is self-component. Hence it must be tolerated (not attacked) by the immune system and functionally eliminated (or killed). After this selective killing, the surviving B cells respond only to foreign antigens, which should be attacked. They are now released into circulation and have a totally different response to meeting their respective antigens. The mature B cells are activated instead of being killed, because the antigen now encountered is alien.

18.9.2 Clonal selection

In 1957, the Australian immunologist F MacFarlane Burnet proposed the clonal selection theory, which quickly gained widespread acceptance. If an organism is exposed to an antigen, it becomes immune to a new infection. Prior to the exposure to a particular antigen, the organism lacks the ability to deal with any toxic effects of that antigen. This ability is acquired during immune response. After the successful defeat of infection, the organism retains the capacity to respond rapidly in the event of a re-infection of same or related type. These features are acquired by clonal selection. Clonal selection theory explains, how antibodies specific to antigens that it has not previously encountered are synthesized. It states that the binding of a particular foreign antigen to an antibody on the surface of a B cell stimulates that cell to divide (**Fig. 18.21**), producing large numbers of the particular B cells (clones), and thus large amounts of the particular antibody that recognizes the foreign antigen. In the pool of immature lymphocytes, the unstimulated B cells and T cells (virgin cells) morphologically appears similar. But on exposure to antigen, a matching B cell antibody binds to the antigen, or a T cell whose receptor can recognize it, gets stimulated to divide, probably by some feedback effect from the surface of the cell, where antibody/receptor–antigen interaction occurs. The stimulated cells then develop into mature B or T lymphocytes, which involve change in morphology, for example, an increase in cell size.

Clonal selection theory also explains how a person can have '**immunological memory**' of the first encounter with an antigen. When an animal first encounters an antigen, it produces a relatively slow primary immune response, because only a small number of its cells have membrane-bound antibody molecules or T cell receptors that can respond to the antigen. The initial expansion of a specific B or T cell population on first exposure to an antigen is called **primary immune response**. However, the next encounter with the same antigen will have rapid enhancement and very high **secondary immune response**, because cells that bind the particular antigen were stimulated to divide by the previous encounter (**Fig. 18.22**). The first stimulation of B or T cells to particular antigen lead not only to mature **antibody-producing plasma cells** or **functional T cells**, but also to expanded clones of long-lived, immature B and T cells called **memory cells**. Memory cells can well respond to the specific antigen in the next encounter, which leads to produce more number of functional B and T effector cells.

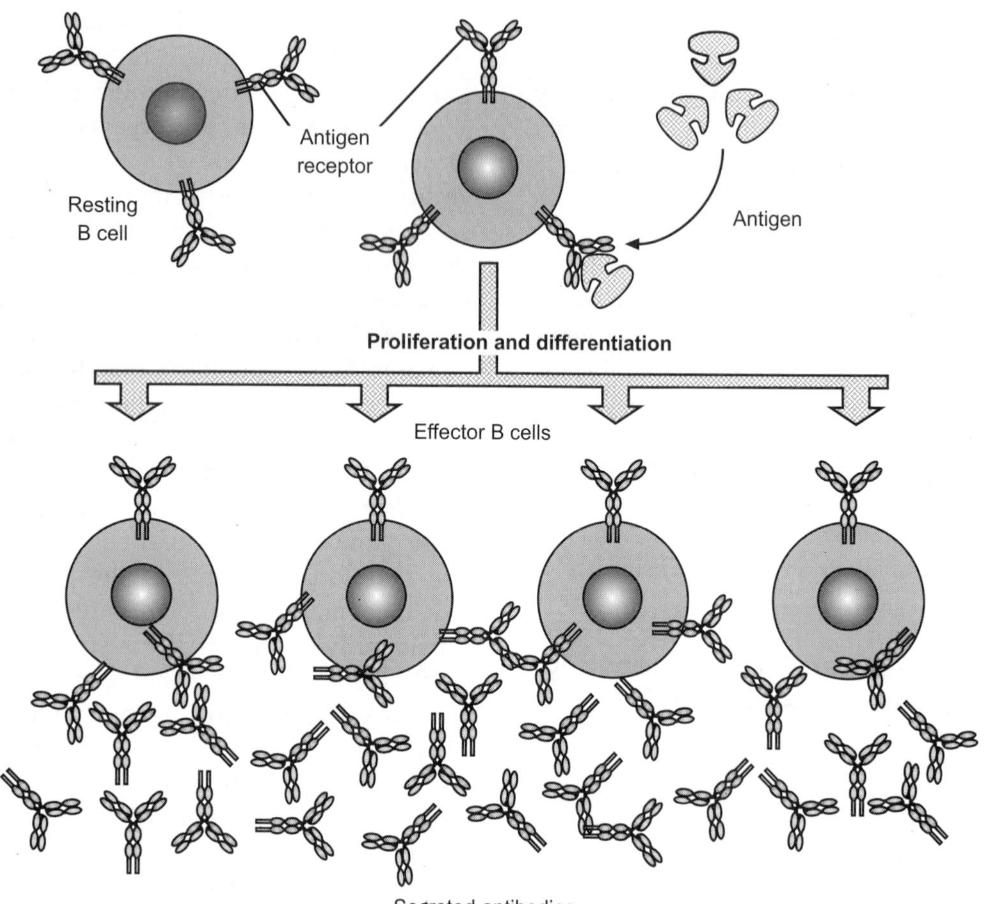

Fig. 18.21 B cell activation. When naive or memory B cells are activated by antigen, they proliferate and differentiate into effector cells. The effector cells produce and secrete antibodies with a unique antigen-binding site, which is the same as that of the original membrane-bound antibodies that served as antigen receptors.

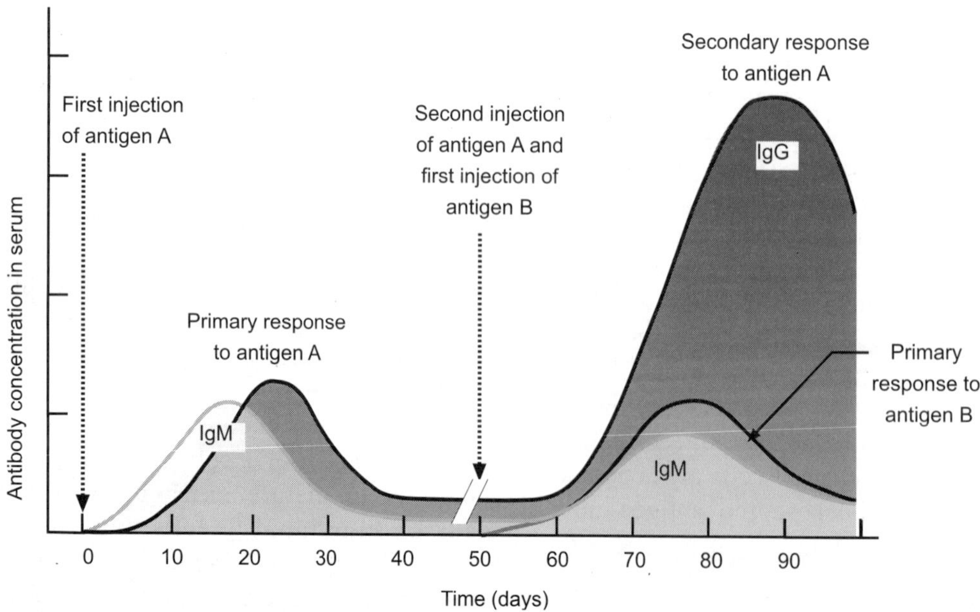

Fig. 18.22 Primary and secondary immune response to first and second dose of antigen. Note that the secondary response is faster and greater than the primary response and is specific for A, indicating that the immune system has specifically 'remembered' encountering antigen A before A.

18.9.3 Antibody secretion

The processes described until now are related to the generation of a highly diverse set of antibody molecules—the key step in the generation of an immune response. The next stage of immune response is the selection of a particular set of antibodies directed against a specific invader. Each immature B cell produced in the bone marrow, expresses a monomeric form of IgM attached to its surface **(Fig. 18.23)**—approximately 10^5 IgM molecules. All these molecules are identical in amino acid sequence and hence, in antigen-binding specificity. Thus, the selection of a particular immature B cell for growth will lead to the amplification of an antibody with a unique specificity. The selection process begins with the binding of an antigen to the membrane-bound antibody.

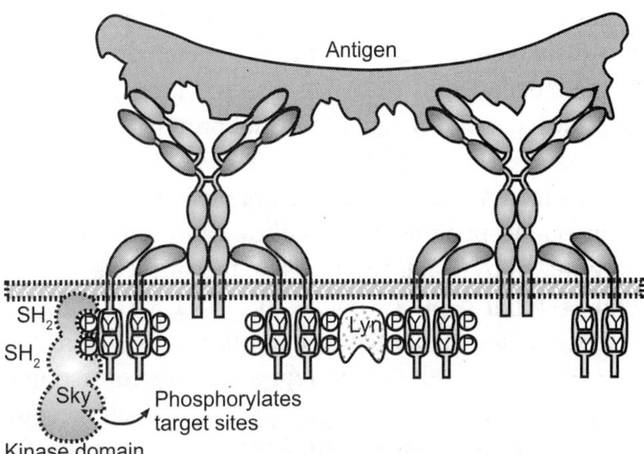

Fig. 18.23 B cell activation. The binding of multivalent antigen links membrane-bound IgM molecules. Which triggers the phosphorylation of tyrosine residues (Y) in the ITAM sequences by protein tyrosine kinases such as Lyn. After phosphorylation, the ITAMs serve as docking sites for Sky, a protein kinase that phosphorylates a number of targets, including transcription factors.

18.10 CELL-MEDIATED RESPONSES

We have discussed the mechanism of humoral immunity, which is mediated by B cells secreting antibody molecules. The target of humoral immunity are the free foreign antigens, i.e., extracellular pathogens, like bacteria and viruses. But they confer little protection against microorganisms that are predominantly intracellular, such as viruses and mycobacteria (that causes tuberculosis and leprosy). These pathogens are shielded from antibodies by the host's cell membrane. To cope with these intracellular pathogens, a different but more subtle strategy, **cell-mediated immunity**, has evolved. The cellular immunity recognizes and destroys cells of the body exhibiting foreign antigens on their surface, regulates and coordinates cell killing and production of antibody. These actions are mostly carried out by the T lymphocytes. T cells, although closely related to B cells, arising from a common precursor cell, do not recognize free antigens. Instead, they recognize only antigens displayed on the surface of other cells of the body.

Major histocompatibility complex (MHC)

Immunological experiments with mice established that ***T cells can recognize antigens only on cells from the same individual***. This identity of an individual is determined by a large complex of genes known as the major histocompatibility complex (MHC). Within the MHC, genes encoding glycoproteins are found on the surface of almost every cell of the body. MHC consists of approximately 15 different genes, most of which are highly polymorphic: over 500 different alleles of MHC genes have been identified **(Table 18.4)**—far more than any other loci in the human genome. The MHC molecules on the surface of cells function as signposts to establish the identity of a given individual. MHC proteins can be subdivided into two major groups: MHC class I and MHC class II molecules.

T cells continually scan the surfaces of all cells and kill those that exhibit foreign markings. The task is very complex since intracellular microorganisms do not leave traces on the surface of their host. Vertebrates, however, have devised an ingenious mechanism, cut and display, to reveal the presence of stealthy intruders. Almost all vertebrate cells exhibit on their surface a sample of peptides derived from the

Table 18.4 Different alleles of MHC class I and class II genes

MHC class II alleles		MHC class I alleles	
Locus	**Number of alleles**	**Locus**	**Number of alleles**
HLA–DRA	2	HLA–A	67
HLA–DRB	179	HLA–B	149
HLA–DQA	18	HLA–C	39
HLA–DQB	29	HLA–E	5
HLA–DPA	8	HLA–F	1
HLA–DPB	69	HLA–G	6
HLA–DMA	4		
HLA–DMB	5		
Total	314		267

digestion of proteins in their cytosol. These peptides are displayed along with integral membrane proteins that are encoded by the MHC genes. More specifically, peptides derived from cytosolic proteins are bound to class I MHC proteins (**Fig. 18.24**). MHC proteins embedded in the plasma membrane tenaciously grip their bound peptides so that they can be touched and scrutinized by T cell receptors on the surface of a killer cell. Foreign peptides bound to class I MHC proteins signal that a cell is infected and mark it for destruction by cytotoxic T cells. A protein assembly consisting of the foreign peptide–MHC complex, the T cell receptor, and numerous accessory proteins triggers a cascade that induces apoptosis (suicidal death) in the infected cell.

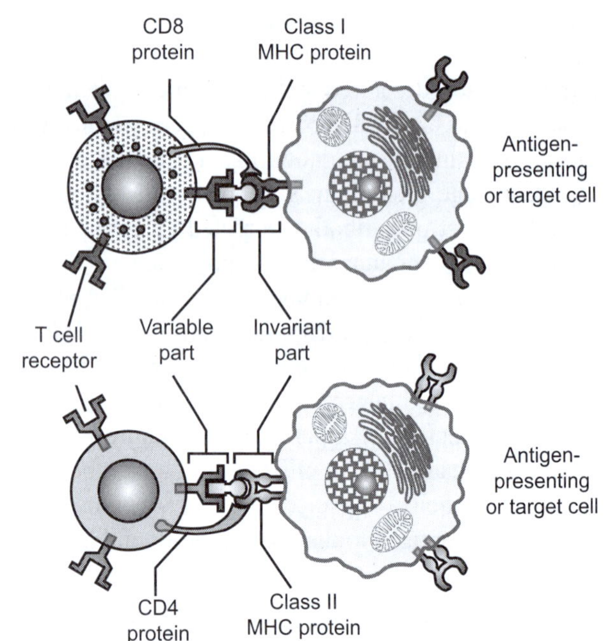

Fig. 18.24 CD4 and CD8 co-receptors on the surface of T cells. Cytotoxic T cells (Tc) express CD8, which recognizes class I MHC proteins, whereas helper T cells (TH) express CD4, which recognizes class II MHC proteins.

Variable and constant regions of T cell receptors

T cells produce one of two types of receptors (TCR). The γ-δ receptor is found on more than 5% of T lymphocytes and synthesized only at an early stage of T cell development. Like Igs, a TCR must recognize a foreign antigen of unpredictable structure. The T cell receptor consists of two 43 kD α (T α) and ß (T ß) chains joined by a disulfide bond (**Fig. 18.25**). Each chain spans the plasma membrane and has a short carboxyl terminal

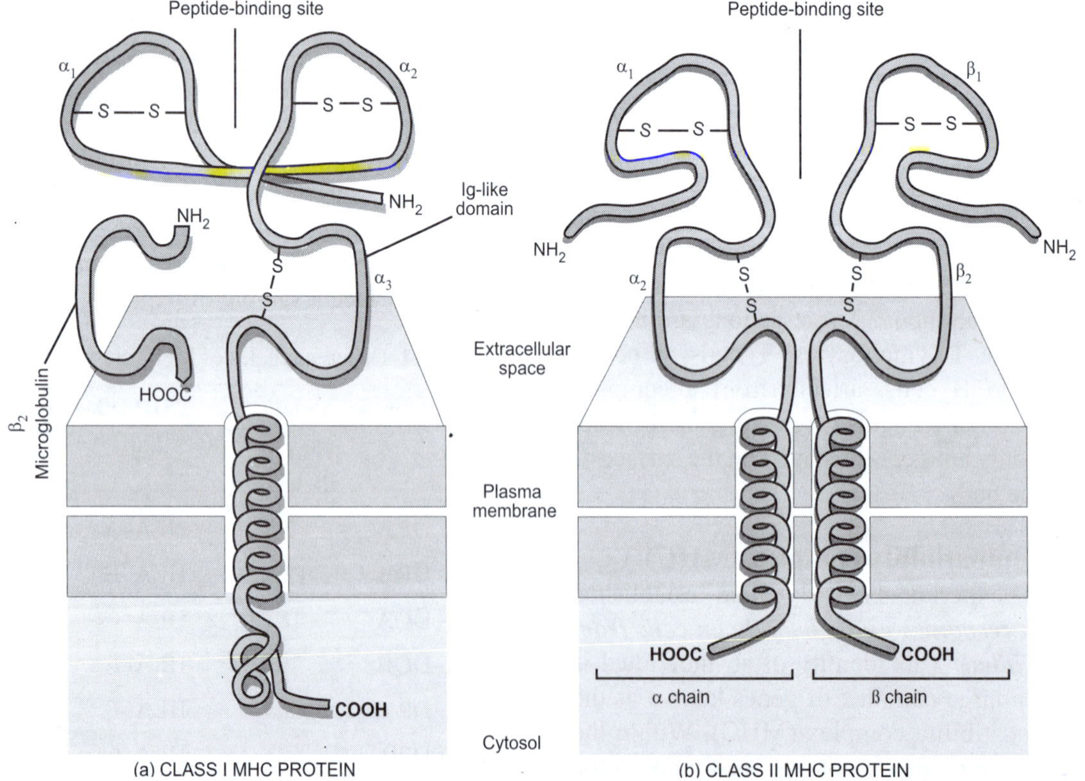

Fig. 18.25 Class I and class II MHC proteins: (a) The α chain of the class I molecules has three extracellular domains (α1, α2 and α3) encoded by separate exons. It is non-covalently associated with a smaller polypeptide chain, β2–microglobulin. (b) In class II MHC proteins, both chains are polymorphic, mainly in the α1 and β1 domains. In both, the two outermost domains interact to form a groove that binds peptide fragments of foreign proteins and presents then to T cells.

region on the cytosolic side. A small proportion of T cells express receptors comprising γand δ in place of αand ß. The T αand T ß chains, like immunoglobulin L and H chains consist of variable and constant regions, and are homologous to the V and C domains of Ig. Assembly of a complete T cell receptor subunit gene entails the recombination of V, D, J and C segments. Furthermore, hypervariable sequences present on the V regions of T α and T ß form the binding site for the epitope. The variable region of T α is encoded by two D-segment genes in addition to 57 V and 13 J-segment genes. Again, the diversity of component genes and the use of slightly imprecise modes of joining them increase the number of distinct T cell receptor proteins formed. At least 10^{12} different specificities arise from combinations of this repertoire of genes.

T cell receptors can recognize a large number of different epitopes. However, all the T cell receptors on a particular T cell have the same specificity. The variable regions of the α and ß chains of the T cell receptor form a binding site that recognizes a combined epitope–foreign peptide bound to a MHC protein. Neither the foreign peptide alone nor the MHC protein alone forms a complex with the T cell receptor. Thus, fragments of an intracellular pathogen are presented in a context that allows them to be detected, leading to the initiation of an appropriate cell-mediated immune response.

Ontogeny of T cells

Leukocytes or white blood cells function in daily housekeeping and defence. All types of blood cells arise from stem cells in bone marrow. T lymphocytes from bone marrow start down a pathway toward a fully differentiated state. When a stem cell migrates from the bone marrow to the thymus, it lacks the cell surface proteins that mediate T cell function, most notably its TCRs. The stem cells proliferate in the thymus to generate a population of T cell precursors. Each of these cells then undergoes the DNA rearrangements that enable it to produce a specific TCR. These cells are then subjected to a complex screening process in the thymus that selects cells with potentially useful T cell receptors (**Fig. 18.26**). In the thymus gland, where helper T (T_H) and killer T cells or cytotoxic T cells (T_C) complete their development.

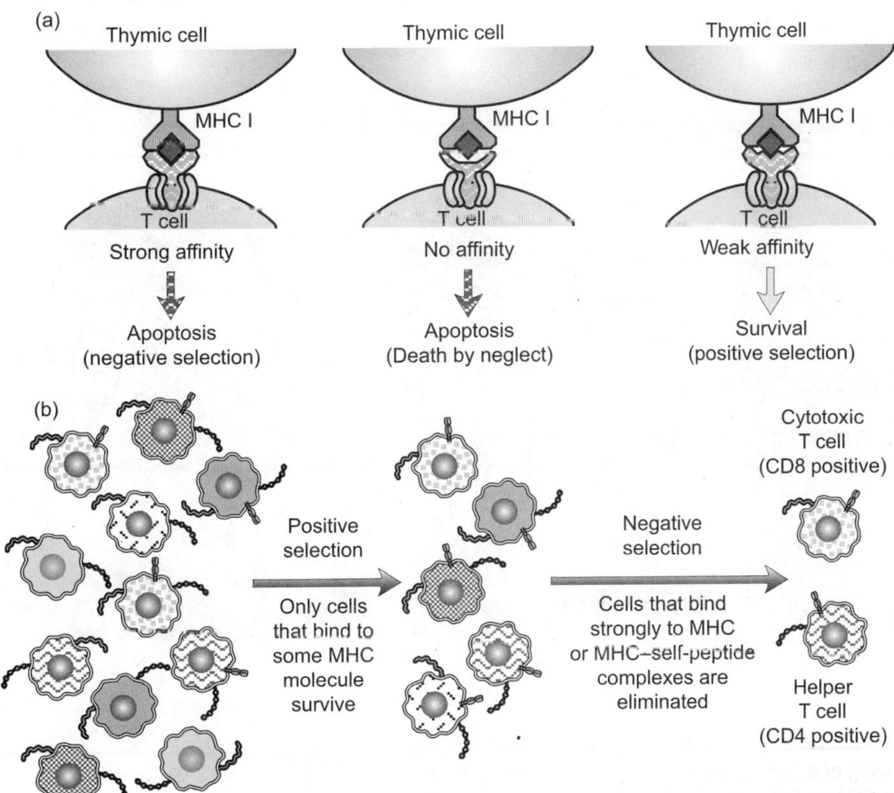

Fig. 18.26 T cell selection: (a) The process of determining the fate of a newly formed T cell in the thymus. (b) In thymus gland, a population of thymocytes is subjected first to positive selection to remove cells that express T cell receptors that will not bind to MHC proteins. The surviving cells are then subjected to negative selection to remove cells that bind strongly to MHC complexes bound to self-peptides.

Cytotoxic T cells

Cytotoxic T cells (T_C cells) have a **CD8 protein**, which is crucial for the recognition of the class I MHC-peptide complex. The abbreviation CD stands for **cluster of differentiation**, referring to a cell surface marker that is used to identify a lineage or stage of differentiation. Each chain in a CD8 dimer contains a domain that resembles an Ig variable domain. CD8 interacts primarily with the relatively constant$_3$ α_3 domain of class I MHC proteins. This interaction further stabilizes the interactions between the T cell and its target. The cytosolic tail of CD8 contains a docking site for Lck, a cytosolic tyrosine kinase akin to Sec. The T cell receptor itself is associated with six polypeptides that form the CD3 complex (**Fig. 18.27**).

Activation of T cells

After leaving the thymus gland the cells circulate in the blood as 'virgin' T cells. These ignore free antigens and unadorned MHC markers on the body's own cells. But they recognize and bind with antigen–MHC complexes on antigen-presenting cells (APCs). The binding of the T cell receptor with the class I MHC-peptide complex and the concomitant binding of CD8 from the T cell with the MHC molecule results in the association of the kinase Lck with the ITAM substrates of the components of the CD3 complex (**Fig. 18.28**).

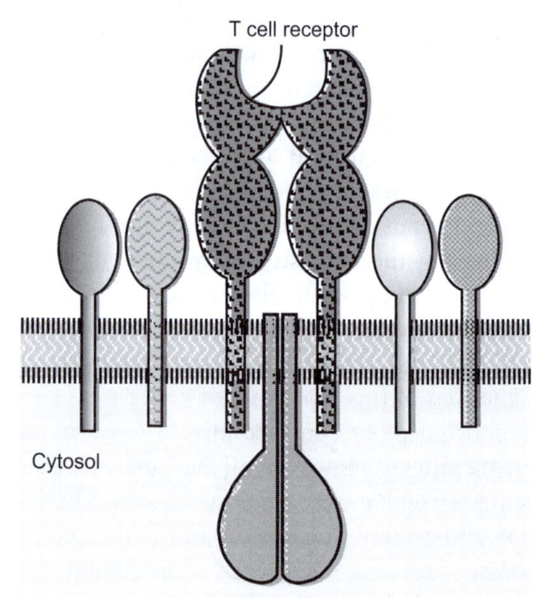

Fig. 18.27 The T cell receptor and its associated CD3 complex. All the CD3 polypeptide chains except zeta chains have extracellular Ig-like domains.

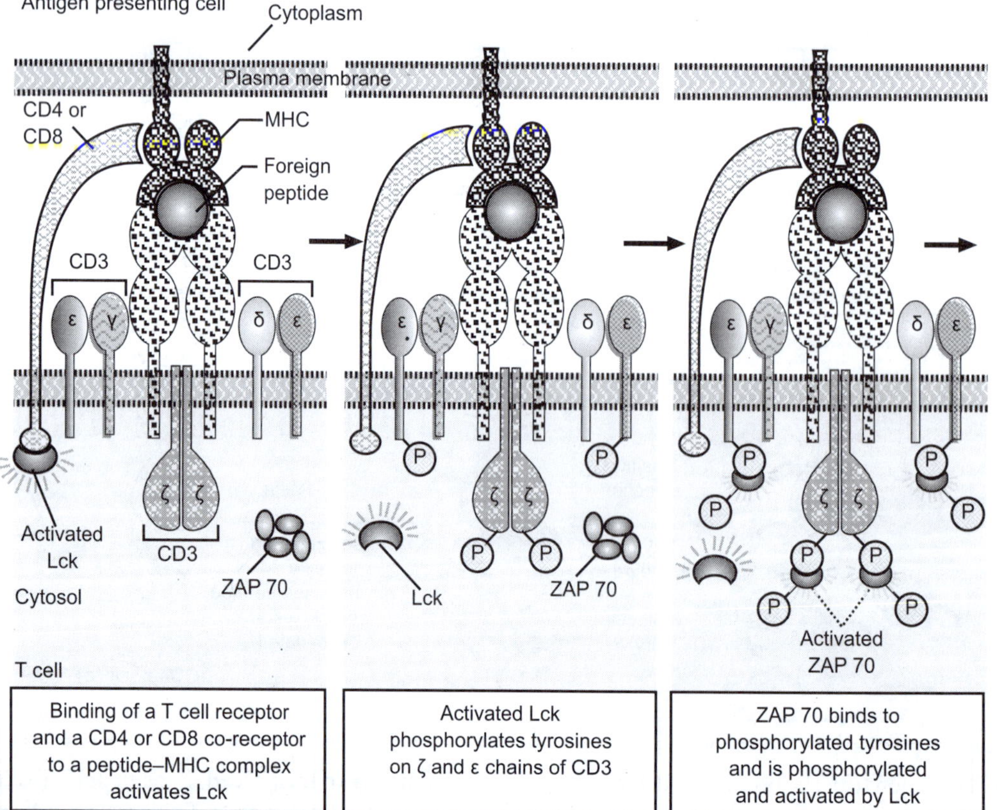

Fig. 18.28 Signalling events initiated by the binding of peptide–MHC complexes with T cell receptors (signal I). (Source: Alberts et al, 2002).

The binding T cells with antigen–MHC complex stimulate them to divide repeatedly and give rise to clones. Their descendents differentiate into subpopulations of effector and memory cells, each with receptors for the antigen. T cell activation has many important consequences. The effector subpopulations of T_H cells secrete interleukins. These secretions fan the repeated cell divisions and differentiation. But the effector T_C cells recognize antigen when it is combined with a particular MHC marker. T_C effector cells destroy infected cells by secreting proteins called **perforins**. These 70 kD proteins form pores (10 nm wide) in target cell plasma membrane. T_C effectors also secrete toxins that disrupt organelles and cause damage on nuclear DNA of their targets (Fig. 18.28). The T_C cells secrete proteases called granzymes into the target cell. These enzymes activate the pathway of apoptosis, leading to the death of target cell and the fragmentation of its DNA. After it has stimulated its target cell to commit suicide, the activated T cell disengages and is stimulated to reproduce. Other T_C cells, including **natural killer cells** (NK cells) recognize tumour cells and kill them on contact.

Helper T cells

Helper T cells (T_H) stimulate the proliferation of specific B lymphocyte and T_C cells, and thereby serve as partners in determining the immune responses that are produced. The importance of T_H cells is known in the case of devastation wrought by AIDS. T_H cells recognize peptides bound to class II MHC molecules. Their helping action is focussed on B cells, macrophages, and dendritic cells. Class II MHC proteins are expressed only on these APC cells, unlike class I MHC proteins, which are expressed on nearly all cells. The peptide presented by class II MHC proteins does not come from the cytosol; it arises from the degradation of proteins that have been internalized by endocytosis, for example, a virus particle that is captured by membrane-bound immunoglobulins on the surface of a B cell (**Fig. 18.29**). This complex is delivered to an endosome, a membrane-enclosed acidic compartment, where it is digested. The resulting peptides become associated with class II MHC proteins, which move to the cell surface. Peptides from the cytosol cannot reach class II proteins, whereas peptides from endosomal compartments cannot reach class I proteins. The association of a foreign peptide with a class II MHC protein signals

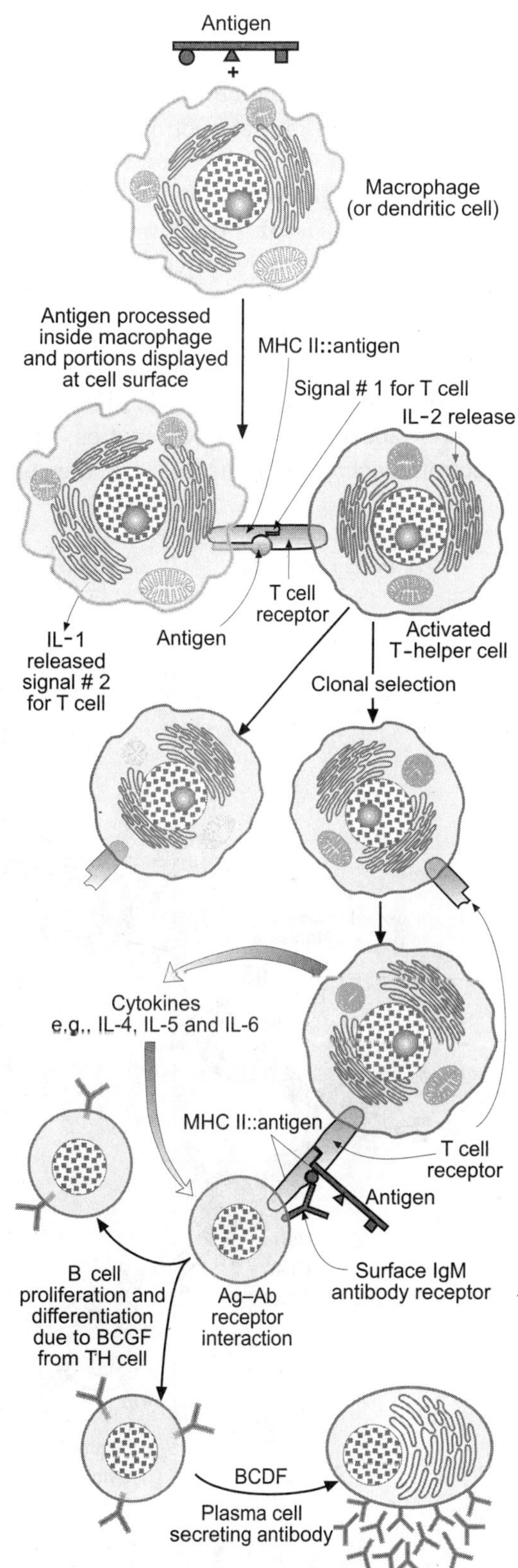

Fig. 18.29 Diagram showing the role of T_H cells in activation of B cell division and production of antibodies by differentiated plasma cell. The activation events begin when macrophage or dendritic cells encounter the antigen. BCGF: B cell growth factors; BCDF: B cell differentiation factors.

that a cell has encountered a pathogen and serves as a call for help. In contrast, association with a class I MHC protein signals that a cell has succumbed to a pathogen and is a call for destruction.

Diversity of MHC proteins

MHC class I and class II proteins, the presenters of peptides to T cells, were discovered because of their role in **transplantation rejection**. A tissue transplanted from one person to an other is usually rejected by the immune system. In contrast, tissues transplanted from one identical twin to another are accepted. Genetic studies revealed that rejection occurs when tissues are transplanted between individuals with different genes in the MHC, a cluster of more than 75 genes playing key roles in immunity. The MHC encodes **class I MHC proteins** and **class II proteins**, as well as **class III proteins** (components of the **complement cascade**) and other proteins that play key roles in immunity. **Figure 18.30** shows various phases of T cell response.

18.11 IMMUNODEFICIENCY DISEASES

The immune system and inter-relationships of its components with other cell types provide the ability to cope with enormous variety of pathogens and other

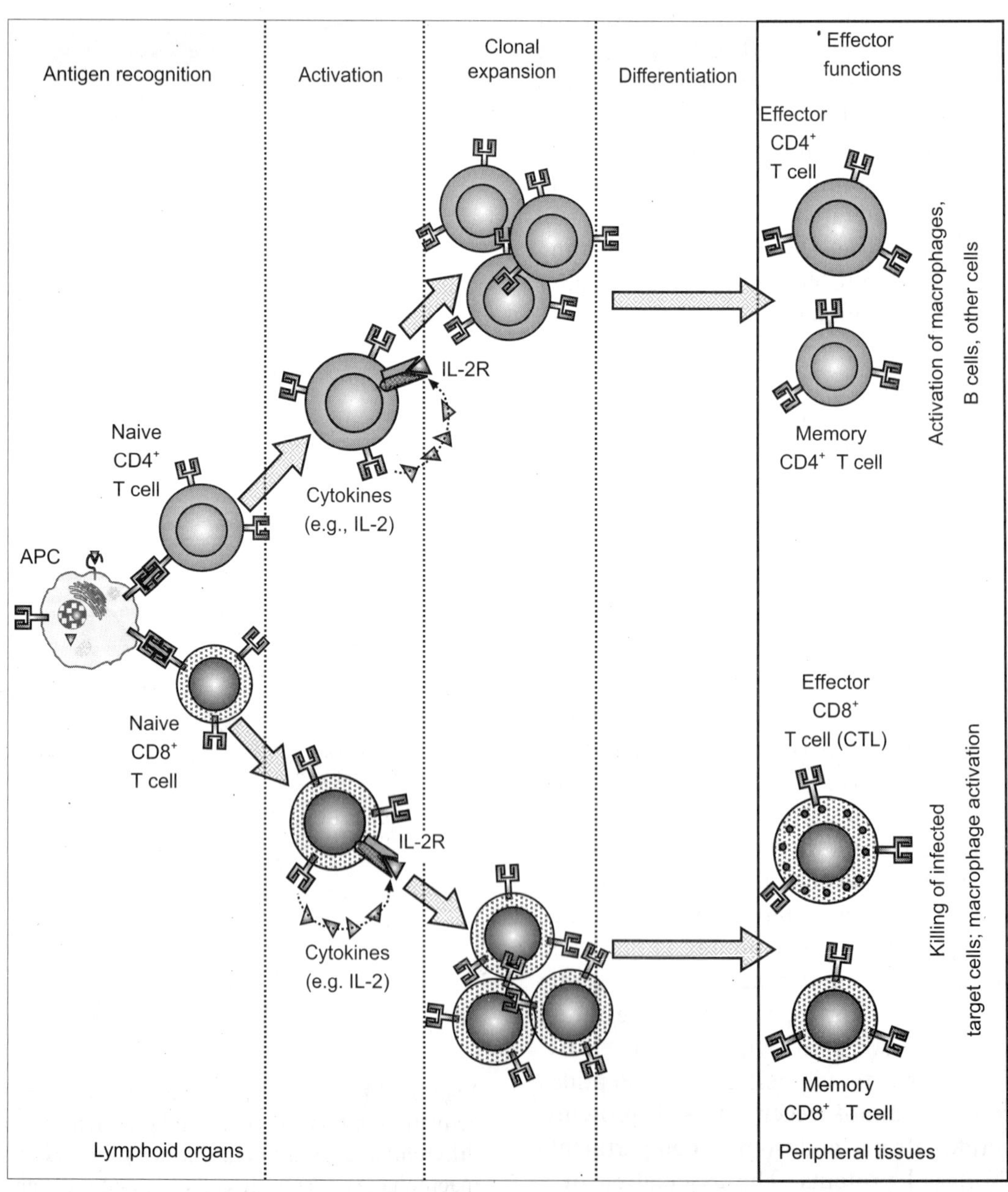

Fig. 18.30 Steps in the activation of T lymphocytes.

cellular abnormalities. It is well established that defects in the development and functions of the immune system (congenital or acquired) result in increased susceptibility to infections and to certain cancers. Disorders caused by defective immunity are called immunodeficiency diseases. Some of these may result from genetic abnormalities in one or more components of the immune system (called congenital or primary immunodeficiencies). Other defects in the immune system may result from infections, nutritional abnormalities, or treatments that cause loss or inadequate function of various components of the immune system (called acquired or secondary immunodeficiencies). In the following sections, a few representative cases are discussed.

18.11.1 Congenital immunodeficiencies

Congenital immunodeficiencies are caused by genetic defects that lead to defective maturation or functions of the different components of the immune system. Congenital immunodeficiencies greatly differ in clinical and pathological manifestations. Some of these disorders include increased susceptibility to infections that may be manifested immediately after birth and may be fatal unless the immunologic defects are corrected. Patients with congenital immunodeficiency are highly susceptible to infections (**Table 18.5**).

Defects in innate immunity

Abnormalities in the two components of innate immunity, phagocytes and the complement system are important causes innate immunodeficiency (**Table 18.6**). For example, chronic granulomatous disease is caused by mutations in the enzyme phagocyte oxidase, which catalyzes the production of microbicidal reactive oxygen intermediates in lysosomes. As a result, neutrophils and macrophages are unable to kill the microbes. To compensate this defect more number of macrophages will be recruited to the site of infection, causing accumulation of phagocytes, which resemble granulomas. Leukocyte adhesion deficiency is caused by mutations in genes encoding integrins or in enzymes required for the expression of ligands for selectins. As a result, blood leukocytes do not bind firmly to vascular endothelium and are not recruited normally to sites of infection. Deficiencies of almost every complement protein, and many

Table 18.5 General features of immunodeficiency diseases

Type of deficiencies	Cytological and laboratory abnormalities	Infectious consequences
B cell	Absent or reduced follicles and germinal centres in lymphoid organs. Reduced serum Ig levels.	Pyogenic bacterial infections
T cell	May be reduced T cell zones in lymphoid organs Reduced DTH reactions to common antigens Defective T cell proliferative responses to mitogens in vitro	Viral and other intracellular microbial infections (e.g., *Pneumocystis carinii*, atypical mycobacteria, fungi) Virus-associated malignancies (e.g., EBV-associated lymphomas)
Innate	Variable, depending on which component of innate immunity is defective	Variable; pyogenic bacterial infections

Table 18.6 Clinical association of primary complement deficiencies

Component	Clinical features
C1 esterase inhibitor	Hereditary angio-oedema Immune complex disease
C1q, C1r, C1s, C4, C2	Immune complex disease, e.g., systemic lupus erythematosus, vasculitis and glomerulonephritis
C3, factor 1, factor H	Pyogenic infection, haemolytic–uremic syndrome
C5, C6, C7, C8, C9	Gonococcal and meningococcal infection
Properdin, factor D	Gonococcal and meningococcal infection

complement regulatory proteins, have been described. The Chediak–Higashi syndrome is an immunodeficiency disease in which the lysosomal granules of leukocytes do not function normally. Consequently, phagocytes and natural killer cells become affected and cells become susceptible to infections.

18.11.2 Acquired immunodeficiencies

Deficiencies of the immune system often develop due to abnormalities acquired later in life. Common causes for serious immunodeficiency in clinical practice are a number of diseases and the therapies used to treat them (**Table 18.7**). The most important of these abnormalities is HIV infection, which will be discussed in the following sections.

Acquired immunodeficiency syndrome (AIDS)

The most widespread and feared immunodeficiency is AIDS (acquired immune deficiency syndrome). AIDS is characterized by the increased susceptibility to viral and microbial infections. In 1981, the first cases of AIDS were recognized. The victims died of rare infections because their immune system was crippled. Two years later Luc Montagnier and coworkers identified the cause. AIDS is produced by the human immunodeficiency virus (HIV). It is estimated that there are more than 42 million HIV-infected individuals worldwide and more than 21 million deaths are attributable to this disease with an annual death toll of more than 3 million. One estimate reported that more than 30% of the population has been infected with HIV, suggesting one of the most devastating afflictions in the history of humankind.

Table 18.7 Important acquired (secondary) immunodeficiency diseases

Causes
Human immunodeficiency virus infection
Protein-calorie malnutrition
Irradiation and chemotherapy treatments, immunosuppressive drugs, bone marrow ablation
Cancer metastases to the bone marrow
Spleenectomy, removal of spleen
Burns
Nephrotic syndrome
Uremia
Chronic inflammatory disease
Persistent infection, e.g., malaria, leprosy
Plasmapheresis

HIV is a retrovirus, enveloped ssRNA virus. An infectious HIV virion enveloped by a lipid bilayer membrane containing two glycoproteins: gp41 spans the membrane and is associated with gp120, which is located on the external face (**Fig. 18.31**). The core of the virus contains two copies of RNA genome and associated tRNAs and several molecules of reverse transcriptase. The genomic RNAs are surrounded by many copies of two proteins called p18 and p24.

Pathogenesis of AIDS: The host cell for HIV is the T_H cell and it can also infect macrophages and dendritic cells. The glycoprotein gp120 molecules

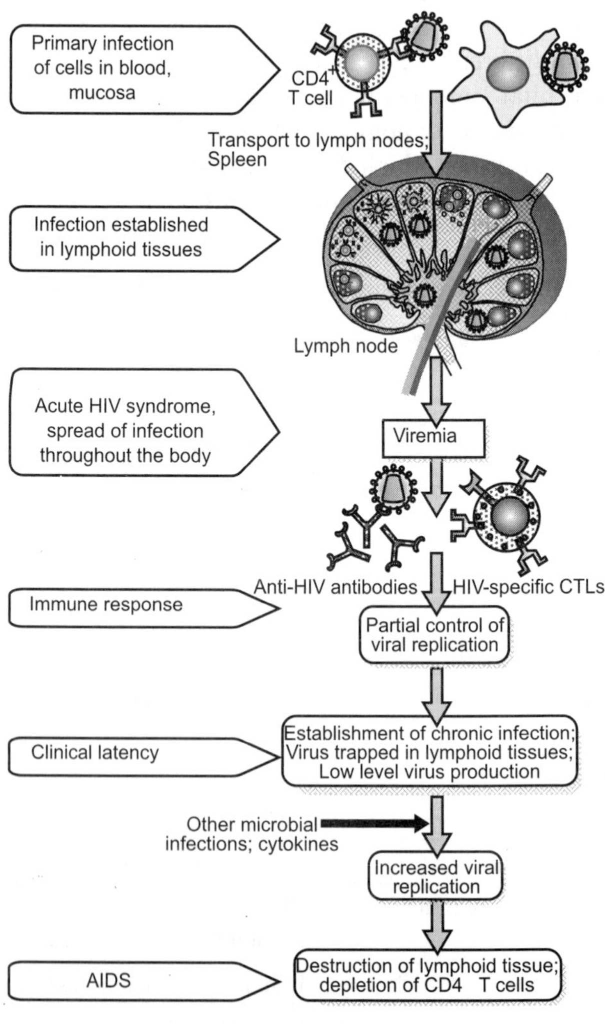

Fig. 18.31 The pathogenesis of HIV. The stages of HIV disease correlated with a progressive spread of HIV from the initial site of infection to lymphoid organs.

on the membrane of HIV bind to CD4$^+$ molecules on the surface of the T_H cell and particular chemokine receptors (CXCR4 and CCR5) on human cells. This interaction allows the associated viral gp41 to insert its amino terminal head into the host cell membrane. The viral membrane and the helper cell membrane fuse, and the viral core is released directly into the cytosol. A DNA copy of the viral RNA is synthesized by the virus reverse transcriptase enzyme and the DNA integrates into the host cell's DNA by the action of the integrase enzyme. The integrated viral DNA is called a *provirus*. The virus can produce a large number of RNAs and then proteins form the core structure, which migrates to the cell membrane, acquires a lipid envelope from the host, and is shed as an infectious viral particle. It is possible for the integrated HIV provirus to remain latent within infected cells for months or years, hidden from the immune system.

Infection by HIV leads to the destruction of T_H cells, because, the permeability of the host plasma membrane is markedly increased by the insertion of viral glycoprotein and the budding virus particles. The influx of ions and water disrupts the ionic balance, causing osmotic lysis of infected cells. AIDS is caused by two major classes of HIV: the common HIV 1 and the much less common HIV 2. The clinical and pathologic manifestations of full-blown AIDS are primarily the result of increased susceptibility to infections and some cancers, as a consequence of immune deficiency. The two most common types of cancers are B cell lymphomas, and tumours of small blood vessels (i.e., Kaposi sarcoma). AIDS patients show defective cytolytic T lymphocyte (CTL) responses to viruses, this is because, the CD4$^+$ are required for full CD8$^+$ CTL responses against many viral antigens.

Therapy and vaccination strategies: The current treatment strategy is aimed at controlling replication of HIV and the infectious complications of the disease. Many drugs that inhibit the activity of viral reverse transcriptase, protease and integrase enzymes have been considerably successful. The treatment, called 'highly active antiretroviral therapy (HAART),' is expensive and its long-term efficacy is not known. Currently, an intensive search for developing effective and affordable vaccines that can cure HIV is ongoing.The treatment should be affordable by all.

18.12 IMMUNOLOGIC TOLERANCE AND AUTO-IMMUNITY

The primary function of the immune system is to protect the host from invasion by foreign organisms. It is capable of reacting to an enormous variety of microbes, but it does not react against each individual's own (self) antigens. This unresponsiveness to self-antigens is also called immunologic tolerance. But the fundamental question is: How does the immune system avoid mounting attacks against the host organisms or distinguish between self and non-self? The selection processes that occur early in the developmental pathways for immune cells kill or suppress those immune cells that react strongly with self-antigens. However, whenever there is failure of these mechanisms of self-tolerance, the immune system can attack the individual's own cells and tissues. Such reactions are called auto-immunity, and the diseases they cause are called auto-immune diseases.

Auto-immunity may result in the production of antibodies against self-antigens or the activation of T cells reactive with self-antigens (**Fig. 18.32**). Autoimmune diseases range from the rare systemic lupus erythematosus (antibodies raised against DNA and RNA) to common diseases like insulin-dependent diabetes mellitus, multiple sclerosis and rheumatiod arthritis. In these illnesses, immune activated against self-antigens result in damage to selective tissues that express the antigen. In many cases, the cause of the generation of self-reactive antibodies or T cells is unclear. However, in other cases, infectious organisms such as bacteria or viruses may play a role. Infection leads to the generation of antibodies and T cells that react with many different epitopes from the infectious organism. If one of these antigens closely resembles a self-antigen, an autoimmune response can result. For example, Streptococcus infections sometimes lead to rheumatic fever owing to the production of antibodies to streptococcal antigens that cross-react with exposed molecules in heart muscle.

18.12.1 Immune responses against tumours

Many studies have shown that adaptive immune system also prevents the outgrowth of transformed cells or destroy abnormal cells before they become invasive tumours. This phenomenon is called

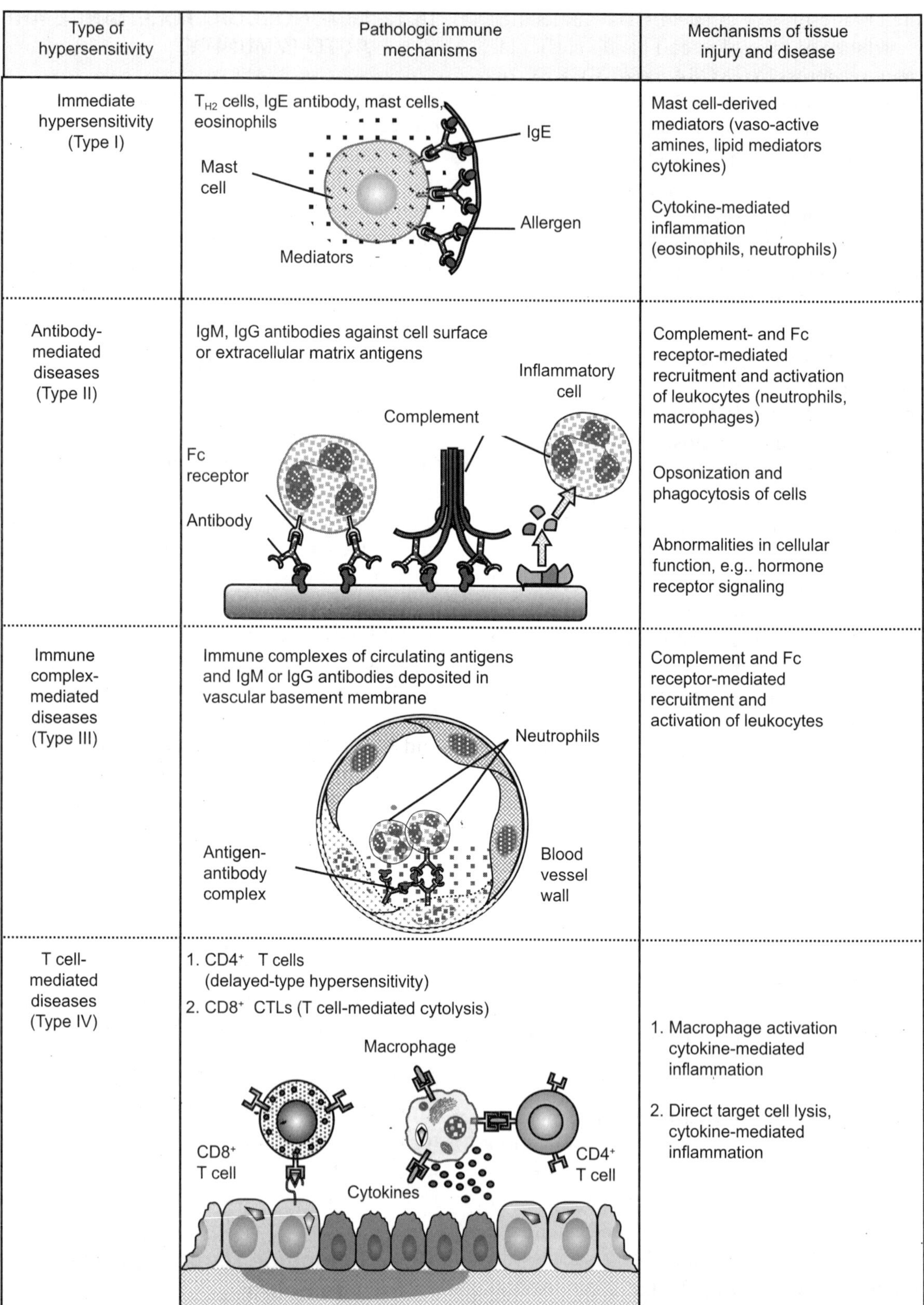

Type of hypersensitivity	Pathologic immune mechanisms	Mechanisms of tissue injury and disease
Immediate hypersensitivity (Type I)	T_{H2} cells, IgE antibody, mast cells, eosinophils	Mast cell-derived mediators (vaso-active amines, lipid mediators cytokines) Cytokine-mediated inflammation (eosinophils, neutrophils)
Antibody-mediated diseases (Type II)	IgM, IgG antibodies against cell surface or extracellular matrix antigens	Complement- and Fc receptor-mediated recruitment and activation of leukocytes (neutrophils, macrophages) Opsonization and phagocytosis of cells Abnormalities in cellular function, e.g.. hormone receptor signaling
Immune complex-mediated diseases (Type III)	Immune complexes of circulating antigens and IgM or IgG antibodies deposited in vascular basement membrane	Complement and Fc receptor-mediated recruitment and activation of leukocytes
T cell-mediated diseases (Type IV)	1. $CD4^+$ T cells (delayed-type hypersensitivity) 2. $CD8^+$ CTLs (T cell-mediated cytolysis)	1. Macrophage activation cytokine-mediated inflammation 2. Direct target cell lysis, cytokine-mediated inflammation

Fig. 18.32 Illustration of four types of hypersensitivity diseases. In each type of hypersensitive reactions different immune effector mechanisms cause tissue damage and disease.

immune surveillance. However, incidences of tumour outgrowth can be seen even in individuals with normal immune system, indicating tumour cells can escape from immune surveillance. **Figure 18.33** shows the different phases of T cell tumour response.

18.12.2 Antibody production by In vitro and genetic engineering

In vitro immunization and fusion of peripheral lymphocytes

A number of laboratories have updated the use of *in vitro* immunization methods. These methods were developed because frequently adequate numbers of antigen-specific B cells are not stimulated during in vivo immunization procedures. Although these methods have some benefits, the main drawbacks are the production of predominantly IgM subclass and the difficulty of standardizing the system. Because human spleenic and tonsillar tissue is largely unavailable, it is preferable to use peripheral blood lymphocytes for production of human monoclonal antibodies. This has been successful only recently, because, procedures have been developed to separate lymphocytes into several subpopulations. These were activated and reconstituted to give a population with a specific B:T cell ratio. The technique is quite intricate, but appears to produce B cells, which are amenable to fusion for production of human monoclonal antibodies.

Monoclonal antibodies

If an animal is injected (immunized) with an antigen, antibodies will be produced in the serum against

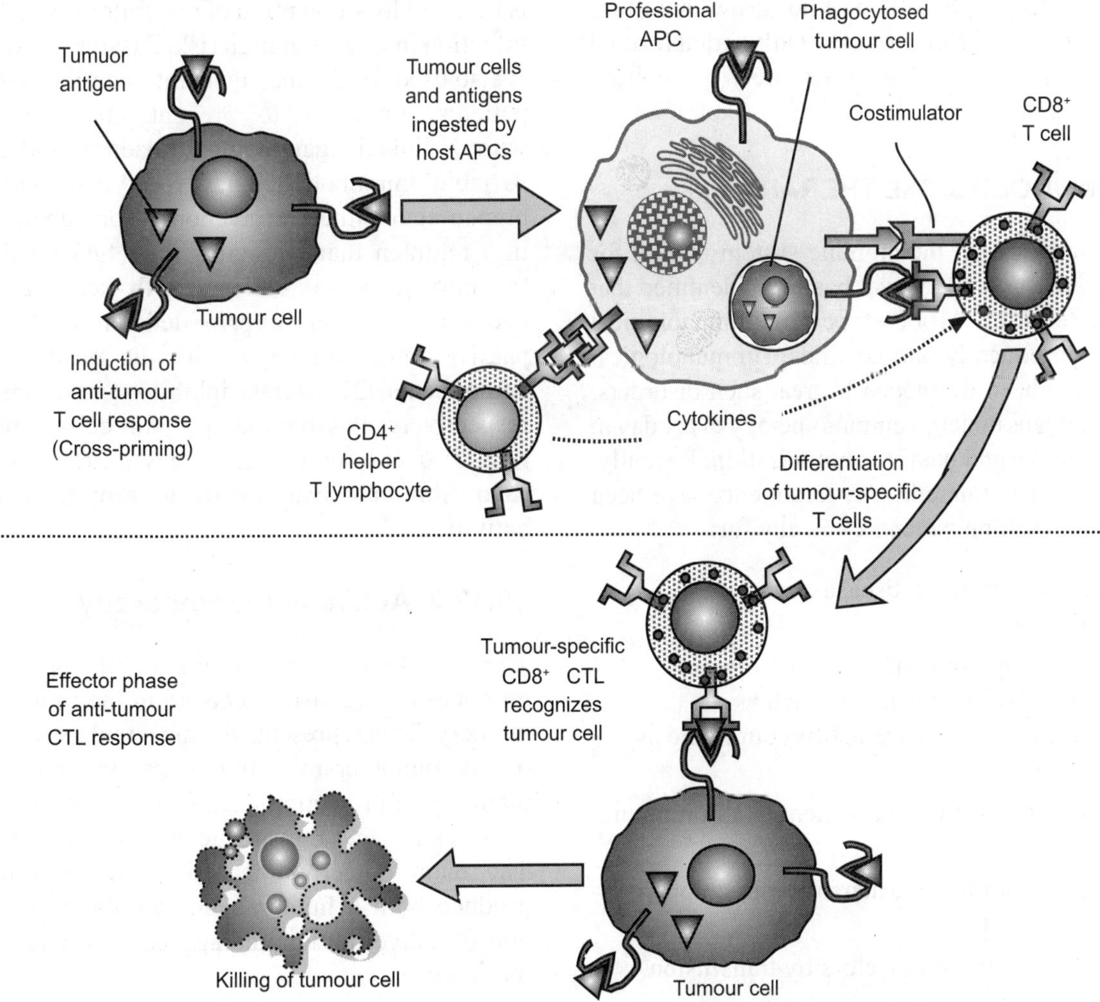

Fig. 18.33 Induction of CD8 T cell response against tumours. CD8 T cell responses to tumours may be induced by cross-priming, in which the tumour cells/antigens are taken up by professional APCs, processed and presented to T cells.

the antigen. From this, an immune serum can be obtained and used for a variety of purposes such as to experimentally detect a particular protein. However, such sera may contain antibodies specific for a wide variety of epitopes of different proteins (polyclonal), so that the antibody of interest will be only a very small proportion of the total. Soon it was realized that a completely specific (pure) antibody, in which every immunoglobulin molecule is specific for the antigen of interest, would be a very powerful tool for molecular analysis.

Cesar Milstein and Georges Kohler from Cambridge were awarded the Nobel Prize for developing this procedure. A given B cell produces only one antibody in terms of binding specificity. If a B cell that produces the wanted antibody could be isolated and grown in culture, the antibody produced would be absolutely pure and therefore absolutely specific (monoclonal). It is possible to isolate such a specific B cell, but B cells will not grow in culture for more than a brief period. Only transformed cancerous cells will grow indefinitely in culture (Channarayappa, 2006).

18.13 IMMUNOLOGICAL THERAPY

Normal functioning of the immune system is vital for human life and survival. Now it has been identified that humans suffer from various types of immunological disorders. Consequently, various kinds of immunological therapies have been developed to treat such disorders. Naturally, humans undergo immunotherapy every day in the form of active and passive immunization. Recently, various types of immunotherapy treatments have been developed for treating patients clinically for:

- Modulation of allergic diseases by desensitization
- modulation of inflammatory diseases by drugs and by biological agents such as monoclonal antibodies against tumour necrosis factor (TNF)
- Suppression of the immune system in autoimmune diseases
- Manipulation of the bone marrow with colony-stimulating factors
- Replacement of missing factors by transfusion
- Treatment of immunodeficiency by bone marrow or stem cell transplants or by direct gene manipulation (gene therapy).

The development of chimeric and humanized monoclonal antibodies made a significant impact on direct use in the treatment of cancer and other diseases.

18.13.1 Passive immunotherapy

Passive immunotherapy refers to the transfer of immunity, usually antibody, to a naive recipient. It can provide instant protection, but does not stimulate the immune system directly by an antigen, so there is no long lasting memory. Passive transfer of antibody will therefore provide protection for no more than 6 months. However, it has been used in conjunction with active immunotherapy to provide immediate protection while awaiting the recipient's response to the active component. Passive immunization is commonly used for protection against tetanus and rabies after exposure. A monoclonal antibody, **palivizumab**, is licensed for prevention of respiratory syncitial virus infection in infants at high risk. Passive immunization is also used for the management of antibody-deficient patients. In order to prevent infection, patients with X-linked agammaglobulinaemia and common variable immunodeficiency receive pooled normal human IgG either intravenously or subcutaneously in a regimen that maintains their IgG levels within the normal range. Patients with hereditary angio-oedema and α_1-antitrypsin deficiency also receive passive immunotherapy with the purified missing components, C1-esterase inhibitor and α_1-antitrypsin, respectively. Passive therapy with cells is used only rarely; occasionally, granulocyte transfusions are used from matched donors to provide phagocytic activity.

18.13.2 Active immunotherapy

The most desirable form of immunotherapy is one that produces a long lasting change in the immunological memory. This represents the most widely used method of immunotherapy. Immunization provides the mainstay of protection against infectious diseases and historically, the Edward Jenner developed method. The main objective of active immunization is to produce both a humoral and cellular response. The aim of active immunotherapy is to produce antibody that will:

- Opsonize bacteria and facilitate phagocytosis
- Bind complement, leading to direct bacterial lysis

- Bind to and neutralize bacterial toxins
- Prevent uptake of viruses into cells by blocking surface receptors
- Enhance phagocytosis of viral particles, preventing dissemination.

Vaccines are developed using either related non-pathogenic organisms or organisms modified to reduce or abolish their capacity to produce disease, a process of **attenuation**. In order to improve vaccines, the antigenic component is often coupled with additional chemical agents that reduce the solubility of the vaccine, ensuring that the release of the active agent is slower (a depot injection) and with agents to increase the immunological response, i.e., **adjuvants** (e.g., alum). It is usual for more than one dose of a vaccine to be required, as the first dose has a priming effect, leading to the production mainly of IgM, while subsequent booster doses lead to the rapid and persistent production of IgG.

Genetic engineering has been used specifically to produce vaccines. Large viruses such as Vaccinia can be used as vectors to carry DNA for antigens of other microbes. Similarly, genetically modified *Salmonella* has been used to target antigens to mucosal macrophages. Recently, many attempts were made to use purified DNA directly as a vaccine, adsorbed on gold particles. On injection into the muscle, this may lead to production of specific proteins, leading to an immune response against the proteins.

Immunotherapy for cancer

Recently, a lot of emphasis has been given to the development of vaccines that will prevent or treat cancer. A major problem has been that immune response to tumours is usually poor. Approaches have included vaccination against pathogens known to be implicated in tumourigenesis, such as papilloma-viruses in cervical cancer, and the development, through genetic techniques, of vaccine against tumour-specific antigens, often coupled with genes for molecules such as IL-2, IL-6 and granulocyte-macrophage colony-stimulating factor, which stimulate the immune response.

Immunotherapy for allergy and for auto-immune diseases

Immunotherapy treatment works by inhibiting immune response. This could involve re-establishing lymphocyte non-responsiveness (i.e., tolerance) to the stimulating allergens or autoantigens, boosting suppressive mechanisms, or diverting the immune response into less damaging pathways. Allergic desensitization is a form of antigen-specific immunotherapy. In this case, the purified allergen is injected subcutaneously on a weekly basis in increasing doses to patients with severe allergies. This leads to a reduction in clinical allergen reactivity that is specific for the allergen injected.

Immunomodulatory therapy

A number of drugs have been used to modulate the immune system for successful transplantation. However, the same drugs are also used, usually in lower doses, to modulate unwanted immune responses in the context of autoimmune diseases. For example, **corticosteroid** therapy has both immunomodulatory and anti-inflammatory properties. Corticosteroids bind within the cell to high affinity glucocorticoid receptors and are transported to the nucleus where they have a direct effect on nuclear transcription. Thus, they prevent the production of essential cytokines and chemokines, reducing lymphocyte activation and proliferation. The effect is not antigen-specific and corticosteroids also act on macrophages and neutrophils to reduce the inflammatory response. However, corticosteroids are lymphocytotoxic and have myriad side effects, altering fat and glucose metabolism, suppressing adrenal function, causing osteoporosis and raising blood pressure. Other immunomodulators such as methotrexate (a folate antagonist), azathioprine (which interferes with nucleic acid synthesis), cyclosporin and tacrolimus (inhibits of calcineurin) and cyclophosphamide (a lymphocytotoxic drug, cross-links DNA and prevents replication).

Cytokines, interferons and colony-stimulating factors

Colony-stimulating factors (CSFs) have been used clinically for some time to treat the cytopenia associated with chemotherapy for malignant disease. The use of CSFs such as GM-CSF and granulocyte CSF (G-CSF) can abolish or reduce the neutropenia and have a beneficial effect in preventing secondary infection. α-Interferon (IFN-α) along with the antiviral drug ribavirin, is used clinically as an antiviral agent as part of the treatment of viral hepatitis, particularly hepatitis C. IFN-γ is used as an adjunct to antibiotic therapy in the treatment of infections in chronic granulomatous disease, which is a potent immuno-stimulator. IFN-ß is used in the management of some cases of multiple sclerosis, although its full efficacy and its therapeutic mode of action are unclear. IL-2 has been used as an adjunctive therapy in the treatment of renal tumours and melanoma, and also in HIV infection. Imiquinod, a new immunostimulatory drug, is recommended for topical treatment of ano-genital warts caused by papillomavirus, and has also proved effective against various types of skin cancer.

Anti-inflammatory and anti-allergic therapy

Although corticosteroids are the most common anti-inflammatory drugs, they are being replaced by newer drugs due to their side effects. Prostaglandins (PG) and leukotrienes are important inflammatory mediators. PGE_2 causes vasodilatation and bronchodilatation, while PGF_{2a} cases bronchoconstriction. Leukotrienes have similar effects to histamine, causing vasodilatation, increased vascular permeability and hypotension. The non-steroidal anti-inflammatory drugs (NSAIDs) work by inhibition of the enzyme cyclooxygenase (COX) to prevent the production of prostaglandins from arachidonic acid. Drugs such as **aspirin** are non-selective inhibitors, but newer drugs such as celecoxib are more selective inhibitors of COX-2, and work preferentially on inflammatory cells while leaving the gastric protective mechanism unaffected.

Anti-allergic drugs: The mainstay of anti-allergic therapy is the use of antihistamines, which block the H_1-histamine receptors on blood vessels and smooth muscle. Chlorphenamine readily crosses the blood-brain barrier and causes marked sedation, but cetrizine and fexofenadine do not readily cross the blood-brain barrier and are therefore less sedative and more potent blocking agents and have longer durations of action. Drugs that interfere with histamine release from mast cells are also used. These include membrane stabilizers such as sodium cromoglicate. These drugs tend to be soluble and are therefore not adsorbed from the gut. Drugs that block cellular calcium channels, such as nifedipine or verapamil, also block histamine release.

Exercises

1. What are the three lines of defence against infection?
2. Explain with a diagram, the major white blood cells.
3. What are the main differences between natural immunity and acquired immunity?
4. Explain the steps involved in activation of compliment determination.
5. Briefly explain the different types of immune activation.
6. Explain the mechanism of antigen processing.
7. Illustrate with diagrams the basic structure of IgG.
8. Explain the steps involved in the production of antibody polypeptide chains.
9. What are the different mechanisms involved in generation of antibody diversity?
10. How are potentially self-reacting lymphocytes removed?
11. Explain briefly the major histo-compatible (MHC I and II) types?
12. Draw a diagram of the ontogeny of T cell.
13. Explain how HIV is related to immuno-modulation.
14. What role does the immune system play in cancer prevention?
15. Explain the different steps involved in genetically engineered antibody production.

Summary

- Animals are subject to infection by viruses, bacteria, protists, fungi and animal parasites.
- During evolution, most organisms have developed several mechanisms that allow them to recognize and ***destroy these infectious agents*** or ***inhibit their invasion***.
- Vertebrates have developed a complex but highly efficient defence system (***immune system***) to guard against a myriad of agents.
- The outer surface of the body and the linings of its internal tracts provide an excellent barrier to penetration by viruses, bacteria and parasites.
- The peripheral lymphoid organs, which consist of the lymph nodes, the spleen, and the mucosal and cutaneous immune systems, facilitate concentration of antigens.
- APCs and lymphocytes interact efficiently with each other to establish ***adaptive immune response***.
- When physical barriers and inflammation are not enough to overwhelm an invader, an infection may become well established.
- Hundreds and thousands of white blood cells (***B and T lymphocytes***) are formed and initiate war against the invaders.
- Blood also contains a group of soluble proteins called ***complement*** that bind to extracellular pathogens, triggering their destruction.
- The complement system may be activated by microbes in the absence of antibody, as part of the innate immune response to infection, and by antibodies attached to microbes, as part of adaptive immunity.
- In response to infections, macrophages and others, cells secrete proteins called cytokines that mediate many of the cellular reactions of innate immunity.
- ***Cytokines*** are soluble proteins that mediate immune and inflammatory reactions. They are responsible for communication between leukocytes and between leukocytes and other cells.
- Another type of innate antiviral response is initiated within the infected cell itself. Virus-infected cells produce substances called ***type 1 interferons*** that are secreted into the extracellular space, where they bind to the surface of non-infected cells rendering them resistant to subsequent infection.
- The immune system comprises two parallel but interrelated systems.
  - Production of antibodies, which are soluble proteins. They combine with the foreign antigen resulting in protective events called ***antibody-mediated response***.
  - Recognition of abnormal cells in the body by the cytotoxic T cells. This is known as ***cell-mediated immunity***.
- To cope with intracellular pathogens, a different cell-mediated immunity has evolved. The cellular immunity recognizes and destroys cells of the body exhibiting foreign antigens on their surface. These actions are carried out by T lymphocytes.
- The binding T cells with antigen–MHC complex stimulate them to divide repeatedly and give rise to *clones*.
- Normal functioning of the immune system is vital for human life and survival.
- Humans suffer from various types of immunological disorders. Consequently, various immunological therapies have been developed to treat such disorders.
- ***Passive immunotherapy*** refers to the transfer of immunity, usually antibody, to a naïve recipient, which provides instant protection.
- ***Active immunization***, on the other hand, produces both a humoral and cellular response.
- A number of drugs have been used to modulate the immune system for successful transplantation.
- ***Corticosteroids*** prevent the production of essential cytokines and chemokines, reducing lymphocyte activation and proliferation.
- ***Colony-stimulating factors (CSFs)*** have been used clinically to treat the cytopenia associated with chemotherapy for malignant disease.
- ***Prostaglandins*** (PG) and ***leukotrienes*** are important inflammatory mediators. The mainstay of anti-allergic therapy is the use of ***antihistamines***, which block the H1–histamine receptors on blood vessels and smooth muscle.

Further Reading

Books

Abbas AK and Lichtman AH (2004), *Basic Immunology: Functions and Disorders of the Immune System*, 2nd edition, Saunders.

Abbas AK, Lichtman AH and Pober JS (2003), *Cellular and Molecular Immunology*, 5th edition. Saunders.

Albert B, Johnson A, Lewis J, Raff M, Roberts K and Walter P (2002), *Molecular Biology and the Cell*, 4th edition, Garland Science.

Channarayappa (2006), *Molecular Biotechnology: Principles and Practices*, Universities Press, India and CRC Press, USA.

Goldsby RA, Kindt TJ and Osborne BA (2000), *Kuby Immunology,* 4th edition, WH Freeman and Company.

Jordan RL (1990), Strategy and techniques for the production of monoclonal antibodies; monoclonal antibody applications for viruses. In Hampton R, Ball E and De Boer S, (eds), *Serological methods for detection and identification of viral and bacterial plant pathogens,* APS Press.

Todd I and Spickett G (2005), *Immunology: Lecture Notes,* 5th edition, Blackwell Publishing Ltd.

References

Ada GL and Nossal G (1987), The clonal selection theory, *Sci. Am.,* 257(2):62–9.

Buckley RH (2000), Advances in immunology: Primary immunodeficiency diseases due to defects in lymphocytes, *New England Journal of Medicine*, 343:1313–24.

Burton DR (1990), Antibody: The flexible adaptor molecule, *Trends Biochem. Sci.* 15:64–69.

Collins M (1991) Death by a thousand cuts, *Curr. Biol.*, 1:140–2.

Davies DR, Padlan EA and Sheriff S (1990), Antibody–antigen complexes, *Annu. Rev. Biochem.*, 59:439–73.

Davis MM (1990), T cell receptor gene diversity and selection, *Annu. Rev. Biochem.*, 59:475–96.

Davis MM (2003), Dynamics of cell surface molecules during T cell recognition, *Annu. Rev. Biochem.*, 72:717–42.

Esser C and Radbruch A (1990), Immunoglobulin class switching molecular and cellular analysis, *Annu. Rev. Immunol.*, 8:717–35.

Fauci AS (1988), The human immunodeficiency virus: Infectivity and mechanisms of pathogenesis, *Science,* 239:617–22.

Gallo RC and Montagnier L (1988), AIDS in 1988, *Sci. Am.*, 259(4):41–8.

Garcia KC, Teyton L and Wilson IA (1999), Structural basis of T cell recognition, *Annu. Rev. Immunol.*,17:369–97.

Guermonprez P, Valladeau J, Zitvogel L, Thery C and Amigorena S (2002), Antigen presentation and T cell stimulation by dendritic cells, *Annu. Rev. Immunol.,* 20:621–67.

Hazenberg MD, Harmann D, Schuitemaker H, and Miedema F (2000), T cell depletion in HIV-1 infection: $CD4^+$ T cells go out of stock, *Nature Immunology*, 1:285–89.

Hennecke J and Wiley DC (2001), T cell receptor–MHC interactions up close, *Cell,* 104:1–4.

Honjo T and Habu S (1985), Origin of immune diversity: Genetic variation and selection, *Annu. Rev. Biochem.*, 54:803–30.

Janeway C and Medzhitov (2002), Innate immune recognition, *Annu. Rev. Immunol.*, 20:197–216.

Janeway CJ (1992), The T cell receptor as a multi-component signalling machine: CD4/CD8 coreceptors and CD45 in T cell activation, *Annu. Rev. Immunol.*, 10:645–74.

Kwong PD, Wyatt R, Robinson J, Sweet RW, Sodroski J and Hendrickson WA (1998), Structure of an HIV gp120 envelope glycoprotein in complex with the CD4 receptor and a neutralizing human antibody, *Nature,* 393:648–59.

Lewis SM and Wu GE (1997), The origins of V(D)J recombination, *Cell,* 88:159–62.

Nossal GJV (1993), Life, death and the immune system, *Sci. Am.*, 269(3):53–62.

Pardoll D (2003), Does the immune system see tumours as foreign or self? *Annu. Rev. Immunol.,* 21:807–39.

Robertson M (1999), Antigen presentation, *Curr. Biol.*, 9:R829–830.

Silverstein AM (2002), The clonal selection theory: What it really is and why modern challenges are misplaced, *Nature Immunol.*, 3:793–96.

Takeda K, Kaisho T and Akira S (2003), Toll-like receptors, *Annu. Rev. Immunol.*, 21:335–76.

Tonegawa S (1985), The molecules of the immune system, *Sci. Am.*, 253(4):122–31.

Tonegawa S (1988), Somatic generation of immune diversity (Nobel lecture in physiology or medicine, 1987), *In Vitro Cell Dev. Biol.,* 24:253–65.

Williams AF and AN Barclay (1988), The immunoglobulin superfamily–domains for cell surface recognition, *Annu. Rev. Immunol.,* 6:381–406.

Wong P and Pamer EG (2003), CD8 T cell responses to infectious pathogens, *Annu. Rev. Immunol.,* 21:29–70.

19 NON-CELLULAR LIFE FORMS

Learning objectives

After reading this chapter, you will be able to:

- List the various forms of non-cellular organisms
- Understand the classification, characteristics and pathognomic features of plasmids, phages, viruses and viroids

19.1 INTRODUCTION

So far we have studied the structure, functions, and various biological processes related to cellular systems. However, there are many other molecular structures, which show some kind of life activities but exist without cells. These are called as **non-cellular** or **acellular** life forms (**Acytota** or **Aphanobionta**).

Non-cellular life (i.e., without a cellular structure) forms were discovered when it was found that the large and complex Mimivirus could make some proteins that are in turn involved in the synthesis of proteins. This discovery suggests the possibility that some viruses may have evolved from earlier forms that could produce proteins independent of a host cell. If so, there may at one time have been a viral domain of life. Also, since most viruses have not been isolated and studied, it is theoretically possible that there may still be some unidentified viruses that can produce their own proteins without the aid of a host cell. It is not clear that all small viruses have originated from more complex viruses by means of genome size reduction. A viral domain of life may only be relevant to certain large viruses such as nucleocytoplasmic large DNA viruses (NCLDVs) like the Mimivirus.

When considering non-cellular life, besides viruses, other forms might also be included, like viroids, satellites, plasmids, phagemids, cosmids, transposons and prions. However, in this chapter the discussion is restricted only to plasmids, bacteriophages, viruses, and viroids.

19.2 PLASMIDS

Plasmids occur naturally in most bacteria and are sometimes found in eukaryotic organisms, yeast, protozoa, and plants. In humans, certain viruses (including human papilloma virus and Epstein–Barr virus) can also exist as plasmids within the cell nuclei. However, plasmids have been most intensively studied in bacteria. The term 'plasmid' was coined by the American molecular biologist Joshua Lederberg in 1952. Plasmid is an extra-chromosomal DNA molecule separated from the chromosomal DNA and capable of autonomous replication. In a majority of the cases, they are covalently closed and supercoiled circles of dsDNA. The size of a plasmid ranges from 1 kbp to over 400 kbp. There can be one to hundreds of the same plasmid in a single cell, or for certain artificial plasmids. Even thousands of copies, Plasmids can be a part of the mobilome, since they are often associated with conjugation, a mechanism of horizontal gene transfer.

Plasmids often contain genes or gene cassettes that confer a selective advantage to the bacterium harbouring them, such as the ability to make it antibiotic-resistant. The genes carried by plasmids represent a mobile reservoir of genetic information that can endow the host bacterial cell with the ability to grow in otherwise hostile environments. Every plasmid contains at least one DNA sequence that serves as an origin of replication (ori), which enables the plasmid DNA to be duplicated independently from the chromosomal DNA. The plasmids of most bacteria are circular (**Fig. 19.1**), but linear plasmids are also known, which superficially resemble the chromosomes of most eukaryotes.

19.2.1 Classification of plasmids

Plasmids are just composed of simple nucleic acids and live in compatible host cells. So far, more than 1000 different kinds of bacterial plasmids have been identified, all sharing a circular DNA structure and the ability to replicate. Plasmids can be classified in a number of ways.

According to phenotype

Although a simple method, because of the abundance of mobile elements and the tendency for plasmids to undergo recombination, it is difficult to devise a methodical system of nomenclature for plasmids based on phenotype alone. Many eukaryotic plasmids lack a phenotype and for others, the phenotype arises from the behaviour of the plasmid, rather than from any encoded function. However, it is possible for plasmids of different types to co-exist in a single cell (seven different plasmids have been found in *E. coli*). But related plasmids are often incompatible, as that only one of them survives in the cell line, due to the regulation of vital plasmid functions. Therefore, plasmids can be assigned into compatibility groups.

- ***F plasmids (Fertility plasmids)***: The differences in sexual behaviour of bacteria, between donors and recipients were found to be caused by an extra-chromosomal DNA plasmid called F, the sex or fertility factor. Donor cells, containing F (called F^+), could transmit this factor in high frequency to recipient cells (F^-), which become, in turn, donor cells. F plasmids contain *tra* genes and therefore are capable of transferring plasmid DNA through conjugation.
- ***R plasmids (Resistance plasmids)***: These were first detected in Japan in 1957 when strains of *Shigella* were found resistant to many of the antibiotics used during a dysentery epidemic. They contain genes that can build to resistance antibiotics, or poisons. These differ from F plasmids in lacking the ability to integrate themselves stably into the bacterial chromosomes. Examples are R1, R46, RK6 and multiple resistance plasmids.

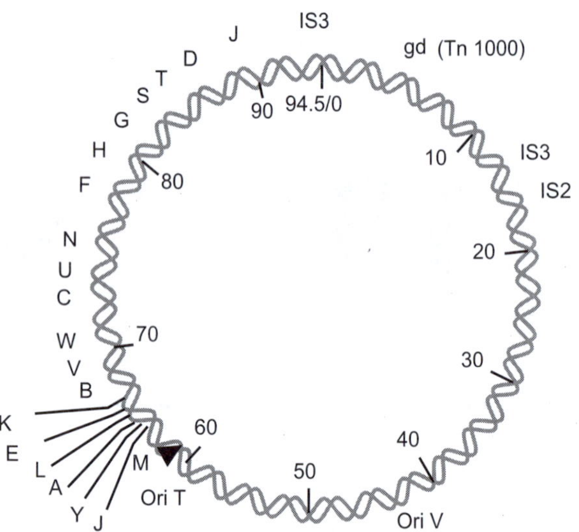

Fig. 19.1 A map of the sex plasmid F. Points are given in kilobases. The capital letters refer to the midpoints of the locations of the corresponding *tra* genes. The insertion sequences, γδ (Tn 1000), IS2, and IS3, are shown shaded. The OriT shows the location of the origin for transfer replication, as well as the direction of transfer. OriV shows the location of the origin for vegatative (non-transfer) replication.

- ***Col plasmids***: These contain genes that code for bacteriocins (colicins), proteins that can kill susceptible bacteria. In *E. coli* strains, different colicins are found (E1, E2, E3, I, K, AgK84, CloDF13, ColE1-K30, ColV, I-K94), each distinguished by specificity for particular susceptible bacterial strains through recognition of special receptor sites on the cell surface. Colicinogenic cells (Col^+) are cells carrying the colicinogenic factor in inhibited or repressed form and are immune to death by colicins of their own type and capable of transferring Col^- cells into Col^+.
- ***Degradative plasmids***: They enable the digestion of unusual substances, e.g., toluene or salicylic acid, e.g., Lac, TOL, Cit.
- ***Virulence plasmids***: These turn the bacterium into a pathogen, e.g., Ent plasmids, Ti, Ri.
- ***Cryptic***: Low molecular weight circular DNA molecules have been reported in many genera of bacteria for which no biological role has been established. For example 2μ, *Mauriceville*.
- ***Recombinant plasmids***: These are constructed *in vitro* from parts of naturally occurring plasmids. *Phagemids*, are chimeric cloning

vectors, containing one or two replication from fl phages; therefore, they replicate like plasmid and upon infection with a helper phage, produce and package ssDNA. *Cosmids* are plasmids that contain the cohesive-end site (*cos*) λ which allows their packaging into λ capsids after they have been ligated to large pieces of DNA. *Phasmids*, are chimeric vectors composed of bacteriophage and plasmid genes. *Shuttle vectors* contain two origins of replication, from very different plasmids or phages, which allow them to replicate in other bacteria in addition to *E. coli*. Some commercially available recombinant vectors are derivatives of pBR322, pML31, pBluescriptII and so on. *Plasmid expression vectors*, a large number of plasmid vectors have been constructed that contain powerful promoters capable of generating large amount of mRNA from cloned genes. In most cases, plasmid expression vectors are designed to express foreign proteins that are not linked to any prokaryotic sequences; more commonly fusion proteins that are encoded partly by the vector and partly by an open reading frame in the cloned segment of the foreign DNA.

According to structure

Although a majority of the plasmids exist as double-stranded closed circles of DNA, a number of single-stranded circular DNA plasmids have been identified in *Streptomyces* and *Clostridium* species, and linear dsDNA plasmids have been isolated from several bacterial and eukaryotic sources (e.g., *Borrelia hermsii*). RNA plasmids also exist. Viroids are specialized circular ssRNA plasmids, which carry no genes. Some bipartite, linear, dsRNA elements found in yeast also conform to the definition of a plasmid. These are known as killer factors because they confer a killer phenotype upon the host.

According to intrinsic properties

Classification based on intrinsic properties (transfer, replication and maintenance) is another way of grouping plasmids, by their ability to transfer to other bacteria. Conjugative plasmids contain *tra* genes, which perform the complex process of conjugation, the transfer of plasmids to another bacterium. Non-conjugative plasmids are incapable of initiating conjugation; hence they can only be transferred with the assistance of conjugative plasmids, by 'accident'. An intermediate class of plasmids carry only a subset of the genes required for transfer. They can 'parasitize' a conjugative plasmid, transferring at high frequency only in its presence. Plasmids are now being used to manipulate DNA and may possibly be a tool for curing many diseases.

19.2.2 Plasmid transfer

From the evolutionary point of view, perhaps no plasmid-borne functions are more important than those permitting plasmid transfers from cell to cell, thereby mediating the flow of genes through a bacterial population. Plasmids have the ability to promote the transfer of themselves and other plasmids from one cell to another, termed conjugative ability. **Conjugation** is defined as the unidirectional transfer of genetic information between cells by cell-to-cell contact. As such, it is not restricted to plasmids, but can occur with any DNA so long as the critical elements are present in the cell. The term 'unidirectional' refers to the fact that a copy of the plasmid is transferred from one cell, termed the 'donor', to another cell, termed the 'recipient'. The prototype conjugative plasmid of *E. coli* is the F plasmid (sex or fertility factor), a large circular plasmid of approximately 95 kbp present in 1–2 copies per cell. Cells carrying the F plasmid are called males (F^+), and are capable of mating with cells lacking the F plasmid (females, F^-). Conjugation is a complex process involving the gene products of a large gene complex within the plasmid (**Fig. 19.2**).

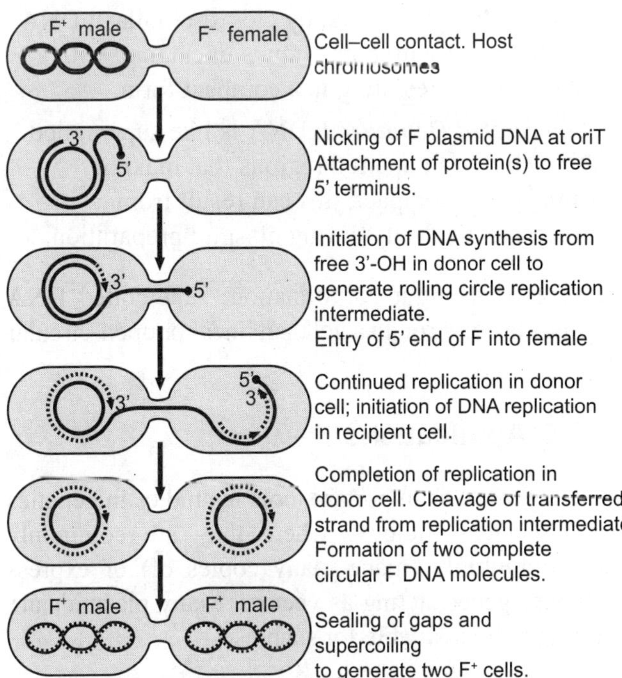

Fig. 19.2 Model for transfer of F plasmid DNA from F^+ male (donor) cell and F^- female (recipient) cell using a rolling circle replication model.

19.2.3 Plasmid DNA replication

Replication and maintenance are totally intrinsic properties applicable to all plasmids. They replicate autonomously, but may be coupled to the replication of the host genome and may be influenced by the host and its environment. Plasmids must be able to sense the division rates of their hosts and alter their own replication rates accordingly. Plasmids, however, require use of the host-encoded replication machinery and encode very few factors of their own required for the initiation of replication and its regulation.

19.2.4 Plasmid conformations

Plasmids may exist in different conformations. Their DNA may appear in one of five conformations, which (for a given size) run at different speeds in a gel during electrophoresis. The conformations are listed below in order of electrophoretic mobility (speed for a given applied voltage) from slowest to fastest:

- ***Nicked open-c*** has free ends, either because both strands have been cut, or because the DNA was linear *in vivo*.
- ***Relaxed Circular***: DNA is fully intact with both strands uncut, but has been enzymatically 'relaxed' (supercoils removed).
- ***Supercoiled*** (or Covalently closed-circular): DNA is fully intact with both strands uncut, and with a twist built in, resulting in a compact form.
- ***Supercoiled Denatured***: DNA is like supercoiled DNA, but has unpaired regions that make it slightly less compact; this can result from excessive alkalinity during plasmid preparation.

Because of its tight conformation, supercoiled DNA migrates faster through a gel than linear or open-circular DNA.

19.2.5 Applications

Plasmids serve as important tools in studies in genetics and molecular biology, where they are commonly used to multiply (make many copies of) or express particular genes, acting as vectors. Many plasmids are commercially available for such uses.

Plasmid vectors

Generally, plasmids used in genetic engineering are called vectors. They are used to transfer genes from one organism to another, which contain a genetic marker conferring a phenotype that can be selected for or against. Most also contain a multiple cloning site (MCS, or polylinker), which is a short region containing several commonly used restriction sites allowing the easy insertion of DNA fragments at this location. The insertion of the target DNA inactivates the marker gene. This is the basis of the selection of the recombinant and this process is known as **insertional inactivation**.

Gene cloning

The gene to be cloned is inserted into copies of a plasmid, which contains genes that make cells resistant to particular antibiotics. Next, the plasmids are inserted into bacteria by a process called transformation. Then, the bacteria are exposed to the particular antibiotics. Only bacteria, which take up copies of the plasmid, survive in the antibiotic, since the plasmid makes them resistant. In particular, the protecting genes are expressed (used to make a protein) and the expressed protein breaks down the antibiotics. In this manner, the antibiotics act as a filter to select only the modified bacteria. Now these bacteria can be grown in large amounts, harvested and lyzed. A large number of plasmid clones can be isolated from cell lysate.

Production of recombinant proteins

Another major use of plasmids is for making large amounts of proteins. In this case, researchers grow bacteria containing a plasmid harbouring the gene of interest. Just as the bacteria produce proteins to confer antibiotic resistance, they can also be induced to produce large amounts of proteins from the inserted gene. This is a cheap and easy way of mass-producing a gene or the protein, for example, insulin or even antibiotics.

Plasmid-free strains

In the absence of selective pressure, having the plasmid is irrelevant to the host, and the plasmid is genetically dispensable. Under these circumstances, the plasmid could conceivably represent a genetic burden to the cell and would, therefore, likely be lost from populations. In research laboratories, sometimes, it is necessary to isolate plasmid-free derivative from a strain currently containing a plasmid. The procedure is termed **curing**, and can happen:

- Spontaneously (perhaps replicating isolated colonies if the plasmid confers a scorable phenotype)

- Following an enrichment (again, if the plasmid confers a growth phenotype)
- By selecting a different, but incompatible, plasmid into the cell
- By treatment with elevated temperature or chemicals such as acridines, ethidium bromide, sodium dodecyl sulphate (SDS) and novobiocin (since the first two chemicals are known as mutagens, they should be used with restraint).

19.3 BACTERIOPHAGE

A bacteriophage (from 'bacteria' and Greek, *phagein*, 'to eat') is one of a number of viruses that infect bacteria. The term is commonly used in its shortened form, phage. Typically, bacteriophages (or phage) consist of an outer protein hull enclosing genetic material usually ssRNA (single-stranded RNA), dsRNA, ssDNA, or dsDNA between 5 bp and 500 kbp long with either circular or linear arrangement. Phages are much smaller than the bacteria they destroy—usually between 20 nm and 200 nm in size (**Fig. 19.3**).

Phages are estimated to be the most widely distributed and diverse entities in the biosphere. They are ubiquitous and can be found in all reservoirs populated by bacterial hosts, such as soil or the intestine of animals. One of the densest natural sources for phages and other viruses is seawater, where up to 10^9 virions per millilitre have been found at the surface, and up to 70% of marine bacteria may be infected by phages. They are also found in drinking water and in some foods, including fermented vegetables and meats e.g., pickle, salami, where they serve the function of controlling the growth of bacteria.

They have been used for over 60 years as an alternative to antibiotics in the former Soviet Union and eastern Europe. They are now seen as a hope against multidrug-resistant strains of many bacteria. However, in the case of methicillin-resistant *Staphylococcus aureus* (MRSA), an infecting phage causes the bacteria to become more virulent and difficult to contain.

19.3.1 Discovery of phages

Since ancient times, there have been documented reports of river water having the ability to cure infectious diseases, such as leprosy. In 1896, Ernest Hanbury Hankin reported that something in the waters of the Ganga and Yamuna in India had marked antibacterial action against *Vibrio cholerae,* a cholera-causing bacteria and could pass

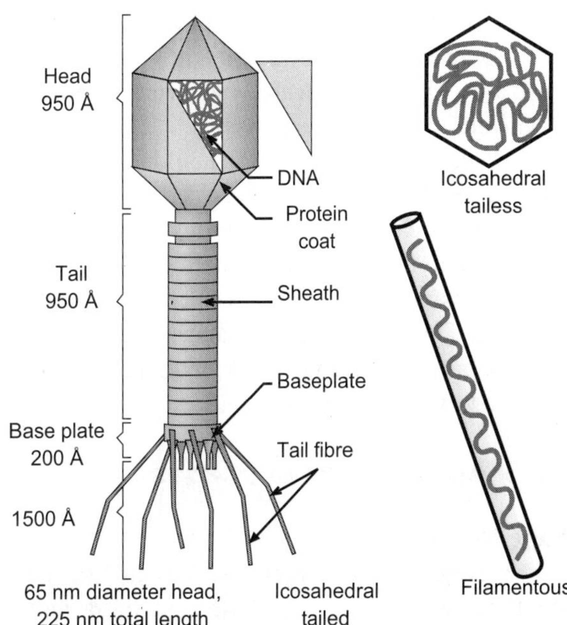

Fig. 19.3 Three basic phage structures (not drawn to scale).

through a very fine porcelain filter. In 1915, British bacteriologist Frederick Twort, superintendent of the Brown Institution of London, discovered a small agent that infected and killed bacteria. Independently, French-Canadian microbiologist Félix d'Hérelle, working at the Pasteur Institute in Paris, announced on September 3 1917 that he had discovered 'an invisible, antagonistic microbe of the dysentery bacillus', a virus parasitic on bacteria.' D'Hérelle called the virus a bacteriophage or bacteria-eater.

Hankin's discovery of antibacterial action in the waters of the Ganga and Yamuna against *Vibrio Cholerae*, has substantiated the Hindu religious belief that a dip or a drink of water of the Ganga could cure antibiotic-resistant bacterial infection. The waters of Ganga when added in adequate amount, to water from from other sources caused the bacteriophage in the latter to quickly multiply and clean thus the water. This is why since ancient times Indians have been carrying Ganga water back home to clean their local water resources. People practice it even today, except that most of them do not know the scientific basis.

19.3.2 Bacteriophage in the environment

Phages are found to be more abundant in the water column of freshwater and marine habitats than previously thought and they can cause significant mortality of bacterioplankton. Phages are not only the most abundant biological entities but probably also the most diverse ones.

A broad number of food products, chemicals, and biotechnology products are manufactured industrially by large-scale bacterial fermentation of various organic substrates. The risk that bacteriophage contamination could rapidly bring fermentation to a halt and cause economic setbacks is a serious threat in these industries. The relationship between bacteriophages and their bacterial hosts is very important in the food fermentation and dairy industries. Defence strategies and systems to curtail the propagation and evolution of phages have been developed over the decades.

19.3.3 Applications of phages

Phage therapy

Felix d'Herelle was highly successful in treating dysentery in rural France, cholera in India and later cured diseases like typhoid, bubonic plague, wound infections, avian typhosis and hemorrhagic septicemia in buffalo using phage therapy, whereas other early attempts to treat infections with phages gave mixed results. Antibacterial use of phages continued in the former Soviet Union, with some success. Only recently, has this technology again gained popularity in the United States, due to emergence of many antibiotic-resistant strains of bacteria. The manner in which phages are applied depends on the type of condition from which a patient suffers. They can be applied topically, during surgery or orally. They can also be applied by means of injection. The nature of the immune system (that automatically fights against any viruses introduced into the bloodstream or lymphatic system) means that injected phages are only used on rare occasions, such as when no other method of application would be effective. Some benefits of phage therapy are given below:

- Phages are highly specific, whereas antibiotics kill all bacteria without specificity; beneficial bacteria (e.g., *E. coli* in the intestinal tract) that perform crucial functions, are also affected by antibiotics and harmful pathogens can then grow more easily. Secondary infections like those by the *Pseudomonas* species or *Clostridium difficile* cause severe diarrhea and colon infections.
- Phages specifically target the harmful bacterium, eliminate it, and leave the beneficial bacteria intact.
- Bacteria eventually grow resistant to antibiotics, but phages are able to change and adapt in ways that make them a constant threat to bacteria. All viruses change and adapt themselves in time, and phages are no exception.
- Phages cannot cause disease to humans, animals, or plants; they can only kill the harmful bacteria.
- For almost all known bacterial species there exists one or more phages specific to that species
- Due to their exponential rate of self-replication, usually a small dose of phages is sufficient for curing a bacterial infection. Phages can penetrate deep into an infection and destroy whole of the particular bacterium growing there. Antibiotics, on the other hand, often encounter difficulty penetrating deep bacterial infections and delivery of the antibiotic to all of the bacteria then becomes a significant obstacle. This obstacle rarely exists in the case of phages.
- In addition to being self-replicating, phages are also self-limiting. When all of the bacteria are infected with phages, their numbers start to decline and the number of phages also decreases. Phages require their specific bacterium in order to exist and, in the absence of that specific bacterium, are eliminated rapidly.
- Preparations of phages are highly stable and can be dispersed in any media. They can be stored for long periods, and have a low cost of production.
- Bacteriophage therapy is currently being used to treat post-burn bacterial infections. Within 24 hours, burn patients can start suffering from opportunistic bacterial attacks. As an alternative to treating post-burn bacterial infections by antibiotics, phages have been in use in certain parts of the world, such as Georgia (a former Soviet republic and now a member of the Commonwealth of Independent States (CIS)) and Poland, and this approach has now been more widely recognized. Results have shown that bacteriophage therapy has an 80% success rate against *Enterococcus* infections and up to 90% against *Staphylococcus aureus*, *Pseudomonas aeruginosa*, *E. coli* and *Klebsiella pneumoniae*. *P. aeruginosa* is the most common pathogen for post-burn infection, and it is known to be notoriously resistant to a variety of antibiotics.
- In addition to treating human illnesses, phages are being used to treat illnesses in livestock. A bacteriophage which is highly active and

rapidly lytic in vitro and which attaches to the K1 capsule antigens of bacteremic strains of *E. coli* has been very effective in preventing and treating septicemia and cerebritis or meningitis in chicken. The bacteriophage treatment was shown to be more effective than multiple doses of antibiotic. Another practical use of phages is for bacterial identification through a process called **phage typing**, which is the use of patterns of sensitivity to a specific battery of phages to precisely identify microbial strains.

- Enzymes from phages have also been isolated and shown to have antibacterial qualities. One such enzyme is PlyG, which is a type of bacteriophage lysin. The lysins, produced by the phage, translocate from the bacterial cytoplasm into the cell wall matrix, where they rapidly hydrolyze covalent bonds essential for peptidoglycan integrity, causing bacterial lysis and concomitant release of progeny phages. PlyG is produced by the phage gamma, which infects *Bacillus anthracis*, the bacterium that causes anthrax. Isolated PlyG has been shown to kill *B. anthracis* in vitro and in vivo.
- Selecting new phages to combat an infection is a much quicker process than finding new antibiotics, which sometimes can take up to several years.
- Phages replicate at the site of infection, but antibiotics are eliminated from the body after they are metabolized.

Disadvantages of phage therapy: The sophisticated ability of phages to destroy their bacterial hosts can also have negative commercial impact; phage contaminants occasionally spread havoc and financial disaster for the various fermentation industries that depend on bacteria, such as cheese production and fermentative synthesis of chemicals. Nonetheless, the use of phages may soon replace current antibiotics in the treatment of bacterial infections.

Other areas of bacteriophage use

In August, 2006 the United States Food and Drug Administration (FDA) approved using phages on cheese to kill the *Listeria monocytogenes* bacteria, giving them GRAS status (Generally Recognized as Safe). In July 2007, the same phages were approved for use on all food products. Government agencies in the West have for several years been looking to Georgia and the former Soviet Union for help with exploiting phages for counteracting biological weapons and toxins e.g., anthrax, botulism. Other uses include spray application in horticulture for protecting plants and vegetable produce from decay and the spread of bacterial diseases. Other applications for phages are as biocide e.g., hospitals and as a preventive treatment for catheters and medical devices prior to use in clinical settings. Technology is now available for phages to be applied to dry surfaces, e.g., uniforms, curtains and sutures for surgery.

19.3.4 Classification of phages

Phages are classified by the International Committee on Taxonomy of Viruses (ICTV) based on morphology and nucleic acid (**Table 19.1**).

Table 19.1 ICTV classification of phages

Order	Family	Morphology	Nucleic acid
Caudovirales, Myoviridae	Non-enveloped	Contractile tail, linear	dsDNA
Siphoviridae	Non-enveloped	Long non-contractile tail, linear	dsDNA
Podoviridae	Non-enveloped	Short non-contractile tail, linear	dsDNA
Tectiviridae	Non-enveloped	Isometric, linear	dsDNA
Corticoviridae	Non-enveloped	Isometric, circular	dsDNA
Lipothrixviridae	Enveloped	Rod-shaped, linear	dsDNA
Plasmaviridae	Enveloped	Pleomorphic, circular	dsDNA
Rudiviridae	Non-enveloped	Rod-shaped, linear	dsDNA
Fuselloviridae	Non-enveloped	Lemon-shaped, circular	dsDNA
Inoviridae	Non-enveloped	Filamentous, circular	ssDNA
Microviridae	Non-enveloped	Isometric, circular	ssDNA
Leviviridae	Non-enveloped	Isometric, linear	ssRNA
Cystoviridae	Enveloped	Spherical, segmented	dsRNA

Note: ds = double-stranded, and ss = single-stranded.

19.3.5 Replication

Phages may have a lytic cycle or a lysogenic cycle (**Fig. 19.4**), but a few viruses are capable of carrying out both. With lytic phages such as the T4 phage, bacterial cells are broken open (lyzed) and destroyed after immediate replication of the virion. As soon as the cell is destroyed, the new phage viruses can find new hosts. Lytic phages are most suitable for phage therapy.

In contrast, the lysogenic cycle does not result in immediate lyzing of the host cell. Those phages able to undergo lysogeny are known as temperate phages. Their viral genome will integrate with the host DNA and replicate along with it fairly harmlessly, or may even become established as a plasmid. The virus remains dormant until host conditions deteriorate, perhaps due to depletion of nutrients, and then the endogenous phages (known as **prophages**) become active. At this point, they initiate the reproductive cycle resulting in lysis of the host cell. As the lysogenic cycle allows the host cell to continue to survive and reproduce, the virus is reproduced in all of the cell's offspring.

Sometimes prophages may provide benefits to the host bacterium while they are dormant by adding new functions to the bacterial genome in a phenomenon called lysogenic conversion. A famous example is the conversion of a harmless strain of *Vibrio cholerae* by a phage into a highly virulent one, which causes cholera. This is why temperate phages are not suitable for phage therapy.

19.3.6 Infection process of bacteriophages

Attachment and penetration

Now we know that the lifecycle of a phage begins when a phage particle adsorbs (attaches) to the surface of a susceptible bacterium (**Fig. 19.4**). To enter a host cell, phages attach to specific receptors on the surface of bacteria, including lipopolysaccharides, teichoic acids, proteins or even flagella. This specificity means that a bacteriophage can only infect certain bacteria bearing receptors that they can bind to, which in turn determines the phage's host range. As phage virions do not move independently, they must rely on random encounters with the right receptors when in solution (blood, lymphatic circulation, irrigation, soil, water and so on.).

Complex phages use a syringe-like motion to inject their genetic material into the cell. After making

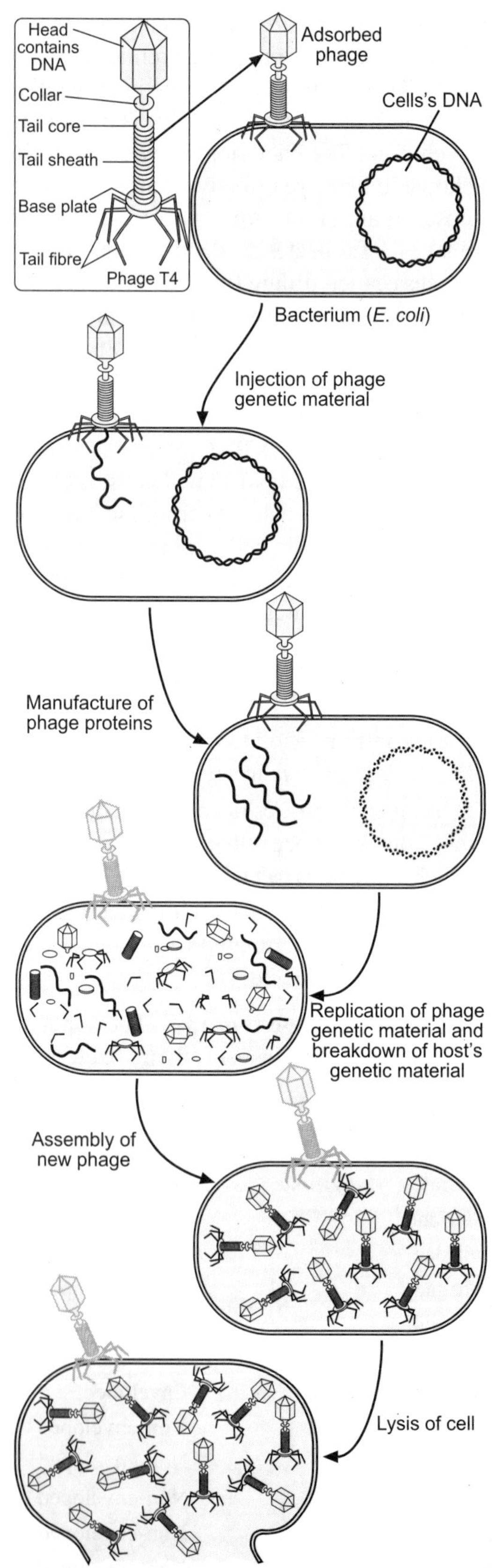

Fig. 19.4 The viral life cycle, using T4 infection of *E.coli* as an example.

contact with the appropriate receptor, the tail fibres bring the base plate closer to the surface of the cell. Once attached completely, the tail contracts, possibly with the help of ATP present in the tail (as shown by Prescott, 1993), injecting genetic material through the bacterial membrane.

Synthesis of proteins and nucleic acid

Within minutes, bacterial ribosomes start translating viral mRNA into protein. For RNA-based phages, RNA replicase is synthesized early in the process (**Fig. 19.5**). Proteins modify the bacterial RNA polymerase so that it preferentially transcribes viral mRNA. The host's normal synthesis of proteins and nucleic acids is disrupted, and it is forced to manufacture viral products instead. These products go on to become helper proteins, which help assemble the new virions, or proteins involved in cell lysis.

Phage assembly

In the case of the T4 phage, the construction of new virus particles involves the assistance of helper proteins. The base plates are assembled first, with the tails being built upon them afterwards. The head capsids, constructed separately, will spontaneously assemble with the tails. The DNA is packed efficiently within the heads. The whole process takes about 15 minutes.

Release of virions

Phages may be released via cell lysis or by host cell secretion. In the case of the T4 phage, in just over 20 minutes after injection, more than 300 phages are be released via lysis within a certain time-scale. This is achieved by an enzyme called endolysin, which attacks and breaks down the peptidoglycan. In contrast, 'lysogenic' phages do not kill the host but become long-term parasites and make the host cell continually secrete more new virus particles. The new virions bud off the plasma membrane, taking a portion of it with them to become enveloped viruses possessing a viral envelope. All released virions are capable of infecting a new bacterium.

19.4 VIRUSES

The simplest and most primitive infectious molecule is the virus (L., venom or poisonous fluid). Although, viruses do not fulil the criteria for being a living organism (lack cellular organization) they are intimately associated with living things and play a very significant role in biological systems. Virus as a minimalist life form and acellular in nature, exists as obligatory intracellular parasite. They neither belong to prokaryotes nor eukaryotes. They exhibit properties of life only when present inside cells of other organisms. The host cell can be any cell type including bacterial, fungal, protozoan, animal and plants. Inevitably, viruses are responsible for many kinds of diseases in almost all the cellular systems.

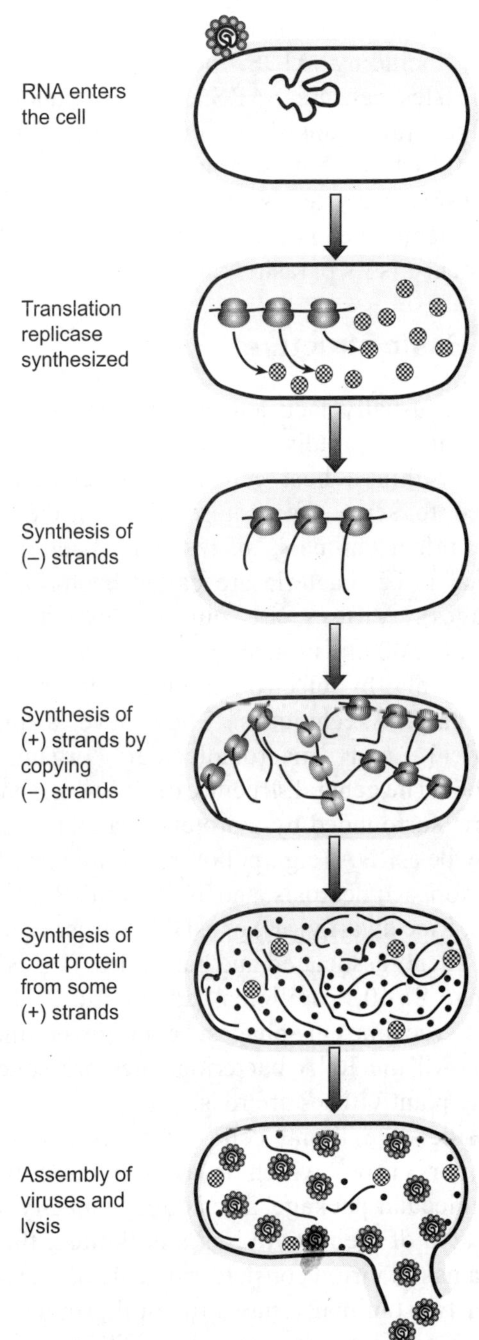

Fig. 19.5 Schematic diagram of the life cycle of a RNA phage.

Viruses occur in a wide variety with different shapes, sizes, and constructions, but all of them share certain common properties, i.e., they cannot reproduce unless present within a host cell. The key to the success of virus is its parasitic nature. It borrows functions from its host cell. A diverse number of viruses are also known to cause devastating diseases in plants. In fact, the first virus to be studied was the TMV (Wendell Stanley, 1935). Since then, a large group of plant and animal viruses have been isolated and studied. Viruses are responsible for dozens of human diseases, including AIDS, polio, influenza, cold sores, measles, cancers, SARS and H1N1 pandemics. Viruses generally consist of an inner core of genetic material (DNA or RNA) surrounded by outer protein coat (**capsid**). Very few virus species are enveloped within a membrane. The key to the virus' simplicity and existence is its parasitic nature.

19.4.1 Basic structure

As viruses usually lack any machinery to catalyze metabolism or protein synthesis, they can only multiply within a host cell whose metabolism is subverted to serve the requirements of the virus. Viruses infect animals, plants and bacteria and those that infect bacteria are called bacteriophages (or phages). Viruses are quite a varied group, ranging 30–300 nm in size. Therefore, they can be observed only by electron microscopy and x-ray crystallography techniques. They have a regular geometrical and macromolecular organization (**Fig. 19.6**). The general structure of a virus is a nucleic acid core, surrounded by a protein coat. The nucleic acid may be ssRNA (e.g., poliovirus, influenza, MS2, satellite tobacco necrosis, and TMV) or dsRNA (e.g., reovirus, Rice dwarf virus) or ssDNA viruses (gemini viruses, adeno-associated and parvovirus) or dsDNA (polyoma, SV40, papilloma, adenovirus 2, herpes simplex, vaccinia, T–phages, cauliflower mosaic viruses). All the RNA bacteriophages are spherical but many plant viruses are rods.

Outside of a living cell, the virus exists as a particle, or virion, which is a little more than a macromolecular package. The coding capacity of the nucleic acid of the naked viruses is limited; thus the coat of a naked virion consists primarily of one (plant virus) or two (animal virus) structural proteins. This building block is called a **protomer**. The protein coat protects the nucleic acid from damage and may confer a specific host range on the potential infectivity of the particle. The capsid—nucleic acid unit— is called a nucleocapsid. The coat may consist of many identical molecules of protein (capsomeres) or may be a complex structure such as one shown in **Fig. 19.6**. The capsomeres are of different shapes such as the hollow prism, hexagonal, pentagonal, lobular, or any other shape. The specific arrangement of capsomeres in the capsid determines the shape of a virion.

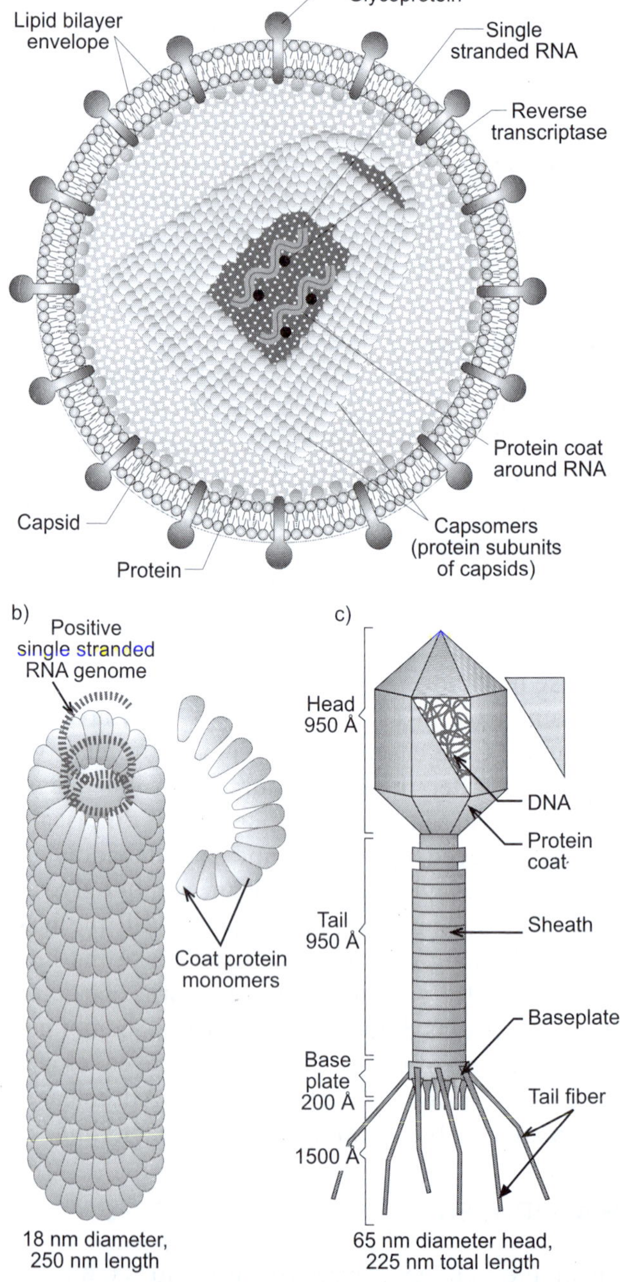

Fig. 19.6 Body plans of viruses: (a) Structure of HIV virus, (b) Morphology of tobacco mosaic virus (TMV), and (c) Complex viruses, such as the T-even bacteriophages (not drawn to scale) (**See Plate 16**).

Non-enveloped viruses

Many viruses have a capsid whose subunits are organized into a polyhedron, that is, a structure having planar faces. Most viruses generally have the following three different types of symmetry.

- ***Icosahedral symmetry:*** A particularly common polyhedral shape of viruses is the 20-sided icosahedron. The capsomers are organized in several ways to form an icosahedral (20-faced) capsid. Icosahedral symmetry arises in order to allow the virus to be in a state of minimum energy. An icosahedral capsid comprises both pentamers (capsomers with five structural units) and hexamers (capsomers with six structural units). In an icosahedral virus the minimum number of capsomers is 12 or multiples thereof.
- ***Helical or cylindrical symmetry:*** The rod-shaped helical capsid of viruses such as tobacco mosaic virus (TMV), bacteriophage M13 and influenza virus, consists of numerous identical capsomers arranged into a helix because they are thicker at one end than the other (tapered end).
- ***Complex symmetry:*** This group belongs to the viruses with complex shaped capsids, which are of two shapes, those without identifiable capsids (e.g., pox viruses) and those with tadpole-shaped structures in which each part has a different sort of symmetry (e.g., T-even phages of *E. coli*). Some viruses such as rabies virus are bullet-shaped.

Enveloped viruses

Some viruses such as for herpes, influenza, mumps and Semliki forest virus are surrounded by a 100 to 150Å thick spiked membrane. This membrane contains a lipid bilayer of plasma membrane of the host cell from which the virus-specific protein molecules or spikes project. The enveloped virions do not have well-characterized structures. In these, the nucleocapsid is encased in a loose membranous envelope. The shape of the virion is variable because the envelope is not rigid. The envelope is derived from the cell membrane, but usually contains virus-coded proteins. Many virions contain virus-encoded enzymes not present in the host cell. Sometimes these enzymes are nucleic acid replicases or are factors needed for adsorption of the virion to the host cell.

19.4.2 Classification of viruses

Viruses are not named according to the method of binomial nomenclature like other organisms. Viruses tend to be named in a random fashion according to the disease caused, the host organism, based on replication strategy, nature of genetic material, or some coded system (e.g., T1, T2, P1 phages).

Eukaryotic viruses

The purpose of a virus infection is genome replication and the production of many progeny virions. Viruses can be classified with respect to their structure or their host cell, gene expression and replication. The universal Baltimore classification (devised in 1971) divided viruses into five families based solely on the replication mechanisms known then (**Fig. 19.7**). However, more categories were needed to include recent discoveries (**Table 19.2**). In the case of eukaryotic viruses, there are four morphological types such as naked icosahedral, naked helical, enveloped icosahedral, and enveloped helical.

Based on replication strategy

The complete virus particle is called a virion and generally contains only one type of nucleic acid, which may be ssRNA or dsRNA or ssDNA or dsDNA. The strategy for genome replication depends on the type of nucleic acid, its strandedness and its sense. On transcription of DNA, only one strand gives rise to mRNA. The mRNA-like strand is given the designation 'plus (+) strand'. The complementary strand is called the 'minus (–) strand'.

19.4.3 Life cycle of viruses

Viral lifecycles are so varied that no single mechanism can account for any one of the stages listed below (**Fig. 19.8**). The lifecycle of all virulent viruses consists of the following stages:

- adsorption of the virus on the host cell
- penetration of viral nucleic acid into the cell
- development of virus-specific functions, alteration of cell functions, replication of the nucleic acid, and synthesis of other virus constituents
- assembly of progeny virus particles
- release of virus particles from the cell.

19.4.4 Animal DNA viruses

DNA viruses, like some RNA viruses, do not carry special enzymes along with the genome. Some DNA viruses are considerably more complex and have

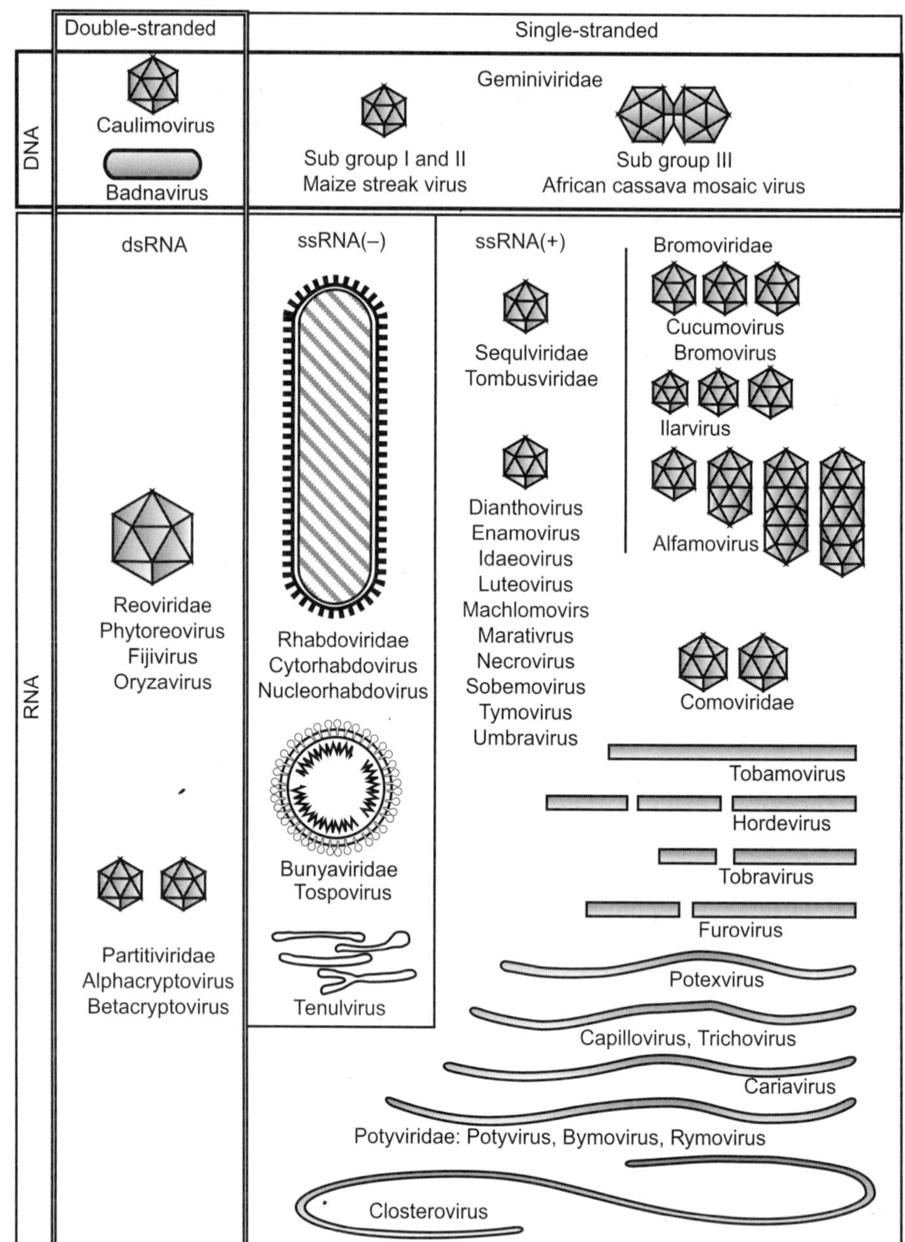

Fig. 19.7 Families and genera of plant viruses. Drawn based on the Sixth report of the International Committee on Taxonomy of Virusus (ICTV).

evolved with more than 100 genes, which give them more versatility. The large animal virus such as herpes simplex virus (HSV) can code for more than 100 proteins, and, as with the large phages (e.g., T4) many of these proteins duplicate the functions of host proteins.

Treatment: In general, viral diseases cannot be treated with antibiotics, because the virus uses the cell's machinery to replicate itself. Thus, any interference with virus metabolism automatically affects the infected host. However, because the larger viruses substitute their own enzymes (e.g., DNA polymerase and thymidine kinase) for those of the cell, it is possible to find some chemicals to which only the viral enzymes are susceptible. Acyclovir [9-(2-hydroxymethyl)-guanine] is one such chemical, which induces thymidine kinase (in contrast to the host enzyme). The triphosphate inhibits the virally coded DNA polymerase. Acyclovir is used to treat herpes infections but cannot bring about a complete cure as the virus exhibits a latent phase in which the thymidine kinase is inactive. In a similar manner, infections with HIV can be treated with azidothymidine, which is phosphorylated and incorporated by the virally coded reverse transcriptase

Table 19.2 Some important animal viruses used in molecular biology research

Virus family	Type of nucleic acid	Representative virus	Genome size (kbp)	Tumour forming
Picornavirus	ssRNA	Polio virus	7–8	No
Orthomyxovirus	ssRNA	Influenza	14	No
Paramyxovirus	RNA	Measles	16–20	-
Rhabdovirus	ssRNA	Vesicular stomatitis	-	No
Togavirus or arbovirus	ssRNA	Sindbis, Semliki forest, eastern equine encephalitis, rubella	12	-
Retrovirus	ssRNA	Various leucosis and leukemia, Rous sarcoma, mouse mammary tumour	9	Yes
Flaviviruses	RNA	Yellow fever virus	10	-
Adenovirus	dsDNA	Adenovirus	36	Some
Hepadnavirus	DNA	Hepatitis B virus	3.2	-
Herpesvirus	dsDNA	Herpes zoster (chicken pox) Herpes simplex I (cold sores)	120–200	Some
Poxvirus	dsDNA	Vaccinia, fibroma	130–280	Some
Reovirus	dsRNA	Reovirus	-	No
Polyoma	dsDNA	Polyoma, SV40	-	Yes
Papilloma	dsDNA	Shope papilloma	5–8	Yes

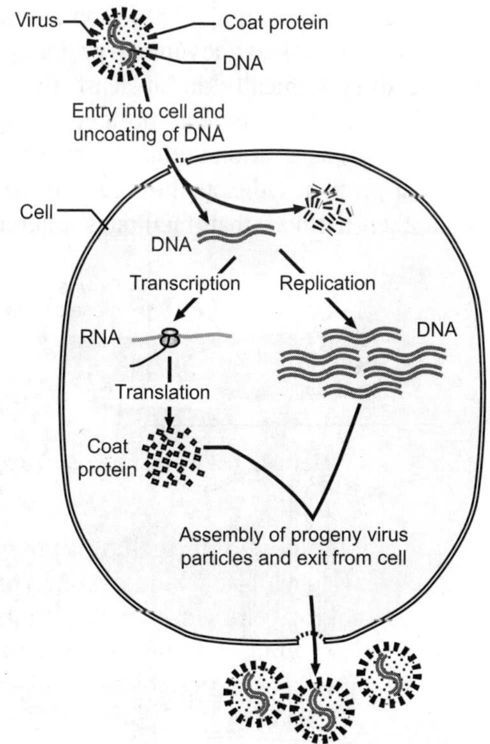

Fig. 19.8 A generalized scheme of viral life cycle. The virus in the diagram shown consists of a small double-stranded genomic DNA, and only expression of a single viral coat protein has been shown. Viruses contain many genes required for replication, asssembly, cellular movement, vector specificity and so on.

leading to production of defective viral DNA. Other approaches for therapy are immonological and vaccine therapies are now available.

19.4.7 Plant viruses

Plant viruses have very compact genomes. An archetypal minimal virus genome would contain genes to encode a replication protein (RP), a movement protein (MP) and a coat protein (CP). If the virus is transmitted by the vectors then the virus needs to have a vector transmission (VP) gene. Generally, viral genes are arranged linearly in the order starting from 5' end replication, movement, coat protein (e.g., TMV). With this arrangement, each gene's function as early or late is fixed; however, many variations can be seen on this basic theme, which are required for adaptations to multiplication requirements. Of all the plant viruses studied so far, 94% have RNA genomes, and only 6% use DNA as their genetic material. The majority (about 82%) of RNA viruses have linear positive-sense ssRNAs. It is said to be **positive-sense RNA**, when the genomic RNA can be translated directly by ribosomes (as if it were mRNA). If a complementary strand must be made before translation can take place, the RNA is said to be **negative sense**.

During translation of viral RNAs, host cell ribosomes move along mRNA, starting at the 5' end, until they come across the first usable start codon (generally AUG) and stop at the first stop codon it comes to in the frame. Any start codons following this stop codon are ignored; thus, no downstream genes will be expressed. Plant viruses have developed several different strategies to work around this limitation (**Fig. 19.9**):

- ***Creation of subgenomic RNA(s)***: The viral RNA can have promoter structures, which the viral RNA-dependent RNA polymerase can transcribe as additional subgenomic RNAs, each having a start codon close to its 5' end.
- ***Multipartite genomes:*** The viral genome is divided into two or more RNAs that are packaged in one or more virus particles. In most cases, all particles need to be present in a cell to be infectious.
- ***Polyproteins:*** Some viral RNAs are first transcribed into a single, long polyprotein, which is later processed into many functional proteins by virus-encoded proteinases.
- ***Read through:*** Some proteins are translated by suppression of the first stop codon. This results in a fusion protein that can have a different function at its C-terminal end.
- ***Translational frameshift***: In a few cases, it has been shown that ribosomes change to a different reading frame during translation, resulting in a new protein.

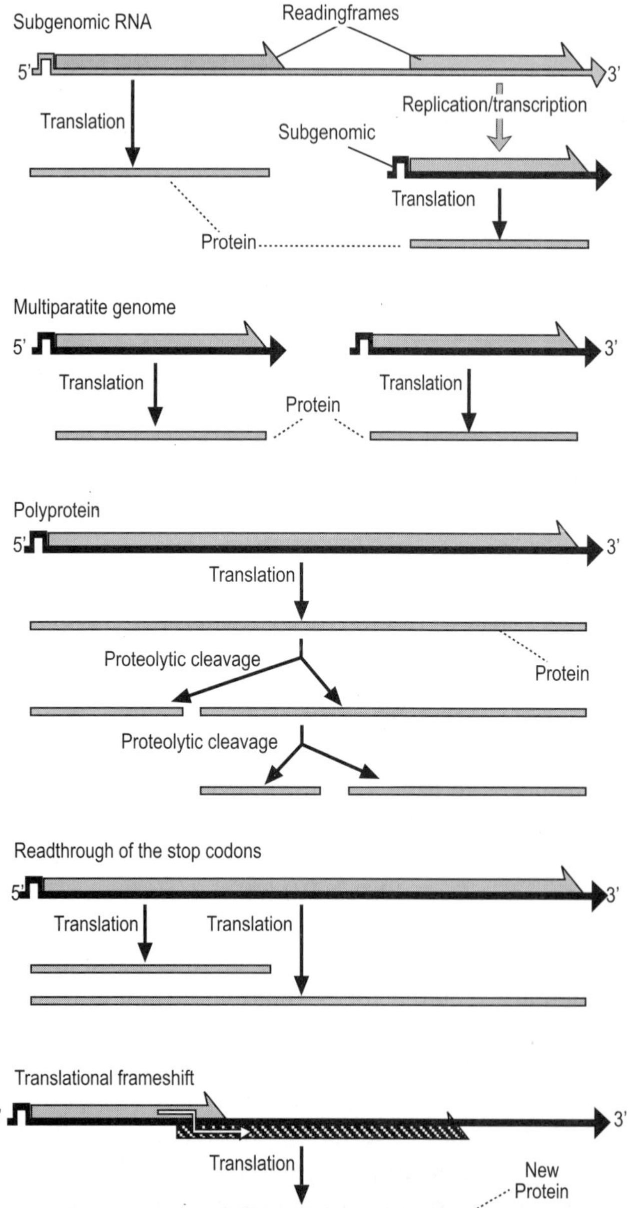

Fig. 19.9 Strategies for the expression of compact viral genomes.

Plant viruses move intercellularly only after they enter the host organism's symplasm. The cytoplasm of each cell is connected to its neighbours by plasmodesmata (**Fig. 19.10**). Viruses can move through the plasmodesmata, only after widening the pore by plasmodesmatal manipulation. They contain special proteins to do this, called movement proteins (MP). Viruses spread systemically in at least two ways: (i) as nucleic acid–movement protein complexes (e.g., TMV); or (ii) as a virion within a tubule made by the movement protein. Other than these two methods, there is another method that facilitates intercellular

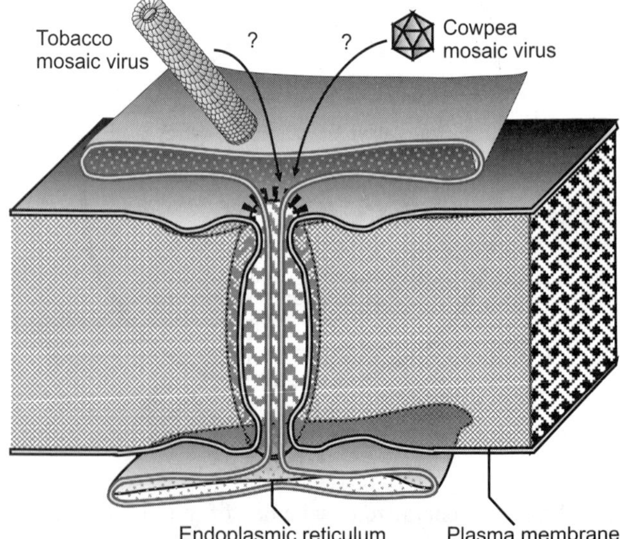

Fig. 19.10 Intercellular movement of virus through plasmodesmata. For the viruses to fit through, the plasmodesmata must widen and the viral nucleic acid must stretch.

movement. This morphogenetic potentiality is seen in a family of plant viruses; the best known of this type is a member of cowpea mosaic virus. When the movement protein and whole virions, of these viruses are injected into protoplasts, the plasma membrane is induced to build external tubular extensions; through which whole virions move from one cell to other.

19.5 SUBVIRAL AGENTS

Subviral agents are non-cellular organisms containing a genome but lacking one of the essential features of a virus (**Table 19.3**). The common property of viruses and subviral agents, which distinguishes them from other self-replicating genetic elements (e.g., plasmids, transposable elements and cell organelles) is their existence in a stable extracellular form, that allows them to be infectious in their own right.

19.5.1 Viroids

Viroids are much smaller than viruses and consist of a short circular RNA molecule (240–600 nt), with considerable secondary structure. The term viroid was coined by TO Diener (1971), who reported that potato spindle-tuber disease is caused by an infectious agent consisting of a small circular RNA molecule that totally lacks a protein coat, which he called **potato spindle tuber viroids** or **PSTV** (**Table 19.4**). Viroids genome does not contain AUG codon (no protein synthesis), but they are replicated autonomously. They infect higher plants causing diseases such as potato spindle tuber disease (PSTV) and cadang cadang (**Fig. 19.11**), a disease of the coconut palm. Variants exist (called virusoids and satellite RNAs) in which the circular form or a linear version is encapsidated with the intervention or a helper virus.

Fig. 19.11 Structure of the coconut cadang cadang viroid (246 nucleotides). The loops represent unpaired regions and the cross-lines, paired bases.

Table 19.3 Different classes of viruses and subviral agents

Agent	Distinguishing properties
Virus	Viral genomes can be either DNA or RNA. The genome is enclosed in a protein coat, which mediates infection and the code of coat proteins and other functions required for infection cycle.
Viroid	RNA genome; it is naked, infection is mediated mechanically, genome encodes no proteins.
Satellite virus	Viral replication is dependent on co-infection by a second 'helper' virus of a different type, which supplies missing functions in trans. The satellite virus and helper viruses are separate entities. They encode their own capsids and infect the cell independently.
Satellite nucleic acid	A DNA or RNA genome with virus-like coding properties but lacking the ability to encode its own capsid and depends on helper virus for the late functions. In plants, viroid-like RNA is packaged as a stowaway in the capsid of its helper virus which helps in infection and replication functions. The helper virus is independent of the satellite.
Virusoid	RNA genome found in plants, structurally and functionally similar to a viroid, and packaged into a viral capsid, but actually part of the virus genome and therefore not expendable like satellite RNA
Virino	Theoretical infectious particle, which has a non-coding nucleic acid genome packaged in a protein coat derived from the host.
Prion	Proteineceous infectious particle resistant to treatments that destroy nucleic acids. Commonly taken to mean an infectious agent composed entirely of a host-encoded protein which can change to a pathogenic and 'self-replicating' conformation.

Table 19.4 Primary and secondary structure of the PSTV viroid

1	CGGAACUAAA	CUCGUGGUUC	CUGUGGUUCA	CACCUGACCU	CCUGAGCAGA	AAAGAAAAAA
61	GAAGGCGGCU	CGGAGGAGCG	CUUCAGGGAU	CCCCGGGGAA	ACCUGGAGCG	AACUGGCAAA
121	AAAGGACGGU	GGGGAGUGCC	CAGCGGCCGA	CAGGAGUAAU	UCCCGCCGAA	ACAGGGUUUU
181	CACCCUUCCU	UUCUUCGGGU	GUCCUUCCUC	GCGCCCGCAG	GACCACCCCU	CGCCCCCUUU
241	GCGCUGUCGC	UUCGGCUACU	ACCCGGUGGA	AACAACUGAA	GCUCCCGAGA	ACCGCUUUUU
301	CUCUAUCUUA	CUUGCUUCGG	GGCGAGGGUG	UUUAGCCCUU	GGAACCGCAG	UUGGUUCCU

Infection of plant cells by viroids is dependent on mechanical damage of the host plant. Viroids differ from RNA viruses in three major aspects: size (about 250–1700 bases); the genome does not encode any proteins and they are not encapsidated. About 30 distinct types of viroids have been identified: some generate characteristic disease symptoms and others are cryptic. All viroids undergo extensive intermolecular base-pairing to form rigid rod-like structures (hammer head structure with self-cleaving properties) with conserved secondary structures, and two viroid 'families' have been recognized on this basis. In mixed infections, frequent recombination occurs between homologous domains of individual viroids to generate novel sequence variants with different pathogenic properties.

Viroid replication requires host-encoded RNA polymerase, which may have weak RNA replicase activity. Replication is via rolling circle mechanism to produce first (–) strand concatemers (which may or may not be circularized) followed by (+) strand concatemers, which are self-cleaved and ligated to regenerate the infectious RNA. How single-stranded circular genomes are released from the double-stranded concatemers is poorly understood, although the avocado sun blotch viroid can catalyze self-cleavage, and several of the structural features of viroids are also found in group I introns, suggesting autocatalytic monomerization in some species and perhaps host-dependent cleavage in others (e.g., ribozymes, self-splicing introns). It is possible that the pathogenic effect of these RNA molecules lies in their ability to catalyze cleavage of other RNAs (e.g., mRNAs).

The only human disease known to be caused by a viroid is hepatitis D (delta). Hepatitis delta virus is a satellite of human hepatitis B virus.For hepatitis D to occur there must be simultaneous infection of a cell with both the hepatitis B virus and the hepatitis D viroid.

19.5.2 Satellite virus

Satellite viruses are subviral agents, which depend on a helper virus for successful infection and propagation. There are two types of satellite viruses: one which **encodes its own coat protein** (can infect without help) and **satellite nucleic acids**, which cannot. Satellite viruses include the eukaryotic *Dependovirus* genus (e.g., adeno-associated viruses) and several ssRNA plant viruses with a diverse range of helper viruses. Satellite nucleic acids can be divided into two classes. Some behave like viruses, but require help with late functions and hence utilize the capsid proteins of the helper virus (e.g., bacteriophage P4, which uses the capsid proteins of bacteriophage P2, also killer factors). Other satellites, found only in plants, resemble viroids in structure and have no coding function, but hitchhike within the capsid of a helper virus to mediate infection. These satellites depend on the helper virus for infection and replication, but the associated viruses are independent of the satellite. Viral infections can occur with or without the satellite. It can influence the symptoms produced in the infected plant and reduce the yield of the helper virus by competing for cellular components during replication. Cucumber mosaic virus (CMV), for example, causes mild symptoms in infected tomato plants, but in the presence of CMV-associated satellite RNA causes lethal necrosis. In other plants, however, the presence of the satellite RNA reduces the severity of the symptoms. In some viruses, viroid-like elements represent a permanent component of the genome. For example, the velvet tobacco mottle virus has a bipartite RNA genome consisting of RNA1 and RNA2, the latter resembling a viroid but actually being essential for viral infection and replication; in this situation the RNA2 is known as a **virusoid**. Conversely, in the closely related Lucerne transient streak virus, RNA2 is a non-essential satellite.

Exercises

1. What are plasmids and how are they classified?
2. Explain briefly the applications of plasmids.
3. What are bacteriophages?
4. Explain briefly the uses of phages.
5. Draw the basic structure of different viruses.
6. Explain the different steps of viral infection in plants with a diagram.
7. Describe the nature and mode of infection of viroids.

Summary

- Non-cellular life was discovered when it was found that the large and complex Mimivirus could make some proteins that are involved in the synthesis of proteins.
- When considering non-cellular life, besides viruses, other forms, such as ***viroids, satellites, plasmids, phagemids, cosmids, transposons and prions***, might also be included.
- Plasmids occur naturally in most bacteria and are sometimes found in eukaryotic organisms, yeast, protozoa, and plants. In humans, certain viruses can also exist as plasmids within cell nuclei. A plasmid is an extra-chromosomal DNA molecule, capable of autonomous replication.
- Plasmids often contain genes that confer a selective advantage to the bacterium harbouring them (e.g., antibiotic-resistant). They are composed of ***simple nucleic acids*** and ***live in compatible host cells***. Their DNA may exist in different conformations: ***linear, nicked, relaxed circular, supercoiled and denatured***. Plasmids serve as important tools in genetics and molecular biology, where they are commonly used as cloning vectors to multiply or express particular genes. Another major use of plasmids is to make large amounts of proteins.
- A ***bacteriophage*** is one of a number of viruses that ***infect bacteria***. The genetic material can be ssRNA, dsRNA, ssDNA, or dsDNA between 5 kbp and 500 kbp long with either ***circular*** or ***linear arrangement***.
- Phages are estimated to be the most widely distributed and diverse entities in the biosphere. They are ubiquitous and can be found in all reservoirs populated by bacterial hosts.
- The relationship between bacteriophages and their bacterial hosts is very important in the context of the food fermentation and dairy industry.
- Use of phages as antibacterial agents was highly successful in treating many diseases and still continues in many countries. The manner in which phages are applied depends on the type of condition from which a patient suffers. The use of phages to destroy their bacterial hosts can also have negative commercial impact; mainly in various fermentation industries that depend on bacteria.
- The simplest and most primitive infectious molecule is the ***virus***. It has a ***minimalist life form*** and ***acellular*** in nature, exist as ***obligatory intracellular parasite***. They exhibit properties of life only when present inside cells of other organisms.
- Viruses infect animals, plants and bacteria and range 30–300 nm in size. The general structure of a virus is a ***nucleic acid core (RNA or DNA)***, surrounded by ***a protein coat***.
- Some viruses such as herpes ***virus***, influenza ***virus***, mumps ***virus*** and Semliki forest virus are surrounded by a 100–150Å- thick spiked membrane. This membrane contains lipid bilayer of plasma membrane of the host cell from which the virus-specific protein molecules or spikes project.
- The purpose of a virus infection is genome replication and the production of many progeny virions. Viral lifecycles are so varied that no single mechanism can account for any one of the stages. In general, ***viral diseases cannot be treated with antibiotics***.
- ***Subviral agents*** are non-cellular organisms, for example, ***viroids*** are much smaller than viruses and consist of a short circular RNA molecule (240–600 nt), with considerable secondary structure. Viroid genome does not contain AUG codon, but is replicated autonomously.
- Viroids infect higher plants causing diseases such as potato spindle tuber disease (PSTV) and cadang cadang, a disease of the coconut palm. Variants exist (called virusoids and satellite RNAs) in which the circular form or a linear version of it encapsidated with the intervention or a helper virus.
- The only ***human disease*** known to be ***caused by a viroid*** is ***hepatitis D (delta)***. Hepatitis delta virus is a satellite of human hepatitis B virus. For hepatitis D to occur there must be simultaneous infection of a cell with both the hepatitis B virus and the hepatitis D viroid.
- Satellite viruses are subviral agents, which depend on a helper virus for successful infection and propagation. There are two types of satellite viruses: with or without own coat protein and satellite nucleic acids, which do not.
- Viral infections can occur with or without the satellite. Cucumber mosaic virus, for example, causes mild symptoms in infected tomato plants, but in the presence of CMV-associated satellite RNA, lethal necrosis results.

Further Reading

Books

Becker WM, Kleinsmith LJ and Hardin J (2000), *The World of the Cell*, 4th edition, Benjamin Cummings, New York.

Cappuccino JG and Sherman N (1999), *Microbiology: A Laboratory Manual,* Benjamin Cummings.

Klein DW, Prescott LM, Harley J (1999), *Microbiology*, WCB/McGraw-Hill, Boston.

Madigan MT, Martinko JM (2005), *Brock Biology of Microorganisms*, 11th edition, Pearson Prentice Hall, Upper Saddle River, NJ.

Matthews REF (1991), *Plant Virology*, Academic Press, San Diego.

McGrath S and van Sinderen D, (eds), (2007), *Bacteriophage: Genetics and Molecular Biology*, 1st edition, Caister Academic Press.

Russell DW, Sambrook J (2001), *Molecular Cloning: A Laboratory Manual,* Cold Spring Harbor Laboratory Press, Cold Spring Harbor, New York.

Starr C and Taggart R (1995), *Biology, The Unity and Diversity of Life*, Wadsworth Publishing Company, A Division of International Thomson Publishing Inc.

Tooze J (1981), *The Molecular Biology of Tumour Viruses,* Cold Spring Harbor Laboratory Press, Cold Spring Harbor, New York.

References

Aguzzi A and Weissmann C (1997), Prion research: The next frontiers, *Nature,* 389:795.

Ahmad S (2002), Treatment of post-burn bacterial infections by bacteriophages, specifically ubiquitous *Pseudomonas* spp. notoriously resistant to antibiotics. *Medical Hypotheses*, 58(4):327–31.

Barrow P, Lovell M and Berchieri A (1998), Use of lytic bacteriophage for control of experimental *Escherichia coli* septicemia and meningitis in chickens and calves, *Clin. Diagn. Lab. Immunol.* 5(3):294–98.

Bishop JM (1982), Cellular oncogenes and retroviruses, *Annu. Rev. Biochem.* 52:301.

Caughey B and Chesebro B (1997), Prion protein and the transmissible spongiform encephalopathies, *Trends Cell Biol.* 7:56–62.

Channarayappa (1992), Resistance mechanism of *Lycopersicon* species to the vector *Bemisia tabaci* and ultrastructural changes in leaf curl virus-infected tomato plants, PhD Thesis submitted to University of Agricultural Sciences (UAS), Bangalore. India.

Collmer C and Howell S (1992), Role of satellite RNA in the expression of symptoms caused by plant viruses, *Annu. Rev. Phytopathol.* 30:419–42.

Diener TO (1991), The frontiers of life: The viroids and viroid-like satellite RNAs, In: Marmorosch K, (ed.), *Viroids and Satellites: Molecular Parasites at the Frontiers of Life* CRC Press.

Duckworth D and Gulig P (2002), Bacteriophage: potential treatment for bacterial infections, *Biodrugs,* 16(1):57–62.

Gerdes K, Rasmussen PB, Molin S (1986), Unique type of plasmid maintenance function: Post-segregational killing of plasmid-free cells, *Proc. Natl. Acad. Sci.,* 83 (10).

Gilbertson R and Lucas WJ (1996), How do viruses traffic on the 'vascular highway'? *Trends in Plant Science*, 1:260–65.

Hynes R (1980), Cellular location of viral transformation proteins, *Cell*, 21:601.

Kajander EO and Ciftcioglu N (1998), Nanobacteria: An alternative mechanism for pathogenic intra- and extracellular calcification and stone formation, *Proc. Natl. Acad. Sci.*, 95:8274–79.

Katz RA and Skalka AM (1994), The retroviral enzymes, *Annu. Rev. Biochem.*, 63:133–73.

Lederberg J (1952), Cell genetics and hereditary symbiosis, *Physiol. Rev.* 32 (4): 403–30.

Mitra S (1980), DNA replication in viruses, *Annu. Rev. Genetics*, 14:347.

Moffat AS (1994), Mapping the sequence of disease resistance, *Science,* 265:1804–5.

Nowak MA and McMichael AJ (1995), How HIV defeats the immune system, *Sci. Am.*, 273:42–9.

Tabler M and Tsagris M (2004), Viroids: petite RNA pathogens with distinguished talents, *Trends Plant Sci.*, 9: 339–48.

Varmus HE (1982), Form and function of retroviral proviruses, *Science,* 216:812.

20 Tools for the Study of Cells

Learning objectives

After reading this chapter, you will be able to:

- List out various techniques used in studying cell structure
- Distinguish between the unique features of each technique
- Identify different organisms used in studying cellular structure

20.1 INTRODUCTION

The study of cells (or cell biology) is basically concerned with understanding the cell structure and functions. It overlaps with many other areas of biology, cytology, genetics, molecular biology, biochemistry and uses various techniques that have been developed in those fields. The principles and experimental approaches that have been drawn from all of these have been used to dissect and elucidate the cell structure and functions. In this chapter, the discussion is limited to some important techniques and methodologies that have greater impact or sensitivity.

20.2 SIZE RELATIONSHIPS

The important property of living organisms is their size. The size of an organism (e.g., microorganisms) ranges from the relatively large (100 μm) down to the tiny viruses (0.01 μm), which cannot be seen by the naked eye. Most bacteria are about 1 μm to 0.5 μm in length and it can reach even up to 20 μm. The size of virus is often expressed in **nanometres** (nano) to avoid decimals of micrometres, whereas, the sizes of eukaryotes ranges from micrometres to metres, for example, yeast (8 μm) and human neurons up to seven feet (**Fig. 20.1**). To visualize particles too small to be seen with the naked eye, the study of cells and their organelles requires use of sophisticated equipments and specialized

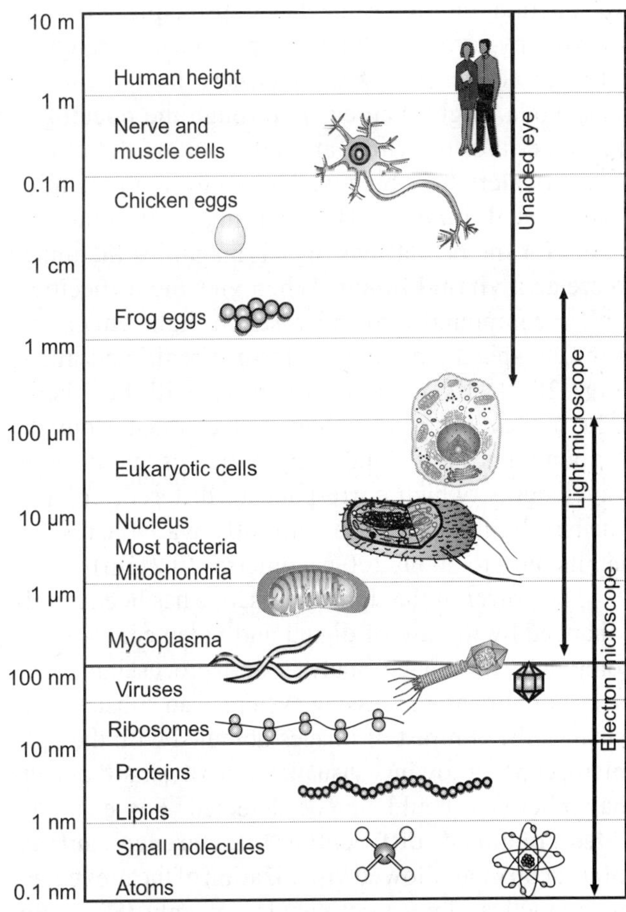

Fig. 20.1 Size comparison among various living things.

techniques. A cell biology laboratory thus, must have a good microscope. Therefore, it is necessary to know the working procedure and some basic principles of microscopy.

20.3 MICROSCOPY

The microscope is a unique instrument that made studies of cell biology possible. Indeed, the very discovery of cells arose from the development of the microscope. Robert Hooke coined the term "cell" following his observations of a piece of cork with a simple light microscope in 1665. Generations of biologists followed Hooke with steady improved instruments. Now probably more than a dozen types of microscopes with varied range of resolution and magnifications are available.

Light microscopy

The light microscope is the basic microscopic system used in microbiology. It works by bending (refracting) light rays. This instrument is also called **bright-field microscope**, or **compound microscope**. All operate on the same basic principle. In light microscopy, visible light is projected through a substage condenser, which focusses the light into a sharp cone. Light then enters through the opening in the stage, into the slide, and bounces off the object. Next, it enters the objective lens to form a magnified image (**real image**). The real image becomes an object for the ocular lens, which magnifies the image to create a **virtual image**. When viewing a specimen with a compound light microscope, light emanating from the specimen bends to form its enlarged image (**Fig. 20.2**). Its internal structures will be visible only if they differ in colour and contrast from the surroundings (**Fig. 20.3**). Light travels at different wavelengths, which correspond to different colours. The best light microscopes currently available resolve details only to about 200 nanometres (0.2 μm).

The power of the light microscope has been highly enhanced by the use of digital and video cameras and computers for image analysis and processing. Such electronic image processing systems can substantially enhance the contrast of images obtained with the light microscope, allowing visualization of small objects that otherwise could not be detected. For example, **video-enhanced differential interference-contrast microscopy** has allowed visualization of the movement of organelles along microtubules of only 0.025 μm. To use the microscope efficiently and with minimal

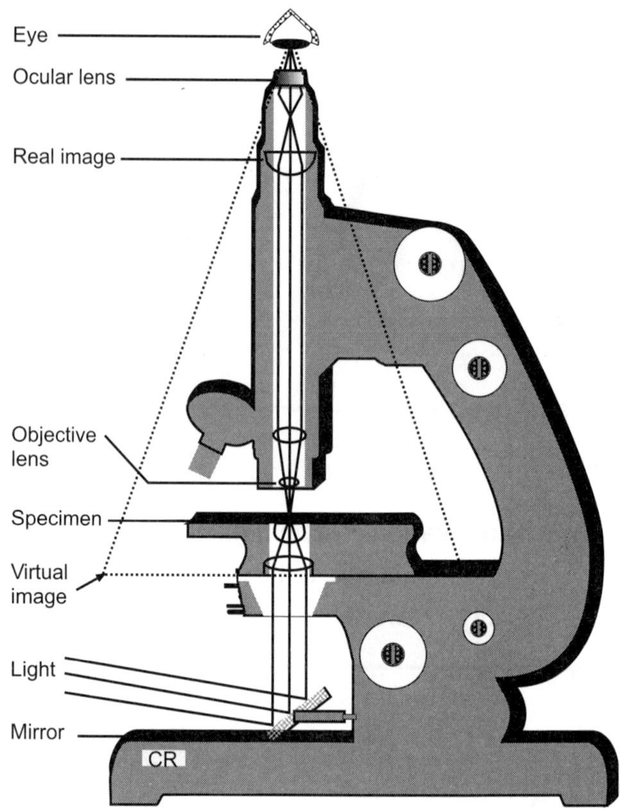

Fig. 20.2 Sectional diagram through a compound light microscope; that is, a microscope that has both an objective and ocular lenses.

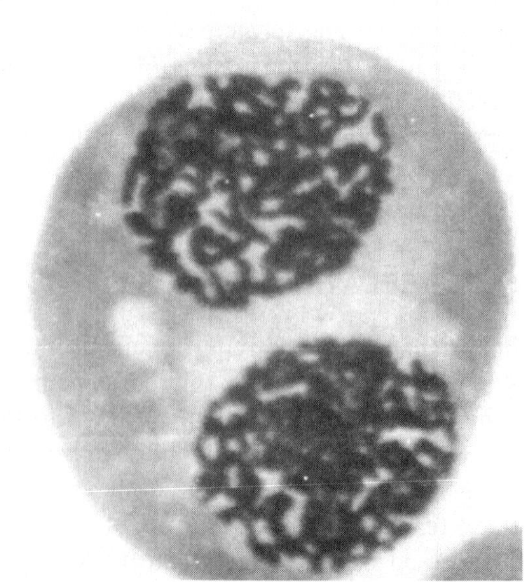

Fig. 20.3 Light microscopic photograph showing a binucleated animal cell stained with Giemsa stain and observed at 1000× magnification (oil-immersion). Pink, thread-like structures in a nucleus represent chromosomes and bluish background is cytoplasm enclosed in a plasma membrane (Channarayappa, 1990).

frustration, one should understand the basic principles of microscopy, which includes magnification, resolution, numerical aperture, illumination, and focussing.

Dark-field microscopy

Here, the specimen background remains dark, and only the object is illuminated (**Fig. 20.4**). This is accomplished by equipping the light microscope with a special kind of condenser that transmits a hollow cone of light from the source of illumination. A special condenser mounted under the stage of the dark-field microscope scatters the light and causes it to hit the object from different angles. Some light bounces off the object into the lens to mark the object visible, but the surrounding area appears dark because it lacks background light. Dark-field microscopy helps in the diagnosis of diseases caused by spiral bacteria because these organisms are near the limit of resolution and do not stain well.

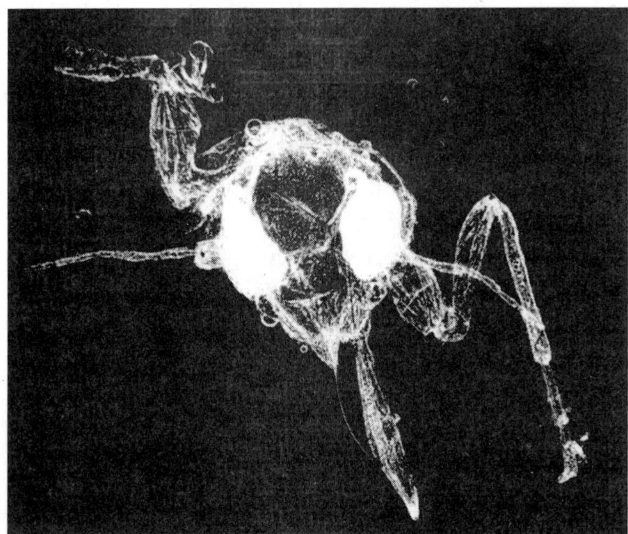

Fig. 20.4 Whitefly (*Bemisia tabaci*) vector head is observed under dark-field microscope. Insect mouth parts and stylet can be distinctly seen (Channarayappa, 1992).

Phase-contrast microscopy

It is extremely valuable for studying living unstained cells that is widely used in applied and theoretical studies. This microscope makes highly transparent objects more visible. Phase-contrast microscope converts differences in refractive index into differences in intensity, which are visible to the eye. This is accomplished by separating the light that enters the objective lens from the diffracted light emanating from the specimen and causing light rays from these two sources to interfere with one another. The relative brightness or darkness of each part of the image reflects the way in which the light from that part of the specimen interferes with the direct light.

Special microscope parts are used in phase-contrast microscopy. A series of special condensers and filters split a light beam and throw the light rays slightly out of phase. The separated beams of light then pass through and around microscopic objects, and small differences in the densities of the objects show up as different degrees of brightness and contrast. With the phase contrast microscope, researchers can see cells alive and unstained. Internal details not visible with the light microscope may also be observed.

Fluorescence microscopy

Fluorescence microscopy is a widely used and very sensitive method for studying the intracellular distribution of molecules. The technique has been applied to identify subcellular structures and is a mainstay of modern cell biology. The cells are coated with a fluorescent dye, such as **fluorescein**, then, illuminated with ultraviolet light energy. The energy excites electrons in the dye, and they move to higher energy levels. However, the electrons quickly drop back to their original energy levels and give off the excess energy as visible light. Consequently, the coated objects appear to fluoresce.

The special features of fluorescence microscopy with respect to illumination of the specimen are shown in **Fig. 20.5**. The function of the exciter

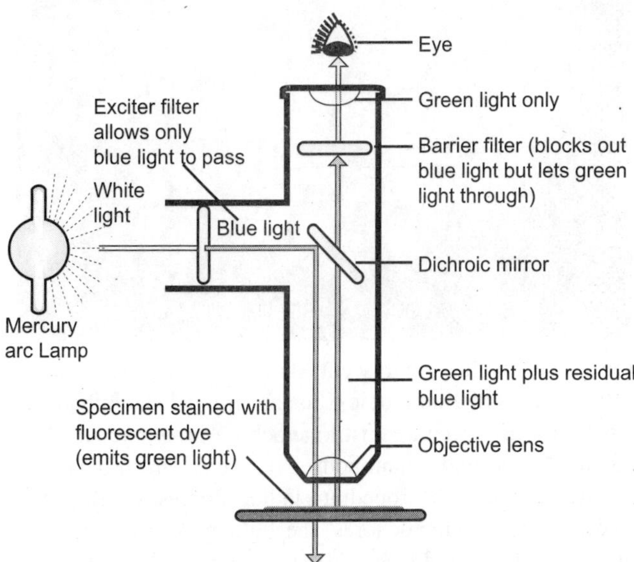

Fig. 20.5 The fluroscence microscope.

filter is to remove all but the blue light; the barrier filter blocks out blue light and allows green light (or other light emitted by the fluorescing specimen) to pass through and reach the eye. Barrier filters are selected on the basis of the dye used. It is possible to chemically combine fluorescent dyes with antibodies (**Fig. 20.6**). Antibodies to which a fluorescent dye is attached are referred to as labelled antibodies. Thus, labelled antibodies can be mixed with a suspension of specimen and then the preparation examined by fluorescent microscopy. This procedure is known as the fluorescent antibody technique or immunofluorescence. An important application of fluorescent microscopy is in the fluorescent antibody technique used to identify an antigenic region (cell, macromolecule).

Confocal microscopy

Confocal microscopy combines fluorescence microscopy with electronic image analysis to obtain images with increased contrast and detail. A tiny light source is supplied by a laser, is focussed on the specimen at a particular depth. The emitted fluorescent light is then collected using a detector, such as a video camera. The light is first passed through a pinhole aperture called a confocal aperture, placed at precisely the point where light emitted from the chosen depth of the specimen comes to a focus (**Fig. 20.7**). Consequently, only light emitted from the plane of focus is able to reach the detector. Scanning across the specimen generates a 2D image of the plane of focus, a much sharper image than that obtained with standard fluorescence microscopy. The series of images obtained at different depths can be used to reconstruct a 3D image of the sample.

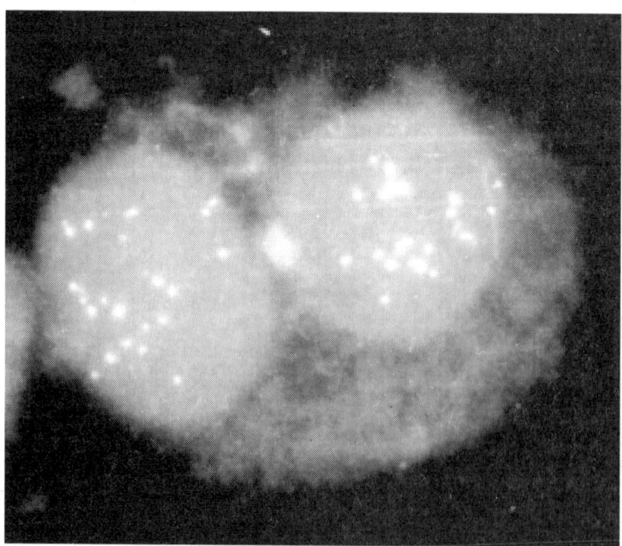

Fig. 20.6 Immunofluorescent staining technique. Animal cells were CREST antibodies (human antinuclear antibodies), which are produced against kinetochore proteins and FITC conjugated secondary antibodies. The nuclear chromatin is counterstained with propedium iodide fluorescent dye. The small spots are kinetochores, the light grey background is nuclear material and the dark grey background is cytoplasm (Channarayappa, 1990).

Multiphoton excitation microscopy

Multiphoton excitation microscopy is used to observe living cells very effectively. The specimen is illuminated with a wavelength of light such that

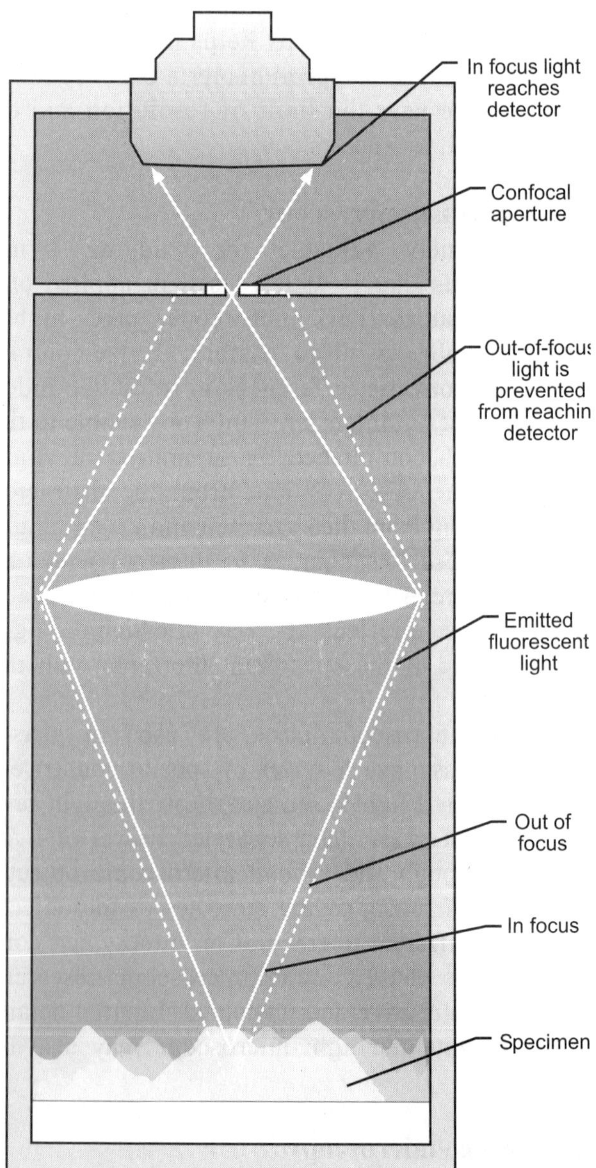

Fig. 20.7 Confocal microscopy.

excitation of the fluorescence dye requires the simultaneous absorption of two or more photons. The probability of two photons simultaneously exciting the fluorescent dye is only significant at the point in the specimen upon which the input laser beam is focussed, so fluorescence is only emitted from the plane of focus of the input light. This highly localized excitation automatically provides 3D resolution, without the need for passing the emitted light through a pinhole aperture, as in confocal microscopy. Moreover, it also allows 3D imaging of living cells.

Electron microscopy

Because of the limited resolution of the light microscope, analysis of a the details of cell structure has required the use of more powerful microscopic technique viz., **electron microscopy**. Two types of electron microscopy—transmission and scanning—are widely used to study the cells.

Transmission Electron Microscope (TEM), employs electrons with wavelengths of about 0.004 nm. In TEM, a magnetic field acts as a lens; it diverts electrons along defined paths and channels them to a focal point (**Fig. 20.8**). To observe fine details of biological samples, cells are sliced extremely thin. Then, electrons scatter in patterns that correspond to the density of different structures. Dense cell parts will be the darkest areas in the final image formed. Most cell structures are somewhat transparent to electrons, so they must first be stained with heavy metal dyes (**Fig. 20.9**). With the electron microscope, magnifications up to 250,000X are commonly obtained and the resolution of 0.5 nm can be achieved, compared with 200 nm for the light microscope.

Scanning electron microscope (SEM) allows the researchers to see the surfaces of objects in the natural state and without sectioning. The specimens are placed in the vacuum chamber and covered with a thin coat of gold to increase electrical conductivity and decrease blurring. A narrow beam of electrons is used in scanning electron microscopes (**Fig. 20.10**). The beam moves back and forth across a specimen surface. The metal coating responds by emitting some of its own electrons (secondary electrons). These electrons that are generated from this surface are collected by a scintillation crystal, which converts each electron impact into a flash of light. Each light flash inside the crystal is amplified by a photomultiplier and used to build up an image on a fluorescent screen. Equipment similar to a television camera detects the emission patterns, and an image is formed. SEM magnification is limited to about 75,000 to 100,000 times. The SEM approaches the resolution as high as 5–20 nm. However, its images have fantastic depth of focus (**Fig. 20.11**).

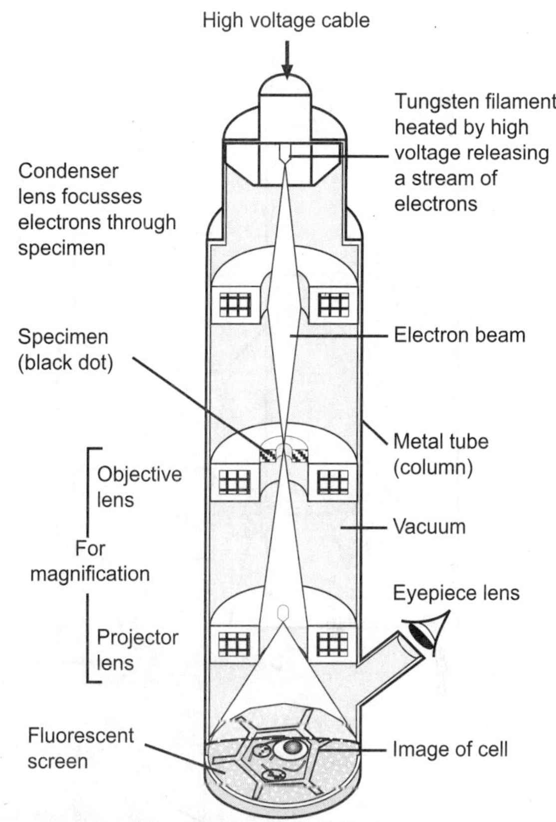

Fig. 20.8 Pathway of the transmission electron microscope.

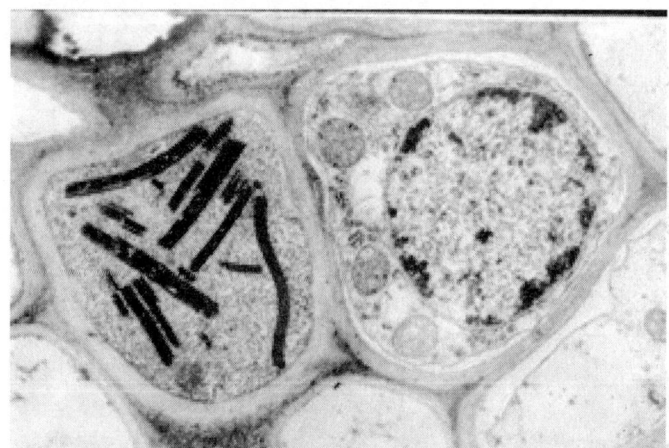

Fig. 20.9 TEM of a cross section of tomato leaf vein showing two phloem bundle cells. The cell on the left contains viral aggregates and the cell on the right shows a nucleus, nuclear envelope, mitochondria, and cell wall (Channarayappa, 1992).

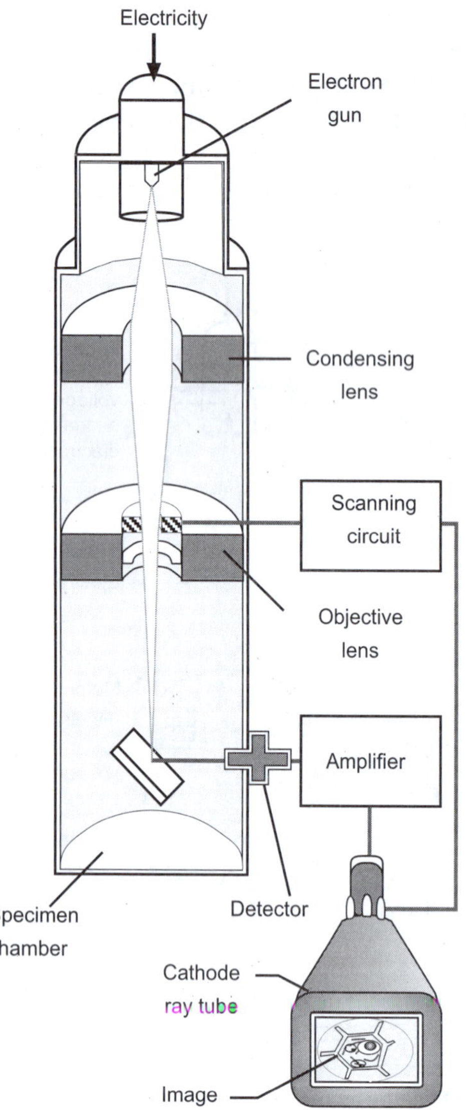

Fig. 20.10 Pathway of the scanning electron microscope.

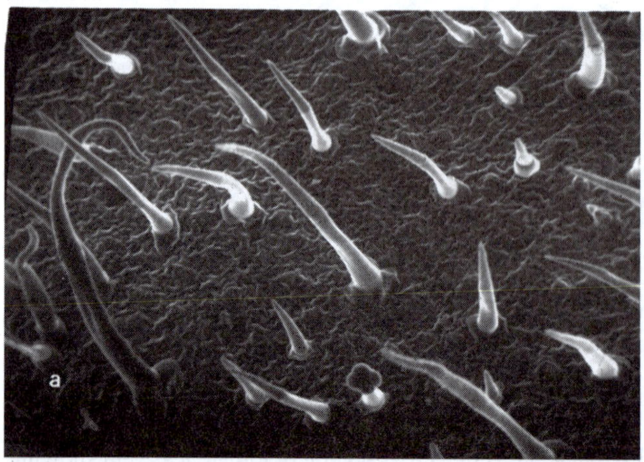

Fig. 20.11 Scanning electron image of tomato leaf surface. SEM image showing the a three dimensional structures of trichomes (Channarayappa, 1992).

Scanning tunnelling and atomic force microscopy

Recently, a new kind of microscope with very high resolving power has been developed: scanning tunnelling microscopy and atomic force microscopy. The former uses a very fine, electrically charged metal tip, which is scanned across the sample. As electrons leak (tunnel) between the tip and the surface supporting the sample, the resistance they encounter varies according to the height of microscopic objects lying on the surface. The resulting fluctuations in current produce a video display of the surface with a resolution comparable to that of an EM. In atomic force microscopy, an extremely sharp tip is either dragged or trapped back and forth across the sample, and its up-and-down motion is detected by the deflection of a laser beam reflected off the cantilever, which holds the tip. This motion is greatly amplified to give a contour map of the object. Both of these techniques have the enormous advantage over electron microscopy that wet, even immersed, sample can be studied. Both techniques are still under improvement and have fascinating potential.

20.4 SPECTROSCOPY

Spectroscopy is the use of the absorption, emission, or scattering of electromagnetic radiation by matter to qualitatively or quantitatively study of the matter or to study physical processes such as:

- ***Absorption***: A transition from a lower level to a higher level with transfer of energy from the radiation field to an absorber, atom, molecule, or solid
- ***Emission***: A transition from a higher level to a lower level with transfer of energy from the emitter to the radiation field; if no radiation is emitted, the transition from higher to lower energy levels is called non-radiative decay.
- ***Scattering***: Redirection of light due to its interaction with matter; scattering might or might not occur with a transfer of energy, i.e., the scattered radiation might or might not have a slightly different wavelength compared to the light incident on the sample.

The matter can be atoms, molecules, atomic or molecular ions, or solids. The interaction of radiation with matter can cause re-direction of the radiation and/or transitions between the energy levels of the atoms or molecules.

A variety of techniques have been developed with the principle of spectroscopy, i.e., study of interaction between electromagnetic radiation and the substance. Electromagnetic spectrum (light, heat, microwaves, infrared, X-rays) have electromagnetic waves with average speed of $3x10^8$ m/sec. These waves are composed of two components, such as electric field and magnetic field, which oscillate perpendicular to each other.

When an atom is exposed to light, its electrons move from ground state to the excited state with the absorption of light or electromagnetic radiation. The excited state of electron, lasts for a short period of a time and soon it tend to returns to the ground state. During this return/drop the atom emits radiation of certain specific wavelength (energy) which is utilized in spectroscopic analysis. Besides absorption or emission of light, atoms of certain substance undergo changes in electronic or nuclear properties, when exposed to electromagnetic radiation. The study of absorption or emission properties or change in nuclear structure provides a lot of information depending on the kind of molecule analyzed.

Mass spectrometry

Mass spectrometers (MS) are analytical instruments used primarily to measure the masses of molecules, determine chemical formulas and molecular structure, and identify unknown substances. Mass spectrometers accomplish these feats by converting the substances in a sample into positively charged (cations and radical cations), gaseous ions, which are accelerated through a curved tube toward a negatively charged plate. This is achieved by a very low concentration of sample molecules allowed to leak into the ionization chamber (which is under a very high vacuum) where they are bombarded with a high-energy electron beam having the molecules fragment and the positive ions produced are accelerated through a charged array into an analyzing tube. As the ions pass through the tube, they are subjected to a magnetic field that causes them to separate from one another according to their molecular mass or according to their mass-to-charge (m/z) ratio. The ions strike an electronic detector located at the end of the tube. Smaller ions travel faster and strike the detector more rapidly than larger ions. The input to the detector is converted into a series of peaks of ascending m/z ratio.

The output of the mass spectrometer shows a plot of relative intensity versus the mass-to-charge ratio (m/z). The most intense peak in the spectrum is termed the base peak and all others are reported relative to its intensity. The peaks themselves are typically very sharp, and are often simply represented as vertical lines.

Nuclear magnetic resonance spectroscopy

Nuclear Magnetic Resonancee Spectroscopy (NMR) is a technique used for detecting interaction between the nuclei of an atom with the magnetic field of the electromagnetic radiation. The types of information accessible via high resolution NMR include:

- functional group analysis (chemical shifts)
- bonding connectivity and orientation (J coupling)
- through space connectivity (Overhauser effect)
- molecular conformations, DNA, peptides and enzyme sequence and structure
- chemical dynamics (Line shapes, relaxation phenomena).

20.5 FLOW CYTOMETER

The flow cytometer is based on the principle that a beam of light (usually laser light) of a single wavelength is directed onto a hydrodynamically focused stream of fluid. A number of detectors are aimed at the point where the stream passes through the light beam; one in line with the light beam (Forward Scatter or FSC) and several perpendicular to it (Side Scatter, SSC and one or more fluorescent detectors). Each suspended particle passing through the beam scatters the light in some way, and fluorescent chemicals found in the particle or attached to the particle may be excited into emitting light at a lower frequency than the light source. This combination of scattered and fluorescent light is picked up by the detectors and by analyzing fluctuations in brightness at each detector (one for each fluorescent emission peak) it is possible to extrapolate various types of information about the physical and chemical structure of each individual particle. FSC correlates with the cell volume and SSC depends on the inner complexity of the particle (i.e., shape of the nucleus, the amount and type of cytoplasmic granules or the membrane roughness). Some modified flow cytometers have eliminated the need for fluorescence and use only light scatter for measurement and others form images of each cell's fluorescence, scattered light, and transmitted light.

Fluorescence-activated cell sorting

Fluorescence-activated cell sorting (FACS) is a specialized type of flow cytometry generally for sorting a heterogeneous mixture of cells into two or more containers, one cell at a time, based upon the specific light scattering and fluorescent characteristics of each cell. It is a very useful instrument as it provides fast, objective and quantitative recording of fluorescent signals from individual cells as well as physical separation of cells of particular interest.

The instrument works by sending cell suspension in the centre of a narrow, rapidly flowing stream of liquid. The flow is arranged so that there is a large separation between cells relative to their diameter. A vibrating mechanism causes the stream of cells to break into individual droplets. The system is adjusted in such a way that there is a low probability of a droplet having more than one cell. Just before the stream breaks into droplets the flow passes through a fluorescence measuring station where the fluorescent character of interest of each cell is measured. An electrical charging ring is placed just at the point where the stream breaks into droplets. A charge is placed on the ring based on the immediately prior fluorescence intensity measurement and the opposite charge is trapped on the droplet as it breaks from the stream. The charged droplets then fall through an electrostatic deflection system that diverts droplets into containers based upon their charge.

Uses of flow cytometery

Flow cytometery instrument can be used for many purposes.

- Cell volume and morphological complexity of cells
- Cell pigments such as chlorophyll, xanthophyll, or phyco-erythrin
- DNA (cell cycle analysis, cell kinetics, proliferation) and RNA content and kinetics
- Chromosome analysis and sorting (library construction, chromosome paint)
- Protein expression and localization in a cell
- Transgenic products *in vivo*, particularly the green fluorescent protein or related fluorescent proteins
- Antigens of cell surface (cluster of differentiation (CD) markers), intracellular (various cytokines, secondary mediators) and nuclear origin
- Enzymatic activity
- pH, intracellular ionized calcium, magnesium, membrane potential
- Membrane fluidity
- Apoptosis (quantification, measurement of DNA degradation, mitochondrial membrane potential, permeability changes, caspase activity)
- Cell viability
- Monitoring electropermeabilization of cells
- Oxidative burst
- Characterizing multidrug resistance (MDR) in cancer cells.

Applications of flow cytometery

It has broad application in medicine (especially in transplantation, hematology, tumour immunology and chemotherapy, genetics and sperm sorting in IVF). In marine biology, the auto-fluorescent properties of photosynthetic plankton can be exploited by flow cytometry in order to characterize abundance and community structure. In protein engineering, flow cytometry is used in conjunction with yeast display and bacterial display to identify cell surface-displayed protein variants with desired properties.

20.6 X-RAY CRYSTALLOGRAPHY

X-ray crystallography (or x-ray diffraction) is a scattering process; crystalline materials scatter x-rays at well-defined angles. If the wavelength of the incident x-rays is known, this allows calculation of the distances between planes of atoms within the crystal. The intensities of the scattered x-rays give information about the atomic positions and allow the arrangement of the atoms within the crystal structure to be calculated.

20.7 SUBCELLULAR FRACTIONATION

Most cells contain a variety of different subcellular organelles. In order to study the detailed functions and structure of organelles, it is very important to isolate the organelles of eukaryotic cells in a form that can be used for biochemical studies. This is usually accomplished by **differential centrifugation**—a technique developed by Albert Claude, Christian de Duve, and their colleagues in the 1940s and 1950s to separate the components of cells on the basis of their size and density.

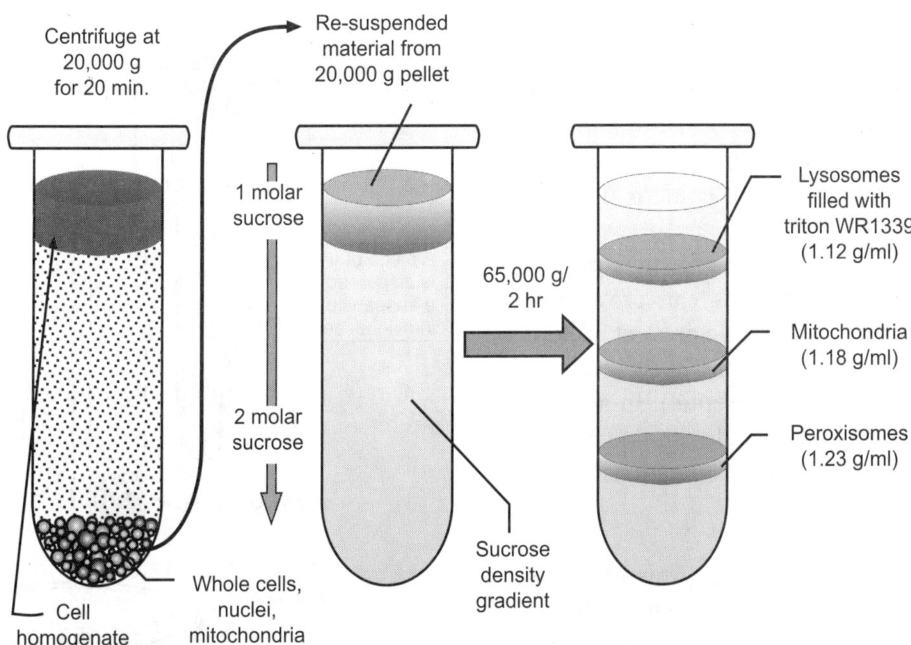

Fig. 20.12 Purification of subcellular fractions by densiy gradient equilibrium centrifugation. In this example, the medium is composed of a continuous sucrosedensity gradient, and the different organelles sediment until they reach a place in the tube equal to their own density, where they form bands.

The first step in subcellular fractionation is the disruption of the plasma membrane under conditions that do not destroy the internal components of the cell. The broken cell is then fractionated into its components by a series of centrifugations in an **ultracentrifuge**, which rotates samples at very high speeds to produce forces up to 500,000 times greater than gravity (i.e., >500,000g). This force causes cell components to move towards the bottom of the centrifuge tube and form a pellet at a rate that depends on their size and density, with the largest and heaviest structures sedimenting most rapidly. The fractions obtained from differential centrifugation correspond to the enriched but not still completely pure. A degree of purification can be achieved by **density gradient centrifugation**, in which organelles are separated by sedimentation through a gradient of a dense substance, such as sucrose. The particles of different size sediment through the gradients at different rates move as discrete bands (**Fig. 20.12**).

20.8 CELL CULTURE TECHNIQUES

The success of cellular studies depends on how readily cells can be grown and manipulated in the laboratory. Learning to grow cells outside the organism, that is, in **cell culture**, has proved to be one of the most valuable technical achievements in the study of biology. Cultured cells can be obtained in large quantities; most cultured cells contain only a single type of cell; many different cellular activities, such as endocytosis and macromolecular biosynthesis, can be readily studied in cell cultures; cells can differentiate in culture, and cultured cells respond to treatment with drugs, hormones, growth factors, and other active substances. Although, growing plant and animal cells is more difficult than the culture of bacteria or yeasts, a wide variety of animal and plant cells have been successfully grown and manipulated in culture. Such in vitro cell culture systems have enabled scientists to study cell growth and differentiation, as well as to perform genetic manipulations required to understand gene structure and function.

The early tissue culturists employed media containing a great variety of unknown substances (not completely defined). In vitro cell growth was accomplished by adding fluids obtained from living systems (e.g., lymph, blood serum, or embryo homogenates). Cells require a considerable variety of nutrients, hormones, growth factors, and co-factors to remain healthy and proliferate. Even today, most cell culture media contain large amounts of serum. However, one of the primary goals of cell culturists has been to develop defined, serum-free media that supports the growth of cells. By using serum-free media many cell lines have been successfully grown in vitro. The composition of these chemically defined media, is relatively complex; they consist of a mixture of nutrients and vitamins, together with a variety of purified proteins, including insulin, epidermal growth factor, and transferrin.

Culture medium is rich in nutrients, thus medium contaminated with micro-organisms can grow faster and exhaust nutrients very quickly. To prevent bacterial contamination, cell culture operations must be carried out under sterile conditions (i.e., aseptic). In addition, the medium should be supplemented with one or more antibiotics to prevent microbial contaminations.

Animal cells

In 1955, Harry Eagle studied the growth of two established cell lines (e.g., malignant tumour cells from patient, Henrietta Lacks, called HeLa cells and a mouse fibroblast line called L cells) for the development of nutrient medium that support the growth of cells in in vitro. He was able to grow these cells in a medium consisting of a mixture of salts, carbohydrates, amino acids, and vitamins, supplemented with serum protein. The medium developed by Eagle is still the basic medium used for animal cell culture.

Three kinds of animal cell cultures are in common use: the **primary culture**, the **established cell line** and the **embryonic cell culture**. Animal cell cultures are initiated by the dispersion of a piece of tissue into a suspension of its component cells, which is then added to a culture dish containing nutrient media. Most animal cell types such as fibroblasts and epithelial cells attach and grow on the plastic surface of dishes. Under appropriate conditions, many specialized cells types can also be grown in culture. The initial cell cultures established from a tissue are called **primary cultures** (**Fig. 20.13**). For example, human fibroblasts can usually be cultured for 50 to 100 population doublings, after which they stop growing. In contrast, embryonic stem cells and cells derived from tumours frequently proliferate indefinitely in cultures. **Embryonic stem cells** are very versatile, since they can differentiate into all cell types present in adult organisms. In addition, a number of immortalized rodent cell lines have been isolated from cultures of normal fibroblasts.

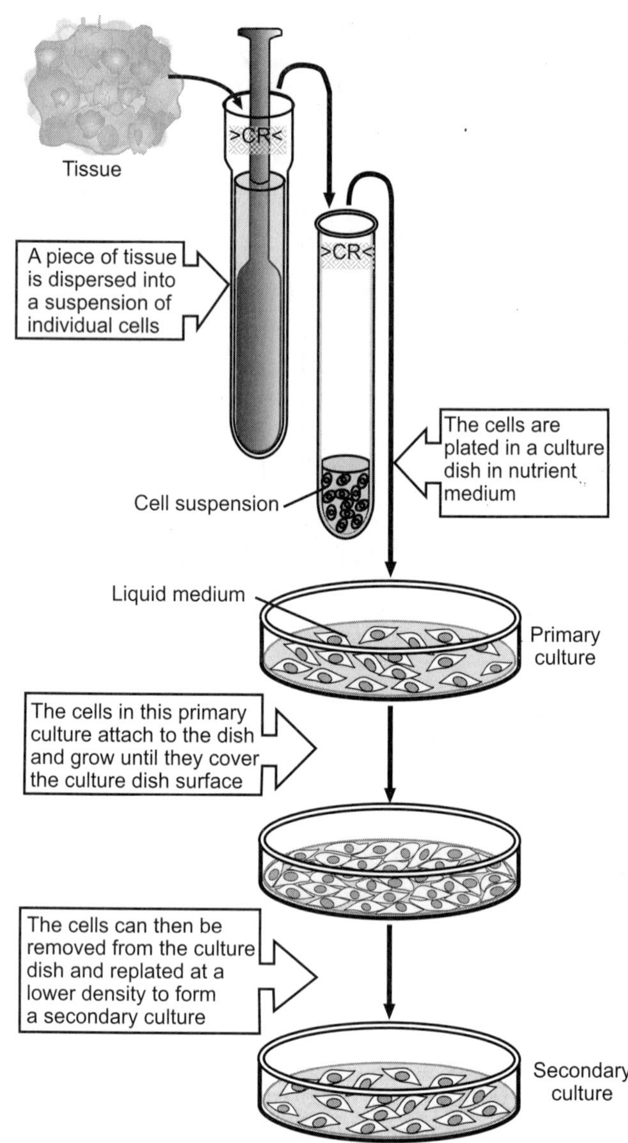

Fig. 20.13 Steps involved in animal cell culture.

Embryonic stem cells

The cells of early embryos are unique in their ability to proliferate and differentiate into all of the types of cells that make up the tissues and organs of adult animals. Embryonic stem cells were first cultured from mouse embryos. These can be propagated indefinitely in culture and if reintroduced into early embryos, are able to give rise to cells in all tissues of animals. The establishment of embryonic stem cell lines has had a major impact on studies of mouse genetics and development as well as opening new possibilities for the treatment of a variety of human diseases. In 1998, two groups of researchers developed the first lines of human embryonic stem cells. Because of the proliferative and differentiative capacity of these cells, they offer the hope of providing new therapies for the treatment of a variety of diseases.

Therapeutic cloning

Animal cloning by somatic cell nuclear transfer (nucleus of a cell transferred into an unfertilized enucleated egg), together with the properties of embryonic stem cells, opens the possibility of therapeutic cloning (**Fig. 20.14**). In therapeutic cloning, a nucleus from an adult human cell would be transferred to an enucleated unfertilized egg, which would then be used to produce an early embryo in culture. These cells are totipotent and could be used to produce an early cloned embryo to generate appropriate types of differentiated cells for transplantation therapy. This technique has many advantages such as production of embryonic stem cells that would be identical to the recipient of the transplant, who was the donor of the adult somatic cell nucleus. This bypasses the barrier of the immune system in rejecting the transplanted tissue. This method can be combined with gene transfer to correct a variety of inherited genetic deficiencies. Therapeutic cloning has enormous potential to treat a wide variety of devastating disorders for which stem cell transplantation therapy could be applied, including Parkinson's disease, Alzheimer's disease, diabetes, and spinal cord injuries. In spite of having many advantages, many challenges remain in exploiting the benefits of therapeutic cloning to humans. Thus, continuing research in this area is

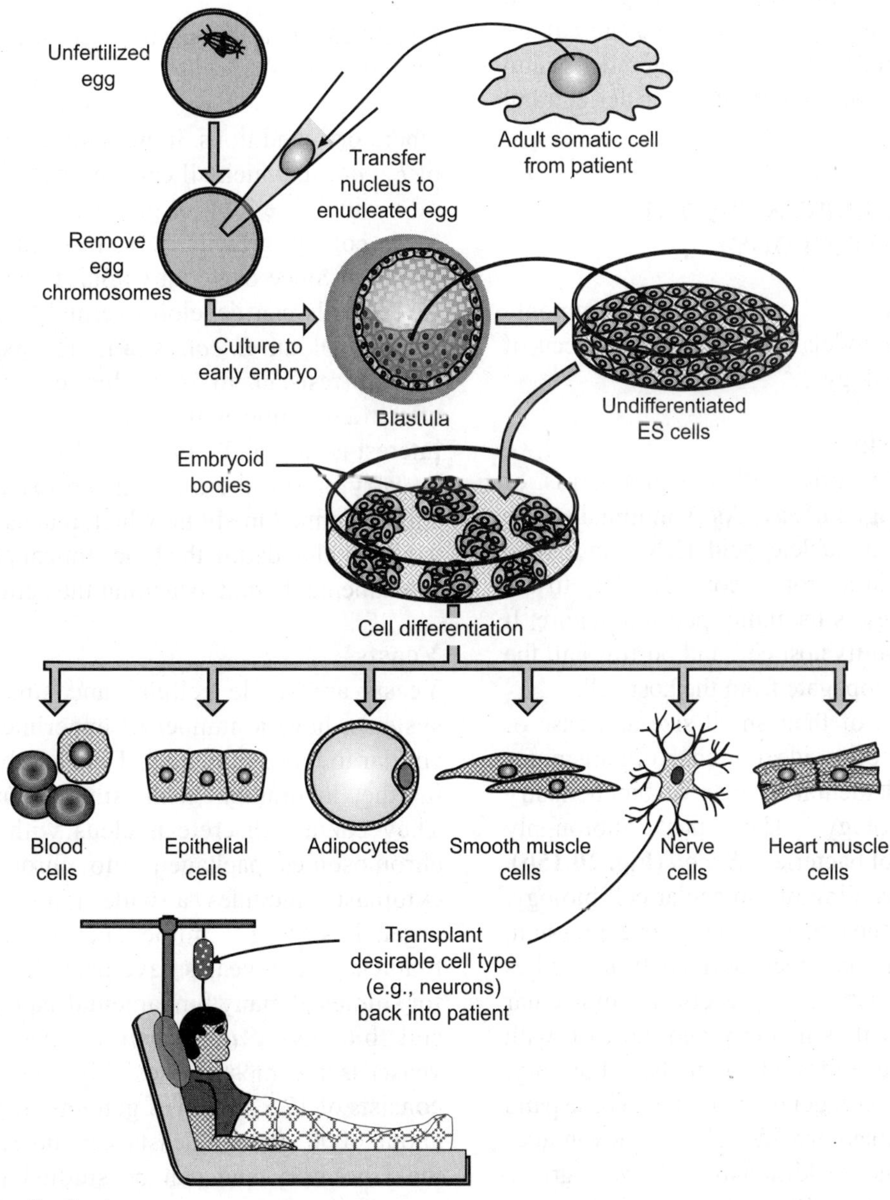

Fig. 20.14 Therapeutic cloning.

highly promising and provides an opportunity to develop new treatment approaches to a broad range of human diseases.

Plant cell culture

Plant cells can be cultured in nutrient media containing appropriate growth regulatory molecules. Plant growth regulators are small molecules when provided with appropriate mixtures of this growth regulatory and other nutrients, many types of plant cells proliferate in culture, producing a mass of undifferentiated cells called a callus. Plant cells are totipotent (a single cell can regenerate into a complete plant), can give rise to different cell types in cultures, provided appropriate hormones and nutrients are added. The ability to produce a new plant from a single cell that has been manipulated in culture makes it easy to introduce genetic alterations into plants, opening important possibilities for agricultural genetic engineering.

20.9 MODEL ORGANISMS USED IN MOLECULAR BIOLOGY

Different types of cells and organisms are commonly used as experimental models to study various aspects of cell and molecular biology.

Viruses and bacteria

The simplest life form (often called as particle) is, of course, the virus (**Fig. 20.15a**). As a minimalist life form, it consists of a nucleic acid (DNA or RNA) inner core enclosed in a protein coat. The key to the virus' survival strategy is its highly parasitic nature. It absolutely depends on its host cell and borrows all the functions needed to propagate from the host cell.

Bacteria, because of their small size and ease of culture, have become an ideal model organism for studying many fundamental aspects of biochemistry and molecular biology. The most thoroughly investigated species of bacteria is *E. coli* (**Fig. 20.15b**). Much of our understanding of molecular cell biology, on DNA replication, the genetic code, gene expression, and protein synthesis has been derived from studies of *E. coli*. This organism is very useful for molecular biologists because of its simplicity and the ease with it can be propagated and studied in the laboratory. The smaller size of *E. coli* genome (4.6 mega base pairs and contains 4000 genes) provides obvious advantages for genetic analysis. Additionally, *E. coli* grows rapidly (it divides every 20 minutes) under normal

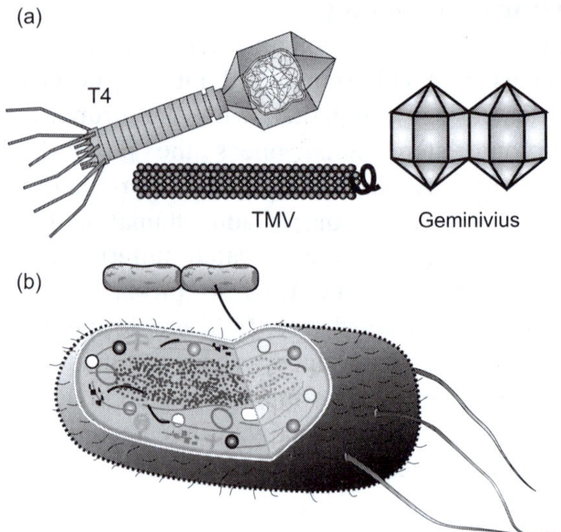

Fig. 20.15 Model organisms used in molecular biology: (a) Virus and (b) *E. coli*.

laboratory conditions. Importantly, a clonal population of *E. coli*, in which all cells are derived from division of a single cell of origin, can be readily isolated as a colony grown on semi-solid agar-containing medium. Since bacterial colonies containing as many as 10^8 cells can develop overnight, selecting genetic variants of an *E. coli* strain, for example, mutants that are resistant to an antibiotic or transformed cell after transfection with a vector containing a gene of interest is easy and rapid. Bacteria also have an ability to carry out all the biosynthetic reactions using very simple defined medium, which makes them extremely useful in elucidating the biochemical pathways that are fundamental to understanding the cellular functions.

Yeasts

Yeasts are single cellular and simplest eukaryotic systems, have a number of experimental advantages similar to those of *E. coli*. They can be grown rapidly in the laboratory with simple culture medium. They contain discrete nucleus with multiple linear chromosomes packaged into chromatin, and their cytoplasm includes a wide range of intracellular organelles (e.g., mitochondria and cytoskeletal filament. Thus, yeasts have provided a crucial model for studies of many fundamental aspects of eukaryotic cell biology. *Saccharomyces cerevisiae*, (bakers, yeast) is the most extensively studied (**Fig. 20.16**), consists of 12×10^6 bp of genomic DNA and contains about 6000 genes. Yeasts can be readily grown in the laboratory and can be studied by many of the same molecular genetic approaches that have proved

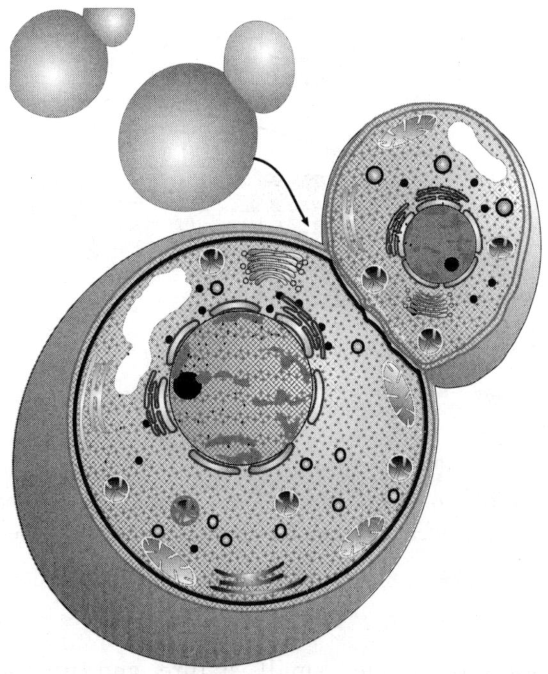

Fig. 20.16 Yeast (*Saccharomyces cerevisiae*), a primitive eukaryote, has been used as a model organism.

so successful with *E. coli*. Yeast cells undergo cell division at every 3 hours and they can be easily grown as colonies from a single cell. The existence of haploid and diploid cells in *S. cerevisiae* populations facilitated various kinds of genetic analysis. Thus, similar to bacteria, yeasts can be used for a variety of genetic manipulations. Yeast mutants have been important in understanding many fundamental processes in eukaryotes and the general features of yeast cell structure and function can be applied to all eukaryotic cells.

Nematode

The nematode *Caenorhabditis elegans* (**Fig. 20.17**) has several notable features that make it one of the most widely used models for studies of animal development and cell differentiation. *C. elegans* has a rapid life cycle. It can be cultured on petri-dishes and fed a simple diet of bacteria. They grow well at a range of temperatures. At 25°C, fertilized embryos complete development in 12 hours and hatch into free-living animals capable of complex behaviours. The first stage juvenile (L1) passes through four juvenile stages (L1–L4) over the course of 40 hours to become a sexually mature adult. The adult hermaphrodite can produce up to 300 self-progeny over the course of about 4 days, or can be mated with rare males to produce up to 1000 hybrid progeny (**Fig. 20.17**). The adult lives for about 15 days, but under stressful conditions the L1 stage animal can enter an alternative developmental stage called **dauer**, which is resistant to environmental stresses and can survive for months.

Genome of *C. elegans* (~100 mbp) is larger than those of unicellular eukaryotes and is simpler and more manageable than the genome of most animals. *C. elegans* genome sequence has been determined, revealing that the genome of *C. elegans* contains approximately 19,000 genes. It has a simple body plan. The prominent organ in the adult hermaphrodite is the gonad, which contains the proliferating and differentiating germ cells, fertilization chamber and uterus for temporary storage of young embryos. The entire adult worm contains only 959 somatic cells, plus 1000 to 2000 germ cells. This worm can be conveniently grown and subjected to genetic manipulations in the laboratory. Due to its cellular simplicity, the detailed microscopic observations on embryonic origin and lineage of all the cells in the adult worm, has been extensively studied. Genetic studies have also been made in mutants that carry mutations for important genes. Interestingly, similar genes have also been shown to work in complex animals (including humans)

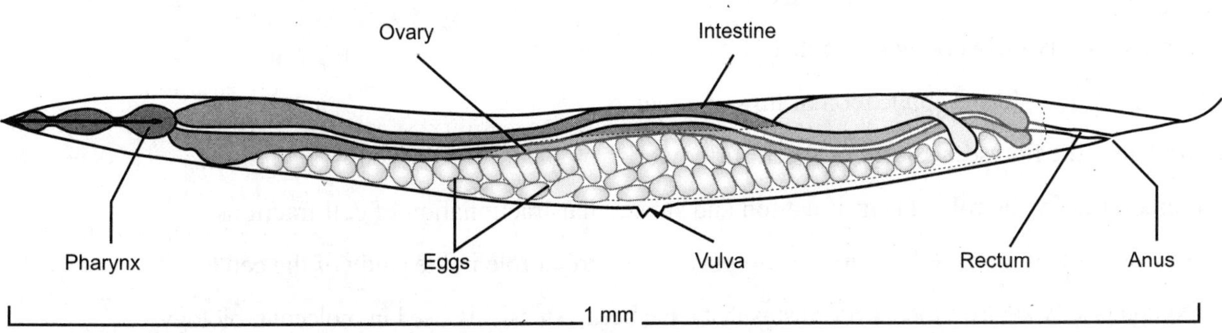

Fig. 20.17 The female nematode *Caenorhabditis elegans*.

making *C. elegans* an important model for studies of animal development.

Drosophila melanogaster

The fruit fly, *Drosophila melanogaster* (**Fig 20.18**) has been a crucial model organism in developmental biology. The genome of *Drosophila* contains only about 14,000 genes. It can be easily maintained and bred in the laboratory, and the short reproductive cycle (about 2 weeks) makes it a very useful organism for genetic experiments. Many fundamental concepts of genetics, such as the relationship between genes and chromosomes, were derived from studies of *Drosophila* early in the 20th century, particularly, TH Morgan and his colleagues. The detailed studies in *Drosophila* have revealed many genes that control development and differentiation, and current methods of molecular biology have allowed the functions of these genes to be analyzed in detail. These studies also led to advancement in understanding of the molecular mechanisms that govern animal development, particularly with respect to formation of the body plan of complex multicellular organisms.

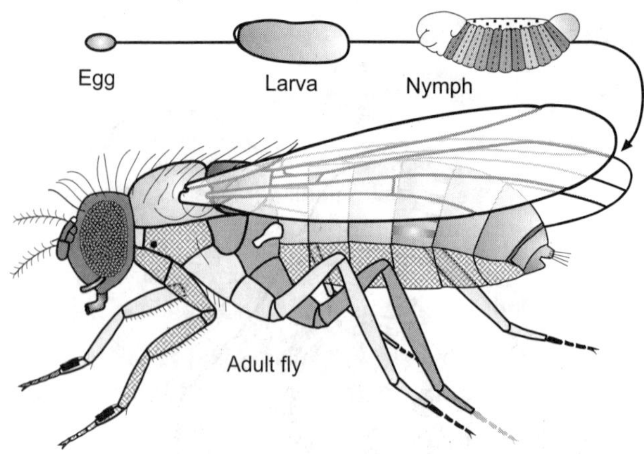

Fig. 20.18 *Drosophila melanogaster*, fruit fly, has been used as a model organism in genetics and developmental biology studies.

Arabidopsis thaliana: A model plant

The small flowering plant *Arabidopsis thaliana* has been widely used as a model plant to study the molecular biology of plants. It is a small, dicotyledonous weed that has become the model species for most aspects of plant biology. It has a short life cycle, small stature and produces a large number of offspring, making it ideally suited for genetic and mutational analysis. It is also easily transformed and, most importantly for genome sequencing, it has one of the smallest genomes and less non-coding sequences in higher plants (125 mbp). *A. thaliana* was therefore chosen to be the first plant to have its complete genome sequenced. These studies have led to the identification of genes involved in various aspects of plant development.

Exercises

1. Describe the working principle of a light microscope with a diagram.
2. How does the electron microscope differ from the light microscope?
3. How does the transmission electron microscope differ from the scanning electron microscope?
4. What is spectroscopy and how is it used?
5. Explain briefly the working principle of a mass spectrometer.
6. Give the principle and application of flow cytometers
7. How does the florescence-activated cell sorter work?
8. Explain briefly the role of centrifugation and subcellular fractionation of cell fractions.
9. How does the in vitro cell culture system play an important role in the study of the cell?
10. Explain briefly any two model-organisms that have been extensively used in molecular biology.

Summary

- The lowest and most simple form of life is a ***cell***.
- Some organisms are single cellular and others are multicellular in nature.
- Even the simplest unicellular organisms exhibit all "the hallmark properties" of life, indicating the cell is the ***fundamental unit of life***.
- The most intensively studied area of biological sciences was dissecting the cell, revealing the composition and their organization and uncovering the architecture of cellular interiors.
- The study of cells (or cell biology) is basically concerned with understanding the cell structure and functions. Thus, the study of the cell requires various techniques that have been developed in various fields.
- The ***microscope*** is a unique instrument which has made studies of cell biology possible. Currently, more than a dozen kinds of microscopes with varied range of resolution and magnifications are available.
- Some other recently developed technique that are used extensively in cell biology are different types of ***spectroscopy***, ***x-ray crystallography***, ***flow cytometer*** and so on.
- In order to study the composition and functions of cells it requires ***subcellular fractionation of cells*** by using different types of centrifuges.
- The success of cellular studies depends on how readily cells can be grown and manipulated in the laboratory.
- When provided with the appropriate conditions most animals and plant cells can live, multiply, and even express differentiated properties in a tissue-culture dish.
- The cells can be monitored in culture continuously or analyzed biochemically, cytologically and also by molecular studies.
- Using ***cell cultures*** various manipulations of cells can be done, including somatic cell fusion, embryonic cell differentiation, nuclear transfer, therapeutic cloning can be done.
- Different types of cells and organisms have been used as experimental models to study various aspects of cell and molecular biology.
- ***E. coli, yeast, nematodes, Drosophila*** and ***Arabidopsis thaliana*** are commonly used experimental models.

Further Reading

Books

Brown TA (2002), *Genomes*. 2nd edition, BIOS Scientific, Oxford, UK.

Channarayappa (2006), *Molecular Biotechnology: Principles and Practices*, Universities Press, India and CRC Press, USA.

Drury RAB, Wallington EA and Cameron R (1967), *Carleton's Histological Techniques*. Oxford University Press.

Hartwell L, Hood L, Goldberg ML, Reynolds AE, Silver LM and Veres RC (2003), *Genetics: From Genes to Genomes*, 2nd edition, McGraw-Hill, New York.

Heywood VH, (ed.), (1971), *Scanning Electron Microscopy: Systematic and Evolutionary Applications*, Academic Press Inc., London.

Loken MR (1990), Immunofluorescence Techniques in Flow Cytometry and Sorting, JohnWiley & Sons.

Pelczar MJ, Chan ECS, and Krieg NR (1993), *Microbiology*, Tata McGraw-Hill, New Delhi.

Watt IM (1985), *The Principles and Practices of Electron Microscopy*, Cambridge University Press, Cambridge.

Wischnitzer S (1990), *Introduction to Electron Microscopy*, Pergamon Press Inc.

References

Channarayappa (1993), Resistance mechanism of *Lycopersicon* species to the vector *Bemisia tabaci* and Ultrastructural changes in leaf curl virus infected tomato leaves, PhD Thesis Submitted to the University of Agricultural Sciences, Bangalore India.

Channarayappa (1990), Cytogenetic analysis of clastogens and aneuploidogens by immunofluorescent and micronucleus methods, PhD thesis submitted to the West Virginia University, Morgantown, West Virginia, USA. W Microfilm 3365.

Cohen J and M Hogan (1994), The new genetic medicines, *Sci. Am.*, December, 76–82.

Index

D

E

G

H

I

N

O

P

Q

R

T